STRONG FIELD PHYSICS

This book presents the foundational physics underlying the generation of high-intensity laser light and its interaction with matter. Comprehensive and rigorous, it describes how the strong electric and magnetic fields of a high-intensity light pulse can shape the nonlinear dynamics of all forms of matter, from single electrons up to atomic clusters and plasmas. Key equations are derived from first principles and important results are clearly explained, providing readers with a firm understanding of the fundamental concepts that underlie modern strong field physics research. The text concludes with suggestions for further reading, along with an extensive reference list. Effective as both an educational resource and as a reference text, this book will be invaluable to graduates and researchers across the atomic, molecular and optical (AMO) and plasma physics communities.

TODD DITMIRE is Professor of Physics at the University of Texas at Austin, and is Director of the Center for High Energy Density Science. He experimentally studies high-intensity laser–matter interactions and develops ultrahigh peak power lasers. He is Fellow of the American Physical Society and cofounder of the laser company National Energetics.

STRONG FIELD PHYSICS

Ultra-Intense Light Interaction with Matter

TODD DITMIRE
The University of Texas at Austin

CAMBRIDGE
UNIVERSITY PRESS

Shaftesbury Road, Cambridge CB2 8EA, United Kingdom

One Liberty Plaza, 20th Floor, New York, NY 10006, USA

477 Williamstown Road, Port Melbourne, VIC 3207, Australia

314–321, 3rd Floor, Plot 3, Splendor Forum, Jasola District Centre,
New Delhi – 110025, India

103 Penang Road, #05–06/07, Visioncrest Commercial, Singapore 238467

Cambridge University Press is part of Cambridge University Press & Assessment,
a department of the University of Cambridge.

We share the University's mission to contribute to society through the pursuit of
education, learning and research at the highest international levels of excellence.

www.cambridge.org
Information on this title: www.cambridge.org/9780521760836

DOI: 10.1017/9781139027984

© Todd Ditmire 2025

This publication is in copyright. Subject to statutory exception and to the provisions
of relevant collective licensing agreements, no reproduction of any part may take
place without the written permission of Cambridge University Press & Assessment.

When citing this work, please include a reference to the DOI 10.1017/9781139027984

First published 2025

A catalogue record for this publication is available from the British Library

A Cataloging-in-Publication data record for this book is available from the Library of Congress

ISBN 978-0-521-76083-6 Hardback

Additional resources for this publication at www.cambridge.org/9780521760836

Cambridge University Press & Assessment has no responsibility for the persistence
or accuracy of URLs for external or third-party internet websites referred to in this
publication and does not guarantee that any content on such websites is, or will remain,
accurate or appropriate.

Contents

Preface		*page* ix
1	**Introduction to Strong Field Physics**	**1**
1.1	Definition of Strong Field Physics	1
1.2	Historical Overview	2
1.3	Outline of This Book	9
1.4	Units, Variables and Mathematical Notation	10
2	**Strong Field Generation by High-Intensity Lasers**	**12**
2.1	Types of Lasers Used in Strong Field Physics and Chirped Pulse Amplification	12
2.2	Common CPA Laser Gain Materials	16
2.3	Femtosecond Oscillators and CPA Front Ends	27
2.4	Pulse Stretchers and Compressors	35
2.5	CPA Amplifiers and Amplifier Design Issues	47
2.6	Beam Focusability and Adaptive Optics	62
2.7	Other Laser Technologies Used in Strong Field Physics Studies	66
3	**Strong Field Interactions with Free Electrons**	**70**
3.1	Introduction and Definitions of the Fields	70
3.2	Dynamics of Electrons in Fields at Subrelativistic Intensities	72
3.3	Dynamics of Electrons in Fields at Relativistic Intensities	78
3.4	Radiation Emitted by Electrons Driven in a Strong Laser Field	91
3.5	Ejection of Free Electrons from the Focus of an Intense Laser	110

	3.6	Intense Laser Interactions with Relativistic Electron Beams	128
	3.7	Quantum Mechanical Description of a Free Electron in a Strong Laser Field	136
4	Strong Field Interactions with Single Atoms		140
	4.1	Introduction	140
	4.2	Multiphoton Ionization and Lowest-Order Perturbation Theory	143
	4.3	The Strong Field Approximation	166
	4.4	Tunnel Ionization	183
	4.5	Nonsequential Double Ionization and Electron Rescattering	204
	4.6	Multiple Ionization of Multielectron Atoms and Ions	216
	4.7	Above-Threshold Ionization	223
	4.8	Ionization Stabilization	248
	4.9	Relativistic Effects	258
5	Strong Field Interactions with Molecules		262
	5.1	Introduction	262
	5.2	Strong Field Ionization of Molecules	262
	5.3	Molecular Alignment in Strong Fields	283
	5.4	Molecular Bonds in Strong Fields	292
	5.5	Coulomb Explosion of Multiply Ionized Molecules	309
6	Strong Field Nonlinear Optics		316
	6.1	Introduction to High-Harmonic Generation	316
	6.2	Single-Atom Physics of High-Harmonic Generation	321
	6.3	Propagation and Phase-Matching in High-Harmonic Generation	350
	6.4	Attosecond Pulse Generation	376
7	Strong Field Interactions with Clusters		390
	7.1	Introduction to Clusters in Laser Fields and the Nanoplasma Model	390
	7.2	Electron Dynamics and Cluster Ionization	402
	7.3	Laser Absorption and Heating of Cluster Nanoplasmas	434
	7.4	Cluster Explosions	463
	7.5	Laser Absorption Dynamics in an Exploding Cluster	482
	7.6	Bulk Cluster Plasma Target Effects	489
8	Strong Field Interactions with Underdense Plasmas		495
	8.1	Introduction to Light Propagation in Underdense Plasma	495
	8.2	Plasma Formation and Heating Mechanisms	498

	8.3	Beam Propagation of Intense Laser Pulses through Underdense Plasma	511
	8.4	Propagation in Underdense Plasma in the Relativistic Regime	535
	8.5	Temporal Phase and Amplitude Modulation of Intense Pulses in Plasma	583
	8.6	Plasma Instabilities Driven by Intense Laser Pulses	590
	8.7	Laser Plasma Electron Acceleration	609
	8.8	X-ray Production by Betatron Oscillations in Wakefield Ion Cavities	643
9	Strong Field Interactions with Overdense Plasmas		659
	9.1	The Structure of Solid Target Plasmas	660
	9.2	Ponderomotive Force Effects on Solid Target Plasmas	692
	9.3	Collisionless Absorption	708
	9.4	Hot Electron Production and Transport	733
	9.5	High-Harmonic Generation from Solid-Density Plasma	747
	9.6	Ion Acceleration	754
	9.7	Solid Target X-ray Production	771
	9.8	Strong Magnetic Field Production	775
	9.9	Relativistic-Induced Transparency	780
	9.10	High-Intensity Laser Solid-Target Effects and Applications	784

Appendix   List of Symbols   788

Bibliography   803
Index   847

Further resources are available online at www.cambridge.org/ditmire

Preface

This book concerns itself with presenting the foundational physics underlying how ultra-intense light interacts with various forms of matter. This field of study is often referred to using the elegant term strong field physics. What defines "high intensity" is not exact but, generally speaking, the strong field regime begins at an intensity approaching 10^{14} W/cm^2, as I will argue in Chapter 1. This topic has, at the time of this writing, led to two recent Nobel Prizes in Physics, the 2018 prize given to Strickland and Mourou for chirped pulse amplification, and the 2023 prize presented to L'Huillier, Agostini, and Krausz for attosecond pulse physics.

My goal in writing this book (which I admit has now grown into a tome) was motivated by what I felt was a need for scientists entering the field (such as new graduate students) to have a book which attempted to cover in a comprehensive manner most of the foundational physics underlying strong-field physics and what happens when intense light irradiates matter. The result, which was written over a number of years (more years than it should have taken and more than I am willing to admit), is a textbook which I hope the reader will find an important reference on the expanse of topics that comprise a strong field physics.

In fact, the field has an interesting breadth. On one hand, much of strong field physics concerns itself with atoms and molecules and, as such, research is pursued within the Atomic-Molecular Optical (AMO) physics community. The term "strong field physics" itself is most often employed in the context of AMO research. However, the study of high-intensity laser–matter interactions inevitably evolves in many regimes to study interactions with plasmas, a topic of broad interest which is pursued by many plasma physicists around the world. This book is intended to address in detail the physics explored by both AMO and plasma physics communities. While these subfields often overlap, for the most part they represent two fairly distinct research areas. I hope that the present work will be found useful by students studying in either one of these complementary areas and that it might inform students in one about the role that effects from the other play in high-intensity light phenomena. So while there are a number of very good textbooks centered either on the atomic or the plasma aspect of high-intensity light–matter interactions, to my knowledge, there are no books that cover both expansively. My hope is that this book partially fills that gap.

My intentions with this book are also motivated by another bias I hold. My feeling is that, as a subfield of physics (or any science) matures, there comes a point at which students of that field must focus on recent publications in the field and can be overwhelmed if they need to look back in the literature to read scores of foundational publications to attain the needed background to understand modern research in that field. Students in such a mature field benefit from the availability of a library of textbooks that give them the foundations to understand the findings of past research without having to delve into the original papers that laid those foundations. I, for example, in doing research on new laser technologies would be doomed if I had to start with the publication by Maiman on the first laser and work through the literature from there. I have the benefit of a wealth of excellent laser physics textbooks like that of Siegman or that by Milonni and Eberly to lean on. Strong-field physics has reached that state of maturity and breadth. While there are now several recent excellent textbooks on high-intensity laser–matter interaction physics, I felt that there was still a place for a comprehensive book that was devoted to treating broad foundational material to complement those other textbooks. I hope that the reader finds that this book plays an important tool in their arsenal on this subject to aid them as a broad reference in their research. So, while I would never presume that this book will ever have the enduring importance and authority of Siegman's lasers textbook, in a word, my aspiration for this book is that it becomes in some way the "Siegman" of strong-field physics.

It is one of my characteristics, I suppose, as an experimentalist and not a theorist, that I am often frustrated when an important equation or result is merely stated without any derivation or is not based on more fundamental results. I have attempted with some energy to develop the most important equations meant to describe physics considered in the book from as far back in first principles as possible. This often means that the result has a heritage which can be traced at some point in the book back to basic equations like Schrödinger's equation, Maxwell's equations, a system's Hamiltonian or Lagrangian, the plasma fluid or kinetic equations, Lorentz transforms or something of similar foundational importance. While I must apologize to the reader that this pedagogical approach often leads to lengthy derivations in the text, I do hope that they will find it helpful to see an important result traced back to fundamental undergraduate or graduate physics. I personally find that such detailed derivations aid me in understanding which physics is and which is not included in an equation and it lays bare the approximations which go into that equation. I admit I have had to punt on this rigorous goal in some cases because of the length of the text. When I have found myself in this quandary I have tried to perform phenomenological or intuitive derivations of the formulae, or at least derive approximate forms of the more accurate formulae, which are then presented.

It goes almost without saying, that a research area such as strong field physics advances at a remarkable pace, driven, in this case, by the rapid advances in laser technology which have occurred since the early 1980s. As with all vibrant science fields, new experiments on the many facets of strong field physics are published on a daily basis. Consequently, I have striven, whenever possible, to avoid a discussion of recent developments which are still undergoing ongoing research for validation, not well explained or somewhat controversial. While often topics such as these represent a particularly exciting aspect of modern strong field physics research, to do justice to such topics would demand a thorough discussion of recent experiments along with a comparison of the theories and simulations published to explain the results. Doing this would cause such a textbook to degenerate quickly into a review which would be out of date almost immediately. Instead, in the interest of creating that broad foundational tool I just mentioned, I have steered clear of these topics and chosen to present and explain the aspects of the field which are well-researched and widely accepted as canon.

It is also tempting, particularly as an experimentalist, to reproduce many published data to elucidate the various topics considered in this book. I have resisted this temptation. In my judgment, reproducing published data from the literature mandates at least a basic description of the specific experimental conditions in which such data were acquired, since often the detailed features of data in strong field experiments depend critically on the parameters and particulars of an experiment. Recounting these experimental specifics in the detail that they usually demand would, no doubt, distract the reader from the more fundamental physics underlying those data. So instead of reproducing published data, I have chosen to present generic cartoons of data. This eliminates the need for listing distracting experimental details and allows me to highlight the salient feature of the effect. Hopefully, the reader will forgive my active avoidance of presenting actual experimental results and finds the more schematic illustrations educational. If the reader desires more detailed information on how experimental data have been used to prove the effects we discuss, they will hopefully explore these results through the extensive list of references given, which are by no means comprehensive but are hopefully complete enough to launch an interested student of the field into a fruitful literature search on results in that area.

I have attempted not to presuppose too much advanced expertise on the part of the reader to make it accessible to graduate students beginning their careers in this field, though knowledge of quantum mechanics and electromagnetism at the upper undergraduate or first-year graduate school level is a must. Some previous knowledge or course work in optics and laser physics would probably be of aid, and exposure to introductory plasma physics is probably necessary, particularly in the later chapters. To assist in deficiencies of background in the latter areas I have attempted to give brief introductions to relevant plasma physics and Gaussian optics in Appendices A and B (available at www.cambridge.org/9780521760836).

The book is organized into a modest number of chapters, each of which is something of a "mini-textbook" on high-intensity interactions with a particular class of target. After a historical overview and a chapter on the high-intensity laser technology used to study strong field phenomena, there follow seven chapters detailing the physics of increasingly larger (and more complex) systems in a high-intensity light field. At first, I consider the simplest possible target, a single electron, and explore at some length the dynamics of this point particle in fields at which the electron motion is relativistic (and, hence, very nonlinear). This is followed by two chapters on microscopic, multielectron systems, namely atoms and then molecules, in which a quantum description of strong field interactions must be developed. The next chapter, Chapter 6, builds on the models developed in Chapters 4 and 5 to elucidate the physics of high-order harmonic generation. The second half of the book is composed of three chapters which tackle more complex systems, namely mesoscopic scale atomic clusters, followed by chapters on interactions with underdense (gaseous) plasmas and overdense (solid target) plasmas. These final three chapters rely heavily on fluid models of plasma physics, supported by kinetic theory results to handle the effect of the many collisions that particles in plasmas experience. While my intention to present the material in terms of these mini-textbook chapters is that each could stand alone as a reference aid, later chapters will occasionally draw on results from earlier chapters (e.g., quantum ionization models from Chapter 4 are used liberally in Chapters 5 and 6, and the plasma chapters draw on the single electron dynamic results of Chapter 3). Consequently, the book does present a monolithic body of material.

Finally, I must acknowledge the shoulders of the giants upon which I stand. I have been gifted in my career to work with, and learn from, some of the pioneers of this field. First, I acknowledge the mentorship and guidance of the late Howard Powell, who was a pioneer in the development of the high-energy laser technology which has made this field possible. Next, I must thank the brilliant experimentalists and theorists I have had the opportunity to work with and be educated by. Donna Strickland (2018 Nobel Laureate in Physics) taught me in the lab as a beginning graduate student at Lawrence Livermore National Laboratory (LLNL) how to align a laser. I treasure her mentorship at such an early point in my career and value her friendship. Two years later another famous strong field physicist came to work in our lab at LLNL (directed by my advisor Mike Perry). Anne L'Huillier, 2023 Nobel Laureate in Physics, worked with me and my coworker Kim Budil (currently director of LLNL) on high harmonics with the LiSAF laser I had built. I am proud to have been able to contribute even just a small part to Anne's pioneering and monumental career of work. Over the years I have had the pleasure of working with giants of this field like Phil Bucksbaum, Roger Falcone, Lou DiMauro, Rick Freeman, Toshi Tajima, Gerard Mourou, Mike Downer, Henry Hutchinson, Scott Wilks, Farhat Beg and many others. I thank them all for what they have taught me and for being good friends and colleagues.

1
Introduction to Strong Field Physics

1.1 Definition of Strong Field Physics

Strong field physics (or "high field physics" in some of the literature) refers to phenomena that occur during the interaction of intense electromagnetic waves with matter of various forms. While it is possible to create "strong" static electric or magnetic fields in the laboratory, by far the highest field strengths that can be produced in the lab occur in the electromagnetic fields of a focused high-power laser pulse. For example, while the strongest DC electric fields produced in the lab rarely exceed one MV per cm (limited by the ability of materials to withstand plasma breakdown at high fields), a petawatt-class high-intensity laser can produce oscillatory electric fields with values of over one TV per cm, nearly six orders of magnitude greater than any laboratory-produced DC field. The highest DC magnetic fields produced by laboratory magnets are around one megagauss, whereas a focused petawatt-class laser can produce an oscillatory magnetic field with peak values of many gigagauss.

The interaction of such high-intensity focused electromagnetic radiation with matter can lead to exotic physics. While strong field interactions have been accessed with microwave radiation (Gallagher 1992), traditionally, strong field physics has been studied with intense optical and near-infrared (IR) pulses generated by high-intensity lasers. These interactions occur in a regime in which the electric field of the optical wave dominates the motion and dynamics of electrons subject to these fields. They are characterized by interactions that are often highly nonlinear. At the highest intensities that are accessible, the motion of electrons can become relativistic during each optical cycle, and the magnetic field of the light pulse becomes important in affecting the motion of electrons in the field.

There is no generally accepted definition for when an electromagnetic field is high enough amplitude to enter the "strong field" regime, and, to a certain extent, the threshold for strong field physics will depend on the particular situation. However, to guide the reader it is interesting to ask at what focused intensity might we expect to encounter strong field phenomena. There are two ways to look at this question.

From the standpoint of the interaction of an intense laser field with a free atom, it is fair to say that the strong field regime is entered when the light field is intense enough that perturbation theory breaks down in the quantum mechanical description of the interaction with the bound electrons. This breakdown occurs when the light intensity is high enough that the peak electric field of the wave $E_0 = \sqrt{8\pi I/c}$ (where I is the light intensity and c is the speed of light) approaches the atomic unit of electric field $E_a = e/a_B^2 = 5.1 \times 10^9$ V/cm, the field felt by an electron in a hydrogen atom (where a_B is the Bohr radius and e is the charge of the electron). Light acquires this electric field at an intensity of 3.5×10^{16} W/cm^2. Because perturbation theory relies on the convergence of a perturbation series, in practice, nonperturbative effects become manifest at fields that are a few percent of this value. Consequently, strong field effects in focused laser interactions with atoms become evident at intensities above about 10^{14} W/cm^2.

Alternatively, from the standpoint of laser light interacting with the free electrons in a plasma, we might argue that the strong field regime will be entered when the laser can drive oscillations of the free electrons with an energy that is comparable to or exceeds the thermal energy of the electrons in that plasma. When this occurs, the laser field dominates the bulk motion of the plasma electrons. Since plasmas begin to form at electron temperatures comparable to atomic ionization potentials, many of the plasmas encountered in the lab have temperatures of ~10 eV up to a few keV. Since the quiver energy of an electron in a near-infrared field acquires a value of a few eV at intensities just under 10^{14} W/cm^2, we can fairly say that the strong field regime of laser plasma interactions is entered at focused intensity above 10^{14} W/cm^2.

With these two alternative views of laser–matter interaction, we will take as a starting point for this book that strong field physics is accessed at laser intensities above 10^{14} W/cm^2 (or in a few cases just below this). These days, such intensities are considered quite modest and can be produced with rather compact, tabletop lasers. The upper end of our realm of study is limited only by the experimental ability to create and focus very high-peak-power lasers. At the time of this writing, the highest intensities produced have been in the vicinity of a few times 10^{22} W/cm^2 and there are lasers which will soon produce intensity one order of magnitude higher.

Describing strong field phenomena over an intensity window spanning nine orders of magnitude is daunting. However, many of the theoretical techniques for describing these interactions are applicable over a wide range of intensity. Only when the motion of electrons becomes relativistic in the laser field (an effect which occurs at intensity around 10^{18} W/cm^2 in most near-IR fields) do the theoretical descriptions require amendment. Such intensities are also characterized by high magnetic fields and optical forces. For example, in a pulse with intensity of 10^{18} W/cm^2, an intensity fairly easily accessed by modern tabletop ultrafast lasers, the peak electric field is 3×10^{10} V/cm and the optical magnetic field is 100 MG. The light pressure, I/c, is ~0.3 Gbar. At the time of the writing of this book, the highest-peak-power lasers can reach 100 TV/cm fields and >10 Tbar light pressures.

1.2 Historical Overview

Theoretical considerations of how intense light interacts with matter are not particularly new and predate the invention of the laser. High-intensity light excitation of a multi-photon process was considered as early as 1931 when Maria Goeppert-Mayer discussed two-photon absorption in her PhD thesis (Goeppert-Mayer 1931). It was not until after the demonstration of the laser, however, that a true multiphoton process was observed in the laboratory when, in 1961, Peter Franken working at the University of Michigan observed second harmonic conversion of a ruby laser pulse in a quartz crystal (Franken *et al.* 1961). It was the rapid development in laser technology in the 1960s and early 1970s that led to a rapid increase in peak power enabling true strong field physics studies. Q-switching (McClung and Hellwarth 1962) and mode-locking (Demaria *et al.* 1966) were technological advancements that permitted the construction of lasers with peak power over 1 GW and focusable intensity entering the strong field regime.

How a strong field interacts with a free electron in the relativistic regime was a topic studied by a number of authors in the early 1960s (often in an astrophysical context), and a reasonable date for the dawn of the field of strong field physics can be marked by the 1964 theoretical publication by Brown and Kibble on the relativistic dynamics of a free electron in a laser field (Brown and Kibble 1964). However, the theoretical study of strong field physics really began in earnest with a classic paper by Russian physicist L. V. Keldysh in the following year (Keldysh 1965). In this paper, the rate of ionization of an atom or ion in a strong laser field was first derived with a nonperturbative quantum theory. At the time, this work was largely ignored in the West. Keldysh's theoretical work amazingly predicted phenomena such as tunnel ionization and high-order above-threshold ionization, effects which became the focus of strong field physics experiments a decade later and have been at the center of strong field research for many years.

The first real experimental observation of a nonperturbative strong field physics effect occurred in the groundbreaking experiment of Agostini *et al.* in 1979 at the Saclay lab in France (Agostini *et al.* 1979). This work was one of the pioneering discoveries that led to the 2023 Nobel Prize in Physics. Their experiment observed, for the first time, truly nonperturbative multiphoton effects in laser–atom interactions by examining photoelectron production from intense six-photon ionization of Xe atoms at intensity up to 4×10^{13} W/cm^2. They found that electrons were ejected during ionization with energy higher than that expected from absorption of the minimum number of photons needed for ionization. At the highest intensities studied in that experiment, the ejected electrons were emitted in a number of energy peaks separated in energy by one-photon quanta with almost equal electron yield over the first four or five peaks, an effect that was coined "above threshold ionization" (ATI) shortly after its observation (Fabre *et al.* 1982). Despite Keldysh's prediction of this very effect 15 years earlier, this experiment was greeted with surprise by the atomic physics community at the time, as it was expected that these higher-order peaks would be emitted with exponentially decreasing amplitude as predicted by lowest-order perturbation theory. This observation of nonperturbative effects sparked a long campaign of experiments and theoretical work on strong laser field ionization of atoms and ions that continues to this day.

The early theoretical work of Keldysh was subsequently elaborated on by two authors in the 1970s and 1980s, F. Faisal and H. Reiss (see in Faisal's book of 1987; and Reiss 1980), leading to so-called Keldysh–Faisal–Reiss or KFR theories, which represent the primary basis for analytic strong field theory in atoms to the present. Experimental work in strong field physics exploded at about the same time, propelled at first by the development of mode-locking, permitting production of picosecond and, shortly thereafter, femtosecond laser pulses which could then be amplified with broadband dye laser amplifiers (Shank and Ippen 1974). However, the field really skyrocketed with the revolutionary development of chirped pulse amplification (CPA) in 1985 by Gerard Mourou and Donna Strickland and at the University of Rochester, a discovery which led to the award of the 2018 Nobel Prize in Physics to Mourou and Strickland (Strickland and Mourou 1985). This laser technology advance permitted construction of tabletop-scale lasers with powers well in excess of 1 terawatt, and such systems quickly proliferated around the world, particularly in the US, France and Britain. It was soon realized that this technology might lead to lasers with power above 1 petawatt (Maine et al. 1987).

Work on atomic ionization made possible by the development of CPA and inspired by a number of Russian theoretical works that followed Keldysh's initial paper (Perelomov et al. 1966) showed that strong field ionization could often be described by tunneling theories (August et al. 1991). This intellectual leap has inspired much of the modern understanding of strong field ionization and a number of other strong field phenomena in atoms, ions and molecules.

Those early experiments in strong field multiphoton ionization were followed by the first observation of nonperturbative nonlinear optical phenomena in high-order harmonic generation at the University of Illinois, Chicago in 1987 (McPherson et al. 1987). This group found that interactions of an intense laser pulse with a gas of atoms led to emission of a range of high harmonics of the laser frequency. The initial observation of high harmonics was striking in that a span of harmonics were emitted which extended to high orders with almost constant intensity, completely at odds with lowest order perturbation theory. This early observation of high-harmonic generation in Chicago was followed up by experiments from the Saclay group in France which observed a number of curious trends in the character of the harmonic spectra and confirmed the formation of a "plateau" in the harmonic spectrum over a large number of harmonic orders (L'Huillier and Balcou 1993). This line of work by Anne L'Huillier led to her 2023 Nobel Prize in Physics.

In fact, not long after the first high harmonic experiments in the mid 1990s very high nonlinear orders, >100, were reported (Chang et al. 1997), resulting in the production of coherent light well into the soft X-ray region. These experiments demonstrated, in a dramatic manner, that at the intensities now available, quantum mechanical multiphoton processes with hundreds or even thousands of photons were possible. High-harmonic generation (or just "HHG") continues to be studied actively. One of the most mysterious aspects of these high harmonic studies was explained nearly simultaneously in 1993 by Ken Kulander at Lawrence Livermore National Laboratory (Kulander et al. 1993) and Paul Corkum at The National Research Council Canada (Corkum 1993). Both surmised that the extent of the so-called harmonic plateau could be explained by a relatively simple quasi-classical model of the laser-driven electrons in an ionizing atom. This "simple-man's" semiclassical treatment is now the basis for much of our understanding of strong field ionization, above-threshold ionization, and high-harmonic generation.

For a time, a push to produce the shortest wavelengths possible by high-harmonic generation drove the research field. However, research in high-harmonic generation was reenergized by a number of remarkable proposals in the mid and late 1990s that suggested that the broad, coherent spectrum of the high harmonic comb could lead to generation of pulses with duration under 1 fs, in the attosecond regime. The first attosecond pulse was demonstrated in the lab (with pulse duration of 650 as) by Ferenc Krausz et al. in Vienna (Hentschel et al. 2001). This work led to Krausz's share of the 2023 Nobel Prize in Physics. This demonstration has essentially spawned an entire subfield of ultrafast research devoted to generating and using ever-shortening XUV pulses produced by controlled HHG (Krausz and Corkum 2002). Pulses under 100 as are now routinely generated and characterized in labs around the world.

Another important development in strong field atomic physics arose in 1992 with the experimental observation at Lawrence Livermore National Laboratory of nonsequential double ionization of He atoms in femtosecond 800 nm pulses at intensity around 10^{15} W/cm^2 (Fittinghoff et al. 1992). (In fact, the experimental signature of this effect was apparent in ionization experiments performed in France as early as the mid 1980s (L'Huillier et al. 1982).) This observation challenged the long-held assumption that intense lasers interacted almost solely with one electron (the most loosely bound electron) at a time, resulting in sequential, uncorrelated stripping of the electrons from the atom as the light intensity increased. Details of this effect were illuminated in a classic, careful experiment performed by Walker and DiMauro shortly after the report of Fittinghoff (Walker et al. 1994). Though innumerable theoretical and computational studies have been performed to understand this multielectron physics, it turns out that the simple quasi-classical model of Kulander and Corkum describes this effect for the most part. The recollision of a tunnel-ionized electron driven back into the parent ion by the oscillating laser field can liberate a second electron. This phenomenon is now well understood, and there are a host of experimental results in the literature describing the various nuances of this effect.

The vast majority of effort in strong field studies in the atomic regime has focused on ionization and high-harmonic generation from single atoms. However, a number of fascinating effects of strong field interactions with small molecules have been investigated since the 1980s. Early studies examined the fate of molecular bonds in modest laser intensity, exploring the so-called bond softening that results from the interaction of the laser field with the bonding electrons. A particularly puzzling observation made in the late 1980s in studies of the explosion of diatomic molecules multiply ionized by an intense laser was the topic of much discussion in the strong field community for a few years in the 1990s. Experiments on the Coulomb explosion of diatomic molecules were initially inexplicable: the energies of the ions ejected from the Coulomb explosion almost always seemed to occur at an ion separation somewhat larger than the equilibrium separation of the molecular bond, an effect which appeared to be largely independent of laser pulse duration (Schmidt et al. 1994). This mystery was solved later in the 1990s when it was realized that molecules preferentially ionize when the nuclear separation increases to a so-called "critical separation" (Chelkowski and Bandrauk 1995; Posthumus et al. 1995).

To a certain extent, strong field ionization of molecules continues to confound investigators to some degree. Molecules almost always tend to show ionization rates that are substantially lower than atoms with similar ionization potentials (Talebpour *et al.* 1996). This effect seems to be partially understood in the context of modern tunneling and strong field approximation theories that account for the extended, nonspherical nature of the wavefunction of bonding and antibonding molecular orbitals. However, to a large extent, the specifics of this effect are not yet understood and anomalous behavior in molecular ionization continues to be observed. It would seem that multielectron physics is much more important in molecular strong field interactions than it is in atoms, and this has challenged modern computational simulations of the interactions.

For a time starting in 1993 with inexplicable results out of the Rhodes group at the University of Illinois Chicago (McPherson *et al.* 1993), there was a period of interest in the study of strong field interactions of laser pulses with clusters of atoms. The first anomalous experimental signature was copious X-rays emitted when gas jets that carried these large clusters were irradiated at intensity $>10^{16}$ W/cm^2. It was soon realized by Ditmire and coworkers at Lawrence Livermore National Laboratory (hereafter denoted as LLNL) that, in fact, these small clusters actually formed small "nanoplasmas" under intense irradiation that could efficiently absorb the incoming laser light and lead to bright X-ray emission and fast ejection of highly charged ions (Ditmire *et al.* 1995). For the most part, intense near-IR laser pulse interactions with clusters are now largely understood in the context of this nanoplasma model, and, at this time, most research on clusters has shifted to study of these clusters in intense XUV and X-ray pulses. A number of exotic phenomena have been observed in these strong field cluster interactions including the production of D_2 nuclear fusion neutrons in a gas of deuterium clusters (Ditmire *et al.* 1999).

While the study of strong field physics had its origins in the study of atomic ionization, it was also realized early on that strong field interactions with plasmas would manifest unique effects, not only through the ionization of atoms and ions in the plasma but in the collective motion of the plasma electrons driven by the strong forces of an intense laser pulse. Strong field studies in plasmas paralleled the atomic and molecular physics studies of the 1980s, 1990s and 2000s.

One of the first pioneering proposals for exploiting strong field interactions in underdense plasmas came with the classic paper of Tajima and Dawson in 1979 in which they proposed accelerating electrons in the wake of a plasma wave set up by the passage of an intense laser pulse through the plasma (Tajima and Dawson 1979). Using the ponderomotive forces of the intense light field, they surmised that very high accelerating gradients could be created in the traveling wave behind a laser pulse which could accelerate electrons to very high (GeV) energies over distances of only a few centimeters. This proposal sparked a vigorous research effort into this so-called "wakefield acceleration." At the time of this writing nonlinear plasma waves produced by femtosecond pulses at intensity $>10^{19}$ W/cm^2 propagating through ionized gases have accelerated electrons to energies of ~ 10 GeV over lengths of only a couple of centimeters (Gonsalves *et al.* 2019). The acceleration in highly nonlinear waves, now referred to as bubble acceleration, has been exploited in experiments around the world.

Another flurry of activity in this field was initiated around 1990 when it was proposed by Burnett and Corkum that strong field ionization of atoms in a gaseous target might be able to set up the conditions for gain in the soft X-ray region (Burnett and Corkum 1989). Though a robust, high-gain, field-ionized recombination X-ray laser was never really realized experimentally, this line of research has led to demonstration of a number of compact femtosecond laser-driven XUV and soft X-ray lasers based on other schemes. About the time that the community was exploring intense ultrafast production of plasmas in gas targets, a remarkable and completely unexpected observation was made at the University of Michigan when intense femtosecond pulses were weakly focused in air (Braun et al. 1995). They observed an extremely long plasma filament produced by the laser, extending for many meters, down the hall from the laser lab. This effect was, at first, unexplained but is now understood to result from an interplay between self-focusing of the intense pulses in the air and refraction of the production of the plasma, yielding a "moving" focus for different slices of the pulse. Many subtle aspects of this visually arresting effect have been elucidated in experiments over the past 25 years.

Strong field interactions in underdense plasma have led to the observation of other nonlinear phenomena, particularly when laser intensities have entered the relativistic regime. These phenomena include relativistic self-focusing, self-phase modulation and nonlinear forward-directed Raman scattering in gaseous plasma targets. Furthermore, exotic effects such as betatron X-ray emission from oscillating electrons in nonlinear plasma wakes have been observed in these experiments.

Finally, studies of high-intensity laser interactions with solid targets go back to the earliest days of the laser. After "giant" laser pulses were produced by Q-switching in the early 1960s, studies of the explosion of plasmas created by irradiation of a solid target were published. Experiments in the strong field regime began in earnest with the availability of joule-class, subpicosecond lasers from the invention of CPA in the mid 1980s. Most early studies were essentially phenomenological experimental studies of the X-rays produced by these solid target plasmas. However, studies of high-intensity laser interactions with solid targets saw an enormous increase in activity in the early 1990s with the proposal of the so-called fast ignition concept. This idea came soon after a remarkable initial study by a group at Stanford (Kmetec et al. 1992), followed by experiments at Berkeley and LLNL in the early 1990s that showed that very efficient generation of MeV "hot" electrons and MeV photons accompanied irradiation of solids at intensity approaching 10^{18} W/cm^2. Study of the collisionless absorption mechanisms that lead to these multi-MeV hot electrons has been the topic of many high-intensity laser-plasma studies for 20 years since the observation of efficient hot electron production in the 1990s.

The fast ignition concept inspired by this research was forwarded by Max Tabak and coworkers at LLNL (Tabak et al. 1994). Tabak's idea suggested that an intense picosecond laser could produce a high-energy (many joules) burst of multi-MeV electrons which could be injected into the compressed fuel of an inertial confinement fusion (ICF) implosion. These hot electrons could serve to heat the compressed fuel and ignite it, triggering a fusion ignition burn.

The fast ignition proposal energized the high-intensity laser-plasma community and essentially led to 20 years of extensive research on the production and propagation of hot electrons in solid targets irradiated at relativistic intensity. At the time of the writing of this book, it is generally believed that the hot electron generation efficiency and behavior of the high-peak-current hot electron bunches in the compressed plasma of an ICF experiment are not favorable for achieving ignition with any reasonable amount of short pulse laser energy (say <200 kJ). Nonetheless, the worldwide research on high-intensity laser production of hot electrons has led to a comprehensive understanding of how intense laser pulses couple to electrons in an overdense plasma and how such electrons propagate. This has resulted in many studies of exotic phenomena such as kilotesla magnetic field generation and positron production in the laser plasma (Cowan et al. 1999).

Fast ignition and hot electron research also propelled the development of high-peak-power laser technology as it was clear early on that fast ignition would demand picosecond pulsed lasers with peak power over 1 PW. The late 1990s saw the demonstration of the first petawatt laser at LLNL on the NOVA ICF laser, a project motivated in large part by the fast ignition idea (Perry et al. 1999). This first demonstration of a petawatt of peak power has led to a proliferation of PW peak power lasers around the world, and an explosion of strong field research at relativistic intensities. After the large aperture grating development at LLNL which enabled that first PW laser on NOVA, a handful of large Nd:glass-based PW lasers emerged at national laboratories such as the Rutherford Appleton Lab in the UK and the ILE in Osaka, Japan. The past 25 years have seen numerous additional PW lasers at scales from 40 J to 500 J see completion; currently dozens of laser labs around the world operate lasers with power $\sim$1 PW. In fact, a number of lasers at peak power near or in excess of 10 PW have now been constructed.

A particularly active research area which spun out of the research on fast ignition is in proton and other ion acceleration. The PW laser on NOVA at LLNL made a surprising observation of multiple tens of MeV proton ejection from thin metal targets at intensity above 10^{20} W/cm^2 in some of that laser's early experiments (Snavely et al. 2000). Fast ion ejection from pulsed laser irradiation of solid target plasmas was, in fact, a well-known phenomenon, with observations dating back to the earliest laser plasma experiments in the early 1960s. However, the LLNL PW results were remarkable in the high energy of the observed protons (>10 MeV) and the efficiency with which these protons were produced (with $\sim$10 percent of the total laser energy emerging in the pulse of fast protons from the back of the target). After a short period of controversy about the mechanism for this surprising hot proton ejection, it is now well established that these protons arise from the hot electrons produced in a solid target interaction and the sheath fields these electrons produce as they attempt to exit out the back side of a thin foil. This so-called target normal sheath acceleration (TNSA) is now well understood and well characterized.

The broad proton spectra that are produced by TNSA has prompted, since 2000, a vigorous research effort into alternate ion acceleration mechanisms which might yield higher proton energies than TNSA and which could produce nearly monoenergetic bursts of MeV protons or heavier ions. This research has been much of the impetus to understand solid density plasma interactions at highly relativistic intensity where radiation pressure becomes important. Currently research into ion acceleration by PW-class lasers is among the most active areas of strong field plasma physics research worldwide. The fast electrons produced in PW-class laser interactions with solids led to the triggering of various nuclear reactions in the targets (Cowan et al. 2000). Such laser-induced nuclear reactions have been one of the many byproducts of these interactions which have presented the promise of laser-driven nuclear applications (like deployment of compact neutron sources or radioactive transmutation). Applications like this drive much of modern research in the field. Laser-driven ion acceleration has led to something of a new renaissance in strong field laser-plasma studies and a rebirth in interest in fast ignition. It is now thought that the production of intense proton bursts could be an alternate, viable way to fast ignition (Roth et al. 2001). At this writing this research avenue is again gaining momentum because of the recent demonstration of fusion ignition at the LLNL NIF ICF facility and the promise of laser-driven fusion energy (Abu-Shawareb et al. 2002).

The past two decades the past two decades have seen strong field research advance in many ways. Particularly exciting progress continues in the manipulation of attosecond extreme ultraviolet pulses via high-harmonic generation. Relativistic intensities are now regularly generated by many labs and laser–plasma interactions in this regime are being better understood all the time. Dramatic progress has been made in laser-driven plasma acceleration of electrons since 2010. In fact, relativistic effects, where the change in the mass of the electron which comes from its acceleration by the laser field to velocity approaching the speed of light within one optical cycle are now regularly seen in laser plasma experiments at intensity above 10^{20} W/cm^2. This electron mass increase leads to exotic plasma physics such as relativistic plasma transparency where a normally optically opaque dense plasma becomes transparent to the laser light because of the electron mass shift, or relativistic self-focusing where the optical properties of the plasma are changed by this mass change, focusing the laser light. In fact, QED effects might start playing a role in laser plasma interactions at the extreme intensities now attainable. At the time of writing this book, the experimentally obtainable intensity frontier is approaching 10^{23} W/cm^2. Intensities exceeding 10^{24} W/cm^2 should be reached within the next decade.

1.3 Outline of This Book

This book is intended to introduce many of the fundamental concepts underlying modern strong field physics research. These concepts span descriptions of intense light interactions with single electrons, individual atoms, ensembles of atoms in molecules and clusters, and many charged particles in plasmas. This book does not represent a comprehensive review of modern strong field physics research and is not a survey of recent results in the field. No attempt is made to discuss specific experimental results that confirm the phenomena presented (though citations to such work are often given). Instead, the basic phenomena underlying the more complex effects observed in strong field physics will be discussed, and the basic equations needed to describe these high field effects will be derived. If a more detailed review of the various aspects of strong field physics is desired, there have been a number of excellent review articles published in recent years and a number of focused textbooks. A listing of some of these complementary books can be found in the bibliography.

The book begins with a comprehensive review of the technology employed to access strong field physics. Then, the text, in a series of chapters, marches through discussions of high-intensity laser pulse interactions with systems of increasing complexity. This begins with an examination of strong field interactions with free electrons, in which the considerations are essentially classical but demand relativistic dynamics descriptions. This is followed by discussions on interactions with atoms and then small molecules, topics which mandate a dive into quantum descriptions of the interactions, though I attempt to utilize quasi-classical models when I can. Building on these quantum strong field models, the book then turns to a key aspect of strong field nonlinear optics, generation of high-order harmonics of the laser field in gases of atoms and molecules. After this, the book essentially moves to many-body systems in which understanding relies heavily on concepts in plasma physics. This set of chapters begins with a discussion of high-intensity interactions with microscopic clusters of atoms. The chapters then include a consideration of strong field interactions with macroscopic scale plasmas, first with underdense plasmas which are low enough density to allow light propagation within them and then concluding with a chapter on interactions with solid density plasmas which are overdense and reflect the light wave. This ordering of chapters is designed so that models of the more complex systems can be built from the physics explored in earlier chapters on smaller systems. I attempt to write each chapter such that it can essentially stand alone as a comprehensive description of that aspect of strong field phenomena; however, the utilization of models from earlier chapters as building blocks for models describing the more complex systems does mean that the reader will benefit from a sequential study of the material in the various chapters.

1.4 Units, Variables and Mathematical Notation

For the entirety of the book, I use CGS units, unless otherwise stated. When mixed units are employed, those units will be listed after the variable in brackets (e.g. intensity, I [W/cm^2]).

All scalar variables will be denoted by symbols in italic text. Vectors (traditional 3-vectors) will be given as bold-faced text symbols (e.g. momentum, **p**). When a vector is considered, if the same symbol is employed in nonboldface, italicized text, this will be implied to mean the magnitude of that vector (e.g. $|\mathbf{p}_0| = p_0$). When 4-vectors are denoted, they will be given by italicized nonboldface text symbols with Greek symbol subscript/superscript counters (e.g. the four-momentum is p_μ with contravariant counterpart p^μ).

I will attempt to hold to as many of the widely used conventions in naming variables as possible. Because of the frequent use of a particular symbol in different senses in the literature (e.g. γ can denote the relativistic factor in special relativity, the Keldysh parameter, or the ratio of specific heats in thermodynamics), we will make liberal use of subscripts in naming many variables. To aid the reader with the bewildering array of variable symbols, an extensive (though incomplete) list of variable symbols is provided in the end-of-book Appendix. Because of the extensive employment of variables describing the laser field

1.4 Units, Variables and Mathematical Notation

throughout all chapters in the book, to minimize clutter, the symbols used to describe the frequency, wavelength and wavenumber of the incident laser will always be written without a subscript (i.e. the laser's wavelength, wavenumber, frequency and angular frequency will always be written λ, k, ν, ω). The mass of the electron will always be denoted by an unsubscripted m; the masses of other particles will be written with a descriptive subscript (e.g. mass of the proton: m_p). Generally speaking, when any frequency is denoted with an $\omega_{whatever}$ it is meant to denote an "angular" frequency which is $2\pi \times$ a cycle frequency, which will always be denoted as $\nu_{whatever}$.

Inevitably I will make approximations in our discussions. I will employ the $=$ symbol to mean exactly equal. When $\cong$ is employed it will mean that the two sides are approximately equal and the approximation made is meant to be rather accurate, which means the reader should interpret the sides of the equation as nearly quantitatively equivalent (e.g. $a_0^2 + 1 \cong a_0^2$ for large a_0). When the symbol $\approx$ is used, the implication is that the two sides of the equation are only roughly equal, perhaps to no better than an order of magnitude. Finally, when $\sim$ is used to relate two mathematical expressions, it means that the left-hand quantity simply scales as the right-hand side, with prefactors dropped for illustrative purposes (e.g. $energy \sim v^2$). Hence the two sides will not necessarily be numerically equal to the same order of magnitude or even have the same units.

2
Strong Field Generation by High-Intensity Lasers

2.1 Types of Lasers Used in Strong Field Physics and Chirped Pulse Amplification

To access focused intensities of over 10^{13} W/cm^2, the parameter range of interest in this book, laser pulses with peak power of over $\sim$1 GW are typically required. With modern short pulse laser technology this power level can be accessed with a rather modest laser. In fact, the state of the art in high-power lasers at the time of this writing is lasers with powers up to 10 PW (10^{16} W). Since much of the strong field phenomena we will examine will be bounded by the parameters that can be accessed experimentally (wavelength, pulse duration, focal spot size, etc.), it is of some benefit to take a closer look at the technology and underlying laser physics used in strong field physics experiments.

If a typical high-peak-power laser is constructed so that the phase front of the beam is well controlled, it can typically be focused to a spot size of only a few wavelengths. For a near-infrared (near-IR) laser this means that focal spot sizes of <3 μm are routinely achievable. A 1 GW laser focused to 10 μm yields a peak intensity of $\sim 10^{15}$ W/cm^2, while a 1 PW laser focused to the same spot size yields focused intensity of $\geq 10^{21}$ W/cm^2. Consequently, the intensity range that we will consider in our discussions will range from 10^{13} W/cm^2, the intensity at which strong field phenomena really begin to manifest, up to a few times 10^{22} W/cm^2, the extreme irradiance currently accessible by the highest-power lasers (Bahk *et al.* 2004; Tiwari *et al.* 2019).

The wavelengths used in strong field interactions are determined by the laser materials available for construction of these lasers and the ability to frequency multiply those wavelengths by nonlinear optics crystals. Though recent developments in X-ray free-electron lasers have extended strong field physics to the very short-wavelength range, we will for the most part confine our discussions to interactions with wavelengths commonly available from traditional high-peak-power lasers. This essentially means, as we discuss in Section 2.2, that interactions at wavelengths between 248 nm and 1,100 nm will dominate our discussions. Longer wavelengths are occasionally encountered in some experiments ranging out to 10.6 μm.

The architecture of lasers used in strong field physics is varied, with differences leading to unique laser pulse characteristics, including pulse duration, pulse shape, pulse contrast and focusability. With the exception of a few other architectures (which we briefly discuss

2.1 Types of Lasers Used in Strong Field Physics and Chirped Pulse Amplification

at the end of this chapter), by far the most common approach to construction of a high-peak-power laser is the technique of chirped pulse amplification.

2.1.1 Chirped Pulse Amplification Architecture

Chirped pulse amplification, or simply CPA, allows amplification of pulses with duration well under 1 ps, typically down to <20 fs in the shortest pulse systems, to energies above 1 mJ per pulse and, in the largest CPA lasers, up to energies of many hundreds of joules. Therefore, TW peak powers are easily attainable, even in modest "table top" configurations at repetition rates in the kilohertz range. The challenge of amplifying ultrashort pulses to such extreme peak powers lies in the fact that pulses of high power interact with the laser gain medium nonlinearly. This nonlinear interaction, which we will explore in more detail later in this chapter, manifested primarily by the nonlinear variation of the index refraction of a material when very intense light passes through that material, leads to a number of deleterious effects which ultimately clamp the maximum power that can be propagated in a laser amplifier. The most significant limitation is self-focusing of the pulse resulting from the nonlinear increase in refractive index spatially where the pulse is at its highest intensity.

The mitigation of these nonlinear effects by the technique of CPA was first demonstrated at the University of Rochester by Strickland and Mourou in 1985 (Strickland and Mourou 1985). This result was soon followed by the first demonstration of pulses with over 1 TW of power on a "tabletop" scale laser by the same group (Maine et al. 1988). This first TW laser delivered laser pulses of wavelength near 1053 nm (the near-infrared) with pulse duration of about 1 ps with $\sim$1 J of energy. Since then, the CPA technique has been implemented in a wide range of laser gain materials and has scaled to the PW level in a number of laboratories (Perry et al. 1999). Furthermore, the CPA technique has been widely utilized in a class of high-repetition-rate, lasers which can generally deliver sub-100 fs pulses with energy of over 1 mJ at >1 kHz repetition rate.

The essence of CPA is to circumvent the limits of nonlinear effects in the amplifier by amplifying an ultrafast pulse after it has been stretched temporally by a large factor. By stretching a femtosecond pulse by 10,000 or more, the peak power of the amplified pulse is reduced by an equivalent factor and the nonlinear effects which would normally lead to catastrophic failure of the amplifier material are avoided. This stretched pulse is then recompressed to nearly the initial pulse duration after amplification, retrieving a fs pulse with high energy. The CPA technique stretches the pulse by inducing a large amount of group delay across the pulse's spectrum, leading to a nearly linear sweep of frequency in time, a situation in which the pulse is said to be frequency "chirped." The CPA technique, therefore, requires careful manipulation of the spectral phases across the bandwidth of the laser pulse to insure that this phase can be reversed and the pulse is compressed with high fidelity. This approach also demands that the broad bandwidth of the spectrum be retained throughout amplification so that a short pulse can be recreated after compression.

At the most basic level, the architecture of a CPA laser has four distinct elements. The integration of these four elements in a CPA laser is illustrated in Figure 2.1. First, there is

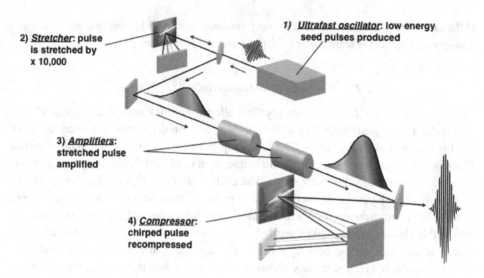

Figure 2.1 Schematic illustrating the four-part architecture of a CPA laser system.

an ultrafast oscillator delivering the initial short pulses at low energy. This oscillator is a mode-locked laser which will deliver from a few picojoules to 1 µJ of energy per pulse, usually at high (>MHz) repetition rate. The next element can be some preamplification or pulse temporal contrast control optics, though in the most basic CPA design the second element is a pulse stretcher which induces enough dispersion on the seed pulse from the oscillator to stretch the pulse to a duration of usually 100 ps to 1 ns. The vast majority of pulse stretchers use a grating combined with a sequence of flat and focusing optics to impart the large spectral dispersion needed. The grating spreads the spectral bandwidth of the ultrafast seed pulse spatially, and this wavelength spread is recombined with an imaging telescope in such a way that the long-wavelength components of the spectrum precede the short-wavelength components (in other words, give a positive chirp to the pulse).

The third element of the CPA laser is the sequence of laser amplifiers which boost the energy of the stretched seed, often only a few picojoules after traversing the stretcher, to the desired final energy. The gain material chosen for this stage of the system must amplify a wide enough bandwidth that short pulses can be recovered after amplification by a factor of typically 10^6–10^{12}. Finally, the amplified pulse is propagated into the fourth element of the system, the pulse compressor. Like the stretcher, the compressor utilizes gratings to recompress the amplified stretched pulse, except that this element must impart negative chirp to undo the positive chirp of the stretcher. The compressor and stretcher must, in principle, be nearly exactly matched in oppositely signed dispersion, though in actuality, frequency-dependent phase imparted to the pulse as it passes through the dispersive optical elements of the amplifier chain must also be compensated to retrieve nearly complete recompression.

2.1 Types of Lasers Used in Strong Field Physics and Chirped Pulse Amplification

2.1.2 Classes of CPA Lasers and Overriding Design Issues

While there are many kinds of CPA systems, there are a number of issues and critical design challenges common to all CPA lasers.

The most important aspect of the CPA laser design is that the laser intensity in the chain be lowered sufficiently that nonlinear effects do not lead to optical damage in the chain. This will lead to a condition on the minimum pulse duration to which the pulse can be stretched. To reach the saturation fluence of the common CPA amplifier materials (like Ti:sapphire and Nd:glass), which is around 1–10 J/cm^2, this requirement almost always demands that the stretched pulse be lengthened to a duration of at least 100 ps.

Next, it is critical that the chirp of the pulse imparted by the stretcher and the amplifier optics be very closely compensated by the compressor. This is particularly challenging for CPA lasers operating with laser pulses of duration below ∼50 fs because of their very large ($\Delta\lambda > 20$ nm) bandwidth. This issue often demands sophistication in the design of the stretcher and implementation of additional passive or active phase elements in the system. It is also critical that gain narrowing, the selective amplification of a portion of the seed pulse's spectrum because of the finite gain bandwidth of the amplifier material, be minimized to retain the broad spectrum of the initial short pulse. Furthermore, it is often quite important that the spatial profile and phase front of the amplified pulse be well maintained so that the amplified pulse can be focused to a near diffraction-limited spot after recompression. This requirement places demands on the amount of nonlinear phase that can be tolerated, the quality of the optics used and other factors such as thermal management and heat removal in high-average-power systems.

Specific kinds of CPA laser each have their own set of additional challenges, and we attempt to address some of them in the sections that follow. While there have been a wide range of CPA implementations and designs, generally speaking, modern CPA lasers tend to fall into one of three categories:

a) *High-repetition-rate, low-energy femtosecond CPA lasers.* This category is the most widespread in the science community worldwide, used extensively in chemistry and atomic physics, and many such systems are commercially available. This class of CPA lasers is almost always based on Ti:sapphire. They typically deliver pulses with about 1 mJ of energy and are recompressed to pulse durations of 100–30 fs (the reasonable bandwidth limit of Ti:sapphire). They operate at high repetition rate, usually at 1 kHz (e.g. Rudd et al. 1993; Fu et al. 1997; Adachi et al. 2008), though higher-average-power systems can be pumped to operate at repetition rates over 10 kHz (Backus et al. 2001). A subclass of this genre of CPA lasers consists of those pumped by a continuous-wave pump source and operate at the inverse of the radiative lifetime of Ti:sapphire, around 200 kHz (Norris 1992) at pulse energies of a few μJ.

b) *Multi-TW tabletop CPA systems.* These systems are the most commonly used in high field physics research and are also usually based on Ti:sapphire, though a significant fraction utilize Nd:glass or optical parametric CPA (OPCPA; see Section 2.2.6). They typically operate at 10 Hz if they employ Ti:sapphire, lower if glass is the amplifier material. Pulse energies typically are 100 mJ to 10 J and the amplified pulse duration is in the 30–100 fs range in sapphire, 300 fs – 1 ps in glass.

c) High-energy CPA lasers. Scaling of CPA to high energy, in the regime of 10 J to 1 kJ has been performed in few labs worldwide, mainly for high-intensity laser–plasma interaction studies. These systems almost always use Nd:glass as the final amplifier because glass amplifiers can be fabricated at large aperture (>10 cm) with high quality; however, they often use Ti:sapphire or OPCPA (Perry *et al.* 1999; Ross *et al.* 1997) as a broadband front-end amplifier. These lasers are single shot, meaning they fire on the order of once every 30 min to 1 hr to allow a cool down of the glass amplifiers. Their pulse durations are typically longer than about 500 fs because of the bandwidth limitation in glass, though hybrid examples with shorter pulses, down to 100 fs, have been demonstrated (Gaul *et al.* 2010).

2.2 Common CPA Laser Gain Materials

Because the design architecture of CPA lasers is heavily dependent on the capabilities of the laser gain media, we next turn to a survey of these media. While many properties are desirable for laser materials in CPA, such as optical quality at large aperture, good thermal heat conductivity or high peak gain cross section, the single most important aspect of a laser material is the width of the gain bandwidth. Broad gain bandwidth is critical to maintain a broad spectrum and hence a short optical pulse upon compression. It is this requirement which drives the selection of the materials discussed in the following sections and excludes otherwise excellent, widely used laser materials such as Nd:YAG in CPA. Table 2.1 summarizes the various important parameters of many laser materials used in CPA.

2.2.1 Ti:Sapphire

By far the most widely used gain material for CPA lasers is Ti-doped sapphire, $Ti^{3+}:Al_2O_3$ (Moulton 1986). Ti:sapphire is one of a class of materials known as vibronic laser materials, which achieve a very broad gain bandwidth through a coupling of the stimulated emission of the ions in the crystal lattice with the phonons of the lattice (Wall and Sanchez 1990). The electronic structure of Ti^{3+} is in fact very simple because there is only one electron left in the 3d shell, which interacts with the surrounding six oxygen atoms in the octahedral structure of the sapphire. This energy level diagram is shown in Figure 2.2a. Three of the five angular momentum states of the 3d electron do not point directly at the neighboring oxygen atoms, yielding a triplet lower level labeled ^3T. The other two angular momentum states do point at the oxygen atoms and hence yield a doublet state with higher energy, the ^2E state. The lowest energy configuration of the Ti ion in this excited ^2E state has the ion displaced slightly with respect to the oxygen atoms. The energy difference between the ground state and the excited state corresponds roughly to green photons (~2 eV), which is where Ti:sapphire predominantly absorbs. The

Table 2.1 Optical parameters for many common CPA laser materials

	Nd:glass Phosphate	Nd:glass Silicate	Ti:sapphire	Cr:LiSAF	Alexandrite	Yb:glass phosphate	YbLYAG	KrF
Peak Wavelength	1,054 nm	1,061 nm	800 nm	850 nm	755 nm	1,010 nm	1,030 nm	240 nm
Peak Cross section	3.7×10^{-20} cm^2	2.7×10^{-20} cm^2	4.1×10^{-20} cm^2	5×10^{-20} cm^2	1.0×10^{-20} cm^2	5×10^{-21} cm^2	2.1×10^{-20} cm^2	2.5×10^{-16} cm^2
Sat. Fluence at Peak	5.1 J/cm^2	7.0 J/cm^2	0.6 J/cm^2	4.7 J/cm^2	26 J/cm^2	40 J/cm^2	9.2 J/cm^2	0.003 J/cm^2
Fluorescent Lifetime	347 µs	330 µs	3.2 µs	67 µs	260 µs	2,000 µs	967 µs	7 µs
Gain Linewidth	20 nm	28 nm	230 nm	180 nm	100 nm	50 nm	12 nm	2 nm
Thermal Conductivity	0.7 W/m-k	1.4 W/m-k	33 W/m-k	3.1 W/m-k	23 W/m-k	0.7 W/m-k	11.2 W/m-k	—
Refractive index	1.50	1.56	1.76	1.41	1.74	1.50	1.83	1
Refractive index dn/dT	-6.8×10^{-6}/°C	-2.9×10^{-6}/°C	-1.3×10^{-6}/°C	-4×10^{-6}/°C	-8.0×10^{-6}/°C	-6×10^{-6}/°C	-7.8×10^{-6}/°C	—

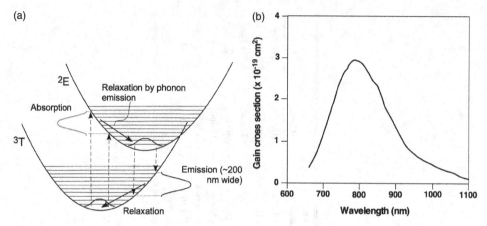

Figure 2.2 (a) Energy level diagram of Ti^{3+} doped in sapphire showing the broad emission that results from vibronic coupling to phonons. (b) Emission gain cross section as a function of wavelength of T:sapphire.

displacement of the lowest energy state of the ^2E level along with the fact that this electron state couples directly to the vibrational modes of the Ti ion lead to a nearly ideal four-level laser system as Figure 2.2a illustrates. Absorption of a green photon by the Ti ion in its ground state populates a range of vibrational states of the displaced ^2E state, yielding a very broad absorption band. Very fast decay of the Ti ion to its lowest vibrational level through coupling to the surrounding lattice leads to rapid population of the upper laser level. This inversion leads to gain with photons of wavelength around 800 nm, but again, because the lasing can occur to a range of vibrational levels in the lower state the emission bandwidth is also very broad. Finally, rapid depopulation of the lower laser level occurs by relaxation of the Ti ion back to its central equilibrium position and transfer of energy to the surrounding lattice.

While a number of laser materials exhibit similar vibronic broadening, Ti:sapphire is by far the most successful such tunable solid-state vibronic laser material, exhibiting gain over a spectral region ranging from 650 nm to greater than 1,000 nm. The wavelength peak of the gain curve of Ti:sapphire, whose gain spectrum is reproduced in Figure 2.2b, is at 800 nm, but lasing action is possible from 700 nm to $\sim$1 μm, a fact which supports amplification of extremely short pulses. (100 nm bandwidth yields pulses of $\sim$10 fs at 800 nm.)

The broad bandwidth of Ti:sapphire permits its use in CPA at a range of wavelengths. In fact, CPA in Ti:sapphire at a wavelength of 1,054 nm (as a front end for a Nd:glass CPA laser) has been demonstrated, though the gain cross section in this spectral region is very low and requires active suppression of spontaneous emission of the system near its gain peak. The vast majority of Ti:sapphire CPA lasers operate at 800 nm, where gain bandwidth of almost 200 nm is accessible. In fact, the duration of compressed pulses in a Ti:sapphire CPA laser is usually limited by the bandwidth of the other optics used (such as polarizers, mirrors and compressor optics). Ti:sapphire CPA lasers can routinely be built that operate with duration under 50 fs and, with the implementation of a variety of spectral shaping technique, amplified compressed pulses below 20 fs have been demonstrated (Barty et al. 1996).

Ti:sapphire also has extremely favorable thermal properties, making it suitable for use in high-repetition-rate systems. The various relevant properties of Ti:sapphire are summarized in Table 2.1, showing that the thermal conductivity is 33 W/m-K at room temperature and even higher when cryogenically cooled. High-average-power CPA systems have been demonstrated that utilize this advantage of Ti:sapphire. The most significant limitation of Ti:sapphire is that its upper state lifetime is only 3.2 μs. This effectively prohibits this use of this material in flashlamp-pumped configurations, because typical flashlamp pulse durations are at least 100 μs or longer in duration. Consequently, Ti:sapphire is usually pumped by another, nanosecond-duration laser. Because the absorption cross section of Ti:sapphire peaks near 500 nm, a natural choice for efficient pumping of Ti:sapphire is to frequency double the 1,064 nm output of a Q-switched Nd:YAG laser to a wavelength of 532 nm (or, at high energy, the frequency doubling of a Nd:glass laser). Virtually all Ti:sapphire-based CPA lasers employ this architecture.

Using this laser pumping, high gain can usually be achieved in a short crystal. The gain cross section of Ti:sapphire is high, 4.1×10^{-19} cm^{-3} for polarization oriented parallel to the c-axis (Ti:sapphire is a uniaxial birefringent crystal), yielding a saturation fluence of 0.6 J/cm^2. Small signal gain of ~3 per pass can be achieved simply in a 1-cm-long Ti:sapphire crystal doped with Ti at 0.1% by weight when pumped at around 1 J/cm^2.

2.2.2 Nd:glass

The first CPA lasers built in the mid to late 1980s used Nd:glass as the amplifier material. Nd:glass is a four-level laser material (Yamanaka et al. 1981) which exhibits gain in the near-infrared (IR) at a wavelength around 1 μm (Caird et al. 1991). Nd can be doped either in silicate glass or phosphate-based glass, the latter almost exclusively used in modern high-energy lasers. Like Nd:YAG, lasing occurs with the 4f electron of Nd^{3+} between the $^4F_{3/2}$ and $^4I_{11/2}$ states. Unlike YAG, however, this line is inhomogeneously broadened by interaction with the crystal field of the surrounding amorphous glass host material, lowering the effective peak cross section but permitting amplification over 20–30 nm of bandwidth and hence supporting short pulse amplification. The Nd^{3+} ion exhibits broad absorption bands around 600, 750, 800 and 860 nm, the latter two amenable to pumping with AlGaAs laser diodes.

Nd:glass remains a commonly used material in high-energy CPA lasers systems, mainly because large-aperture, high-optical-quality rods and disks can be fabricated easily in glass. It is, for example, straightforward to manufacture high-quality Nd:glass rods with diameter up to 70 mm, and Nd:glass disks with clear aperture up to ~50 cm have been demonstrated. Typically, any CPA lasers with energy above 200 J almost always use Nd:glass as the final amplifier material. In addition to the advantages offered by high-quality materials at large aperture, the long upper state lifetime of Nd:glass (typically around 330 μs for most phosphate glasses) and its various absorption bands between 500 and 800 nm permit easy flashlamp pumping. Flashlamp pumping is used for most all large-aperture (>20 cm)

Nd:glass amplifiers in CPA lasers. Flashlamp-pumped Nd:glass has been used to scale CPA to greater than 1 PW, employed in lasers with energy of >200 J in pulse duration of 200–450 fs (Perry et al. 1999).

The gain properties of Nd:glass are well suited to most CPA architectures. The gain cross section of the most commonly used Nd:glasses (such as Schott LG-760 or Kigre Q-88) is around $3.5 - 4.0 \times 10^{-20}$ cm^2, which means that the saturation fluence in Nd:glass ($h\nu/\sigma$, the photon energy divided by the gain cross section) is around 5–6 J/cm^2. This is comparable to the damage fluence of most glasses when irradiated with pulses of around 1 ns duration, the typical stretch duration of most CPA lasers. Consequently, Nd:glass can be operated in CPA lasers at extraction fluences which maximize its good energy storage properties. Furthermore, Nd:glass has a relatively modest nonlinear refractive index (n_2) equal to about 1.0×10^{-13} esu for most phosphate glasses (though higher in silicate), a fact which aids in minimizing B-integral in these laser systems (whose deleterious effects will be described later).

The most commonly used Nd:glass amplifier materials are either silicate based (SiO$_2$) or phosphate based (P$_2$O$_5$). Phosphate glasses have gain centered near 1054 nm while silicate glasses have a gain peak a little further to the red, around 1,061 nm. The common practice in almost all modern Nd:glass CPA lasers is to use phosphate glass because it exhibits a higher cross section than silicate glasses ($\sim 4 \times 10^{-20}$ versus $\sim 2.7 \times 10^{-20}$ in silicate) so it is easier to achieve reasonable gain in flashlamp-pumped applications.

Perhaps the greatest drawback of using Nd:glass in CPA lasers is that glass exhibits a somewhat narrow gain spectrum, a fact which limits the shortest pulses possible. While Nd:glass has a broader gain spectrum than neodymium doped in other common crystalline hosts such as YAG and YLF (whose narrow gain spectrum eliminates these otherwise excellent gain materials for ultrafast CPA applications), it is still rather limited when compared to the other commonly used gain material in CPA, Ti:sapphire. The gain spectra for two typical phosphate and silicate glasses are illustrated in Figure 2.3. As this figure illustrates, the gain bandwidth for phosphate glass is around 20 nm (FWHM) and a little

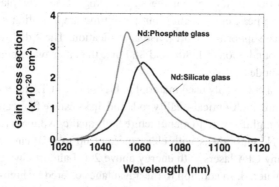

Figure 2.3 Emission gain cross section as a function of wavelength of the two most commonly used Nd:glasses: phosphate and silicate glass. These specific cross sections are for Schott laser glass LG-750 (phosphate) and LG680 (silicate).

broader (~25 nm) in silicate. Because of the gain narrowing of the seed pulse spectrum when this cross section spectral profile is exponentiated during unsaturated amplification, Nd:glass CPA systems, which use the Nd:glass to yield the majority of the gain (typically a gain factor of 10^6–10^{10}), usually yield amplified CPA pulses with less than 4–5 nm of bandwidth. This results in transform-limited pulse durations no shorter than ~500 fs.

The mixing of phosphate and silicate glasses can be used to broaden the effective gain spectrum to a certain degree. This is a technique that has been used to achieve pulse duration below 300 fs in some predominately glass amplifier chains and as low as 130 fs in a hybrid Nd:glass/OPCPA chain (Gaul et al. 2010).

The second major drawback of using Nd:glass in CPA lasers is the rather poor thermal properties of glass, which has an order-of-magnitude lower thermal conductivity than crystals (see Table 2.1). Glass has a thermal conductivity of ~0.5 W/m-K, far less than 14 W/m-K of Nd:YAG (Koechner 2006). Consequently, flashlamp-pumped Nd:glass, a pumping technique which generates significant heat from the inefficiency inherent in broadband optical illumination for populating the upper laser level, has a limited repetition rate. For example, a ~1 cm diameter Nd:glass rod will usually be limited to a repetition rate of ~1 shot/30 seconds while large-aperture Nd:glass disk amplifiers require 30 minutes to 1 hr to cool sufficiently between shots to maintain high beam quality. Active cooling of slabs is possible to permit repetition rates approaching 10 Hz in some cases when pumped by laser diodes.

Because of these two major limitations of Nd:glass, most modern high-energy CPA lasers utilize a different amplification approach in the front end of the laser chain, boosting the pulse energy of the stretched pulse up to a few mJ to a joule before injecting these pulses into the Nd:glass power amplifiers. A commonly implemented architecture for Nd:glass CPA lasers utilize a broadband front end which uses Ti:sapphire operating in the wings of the gain curve out at $\lambda = 1$ μm (Perry et al. 1999) or OPCPA (Gaul et al. 2010). This approach has the advantage of providing a high-bandwidth seed to the Nd:glass, somewhat ameliorating the effects of gain narrowing.

2.2.3 Yb-Based Materials

Another rare earth lasing ion of interest for CPA (as of the writing of this book) is Yb^{3+} doped in materials such as YAG or fluorophosphate glass. Yb is an interesting ion because it has a very simple electron structure: it has only one vacancy in the 4f shell so it has only two bound levels, the $^2F_{7/2}$ ground state and the $^2F_{5/2}$ excited state. Further splitting of this upper level allows Yb to lase at room temperature as a quasi 3-level laser system with absorption lines at 941 nm and 970 nm, and lasing at a wavelength near 1,030 nm to a lower level only 0.08 eV above the ground state. Thermal population of this lower laser level makes the system act like a 3-level laser, though cryogenic cooling of Yb-doped material can remove the thermal population of the lower level and allow 4-level lasing.

Yb^{+3} doped in fluorophosphate glass is interesting because this glass can be efficiently pumped by diodes at a wavelength of 970 nm, making it promising for high-energy (~100 J), high-repetition-rate (>1 Hz) petawatt-class lasers (Keppler et al. 2011).

Unfortunately Yb:glass has a very low gain cross section ($\sim 1 \times 10^{-20}$ cm^2), but it does have excellent energy storage characteristics (with a 2 ms upper state lifetime.) With lasing near 1030 nm, it has a gain bandwidth of $\sim$70 nm, making 100 fs pulse amplification possible.

Yb doped in YAG also exhibits some gain bandwidth ($\sim$2 nm) which allows pulses of a few picoseconds to be amplified (Klingebiel et al. 2011; Schneider et al. 2014). This has led to a class of lasers, often based on thin disks, that take advantage of the good thermal characteristics of YAG, the diode pumpability of the Yb^{3+} ion and the small quantum defect associated with this pumping at 970 nm to allow CPA at a few picoseconds with rather high average power (say, a few tens of millijoules at a few kilohertz).

2.2.4 Cr:LiSAF, Alexandrite and Other Lesser Used Materials

A number of other solid-state laser materials have been explored for use in CPA. One of the principal motivations for these developments has been to find a material that exhibits the broad gain bandwidth of Ti:sapphire but which also has the same favorable pumping options as Nd:glass. Other materials have been explored as a means to enable efficient diode pumping of the CPA amplifiers with the hopes of amplifying to high energies at repetition rates much higher than those possible with flashlamp-pumped Nd:glass. Many of these alternate laser materials are, like Ti:sapphire, vibronic lasers, yielding broad bandwidth, with Cr^{3+}-doped materials being the most heavily investigated. In fact there are scores of broadband Cr-based materials that have been explored in femtosecond oscillators at various near-IR wavelengths, however, most have never been implemented in CPA lasers used in real strong field physics research. Cr-based materials have the advantage that they absorb pump light in three broad spectral bands centered near 300 nm, 450 nm and 650 nm, which can be easily matched to flashlamp output spectrum. What is more, their upper state lifetime is typically 70–300 μs. The two materials which have seen some success in CPA implementations are alexandrite and Cr:LiSAF.

Alexandrite (Walling et al. 1979), the common name for chromium-doped chrysoberyl (Cr^{3+}:BeAl$_2$O$_4$), has been demonstrated in CPA lasers operating with pulse width down to 200 fs (Pessot et al. 1989; Hariharan et al. 1996). Alexandrite has a 260 μs lifetime, so it is suited for flashlamp pumping and can be fabricated in rods of 1–2 cm diameter. It has about a 120 nm gain cross section bandwidth (FWHM) centered around a gain peak at 780 nm. While Alexandrite exhibits good thermal and reasonable mechanical properties, the principal limitation to its broader implementation in CPA lasers is its rather low gain cross section of only 1×10^{-20} cm^2, making compact, high-gain amplifier chains extremely challenging.

A somewhat more successful material for CPA applications has been Cr^{3+}:Li SrAlF$_6$ (known commonly as LiSAF). This is a material specifically engineered as a broadband gain material (Payne et al. 1994). It has an upper state lifetime of 67 μs which is marginally long enough for efficient flashlamp pumping with properly designed short pulse forming networks (Ditmire and Perry 1993; White et al. 1992). Its greatest draw is that it can be flashlamp or diode pumped and exhibits gain centered at 850 nm with bandwidth of 180 nm, nearly as broad as the 230 nm gain bandwidth of Ti:sapphire. LiSAF has a peak gain cross section comparable to Nd:glass (5×10^{-20} cm^{-2}) so high-gain rods are possible and good energy extraction by saturation can be performed. Unfortunately, scaling of LiSAF to larger aperture and higher energy, where laser pumping Ti:sapphire is very challenging, has been hindered by the fact that large-aperture LiSAF crystals are usually marred by optical distortions from internal crystal stresses. Furthermore, LiSAF has other mechanical weaknesses, being a soft crystal which is prone to dissolve in coolant water over time if the coolant pH is not carefully held near 7.0. LiSAF still holds promise as a broadband CPA material because it can be diode pumped near 670 nm by AlGaInP diodes.

Finally, it is worth mentioning that Er-doped materials, glass in particular, have been explored for CPA. Erbium is a rare earth ion that lases at a wavelength near 1.5 μm and has found its greatest use in Er-doped fiber ultrafast oscillators. However, at the time of this writing no published strong field studies with lasers based on this material have been published.

2.2.5 Excimer Lasers

Excimer lasers are based on gain in discharge-pumped gases of mixtures of a noble gas, like Kr (or Ar and Xe) with fluorine or chlorine and helium. An inversion can be developed by producing a bound state molecule between the noble gas and F or Cl in an excited electronic state. This excited excimer molecule radiatively decays to the ground state, which is naturally unbound. Rapid separation of the excimer components leads to 4-level laser operation. Because the lower laser level energy is strongly dependent on atomic separation of the two excimer atoms, the excited excimer state can emit over a rather broad range of wavelengths, and the excimer laser exhibits bandwidth suitable for amplification of femtosecond pulse. The most widely used excimer for ultrashort applications is KrF which lases in the ultraviolet at 248 nm with about 2 nm of bandwidth. Excimers exhibit high wall-plug efficiency from electrical power to laser power because they are pumped directly by the high-energy electrons from a cathode in a discharge.

While not widely pursued, CPA in excimer lasers has been demonstrated (Divall et al. 1996). The highest-energy demonstration of such a CPA laser was in KrF. The principal motivation for pursuing CPA in KrF is that gain in this medium resides in the UV, at 248 nm, much shorter than is accessible with solid state laser materials. This short wavelength is desirable for many laser-plasma applications of CPA. The architecture of a KrF CPA laser involves the amplification of stretched pulses at 746 nm in a dye system or in Ti:sapphire followed by frequency tripling of the pulse to 248 nm. This is seeded into a large e-beam pumped KrF cell yielding many joules of stretched pulse energy at 248 nm. The largest such KrF CPA laser delivered 2 J pulses with compressed pulse duration of 200 fs (Ross et al. 1994).

KrF as a CPA amplifier has numerous drawbacks which has quelled widespread implementation, despite the attractiveness of the short wavelength in many applications. The saturation fluence of KrF is extremely low (0.003 J/cm^2), mandating very large-aperture amplifiers for even modest energy. Furthermore, compression gratings at this short

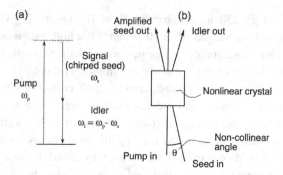

Figure 2.4 Diagram illustrating how OPCPA works. In (a) is the energy diagram showing pump, seed and idler photons. The geometry of the three waves in shown in (b).

wavelength are quite inefficient, so throughput in the compressor is reduced to about 25%, well below the 50–90% possible in the near-IR.

2.2.6 Optical Parametric CPA

An alternative approach to amplifying broad-band laser pulses in CPA emerged in the late 1990s and has become a principal building block of almost all modern CPA lasers. This approach relies on parametric amplification (Cerullo and De Silvestri 2003) of the chirped pulse in nonlinear crystals, a technique that has come to be called optical parametric chirped pulse amplification (OPCPA). Instead of amplifying the seed pulse in a broadband energy-storing laser medium, the chirped pulse is instead copropagated with a second, shorter-wavelength laser pulse in a nonlinear crystal. Through the process of parametric amplification, the chirped seed pulse is amplified (Dubietis et al. 1992; Ross et al. 1997; Ross et al. 2000). This process is schematically illustrated in Figure 2.4 where the chirped seed pulse is the signal and, as the figure shows, a second idler beam at a frequency of $\omega_{idler} = \omega_{pump} - \omega_{signal}$ is produced.

This parametric process relies on the crystal being cut so that the three beams are nearly phase matched, such that

$$\Delta k = \left| \mathbf{k}_p - \mathbf{k}_s - \mathbf{k}_i \right| = 0, \qquad (2.1)$$

where $\mathbf{k}_p$, $\mathbf{k}_s$ and $\mathbf{k}_i$ are the wavenumber vectors of the pump pulse, the amplified seed signal and the produced idler beam, respectively. In practice it is not possible to achieve perfect phase-matching for all wavelengths, but in many crystals it is possible to achieve $\Delta k \approx 0$ for a rather broad range of wavelengths leading to very good amplification bandwidth for broadband CPA pulses. In fact, it has been shown that bandwidths sufficient for recompression of pulses with duration near 10 fs are possible with OPCPA. Gain bandwidth of $\sim$100 nm is routinely possible, for example, in the parametric amplification of pulses with wavelength near 1 μm and pump wavelength near the chirped pulse's second harmonic ($\sim$532 nm) in commonly used nonlinear crystals such as KDP and BBO (Eimerl et al. 1987).

2.2 Common CPA Laser Gain Materials

In addition to the advantage of offering very broad bandwidth (Zheng et al. 2009), OPCPA also has the advantage that very high gain per stage is possible. The gain of a single OPCPA crystal is given by

$$G_{OPA} = 1 + \gamma_{OPA}^2 l^2 \frac{\sinh^2\left[\sqrt{\gamma_{OPA}^2 l^2 - \Delta k^2 l^2/4}\right]}{\gamma_{OPA}^2 l^2 - \Delta k^2 l^2/4}, \qquad (2.2)$$

where l is the crystal length, Δk is the phase mismatch defined in Eq. (2.1) (which varies with wavelength and hence defines the gain bandwidth) and γ_{OPA} is the optical parametric amplification gain coefficient given by

$$\gamma_{OPA} = 4\pi d_{eff} \sqrt{\frac{2\pi I_0}{n_p n_s n_i c \lambda_s \lambda_i}}. \qquad (2.3)$$

Here I_0 is the intensity of the pump beam, d_{eff} is the effective nonlinear polarization of the crystal used and the n_xs and λ_xs are the refractive indices and wavelengths of the signal, idler and pump pulses. For example, when Beta-Barium Borate (BBO) is employed and pumped at an intensity of 500 MW/cm^2 by a 532 nm pulse, a single, 2 cm long crystal will yield a gain of ~100,000 at 1054 nm. Because of the high gains possible, OPCPA offers a compact alternative to the regenerative amplifier stage.

The most common crystals for OPCPA tend to be Beta-Barium Borate (BBO), potassium dihydrogen phosphate (KDP) and lithium triborate (LBO). BBO is perhaps the most commonly used crystal in OPCPA applications, as it has a large d_{eff} (2.1 × 10^{12} m/V for Type I amplification of 1054 nm pulses by a 532 nm pump) when compared to the other crystals, such as KDP (where d_{eff} is only about 0.29 × 10^{12} m/V for Type I amplification of the same wavelengths). Beta-Barium Borate can, therefore, yield high gain when pumped by about 200–500 MW/cm^2 while similar gains can only be achieved in KDP at pump intensities in excess of 1 GW/cm^2. The clear aperture over which high optical quality can be retained during manufacture of BBO crystals is, however, limited to about 3 × 3 cm, while KDP can be grown with very high quality to very large apertures (>20 cm). Lithium triborate can also be produced at the 10 cm aperture. Consequently LBO has become the crystal most likely used for high-energy (>1 J) OPCPA applications in the near-IR. The damage threshold for both BBO and KDP at 532 nm is around 5 GW/cm^2 for 10 ns pulses, so the high pump fluences required for crystals like KDP and LBO can push up against damage issues unless shorter pump pulses are used.

Broadband amplification in OPCPA is easily accomplished when deployed in the colinear, nearly degenerate configuration, that is, when the pump is nearly at the second harmonic of the seed and the pump and seed cross at a shallow angle (say ≤1°). This is easily achieved, for example, when a frequency-doubled Nd:YAG laser operating at 532 nm amplifies pulses near 1 μm, the situation needed for using OPCPA as a broadband preamplifier in a Nd:glass laser chain. The gain spectrum for such a situation in BBO is illustrated in Figure 2.5a. When a nondegenerate configuration is desired, such as amplifying 800 nm pulses with a 532 nm pump, in general, colinear amplification will not yield broadband

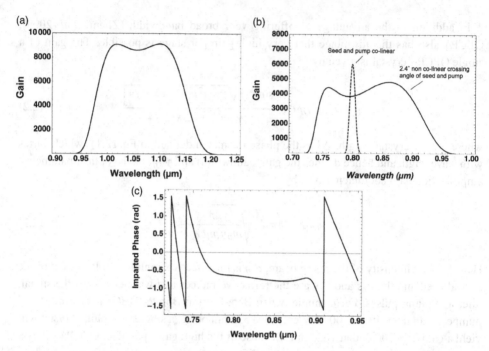

Figure 2.5 (a) Optical parametric amplification gain of a BBO crystal near a wavelength of 1 μm pumped nearly collinearly. This gain is from a 5 mm thick OPCPA crystal pumped at 3 GW/cm^2 by a 532 nm pulse oriented 1° off the beam path of the seed beam. (b) OPA gain of the same 5 mm thick BBO crystal cut for phase-matching near 800 nm. The solid line shows the broadband gain that can be achieved around 800 nm by pumping at a noncollinear angle (in the crystal) of 2.4°. The dashed line shows that the gain is narrowband if the pump beam is propagated colinearly to the seed beam. (c) Phase imparted to the seed as a function of wavelength by amplification in the noncollinear geometry illustrated in part (b).

amplification. This is illustrated in Figure 2.5b. Instead it is necessary to resort to a noncollinear geometry between pump and seed, and tune the crossing and phase-matching angles to find an inflection in the phase-matching curve that yields broad band amplification. For example, such broadband, noncollinear amplification of 800 nm pulses with 532 nm pump pulses can be achieved in Type-I BBO (pump is extraordinary, seed and idler are ordinary waves) when the two beams are crossed at an angle of about 2.4° and the crystal is cut 23.8° from the crystal c-axis (see Figure 2.5b).

Practical implementation of OPCPA demands certain requirements, some of which make OPCPA challenging when compared to traditional laser amplifiers. The most pressing design consideration arises from the fact that the gain is very nonlinear with pump laser intensity. This requires that the pump laser beam not have any short-time-scale intensity oscillations on the pulse, which might be greatly amplified and imprinted on the chirped seed pulse. A Q-switched oscillator will usually have temporal structures arising from mode beating in the resonator cavity of the pump laser. Consequently, OPCPA pump lasers almost always must be injection-seeded to force lasing on a single longitudinal mode.

A related problem arises from the natural temporal pulse shape of the pump laser pulse. Even an injection-seeded Q-switched pump laser will have pulse shape which rises and falls with a few ns times scale, often with a pulse shape approximating that of a Gaussian. Because of the nonlinear nature of the parametric amplification, a copropagating seed pulse will see higher gain at the central, high-intensity temporal region of the pump while the wings of the seed will see lower gain. This has the effect of shortening the chirped pulse which in turn narrows the spectrum of the pulse because of the one-to-one correspondence between spectrum and time. This effective gain narrowing can be ameliorated if the pump pulse is much longer (at least $\times$ 5) than the seed pulse. This, however, results in poor pump-to-seed conversion efficiency because of the great mismatch in pulse duration and the inability to extract energy from the pump over its entire temporal profile. Completely eliminating this effect requires that the pump pulse be flat in time, requiring often challenging pulse shaping of the pump.

A mismatch in pump and seed pulse durations can also result in unwanted spontaneous amplification (ASE). Spontaneous amplification and parasitic lasing can be a particular problem in OPCPA because the gain in a single crystal can be so large (4 or even 5 orders of magnitude). Even with an anti-reflection coating lowering reflection to 0.1%, a gain of 10^4 can easily produce parasitic lasing within a crystal. The approach now used in the newest CPA lasers is to eliminate this ASE by utilizing picosecond pump pulses in the first stages of gain, a technique detailed in the following section. In addition to these temporal considerations, the nonlinear nature of OPCPA also demands very high-quality spatial profiles on the pump beam. This is crucial not only to circumvent potential damage but also to ensure that any spatial perturbations on the pump are not amplified and imparted to the seed. Again, optimum pumping efficiency really requires the pump pulse to be close to flat top in space to maintain uniform gain over the entire spatial profile of the seed.

The parametric amplification process in OPCPA also results in the introduction of a phase term onto the amplified pulse, This phase term can be significant and must be corrected in the dispersion compensation of the entire CPA chain. Figure 2.5c illustrates the phase imparted as a function of wavelength during amplification of 800 nm pulses pumped by 532 nm pulses in BBO.

2.3 Femtosecond Oscillators and CPA Front Ends

2.3.1 Mode-Locking

All CPA systems need an initial seed pulse. This pulse must be at least as short as the duration desired after amplification and compression, or at least contain the necessary bandwidth with well-controlled phase profile. The seed laser pulses must also be matched to the wavelength of the gain material to be employed in the CPA amplifier chain. In practice, the short pulses are generated inside a laser oscillator and subsequently amplified. The most common technique for generating the ultrashort pulses needed for CPA chains is mode-locking (Jain *et al.* 1992).

The general idea of mode-locking is illustrated in Figure 2.6. A mode-locked oscillator is constructed like most other laser oscillators with two cavity end mirrors and gain medium between these mirrors (along with other transmissive optics). Such an optical cavity can support light wavelengths such that an integer multiple of half of the wavelength is equal to the cavity length, L. Since all cavities have some materials inserted (like the gain medium), we can define these longitudinal modes of the cavity in terms of the effective cavity length, L_c such that

$$L_c = L + \sum_i l_i(n_i - 1), \tag{2.4}$$

where l_i is the length of the various i materials with refractive index n_i inserted into the cavity. The frequencies of the longitudinal modes of the cavity then are

$$v_j = j \frac{c}{2L_c}, \tag{2.5}$$

where j is an integer.

When an oscillator is made to lase with a gain element that has a broad wavelength spectrum, a large number of longitudinal modes will lase at any one time (unless special steps are taken to *suppress* all but one mode). For example, a typical Ti:sapphire oscillator, the most common kind of seed laser employed in modern CPA lasers, with mirror separation on the order of $L \sim 1\text{m}$ and the wavelength centered around $\lambda = 800$ nm, has over 300,000 possible longitudinal modes within the large gain bandwidth of Ti:sapphire. The frequency spacing between two adjacent modes is given by $\Delta v_j = c/2L_c$.

Generally speaking, in any oscillator, each mode can lase separately from each other with no fixed phase relation between them. This is the situation described schematically at the top of Figure 2.6. This situation will result in a CW output wave with random intensity fluctuations caused by interferences between multiple modes, each of slightly different wavelength. If, however, all of the modes can be forced to lase with well-defined phase between them, they yield an electric field structure inside the lasing cavity that is of the form

$$E(t) = \sum_j E_j e^{i2\pi v_j t}, \tag{2.6}$$

where the E_j components describe the lasing amplitude of the jth longitudinal modes and will be related to the spectral shape of the gain medium in the cavity. We recognize this sum of equally phased waves as a Fourier sum which can yield a repeating spike in electric field which propagates back and forth in the laser cavity. This situation in which all modes are forced to have the same phase and sum to a short temporal laser pulse spike is known as mode-locking.

The pulse pattern described by this Fourier series repeats after every round-trip time in the resonator cavity $= c/2L_c$. One of the mirrors, the output coupler, leaks a small fraction of the pulse energy every time the pulse bounces off that mirror and generates repetitive pulses outside the oscillator at a repetition rate $v_{out} = c/2L_c$, which is identical to the frequency spacing of adjacent longitudinal modes in the cavity. The number of modes,

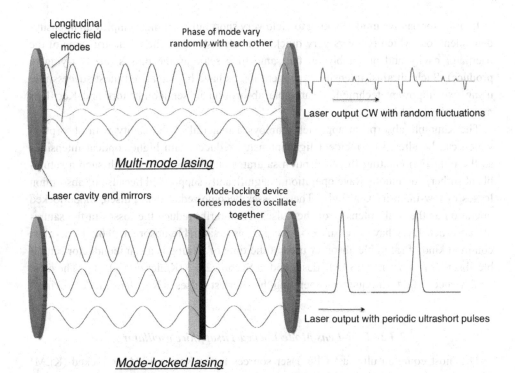

Figure 2.6 Illustration of how mode-locking works to yield a train of ultrashort pulses.

N, that oscillate simultaneously determines the mode-locked bandwidth, $N\Delta v_j$, and the minimum pulse duration $\tau \approx 1/N\Delta v_j$.

In this way, ultrafast laser pulses can be produced. They emerge from the oscillator cavity at a repetition frequency typically ~100 MHz for ~1 m length cavities. There are various methods for locking the phases of these modes. A complete survey of mode-locking techniques is beyond the scope of this chapter. Generally, however, there are two approaches to achieve mode-locking: a short pulse and hence locked modes can be forced by introducing an active time windowing in the cavity, usually referred to as active mode-locking, or by designing the oscillator cavity so that it becomes energetically favorable to lase with the modes locked, known as passive mode-locking. The first technique can be thought of as modulating a loss in time and taking advantage of the fact that when the modes are locked a short time pulse results. The second technique can be performed through introducing an intensity-dependent loss and taking advantage of the fact that when the modes are locked an intense spike is formed in the oscillator cavity. Active mode-locking, can be achieved by modulating the loss in the cavity temporally with acousto-optic modulation, electro-optic modulation, or synchronously pumping with a pulsed source. However, the shortest pulses have been produced with passive mode-locking, and it is passively mode-locked lasers which are most common in CPA constructions.

In order for passive mode-locking to yield very short pulses, it must employ an intensity-dependent loss which recovers very quickly (much faster than the round-trip time of the oscillator cavity and preferably on the same time scale as the pulses one is trying to produce). Such ultrafast intensity-dependent losses have been realized most successfully using two important techniques: saturable absorption in semiconductors and Kerr-lens mode-locking (KLM).

The saturable absorption approach employs a material in the cavity with absorption which can be altered by incident light intensity, reduced with higher optical intensities as the transition creating the absorption saturates and hence drops. With such a saturable absorber, continuous wave operation is significantly suppressed because of absorption losses of low-intensity CW light. These losses become reduced in pulsed, mode-locked operation as the peak intensity of the pulses temporarily reduce the losses in the saturable absorber. Dyes have been used in the past as a saturable absorber, though the most common kind of saturable absorber used at the time of writing is a semiconductor saturable absorber mirror or "SESAM" described in Section 2.3.3 (Keller *et al.* 1996). The most widely successful technique for generating the shortest pulses is Kerr lens mode-locking.

2.3.2 Kerr-Lens Mode-Locked Ti:sapphire oscillators

The most common ultrafast CPA laser sources is the Kerr-lens Mode-locked (KLM) Ti:sapphire oscillator (Brabec *et al.* 1992; Spielmann *et al.* 1994). The design of a typical KLM Ti:sapphire laser is illustrated in Figure 2.7. The basis for operation of such a laser is to rely on the self-focusing that occurs in the sapphire gain medium by intense pulses when mode-locked. KLM is obtained through the a nonlinear optical lensing effect at an intracavity focus, where the index of refraction $n = n_0 + \gamma_{NL} I(r, t)$ is modified by a dependence on the spatial and temporal intensity profile $I(r, t)$. The Gaussian-shaped spatial profile of the lasing spatial mode in the cavity (common to stable resonator cavities) leads to a self-focusing effect in materials with nonlinear index n. The laser resonator is typically adjusted

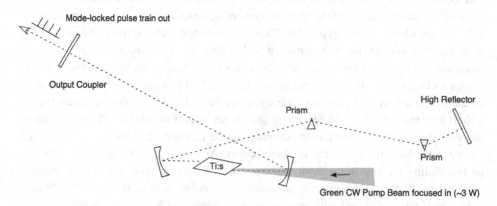

Figure 2.7 Optical diagram of a typical Kerr-lens mode-locked Ti:sapphire oscillator.

towards the edge of its geometric stability limits, such that the mode size is smaller for the high-intensity pulsed operation than in CW operation. This leads to higher losses of CW lasing when an aperture is introduced and lower gain by the CW mode because the gain in the Ti:sapphire is spatially localized by the pump laser. An intra-cavity aperture suppresses the CW lasing and pushes the laser into mode-locking. Kerr-lensing happens on the femtosecond time scale and is inherently broadband, making it suitable for the very short (<20 fs) oscillators possible with Ti:sapphire. Pulses from Ti:sapphire oscillators have been generated with duration as short as 5 fs (Jung et al. 1997; Ell et al. 2001), which is less than two optical cycles at the center wavelength of 800 nm.

As illustrated in Figure 2.7 a typical Ti:sapphire oscillator has a Ti:sapphire crystal situated inside a subresonator composed of focusing mirrors pumped by a secondary "green" laser, such as a frequency-doubled diode-pumped solid state laser (e.g. $Nd:YVO_4$) or an argon-ion laser. The oscillator cavity is further composed of an output coupler and some form of dispersion compensation. In the case of the laser illustrated in Figure 2.7, pulse broadening occurs because the positive group velocity dispersion of materials in the cavity is compensated by two prisms situated so that the blue end of the spectrum travels a shorter distance than the red end. Balancing this geometric negative dispersion with positive dispersion in the cavity materials can yield a very short pulse directly out of the cavity through the output coupler. Average output powers of 0.2–1 W are typical at repetition rates of 25–150 MHz when green pump powers of 3–5 W are employed.

2.3.3 Yb- and Er-Based Mode-Locked Fiber Lasers

A second class of mode-locked oscillators finding widespread use in modern CPA lasers are lasers based on lasing in a doped fiber passively mode-locked with a fast saturable absorber (Ortac et al. 2007; Fernmann and Hartl 2009; Limpert et al. 2011). There are innumerable variations of designs of this class of laser (Zhang et al. 2014). Early systems were based on Er-doped fibers lasing at 1.5 μm, but more recent development of Yb-based fiber lasers producing pulses at wavelength near 1 μm have made these mode-locked fiber systems very attractive as a robust seed for CPA lasers. As already discussed, Yb is attractive because it can be easily pumped by laser diodes at wavelength near 970–980 nm. These lasers can be mode-locked in a number of ways, but a common method is to use a SESAM at the back end of the fiber cavity (Saraceno et al. 2012).

One simple example of how such an oscillator can be constructed is shown in Figure 2.8a (Fermann and Hartl 2013). A glass fiber doped with Yb is injected with diode light at 980 nm with a wavelength division coupler, creating an inversion in the Yb fiber. Because of the dispersion of the fiber there needs to be some means of reversing this pulse broadening from dispersion. One common way is to use a chirped fiber Bragg grating on one end of the fiber laser cavity. This is a fiber with a modulated refractive index that reflects light based on constructive interference from scattering off the index modulation. By varying the modulation wavelength as a function of position in the fiber, different wavelengths get reflected at different depths and thereby reverse the frequency chirp imparted by the fiber.

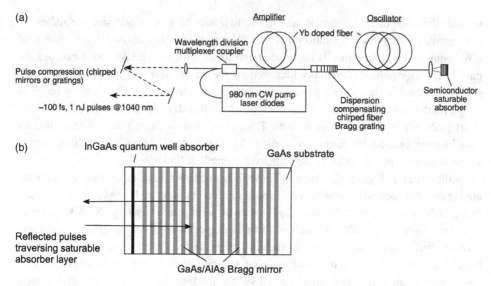

Figure 2.8 (a) Optical schematic of a dispersion compensated, modelocked Yb fiber oscillator. (b) The design of a semiconductor saturable absorber mirror (SESAM).

Mode-locking can be achieved by the introduction of a SESAM at the other end of the cavity (Keller *et al.* 1996). How such a mirror works is shown in Figure 2.8b. Reflectivity comes from a Bragg mirror stack of alternating semiconductors grown on a GaAs substrate. The reflected light passes through a very thin layer composed of an InGaAs semiconductor. The effective absorption wavelength of this layer can be selected by varying the bandgap of the semiconductor by varying the amount of In doped into the GaAs. The mirror is operated as an antiresonant Fabry–Perot cavity. Absorption at the Yb laser wavelength saturates when high-intensity spikes are present so that the laser is only above threshold for lasing when it is mode-locked. The saturation fluence of most SESAM absorbing layers is about 100 μJ/cm^2.

The output power of the laser can be boosted by passing the output through an amplifier fiber. Ultrashort pulses are retrieved by compressing external to the amplifier fiber, often with chirped mirrors. Output parameters vary widely with device and design but typically 100 fs pulses can be produced at 1,030–1,050 nm wavelengths with output powers of 100 mW to a few watts.

2.3.4 Carrier-Envelope Phase Control of Femtosecond Oscillators

Mode-locking yields pulses that are a visible or near-IR electromagnetic (EM) wave underneath some carrier envelope with an amplitude shape determined by the bandwidth of the gain medium and details of the mode-locking technique. Note, however, that the phase of the wave underneath this pulse carrier envelope is not uniquely determined. Practically speaking, in many strong field experiments this phase is irrelevant and no attention is paid to it. However, as we will see at various points later in this book, when pulses of only a

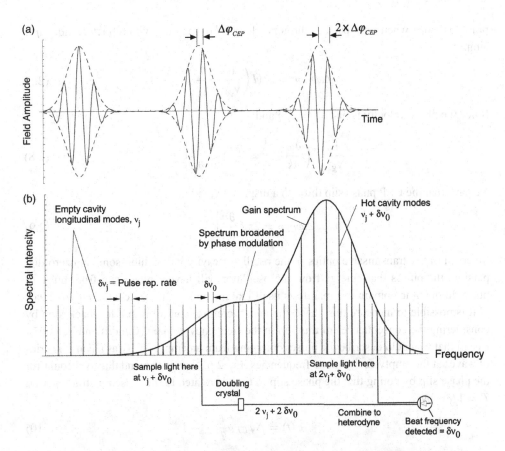

Figure 2.9 (a) Time domain plot of pulses from a mode-locked oscillator showing how the difference between phase velocity and group velocity in the intercavity materials leads to a phase slip of $\Delta\varphi_{CEP}$ in subsequent pulses emerging from the oscillator. (b) Frequency domain plot of the longitudinal modes in an oscillator cavity and the bandwidth supported by the lasing medium. The shift of longitudinal modes by a frequency $\delta\nu_0$ results from the pulse-to-pulse carrier envelope phase slip. The bottom of the figure shows how comparing light from frequency doubling of spectrum near ν_j with light near $2\nu_j$ yields a heterodyne beat equal to the phase slip offset $\delta\nu_0$.

few optical cycles are employed (say ≤10 fs), the value of this so-called carrier-envelope phase (or CEP) can play an important role (Jones *et al.* 2000). This is particularly true in attosecond pulse generation. This can be seen roughly in the three pulses depicted in Figure 2.9, where it is clear that the actual peak electric field value depends on the CEP of the wave underneath the pulse envelope.

In fact, the CEP phase will vary from one pulse to the next in the pulse train of a single mode-locked oscillator. This is a consequence of the difference between the phase velocity of a wave, v_φ, and the group velocity of the light pulse, v_g, through a material which exhibits spectral dispersion. The phase slip of a wave of frequency ν with respect to its pulse envelope when it propagates through a length l of medium with refractive index η is simply

$$\Delta\varphi_{CEP} = 2\pi\nu l \left(\frac{1}{v_\varphi} - \frac{1}{v_g}\right). \quad (2.7)$$

Since the phase velocity is just $v_\varphi = c/\eta$ and

$$\frac{1}{v_g} = \left(\frac{\partial\omega}{\partial k}\right)^{-1} = \frac{\eta}{c} - \frac{1}{v}\frac{\partial\eta}{\partial\lambda}\bigg|_{\lambda_0}, \quad (2.8)$$

we have that the CEP phase slip through a dispersive medium is

$$\Delta\varphi_{CEP} = 2\pi l \frac{\partial\eta}{\partial\lambda}\bigg|_{\lambda_0}. \quad (2.9)$$

Since all of the transmissive optics in the oscillator cavity will exhibit some nonzero dispersion, the pulses that emerge from the oscillator will have a phase slip from pulse to pulse. In the time domain this manifests itself like the pulse train depicted in Figure 2.9a.

It is possible to measure this phase slip and, therefore, control it in a feedback loop by considering the consequences of this slip in the frequency domain (Cundiff and Ye 2005). Recall that we could describe the pulses emerging from an empty cavity as a Fourier series of waves at the empty cavity mode frequencies, Eq. 2.6. Now we amend this to account for the phase slip by noting that the phase slip in time is related to the pulse repetition period $T = 1/\delta v_j$ by

$$\varphi(t) = \Delta\varphi_{CEP}\frac{t}{T}. \quad (2.10)$$

So the electric field pulse train emerging from the oscillator must be multiplied by $e^{i\varphi(t)}$ which means that the Fourier representation of this pulse train is

$$E(t) = \sum_j E_j e^{i2\pi v_j t} e^{i\Delta\varphi_{CEP}\delta v_j t}$$

$$= \sum_j E_j e^{i2\pi(v_j + \Delta\varphi_{CEP}\delta v_j/2\pi)t}. \quad (2.11)$$

The pulse-to-pulse carrier envelope phase slip is actually just a shift on the frequencies of the empty cavity modes of magnitude $\delta v_0 = \Delta\varphi_{CEP}\delta v_j/2\pi$. This is illustrated in the frequency domain in Figure 2.9b. The mode-locked pulse train is actually a sum of shifted modes.

The size of this frequency shift, and hence the magnitude of the pulse-to-pulse CEP phase slip can be measured using a so-called f-2f interferometer. The principal of this measurement is illustrated at the bottom of the frequency spectrum plot of Figure 2.9b. In practice, the spectrum width can be broadened by self-phase modulation in a modulated fiber so that the pulses have spectral components in the red wing at a frequency given by $v_j + \delta v_0$ and spectral components in the blue wing at one octave higher frequency, near the second harmonic of the red component at $2v_j + \delta v_0$. The red wing of the spectrum can

2.4 Pulse Stretchers and Compressors

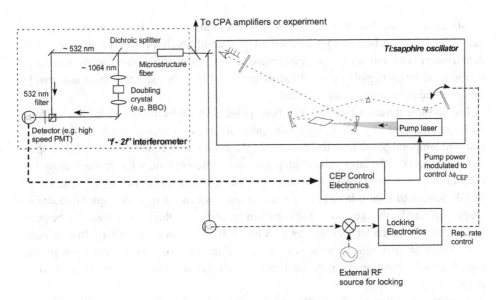

Figure 2.10 Typical optical setup for stabilizing the carrier-envelope phase of a mode-locked Ti:sapphire oscillator. The CEP phase slip is measured with the "f-2f" interferometer yielding a heterodyne beat frequency that can be locked in a feedback loop. The CEP phase slip is controlled by small variation of the intensity of the pulse in the cavity through tuning of the pump power. The repetition rate of the laser is also locked to an external RF frequency by rotating a mirror in the spatially dispersed part of the cavity to induce group delay.

be frequency doubled in a nonlinear crystal to a frequency of $2v_j + 2\delta v_0$. Heterodyning this light with the blue-wing spectral component yields a beat frequency in the RF regime equal to $\delta v_0 = \Delta\varphi_{CEP}\delta v_j/2\pi$, yielding a direct measurement of the CEP slip.

An example of how this can be realized in the lab on a mode-locked Ti:sapphire laser illustrated in Figure 2.10 (Cundiff *et al.* 2008). Typically some of the power of the output pulse train is sampled and sent into a microstructured fiber to broaden the spectrum through self-phase modulation. This yields an octave-spanning spectrum with frequency components at both $v_j + \delta v_0$ and $2v_j + \delta v_0$. The low-frequency component, which might be around 1,064 nm, is split with a dichroic mirror and frequency doubled in a BBO crystal. This is then recombined with the 532 nm component of the femtosecond pulse spectrum on a fast photomultiplier (PMT) to yield an RF signal that can be minimized in a feedback loop. One common way to adjust the CEP is to vary slightly the power in the cavity by varying the pump laser power. A separate loop controls the repetition rate of the cavity by rotating a mirror in the spatially dispersed end of the cavity to induce slight group delay.

2.4 Pulse Stretchers and Compressors

2.4.1 Basic Principles and the Two-Grating Pulse Compressor

One of the key elements of the CPA architecture is the stretching of the seed pulse prior to amplification and recompression of the pulse after amplification. This mandates the construction of two optical elements, the first that induces a very large amount of spectral dispersion (the stretcher) on a laser pulse and the second which reverses this dispersion (the compressor). The pair must also compensate, as well as possible, the dispersion induced on the pulse by propagation through the material and any other spectral phase acquired by the stretched pulse during amplification.

The shortest duration possible in a laser pulse is given by its transform-limited form which is determined by its bandwidth and pulse shape (such a transform limited pulse has constant temporal phase over the duration of the pulse). Dispersion causes different spectral components of the laser pulse to propagate with different time delays through an optical component or system. CPA systems work by adding a large amount of mostly linear positive dispersion to a short laser pulse (meaning low frequencies precede higher frequencies or, put differently, the red part of the spectrum precedes the blue). This broadens the pulse in time by a large factor, perhaps by a factor of 10^3–10^4 over the initial ultrafast pulse duration, to lower significantly the peak power and intensity of the pulse during amplification. After amplification this positive dispersion is reversed to retrieve a short, high-power pulse.

The basic building block of CPA, then, is an optical construction that can induce an enormous amount of negative dispersion, a design first proposed by Treacy (Treacy 1969). Treacy realized that a pair of plain gratings with their faces and grooves parallel produces a time delay for broadband light that is an increasing function of wavelength. This grating configuration has become the standard design for pulse compressors in CPA. Figure 2.11 shows such a pair of gratings. Typically each grating has a groove density $N = 1/\delta_{gr} \sim 600$–$1,800$ lines/mm. The gratings are separated by a distance d (which might be of the order of 1–2 m) which means that the normal separation of the planar faces of the gratings is $b = d/\cos[\theta_d(\lambda_0)]$ where $\theta_d(\lambda_0)$ is the diffracted angle of the light at

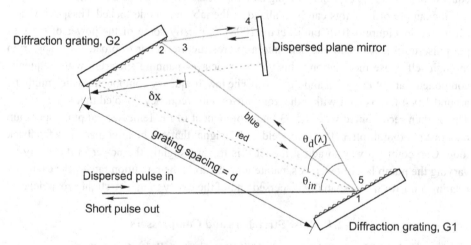

Figure 2.11 Diagram of the widely employed Treacy two-grating pulse compressor. This device imparts large amounts of negative group delay dispersion on a chirped pulse to compress it to nearly its transform limit time duration.

the central wavelength of the incoming broadband pulse. A laser pulse incident on the first grating at angle θ_{in} gets diffracted according to the grating equation

$$\sin\theta_{in} + \sin\theta_{out} = \frac{m_d \lambda}{\delta_{gr}}, \qquad (2.12)$$

where the diffraction order m_d is -1 (or 1 in some systems) for efficient first-order diffraction. The second grating reverses the angular chirp such that different wavelengths propagate parallel again. The pulse is then usually passed back through the two gratings to undo the spatial spread of wavelengths and induced further negative dispersion.

A simple inspection of Figure 2.11 illustrates how the large group delay spread is induced on the spectrum: the red component of the spectrum gets diffracted to a larger angle than the blue components by the first grating, which causes it to travel a greater distance back and forth through the system than the blue components. A simple estimate for the extent of the group delay induced by such a grating pair can be made by noting that the additional distance the red component travels in and out of the grating pair is roughly $4 \times \delta x$, as noted in the figure. We can further simplify our assessment by assuming that the pulse sent into the grating pair is incident near the Littrow angle, namely that angle such that $\theta_{in} \approx \theta_d(\lambda_0) = \gamma_{Lit}$. The grating equation becomes

$$\sin\gamma_{Lit} \cong \frac{\lambda}{2\delta_{gr}}, \qquad (2.13)$$

which means that if the incident pulse has a spread in wavelength of $\Delta\lambda$, then the angular spread of the spectrum on the second grating (G2 in Figure 2.11) is

$$\Delta\gamma \cong \frac{\Delta\lambda}{2\delta_{gr}\cos\gamma_{Lit}}. \qquad (2.14)$$

If the spatial spread of the colors on G2 is roughly $d\Delta\gamma$ and the extra path difference for the red component is $\Delta x \approx \Delta\gamma \tan\gamma_{Lit}$, then the group delay spread of the pulse's spectrum is

$$\Delta\tau \cong \frac{4\delta x}{c} \cong \frac{2d\Delta\lambda}{c\delta_{gr}}\frac{\sin\gamma_{Lit}}{\cos^2\gamma_{Lit}}. \qquad (2.15)$$

Putting in some typical numbers, Eq. (2.15) predicts that a compressor that is built of gratings with 1,800 l/mm groove density separated by 1 m will yield ~380 ps on a 20 nm spectrum at 800 nm wavelength (typical numbers for a 50 fs Ti:sapphire laser). This is about the stretched pulse duration (200–500 ps) that is typically used in many Ti:sapphire systems.

While in principle it would be possible to use such a two-grating system to stretch a pulse by imparting negative dispersion for CPA amplification; in essentially all CPA lasers this configuration is used to compress a pulse that has initial positive dispersion induced for the stretch prior to amplification. This is because this configuration has no transmissive or focusing reflective optics which might be prone to damage or unwieldy apertures. It also requires fewer optical elements and is easier to be scaled in size. So it is best suited for compressing pulses after they have been amplified to high energy where the beam aperture is larger and demands large-aperture reflective optics. Virtually all compressors in CPA are based on this configuration. (Though sometimes four gratings are employed instead of two where the pulse is not passed back through the first two gratings to remove the spatial dispersion but is instead passed through two additional identically arranged gratings in a mirror image configuration after being dispersed by the first two gratings. This is sometimes needed for very large-aperture amplified beams.) Since the Treacy configuration is used to compress the pulse, a different design is required to induce positive dispersion and stretch the pulse at the front end of a CPA chain.

2.4.2 The Basic Two-Grating, Two-Lens Stretcher

It turns out that the configuration that is conjugate to the Treacy compressor of Figure 2.11 was proposed in the mid 1980s by Martinez (Martinez 1987). His proposed configuration is shown in Figure 2.12. Martinez showed that a pair of antiparallel gratings with a unity magnification telescope between them is a complementary system to the grating pair compressor in all dispersion orders, as long as optical aberrations are negligible. As can be seen in Figure 2.12, in this configuration the opposite is true for the path lengths traveled by the various wavelengths: now the red portion of the spectrum traverses a shorter distance than the blue. The magnitude of the imposed positive group delay can be easily seen if one notes that the one-to-one telescope between the gratings projects an inverted image (G1') of the first grating (G1) behind the second grating (G2), which makes the system equivalent to an effective negative grating distance $d = 2f - d_2 - d_1$, where d_i are the spacings between the two gratings and the nearest telescope lens. Hence the dispersion terms are perfectly matched in all orders apart from the sign. So the temporal spread for a given bandwidth estimated in Eq. (2.15) yields an approximate estimate for

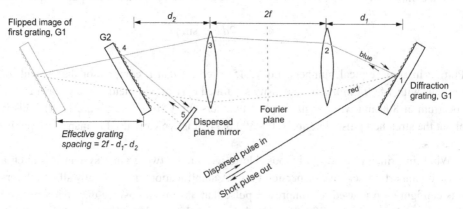

Figure 2.12 Optical diagram of the Martinez two-grating lens-based pulse stretcher. This device works in the opposite manner to the two-grating compressor by using a 2f lens pair to create an inverted image of the first grating and thereby impart large positive group delay dispersion on the pulse.

the positive group delay of the Martinez configuration if $d = 2f - d_2 - d_1$ is used in this equation for the grating spacing.

This concept is used in most CPA designs for the stretcher, and early CPA systems did indeed use lenses in the telescope. However, chromatic aberrations in the lenses impart higher-order dispersion terms and spatial chirp which is hard to correct when particularly broad spectra are propagated through the system. Consequently the lens-based stretcher is only useful for systems with ultimate pulse durations of longer than about 200 fs (bandwidths of under 5 nm). The problem of chromatic aberrations can be circumvented if the lenses are replaced by curved reflective optics.

2.4.3 All-Reflective Optics Stretcher Design

The technique used in stretchers for most CPA systems amplifying sub-100 fs pulses is to use only reflective optics. Aberrations from the lenses in the basic Martinez stretcher design tend to produce phase terms which are hard to cancel in most systems. They also impart spatially dependent phase distortions to the beam. To avoid chromatic and other aberrations altogether, the lenses in the stretcher can be replaced by reflective mirrors (i.e. a focusing mirror replaces the lenses).

The first design step in most modern CPA stretchers is to use only one grating in a folded configuration. Note in Figure 2.12 that the left half of the stretcher is merely a mirror image of the right half. Therefore a flat mirror can be placed at the midpoint of the 2f telescope and the beam, with small vertical offset as it passes through the system traverses a path like that in 2.12 but with four hits on one grating (instead of two hots each on two gratings). Next, the single lens in this configuration is then replaced with a curved optic (or optics).

Using a reflective mirror instead of a lens does lead to design challenges, essentially resulting from the geometric problems of placing a large-aperture mirror near the grating. This problem is particularly significant in high-bandwidth stretchers (those passing >50 nm) because the aperture of the optics must be large to capture the wide spatially dispersed bandwidth of the pulse. For example, if the first lens in the Martinez stretcher is replaced by a mirror reflecting at 0° the first reflection of the curved mirror would then be blocked by the grating. Several solutions have been developed to overcome this problem. A typical all-reflective pulse stretcher, one first proposed by Banks and Perry (Banks et al. 2000) is shown in Figure 2.13a. In such a stretcher the dispersed beam after striking the grating for the first time strikes a spherical curved mirror on axis and then reflects from a reflective stripe displaced slightly vertically located either by coating a reflective strip on the grating directly or (as illustrated in Figure 2.13a) by placing a thin flat mirror just in front of the grating. This reflective strip send the spatially dispersed beam to a flat mirror that rests at the mid-point of the telescope to fold the stretcher. The beam folds back on itself with slight vertical displacement. A vertical retro mirror pair at the output displaces the dispersed beam for its second round trip through the system to yield the temporally dispersed stretched pulse. Residual aberrations arise from the fact that the grating is not in the focal plane of the mirror and the majority of the rays strike the surface off-axis. These can be reduced by a second pass through the system with an image inversion between the passes.

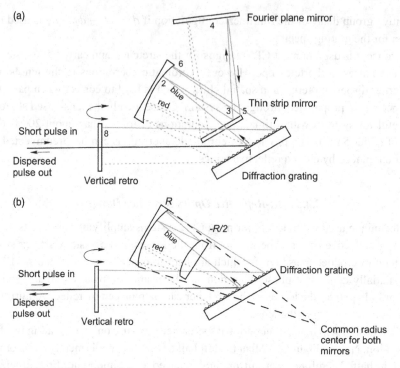

Figure 2.13 (a) Example of the implementation of the Martinez pulse stretcher with all-reflective optics to eliminate chromatic aberration from lenses (often called the Perry–Banks stretcher). (b) Optical schematic of an all-reflective pulse stretcher using two curved reflective optics in the Offner triplet configuration to eliminate spherical aberrations that are present in the stretcher of Figure 2.15a.

Such a design is useful for pulses with width down to about 30 fs. Even greater bandwidths demand further aberration compensation. One widely used design for CPA lasers striving to yield compressed pulses under 30 fs is illustrated in Figure 2.13b. Here the single curved spherical mirror forming the two telescope optics is replaced by two curved mirrors arranged in the so-called Offner triplet configuration (Cheriaux 1996). This combination uses two concentric spherical mirrors, one concave and one convex. The symmetry of this arrangement allows only for symmetrical aberrations (i.e. spherical, astigmatism), but they will cancel between the two mirrors if the radius of curvature ratio is 2:1 and opposite in sign.

2.4.4 Compressor Design

The final element of a CPA system is the compressor, which as we have already discussed is almost always the two-grating Treacy configuration already discussed. Some practical aspects of the compressor design to bear on performing strong field experiments and bear mentioning. Because the final compressed pulse is high energy it must be expanded to bring it below the damage threshold of the gratings. For pulses near 800 nm wavelength gold-coated gratings must be used, which damage at about 300 mJ/cm^2 on the grating surface. If pulses close to 1 μm wavelength are being compressed then there are high-quality multilayer dielectric (MLD) gratings available which have damage threshold closer to 1 J/cm^2 (Perry et al. 1995; Shore et al. 1997). Gold gratings have diffraction efficiency such that roughly 60–65% of the pulse energy can be transmitted through the compressor, while MLD gratings permit close to 90% transmission of the pulse energy.

Because the compressed pulses in either case will have a beam intensity of roughly 1 TW/cm^2 it is impossible to propagate the pulses through any significant length of air without nonlinear effects degrading the beam quality. Therefore it is usual to construct the compressor in a vacuum tank and propagate the pulse to the experiment in vacuum. Very large-aperture beams can be potentially compressed by coherently "tiling" multiple gratings (Qiao et al. 2008), though this is uncommon at the time of this writing.

2.4.5 Phase Control in CPA

One of the most challenging aspects to a successfully designed CPA laser is the management of spectral dispersion (Backus et al. 1998). This refers to the frequency-dependent phase that gets imparted to the laser pulse as it propagates through the system. Such dispersion arises from materials through which the pulse propagates which have refractive indices that vary with wavelength or when optical components cause different wavelengths to travel through different path lengths. As such, dispersion can generally be divided into material dispersion that arises from propagation through material (glass, laser gain media, air, etc ...) and into geometric dispersion that results from propagating along different optical paths within a system. It is very large dispersion which is imparted to a pulse by the stretcher to lengthen the pulse for amplification in CPA and the reversal of this dispersion by the compressor to retrieve an ultrashort pulse.

The essence of the design challenge in CPA is to impart as little net total spectral dispersion to the pulse's spectrum as possible to allow delivery of a pulse that is very short in time (Fittinghoff et al. 1998). This can be a challenge because of the very large group delay dispersion that must be imparted to the initial pulse to stretch it for safe amplification and the large amount of materials that must be traversed to amplify the pulse in the laser chain. Ultimately all of the dispersion picked up by the pulse should be largely canceled by the compressor and phase compensation devices. The extent to which spectral dispersion affects the final pulse duration and shape can be, to a reasonable approximation, analyzed in the frequency domain. (This neglects the effects of gain saturation and self-phase modulation in the time domain, but is of reasonable accuracy for most laser designs.)

To make such a frequency domain analysis we can approximate the effects of dispersion by considering an initial laser pulse in time, $E_{in}(t)$, which is composed of an oscillating carrier frequency, characterized by that pulse's central frequency ω_0, or it is central wavelength λ_0, underneath a slowly varying pulse envelope, $s_{in}(t)$. This might describe an initial femtosecond pulse to be used as a seed pulse for the CPA chain. The pulse field spectrum can be found by simple Fourier transform of this pulse, which, if we use complex notation to describe the laser electric field, is

$$\tilde{s}_{in}(\omega) = \frac{1}{2\pi} \int_{-\infty}^{\infty} s_{in}(t) e^{-i\omega_0 t} e^{i\omega t} \, dt. \quad (2.16)$$

(Note: in this subsection we will temporarily forgo with the convention used in most of this book in which the central carrier frequency and wavelength of the laser pulse is denoted by unsubscripted variables. Instead we will for the moment denote the center frequency and wavelength of the laser's spectrum by ω_0 and λ_0, using the unsubscripted variables to denote the variation of that parameter underneath the spectrum of the pulse. This notation is a bit more in keeping with the usual ultrafast laser literature.) The effects of all the dispersion imparted to the pulse summed over all components and geometrically dispersing elements (like pulse stretcher and compressor) can be simulated by multiplying by a net spectral phase function:

$$\tilde{s}_{fin}(\omega) = \tilde{s}_{in}(\omega) e^{i\varphi(\omega)}. \quad (2.17)$$

The final pulse that results at the end of the chain is the inverse Fourier transform of this phase-modulated spectrum

$$E_{fin}(t) = \int_{-\infty}^{\infty} \tilde{s}_{in}(\omega) e^{i\omega t} \, d\omega. \quad (2.18)$$

Ideally the spectral phase of the laser chain would sum to zero so that the initial laser field shape and duration (amplified in amplitude of course) is retrieved at the end of the chain. In practice, this is rarely possible so the best that can be done is to minimize the residual phase to yield a pulse near to the initial pulse duration and shape.

To quantify the impact of various elements in the laser chain on the residual phase it is helpful to expand the spectral phase in a Taylor series centered around the central laser frequency, ω_0:

$$\varphi(\omega) = \varphi_0 + \varphi_1(\omega - \omega_0) + \varphi_2(\omega - \omega_0)^2 + \varphi_3(\omega - \omega_0)^3 + \varphi_4(\omega - \omega_0)^4 + \ldots \quad (2.19)$$

The various coefficients in this polynomial expansion have a physical interpretation. The first is simply a constant and represents a global phase shift. The other constants are

$$\varphi_1 = \left.\frac{\partial \varphi(\omega)}{\partial \omega}\right|_{\omega=\omega_0} \quad \text{Group delay,} \quad (2.20a)$$

$$\varphi_2 = \frac{1}{2} \left.\frac{\partial^2 \varphi(\omega)}{\partial \omega^2}\right|_{\omega=\omega_0} \quad \text{Group delay dispersion,} \quad (2.20b)$$

$$\varphi_3 = \frac{1}{6} \left.\frac{\partial^3 \varphi(\omega)}{\partial \omega^3}\right|_{\omega=\omega_0} \quad \text{Third-order dispersion,} \quad (2.20c)$$

$$\varphi_j = \frac{1}{j!} \left.\frac{\partial^j \varphi(\omega)}{\partial \omega^j}\right|_{\omega=\omega_0} \quad j\text{th-order dispersion.} \quad (2.20d)$$

The group delay term is simply phase acquired by the pulse as it propagates. It is proportional to the path length traversed by the pulse. The next term, quadratic in frequency, is

2.4 Pulse Stretchers and Compressors

the group delay dispersion. It is nonzero if the pulse travels in elements such that different wavelengths travel different distances (like the spectral components in a grating stretcher or compressor) or through material that has a wavelength-dependent refractive index. This phase term imparts a linear sweep in frequency on the pulse. The next term is the third-order dispersion followed by higher orders. It is rarely necessary to consider terms above the fifth order for almost all CPA lasers (Kane and Squier 1997).

A good illustration of the effects of the quadratic term (the group delay dispersion or GDD) can be seen if we assume that the initial pulse is a well-defined Gaussian in time. This pulse is characterized by its $1/e^2$ pulse duration, τ, or more useful is the final full width at half maximum (FWHM) time duration of the pulse:

$$s_{in}(t) = E_0 \exp\left[-t^2/\tau^2\right] = E_0 \exp\left[-2\ln 2 t^2/\Delta t_{FWHM}^2\right]. \tag{2.21}$$

Fourier transform of this shows that the field spectrum is also a Gaussian:

$$\tilde{s}_{in}(\omega) = \frac{1}{2\pi} \int_{-\infty}^{\infty} \exp\left[-t^2/\tau^2\right] e^{-i\omega_0 t} e^{i\omega t} dt$$

$$= \tilde{s}_0 \exp\left[-(\omega - \omega_0)^2 \tau^2/4\right]. \tag{2.22}$$

The FWHM width of this spectrum in frequency is simply

$$\Delta \nu_{FWHM} = \frac{\sqrt{2\ln[2]}}{\pi \tau}. \tag{2.23}$$

We say that this pulse is transform-limited. It has no residual spectral phase so its pulse duration is as short as it can be for the given spectral bandwidth. The product of the pulse duration and spectral bandwidth (an uncertainly relation) yields the well-known result for a transform-limited Gaussian pulse; the time-bandwidth product of this perfect Gaussian pulse is 0.44,

$$\Delta \nu_{FWHM} \Delta t_{FWHM} = \frac{2\ln[2]}{\pi} = 0.44. \tag{2.24}$$

Now imagine that some second-order phase (GDD) is imparted to the pulse. This phase function is multiplied on the spectrum in the spectral domain (i.e. Eq. 2.17) and inverse Fourier transformed to yield the electric field:

$$E'_{fin}(t) = \int_{-\infty}^{\infty} \tilde{s}_0 \exp\left[-(\omega-\omega_0)^2 \tau^2/4\right] \exp\left[i\varphi_2(\omega-\omega_0)^2\right] e^{i\omega t} dt$$

$$= E'_0 \exp\left[-\frac{t^2}{\tau^2(1+16\varphi_2^2/\tau^4)}\right] \exp\left[-i\omega_0 t - i\frac{4\varphi_2 t^2}{\tau^4(1+16\varphi_2^2/\tau^4)}\right]. \tag{2.25}$$

Clearly the pulse duration has been lengthened, so the new duration is

$$\Delta t'_{FWHM} = \sqrt{2\ln[2](1+16\varphi_2^2/\tau^4)} \tau > 0.44 \tag{2.26}$$

with time-bandwidth product larger than 0.44. Furthermore, we see that the instantaneous frequency in the pulse, $\omega = \partial\varphi/\partial t$, exhibits a linear sweep of the form $\omega = \omega_0 + b_c t$, where the chirp parameter is

$$b_c = \frac{4\varphi_2}{\tau^4 \left(1 + 16\varphi_2^2/\tau^4\right)}. \tag{2.27}$$

In CPA, the chirp placed on the pulse in this manner is often cited in terms of the number of picoseconds of stretch per nanometer of bandwidth. Typical values for a CPA system might be 20–40 ps/nm in Ti:sapphire systems and $\sim$100–200 ps/nm for Nd:glass-based lasers. The various orders of phase imparted to a pulse as it travels through a piece of transparent material of length l and refractive index $\eta(\lambda)$ are easily calculated. They are, to fourth order,

$$\varphi_1^{(mat)} = \left.\frac{\partial\varphi}{\partial\omega}\right|_{\omega=\omega_0} = \left.\frac{\partial\lambda}{\partial\omega}\frac{\partial}{\partial\lambda}\left(\frac{2\pi}{\lambda}\eta(\lambda)l\right)\right|_{\lambda=\lambda_0}$$

$$= \frac{\eta(\lambda_0 l)}{c} - \frac{\lambda_0 l}{c}\left.\frac{\partial\eta}{\partial\lambda}\right|_{\lambda=\lambda_0} \tag{2.28a}$$

$$\varphi_2^{(mat)} = \left.\frac{1}{2}\frac{\partial\varphi_1}{\partial\omega}\right|_{\omega=\omega_0} = -\frac{\lambda^2}{4\pi c}\left.\frac{\partial\varphi_1(\lambda)}{\partial\lambda}\right|_{\lambda=\lambda_0}$$

$$= \frac{\lambda_0^3 l}{4\pi c^2}\left.\frac{\partial^2\eta}{\partial\lambda^2}\right|_{\lambda=\lambda_0} \tag{2.28b}$$

$$\varphi_3^{(mat)} = \left.\frac{1}{3}\frac{\partial\varphi_2}{\partial\omega}\right|_{\omega=\omega_0} = -\frac{\lambda^2}{6\pi c}\left.\frac{\partial\varphi_2(\lambda)}{\partial\lambda}\right|_{\lambda=\lambda_0}$$

$$= -\frac{\lambda_0^4 l}{8\pi^2 c^3}\left(\left.\frac{\partial^2\eta}{\partial\lambda^2}\right|_{\lambda=\lambda_0} + \frac{\lambda_0}{3}\left.\frac{\partial^3\eta}{\partial\lambda^3}\right|_{\lambda=\lambda_0}\right) \tag{2.28c}$$

$$\varphi_4^{(mat)} = \left.\frac{1}{4}\frac{\partial\varphi_3}{\partial\omega}\right|_{\omega=\omega_0} = -\frac{\lambda^2}{8\pi c}\left.\frac{\partial\varphi_3(\lambda)}{\partial\lambda}\right|_{\lambda=\lambda_0}$$

$$= -\frac{\lambda_0^5 l}{16\pi^3 c^4}\left(\left.\frac{\partial^2\eta}{\partial\lambda^2}\right|_{\lambda=\lambda_0} + \frac{2\lambda_0}{3}\left.\frac{\partial^3\eta}{\partial\lambda^3}\right|_{\lambda=\lambda_0} + \frac{\lambda_0^2}{12}\left.\frac{\partial^4\eta}{\partial\lambda^4}\right|_{\lambda=\lambda_0}\right) \tag{2.28d}$$

Notice that if the refractive index does not vary with wavelength the group velocity of the light, v_g, related to the group delay simply by $\varphi_1 = l/v_g$, is just the same as the phase velocity $v_\varphi = c/\eta$ and all dispersion terms vanish. Only if the material exhibits a refractive index with some curvature do the dispersion terms become nonzero (the case for essentially all transmissive optics used in lasers). These terms are easily calculated using published Sellmeier equations for the optical materials in the laser chain.

Calculating these dispersion coefficients for the stretcher and compressor are a bit more subtle. Because of aberrations imposed by focusing optics it is usually necessary to find them numerically for most stretcher designs by some technique such as ray tracing. For the two-grating Treacy compressor (and an ideal aberration free stretcher) it is possible to find them analytically. To do this, one notes that group delay is simply the physical propagation length of a particular wavelength which can be found from the geometry of the grating pair. There are a variety of ways of deriving and presenting the phase terms. Following Treacy, consider a plane drawn between the two gratings perpendicular to the input beam axis. The diffracted ray propagates a slant distance from G1 to G2 of $\chi(\lambda) = b/\cos[\gamma_{Lit} - \theta(\lambda)]$ (recalling that b is the normal distance between the faces of the parallel gratings, γ_{Lit} is the input incidence angle and θ is the diffracted ray angle). Therefore the total propagation length for a given wavelength from the perpendicular plane at input to output is $z(\lambda) = \chi(\lambda)(1+\cos[\theta(\lambda)])$. The wavelength dependence of this is just given by the grating equation, $\sin\gamma + \sin[\theta(\lambda)] = \lambda/\delta_g$, and the group delay is $\varphi_1^{(comp)} = z(\lambda)/c$. Differentiation of this group delay in a manner similar to that performed earlier for materials leads to the various dispersion orders. These calculations are tedious and will not be reproduced here. The various phase terms, calculated for a complete double-pass trip through the grating pair, presented in terms of the diffracted angle of the central laser wavelength $\theta_0 = \theta(\lambda_0)$ are

$$\varphi_2^{(comp)} = -\frac{\lambda_0^3 d}{2\pi c^2 \delta_g^2 \cos^2\theta_0} \tag{2.29a}$$

$$\varphi_3^{(comp)} = \frac{\lambda_0^4 d}{4\pi^2 c^3 \delta_g^2 \cos^2\theta_0}\left(1 + \frac{\lambda_0}{\delta_g}\frac{\sin\theta_0}{\cos^2\theta_0}\right) \tag{2.29b}$$

$$\varphi_4^{(comp)} = -\frac{\lambda_0^5 d}{32\pi^3 c^4 \delta_g^2 \cos^2\theta_0} \times$$
$$\left\{4 + 8\frac{\lambda_0}{\delta_g}\frac{\sin\theta_0}{\cos^2\theta_0} + \frac{\lambda_0^2}{\delta_g^2}\left[1 + \left(6 + 5\tan^2\theta_0\right)\tan^2\theta_0\right]\right\}. \tag{2.29c}$$

These equations can be used to calculate the dispersion orders of an ideal aberration free stretcher as well if the grating separation distance, d, is set to a negative number. Note that the fourth- and second-order terms always have the same sign and the third-order phase always has the opposite sign. The stretched pulse duration that such a grating pair compresses can be estimated from the GDD of the system, Eq. (2.29a), and Eq. (2.26), noting that the GDD term is large so we can say (since $\Delta\nu \approx (c/\lambda^2)\Delta\lambda$) that

$$\Delta\tau \cong \sqrt{32\ln[2]}\frac{\varphi_2}{\tau} = 4\pi\varphi_2\Delta\nu_{FWHM}$$
$$\cong \frac{2\lambda_0 d}{c\delta_g^2 \cos^2\theta_0}\Delta\lambda. \tag{2.30}$$

This result differs from our simple geometric result of Eq. (2.15) when Littrow is assumed for the diffracted angle ($\theta_0 = \gamma_{Lit}$), by a factor of $\lambda_0/\delta_g \sin\gamma_{Lit}$ – a factor close to unity.

This analysis shows that in principle, the compressor perfectly compensates the induced spectra phase to all orders if the grating incidence angle and effective separation are identical in stretcher and compressor. In practice, this perfect compensation is not possible because the pulse must traverse some amount of dispersive material in the amplifier chain. Consequently, phase control in a CPA laser becomes a design exercise in attempting to fulfill the following conditions simultaneously:

$$\varphi_2^{(stretch)} + \varphi_2^{(mat)} + \varphi_2^{(comp)} = 0 \qquad (2.31a)$$
$$\varphi_3^{(stretch)} + \varphi_3^{(mat)} + \varphi_3^{(comp)} = 0 \qquad (2.31b)$$
$$\varphi_4^{(stretch)} + \varphi_4^{(mat)} + \varphi_4^{(comp)} = 0. \qquad (2.31c)$$

Equation (2.31a) is easy to fulfill; the grating separation of the stretcher or compressor is merely changed to compensate for the GDD from the laser chain materials. The second condition, zeroing third-order dispersion (TOD), is achieved almost as easy by noting that TOD and GDD vary differently with diffraction angle, so tuning the input angle of the stretcher or compressor presents a separate parameter by which the TOD and GDD can be independently reduced to zero. Compensation of dispersion up to third order is usually adequate to achieve good recompression fidelity in laser pulses that are 50 fs or longer. (Phase control demands on these longer pulses are more demanding, however, if high temporal contrast is needed, as discussed later.)

Shorter pulses usually demand at least partial compensation if not complete zeroing of the fourth-order dispersion (FOD). Doing this is more challenging. One solution is to vary the amount of dispersive material is introduced into the chain, a technique that essentially presents a third "tuning knob" to zero all three dispersion equations in (2.31). Introducing a judiciously chosen amount of material in the low-energy section of the laser can often minimize FOD to a manageable level, though this technique is not always convenient and introduces other problems (like nonlinear phase modulation effects). In many CPA designs, incomplete compensation of FOD means that one must find a compromise in choice of GDD to compensate partially for the phase distortion imposed by FOD over the central part of the spectrum.

Other techniques for phase control through the fourth order have been implemented in many CPA systems (Lemoff and Barty 1993), including using prisms in the stretcher, employing mixed groove-density stretcher/compressor gratings or introducing chirped mirrors. In most CPA systems it is quite tedious and sometimes not possible to correct all significant phase terms, particularly in high-bandwidth systems. The spectral phase can vary from small alignment drifts off many optics in the laser, especially the stretcher and compressor angles. Phase control devices have been developed which can be inserted in the chain to correct phase terms and to control the temporal shape of the pulse. Such devices are of particular utility if the CPA laser is to be used in an application where details of the pulse shape and phase profile are important (such as some attosecond pulse generation experiments or in physical chemistry). Two commonly used pulse shapers that can modulate the phase and the amplitude of the pulse in a CPA system are liquid crystal spatial light modulators (SLM) and acousto-optic programmable dispersive filters (AOPDF). The first employs a linear array SLM in the Fourier plane at the mid-point of the imaging telescope in a stretcher to control the phase as a function of wavelength. In an AOPDF (Tournois 1997) the spectral phase of an acoustic wave is directly transferred to the spectral phase of an optical pulse in an optical crystal such as KDP. The acoustic pulse is copropagated in the crystal with the chirped pulse. The phase and the amplitude of the optical pulse are controlled by the phase and amplitude of the acoustic wave.

2.5 CPA Amplifiers and Amplifier Design Issues

2.5.1 Typical Modern CPA Architectures

While there are innumerable laser design architectures that have been successfully implemented in CPA systems, most CPA lasers have a few common design elements. It is, therefore, informative to consider briefly a few example CPA laser designs and then to examine a few of the salient elements of many of these designs.

Figure 2.14 illustrates an example of an early typical CPA laser based on Ti:sapphire delivering compressed pulses of about 1 J with pulse duration of ~30–50 fs at a repetition rate of 10 Hz. (Many such systems have been published. A few good examples can be found in Kmetec et al. (1991), Barty et al. (1994), Chambaret et al. (1996), Bonlie et al. (2000), Pittman et al. (2002) and Aoyama et al. (2003). The design uses a mode-locked Ti:sapphire oscillator to generate the initial 20 fs pulses at a wavelength of 800 nm. These pulses, emerging from the oscillator at a repetition rate of the mode-locked laser cavity near

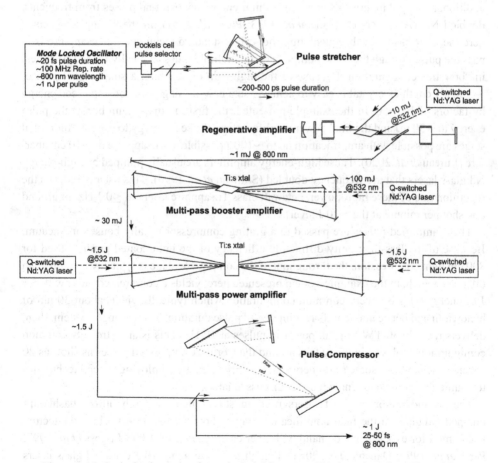

Figure 2.14 Optical schematic of a typical multiterawatt CPA laser based on Ti:sapphire amplification.

100 MHz, each have an energy of about 1 nJ. Pulses are selected by a Pockels cell pulse slicer at 10–1,000 Hz and the pulses are sent into a reflective optic pulse stretcher which chirps the pulses to a duration of ~200–500 ps. These stretched pulses are then injected into a regenerative amplifier (or "regen," an element discussed in greater detail in the next section). The regenerative amplifier employs a Ti:sapphire crystal pumped by pulses from a Q-switched Nd:YAG laser delivering typically 10–30 mJ pulses of 532 nm light at 10 Hz. The seed pulse passes through the laser-pumped sapphire roughly 20 times, boosting its energy from about 100 pJ to ~1 mJ. (In new systems the regen is usually replaced by some kind of OOPCPA architecture, discussed below.) It is common to operate the laser up to this energy level at repetition rate of 1 kHz, though subsequent energy-boosting amplifiers will usually be reduced to 10 Hz (Yamakawa and Barty 2000). It is also possible at this point to compress the pulse and insert some kind of nonlinear pulse contrast-enhancing element, such as an XPW cleaner described in Section 2.5.8.

After the regenerative amplifier, the pulse is further boosted in energy by allowing its beam size to increase by propagation or insertion of a telescope and then passing it through additional Ti:sapphire crystals pumped by high-energy nanosecond pulses from frequency doubled Nd:YAG lasers (Leblanc et al. 1993; Ple et al. 2007). In the example presented here the laser has two subsequent amplifiers. The standard technique in Ti:sapphire is to pass the pulse through the same pumped crystal multiple times (typically four to six times) in "bow-tie" configuration. Because of the high gain of sapphire, a small signal gain of 3–5 in a 1 cm thick crystal is easy to achieve, so a bow-tie stage can boost the seed energy by factors of 10 – 50. In the example presented, the first multipass amp boosts the pulse energy to ~30 mJ and the final power amplifier raises the seed energy to ~1.5 J. Additional stages are possible with amplification up to ~100 J possible in Ti:sapphire of ~10 cm aperture (Lureau et al. 2020). These high-energy amplifiers are usually pumped by high-energy Nd:glass lasers that are frequency doubled (Sullivan et al. 1996). This usually reduces the repetition rate of these high-power, petawatt class Ti:sapphire lasers to ≤ 0.1 Hz (or around one shot per minute at the >20 J level).

These amplified pulses are passed to a grating compressor usually housed in vacuum. Because of the less-than-unity diffraction efficiency of the gold-coated gratings used for 800 nm light compression, the throughput of the compressor stage is typically no more than 60–65%, which, in the example system presented here, yields a compressed pulse with ~1 J of energy. This is a very common configuration which typically yields about 30 nm of bandwidth and hence about 30 fs recompressed pulse duration. This example system, then, delivers roughly 30 TW of peak power in pulses at 10 Hz. This is an extremely common configuration deployed in many labs around the world. (Compressed pulses as short as 20 fs can be achieved in some mid-energy Ti:sapphire lasers by employing special techniques to counteract gain narrowing, an effect discussed later).

The second major class of lasers used in strong field research utilize flashlamp-pumped Nd:glass in the final amplification stages (Powell et al. 1990). This architecture is common for experiments demanding higher energy, say 100–1,000 J (Ross et al. 1997; Perry et al. 1999; Danson et al. 2004), though there are many 10 J class Nd:glass lasers employed worldwide. Figure 2.15 presents a generic, typical modern high-energy Nd:glass

2.5 CPA Amplifiers and Amplifier Design Issues

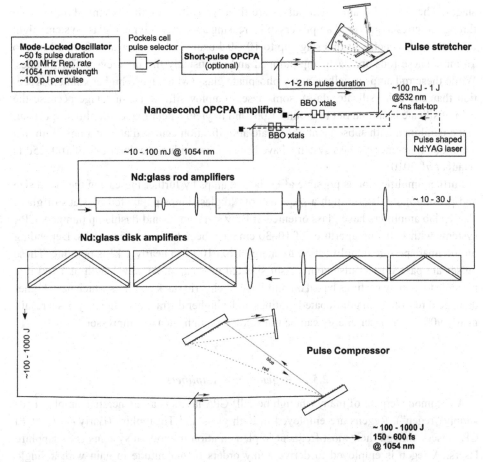

Figure 2.15 Optical schematic of a typical high-energy CPA laser based on OPCPA seeding Nd:glass power amplifiers.

CPA laser. As with the Ti:sapphire system, low-energy ultrafast pulses are generated in a mode-locked oscillator, in this case with center wavelength near the gain peak of phosphate glass, 1,053 nm. These pulses are pulse-selected with a Pockels cell to a few Hz rep. rate. At this point it is possible to boost the energy of the pulses to the millijoule level before significant stretching using a picosecond short pulse OPCPA module for enhanced pulse contrast. This element is proliferating in the newest high-energy CPA lasers (including in Ti:sapphire systems).

The pulse at this point is stretched, typically to duration above 1 ns, to allow high fluence energy extraction in the Nd:glass amps. After stretching, it is now common to perform the initial amplification to the 10 mJ to 1 J level in a series of Ti:sapphire (Rouyer et al. 1993) or, more commonly, OPCPA amplifiers pumped by pulse-shaped, frequency doubled Nd:YAG lasers. This OPCPA stage helps reduce prepulses and helps maintain broad bandwidth prior to final amplification in the moderate-bandwidth Nd:glass stages. These OPCPA-amplified pulses are then typically spatially expanded and passed through a series of glass rod amplifiers of increasing aperture. A typical glass system might boost the pulse energy at this stage up to 10–30 J using rods in apertures from 10 mm to 70 mm. These rods are flashlamp pumped and can usually fire once every few minutes. While these rod amps usually employ phosphate glass (which has a higher gain cross section than the older silicate glass), some lasers employ silicate at this stage because the gain of silicate is shifted out to around 1,061 nm to yield a net broader amplified spectrum when combined with subsequent phosphate amplification centered at 1,053 nm. With this technique high-energy glass systems have been shown with pulse durations of 100–250 fs (Gaul et al. 2010).

Further amplification is possible after the rod amps by further increasing the beam size and passing the pulse through a sequence of Nd:glass amps populated by slabs of glass. These slab amplifiers have glass oriented at Brewster's angle and flashlamp pumped. CPA systems with slab amp apertures of 10–30 cm have been built at various labs. Depending on the final aperture, amplification to energy from 100 J to nearly 1 kJ is possible. These pulses are passed to a pulse compressor situated in vacuum. At wavelength of 1 μm it is possible to employ gratings based on multilayer dielectric stack coatings which have higher damage threshold than gold-coated gratings and a higher diffraction efficiency. As a result, nearly 90% of the laser energy can be transmitted though such a compressor.

2.5.2 Regenerative Amplifiers

A common element of many (though not all) CPA lasers is a regenerative amplifier (or simply "regen"). Regens are employed in both glass and Ti:sapphire (Barty et al. 1996) CPA lasers, though their most frequent implementation in modern systems is in sapphire lasers. A regen is employed to derive many orders of magnitude in gain with a single gain element by passing the pulse numerous times through the amplifier material. This technique often yields a net gain of over 10^6 in a single stage, achieved by passing a gain material with small-signal, single-pass gain of ∼3–10 as many as 15–30 times. A typical regen is illustrated in Figure 2.16 where a Ti:sapphire-based regen is shown.

A regen is usually designed as a stable resonator cavity which amplifies a TEM_{00} mode (Murray and Lowdermilk 1980). This has the advantage of maintaining the good beam quality of the seed on every pass and yielding on output a high-quality, Gaussian spatial profile which can be relay imaged through the remainder of the CPA amplifier chain. Most regens work with two end mirrors, with radii of curvature R_1 and R_2 separated by length L, which satisfy the usual resonator stability criterion

$$0 < \left(1 - \frac{L}{R_1}\right)\left(1 - \frac{L}{R_2}\right) < 1. \tag{2.32}$$

Since a large mode is desired in the optics, a regen is usually designed so that the damage-prone, sensitive optics (such as Pockels cells and gain material) are located in the regions of the resonator cavity where the mode is largest (usually near the end mirrors).

2.5 CPA Amplifiers and Amplifier Design Issues

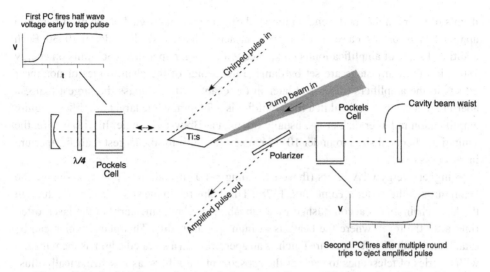

Figure 2.16 Optical layout for a typical regenerative amplifier based on Ti:sapphire amplification.

The other important element of a regen is the optical switch (or switches) which captures a single pulse for multiple passes and then expels it when full amplification has been completed. This is most often accomplished with a combination of polarizers and Pockels cells driven by ns-scale, high-voltage pulses. Since retention of broad bandwidth is of great importance in CPA, and the regen has the disadvantage that bandwidth-limiting optics such as polarizers are passed multiple times, it is preferred to design the regen with as few optics as possible. This can be achieved in practice with modern HV switching technology which allows a single Pockels cell to be switched on and off two times within a couple of hundred ns to perform both pulse capture and switching out with the same cell (as opposed to the more simple design using two quarter-wave cells shown in Figure 2.16). It is furthermore possible to place spectral filters that preferentially pass the wings of the spectrum to counteract gain narrowing during amplification and permit broader band amplification.

Because of the bandwidth limitation that regens impose through the need for bandwidth-limited optics, the shortest pulse CPA systems (those operating with durations below 30 fs) usually forgo with a regen and supplant it with a high-gain multipass amplifier, or OPCPA technology. This also avoids the necessity to compensate for meters of material dispersion that can easily accumulate in regens, though with the trade-off of losing the spatial profile filtering of a regen. Without employing intercavity spectral filtering, most Ti:sapphire regens pass about 30 nm of bandwidth, sufficient for compression to about 30 fs.

2.5.3 Relay Imaging and Spatial Filtering

Subsequent to amplification in the regen (or other high-gain initial amplifier element like an OPCPA stage), higher-energy (>100 mJ) CPA systems require power amplifiers. The details of their amplifiers depend on the kind of gain material used, but the most common approach is to construct a series of amplifiers, each with a gain of the order of 10–30. Each additional stage of amplification is conducted with a larger aperture. The limits on energy extraction from any stage are set by damage thresholds of the chain optics or nonlinear effects in the amplifiers (discussed later in the chapter). In Nd:glass the largest realistic aperture that can be obtained in a rod amplifier is ~70 mm, while larger amplifiers require amplification in larger disk amps, usually oriented at Brewster's angle. In Ti:sapphire, the limitation of crystal sizes to under 10–15 cm sets the limit for the largest realistic aperture in these systems.

In higher-energy CPA lasers (those exceeding ~1 J) it is desirable to relay-image the beam through the system (Hunt et al. 1978). This helps retain the smoothest beam through the laser chain and is accomplished by establishing a relay plane early in the laser, often right after the regen where the beam is of high spatial quality. This plane, which can be established with a "soft" aperture (such as an aperture with a serrated edge), is then imaged with a series of telescopes to each of the subsequent amplifiers, as is schematically illustrated in Figure 2.17. This technique is critical in most high-energy glass lasers because of the high fluences one must operate at to achieve good energy extraction. This relay of

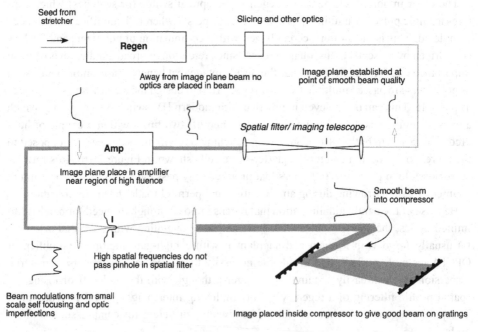

Figure 2.17 Illustration of how relay imaging and spatial filtering can be used to optimize energy extraction from flashlamp-pumped amplifiers by propagating a nearly flattop spatial beam profile. The flattop beam is relay-imaged through the system by a series of telescopes that throw the image of a smooth flattop beam to the output of each amplification stage. Small-scale intensity fluctuations are removed at each stage by spatially filtering the beam with pinholes in each telescope.

the initial image is designed to place the image at the highest fluence, most damage-prone optics, including the final grating in the compressors. The relay telescopes will usually have to be focused in vacuum. The beam can be spatially cleaned by the insertion of pinholes at the focus. Because the focus of a beam essentially represents the 2D Fourier transform of the beam being focused, high-spatial-frequency modulations on the beam or phase aberrations will focus to large radii.

Passing through a pinhole at this far field plane removes these high-frequency modulations. A good rule of thumb in designing such spatial filters is to place a pinhole which is roughly 10 times the diffraction limit of the focused beam size at the telescope focal plane.

2.5.4 Gain Saturation and Efficient Energy Extraction

As with any laser amplifier system, efficient laser amplification in CPA laser system is best achieved by saturated amplification in the laser amplifier (Frantz and Nodvik 1963). This occurs when the extracted pulse fluence approaches or exceeds the material's saturation fluence, $h\nu/\sigma$. This fluence is about 1 J/cm^2 in Ti:sapphire, and most CPA systems employing Ti:sapphire, for example, work with amplified seed pulse fluences very near this value. Approaching the saturation fluence is somewhat more difficult in Nd:glass because of nonlinear effects in the amps and damage threshold constraints of the chain optics. The standard approach in CPA is to stretch the pulse sufficiently to allow operation near the saturation fluence while staying safely below the damage threshold of optics, a fluence which typically increases as the square root of the pulse duration. As a consequence, most Ti:sapphire lasers operate with stretched pulses only 200–500 ps while the higher-saturation fluence glass lasers work better with pulses that are longer (realistically 1–2 ns).

Gain saturation can have a modest effect on the amplified pulse spectrum. Because the seed pulse is chirped, a pulse undergoing saturated amplification sees higher gain near the beginning of the pulse, which corresponds to the long-wavelength portion of the spectrum of a positively chirped seed pulse. This will tend to pull the amplified spectrum toward the red since most CPA lasers operate with positive chirp. This can have a modest impact on the amplified bandwidth and hence the pulse duration upon compression.

2.5.5 Gain Narrowing and Bandwidth Considerations

A more important consideration in keeping the broad bandwidth of pulses in CPA lasers is to minimize the narrowing associated with the finite bandwidth of the gain media. As mentioned, gain narrowing in Nd:glass is the most significant factor in determining the pulse width of the recompressed pulse. Gain narrowing is a consequence of the fact that the spectrum of the amplified seed pulse, in the limit of small signal, unsaturated, amplification is

$$S_{amp}(\lambda) = S_{seed}(\lambda) \exp\left[\Delta n \sigma_0(\lambda) l\right], \tag{2.33}$$

where S_{amp} and S_{seed} are the seed and amplified spectra, respectively, Δn is the gain inversion density, l is the gain medium propagation length and $\sigma_0(\lambda)$ is the

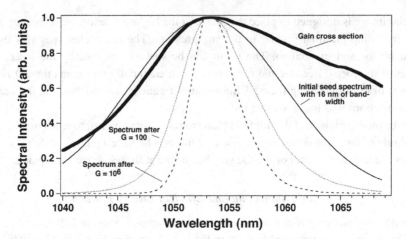

Figure 2.18 Calculation of gain narrowing in unsaturated Nd:phosphate glass amplifiers of a laser pulse with 16 nm of initial bandwidth. Narrowing gains of 100 and 1,000,000 are shown.

wavelength-dependent cross section. Because $\sigma_0(\lambda)$ will be peaked at one wavelength with some finite width (as illustrated, for example, in Nd:glass in Figure 2.3), the exponentiation of the gain cross section leads to an amplified spectrum which is narrower than the seed spectrum. Figure 2.18 illustrates the spectrum of a transform-limited 100 fs pulse centered at a wavelength of 1,053 nm amplified in Schott LG-750 Nd:glass (used in the NOVA laser) by gain factors of 100 and 10^6. Because of the time-bandwidth requirement on the pulse duration, which were stated in Eq. (2.24) for a Gaussian shaped pulse, the narrowing of the spectrum directly affects the best recompressed bandwidth possible.

The gain narrowing that a seed pulse will experience in a CPA chain in the absence of gain saturation can be estimated using Eq. (2.33) and assuming that the seed and amplified spectra are Gaussians as well as the spectral shape of the gain cross section (all not completely accurate approximations). Taylor-expanding the gain cross section spectrum in the exponent leads to the following formula for estimating the extent of gain narrowing:

$$S_{amp}(\lambda) = S_{seed}(\lambda) \exp\left[\Delta n l \sigma_0 e^{-(\lambda-\lambda_0)^2/\Delta\lambda_{gain}^2}\right] \quad (2.34)$$

$$S_{amp} \exp\left[-\frac{(\lambda-\lambda_0)^2}{\Delta\lambda_{amp}^2}\right] = S_{seed} \exp\left[-\frac{(\lambda-\lambda_0)^2}{\Delta\lambda_{seed}^2}\right]$$
$$\times \exp\left[\Delta n l \sigma_0 \left(1 - \frac{(\lambda-\lambda_0)^2}{\Delta\lambda_{gain}^2} + \cdots\right)\right]$$

$$\cong S_{seed} G_0 \exp\left[-(\lambda-\lambda_0)^2 \frac{\Delta\lambda_{gain}^2 + \Delta\lambda_{seed}^2 \ln[G_0]}{\Delta\lambda_{seed}^2 \Delta\lambda_{gain}^2}\right] \quad (2.35)$$

$$\Delta\lambda_{amp} \approx \frac{\Delta\lambda_{seed} \Delta\lambda_{gain}}{\sqrt{\Delta\lambda_{gain}^2 + \Delta\lambda_{seed}^2 \ln[G_0]}}. \quad (2.36)$$

Here $\Delta\lambda_{amp}$ is the amplified bandwidth, $\Delta\lambda_{seed}$ is the seed bandwidth, and $\Delta\lambda_{gain}$ is the gain cross section bandwidth. G_0 is the total small-signal gain of the amplifier system at the spectral peak of the gain. Equation (2.36) illustrates, for example, that if a seed pulse at 1,054 nm wavelength with sufficient bandwidth for a 100 fs transform-limited Gaussian pulse (16.2 nm) is amplified in LG-760 Nd:glass, which has a gain cross section bandwidth of about 19.5 nm, by a gain of 10^6 the resulting pulse will retain only 4.9 nm of bandwidth, broadening the recompressed pulse to at least 320 fs. If G_0 is very large, the amplified pulse bandwidth becomes independent of the seed bandwidth and approaches that bandwidth of the gain material divided by $(\ln[G_0])^{1/2}$.

It should also be mentioned that, because the seed pulse is chirped, its pulse duration is directly related to its bandwidth. Gain narrowing therefore has the secondary effect of shortening the stretched pulse during amplification. This can have deleterious effects as the pulse propagates through the amplifiers gaining energy. The damage threshold of most optics decreases as the square root of the pulse duration and the effective intensity resulting in B-integral accumulation increases with this pulse shortening.

It is also important to note that other factors can narrow the seed pulse's spectral bandwidth in a CPA laser, with the accompanying effects discussed earlier. The limited bandwidth of optics employed in the CPA system can set an upper limit on the effective pulse bandwidth which can be amplified. This is most often a problem in CPA lasers operating with substantial spectral bandwidth, such as Ti:sapphire lasers designed to amplify recompressed pulses with duration under 50 fs. These systems often require bandwidth in excess of 30 nm, which is taxing for many optics, particularly thin-film polarizers, which are frequently utilized in regenerative amplifiers and pulse-slicing (contrast-enhancing) modules.

2.5.6 Nonlinear Self-Focusing, Self-Phase-Modulation and the B-integral

Avoiding nonlinear phase accumulation by an intense pulse as it is amplified in laser gain media is one of the principal motivations for the CPA technique. This nonlinear phase is a consequence of the fact that the refractive index of any material in fact has a dependence on intensity often written as

$$\eta = \eta_0 + \gamma_{NL} I, \qquad (2.37)$$

where γ_{NL} is related to the commonly cited nonlinear refractive index, η_2 (or usually n_2 in the notation of the nonlinear optics literature which defines it by $n = n_0 + n_2 |E_0|^2$) through the relation $\gamma_{NL} = 40\pi \eta_2/c\eta_0$. This intensity dependence to the refractive means that particularly intense pulses will alter the refractive index in a significant way, increasing the index in regions of the pulse with the highest intensity. The extent to which this nonlinear change in refractive index is important in a CPA chain can be quantified by calculating the total integrated phase induced by the nonlinear term, a quantity called the B-integral:

$$B = \frac{2\pi}{\lambda} \int_0^l \gamma_{NL} I(z)\, dz, \qquad (2.38)$$

Table 2.2 *Summary of the nonlinear refractive indices of some common CPA optical materials*

Material	Nonlinear refractive index, $\gamma_{NL}(\text{cm}^2/\text{W})$
Phosphate glass	2.9×10^{-16}
Silicate glass	3.9×10^{-16}
Sapphire (o)	2.9×10^{-16}
Sapphire (e)	3.1×10^{-16}
BK-7 glass	3.4×10^{-16}
Fused silica	2.5×10^{-16}
KDP (o)	2.1×10^{-16}
KDP (e)	2.2×10^{-16}
MgF$_2$	0.76×10^{-16}
LiF	0.70×10^{-16}

where l is the total propagation length through a material with γ_{NL}, and $I(z)$ is the intensity of the pulse as a function of propagation position (unchanging for a collimated beam in most materials but growing as a pulse propagates through an amplifier, for example.)

The values of γ_{NL} for a variety of optical materials commonly employed in CPA are summarized in Table 2.2. These values suggest that B will be significant (i.e. $\sim$1) when pulses of about 5 GW/cm^2 propagate through 10 cm of material. This tends to set the pulse duration that the seed pulse must be stretched in CPA as we have discussed previously. For example, if an amplifier is designed to operate with an output fluence of 1 J/cm^2, this B-integral requirement mandates pulses with duration of >200 ps.

The accumulation of significant B in an amplifier chain can have three important deleterious consequences in CPA amplifiers. First, as in any high-power laser amplifier, nonlinear phase can lead to a phenomenon known as self-focusing. In regions of the laser beam's spatial profile where the intensity is highest, the slightly higher refractive index results in concave curved phase fronts, leading to focusing. A focused beam becomes more intense, leading to further self-focusing ending in damage to the optical material. While this can occur for the entire laser beam, as most beams will have higher intensity toward the center of the beam, this self-focusing effect tends to be most catastrophic for small intensity perturbations on an otherwise smooth beam. This effect is known as small-scale self-focusing and is the ultimate limit to the peak power that can be extracted from a laser amplifier. It has been shown that the spatial scale of the fastest growing perturbation in small scale self-focusing tends to grow with B as e^{2B}. So this suggests that the total B-integral must be such that $\Sigma B \leq 1 - 2$ to tame this effect. The usual way of preventing small-scale intensity perturbations from growing to levels that damage the optical components is to insert spatial filters in the chain at intervals along the propagation. It is usually a good design goal to keep $B \leq 1$ between each spatial filter in the amplifier chain.

B-integral accumulation in a CPA amplifier chain has a second consequence. Because of the temporal intensity dependence of the chirped pulse, there will be a nonlinear phase

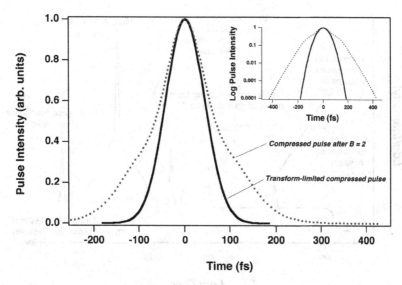

Figure 2.19 Calculation showing the effects that B-integral of 2 has on the recompression of a chirped pulse with initially a 50 fs Gaussian transform limit.

profile imparted to the time profile of the pulse. Since the seed pulse is chirped and there is a one-to-one correspondence of time to wavelength, this time variation of phase results in phase imparted directly on the pulse spectrum (Perry et al. 1994). This phase is not easily compensated by the stretcher/compressor pair, alone, and will tend to distort, and broaden the compressed pulse. An example of pulse broadening on a compressed pulse with a Gaussian spectrum after it has experienced a B-integral of 2 is illustrated in Figure 2.19. As this calculation indicates, it is usually best to keep the total B-integral through the entire CPA chain to ≤ 1.

The third negative effect that B-integral has is in inducing prepulses on the main laser pulse. This effect is discussed in more detail in the next section.

2.5.7 Pulse Temporal Contrast Control

In many applications of CPA lasers, it is necessary to focus the high-power pulse to a high intensity, frequently above 10^{18} W/cm^2 and occasionally up toward 10^{22} W/cm^2 with the highest-power systems. It is often desirable for such an intense pulse to interact with a target, solid, gas or otherwise, that has been largely undisturbed until the subpicosecond pulse rises quickly to its peak intensity. Unfortunately, many processes in a CPA amplifier chain result in pulses which contain a nonnegligible amount of energy in a temporal window, well ahead (many ps to ns) of the rise of the main pulse, which would otherwise rise from negligible intensity to its peak in only a few hundred femtoseconds in most cases. Because most materials break down into a plasma state, and begin to expand, when illuminated with intensity as low as 10^{12} W/cm^2 (depending on material and pulse duration), it is obvious that any such prepulses can perturb the target, even if they contain only

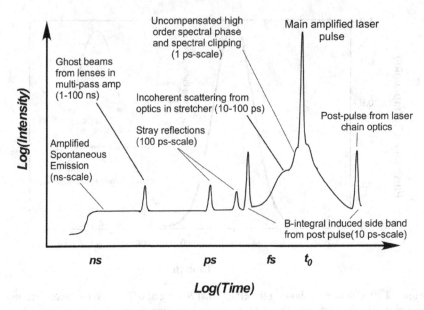

Figure 2.20 Schematic illustration of the various types of prepulses that can arise in CPA lasers along with their approximate time scale before the main pulse.

$10^{-5}-10^{-10}$ of the main pulse's peak intensity. Consequently, suppression of prepulses in CPA lasers is a major concern for many designs that will be used in experiments.

There are many reasons for the formation of prepulses, and there have been a number of methods published to address some of these problems. Figure 2.20 shows some of the zoo of prepulses that can arise and the sort of time scales that they appear in. While the formation of prepulses is strongly dependent on the details of the laser design, there are a few well-known reasons for the formation of prepulses in CPA lasers.

1) Amplified Spontaneous Emission. Any gain medium, when excited to a state of inversion, can spontaneously emit radiation which can be subsequently amplified in the gain medium prior to the passage of the main, chirped seed pulse. This amplified spontaneous emission (ASE) yields a long pedestal many nanoseconds before the main pulse. The ASE pedestal can be suppressed to a large extent by optical switching with Pockels cells; however, the nanosecond time-scale rise of most high-voltage pulse slicers results in a prepulse of at least a few hundred picoseconds duration. In Ti:sapphire amplifiers the ASE level is often about 10^{-6} that of the peak power of the main pulse. Typically a regenerative amplifier is the source of much of this ASE. There are techniques to suppress this ASE prior to amplification in a power amplifier chain, as discussed later in Section 2.5.10. OPCPA amplifiers also exhibit ASE but only over the time scale of the pump pulse when gain is present. Therefore, as described in Section 2.5.11 nanosecond time scale ASE can be largely eliminated if much of the system gain comes from an OPA stage that uses a pump that is only picoseconds in duration.

2) Nonlinear side band formation: It is quite common for a laser system to generate *post-pulses* in any amplifier chain. For example, any flat transmissive optic, such as a waveplate or Pockels cell, can generate a postpulse through a back-and-forth double bounce of a small fraction of the main pulse. Usually such postpulses are of no consequence to any strong field experiment; however, because the main pulse (and hence the replica postpulse) is chirped and long in duration in the CPA amplifier chain, they usually overlap. If these pulses then experience some nonlinear phase (i.e. significant B-integral) they can essentially mix, producing replica temporal sidebands both after *and* before the main pulse. Upon compression, these nonlinear sidebands also compress and result in prepulses of short duration at points ahead of the main pulse corresponding to the temporal delay of the initial postpulse. Because it is often very difficult to eliminate all postpulses in a system, and all CPA lasers must be designed with some intrinsic B-integral, these postpulses are notoriously difficult to eliminate.

3) Spectral clipping and uncompensated spectral phase: Both the pulse stretcher and compressor have finite aperture optics which means that, in practice, they have a finite window to pass the broad bandwidth of the pulse. This window can often result in a hard cut to the spectrum. When a sharp edge in the spectrum of a pulse is Fourier transformed, side lobes appear in the transform-limited pulse profile. These lobes can be significant and act as effective prepulses in the time window of a few picoseconds. In practice, the spectral transmission window of the stretcher should be 3–4 times larger than the spectral width, FWHM, of the final pulse spectrum. The sharpness of the spectral edge reduces with an increasing beam size so the spectral window in the compressor (where the beam will be much bigger than in the stretcher) can be smaller. Another cause of prepulse in the time window of a few hundred femtoseconds is uncompensated spectral phase (such as third-order dispersion). While such phase may be reduced to the point that the main pulse compresses to nearly its original duration, a small amount of residual phase can lead to pulse distortion in the wings at intensity $\sim 10^{-2}$ of the main pulse and yield a pulse that has much higher intensity in the wings than a pure Gaussian or sech^2 pulse. Only further phase compensation (often with a phase control device as discussed earlier) can eliminate this source of prepulse.

4) Scattering and ghost beams in multipass amplifiers. Many CPA architectures employ multiple passes through an amplifier to enhance energy extraction and give high gain in a single stage. It is possible, then, for scattering from optical surfaces on an early pass to send laser energy back into the output line well before the main pulse. Typically with good high-quality antireflection coatings on optics and the fact that these backscattered beams do not compress or focus as well as the main pulse, these ghost beam prepulses are quite low-intensity, perhaps $<10^{-8}$ of the main pulse. However, they do occur many nanoseconds before the main pulse and can be a huge concern in some strong field experiments. Removing lenses from the multipass architecture (replaced with all off-axis parabolic focusing optics) and careful spatial baffling can greatly reduce the number and intensity of these ghost beam prepulses. However, the only way to eliminate them completely is to employ an amplifier design that is solely single pass through all amplifiers.

5) *Incoherent scattering from optics in the stretcher.* Sometimes a broad wing of prepulse in the ~100 ps time window is observed. This can sometimes be attributed to scattering of light in the stretcher where different frequencies are taking different group delay paths. This effect seems to be a consequence of the grating itself and is variable depending on the grating quality and type.

2.5.8 Cross-Polarized Wave Generation

There have been a number of techniques implemented to clean temporally amplified pulses in CPA. Most of these techniques have aimed at reducing the level of ASE background which plagues Ti:sapphire lasers in particular. The general idea is to conduct much of the CPA pulse amplification up to the millijoule level, which typically constitutes the bulk of the gain factor (from say 100 pJ to 1 mJ) and hence produces most of the ASE. Then pulses are compressed and cleaned with some nonlinear optical element. One commonly implemented technology in many CPA front ends is the cross-polarized wave generation (XPW) technique (Jullien *et al.* 2009; Ricci *et al.* 2013). This element is often inserted after a regen in a Ti:sapphire system. Pulses must be compressed at this stage and passed through the XPW system. After this, the temporally clean pulses are restretched and further amplified to full energy.

XPW is a four-wave degenerate nonlinear process, akin to the Kerr effect in a medium (where the refractive index is locally changed by the intense field). This process, however, can only take place in a medium which exhibits an anisotropy in its third-order nonlinearity. BaF_2 is the most common material used with this property. The nonlinear process can be thought of as three linearly polarized photons interacting to produce a single photon at the same frequency but with polarization perpendicular to the input three photons: $\omega^{(\perp)} = \omega^{(\parallel)} + \omega^{(\parallel)} - \omega^{(\parallel)}$. As a result, a perpendicularly polarized pulse is produced in the crystal with intensity profile that is essentially the third power of the input pulse. Consequently ASE and other prepulses can be reduced from the 10^{-5} or 10^{-6} level to many orders of magnitude greater contrast.

A typical implementation of XPW at the mJ level is shown in Figure 2.21. The XPW-generating crystals are located between two crossed polarizers which essentially reject the

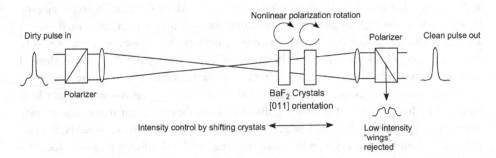

Figure 2.21 Optical schematic of how crossed polarized wave generation (XPW) can be used to clean temporally an ultrafast laser pulse.

incoming linearly polarized pulse in the absence of any XPW polarization rotation. The pulse is focused to increase its local intensity in the XPW crystals with the location of the crystals scanned after the focus to find the optimum intensity for maximum polarization rotation efficiency. Typically two BaF_2 crystals in either Z-cut or holographic cut orientation are used. Two crystals improve the efficiency of passing the main pulse; single crystals saturate at about 10 percent efficiency while two crystals can be optimized to yield nearly 30 percent passage. The unrotated weak fields of the prepulse are rejected by the second polarizer, though the ultimate contrast that can be achieved is determined by the extinction of the polarizer employed so, in practice, contrast enhancement of three to four orders of magnitude is typical. This process has an auxiliary benefit in that it shortens the main pulse by a factor of about $3^{1/2}$ (and hence broadens the usable bandwidth by the same factor) because of the I^3 dependence of the process.

2.5.9 Picosecond OPA Front Ends

Another contrast enhancement approach, which has found widespread application in many modern CPA systems, eliminates the nanosecond ASE prepulse by never generating it in the first place. This technique involves boosting the CPA seed pulse to the mJ level by using parametric amplification pumped by laser pulses that are only 20 ps or shorter, so called short pulse- or picosecond OPCPA (SP-OPCPA) (Tavella et al. 2005; Ahmad et al. 2009). The idea is that in parametric amplification, there is no energy storage medium which can decay by spontaneous emission so ASE can only occur in a time window over which the amplifier crystal is pumped by the pump laser pulse. If a pulse of only 10 or 20 ps is employed, there can be no ASE outside this window, making ASE contrast from a few picoseconds in front of the main pulse forward essentially infinite.

The challenge of this approach is that because of the short duration of the pump pulse, precise ($\sim$1 ps) optical synchronization must exist between the pump and seed pulse. This optical precision must usually be achieved by deriving the pump and seed pulses from the same mode-locked laser source. An example of SP-OPCPA implementation is shown in Figure 2.22. The pulse train of a mode-locked laser operating at the seed pulse wavelength (800 nm say for a Ti:sapphire amp chain, 1 µm for a glass-based system) is split with the majority of the train going to seed the pump laser. Because of the much longer optical path delay in the pump laser, the actual pulse chosen for pump laser amplification is selected from the mode-locked pulse train a few ns before the pulse selected for the seed. (In practice, subpicosecond synchronization between pulses so close together in the pulse train is still maintained.)

If the seed laser is at 800 nm, it is necessary to broaden the spectrum so that there is some spectral component out at 1 µm for seeding the Nd:YAG- or YLF-based seed laser. This can be done in a micro-structured fiber (akin to how this is performed for CEP stabilization). The seed to the pump is amplified in a YAG or YLF system typically composed of a regen and a power amplifier. YAG can support pulses with duration around 20 ps while YLF can yield slightly shorter pulses around 8 ps. These amplified pulses are frequency doubled to 0.5 µm and sent, with appropriate time delays, into short parametric

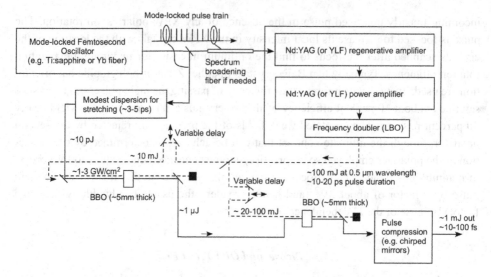

Figure 2.22 Optical schematic of a typical short-pulse, picosecond OPCPA front end. Pulses from the same mode-locked pulse train are used to seed a picosecond-pulse-duration pump laser and the ultrafast beamline, ensuring subpicosecond time synchronization. Gain narrowing in the YAG (or YLF) pump source leads to a frequency-doubled pump pulse that is ~20 ps (or ~8 ps in YLF). This pulse is used to pump OPA amplifiers that amplify the broadband seed. Spontaneous amplification outside the window of the picosecond pump pulses is therefore completely eliminated.

amplifier crystals. 5-mm-thick BBO pumped at 1–3 GW/cm² is typical. The seed beam line on the other hand has the selected seed pulse slightly stretched (to perhaps 3–5 ps) to derive good temporal overlap with the pump pulse and then it is passed through a series of BBO amplifiers operated with broad bandwidth pumped by the picosecond green pulses. Gain of 1,000–10,000 per crystal can be derived, yielding mJ seed pulses with two or three BBO stages. These pulses can be compressed (often with chirped mirrors) or they can be further stretched for amplification in the power amplifiers of the rest of the CPA laser chain. This technique has been shown to yield pulse contrast of better than 10^{-10} over a window from a few picoseconds forward of the main pulse.

2.6 Beam Focusability and Adaptive Optics

2.6.1 CPA Beam Focusing Considerations

Once the CPA pulses are compressed, generating strong fields requires focusing those pulses to a small spot. For systems with modest peak power (say up to 1 TW) it is possible to focus the pulses with a simple plano-convex lens. Spherical aberrations in such an optic, however, limit high-quality focusing to f/#s (where the f/# is defined as the ratio of the focal length of the optic to a suitably defined beam diameter) of greater than about f/20. Furthermore, nonlinear effects (B-integral) in the lens can impart spatial phase that

2.6 Beam Focusability and Adaptive Optics

degrades the focus or causes filamentation of the beam and optic damage. These nonlinear effects in the lens can be mitigated to some extent by employing a lens manufactured from a low-n_2 material like MgF or LiF (see Table 2.2). Lenses are widely employed in strong-field high-harmonic generation experiments that work best with long focal lengths. However, to derive the highest focal intensities, with f/#s from 10 down to 1, it is typical to employ parabolic shaped off-axis focusing mirrors. These optics are often very sensitive to alignment. Such focusing is almost always performed in a vacuum chamber directly linked to the vacuum chamber housing the pulse compressor.

The tightest focal spots and hence highest intensities that can be achieved are with beams that have an absolutely flat phase front across the beam. In that case, a beam with a Gaussian spatial profile (a TEM$_{00}$ spatial mode) will focus (with a perfect, aberration-free focusing optic) to a $1/e^2$ spot size (diameter) of 4 $(\lambda/\pi) f/\#$ if the f/# is defined as $f/2w_0$, w_0 being the radius of the $1/e^2$ intensity point of the beam incident on the focusing optic. In many large CPA systems the beam passed from the compressor will be closer to a flat-top beam, more accurately described by a high-order super-Gaussian. In that case, the far-field pattern is better described by the Airy pattern which has a diameter defined at the first radial minimum of the focal distribution equal to 2.44 $\lambda f/\#$. This suggests that in most strong field physics experiments the best focal spot size that can possibly be achieved is a few wavelengths in diameter. So a 1 TW laser might reach peak intensity of $\sim 1 \times 10^{19}$ W/cm^2 when focused to a radius of 2 wavelengths, and a PW laser can reach intensity exceeding 10^{22} W/cm^2.

In practice, achieving such tight focal spots is experimentally very challenging. They require active correction of the wavefront as we will discuss later in the chapter. As a result, typical numbers for usual CPA lasers are at peak intensities about an order of magnitude below these ideal values. Actual beams are imperfect, having phase aberrations or a combination of spatial modes. While it is common in laser physics to quantify this non-ideal focusing behavior by the M^2 factor (Sasnett 1989), it is more common and more useful from the standpoint of a strong field experiment, however, to quantify the focal spot intensity in terms of the Strehl ratio S. S is a factor ≤ 1 equal to the ratio of the peak focused intensity to the focused intensity of an ideal beam with no phase aberrations at all (diffraction limit). For example, $S = 0.25$ means that the peak intensity achievable is one-fourth the intensity of the same beam profile with a perfectly flat phase front focused with a perfect optic.

The degradation of the beam's Strehl ratio from phase front modulation has many sources in a CPA laser system. In reality, an optical system such as a CPA amplifier chain always has some aberrations that lead to an inhomogeneous phase front and degrade the focus quality. Aberrations accumulate throughout a CPA laser from surface and bulk errors in optics and from misalignment. Surface wavefront errors arise, for example, from polishing errors or manufacturing limitations, or from stress deformations in optics due to mounting stresses or sag from gravity. Wavefront errors can accumulate in transmissive optics because of material anisotropies, and thermal- or stress-induced variations of the index of refraction. Thermal effects are particularly prevalent in the amplifiers which must be pumped to drive an inversion and inevitably are heated. Chromatic aberrations in lenses can result in wavefront errors with transverse spread of the focus and can result in longitudinal focusing errors as well. Slight misalignment of gratings in the pulse compressor can degrade the wavefront. Finally, nonlinear optical effects upon transmission of the amplifiers and other optics (again B-integral induced self-focusing effects) can further degrade the beam wavefront quality.

The mathematics of quantifying the effects of wavefront aberrations have been extensively studied. It is worth here making a few comments on the characterization of these aberrations in the context of CPA laser beams. The final (usually round) beam can be characterized by a wavefront phase function of the form $\varphi_{WF}(\rho, \theta)$, where ρ is the normalized radius in the beam (varying from 0 to 1 at the edge of the beam) and θ is the azimuthal angle in the round beam. A perfect beam would have $\varphi_{WF}(\rho, \theta) = constant$. If uniform intensity is assumed over the beam, simple diffraction theory tells us that the Strehl ratio is simply

$$S = \left|\left\langle e^{i\varphi_{WF}(\rho,\theta)} \right\rangle\right|^2, \quad (2.39)$$

where the brackets denote an average over the beam front. Clearly it is desirable in a CPA laser to keep φ_{WF} as constant as possible over a beam to yield a small high-intensity focus.

It is common to describe an aberrated wavefront in terms of Zernike polynomials, $Z_i^j(\rho, \theta)$, in the form of an expansion over these polynomials:

$$\varphi_{WF}(\rho, \theta) = \sum_{i,j} c_i^j Z_i^j(\rho, \theta), \quad (2.40)$$

where the c_i^j are the amplitudes of the aberrations described by that Zernike polynomial and the sum is over radial order, i, and angular frequency, j. For example, keeping the first eight terms of a Zernike polynomial expansion, the wavefront of a beam could then be characterized as

$$\varphi_{WF}(\rho, \theta) = c_0 + c_1 \rho \cos\theta + c_2 \rho \sin\theta + c_3(\rho^2 - 1) + c_4 \rho^2 \cos 2\theta$$
$$+ c_5 \rho^2 \sin 2\theta + c_6(3\rho^2 - 2)\rho \cos\theta + c_7(3\rho^2 - 2)\rho \sin\theta \quad (2.41)$$
$$+ c_8(6\rho^4 - 6\rho^2 + 1) + \ldots,$$

where these first eight terms describe the following phase aberrations: piston, x-tilt, y-tilt, focus, astigmatism at 0 degrees, astigmatism at 45 degrees, coma with x-tilt, coma with y-tilt and spherical aberration, respectively. A useful way to approximate the Strehl that might be expected from an uncorrected sequence of aberrations from a laser chain is to estimate from the quality of the optics in the chain the root-mean-square deviation of the wavefront phase

$$\sigma_{WF}^2 = \left\langle \left(\varphi_{WF}(\rho, \theta) - \bar{\varphi}_{WF}\right)^2 \right\rangle, \quad (2.42)$$

where $\bar{\varphi}_{WF}$ is the average of the wavefront phase over the beam. The Strehl ratio can be estimated from this single number by the Ruze formula in antenna theory (Born and Wolf 1980):

$$S \approx e^{-\sigma_{WF}^2}. \quad (2.43)$$

It is usually desirable to construct CPA lasers such that the Strehl ratio is higher than about 0.5 (which is roughly equivalent to saying that the CPA beam can be focused to a spot size about a square root of two times its diffraction-limited beam size). Strehl ratios ~0.8 or 0.9 are typically possible with corrective adaptive optics in most modern CPA chains.

2.6.2 Adaptive Optic Technologies Used in CPA

Strehl ratio and peak intensity of a CPA system can often be enhanced by the deployment of an adaptive optic to correct for many of the aberrations imposed on the pulse. This technique, drawn in large part from observational astronomy, is most critical in CPA amplifier chains working at laser aperture such as PW-class Nd:glass lasers, where maintenance of beam quality adequate for good focusing is not possible over the large optics employed. Corrective adaptive optics are also quite helpful on systems that generate considerable thermal aberrations because of elevated repetition rate, such as Ti:sapphire amps of a few cm aperture operated at >1 Hz (Planchon et al. 2005). In that case, it is desirable to have some sort of active correction on the wavefront aberrations that includes the insertion of an adaptive optic (deformable mirror), some wavefront sensor near the end of the chain to diagnose the wavefront aberration and some kind of feedback loop electronics. Adaptive optics are a technology that aims at correcting the aberrated wavefront of an optical system to improve its imaging or focusing quality to the transform limit (Nees et al. 1998). Adaptive optics have been used in astronomical telescopes to correct atmospheric distortions and to correct the human cornea and lens in retinal imaging. In using adaptive optics, the wavefront distortions are first measured across the laser beam with some sort of wave-front sensor and then corrected with an optic which can impart spatial phase over the aperture of the beam, usually done with a deformable mirror or a liquid crystal array. Adaptive optics has been used over the last decade to improve the focusability of many high energy CPA laser pulses.

The most common wavefront sensors used in CPA are the Shack–Hartmann sensor (Hardy 1978) and shearing interferometers (Rimmer and Wyant 1975; Chanteloup et al. 1998). The Shack–Hartmann sensor is formed with an array of micro-lenses, each with the same focal length ("lenslets") that focus the beam on a CCD sensor. The tilt of the local wavefront across each lenslet can be determined from shift of the focal spot with respect to a reference. The wavefront is then reconstructed from the discrete step of tilts over each lenslet. A typical Shack–Hartmann sensor for CPA laser wavefront sensing might have a lenslet array with $\sim 10^3$ lenses that focus on a megapixel CCD sensor which can resolve up to $\lambda/100$ wavefront errors. Figure 2.23 illustrates how such a wavefront sensor used in conjunction with a deformable mirror can improve the quality of the wavefront phase and lead to high Strehl ratios under tight focusing. Here a Shack–Hartmann sensor diagnoses the wavefront of the amplified laser beam by analyzing the leakage through a beam splitter. A deformable mirror with feedback from the wavefront sensor corrects the wavefront before the final focusing optic.

Several different kinds of technology can be used in the deformable mirror. Technologies used frequently include segmented mirrors, bimorph and monomorph mirrors, and

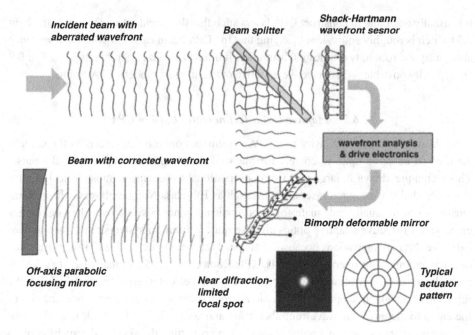

Figure 2.23 Illustration of how a spatially aberrated beam can be corrected to a beam focusable to near the diffraction limit by employing a wavefront sensor and a deformable mirror to correct these aberrations (figure courtesy of E. Gaul).

micro-electro-mechanical systems (MEMS). Segmented mirrors use discrete segments, each controlled by three piezoelectric actuators. The mirror surface can be made of a thin faceplate rather than individual sections on an array of actuators to avoid discontinuities. Bimorph mirrors consist of two piezoelectrics that are oppositely polarized and then bonded together. The front mirror surface and the back surface are grounded and electrodes are sandwiched between the two wafers. Voltage applied to the electrodes induces local bending of the piezo. MEMS deformable mirrors have high spatial resolution, fast response time but limited size, and damage fluence so they have not found widespread use in high energy CPA lasers. High quality optics in CPA lasers lead to wavefront distortion with only a few significant aberration terms. Strehl ratios >0.6 are readily achieved after correcting the wavefront by deformable mirrors with tens of actuators.

2.7 Other Laser Technologies Used in Strong Field Physics Studies

We conclude this chapter by noting that lasers based on technologies other than CPA have been used in strong field physics studies over the years, particularly in the early days of strong field research before the invention or widespread proliferation of CPA. These technologies are sometimes employed for their ability to access unique parameter ranges such as far IR or some visible wavelengths. We briefly mention here three such technologies finding application in strong field research in the past or currently in a small number of labs.

2.7 Other Laser Technologies Used in Strong Field Physics Studies

2.7.1 Picosecond CO_2 Lasers

An interesting alternative approach to producing strong fields for experiments involves amplifying picosecond pulses directly in CO_2 amplifiers (Corkum 1985). CO_2 lasers have an interesting place in strong field research: they lase at a far IR wavelength of 10.6 μm. Because of the wavelength squared dependence of the ponderomotive energy of a quivering electron in a laser field, this long wavelength therefore accesses quite high ponderomotive energies even when the focused intensity is not at the extreme. Such lasers can yield up to about 10 terawatts of peak power by delivering multijoule pulse energies in pulses of down to 3 ps in duration (Tochitsky et al. 2012).

CO_2 lases between two vibrational levels, excited in an electrical discharge. A gas mixture of CO_2, N_2 and He is employed in the discharge. Free electrons in the discharge tend to excite the first vibrational level in N_2, which is nearly resonant with the asymmetric vibrational mode of the CO_2. This vibrational mode it excited by colliding with those N_2 molecules. The CO_2 then lases on an inversion between two excited vibrational levels. Collisions with the light He atoms in the gas depopulate the lower laser vibrational levels to make the system work as a 4-level laser. Each vibrational level has a ladder of rotational levels and the system will lase between a whole range of rho-vibrational modes, leading to a gain spectrum that is a broad comb of discrete lines. If this system is then operated at high pressures (say, above 10 atm), pressure broadening of the sharp rho-vibrational laser lines leads to a nearly smooth broad gain spectrum with about 1 THz (∼0.3 μm) of gain bandwidth. This is broad enough to support amplification of pulses as short as 3 ps.

In a typical realization of a picosecond CO_2 laser for strong field plasma research, like the laser deployed at UCLA (Haberberger 2010), a 10.6 μm picosecond seed pulse is generated by taking a long (500 ns) 10 μm pulse and gating it in a CS_2 cell with a picosecond pulse from a small Nd:glass CPA laser by the nonlinear Kerr effect. The short 1 μm pulse from the glass laser produces a transient polarization rotation for the 10.6 μm pulse so that a polarizer rejects most of the long pulse and passes just that short segment rotated by the short glass laser pulse in the Kerr cell. This seed pulse is first amplified in a high-pressure CO_2 discharge situated in a regenerative amplifier cavity which is output coupled and hence produces a train of pulses at the ∼10 ns roundtrip time of the regen cavity. The pulse is amplified to about 10 mJ and a single pulse from the train is selected with a Pockels cell. Finally the pulses are amplified at large aperture (∼20 cm) in a second high-pressure CO_2 discharge power amplifier under multipass up to an energy approaching 50 J. The final pulse duration is about 3 ps (yielding nearly 15 TW of peak power).

2.7.2 Amplified Picosecond Nd:YAG Lasers

Pulses of about 40–100 ps can be amplified in Nd:YAG amps directly (Noom et al. 2013). YAG is an otherwise brilliant laser gain material: it has high gain cross section it has good thermal conductivity, and it can be manufactured to rods of up to 20 mm diameter. It can be flashlamp-pumped at repetition rate of 10 Hz at rod sizes up to 16 mm with ease. Pulses of ∼50 ps can be generated in an actively mode-locked YAG oscillator. These pulses can then be amplified directly in YAG in a regenerative amplifier followed by amplification in a laser chain of YAG rods of increasing aperture up to as high as $\sim$1 J. Such pulses can be frequency-doubled with good efficiency in KDP or LBO to yield high-power 532 nm pulses.

This architecture is used as a pump laser for a number of other sources, such as the SP-OPCPA front end described earlier or the femtosecond dye laser chain described in the next subsection. Such YAG lasers have also been used directly in strong field experiments particularly when visible photons were desired from doubled or tripled 1064 nm pulses. They can access intensity above 10^{14} W/cm^2 when focused with thin lenses.

2.7.3 Amplified Colliding Pulse Modelocked Dye Lasers

Prior to the development of Ti:sapphire as an ultrafast laser amplifying medium in the late 1980s, by far the most common way of generating ultrafast laser pulses was with tunable dye laser media (Schafer et al. 1966). While dyes are no longer much used in strong field research (with a few rare exceptions when wavelengths in the $\sim$600 nm range are desired), many of the first strong field physics experiments were conducted with amplified dye lasers, so it is worth mentioning this technology here. Laser dyes are long-chain organic molecules which typically have optically active benzene rings attached. The optically active electrons reside in a singlet ground state; they usually have very large dipole moments and therefore large gain cross sections. The dyes are dissolved in a solvent (at $10^{-3} - 10^{-4}$ molar) to create the laser gain medium. Because of the complexity of the molecule, the electronic transitions are coupled to a large number of vibrational modes which smear the relevant laser transitions into a broad band, often used to create a tunable narrowband laser but which are also suitable for amplification of femtosecond laser pulses.

Dyes typically exhibit gain in the 500–700 nm visible spectral region. Bandwidths of $\geq$40 nm are common, so sub-20 fs pulses are in principle possible from a dye. Because of a shift in the equilibrium spacing of the molecule's atoms in an excited electronic state the dye can be made to lase as a broadband four-level laser (much like Ti:sapphire discussed earlier) with upper-state lifetime of a few tens of nanoseconds. The very short lifetime demands that when used as a pulse amplifier, the dye typically must be pumped by subnanosecond pulses from a frequency doubled Nd:YAG laser. Upon pumping to an inversion, some long-lived ($\sim$100 μs) triplet states in the dye are produced which lead to absorption at the laser wavelength, mandating that the dye be flowed in a jet or glass cell to remove molecules in these triplet states from the gain volume.

The typical architecture of a high-peak-power dye laser used in strong field physics research starts with a colliding pulse mode-locked (CPM) dye laser pumped by a CW green beam like that generated by an Ar ion laser. Such a CPM laser is arranged in a ring cavity with a dye jet gain medium and an intercavity saturable absorber (usually a second dye jet); the ring cavity permits the formation of a counter propagating pulse in the cavity whose overlap in the saturable absorber element aids in achieving higher transient intensity there. (Mode-locking can also be achieved in dye oscillators by synch-pumping

2.7 Other Laser Technologies Used in Strong Field Physics Studies

the dye by a MHz train of green pulses from a frequency-doubled, mode-locked Nd:YAG laser.) The CPM laser produces ~ 100 fs pulses, which can then be subsequently amplified in a chain of dye flowing cells to energy up to ~ 1 mJ (yielding ~ 10 GW peak powers, more than adequate to reach focused intensity above 10^{14} W/cm^2). These amplifiers are usually pumped by frequency-doubled pulses from an amplified, mode-locked Nd:YAG laser like that described in the previous section (Perry et al. 1989).

3
Strong Field Interactions with Free Electrons

3.1 Introduction and Definitions of the Fields

We begin our study of strong laser field interaction with matter by first considering the interaction of strong EM fields with individual electrons. Practically, this means that the free electrons subject to a laser pulse are distant enough from other charged particles that we can ignore the fields and potentials from those other particles. Experimentally, such a situation is realized when the laser pulse is focused into a very diffuse gas (of density, say, less than $\sim 10^{15}$ cm^3) in which electrons have been liberated by ionization. The situation is also realized when an intense laser pulse interacts with a beam of relativistically accelerated electrons (a situation which we will treat explicitly in Section 3.6). While of interest in their own right, the phenomena we will examine in this context and the results we derive will inform our subsequent considerations of more complex systems, such as strong field interactions with atoms and plasmas.

For the majority of this chapter, we will consider the electron as a point-like classical particle. For free electrons, this is an excellent assumption, as quantum effects play almost no role in the dynamics of the electron motion. That this is so can be seen by noting that the de Broglie wavelength (h/p) of the free electron is small compared to the spatial scales of importance in the problem. Since we ignore in this chapter the confining potential of any atom or ion, the main spatial scale of interest is the electron oscillation amplitude, which, as we will see in the next section, exceeds the de Broglie wavelength of an oscillating electron in the laser field at intensities above about 10^{13} W/cm^2.

As discussed in Chapter 1, because of the high intensities involved, it is an excellent approximation to treat the laser radiation as a classical field, ignoring the quantum character of the light because of the enormous number of photons involved in the interactions. Only in Section 3.6 will we need to consider the quantum nature of the incident radiation because of the large Doppler upshift in photon energy experienced by an electron in a relativistic beam.

In this chapter we will first examine the time-dependent trajectory and energy of electrons in an infinite plane-wave field. We will start by considering these dynamics at modest field intensities where special relativity can be ignored. Though the derivations in this regime are essentially trivial, this will give us good physical insight into the laser/electron

3.1 Introduction and Definitions of the Fields

interaction dynamics. After that, we will tackle the problem of finding the electron dynamics at higher intensities where relativistic effects become important. Next, we will examine the radiation that is reemitted by the free electron as it is driven by the laser field, spending particular attention to the high-order harmonic radiation that results from the highly anharmonic motion of relativistic electrons at high intensity. We will then turn to the situation in which the laser field is no longer treated as an infinite plane-wave but is instead confined spatially by a focus. This will lead to an understanding of how electrons are ejected by the intensity gradient in a focus. The specifics of all of these phenomena will then be considered in the special case that the electrons irradiated by the laser are delivered with initial high energy from an accelerator. Finally, in the last section of this chapter we will develop the quantum mechanical description of the wavefunction of a free electron in the laser field; a derivation performed in anticipation of the ionization physics of atoms and ions we will consider in the next chapter.

3.1.1 Description of the Fields

Before proceeding, we must first discuss how we will describe the fields and relate these fields to observables like intensity. As mentioned, we treat the fields classically and forgo with a quantum description. This is an excellent approximation for the intensities that we are considering. This can be seen by noting that a classical description of the fields is appropriate when there are many photons in the volume of one cubic wavelength. Even at the rather modest intensity of 10^{13} W/cm^2 (at least modest by the standards of the phenomena we are going to consider), the photon density is $I\lambda^3/\hbar\omega c \approx 2 \times 10^9$ photons/cubic wavelength (with wavelength of 1 μm). This very large photon density justifies a classical description of the laser field, even when a quantum description of the incident matter is needed.

For the vast majority of the time, it is adequate to describe the incident laser as a coherent, single-mode field which permits easy mathematical description of the electric and magnetic fields. The electric field for a single-mode laser pulse of well-defined wavelength is

$$\mathbf{E}(\mathbf{x}, t) = \mathbf{E}_0 s(\mathbf{x}, t) \cos(\omega t - kz), \tag{3.1}$$

where $\mathbf{x} = x\hat{\mathbf{x}} + y\hat{\mathbf{y}} + z\hat{\mathbf{z}}$ is the position vector. The $\mathbf{E}_0$ is the peak electric field strength and includes any polarization description. (The reader is to be reminded that throughout the book symbols used to describe the frequency, wavelength and wavenumber of the incident laser will always be written without a subscript, that is, the laser's wavelength, wavenumber, frequency and angular frequency will always be written λ, k, ν, ω.) We will, by convention, treat the laser field as propagating along the z-axis. When the field is linearly polarized, we shall assume that the electric field points along the x-axis. The function $s(\mathbf{x}, t)$ describes the temporal envelope of the laser pulse and any spatial focusing if we do not consider an infinite plane-wave.

By the same token, we write the magnetic field of the laser as

$$\mathbf{B}(\mathbf{x}, t) = \mathbf{B}_0 s(\mathbf{x}, t) \cos(\omega t - kz), \tag{3.2}$$

which is related to the electric field by the fact that $E_0 = B_0$ and $\mathbf{E}_0 \perp \mathbf{B}_0$. Therefore, for linearly polarized fields the B-field is pointed along the y-axis. These fields are related to the intensity of the light by the Poynting vector, $\mathbf{S}$:

$$\mathbf{S} = \frac{c}{4\pi} \mathbf{E} \times \mathbf{B} = \frac{c}{4\pi} |\mathbf{E}|^2 \,\hat{\mathbf{k}}. \tag{3.3}$$

The intensity, I, is usually defined as the magnitude of the Poynting vector averaged over a full optical cycle, T, which relates the intensity to the peak electric field value by

$$I = |\langle \mathbf{S} \rangle_t| = \frac{c}{8\pi} E_0^2. \tag{3.4}$$

(For light in a medium of dielectric constant ε, this relation is amended to read $I = c\sqrt{\varepsilon} E_0^2 / 8\pi$.) This gives us a convenient relationship between peak field and intensity in practical units:

$$I\left[W/cm^2\right] = 1.32 \times 10^{-3} \left|E_0 \left[V/cm\right]\right|^2. \tag{3.5}$$

It will frequently be convenient to work with the potentials, Φ and $\mathbf{A}$, instead of directly working with the electric and magnetic fields. These are related to $\mathbf{E}$ and $\mathbf{B}$ in the usual way:

$$\mathbf{E} = -\frac{1}{c}\frac{\partial \mathbf{A}}{\partial t} - \nabla \Phi, \tag{3.6a}$$

$$\mathbf{B} = \nabla \times \mathbf{A}. \tag{3.6b}$$

We will typically work in the Coulomb gauge (often called the "radiation gauge"), which allows us to set

$$\nabla \cdot \mathbf{A} = 0. \tag{3.7}$$

3.2 Dynamics of Electrons in Fields at Subrelativistic Intensities

We now turn to a derivation of the trajectories of an electron in a field in which the laser intensity is low enough that relativistic effects can be ignored. At what intensity this simplification breaks down will be derived in the following section.

3.2.1 Lagrangian and Hamiltonian of a Free Electron in an EM Wave

When special relativity is excluded, the electron motion can be derived in a straightforward manner from the usual Newtonian equations of motion. However, since we will find it advantageous to work with potentials, we derive some insight by developing our derivation with the Lagrangian formalism. Even though the potential of a charged particle in an electromagnetic wave is velocity dependent, it is possible to write the Lagrangian for that particle in terms of the scalar and vector potentials. From the discussion in Goldstein (1980, p. 21), we can write the Lagrangian for a charged particle of charge q in an electromagnetic field in terms of independent generalized coordinates $\mathbf{x}$ and $\mathbf{v}$ (which we can select to be Cartesian position and velocity) as

3.2 Dynamics of Electrons in Fields at Subrelativistic Intensities

$$L(\mathbf{x}, \mathbf{v}, t) = \frac{1}{2}mv^2 - q\Phi + \frac{q}{c}\mathbf{A} \cdot \mathbf{v}. \qquad (3.8)$$

(Here we remind the reader that throughout this book the mass of the electron will always be denoted by an unsubscripted m; the masses of other particles will be written with a descriptive subscript, for example mass of the proton: m_p.)

We find the equations of motion from the usual Euler–Lagrange equations

$$\frac{d}{dt}\frac{\partial L}{\partial v_i} - \frac{\partial L}{\partial x_i} = 0, \qquad (3.9)$$

where the subscript i references Cartesian direction: $x_i = x, y, z$ and $v_i = v_x, v_y,$ and v_z. Working with the Lagrangian has the advantage of allowing us to derive certain constants of motion. First, we have the canonical momentum that is conjugate to the coordinate x_i:

$$P_i = \frac{\partial L}{\partial v_i}. \qquad (3.10)$$

The reader should recall that the canonical momentum is, in general, not necessarily equal to the kinetic momentum $P_i \neq mv_i$. So, we will typically denote the canonical momentum with a capital P and kinetic momentum by a lowercase p ($= mv$ in the nonrelativistic case). The value in finding the canonical momentum is that P_i is a conserved quantity if the Lagrangian does not explicitly contain the conjugate variable, x_i. (The Langrangian is said to be cyclic in that variable.)

For an electron with $q = -e$, the Lagrangian of Eq. (3.8) yields the canonical momenta for an electron in the electromagnetic wave:

$$P_i = mv_i - \frac{e}{c}A_i. \qquad (3.11)$$

P_i, then, is the kinetic momentum of the electron adjusted by a term varying as the component of the vector potential of the EM wave in that direction.

We can also find the total energy of the electron in the wave by finding the Hamiltonian. Formally, the Hamiltonian is related to the Lagrangian and the particle's canonical momentum by the equation

$$H(\mathbf{x}, \mathbf{p}, t) = \dot{\mathbf{x}} \cdot \mathbf{P} - L(\mathbf{x}, \dot{\mathbf{x}}, t). \qquad (3.12)$$

In Hamiltonian mechanical theory, the canonical equations of the Hamiltonian tell us that (Goldstein 1980, p. 356)

$$\frac{dH}{dt} = -\frac{\partial L}{\partial t}, \qquad (3.13)$$

which, of course, indicates that if t does not appear explicitly in L, then the Hamiltonian and the total energy will be constant in time. The Hamiltonian for an electron in the laser field can be found using Eqs. (3.8) and (3.12) along with the fact that $\mathbf{p} = \mathbf{P} + e\mathbf{A}/c$ to yield

$$H = \dot{\mathbf{x}} \cdot \mathbf{P} - \tfrac{1}{2}m\dot{x}^2 - e\Phi + \frac{e}{c}\mathbf{A} \cdot \dot{\mathbf{x}},$$

$$H = \mathbf{P}\cdot\mathbf{p} - \frac{\mathbf{p}\cdot\mathbf{p}}{2m} - e\Phi + \frac{e}{mc}\mathbf{A}\cdot\mathbf{p},$$

$$H = \frac{1}{2m}\left(\mathbf{P} + \frac{e}{c}\mathbf{A}\right)^2 - e\Phi. \tag{3.14}$$

This form of the Hamiltonian, written in terms of the canonical momentum, will become particularly useful later when we turn to a quantum mechanical description and the first quantization principle where it is the canonical momentum conjugate to the position variables which will be replaced with the gradient operator. From the classical standpoint, H is simply the sum of kinetic ($= p^2/2m$) and electrostatic potential energy.

3.2.2 Motion of a Free Electron in a Plane-Wave

Now we are armed to find the equations of motion for the electron. For a free electron in vacuum we have that $\nabla\Phi = 0$ so that the E-field is simply

$$\mathbf{E} = -\frac{1}{c}\frac{\partial \mathbf{A}}{\partial t}. \tag{3.15}$$

For the time being we consider a monochromatic, linearly polarized, infinite plane-wave so that $s(t) = 1$ in the definition of the fields. Since we will work predominantly with the vector potential we will define the laser field as

$$\mathbf{A} = A_0 \sin(\omega t - kx)\,\hat{\mathbf{x}}, \tag{3.16}$$

which by (3.6) means that the electric and magnetic fields are

$$\mathbf{E} = -E_0 \cos(\omega t - kx)\,\hat{\mathbf{x}}, \tag{3.17a}$$
$$\mathbf{B} = -B_0 \cos(\omega t - kx)\,\hat{\mathbf{y}}. \tag{3.17b}$$

The Euler–Lagrange equations yield the equations of motion for the electron, which are

$$\frac{d}{dt}\left(mv_x - \frac{e}{c}A_x\right) = 0, \tag{3.18}$$

$$\frac{d}{dt}(mv_x) + \frac{e}{c}\frac{\partial A_x}{\partial z}v_z = 0. \tag{3.19}$$

Equation (3.18) is simply a statement of the conservation of transverse canonical momentum (since the Lagrangian with vector potential (3.16) is cyclic in the transverse coordinates, x and y).

Of course, these equations could just as easily have been derived from Newton's equation with electric field and Lorentz forces:

$$\frac{d\mathbf{p}}{dt} = -e\mathbf{E} - e\frac{\mathbf{v}}{c}\times\mathbf{B}. \tag{3.20}$$

For the moment we can ignore the second term in Eq. (3.19) as being small because of the v_x/c factor, which means simply that the electron velocity in z remains at its initial value before arrival of the laser, $v_z = v_{z0}$; the laser field does not affect the electron in the direction of laser propagation. Integrating (3.18) yields the transverse oscillatory velocity of the electron along the x-axis in the field

3.2 Dynamics of Electrons in Fields at Subrelativistic Intensities

$$mv_x = \frac{e}{c}A_x + \text{const}, \tag{3.21}$$

$$v_x = \frac{e}{mc}A_0 \sin(\omega t - kz) + v_{x0}, \tag{3.22}$$

which then lets us integrate one more time to find the explicit trajectory of the electron along the x-direction:

$$x = \frac{e}{mc\omega}A_0 \cos(\omega t - kz) + v_0 t + x_0. \tag{3.23}$$

Equations (3.22) and (3.23) illustrate that in the field, aside from any initial velocity, the electron executes purely harmonic motion along the axis of the field's polarization at the incident laser's frequency. The maximum velocity attained by the electron during its oscillation (assuming that the electron has no initial velocity when the field is turned on) is

$$v_{osc} = \frac{eA_0}{mc} = \frac{eE_0}{m\omega}. \tag{3.24}$$

This peak oscillation velocity is proportional to the incident peak field strength times the wavelength of the field. Equation (3.23) shows that the trajectory of the quivering electron is in phase with the oscillating electric field and has an oscillation amplitude of

$$x_{osc} = \frac{eA_0}{mc\omega} = \frac{eE_0}{m\omega^2}. \tag{3.25}$$

This oscillation amplitude can be a good fraction of the laser wavelength, even for modest incident laser intensity. For example, a 1,053 nm wavelength laser focused to an intensity of 10^{17} W/cm^2 will drive an electron with an amplitude of over 40 nm.

Another important finding from this simple analysis is that if an electron is irradiated by a finite-duration laser pulse, as long as the pulse intensity rises and falls adiabatically, the laser imparts no net momentum to the electron. This can be seen by returning to Eq. (3.18) and multiplying A_0 in Eq. (3.16) by a finite intensity envelope, $s(t)$, such that $A_0 \to A_0 s(t - z/c)$. This choice of argument for s ensures that the pulse envelope travels in the positive z-direction with the wave. As discussed in Chapter 2, a typical temporal envelope might be a Gaussian pulse such that $s(t) = \exp[-2\ln 2 t^2/\Delta t_{FWHM}^2]$. As long as

$$\frac{ds}{dt} \ll \omega, \tag{3.26}$$

which is equivalent to requiring that the pulse duration is long compared to the optical period $\Delta t \gg T = 2\pi/\omega$, then we can ignore this temporal derivative of the pulse envelope and Eq. (3.18) returns Eq. (3.22) with the simple resubstitution of $A_0 \to A_0 s(t)$. The oscillation velocity now tracks the amplitude of the laser pulse envelope,

$$v_{osc} = \frac{eA_0 s(t)}{mc}, \tag{3.27}$$

and clearly returns to zero as the pulse intensity falls to zero. The laser imparts no momentum to the electron during the pulse interaction with that electron. It can be said that any energy given to the electron by the rise of the laser pulse is given back to the field by the electron as the pulse falls back to zero.

3.2.3 Average Kinetic Energy of the Electron: The Ponderomotive Energy

It is now natural to ask what the average kinetic energy of the oscillating electron is in this laser field. The instantaneous kinetic energy, ε_K, of the electron is

$$\varepsilon_K = \frac{e^2 A_0^2}{2mc^2} \sin^2(\omega t - kz). \tag{3.28}$$

We define a quantity usually called the "pondermotive energy" as the average energy the electron acquires over a full laser cycle $U_p = \langle \varepsilon_K \rangle_{cycle}$. This energy is

$$U_p = \frac{e^2 A_0^2}{4mc^2} = \frac{e^2 E_0^2}{4m\omega^2}. \tag{3.29}$$

The ponderomotive energy scales with laser intensity and wavelength as $U_p \sim I\lambda^2$, so longer-wavelength laser pulses exhibit higher electron ponderomotive energy. In practical units, a convenient formula for U_p can be written as

$$U_p[eV] = 9.33 \times 10^{-14} I[W/cm^2] \lambda[\mu m]^2. \tag{3.30}$$

The ponderomotive energy will frequently set the energy scale at which some strong field interaction physics occurs. As Eq. (3.30) illustrates, U_p is ~1 eV, the energy of a near-IR photon, at an intensity of around 10^{13} W/cm^2 (again assuming a 1 μm laser wavelength). This is another reflection of the fact that at intensities above this value it is appropriate to think of the field classically instead of as a flux of individual 1 eV photons, since many photons must participate if the electron is to acquire ponderomotive energy of many eV. At an intensity of around 10^{17} W/cm^2 the electron ponderomotive energy is around 10 keV. At this point the laser field dominates the motion of electrons even in high-temperature plasmas (the temperature of a fusion plasma such as that found in a solar interior is of the order of 10 keV).

U_p is frequently referred to as the "ponderomotive potential" in much of the literature. This term is misleading and will be avoided in this book. U_p is not a potential in the rigorous sense in that it is not derived from the gradient of a force. Instead, the ponderomotive energy is a useful approximation which relies on the electron experiencing many field oscillations at nearly the same field strength. As we shall see, when the electron travels through a focused laser field at a velocity high enough that it samples a range of intensities over the course of one oscillatory period, the concept of a cycle-averaged ponderomotive energy becomes meaningless.

3.2.4 Dynamics of the Electron in a Circularly Polarized Field

We can now consider the trajectory of an electron in a circularly polarized field. We define such a field in terms of the vector potential in the following way:

$$\mathbf{A} = \frac{1}{\sqrt{2}} A_0 \sin \omega \tau \hat{\mathbf{x}} \pm \frac{1}{\sqrt{2}} A_0 \cos \omega \tau \hat{\mathbf{y}}, \tag{3.31}$$

which is chosen to ensure that our previous relationships between A_0, E_0 and intensity are still valid. The sign between the terms is chosen to select between right circular and left circular polarization. As before, the Lagrangian remains cyclic in x and y so P_x and P_y are both conserved. As before, $v_z = 0$ and the perpendicular velocity $\mathbf{v}_\perp = v_x\hat{\mathbf{x}} + v_y\hat{\mathbf{y}}$ reflects conservation of P_x and P_y.

$$\mathbf{v}_\perp = \frac{eA_0}{\sqrt{2}mc}\left(\sin\omega\tau\hat{\mathbf{x}} \pm \cos\omega\tau\hat{\mathbf{y}}\right) + \mathbf{v}_0. \tag{3.32}$$

The resulting electron trajectory is

$$\mathbf{x} = \frac{eA_0}{\sqrt{2}mc\omega}\left(\cos\omega\tau\hat{\mathbf{x}} \mp \sin\omega\tau\hat{\mathbf{y}}\right) + \mathbf{v}_0 t, \tag{3.33}$$

which means that the electron traces a circle with radius, r,

$$r = \frac{eA_0}{\sqrt{2}mc\omega}. \tag{3.34}$$

3.2.5 Residual Drift Energy of Electrons Born in the Field

With Eq. (3.27) we showed that if an electron is subject to a laser pulse that rises and falls slowly, the oscillatory velocity falls to zero at the end of the pulse and the electron retains only the velocity that it had prior to the pulse arrival. This will not be true if we take as the initial condition the electron beginning its oscillation in the field at some arbitrary phase. This situation is an excellent approximation to that in which a free electron is injected in the field by the near instantaneous ionization of an atom, ion or molecule in the laser field (a process which we will explore in depth in the next chapter). We will often refer to this process by stating that the electron is "born" in the field at a well-defined phase $\varphi_0 = \omega t_0 - kz_0$. In the case of optical ionization we will often approximate the initial conditions of the electron as being born into the field at rest.

When this electron is injected into the field, its velocity will not in general return to zero after the pulse has fallen to zero intensity. We refer to this remaining electron velocity as residual drift energy. This energy can be found if we return to Eq. (3.22) for the velocity of the electron and impose the condition that the velocity at the time and place of the electron's birth in the field, t_0 and z_0, is zero. For linearly polarized light, the velocity along x is

$$v_x = \frac{e}{mc}A_0 s(t - z/c)\sin(\omega t - kz) - \frac{e}{mc}A_0^{(0)}\sin\varphi_0, \tag{3.35}$$

where we have accounted for the fact that the peak vector potential varies in time over the laser pulse, $s(t)$. The $A_0^{(0)}$ is the amplitude of the vector potential at the time that the electron is born. As $s(\infty) \to 0$ the electron retains some residual velocity, which is from Eq. (3.35),

$$v_{res} = \frac{e}{mc}A_0^{(0)}\sin\varphi_0, \tag{3.36}$$

in the direction of the laser polarization. If the phase of birth is $\varphi_0 = 0$, then Eq. (3.16) shows that the electron is born at the peak of the electric field and it ends up with no residual drift velocity after the pulse has passed. The residual drift energy of the electron is

$$U_{res} = \frac{e^2}{2mc^2}\left(A_0^{(0)}\right)^2 \sin^2 \varphi_0. \tag{3.37}$$

This can be written in terms of the ponderomotive energy seen by the electron at the time of its introduction into the field as

$$U_{res} = 2U_p \sin^2 \varphi_0. \tag{3.38}$$

So, electrons born near the peak of the electric field oscillation acquire almost no residual drift energy, but if electrons are born near the zero of electric field, they will acquire up to twice the ponderomotive energy as residual drift kinetic energy.

This observation is simply another way to state conservation of canonical momentum. If the electron is injected into field at rest near the peak of the electric field, the canonical momentum, as calculated by Eq. (3.11), is zero. When the field amplitude ramps down to zero, the electron retains no kinetic momentum. However, if the electron is born near a zero in electric field, the vector potential is near a maximum and the electron acquires some canonical momentum, even if its initial kinetic momentum is zero. Because this transverse canonical momentum must be preserved, as the vector potential amplitude drops, this canonical momentum is transferred into transverse kinetic momentum.

When an electron is born into a circularly polarized field, the situation is different. Equation (3.32) indicates that if the electron is born at rest,

$$\mathbf{v}_{\perp res} = \frac{e}{\sqrt{2}mc} A_0^{(0)} \left(\sin \varphi_0 \hat{\mathbf{x}} \pm \cos \varphi_0 \hat{\mathbf{y}}\right), \tag{3.39}$$

which yields for the residual drift energy

$$\varepsilon_{res} = \frac{e^2}{4mc^2}\left(A_0^{(0)}\right)^2 \left(\sin^2 \varphi_0 + \cos^2 \varphi_0\right), \tag{3.40}$$

$$\varepsilon_{res} = U_p. \tag{3.41}$$

Now there is no phase at which the residual drift energy is zero but instead the electron always acquires U_p worth of drift energy. From the standpoint of canonical momentum conservation, Eq. (3.31) shows that at every phase in a circularly polarized field an electron born at rest acquires some canonical momentum so it always ends up with some drift energy equal to the ponderomotive energy at the time of injection. The direction of that residual drift velocity is perpendicular to the direction of the electric field at the time the electron is born.

3.3 Dynamics of Electrons in Fields at Relativistic Intensities

Next we will consider a more complex and subtle problem, which is to find the trajectories of electrons when the laser field is strong enough that the electron velocity becomes relativistic. We first ask, when is a relativistic treatment necessary? It makes sense that relativistic dynamics will need to be calculated when the previously calculated nonrelativistic oscillation velocity approaches c, $v_{osc}/c \to 1$. We can define a dimensionless parameter that is the ratio v_{osc}/c:

3.3 Dynamics of Electrons in Fields at Relativistic Intensities

$$a_0 \equiv \frac{v_{osc}}{c} = \frac{eA_0}{mc^2}. \tag{3.42}$$

The a_0 is usually referred to as the normalized vector potential and its value gives an indication of how relativistic the electron motion will be in the laser field. When $a_0 \ll 1$, neglect of relativity is justified; however, when a_0 approaches 1, clearly a relativistic treatment of the electron motion is needed and the results of Section 3.2 will no longer be accurate. In practical units we can write the normalized vector potential as

$$a_0 = \frac{eE_0}{mc\omega} = 8.55 \times 10^{-10}\sqrt{I\left[W/cm^2\right]}\lambda\,[\mu m]. \tag{3.43}$$

Here, a_0 will be 1 in a laser with wavelength near 1 μm when the intensity is 1.4×10^{18} W/cm^2. So for most near-IR lasers, a relativistic treatment of the electron motion in the field is needed at intensities above 10^{17} W/cm^2. Above 10^{18} W/cm^2 the electron motion is strongly relativistic and, as we will see, deviates sharply from the harmonic motion described in the previous section.

3.3.1 Relativistic Equations of Motion

This problem has been addressed by a number of authors over the years (Kibble 1966; Eberly and Sleeper 1968; Gunn and Ostriker 1971; Bardsley et al. 1989). While the solution for the electron trajectory in a relativistic field can be found elegantly with the Hamilton–Jacobi formalism, we will approach the problem as we did in the nonrelativistic limit with the somewhat more intuitive Lagrangian technique. We will adapt the usual definitions of special relativity, with $\beta = v/c$ and $\gamma = (1 - \beta^2)^{-1/2}$. The relativistic Lagrangian for a charged particle in the electromagnetic field is (Goldstein 1980, p. 322)

$$L(\mathbf{x}, \mathbf{v}) = -\frac{mc^2}{\gamma} + \frac{q}{c}\mathbf{A}\cdot\mathbf{v} - q\Phi. \tag{3.44}$$

We shall ignore radiation reaction throughout this analysis, an assumption which we will revisit again in the next section. Again, since we consider a negatively charged electron in vacuum, we have that $\nabla\Phi = 0$, which means that the Lagrangian we seek is

$$L(\mathbf{x}, \mathbf{v}) = -mc^2\sqrt{1 - \beta^2} + \frac{q}{c}\mathbf{A}\cdot\mathbf{v}. \tag{3.45}$$

Once more it will be helpful to find the canonical momenta conjugate to the position coordinates represented by this Lagrangian,

$$P_i = \frac{\partial L}{\partial v_i}, \tag{3.46}$$

which, when applied to (3.45), give us the following:

$$\mathbf{P} = \gamma m\mathbf{v} - \frac{e}{c}\mathbf{A}. \tag{3.47}$$

As in the nonrelativistic case we again find that the canonical momentum is simply the kinetic momentum plus e/c times the vector potential,

$$P_i = p_i - \frac{e}{c}A_i, \tag{3.48}$$

identifying the relativistic linear momentum with $\gamma m \mathbf{v}$. Also note that we can find the relativistic Hamiltonian from the Lagrangian with

$$H = \mathbf{P} \bullet \mathbf{v} - L(\mathbf{x}, \mathbf{v}) \tag{3.49}$$

yielding the well-known formula

$$H = \sqrt{p^2c^2 + m^2c^4} = \sqrt{c^2\left(\mathbf{P} + \frac{e}{c}\mathbf{A}\right)^2 + m^2c^4} \tag{3.50}$$

which is alternatively expressed as

$$H = \gamma mc^2 = mc^2\sqrt{1 + \frac{p^2}{m^2c^2}}. \tag{3.51}$$

We will begin with a consideration of an electron in a linear polarized field, though we will find it convenient from this point forward to work with the proper time of the electron, τ, where

$$\tau = t - \frac{z}{c}, \tag{3.52}$$

noting that t here represents the laboratory time, namely the time experienced by the stationary observer in the laboratory frame. The laser field for a linearly polarized infinite plane-wave can be written as

$$\mathbf{A}(\tau) = A_0 \sin \omega \tau \, \hat{\mathbf{x}}. \tag{3.53}$$

The sign of z/c in the proper time is chosen to ensure that the wave moves in the $+z$ direction. We also note at this point that while both ω and τ must be transformed from the lab frame to the electron rest frame, the product $\omega\tau$, which is simply the phase of the wave, is Lorentz frame invariant. This makes sense given that phase is just a counting of the peaks and troughs of a wave which must remain the same independent of reference frame.

3.3.2 Electron Trajectory in a Relativistic Field

For simplicity let us consider the x and z motion of the electron separately and apply the Euler–Lagrange equation, Eq. (3.9) to Eq. (3.45), to derive the equations of motion. First, evaluating the derivatives in the x-direction and noting that L does not explicitly depend on x so that the right-hand side of (3.9) is zero gives us for the x-motion the first constant of motion

$$\frac{d}{dt}\left(\frac{mc\beta_x}{\sqrt{1-\beta^2}} - \frac{e}{c}A_x\right) = 0. \tag{3.54}$$

Using the definition of relativistic momentum

$$\frac{mc\beta_x}{\sqrt{1-\beta^2}} = \gamma m v_x = p_x \tag{3.55}$$

yields the result for the momentum in the x-direction:

$$p_x(\mathbf{x}, t) = \frac{e}{c} A_x + p_{x0}. \tag{3.56}$$

A similar calculation using (3.9) with the z-axis variable gives

$$\frac{d}{dt}\left(\frac{mc\beta_z}{\sqrt{1-\beta^2}}\right) = \frac{dp_z}{dt} = -e\beta_x \frac{\partial A_x}{\partial z}. \tag{3.57}$$

Finally we need to consider the energy of the electron whose change we can easily find with the help of Eqs. (3.13) and (3.45):

$$\frac{d}{dt}\left(\gamma mc^2\right) = -e\boldsymbol{\beta} \cdot \frac{\partial \mathbf{A}}{\partial z}. \tag{3.58}$$

We now can write the relevant equations of motion in terms of the proper time:

$$p_x = \frac{e}{c} A_0 \sin\omega\tau + p_{x0} \tag{3.59}$$

$$\frac{dp_z}{dt} = -\frac{e}{c} \beta_x A_0 \omega \cos\omega\tau \tag{3.60}$$

$$mc^2 \frac{d\gamma}{dt} = -e\beta_x A_0 \omega \cos\omega\tau. \tag{3.61}$$

Equations (3.59)–(3.61) are just manifestations of the more familiar equations of motion for momentum and energy:

$$\frac{d\mathbf{p}}{dt} = -e\mathbf{E} - e\frac{\mathbf{v}}{c} \times \mathbf{B}, \quad \frac{dH}{dt} = \mathbf{v} \cdot \mathbf{F} = -e\mathbf{v} \cdot \mathbf{E}.$$

Combining Eqs. (3.60) and (3.61) to eliminate the vector potential

$$c\frac{dp_z}{dt} = -eA_0\beta_x\omega\cos\omega\tau = \frac{d}{dt}\left(\gamma mc^2\right) \tag{3.62}$$

yields the next constant of motion:

$$\frac{d}{dt}(p_z - \gamma mc) = 0, \tag{3.63}$$

which integrates to yield a useful relationship between the electron's longitudinal momentum and relativistic gamma factor

$$p_z - \gamma mc = p_{z0} - \gamma_0 mc. \tag{3.64}$$

Here p_{z0} and γ_0 are the longitudinal momentum and gamma factor of the electron before the turn-on of the laser field.

Let us consider the specific case of an electron that is initially at rest prior to the arrival of the intense laser pulse so that $p_{z0} = 0$ and $\gamma_0 = 1$. In that case, Eq. (3.64) gives for the longitudinal momentum of the electron

$$p_z = (\gamma - 1)mc, \tag{3.65}$$

which, when combined with the expression for the relativistic gamma factor,

$$\gamma^2 = 1 + \frac{p^2}{m^2c^2} = 1 + \frac{p_x^2}{m^2c^2} + \frac{p_z^2}{m^2c^2} = 1 + \frac{2p_z}{mc} + \frac{p_z^2}{m^2c^2}, \quad (3.66)$$

yields a relationship between the longitudinal momentum and the momentum along the field polarization. Equating the second terms of the equality in (3.66) relates p_x and p_z by

$$p_z = \frac{p_x^2}{2mc}. \quad (3.67)$$

Equation (3.67) combined with Eq. (3.56) solves the problem for the electron's momentum in terms of the proper time (assuming $p_{x0} = 0$). The components of the transverse electron momentum found in (3.56) can be written in terms of the normalized vector potential as

$$p_x = a_0 mc \sin \omega\tau, \quad (3.68)$$

$$p_y = 0. \quad (3.69)$$

Equation (3.67) can now be utilized with the solution for p_x to yield

$$p_z = \frac{e^2 A_0^2}{2mc^3} \sin^2 \omega\tau, \quad (3.70)$$

which can be written with the help of a trigonometric identity, $(\sin^2 a = (1 - \cos 2a)/2)$, in the more suggestive form

$$p_z = \frac{1}{4} a_0^2 mc \left(1 - \cos(2\omega\tau)\right). \quad (3.71)$$

The electron undergoes sinusoidal motion in the x-direction with respect to the proper time. However, unlike the nonrelativistic solution, we now see that the electron's momentum in the z-direction is not zero but is finite and positive.

Next, we desire expressions for the position of the electron as a function of proper time. Since we have so far found the momentum as a function of proper time, we write

$$\mathbf{p} = \gamma m \frac{d\mathbf{x}}{dt} = \gamma m \frac{d\mathbf{x}}{d\tau} \frac{d\tau}{dt}. \quad (3.72)$$

Return to the definition of the proper time to relate its differential element to a differential element of laboratory time,

$$\frac{d\tau}{dt} = \frac{d}{dt}(t - z/c) = 1 - \frac{v_z}{c}, \quad (3.73)$$

which, when Eq. (3.65) is employed and the relationship between momentum and velocity is recalled, $\gamma m v_z = p_z$, give us the necessary relationship between differentials of proper and laboratory time

$$\frac{d\tau}{dt} = 1 - \frac{p_z}{\gamma mc} = \frac{1}{\gamma}. \quad (3.74)$$

3.3 Dynamics of Electrons in Fields at Relativistic Intensities

This is just a statement of time dilation for the electron moving in the lab frame. Equation (3.72) gives us an equation which we can easily integrate with respect to proper time:

$$\mathbf{p} = m\frac{d\mathbf{x}}{d\tau}. \tag{3.75}$$

Employing our solution (3.68) for the transverse momentum as a function of proper time gives us the electron's transverse trajectory:

$$x(\tau) = \int_0^\tau a_0 c \sin\omega\tau' \, d\tau',$$

$$x(\tau) = \frac{a_0 c}{\omega}(1 - \cos\omega\tau). \tag{3.76}$$

A similar integration of (3.71) yields the longitudinal electron trajectory

$$z(\tau) = \frac{1}{4}a_0^2 c\tau - \frac{a_0^2 c}{8\omega}\sin(2\omega\tau). \tag{3.77}$$

Equations (3.68)–(3.71) and (3.76)–(3.77) essentially solve the problem, at least as far as we can, analytically. These equations would seem to give the desired result of the electron trajectory in the laboratory frame; however, this solution is slightly deceiving, as it does not yield an explicit expression for $\mathbf{p}(t)$ and $\mathbf{x}(t)$. These are only *implicit* solutions for the momentum and position as functions of proper time: they depend on τ, which in turn depends on position, z. In fact, an analytic solution for the explicit trajectory of the electron is not possible, and we must rely on numerical solutions of the equations of motion, Eqs. (3.54) and (3.57), to derive the actual electron position versus laboratory time. Nonetheless, these solutions will give us insight into the relativistic motion of the electron and will prove valuable when we consider the radiation produced from the laser-driven electron.

3.3.3 Longitudinal Drift Velocity of the Electron

While Eqs. (3.76) and (3.77) would suggest that the electron motion is harmonic, this is not true. The electron motion is only harmonic with respect to the proper time, not with respect to a laboratory observer. In fact, the motion is not periodic in space. Figure 3.1 shows the trajectory of an electron in a 1 μm wavelength field with three different values of a_0 during a laboratory time interval of 20 fs. The electron motion is strongly anharmonic and exhibits increasing transverse amplitude with increasing laser field strength. It is interesting to note that the number of transverse oscillations over the fixed time interval of the calculation decreases with increasing a_0, a consequence of the differing rates of laboratory and proper time for the electron. Particularly striking in this plot is that the electron acquires a time-averaged motion along the z-axis, a manifestation of the constant first term for p_z seen in equation (3.71). This drift velocity increases with increasing intensity (a_0), though the motion along z in one cycle is already a fair fraction of the x-axis oscillation amplitude at an intensity as low as 3×10^{17} W/cm^2 ($a_0 = 0.5$). This forward drift motion is a consequence of the laser's magnetic field: as the electron is driven along x by the

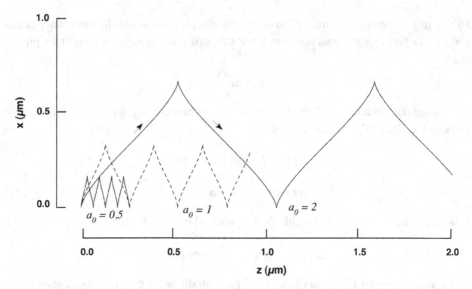

Figure 3.1 Trajectories of an electron driven by a field with $\lambda = 1{,}054$ nm for three different values of a_0 over a laboratory time interval of 20 fs.

electric field, the $\mathbf{v} \times \mathbf{B}$ force drives the electron in the $+z$-direction. Because the transverse, electric field-driven velocity oscillates at $\pi/2$ out of phase with $\mathbf{E}$ and $\mathbf{B}$, there is a periodic acceleration and stopping deceleration by the B-field during the course of an E-field cycle.

We can use our solution for the z component of the electron's motion with respect to proper time, Eq. (3.77), to find this time-averaged longitudinal drift velocity. Note that the electron trajectory in z has a drift term and an oscillatory term which averages to zero over many optical cycles: $z(\tau) = z_d + z_{osc}$. The z component of the velocity, then, must also have a constant drift component and an oscillatory component: $v_z(\tau) = v_d + v_{osc}$. The drift component is then found from (3.77):

$$\begin{aligned} v_d &= \frac{dz_d}{dt} = \frac{d}{dt}\left(\frac{1}{4}a_0^2 c\tau\right) \\ &= \frac{1}{4}a_0^2 c\,(1 - v_z/c) \\ &= \frac{1}{4}a_0^2 c\left(1 - \frac{v_d + v_{osc}}{c}\right) \end{aligned} \quad (3.78)$$

Since we are averaging over an optical cycle to find the time-averaged drift velocity, $\langle v_{osc}\rangle_T = 0$, this lets us solve for the drift velocity to find

$$\frac{v_d}{c} = \frac{a_0^2}{4 + a_0^2}. \quad (3.79)$$

3.3 Dynamics of Electrons in Fields at Relativistic Intensities

This longitudinal electron drift velocity is 50 percent c at $a_0 = 2$ (intensity of 5×10^{18} W/cm^2 in a 1 μm laser field).

3.3.4 Electron Energy in the Relativistic Field

We can also use our implicit solutions for the electron momentum to find the energy of an electron in the laser field (for an electron initially at rest). The total energy of the electron (rest energy plus kinetic energy) is found with the help of Eq. (3.67):

$$U = \gamma mc^2 = mc^2 \sqrt{1 + \frac{p_x^2 + p_z^2}{m^2 c^2}}$$

$$= mc^2 \sqrt{1 + \frac{2p_z}{mc} + \frac{p_z^2}{m^2 c^2}}$$

$$= mc^2 \left(1 + \frac{p_z}{mc}\right). \tag{3.80}$$

Equation (3.80) with (3.70) then tells us that the electron energy as a function of proper time is

$$U = mc^2 \left(1 + \frac{1}{2} a_0^2 \sin^2 \omega\tau\right), \tag{3.81}$$

which yields a cycle-averaged electron energy of

$$\langle U \rangle = mc^2 \left(1 + a_0^2/4\right). \tag{3.82}$$

This energy is the rest energy of the electron plus a term which is identical to the non-relativistic ponderomotive energy result of Eq. (3.29). This total energy results from a combination of transverse oscillation velocity and longitudinal drift velocity.

3.3.5 Electron Trajectory in a Circularly Polarized Field

Consider for a moment the trajectory of the electron when the field is circularly polarized, described by Eq. (3.31). Conservation of transverse canonical momentum, for an electron initially at rest, gives us a result similar to Eq. (3.59):

$$\mathbf{p}_\perp = p_x \hat{\mathbf{x}} + p_y \hat{\mathbf{y}} = \frac{e}{c} \mathbf{A}(t), \tag{3.83}$$

which yields with Eq. (3.31)

$$\mathbf{p}_\perp = \frac{1}{\sqrt{2}} a_0 mc \left(\sin \omega\tau \hat{\mathbf{x}} \pm \cos \omega\tau \hat{\mathbf{y}}\right). \tag{3.84}$$

Reasoning along the lines leading to Eq. (3.67) lets us write the relationship between transverse and longitudinal momenta:

$$p_z = \frac{p_\perp^2}{2mc}, \tag{3.85}$$

which when combined with Eq. (3.84) shows us that the longitudinal momentum of the electron is constant in time:

$$p_z = \frac{1}{4}a_0^2 mc. \qquad (3.86)$$

Equation (3.75) lets us retrieve the electron trajectories in position:

$$x = \frac{a_0 c^2}{\sqrt{2}\omega}(1 - \cos\omega\tau), \qquad (3.87)$$

$$y = \frac{a_0 c^2}{\sqrt{2}\omega}\sin\omega\tau, \qquad (3.88)$$

$$z = \frac{1}{4}a_0^2 c\tau. \qquad (3.89)$$

The electron does not exhibit the oscillatory behavior in the z-direction seen when irradiated by linear polarization. Instead, the electron flies with constant speed along z, with a velocity which is exactly the same as the average drift velocity in a linearly polarized field, given by (3.79). We see that now the electron traces a corkscrew trajectory in space, with the direction of helicity determined by the direction of polarization rotation. The radius of this corkscrew motion is

$$r = \frac{a_0 c}{\sqrt{2}\omega}. \qquad (3.90)$$

Figure 3.2 shows the trajectories of an electron subject to linear and circular polarization in fields with $a_0 = 2$.

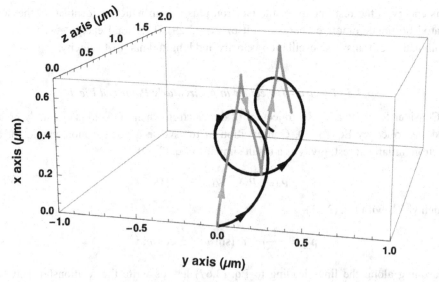

Figure 3.2 Trajectories of an electron driven by a field with $\lambda = 1,053$ nm for $a_0 = 2$ when the field is linearly polarized (gray line) and circularly polarized (black line) over a laboratory time interval of 20 fs.

3.3.6 Residual Drift Energy and Forward Motion of the Electron in a Finite Duration Pulse

We can now ask the question we posed in the previous section: what energy does an electron retain if the laser field amplitude is adiabatically ramped back down to zero. Again we consider what this energy will be for an electron ejected, or "born" into the field at an arbitrary phase of the field, $\varphi_0 = \omega \tau_0$ at rest. Returning to the case of linear polarization, now the solutions for the momenta are needed subject to the initial conditions that the electron is born with zero momentum at this arbitrary phase. We can, along the lines of our nonrelativistic treatment, write the laser field amplitude as slowly varying with proper time, $a_0(\tau)$, with the electron being injected at a field amplitude written as $a_0(0)$. Equation (3.59) gives the transverse momentum

$$p_x(\tau) = a_0(\tau)mc \sin \omega\tau - a_0(0)mc \sin \varphi_0, \tag{3.91}$$

and Eq. (3.67) delivers the longitudinal momentum

$$p_z(\tau) = \frac{mc}{2} (a_0(\tau) \sin \omega\tau - a_0(0) \sin \varphi_0)^2. \tag{3.92}$$

With Eq. (3.80) and cycle averaging, we have for the total energy of the electron

$$\langle U \rangle = mc^2 \left(1 + \frac{a_0^2}{4} + \frac{a_0(0)^2}{2} \sin^2 \varphi_0 \right), \tag{3.93}$$

which means that after the laser pulse intensity falls to zero, the electron retains residual kinetic energy equal to

$$\langle \varepsilon_K \rangle_{res} = \frac{1}{2} a_0(0)^2 mc^2 \sin^2 \phi_0. \tag{3.94}$$

Again, the residual drift energy of the electron is zero if it is born in the field at the peak of an electric field oscillation and is maximum when born at a null in the electric field. It is interesting to note that this result is identical to the nonrelativistic residual drift energy that we found in Eq. (3.37).

If the electron is present before the arrival of the laser pulse and has no initial laboratory frame momentum, $a_0(0) = 0$, which means that the electron must return to rest after the pulse falls to zero. However, once the pulse field falls back to zero, the electron will have moved a finite amount along the direction of the EM field propagation. To examine this in more detail, we will change notation slightly and explicitly treat the laser pulse envelope with a separate function, which is equal to one at its peak,

$$\mathbf{A}(\tau) = A_0 S(\tau) \sin \omega\tau \hat{\mathbf{x}}, \tag{3.95}$$

where A_0 now is the vector potential at the peak of the laser pulse, where the function S is equal to one. The condition that we have assumed, that the laser pulse rises and falls slowly can be examined explicitly by looking at the energy gain of the electron in the field, given by Eq. (3.58) including the pulse envelope

$$mc^2 \frac{d\gamma}{dt} = -e\boldsymbol{\beta} \cdot \frac{\partial \mathbf{A}}{\partial t}$$

$$= -e\beta_x \omega A_0 S(\tau) \cos \omega \tau - e\beta_x \omega A_0 \sin \omega \tau \frac{\partial S(\tau)}{\partial t}. \tag{3.96}$$

We see that we retrieve our previous results if we can ignore the second term on the right-hand side of this equation, which is the same as saying that

$$\frac{\partial S}{\partial t} \ll \omega \quad \Rightarrow \quad \Delta t \gg \omega^{-1} \sin \varphi_0, \tag{3.97}$$

where Δt is the pulse width of the envelope $S(\tau)$. Practically speaking, if the laser wavelength is ~ 1 μm, this condition holds for pulses with >5 fs duration, true for virtually all state-of-the art lasers. If we ignore that second term we can rewrite the solution for the longitudinal momentum, Eq. (3.70), with the pulse envelope explicitly included:

$$p_z = \frac{1}{4} a_0^2 mcS^2(\tau)(1 - \cos(2\omega\tau)). \tag{3.98}$$

This allows us to find the total motion of the electron along the $+z$ direction during the time history of the pulse. Recalling Eq. (3.75), the electron is pushed a distance

$$\Delta z = \frac{1}{m} \int_{-\infty}^{\infty} p_z \, d\tau$$

$$= \frac{1}{4} a_0^2 c \int_{-\infty}^{\infty} S^2(\tau) d\tau \tag{3.99}$$

by the laser pulse.

For the sake of a concrete example, suppose that the laser pulse has a temporal envelope described by a Gaussian

$$S(\tau) = \exp\left[-2 \ln 2 \tau^2 / \Delta t_{FWHM}^2\right]. \tag{3.100}$$

Integrating (3.99) with the pulse envelope (3.100) yields for the forward motion of the electron in the pulse

$$\Delta z = \frac{\pi^{1/2}}{8\sqrt{\ln 2}} a_0^2 c \Delta t. \tag{3.101}$$

To give a numerical example, if the pulse is 100 fs in duration, this forward motion is about 10 μm for an $a_0 = 1$. So a strongly relativistic pulse, with intensity above, say 10^{19} W/cm² where $a_0 \sim 3$–10 will push a free electron a macroscopic distance of a fair fraction of a millimeter. This distance is often comparable to or greater than the Rayleigh range of the laser focus. We will return to the question of electron ejection from the focus of a laser in Section 3.5.

Now if the electron does not come to rest as $S(\tau)$ falls to zero because it was injected into the field at a nonzero canonical momentum it retains residual energy, as we have seen, which can be written in terms of the pulse envelope and the peak normalized vector potential

$$\varepsilon_{res} = mc^2 \left(1 + \frac{a_0^2 S(\tau_0)^2}{2} \sin^2 \omega \tau_0\right). \tag{3.102}$$

3.3 Dynamics of Electrons in Fields at Relativistic Intensities

The electron will be traveling at some angle with respect to the laser propagation direction. In the nonrelativistic case, we found that the electron drifted in a direction parallel to the laser electric field, but now we might expect there to be residual drift momentum in the z-direction at higher fields. We can find the electron's residual drift angle with respect to the laser propagation direction, θ_r, from the momentum components. If the energy is written

$$\varepsilon_{res} = \gamma_{res} mc^2, \qquad (3.103)$$

then the z component of the residual drift momentum is

$$p_{res} = (\gamma_{res} - 1)mc. \qquad (3.104)$$

The tangent of the electron residual drift angle is the ratio of the transverse and longitudinal momentum components

$$\tan \theta_r = \frac{p_{xres}}{p_{zres}}$$
$$= \frac{\sqrt{2mcp_{zres}}}{p_{zres}}, \qquad (3.105)$$

yielding for this residual drift angle in terms of the residual electron energy

$$\theta_r = \tan^{-1}\left[\sqrt{\frac{2}{\gamma_{res} - 1}}\right]. \qquad (3.106)$$

Written in terms of the value of a_0 and the field phase at the point the electron is injected into the field,

$$\theta_r = \tan^{-1}\left[\frac{2}{a_0(0) \sin \varphi_0}\right]. \qquad (3.107)$$

At high a_0 this relationship implies that the residual drift angle scales as $\theta_r \sim 2/a_0(0)$.

Equation (3.106) is simply a reflection of the momentum imparted to the electron by the photons it had to absorb to acquire its residual drift energy. We can see this if we note that N photons are required to impart the residual kinetic energy given to the electron by the passing laser pulse:

$$K_{res} = \varepsilon_{res} - mc^2 = (\gamma_{res} - 1)mc^2$$
$$= N\hbar\omega. \qquad (3.108)$$

These N photons deliver momentum along their propagation direction, which is

$$p_z = N\hbar k = N\hbar \frac{\omega}{c}. \qquad (3.109)$$

Using the definition for θ_r, we have

$$\frac{p^2}{p_z^2} = \frac{p_x^2 + p_z^2}{p_z^2} = \tan^2 \theta_r + 1. \qquad (3.110)$$

Conservation of energy with the absorption of N photons tells us that

$$mc^2 \left(1 + \frac{p^2}{m^2 c^2}\right)^{1/2} = N\hbar\omega + mc^2, \quad (3.111)$$

$$p^2 = \left(\frac{N\hbar\omega}{c}\right)^2 + 2N\hbar\omega m,$$

Combining Eqs. (3.109) and (3.111) to eliminate the momentum terms in (3.110) gives

$$\tan^2 \theta_r + 1 = \left(\frac{c}{N\hbar\omega}\right)^2 \left(\frac{N^2 \hbar^2 \omega^2}{c^2} + 2N\hbar\omega\right), \quad (3.112)$$

$$\tan \theta_r = \sqrt{\frac{2mc^2}{N\hbar\omega}}, \quad (3.113)$$

which, with (3.108), returns our previous result

$$\tan \theta_r = \sqrt{\frac{2}{\gamma_{res} - 1}}. \quad (3.114)$$

3.3.7 Effective Electron Mass in the Strong Field

Before turning to the question of what radiation the electron re-radiates under the laser drive, it is interesting to look at defining an effective mass in the field. Because of the high velocities acquired in a strong field, the electron mass effectively increases when compared to its rest mass, an observation first made by Brown and Kibble (1964). Viewing the electron mass as effectively increased due to the laser field drive will be particularly useful when we turn to the physics of relativistically driven plasmas. The question we want to answer is, what is the effective electron mass when we think in terms of a cycle averaged energy and electron momentum (in linear polarization)?

To ascertain this, we start with the usual equation for the electron energy

$$U^2 = p^2 c^2 + m^2 c^4 \quad (3.115)$$

and recall that in the field, the cycle averaged energy of an electron initially at rest in the laser field is

$$\langle U \rangle = mc^2 \left(1 + a_0^2/4\right). \quad (3.116)$$

Equations (3.68)–(3.71) tell us that the cycle-averaged momentum is

$$\langle p \rangle = \frac{1}{4} a_0^2 mc. \quad (3.117)$$

We can then ask what effective momentum will lead to the fulfillment of Eq. (3.115) in the sense of cycle-averaged quantities. While not rigorous, we can with simple intuition write

$$\langle U \rangle^2 - \langle p \rangle^2 c^2 = m_{eff}^2 c^4, \quad (3.118)$$

3.4 Radiation Emitted by Electrons Driven in a Strong Laser Field

which, with the substitution of the cycle averaged quantities,

$$m^2 c^4 \left(1 + a_0^2/2 + a_0^4/16\right) - m^2 c^4 a_0^4/16 = m_{eff}^2 c^4, \qquad (3.119)$$

let us solve for the desired cycle-averaged effective electron mass:

$$m_{eff} = m\sqrt{1 + a_0^2/2}. \qquad (3.120)$$

So we can often think of an ensemble of free electrons driven in a strong laser field as if they were a collection of "heavy" electrons with mass increased by the factor written in (3.120). This is admittedly a "hand waving" treatment of the situation and not rigorous; it does, however, often give us some insight into the nature of electronic matter subject to relativistic intensities, which will often act like a gas of electrons with increased mass.

3.4 Radiation Emitted by Electrons Driven in a Strong Laser Field

Clearly the acceleration the electrons undergo in the strong field must in turn lead to the emission of electromagnetic radiation. Now that we have found the trajectory of the electron in the field, we are in a position to calculate the spectrum and angular emission of this reemission. If the intensity of the field is nonrelativistic, the electron motion is harmonic and the pattern of the scattered radiation is easily calculated (Sarachik and Schappert 1970; Esarey et al. 1993; Castillo-Herrera et al. 1993; Hartemann 1998).

3.4.1 Radiation from an Electron in a Subrelativistic Field

We can easily find the radiated emission from a sinusoidally driven electron with the famous Larmor formula:

$$\mathbf{E}_s = \frac{e}{c} \left[\frac{\hat{\mathbf{n}} \times \hat{\mathbf{n}} \times \dot{\boldsymbol{\beta}}}{R} \right]_{ret}. \qquad (3.121)$$

The geometry of the emission is depicted in Figure 3.3. Here $\dot{\boldsymbol{\beta}}$ is the acceleration of the electron normalized to c, and $\hat{\mathbf{n}}$ is the unit vector pointing from the electron to the observation point. The entire quantity in the bracket is evaluated at the retarded time $t_{ret} = t - R/c$ where R is the distance from emitter to observer. As Figure 3.3 illustrates, we can write $\hat{\mathbf{n}} \times \hat{\mathbf{n}} \times \dot{\boldsymbol{\beta}} = \dot{\beta} \sin \theta \hat{\mathbf{n}}$. What we observe from the oscillating electron is the radiated power, which is given by the Poynting vector

$$\mathbf{S} = \frac{c}{4\pi} \mathbf{E} \times \mathbf{B}$$
$$= \frac{c}{4\pi} |\mathbf{E}_s| \hat{\mathbf{n}}. \qquad (3.122)$$

The power radiated by the electron per unit solid angle is

$$\frac{dP}{d\Omega} = \frac{cR^2}{4\pi} E_s^2. \qquad (3.123)$$

92   *Strong Field Interactions with Free Electrons*

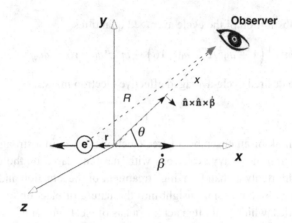

Figure 3.3 Geometry of scattered radiation emission from a laser-driven electron.

Equation (3.22) tells us that at nonrelativistic intensity the electron acceleration is simply (ignoring any phase shift due to z position)

$$\dot{\beta} = \frac{eE_0}{mc} \cos \omega t, \tag{3.124}$$

which when substituted into Eq. (3.121) gives the magnitude of the electric field of the scattered radiation

$$E_s = \frac{e^2}{mc^2 R} \cos \omega t \, \sin \theta. \tag{3.125}$$

The observed time-average of the power radiated per unit solid angle then becomes

$$Q(\Omega) = \left\langle \frac{dP}{d\Omega} \right\rangle = \frac{e^4 E_0^2}{8\pi m^2 c^3} \sin^2 \theta. \tag{3.126}$$

We have retrieved the well-known Thomson scatter emission pattern, which peaks in the direction perpendicular to the direction of the laser polarization. The radiation emitted by the quivering electron is at the same frequency as the driving laser frequency with intensity proportional to the driving laser intensity, consequences of the purely harmonic motion of the electron at modest intensity.

3.4.2 Radiation Emission from Electrons in Relativistic Intensity Fields

This trivial result at modest intensity is significantly altered when the field intensity is high enough that the electron motion becomes relativistic (i.e. a_0 approaches or exceeds 1). As we have seen, the electron motion becomes highly anharmonic, so it follows that the electron emission is likely to include higher-order harmonics of the laser radiation. What is more, because the force drives drift motion of the electron in the direction of the laser propagation, the electron motion is not even periodic. So, at first look, it is unclear what it even means when we say that the electron emits high harmonics of the incident field. The

3.4 Radiation Emitted by Electrons Driven in a Strong Laser Field

forward drift motion of the electron will lead to Doppler shifts of the incident radiation in the frame of the electron and shifts of emitted radiation back from the frame of the emitting electron into the lab frame.

There are essentially two ways to attack this problem analytically. The first, which was first presented in a classic paper by Sarachik and Schappert (1970), involves Lorentz transforming into a frame in which the electron motion *is* periodic and calculating the harmonic emission of the periodically oscillating electron. One then Lorentz transforms the radiation back into the observer's frame. This approach requires careful and subtle transformation of various variables between frames and special attention to time intervals in the frame of the emitting electron, the frame of periodic motion and the observer's frame. The second approach involves simple solution of the radiation equations with electron trajectories calculated in the laboratory frame, an approach explored in considerable detail in the excellent work of Esarey, Ride and Sprangle (Esarey *et al.* 1993). Ultimately it is this second approach which we will use to solve the problem, as it most readily yields equations which can be numerically solved to derive the emitted spectrum and angular distribution of the scattered radiation. We will, however, explore the first approach for a bit as it gives us insight into the nature of the emitted radiation spectrum and forward folding of the spatial distribution as we increase the incident intensity.

Before proceeding it is likely helpful to remind the reader of the relativistic mechanics of four-vectors and Lorentz transforms. We can express certain quantities as four vectors, which we will write as

$$x_\mu = (ix_0, x_1, x_2, x_3) = (ix_0, \mathbf{x}). \tag{3.127}$$

Any four-vector will transform into a frame moving with velocity β and Lorentz factor γ according to the well-known equations

$$x'_0 = \gamma(x_0 - \boldsymbol{\beta} \cdot \mathbf{x}), \tag{3.128a}$$

$$\mathbf{x}' = \mathbf{x} + \frac{\gamma - 1}{\beta^2}(\boldsymbol{\beta} \cdot \mathbf{x})\boldsymbol{\beta} - \gamma \boldsymbol{\beta} x_0. \tag{3.128b}$$

Some of the four-vectors we will encounter in our analysis include the position four-vector,

$$x_\mu = (ict_0, x, y, z), \tag{3.129}$$

the momentum four-vector,

$$p_\mu = (iU/c, p_x, p_y, p_z), \tag{3.130}$$

where U is the particle energy, the wavenumber,

$$k_\mu = (i\omega/c, k_x, k_y, k_z) \tag{3.131}$$

and the potential four-vector

$$A_\mu = (i\Phi, A_x, A_y, A_z). \tag{3.132}$$

Recall that the absolute value of any four vector, which we write as $x_\mu x^\mu$, where we use the usual convention of summing over repeated indices, is invariant under Lorentz transform. Since we can write the normalized vector potential in terms of the potential four-vector as

$$a_0 = \frac{e}{mc^2}\sqrt{A_\mu A^\mu}, \tag{3.133}$$

we see that a_0 is frame invariant. This is, of course, not true of the electric field strength or field frequency.

3.4.3 Radiation Emission from an Electron in the Periodic Rest Frame

We will gain insight into the emission of radiation in the relativistic intensity regime if we examine the problem from the viewpoint of a frame in which the electron motion is periodic. This will allow us to see more clearly why the emission contains higher harmonics, what frequency those harmonics are actually observed as back in the laboratory frame and what effect the forward drift motion of the electron has on the angular distribution of the emission in the lab frame. In this analysis we will loosely follow the derivation of Sarachik and Schappert (Sarachik and Schappert 1970). The key is to find a frame which moves with the forward drifting electron at a velocity such that the average momentum of the electron is zero. Of course, The frequency of the drive laser is Doppler-shifted in this frame. Then, the emission of radiation will be at harmonics of this frequency; a Lorentz transform back into the lab frame will show us what the frequency of the emitted radiation will be as a function of angle with respect to the laser propagation direction.

The zero average momentum frame is often termed the R-frame (the "rest frame," though the electron is not at rest in this frame, its motion is simply periodic with no net drift motion). The velocity of this frame is simply the drift velocity which we have already derived in Eq. (3.79). The frequency of the laser in this frame is easily found by Eq. (3.128a) since the frequency is the first component of the wavenumber four-vector. Since the laser co-propagates in parallel with the drifting electron, the transform is

$$x'_0 = \gamma(x_0 - \beta_d x_1). \tag{3.134}$$

Because the frame we seek moves with the electron drift velocity, $\beta = \beta_d$, which leads, with the result found in (3.79), to a gamma factor for this drift frame in terms of the normalized vector potential

$$\gamma = \gamma_d = (1 - \beta_d^2)^{-1/2}$$
$$= \frac{1}{\sqrt{1 - \left(\frac{a_0^2}{4 + a_0^2}\right)^2}}$$
$$= \frac{1 + a_0^2/4}{\sqrt{1 + a_0^2/2}}. \tag{3.135}$$

The driving laser frequency in the lab frame is then Doppler down-shifted to ω_R according to Eq. (3.134):

3.4 Radiation Emitted by Electrons Driven in a Strong Laser Field

$$\omega_R = \omega\gamma_d(1-\beta_d) = \omega\frac{1+a_0^2/4}{\sqrt{1+a_0^2/2}}\left(1-\frac{a_0^2/4}{1+a_0^2/4}\right),$$

$$\omega_R = \frac{\omega}{\sqrt{1+a_0^2/2}}. \tag{3.136}$$

We can now find the trajectory of the electron is this frame by simply transforming our previously derived lab frame solutions in terms of proper time into this R-frame. The momenta of an electron moving in a linearly polarized field were given in Eqs. (3.68)–(3.70) which transform as the components of the momentum four-vector into the R frame,

$$p_z^{(R)} = \gamma_d(p_z - \beta_d U/c), \tag{3.137}$$

$$\mathbf{p}_\perp^{(R)} = \mathbf{p}_\perp, \tag{3.138}$$

leading to an expression of the transverse momentum in the R-frame in terms of the R-frame proper time

$$p_x^{(R)} = a_0 mc \sin\omega_R\tau_R. \tag{3.139}$$

To find τ_R we note that the phase of a wave is a Lorentz invariant since the phase amounts to a simple counting of peaks and troughs in a wave which cannot change from frame to frame

$$\varphi = \omega\tau = \omega_R\tau_R, \tag{3.140}$$

so we can say

$$\tau_R = \tau\sqrt{1+a_0^2/2}. \tag{3.141}$$

A transform also yields the R-frame momentum in the z-direction:

$$p_z^{(R)} = \gamma_d(p_z - \beta_d U/c), \tag{3.142}$$

which can be written with the help of Eqs. (3.80), (3.79), (3.71) and (3.135) as

$$p_z^{(R)} = \gamma_d(p_z(1-\beta_d) - \beta_d mc), \tag{3.143}$$

$$p_z^{(R)} = \frac{1}{\sqrt{1+a_0^2/2}}\frac{a_0^2 mc}{4}\cos 2\omega_R\tau_R. \tag{3.144}$$

As expected, the momenta average to zero. Integration with respect to the proper time, accounting for the transform of proper time to the R-frame, Eq. (3.141), yields the electron trajectory:

$$\mathbf{x} = \frac{1}{m\sqrt{1+a_0^2/2}}\int \mathbf{p}^{(R)}\,d\tau_R, \tag{3.145}$$

$$x = \frac{1}{\sqrt{1+a_0^2/2}}\frac{a_0 c}{\omega_R}(1-\cos\omega_R\tau_R), \tag{3.146}$$

$$z = \frac{1}{(1+a_0^2/2)}\frac{a_0 c}{8\omega_R}\sin(2\omega_R\tau_R). \tag{3.147}$$

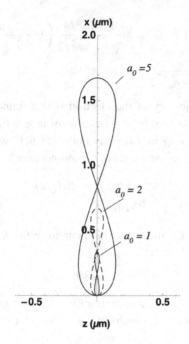

Figure 3.4 Trajectories of an electron driven by a field with $\lambda = 1{,}054$ nm in the R-frame for three different values of a_0.

This solution describes the motion of a "figure-8" as illustrated in Figure 3.4 where the trajectories of electrons driven at various a_0 are plotted. Notice that at the extrema of the x-motion of the electron, the electron is always moving in the positive z-direction. The width of the figure-8 increases with increasing field strength. Comparison of Eqs. (3.146) and (3.147) illustrates that the aspect ratio of this figure-8 motion

$$\frac{x_{\max}}{z_{\max}} = \frac{8\sqrt{1 + a_0^2/2}}{a_0} \tag{3.148}$$

and approaches a maximum value for $a_0 \to 0$ as

$$\frac{x_{\max}}{z_{\max}} = 4\sqrt{2}. \tag{3.149}$$

We can perform a similar calculation for the R-frame trajectory of an electron irradiated by circular polarization. We have found that the electron drifts in the z-direction with the same velocity in this polarization as does an electron subject to linear polarization. What we find is that

$$\mathbf{p}_\perp^{(R)} = \mathbf{p}_\perp \tag{3.150}$$

$$p_z^{(R)} = 0, \tag{3.151}$$

which yield for the electron trajectory a stationary circle in the x–y plane

$$\mathbf{x}_\perp = \frac{1}{\sqrt{1 + a_0^2/2}} \frac{a_0 c}{\sqrt{2}\omega_R} \left[(1 - \cos\omega_R \tau_R)\hat{\mathbf{x}} \pm \sin\omega_R \tau_R \hat{\mathbf{y}} \right] \tag{3.152}$$

with a radius of

$$r = \frac{1}{\sqrt{1+a_0^2/2}} \frac{a_0 c}{\sqrt{2}\omega_R}. \tag{3.153}$$

Equations (3.146) and (3.147) show that the electron motion is indeed periodic in the R-frame. They also seem to suggest that the motion of the electron is harmonic, oscillating at the fundamental frequency of the drive (Doppler shifted, of course, into the R-frame) in the polarization direction, and oscillating at the second harmonic of ω_R in the longitudinal direction. So it would appear that the scattered radiation will have spectral components only at the first and second harmonics. This picture is, of course, incorrect because the electron oscillates at these two frequencies only with respect to the proper time. The motion of the electron in the R-frame observer's time (the time of an observer moving with the electron at speed β_d) is, in fact, anharmonic because the electron experiences longitudinal motion at the same time as it undergoes transverse motion. This gives rise to an entire comb of high-order harmonics scattered at integer multiples of the R-frame driving frequency $\omega_R^{(scat)} = q\omega_R$.

It is easy to see that a range of harmonics are emitted by a simple analysis of the electron trajectory if we notice that the harmonic motion of the electron in x scales something like

$$x \sim x_{osc} e^{i\omega t - ikz}$$

and in z the motion scales like

$$z \sim z_{osc} e^{i2\omega t - i2kz}.$$

If we assume for the moment that the electron motion in the z-direction is small with respect to the drive wavelength, we Taylor-expand the x and z motion with respect to z:

$$x \sim x_{osc} e^{i\omega t} \left(1 - ikz + \frac{k^2 z^2}{2} - \ldots\right) \tag{3.154}$$

$$z \sim z_{osc} e^{i2\omega t} \left(1 - i2kz + 2k^2 z^2 - \ldots\right). \tag{3.155}$$

When we substitute (3.155) back into the expansion for z for both axes of motion we find that the x and z trajectories look something like

$$x \sim x_{osc} \left(e^{i\omega t} - ikz_{osc} e^{i3\omega t} - \frac{k^2 z^2}{2} e^{i5\omega t} - \ldots\right), \tag{3.156}$$

$$z \sim z_{osc} \left(e^{i2\omega t} - ikz_{osc} e^{i4\omega t} + 2k^2 z_{osc}^2 e^{i6\omega t} - \ldots\right). \tag{3.157}$$

It is now apparent that the x motion actually contains a comb of harmonics at odd integer multiples of the drive frequency, starting with the fundamental frequency (the standard linear Thomson scattering), while the z motion contains a comb of higher harmonics at even integer multiples, starting at the second harmonic. This analysis shows that, since we expect no radiation in the direction parallel to the electron's motion, any harmonic emission we observe should have nulls in the emission along the laser polarization direction if the harmonic is odd and should be zero in the axis of light propagation if the harmonic is even.

3.4.4 Transformation of Radiation in the R-Frame Back to the Lab Frame

To find the spectrum of emitted radiation back in the laboratory frame we Lorentz transform to a frame moving the other direction at the drift velocity so that $\beta = -\beta_d$. The frequency of the qth harmonic of ω_R is related to the qth harmonic frequency in the lab frame, ω_q, by a transform

$$q\omega_R = \omega_q \gamma_d (1 - \beta_d \cos\theta) \tag{3.158}$$

where θ is the angle between the propagation direction of the laser and the observation angle in the lab frame. This gives in terms of a_0:

$$\omega_q = q\omega \frac{1}{\sqrt{1+a_0^2/2}} \frac{\sqrt{1+a_0^2/2}}{1+a_0^2/4} \left(1 - \frac{a_0^2/4}{1+a_0^2/4}\cos\theta\right)^{-1}$$

$$= q\omega \left(1 + \frac{a_0^2}{4}(1-\cos\theta)\right)^{-1} \tag{3.159}$$

or

$$\omega_q = \frac{q\omega}{1 + (a_0^2/2)\sin^2(\theta/2)}. \tag{3.160}$$

So, when we speak of harmonic emission from the driven electron in the lab frame, we must refer to Eq. (3.160) which shows that the actual frequency of observed scattered radiation is direction dependent. In fact, if $a_0 \geq 2^{1/2}$ then *all* frequencies between one harmonic and the next are emitted depending on observation angle. Consequently one must be careful when speaking of the emission of a specific harmonic from the electron. At high a_0, emission at a given wavelength at a given angle may contain contributions from a number of different "harmonics." This spectral shifting we can see resulted from the Doppler downshift of the laser to low frequency when we transformed into the R-frame and the angle-dependent Doppler shift of the emitted radiation upon transform back into the lab frame. In the direction of laser propagation these Doppler shifts cancel, and the observed radiation is at the true qth harmonic of the drive wavelength.

We can also derive some insight into the angular distribution of the emitted radiation through this R-frame analysis. Consider the power emitted by the electron per unit solid angle in the R-frame

$$\frac{dP_R^{(q)}}{d\Omega_R} = \text{Power radiated per unit solid angle in } R\text{-frame.} \tag{3.161}$$

For the first harmonic, this will just be the usual Thomson scatter pattern, Eq. (3.126), in the R-frame. Since energy transforms as the 0th component of the momentum four-vector we can write the Lorentz transform between L-(lab) and R-frames such that

$$dP_L^{(q)} dt_L = \gamma_d dP_R^{(q)} dt_R (1 + \beta_d \cos\theta)_R \tag{3.162}$$

where dt_L and dt_R are time intervals for observers in the lab and R-frames. We can transform the emission angle in the R-frame to the lab frame by writing Lorentz transforms for distances in the two frames:

3.4 Radiation Emitted by Electrons Driven in a Strong Laser Field

$$x_L = \gamma_d x_R (1 + \beta_d \cos \theta_R); \quad x_R = \gamma_d x_L (1 - \beta_d \cos \theta_L) \tag{3.163}$$

which yield for the relationship between angles in the two frames

$$\cos \theta_R = \frac{\cos \theta_L - \beta_d}{1 - \beta_d \cos \theta_L}. \tag{3.164}$$

This relationship lets us relate the differential elements of solid angle in the two frames

$$\frac{d\Omega_R}{d\Omega_L} = \frac{d(\cos \theta_R)}{d(\cos \theta_L)} = \frac{1}{1 - \beta_d \cos \theta_L} \frac{\beta_d \cos \theta_L - \beta_d^2}{(1 - \beta_d \cos \theta_L)^2}$$

$$= \frac{1 - \beta_d^2}{(1 - \beta_d \cos \theta_L)^2}, \tag{3.165}$$

which in terms of incident a_0 is

$$\frac{d\Omega_R}{d\Omega_L} = \frac{1 + a_0^2/2}{1 + (a_0^2/2)\sin^2(\theta_L/2)}. \tag{3.166}$$

We also need the ratio of the time intervals in the two frames. If we denote with a prime the time interval for the quivering electron, the retarded time of the electron can be written in terms of the observer's time interval Δt:

$$\Delta t' = \Delta t + \hat{\mathbf{n}} \cdot \Delta \mathbf{x}/c. \tag{3.167}$$

The total power scattered by the electron is a quantity measured over many optical cycles, so we know that in the R-frame, $\langle \Delta \mathbf{x} \rangle = 0$, which allows us to deduce that in this frame, on average over many cycles the electron's and observer's time intervals are the same $\Delta t'_R = \Delta t_R$. In the lab frame the electron drifts over a length $\langle \Delta \mathbf{x} \rangle = c \beta_d \Delta t$, so the time intervals in the lab frame are related by

$$\Delta t_L = \Delta t'_L (1 - \beta_d \cos \theta_L). \tag{3.168}$$

Time dilation for the electron requires that $\Delta t'_L = \gamma_d \Delta t'_R$, so the observer's time intervals are related by this expression:

$$\frac{\Delta t_L}{\Delta t_R} = \gamma_d (1 - \beta_d \cos \theta_R). \tag{3.169}$$

Combining Eqs. (3.162), (3.164), (3.165) and (3.169) yields the relationship between the scattered radiation pattern in the R- and lab frames:

$$\frac{dP_R^{(q)}}{d\Omega_L} = \frac{dP_R^{(q)}}{d\Omega_R} \gamma_d (1 + \beta_d \cos \theta_R) \frac{dt_R}{dt_L} \frac{d\Omega_R}{d\Omega_L} \tag{3.170}$$

$$\frac{dP_L^{(q)}}{d\Omega_L} = \frac{(1 + a_0^2/2)^2}{(1 + (a_0^2/2)\sin^2(\theta_L/2))^4} \frac{dP_R^{(q)}}{d\Omega_R}. \tag{3.171}$$

Equation (3.171) is an interesting if not unexpected result. It shows that any emitted radiation in the R-frame is effectively folded toward the direction of laser propagation as a_0 is increased. This is a direct consequence of the forward-directed drift motion of the electron at high intensity. It is this relativistic physics which causes synchrotron radiation from highly relativistic electrons to emerge from the machine in a nearly collimated forward-directed beam. The extent of this forward folding of the emitted harmonics can be seen if we ask at what angle is the half maximum of the transformation function multiplying the R-frame distribution in (3.171). If the half-width emission angle in the lab frame is small, we see that the angular half width of this folding function is, for large a_0,

$$\frac{1}{2} \approx \frac{1}{\left(1 + (a_0^2/2)\theta_{HW}^2\right)^4}$$

$$\theta_{HW} \approx \frac{\sqrt{8(2^{1/4} - 1)}}{a_0} = \frac{1.2}{a_0}. \tag{3.172}$$

So, for example, a $\lambda = 1$ μm laser focused to an intensity of 10^{20} W/cm^2 ($a_0 \cong 10$) will drive harmonic emission from free electrons which is emitted largely in a forward cone of roughly 100 mrad in extent. This emission cone is essentially $1/\gamma_d$, which can be seen from Eq. (3.164) by setting $\theta_R = 90°$:

$$\cos\theta_{HW} = \beta_d; \quad \sin\theta_{HW} = \sqrt{1 - \beta_d^2} = \frac{1}{\gamma_d} = \frac{\sqrt{1 + a_0^2/2}}{1 + a_0^2/4}. \tag{3.173}$$

3.4.5 Scattered Radiation Spectrum and Angular Distribution of the Scattered Radiation

With this intuition on the characteristics of the spectrum and angular distribution of the scattered radiation at relativistic intensity, we can turn finally to analytic solution of the problem. Here we will forgo the Sarachik and Schappert approach in solving the radiation equations in the lab frame. In this analysis we follow closely the derivation of Esarey, Ride and Sprangle (Esarey *et al.* 1993). As we have done all along, we will continue to ignore any effects of radiation reaction on the motion of the electrons, an assumption we will justify at the end of the derivation.

As with most radiation problems, we begin with the Liénard–Wiechert potentials for the fields from moving charges (Jackson 1975, p. 657)

$$\Phi(\mathbf{x}, t) = \left[\frac{q}{(1 - \boldsymbol{\beta} \cdot \hat{\mathbf{n}})R}\right]_{ret}, \tag{3.174}$$

$$\mathbf{A}(\mathbf{x}, t) = \left[\frac{q}{(1 - \boldsymbol{\beta} \cdot \hat{\mathbf{n}})R}\boldsymbol{\beta}\right]_{ret}, \tag{3.175}$$

where the times are evaluated at the retarded time of the moving charge. As Jackson shows, by carefully taking derivatives to find the electric field, these potentials yield for the emitted field of the moving charge in the geometry outlined in Figure 3.3

$$\mathbf{E}(\mathbf{x}, t) = q\left[\frac{\hat{\mathbf{n}} - \boldsymbol{\beta}}{\gamma^2(1 - \boldsymbol{\beta} \cdot \hat{\mathbf{n}})^3 R^2}\right]_{ret} + \frac{q}{c}\left[\frac{\hat{\mathbf{n}} \times [(\hat{\mathbf{n}} - \boldsymbol{\beta}) \times \dot{\boldsymbol{\beta}}]}{\gamma^2(1 - \boldsymbol{\beta} \cdot \hat{\mathbf{n}})^3 R}\right]_{ret}. \tag{3.176}$$

The first term is usually referred to as the velocity field, which is the changing electrostatic fields dropping off as R^{-2} and does not concern us here; the second term is normally called

3.4 Radiation Emitted by Electrons Driven in a Strong Laser Field

the acceleration field and is the term which gives a propagating electromagnetic wave. We are after the power radiated by the electron as a function of angle, which (3.122) tells us is

$$\chi = \left(\frac{c}{4\pi}\right)^{1/2} [R\mathbf{E}]_{ret}, \qquad (3.177)$$

making sure to evaluate the fields at the appropriate retarded time now that relativistic particle velocities are involved. The spectrum of the emission is just the Fourier transform of the temporal emission

$$\tilde{\chi}(\omega_R) = \mathfrak{F}[\chi(t)]. \qquad (3.178)$$

These two quantities are related by Parseval's theorem which states that the total energy emitted integrated over all time must be equal to the total energy emitted integrated over all radiated frequencies (ω_r):

$$\frac{dW}{d\Omega} = \int |\chi(t)|^2 dt; \quad \frac{dW}{d\Omega} = \int |\tilde{\chi}(\omega_r)|^2 d\omega. \qquad (3.179)$$

We are ultimately after the intensity radiated per solid angle per unit frequency. Since

$$\frac{dW}{d\Omega} = \int_0^\infty \frac{d^2 I(\omega, \hat{\mathbf{n}})}{d\omega_r d\Omega} d\omega_r, \qquad (3.180)$$

the integral over all frequencies (positive and negative) in (3.179) lets us say that the intensity distribution we seek is

$$\frac{d^2 I(\omega, \hat{\mathbf{n}})}{d\omega_r d\Omega} = 2 |\tilde{\chi}(\omega_r)|^2. \qquad (3.181)$$

The Fourier transform of the acceleration field

$$\tilde{\chi}(\omega_r) = \left(\frac{e^2}{8\pi^2 c}\right) \int_{-\infty}^{\infty} \left[\frac{\hat{\mathbf{n}} \times [(\hat{\mathbf{n}} - \boldsymbol{\beta}) \times \dot{\boldsymbol{\beta}}]}{\gamma^2 (1 - \boldsymbol{\beta} \cdot \hat{\mathbf{n}})^3}\right]_{ret} e^{i\omega_r t} dt \qquad (3.182)$$

must be found in terms of retarded time $t' = t - R/c$. Since the observation point is far away, we can say the distance from emitter to observer is $R \cong x - \hat{\mathbf{n}} \cdot \mathbf{r}$, where x is the distance from the central point of the particle's trajectory to the observation point, and $\mathbf{r}$ is the location of the electron with respect to this central point (see Figure 3.3).

We can relate the differential element of retarded time interval to the observers time interval by

$$dt = dt' + d(x - \hat{\mathbf{n}} \cdot \mathbf{r})/c = (x - \hat{\mathbf{n}} \cdot \boldsymbol{\beta}) dt', \qquad (3.183)$$

which permits us to evaluate Eq. (3.182),

$$\tilde{\chi}(\omega_r) = \left(\frac{e^2}{8\pi^2 c}\right) \int_{-\infty}^{\infty} \frac{\hat{\mathbf{n}} \times [(\hat{\mathbf{n}} - \boldsymbol{\beta}) \times \dot{\boldsymbol{\beta}}]}{(1 - \boldsymbol{\beta} \cdot \hat{\mathbf{n}})^2} e^{i\omega_r(t' + R/c)} dt', \qquad (3.184)$$

and find our desired intensity distribution

$$\frac{d^2 I(\omega, \hat{\mathbf{n}})}{d\omega_r d\Omega} = \frac{e^2}{4\pi^2 c} \left| \int_{-\infty}^{\infty} \frac{\hat{\mathbf{n}} \times [(\hat{\mathbf{n}} - \boldsymbol{\beta}) \times \dot{\boldsymbol{\beta}}]}{(1 - \boldsymbol{\beta} \cdot \hat{\mathbf{n}})^2} e^{i\omega_r(t' + \hat{\mathbf{n}} \cdot \mathbf{r}/c)} dt \right|^2. \qquad (3.185)$$

As Jackson shows, this formula can be integrated by parts, noting that

$$\frac{\hat{n} \times [(\hat{n} - \beta) \times \dot{\beta}]}{(1 - \beta \cdot \hat{n})^2} = \frac{d}{dt} \frac{\hat{n} \times \hat{n} \times \beta}{1 - \beta \cdot \hat{n}}$$

to give

$$\frac{d^2 I(\omega, \hat{n})}{d\omega_r d\Omega} = \frac{e^2 \omega_r^2}{4\pi^2 c} \left| \int_{-\infty}^{\infty} \hat{n} \times \hat{n} \times \beta e^{i\omega_r(t' + \hat{n} \cdot r/c)} dt \right|^2, \quad (3.186)$$

where we have omitted primes for clarity.

Following the approach of Esarey et al. (Esarey et al. 1993), we will assume that the irradiation of the electron is over a finite time interval, T, whose emission is written for brevity as

$$\xi = \frac{d^2 I(\omega, \hat{n})}{d\omega_r d\Omega} \bigg|_{\text{interval}=T} \quad (3.187)$$

$$\xi = \frac{e^2 \omega_r^2}{4\pi^2 c} \left| \int_{-T/2}^{T/2} \hat{n} \times \hat{n} \times \beta e^{i\omega_r(t' + \hat{n} \cdot r/c)} dt \right|^2. \quad (3.188)$$

This equation is essentially the same as saying that the electron sees a finite laser pulse, which is flat-top in time. Certainly we could have included a realistic laser intensity envelope in Eq. (3.188), but this just leads to further complications in the algebra with no additional physical insight. (The consequences of this simplification are merely on the spectral shape of each individual harmonic, as we shall see.) We will also consider linear polarization though, circular polarization leads to some simplifications of the algebra.

It is convenient to work in polar coordinates which allow us to say that in the far field approximation, $\hat{n} = \hat{r}$, where θ is the angle measured from the propagation axis of the laser ($\hat{z}$) and φ is the angle with respect to the laser polarization axis ($\hat{x}$). We therefore have

$$\hat{r} = \sin\theta \cos\varphi \hat{x} + \sin\theta \sin\varphi \hat{y} + \cos\varphi \hat{z}, \quad (3.189a)$$
$$\hat{\theta} = \cos\theta \cos\varphi \hat{x} + \cos\theta \sin\varphi \hat{y} + \sin\theta \hat{z}, \quad (3.189b)$$
$$\hat{\varphi} = -\sin\varphi \hat{x} + \cos\varphi \hat{y}. \quad (3.189c)$$

That the electron moves in only the x- and z-directions lets us write

$$\hat{n} \times \hat{n} \times \beta = -\beta_\theta \hat{\theta} - \beta_\psi$$
$$= (-\cos\theta \cos\varphi \beta_x + \sin\theta \beta_z)\hat{\theta} + \sin\varphi \beta_x \hat{\varphi}, \quad (3.190)$$
$$\hat{n} \cdot r = \sin\theta \cos\varphi x + \cos\theta z. \quad (3.191)$$

The term in (3.190) sets the polarization of the scattered emission, which will have $\hat{\theta}$ and $\hat{\varphi}$ components. The intensities of these two polarization components add, which gives, from substitution of (3.190) and (3.191) in (3.188),

3.4 Radiation Emitted by Electrons Driven in a Strong Laser Field

$$\xi = \frac{e^2\omega_r^2}{4\pi^2 c}\left\{\left|\int_{-T/2}^{T/2}(-\cos\theta\cos\varphi\beta_x + \sin\theta\beta_z)\right.\right.$$

$$\left.\times \exp\left[i\omega_r(t - \sin\theta\cos\varphi x/c - \cos\theta z/c)\right]dt\right|^2$$

$$\left.+ \left|\int_{-T/2}^{T/2}\sin\varphi\beta_x \exp\left[i\omega_r(t - \sin\theta\cos\varphi x/c - \cos\theta z/c)\right]dt\right|^2\right\}. \quad (3.192)$$

Utilizing the electron trajectories as a function of proper time found in Eqs. (3.76) and (3.77), the product of the velocity and lab time intervals can be found easily:

$$\beta_i dt = \frac{1}{c}\frac{dx_i}{d\tau}d\tau. \quad (3.193)$$

Introduction of (3.193) into (3.192) will allow us to integrate over proper time (which is the only solution we have for the trajectory of the electrons).

First, let's look at the phase term in (3.192), which we denote $\eta(\tau)$ and insert the electron trajectories

$$\eta(\tau) = \omega_r(\tau - z/c) - \sin\theta\cos\varphi x/c + \cos\theta z/c, \quad (3.194)$$

$$\eta(\tau) = \left(1 + (1-\cos\theta)a_0^2/4\right)\omega_r\tau + \frac{\omega_r}{\omega}a_0\sin\theta\cos\varphi\cos\omega\tau$$

$$- \frac{\omega_r}{\omega}(1-\cos\theta)(a_0^2/8)\sin 2\omega\tau. \quad (3.195)$$

The scattered radiation now is

$$\xi = \frac{e^2\omega_r^2}{4\pi^2 c}\left\{\left|\int_{-\tau_0/2}^{\tau_0/2}\left(-\cos\theta\cos\varphi\frac{dx}{d\tau} + \sin\theta\frac{dz}{d\tau}\right)e^{i\eta(\tau)}d\tau\right|^2\right.$$

$$\left.+ \left|\int_{-\tau_0/2}^{\tau_0/2}\sin\varphi\frac{dx}{d\tau}e^{i\eta(\tau)}d\tau\right|^2\right\}, \quad (3.196)$$

where $\omega\tau_0/2\pi$ is the number of cycles in the pulse. This becomes with differentiation of (3.76) and (3.77)

$$\xi = \frac{e^2\omega_r^2}{4\pi^2 c}\left\{\left|\int_{-T/2}^{T/2}\left(-a_0\cos\theta\cos\varphi\sin\omega\tau + \frac{a_0^2}{4}\sin\theta(1-\cos 2\omega\tau)\right)e^{i\eta(\tau)}d\tau\right|^2\right.$$

$$\left.+ \left|\int_{-T/2}^{T/2}a_0\sin\varphi\sin\omega\tau e^{i\eta(\tau)}d\tau\right|^2\right\}. \quad (3.197)$$

Equation (3.197) with phase given by (3.195) is suitable for straightforward numerical integration to find the entire spectrum of radiation as a function of radiated frequency and angle. For example, Figure 3.5 shows the spectra calculated numerically with Eq. (3.197) at an observation angle of 90° with respect to both the axis of laser propagation and laser polarization for three increasing values of laser intensity.

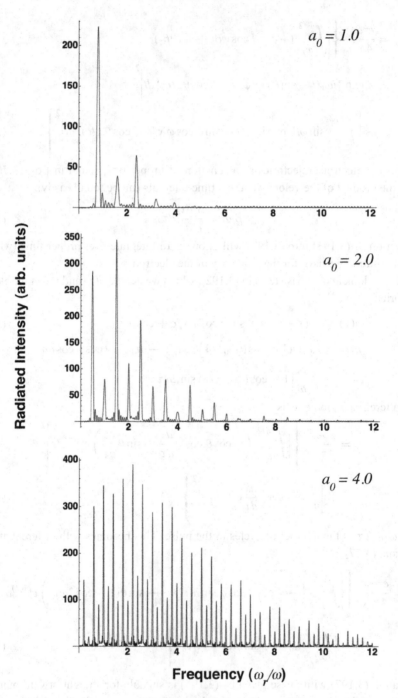

Figure 3.5 Calculated scattered radiation emission from an electron observed in the y-direction ($\theta = 90°$ and $\varphi = 90°$) for three different values of a_0.

3.4 Radiation Emitted by Electrons Driven in a Strong Laser Field

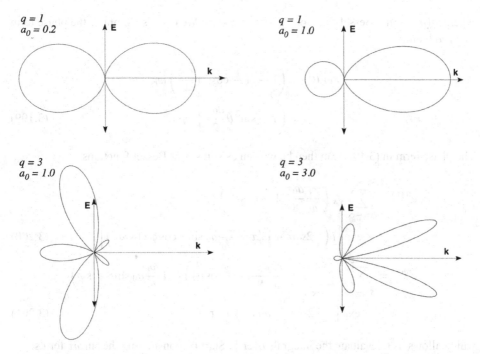

Figure 3.6 Calculated scattered radiation emission patterns for two harmonic orders in the x–z plane from an electron driven by two different values of a_0.

The spectrum does indeed emerge as a comb of discrete peaks, though the order positions are shifted to lower frequency than integer multiples of the laser fundamental, a consequence of the Doppler downshift of the laser in frame of the forward-drifting electron. The number of harmonics emitted increases dramatically as the field strength is increased; in fact, the strongest harmonic peak moves to orders higher than $q = 1$ at $a_0 > 1$. The Doppler downshift of the harmonics also becomes more substantial at higher field strength.

Figure 3.6 shows the angular distribution of $q = 1$ and $q = 3$ emission in the xz plane (the plane of laser polarization). Even though a single harmonic is plotted, it should be recalled that the frequency of emission changes with angle in each of these plots. At low intensity, we observe the usual Thomson scatter emission pattern, which is Lorentz-folded into the propagation direction as a_0 is increased. The $q = 3$ pattern exhibits a number of lobes with a zero in the emission pattern, which also is folded forward at the higher field strength, in the direction of laser propagation.

We can make some further analytic progress by introducing the Jacobi–Anger formula to equation (3.197)

$$e^{ib\sin\sigma} = \sum_{s=-\infty}^{\infty} J_s(b) e^{is\sigma}, \tag{3.198}$$

where $J_s(b)$ are the Bessel functions of the first kind. We can also simplify the phase term by introducing

$$\omega_r'(\theta) \equiv \left(1 + (1 - \cos\theta)\frac{a_0^2}{4}\right)\omega_r$$
$$= \left(1 + \sin^2\theta \frac{a_0^2}{2}\right)\omega_r. \tag{3.199}$$

The phase term in (3.196) can then be written as sums over Bessel functions

$$e^{i\eta(\tau)} = \sum_{s=-\infty}^{\infty} J_s\left(\frac{\omega_r}{\omega}\frac{a_0^2}{8}(1 - \cos\theta)\right)$$
$$\times \exp\left[i\left(-2s\omega\tau + \omega_r'\tau + \frac{\omega_r}{\omega}a_0\sin\theta\cos\varphi\cos\omega\tau\right)\right]. \tag{3.200}$$

$$e^{i\eta(\tau)} = \sum_{s=-\infty}^{\infty}\sum_{q'=-\infty}^{\infty} J_s\left(\frac{\omega_r}{\omega}\frac{a_0^2}{8}(1 - \cos\theta)\right)J_{q'}\left(\frac{\omega_r}{\omega}a_0\sin\theta\cos\varphi\right)$$
$$\times \exp\left[i\left(-2s\omega\tau + q'\omega\tau + \omega_r'\tau\right)\right] \tag{3.201}$$

which allows us to evaluate the integrals over τ. Start by considering the $\sin\omega\tau$ terms:

$$\int_{-T/2}^{T/2} \sin\omega\tau \exp\left[-i(2s\omega + q'\omega - \omega_r')\tau\right] d\tau$$
$$= \int_{-T/2}^{T/2} -\frac{i}{2}\{\exp\left[-i\left((2s + q' - 1)\omega - \omega_r'\right)\tau\right] d\tau$$
$$- \exp\left[-i\left((2s + q' + 1)\omega - \omega_r'\right)\tau\right] d\tau\}. \tag{3.202}$$

In the first term of the right-hand side of (3.202), we make the replacement $2s + q' - 1 = q$ and in the second term set $2s + q' + 1 = q$. Similarly, in the second term we can replace

$$1 - \cos 2\omega\tau = \frac{1}{2}\left(2 - e^{i2\omega\tau} - e^{-i2\omega\tau}\right) \tag{3.203}$$

in which the resulting three exponential terms are replaced with $2s + q' = q$ and $2s + q' \pm 2 = q$, respectively. Then the integral over proper time gives the spectral dependence

$$\int_{-T/2}^{T/2} \exp\left[i(\omega_r' - q\omega)\tau\right] d\tau = 2\frac{\sin\left[(\omega_r' - q\omega)T/2\right]}{\omega_r' - q\omega}. \tag{3.204}$$

With these substitutions we see that the harmonic spectrum can be written

$$\xi = \frac{e^2\omega_r^2 T^2}{4\pi^2 c}\sum_{q=1}^{\infty} \text{sinc}^2\left[(\omega_r' - q\omega)T/2\right]\left[A_q(\theta,\varphi) + B_q(\theta,\varphi)\right], \tag{3.205}$$

3.4 Radiation Emitted by Electrons Driven in a Strong Laser Field

where the intensity factors are

$$\mathcal{A}_q(\theta,\varphi) = \frac{1}{4}a_0^2 \cos^2\theta \cos^2\varphi \left[\sum_{s=-\infty}^{\infty} J_s(\alpha_1)\left(J_{q-2s+1}(\alpha_2) - J_{q-2s-1}(\alpha_2)\right)\right]^2$$

$$+ \frac{1}{64}a_0^4 \cos^2\theta \left[\sum_{s=-\infty}^{\infty} J_s(\alpha_1)\left(J_{q-2s}(\alpha_2) - J_{q-2s+2}(\alpha_2) - J_{q-2s-2}(\alpha_2)\right)\right]^2$$

(3.206a)

$$\mathcal{B}_q(\theta,\varphi) = \frac{1}{4}a_0^2 \sin^2\varphi \left[\sum_{s=-\infty}^{\infty} J_s(\alpha_1)\left(J_{q-2s+1}(\alpha_2) - J_{q-2s-1}(\alpha_2)\right)\right]^2 \quad (3.206b)$$

and the arguments of the Bessel functions are

$$\alpha_1 = \frac{\omega_r}{\omega}\frac{a_0^2}{8}(1-\cos\theta), \tag{3.207a}$$

$$\alpha_2 = \frac{\omega_r}{\omega}a_0 \sin\theta \cos\varphi. \tag{3.207b}$$

While it appears that these substitutions have made no progress in simplifying the scattered radiation distribution, they do show that the spectrum can be written as an explicit sum over harmonic orders. The factor $\mathcal{A}_q(\theta,\varphi) + \mathcal{B}_q(\theta,\varphi)$ can be interpreted as the intensity and angular distribution of the qth harmonic. The spectrum of each term in the q sum is a sharply peaked sinc^2 function centered at a frequency given by

$$\omega_r' - q\omega = 0. \tag{3.208}$$

Examination of Eq. (3.199) indicates that the frequency of each qth harmonic peak is just the harmonic frequency that we derived in Eq. (3.160) from R-frame Doppler shifting arguments. The shape of the resonance function, a sinc^2 function, is simply an artifact of assuming constant irradiation intensity over a finite time interval, T. In reality, the spectral shape of each harmonic is determined by the temporal shape of the driving laser pulse.

We are now able to use the intensity factors of (3.206) to plot the angular distribution of each harmonic, recognizing that the frequency of emission of that harmonic varies according to observation angle as described by Eq. (3.160) or (3.199). At arbitrary angles this form is not much easier than numerical integration of (3.197) because of the need to evaluate the infinite sums. However, Eqs. (3.205)–(3.207) can let us evaluate easily the shape of the spectrum at certain interesting angles. First consider emission in the forward direction parallel to the laser propagation, $\theta = 0$. Equation (3.207) shows that both α factors are zero in this case. Since $J_\sigma(0) = \delta_\sigma$ the intensity factors become

$$\mathcal{A}_q(\theta,\varphi) + \mathcal{B}_q(\theta,\varphi) = \frac{a_0^2}{4}\left[\sum_{s=-\infty}^{\infty} J_0(0)\left(J_{q+2s+1}(0) - J_{q+2s-1}(0)\right)\right]^2$$

$$= \frac{a_0^2}{4}\left(J_{q+1}(0) - J_{q-1}(0)\right)^2$$

$$= \frac{a_0^2}{4}\left(\delta_{q+1} - \delta_{q-1}\right)^2. \tag{3.209}$$

So directly in the forward direction *only* the fundamental is emitted, and, as Eq. (3.160) told us, this emission is right at the initial laser frequency.

A more interesting case is that of the backscattered radiation. This is of particular importance, as we will see, in experiments in which a femtosecond laser irradiates an electron beam head-on with the intention of producing backscattered radiation upshifted by the Lorentz factor of the beam (see Section 3.6). In the backscattered case $\alpha_\perp = 0$, and

$$\alpha_1 = \frac{a_0^2}{4}\frac{\omega_r}{\omega} = q\frac{a_0^2}{4(1+a_0^2/2)}, \qquad (3.210)$$

which reduce the amplitude factors to

$$\mathcal{A}_q(\theta,\varphi) + \mathcal{B}_q(\theta,\varphi) = \frac{a_0^2}{4}\left[J_{(q+1)/2}\left(\frac{\omega_r}{\omega}\frac{a_0^2}{4}\right) - J_{(q-1)/2}\left(\frac{\omega_r}{\omega}\frac{a_0^2}{4}\right)\right]^2. \qquad (3.211)$$

At low a_0 we see that

$$\mathcal{A}_q(\theta,\varphi) + \mathcal{B}_q(\theta,\varphi) \xrightarrow{a_0 \to 0} \frac{a_0^2}{4}\left[J_{(q+1)/2}(0) - J_{(q-1)/2}(0)\right]^2$$

$$= \frac{a_0^2}{4}\left[\delta_{q+1} - \delta_{q-1}\right]^2, \qquad (3.212)$$

so the spectrum reduces to the simple Thomson scatter result with only $q = 1$ emission. At higher a_0, the spectrum of harmonics has a shape that goes as

$$S_q(\theta,\varphi) \sim q^2\left[J_{(q+1)/2}\left(q\frac{a_0^2}{4(1+a_0^2/2)}\right) - J_{(q-1)/2}\left(q\frac{a_0^2}{4(1+a_0^2/2)}\right)\right]^2. \qquad (3.213)$$

The amplitudes of each q harmonic is plotted from this equation for a range of a_0 values in Figure 3.7, illustrating again that the number of higher harmonics increases dramatically as a_0 increases. Using Eq. (3.211), average harmonic number as a function of a_0 is plotted. This plot illustrates that the average nonlinear order increases roughly as $q_{ave} \cong a_0^{2.9}$. These trends have been confirmed in experiments (Chen et al. 1998).

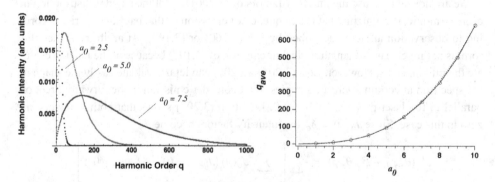

Figure 3.7 Plot of the backscattered harmonic intensity from a free electron for three different values of a_0. As a function of harmonic order (left). On the right is a calculation of the average emitted harmonic order as a function of normalized vector potential. The dashed line is a fit with $q_{ave} = a_0^{2.86}$.

3.4.6 The Potential Effects of Radiation Reaction on the Electron

Before concluding this section, let us take a very brief look at the amount of power radiated by the electron to determine if the neglect of radiation reaction is valid (in the classical limit). The total power radiated by a charge is given by the relativistic Larmor formula (Jackson 1975, p. 660)

$$P_{rad} = -\frac{2}{3}\frac{e^2}{m^2c^3}\left(\frac{dp_\mu}{d\tau}\frac{dp^\mu}{d\tau}\right) \tag{3.214}$$

or, in terms of momentum and energy explicitly,

$$P_{rad} = \frac{2}{3}\frac{e^2}{m^2c^3}\left[\left(\frac{d\mathbf{p}}{d\tau}\right)^2 - \frac{1}{c^2}\left(\frac{dU}{d\tau}\right)^2\right]. \tag{3.215}$$

From the trajectories of an electron in a linearly polarized field, the derivatives of the four-momenta are

$$\frac{d\mathbf{p}}{d\tau} = a_0 mc\omega \cos\omega\tau\, \hat{\mathbf{x}} + \frac{1}{2}a_0^2 mc\omega \sin 2\omega\tau\, \hat{\mathbf{z}}, \tag{3.216}$$

$$\frac{dU}{d\tau} = mc^2\omega \frac{d}{d\tau}\left(1 - \frac{p_z}{mc}\right)$$

$$= \frac{1}{2}a_0^2 mc^2\omega \sin 2\omega\tau. \tag{3.217}$$

The total radiated power as a function of proper time then is

$$P_{rad} = \frac{2}{3}\frac{e^2\omega^2}{c}a_0^2 \cos^2\omega\tau. \tag{3.218}$$

To estimate if this power loss is significant, we calculate the energy loss by the electron by integrating the power loss over one complete optical cycle

$$U_{loss} = \int_0^{2\pi/\omega} P_{rad}\,d\tau$$

$$= \frac{2\pi}{3}\frac{e^2\omega}{c}a_0^2. \tag{3.219}$$

We can compare that energy loss to the average energy of the electron in the field and find that

$$\frac{U_{loss}}{\langle U \rangle} = \frac{2\pi}{3}\frac{e^2\omega}{mc^3}\frac{a_0^2}{1 + a_0^2/4}. \tag{3.220}$$

This ratio is $\sim 10^{-7}$ at high a_0, so it seems that energy losses by radiation are relatively unimportant in the overall trajectory of the electron.

It turns out that the inclusion of radiation reaction does have some interesting consequences (Macchi 2013, pp. 17–21). The most intuitive way to think of radiation reaction consequences is as applying a kind of friction force to the motion of the electron (and this physics is often referred to as radiation friction). Consider an electron with some initial velocity undergoing gyromagnetic circles in an externally imposed magnetic field.

We know, of course, from the centripetal acceleration of that electron that it will radiate an EM field. Since energy must be conserved, the electron loses kinetic energy (even though the magnetic field itself does no work on the electron) and will ultimately spiral in toward its center of circular motion. This picture implies that the electron motion could be modeled as if it was feeling a friction force, slowing it down. In fact there is no completely accepted form for this radiation reaction friction. The most commonly used approximate equation to quantify radiation friction for relativistic electrons is the Landau–Lifshitz formula (Landau and Lifshitz 1994, Sec. 76). Implementation of this classical field formula indicates that radiation friction is largely a negligible effect for free electrons accelerated by laser fields with intensity at least up to 10^{23} W/cm^2. There is, in fact, an analytic solution for the motion of an electron in a plane-wave laser field subject to the Landau and Lifshitz approximation to the radiation friction, though we will not explore it here (Di Piazza 2008).

An interesting consequence of the radiation friction force is that its inclusion breaks the conservation of canonical momentum in the laser field. Recall that it was this conservation that led to the conclusion that a free electron could acquire no net energy from a passing plane-wave pulse. The oscillating electron merely gave up its oscillation energy back to the field as long as the plane-wave's pulse rose and fell adiabatically (over many wavelengths). The inclusion of a radiation friction term breaks this conservation law and a free electron can indeed acquire some energy from a passing plane-wave pulse. Again, this effect is so small as to be negligible in the considerations made in this chapter, and our neglect of the effect is completely justified and accurate.

We must also note that the radiation friction force approximated in the Landau–Lifshitz formula ignores any quantum effects of the radiation. This complication is beyond the scope of the treatment here and demands a theoretical treatment within the rubric of a fully QED consistent theory. It should be said that once the emitted radiation reaches photon energies that are a significant fraction of the electron rest mass, our conclusions are incorrect and quantized effects in the radiation reaction can play a significant role in the electron dynamics.

3.5 Ejection of Free Electrons from the Focus of an Intense Laser

Until now we have considered the electron dynamics in an infinite plane-wave. Of course, actual strong field physics experiments are performed with a focused laser beam that has finite spatial extent and consequently a gradient in intensity. This gradient in field amplitude can lead to time-averaged forces on the electron that can push the electron from regions of high intensity toward those of lower intensity (Bucksbaum et al. 1987; Hartemann et al. 1995; Monot et al. 1993; Quesnel and Mora 1998; Mulser and Bauer 2010, p. 193). This phenomenon is often referred to as ponderomotive acceleration or ponderomotive scattering.

A naïve view of electron scattering in a focused laser field might argue that the only forces the electron experiences outside of its quiver motion is through spontaneous Thomson scattering, which has a very low cross section equal to 0.7×10^{-24} cm^2. In a 100 fs pulse at an intensity of, say, 10^{16} W/cm^2, the photon fluence is $\sim 6 \times 10^{21}$ photons/cm^2 so each electron has less than a 1% probability of even scattering one photon, each of which would impart a momentum of ~ 1 eV/c. In actuality, the electron experiences a coherent laser beam so that this minuscule spontaneous scattering is dominated by stimulated scattering. One can think of the scattered photon as entering a coherent mode of the laser beam, which means that the scattering rate is enhanced by a factor equal to the number of photons in the mode, an enormous number at high intensity. Consequently, field scattering from the electron can impart a significant momentum to the electron. Again, this view of many photons participating in the scattering process justifies our classical description of the fields.

To study this effect quantitatively, we will need a description of the laser field near the focus. In practice most real laser beams have some amount of multiple spatial modes in their spatial distribution so the actual intensity distribution near a focus is rather complicated. It is often a fairly good approximation to treat the laser field near focus as that of a perfectly coherent, single TEM_{00} spatial mode beam. Solution of the paraxial wave equation, performed in detail in Appendix A, leads to an analytic solution to the electric field near the laser focus, which is Gaussian in the transverse profile, with a spatially varying phase. For linearly polarized light, this single-mode focal distribution can be written

$$E_x = E_0 \frac{w_0}{w(z)} \exp\left[-\frac{r^2}{w(z)^2}\right] \cos\left[\omega\tau + \tan^{-1}(z/z_R) - \frac{zr^2}{z_R w(z)^2}\right], \quad (3.221)$$

where the axially dependent $1/e^2$ radius is given by

$$w(z) = w_0 \sqrt{1 + \frac{z^2}{z_R^2}}. \quad (3.222)$$

Here

$$z_R = \frac{\pi}{\lambda} w_0^2 \quad (3.223)$$

is the so-called Rayleigh range of the focus, which quantifies the longitudinal extent over which the focal spot remains near the diffraction-limited spot size of w_0. At $|z|$ values greater than z_R diffraction causes spreading in the beam profile. The $\tan^{-1}$ term in the phase of the field is the well-known Guoy phase shift which leads to the phase shift of $180°$ as the field passes through focus. The reader is to be reminded that this solution for a focused laser field is simply an approximation; it represents a solution to the paraxial wave equation which is only an approximation to the full wave equation. (More on this later in this section and in Appendix A.)

3.5.1 Time-Averaged Ponderomotive Force on an Electron in an Intensity Gradient

Let us now consider the motion of an electron in a field like that described by Eq. (3.221). We begin our discussion in the nonrelativistic regime (i.e. at intensity below about 10^{17} W/cm^2 of near-IR lasers). As we now know, the electron will experience harmonic motion along the x-direction because of the oscillating electric field. In the absence of a

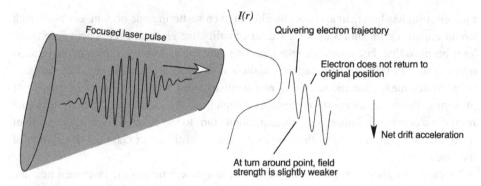

Figure 3.8 Qualitative description of why an electron experiences a time-averaged force toward lower intensity in a laser field gradient.

field gradient we have found that the electron motion is periodic and returns to its initial position after one complete optical cycle. If the electron is instead in a field with an intensity gradient, the situation is akin to that depicted in Figure 3.8. As the electron quivers in one direction, it can move to a region of lower (or higher) field strength. If it moves, say, to lower field strength, when the phase of the field reverses the sign of E and drives acceleration of the electron in the other direction, the magnitude of this force will be somewhat less than the force the electron felt on the previous half cycle that drove it out to lower intensity. The result is that the electron does not quite return to its previous starting point for that cycle. Consequently the electron will "drift" toward lower intensity on every cycle. If the laser pulse is on long enough, the electron will ultimately be forced out of the focus by this time-averaged drift force.

We can see why this occurs in a linearly polarized field by assuming that the gradient in intensity is weak and examining the electron trajectory in this gradually varying field. In the absence of relativistic effects the electron is subject to

$$m\dot{\mathbf{v}} = -e\mathbf{E}(\mathbf{x}) \qquad (3.224)$$

in a field with a field gradient which could be described as a Gaussian (though need not be). If E_1 is the slowly varying field amplitude with radius,

$$\mathbf{E}(\mathbf{x}) = E_0 e^{-r^2/w_0^2} \cos\omega\tau \, \hat{\mathbf{x}} = E_1(r)\cos\omega\tau \, \hat{\mathbf{x}}, \qquad (3.225)$$

we can Taylor-expand the field about some point at radius r_0, as long as the variation of E_1 is slow. Practically speaking, this will require that the quiver amplitude is small compared to the field gradient $E_1/\partial_x E_1 \sim w_0 \gg x_{osc}$. We can treat the spatially varying field with the first two terms of this expansion:

$$\mathbf{E}(\mathbf{x}) = E_1(r_0)\cos\omega\tau \, \hat{\mathbf{x}} + \frac{\partial E_1(r_0)}{\partial x}(x-x_0)\cos\omega\tau \, \hat{\mathbf{x}} + \ldots. \qquad (3.226)$$

Since each term in the field expansion is smaller than the previous one, we treat the motion of the electron in a perturbation expansion in transverse velocity, $v_x = v_{0x} + v_{1x} + \ldots$,

3.5 Ejection of Free Electrons from the Focus of an Intense Laser

and position $x = x_0 + x_1 + \ldots$. The solution to the zeroth-order terms results from the first term in the expansion of (3.226) and we retrieve our previous plane-wave solutions for transverse velocity and position

$$v_{0x} = -\frac{eE_1(r_0)}{m\omega} \sin \omega \tau, \tag{3.227}$$

$$x_0 = \frac{eE_1(r_0)}{m\omega^2} \cos \omega \tau. \tag{3.228}$$

Using standard perturbation theory methods, we find the solution to the first-order term of the electron velocity by equating the second terms in the velocity and field expansion of (3.224) and inserting the zeroth-order solution of the position into the first-order term. This gives for dv_{1x}/dt

$$\begin{aligned} m\frac{dv_{1x}}{dt} &= -ex\frac{\partial E_1(r_0)}{\partial x} \cos \omega \tau \\ &= -\frac{e^2}{m\omega^2} E_1(r_0) \frac{\partial E_1(r_0)}{\partial x} \cos^2 \omega \tau \\ &= -\frac{e^2}{2m\omega^2} \frac{\partial E_1(r_0)^2}{\partial x} \cos^2 \omega \tau. \end{aligned} \tag{3.229}$$

We then define an effective cycle-averaged force on the electron, f_p, by averaging Eq. (3.229) over a complete electron oscillation,

$$F_p \equiv \left\langle m\frac{dv_1}{dt} \right\rangle, \tag{3.230}$$

yielding

$$F_p = -\frac{e^2}{4m\omega^2} \frac{\partial}{\partial x} E_1(r_0)^2. \tag{3.231}$$

This cycle-averaged force is proportional to the negative of the gradient (in the x direction this case) of the square of the electric field amplitude, a quantity which is, of course, proportional to the local intensity. This cycle-averaged force is usually called the ponderomotive force. This view is only really valid if the gradient is slow when compared to the single-cycle oscillation amplitude of the electron. If this is not true (which can occur for very small focal spots or very strong fields) the perturbation expansion employed here is invalid.

We have already seen that the cycle-averaged ponderomotive energy, U_p, of an electron in a laser field is proportional to intensity. This result then leads us to consider the ponderomotive energy as some kind of effective potential energy, whose gradient leads to the ponderomotive force. This phenomenological view would suggest that we could write the ponderomotive force as

$$\mathbf{F}_p = -\nabla U_p. \tag{3.232}$$

This equation can be written with the help of Eq. (3.29) as

$$\mathbf{F}_p = -\frac{e^2}{4m\omega^2} \nabla E_1^2 = -\frac{e^2}{4mc^2} \nabla A_1^2. \tag{3.233}$$

Equation (3.233) is, in fact, the correct form of the cycle-averaged ponderomotive force (Quesnel and Mora 1998). For this reason U_p is often referred to in the literature as the "ponderomotive potential." Of course, U_p is not, rigorously speaking, a true potential; it is a cycle-averaged quantity. We will refrain from using this term and simply refer to the ponderomotive energy in the remainder of this book.

3.5.2 Role of Longitudinal Fields in the Correct Description of the Ponderomotive Force

Equation (3.233) appears to have roughly the same form as Eq. (3.231). However, a notable discrepancy arises when we compare (3.233) to (3.231). Consider an electron in a cylindrically symmetric laser focus at arbitrary x and y position (like that pictured in Figure 3.9). Equation (3.231) predicts that the electron feels a force in the x-direction, the direction of the electric field polarization. However, if we are to view the ponderomotive force as negative the gradient in ponderomotive energy and hence negative the gradient in intensity, the force the electron feels should be *radial*. Clearly (3.233) makes a different prediction than (3.231). Equation (3.231) would suggest that electrons ejected from the focus will acquire energy directly along the laser polarization no matter where in the focus they initially reside, leading to an anisotropic electron distribution outside of the focus. On the other hand, Eq. (3.233) predicts that electrons ejected from the focus will exit in a direction along a radius.

If we consider the field at focus to be described by a distribution like (3.221) it is difficult to see how the electron could feel any force at all in the y-direction as predicted by (3.233). If we include the magnetic field, which for a field like (3.221) is $\mathbf{B} \parallel \hat{\mathbf{y}}$, the problem is not solved. This B-field gives a force along the z-direction since $\mathbf{v}_{osc} \times \mathbf{B} \parallel \hat{\mathbf{z}}$. So the inclusion of the Lorentz force cannot explain the y-directed force that Eq. (3.233) would imply.

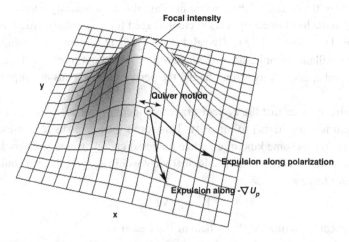

Figure 3.9 Two possible ponderomotively driven paths of an electron in a laser focus. The path of expulsion along the ponderomotive energy gradient is the path actually driven by the field.

3.5 Ejection of Free Electrons from the Focus of an Intense Laser

The solution to this puzzle rests with the fact that a field distribution like Eq. (3.221) is not the correct distribution for an electric field near a focus (Davis 1979; Cicchitelli et al. 1990). In fact, a Gaussian distribution with purely transverse fields does not satisfy Maxwell's equations. Maxwell's equations in vacuum demand that $\nabla \cdot \mathbf{E} = 0$, which is exactly satisfied for an infinite plane-wave. However, any field written with E directed along x at all points in space with an amplitude gradient in x has

$$\frac{\partial E_x}{\partial x} \neq 0, \tag{3.234}$$

which must mean that the electric field of the Gaussian distribution has nonzero divergence

$$\nabla \cdot \mathbf{E}_{Gaussian} \neq 0. \tag{3.235}$$

In reality, the field near a focus *must* contain nonzero E_y and E_z components to fulfill the requirements of Maxwell's equations in vacuum. Because we must also have $\nabla \cdot \mathbf{B} = 0$ we need to draw the same conclusion about the magnetic field. This finding is not surprising when we recall that (3.221) results from the solution of the paraxial wave equation and, therefore, must only be an approximate solution for the field near focus.

Cicchitelli, Hora and Postle (Cicchitelli et al. 1990) first showed that the additional field components near a Gaussian focus were necessary to describe correctly the ponderomotive motion of an electron in that focus. In reality, these additional field components are small compared to the transverse field oriented along the polarization direction. They can be expanded in terms of the small parameter

$$\kappa_f = \frac{1}{kw_0}$$

and, as we will see, only fields up to the first-order correction in κ are needed to derive the correct form of the ponderomotive force.

It is possible to estimate the size of these forces from the zero divergence requirement. If we treat the Gaussian beam as the zeroth-order solution for the field, we can use the divergence condition to find the z component of the electric field to first order. We can write

$$\frac{\partial E_x}{\partial x} + \frac{\partial E_z}{\partial z} = 0 \tag{3.236}$$

if we ignore the y component of E, which is justified as E_y will be second order in κ since

$$\frac{\partial E_z}{\partial z} \sim \frac{1}{k}\frac{\partial E_y}{\partial y}.$$

We can then say that

$$E_z \cong -\int \frac{\partial E_x}{\partial x} dz, \tag{3.237}$$

which can be evaluated if we approximate E_x by the zeroth-order Gaussian field:

$$\frac{\partial E_x}{\partial x} \cong -E_0 \frac{w_0}{w} \frac{2x}{w^2} \exp\left[-r^2/w^2\right] \cos(\omega t - kz). \tag{3.238}$$

Integrating this (and saying that $w^3 \cong w^2 w_0$ near a focus) gives us the approximate first-order expression for the longitudinal electric field near a focus,

$$E_z \cong -2E_0 x \frac{w_0}{w^2} \kappa_f \exp\left[-r^2/w^2\right] \sin(\omega t - kz). \tag{3.239}$$

This field component is first order in the small parameter κ, and is 90° out of phase with the transverse field, E_x.

A rigorous derivation of the additional field components has been presented by a number of authors (Davis 1979; Cicchitelli et al. 1990; Quesnel and Mora 1998). (The outlines of this derivation are discussed in Appendix A.) Cicchitelli et al., for example, showed that the correct form of the first-order longitudinal field with the correct phase is

$$E_z = -2E_0 x \frac{w_0}{w^2} \kappa_f \exp\left[-\frac{r^2}{w^2}\right] \sin\left(\omega t - kz - 2\tan^{-1}\frac{z}{z_R} - \frac{zr^2}{z_R w^2}\right). \tag{3.240}$$

Since the magnetic field is also subject to the zero divergence requirement, an identical analysis leads to a similar result for the longitudinal component of the B-field

$$B_z = -2E_0 y \frac{w_0}{w^2} \kappa_f \exp\left[-\frac{r^2}{w^2}\right] \sin\left(\omega t - kz - 2\tan^{-1}\frac{z}{z_R} - \frac{zr^2}{z_R w^2}\right). \tag{3.241}$$

We are now in a position to calculate the ponderomotive force with the inclusion of these longitudinal fields. We will perform a similar perturbative analysis with the additional fact that there are z components of the fields which are first order in a small parameter. Starting with the usual equation of motion for the electron

$$\frac{d\mathbf{p}}{dt} = -e\mathbf{E} - \frac{e}{c}\mathbf{v} \times \mathbf{B}, \tag{3.242}$$

we can generally describe the spatially dependent field in terms of oscillation phase $\varphi = \omega\tau$:

$$\mathbf{E} = \mathbf{E}_1(\mathbf{x}) \cos\varphi. \tag{3.243}$$

As before, the field can be Taylor-expanded in position about some point which we will simply denote as position 0:

$$\mathbf{E} = [\mathbf{E}_1(0) + \mathbf{x} \cdot \nabla \mathbf{E}_1(0) + \ldots] \cos\varphi. \tag{3.244}$$

Again, this construction is valid only for slowly varying field amplitudes to permit the expansion to converge, mandating that $\mathbf{x} \bullet \nabla E_1(0)$ be small with respect to $E_1(0)$.

Writing a perturbation expansion for the electron's momentum and position, being cognizant of the possibility for motion in all three directions

$$\mathbf{p}(t) = \mathbf{p}_0 + \mathbf{p}_1 + \ldots$$
$$\mathbf{x}(t) = \mathbf{x}_0 + \mathbf{x}_1 + \ldots.$$

Since the zeroth-order field includes only the electric field along the x polarization direction, the zeroth-order quiver motion solution of the electron is the same as before:

$$\frac{d\mathbf{p}_0}{dt} = -e\mathbf{E}_1(0) \cos\varphi, \tag{3.245}$$

3.5 Ejection of Free Electrons from the Focus of an Intense Laser

$$\mathbf{p}_0 = \frac{e}{\omega} E_1 \sin\varphi \hat{\mathbf{x}}, \quad (3.246)$$

$$\mathbf{x}_0 = \frac{e}{m\omega^2} E_1 \cos\varphi \hat{\mathbf{x}}. \quad (3.247)$$

Now the first-order terms have elements which include E_z and B_z, so the equation for the first-order drift momentum utilizes (3.247) to give

$$\frac{d\mathbf{p}_1}{dt} = -e\mathbf{x}_0 \cdot \nabla E_x \hat{\mathbf{x}} - e\mathbf{x}_0 \cdot \nabla E_z \hat{\mathbf{z}} - \frac{e}{mc} \mathbf{p}_0 \times \mathbf{B}. \quad (3.248)$$

As we have discussed, we can use one of Maxwell's equations to estimate the size of the longitudinal B-field (see Appendix A):

$$B_z \cong -\int \frac{\partial B_y}{\partial y} dz. \quad (3.249)$$

In terms of these explicit field components, the equation for the first-order drift momentum is

$$\frac{d\mathbf{p}_1}{dt} = -\frac{e}{m\omega^2} E_1 \cos\varphi \frac{\partial}{\partial x} (E_1 \cos\varphi) \hat{\mathbf{x}} - \frac{e}{m\omega c} E_1 \sin\varphi \left(-B_z \hat{\mathbf{y}} + B_y \hat{\mathbf{z}}\right). \quad (3.250)$$

Assuming again that the zeroth-order solution for the B-field is just the transverse field along y, we can say that $B_y = E_1 \cos\varphi$, which lets us integrate for B_z to get

$$B_z \cong -\int \frac{\partial}{\partial y} E_1 \cos\varphi \, dz = -\frac{1}{k} \frac{\partial}{\partial y} E_1 \sin\varphi. \quad (3.251)$$

Here we have ignored the slow z variation of the field amplitude E_1 with respect to the wavelength-scale variation of the field. Putting (3.251) into (3.250) leads to an expression for the first-order drift force, which is

$$\frac{d\mathbf{p}_1}{dt} = -\frac{e^2}{2m\omega^2} \cos^2\varphi \frac{\partial E_1^2}{\partial x} \hat{\mathbf{x}} - \frac{e^2}{2m\omega^2} \sin^2\varphi \frac{\partial E_1^2}{\partial y} \hat{\mathbf{y}} \\ - \frac{e^2}{m\omega c} E_1^2 \cos\varphi \sin\varphi \hat{\mathbf{z}}. \quad (3.252)$$

The first term on the right-hand side of (3.252) comes from the gradient in E_x (as we found previously), but now there are two additional terms which yield forces along y and z. The second term (y component of the force) has appeared due to the inclusion of B_z in the Lorentz force and now has the form of the gradient of the field amplitude squared along the y-axis. The third term, pointing along z, arose from the inclusion of the force from B_y.

Consider the z term. We can write

$$\frac{\partial}{\partial z} (E_1 \cos\varphi) = -kE_1 \sin\varphi + \cos\varphi \frac{\partial E_1}{\partial z}, \quad (3.253)$$

which lets us say that the z component of the drift force is

$$\left(\frac{d\mathbf{p}_1}{dt}\right)_z = -\frac{e^2}{m\omega c} \frac{E_1}{k} \cos\varphi \left(\frac{\partial E_1}{\partial z} \cos\varphi - \frac{\partial}{\partial z} (E_1 \cos\varphi)\right), \quad (3.254)$$

or the total first-order drift force is

$$\frac{d\mathbf{p}_1}{dt} = -\frac{e^2}{2m\omega^2}\left(\cos^2\varphi\frac{\partial E_1^2}{\partial x}\hat{\mathbf{x}} + \sin^2\varphi\frac{\partial E_1^2}{\partial y}\hat{\mathbf{y}} + \cos^2\varphi\frac{\partial E_1^2}{\partial z}\hat{\mathbf{z}}\right)$$
$$-\frac{e^2}{m\omega^2}E_1\cos\varphi\frac{\partial}{\partial z}(E_1\cos\varphi)\hat{\mathbf{z}}.$$
(3.255)

The last term in Eq. (3.255) will average to zero when we average this force over a complete optical cycle. Performing this cycle average delivers the result we seek, which is the true ponderomotive force

$$\left\langle\frac{d\mathbf{p}_1}{dt}\right\rangle_T = \mathbf{F}_p = -\frac{e^2}{4m\omega^2}\left(\frac{\partial E_1^2}{\partial x}\hat{\mathbf{x}} + \frac{\partial E_1^2}{\partial y}\hat{\mathbf{y}} + \frac{\partial E_1^2}{\partial z}\hat{\mathbf{z}}\right)$$
$$= -\frac{e^2}{4m\omega^2}\nabla E_1^2 = -\nabla U_p.$$
(3.256)

Equation (3.256) reproduces our heuristic result that the cycle-averaged force a quivering electron feels is proportional to the gradient in the ponderomotive energy. This derivation illuminates the origin of the various components of the ponderomotive force. As before, the x component is due simply to a gradient in the x component of the electric field, which is sampled by the quiver motion of the electron in the x-direction. The y component of force arises from the presence of the small z component of the B-field. B_z is almost exactly in phase with the zero-order oscillating velocity, v_x. The continuous application of the Lorentz force from the $v_x B_z$ term yields a component of force in the y-direction. It is curious to note that the longitudinal electric field, E_z, plays almost no role in the drift motion of the electron in this case, a fact that has been confirmed through numerical simulations (Quesnel and Mora 1998).

3.5.3 Ponderomotive Ejection from the Focus with Long and Short Laser Pulses

The electron subject to intense irradiation will tend to feel a cycle-averaged drift force in the radial direction which drives the electron out of the focal region. If the laser pulse has some temporal envelope, in general it will be necessary to solve the drift equations of motion subject to the time-varying ponderomotive force, particularly if the free electron feels that force as the pulse intensity rises up from zero on the leading edge of the pulse. We can make some simple estimates for the role that the laser pulse duration will play by estimating the time it takes to eject an electron from the focus.

In a focus with radius w_0 the radial ponderomotive force is roughly

$$m\ddot{x} = -\nabla U_p \approx -U_p(0)/w_0,$$
(3.257)

where we denote the ponderomotive energy at the peak of the distribution by $U_p(0)$. Subject to this force, the time to eject the electron from the focus is roughly determined by the time it takes the electron to be driven a focal radius

$$x \sim \frac{U_p(0)}{mw_0}t_{ej}^2.$$
(3.258)

3.5 Ejection of Free Electrons from the Focus of an Intense Laser

So we would expect that the electron will be completely expelled from the focus during the laser pulse if the laser pulse duration is longer than this ejection time. This gives us a rough estimate for the "long pulse regime":

$$\Delta t_p > w_0 \sqrt{\frac{m}{U_p(0)}}. \quad (3.259)$$

As an example, if a 1 μm laser is focused to a spot size of about 10 μm to give an intensity of 10^{16} W/cm^2, the long pulse regime is entered for pulses of longer than about 1 ps.

If the electron is born into the field by, say, ionization of an atom or ion, electrons in the long laser pulse regime will "roll" down the ponderomotive hill and exit the focus with kinetic energy essentially equal to the ponderomotive energy at the time of the electron's birth. This kinetic energy is in addition to any energy that the electron might acquire at the moment of birth (moment of ionization), a situation which we will consider in the next chapter. If, on the other hand, the pulse is short in duration, the pulse will turn off before the electron has completely acquired the ponderomotive energy by ejection from the focus. This ponderomotive acceleration plays an important role in energy observed in electrons formed in a laser focus by field ionization since any experiment necessarily observes the electron energy at a point well outside the focal volume.

To illustrate this, Figure 3.10 shows the energy of an electron born in a $\lambda = 1$ μm laser field with zero canonical momentum (which occurs when the birth is near the peak of an

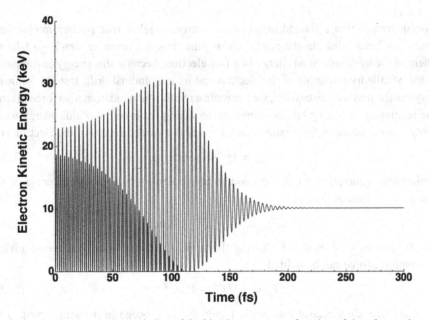

Figure 3.10 Numerical calculation of the kinetic energy as a function of time for an electron born near the center of a laser focus. The laser field has $\lambda = 1054$ nm focused to a spot with $w_0 = 3$ μm and peak intensity of 1×10^{17} W/cm^2 (peak $U_p = 10$ keV). The laser pulse was 1 ps in duration and the electron injected at the peak of the electric field so that it has no initial canonical momentum.

electric field cycle), as a function of time in a focused laser field. The pulse is long, so the electron, which is born within 0.1 μm of the peak of the field focused to a spot of $w_0 = 3$ μm with peak intensity of 10^{17} W/cm^2, first experiences oscillations with average energy of 10 keV, the ponderomotive potential of this intensity. As time goes on, the electron exits the focus and ultimately ends up with a constant directed kinetic energy equal to its initial cycle-averaged oscillation energy.

3.5.4 The Ponderomotive Force at Relativistic Intensity

The cycle-averaged ponderomotive force in a laser focus at relativistic intensity is a rather more complicated concept and requires greater subtlety in its treatment. A number of authors over the years have published derivations of a relativistic ponderomotive force (Schmidt and Wilcox 1973; Bauer et al. 1995; Startsev and McKinstrie 1997; Quesnel and Mora 1998; Bituk and Fedorov 1999; Kaplan and Pokrovsky 2005). The majority of these formulae were cast in terms of the derivative of the momentum with respect to the implicit proper time, which makes them of limited use for actually finding electron trajectories. In fact, the ponderomotive force concept itself can become questionable in the relativistic regime, as the electron oscillation amplitude increases and the laser is focused more tightly to reach these intensities. The ponderomotive force relies on the idea of cycle averaging: yet we have seen that an electron at relativistic intensity experiences fewer oscillations for a given laboratory time interval as the field strength is increased (recall the plots of Figure 3.1).

Nonetheless, within a limited range of parameters, a relativistic ponderomotive force formula can have value. In the relativistic regime it is not accurate simply to take the gradient of the cycle-averaged energy of a free electron, because this energy encompasses both the oscillatory quivering of the electron and its longitudinal drift. Instead, following the logic in the previous discussion, we can write a relativistic ponderomotive force in terms of the relativistic quivering kinetic energy ε_K by taking the gradient of this energy with a suitably chosen time-averaged gamma factor. Since the kinetic energy of the electron is

$$\varepsilon_K = (\gamma - 1)\, mc^2, \tag{3.260}$$

the relativistic generalization of the ponderomotive energy is a suitable average of this quantity over a laser period, T,

$$\langle \varepsilon_K \rangle_T \equiv U_p = (\bar{\gamma} - 1)\, mc^2, \tag{3.261}$$

where $\bar{\gamma}$ is to be determined. Following the same reasoning as earlier, the relativistic ponderomotive force can be written

$$\mathbf{F}_p = -\nabla (\bar{\gamma} - 1)\, mc^2. \tag{3.262}$$

To find the cycle-averaged relativistic factor, $\bar{\gamma}$, we proceed in the same spirit as the nonrelativistic derivation, in which we say that the momentum has an oscillatory component in the transverse direction, which we denote as $\mathbf{p}_0$, and a slow drift component,

3.5 Ejection of Free Electrons from the Focus of an Intense Laser

denoted $\mathbf{p}_1$. We work with the transverse momentum, $\mathbf{p}_\perp = p_x\hat{\mathbf{x}} + p_y\hat{\mathbf{y}}$, and utilize conservation of transverse canonical momentum to find the zeroth-order momentum

$$\frac{d}{dt}\left(\mathbf{p}_{\perp 0} - \frac{e}{c}\mathbf{A}_\perp\right) = 0. \tag{3.263}$$

This allows us to write the electron momentum in terms of the laser's vector potential as

$$\mathbf{p}_\perp = \mathbf{p}_{\perp 0} + \mathbf{p}_{\perp 1} = \frac{e}{c}\mathbf{A}_\perp + \mathbf{p}_{\perp 1}. \tag{3.264}$$

Recalling our previous derivation, conservation of energy gave us Eq. (3.63). For reasons we shall discuss later, we can make a reasonable assumption that at intensities that are not too high, $p_z - \gamma mc$ varies slowly compared to the fast transverse oscillations; we can say that this quantity is nearly the same as its cycle-averaged quantity:

$$p_z - \gamma mc \cong \langle p_z - \gamma mc \rangle_T. \tag{3.265}$$

It follows that we can equate $p_z - \gamma mc$ with the slow drift component, which we write as

$$p_z - \gamma mc \cong p_{z1} - \overline{\gamma} mc. \tag{3.266}$$

The calculation of γ^2 for the electron with explicit substitution of fast oscillatory and slow drift terms is

$$\gamma^2 = 1 + \frac{1}{m^2c^2}\left(p_\perp^2 + p_z^2\right)$$

$$= 1 + \frac{1}{m^2c^2}\left(p_{\perp 1}^2 - 2\frac{e}{c}\mathbf{p}_{\perp 1}\cdot\mathbf{A}_\perp + \frac{e^2}{c^2}A_\perp^2 + (p_{z1} - \overline{\gamma}mc + \gamma mc)^2\right), \tag{3.267}$$

which becomes

$$-m^2c^2 = p_{\perp 1}^2 + p_{z1}^2] - 2\frac{e}{c}\mathbf{p}_{\perp 1}\cdot\mathbf{A}_\perp + \frac{e^2}{c^2}A_\perp^2 + \overline{\gamma}^2 m^2 c^2 \\ + 2\gamma mc p_{z1} - 2\overline{\gamma} mc p_{z1} - 2\overline{\gamma}\gamma m^2 c^2. \tag{3.268}$$

We now cycle-average both sides of (3.268) so that $\gamma \to \overline{\gamma}$ and

$$\langle \mathbf{p}_{\perp 1}\cdot\mathbf{A}_\perp \rangle = 0. \tag{3.269}$$

Defining $\mathbf{A}_1$ as the slow spatially varying amplitude of the oscillating vector potential

$$\mathbf{A}_\perp = \mathbf{A}_1(\mathbf{x})\sin\varphi \quad\Rightarrow\quad \left\langle A_\perp^2\right\rangle_T = \frac{1}{2}A_1^2$$

and noting that $p_\perp^2 + p_{z1}^2 = p_1^2$ lets us solve Eq. (3.268) for $\overline{\gamma}$ to give

$$\overline{\gamma} = \sqrt{1 + \frac{p_1^2}{m^2c^2} + \frac{e^2}{2m^2c^4}A_1^2}. \tag{3.270}$$

We can use this result to find the relativistic ponderomotive force

$$\mathbf{F}_p = -mc^2\nabla\overline{\gamma}. \tag{3.271}$$

Taking the gradient of Eq. (3.270) finally gives us

$$\mathbf{F}_p = -\frac{e^2}{4\bar{\gamma}mc^2}\nabla A_1^2, \qquad (3.272)$$

or in terms of the electric field:

$$\mathbf{F}_p = -\frac{e^2}{4\bar{\gamma}m\omega^2}\nabla E_1^2. \qquad (3.273)$$

Equation (3.273) reduces to the nonrelativistic result, Eq. (3.256), when $\bar{\gamma} \to 1$. In terms of normalized vector potential

$$\bar{\gamma} = \sqrt{1 + \frac{p_1^2}{m^2c^2} + \frac{a_0^2}{2}}, \qquad (3.274)$$

and the ponderomotive force is

$$\mathbf{F}_p = -\frac{mc^2}{4\bar{\gamma}}\nabla a_0(\mathbf{x})^2. \qquad (3.275)$$

It is possible to perform a more rigorous derivation of Eq. (3.272) using an approach like that employed in Eqs. (3.242) to (3.256), though we will not reproduce such an analysis here. (See Quesnel and Mora (1998) for details.) The time-averaged result in Eq. (3.274), incidentally, is the same as our hand-waving discussion of the effective time-averaged mass increase of the electron in a strong field found in Eq. (3.120) if we ignore the slow drift component of the momentum (which is usually a good approximation, for example, in a plasma where the space charge restoring force of the heavy ions restricts the drift motion of the electrons).

The relativistic ponderomotive force can be used to calculate the cycle-averaged trajectory of a relativistic electron in the laser focus. As long as the assumptions made to derive this force are valid, it accurately represents the forces an electron feels from all components of the field near focus up to first order in κ. Figure 3.11 illustrates three such calculations for a moderately relativistic intensity (peak $a_0 \approx 2$). This figure shows the trajectories of three electrons in a laser focus initially at rest before the arrival of the 100 fs pulse (see figure caption for the other parameters of the calculation). The exact trajectory of the electron using the full equations of motion with fields up to the first order in κ (which include z components of $\mathbf{E}$ and $\mathbf{B}$) is shown. The trajectory predicted by Eq. (3.273) for these three electrons is also shown. Clearly the intense pulse ejects all electrons in the radial direction. The relativistic cycle-averaged ponderomotive force does an adequate job in these cases of predicting the trajectory of the oscillation center of each electron. Also note that the electrons are driven in the $+z$-direction from the force produced by the gradient of the rising pulse intensity in time.

The electrons gain radial kinetic energy from the ponderomotive scattering. Figure 3.12 shows the energy of one electron from Figure 3.11 (the electron at $x = y = 2$ μm) as a function of time. As the pulse ramps up, the oscillation energy builds; as the ponderomotive force accelerates the electron radially, this oscillation energy is converted to directed

3.5 Ejection of Free Electrons from the Focus of an Intense Laser

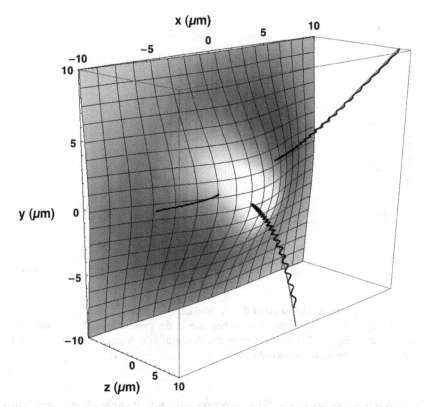

Figure 3.11 Numerical calculation of the trajectory of three electrons at three different initial positions in a laser focus. The laser field has $\lambda = 1{,}057$ nm focused to a spot with $w_0 = 10$ μm and peak intensity of 4×10^{18} W/cm^2 (peak $a_0 = 1.8$). The laser pulse was 100 fs in duration and the electrons are all initially at rest prior to the arrival of the laser pulse. The black trajectories are the solution of the full equations of motion, and the gray trajectories are for the same electrons using the cycle-averaged relativistic ponderomotive force. The focal spot profile is superimposed in the background to illustrate the intensity distribution.

kinetic energy. The ponderomotive energy from Eq. (3.270) corresponding to the cycle-averaged trajectory is also plotted in 3.12 as a dashed gray line. In this moderately relativistic intensity regime the ponderomotive force approximation is accurate at predicting the average trajectory of the electron exiting the focus. As the pulse intensity rises temporally, the electron energy rises. This oscillatory energy gets converted to directed energy as the electron exits the focus. This is reflected in the way that the cycle-averaged result (gray dashed line) tracks the oscillations of the full calculation as the oscillation energy wanes and converts to direct energy. It is interesting to note, however, that the ponderomotive energy of Eq. (3.274) is not the same during the oscillations of the laser pulse as the actual average energy of the electron of the full trajectory. The oscillating energy (solid line in Figure 3.12) has an average that is well above the ponderomotive energy depicted by the dashed gray line. Equation (3.274) is obviously not the same as Eq. (3.82).

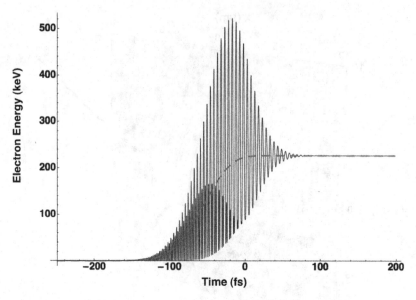

Figure 3.12 Numerical calculation of the electron energy from the electron in Figure 3.10 that starts at $x = y = +2$ μm. The dashed line is the cycle-averaged ponderomotive energy given by Eq. (3.274) from the same electron with its trajectory calculated with the cycle-averaged relativistic ponderomotive force.

The cycle-averaged energy of the actual electron trajectory is more closely approximated by Eq. (3.82) even though the ponderomotive gamma of Eq. (3.274) yields the more correct approximation to the ejected electron trajectory.

3.5.5 Breakdown of the Ponderomotive Force Approximation and Electron Ejection at Highly Relativistic Intensity

As we have already cautioned, care must be exercised when applying the concept of the ponderomotive force to determine the trajectory and energy of an electron, particularly at high intensity in tightly focused beams. $\mathbf{F}_p$ is a cycle-averaged quantity that applied in the case of weak field amplitude gradients. To illustrate just how dramatically this concept can break down in real strong field interactions, we show in Figure 3.13 the numerically calculated trajectory of an electron born in a laser field with a peak intensity of 10^{21} W/cm^2 ($a_0 \cong 30$) very close to the center of the focal spot (slightly displaced in the $+x$-direction). The electron is placed in the field at the peak of the electric field oscillation (which simulates the ionization of an electron via tunnel ionization as described in Chapter 4). The trajectory calculated using the actual fields (up to first order in κ) is shown in black while the trajectory calculated using the relativistic ponderomotive force is shown in gray. To the right in Figure 3.13 is the electron kinetic energy as a function of time for the exact solution (black) and the cycle-averaged ponderomotive force solution. Obviously, the ponderomotive force solution looks nothing like the actual trajectory of the electron. Subject only to

3.5 Ejection of Free Electrons from the Focus of an Intense Laser

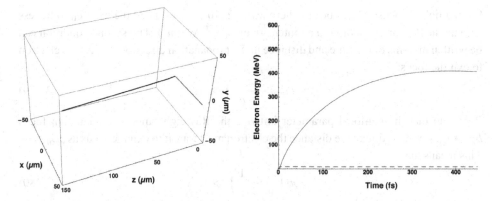

Figure 3.13 Numerical calculation of the trajectory of an electron injected at the peak of the electric field very near the center of a $w_0 = 10$ μm laser focus. The laser field has $\lambda = 1{,}054$ nm and is focused to a peak intensity of 1×10^{21} W/cm^2. The actual electron trajectory (in black) is compared to the trajectory predicted by Eq. (3.275) (in gray) in the plot on the left. The energy of the electron is plotted on the right, with the erroneous cycle-averaged energy plotted in this panel as a dashed gray line.

the cycle-averaged ponderomotive force, the electron is immediately ejected directly in the radial direction (in this case toward $+x$ since that is the direction we displaced the electron). As we would expect, the relativistic ponderomotive energy of ~10 MeV is simply converted to directed kinetic energy, and the electron exits the focal spot with almost 10 MeV of energy (see the gray dashed line) in the cycle-averaged approximation. The electron acquires no forward directed momentum because there is no ponderomotive gradient in this direction at the peak of a laser pulse at best focus. In reality, the electron is ejected largely in the forward direction at a velocity near c and the electron acquires many hundreds of MeV of energy from the field as it surfs along with the laser pulse.

This figure gives us a hint as to why the ponderomotive approximation breaks down. It is a cycle-averaged quantity, yet it is clear that for the electron in Figure 3.13, less than one complete optical cycle is experienced in the electron's frame before it exits the focal volume. To be more explicit about the range of validity of the ponderomotive approximation, we demand that the field amplitude gradient be small with respect to the electron oscillation amplitude,

$$x_{osc} \left| \frac{E}{\nabla E} \right| \ll 1, \tag{3.276}$$

and that the electron experience many optical cycles in its moving frame, which means that

$$\omega \tau_{exit} \gg 1. \tag{3.277}$$

The first of these conditions mandates that the ponderomotive approximation is valid when

$$\frac{a_0 c}{\omega w_0} \ll 1. \tag{3.278}$$

So in a tightly focused laser beam where $w_0 \sim 2\text{--}10 \times \lambda$, we need the peak a_0 to be less than about 10 (corresponding to an intensity of $\sim 10^{20}$ W/cm^2). The second condition can be written in terms of the time and distance in the propagation direction it takes the electron to exit the focus:

$$\omega \left(t_{exit} - \frac{z_{exit}}{c} \right) \gg 1. \tag{3.279}$$

Recalling that the confocal parameter is twice the Rayleigh range of a Gaussian focus $b = kw_0^2$, we can estimate the distance the electron must transit to exit the focus as $z_{exit} \approx b$. This means that

$$\omega b \left(\frac{1}{v_z} - \frac{1}{c} \right) \gg 1, \tag{3.280}$$

or since the longitudinal velocity can be assumed to be near c and we can approximate $\omega/v_z \approx \omega/c = k$, the condition that the electron experiences many oscillations is

$$1 - \frac{v_z}{c} \gg \frac{1}{k^2 w_0^2} = \kappa^2. \tag{3.281}$$

Estimating the average longitudinal velocity by the drift velocity yields a more stringent condition on the maximum a_0 in which the ponderomotive force approximation is valid:

$$1 - \frac{a_0^2}{4 + a_0^2} \gg \kappa^2$$

$$a_0 \lesssim 2\sqrt{1 - \kappa^2}. \tag{3.282}$$

This condition indicates that in most foci, the ponderomotive approximation will begin to break down when the peak a_0 exceeds 1 by only a bit. This is because at high a_0 an electron quickly acquires forward velocity near c and the electron tends to surf with the field and not really oscillate many times in it. Note that this condition tends not to be quite as stringent for free electrons that are already present before the arrival of the laser pulse, because in that situation, the electron will tend to be ejected from the laser focus on a time scale before the main, most intense part of the pulse arrives (see Figure 3.11). This condition is not nearly as stringent in a plasma because space charge forces can serve to prevent the rapid expulsion from the focus and the ponderomotive approximation is valid to significantly higher a_0.

We have determined that when the ponderomotive force approximation is valid, the electron can be seen as rolling down the ponderomotive energy hill and acquires kinetic energy in the radial direction that is near that of the peak ponderomotive energy experienced. This energy estimate is completely wrong for an electron born into a focus at moderate to high field strength. On the contrary, in a situation like that depicted in Figure 3.13 at high a_0, the electron will acquire energy from the laser field nearly continuously during ejection and it will be ejected at a shallow angle which is determined by its rather complicated dynamics in the 3D field distribution near focus. (This ejection angle can be quite different than the plane-wave electron flight angle of Eq. (3.106) (Maltsev and Ditmire 2003).) The energy gain in this strongly relativistic ejection case can be written

3.5 Ejection of Free Electrons from the Focus of an Intense Laser

$$\frac{d}{dt}\gamma mc^2 = -e\mathbf{E} \cdot \mathbf{v}. \tag{3.283}$$

As long as the electron surfs with the field in phase, it will continuously pick up energy from the field. If θ_{ej} is the angle with respect to the laser propagation axis that the electron flies out of the focus, it will be in phase with the field as long as

$$\frac{v_z}{c}\cos\theta_{ej} < 1. \tag{3.284}$$

In linear polarization, the energy gain is

$$\frac{dU}{dt} = -e(E_x v_x + E_z v_z). \tag{3.285}$$

Near the laser focus, the electric field components are

$$E_x \cong E_0 \cos(\omega\tau + \varphi_G), \tag{3.286}$$

$$E_z \cong -\frac{2\kappa x E_0}{w}\sin\left(\omega\tau + \varphi_G - \tan^{-1} z/z_R\right), \tag{3.287}$$

where φ_G is the usual Gaussian focus phase term. Since $v_z t \approx c$ and $x = v_x t$,

$$\frac{dU}{dt} \approx -e v_x E_0 \left(\cos\varphi_G + 2\frac{\kappa ct}{w}\sin\varphi_G\right). \tag{3.288}$$

When the term in the parentheses of (3.288) goes to zero, the electron stops gaining energy and starts losing it back to the field.

We can proceed with some additional rough approximations. The total energy gain of the ejected electron is

$$\gamma_{ej} \approx -\frac{e}{mc^2}\int \mathbf{E} \cdot \mathbf{v}\, dt. \tag{3.289}$$

Though we do not know the actual ejection angle of the electron, we know it must not be too different from the angle predicted by photon momentum conservation in a plane-wave which tells us that $\theta_{ej} \approx \sqrt{2/(\gamma_{ej}-1)}$. We then estimate the electron velocity components as $v_z \approx c$, $v_x \approx c\theta_{ej}$ and assume that the electron ejection is largely in the z-direction, which lets us write

$$\gamma_{ej} \approx -\frac{e}{mc^2}\int E_x v_x\, dt$$

$$= -\frac{e}{mc^2}\int \frac{E_0 \cos\varphi_G}{\sqrt{1+z^2/z_R^2}}\theta_{ej}\, dz. \tag{3.290}$$

Since the electron at high a_0 stays in phase with the field over distance equal to a Guoy phase shift, we can say that $\cos\phi_G \approx -1$ for an electron born near the peak of an electric field oscillation. These approximations finally give the estimate for the energy gained by the electron from the field,

$$\gamma_{ej} \cong \frac{eE_0\theta_{ej}}{mc^2}\int_0^{z_R} \frac{1}{\sqrt{1+z^2/z_R^2}}\, dz,$$

$$\gamma_{ej} \cong \frac{eE_0\theta_{ej}z_R}{mc^2}\sinh(1) = 0.88\frac{eE_0\theta_{ej}z_R}{mc^2}. \tag{3.291}$$

Noting that $\theta_{ej} = \sqrt{2/\gamma_{ej}}$ tells us that

$$\gamma_{ej} \approx (a_0 k z_R)^{2/3} \tag{3.292}$$

The ejected electron energy is proportional to two-thirds power of the incident peak field times the Rayleigh range which means, not surprisingly, that the electron energy depends not only on the peak field strength but also on the focal geometry. This focal geometry dependence was completely absent in the ponderomotive force picture of the electron ejection. This is a simple approximate formula which can give a sense of the energy expected in the ejection of electrons born in a highly relativistic field. Equation (3.292) predicts that the electron ejected in the situation considered in Figure 3.13 would be 720 MeV, which is in the order of magnitude of the actual simulated ejection energy of $\sim$400 MeV.

3.6 Intense Laser Interactions with Relativistic Electron Beams

While we have considered strong field interactions with electrons initially at rest, the results we derived in this chapter can be quite easily applied to situations in which the initial electron momenta are not zero by a simple change of initial conditions. The situation most often encountered that leads to this is when an intense short laser pulse is focused into the beam of previously accelerated electrons. These electrons can be accelerated by a linac or by another laser pulse, possibly in a plasma. One of the principal motivations for performing this experiment is that the photons that are scattered by a bunch of relativistic electrons will be greatly Doppler upshifted in frequency. This upshift represents an interesting way to produce femtosecond pulses of x-ray photons. This process is essentially an example of the linear (and nonlinear) Thomson scattering that we have already considered with the difference that the initial conditions involve electrons that are already at relativistic velocity in the laboratory frame. This process is often referred to as inverse Compton scattering.

3.6.1 Inverse Compton Scattering by a Relativistic Electron Beam

Classic Compton scattering involves the scattering of a high-energy photon by an electron initially at rest. If the photon momentum is large, this scattering results in the recoil of the electron and a Doppler downshift of the scattered photon. When a laser pulse irradiates a bunch of relativistic electrons, the scattered photon can, instead, be Doppler upshifted in frequency. We have up until now considered laser photons which have photon energy way below the electron's rest energy, which is one reason we have been able to ignore the quantum nature of the light and treat the laser field classically.

Continuing for the moment to ignore the photon's momentum, consider a beam of electrons from an accelerator which has initial energy and velocity γ_0 and β_0. To find the frequency of any scattered photon, we Lorentz-transform the incident field into the rest frame of the electron. If the incoming laser propagates on an axis which is at an angle α with respect to the incoming electron's axis (with $\alpha = 0$ denoting a head-on collision), the frequency of the field in the rest frame of the electron beam is

3.6 Intense Laser Interactions with Relativistic Electron Beams

$$\omega_R = \gamma_0 \omega \left(1 + \beta_0 \cos \alpha \right). \tag{3.293}$$

This frequency is then related to the scattered light frequency, ω_s, by an inverse Lorentz transform

$$\omega_R = \gamma_0 \omega_S \left(1 - \beta_0 \cos \theta \right), \tag{3.294}$$

where the angle θ is again measured with respect to the incoming electron's axis in the lab frame (so $\theta = 0$ would correspond to 180° back scattering of the photon).

The frequency of light scattered by the electron is

$$\omega_S = \frac{1 + \beta_0 \cos \alpha}{1 - \beta_0 \cos \theta}. \tag{3.295}$$

If we consider a head-on collision between the light pulse and a strongly relativistic electron beam ($\beta_0 \cong 1$) followed by 180° backscatter, the scattered photon is upshifted by

$$\omega_S = \omega \left(\frac{1 - \beta_0}{1 + \beta_0}\right) = \gamma_0^2 (1 + \beta_0)^2 \omega$$

$$\cong 4\gamma_0^2 \omega \tag{3.296}$$

So a modest RF linac producing electrons with a 50 MeV electron beam from a linac would upshift 1 eV photons to the hard X-ray region of photon energy ~40 keV.

What is more, the vast majority of the scattered photons will emerge from the intersection of the electron bunch and laser pulse with narrow angular divergence. The angles in the lab and rest frame of the electrons are related by

$$\cos \theta_R = \frac{\cos \theta_L - \beta_0}{1 - \beta_0 \cos \theta_L}, \tag{3.297}$$

so that even if a photon is scattered into an angle at 90° with respect to the beam axis $\theta_R = \pi/2$, the emergence angle of that upshifted photon in the lab frame is given by

$$\cos \theta_L = \beta_0,$$

$$\sin \theta_L = \sqrt{1 - \beta_0^2} = \frac{1}{\gamma_0}, \tag{3.298}$$

which means that the pulse of scattered photons will be directed in a narrow cone of angle $\theta_L \sim 1/\gamma_0$ in the lab frame. Consequently, the X-ray pulse produced can be quite collimated about a small angle in the direction of the electron beam propagation if the electron beam is at high energy, independent of the incident laser pulse angle. For example, if laser photons are scattered by a beam of 200 MeV electrons ($\gamma_0 = 390$) the scattered X-ray beam will have a divergence of 5 mrad, about as collimated as a typical laser pointer.

3.6.2 Effects of Photon Momentum and Strong Fields on the Scattered Photons

A complication does arise when the electron beam is strongly relativistic. While the laser photon energy is well below the electron rest energy in the lab frame, which permitted us to ignore any effects of electron recoil in our analysis of Section 3.4, the photon energy in the

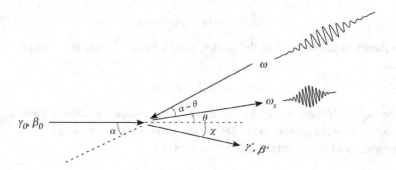

Figure 3.14 Geometry for nonlinear inverse Compton scattering of a laser by a relativistic electron.

rest frame of the electron beam may be rather large and the effects of electron recoil from the finite momentum of the incoming photon can no longer be neglected. This represents the true Compton scattering regime. For example, in the 50 MeV electron beam we considered, this effect is probably negligible. Since $\omega_R \cong 2\gamma_0 \omega$ for head-on laser incidence, the photon energy in the electron's rest frame is around 200 eV, still well below mc^2. But in experiments like those performed at the SLAC linac (Meyerhofer 1997), the initial electron beam was at 47 GeV. In that case the photon energy in the electron's rest frame can be as high as $\sim$200 keV, which is a large fraction of mc^2.

We can find the frequency of the scattered photon by performing a kinematics analysis in the lab frame, accounting for the momenta of the incoming and outgoing photons. With the scattering geometry depicted in Figure 3.14, conservation of 4-momentum is

$$p_\mu + k_\mu = p'_\mu + k'_\mu. \tag{3.299}$$

Unprimed quantities refer to parameters before the collision of the electron and photon; primed quantities give parameters of the scattered photon and recoiling electrons. We will consider the more general case in which multiple photons can be scattered via a nonlinear scattering process with multiphoton order q. Equation (3.299) yields three equations

$$\gamma_0 mc^2 + q\hbar\omega = \gamma' mc^2 + \hbar\omega_S, \tag{3.300}$$

$$\gamma_0 mc\beta_0 - q\hbar\frac{\omega}{c}\cos\alpha = \gamma' mc\beta' \cos\chi + \hbar\frac{\omega_S}{c}\cos\theta, \tag{3.301}$$

$$-q\hbar\frac{\omega}{c}\sin\alpha = \gamma' mc\beta' \sin\chi + \hbar\frac{\omega_S}{c}\sin\theta. \tag{3.302}$$

After considerable algebra to eliminate χ, γ' and β', along with the use of the trigonometric identity $\cos\alpha \cos\theta + \sin\alpha \sin\theta = \cos(\alpha - \theta)$, we find that

$$\omega_S = \frac{q\gamma_0(1 + \beta_0 \cos\alpha)\omega}{\gamma_0(1 - \beta_0 \cos\theta) + (q\hbar\omega/mc^2)(1 + \cos(\alpha - \theta))} \tag{3.303}$$

This equation is a more general form of our result, Eq. (3.295), that accounts for potentially q scattered photons with the addition of a term in the denominator. This second term in the denominator is a consequence of the momentum of these q incoming photons with the prefactor given as

$$\frac{P_{recoil}}{mc} = \frac{q\hbar\omega}{mc^2}\omega_S. \tag{3.304}$$

So what happens when the incident laser is intense with a_0 approaching or exceeding 1? We know that an electron at rest acquires momentum from the laser field in the direction of laser propagation. This strong field effect, then, is like an addition to the photon momentum. We need to return to our previous electron momentum analysis but apply different initial conditions to find the additional momentum that a strong field will impart to the electron, and add this to the photon recoil momentum described by Eq. (3.304). The initial electron momentum is $p_0 = \gamma_0 mc\beta_0$ which we can decompose into components parallel and perpendicular to the laser propagation direction:

$$p_{0\parallel} = -\gamma_0 mc\beta_0 \cos\alpha, \tag{3.305a}$$

$$p_{0\perp} = -\gamma_0 mc\beta_0 \sin\alpha. \tag{3.305b}$$

Equation (3.63) is still a valid constant of motion

$$\frac{d}{dt}(p_\parallel - \gamma mc) = 0, \tag{3.306}$$

which when integrated yields

$$p_\parallel = (\gamma - \gamma_0)mc - p_0 \cos\alpha. \tag{3.307}$$

Conservation of transverse canonical momentum gives us in the perpendicular direction

$$p_\perp = -\gamma_0 mc\beta_0 \sin\alpha + a_0 mc \sin\omega\tau. \tag{3.308}$$

Introducing these momentum components into the relativistic factor of the quivering electron in the laser field lets us write

$$\gamma^2 = 1 + \frac{1}{m^2c^2}\left(p_\perp^2 + p_\parallel^2\right)$$

$$= \left(\frac{p_\parallel}{mc} + \gamma_0 + \gamma_0\beta_0\cos\alpha\right)^2, \tag{3.309}$$

from which we can solve for the parallel momentum imparted by the laser field:

$$p_\parallel = \gamma_0 mc\beta_0 \cos\alpha + \frac{a_0^2 mc \sin^2\omega\tau}{2\gamma_0(1+\beta_0\cos\alpha)} - \frac{\beta_0 a_0 mc}{1+\beta_0\cos\alpha}\sin\alpha\sin\omega\tau. \tag{3.310}$$

For the purposes of scattering radiation and the electron recoil momentum, we need the cycle-averaged parallel momentum

$$\langle p_\parallel \rangle = \gamma_0 mc\beta_0 \cos\alpha + \frac{a_0^2 mc}{4\gamma_0(1+\beta_0\cos\alpha)}. \tag{3.311}$$

The first term is obviously the initial parallel momentum of the electron so we can say that the momentum imparted by the laser field is

$$P_{recoil}^{(SF)} = \frac{a_0^2 mc}{4\gamma_0(1+\beta_0\cos\alpha)}. \tag{3.312}$$

Adding this to the photon momentum gives us a formula for the scattered radiation frequency from the scattering of q photons, corrected for strong field effects:

$$\omega_S = \frac{q\gamma_0 \left(1 + \beta_0 \cos\alpha\right) \omega}{\gamma_0 \left(1 - \beta_0 \cos\theta\right) + \left[\frac{q\hbar\omega}{mc^2} + \frac{a_0^2 mc}{4\gamma_0(1+\beta_0 \cos\alpha)}\right](1 + \cos(\alpha - \theta))}. \quad (3.313)$$

Note that if the photon momentum is negligible, $\gamma_0 q\hbar\omega \ll mc^2$, and the electron is initially at rest, we recover our original scattered radiation result, Eq. (3.160). This strong field correction not only affects the scattered radiation frequency, irradiation of a beam at high a_0 will also shift the scattered spectrum away from the first harmonic toward higher multiphoton orders. Recall that the scattered radiation tends to have an average multiphoton order that increases as $\sim a_0^3$. This will also happen in the scattering from an electron beam.

It is also possible to ask what the energy of the scattered electron will be, since this is a potential observable of this effect. For head-on scattering by a highly relativistic electron beam, we can use the fact that $\gamma_0^2 (1 - \beta_0) = 1/2$ to say that the photon energy is

$$\omega_S(\theta = 0) \cong \frac{2q\gamma_0^2 \left(1 + \cos\alpha\right) \omega}{1 + \left(2\gamma_0 q\hbar\omega/mc^2 + a_0^2/4\right)(1 + \cos\alpha)}, \quad (3.314)$$

which means that from conservation of energy, the minimum energy of scattered electrons must be

$$\gamma_{\min} \cong \frac{\gamma_0 \left(1 + a_0^2/4\right)}{1 + \left(4\gamma_0 q\hbar\omega/mc^2 + a_0^2/2\right)}. \quad (3.315)$$

So, strong field effects will manifest themselves in the minimum energy of the electrons scattered out of the relativistic beam by the laser (Meyerhofer 1997). Of course, the energy of the scattered electrons is further reduced at high a_0 irradiation because multiphoton scattering becomes much more probable.

3.6.3 Electron Beam Emittance and Other Nonideal Effects on Scattered Photon Beam Bandwidth

If inverse Compton scattering is utilized to generate a beam of X-rays, often the bandwidth, or more specifically, the brightness $d^2I/d\omega d\Omega$, is an important parameter. There are a number of factors which will broaden the spectrum of the scattered photon beam. First, the intrinsic spread in emission angles is the main cause of bandwidth. Any real electron beam has a finite brightness which is usually characterized by its normalized emittance. The transverse emittance of a beam is $\varepsilon_e = \gamma_0 r_b \theta_b$, where r_b is the average beam radius and θ_b is the average electron angular spread. Typical electron beams from RF linacs with range from 1 π mm-mrad for the best state-of-the-art accelerators with photo injectors, up to many tens of millimeters – mrad. The finite transverse emittance will lead to a spread of angles in the electron bunch as it interacts with the laser. This leads to a spread in α angles and a commensurate spread in scattered photon energy. There is also the energy spread of any real beam that further contributes to this effect. The beam will

3.6 Intense Laser Interactions with Relativistic Electron Beams

also necessarily have some finite longitudinal emittance which leads to an electron energy spread $\Delta\gamma_0\gamma_0 = \varepsilon_L^2/2r_b^2$.

Other nonideal effects can cause a spread in photon spectral bandwidth. The laser itself has a spread in bandwidth. This was manifested in our scattered spectral analysis with the sinc^2 function of Eq. (3.205). A transform limited pulse leads to an incident bandwidth spread of $\Delta\omega/\omega \sim 1/T$. Finally, in the strong field regime the presence of a range of intensities, both temporally and spatially, will lead to a spectral broadening through the range of a_0^2 values in the interaction.

3.6.4 Pair Production During Multiphoton Scattering in a Relativistic Beam

At high enough field strengths it is well known that electron–positron pairs can be produced in the vacuum (Holstein 1999; Melissinos 2008). This process can be thought of as the tunneling of positrons from the negative energy states predicted by the Dirac equation into real, positive energy states. Such tunneling is made possible by the imposition of a very large electric field. Before considering this process from the laser field in the interaction of an electron beam, let's first examine it in the simple situation of a strong static fielding vacuum.

The pair production process can be envisioned by examining Figure 3.15, which shows the deformation of the potential of charge particles in the Dirac sea due to the imposition of a strong electric field. The rate at which charge particle pairs tunnel through the $2mc^2$ energy barrier separating the negative and positive energy states in vacuum can be estimated by the standard quasi-classical WKB approximation. If a static field of strength E_0 is imposed, the barrier width is related to the field by

$$eE_0 s = 2mc^2. \tag{3.316}$$

The WKB approximation says that we can estimate the rate that charged particles tunnel through this barrier (at a speed of c) by a rate given by

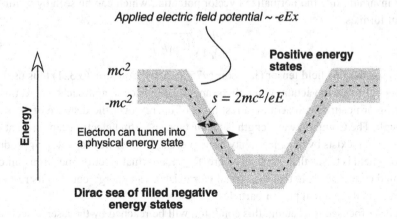

Figure 3.15 Schematic explaining how electron–positron pairs can be produced in vacuum by a strong electric field by tunneling from negative energy states.

$$W_{e^+e^-} \approx 2\rho \frac{c}{s} e^{-2\xi}, \qquad (3.317)$$

where ρ is some appropriate density of states and the barrier penetration factor is (Schiff 1968, p. 271)

$$\xi = \frac{1}{\hbar} \int \sqrt{2m(V(x) - \varepsilon)} dx. \qquad (3.318)$$

Approximating the potential in the barrier region by a linear variation with width given by s gives for this penetration factor

$$\xi = \frac{1}{\hbar} \int_0^s \sqrt{2m(2mc^2(1 - x/s))} dx$$
$$= \frac{4}{3} mcs = \frac{8}{3} \frac{m^2 c^3}{e\hbar} \frac{1}{E_0}. \qquad (3.319)$$

We can define at this point the Schwinger field, E_S:

$$E_S = \frac{m^2 c^3}{e\hbar} = 1.3 \times 10^{16} \text{ V/m}, \qquad (3.320)$$

which is the field strength required for an electron to gain its rest mass in energy over a distance of one Compton wavelength. The approximate rate of pair production becomes

$$W_{e^+e^-} \approx \rho \frac{e}{mc} E_0 \exp\left[-\frac{16}{3} \frac{E_S}{E_0}\right]. \qquad (3.321)$$

This production rate is extremely nonlinear with increasing field strength. In fact, as we will see in the next chapter, the production of free electrons by any tunneling process (such as from the binding potential of an atom or ion) typically exhibits a dependence with field strength like $W \sim \exp[-a/E_0]$. Our simple approximate model somewhat overestimated the prefactor in the exponent. A more exact calculation of the pair production rate tells us that the rate is actually $W_{e^+e^-} = \alpha_{FS} E_0^2/\pi^2 \exp[-\pi E_S/E_0]$ (Schwinger 1951).

We define an appropriate dimensionless parameter, $\Upsilon \equiv E_0/E_S$. Υ, it turns out, is a Lorentz invariant, like the normalized vector potential, which can be seen by writing it in covariant form as

$$\Upsilon = \frac{e\hbar}{m^3 c^5} \sqrt{\langle F_{\mu\nu} p^\nu \rangle}, \qquad (3.322)$$

where $F_{\mu\nu}$ is the EM field tensor (Jackson 1975, p. 550). Equation (3.321) tells us that the pair production rate in vacuum will only be appreciable once Υ approaches 1. At this field, an electron or positron can acquire a rest mass of energy over the distance of a Compton wavelength. The Compton wavelength is the distance a virtual electron can travel at c during the period it exists by the uncertainty principle. So, another way to view pair production by a strong field is to say that in a field where $\Upsilon \geq 1$, a virtual electron–positron pair can be accelerated on a time scale that the virtual pair exists to an energy equal to its rest energy and thereby make it a real pair of particles.

For a laser focused in vacuum, this condition will be reached by the laser's electric field at an intensity of 2.2×10^{29} W/cm^2, which is well beyond any current or near future

3.6 Intense Laser Interactions with Relativistic Electron Beams

laser capability. In fact, even if such an intense laser field could be made by the focus of a single beam in vacuum, momentum conservation would prohibit the production of the pair. Momentum conservation demands the collision of either two or more photons or photons with a real particle. Consequently, while pair production is not possible with real world strong field lasers in vacuum alone, the conditions for pair production can be achieved when a laser collides with a relativistic beam of electrons.

One way to understand this is that while the electric field of the laser is not sufficient in the lab frame to create a pair, the electric field of the laser field is Lorentz boosted in the frame of the electron. The field in the electron's frame is $E' = \gamma_0 \beta_0 E_0$. For highly relativistic beams this boosting factor can be very large (for example $\gamma_0 = 9 \times 10^4$ in the SLAC experiment referred to earlier (Meyerhofer 1997)). So, a pair can be produced in the electron rest frame via a collision of the strong laser field with the electron (which is now present to help conserve momentum in the pair-producing event). While this direct process is possible, in a beam there turns out to be a much greater probability of producing pairs through a two-step process. We know that the electrons can scatter photons with a significant Doppler boost in the photon energy. It is then possible to produce a pair in the following multiphoton process,

$$\hbar \omega_\gamma + q \hbar \omega \rightarrow e^- + e^+,$$

in which q laser photons collide with a single high-energy backscattered photon from the electron beam. The pair is then produced in the center-of-mass frame of the colliding photons, assuring conservation of momentum.

It is straightforward to find the COM frame, which we will do assuming that the laser collides head-on with the scattered, high-energy photon. The momentum of q_p laser photons is

$$p_L = q_p \frac{\hbar \omega}{c}, \qquad (3.323)$$

while that of the scattered photon, which we assume occurs by 180° backscatter for simplicity, is

$$p_s = \frac{\hbar \omega_s}{c} \cong 4\gamma_0^2 \frac{\hbar \omega}{c}. \qquad (3.324)$$

We Lorentz transform to a frame which is moving with the COM of the photons

$$p'_L = \gamma_{CM} q_p \frac{\hbar \omega}{c} (1 + \beta_{CM}) \quad p'_s = \gamma_{CM} 4 \hbar \gamma_0^2 \frac{\omega}{c} (1 - \beta_{CM}) \qquad (3.325)$$

so that $p'_L = p'_s$, which means that

$$4\gamma_0^2 (1 - \beta_{CM}) = q_p (1 + \beta_{CM})$$

$$4\gamma_0^2 \left(1 - \beta_{CM}^2\right) = q_p (1 + \beta_{CM})^2 \cong 4 q_p,$$

so the Lorentz factor of the COM frame for the photon collision is

$$\gamma_{CM} = \gamma_0 / q_p. \qquad (3.326)$$

To find the field components in the COM frame, we note that the fields are components of the EM field tensor F_{ab} and therefore transform as (Jackson 1975, p. 552)

$$F'_{\mu\nu} = \frac{\partial x'_\mu}{\partial x_\delta} \frac{\partial x'_\nu}{\partial x_\varepsilon} F_{\delta\varepsilon}. \quad (3.327)$$

This lets us write

$$E'_x = \gamma_{CM} \left(E_x + \beta_{CM} B_y \right) \quad (3.328)$$

or

$$E'_0 = 2\gamma_{CM} E_0. \quad (3.329)$$

High enough γ_{CM} will increase the field amplitude and push Y toward the pair production threshold of $Y \sim 1$. The threshold set by conservation of energy to produce a pair in the COM frame is

$$\hbar \omega'_s = \hbar q_p \omega' = mc^2. \quad (3.330)$$

Lorentz transform of the scattered photon energy into the COM frame means that $\omega'_s = \omega_s/2\gamma_{CM}$, where the backscattered photon energy in the lab frame is $\omega_s = 4\gamma_0^2 q\omega$. By the same token, the energy of the laser photons in the COM frame are $\omega' = 2\gamma_{CM}\omega$.

Equation (3.330) says that

$$q_p \hbar^2 \omega_s \omega = m^2 c^4, \quad (3.331)$$

which lets us write the threshold minimum laser photon energy needed for pair production in an electron beam with energy $\gamma_0 mc^2$:

$$\hbar \omega_{thresh} = \frac{mc^2}{2\gamma_{CM}\sqrt{qq_p}}. \quad (3.332)$$

Here q is the number of photons producing the scattered photon and q_p is, again, the number of laser photons participating in the pair-producing collision. So it is apparent that this process scales rather strongly with laser intensity because the average multiphoton orders scale with a_0 like $\langle q_p \rangle \sim \langle q \rangle \approx a_0^{2.8}$. In a given electron beam, a stronger laser field lowers the photon energy threshold for pair production.

3.7 Quantum Mechanical Description of a Free Electron in a Strong Laser Field

Up to this point in this chapter, we have treated the free electron as a classical point particle subject to the forces imposed on it by the fields of the electromagnetic wave. As we look forward to the next chapter where we will be concerned with the electron as it transitions from being bound in an atom or ion to a free state in the continuum, we will need a quantum mechanical description of the free electron. Our goal will be to find wavefunctions that are eigenfunctions of the Schrödinger equation containing the laser field terms. We will confine our discussion in this case to nonrelativistic intensities, though relativistic wavefunctions are analytically derivable (Fedorov 1997, p. 42).

3.7 Quantum Mechanical Description of a Free Electron

We require the Schrödinger equation in a form that incorporates the strong field. We have already laid the groundwork for finding the form we need. Recalling Eq. (3.14) where we found the electron Hamiltonian in terms of the canonical momentum conjugate to the position variables

$$H = \frac{1}{2m}\left(\mathbf{P} + \frac{e}{c}\mathbf{A}\right)^2 - e\Phi, \qquad (3.333)$$

we make the usual operator substitutions for the first quantization

$$H \to i\hbar\frac{\partial}{\partial t} \qquad \mathbf{P} \to -i\hbar\nabla.$$

Letting the Hamiltonian operate on the electron wavefunction yields

$$i\hbar\frac{\partial \psi}{\partial t} = -\frac{\hbar^2}{2m}\nabla^2\psi - \frac{i\hbar e}{2mc}\nabla\cdot(\mathbf{A}\psi) - \frac{i\hbar e}{2mc}\mathbf{A}\cdot\nabla\psi + \frac{e^2}{2mc^2}A^2\psi. \qquad (3.334)$$

Since $\nabla\cdot\mathbf{A}\psi = \psi\nabla\cdot\mathbf{A} + \mathbf{A}\cdot\nabla\psi = \mathbf{A}\cdot\nabla\psi$, we can simplify by utilizing our choice of the Coulomb gauge (radiation gauge) where we set $\nabla\cdot\mathbf{A} = 0$. This yields what is usually called the $\mathbf{A}\cdot\mathbf{p}$ form of the interaction Hamiltonian:

$$i\hbar\frac{\partial\psi}{\partial t} = -\frac{\hbar^2}{2m}\nabla^2\psi + \left(-\frac{i\hbar e}{2mc}\mathbf{A}\cdot\nabla\psi + \frac{e^2}{2mc^2}A^2\right)\psi. \qquad (3.335)$$

While it does not include the effects of electron spin, this form of the Hamiltonian includes the effects of both electric and magnetic field on the wavefunction.

Next we make the common dipole approximation simplification where we take advantage of the fact that the laser wavelength is much longer than the electron excursion distance (obviously an approximation which fails at higher intensities)

$$e^{i\mathbf{k}\cdot\mathbf{x}} = 1 + i\mathbf{k}\cdot\mathbf{x} + \dots$$
$$\cong 1,$$

which lets us write simply for the vector potential $\mathbf{A}(x,t) \to A_0 \sin\omega t \hat{\varepsilon}$. The polarization of the laser is contained in $\hat{\varepsilon}$. Since $\mathbf{B} = \nabla\times\mathbf{A}$, making this dipole approximation simplification for $\mathbf{A}$ and removing the z dependence is tantamount to setting the magnetic field to zero.

Equation (3.335) is the equation we seek. It contains two terms on the right, the second term on the right in this equation, containing the field is called the interaction Hamiltonian and will usually be denoted H_I.

3.7.1 Conversion to the Length Form of the Interaction Hamiltonian

In the context of the dipole approximation, we can transform the interaction Hamiltonian to a more familiar form. We do this by performing a gauge transformation on the wavefunction in (3.335):

$$\chi = \exp\left[i\frac{e}{\hbar c}\mathbf{x}\cdot\mathbf{A}\right]\psi. \qquad (3.336)$$

This transformation has no effect on the probability distribution of the electron which depends only on the absolute squared value of either wavefunction: $|\psi|^2 = |\chi|^2$. Inserting (3.336) into (3.335)

$$\frac{e}{c}\mathbf{x} \cdot \frac{\partial \mathbf{A}}{\partial t}\chi + i\hbar\frac{\partial \chi}{\partial t} =$$
$$-\frac{\hbar^2}{2m}\nabla \cdot \left(-i\frac{e}{\hbar c}\mathbf{A}\chi \exp\left[-i\frac{e}{\hbar c}\mathbf{x}\cdot\mathbf{A}\right] + \nabla\chi \exp\left[-i\frac{e}{\hbar c}\mathbf{x}\cdot\mathbf{A}\right]\right)\exp\left[i\frac{e}{\hbar c}\mathbf{x}\cdot\mathbf{A}\right]$$
$$-i\frac{\hbar e}{mc}\mathbf{A}\cdot\left(-i\frac{e}{\hbar c}\mathbf{A}\chi + \nabla\chi\right) + \frac{e^2 A^2}{2mc^2}\chi \tag{3.337}$$

yields

$$i\hbar\frac{\partial \chi}{\partial t} = -\frac{\hbar^2}{2m}\nabla^2\chi - \frac{e}{c}\mathbf{x}\cdot\frac{\partial \mathbf{A}}{\partial t}\chi. \tag{3.338}$$

From Eq. (3.6a) we can rewrite (3.338) in terms of the electric field directly

$$i\hbar\frac{\partial \chi}{\partial t} = -\frac{\hbar^2}{2m}\nabla^2\chi - e\mathbf{x}\cdot\mathbf{E}\chi. \tag{3.339}$$

This equation is termed the "length form" of the Schrödinger equation. We will tend to work in the $\mathbf{A} \cdot \mathbf{p}$ form, though either form is acceptable in the dipole approximation, the choice usually being driven by mathematical or computational convenience.

3.7.2 Volkov States of the Free Electron in a Laser Field

We now seek an eigenfunction of Eq. (3.335). If the field is zero, $H_I = 0$, and we have simply

$$i\hbar\frac{\partial \psi}{\partial t} = -\frac{\hbar^2}{2m}\nabla^2\psi. \tag{3.340}$$

The solution to this is just the plane-waves

$$\psi = \alpha \exp\left[i\mathbf{k}_e \cdot \mathbf{x} - i\varepsilon_{k_e} t/\hbar\right]. \tag{3.341}$$

This wave function can either be normalized in a box of length L, so that $\alpha = L^{-3/2}$ or by delta function normalization with $\alpha = (2\pi)^{-3/2}$. The ε_{k_e} is the kinetic energy of the electron

$$\varepsilon_{k_e} = \frac{\hbar^2 k_e^2}{2m}; \tag{3.342}$$

$k_e = p/\hbar$ is the wavenumber of the electron.

In the spirit of this free space plane-wave solution, we seek a solution of Eq. (3.335) of the form

$$\psi = a\exp\left[i\varsigma(t) + i\mathbf{k}_e \cdot \mathbf{x}\right], \tag{3.343}$$

3.7 Quantum Mechanical Description of a Free Electron

where a is some suitable normalization factor. Substitution of this wavefunction into the Schrödinger equation yields

$$-\hbar \frac{\partial \varsigma}{\partial t} a \exp\left[i\varsigma(t) + i\mathbf{k}_e \cdot \mathbf{x}\right] = \left(\frac{\hbar^2 k_e^2}{2m} + \frac{\hbar e}{mc} \mathbf{A} \cdot \mathbf{k} + \frac{e^2}{2mc} A^2\right) a \exp\left[i\varsigma(t) + i\mathbf{k}_e \cdot \mathbf{x}\right]. \quad (3.344)$$

We see that in defining $\mathbf{k}_e$ by Eq. (3.343) we must associate $\hbar \mathbf{k}_e$ with the canonical momentum of the electron. Solving for ς gives us

$$\varsigma(t) = -\frac{1}{\hbar} \int \left(\frac{\hbar^2 k_e^2}{2m} + \frac{\hbar e}{mc} \mathbf{A} \cdot \mathbf{k} + \frac{e^2}{2mc} A^2\right) dt. \quad (3.345)$$

Recalling the dipole approximation for $\mathbf{A}$, this solution finally gives for the wavefunction in the laser field

$$\psi = a \exp\left[-i(\varepsilon_{k_e} + U_p)\frac{t}{\hbar} + i\frac{e}{mc\omega} A_0 \hat{\varepsilon} \cdot \mathbf{k}_e \cos \omega t - i\frac{U_p}{2\hbar\omega} \sin(2\omega t) + i\mathbf{k}_e \cdot \mathbf{x}\right] \quad (3.346)$$

where we have inserted the usual nonrelativistic ponderomotive energy, $U_p = e^2 A_0^2 / 4mc^2$.

This wavefunction is known as a Volkov state. It describes a free electron subject to an oscillating field in the dipole approximation. It has the form of a plane-wave, in which the kinetic energy of the electron has added to it the cycle-averaged ponderomotive energy. It also has the position, $\mathbf{x}$, shifted by a term that goes as the oscillating position of the electron $(eA_0/mc\omega) \cos \omega t$ (recall Eq. (3.23)).

Turning to the length form of the interaction Hamiltonian, we use the transformation of Eq. (3.336) to find the Volkov state in the length gauge

$$\psi = a \exp\left[-i(\varepsilon_{k_e} + U_p)\frac{t}{\hbar} + i\frac{e}{mc\omega} A_0 \varepsilon \cdot \mathbf{k}_e \cos \omega t \right. \\ \left. - i\frac{U_p}{2\hbar\omega} \sin(2\omega t) + i\left(\mathbf{k}_e + \frac{e}{\hbar c} A_0 \sin \omega t \, \varepsilon\right) \cdot \mathbf{x}\right]. \quad (3.347)$$

4
Strong Field Interactions with Single Atoms

4.1 Introduction

How an electron is freed from an atom by photoionization and what kinetic energy it retains after ionization are among the central questions of modern physics. The photoelectric effect is the simplest manifestation of this phenomenon; it describes the photoionization process in a solid when the incident photon energy is greater than the work function of the material. This law is formally justified quantum mechanically by first-order perturbation theory and describes the absorption of a single photon by bound electrons in the target, resulting in a transition of that electron into the free continuum. The kinetic energy of the outgoing electron is the difference between the photon energy and the work function, or the ionization potential when individual atoms are considered. This description is appropriate in incoherent as well as coherent light sources but only when the incident intensity is weak enough to justify the use of Fermi's Golden Rule.

In this chapter we turn to the question of how atoms are ionized in light fields in which the photon energy is typically less than the atom's ionization potential but the intensity is high. While there are a range of physical phenomena which can occur in atoms at high intensity, we will confine our considerations to the widely studied physics of electron ionization from a single atom or ion, and the acquisition of energy by that liberated electron from the light field during the ionization process. (We will use the term "atom" in this chapter to refer not only to the neutral particle but partially stripped ions as well.) Our principal focus will be on the rate that electrons are liberated from a particular atomic species as a function of laser parameters such as intensity. We will also examine the energy spectra and angular distributions of the ionized electrons. The near-IR, optical and near-UV laser sources used for almost all strong field physics studies have a photon energy which is considerably less than the ionization potential of most atoms, so unlike the simple photoelectric effect, it is inevitable that multiple photons must participate in the ionization process. Just how many photons participate is an interesting question.

Unlike our discussions in the previous chapter, where it was completely adequate to treat the electron as a point charge propagating in a classically described field, we will now need to address directly the quantum mechanical character of the electron because of its confinement initially in the potential of the atom. On the other hand, it will be largely unnecessary

4.1 Introduction

to utilize special relativity in our discussions, even at intensities above 10^{18} W/cm^2, where a relativistic treatment is critical when considering a free electron. The reason for this is that during the ionization of atoms in the laser field, the electron velocity is rarely much above that of its initial velocity when bound in the atom. This velocity is well below c for all but the most highly charged ions. It is true that once the electron has been liberated in a relativistic intensity field, its subsequent motion must be treated relativistically, relying on the physics discussed in the previous chapter, but the rate of ionization up to that point can usually be accurately described with nonrelativistic theories. At the end of the chapter we will look briefly at the effects that relativity has on the ionization dynamics.

4.1.1 The Single Active Electron Approximation

When the ionization of any atom larger than a hydrogen atom is considered, the problem of calculating the ionization dynamics of multiple electrons can quickly devolve into a hopelessly complicated situation. Not only does the problem involve a multibody quantum mechanical ensemble, but the effects of correlation between the Fermionic electrons must also be included. While modern computation techniques do allow large-scale simulation of two electrons in a binding potential (at the time of this writing), clearly some approximation is needed to deal with the complications posed by multielectron atoms.

A viable path forward is to employ an approach known as the single active electron (SAE) approximation. The essence of this approximation involves treating the strong laser field as if it interacts with only one electron at a time, the electron which is most weakly bound with an energy given by the ionization potential of that particular atomic (or ionic) species (Protopapas *et al.* 1997). The remaining electrons in the atom are assumed to be effectively frozen and serve to create the effective potential in which the outermost active electron is bound. The dynamics of the electron in the strong field are then solved using this static effective potential. The effects of electron correlation in the strong field are effectively ignored – with the exception of the inclusion of correlation effects in the calculation of the initial, field-free, quantum wavefunction, which is the usual given initial condition for most calculations (Becker and Rottke 2008). The SAE approach has been widely employed in strong field theories because it reduces the multielectron problem to the much more tractable problem of a single bound particle (Burnett *et al.* 1993).

One consequence of the SAE picture is that the ionization of a multielectron atom is viewed as purely sequential. The strong laser field first removes one electron before affecting the second, more tightly bound electron. Further ionization of additional electrons occurs in a stepwise fashion. Therefore, an atom subject to an intense laser field will experience single-electron ionization at some intensity during the laser pulse, followed by subsequent ionization events at higher intensity later in the pulse if additional, more tightly bound electrons are present. This process of sequential multielectron ionization was established in early experiments (Perry *et al.* 1988a; Mainfray and Manus 1991; Augst *et al.* 1991), though a notable exception to this picture arises, which we will address in Section 4.5.

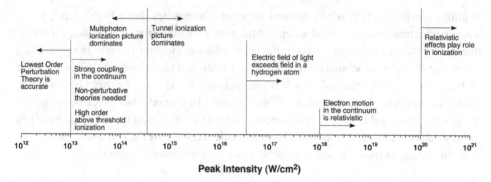

Figure 4.1 Dominant physics regimes of strong-field atomic interactions as the peak laser intensity increases (assuming near-IR wavelengths).

4.1.2 Nonlinear Physics Regimes in Strong Field Interactions with Atoms

Because many photons participate during strong field irradiation, the physics is inevitably highly nonlinear. This nonlinearity leads to a wealth of interesting effects. Among these is the fact that the ionization rate of atoms in a strong field is very strongly dependent on the incident light intensity. Combined with the temporal and spatial variation in intensity that accompanies any real focused laser pulse, this nonlinear physics leads to many complications. The nature of the nonlinearity evolves as the incident intensity increases. Figure 4.1 attempts to summarize some of the different physical regimes that are entered as the intensity is increased (Burnett *et al.* 1993).

Multiphoton effects have been well known for decades and were observed shortly after the invention of the laser once intensities of $>10^8$ W/cm^2 were obtainable in the laboratory. Multiphoton physics can be, at modest intensity, explained by the use of perturbation theory, as is employed in the single-photon photoelectric effect, up to intensities of about 10^{13} W/cm^2 (dependent on laser wavelength). At around 10^{13} W/cm^2 the laser field begins to modify the bound energy levels of the field-free atom. Climbing up in intensity by another order of magnitude, multiphoton processes start to occur with high probability in the continuum. At this point, description of the physics demands so-called nonperturbative theories. This regime can be thought of as the field significantly modifying the continuum. At intensity above 10^{16} W/cm^2, the field becomes comparable to an atomic unit of field and, therefore, the laser field tends to overwhelm the atomic structure. At intensities at and above this point, the laser field significantly distorts the binding potential of the atom and the concept of bound states begins to break down. Ionization can then be described through the process of tunnel ionization by an adiabatically varying field. Ultimately, at intensities above 10^{18} W/cm^2 relativistic descriptions of the laser interaction with the ionized electron will be needed.

Through this chapter we will address each of these nonlinear regimes in turn. We will start with a consideration of multiphoton physics, which we will initially describe by lowest-order perturbation theory. We argued in the introduction that the intensity at which

4.2 Multiphoton Ionization and Lowest-Order Perturbation Theory

perturbation theory breaks down could be considered the threshold for strong field physics. Therefore an understanding of multiphoton perturbation theory will give us insight into the physics we can expect as this theory falters with increasing intensity. Next we will turn to a widely used approximation which permits quantitative calculation at intensities where the laser field is no longer a small perturbation. Once we have developed this so-called strong field approximation, we will be in a position to look at atomic ionization at high intensity in terms of electron tunneling, a picture which will allow understanding ionization to high intensities. Next we will consider physics beyond the SAE approximation. Before concluding with a look at more exotic aspects of strong field atomic physics, we will examine in detail some of the experimental observables which result from these various regimes of strong field ionization including multiply charged ion production and complex free electron energy spectra and angular distributions.

4.2 Multiphoton Ionization and Lowest-Order Perturbation Theory

In the context of the SAE picture, one way to view the ionization of the atom and the freeing of a single electron into the continuum is to consider the most weakly bound electron as absorbing a number of photons through dipole transitions from the incident field. During this multiphoton ionization process (or MPI), enough photons must be absorbed to overcome the binding energy of the atom (or ion), the ionization potential, I_p (Perry et al. 1988b). A schematic illustration of the single-photon ionization process and the MPI generalization of this process are shown in Figure 4.2.

Because we are focusing for the most part on strong field laser sources with wavelengths between about 1,100 nm and 200 nm (photon energies of $\sim$1 eV to 6 eV), we are talking about processes requiring at least two or three photons for even the most weakly bound atoms (like Li) subject to UV fields. More typical multiphoton orders might be 7–10 photons for near-IR irradiation of noble gas atoms, or many more photons for ionization of highly charged ions. The number of photons absorbed to ionize the atom must be the

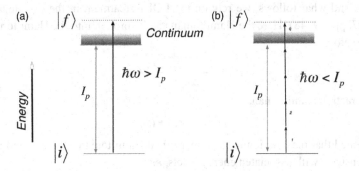

Figure 4.2 Schematic picture of single-photon ionization (a) and multiphoton ionization (b), where the atom absorbs a number of photons from the intense laser field, freeing the most weakly bound electron into the continuum.

minimum number needed to overcome the binding potential plus any field-induced shift of that binding potential, though in practice many more photons can participate in sufficiently intense laser fields.

4.2.1 MPI at Modest Intensities: Treating the Laser Field as a Perturbation

A quantitative study of the MPI process is best begun from the standpoint of lowest-order perturbation theory (LOPT). For the moment we will consider ionization of the atom by the absorption of the minimum number of photons needed to overcome the ionization potential. Ignoring for the moment any shifts in energy of an electron in the continuum that might arise from the presence of the laser field itself, the number of photons needed for ionization can be succinctly stated as:

$$q_{\min} = Integer\left[\frac{I_p}{\hbar\omega} + 1\right]. \tag{4.1}$$

If the laser electric field is not too strong, we can treat it as a perturbation. What it means to be "not too strong" is a question that we will answer quantitatively in a moment. However, we can say qualitatively that the electric field should be well below the field felt by an electron bound in the atom to be irradiated. In hydrogen, this electric field is roughly $E_a = e/a_B{}^2 = 3.5 \times 10^9$ V/cm (where a_B is the Bohr radius = 5.29×10^{-9} cm). Since a field of 5×10^8 V/cm corresponds to an intensity of around 3×10^{14} W/cm^2, we might expect, then, that this approach will fail at intensities much greater than 10^{14} W/cm^2 (though ultimately we will see that that breakdown point will depend on laser frequency).

Consider first the evolution of the wavefunction of a single electron in the effective binding potential of the atom $V(\mathbf{x})$ subject to a laser field in the dipole approximation. It will be convenient for the time being to work with the length form of the EM interaction Hamiltonian term (see Eq. (3.340)), which means that the wavefunction of the active electron is subject to the following Schrödinger equation:

$$i\hbar\frac{\partial\psi}{\partial t} = -\frac{\hbar^2}{2m}\nabla^2\psi + V(\mathbf{x})\psi - e\mathbf{x}\bullet\mathbf{E}(t)\psi. \tag{4.2}$$

With this and what follows, we rely on the LOPT treatment in the excellent book by Faisal (1987, pp. 29–48). Here the Hamiltonian can be broken into the Hamiltonian of the unperturbed atom,

$$H_0 = -\frac{\hbar^2}{2m}\nabla^2 + V(\mathbf{x}), \tag{4.3}$$

and the interaction Hamiltonian,

$$H_I = -e\mathbf{x}\bullet\mathbf{E}(t). \tag{4.4}$$

It is assumed that the solution to the unperturbed Hamiltonian is known and that it has known eigenstates with associated energy levels, ε_j:

$$H_0|j\rangle = \varepsilon_j|j\rangle. \tag{4.5}$$

The interaction Hamiltonian is considered a small perturbation on H_0.

4.2 Multiphoton Ionization and Lowest-Order Perturbation Theory

Our goal here is to find a rate of ionization, so we need to find a propagation operator which, when acting on the initial wavefunction of the electron prior to the arrival of the laser field, $|i\rangle$, propagates the evolution of that wavefunction forward in time under the influence of the interaction Hamiltonian. The $|i\rangle$ is typically the ground state of the atom or ion being irradiated (though it need not be). We define this evolution operator as

$$\psi_i(t) = \hat{u}(t)|i\rangle, \tag{4.6}$$

which permits us to find the amplitude of the wavefunction in some final state $|f\rangle$ (which is presumably some continuum state if we want to know the rate of ionization) by projecting onto that final state. This transition amplitude is

$$A'_{fi} = \langle f|\psi_i(t)\rangle = \langle f|\hat{u}(t)|i\rangle, \tag{4.7}$$

and the probability of making a transition from $|i\rangle$ to $|f\rangle$ is just the absolute square of this amplitude. Introduce a simple transformation on the wavefunction

$$\chi(t) = \exp[iH_0 t/\hbar]\psi(t), \tag{4.8}$$

which, when substituted into the Schrödinger equation (4.2), yields

$$i\hbar\frac{\partial}{\partial t}(e^{-iH_0 t/\hbar}\chi) = H_0 e^{-iH_0 t/\hbar}\chi + H_I e^{-iH_0 t/\hbar}\chi$$

$$i\hbar\frac{\partial \chi}{\partial t} = e^{iH_0 t/\hbar} H_I e^{-iH_0 t/\hbar}\chi, \tag{4.9}$$

leading us to write the modified Hamiltonian in the interaction picture for the wavefunction evolution:

$$i\hbar\frac{\partial \chi}{\partial t} = \hat{H}_I \chi. \tag{4.10}$$

Equation (4.10) is trivially solved:

$$\chi(t) = -\frac{i}{\hbar}\int H_I \chi\, dt. \tag{4.11}$$

Applying the boundary condition that the wavefunction begin in $|i\rangle$ prior to the turn-on of the laser field and the condition that laser intensity starts at 0 at $t = -\infty$, this solution can be written:

$$\chi(t) = |i\rangle - \frac{i}{\hbar}\int_{-\infty}^{t} \hat{H}_I(t')dt'. \tag{4.12}$$

To find the transition amplitude (4.7) we can assume that the final state is an eigenstate of the unperturbed Hamiltonian, H_0, which gives for the amplitude after transforming the wavefunction

$$A'_{fi} = e^{-i\varepsilon_f t/\hbar}\langle f|\chi(t)\rangle. \tag{4.13}$$

Since a phase term has no impact on the observed transition probability, for simplicity we work with the modified transition amplitude and modified evolution operator

$$A_{fi} = \langle f|\chi(t)\rangle = \langle f|u(t)|i\rangle. \tag{4.14}$$

The problem of solving the Schrödinger equation has been couched in terms of finding the evolution operator $u(t)$:

$$\chi(t) = u(t)|i\rangle. \qquad (4.15)$$

It is at this point that we must use some approximate method to find the solution for this evolution operator from the Schrödinger equation. Eliminating the wavefunction by inserting Eq. (4.15) into Eq. (4.12),

$$u(t)|i\rangle = |i\rangle - \frac{i}{\hbar}\int_{-\infty}^{t} \hat{H}_I(t')u(t')|i\rangle\, dt', \qquad (4.16)$$

and "un-projecting" the evolution operator from the initial state yields an implicit equation for $u(t)$:

$$u(t) = 1 - \frac{i}{\hbar}\int_{-\infty}^{t} dt'\, \hat{H}_I(t')u(t'). \qquad (4.17)$$

A perturbation expansion develops if we solve this equation by iteration. Repeated insertion of (4.17) back into the $u(t)$ term on the right-hand side of (4.17) yields

$$u(t) = 1 - \frac{i}{\hbar}\int_{-\infty}^{t} dt'\, \hat{H}_I(t') + \left(\frac{i}{\hbar}\right)^2 \int_{-\infty}^{t} dt' \int_{-\infty}^{t'} dt''\, \hat{H}_I(t'')\hat{H}_I(t') + \ldots. \qquad (4.18)$$

The evolution operator can be expressed as a sum of terms of order q,

$$u(t) = 1 + \sum_{q=1}^{\infty} u_q(t), \qquad (4.19)$$

where the qth-order evolution operator term is expressed in terms of the interaction Hamiltonian as

$$u_q(t) = \left(-\frac{i}{\hbar}\right)^q \int_{-\infty}^{t} dt_1 \hat{H}_I(t_1) \int_{-\infty}^{t_1} dt_2 \hat{H}_I(t_2) \ldots \int_{-\infty}^{t_{q-1}} dt_q \hat{H}_I(t_q). \qquad (4.20)$$

4.2.2 Physical Interpretation of the Perturbation Expansion

To understand the physical significance of the qth-order term of the evolution operator, recall that the transition amplitude from the initial atomic state to the final (likely continuum) state is governed by the sum of these terms:

$$A_{fi} = \langle f|1 + u_1 + u_2 + \ldots |i\rangle. \qquad (4.21)$$

Since we are interested in actual transitions to different states, and, of course, $\langle f|i\rangle = 0$, the first term in the sum that could actually lead to a real transition is u_1:

$$A_{fi}^{(1)} = \langle f|u_1(t)|i\rangle. \qquad (4.22)$$

The interaction Hamiltonian in the length form is

$$\hat{H}_I(t) = e^{iH_0 t/\hbar} e\mathbf{x} \cdot \mathbf{E} e^{-iH_0 t/\hbar}. \qquad (4.23)$$

4.2 Multiphoton Ionization and Lowest-Order Perturbation Theory

For now we assume that the incident laser is linearly polarized and, as before, we take the x-axis as the polarization axis. Introduction of the electric field of the laser in the dipole approximation,

$$\mathbf{E} = \frac{E_0}{2}(e^{-i\omega t} + e^{i\omega t})\hat{\mathbf{x}}, \quad (4.24)$$

gives us an expression for the first-order transition amplitude

$$A_{fi}^{(1)} = \langle f | \left(-\frac{i}{\hbar}\int_{-\infty}^{t} dt'\, e^{iH_0 t'/\hbar} ex \frac{E_0}{2}(e^{-i\omega t'} + e^{i\omega t'})e^{-iH_0 t'/\hbar}\right) |i\rangle. \quad (4.25)$$

Since we will choose the initial and final states to be eigenstates of the unperturbed atomic Hamiltonian, H_0, we can reverse the order of integration and write the first-order transition amplitude as two terms:

$$\begin{aligned} A_{fi}^{(1)} = & -\frac{i}{\hbar}\int_{-\infty}^{t} dt' \exp\left[i\left(\frac{\varepsilon_f}{\hbar} - \frac{\varepsilon_i}{\hbar} - \omega\right)t'\right]\frac{eE_0}{2}\langle f|x|i\rangle \\ & -\frac{i}{\hbar}\int_{-\infty}^{t} dt' \exp\left[i\left(\frac{\varepsilon_f}{\hbar} - \frac{\varepsilon_i}{\hbar} + \omega\right)t'\right]\frac{eE_0}{2}\langle f|x|i\rangle \end{aligned}. \quad (4.26)$$

Our interest is in the transition amplitude on a time scale much longer than the atomic time scale, so it is appropriate to let $t \to \infty$ in the upper limit of the integrals. Recalling the Fourier expression for the Dirac delta function

$$\delta(z) = \frac{1}{2\pi}\int_{-\infty}^{\infty} e^{izt}dt \quad (4.27)$$

removes the integrals, delivering for the first-order transition amplitude

$$A_{fi}^{(1)} = -\frac{i}{\hbar}\pi eE_0 \langle f|x|i\rangle \delta\left(\frac{\varepsilon_f}{\hbar} - \frac{\varepsilon_i}{\hbar} - \omega\right) - \frac{i}{\hbar}\pi eE_0 \langle f|x|i\rangle \delta\left(\frac{\varepsilon_f}{\hbar} - \frac{\varepsilon_i}{\hbar} + \omega\right). \quad (4.28)$$

The physical interpretation of this expression for A_{fi} is clear if we examine the energy conservation imposed by the delta functions. The first term demands a transition to a final state with energy such that $\varepsilon_f - \varepsilon_i = \hbar\omega$, which would correspond to the absorption of one photon by the atomic electron from the initial state of energy ε_i to a state of energy ε_f. Of course, if there is no state near this final energy, the transition amplitude approaches zero. The second term is nonzero only if $\varepsilon_f - \varepsilon_i = -\hbar\omega$, a statement of energy conversation when one photon is emitted. These two terms in the transition amplitude, which we can now associate with the absorption and emission of one photon, respectively, are schematically illustrated in Figure 4.3.

Note that the transition amplitude is proportional to the dipole matrix element between the initial and final states $\langle f|x|i\rangle$, and the electric field amplitude E_0, which, of course, means that the transition probability, $\left|A_{fi}^{(1)}\right|^2$, will be proportional to the laser intensity. We have simply reproduced the first-order perturbation transition rate (which leads to Fermi's Golden Rule). This result tells us to interpret the various u_q terms in the perturbation expansion for $u(t)$ of (4.21) as transition processes involving q photons.

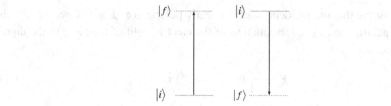

Figure 4.3 Illustration of the two first-order perturbation theory terms, corresponding to the absorption or emission of one photon.

This interpretation is reinforced by considering the second-order transition amplitude term, which is

$$A_{fi}^{(2)} = \langle f | u_2(t) | i \rangle$$
$$= \left(-\frac{i}{\hbar}\right)^2 \int_{-\infty}^{t} dt' \int_{-\infty}^{t'} dt'' \, \langle f | \hat{H}_I(t'') \hat{H}_I(t') | i \rangle. \quad (4.29)$$

The interaction with the field now effectively appears twice in the double application of the interaction Hamiltonian, implying that two photons participate in the transition process represented by this term. Evaluating this term is more difficult than the first-order term. Evaluation is made possible by inserting a complete set of states between the two interaction Hamiltonian operators. If we consider the j stationary eigenstates of the unperturbed Hamiltonian, we know that a sum over a complete set of states yields the unity operator:

$$\int \sum_j |j\rangle \langle j| = 1. \quad (4.30)$$

Here the combination of the integral and summation signs means that this sum is an actual sum over discrete bound states plus an integral over unbound continuum states. Inserting this complete set of states between the interaction Hamiltonian operators gives

$$A_{fi}^{(2)} = \left(-\frac{i}{\hbar}\right)^2 \int_{-\infty}^{t} dt' \int_{-\infty}^{t'} dt'' \, \langle f | \hat{H}_I(t'') | j \rangle \langle j | \hat{H}_I(t') | i \rangle. \quad (4.31)$$

Using Eq. (4.5) and the definitions in Eqs. (4.10) and (4.24), the interaction Hamiltonian matrix elements can be written

$$\langle j' | \hat{H}_I | j \rangle = \frac{eE_0}{2} e^{i(\varepsilon_j - \varepsilon_{j'} - \hbar\omega)t/\hbar} \langle j' | x | j \rangle + \frac{eE_0}{2} e^{i(\varepsilon_j - \varepsilon_{j'} + \hbar\omega)t/\hbar} \langle j' | x | j \rangle. \quad (4.32)$$

Combined with integration over the first time-integral

$$\int_{-\infty}^{t'} dt'' \exp\left[i \left(\varepsilon_j/\hbar - \varepsilon_i/\hbar \pm \omega\right) t''\right] = -i \frac{\exp\left[i \left(\varepsilon_j/\hbar - \varepsilon_i/\hbar \pm \omega\right) t'\right]}{(\varepsilon_j/\hbar - \varepsilon_i/\hbar \pm \omega)}, \quad (4.33)$$

we have, finally, for the second-order transition amplitude, the sum of four terms:

$$A_{fi}^{(2)} = -i \left(-\frac{i}{\hbar}\right)^2 2\pi\hbar^2 \left(\frac{eE_0}{2}\right)^2 \int \sum_j \left[\frac{\langle f | x | j \rangle \langle j | x | i \rangle}{(\varepsilon_j - \varepsilon_i + \hbar\omega)} \delta\left(\varepsilon_j - \varepsilon_i - 2\hbar\omega\right)\right.$$

4.2 Multiphoton Ionization and Lowest-Order Perturbation Theory

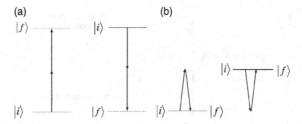

Figure 4.4 Illustration of the four second-order perturbation theory terms; the first corresponds to the absorption of two photons, the second to emission of two photons and the third and fourth correspond to the absorption and emission of one photon each, returning the electron to its original energy state.

$$+ \frac{\langle f|x|j\rangle \langle j|x|i\rangle}{(\varepsilon_j - \varepsilon_i + \hbar\omega)} \delta(\varepsilon_j - \varepsilon_i + 2\hbar\omega)$$

$$+ \left(\frac{\langle f|x|j\rangle \langle j|x|i\rangle}{(\varepsilon_j - \varepsilon_i - \hbar\omega)} + \frac{\langle f|x|j\rangle \langle j|x|i\rangle}{(\varepsilon_j - \varepsilon_i + \hbar\omega)} \right) \delta(\varepsilon_f - \varepsilon_i) \bigg]. \quad (4.34)$$

The interpretation of these four terms is made clear by the energy conservation imposed by the delta functions. The two-photon transitions of these four terms are illustrated schematically in Figure 4.4. The first term is the amplitude for absorption of two photons (Figure 4.4a) and the second is the corresponding term for emission of two photons (Figure 4.4b). The third and fourth terms represent transitions that must end in the initial state (or a state degenerate with the initial energy) and can be seen as the absorption and emission of photons (depicted in Figure 4.4c and d).

Not surprisingly, the amplitude is proportional to the square of the incident field amplitude, meaning the transition probability of this two-photon process scales as the square of intensity. The function for the transition amplitude is now more complex than the single photon case with a sum over *all* possible intermediate states. From the standpoint of MPI, it is the term like the first one in Eq. (4.34) in which we are most interested as the one leading to ionization in the various perturbation orders, if energy conservation will permit the final state to be in the continuum. (The third and fourth terms in Eq. (4.34) will have another useful interpretation which we will return to in a few pages.)

4.2.3 The LOPT Multiphoton Ionization Transition Rate and Cross Section

Repeated iteration to find the qth-order transition amplitude leads to a sequence of terms like those in (4.34), one of which will correspond to absorption of q photons. The process in question is illustrated in Figure 4.5. Continuing the logic that led to Eq. (4.34), including the repeated insertion of a sum over a complete set of states between pairs of the interaction Hamiltonian operator leads naturally to the qth-order q-photon transition amplitude, which can be written:

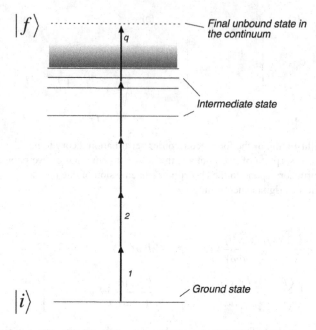

Figure 4.5 Schematic picture of the lowest-order perturbation expansion term corresponding to the absorption of q photons, freeing a single electron into the unbound continuum.

$$A_{fi}^{(q)} = (-i)^{q+1} 2\pi \left(\frac{eE_0}{2}\right)^q \delta\left(\varepsilon_j - \varepsilon_i - q\hbar\omega\right) \int \sum_{j_1} \int \sum_{j_2} \cdots \int \sum_{j_{q-1}}$$

$$\times \frac{\langle f|x|j_{q-1}\rangle \cdots \langle j_2|x|j_1\rangle \langle j_1|x|i\rangle}{\left(\varepsilon_{j_{q-1}} - \varepsilon_i - (q-1)\hbar\omega\right) \cdots \left(\varepsilon_{j_2} - \varepsilon_i - 2\hbar\omega\right)\left(\varepsilon_{j_1} - \varepsilon_i - \hbar\omega\right)} \quad (4.35)$$

For a given atom, finding the LOPT multiphoton ionization probability comes down to seeking the lowest number of photons, q, which lead to the electron acquiring enough energy to overcome the binding ionization potential. Once that number of photons has been ascertained, the probability that the atom will have been ionized by the incident laser pulse is

$$P_{fi} = \left|A_{fi}^{(q)}\right|^2. \quad (4.36)$$

What we are really after is not the probability of ionization but rather an ionization *rate*. We can denote this rate as probability of ionization per unit time and formally write the ionization rate for q-photon ionization $W_{fi}^{(q)}$ as

$$W_{fi}^{(q)} \equiv \frac{P_{fi}}{\Delta t} \quad (4.37)$$

and let Δt go to infinity. This formula is easily manipulated, realizing that the probability is related to the square of the energy-conserving Dirac delta function. A delta function can be written

4.2 Multiphoton Ionization and Lowest-Order Perturbation Theory

$$\delta(z) = \lim_{\Delta t \to \infty} \frac{1}{2\pi} \int_{-\Delta t/2}^{\Delta t/2} \exp[izt] dt$$

$$= \frac{1}{2\pi} \lim_{\Delta t \to \infty} \Delta t \quad (\text{when } z = 0). \tag{4.38}$$

So the ionization rate for LOPT q-photon MPI can be concisely written as:

$$W_{fi}^{(q)}(\mathbf{k}_e) = \frac{2\pi}{\hbar} \left(\frac{e}{2}\right)^{2q} E_0^{2q} \delta(\varepsilon_f - \varepsilon_i - q\hbar\omega) \left|x_{fi}^{(q)}(\mathbf{k}_e)\right|^2, \tag{4.39}$$

where $x_{fi}^{(q)}(\mathbf{k}_e)$ is the generalized multiphoton dipole matrix element equal to the second line of Eq. (4.35). This rate is written as an explicit function of the electron's final momentum $\hbar\mathbf{k}_e$, to remind us that this is a differential rate for outgoing electrons with varying direction and momentum. We also reiterate that this ionization rate exhibits the hallmark of a q-photon process; it is proportional to the incident intensity to the qth power.

For practical calculations, we need to deal with the fact that, when considering ionization, the final state is actually within a continuum of free states characterized by some density of states. The form of the final state will depend, to some extent, on the problem considered, though the simplest approximation for the form of the state is a plane-wave (which ignores the influence of the binding potential or incident radiation). This is essentially the multiphoton equivalent of the Born approximation. If we use the convention that the final state is delta function-normalized, a pure plane-wave would be written

$$|f\rangle = \frac{1}{(2\pi)^{3/2}} e^{i\mathbf{k}_e \cdot \mathbf{x}}, \tag{4.40}$$

and then we can find the final ionization rate by multiplying (4.39) by a density of states and integrating over energy near the outgoing electron's energy, a procedure which gives

$$W^{(q)}(\hat{\mathbf{k}}_f) = \int W_{fi}^{(q)}(\mathbf{k}_e) \rho_e d\varepsilon, \tag{4.41}$$

where the density of states is (Faisal 1987, p. 49)

$$\rho_e = \frac{m k_e}{\hbar^2}. \tag{4.42}$$

In the MPI situation considered so far, where we ignore any shift of the continuum limit in the strong field, the final electron momentum is found from simple energy conservation,

$$\frac{(\hbar k_f)^2}{2m} = \varepsilon_f = q\hbar\omega - I_p, \tag{4.43}$$

to give, for the final differential ionization rate,

$$W^{(q)}(\hat{\mathbf{k}}_f) = 2\pi \frac{m^{3/2} \varepsilon_f}{\hbar^4} \left(\frac{e}{2}\right)^{2q} \left|x_{fi}^{(q)}(\hat{\mathbf{k}}_f)\right|^2 E_0^{2q}. \tag{4.44}$$

The total ionization rate is found by integrating over all outgoing angles for the electron:

$$w^{(q)} = \int W^{(q)}(\hat{\mathbf{k}}_f) \sin\theta \, d\theta \, d\varphi. \tag{4.45}$$

The MPI literature often couches results based on LOPT in terms of a generalized multiphoton cross section. Here the incident photon flux of the laser (photons per cm^2 per second) is the relevant parameter and is related to the incident intensity easily as

$$\Phi_{\hbar\omega} = \frac{I}{\hbar\omega} = \frac{cE_0^2}{8\pi\hbar\omega}, \qquad (4.46)$$

which means that the rate that atoms are ionized by MPI is

$$w^{(q)} = \sigma^{(q)}\Phi_{\hbar\omega}^q, \qquad (4.47)$$

where the multiphoton ionization cross section is

$$\sigma^{(q)} = \int 2\pi \frac{\sqrt{m^3 \varepsilon_f}}{\hbar^4} \left(\frac{2\pi e^2 \hbar\omega}{c}\right)^q \left|x_{fi}^{(q)}(\hat{\mathbf{k}}_e)\right|^2 d\Omega. \qquad (4.48)$$

This quantity does not have the units of a regular cross section but instead has units of cm^{2q} s^{q-1}.

4.2.4 Intensity Validity Range for the LOPT Description of MPI

This analysis now puts us in a position to evaluate the regime in which the LOPT picture is quantitatively accurate in describing the multiphoton ionization rate. Since this rate was derived from a perturbation expansion, we expect that the accuracy of the rate will break down when higher-order terms in the expansion become comparable in size to the lowest-order term. For ionization to a given energy, the term corresponding to the lowest-order term is illustrated in Figure 4.6. The next-order term in the perturbation expansion that will yield the same outgoing electron energy and hence will contribute to the MPI rate in question is the term in which q photons are absorbed and then one additional photon is absorbed followed by emission of the $q+2$ photon. This $q+2$ photon process is compared to the lowest-order process in Figure 4.6.

The term in the perturbation series for the transition amplitude which corresponds to the $q+2$ process depicted in Figure 4.2 can be written from our previous analysis. It is

$$A_{fi}^{(q+2)} = (-i)^{q+3} 2\pi \left(\frac{eE_0}{2}\right)^{q+2} \delta(\varepsilon_j - \varepsilon_i - q\hbar\omega) \int \sum_{j_1} \int \sum_{j_2} \cdots \int \sum_{j_q} \int \sum_{j_{q+1}}$$

$$\times \frac{\langle f|x|j_{q-1}\rangle \ldots \langle j_2|x|j_1\rangle \langle j_1|x|i\rangle}{\left(\varepsilon_{j_{q+1}} - \varepsilon_i + (q+1)\hbar\omega\right)\left(\varepsilon_{j_q} - \varepsilon_i - q\hbar\omega\right)\ldots\left(\varepsilon_{j_1} - \varepsilon_i - \hbar\omega\right)}. \qquad (4.49)$$

Now, the probability for ionization to this energy channel is the square of the sum of the lowest-order term and the next-higher order contributing term in the transition amplitude

$$P_{fi} = \left|A_{fi}^{(q)} + A_{fi}^{(q+2)}\right|^2$$

$$= \left|A_{fi}^{(q)}\right|^2 + A_{fi}^{(q)}\left(A_{fi}^{(q+2)}\right)^* + \left(A_{fi}^{(q)}\right)^* A_{fi}^{(q+2)} + \left|A_{fi}^{(q+2)}\right|^2. \qquad (4.50)$$

4.2 Multiphoton Ionization and Lowest-Order Perturbation Theory

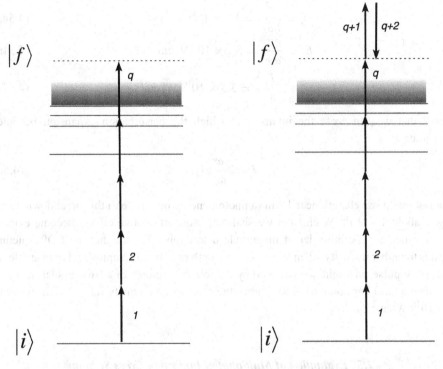

Figure 4.6 Illustration of the lowest-order term yielding multiphoton ionization compared to the next-higher term contributing to ionization rate into that free energy channel. The second term corresponds to the absorption of $q + 1$ photons accompanied by emission of one photon, yielding an outgoing free electron with the same energy as the lowest-order perturbation expansion term.

Consider the second and third terms in Eq. (4.50). It is a fair order-of-magnitude approximation to take

$$\langle j_a | x | j_b \rangle \approx a_B, \tag{4.51}$$

where a_B is the usual Bohr radius (= 51 pm). The magnitude of the third and fourth terms in the ionization probability can be approximated as

$$A_{fi}^{(q)} A_{fi}^{(q+2)} \approx \left(\frac{eE_0 a_B}{2\omega} \right)^2 A_{fi}^{(q)}. \tag{4.52}$$

These higher-order terms become comparable to the lowest-order term when

$$2 \left(\frac{eE_0 a_B}{2\omega} \right)^2 \approx 1. \tag{4.53}$$

Introducing the atomic frequency, atomic electric field, and corresponding "atomic unit of intensity,"

$$\omega_a \equiv \frac{e^2}{\hbar a_B} = 4.1 \times 10^{12} \text{ s}^{-1}, \qquad (4.54\text{a})$$

$$E_a \equiv \frac{e}{a_B^2} = 5.1 \times 10^9 \text{ V/cm}, \qquad (4.54\text{b})$$

$$I_a \equiv \frac{cE_a^2}{8\pi} = 3.5 \times 10^{16} \text{ W/cm}^2, \qquad (4.54\text{c})$$

means that we can write the intensity at which the perturbation expansion becomes inaccurate as

$$I = 2\frac{\omega^2}{\omega_a^2} I_a. \qquad (4.55)$$

In a laser with wavelength near 1 µm (a photon energy of ∼1.2 eV) this breakdown intensity is about 1×10^{14} W/cm² (as we shall see, nonperturbative effects become evident at intensities at least an order of magnitude below this). We see that the LOPT picture applies to higher intensity when shorter-wavelength radiation is employed. For example, in a near UV pulse, as might be generated by the fourth harmonic of a Nd:glass laser, at 263 nm (or the third harmonic of Ti:sapphire), the breakdown intensity for LOPT is closer to 2×10^{15} W/cm².

4.2.5 Evaluation of Multiphoton Ionization Cross Sections

While the derivation of the multiphoton cross section in the framework of LOPT is straightforward, the evaluation of these cross sections is not. The principal difficulty rests in the need to evaluate the sum of dipole matrix elements over the complete set of atomic and continuum states. It is not enough to know the amplitude of these matrix elements but the phases are needed as well. There has been considerable work in the literature on the evaluation of MPI cross sections using Eq. (4.35) and we will not seek to summarize the various methods that have been developed. (Many of the most often used methods for evaluating the MPI cross section are discussed in the excellent book on the subject by Faisal (1987).)

One simple method, however, can be employed which can give reasonable cross sections with a minimum of computation. The method is commonly known as the "truncated summation method" and relies on the fact that, in the various sums over atomic states, typically a small number of terms contribute predominantly to the cross section because of the $\varepsilon_{j_q} - \varepsilon_i - q\hbar\omega)^{-1}$ denominator terms. Those matrix elements, which are closest to a q-photon resonance, will be the terms contributing the most to the sum. We can, therefore, separate out those dominant terms and approximate the remainder of the sum with an appropriately chosen average effective state energy. Each sum over a complete set of states can then be written as

$$\int \sum_j \frac{|j\rangle\langle j|}{\varepsilon_j - \varepsilon_i - q'\hbar\omega} \cong \sum_{j=1}^{\mu} \frac{|j\rangle\langle j|}{\varepsilon_j - \varepsilon_i - q'\hbar\omega} + \int \sum_{j=\mu+1}^{\infty} \frac{|j\rangle\langle j|}{\langle\varepsilon_j\rangle - \varepsilon_i - q'\hbar\omega}, \qquad (4.56)$$

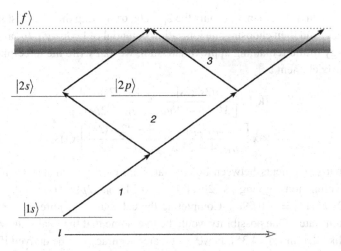

Figure 4.7 Three-photon ionization of a hydrogen atom by near-UV pulses. This illustrates the various photon absorption channels possible with three photons and the angular momentum possibilities.

where the states 1 to μ are selected as the principal contributors to that particular sum, and the remaining terms are approximated with an effective average energy $\langle \varepsilon_i \rangle$. The choice of this average effective energy depends on the situation but it is typically appropriate to choose the ionization energy. Using the fact that the initial sum is over a complete set of states, we can write the approximation term (the second term on the right-hand side in (4.56)) as

$$\frac{1 - \sum_{j=1}^{\mu} |j\rangle\langle j|}{\langle \varepsilon_j \rangle - \varepsilon_i - q'\hbar\omega}. \tag{4.57}$$

Putting each sum in this form reduces the problem to calculating the dipole matrix element between a given initial state and the μ states selected as the most important contributors to the sum.

To illustrate this method, and to explore the variation of the MPI cross section with incident wavelength, consider as an example the three-photon ionization of hydrogen. Such an MPI process would be important with near UV laser pulses in the 200–250 nm range, as might be produced by a KrF laser, the fourth harmonic of an Nd:glass laser, or the third harmonic of Ti:sapphire. This MPI process is illustrated in Figure 4.7 where energy is plotted vertically and the change in angular momentum possible with a linearly polarized pulse is indicated in the horizontal direction. This three-photon diagram suggests that the main contributing term to the one-photon sum will arise from the 2p state and the largest contributing term to the two-photon sum will be the 2s state. In fact, for wavelengths near 245 nm, the two-photon transition is nearly resonant with that 2s state.

Truncating the first sum to include only the 2p state, truncating the second sum to include only the 2s state, and then choosing the ionization potential of hydrogen, I_H, as the average effective energy of the remaining approximate sum, we can write the three-photon dipole transition matrix element as

$$x_{fi}^{(3)} = \langle \mathbf{k}_e | x \left[\frac{|2s\rangle\langle 2s|}{3I_H/4 - 2\hbar\omega} + \frac{1 - |2s\rangle\langle 2s|}{I_H - 2\hbar\omega} \right]$$
$$\times x \left[\frac{|2p\rangle\langle 2p|}{3I_H/4 - 2\hbar\omega} + \frac{1 - |2p\rangle\langle 2p|}{I_H - 2\hbar\omega} \right] x |1s\rangle . \quad (4.58)$$

The dipole matrix elements between bound states are easily calculated from the known hydrogen wavefunctions, giving us $\langle 2p|x|1s\rangle = 0.744 a_B$, $\langle 2p|x^2|1s\rangle = 0$, $\langle 2s|x|2p\rangle = -3a_B$, and $\langle 2s|x^2|1s\rangle = -0.99 a_B^2$. Completing the calculation requires a formula for the final continuum states. One possibility would be to assume that the final state was simply a plane-wave, like that in Eq. (4.35), however, greater accuracy can be derived if we use the known Coulomb continuum state (Becker and Faisal 1999) so that we have accounted for the influence of the remaining charged hydrogen ion on the outgoing electron. This state can be written in terms of the Kummer confluent hypergeometric function $_1F_1(a;b;c)$ as

$$|\mathbf{k}_e\rangle = \left(\frac{8\pi^3}{k_e} \right)^{1/2} \sum_{l,m} C_{k_l} i^l \exp\left[-i\left(\varphi_l + k_e r\right)\right]$$
$$\times r_1^l F_1 \left(i/a_B k_e + l + 1; 2l + 2; 2i k_e r \right) Y_{lm}(\hat{\mathbf{r}}) Y_{lm}(\hat{\mathbf{k}}) \quad (4.59)$$

where the Coulomb phase shift is $\varphi_l = \arg[\Gamma(l + 1 - i/a_B k_e)]$ and the normalization constant is

$$C_{k_e l} = \frac{2^{l+1}}{(2l+1)!} k_e^l \left(\frac{\Pi \left(s^2 + 1/a_B^2 k_e^2\right)}{1 - e^{-2\pi/a_s k_e}} \right)^{1/2} . \quad (4.60)$$

For our purposes we need only retain the $l = 1$ and $l = 3$ terms in this state sum, as these are the only two allowed angular momentum states for the final state, shown in Figure 4.7.

Using this final state wavefunction and integrating over all possible outgoing electron angles, simple numerical calculation yields for the three-photon cross section the curve plotted in Figure 4.8. This curve illustrates the general trend as a function of incident laser wavelength typically observed in MPI cross sections found from LOPT, namely a dramatic increase in the cross section at wavelengths near a multiphoton resonance with an intermediate level. In the present case, there is a two-photon resonance with the 2s state leading to a large enhancement of ionization yield near 244 nm. In the framework considered so far, at this resonant wavelength the ionization rate becomes infinite, so other techniques are needed to evaluate the ionization rate near such a multiphoton resonance. (Of course, in a circularly polarized pulse this peak in the MPI cross section would be absent because a two-photon resonance would not be allowed between the 1s and 2s states by angular momentum selection rules.)

We can use this calculation to estimate the intensity needed to reach significant ionization. Off-resonance, the three-photon cross section is about 10^{-84} cm^6 s^2. If we have a

4.2 Multiphoton Ionization and Lowest-Order Perturbation Theory

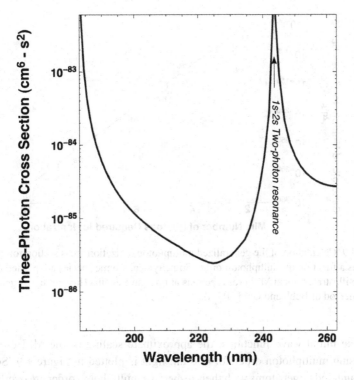

Figure 4.8 Calculation of the three-photon MPI cross section for hydrogen as a function of laser wavelength.

pulse of width $\Delta\tau \sim 100$ fs we get strong ionization when the rate times the pulse duration approaches 1: $\Delta\tau\,\sigma_3(I/h\nu)^3 \sim 1$. This condition is satisfied for intensity of around 1.5×10^{14} W/cm^2. If the pulse is longer, say 100 ps, this ionization saturation intensity is reduced by an order of magnitude. Our calculation is consistent with experimental observations.

A coarse order-of-magnitude estimate of how the multiphoton ionization cross section away from resonances scales with increasing multiphoton order can be gleaned by taking the truncated summation method one step further. If all intermediate states are ignored, the summation over explicit states in (4.56) is ignored and the entire cross section is found with the second, average energy state term in this equation. This approach, first employed by Bebb and Gold (1966), reduces the multiphoton dipole matrix element to a simple, single matrix element calculation of the form

$$x_{fi}^{(q)} \approx \frac{\langle \mathbf{k}_e | x^q | i \rangle}{\prod_{\mu=1}^{q-1}\left(\langle \varepsilon_j \rangle - \varepsilon_i - \mu\hbar\omega\right)}. \tag{4.61}$$

Since all intermediate states are ignored in such a calculation, the choice of the average final state energy is more appropriately an energy among these excited states. Bebb and Gold (1966) suggested using the first excited state energy. Utilizing (4.61) and

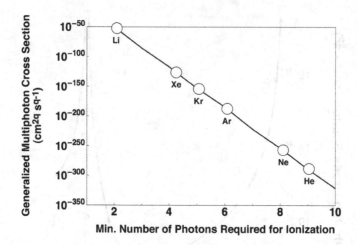

Figure 4.9 Calculation of the generalized multiphoton ionization cross section for 400 nm pulses as a function of multiphoton order. Some typical atomic species are plotted on this curve to illustrate typical MPI cross sections at this wavelength. This curve was generated by the method of Bebb and Gold (1966).

hydrogen-like initial wave functions, the approximate scaling of the MPI cross section with increasing multiphoton order for 400 nm light is plotted in Figure 4.9. Some commonly encountered target atoms with their respective multiphoton order are overlaid on this plot.

4.2.6 Resonances in Multiphoton Ionization

Our simple calculation of the three-photon MPI cross section for hydrogen illustrates one of the complications in assessing multiphoton ionization rates: what is the ionization rate when an intermediate subset of photons is very close to resonance with an excited state of the atom? Such a situation is depicted in Figure 4.10, where there is a q_1-photon resonance with an intermediate state in the q-photon MPI of this atom. Clearly, the ionization rate as calculated by the LOPT approach will yield unrealistically high rates because of the $\varepsilon_1 - \varepsilon_i - q_1 \hbar \omega$ term in the denominator.

This resonantly enhanced MPI situation (REMPI) has been widely considered in the literature (Perry and Landen 1988) and can lead to rather rich and complicated physics (Lambropoulos and Tang 1992). Since many of the most important effects of REMPI occur at modest intensity ($\sim 10^{11}$ W/cm^2), we will not dwell on it in any detail here, other than to outline the techniques that can be used to assess the ionization rate when an intermediate resonance occurs. (Such resonances play a role in the photoelectron spectrum produced in strong field ionization, so we will require a little background to understand that phenomenon later in this chapter.) Generally, the presence of an intermediate resonance greatly enhances the ionization rate and leads to a deviation in the intensity scaling of the rate away from the I^q scaling characteristic of nonresonant MPI (sometimes dramatically).

4.2 Multiphoton Ionization and Lowest-Order Perturbation Theory

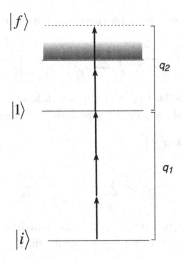

Figure 4.10 Multiphoton ionization in an atom in which there is an intermediate resonance. In this case q_1 photons are nearly resonant with the level $|1>$, which can then be ionized, releasing an electron into the continuum by q_2 photons.

The importance of REMPI in our understanding of strong field atomic ionization is increased by a physical effect we have yet to consider. So far, we have ignored any effect of the strong field on the atomic structure and have assumed that the positions of the atomic levels remain unchanged when the atom is subject to the strong EM field. This is a gross approximation that cannot hold at high intensities, where the field strength begins to approach the atomic field strength. The first signature of the field effect on atomic structure is to shift the position of the field-free eigenstates in energy, a process known as the AC Stark shift. The AC Stark shift can move the position of energy levels around, so that even if the MPI is nonresonant at first, energy levels can be shifted into resonance at a particular intensity, making REMPI more prevalent and more important than it would seem at first (Protopapas *et al.* 1997).

In fact, we have already performed the calculation needed to find the size of the AC Stark shift in an atom within the framework of perturbation theory. The first perturbation term which will contribute to the shift of an energy level is the second-order (two-photon) processes which have the same initial and final states. These processes are pictured in the bottom two diagrams of Figure 4.4 and correspond to the final two terms in Eq. (4.34). The lowest-order perturbation amplitude contributing to the AC Stark shift can be written from (4.34) as

$$A_{ii}^{(2)} = -i\left(-\frac{i}{\hbar}\right)^2 2\pi\hbar^2 \left(\frac{eE_0}{2}\right)^2$$
$$\times \int \sum_j \left(\frac{\langle i|x|j\rangle\langle j|x|i\rangle}{(\varepsilon_j - \varepsilon_i - \hbar\omega)} + \frac{\langle i|x|j\rangle\langle j|x|i\rangle}{(\varepsilon_j - \varepsilon_i + \hbar\omega)}\right)\delta\left(\varepsilon_f - \varepsilon_i\right). \quad (4.62)$$

Energy is conserved so we can replace the delta function using Eq. (4.38) to yield

$$A^{(2)}_{ii}(t) = i\frac{e^2 E_0^2}{2} \int \sum_j \frac{(\varepsilon_j - \varepsilon_i)|x_{ji}|^2}{(\varepsilon_j - \varepsilon_i)^2 - (\hbar\omega)^2}\frac{t}{\hbar}. \tag{4.63}$$

To simplify notation, we introduce $x_{ji} \equiv \langle j|x|i\rangle$, which lets us write the evolution operator of Eq. (4.19) acting to second order on the unperturbed stationary wave function:

$$\begin{aligned}\psi_i(t) &= \left(1 + A^{(2)}_{ii}\right)\psi_i(0) \\ &= \left(1 + i\frac{e^2 E_0^2}{2}\int \sum_j \frac{(\varepsilon_j - \varepsilon_i)|x_{ji}|^2}{(\varepsilon_j - \varepsilon_i)^2 - (\hbar\omega)^2}\frac{t}{\hbar}\right)\psi_i(0).\end{aligned} \tag{4.64}$$

If we couch the wavefunction in terms of an effective (small) energy shift of the stationary eigenstate of the atom, we can write for this stationary state

$$\begin{aligned}\psi_i(t) &= \psi_i(0)e^{-i(\varepsilon_i + \Delta\varepsilon_s)t/\hbar} \\ &\cong (1 - i\Delta\varepsilon_s t/\hbar)\psi_i(0)e^{-i\varepsilon_i t/\hbar}.\end{aligned} \tag{4.65}$$

Comparison of Eqs. (4.64) and (4.65) leads us to conclude that the energy shift of the level in the laser field is

$$\Delta\varepsilon_S = -\frac{e^2 E_0^2}{2\hbar}\int \sum_j \frac{\omega_{ji}|x_{ji}|^2}{\omega_{ji}^2 - \omega^2}, \tag{4.66}$$

where we have introduced $\omega_{ji} = (\varepsilon_j - \varepsilon_i)/\hbar$.

The key conclusion from Eq. (4.66) is that the AC Stark shift of the various atomic levels is different depending on how deeply bound they are. This means that the energy *difference* between levels will shift with increasing incident intensity. So level spacings that may be nonresonant with the field initially can be shifted *into* resonance by the Stark shift. Deeply bound levels typically have $\omega_{ij} \gg \omega$ in near-IR or optical wavelength laser fields and are, therefore, weakly affected by the Stark shift while higher-lying levels are more significantly shifted. For example, the AC Stark shift of the 1s level of hydrogen is a well-known result:

$$\Delta\varepsilon_S(1s) = -\frac{9}{8}E_0^2 a_B^3. \tag{4.67}$$

The Stark shift of levels near the continuum can be estimated from Eq. (4.66) by taking the opposite limit in the denominator, $\omega \gg \omega_{ij}$ and approximating the transition energy in the numerator with the atomic transition frequency, $\omega_{ij} \approx e^2/2\hbar x_{ij}$; $x_{ij} \approx a_B$. High-lying levels, then, will be roughly shifted by

$$\Delta\varepsilon_s \approx e^2 E_0^2/4m\omega^2 = U_p. \tag{4.68}$$

The upper levels will be shifted upward in energy by an amount given simply by the ponderomotive energy. This shift applies not only to the positions of upper Rydberg levels but to the position of the continuum limit as well (more on this in the next section). To illustrate the point, if we considered a 1 μm laser field focused to peak intensity near 10^{14} W/cm^2,

Eq. (4.68) suggests that the atomic ground state may be shifted by only 0.009 eV, but that the upper states (and level of the ionization limit) are shifted by as much as 10 eV. So the positions of resonances can be shifted by energies of many eV, well over the energy of photons employed in most experiments. This analysis also justifies why it is accurate to use unperturbed atomic ground states in the calculation of strong field ionization rates in near-IR or optical laser fields.

The AC Stark shifting of upper energy levels has a number of important consequences in the strong field ionization in intensities in the vicinity of 10^{13}–10^{14} W/cm^2. Figure 4.11 is a cartoon of the effective nonlinear order of the ionization rate ($w \sim I^q$) as a function of the detuning of a laser frequency near a q_1-photon resonance. Far from the resonance, the

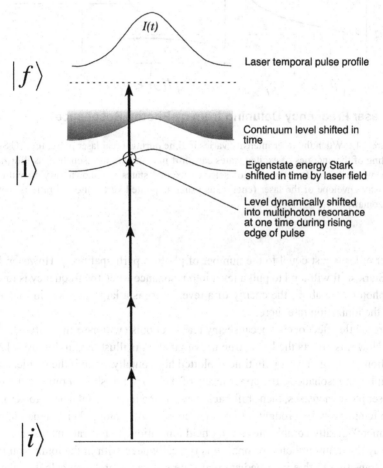

Figure 4.11 The effective nonlinearity of ionization, q' such that $w \sim I^{q'}$ as a function of frequency detuning near a multiphoton resonance. The sharp increase in nonlinearity near the resonance can be attributed to AC Stark shifting of the level into resonance with increasing intensity when the incident field is just slightly away from the optimum resonance frequency.

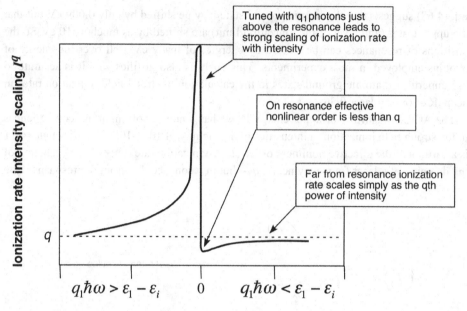

Figure 4.12 When the laser intensity varies in time during a real laser pulse, the AC Stark shifting of the energies of excited states can shift into and out of resonance, as illustrated here, where the shifted energies of selected excited states are shown varying with the intensity envelope of the laser (energy, as usual, is plotted vertically, and time is plotted horizontally).

nonlinear order is just equal to the number of photons participating, q. However, because the AC Stark shift will tend to pull a level into resonance when the frequency is tuned such that q_1 photons are above the energy of a level, there is a large increase in the nonlinear order of the ionization rate there.

A more subtle effect occurs because any real laser pulse will have intensity which varies in time. How this affects the level structure of an atom is illustrated in Figure 4.12, where the position of energy levels with time is plotted horizontally. Even if the incident field has no multiphoton resonances, the upper levels of the atom will shift in time as the intensity of the laser pulse increases, then shift back down as the intensity falls back to zero. A level can then temporarily be brought into resonance with q_1 photons. This "dynamic resonant enhancement" greatly complicates strong field ionization at moderate intensities.

The way that a multiphoton resonance is typically dealt with in the ionization rate is to treat the transition to this intermediate real state separately and then calculate the ionization rate by q_2 photons out of that excited state into the continuum (such that $q_1 + q_2 = q$). The population in the resonant level is determined by the multiphoton generalization of Rabi oscillations (Milonni and Eberly 2010, p. 191), which are most conveniently calculated by the density matrix formalism and the multiphoton version of the optical Bloch

4.2 Multiphoton Ionization and Lowest-Order Perturbation Theory

equations (Lambropoulos and Tang 1992). The slowly varying density matrix elements, $\tilde{\rho}_{mn}$, are calculated in the usual way from the density matrix, ρ_{mn} by $\tilde{\rho}_{mn} = \rho_{mn}(t)e^{-iq_1\omega t}$. The population density of the mth state is just $\tilde{\rho}_{mm}$. If the 0th state is the ground state and we have only one resonant level, denoted with index 1, these elements evolve by the usual optical Bloch equations as (where the Γ_i factors denote decay rates from a state)

$$\frac{d\tilde{\rho}_{00}}{dt} = \operatorname{Im}[\Omega_{01}\tilde{\rho}_{10}] - \Gamma_0\tilde{\rho}_{00} \tag{4.69a}$$

$$\frac{d\tilde{\rho}_{11}}{dt} = -\Gamma_1\tilde{\rho}_{11} - \operatorname{Im}[\Omega_{01}\tilde{\rho}_{10}] \tag{4.69b}$$

$$\frac{d\tilde{\rho}_{10}}{dt} = \frac{i}{2}\Omega_{01}(\tilde{\rho}_{11} - \tilde{\rho}_{00}) + i\Delta\omega\tilde{\rho}_{10} - \frac{\Gamma_1}{2}\tilde{\rho}_{10}. \tag{4.69c}$$

The multiphoton Rabi frequency is

$$\Omega_{01} = x_{10}^{(q_1)} \frac{(eE_0)^{q_1}}{\hbar}, \tag{4.70}$$

which is proportional to the multiphoton dipole matrix element

$$x_{10}^{(q)} = (-i)^{q+1}\left(\frac{1}{2}\right)^q \int \sum_{j_1} \int \sum_{j_2} \cdots \int \sum_{j_{q-1}}$$

$$\times \frac{\langle 1|x|j_{q-1}\rangle \ldots \langle j_2|x|j_1\rangle \langle j_1|x|0\rangle}{(\varepsilon_{j_{q-1}} - \varepsilon_0 - (q-1)\hbar\omega) \ldots (\varepsilon_{j_2} - \varepsilon_0 - 2\hbar\omega)(\varepsilon_{j_1} - \varepsilon_0 - 1\hbar\omega)}. \tag{4.71}$$

The decay rates out of the two levels can be taken to be the nonresonant multiphoton ionization rate out of these levels so that

$$\Gamma_0 = \sigma^{(q)}\Phi_{\hbar\omega}{}^q, \tag{4.72a}$$

$$\Gamma_1 = \sigma^{(q_2)}\Phi_{\hbar\omega}{}^{q_2}. \tag{4.72b}$$

The field detuning needs to account for the AC Stark shift of the two levels considered and can be written $\Delta\omega \equiv q_1\omega - \omega_{10} - \Delta\varepsilon_S(1)/\hbar + \Delta\varepsilon_S(0)/\hbar$ where the Stark shift of the ith level is $\Delta\varepsilon_S(i)$. The ionization probability, then, is found by calculating the probability density left in the two resonant states:

$$P_{ion}(t) = 1 - \tilde{\rho}_{00} - \tilde{\rho}_{11}. \tag{4.73}$$

This theoretical approach to REMPI has been successfully applied to a wide range of experiments. The practical consequences of REMPI in the observed ionization rate as a function of incident laser wavelength is illustrated in Figure 4.13, which is an adaptation of data on UV ionization of Kr. At lower intensities ($<10^{13}$ W/cm^2) resonances in Kr lead to ionization rate enhancements at sharp, well-defined wavelengths (top plot). However, as the intensity approaches the strong field regime near 10^{14} W/cm^2, the dynamic AC Stark shift smears these sharp resonances out to yield broad frequency structure. Ultimately, at high enough peak ponderomotive energies, all wavelengths will be resonant for some period of time.

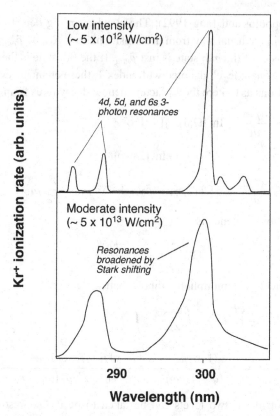

Figure 4.13 Example of resonant MPI rates as a function of wavelength in Kr irradiated by UV pulses. Top plot shows fine atomic structure from three-photon resonances with various levels around the 4d, 5d and 6s manifolds. The bottom plot shows how the structure washes out at higher intensity because of dynamic Stark shifting of these resonances in the time-varying pulse. Plots were generated based on data of Perry and Landen (1988).

4.2.7 MPI beyond LOPT: Strong Coupling in the Continuum

Investigating MPI in the LOPT framework, has limitations and is of limited quantitative use in the strong field regime. We have already seen that LOPT breaks down at intensities much above 10^{13} W/cm^2 in near-IR fields or above $\sim 10^{14}$ W/cm^2 in UV laser pulses. Nonetheless, the perturbation theory picture can give us insight into multiphoton physics even beyond the quantitative limits of LOPT. We can, for example, get a qualitative sense of the nature of strong field ionization at intensities beyond the applicability limit of LOPT by considering the perturbation expansion to even higher orders.

The transition amplitude of an electron to the continuum as found from the infinite series expansion of the propagation operator is, after all, formally an exact solution of the Schrödinger equation for an electron bound in an atomic potential subject to a laser field, as long as the series converges. It was our truncation of the perturbation series at the lowest-order term yielding ionization that was the first significant approximation. If we

4.2 Multiphoton Ionization and Lowest-Order Perturbation Theory

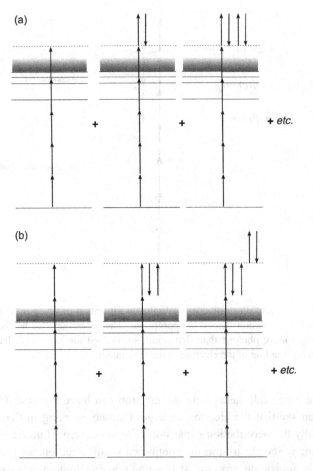

Figure 4.14 Illustration of higher-order MPI terms which can become important in the perturbation expansion as the laser intensity becomes large. **(a)** are terms which contribute to the transition amplitude of the minimum electron energy ionization channel. **(b)** illustrates terms which will contribute to additional ionization channels, which will have higher outgoing electron energies.

consider terms beyond this lowest order, the higher-order terms in the perturbation expansion correspond to multiphoton processes like those depicted in Figure 4.14. First, the lowest energy ionization channel (that treated by the lowest-order term) will be augmented by additional terms involving two, or four or more photons (like the diagrams shown in the top row of Figure 4.14). Furthermore, this is not the only ionization channel possible. Higher electron energies in the continuum are also possible, represented by higher terms in the perturbation expansion like those depicted in the bottom row of 4.14.

In a sense, the situation depicted by Figure 4.15 develops. As the incident intensity gets larger, the ionizing electron begins to undergo numerous transitions in the continuum. As the photon density increases, these continuum transitions are possible because they

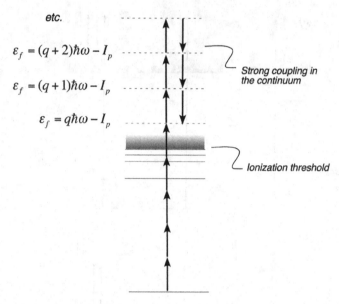

Figure 4.15 Schematic illustration of multiphoton ionization of an atom at an intensity beyond which perturbation theory is appropriate. The ionization can be thought of as involving many more photons than that minimum needed for ionization, leading to an effective strong coupling of the electron in the continuum.

are driven on a time scale faster than the electron can leave the vicinity of the parent nucleus. We can say that the electron undergoes strong coupling in the continuum. Of course, eventually the perturbation expansion fails to converge, but the picture of many photons coupling to the electron in the continuum is still a reasonable picture of what is happening in the ionization process. At this point, a very large number of photons will participate in the ionization process and we expect that the yield scaling with intensity will deviate significantly from the LOPT prediction. Obviously a different theoretical approach is needed in this higher intensity regime.

4.3 The Strong Field Approximation

Calculation of ionization rates of atoms in the high-intensity, "nonperturbative" regime beyond LOPT is one of the most widely pursued efforts in strong field physics since the late 1970s, with history well summarized in Agostini and DiMauro (2008) and Popruzhenko (2014). Numerous theoretical approaches are possible. Direct numerical integration of the Schrödinger equation is one possibility (Kulander *et al.* 1993; Burnett *et al.* 1993; Peng *et al.* 2003; Chirila and Potvliege 2005). Floquet methods are another (Potvliege and Shakeshaft 1992; Burnett *et al.* 1993; Potvliege and Vucic 2009), and one which we will visit in the next chapter. Both of these approaches, however, ultimately demand numerical computations; an analytic theory of strong field ionization is desirable to give us insight into the physical process at work.

4.3 The Strong Field Approximation

The most widely used analytic approach to the strong field ionization problem was first presented by Keldysh in his pioneering paper (Keldysh 1965). The Keldysh approach was later expounded upon by Faisal (1973) and Reiss (1980), which led to what generally became known as the Keldysh–Faisal–Reiss (KFR) theory. Since the early 1980s, the KFR theoretical framework has been improved and adapted to numerous problems in strong field physics (Brandi et al. 1981; Karnakov et al. 2009). The general theoretical approach has come to be known simply as the Strong Field Approximation, or SFA.

The essential element of the SFA is to continue to rely on a perturbation theory approach to the strong field problem, but instead of treating the light field as a weak perturbation on the atomic binding potential, the SFA treats the unperturbed Hamiltonian as that containing the strong light field and assumes, on the other hand, that the binding potential of the atom is the small perturbation. This central approximation is the core of Keldysh's original approach. Furthermore, the SFA assumes that the population density in the ground states undergoes negligible depletion during the ionization process.

Frankly, both of these approximations lack rigorous justification. At the end, they lead to results which are often not particularly quantitatively accurate, though, with corrections, the SFA can yield accurate results in some regimes (such as ionization at high intensities in long wavelength pulses). Despite the dubious rigor and accuracy of these approximations, the SFA elegantly extends the LOPT MPI approach to higher intensities and gives excellent insight into strong field behavior when many photons participate. It bridges the picture of multiphoton processes to the quasi-classical picture of tunnel ionization, and it presciently predicts high-order, above-threshold ionization. As such, the SFA is a superb theoretical framework for understanding strong field ionization at high intensity and is widely employed in the literature. The SFA is often called a nonperturbative theory, though, strictly speaking, this is not correct since a perturbation expansion is employed. It is, however, nonperturbative with respect to the driving field, and, therefore, is applicable, in principle, to very high intensity.

4.3.1 The SFA Perturbation Expansion and the Keldysh Approximation

It is possible to derive the SFA in a concise manner using S-matrix formalism; see Becker and Faisal (2005) and Reiss (1980). We will, however, forgo this formalism, sacrificing its elegance for a slightly more intuitive approach, which closely follows the derivation of Milonni and Sundaram (1993). We proceed based on the two central approximations of the SFA: (1) the atomic potential is treated as a weak perturbation and (2) the ground state of the atom is negligibly depleted. Based on approximation (1), it makes sense to write the Hamiltonian of the irradiated electron

$$H = \frac{\hat{P}^2}{2m} + H_I + V(\mathbf{x}) \tag{4.74}$$

with the usual momentum operator $\hat{\mathbf{P}} = -i\hbar\nabla$. Though Keldysh initially employed the length form of the interaction Hamiltonian, we will follow the work of most subsequent authors and employ the $\mathbf{A} \cdot \mathbf{p}$ form. This choice of interaction Hamiltonian admittedly leads to gauge invariance problems with the theory, but this turns out to be of no real practical consequence (Bauer 2008; Popruzhenko 2014). The interaction Hamiltonian in the dipole approximation, then, is

$$H_I = -i\frac{\hbar e}{mc}\mathbf{A}\cdot\nabla + \frac{e^2}{2mc^2}A^2, \quad (4.75)$$

which lets us write the unperturbed Hamiltonian as

$$H_0 = \frac{\hat{p}^2}{2m} + H_I. \quad (4.76)$$

(Note that H_0 is different than when we considered traditional perturbation theory and is now time dependent.) Following the same perturbation expansion approach described in the previous section, the transition amplitude to the final (time-varying) free state, $f(t)$, is in terms of the time-propagated initial eigenstate

$$A_{fi}(t) = \langle f(t)|\,\psi(t)\rangle = \langle f(t)|\,e^{-iH_0 t/\hbar}\,|\chi(t)\rangle. \quad (4.77)$$

We have explicitly noted that the final state varies in time because the electron is born into a strong oscillating field. Recalling Eq. (4.15), the transition amplitude can be written in terms of a time propagator, $u(t)$, and the initial state of the atom

$$A_{fi}(t) = \langle f(t)|\,e^{-iH_0 t/\hbar} u(t)\,|i\rangle. \quad (4.78)$$

As before, recalling Eq. (4.9), we can write the modified perturbing Hamiltonian term, but now the perturbation arises from the atomic binding potential $V(\mathbf{x})$

$$\hat{V}(\mathbf{x},t) = e^{iH_0 t/\hbar}V(x)e^{-iH_0 t/\hbar}. \quad (4.79)$$

Note that even though the perturbing potential is time independent, the modified perturbing operator is time dependent. Recalling the sequence of steps we used to derive the LOPT transition amplitude, Eqs. (4.14) to (4.18), we can write the SFA perturbation series for the propagation operator

$$u(t) = 1 - \frac{i}{\hbar}\int_{-\infty}^{t} dt'\,\hat{V}(t') + \left(\frac{i}{\hbar}\right)^2\int_{-\infty}^{t} dt'\int_{-\infty}^{t'} dt''\,\hat{V}(t'')\hat{V}(t') + \ldots. \quad (4.80)$$

The second term in this series yields the first-order transition amplitude

$$A_{fi}^{(1)}(t) = -\frac{i}{\hbar}\int_{-\infty}^{t} dt'\,\langle f(t)|\,e^{-iH_0 t/\hbar}\hat{V}(t')\,|i\rangle. \quad (4.81)$$

Now we have the exponentiated unperturbed Hamiltonian terms, H_0, with which to deal. We have already found the eigenstates for this Hamiltonian describing an electron in free space subject to an oscillating light field; they are the Volkov wavefunctions for a free electron in the laser field derived in Chapter 3. This makes the operation on the final state straightforward if we choose this state to be a Volkov state, which satisfies

$$H_0\,|\psi_V(t)\rangle = \varepsilon_f\,|\psi_V(t)\rangle. \quad (4.82)$$

4.3 The Strong Field Approximation

Recall that the Volkov states, because they are eigenfunctions of H_0 (and they depend on electron canonical momentum $\hbar \mathbf{k}_e$, though we have not explicitly stated it here), represent a complete set of states. The choice of a Volkov state for the final state of the ionized electron, $f(t) = \psi_V(t)$, is one of the key steps in the SFA as first derived by Keldysh. It is a significant approximation, as this choice means that the effects of the Coulomb force from the residual parent ion are completely ignored.

Using the completeness of the Volkov states we can, in Eq. (4.81), insert the identity

$$\int d^3 \mathbf{k}'_e \, |\psi_V(\mathbf{k}'_e, t')\rangle \langle \psi_V(\mathbf{k}'_e, t')| = 1, \quad (4.83)$$

use the fact that $\langle \psi_V(\mathbf{k}_e, t) | \psi_V(\mathbf{k}'_e, t') \rangle = \delta(\mathbf{k}_e - \mathbf{k}'_e)\delta(t - t')$, and use Eq. (4.79), to yield for the first-order transition amplitude

$$A_{fi}^{(1)}(t) = -\frac{i}{\hbar} \int_{-\infty}^{t} dt' \, \langle \psi_V(t') | V(\mathbf{x}) e^{-iH_0 t'/\hbar} | i \rangle. \quad (4.84)$$

Next we need to deal with the term $e^{-iH_0 t'/\hbar} |i\rangle$. The initial state, which is presumably the unperturbed ground state of the atom or ion, is not an eigenfunction of H_0. So we could expand the state subject to this operator in a series of complete states:

$$|\psi_i(t)\rangle = e^{-iH_0 t/\hbar} |i\rangle = \sum_j a_j e^{-i\varepsilon_j t/\hbar} |j\rangle. \quad (4.85)$$

Using the second assumption on which the SFA is based, we can say that the atom has a small probability of leaving the ground state (a dubious assumption, which we use anyway). In that case, all terms other than those proportional to the initial ground state in the sum of (4.85) are dropped, letting us say

$$|\psi_i(t)\rangle \cong e^{iI_p t/\hbar} |i\rangle. \quad (4.86)$$

Now the transition amplitude is

$$A_{fi}^{(1)}(t) \cong -\frac{i}{\hbar} \int_{-\infty}^{t} dt' \, \langle \psi_V(t') | V(\mathbf{x}) | \psi_i(t') \rangle. \quad (4.87)$$

Since ψ_i is an eigenfunction of $\hat{P}^2/2m + V(\mathbf{x})$, it satisfies the Schrödinger equation of the free atom

$$\left(\frac{\hat{P}^2}{2m} + V(\mathbf{x}) \right) \psi_i(t) = i\hbar \frac{\partial \psi_i(t)}{\partial t}, \quad (4.88)$$

so, Eq. (4.87) can be rewritten

$$A_{fi}^{(1)}(t) \cong -\frac{i}{\hbar} \int_{-\infty}^{t} dt' \, \langle \psi_V(t') | \left(i\hbar \frac{\partial}{\partial t} - \frac{\hat{P}^2}{2m} \right) | \psi_i(t') \rangle. \quad (4.89)$$

Integration by parts and use of the Schrödinger equation to replace the time derivative of the Volkov state with the operator, H_0, gives

$$A_{fi}^{(1)}(t) \cong -\frac{i}{\hbar}\left[i\hbar\left\langle\psi_V(t')\psi_i(t')\right\rangle\bigg|_{-\infty}^{t} + \int_{-\infty}^{t} dt'\left\langle\psi_V(t')\right|\left(H_0 - \hat{P}^2/2m\right)\left|\psi_i(t')\right\rangle\right]. \tag{4.90}$$

The first term in (4.90) is bounded as $t \to \infty$ and can be ignored. Using Eq. (4.76) gives our final result for the first-order SFA transition amplitude

$$A_{fi}^{(1)}(t) = -\frac{i}{\hbar}\int_{-\infty}^{t} dt'\left\langle\psi_V(t')\right|H_I(t')\left|\psi_i(t')\right\rangle. \tag{4.91}$$

The transition amplitude as written in Eq. (4.91) is known as the Keldysh approximation. It describes the probability amplitude for a transition from the ground state of the atom to a free Volkov state with momentum, $\hbar k_e$. Note that the field enters into this transition amplitude in two ways: through the interaction Hamiltonian and in the final, Volkov state. All atomic physics is contained simply in the initial state of the atom; no intermediate states come into play. It is curious to note that if we replace the Volkov final state in (4.91) with a simple free electron continuum state, we actually retrieve the result of first-order perturbation theory. However, despite this appearance of similarity, the SFA is NOT a first-order expansion in the applied field; it is the first-order expansion with respect to the atomic binding potential and, as such, incorporates the laser field to all orders.

The SFA as embodied in Eq. (4.91) does have a number of significant weaknesses. One we have already commented on is the neglect of depletion of the ground state. In virtually any real ionization situation of interest this approximation does not hold. Treating the initial state as an unperturbed eigenstate becomes a poor approximation when the ponderomotive energy of the field approaches and exceeds the binding potential of the atom because the field begins to affect that initial ground state at that point. This problem means that there is an upper limit to the intensity in which (4.91) is valid, though in practice this does not cause major inaccuracies for relevant experimental parameters.

Most significant, however, is the treatment of the outgoing wavefunction as a pure Volkov wave and ignoring completely the influence of the Coulomb force of the residual ion. This approximation has the gravest consequences. In fact, neglect of the Coulomb force means that Eq. (4.91) tends to underestimate the actual ionization rate in many circumstances by one to two orders of magnitude (at least as determined by Floquet theory computational calculations of ionization) (Becker et al. 2001). The need for a Coulomb correction tends to be less important in longer wavelength fields, where the ponderomotive energy is more dominant, reducing the effect of $V(\mathbf{x})$ on the outgoing electron. It also tends to be less of an issue in circularly polarized fields (which we address in a few sections), largely because the increase in angular momentum associated with the absorption of many photons pushes the wavefunction of the outgoing electron away from the nucleus, decreasing the importance of the Coulomb force. Consequently, the SFA finds its most accurate applications in those two situations. In fact, the issue of the Coulomb effect on the outbound electron can be

4.3 The Strong Field Approximation

addressed with reasonable accuracy by a simple correction to the final wavefunction using the WKB approximation (Krainov 1997; Popov 2004). This we will present shortly.

Despite these shortcomings, Eq. (4.91) contains a remarkable amount of physics in a shockingly concise, simple formula. The SFA in this simple form has been used to predict photo-electron spectra and ionization rates from a number of noble gases with a range of incident wavelengths, with particular success in circularly polarized fields. Most importantly, despite its quantitative inaccuracies, Eq. (4.91) will provide an elegant bridge from modest intensity to high intensity and will give us insight into what happens when a large number of photons participate in the ionization of the atom.

4.3.2 Evaluation of the SFA Ionization Transition Amplitude

To evaluate Eq. (4.91) we proceed with the velocity form of the interaction Hamiltonian. The appropriate form of the Volkov wavefunction (with delta-function normalization) was found in Section 3.7.2. Recalling Eqs. (3.343)–(3.346):

$$\psi_V = \frac{1}{(2\pi)^{3/2}} \exp\left[-\frac{i}{2m\hbar}\int\left(\hbar\mathbf{k}_e + \frac{e}{c}\mathbf{A}(t)\right)^2 dt - i\mathbf{k}_e\cdot\mathbf{x}\right]$$

$$= \frac{1}{(2\pi)^{3/2}} \exp\left[-i\left(\varepsilon_f + U_p\right)\frac{t}{\hbar} + i\frac{e}{mc\omega}A_0\hat{\varepsilon}\cdot\mathbf{k}_e\sin\omega t - i\frac{U_p}{2\hbar\omega}\sin 2\omega t - i\mathbf{k}_e\cdot\mathbf{x}\right]. \quad (4.92)$$

(We have shifted the phase of the field by $\pi/2$ from Eq. (3.346) for future convenience.) Leveraging the properties of this Volkov final state, we can make the following manipulations with the interaction Hamiltonian:

$$-H_I|\psi_V\rangle = \left(i\hbar\frac{\partial}{\partial t} + \frac{\hbar^2}{2m}\nabla^2\right)|\psi_V\rangle$$

$$= \left(i\hbar\frac{\partial}{\partial t} + \frac{\hbar^2 k_e^2}{2m}\right)|\psi_V\rangle = \left(i\hbar\frac{\partial}{\partial t} + \varepsilon_f\right)|\psi_V\rangle. \quad (4.93)$$

Following the same sequence that the KFR papers follow, we now Fourier-decompose the Volkov wavefunction, using the Jacobi–Anger identity, which we first employed in Chapter 3, which states that

$$\exp[iz\sin\theta] = \sum_{n=-\infty}^{\infty} J_n(z)\exp[in\theta] \quad (4.94)$$

and means that the Volkov state, (4.92), can be written in terms of a double sum over Bessel functions

$$\psi_V = \frac{1}{(2\pi)^{3/2}} \exp\left[-i\left(\varepsilon_f + U_p\right)\frac{t}{\hbar} - i\mathbf{k}_e\cdot\mathbf{x}\right]$$

$$\times \sum_{r=-\infty}^{\infty} J_r\left(\frac{eA_0 k_{ex}}{mc\omega}\right)\exp[ir\omega t] \sum_{n=-\infty}^{\infty} J_n\left(\frac{U_p}{2\hbar\omega}\right)\exp[in\omega t], \quad (4.95)$$

where we have introduced the component of the outgoing electron momentum along the polarization vector $\hat{\varepsilon} \cdot \mathbf{k}_e = k_{ex}$. Shifting indices in the sum with $q \equiv r - 2n$ gives for the Volkov state:

$$\psi_V = \frac{1}{(2\pi)^{3/2}} \sum_{q=-\infty}^{\infty} \exp\left[-i\left(\varepsilon_f + U_p - q\hbar\omega\right)\frac{t}{\hbar} - i\mathbf{k}_e \cdot \mathbf{x}\right] \quad (4.96)$$

$$\times \sum_{n=-\infty}^{\infty} J_{q+2n}\left(\frac{eA_0 k_{ex}}{mc\omega}\right) J_n\left(\frac{U_p}{2\hbar\omega}\right)$$

Equation (4.93) with (4.96) gives

$$-H_I |\psi_V\rangle = (U_p - q\hbar\omega) |\psi_V\rangle, \quad (4.97)$$

which allows us now to evaluate the transition amplitude explicitly:

$$A_{fi}^{(1)}(t) = -\frac{i}{\hbar} \int_{-\infty}^{t} dt' \, (U_p - q\hbar\omega) \langle \psi_V(t') \psi_i(t') \rangle \quad (4.98)$$

$$A_{fi}^{(1)}(t) = -\frac{i}{\hbar} \int_{-\infty}^{t} dt' \sum_{q=-\infty}^{\infty} (U_p - q\hbar\omega) \exp\left[-\frac{i}{\hbar}\left(\varepsilon_f + U_p + I_p - q\hbar\omega\right)\right]$$

$$\sum_{n=-\infty}^{\infty} J_{q+2n}\left(\frac{eA_0 k_{ex}}{mc\omega}\right) J_n\left(\frac{U_p}{2\hbar\omega}\right) \langle e^{i\mathbf{k}_e \cdot \mathbf{x}} | i\rangle. \quad (4.99)$$

The projection of the plane-wave part of the final state on the initial state is simply the Fourier transform of the initial state, which we write

$$\langle e^{i\mathbf{k}_e \cdot \mathbf{x}} | i\rangle = \frac{1}{(2\pi)^{3/2}} \int \psi_i(0) e^{i\mathbf{k}_e \cdot \mathbf{x}} d^3\mathbf{x}$$

$$\equiv \phi(\mathbf{k}_e). \quad (4.100)$$

As we did before in our LOPT analysis, we extend the time integration to infinity to give the explicit transition amplitude in the velocity gauge:

$$A_{fi}^{(1)} = -2\pi i \sum_{q=q_{min}}^{\infty} (U_p - q\hbar\omega) \left[\sum_{n=-\infty}^{\infty} J_{q+2n}\left(\frac{eA_0 k_{ex}}{mc\omega}\right) J_n\left(\frac{U_p}{2\hbar\omega}\right)\right] \phi_i(\mathbf{k}_e)$$

$$\times \delta\left(\varepsilon_f + U_p + I_p - q\hbar\omega\right). \quad (4.101)$$

Because of the presence of the delta function, the first sum is taken from the first integer, q_{min}, that fulfills the condition $\varepsilon_f = q\hbar\omega - U_p - I_p$. The summed function in brackets of Eq. (4.101) is the weighting function for the various ionization channels and is referred to by some authors as the generalized Bessel function of three arguments (Faisal 1973):

$$J_q(a,b) \equiv \sum_{n=-\infty}^{\infty} J_{q+2n}(a) J_n(b). \quad (4.102)$$

Reiss (1980) discusses the properties of this function in an appendix to his original article on the SFA.

4.3 The Strong Field Approximation

To conclude this evaluation of the Keldysh amplitude, we need to find the ionization rate, which, as we discussed in Section 4.2, is

$$W_{fi} = \lim_{\tau \to \infty} \frac{|A'_{fi}|^2}{\tau}, \qquad (4.103)$$

so we use the property of the energy-conserving delta function

$$\delta(z) = \lim_{\tau \to \infty} \tau \quad \text{when } z = 0, \qquad (4.104)$$

and then, after multiplying by the density of states in the continuum, we integrate over final electron energy to get for the ionization rate

$$W_{fi}^{(SFA)}(\hat{\mathbf{k}}_e) = 2\pi \frac{m}{\hbar^3} \sum_{q=q_{min}}^{\infty} k_f(q) \left(U_p - q\hbar\omega\right)^2$$

$$\times \left[\sum_{n=-\infty}^{\infty} J_{q+2n}\left(\frac{eA_0 k_{ex}}{mc\omega}\right) J_n\left(\frac{U_p}{2\hbar\omega}\right) \right]^2 |\phi_i(\hat{\mathbf{k}}_e)|^2. \qquad (4.105)$$

The final electron momentum is imposed by the energy-conserving delta function and is

$$k_f(q) = \frac{\sqrt{2m(q\hbar\omega - U_p - I_p)}}{\hbar}. \qquad (4.106)$$

The total ionization rate is found from this differential rate by integrating over all possible directions for the outgoing electron:

$$w^{(SFA)} = \int W_{fi}^{(SFA)}(\hat{\mathbf{k}}_e) \sin\theta \, d\theta \, d\varphi. \qquad (4.107)$$

The SFA ionization rate, as embodied in Eq. (4.105), contains a remarkable amount of physics and gives us profound insight into the nature of strong field ionization. The most significant result is that the total ionization rate is given by the sum over index q, each term of which can be associated with the absorption of q photons, proven by examination of the delta function in Eq. (4.101). As we suspected from our discussion of Section 4.2.7, in the strong field regime the ionization rate exhibits strong coupling in the continuum and is a result of the absorption of many photons, including absorption pathways that involve absorption of more photons than the minimum required for ionization. In fact, the SFA calculation tells us that ALL photon absorption pathways contribute, from the absorption of the minimum number of photons needed to free the electron all the way up to an infinite number of photons. The contribution of each of these absorption channels to the ionization rate is weighted by the square of the generalized Bessel function, which tends to make the first few terms the most significant. However, as the intensity increases (and A_0 increases), the Bessel terms at higher photon number become more significant (see the plot in Figure 4.16 of the magnitude of each term for ionization of He). The SFA is an elegant illustration of the strongly coupled continuum physics discussed qualitatively in Section 4.2.7.

Each term in the sum associated with the absorption of a different number of photons results in a different final state energy for the electron. The next significant finding of

(4.105) is that the energy of the electron is equal to the number of photons absorbed, minus the ionization potential AND minus the ponderomotive energy, $\varepsilon_f = q\hbar\omega - U_p - I_p$. In fact, the minimum number of photons that can participate in the ionization is the number of photons which gives the electron enough energy to overcome not only the binding potential of the atom, but the ponderomotive energy of the electron once it makes it into the continuum. This is not surprising in the context of our discussion of Section 4.2.6 (see Figure 4.12 in particular). We have already seen that the position of the ionization continuum is shifted up in energy by an amount equal to the ponderomotive energy. We can view this physics slightly differently. The amount of energy that the electron must absorb from the laser field to become free is the energy needed to overcome the binding energy of the atom *plus* the energy that that electron must have when it has been liberated into the vacuum where it oscillates with an energy equal to the ponderomotive energy. The only free states available to the electron are those in which it possesses time-averaged kinetic energy equal to the ponderomotive energy. Of course, the electron did not have this "wiggle" energy initially when it was confined in its deeply bound state. It needs to acquire the energy from the field during the ionization process.

It is also interesting to note that the entirety of the atomic physics in the SFA is contained completely in the Fourier transform of the initial bound state. As we have already observed, the SFA effectively ignores the effects of intermediate states on the ionization process, in stark contrast to the predictions of perturbation theory when treating the laser field as the perturbation. This arises from our treatment of the binding potential as a first-order perturbation, an approximation which essentially led to us throwing out any internal structure of the atom. The fact that the ionization rate of each multiphoton channel is proportional to the square of the Fourier transform of the initial electron bound wavefunction has a physical meaning that arises from conservation of momentum. The 3D spatial Fourier transform essentially tells us the spectrum of momenta that the bound electron has initially in the atom. Since, in the dipole approximation, the laser field can impart no momentum (and even outside this approximation its magnitude is insignificant) the *only* momenta that the outgoing electron can have are determined by the spectra of momenta available to the electron in its initial state.

Ultimately, the SFA in the form presented in Eq. (4.105) is not particularly accurate for linearly polarized fields, for reasons already discussed. It ignores the Coulomb field on the outgoing electron, and it ignores the effects of resonances with intermediate states. Furthermore, its accuracy is degraded when ionization depletion of the atom becomes significant, and it has the pitfalls of any first-order perturbation theory. Nonetheless, the essence of the physics contained in this formula has been well established in experiment and more sophisticated theories. The participation of many photon channels in ionization, the ponderomotive shift of the continuum, and the prediction of electrons leaving the ionized atom with a range of final energies are all well-established aspects of strong field ionization. The SFA prediction also gives reasonable answers for the scaling of ionization yield with increasing intensity, even if it underestimates the total magnitude of the ionization rate when uncorrected for the Coulomb force impact on the outgoing electron. The SFA also plays another important role in our understanding of strong field ionization: it bridges the

4.3 The Strong Field Approximation

view of the laser as driving ionization by many photons (multiphoton absorption) in the weak field regime, with a view of the laser as a slowly varying, purely classical field at high fields, as we shall demonstrate.

4.3.3 The SFA in the Weak Field Limit

Briefly consider the SFA ionization rate in the limit of weak fields. Looking at the Bessel function arguments, it is clear that the "weak field limit" in this case means that

$$\frac{eA_0 k_{ex}}{m\omega c} \ll 1 \tag{4.108}$$

or, equivalently,

$$v_{osc} \frac{k_{ex}}{\omega} \ll 1,$$

$$\sqrt{U_p \varepsilon_f} \ll \hbar\omega. \tag{4.109}$$

If the final electron energy is on the order of the photon energy, condition (4.109) indicates that the weak field limit is appropriate when the ponderomotive energy is much less than the photon energy. In near-IR fields (where the photon energy is 1–2 eV), this condition implies that the intensity should be below $\sim 10^{13}$ W/cm^2, not too different from our estimation for the validity regime for LOPT.

Taylor expansion of the first Bessel function yields

$$J_q(x) = \sum_{m=0}^{\infty} \frac{(-1)^m}{m!(m+q+1)!} \left(\frac{1}{2}x\right)^{2m+q}. \tag{4.110}$$

When the argument is small, all Bessel functions but J_0 approach zero, so we can ignore all but the $n = 0$ term in (4.105) and the $m = 0$ term in (4.110). This means that the first Bessel function in (4.105) is

$$J_q(x) \cong \frac{1}{(q+1)!}(x/2)^q, \tag{4.111}$$

and $J_0(x) \cong 1$. The SFA ionization rate then scales as

$$W_{fi}^{(SFA)} \sim A_0^{2q_{min}} \sim I^{q_{min}} \tag{4.112}$$

in weak fields. This intensity scaling of ionization yield is identical to the intensity scaling of the LOPT prediction, a satisfying result.

4.3.4 The SFA in the Limit of Many Photons

Of greater interest to us is what the SFA tells us at high intensity, which we now know corresponds to a regime in which many more photons than just q_{min} participate in the ionization of the atom. To make analytic progress in this high-intensity regime, we follow closely the original analysis of Keldysh (1965). Ultimately, we will replace the infinite sum over multiphoton orders with a continuous integration over q, but first, we must deal with the infinite sum in the multiphoton weighting factor. The following identity can be introduced:

$$\sum_{n=-\infty}^{\infty} J_{q+2n}(a) J_n(b) = \frac{1}{2\pi} \int_{-\pi}^{\pi} \exp[iq\vartheta + ia \sin \vartheta + ib \sin 2\vartheta] \, d\vartheta, \qquad (4.113)$$

which gives

$$\sum_{n=-\infty}^{\infty} J_{q+2n}\left(\frac{eA_0 k_{ex}}{m\omega c}\right) J_n\left(\frac{U_p}{2\hbar\omega}\right) = \\ \frac{1}{2\pi} \int_{-\pi}^{\pi} \exp\left[iq\vartheta + i\frac{eA_0 k_{ex}}{m\omega c} \sin \vartheta + i\frac{U_p}{2\hbar\omega} \sin 2\vartheta\right] d\vartheta \qquad (4.114)$$

This integral can be evaluated approximately using the method of steepest descent (often called the saddle point method). When evaluating an integral of the form

$$\int \exp[Af(z)] \, dz$$

where A is large, the contour of integration (from $-\pi$ to π along the real axis in our case) can be deformed by Cauchy's Theorem to pass through the saddle points (maxima) of the integrand in the imaginary plane. The integration can then be approximated, since the dominant contribution to the integral will come in the vicinity of these saddle points. The saddle points are found from $f'(z_0) = 0$, which, in our case, yields the condition

$$f' = q + \frac{eA_0 k_{ex}}{m\omega c} \cos \vartheta + \frac{U_p}{\hbar\omega} \cos 2\vartheta = 0. \qquad (4.115)$$

Solving Eq. (4.115) yields four saddle points,

$$\cos \vartheta_0 = -\frac{e\hbar A_0 k_{ex}}{4mcU_p} \pm \sqrt{\left(\frac{e\hbar A_0 k_{ex}}{4mcU_p}\right)^2 - \frac{q\hbar\omega}{2U_p} + \frac{1}{2}}, \qquad (4.116)$$

two above the real axis and two below. For algebraic convenience we will, at this point, introduce the following factor:

$$\gamma_K \equiv \sqrt{\frac{I_p}{2U_p}}. \qquad (4.117)$$

This parameter will turn out to have an important physical significance, but for the moment we just note that it is proportional to the laser frequency and the inverse of the field amplitude:

$$\gamma_K = \sqrt{2mI_p}\omega/eE_0. \qquad (4.118)$$

The saddle points are

$$\cos \vartheta_0 = -\frac{\gamma_K \hbar k_{ex}}{\sqrt{2mI_p}} \pm i\gamma_K \sqrt{1 + \frac{\hbar^2 k_{e\perp}^2}{2mI_p}}. \qquad (4.119)$$

4.3 The Strong Field Approximation

The saddle point method involves approximating the integral by expanding $f(z)$ around the saddle points to second order and saying that the integration over the contour is approximately (Zwillinger 1992, p. 230)

$$\int_{C'} \exp[Af(z)]\, dz \cong \sum_{\substack{\text{saddle} \\ \text{points} \\ \text{on } C'}} \sqrt{\frac{2\pi}{A\left|f''(z_0^{(i)})\right|}} e^{Af(z_0^{(i)})}. \qquad (4.120)$$

Let us assume that the outgoing electron energy is not too large so $I_P \gg \varepsilon_f$. We then pass the integration contour through the two saddle points above the real axis, which are approximately

$$\cos\vartheta_1 = -\frac{\gamma_K \hbar k_{ex}}{\sqrt{2mI_p}} + i\gamma_K, \quad \cos\vartheta_2 = \frac{\gamma_K \hbar k_{ex}}{\sqrt{2mI_p}} + i\gamma_K. \qquad (4.121)$$

For our specific integral we have

$$f(\vartheta) = iq\vartheta + i\frac{k_{ex}\sqrt{2I_p/m}}{\omega\gamma_K}\sin\vartheta + i\frac{I_p}{4\hbar\omega\gamma_K}\sin 2\vartheta \qquad (4.122a)$$

$$f''(\vartheta) = -i\frac{k_{ex}\sqrt{2I_p/m}}{\omega\gamma_K}\sin\vartheta - i\frac{I_p}{\hbar\omega\gamma_K}\sin 2\vartheta$$

$$\cong -i\frac{I_p}{\hbar\omega\gamma_K}\sin 2\vartheta. \qquad (4.122b)$$

Note that $\cos\vartheta_1 \cong \cos\vartheta_2 \cong -i\gamma_K$, so

$$\vartheta_i \cong \cos^{-1} i\gamma_K = \frac{\pi}{2} + i\ln\left[\gamma_K + \sqrt{1+\gamma_K^2}\right]$$

$$= \frac{\pi}{2} + i\sinh^{-1}\gamma_K. \qquad (4.123)$$

Since $\sin\vartheta_1 \cong (1+\gamma_K^2)^{1/2}$ and $\sin 2\vartheta_1 = 2\cos\vartheta_i \sin\vartheta_i$, the first and third terms in (4.122a) contribute to the real part of the exponential and the middle term gives the integrand its phase, which is

$$\varphi = \frac{k_{ex}\sqrt{2I_p/m}}{\omega\gamma_K}\sqrt{1+\gamma_K^2}. \qquad (4.124)$$

Finally, since

$$|f''(\vartheta_i)| = \frac{2I_p}{\hbar\omega}\frac{\sqrt{1+\gamma_K^2}}{\gamma_K},$$

we can put all of this together in Eq. (4.120) to yield, with (4.114) and (4.101), the SFA ionization rate:

$$W_{fi}^{(SFA)}(\mathbf{k}_e) = \frac{1}{\hbar}\sum_{q=q_{min}}^{\infty} \delta(\varepsilon_f + U_p + I_p - q\hbar\omega)(U_p - q\hbar\omega)^2 |\phi_i(\mathbf{k}_e)|^2 \frac{\hbar\omega}{I_p}\frac{\gamma_K}{\sqrt{1+\gamma_K^2}}$$

$$\times \exp\left[-2q\sinh^{-1}\gamma_K + \frac{I_p}{\hbar\omega}\frac{\sqrt{1+\gamma_K^2}}{\gamma_K} + \frac{\hbar k_{ex}^2}{\omega m}\frac{\gamma_K}{\sqrt{1+\gamma_K^2}}\right]$$

$$\times \left(1 + (-1)^q \cos\left(\frac{k_{ex}\sqrt{8I_p/m}}{\omega}\frac{\sqrt{1+\gamma_K^2}}{\gamma_K}\right)\right). \tag{4.125}$$

Now we perform the sum over all photon orders, which can be approximated as an integration over energy:

$$\sum_q \to \frac{1}{\hbar\omega}\int d(q\hbar\omega).$$

Energy conservation (imposed by the delta function) leads to the following substitutions in terms of the outgoing electron momentum wavenumber, k_e:

$$(U_p - q\hbar\omega)^2 = \left(\frac{\hbar^2 k_e^2}{2m} + I_p\right)^2, \tag{4.126}$$

$$q = \frac{U_p + \varepsilon_f + I_p}{\hbar\omega} = \frac{I_p}{\hbar\omega}\left(1 + \frac{1}{2\gamma_K^2} + \frac{\hbar^2 k_e^2}{2I_p m}\right). \tag{4.127}$$

To perform the integration of the differential ionization rate over $\mathbf{k}_e$ we need an explicit formula for the Fourier transform of the ground state wavefunction. For illustrative purposes, we can assume a hydrogen-like 1s wavefunction with effective nuclear charge Z:

$$\psi_i(0) = \frac{1}{\sqrt{\pi}}\left(\frac{Z}{a_B}\right)^{3/2} e^{-Zr/a_B}. \tag{4.128}$$

Define in terms of the ionization potential an atomic wavenumber, k_{Ip}, such that

$$I_p = \frac{\hbar^2 k_{Ip}^2}{2m}, \tag{4.129}$$

which means that for the 1s ground state of a hydrogen-like ion of charge Z, $k_p = Z/a_B$. With a wavefunction of this form, the Fourier transform is straightforward; integration by parts twice with respect to r gives

$$\varphi_i(\mathbf{k}_e) = \frac{k_p^{3/2}}{(2\pi)^{3/2}}\int \frac{e^{-k_p r}}{\sqrt{\pi}} e^{i\mathbf{k}_e\cdot\mathbf{x}} d^3\mathbf{x} \tag{4.130}$$

$$\varphi_i(\mathbf{k}_e) = \frac{k_p^{3/2}}{(2\pi)^{3/2}}\iiint \exp\left[ik_e\cos\theta r - k_p r\right] r^2 \sin\theta\, dr d\theta d\varphi$$

$$= -\frac{4}{\sqrt{2\pi}}\frac{k_p^{5/2}}{\left(k_e^2 + k_p^2\right)^2}. \tag{4.131}$$

The integration of (4.125) over k_e is simplified by noting that the argument in the cosine function scales like $\sim (I_p/m)^{1/2} k_e/\omega\gamma_K$. Since we know that many photons participate, this is a rapidly varying function of k_e and we can ignore the cosine term as averaging to zero upon integration. We can also say that most of the ionization rate will be due to ionization pathways nearest threshold where $\varepsilon_f \ll I_p$, so we can set $k_e = 0$ in the prefactor of the exponential in (4.125). (It must be conceded that both of these approximations are not rigorously justified. For example, averaging the cosine term results in ignoring certain quantum interferences.) With these approximations the ionization rate becomes

$$W_H^{(SFA)} \cong \frac{4}{\pi^2 k_{Ip}^3} \frac{I_p}{\hbar} \frac{\gamma_K}{\sqrt{1+\gamma_K^2}} \exp\left[-\frac{2I_p}{\hbar\omega}\left(\left(1+\frac{1}{2\gamma_K^2}\right)\sinh^{-1}\gamma_K - \frac{\sqrt{1+\gamma_K^2}}{2\gamma_K}\right)\right]$$

$$\times \int \exp\left[-\frac{\hbar}{m\omega}\left(\sinh^{-1}\gamma_K k_{e\perp}^2 + \left(\sinh^{-1}\gamma_K - \frac{\gamma_K}{\sqrt{1+\gamma_K^2}}\right)k_{ex}^2\right)\right] d^3k_e,$$
(4.132)

where $k_{e\perp}$ is the electron momentum transverse to the polarization axis. In principle, the integration should be taken over electron energies such that $\hbar^2 k_e^2/2m \geq q_{min}\hbar\omega - U_p - I_p$, though we will extend the lower limit of integration over k_e to 0. This formula, in terms of electron momenta parallel and perpendicular to the polarization axis, can be evaluated in cylindrical coordinates with the integrals:

$$\int_0^\infty k_{e\perp} \exp\left[-\alpha k_{e\perp}^2\right] dk_{e\perp} = \frac{1}{2\alpha}$$
(4.133a)

$$\int_0^\infty \exp\left[-\beta k_{ex}^2\right] dk_{ex} = \sqrt{\frac{\pi}{\beta}}$$
(4.133b)

to yield the final SFA ionization rate for a hydrogen-like 1s state:

$$W_H^{(SFA)} \cong \frac{4}{\sqrt{\pi}} \frac{I_p}{\hbar} \left(\frac{\hbar\omega}{2I_p}\right)^{3/2} \frac{\gamma_K}{\sqrt{1+\gamma_K^2}} \frac{1}{\sinh^{-1}\gamma_K} \frac{1}{\left(\sinh^{-1}\gamma_K - \gamma_K\sqrt{1+\gamma_K^2}\right)^{1/2}}$$

$$\times \exp\left[-\frac{2I_p}{\hbar\omega}\left(\left(1+\frac{1}{2\gamma_K^2}\right)\sinh^{-1}\gamma_K - \frac{\sqrt{1+\gamma_K^2}}{2\gamma_K}\right)\right].$$
(4.134)

Let us consider this formula in the weak field limit. From the definition of γ_K in Eq. (4.118), the weak field limit means that this dimensionless parameter must be $\gg 1$. So "weak field" implies that

$$E_0 \ll \sqrt{2mI_p\omega}/e,$$
(4.135)

which is essentially identical to the condition for the breakdown of LOPT that we derived in Eq. (4.53). In this limit

$$\sinh^{-1}\gamma_K = \ln\left[\gamma_K + \sqrt{1+\gamma_K^2}\right] \cong \ln\gamma_K + \ln 2$$

$$\cong \ln\gamma_K,$$
(4.136)

reducing Eq. (4.134) to

$$W_H^{(SFA)}(\gamma_K \gg 1) \cong \frac{4}{\sqrt{\pi}} \frac{I_p}{\hbar} \left(\frac{\hbar\omega}{2I_p}\right)^{3/2} \ln\left[Z\frac{\omega}{\omega_a}\frac{E_a}{E_0}\right]^{-2} \left(\frac{1}{\gamma_K}\right)^{2(I_p+U_p)/\hbar\omega} \quad (4.137)$$

or

$$W_H^{(SFA)}(\gamma_K \gg 1) \cong \frac{4}{\sqrt{\pi}} \frac{I_p}{\hbar} \left(\frac{\hbar\omega}{2I_p}\right)^{3/2} \ln\left[Z\frac{\omega}{\omega_a}\frac{E_a}{E_0}\right]^{-2} \left(\frac{\omega_a}{ZE_a\omega}\right)^{2\alpha} E_0^{2\alpha}. \quad (4.138)$$

This form of the ionization rate takes on a power law scaling with incident intensity with nonlinear order $\alpha = (I_p+U_p)/\hbar\omega \cong q_{\min}$. Again, the approximate form of the SFA predicts an intensity scaling nearly identical to LOPT in the weak field limit.

It is informative at this point to examine the predictions of the SFA from Eqs. (4.105)–(4.107) and its saddle point approximation (4.134). To illustrate, Figure 4.16 shows a numerical calculation of the ionization of the helium atom by intense 800 nm light. Because He has an ionization potential of 24.6 eV, and the laser has a photon energy of 1.55 eV, at least 16 photons would be required to ionize helium, in the absence of any field-induced continuum shift. Figure 4.16a shows the ionization rate of the various terms contributing to the sum in Eq. (4.105) (integrated over angle), which can be attributed to the absorption of different numbers of photons, for three different intensities. We can see that even at the lowest intensity considered (2×10^{14} W/cm^2) the field shift of the continuum limit has limited ionization to terms involving 25 photons or more. Higher intensities have increasingly higher thresholds for the minimum number of photons that need to be absorbed, reaching as high as 58 for intensity $\sim 10^{15}$ W/cm^2. This shifting of the minimum number of photons needed to higher photon numbers at higher intensity is often called "channel closing" as the lower-order processes become "shut off" by the upward shift of the continuum limit (Eberly et al. 1991; Burnett et al. 1993). We will discuss this in greater detail when we consider the electron energy spectra of a process like this.

Clearly, a LOPT view of the ionization of the deeply bound He atom in this near-IR wavelength is of no validity. At the highest intensity it is apparent that processes with 60–80 photons contribute to the ionization rate. Figure 4.16b plots the total ionization rate for He as a function of intensity as found from the summation of plots like those in Figure 4.16a and, for comparison, as found with Eq. (4.134). In the summed rate, the rate exhibits steps with increasing intensity, a consequence of the sequential closing of the lowest-order channel. At low intensity, the prediction of LOPT with a scaling of $w \sim I^{16}$ is shown for comparison. Since many high-field experiments are performed with pulses with duration of $\sim 10^{-13}$ s, the rate predicted by the SFA indicates that peak intensity around 2×10^{15} W/cm^2 is needed to produce significant (or nearly complete) ionization of He to He$^+$. Experiments indicate that such ionization saturation of He actually occurs at intensity around 5×10^{14} W/cm^2 in 100 fs pulses at 800 nm (Walker et al. 1994). This is reasonable agreement with experiment, though the underestimate of the ionization rate by the SFA in linearly polarized fields is well known and results from the approximations we have already discussed. The rollover in the nonlinear order of ionization rate scaling with intensity seen in Figure 4.16b is well documented in experiment. Indeed, the observed nonlinear order

4.3 The Strong Field Approximation

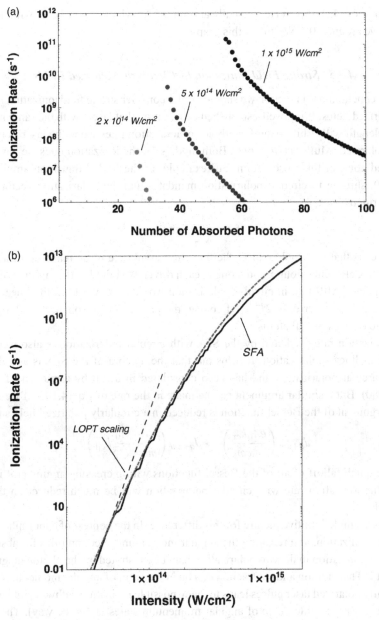

Figure 4.16 (a) The ionization rate for the various multiphoton ionization channels above threshold for MPI of He ($I_p = 24.6$ eV) by 800 nm light at three intensities as calculated by the strong field approximation. A hydrogen-like 1s wavefunction was assumed for the initial wavefunction. (b) Also shown is the total integrated ionization rate calculated by the SFA, integrated over all final electron energies. The solid line shows the sum over all multiphoton ionization channels and the gray dashed line shows the approximation in which an integration over all final energies is performed. The thin dashed line is the intensity scaling predicted by LOPT for 16-photon ionization of He by 800 nm light.

under conditions like this in He is usually around $\sim I^{6-8}$ which is consistent with the slope of the curve around 10^{15} W/cm² in this graph.

4.3.5 Strong Field Ionization by Circularly Polarized Fields

Before concluding this section we should briefly consider strong field ionization in circularly polarized fields. It is a well-established observation that lasers with the same intensity and wavelength will exhibit significantly lower ionization rates when the laser field is circularly polarized (Muller et al. 1992). Both models for the ionization rates we have so far considered confirm this observation. For example, in the LOPT transition amplitude of Eq. (4.30), shifting to circular polarization mandates that the polarization vector should now be written

$$\hat{\varepsilon} \to \frac{1}{\sqrt{2}}(\hat{x} + i\hat{y}).$$

This means that we would expect the transition amplitude to be made up of a sum of dipole matrix elements along x and along y, each driven by a field with amplitude $E_0/2^{1/2}$. So the q-photon MPI rate in circular polarization would be down from the linear polarization rate by the order of $1/2^{q-1}$. Of course, the presence of resonances will affect the actual ratio of rates in real atoms.

The ionization rate predicted by the SFA with circular polarization is also well down from that of linear polarization. It turns out that the algebra of the SFA is actually easier with circular polarization and has been considered in detail by a number of authors (Reiss 1980). But a similar argument can be made in the rate of Eq. (4.105) if we suggest that the argument of the Bessel function is reduced in a circularly polarized field such that

$$J_{q+2m}\left(\frac{eA_0 k_{ex}}{mc\omega}\right) \to J_{q+2m}\left(\frac{e}{mc\omega}\frac{A_0}{\sqrt{2}}k_{e\|}\right).$$

Given the quick falloff of all of the Bessel functions with increasing argument it is obvious that the ionization rate with circular polarization will be much reduced in the SFA framework.

There is a simple intuitive picture for this difference. In the context of a multiphoton picture of this ionization, and recalling the angular momentum selection rules for absorption of a photon, ionization in the two polarizations can be illustrated by the photon diagrams in Figure 4.17. The addition and subtraction of a quantized unit of angular momentum in MPI with linearly polarized laser pulses leads to many possible ionization pathways into the continuum (accessing a wide range of angular momentum states along the way). The steady increase of angular momentum that must accompany absorption of circularly polarized light permits only one MPI pathway. Consequently, MPI beyond, say, two-photon ionization will be significantly faster in a linearly polarized field because of the greater number of ionization channels available to the electron.

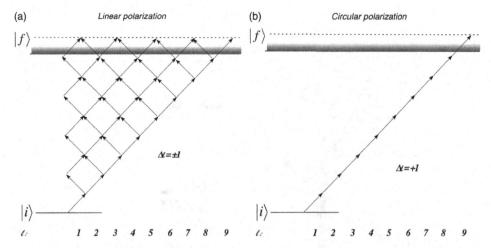

Figure 4.17 Multiphoton ionization of an atom by a linearly and a circularly polarized laser. The angular momentum selection rules enable a large number of ionization channels in a linearly polarized field while only one is allowed in the circularly polarized laser. Consequently, the MPI rate with large numbers of photons is typically significantly higher in linearly polarized fields.

4.4 Tunnel Ionization

So far we have considered a situation in which the outer active electron experiences many oscillations of the sinusoidally varying electric field. It is this periodic oscillation which led us to interpret the ionization in terms of the absorption of many photons, even though we have not explicitly quantized the field. This continued reliance on a classical field is justified; clearly many photons take part in the ionization process, particularly at nonperturbative intensities. However, this view of the field as a classical force acting on the outer (quantum mechanically described) electron of the atom or ion leads us to consider strong field ionization from a completely different viewpoint.

Instead of thinking of the field as a periodically (and rapidly) varying term in the Hamiltonian, we can, alternatively, view the field as distorting the potential that binds the electron to its charged nucleus. This distortion varies in time as the field oscillates. An instantaneous view of the binding potential subject to a strong laser field looks something like that pictured in Figure 4.18. The Coulomb potential binding the electron is distorted by the applied field in such a way that a barrier forms, through which the electron wavefunction can tunnel. When a linearly polarized applied field oscillates, the deformed potential is reversed and there is a window during which the electron can be ionized in the other direction. Because the penetration of the bound electron wavepacket decreases exponentially, the rate of ionization by this tunneling process is expected to be a strongly nonlinear increasing function of applied field strength. This process, tunnel ionization, is a key concept in strong field atomic physics, providing an alternate and complementary view to multiphoton ionization.

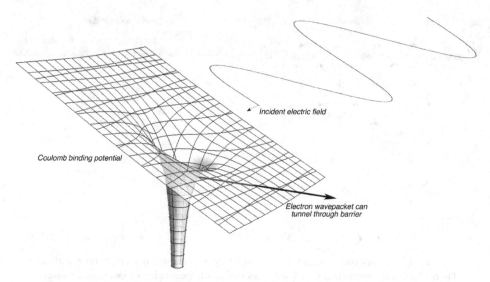

Figure 4.18 Schematic illustration of tunnel ionization in an atom subject to an electric strong field strong enough to deform the binding potential. At sufficiently high field strength, the most weakly bound electron can tunnel through the deformed barrier into the free continuum, as long as the field varies slowly enough for the electron to pass through the barrier before the field changes directions.

4.4.1 Tunneling Time and the Keldysh Parameter

Naturally we now must ask, when is this alternate tunnel ionization view an appropriate picture of strong field ionization? The answer is that this picture is most appropriate when the electron can tunnel through the barrier on a time scale much faster than the oscillating field changes. This limit is often called the adiabatic limit and is most appropriate for low-frequency, long-wavelength fields. To estimate the validity range of tunneling, we need an estimate of the time it takes an electron to tunnel through the barrier.

If the distorted potential of an ion of residual charge Z can be written

$$V(\mathbf{x}) = -\frac{Ze^2}{r} - eE_0 x \cos \omega t, \qquad (4.139)$$

then the width of the barrier through which an electron bound by energy I_p when the laser field is at its maximum can be found from the points on the potential that are lowered to $-I_p$:

$$-\frac{Ze^2}{r} - eE_0 r = -I_p. \qquad (4.140)$$

The points on the distorted potential lowered to the level of the ionization potential are the two solutions of (4.140)

$$r_0 = \frac{1}{2eE_0} \left(I_p \pm \sqrt{I_p^2 - 4Ze^3 E_0} \right), \qquad (4.141)$$

which means that the barrier width is

$$\Delta r_{\text{barrier}} = \sqrt{I_p^2/e^2 E_0^2 - 4Ze/E_0} \cong \frac{I_p}{eE_0}. \tag{4.142}$$

The classical velocity of the bound electron is approximately

$$v \approx \frac{1}{2}\sqrt{\frac{2I_p}{m}}, \tag{4.143}$$

which leads to an estimate for the time to tunnel:

$$\Delta t_b = \frac{\sqrt{2mI_p}}{eE_0}. \tag{4.144}$$

The tunnel ionization picture is appropriate, then, when $\Delta t_b \ll \omega^{-1}$. This condition suggests that we define a parameter

$$\omega \Delta t_b \equiv \gamma_K = \frac{\omega\sqrt{2mI_p}}{eE_0}$$

$$\gamma_K = \sqrt{\frac{I_p}{2U_p}}, \tag{4.145}$$

which is, in fact, the parameter we previously introduced into the equations of the SFA.

The physical significance of this parameter, which is commonly known as the Keldysh parameter, is now evident: when $\gamma_K \ll 1$, the tunneling picture of strong field ionization is appropriate. Equation (4.145) tells us that the tunneling regime is accessed at higher intensities in longer wavelengths. We can see that the weak field regime, the range we previously associated with LOPT and multiphoton ionization, is then characterized by the opposite limit, $\gamma_K \gg 1$. In fact, the Keldysh parameter is a rather soft parameter. Widespread investigations of tunnel ionization experimentally and computationally indicate that the tunneling regime is accessed when γ_K is only somewhat less than 1. For example, in a near-IR field with a wavelength of, say 800 nm, $\gamma_K \sim 1$ in singly charged He ($I_p = 54.4$ eV) at an intensity of $\sim 3 \times 10^{14}$ W/cm^2. Experiments show that the tunneling picture is a rather accurate description of the ionization of He$^+$ to He^{2+} at intensities above $\sim 10^{15}$ W/cm^2 corresponding to $\gamma_K \sim 0.5$.

4.4.2 Exponential Scaling of the Tunnel Ionization Rate

Before we explore the more sophisticated models for tunnel ionization, it is interesting to note that the ionization rate in the tunneling regime tends to exhibit a universal exponential scaling with applied field strength, common across all tunneling models. A quasi-classical WKB model for tunneling tells us the rate a particle's wavefunction leaks through a classically forbidden barrier is proportional to the penetration factor given by Schiff (1966, p. 268):

$$w_{tun} \propto P = \exp\left[-2\int_{r_1}^{r_2} \frac{1}{\hbar}\sqrt{2m(V(r)-U)}\,dr\right], \tag{4.146}$$

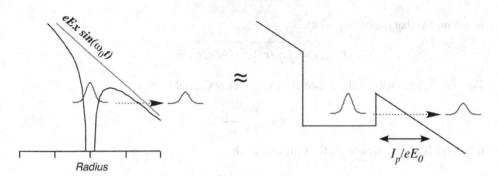

Figure 4.19 A simple one-dimensional picture of the deformed potential in an atom subject to an intense laser field. We can estimate the tunneling rate by treating the barrier as triangular.

where U is the electron energy. A simple estimate for this quantity can be made if we approximate the distorted barrier in a manner pictured in Figure 4.19: we treat the barrier as being shaped approximately by a linear ramp. The penetration factor in (4.146) is then easily evaluated,

$$P \cong \exp\left[-2\int_0^{\Delta r_B} \frac{1}{\hbar}\sqrt{2m|U| - 2meE_0 r}\, dr\right], \qquad (4.147)$$

for barrier width Δr_B. With the approximate barrier width found in (4.142), letting $I_p = U$, integration yields

$$P \cong \exp\left[-\frac{2}{3}\frac{m^{1/2}}{e\hbar}\frac{(2I_p)^{3/2}}{E_0}\right]. \qquad (4.148)$$

An exponential factor of this form is the tell-tale sign of strong field tunneling and will reappear in the tunneling models we now consider.

4.4.3 Tunnel Ionization in a Hydrogen Atom

Ionization by tunneling from the $-1/r$ potential of the hydrogen atom (or hydrogen-like ions) was famously presented in the textbook of Landau and Lifshitz (1991) (amusingly, in the form of the solution to a homework problem). We will find this solution useful as a baseline model for optical tunneling.

First, consider the wavefunction for the ground state of an H-like atom of nuclear charge Z

$$\psi(\mathbf{x}) = \left(\frac{Z^3}{\pi a_B^3}\right)^{1/2} e^{-Zr/a_B}, \qquad (4.149)$$

where $a_B = \hbar^2/me^2$ is the Bohr radius. If we start with the Schrödinger equation in the length form and treat, for the moment, the field as static, we have

$$\frac{\hbar^2}{2m}\nabla^2\psi + \left(\varepsilon + \frac{Ze^2}{r} + eE_0 x\right)\psi = 0. \tag{4.150}$$

This problem is separable in parabolic coordinates. Introducing these coordinates,

$$\xi = r - x, \tag{4.151a}$$
$$\eta = r + x, \tag{4.151b}$$
$$\phi = \tan^{-1}(y/x), \tag{4.151c}$$

the ground state wave function becomes

$$\psi_{1s} = \kappa \exp\left[-\frac{Z}{2a_B}(\xi + \eta)\right] \tag{4.152}$$

with

$$\kappa \equiv \left(Z^3/\pi a_B^3\right)^{1/2}.$$

The separation of the problem into parabolic coordinates will allow us to write the 3D Schrödinger equation in terms of the coordinate η, in a form that looks like the 1D Schrödinger equation, enabling easy application of the WKB approximation to find the tunneling rate.

Some simple assumptions will be necessary. First, we assume that the field is weak enough that the electron emerges from under the barrier at a point $x \gg a_B/Z$. Second, we assume that the energy of the electron is approximately that of the ground state of the hydrogen-like atom in the absence of the field, so that the bound state energy is

$$\varepsilon \cong -\frac{Z^2 e^2}{2a_B} = -\frac{Z^2 m e^4}{2\hbar^2}. \tag{4.153}$$

Furthermore, we assume that we can treat the wavefunction near the nucleus as approximately that of the unperturbed 1s ground state. In the new coordinates the potential felt by the electron is

$$V = -\frac{2Ze^2}{\xi + \eta} - \frac{1}{2}eE_0(\xi - \eta). \tag{4.154}$$

Our interest is in the behavior of the electron's wavefunction as $x \to \infty$, which means that $\xi \to 0$ and $\eta \to \infty$. Since ξ is small in our region of interest, we can say that the wavefunction of the 1s state remains like the unperturbed wavefunction along ξ:

$$\psi_{1s} \sim \exp\left[-\frac{Z\xi}{2a_B}\right]. \tag{4.155}$$

Following the usual procedure for the situation in which the coordinates are separable, we write the wavefunction as a product of functions in terms of the parabolic coordinates

$$\psi_{1s} = f(\xi)g(\eta)e^{im_l\phi}, \tag{4.156}$$

where m_l is the usual magnetic quantum number. When the Schrödinger equation (4.150) is separated into two equations, each depending only on a single parabolic coordinate, we retrieve an eigenvalue equation in terms of separation constant Z_g:

$$\frac{d}{d\eta}\left(\eta\frac{dg}{d\eta}\right) + \left(\frac{m_l^2}{4}\frac{1}{\eta} + \frac{m\varepsilon}{2\hbar^2}\eta + \frac{meE_0}{4\hbar^2}\eta^2\right)g + Z_g\frac{me^2}{\hbar^2}g = 0. \tag{4.157}$$

A similar equation for $f(\xi)$ is derived in terms of its separation constant Z_f (with the exception that the sign of the E_0 term is negative). The separation constants fulfill the condition $Z_g + Z_f = Z$. We are faced with the need to solve for the separation constants; however, we note that in the case of $E_0 = 0$, the g and f equations are identical, which indicates that $Z_g = Z_f = Z/2$. So, if the field is not too large (i.e. smaller than the atomic field) we can rewrite Eq. (4.157) for the η portion of the wavefunction as with $Z_g \approx Z/2$

$$\frac{d}{d\eta}\left(\eta\frac{dg}{d\eta}\right) + \left(\frac{m_l^2}{4\eta} + \frac{m\varepsilon}{2\hbar^2}\eta + \frac{Zme^2}{2\hbar^2} + \frac{meE_0}{4\hbar^2}\eta^2\right)g = 0. \tag{4.158}$$

Making the substitution $\chi(\eta) = \sqrt{\eta}g$ transforms Eq. (4.158):

$$\frac{d^2\chi}{d\eta^2} + \left(\frac{m\varepsilon}{2\hbar^2} + \frac{Zme^2}{2\hbar^2\eta} + \frac{1}{4\eta^2} + \frac{meE_0}{4\hbar^2}\eta\right)\chi = 0. \tag{4.159}$$

We seek a solution, then, for the wavefunction of the form

$$\psi = \kappa\exp\left[-\frac{Z\xi}{2a_B}\right]\frac{\chi}{\sqrt{\eta}}. \tag{4.160}$$

With the first assumption we made, we will demand that this wavefunction approaches that of the unperturbed 1s state near the nucleus. So, we choose a point η_0 such that for $\eta < \eta_0$ the wavefunction is that of the unperturbed 1s state and for $\eta > \eta_0$ the WKB solution for the wavefunction is valid. Since Eq. (4.159) is of the form

$$\frac{d^2u}{dx^2} + k_e^2 u = 0 \tag{4.161}$$

the WKB solution can be written (Schiff 1968, p. 271)

$$u \cong \frac{A}{k_e^{1/2}}\exp\left[\pm\int_{x_1}^{x_2}k_e dx + \frac{i\pi}{4}\right]. \tag{4.162}$$

Our boundary conditions mandate

$$\chi(\eta_0) = \exp\left[-\frac{Z\eta_0}{2a_B}\right], \tag{4.163}$$

which leads us to write, to retain proper normalization,

$$\chi = \sqrt{\frac{\eta_0 k_0}{k_e}}\exp\left[-\frac{Z\eta_0}{2a_B}\right]\exp\left[i\int_{\eta_0}^{\eta_1}k_e d\eta + \frac{i\pi}{4}\right] \tag{4.164}$$

4.4 Tunnel Ionization

with

$$k_e = \sqrt{-\frac{Z^2 me^2}{4\hbar^2 a_B} + \frac{Zme^2}{2\hbar^2 \eta} + \frac{1}{4\eta^2} + \frac{meE_0}{4\hbar^2}\eta}. \quad (4.165)$$

The value of k_e is imaginary under the barrier, and k_0 is the value of the electron momentum wavenumber at the point η_0. To find the ionization rate, what we really seek is the absolute squared value of the wavefunction:

$$|\chi|^2 = \frac{\eta_0 |k_0|}{k_e} \exp\left[-\frac{Z\eta_0}{a_B}\right] \exp\left[-2\int_{\eta_0}^{\eta_1} |k_e| d\eta\right]. \quad (4.166)$$

Following Landau and Lifshitz, we can make some approximations about k_e. Since the region of interest is far from the atom nucleus, $\eta \gg a_B$, in the factor before the exponential; so we can approximate

$$k_e \cong k_1 = \frac{1}{2}\sqrt{\frac{meE_0}{\hbar^2}\eta - \frac{Z^2 me^2}{\hbar^2 a_B}} \quad (4.167)$$

and

$$|k_0| \cong \frac{1}{2}\frac{Zem^{1/2}}{\hbar a_B^{1/2}}. \quad (4.168)$$

We need to find the point η_1 where the electron emerges from the barrier to insert into the upper limit of the integral in (4.166). From Eq. (4.167) this point can be found:

$$0 = \frac{1}{2}\sqrt{\frac{meE_0}{\hbar^2}\eta_1 - \frac{Z^2 me^2}{\hbar^2 a_B}} \quad (4.169)$$

yielding

$$\eta_1 \cong \frac{Z^2 e}{E_0 a_B}. \quad (4.170)$$

Now, in the exponential of (4.166) we return to (4.165) and Taylor-expand k_e with respect to the $1/\eta$ term (since we care about large η to derive the tunneling rate far from the nucleus):

$$|k_e| \cong K_2 + \frac{Zme^2}{4K_2 \hbar^2}\frac{1}{\eta}. \quad (4.171)$$

Define $K_2 = ik_1$ to get for the wavefunction density

$$|\chi|^2 = \eta_0 \frac{Zem^{1/2}}{2\hbar a_B^{1/2}}\frac{1}{k_1} \exp\left[-\frac{Z\eta_0}{a_B}\right] \exp\left[-2\int_{\eta_0}^{\eta_1} K_2 d\eta - \frac{Ze^2 m}{2\hbar^2}\int_{\eta_0}^{\eta_1}\frac{1}{K_2 \eta}d\eta\right]. \quad (4.172)$$

Consider the first integral in the exponential of (4.172):

$$\int_{\eta_0}^{\eta_1} K_2 d\eta = \frac{1}{3}\frac{(Z^2/a_B^2 - meE_0\eta_0/\hbar^2)^{3/2}}{meE_0/\hbar^2}, \quad (4.173)$$

or since η_0 is chosen to be close to the nucleus where we have assumed that the external field does not significantly affect the potential, Taylor expansion gives

$$\int_{\eta_0}^{\eta_1} K_2 d\eta \cong \frac{1}{3}\frac{Z^3 m^2 e^5}{\hbar^4}\frac{1}{E_0} - \frac{Z}{2a_B}\eta_0. \tag{4.174}$$

Evaluation of the second integral in (4.172) gives

$$\int_{\eta_0}^{\eta_1} \frac{1}{K_2 \eta} d\eta \cong -2\frac{\hbar^2}{Zme^2}\ln\left[4\frac{Z^2\hbar^2}{a_B^2 me E_0 \eta_0}\right]. \tag{4.175}$$

These give us for the wavefunction χ

$$|\chi|^2 = 4\frac{Z^3 e}{a_B^2}\frac{1}{E_0}\frac{1}{\sqrt{meE_0\eta/\hbar^2 - Z^2/a_B^2}}\exp\left[-\frac{2}{3}\frac{Z^3 m^2 e^5}{\hbar^4}\frac{1}{E_0}\right]. \tag{4.176}$$

To derive a *rate* of ionization from this result, we think in terms of an electron flux emerging from the barrier, which can be written $|\psi|^2 v_x$ where the wavefunction is evaluated at the point where it emerges from the barrier. To find the total rate that the electron emerges from the atom, we must integrate the flux over an area normal to the direction the electron emerges:

$$w_I = \int |\psi(\eta_1)|^2 v_x \, dA. \tag{4.177}$$

Recalling that $\xi = 0$ far from the nucleus, the electron velocity is easily calculated

$$v_x \approx \sqrt{\frac{2}{m}(\varepsilon - eE_0 x)}$$

$$\approx \sqrt{\frac{2}{m}\left(\varepsilon - \frac{1}{2}eE_0(\xi - \eta)\right)}$$

$$\approx \sqrt{\frac{2}{m}\left(-\frac{Z^2 e^2}{2a_B} + \frac{1}{2}eE_0\eta\right)}. \tag{4.178}$$

In cylindrical coordinates about the axis of the field's polarization, $dA = 2\pi\rho d\rho$ and $\xi\eta = r^2 - x^2 = \rho^2$, which allows us to write for large η

$$d\rho = d\sqrt{\xi\eta} = \frac{1}{2}\sqrt{\frac{\xi}{\eta}}d\eta + \frac{1}{2}\sqrt{\frac{\eta}{\xi}}d\xi$$

$$\approx \frac{1}{2}\sqrt{\frac{\eta}{\xi}}d\xi. \tag{4.179}$$

The total rate of ionization by tunneling becomes

$$w_I = \int_0^\infty |\psi(\eta_1)|^2 v_x 2\pi\rho d\rho$$

$$= \int_0^\infty \kappa^2 \pi \exp[-Z\xi/a_B]|\chi|^2 \sqrt{\frac{eE_0\eta}{m} - \frac{Z^2 e^2}{ma_B}}d\xi, \tag{4.180}$$

which integrates to yield our final result:

$$w_I = 4Z^5 \frac{m^3 e^9}{\hbar^7} \frac{1}{E_0} \exp\left[-\frac{2}{3} \frac{Z^3 m^2 e^5}{\hbar^4} \frac{1}{E_0}\right]. \tag{4.181}$$

This formula yields the instantaneous rate of ionization for a given field strength. In the strong field physics literature it is often called the "simple atom" DC tunnel ionization rate. This equation can be put in a more suggestive form if we introduce $\tilde{E}_0 \equiv E_0/E_a$ and $\tilde{I}_p \equiv I_p/I_H = Z^2$. The rate now becomes

$$w_I = 4\omega_a \frac{\tilde{I}_p^{5/2}}{\tilde{E}_0} \exp\left[-\frac{2}{3}\tilde{I}_p^{3/2} \frac{1}{\tilde{E}_0}\right]. \tag{4.182}$$

The ionization rate is dominated by the exponential term

$$w_{tun} \sim \exp\left[-\frac{2}{3} \frac{Z^3}{\tilde{E}_0}\right], \tag{4.183}$$

which is a factor of the same form as our simple penetration formula, Eq. (4.148). Equation (4.182) demonstrates that the ionization rate will increase very quickly with field, and that the time scale for ionization approaches the characteristic time scale of electron motion in the atom (ω_a^{-1}) when the field strength approaches the binding field of the atom ($\tilde{E}_0/Z^3 \sim 1$).

4.4.4 Cycle-Averaged Simple Atom Tunnel Ionization Rate

In optically induced tunneling, the atom experiences a constant field when the polarization is circular. The rate derived in Eq. (4.182) is the appropriate rate for that form of laser polarization. In a linearly polarized field, however, the atom experiences a periodic increase and decrease of the electric field. To find the appropriate observed rate, it is necessary to average the DC ionization rate over a complete optical cycle. The "cycle-averaged" rate can therefore be written

$$w_{CA} = \frac{1}{2\pi} \int_0^{2\pi} w_I(E_0 \cos\varphi) \, d\varphi. \tag{4.184}$$

Performing this integration with Eq. (4.182) yields

$$w_{CA} = \frac{4}{\pi} Z^5 \omega_a \frac{E_a}{E_0} K_0\left[\frac{2}{3}\left(\frac{I_p}{I_H}\right)^{3/2} \frac{E_a}{E_0}\right], \tag{4.185}$$

where $K_0(a)$ is the modified Bessel function of the second kind. Alternatively, a useful approximation can be made to derive a simpler analytic form of the cycle-averaged ionization rate by noting that the vast majority of the ionization occurs near the peak of the optical cycle. This lets us Taylor-expand the $1/\cos\varphi$ term in the exponential around $\varphi = 0$ and evaluate the integral by ignoring the time dependence in the prefactors to the exponential:

$$w_{CA} = \frac{1}{\pi} \int_{-\pi/2}^{\pi/2} 4\omega_a \left(\frac{I_p}{I_H}\right)^{5/2} \left(\frac{E_a}{E_0 \cos\varphi}\right) \exp\left[-\frac{2}{3}\left(\frac{I_p}{I_H}\right)^{3/2} \frac{E_a}{E_0 \cos\varphi}\right] d\varphi$$

$$\cong \frac{1}{\pi} w_I^{(static\,E)} \int_{-\pi/2}^{\pi/2} 4\exp\left[-\frac{1}{3}\left(\frac{I_p}{I_H}\right)^{3/2} \frac{E_a}{E_o}\varphi^2\right] d\varphi$$

$$\cong \left[\frac{3}{\pi}\left(\frac{I_H}{I_p}\right)^{3/2} \frac{E_0}{E_a}\right]^{1/2} w_I^{(static\,E)} \qquad (4.186)$$

where the limits of the integration have been extended to infinity in the final step, reducing the evaluation to a simple Gaussian integration.

The simple approximate formula of (4.186) illustrates nicely by how much the cycle-averaged rate is reduced from the static field rate. The cycle-averaged rate is clearly reduced with respect to the DC rate (since $E_0/Z^3 E_a \ll 1$ in the tunneling regime) by a significant amount because of the strongly nonlinear decrease in rate over much of the optical cycle during which the field is below its maximum. To illustrate the magnitude of the ionization predicted by these formulae, an example of the ionization rate predicted by the DC and cycle-averaged simple atom tunneling rates for He^+ ions is plotted versus laser intensity in Figure 4.20. Equation (4.185) predicts that the rate ionizing He^+ to He^{2+} will be significant

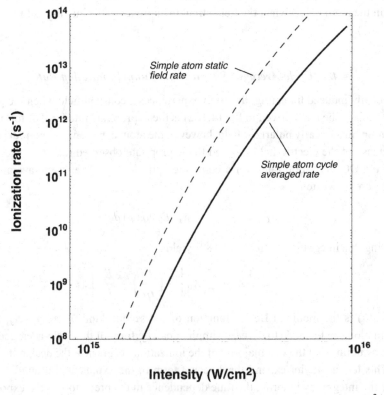

Figure 4.20 Plot of the static simple atom tunneling rate for ionization of He^+ to He^{2+} as a function of electric field strength (dashed line) and corresponding intensity as compared to the cycle-averaged ionization rate.

4.4 Tunnel Ionization

in a ~ 100 fs pulse (i.e. $w_{CA} \sim 10^{13}$ s^{-1}) when the peak intensity of the pulse is around 5×10^{15} W/cm^2. This is consistent with experiments.

4.4.5 The SFA Ionization Rate in the Tunneling Limit

At this point we are faced with an interesting question: the nature of strong field ionization seems to shift from a multiphoton picture to a tunneling picture as the intensity increases and the frequency of the field decreases, but are both effects manifestations of the same physics in two different limits? The answer to this question must, of course, be yes. We have utilized two quite different theoretical formalisms to address these two limits, which suggests that multiphoton and tunneling ionization are rather different processes. However, these two models are simply approximate solutions of the same Schrödinger equation. The unity of these two limiting cases is made elegantly clear in the SFA formalism, as Keldysh (1965) observed in his initial paper on the theory.

We have already seen how the SFA result reduces to a LOPT-type scaling when the field is weak, or namely when $\gamma_K \gg 1$. The SFA also reduces to a formula which is of the form of optical tunneling in the opposite limit, when $\gamma_K \ll 1$. To see this, return to the SFA ionization rate for a hydrogen-like 1s ground state, as we derived in Eq. (4.134). In the tunneling limit, when the Keldysh parameter is small we can say:

$$\sinh^{-1} \gamma_K \approx \gamma_K - \frac{1}{6}\gamma_K^3 + \frac{3}{40}\gamma_K^5 \dots, \qquad (4.187)$$

which means that the term in the exponent of (4.134) is approximately

$$\left(1 + \frac{1}{2\gamma_K^2}\right) \sinh^{-1} \gamma_K - \frac{\sqrt{1+\gamma_K^2}}{2\gamma_K}$$

$$\approx \left(1 + \frac{1}{2\gamma_K^2}\right)\left(\gamma_K - \frac{1}{6}\gamma_K^3 + \dots\right) - \frac{1}{2\gamma_K}\left(1 + \frac{1}{2}\gamma_K^2 + \dots\right)$$

$$\approx \frac{2}{3}\gamma_K + O\left(\gamma_K^3\right). \qquad (4.188)$$

The prefactor to the exponent becomes

$$\frac{\gamma_K}{\sqrt{1+\gamma_K^2}} \frac{1}{\sinh^{-1}\gamma_K} \frac{1}{\left(\sinh^{-1}\gamma_K - \gamma_K/\sqrt{1+\gamma_K^2}\right)^{1/2}}$$

$$\approx \frac{1}{\left((\gamma_K - \gamma_K^3/6) - (\gamma_K + \gamma_K^3/2)\right)^{1/2}}$$

$$\approx \frac{\sqrt{3}}{\gamma_K^{3/2}}. \qquad (4.189)$$

With these approximations and an explicit substitution for the Keldysh parameter, the ionization rate as predicted by the SFA for a 1s state in the tunneling limit becomes

$$w^{SFA}(\gamma_K \ll 1) \cong 4\sqrt{\frac{3}{\pi}\frac{I_p}{\hbar}} \left[\left(\frac{I_H}{I_p}\right)^{3/2} \frac{E_0}{E_a}\right]^{3/2} \exp\left[-\frac{2}{3}\left(\frac{I_p}{I_H}\right)^{3/2} \frac{E_a}{E_0}\right]. \quad (4.190)$$

Comparing this formula with the hydrogen atom tunnel ionization rate derived in the last section, Eqs. (4.182) and (4.186) show that the two rates bear remarkable similarity. Most important, the SFA exhibits exactly the same "tunneling" exponential dependence with the inverse of the field as the simple triangle potential tunneling model of Eq. (4.148) and the simple-atom tunneling model. So we can conclude that the SFA model shows that strong field ionization of an atom reduces to the LOPT scaling at low field strengths and the tunnel ionization scaling at high fields. As Keldysh presciently observed in his original paper, multiphoton and tunnel ionization are merely different manifestations of the same physical effect (Keldysh 1965). We can conclude that the MPI and tunneling pictures are completely compatible and simply represent approximations in two different limits to the actual process of strong field ionization. The Keldysh parameter is seen as the number that quantifies which approximate view is appropriate.

Comparison of Eq. (4.190) and Eqs. (4.182) and (4.186) does highlight one of the largest weaknesses of the SFA in the form we have used. Because the factor $(I_H/I_p)^{3/2}(E_0/E_a)$ is typically less than one (the laser field usually is smaller than the atom's field), the prefactor in the SFA tunneling formula is quite a bit smaller than the prefactor in the simple atom tunneling rate. This under-prediction of the ionization rate by the SFA for linearly polarized light is a well-known problem of the model (Becker *et al.* 2001). Again, it can be principally attributed to the fact that the SFA ignores the effect of the Coulomb binding field on the outgoing electron.

4.4.6 The Coulomb Correction to the SFA Ionization Rate

Inaccuracies in the SFA ionization rates can be corrected to a significant extent by making a simple, WKB correction to the Volkov state used in the SFA derivation to account for the presence of the Coulomb field of the residual nuclear charge (Krainov 1997). Recall that the Volkov state we employed was of the form

$$\psi_V^* = \frac{1}{(2\pi)^{3/2}} \exp\left[\frac{i}{\hbar}\int \frac{1}{2m}\left(\hbar\mathbf{k}_e + \frac{e}{c}\mathbf{A}(t)\right)^2 dt + i\mathbf{k}_e \cdot \mathbf{x}\right]. \quad (4.191)$$

In the spirit of the WKB approximation, we can account in an approximate way for the residual Coulomb field by inserting the Coulomb potential in the energy integral, giving us a Coulomb-corrected version of this Volkov state:

$$\left(\psi_V^{Coul}\right)^* = \frac{1}{(2\pi)^{3/2}} \exp\left[\frac{i}{\hbar}\int \left(\frac{1}{2m}\left(\hbar\mathbf{k}_e + \frac{e}{c}\mathbf{A}(t)\right)^2 + V_{Coul}(r)\right)dt + i\mathbf{k}_e \cdot \mathbf{x}\right]$$

$$= \exp\left[\frac{i}{\hbar}\int V_{Coul}(r) dt\right] \psi_V^* \quad (4.192)$$

4.4 Tunnel Ionization

where

$$V_{Coul}(r) = -\frac{Ze^2}{r}. \tag{4.193}$$

Now put the integral in terms of the momentum, $dt = m\, dr/p_e$. The electron momentum is imaginary under the barrier and is

$$p_e = \sqrt{2m(\varepsilon - V_{tot})}$$
$$\cong \sqrt{2m\left(-\frac{Z^2e^2}{2a_Bn^2} + eE_0r\right)}, \tag{4.194}$$

where we have ignored the Coulomb contribution to the electron energy in this term under the barrier. The Coulomb correction term becomes

$$\exp\left[\frac{i}{\hbar}\int V_{Coul}(r)dt\right] = \exp\left[-\frac{1}{\hbar}\int_{r_i}^{r_f}\frac{Ze^2}{r}\frac{m^{1/2}}{(Z^2e^2/a_Bn^2 + 2eE_0r)^{1/2}}dr\right]. \tag{4.195}$$

We now need to find integration limits for the passage of the electron under the barrier. r_f is the point at which the electron emerges from the barrier, which we already found to be

$$r_f \cong \frac{I_p}{eE_0} = \frac{Z^2e}{2a_Bn^2E_0}. \tag{4.196}$$

The inner limit is more subtle. We need r_i to be away from the atomic ground state. A reasonable choice is the size of the ground state wavefunction

$$r_i \cong \frac{a_Bn}{2Z}. \tag{4.197}$$

Evaluating the integral in (4.195) gives

$$\exp\left[\frac{i}{\hbar}\int V_{Coul}(r)dt\right] = \exp\left[-2\frac{en}{\hbar}\sqrt{a_Bm}\tanh^{-1}\right.$$
$$\left.\left(\sqrt{\frac{Z^2e^2m/a_Bn^2 - 2meE_0r}{Z^2e^2m/a_Bn^2}}\right)\bigg|_{r_i}^{r_f}\right]$$
$$= \exp\left[-2n\left(\tanh^{-1}(0) - \tanh^{-1}\left(\sqrt{1 - \frac{E_0a_B^2n^3}{Z^3e}}\right)\right)\right], \tag{4.198}$$

and since

$$\tanh^{-1}(0) = 0,$$
$$\tanh^{-1}(iz) = -\frac{i}{2}\ln\left(\frac{1-z}{1+z}\right),$$

it reduces to

$$\exp\left[\frac{i}{\hbar}\int V_{Coul}(r)dt\right] \cong \exp\left[-n\ln\left(\frac{E_0 a_B^2 n^3}{2Z^3 e}\right)\right]$$
$$\cong \left(\frac{2Z^3 e}{E_0 a_B^2 n^3}\right)^n. \tag{4.199}$$

Putting this factor in terms of the atom's ionization potential, $I_p/I_H = Z^2/n^2$, gives

$$\exp\left[\frac{i}{\hbar}\int V_{Coul}(r)dt\right] \cong 2^n\left[\left(\frac{I_p}{I_H}\right)^{3/2}\frac{E_a}{E_0}\right]^n. \tag{4.200}$$

This term essentially can be carried through the entire derivation of the SFA. As a result, when the transition amplitude is squared, the corrected ionization rate is

$$W_{SFA}^{Coul} = 2^{2n}\left[\left(\frac{I_p}{I_H}\right)^{3/2}\frac{E_a}{E_0}\right]^{2n} W_{SFA}. \tag{4.201}$$

This correction factor is applicable to the SFA rate at any value of γ_K. Comparisons with numerical simulations of ionization rates indicate that the application of this correction factor essentially boosts the ionization rates for linear polarization to nearly correct levels. The effect of this term is typically to increase the ionization rate by one to two orders of magnitude over the original SFA rate.

4.4.7 Tunnel Ionization Rate for Complex Atoms

Using the Coulomb-corrected SFA ionization rate in the tunneling limit, it is possible to develop a formula for the tunneling rate of multielectron atoms with ground states other than 1s states. This was first done first by Perelomov, Popov and Teren'tev (PPT) in another classic early strong field physics paper which followed shortly after the publication of Keldysh's result. We follow, more or less, the PPT approach in our derivation. Though we will employ the SFA, it is interesting to note that a virtually identical result can be derived using a WKB approach in parabolic coordinates for static field tunneling, similar to the treatment we presented for static field tunneling of a hydrogen atom, and then cycle-averaging that result using approximation (4.191). See Bisgaard and Madsen (2004) for detailed derivation and Lin et al. (2018, p. 94) for discussion.

Recall the differential SFA ionization rate of Eq. (4.130). Our starting point is this equation after approximating the sum over q by an integral and dropping the cosine term as rapidly varying. Generalization of the analysis we performed in Eqs. (4.130)–(4.139) to complex atoms amounts to finding a general form for the Fourier transform of the ground state wavefunction. For simplicity we introduce an approximation for the ground state wavefunction; we consider the wavefunction in the asymptotic region, where $r \gg k_p^{-1}$, where k_p^{-1} is the approximate radial extent of the wavefunction with atomic wavenumber defined by $k_p = (2mI_p)^{1/2}/\hbar$. We can write the asymptotic wavefunction as

$$\psi_i \cong k_p^{3/2} C_{kl}(k_p r)^{\zeta-1} e^{-k_p r} Y_{lm_l}(\theta,\varphi). \tag{4.202}$$

4.4 Tunnel Ionization

A short range potential can be characterized by $\zeta = 0$, an s-state by $\zeta = 1$, a p-state by $\zeta = 2$ and so on. All of the atomic physics is contained in the angular momentum of the spherical harmonic, Y_{lm_l}, and the coefficient C_{kl}. The factor C_{kl} is a normalization coefficient specific to each atom (found by fitting the wavefunction in the asymptotic region); for the ground state of hydrogen $C_{kl} = 2$. This coefficient is tabulated for many atoms (Popov 2004), though a number of approximate formulae for it have been published. A widely used approximation is that derived by Ammosov, Delone and Krainov (ADK) (Ammosov et al. 1986). If we introduce the "effective principal quantum number" in terms of the ionization potential of the charge state in question

$$n^* = \frac{Ze}{\sqrt{2I_p a_B}}, \tag{4.203}$$

the ADK approximation to this coefficient is

$$C_{kl} \approx \left(\frac{2 \exp[1]}{n^*}\right)^{n^*} \frac{1}{\sqrt{2\pi n^*}}. \tag{4.204}$$

With the ground state wavefunction of the form (4.202), the Fourier transform can be written

$$\phi_i(\mathbf{k}_e) = Y_{lm_l}(\hat{\mathbf{k}}_e) \frac{k_p^{3/2} C_{kl}}{(2\pi)^{3/2}} \int_0^{2\pi} \int_0^\pi \int_0^\infty$$
$$\times \exp\left[ik_e \cos\theta - k_p r\right](k_p r)^{\zeta - 1} r^2 \sin\theta \, dr d\theta d\varphi$$

$$\phi_i(\mathbf{k}_e) = Y_{lm_l}(\hat{\mathbf{k}}_e) \sqrt{\frac{2}{\pi}} k_p^{-3/2} C_{kl} \beta_\zeta, \tag{4.205}$$

where we have assumed that the energy of the outgoing electron is small so that we can approximate (4.205) with $k_e \ll k_p$. Here $\beta_\zeta = 1$ for a short-range potential ($\zeta = 0$), $\beta_\zeta = 2$ for s-states, and $\beta_\zeta = 6$, 24 and 120 for p-, d- and f-states, respectively.

For electron motion close to the polarization direction (θ is small), we can approximate the spherical harmonics

$$Y_{lm_l}(\hat{\mathbf{k}}_e) \approx \alpha_{lm_l} \sin^{|m_l|}\theta \, e^{im_l\varphi} \tag{4.206}$$

and, furthermore, approximate the sine function in terms of the transverse outgoing electron momentum

$$\sin\theta \approx \frac{k_{e\perp}}{k_e} \approx \frac{k_{e\perp}}{k_p}. \tag{4.207}$$

This lets us insert a spherical harmonic of the approximate form

$$Y_{lm_l}(\hat{\mathbf{k}}_e) \approx \alpha_{lm_l} \left(\frac{k_{e\perp}}{k_p}\right)^{|m_l|} e^{im_l\varphi} \tag{4.208}$$

with normalization constant

$$\alpha_{lm_l} = \frac{1}{2^{|m_l|} |m_l|!} \left[\frac{(2l+1)}{4\pi} \frac{(l+|m_l|)!}{(l-|m_l|)!}\right]^{1/2}. \tag{4.209}$$

(Just remember that in the context of these equations m_l refers to the angular momentum projection quantum number, not the electron mass which is denoted here as always by m. We have also ignored the sign of $\alpha_{lm_l} \sim -1^{m_l}$ for convenience.)

Insertion of (4.208) into (4.125) leads to integrals similar to those of Eq. (4.132). Following steps like those leading to Eq. (4.134), we integrate the differential ionization rate over outgoing angles using the fact that

$$\int_0^\infty k_{e\perp}^{2|m_l|+1} \exp\left[-ak_{e\perp}^2\right] dk_{e\perp} = \frac{|m_l|!}{2a^{|m_l|+1}}, \qquad (4.210)$$

combined with the observation that in the tunneling limit ($\gamma_K \ll 1$)

$$\left(\sinh^{-1} \gamma_K\right)^{|m_l|+1} \cong \gamma_K^{|m_l|+1} \qquad (4.211)$$

and Eq. (4.205) to yield

$$W_{tun} = 2\sqrt{3\pi}\frac{I_p}{\hbar} C_{kl}^2 \alpha_{lm_l}^2 \beta_\zeta^2 |m_l|! \frac{1}{k_p^{2|m_l|+3}} \left(\frac{\omega m}{\hbar}\right)^{|m_l|+3/2} \frac{1}{\gamma_K^{|m_l|+3/2}} \exp\left[-\frac{2}{3}\left(\frac{2I_p}{\hbar\omega}\right)\gamma_K\right]. \qquad (4.212)$$

With substitution for the Keldysh parameter and recalling that $k_p = (2mI_p)^{1/2}/\hbar$, we can write

$$W_{PPT} = 2\sqrt{3\pi}\frac{I_p}{\hbar} C_{kl}^2 \alpha_{lm_l}^2 \beta_\zeta^2 |m_l|! \left[\left(\frac{I_H}{I_p}\right)^{3/2} \frac{E_0}{E_a}\right]^{m_l+3/2} \exp\left[-\frac{2}{3}\left(\frac{I_p}{I_H}\right)^{3/2}\frac{E_a}{E_0}\right] \qquad (4.213)$$

for the tunnel ionization rate of a complex atom. This result is essentially identical to the original PPT result (Perelomov 1966) who considered tunneling from a short-range potential (i.e. $\beta_\zeta = 1$).

Applying the Coulomb correction factor, (4.201), and explicit use of Eqs. (4.204) and (4.209), we get, finally:

$$W_{PPT/ADK} = \frac{I_p}{\hbar} \beta_\zeta^2 \, 2^{2n*-2|m_l|-1} \sqrt{\frac{3}{\pi}} \left[\left(\frac{2\exp[1]}{n*}\right)^{2n*} \frac{1}{2\pi n*}\right] \left[\frac{(2l+1)(l+|m_l|)!}{|m_l|!(l-|m_l|)}\right]$$

$$\times \left[\left(\frac{I_p}{I_H}\right)^{3/2}\frac{E_a}{E_0}\right]^{2n*-|m_l|-3/2} \exp\left[-\frac{2}{3}\left(\frac{I_p}{I_H}\right)^{3/2}\frac{E_a}{E_0}\right]$$

$$(4.214)$$

This formula is among the most important and often used analytic results in strong field physics. It is a general formula for the tunnel ionization rate of an arbitrary atom or ion in terms of the effective principal quantum number and the angular momentum quantum numbers. It is widely accepted as the standard theory for calculating the rate of tunnel ionization in an atom or ion subject to a linearly polarized field. Experiments have shown this formula to be quite accurate, at least up to intensities of 10^{18} W/cm^2 (Augst et al. 1991), and it is likely accurate to intensities two orders of magnitude above that. In deference to a widely used convention in the literature, we will refer to this ionization rate as the "PPT/ADK ionization rate," though it is often called simply the ADK equation in the literature. This is somewhat unfair since the essential derivation of this ionization rate was made by Perelomov, Popov and Teren'tev (Perelomov et al. 1966) 20 years ahead of the ADK publication on the asymptotic coefficient approximation (Ammosov et al. 1986). (The SFA-based complex atom ionization rate approach of the original PPT analysis has been generalized in recent years to yield formulae that are applicable to $\gamma_K > 1$ and hence extends this tunneling rate to lower-intensity and higher-frequency fields (Popruzhenko et al. 2008). These equations, however, are cumbersome.)

This formula retains the characteristic tunneling exponential factor which we have now seen in every tunneling model we have considered. As usual, this factor tends to dominate the extremely nonlinear scaling of the ionization rate with increasing intensity. To evaluate the ionization rate of any real atom with multiple electrons in a given orbital, one should in principle, sum this rate over all magnetic quantum numbers. In practice, however, when considering low charge states, ionization typically occurs in fields of strength comparable but somewhat less than the atomic field strength. This means that usually

$$\left[\left(\frac{I_p}{I_H}\right)^{3/2} \frac{E_a}{E_0}\right]^{-|m_l|} \ll 1, \qquad (4.215)$$

so the prefactor in Eq. (4.214) essentially insures that the $m_l = 0$ electron will have the highest ionization rate. Therefore, it is usually adequate to calculate the ionization rate with $m_l = 0$ and assume the ionization of subsequent electrons in the formation of higher charge states occurs on a time scale longer than the remaining electrons redistribute themselves in the atom back into $m_l = 0$ states. (Of course, in a multielectron atom, m_l and l are not good quantum numbers, so any discussion differentiating ionization rates with respect to one of these numbers in complex atoms should be treated with caution.)

The reason for this strong preference for ionization of $m_l = 0$ states can be easily seen if one considers the shape of the electron wavefunction for $m_l = 0$ and $m_l = 1$ or -1 (recalling the definition of the quantization axis in (4.207) as along the laser polarization). Figure 4.21 shows, for illustrative purposes, the shape of the wavefunctions of the $l = 1$ (p orbital) electrons with magnetic quantum numbers $m_l = 0$ and the $m_l = \pm 1$ in a potential deformed by the laser's electric field. The $m_l = 0$ state extends initially to the greatest extent along the tunneling axis, so it is not surprising, given the nonlinear nature of the tunneling rate, that it is this m_l state with the highest ionization rate.

A comparison of the ionization rate as a function of intensity as predicted by three of the models we have considered so far for ionization of an He^+ ion to He^{2+} from irradiation by 800 nm light is illustrated in Figure 4.22. This ionization process is well into the tunneling regime for 800 nm light when intensities exceed 10^{15} W/cm². The SFA rate, Eq. (4.134), predicts an ionization rate which is somewhat below the predictions of the PPT/ADK rate and the cycle-averaged simple atom rate. This is a consequence of neglecting the Coulomb correction to the outgoing wavefunction. The PPT/ADK rate, which is generally considered the most accurate, indicates that significant ionization of He to the fully stripped 2+ state will occur in femtosecond pulses for intensities in the vicinity of $5-7 \times 10^{15}$ W/cm².

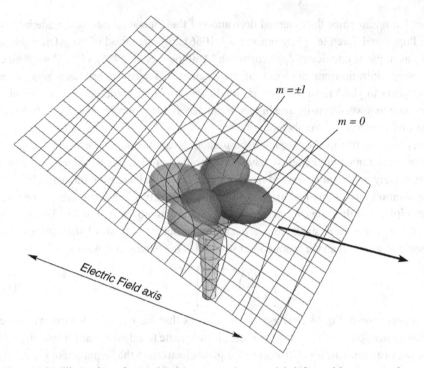

Figure 4.21 Illustration of a *p*-orbital in an atomic potential deformed by a strong laser field, illustrating why the $m = 0$ suborbital tends to have the fastest tunnel ionization rate. The wavefunction extends the furthest along the tunneling axis when $m = 0$.

4.4.8 Barrier Suppression Ionization

Viewing ionization of the electron in a distorted potential like that illustrated in Figure 4.18 leads to the question of what happens as the field is further increased. Ultimately the strength of the field is great enough that the potential is depressed sufficiently to allow the electron to escape directly over the potential barrier, a situation depicted in Figure 4.23. At fields high enough to allow this to happen, we expect that the atom will be completely ionized on a time scale comparable to the characteristic time of electron motion in the atom ($\sim \omega_a^{-1}$). This is commonly known as barrier suppression ionization (BSI).

Calculation of the intensity at which BSI occurs is straightforward. If we assume that the binding potential is that of a bare nucleus so that we can treat it as a pure Coulomb potential (hence ignoring the effects of any more deeply bound electrons), the deformed potential is

$$V(r) = -\frac{Ze^2}{r} - eE_0 r. \qquad (4.216)$$

To find the peak barrier height of this potential, we find its peak value occurring when

$$\left. \frac{\partial V}{\partial r} \right|_{r_p} = 0, \qquad (4.217)$$

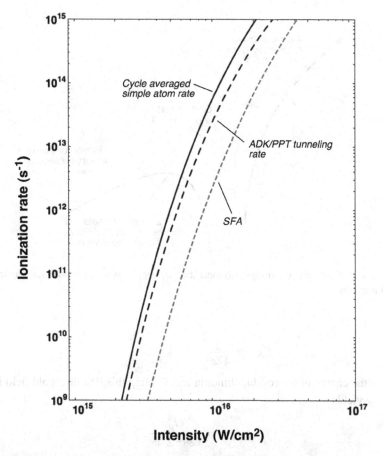

Figure 4.22 Plot of the single-electron atomic ionization rate for the ionization of He$^+$ to He^{2+} versus intensity by 800 nm light for some of the strong-field ionization rate models considered so far in this chapter. (Note that $\gamma_K < 1$ for intensity $>5 \times 10^{14}$ W/cm^2.)

which lets us solve easily for the radial position of the barrier peak:

$$\frac{Ze^2}{r_p} - eE_0 = 0,$$

$$r_p = \sqrt{Ze/E_0}. \tag{4.218}$$

Ionization by barrier suppression occurs when the potential at this peak is equal to the binding energy of the electron in the ground state

$$V(r_p) = -I_p. \tag{4.219}$$

This condition leads directly to the field strength needed for BSI,

$$-I_p = -Ze^2\sqrt{\frac{E_0}{Ze}} - eE_0\sqrt{\frac{Ze}{E_0}}$$

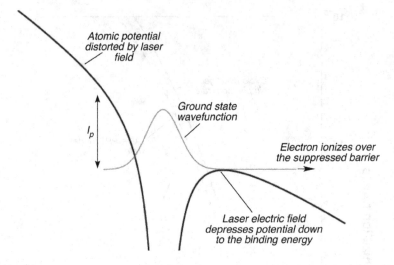

Figure 4.23 Schematic of strong-field ionization by suppression of the potential barrier at high intensity.

$$E_{BSI} = \frac{I_p^2}{4Ze^3}, \qquad (4.220)$$

where Z is the charge of the residual binding ion. Casting this BSI threshold field in terms of intensity yields

$$I_{BSI} = \frac{cI_p^4}{128\pi Z^2 e^6}, \qquad (4.221)$$

which can be written in useful units as

$$I_{BSI}[W/cm^2] = 4.0 \times 10^9 \left(I_p[eV]\right)^4 Z^{-2}. \qquad (4.222)$$

It should be noted that this simple analysis is flawed in that it assumes that the electron motion along the field polarization is independent of motion in the lateral direction. As we have already seen in Section 4.4.3, the hydrogenic ion subject to a static electric field is a problem that is separable in terms of parabolic coordinates. Consequently, the critical field found in (4.220) is inaccurate. A more sophisticated analysis of the true field strength required to suppress completely the binding potential barrier begins with Eq. (4.162) for electron motion along the parabolic coordinate, η, in which the barrier is suppressed; see Mulser and Bauer (2010, p. 276) for a detailed analysis. Such a calculation indicates that a field roughly 2.4 times higher than found by our simple 1D analysis, or intensity 5.5 times higher than that indicated by Eq. (4.221) is needed to suppress fully the binding potential.

Nonetheless, Eq. (4.222) is an extremely convenient, simple formula that permits an estimate for the intensity required to ionize an atom or ion significantly in the tunneling regime. The BSI formula is a surprisingly accurate predictor of the threshold intensity needed for significant tunnel ionization despite its virtually trivial simplicity (Augst *et al.* 1991). For

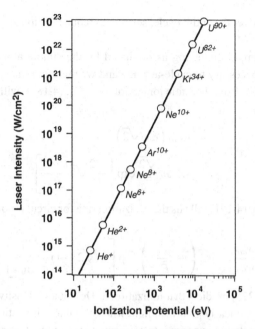

Figure 4.24 Barrier suppression ionization thresholds for various ions as a function of peak intensity.

example, Eq. (4.222) predicts BSI ionization of He$^+$ at an intensity of 9×10^{15} W/cm^2, an intensity which is completely consistent with the PPT/ADK prediction of the rate of ionization of He$^+$ illustrated in Figure 4.22.

We can use this formula as a way to explore easily the intensities required to ionize a wide range of atoms and ions. Figure 4.24 plots the BSI threshold intensity as a function of the ionization potential of an atom or ion. Various ionization products are plotted on this line. This curve illustrates that ionization of rare gas atoms past their valence shell requires intensity $\sim 10^{19}$–10^{20} W/cm^2 and ionization into the K-shell of uranium would require peak intensity in the vicinity of $\sim 10^{23}$ W/cm^2.

4.4.9 Tunnel Ionization by Circularly Polarized Light

We have already seen that MPI is typically slower when the laser field is circularly polarized. The same is true in the tunneling regime. When an atom is subject to a circularly polarized field, it is true that the atom experiences the peak field strength at all times during the optical cycle, not just at the oscillatory peaks as it does in a linearly polarized field. However, for a given intensity, that peak field is reduced in circular polarization:

$$E = E_0/\sqrt{2} \tag{4.223}$$

Because the rate at which an electron tunnels into the continuum is such a strongly non-linear function of the field, this reduction in field more than compensates for the fact that the peak field is felt over the entire cycle, serving to reduce the ionization rate by tunneling significantly.

Consider, for example, tunneling as predicted by the simple atom model. In circular polarization it is unnecessary to cycle-average and we therefore find the rate of ionization in circular polarization from the static ionization rate, Eq. (4.187), with the substitution for the field, Eq. (4.223):

$$w_{cir} = w_{stat}\left(E_0/\sqrt{2}\right)$$
$$= 4Z^5 \omega_a \frac{\sqrt{2}}{\tilde{E}_0} \exp\left[-\frac{2}{3}\sqrt{2}\frac{\tilde{I}_p^{3/2}}{\tilde{E}_0}\right]. \qquad (4.224)$$

Comparison with Eq. (4.191) tells us the ionization rate for circular polarization is reduced by a factor

$$\frac{w_{cir}}{w_{lin}} = \sqrt{2}\left(\frac{\pi}{3}\frac{\tilde{I}_p^{3/2}}{\tilde{E}_0}\right) \exp\left[-(\sqrt{2}-1)\frac{2}{3}\frac{\tilde{I}_p^{3/2}}{\tilde{E}_0}\right]. \qquad (4.225)$$

For example, Eq. (4.225) predicts that ionization of He^+ at an intensity of 5×10^{15} W/cm^2 is 12% of the rate in a linearly polarized field when circular polarization is employed. This effect is well documented in experiments performed in the tunneling regime.

4.5 Nonsequential Double Ionization and Electron Rescattering

Treating strong field atomic ionization by the single active electron approximation would seem to suggest that multiple ionization of an atom is purely sequential. The laser field interacts with one electron at a time, the most weakly bound electron, ejecting it first, with subsequent ionization of more deeply bound electrons occurring later in the laser pulse, presumably at a higher intensity. Is there, however, any physical mechanisms which might lead to simultaneous, or nearly simultaneous, correlated ejection of two (or more) electrons at one time during strong field ionization? By "nearly simultaneous" we mean ejection of multiple electrons in a correlated fashion (from the same parent atom) within a time scale of the order of a laser oscillation. Such a process is usually called nonsequential multiple ionization and can indeed occur. Experimental evidence for such nonsequential, correlated double ionization surfaced as early as 1982 (L'Huillier et al. 1982), though it took at least ten years after this to develop a theoretical understanding of why such a process might take place. Experiments in noble gas atoms have shown that correlated double ionization occurs in linearly polarized near-IR pulses with a probability up to 0.2% of single electron ionization at the same intensity (Fittinghoff et al. 1992; Walker et al. 1994).

The ionization models we have considered so far suggest that such nonsequential double (or multiple) ionization is extremely unlikely. In the multiphoton regime ($\gamma_K > 1$) direct double ionization would require a process such as that depicted in Figure 4.25, the simultaneous absorption of a sufficient number of photons to overcome not just the binding potential of the first electron but also the ionization potential of the next, more deeply bound electron. In principle, perturbation theory does not preclude such a process, it just

4.5 Nonsequential Double Ionization and Electron Rescattering

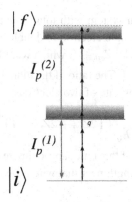

Figure 4.25 Direct multiphoton two-electron nonsequential double ionization.

implies that it is exceedingly improbable. If we say that the MPI nonsequential double ionization (NSDI) process requires q photons to overcome the first electron's ionization potential and s photons to overcome the binding energy of the next more deeply bound electron then, recalling the treatment of Section 4.2.4, LOPT would predict that the rate of the single ionization process, w_1, would compare to the rate of the NSDI process, w_2, as

$$\frac{w_1}{w_2} \sim \left(\frac{I_a}{I}\right)^s. \tag{4.226}$$

Since, in the MPI regime, $I \leq \sim 0.01\, I_a$ and for most atoms subject to optical or near-IR fields $s \geq 4$ or 5, we can say that the direct multiphoton-driven NSDI will have a rate at least eight orders of magnitude below that of the single electron ionization rate.

In the tunneling regime ($\gamma_K < 1$) one can ask, along similar lines of reasoning, if two correlated electrons can tunnel from the atom simultaneously, so-called collective tunneling (Becker *et al.* 2005). A rather nonrigorous, but physically illuminating way to think about such a process is to ask if two electrons, bound by an effective ionization potential given by the sum of the ionization potentials of the two least deeply bound electrons have a tunneling rate that is significant. If we assume for simplicity that such a rate would be given by the same functional form as the single electron rate, we recall that the rate is dominated by the exponential term

$$w_T \sim \exp\left[-\frac{2}{3}\tilde{I}_p^{3/2}\frac{1}{\tilde{E}_0}\right], \tag{4.227}$$

recalling that $\tilde{I}_p = I_p/I_H$ and $\tilde{E}_0 = E_0/E_a$. For illustrative purposes it is a little simpler if we just think in terms of an effective binding charge, Z_{eff}, for the first, most loosely bound electron which will have a tunneling rate which scales as

$$w_T \sim \exp\left[-\frac{2}{3}Z_{eff}^3\frac{1}{\tilde{E}_0}\right]. \tag{4.228}$$

Since we consider a two-electron system with total ionization potential, $I_p = I_p^{(1)} + I_p^{(2)}$, we can estimate the ratio of the single electron to double electron tunneling rate by saying that the second electron is bound by $Z_{eff} + 1$, which would mean that the effective binding potential is $\tilde{I}_p^{(eff)} = 2Z_{eff}^2 + 2Z_{eff} + 1$. The ratio of the tunnel ionization rate of the first single electron to the collective tunneling rate of two electrons would then scale like

$$\frac{w_1}{w_2} \sim \exp\left[\frac{2}{3}\frac{1}{\tilde{E}_0}\left(\left(2Z_{eff}^2 + 2Z_{eff} + 1\right)^{3/2} - Z_{eff}^3\right)\right]. \tag{4.229}$$

To derive a quantitative estimate of this ratio, approximate the relevant field strength by the BSI field, Eq. (4.220), a field strength near that where we expect full ionization of the first electron:

$$\frac{E_{BSI}}{E_a} \approx \left(\frac{Z_{eff}^2 e^2}{2a_B}\right)\frac{1}{4Z_{eff}e^3}\frac{a_B^2}{e} = \frac{Z_{eff}^3}{16}. \tag{4.230}$$

Equation (4.235) suggests that for a neutral atom, with $Z_{eff} \approx 1$ the ratio of one to two electron ionization rates is

$$\frac{w_1}{w_2} \sim \exp\left[\frac{32}{3}\left(5^{3/2} - 1\right)\right] = 1.4 \times 10^{47}$$

while for high Z_{eff} the ratio of ionization rates is

$$\frac{w_1}{w_2} \sim \exp\left[\frac{32}{3}\left(2^{3/2} - 1\right)\right] = 2.9 \times 10^8.$$

We see that even in the case of highly charged ions the "collective tunneling" rate is unlikely to be any higher than seven orders of magnitude below the single electron rate, and for neutral atoms the ratio of rates is far greater.

So these "direct" double ionization rates can be rejected as being insignificant in any practical experimental situation. There are, however, two possible mechanisms for direct NSDI: "shake off" ionization and electron recollision ionization. The latter process plays a far more important role in most real strong field experimental conditions.

4.5.1 Shake-off Double Ionization

Shake-off double ionization is a process which is well known in beta decay and in single-photon photoionization of atoms. When an electron is removed from the atom suddenly (or in the case of beta decay the charge of the binding nucleus is changed suddenly) the remaining bound electron finds itself in an energy state which is no longer an eigenstate of the ion. This means that the original state of the electron is projected onto the complete set of states of the new, more highly charged ion. Since some of the states in this complete set of states will be unbound continuum states, there is a finite probability that the remaining electron will find itself in an unbound state and will itself also become ionized.

In high-energy photoionization of He by X-rays, it is well known that this shake-off mechanism leads to simultaneous double ionization about 1.7% of the time (Dörner et al. 2002). In strong field ionization, the contribution of shake-off is far less important (though in principle still in effect). The reason for this comes from the requirement that the ionization of the first electron must occur "suddenly," that is, instantaneously. To be more concrete, if the remaining electron is to be projected onto the new complete set of states the first electron must have been ejected on a time scale much faster than the relaxation time of the remaining bound electron wavefunction. To zeroth order this electron relaxation time will be on the order of an atomic time scale ($\sim \omega_a^{-1} = 0.024$ fs). While this condition is easily fulfilled in the ejection of the MeV electron in beta decay and in the ejection of an electron by a high-energy X-ray photon, it is not well satisfied in strong field ionization.

The most likely situation for shake-off ionization should occur in the tunneling regime where the first ionized electron is ejected at a well-defined phase of the laser field. In fact, we have already seen that there is some finite time for this electron to tunnel through the barrier, set approximately by the time it takes a classical electron to traverse the width of the suppressed binding potential barrier (recall Figure 4.19). The time scale for removal of an electron by tunneling compared to an atomic relaxation time scale then is

$$\frac{t_{tun}}{t_{atom}} \cong \omega_a t_{tun} = \frac{\omega_a}{\omega} \gamma_K. \tag{4.231}$$

Since the tunneling rate scales in terms of the Keldysh parameter is

$$w_{tun} \sim \exp\left[-\frac{4I_p}{3\hbar\omega}\gamma_K\right],$$

we can say that significant tunneling of the outer least deeply bound electron occurs when

$$\gamma_K \approx \frac{3\hbar\omega}{4I_p} \approx (\# \text{ photons needed for ionization})^{-1}. \tag{4.232}$$

Equations (4.231) and (4.232) indicate that $t_{tun} \approx (3\hbar\omega_a/4I_p)t_{atom} \approx t_{atom}$. Tunneling occurs on about the same time scale as the atomic relaxation time, pushing the process well away from the instantaneous ionization needed to effect shake-off double ionization. Consequently, shake-off double ionization does play a role in strong field NSDI, although it is usually considered a small effect. (Sophisticated two-electron S-matrix calculations of the shake-off double ionization rate indicate that it is about 10^{-4} to 10^{-5} that of the single electron ionization rate (Becker and Faisal 1999; 2005).)

4.5.2 Field-Driven Electron Recollision Dynamics

Correlated dynamics of two or more electrons in strong field ionization are much more prevalent through the process of laser field-driven recollisions (van der Hart and Burnett 2000). This process, which requires linearly polarized fields, results in the double ionization of an atom in the tunneling regime through a process depicted in Figure 4.26. While this can be calculated by employing a second-order correction to the SFA (Milosevic et al. 2007) a simple quasi-classical theory gives much more insight. During tunnel ionization the electron is born into the continuum as it emerges from the barrier of the suppressed binding potential in a strong light field at a well-defined moment and

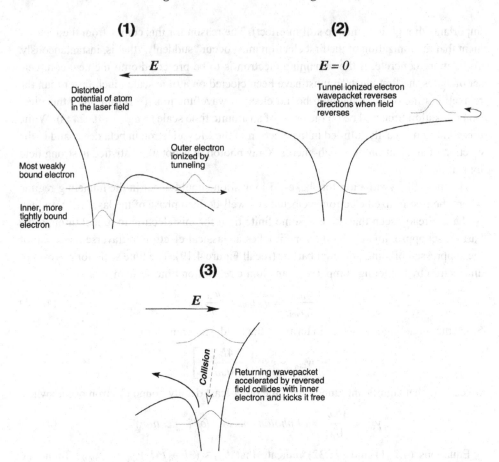

Figure 4.26 Illustration of the quasi-classical three-step recollision process leading to nonsequential double ionization. First an electron is pulled into the continuum by tunneling. During the subsequent propagation of the free electron wavepacket, the electric field of the laser reverses and reverses the trajectory of the free electron. Finally, when the electron returns to the vicinity of the nucleus, one possible outcome if the returning electron has more energy than the binding energy of the next bound electron is an inelastic scattering event which frees that second, more tightly bound electron.

well-defined phase of the laser field. After this point, the free electron can be thought of as a classical point-like particle propagating under the influence of the laser field.

As Figure 4.26 illustrates, once this electron emerges into the continuum (step 1 in Figure 4.26), it freely propagates away from the binding nucleus in the oscillating field. If the electron is "born" into the field in the right range of phases of the field, once the field reverses it drives the electron back to the atom (a process which occurs only if the field is linearly polarized), shown as step 2 of the figure. When the initially ionized electron returns to its parent nucleus, it does so with kinetic energy it has acquired by acceleration in the laser field (step 3). If this energy is sufficient, the electron can inelastically scatter off

4.5 Nonsequential Double Ionization and Electron Rescattering

of the remaining ion, ionizing one of the more deeply bound electrons (or in some cases exciting it to a higher energy level).

In this process the tunnel-ionized electron can eject a second electron within the course of a single optical cycle. This laser-driven recollision model has come to be called the quasi-classical three-step model and forms the framework for understanding many aspects of modern strong field atomic and molecular physics. It represents the dominant process for NSDI and is the basis for understanding a wide range of other strong field phenomena. Clearly, if the field has even a small amount of ellipticity the electron (see Section 3.2.4) the electron will not return to the nucleus as it will be driven not only in the x-direction but in the perpendicular (y) direction as well. The sharp falloff of observed NSDI rates with increased laser field ellipticity is one of the most important experimental confirmations that recollision dynamics are the root of most NSDI.

The kinematics of this three-step quasi-classical model can be easily quantified by considering the trajectory of a free electron after it has been born into the continuum at a well-defined birth time, t_b and birth phase of the field $\varphi_b = \omega t_b$. Assuming that the motion is nonrelativistic ($a_0 \ll 1$), the electron motion along the polarization direction, x, is described simply by

$$\ddot{x} = \frac{eE_0}{m} \cos \omega t. \qquad (4.233)$$

If the electron is born at rest at the location of the nucleus, $x(0) = 0$, then

$$\dot{x} = \frac{eE_0}{m\omega} \sin \omega t - \frac{eE_0}{m\omega} \sin \varphi_b \qquad (4.234)$$

$$x = -\frac{eE_0}{m\omega^2} (\cos \omega t - \cos \varphi_b + \omega t \sin \varphi_b - \varphi_b \sin \varphi_b). \qquad (4.235)$$

Recollision with the nucleus by this liberated electron occurs when it returns to $x = 0$, which means that the phase in the field at which the electron recollides with the nucleus, φ_r, is found from

$$\cos \varphi_r - \cos \varphi_b + \varphi_r \sin \varphi_b - \varphi_b \sin \varphi_b = 0. \qquad (4.236)$$

Once this return phase in found from (4.236), the energy of the returning electron is found with the help of Eq. (4.234) to give

$$U_r = 1/2 m \dot{x} (\varphi_r)^2$$
$$= \frac{e^2 E_0^2}{2m\omega^2} (\sin \varphi_r - \sin \varphi_b)^2$$
$$= 2U_p (\sin \varphi_r - \sin \varphi_b)^2. \qquad (4.237)$$

Numerical solution of Eq. (4.236) for the phase of recollision indicates that the electron only returns to the nucleus for birth phases between 0 and $\pi/2$, namely the electrons that are born in the field during the quarter-cycle after the field peak. Electrons born in the cycle before the field peak never completely return to the vicinity of the nucleus. Figure 4.27 illustrates this by showing electron trajectories along x as a function of time for four birth phases, found from Eq. (4.235). The first three occur at phases $<\pi/2$ and all return to the

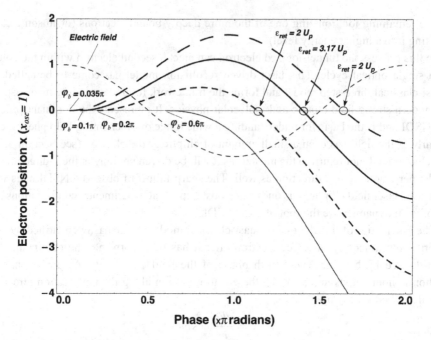

Figure 4.27 Example trajectories of an electron freed by tunneling at various birth phases in the laser's electric field (labeled φ_b). The E-field is illustrated as are four different electron trajectories. The solid line shows the trajectory which yields the highest possible electron energy (3.17 U_p) upon return to the nucleus at $x = 0$. Two other trajectories have return energy of 2 U_p, one being the short trajectory (short dashed line), the other being the long one (long dashed line). A fourth trajectory shows an electron which does not return to the nucleus.

nucleus during the subsequent field half cycle. The fourth trajectory originates in the field at a birth phase $>\pi/2$ and never returns to the nucleus. The magnitude of the kinetic energy of the returning electron is indicated in each case.

The kinetic energy of the recolliding electron, which is of the order of the ponderomotive energy for most birth phases, as a function of its phase of birth, is shown in Figure 4.28. On top of this curve, the magnitude of the field is plotted as is the ionization probability as calculated by the static tunnel ionization formula of Eq. (4.181).

Figure 4.28 tells us a number of interesting things. First, we see that electrons born at the field maximum, where the ionization probability is highest, have return energy of 0. Only later in the field phase, where the ionization probability drops off quickly, does the electron return with finite energy. This figure illustrates that the maximum energy that an electron can acquire in a recollision occurs at a birth phase near $0.1\,\pi$ radians and is equal to 3.17 U_p. This energy factor has become an ubiquitous part of strong field physics and will reappear later when we consider high-order harmonic generation in gases of atoms. Finally, Figure 4.28 illustrates that for all other return energies less than 3.17 U_p, there are two birth phases which yield identical return energies. The trajectories plotted in Figure 4.27, where

4.5 Nonsequential Double Ionization and Electron Rescattering

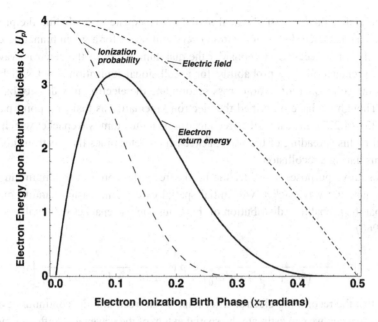

Figure 4.28 Plot of the return energy of an electron as a function of the phase in the field. The amplitude of the field (thin dashed line) at these phases is illustrated as is the probability of tunneling as calculated from the simple atom static field model (long dashed line), illustrating that there is a preponderance of ionization at phases near 0.

two different trajectories, both resulting in a return energy of 2 U_p, are plotted, illuminate the nature of these two same-energy birth phases. Electrons born before 0.1 π, the window in the optical cycle with the higher ionization probability, undergo a longer trajectory prior to recollision than do electrons born after 0.1 π.

The upshot of this analysis is that if the returning electron returns with kinetic energy above the ionization potential of the next most loosely bound electron, it can collisionally ionize that second electron, resulting in nonsequential ionization of two electrons during the same half cycle. This condition for recollision NSDI is met (at least in this simple picture) if the return energy of $3.17 U_p$ at the intensity of ionization of the first electron exceeds the ionization potential of the second electron. This condition is frequently met in many experimental circumstances. For example, ionization of neutral He by laser pulses of 800 nm wavelength becomes significant at intensity $\geq \sim 5 \times 10^{14}$ W/cm^2 as illustrated by Figure 4.16. Equation (3.30) tells us that the ponderomotive energy of electrons at this intensity and wavelength is 30 eV, which means that some of the return energies which reach up to 95 eV exceed the 54.4 eV ionization potential of He$^+$, and we expect recollision NSDI to He^{2+} to occur with some reasonable probability. This is indeed observed in the ionization of He in this intensity range (Walker et al. 1994; Fittinghoff et al. 1992).

4.5.3 Wavepacket Spreading of the Recolliding Electron and Coulomb Focusing

While the simple three-step recollision model describes the basic reasons for NSDI, there are a number of subtleties which determine the magnitude and details of the effect.

For example, the simple quasi-classical model is inadequate in predicting the probability with which a correlated double ionization event will occur for a given tunneling electron. To assess this, it is necessary to consider the quantum nature of the electron wavepacket during its propagation. The probability for a collisional ionization event will be determined by the collisional ionization cross section and the electron flux density back at the nucleus. Though we have described the electron kinematics as that of a point particle in Eqs. (4.233)–(4.237), in reality, the free electron is a quantum wavepacket which spreads in time. It is this spreading of the wavepacket which determines the electron flux back at the nucleus during a recollision.

For illustrative purposes, consider the newly freed electron as a minimum uncertainty Gaussian quantum wavepacket with initial spatial extent δr_0. Basic quantum mechanics tells us that the probability distribution of this Gaussian wavepacket will evolve in time as (Schiff 1968)

$$|\psi(r,t)|^2 = \frac{1}{2\pi \left(\delta r_0^2 + \hbar^2 t^2/4m^2 \delta r_0^2\right)} \exp\left[-\frac{r^2}{2\left(\delta r_0^2 + \hbar^2 t^2/4m^2 \delta r_0^2\right)}\right]. \quad (4.238)$$

If we say that the recolliding electron after it has been freed into the continuum evolves as such a wavepacket, we can estimate the spatial extent of the wavepacket after a time $t_b - t_r$. For example, Figure 4.27 shows that for an electron trajectory resulting in the maximum return energy ($\varphi_b = 0.1\pi$) the return phase is around 1.5π. So the time in the continuum is $\sim 1.4\pi/\omega$. If we say that the initial size of the wavepacket is around one Bohr radius, then the size of the electron wavepacket upon return to the nucleus is

$$\delta r \approx \sqrt{a_B^2 + \hbar^2 (1.4\pi/\omega)^2/4m^2 a_B^2} \quad (4.239)$$

or 2×10^{-7} cm $= 39\ a_B$. The wavepacket has increased by almost a factor of 40 over its original size. To take the example further, the (electron polarization averaged) collisional ionization cross section in an ion such as He$^+$ can be estimated by the so-called Lotz empirical formula (Lotz 1968) which says that

$$\sigma_{CI} \cong 4.4 \times 10^{-14} \frac{\ln\left[\varepsilon/I_p\right]}{\varepsilon I_p} \quad \text{cm}^2 \quad (4.240)$$

if the electron energy, ε, and ionization potential are expressed in eV. If He is tunnel-ionized by an 800 nm laser pulse at intensity of 1×10^{15} W/cm^2, the maximum return energy of recolliding electrons is 3.17×60 eV $= 190$ eV. At this energy the collisional ionization cross section of He$^+$ is around 5×10^{-18} cm^2. The probability that the recolliding electron collisionally ionizes to liberate the second electron is approximately the peak electron fluence $\sim (39\ a_B)^{-2} \approx 2.5 \times 10^{13}$ e/cm^2 times this cross section to yield $\sim 10^{-4}$. In other words, for every tunnel-ionized electron, about 0.01% of the time, a second electron will be nearly simultaneously liberated (Sheehy et al. 1998).

This is a fair order of magnitude estimate for the actual probability of NSDI in an atom like He near ionization saturation of the neutral atom. However, this value is something of an underestimate for the actual probability for recollision NSDI in most atoms

4.5 Nonsequential Double Ionization and Electron Rescattering

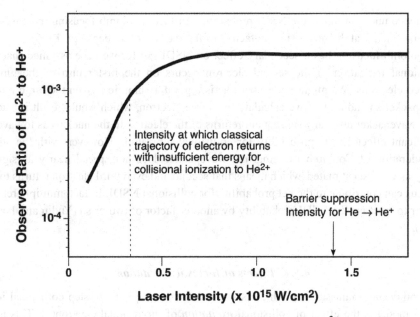

Figure 4.29 Observed probability for double ionization of He to He^{2+} as a function of laser intensity for 800 nm light. Plots generated based on plots in Yudin and Ivanov (2001).

(Yudin and Ivanov 2001). For example, Figure 4.29 shows the observed probability for NSDI as a function of intensity in 800 nm laser pulses. For most intensities where tunnel ionization of the first electron is significant in He ($>5 \times 10^{14}$ W/cm^2) the observed NDSI probability is about an order of magnitude higher than our simple expanding wavepacket estimate.

The origins of this increase in recollision NSDI are complex and rooted in the detailed quantum mechanics of correlated two-electron systems bound in a Coulomb potential. For the most part, however, this discrepancy can be attributed to two effects. The first correction arises from the fact that the wavepacket of the tunnel-ionized electron does not actually expand as if it were a completely free electron. This electron in reality propagates under the influence of the Coulomb field of the residual ion. During the oscillation of the electron wavepacket along x, the Coulomb potential of the ion (ignoring any effects of the confined electron cloud of the still bound electrons) is

$$V_{Coul} \cong -\frac{Ze^2}{r} \approx -\frac{Ze^2}{x}\left(1 - \frac{r_\perp^2}{2x^2}\right), \tag{4.241}$$

where $r_\perp$ is the distance away from the x-axis. While the electron wavepacket oscillates along x, it feels a confining potential in the perpendicular direction from the long-range attraction of the residual Coulomb field. This confining potential serves to restrict the uncertainty-expansion of the recolliding wavepacket, an effect known as Coulomb focusing (Yudin and Ivanov 2001). Consequently, our simple estimate for the spreading of the wavepacket and the recolliding electron flux overestimates the wavepacket spreading and hence underestimates the NSDI probability. In fact, Coulomb focusing increases the electron fluence at the recollision by factors of three or four (Brabec *et al.* 1996).

Coulomb focusing has a secondary effect on NSDI. So far we have assumed that any collisional ionization of the second electron occurs on the first return of the tunnel-ionized electrons. We might argue that this is a good assumption given the large rate of wavepacket spreading that we calculate for a free electron, which would result in such a large wavepacket upon any subsequent returns of the electron to the nucleus as to have no significant effect on the probability for collisional ionization. However, with the added consideration of Coulomb focusing, the wavepacket does not spread nearly as significantly as we first estimated, which means that second and even third electron returns to the nucleus can contribute to the net probability for collisional NSDI. In fact, multiple returns appear to increase the NSDI probability by another factor or two or so (Yudin and Ivanov 2001).

4.5.4 Effects of Inelastic Excitation

Another important subtlety which is ignored in the simple three-step collisional ionization model is the effect of collisional *excitation* of the residual electron(s). This is an inelastic collision channel in addition to collisional ionization. This effect explains an interesting facet of Figure 4.29: that recollision NSDI can occur at intensities below that at which 3.17 U_p is *below* the energy threshold for collisional ionization (Walker *et al.* 1994). A model relying on collisional ionization alone might argue in He, for example, that at intensity below 3×10^{14} W/cm^2 in 800 nm pulses, NSDI cannot occur because the highest energy a returning electron can have is less than the 54.4 eV needed to ionize the residual ion. The experimental observations behind Figure 4.29 indicate that this is not so at those lower intensities.

Again, the reasons for this are complex, and rest in part in the quantum nature of the ionization. For example, any actual returning quantum wavepacket will have a spread of energies, not just the one well-defined return energy that the classical result of Eq. (4.237) indicates. However, the most significant reason for this below-threshold collisional NSDI is inelastic excitation of the bound electron, a process described in Figure 4.30. The returning electron can, instead of ionizing the residual bound electron, collisionally excite it to an upper energy level. This excited electron can then be tunnel or barrier suppression ionized on the subsequent half laser cycle. This process requires the returning electron only to have enough energy to excite the bound electron, not remove it, helping to explain the below-threshold appearance of NSDI at low intensity.

This effect also aids in explaining the higher-than-expected probability for NSDI. At the bottom of Figure 4.30 is plotted the collisional ionization cross section for He$^+$ as a function of electron energy as well as the inelastic excitation cross section and the sum of these cross sections. Clearly, if inelastically excited electrons can be subsequently ionized by the laser field, the effective cross section leading to recollision NSDI is a factor of two to four higher than that of collisional ionization alone. This effect conspires with the Coulomb focusing effects to increase the probability of NSDI in many atoms to $\geq 0.1\%$.

4.5 Nonsequential Double Ionization and Electron Rescattering

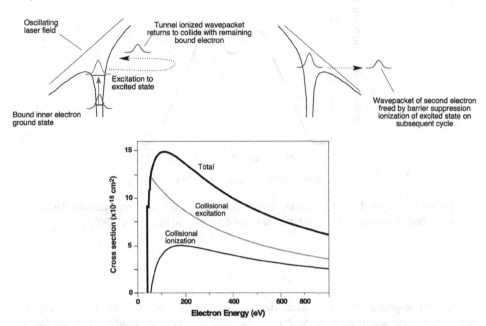

Figure 4.30 Two-step recollision ionization where the recolliding electron excites the bound electron to an upper quantum level (1) which is then subsequently ionized by barrier suppression ionization on the next half cycle (2). The graph on the bottom illustrates the effects of combining the collisional ionization and collisional excitation cross sections in He^+. Cross sections taken from Yudin and Ivanov (2001).

4.5.5 Ion Momenta Distributions

In Section 4.7, we will take up the important topic of electron energy and angular distributions in strong field ionization, but before doing that, it is interesting to note the effects that recollision NSDI has on the momentum distributions of the residual ions. With atoms initially at rest, the resulting ion must retain momentum equal to the momentum of the outgoing electrons in the opposite direction. Conservation of canonical momentum lets us calculate the momentum of the returning tunnel-ionized electron and then, assuming an inelastic collision which yields two electrons with almost zero initial kinetic energy,

$$p_{ion} = \frac{e}{c} A_0 (\sin \varphi_r - \sin \varphi_b) - 2 \frac{e}{c} A_0 \sin \varphi_r$$
$$= -\frac{e}{c} A_0 (\sin \varphi_r + \sin \varphi_b). \tag{4.242}$$

If the maximum rate of NSDI occurs with electrons born near the phase yielding maximum return energy (0.1π), the return phase is $\sim 1.5\pi$ so $\sin \varphi_r + \sin \varphi_b = -0.7$. It follows then that ions that result from recollision NDSI will likely have a two-peak momentum distribution oriented along the laser polarization axis with momenta peaks near

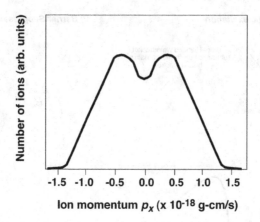

Figure 4.31 Typical ion momentum distribution from recollision nonsequential double ionization of an atom. Plots generated based on data from Weber et al. (2000).

$$p_{ion} \cong \pm 0.7 \frac{e}{c} A_0. \qquad (4.243)$$

This double-peaked ion momentum distribution is indeed what is observed experimentally when ions resulting from NSDI are examined (which is additional proof for the recollision explanation of NSDI). Experimental measurements of ion momentum distributions from ions involved in NSDI look something like the two-peaked distribution plotted in Figure 4.31 (Weber et al. 2000).

4.6 Multiple Ionization of Multielectron Atoms and Ions

Now that we have developed quantitative models for ionization rates, we are in a position to develop a model to describe the number of ions and ion charge states that are to be expected for given laser pulse intensities and pulse durations. Such a treatment is important as it relates the more fundamental quantities like ionization rate with experimental observables such as ion yield and saturation intensity.

4.6.1 Ion Yields Subject to Sequential Ionization

Consider an ensemble of atoms of initial density n_0 each with s bound electrons subject to an intense laser pulse with some temporal intensity envelope such that, $I(t) = I_0 f(t)$. If we care about the total ion yields, we utilize angle and energy integrated ionization rates for the ith charge state, $W_i(t)$ which vary with time through $I(t)$. For the moment, assume that the ionization is purely sequential, which allows us to write the following rate equations for the density of the ith ion charge state:

$$\frac{dn_0}{dt} = -W_0 n_0 \qquad (4.244a)$$

4.6 Multiple Ionization of Multielectron Atoms and Ions

$$\frac{dn_1}{dt} = -W_1 n_1 + W_0 n_0 \quad (4.244b)$$

$$\frac{dn_2}{dt} = -W_2 n_2 + W_1 n_1 \quad (4.244c)$$

$$\vdots$$

$$\frac{dn_s}{dt} = W_{s-1} n_{s-1} \quad (4.244s)$$

It is possible to derive an analytic solution to this set of equations (Chang et al. 1993). We look first at the solution for the density of neutral atoms, which has the simple solution

$$n_0(t) = n_0(0) \exp\left[-\int_{-\infty}^{t} W_0(t') dt'\right]. \quad (4.245)$$

This solution naturally leads us to define an intensity at which significant depletion of the target atom has occurred through ionization. For the ith charge state, this saturation intensity, I_{sat}, can be defined by the condition that

$$\int_{-\infty}^{\infty} W_i\left(I_{sat} f(t)\right) dt = 1, \quad (4.246)$$

which indicates that the saturation intensity is, at least weakly, dependent on the laser pulse duration and pulse shape. For example, if the ionization is multiphoton and described by LOPT such that $W_i = \alpha_i I^q$, the saturation intensity of a species subject to a constant-intensity pulse of duration τ_p is just

$$I_{sat} \cong \left(\frac{1}{\alpha_i \tau_p}\right)^{1/q}. \quad (4.247)$$

Because strong field ionization is usually quite nonlinear in intensity, the actual value of the saturation intensity is usually determined largely by the ionization rate and not very strongly dependent on pulse duration. (I_{sat} is only dependent on the qth root of the pulse duration in the preceding LOPT example.) In the tunneling regime, the BSI intensity, Eq. (4.221), is often a very good estimate for the saturation intensity. For example, the BSI intensity for ionization of He^+ is 8.7×10^{15} W/cm^2, while the saturation intensity for ionization of He^+ by a 50 fs Gaussian pulse found using the PPT/ADK rate in Eq. (4.246) is 8.0×10^{15} W/cm^2.

The formal solution for the first charge state can be derived by substitution of solution (4.245) into Eq. (4.244b):

$$\frac{dn_1}{dt} = -W_1 n_1 + W_0 n_0(0) \exp\left[-\int_{-\infty}^{t} W_0(t') dt'\right]. \quad (4.248)$$

Solution of (4.248) and subsequent resubstitition leads to a series of equations of the form $x'(t) = -wx + y(t)$. Solution of this equation is straightforward and yields for the density of the ith charge state in terms of the time-dependent density of the $(i-1)$th charge state:

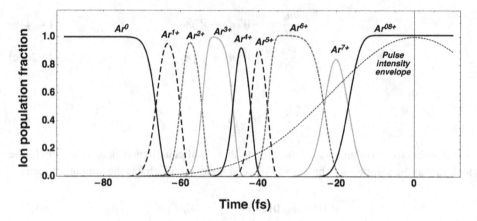

Figure 4.32 Calculated time history of the population fraction of charge states in argon subject to a 50 fs laser pulse at peak intensity of 5×10^{16} W/cm^2 based on the PPT/ADK tunnel ionization rate.

$$n_i(t) = \exp\left[-\int_{-\infty}^{t} W_i(t')dt'\right] \times \int_{-\infty}^{t} \exp\left[-\int_{-\infty}^{t'} W_i(t'')dt''\right] W_{i-1}(t')n_{i-1}(t')dt' \quad (4.249)$$

An example of the time-dependent solution for the ionization of a multielectron atom is illustrated in Figure 4.32. Here the time history of the relative population fraction of various charge states in Ar is calculated using Eq. (4.214) and the PPT/ADK tunneling formula (which, admittedly, is marginally appropriate for the MPI of Ar^0) for atoms subject to a 50 fs FWHM pulse with peak intensity of 5×10^{16} W/cm^2. This calculation illustrates that as the pulse ramps up in intensity on its rising edge, it begins to ionize the neutral Ar atom to 1+ until the neutral is depleted. As the pulse intensity continues to rise, the 2+ ion population grows at the expense of 1+ ions. This process continues until the peak of the pulse is reached, which in this case produces an almost 100% population of Ar^{8+}. Note that atomic shell effects play a role in this history; Ar^{6+} lingers in the pulse longer than its adjacent charge states because it is a closed 3s shell and is harder to ionize, requiring more of an increase in intensity than the more easily ionized open 3p shell electrons. On the other hand, the single 3s electron in the 7+ ion is so weakly bound that 7+ never reaches saturation before it is depleted by ionization to the closed shell of Ar^{8+}.

We can make a useful approximation to the cumbersome solutions of Eq. (4.249) (Chang et al. 1993). An approximate, and more easily calculable form for the population time history of the ith charge state can be derived by noting that, because of the severe depletion of lower charge states, the time history of the ith charge state is dominated by the exponentially decaying tail of the $(i-1)$th charge state. We can approximate this next lower charge state

$$n_{i-1}(t') \cong n_0(0) \exp\left[-\int_{-\infty}^{t'} W_i(t')dt'\right]. \quad (4.250)$$

4.6 Multiple Ionization of Multielectron Atoms and Ions

Since we often care about the population of charges at the end of the laser pulse, this is now easily stated in terms of the total time integrated ionization rates

$$\Gamma_i = \left[-\int_{-\infty}^{\infty} W_i(t')dt' \right] \tag{4.251}$$

to give the following simple solution for the population of each charge state

$$n_0(\infty) = n_0(0) \exp[-\Gamma_0], \tag{4.252a}$$

$$n_i(\infty) \cong n_0(0) (\exp[-\Gamma_i] - \exp[-\Gamma_{i-1}]). \tag{4.252b}$$

It is interesting to note that Γ_i can be calculated analytically for most common ionization rate models and pulse shapes, see Chang et al. (1993), yielding a fully analytic solution for the population density of various charge states.

4.6.2 Focal Volume Integration of Ion Yields

How many ions are actually produced by a focused laser pulse is determined not only by the laser pulse intensity and temporal profile but also by the fact that there is a range of peak intensities in any real laser focus. Quantitatively, the number of ith charge state ions observed in a real ionization experiment requires integration of the rate not only over time but also over the spatial intensity distribution near the focus (Walker et al. 1998)

$$N_i = \int n_i(t = \infty, \mathbf{x}) d\mathbf{x}. \tag{4.253}$$

Describing the intensity near the focus by $I = I_0 f(\mathbf{x})$, it is common to describe the focal distribution of a single spatial mode laser beam by the Gaussian profile we encountered in Section 3.5 (and detailed in Appendix A):

$$I(r,z) = \frac{1}{1+z^2/z_R^2} \exp\left[-\frac{2r^2}{w_0^2 (1+z^2/z_R^2)}\right]. \tag{4.254}$$

With this distribution the total ion yield can be conveniently found by noting that the ionization history will be the same at points in the focus with the same peak intensity. We can therefore consider iso-intensity contours of the focal distribution of Eq. (4.254). Examples of two such contours are shown near a Gaussian focus in Figure 4.33. The total yield of ions, then, is reduced to finding the volume within each iso-intensity contour and then integrating the yield over these volume shells (reducing a 3D integral to a 1D integral). We need to find the volume in the focus that encompasses peak intensities above some intensity I_1. For simplicity we introduce the dimensionless propagation position normalized to the Rayleigh range, $\tilde{z} = z/z_R$, and integrate the volume within the z-dependent radii that delimit the extent of intensity I_1:

$$V(I > I_1) = 2\pi \int_{-z_1}^{z_1} \int_0^{r(z)} r'(z) dr' dz$$

$$= \pi \int_{-z_1}^{z_1} (r(z))^2 dz. \tag{4.255}$$

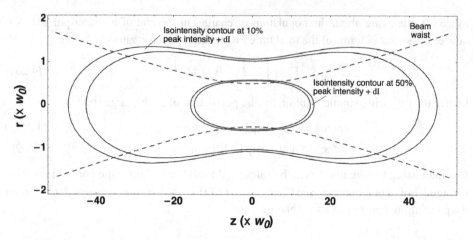

Figure 4.33 Isointensity shells near the focal point of a Gaussian intensity distribution (note difference in scales along r and z).

Using Eq. (4.254), the radius of intensity I_1 is found:

$$(r(z))^2 = -\frac{w_0^2}{2}\left(1+\tilde{z}^2\right)\ln\left[\frac{I_1}{I_0}\left(1+\tilde{z}^2\right)\right]. \tag{4.256}$$

The points along the propagation axis with $r = 0$ where $I = I_1$ are

$$z_1 = \pm z_R\sqrt{\frac{I_0}{I_1} - 1}. \tag{4.257}$$

Substitution of (4.256) and (4.257) into Eq. (4.255) yields for the integrated volume of intensity above I_1

$$V(I > I_1) = \pi w_0^2 z_R\left(\frac{2}{9}\tilde{z}_1^3 + \frac{4}{3}\tilde{z}_1 - \frac{4}{3}\text{Arctan}\tilde{z}_1\right). \tag{4.258}$$

It is now possible to find how the yield of a certain ion charge state will scale with increasing peak laser intensity when that peak intensity is well above the saturation intensity. In this regime, the longitudinal points in the focus encompassing intensity above the saturation intensity for that ion are approximately

$$\tilde{z}_1 = \sqrt{\frac{I_0}{I_{sat}} - 1} \approx \left(\frac{I_0}{I_{sat}}\right)^{1/2}. \tag{4.259}$$

At high intensity the first term in the parentheses of (4.258) dominates the volume which means that the volume over which ionization of the charge state in question is saturated is

$$V(I > I_{sat}) \cong \frac{2}{9}\pi w_0^2 z_R\left(\frac{I_0}{I_{sat}}\right)^{3/2}. \tag{4.260}$$

This relation tells us that if we measure the number of ions being produced in a laser focus at an intensity above the saturation intensity, the ion yield will increase as the peak

4.6 Multiple Ionization of Multielectron Atoms and Ions

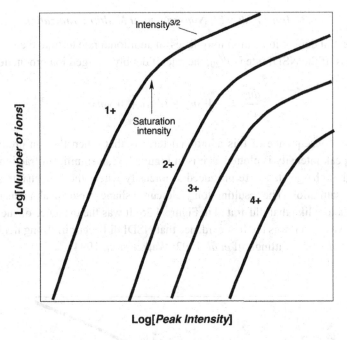

Figure 4.34 Typical behavior of the observed yield of various ion charge states when a dilute gas of atoms is irradiated by laser pulses of increasing intensity and the ionization is purely sequential.

intensity to the 3/2 power. (This assumes that atoms in the target gas are not limited in extent in z and encompass the entire focal volume, as is encountered when a laser is focused into a large target chamber filled with target gas.) As a result of this volume scaling, the measured yield of ions of various charge states as a function of peak laser intensity assumes functional dependence that frequently looks much like the curves illustrated in Figure 4.34. At low peak intensity the production of the 1+ ion begins and the yield scales steeply nonlinear with intensity, reflecting the nonlinearity of the ionization rate with intensity. The slope of this curve on the log-log scale used in Figure 4.34 reflects the nonlinear ionization rate scaling. (For example, if this ion is produced by LOPT MPI of order q, then the slope in this region of the plot is q.)

As the peak intensity increases, the ionization saturates, marked by a rollover in the yield curve. Above this saturation intensity, the number of ions produced scales simply with the volume in the focus where intensity is above the saturation intensity. Above this point the ion yield increases more or less as the 3/2 power of the peak intensity. At the saturation intensity for 1+, a significant number of 2+ ions start to be formed and the yield of this charge state increases nonlinearly up to a peak intensity near the saturation intensity for 2+. This sequential behavior for higher charge states continues, with the relative spacing in intensity between charge states determined by how difficult it is to ionize the next-higher charge state.

4.6.3 Ion Yields with Nonsequential Multiple Ionization

Nonsequential double ionization introduces an additional rate term to the ion population rate equations. If the NSDI rate is W_{20}, the rate of doubly charged ion production becomes

$$\frac{dn_2}{dt} = -W_2 n_2 + W_1 n_1 + W_{20} n_0. \tag{4.261}$$

The practical consequence of this additional term is that when the ion production as a function of peak intensity is plotted as it is in Figure 4.34, a significant fraction of 2+ ions (or next-higher charge state) are produced at intensity where the first charge state is just approaching saturation. The resulting ion yield curve shape versus peak intensity takes on a "knee" structure like that illustrated in Figure 4.35. It was the presence of such knee-like ion yield curves which was the first evidence that NSDI did occur in strong field ionization (L'Huillier et al. 1982; Fittinghoff et al. 1992; Walker et al. 1994).

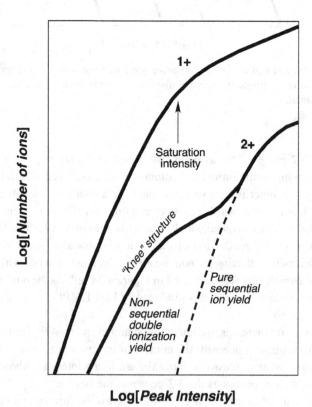

Figure 4.35 Typical behavior of the observed yield of various ion charge states when a dilute gas of atoms is irradiated by laser pulses of increasing intensity when a nonsequential double ionization term is included in the ionization rate equations. The formation of a "knee" in the yield of the higher charge state is because of direct double ionization.

4.7 Above-Threshold Ionization

Though we have not yet explicitly discussed the energies of electrons produced in strong field ionization, it is clear from the various ionization models considered so far that ejected electrons can acquire more energy than that associated with the absorption of the minimum number of photons to overcome the ionization potential. The multiphoton generalization of the photoelectric effect would suggest that ionized electrons have kinetic energy

$$\varepsilon_{\min} = q_{\min}\hbar\omega - I_p. \qquad (4.262)$$

The SFA, for example, indicates that more, sometimes far more, than this minimum number of photons can be absorbed by an atomic electron upon ionization, which inevitably leads to an increase in the kinetic energy of this electron in increments of the photon energy. It is obvious from the SFA ionization rate, Eq. (4.105), that the total ionization rate of an atom is given by a sum over absorbed photons starting from the minimum number of photons needed. Plotting such an SFA rate as is done in Figure 4.16 further reinforces the point. Even in the context of perturbation theory, we have already seen that the truncation of the perturbation series at the lowest-order term corresponding to a multiphoton transition to the continuum is just an approximation. There are higher-order terms in this expansion, some corresponding to absorption of photons beyond q_{min} which would lead to the ejection of electrons with kinetic energy at multiples of the photon energy. What is more, ionization in the tunneling picture does not preclude the electron absorbing more energy from the field.

The experimental observation of electrons with energies above that predicted by Eq. (4.262) has come to be called above-threshold ionization, or ATI (Eberly et al. 1991). This phenomenon was clearly predicted by Keldysh's original derivation of the SFA in 1964 and, even in the context of perturbation theory, is clearly expected as the laser intensity increases. Nonetheless, the atomic physics community was widely surprised when in 1979 electron spectra with kinetic energy peaks above the single peak predicted by Eq. (4.262) were observed during seven-photon ionization of Xe in the pioneering experiment of Agostini et al. (1979). Even at intensity as low as 10^{13} W/cm^2, electron energy spectra something like that illustrated in Figure 4.36c were observed.

This effect was considered a surprise because it was believed that an electron could not absorb additional photons once it had been promoted to the continuum. Our discussions on strong field ionization indicate that this is a naïve and incorrect view. Even though an electron wavepacket can be promoted to the continuum by absorption of q_{min} photons, it does not leave the vicinity of the nucleus immediately, so at sufficiently high photon flux the electron can undergo additional photon-absorbing transitions in the continuum (recall our discussion in Section 4.2.7). Since the first experimental observation of ATI, the literature devoted to both experimental and theoretical study of it has become vast (Eberly et al. 1991). ATI is among the most widely studied phenomena in strong field physics and is now rather well understood (though some subtle features remain incompletely explained).

There have been a number of models forwarded to describe experimentally observed ATI electron energy spectra and angular distributions. These include analytic models such as improvements to the SFA (Reiss 1980) or the model of essential states (Usachenko et al. 2004), and numerical models including computational solution of the time-dependent Schrödinger equation. Each of these models can at times quantitatively match certain specific experiments, often over a limited range of parameters. In reality, the specific details of the shape of any ATI energy spectrum depend, sometimes sensitively, on a variety of experimental parameters including intensity, pulse duration, pulse shape, wavelength, atomic species and even shape of the focal spot. Consequently, a truly predictive model of ATI spectra over a broad range of parameters is difficult. We will not attempt any such detailed theoretical description of the ATI spectra here. Instead we will discuss the qualitative aspects of ATI energy spectra in the context of the ionization models we have already developed (LOPT, SFA, resonantly enhanced MPI, and tunneling) describing widely observed features seen in ATI spectra at various intensity ranges and at various values of the Keldysh parameter.

4.7.1 General Characteristics of ATI Energy Spectra

Figure 4.36 attempts to summarize the generic characteristics of electron ATI spectra generated in optical and near-IR ($\lambda \sim$ 500–1,100 nm) strong field ionization as various windows of laser intensity are entered. (In principle, the trends outlined in this figure also describe ATI by pulses with blue and near UV wavelengths, but since these short wavelengths have lower ponderomotive energies, the relevant intensities for each regime will be somewhat higher.) We will discuss details of the characteristics summarized here in subsequent subsections.

At modest intensities (Figure 4.36a) the ionization is well described by MPI and a LOPT description is adequate. As a result, electrons with the minimum energy required to overcome the binding potential of the atom, described by Eq. (4.262), are observed. As the intensity increases above 10^{12} W/cm^2 (Figure 4.36b) perturbation theory may still be valid but higher-order terms become significant. In this regime higher-order ATI peaks begin to appear in the energy spectrum but, as perturbation theory predicts, with electron yields which decrease rapidly with increasing order. The higher-order peaks are spaced by energy equal to the energy of one laser photon. At intensities above about 10^{13} W/cm^2 the first nonperturbative signatures in the ATI spectrum are seen and the shape of the spectrum begins to take on the form illustrated in Figure 4.36c. A number of ATI orders appear with comparable numbers of electrons, with electron yields increasing for the first few orders before the amplitude of the peaks falls slowly at higher energy.

At intensity of a few times 10^{13} W/cm^2 the ponderomotive energy starts to become comparable to the energy of a single photon (1–2 eV). As a result, the Stark shift of higher energy levels in the atom starts moving these levels in and out of resonance during the rise and fall of the laser pulse, leading to structures in the ATI spectrum over energies which are less than a photon and which are coarsely repeated for higher ATI orders (Figure 4.36d). Finally, at intensity above about 10^{14} W/cm^2 ionization for wavelengths longer than 500 nm typically evolves from a multiphoton character to the tunneling regime ($\gamma_K < 1$), and the spectrum takes on a characteristic shape like that shown in Figure 4.36e. In this

4.7 Above-Threshold Ionization

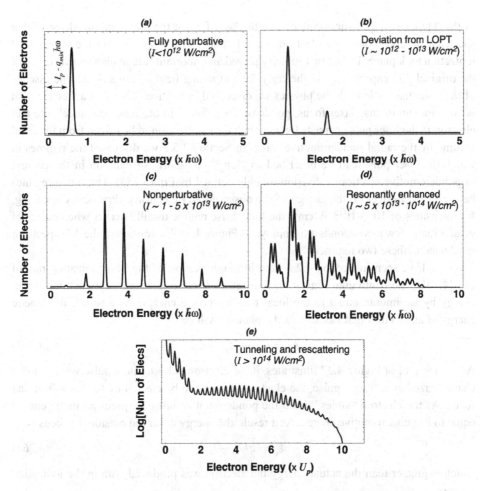

Figure 4.36 Illustration of the qualitative behavior observed in the photoelectron energy spectrum in various intensity regimes, showing the onset of above-threshold ionization, deviations from perturbative behavior (second row), and, finally, evolution to the tunneling limit.

regime, ATI spaced by energy of the photon still persist (at least for pulses longer than a few femtoseconds) because of the periodic nature of tunneling during the course of many wavelengths. The majority of electrons produced have energy between the first ATI peak and $2U_p$ with a more or less decreasing yield with increasing ATI order. In addition, a small number of electrons are ionized with energy between $2U_p$ and $\sim 10U_p$, a consequence of the rescattering of some electrons off of the parent nucleus in the half cycle of the field after they have been liberated by tunneling.

4.7.2 ATI by Laser Pulses in the Long and Short Pulse Regimes

Before we discuss the physics behind these trends in the ATI energy spectrum, it is important now to point out that there can be an important difference between the shape of the ATI energy spectrum produced by the laser field at the atom and the shape of that spectrum observed once the electrons have left the focal region where the strong field ionization took place. In fact this subtlety played an important role in interpreting many of the original ATI experiments in the early days of strong field investigations. The essence of the issue has to do with the physics we discussed in Section 3.5: when a free electron is born into an intense laser focus, ponderomotive forces in the focus can accelerate that electron *if* the laser pulse persists long enough for the electron to be driven from the focal volume by the radial ponderomotive force. In Section 3.5.3 we discussed the regimes in which the laser pulse can be described as "long" or "short" in duration in the context of ponderomotive ejection, a division roughly defined by Eq. (3.253). These two regimes have a distinct effect on the energy of the observed ATI electrons. (Practically speaking, for intensities of 10^{13}–10^{15} W/cm^2, the long pulse regime usually occurs when pulses of greater than a few picoseconds are employed.) Figure 4.37 illustrates how the ATI spectrum is affected in these two regimes.

As we have seen in Section 4.2.6 (and in the derivation of the SFA ionization rate of Eq. (4.101)) the continuum ionization limit of an atom in a strong field is shifted up in energy by an amount equal to the local ponderomotive energy. As a result, the kinetic energy of an electron that has absorbed q photons will be

$$\varepsilon_q = q\hbar\omega - I_p - U_p. \tag{4.263}$$

As the top plot of Figure 4.37 illustrates, if the electron is born into a pulse which can be characterized as a "long" pulse, the electron has time to be ejected by the laser from the focus. As the electron "slides" down the ponderomotive hill it acquires a kinetic energy equal to its ponderomotive energy. As a result, the energy observed outside the focus is

$$\varepsilon'_q = q\hbar\omega - I_p, \tag{4.264}$$

which is greater than the actual energy the electron was produced with in the ionization process.

On the other hand, if the pulse is "short," the situation illustrated in the bottom of Figure 4.37 occurs. The pulse falls to zero intensity before the electron exits the focus and any ponderomotive energy the electron has when it has been freed into the continuum is given back to the field. The observed kinetic energy, then, is simply the actual energy it was born with, given by Eq. (4.263). We can see, then, that the spectrum observed in short laser pulses actually reflects the shape of the ATI energy spectrum of the electrons during the ionization process. Electrons born into a long pulse, however, are observed with the total energy they acquire above the field-free ionization potential of the atom.

4.7.3 ATI in the Perturbative Regime

ATI at modest intensities initially has a spectral shape like that of the plot in Figure 4.36a with higher-order ATI peaks appearing as the intensity increases. Of course, the amplitude of these higher-order ionization terms in the perturbation series scale like $(I_0/I_a)^q$ so we expect the ATI orders in this regime to decrease exponentially with increasing ATI order.

4.7 Above-Threshold Ionization

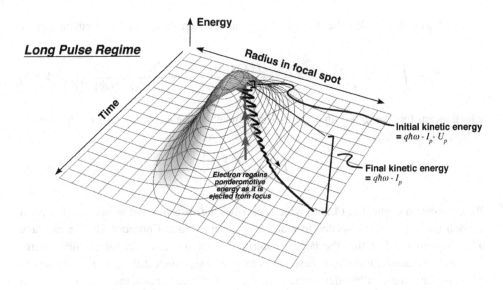

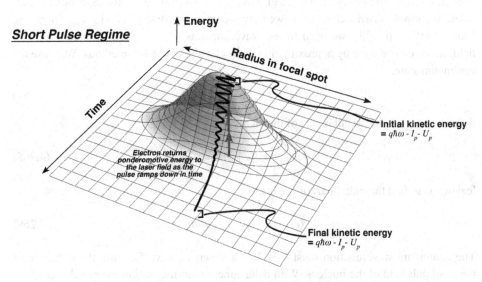

Figure 4.37 Illustration of the consequences on the observed electron energy spectrum when the laser pulse is in the long pulse regime or when it is in the short pulse regime.

It is interesting, then, to ask at what intensity might we expect the amplitude of the second ATI peak to approach that of the first? This is the intensity where a "nonperturbative" kind of ATI spectrum, like that of Figure 4.36c, will start to form. This question can be answered quantitatively by comparing the magnitude of the next-higher term in the perturbation expansion with the lowest-order perturbation term. (For this analysis we follow closely developments made by Fedorov (1997, p. 324).)

Recall from Eq. (4.35) that the qth-order multiphoton dipole matrix element can be written as

$$x_{fi}^{(q)} = \int \sum_{j_1} \cdots \int \sum_{j_{q-1}} \frac{\langle f|x|j_{q-1}\rangle \cdots \langle j_1|x|i\rangle}{(\varepsilon_{j_{q-1}} - \varepsilon_i - (q-1)\hbar\omega) \cdots (\varepsilon_{j_1} - \varepsilon_i - \hbar\omega)}, \qquad (4.265)$$

which means that the $(q+1)$th order matrix element can be written in terms of this lower-order element, giving

$$x_{fi}^{(q+1)} = \int \sum_{j_q} \frac{\langle f|x|j_q\rangle x_{fi}^{(q)}}{(\varepsilon_{j_q} - \varepsilon_i - q\hbar\omega)}. \qquad (4.266)$$

We consider an element in (4.266) which describes the absorption of an additional photon beyond those q photons needed to reach a continuum state. Consequently, we can take advantage of the fact that the largest contribution to the sum over intermediate states will be continuum states since these states have energies such that $\varepsilon_{j_q} - \varepsilon_i$ is closest to $q\hbar\omega$ and have the smallest denominator in the sum. We can approximate, then, this sum over all intermediate states by the integral over continuum states. This requires us to calculate the dipole matrix element between two continuum states, $\langle k_{q+1}|x|k_q\rangle$. Following Fedorov (1997, p. 325), we need to use wavefunctions which account for the Coulomb field, which can be done by approximating these elements by WKB methods. We write the continuum state:

$$|k\rangle = \chi_k(r) Y_{lm}(\theta, \varphi). \qquad (4.267)$$

Since $x = r\cos\theta$, the angular part of the element is

$$\int Y_{10}^* \cos\theta\, Y_{00} d\Omega = \frac{1}{\sqrt{3}}, \qquad (4.268)$$

leaving us to find the radial part of the integral

$$\langle r\rangle = \int \chi_{k_{q+1}}^* r \chi_{k_q} r^2 dr. \qquad (4.269)$$

The continuum wavefunction must look like a spherical wave far from the influence of the Coulomb field of the nucleus. With delta-function normalization the radial part of the WKB continuum states can be approximated,

$$\chi_k = k_\infty \sqrt{\frac{2}{\pi k_e(r)}} \frac{1}{r} \sin\left[\int k_e(r) dr\right], \qquad (4.270)$$

where k_∞ is the electron wavenumber far from the nucleus. If the electron has energy above the ionization threshold of ε_k, and we ignore the centrifugal potential from any electron angular momentum, we can write for the electron wavenumber

$$k_e(r) = \frac{1}{\hbar} \sqrt{2m\left(\varepsilon_k + \frac{Ze^2}{r}\right)} \qquad (4.271)$$

4.7 Above-Threshold Ionization

and

$$k_\infty = \frac{\sqrt{2m\varepsilon_k}}{\hbar}. \tag{4.272}$$

Since the absorption of photons by the electron occurs near the nucleus, the free kinetic energy ε_k, is much smaller than the binding Coulomb potential, allowing us to Taylor-expand the wavenumber about this small energy:

$$k_e(r) \cong \frac{1}{\hbar}\sqrt{2m\frac{Ze^2}{r}} + \frac{1}{\hbar}\sqrt{\frac{mr}{2Ze^2}}\varepsilon_k. \tag{4.273}$$

If we insert Eq. (4.273) into the sine function of (4.270), keep only the first term of (4.273) when inserting into the prefactor of (4.270), and expand the sine function into exponential functions, we can write for the radial part of the matrix element (4.269)

$$\langle r \rangle \cong \frac{2}{\pi} k_\infty k'_\infty \frac{\hbar}{\sqrt{2mZe^2}} \int_0^\infty r^{3/2} \frac{i}{2}$$

$$\times \left(\exp\left[i\frac{1}{\hbar}\sqrt{\frac{m}{2Ze^2}}(\varepsilon_k - \varepsilon_{k'}) \int_0^r r'^{1/2} dr' \right] + c.c. \right) dr$$

$$\cong \frac{i}{\pi\hbar}\sqrt{\frac{2m\varepsilon_k\varepsilon_{k'}}{2e^2}} \int_0^\infty r^{3/2} \left(\exp\left[i\frac{1}{3\hbar}\sqrt{\frac{2m}{Ze^2}}(\varepsilon_k - \varepsilon_{k'}) r^{3/2} \right] + c.c. \right) dr, \tag{4.274}$$

where the primed and unprimed wavenumbers refer to the continuum states before and after absorption of the photon.

This integral can be performed analytically, noting that

$$\int_0^\infty x^{3/2} \exp\left[iax^{3/2}\right] dx = \frac{2}{3}\Gamma\left(\frac{5}{3}\right)(-ia)^{-5/3}, \tag{4.275}$$

where $\Gamma(x)$ is the usual gamma function, and the fact that $(-i)^{-5/3} = \exp[-i\pi/6]$ to yield for the radial matrix element

$$\langle r \rangle \cong \alpha_r i \frac{3^{5/3} \hbar^{2/3}}{2^{1/3} \pi} \left(\frac{Ze^2}{m}\right)^{1/3} \frac{\sqrt{\varepsilon_k \varepsilon_{k'}}}{(\varepsilon_k - \varepsilon_{k'})^{5/3}}. \tag{4.276}$$

Here the numerical cofactor, $\alpha_r = 4\Gamma(5/3)\cos\pi/6 = 1.04$. Recalling that the energy difference between the initial and final continuum states is $\varepsilon_k - \varepsilon_{k'} = \hbar\omega$, we use (4.276) and write for the $(q+1)$th dipole matrix element an integral over all continuum energies

$$x_{fi}^{(q+1)} \cong \int_0^\infty i\alpha_r \frac{\hbar^{2/3}}{\pi} 3\left(\frac{9Ze^2}{2m}\right)^{1/3} \frac{\sqrt{\varepsilon_k(\varepsilon_k + \hbar\omega)}}{(\hbar\omega)^{5/3}} \frac{x_{fi}^{(q)}}{(\varepsilon_k - \varepsilon_i - q\hbar\omega)} \frac{d\varepsilon_k}{\varepsilon_k}. \tag{4.277}$$

To evaluate this integral, we use what is commonly known as the "pole approximation" (Fedorov 1997, p. 337) where, in the integral, we say

$$\lim_{\delta \to 0} \frac{1}{x + i\delta} \approx -i\pi \delta(x). \tag{4.278}$$

Employing this approximation in Eq. (4.277), we get finally

$$x_{fi}^{(q+1)} \cong \alpha_r 3 \left(\frac{9Ze^2\hbar^2}{2m} \right)^{1/3} \frac{x_{fi}^{(q)}}{(\hbar\omega)^{5/3}}, \qquad (4.279)$$

where we have set $\varepsilon_k \approx \varepsilon_k'$.

This result lets us evaluate the ratio of neighboring ATI peaks in the perturbative regime. Using our perturbation theory result for q-photon ionization, Eq. (4.49) and the multiphoton dipole matrix element, Eq. (4.279), we find that the ratio between $(q + 1)$ and q-photon ionization rates is

$$\frac{W^{(q+1)}}{W^{(q)}} \cong \alpha_r^2 9 \left(\frac{9Ze^2\hbar^2}{2m} \right)^{2/3} \frac{1}{(\hbar\omega)^{10/3}} \left(\frac{eE_0}{2} \right)^2. \qquad (4.280)$$

These multiphoton rates will be equal, and neighboring ATI peaks will have comparable magnitude at a critical intensity where the ratio in Eq. (4.280) is ≈ 1. At this point the ATI spectrum will begin to take on a nonperturbative character and will begin to look like the spectrum in Figure 4.36c. This critical intensity (for $Z = 1$ and $\alpha_r \approx 1$) is

$$I_c \cong \frac{1}{18\pi} \left(\frac{2m}{9e^2\hbar^2} \right)^{2/3} \frac{c(\hbar\omega)^{10/3}}{e^2}. \qquad (4.281)$$

In laser fields of optical and near-IR wavelengths, the critical intensity for the onset of nonperturbative ATI spectra predicted by Eq. (4.281) is at a surprisingly modest intensity. In a pulse with wavelength at 527 nm (second harmonic of an Nd-based laser) this critical intensity is 1.5×10^{12} W/cm^2. In an 800 nm pulse it is even lower, being around 4×10^{11} W/cm^2. This analysis explains why electron energy spectra from the ionization of noble gas atoms by optical wavelength pulses are usually observed with many ATI peaks (5–10) of comparable height at intensities as low as $\sim 10^{13}$ W/cm^2, where the field is still nearly two orders of magnitude below an atomic unit of electric field (Eberly et al. 1991). The observation of ATI spectrum of this form with these modest intensities was initially considered a surprise, but we can see that in fact such behavior is completely consistent with the perturbative multiphoton ionization model.

4.7.4 Nonperturbative ATI in the Long Pulse Regime and Peak Suppression

What form the observed nonperturbative ATI energy spectrum takes does depend significantly on whether the experiment is performed with long or short laser pulses. As we discussed in Section 4.7.2, ATI can be thought of as being in the long pulse regime when the laser pulse duration, τ_p, and focal spot size, w_0, fulfill the condition

$$\tau_p \geq w_0 \sqrt{\frac{m}{U_p}}. \qquad (4.282)$$

At the onset of the nonperturbative regime of ATI, peak intensity is around 10^{13} W/cm^2 which means $U_p \approx 0.5 - 1$ eV. This means that the long pulse regime is typically reached in

4.7 Above-Threshold Ionization 231

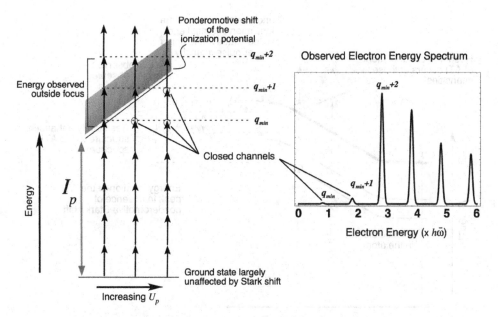

Figure 4.38 Effects of increasing intensity and corresponding ponderomotive continuum energy shift show closing of lowest-order MPI channels as intensity increases. The characteristic ATI spectrum in the multiphoton regime when electrons are observed in the long pulse limit, illustrating the effects of channel closing which suppress the lowest-order ATI energy peaks.

spots of focal radius ~10 µm when pulse durations are ~30 ps or longer. As discussed earlier, under this condition the observed ATI spectrum has the peculiar quality that, because of ponderomotive ejection from the focus, the electron energies observed are those energies given by the difference between the number of photons actually absorbed during ionization and the field-free, vacuum ionization potential of the atom, not the actual energy of the electron at the moment of ionization, which is affected by the upward ponderomotive Stark shift of the ionization continuum.

This effect has a consequence which is summarized in Figure 4.38. As local laser intensity is increased, the ponderomotive Stark shift of the ionization threshold mandates that an increasing number of photons is required to overcome the threshold. But because the observed energy is that of the difference of this many absorbed photons and the unshifted ionization potential, there will be some low ATI orders which appear to be produced, as the energy spectrum reproduced in the right-hand panel of Figure 4.38 illustrates. So in addition to the onset of the production of a range of ATI peaks of comparable peak height, the first few ATI peaks can be reduced to zero amplitude by this channel-closing effect. This observation is usually referred to as peak suppression. The number of ATI orders suppressed depends on how many times the ponderomotive energy is over the photon energy, and the number of suppressed peaks will be roughly $\approx U_p/\hbar\omega$. In real experimental situations, the fact that the focal spot has a range of intensities serves to produce at least a small

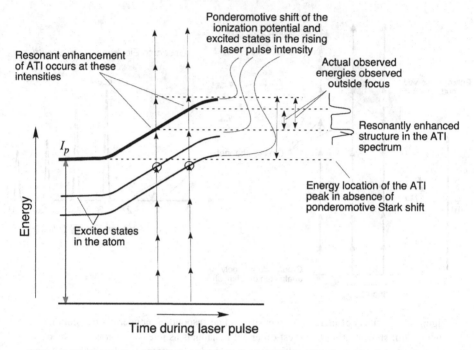

Figure 4.39 Illustration of how the dynamic Stark shift in a time-varying laser pulse can give rise to resonant enhancement of the ionization rate which leads to a predominance of electrons with particular ATI energies (leading to so-called Freeman resonances).

number of electrons at lower ponderomotive energy, so ATI spectra in this regime rarely show 100% suppression of these ATI channels.

4.7.5 Nonperturbative ATI in the Short Pulse Regime

ATI energy spectra produced in short laser pulses evince a rather different trend. In this regime, it is often observed that the Stark shift of higher unoccupied energy levels has a dramatic effect on the shape of the ATI spectrum. Recalling our discussion of Section 4.2.6, we know that when the intensity is sufficient to Stark shift upper energy levels by a good fraction of the photon energy (while largely not affecting the more deeply bound ground state) this Stark shift can resonantly enhance the MPI rate transiently as upper levels are shifted into a multiphoton resonance at different times in the rising and falling laser pulse temporal envelope. Figure 4.39 helps to explain the consequences of this shift.

In the long pulse regime, resonant enhancement of the ionization rate has no notable effect on the shape of the ATI spectrum; any signature of what the actual ponderomotive shift was when an electron was born is erased by the subsequent reacquisition of that ponderomotive energy during ejection from the focus. The situation in short pulses is quite different. As we have discussed, the observed energies of electrons in this case are directly equal to the energy that an electron acquires above the ponderomotively shifted continuum limit at the moment of its ionization. The shifting of certain upper levels into resonance during the laser pulse serves to produce a large number of electrons at specific ponderomotive energies. That means that the net energy of electrons produced in any given ATI order (i.e. a given number of absorbed photons) will reflect the specific ponderomotive shift that corresponds to an enhancement of the ionization rate from a transiently produced resonant enhancement. Therefore, as upper levels shift in and out of resonance during a laser pulse, very specific energy differences between the energy acquired by q-photon absorption and the ponderomotively shifted ionization potential will show up as enhanced peaks in the ATI spectrum. The electron energy that is observed for the time that the ionization rate is increased through such a transient resonant enhancement will be the difference between the energy of the absorbed photons and the ionization potential shifted by the ponderomotive energy at that particular resonance. Different levels will come into resonance at different intensities, producing enhanced electron production at different net energies and a peak in the spectrum at a specific shifted energy. As a result, fine structure like that shown in Figure 4.36d, at an energy scale less than the quantized ATI peaks separated by one photon energy unit, will appear. These resonance substructures in the ATI spectrum are often referred to as Freeman resonances (Freeman and Bucksbaum 1991; Freeman et al. 1992).

In principle, these resonance peaks will repeat in each ATI peak. Figure 4.40 shows an example of how this effect manifests in ATI spectrum resulting from the MPI of Xe atoms by 600 nm laser pulses. At intensity approaching 10^{14} W/cm^2, Stark shifts bring f and p

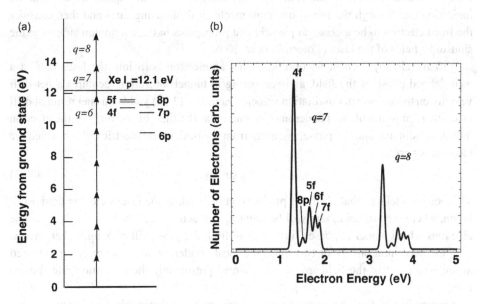

Figure 4.40 Example shape of two ATI electron energy peaks in the short pulse regime when Freeman resonances come into play. This spectrum is similar to that observed when Xe is irradiated by 600 nm pulses near 10^{14} W/cm^2. The diagram in (a) shows the Xe excited states that play a role in the multiphoton resonantly enhanced ATI structure.

energy levels into six-photon resonance. The energy level diagram in Figure 4.40 shows that a Rydberg f-level, like the 7f state, is brought into resonance at the lowest intensity shift first. A resonantly enhanced peak in the ATI spectrum at energy just below the unshifted location of an ATI peak occurs (see the example ATI spectrum in part (b) of Figure 4.40). This is followed by subsequent shifts of 6f, 5f, 8p and 4f levels into six-photon resonance, each in turn as intensity increases. The ATI spectrum exhibits peaks corresponding to the resonances at lower and lower electron energy since the ponderomotively shifted continuum limit moves up as each state moves into resonance at higher intensity. Of course, the magnitude of each resonance peak depends on the magnitude of the multiphoton matrix element connecting it with the ground state as well as the amount of time during the pulse that each level remains in resonance.

4.7.6 ATI in the Quasi-Classical Tunneling Regime

This discussion of ATI has been in the context of a multiphoton ionization picture, as this is usually the relevant picture in the regime where nonperturbative effects begin, optical to near-IR pulses at intensity of 10^{13} to a few times 10^{14} W/cm^2. However, we know that at higher intensity at typical laser wavelengths the Keldysh parameter falls below one and the ionization takes on a tunneling character. What, then, is to be expected of the ionized electron energies in this regime? If we continue to ignore relativistic effects, this question can be answered in the context of the quasi-classical model of tunneling ionization, like that used to explain NSDI (Corkum et al. 1989; Corkum et al. 1992). Again, we can find the ionization rate through the use of quantum mechanical tunneling rates and then consider the freed electron to be a classical particle that propagates in the continuum subject to the sinusoidal fields of the laser (Colosimo et al. 2008).

We have already seen in Section 3.2.5 that an electron born into the laser field at a well-defined phase of the field, as is occurring in tunnel ionization, acquires a net drift velocity on top of its usual oscillation velocity. Equation (3.38) yields this net residual drift velocity, ε, in terms of the ponderomotive energy at the time of the birth of the electron into the continuum and the phase, φ_0, away from the peak of the electric field in which the electron is born:

$$\varepsilon = 2U_p \sin^2 \varphi_0. \tag{4.283}$$

This equation tells us that electrons produced at the peak of the laser's electric field oscillation, where the tunneling rate will be highest, will acquire zero net kinetic energy. The electrons which tunnel at phases further away from the peak will pick up higher energy, with the maximum energy of two times the local ponderomotive energy being imparted at the zero point of the field, $\varphi_0 = \pi/2$ (where presumably the ionization rate falls to zero).

It is possible to find the spectrum of electron energies that might be expected in this quasi-classical ionization picture by combining Eq. (4.283) with a static field tunneling rate like the hydrogenic ion rate of Eq. (4.182) (Burnett and Corkum 1989). Such a rate yields a weighting for the number of electrons produced as a function of the phase of the

4.7 Above-Threshold Ionization

field which combine with the quasi-classical model for residual energy to yield an energy spectrum. If we note that the laser's field can be written

$$E = E_0 \cos\varphi, \tag{4.284}$$

we have with Eq. (4.283) that the relation between laser field strength and liberated electron energy is

$$E = E_0 \left(1 - \frac{\varepsilon}{2U_p}\right)^{1/2}. \tag{4.285}$$

Using the hydrogen-like ion DC tunneling rate

$$w_{sf} = 4\tilde{Z}^5 \omega_a \frac{E_a}{E} \exp\left[-\frac{2}{3}\left(\frac{I_p}{I_H}\right)^{3/2} \frac{E_a}{E}\right], \tag{4.286}$$

the energy distribution of electrons produced by tunneling over an optical cycle is

$$f(\varepsilon)\,d\varepsilon = aw_{sf}(t)\,dt, \tag{4.287}$$

where a is some suitable normalization factor. From Eq. (4.283) we can write

$$d\varepsilon = 4U_p\omega \sin\varphi \cos\varphi\, dt, \tag{4.288}$$

which, when combined with relations (4.285) and (4.286), yield for the electron energy distribution from one optical cycle of tunneling

$$\begin{aligned} f(\varepsilon)d\varepsilon &= a\frac{1}{\sin\varphi \cos^2\varphi} \exp\left[-\frac{2}{3}\left(\frac{I_p}{I_H}\right)^{3/2} \frac{E_a}{E_0 \cos\varphi}\right] \\ &= a\sqrt{\frac{2U_p}{\varepsilon}} \frac{1}{1-\varepsilon/2U_p} \exp\left[-\frac{2}{3}\left(\frac{I_p}{I_H}\right)^{3/2} \frac{E_a}{E_0} \frac{1}{(1-\varepsilon/2U_p)^{1/2}}\right]. \end{aligned} \tag{4.289}$$

An example of electron energy distributions generated by this formula are illustrated in Figure 4.41 for two intensities ionizing He. Not surprisingly, the energy distribution is strongly peaked at zero energy, reflecting the strong preference of tunneling at the peak of the field. The distribution then falls off rapidly toward the maximum electron energy of $2U_p$, a consequence of the exponential factor in the distribution of Eq. (4.289).

This analysis would imply that the ATI spectrum in the tunneling regime is smooth and lacks the peaks we have seen associated with the absorption of additional photons in the continuum. It is important to note that the spectrum of (4.289) is just that generated by a single half optical oscillation. In reality, tunneling will occur over a sequence of oscillations (with the exception of extremely short pulses, a situation considered in Section 4.7.9). This means that the ionization can be thought of as the field periodically pulling off a sequence of wavepackets in one direction by tunneling from the atom, with temporal spacing given by the laser optical period $2\pi/\omega$. If one were then to think about the electron energy spectrum as a Fourier transform of the temporal structure of the tunnel-ionized wavefunction, this periodic structure in the freed wavefunction leads to the generation of peaks in the electron energy spectrum.

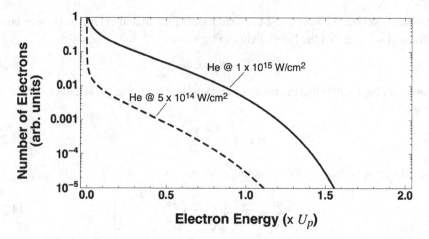

Figure 4.41 Typical electron energy spectra resulting from one cycle of tunnel ionization of an ensemble of He atoms by one laser oscillation cycle in the tunnel ionization regime.

For example, consider the freed wavefunction that has been released by tunneling over j optical cycles. The electron energy spectrum, in terms of the Fourier transform of the freed electron wavefunction, ψ, could be written

$$f(\varepsilon) \propto \left| \int \sum_j \psi(t + j2\pi/\omega) \exp\left[i\varepsilon t/\hbar\right] dt \right|^2. \qquad (4.290)$$

As a simple illustration, if the electron tunnels over two optical cycles the electron energy spectrum is

$$f(\varepsilon) \propto |1 + \exp[i2\pi\varepsilon/\hbar\omega]|^2 \left|\tilde{\psi}(\varepsilon)\right|^2$$
$$\propto (1 + \cos[2\pi\varepsilon/\hbar\omega]) \left|\tilde{\psi}(\varepsilon)\right|^2, \qquad (4.291)$$

where $\left|\tilde{\psi}(\varepsilon)\right|^2$ represents the Fourier transform of the wavepacket liberated during one of the optical cycles, and hence, the single cycle energy spectrum like that of Eq. (4.289). We can see from the two-cycle example that this periodic release of the wavepackets leads to modulation of the single-cycle electron energy spectrum at energy intervals of the photon energy. Therefore, the ATI spectrum in the tunneling regime will exhibit a shape like that of Figure 4.41 with peaks separated by $\hbar\omega$, superimposed on this spectral envelope, a peak spacing just as in the multiphoton regime.

4.7.7 Rescattering Effects in ATI

This is not the complete story of ATI in the tunneling regime, however. We have already seen that the rescattering of tunnel-ionized electrons can lead to inelastic collisions with the parent nucleus and subsequent double ionization of the atom. Tunnel-ionized electrons

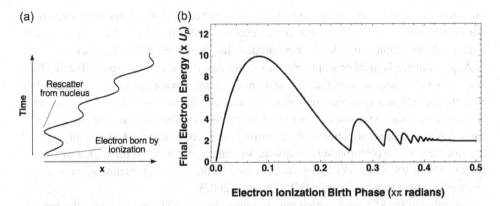

Figure 4.42 Trajectory of an electron undergoing elastic backscattering in a laser field-driven recollision (a). A calculation of the energy of electrons which undergo a single 180° elastic backscattering event from the parent nucleus as a function of the phase in the field at which the electron initially tunnel ionizes is shown at right.

that return to the nucleus can also scatter elastically from the residual Coulomb potential of the parent nucleus. The residual drift energy of tunnel-ionized electrons predicted by Eq. (4.283) assumed that the freed electron oscillated untouched in the laser field, and that the oscillating electron adiabatically gave back any ponderomotive energy to the field when the pulse died away. An elastic Coulomb collision with the parent nucleus can break that adiabaticity. Consequently, if a tunnel-ionized electron does return to the parent nucleus, something that happens only for electrons liberated in the quarter optical cycle *after* the peak of the electric field, there is a finite probability that a scattering collision will break the oscillatory adiabaticity, and the electron will acquire *more* energy than Eq. (4.283) predicts (Chen et al. 2009). This situation is generically illustrated in the left hand panel of Figure 4.42 (Becker et al. 2002).

In fact, the maximum energy that the electron can acquire will occur if that electron underwent a 180° elastic scattering from the nucleus. In that case the electron still has its field-derived oscillatory energy, which as we know can be up to 3.17 U_p for electrons born 18° after the peak of the electric field, but now they are traveling the opposite direction. The plot in part (b) of Figure 4.42 shows a calculation of the final kinetic energy that an electron has as a function of the phase of birth in the field after the field peak if it does undergo such a 180° collision. (Collisions resulting in smaller scattering angles will yield residual energies less than those plotted in Figure 4.42.) As this figure illustrates, tunnel-ionized electrons undergoing 180° scattering can acquire an energy up to $10U_p$ (in fact, 10.007 U_p to be more precise). This occurs for electrons born at a phase 15° after the peak of the electric field, just slightly earlier than the electrons acquiring the maximum recollision energy.

This rescattering leads to a low-amplitude plateau in the ATI spectrum of tunnel-ionized electrons that extends in energy well out past the $2U_p$ energy cutoff of the quasi-classical prediction of Eq. (4.283) to a cutoff in the vicinity of $10U_p$ (Milosevic et al. 2010). Observed ATI spectra in the tunneling regime from linearly polarized pulses typically take on a shape like that illustrated in Figure 4.36e. Low ATI orders fall off quickly with increasing electron energy out to 2 U_p. Above that energy a broad plateau of ATI orders, usually with amplitude about two orders of magnitude below the lowest ATI orders, is seen.

A quantum mechanical description of this recollision physics is in general difficult. However, it is interesting to note that the SFA model does provide a means to evaluate this rescattering ATI in a quantum framework. Remember that the SFA model for ionization we considered was the result of a perturbation expansion, where the atomic potential was treated as the small perturbation to the electron in the strong laser field. We simply utilized the first term in the perturbation expansion. We can, in the spirit of the higher-order perturbation theory we have already explored, return to this SFA perturbation expansion and examine the next term in the series (Becker et al. 2002).

Returning to the SFA propagation operator given in Eq. (4.80), we now take the second-order term in this expansion. That second-order propagation operator is

$$u_2^{(SFA)}(t) = -\frac{1}{\hbar^2} \int_{-\infty}^{t} dt' \int_{-\infty}^{t'} dt'' \hat{V}(t'') \hat{V}(t'). \tag{4.292}$$

Following the procedure we used to derive LOPT, we write the transition amplitude that results from this propagator and insert a complete set of Volkov states between the two atomic potential energy operators to yield

$$A_{fi}^{(2)} = -\frac{1}{\hbar^2} \int_{-\infty}^{t} dt' \int_{-\infty}^{t'} dt'' \int d^3 k' \\ \times \left\langle \psi_V(\mathbf{k}_e) \left| \hat{V}(t'') \right| \psi_V(\mathbf{k}'_e) \right\rangle \left\langle \psi_V(\mathbf{k}'_e) \left| \hat{V}(t') \right| \psi_i \right\rangle \tag{4.293}$$

Utilizing the previously derived Keldysh approximation on the second term in this integral and steps similar to those employed in Eqs. (4.83) to (4.86), we find for the transition amplitude

$$A_{fi}^{(2)}(t) = -\frac{1}{\hbar^2} \int_{-\infty}^{t} dt' \int d^3 k'_e \left\langle \psi_V(\mathbf{k}_e, t') | V(\mathbf{x}) | \psi_V(\mathbf{k}'_e, t) \right\rangle \\ \times \int_{-\infty}^{t'} dt'' \left\langle \psi_V(\mathbf{k}'_e, t'') \left| H_I(t'') \right| \psi_i \right\rangle. \tag{4.294}$$

This amplitude is added to the previously derived first-order SFA transition amplitude, Eq. (4.89), to find the transition probability of the electron reaching a Volkov state in the continuum with electron wakenumber $\mathbf{k}_e$.

Equation (4.294) for the second-order SFA transition amplitude has a remarkably intuitive interpretation now that we understand high-order ATI through the simple quasi-classical rescattering model. Reading the terms in this amplitude from right to left, we can say that the electron begins in its initial state, described by ψ_i. The electron interacts with the laser field through the interaction Hamiltonian and makes a transition to a free state described by the Volkov wavefunction of wavenumber $\mathbf{k}'_e$. The electron then, after some time propagating in the laser field rescatters off of the parent ion potential via the ion potential operator $V(\mathbf{x})$ to a final electron energy wavenumber $\mathbf{k}_e$. The process is

4.7 Above-Threshold Ionization

completely described when the transition amplitude to k_e is integrated over all possible intermediate electron wavenumbers k'_e.

In practice, the 5D integral of Eq. (4.294) is challenging to evaluate. It has been numerically evaluated by some authors and does yield remarkably accurate productions for ATI energies and angular distributions in the high-order ATI rescattering cutoff energy region (Becker *et al.* 1994).

It is also interesting to note that this second-order SFA transition amplitude can be rewritten in another suggestive form. Using Eq. (3.346) for the explicit form of the phase of the Volkov wavefunctions, the second-order SFA amplitude can be expressed in terms of a total phase and a weighting term:

$$A_{fi}^{(2)}(t) = -\frac{1}{\hbar^2} \int_{-\infty}^{t} dt' \int d^3 k'_e \int_{-\infty}^{t'} dt'' M_{fi} \exp\left[iS_{k_{ei}}(k'_e, t_0, t')/\hbar\right]. \quad (4.295)$$

The phase term is

$$S_{k_{ei}}(k'_e, t_0, t') = -\frac{1}{2m}\int_{t'}^{\infty}\left(\hbar k_e + \frac{e}{c}A(t)\right)^2 dt - \frac{1}{2m}\int_{t_0}^{t'}\left(\hbar k'_e + \frac{e}{c}A(t)\right)^2 dt + I_p t_0, \quad (4.296)$$

which is weighted by a term which can be expressed in terms of the initial wavefunction and simple continuum plane-waves of wave numbers k_e and k'_e

$$M_{fi} = \langle k_e|V(r)|k'_e\rangle \langle k'_e|H_I|i\rangle, \quad (4.297)$$

which can be evaluated using the methods described leading up to Eq. (4.93). Inspection of the phase, Eq. (4.296), reveals that it takes the form of a "modified" action for the electron (slightly different than the classical action defined by time integral of the classical Lagrangian). The "modified action" in this case is formed by an electron bound by the ionization potential up to the time t_0. This is followed by action of the electron in the laser vector potential with intermediate mechanical momentum $p'_e = k'_e + eA(t)/c$ over the time interval between being released by the initial tunneling at t_0 to the time of the recollision at t'. Then, finally, the scattered electron acquires action with its final momentum in the field over a time interval from the recollision to infinity. We will find this form for the phase of the recollision amplitude very useful in Chapter 6 where we will consider high-order harmonic generation by recolliding electrons.

(It is also interesting to note that this interpretation of the next term in the SFA transition amplitude as a rescattering term gives us some information about the convergence properties of the SFA perturbation expansion. Since this implies that we can interpret higher-order terms in this expansion in terms of multiple scatterings from the nucleus, it is clear, at least in long-wavelength fields, that wavepacket spreading makes subsequent terms representing multiple returns far less important. We expect, then, that the SFA expansion will converge rapidly and justifies the Keldysh approximation as the dominant term describing ionization. This reasoning also argues that the SFA expansion will converge slowly in very short-wavelength light where the electron wavepacket spends little time in the continuum between rescatterings; many returns will occur and, hence, many terms in the SFA expansion would be needed for an accurate calculation.)

As one final postscript to this description of rescattering ATI, we should note that experimental observations of the ATI plateau in the electron energy range of 2× to ∼10 × U_p are not always like that pictured in Figure 4.36e, where the plateau is characterized by ATI orders on a smooth single-humped envelope. In some atoms and certain intensities, broad oscillatory structure is observed in the shape of the plateau envelope (Wassaf et al. 2003; Paulus et al. 2001). This structure is usually seen to be very dependent on laser pulse intensity and target atom, which implies that ponderomotively shifted resonances or continuum limits play a role. We will not discuss this effect in depth other than noting that the best explanation for it probably arises from electrons which acquire energy from the field but not enough to escape completely the influence of the binding potential. Such a situation arises, for example, if Rydberg levels are ponderomotively shifted into resonance during the laser pulse, the same physics leading to Freeman resonances in the ATI spectrum. If this occurs, some small number of electrons will be temporarily promoted to one of these upper levels. Such an electron will then linger in the vicinity of the nucleus, increasing the probability of recolliding with the nucleus one or more times during optical cycles subsequent to the promotion of the electron. This enhanced rescattering will give rise to intensity-dependent enhancements of electron production at high energy and hence lead to structure in the ATI plateau (Milosevic et al. 2007).

4.7.8 ATI with Circularly Polarized Pulses

When a circularly polarized field performs the ionization, ATI energy spectra exhibit a noticeable difference from spectra generated by linear polarization in otherwise identical conditions (intensity, atomic species etc.). It has already been noted that the ionization rate tends to decrease markedly when circular polarization is employed, so the amplitude of ATI peaks drops markedly when this kind of polarization is used. To understand how circular polarization affects the shape of an ATI spectrum, first consider ATI electron production in the multiphoton ionization regime. If q_m is the minimum number of photons needed to ionize the atom and s is the number of photons absorbed above q_m leading to additional ATI peaks, the selection rules for absorption of circularly polarized light (where $\Delta m_l = \pm 1$) mandate that the final angular momentum of the freed electron is

$$\ell_f = (q_m + s)\,\hbar. \tag{4.298}$$

The effective potential that the electron experiences upon promotion to the continuum includes the Coulomb potential of the parent ion and the centripetal potential of the final angular momentum, so is

$$V_l(r) = -\frac{Ze^2}{r} + \frac{\ell_f(\ell_f + 1)\,\hbar^2}{2mr^2}. \tag{4.299}$$

This high-angular-momentum final state leads to marked suppression of low ATI peak orders and an effective shifting of the highest-amplitude observed ATI order to energy higher than similar focal conditions with a linearly polarized pulse.

4.7 Above-Threshold Ionization

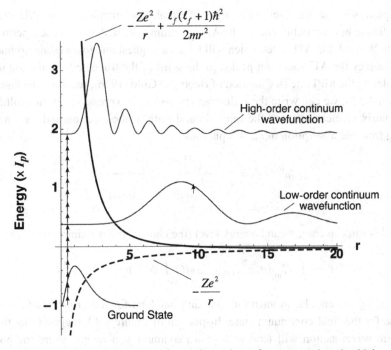

Figure 4.43 Plot of the energy potential near the nucleus of an atom undergoing high-order MPI showing how the formation of the centripetal barrier for electrons acquiring high angular momentum (as occurs in ionization by circularly polarized pulses) pushes the continuum wavefunction of electrons ionized with low MPI orders out to large radii, decreasing the effective overlap of the wavefunction with the initial ground state wavefunction and, therefore, suppressing the MPI rate for those lower orders.

The origin of this can be understood when one considers the shape of the continuum final state wavefunction arising from multiphoton promotion to the continuum with effective potential of Eq. (4.299). This situation is illustrated in the potential energy diagram of Figure 4.43, where the Coulomb potential and ground state wavefunction are plotted as a function of radius around the nucleus and are compared to the radial component of continuum wavefunctions at low and high order in the high angular momentum potential described by Eq. (4.299). (These functions are those of Eq. (4.59).) The high-angular-momentum final continuum states resulting from circularly polarized excitation have a centripetal potential which pushes the continuum state away from the nucleus. This effect is most pronounced for electrons with low energy above the ionization threshold (corresponding to low-order ATI peaks). This wavefunction has very small overlap in radius with the ground state bound near the nucleus, which implies that the multiphoton dipole matrix element will be small. On the other hand, when the electron acquires higher energies above the ionization threshold (corresponding to higher ATI orders, i.e. large s), the effect of the centripetal potential is lessened and the final continuum state is closer to the nucleus. Ionization to the continuum associated with these higher orders has greater overlap with the initial ground state and a larger ionization rate.

To explore semiquantitatively the effects of circular polarization on the ATI spectrum and to estimate how far out in energy the ATI spectrum is shifted, we can use a perturbative approach. We seek the ATI order which will have the highest rate to ascertain around what electron energy the ATI spectrum peaks. In the spirit of the truncated summation method for calculating the MPI rate (the method of Bebb and Gold (1966), recalling the discussion of Section 4.2.5), we can write the following approximate expression for the multiphoton dipole matrix element between the initial bound state of the electron and the final state resulting from the absorption of $q_m + s$ photons

$$x_{fi}^{(q_m+s)} \cong \frac{\langle f | x^{q_m+s} | i \rangle}{\prod_{\eta=1}^{q_m+s-1} (\langle \varepsilon_j \rangle - \varepsilon_i - \eta \hbar \omega)}, \qquad (4.300)$$

where $\langle \varepsilon_j \rangle$ is some average bound energy level (the choice here is unimportant) and

$$\langle f | x^{q_m+s} | i \rangle = \int Y_{l'm'_l}^*(\theta, \varphi) Y_{lm_l}(\theta, \varphi) d\Omega \int \psi_f^*(r) r^{q_m+s} \psi_i(r) r^2 dr. \qquad (4.301)$$

Here l and m_l are angular momentum quantum numbers for the ground state and l', m'_l are those for the final continuum state. Inspection of Figure 4.43 suggests that the final continuum wavefunction will tend to have a maximum at a radius where the potential (4.299) is roughly equal to the free energy of the electron. Ignoring the ponderomotive shift of the ionization potential, the radius at which the final state wavefunction is maximum, r_f, is given by the condition

$$-\frac{Ze^2}{r_f} + \frac{\ell_f (\ell_f + 1) \hbar^2}{2mr_f^2} = (q_m + s) \hbar \omega - I_p = \varepsilon_f. \qquad (4.302)$$

We expect the maximum ionization rate will occur to the ATI channel that leads to a correspondence in radius between the peak of the continuum wavefunction as determined from Eq. (4.302) and the peak of the ground state wavefunction.

We can continue our simple estimate by considering the ground state wavefunction to be hydrogenic and by finding where the product $r^{q_m+s} \psi_i(r)$ peaks. The peak radius of this function, r_p, is given by the condition

$$\frac{\partial}{\partial r} \left(r^{q_m+s} \exp[-Zr/a_B] \right) \bigg|_{r=r_p} = 0$$

$$r_p = \frac{a_B}{Z} (q_m + s). \qquad (4.303)$$

The maximum circularly polarized ATI order then occurs when $r_p \approx r_f$. This condition will give us the maximum ATI order, s_{MAX}. Using Eq. (4.298) and ignoring the Coulomb term, Eq. (4.302) becomes

$$\frac{(q_m + s_{MAX})(q_m + s_{MAX} + 1) \hbar^2}{2mr_p^2} \approx (q_m + s_{MAX}) \hbar \omega - I_p \qquad (4.304)$$

4.7 Above-Threshold Ionization

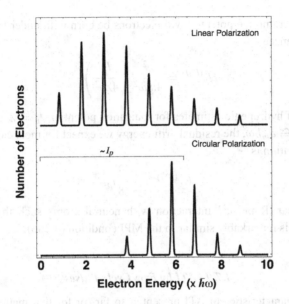

Figure 4.44 Typical nonperturbative ATI energy spectrum for linearly polarized laser pulses and for circularly polarized laser pulses, showing the effective suppression of a number of lower-order (lower-energy) ATI peaks in the case of circular polarization.

Using (4.303) and approximating the ionization potential

$$I_p \cong \frac{Z^2 e^2}{2a_B} = \frac{Z^2 \hbar^2}{2ma_B^2} \approx q_m \hbar\omega, \qquad (4.305)$$

we find finally that we expect the peak ATI order in circularly polarized pulses will be

$$s_{MAX} \cong I_p/\hbar\omega. \qquad (4.306)$$

Since in many situations, many photons are required to ionize the atom (such as 800 nm laser light ionization of argon where $I_p/\hbar\omega \approx 11$) this relation implies that the ATI spectrum, even in weakly nonperturbative intensities will be pushed to orders well above the few orders above threshold. Figure 4.44 illustrates how the shape of an ATI spectrum in mildly nonperturbative intensities differs between circularly polarized and linearly polarized pulses.

This conclusion is also consistent with the quasi-classical model for ATI electron energies produced in the tunneling regime. We have already seen that the electron energy spectrum from tunnel ionization by linear polarization results in a distribution of residual drift energies peaking at zero kinetic energy (Figure 4.41). But we know from Eq. (3.41) that the residual drift energy of all electrons born in a circularly polarized field is just the ponderomotive energy at birth. So, again, ATI electrons born in circular polarization have a spectrum shifted up in energy. The magnitude of the shifted energy can be estimated if we assume that the BSI condition describes when the rate of electrons is significant. If we ask, at what ponderomotive potential will electrons be born with under the BSI condition Eq. (4.220), we can say

$$U_p^{(BSI)} = \frac{e^2}{4m\omega^2}\left(\frac{I_p^2}{4Ze^3}\right)^2 \tag{4.307}$$

or, with the use of hydrogenic estimates for ionization potential, $I_p = Z^2 e^4 m/2\hbar^2$ and the approximation $I_p \cong q_m \hbar\omega$, the residual drift energy we expect for most electrons produced in circular polarization is

$$U_p^{(BSI)} \cong \frac{q_m^2}{128} I_p. \tag{4.308}$$

For optical or near-IR pulses' interaction with neutral atoms such that $q_m \approx 6\text{--}15$, condition (4.308) is remarkably similar to the MPI condition (4.306).

4.7.9 ATI by Few-Cycle Pulses

Many of the characteristics of ATI presented so far (or for that matter, almost all the ionization models so far discussed) have essentially considered ATI electron production in quasi-CW fields. This is a reasonable approximation as long as the pulse duration of the laser is significantly longer than an oscillation period. When the laser pulse is short enough that this condition is not met and is only a few optical cycles in duration, which occurs, say, when near-IR pulses have duration under 10 fs, ATI spectra develop some interesting modifications to the picture already presented (Milosevic *et al.* 2006). First, if the ATI spectrum arises from ionization in the tunneling regime, tunneling may occur only over one cycle, that cycle with the highest field amplitude in the few-cycle pulse. Since structure in the ATI spectrum with peaks separated by energy $\hbar\omega$ can be understood as quantum wavepackets being produced once every optical cycle, this reduction to single-cycle ionization tends to eliminate such photon energy–spaced peaks. The ATI spectra observed from single-cycle pulses is really more of a smooth envelope with a broad plateau extending out to 10 U_p.

An even more interesting feature seen in the rescattering plateaus of ATI from single-cycle pulses arises as a consequence of the phase of the optical wave with respect to the envelope of the laser pulse (Kling *et al.* 2008). As we will discuss in the next section and as is apparent from the discussion of tunneling ATI, electrons produced by strong field ionization are emitted mainly along the laser polarization. For linearly polarized pulses of many optical cycles, clearly the electron spectra observed along the polarization will be of the same energy and amplitude in either direction along the polarization. There are more or less the same number of field oscillations with the electric field pointing in either direction.

In few-cycle pulses that is not the case (Grasbon *et al.* 2003; Liu *et al.* 2004; Milosevic *et al.* 2006). Consider the electron field of a few-cycle Gaussian laser pulse, such that the pulse duration $\tau_p \approx 2\pi/\omega$:

$$E(t) = E_0 \exp\left[-t^2/\tau_p^2\right] \sin\left[\omega t + \varphi_p\right]. \tag{4.309}$$

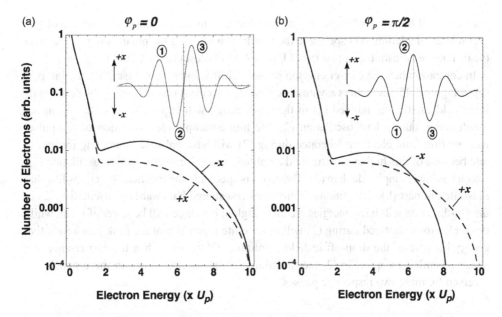

Figure 4.45 Effect on the temporal shape of the electric field for two different phases of the field with respect to the pulse-intensity envelope (inset) in a few-cycle pulse. The effect that this field envelope phase has on the rescattering electron energy spectrum in the two polarization directions when such single-cycle pulses are employed in tunnel ionization. The situation is reversed when phases are shifted by an additional factor of π.

Here we have introduced the carrier-envelope phase, φ_p, describing the phase shift between the optical wave and the peak of the Gaussian pulse-shape envelope. The insets within the two panels of Figure 4.45 show such a pulse (with about 5 fs duration) for two values of this carrier-envelope phase. Obviously the shape of the electric field in time is different in these two cases. Consider the energies of rescattered ATI electrons that we might expect from a pulse with $\varphi_p = 0$, like that in part (a) of Figure 4.45. This pulse has one half optical oscillation with maximum electric field directed along $-x$ and a second optical oscillation of equal field strength oriented along $+x$.

As the electric field amplitude rises, the first significant tunneling rate occurs, at the peak of the field oscillation marked (1) in the figure inset. Electrons born in the quarter cycle after the peak of this oscillation are ejected in the $-x$-direction upon ionization and are brought back to the nucleus, rescattering while subject to optical oscillation (2) where the field is largest in the pulse and driving the electron in the $+x$-direction. Electrons that tunnel ionize during oscillation (2) in the $+x$-direction will rescatter while being driven toward $-x$ during oscillation (3), which has the same amplitude as half oscillation (2) and will presumably produce the same energies in the rescattering process. The differences between ionization in these two oscillations is that the ground state is depleted by ionization during oscillation (2) and by ionization during (1). Therefore when electrons are observed in the two directions $-x$ and $+x$ they will have the same energy in the rescattering plateau, but there will be fewer of these rescattered electrons in the $+x$-direction because of this ground state depletion. ATI spectra in the two directions along the polarization will take on the differences illustrated in part (a) of Figure 4.45 (Milosevic et al. 2006).

In contrast, when the carrier envelop phase is shifted by $\pi/2$, a pulse illustrated in part (b) of Figure 4.45, the field is asymmetric along x. Electrons ionized during the first significant field oscillation, labeled (1) in the $+x$-direction will experience rescattering in the $-x$-direction during half oscillation (2), the highest-amplitude oscillation of the pulse. Rescattering from electrons liberated during (2) will have enhanced tunneling ionization rate because of this higher electric field amplitude but will rescatter during oscillation (3), which has lower amplitude than (2). The result is spectra in the two field directions like that illustrated in part (b) of Figure 4.45; electrons rescattered backwards by oscillation (2) in the $+x$-direction will have energies reaching higher, but there will be fewer of them, while those electrons scattered during (3) will never quite reach 10x of the peak ponderomotive energy because of the drop-off in field amplitude. Of course, when the carrier-envelope phase is shifted by π or $3\pi/2$, the directions of the spectra shown in these panels are reversed for these two respective phases.

4.7.10 Angular Distributions of ATI Electrons

We conclude our discussion of ATI by assessing what the angular distribution of the scattered electrons becomes upon ionization. Generally in linearly polarized fields at nonrelativistic intensity, ATI electrons are ejected in narrow cones along the laser field axis. The angular width of this cone tends to decrease with increasing ATI order, with a notable exception for electrons produced in the rescattering plateau. This general trend can be seen in the various models for ATI already presented.

In the context of perturbative MPI, the distribution of electrons with respect to the laser polarization direction, characterized by angle θ, can be estimated through the Bebb and Gold truncated summation method (Bebb and Gold 1966) we employed in Section 4.7.8. The differential ionization rate is proportional to the square of the multiphoton dipole matrix element for $N = q_m + s$ photon ionization

$$x_{fi}^{(q_m+s)} \sim \langle f | x^{q_m+s} | i \rangle = \langle f | r^N \cos^N \theta | i \rangle, \tag{4.310}$$

which implies that MPI into the s-order ATI peak in the MPI regime will have angular distribution around the polarization axis of

$$\chi_N(\theta) \sim \cos^{2N} \theta. \tag{4.311}$$

Higher ATI orders are more collimated. A similar conclusion can be gleaned from the SFA rate for N-photon ionization. Recalling the differential ionization rate from the SFA found in Eq. (4.105), the angular distribution of a given N-photon ionization channel for spherically symmetric ground state wavefunction is

$$\chi_N(\theta) \sim \left[\sum_{n=-\infty}^{\infty} J_{N+2n}\left[\frac{eA_0 k_{ex}}{m\omega c}\right] J_n\left[\frac{U_p}{2\hbar\omega}\right] \right]^2. \tag{4.312}$$

4.7 Above-Threshold Ionization

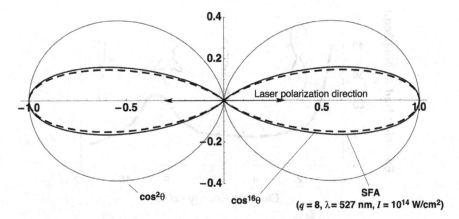

Figure 4.46 Polar plot of the angular distribution of photo-ionized electrons with respect to the laser electric field direction for $q = 8$ electrons as predicted by the SFA (thick solid line). The distribution as predicted by perturbation theory ($\cos^{16}\theta$, dashed line) and a dipole distribution ($\cos^2\theta$, thin line) are also shown for comparison.

For illustrative purposes, the angular distributions of electrons resulting from eight-photon ionization in a 527 nm pulse at intensity of 10^{14} W/cm^2 (parameters in the MPI regime) as predicted by the perturbation model Eq. (4.311) and the SFA, Eq. (4.312), are plotted in Figure 4.46. The distribution predicted by single-photon photo-ionization is also plotted here as a thin grey line to illustrate how much more collimated these multiphoton ATI electrons are.

That electrons are collimated along the field in the tunneling regime is obvious from a picture like that of Figure 4.18. Electrons will tend to tunnel along the narrow channel where the binding potential is depressed by the laser field. The extent of this collimation can be quantified by examining the SFA in the small Keldysh parameter limit. Recalling the differential ionization rate predicted by the SFA for many photons, Eq. (4.132):

$$W^{(SFA)}(\mathbf{k}_e) \sim \exp\left[-\frac{\hbar}{\omega m}\left(\sinh^{-1}\gamma_K k_{e\perp}^2 + \left(\sinh^{-1}\gamma_K - \frac{\gamma_K}{\sqrt{1+\gamma_K^2}}\right)k_{ex}^2\right)\right] \quad (4.313)$$

which in the small Keldysh parameter limit says

$$W^{(SFA)}(\mathbf{k}_e) \sim \exp\left[-\frac{\hbar}{\omega m}\left(\gamma_K k_{e\perp}^2 + \frac{\gamma_K^2}{2}k_{ex}^2\right)\right], \quad (4.314)$$

implying a Gaussian distribution of electron momenta in the transverse and field parallel directions. The angular width of these electrons can be estimated from this distribution:

$$\Delta\theta \approx \Delta k_{e\perp}/\Delta k_{ex}$$
$$\approx \sqrt{\gamma_K/2}. \quad (4.315)$$

As γ_K decreases and the field moves further into the tunneling regime, the tunnel-ionized electrons become more collimated.

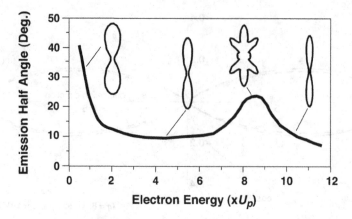

Figure 4.47 Characteristic photoelectron angular distributions at various ATI energies, showing the effects of rescattering in the ATI rescattering plateau (inset). A typical experimental curve of the half maximum of the emitted cone of electrons as a function of energy from tunnel ionization is shown at the bottom, illustrating the initial decrease in the emission angle of electrons as electron energy increases up to $2U_p$ followed by a broadening of the electron emission cone at higher energy due to the fact that these higher-energy electrons result from rescattering from the parent nucleus. Plot generated based on data from Yang *et al.* (1993).

One notable exception to the trend that electron emission becomes more collimated at higher electron energy occurs in the rescattering plateau. Figure 4.47 reproduces the kind of data on ATI angular distributions often observed in the tunneling regime (Yang *et al.* 1993). Here the emission angle of ATI electrons as a function of energy is plotted as a function of electron energy. Characteristic electron distributions throughout this energy range are superimposed on this plot. As the electron energies increase (moving to higher ATI orders), the angular spread of the electrons rapidly decreases, at least in the pure tunneling region up to energy of $2U_p$. At energies above this, where electrons are produced primarily through rescattering, the angular spread of the electrons actually increases, a consequence of the Coulomb scattering of the electrons initially emitted along the laser polarization. "Ears" in the angular distribution are often seen at some energies in the rescattering plateau (as the polar plot of electrons for energy near $8U_p$ illustrates), though the specific shape of these complicated structures depends on intensity and parent atom, implying that the residual electronic structure of the remaining parent ion shapes the distribution of the rescattered electron.

4.8 Ionization Stabilization

It would seem to be so obvious that it almost requires no explicit statement that the ionization rate of an atom or ion increases when the intensity of the illuminating light increases. Every theoretical model we have examined so far confirms this extremely intuitive declaration. This would seem to be a generally true statement: single-photon ionization increases

4.8 Ionization Stabilization

as photon flux increases, MPI increases nonlinearly as photon flux grows, and the tunneling rate of electrons from an atom grows exponentially with increasing electric field strength. However, it turns out that there are certain situations in strong fields where the ionization rate is thought to *decrease* with *increasing* light intensity (Gavrila 2002; Fedorov 1997, p. 361). Such a situation is, at first thought, completely nonintuitive. It has come to be called ionization stabilization.

A number of different kinds of ionization stabilization have been explored in the strong field physics literature. These include interference stabilization, which involves coherent interferences between closely spaced Rydberg levels when both are coupled to the continuum by a strong field (see Fedorov 1997, p. 361 ff. for an extensive treatment of this manifestation of stabilization). A form of stabilization dependent on the pulse shape, known as dynamic stabilization, has also been explored (see the excellent review article by Gavrila (2002) for details on dynamic stabilization). We will forgo a discussion of these special situations and, in the spirit of our previous discussions on strong field ionization, will instead focus on the more fundamental question of whether there are intrinsic properties of an atom in its ground state subject to a monochromatic strong field which would lead to a regime where the ionization rate of that atom decreases with increasing intensity. This more general form of ionization stabilization is often termed quasistationary or adiabatic stabilization. In truth, there is, at this writing, scant experimental evidence for quasistationary stabilization of an atom in a strong field. However, the theoretical basis for such ionization stabilization is rather well established and leads to the conclusion that ionization stabilization is a verifiable strong field phenomenon.

The reason we have yet to encounter the essence of the physics that leads to stabilization is because we have neglected in our treatment of multiphoton ionization the distortion of the ground state wavefunction by the strong field. This we did based on the conclusion that the AC Stark shift of a deeply bound level was negligible. That assessment followed from the logic that "deeply bound" meant that the atomic transition frequencies were much greater than the light frequency, usually an excellent approximation for near-IR and optical laser wavelengths. Based on that assumption, we have used unperturbed atomic ground states to calculate the ionization rate of atoms in a strong field. In short-wavelength fields, however, where the photon energy approaches or exceeds the binding energy of the atom, at high enough intensity the incident field *can* significantly distort the ground state wavefunction. Such distortion can lead to a net reduction in the ionization rate.

4.8.1 The Kramers–Henneberger Hamiltonian

The origin of high-frequency quasi-stationary stabilization can be seen if we transform the Hamiltonian of an atom subject to a strong oscillating field into a frame that moves with a classically oscillating electron (Mostowski and Eberly 1991). Recall the $\mathbf{A} \cdot \mathbf{P}$ form of the atom's SAE Hamiltonian

$$H = \frac{1}{2m}\left(\mathbf{P} + \frac{e}{c}\mathbf{A}\right)^2 + V(\mathbf{x}). \qquad (4.316)$$

We can transform into the oscillating frame through the use of a translation operator, T, of the form of a unitary transformation

$$T = \exp[\mathbf{a} \cdot \nabla] = \exp[-i\mathbf{a} \cdot \mathbf{P}/\hbar]. \tag{4.317}$$

That such an operator translates a wavefunction by the vector $\mathbf{a}$ can be seen by Fourier decomposing the wavefunction

$$\psi(\mathbf{x}) = \frac{1}{2\pi} \int_{-\infty}^{\infty} \tilde{\psi}(\mathbf{k}_e) e^{i\mathbf{k}_e \cdot \mathbf{x}} dx. \tag{4.318}$$

Application of the operator T to this wavefunction yields

$$T\psi(\mathbf{x}) = e^{-i\mathbf{a} \cdot \mathbf{k}_e} \frac{1}{2\pi} \int_{-\infty}^{\infty} \tilde{\psi}(\mathbf{k}_e) e^{i\mathbf{k}_e \cdot \mathbf{x}} dx$$

$$= \frac{1}{2\pi} \int_{-\infty}^{\infty} \tilde{\psi}(\mathbf{k}_e) e^{i\mathbf{k}_e \cdot \mathbf{x} - \mathbf{a}} dx$$

$$= \psi(\mathbf{x} - \mathbf{a}). \tag{4.319}$$

An oscillating free electron has trajectory

$$x(t) = \frac{eA_0}{mc\omega} \cos \omega t = \frac{e}{mc} \int^t A(t') dt', \tag{4.320}$$

which implies that the operator we should apply is of the form

$$T = \exp\left[i \frac{e}{mc\hbar} \int^t \mathbf{A}(t') \cdot \mathbf{P} dt'\right]. \tag{4.321}$$

It is furthermore convenient to remove the ponderomotive term in the Hamiltonian by a second unitary transformation of the form

$$S = \exp\left[-i \frac{e^2}{2mc^2\hbar} \int^t A^2(t') dt'\right]. \tag{4.322}$$

Application of these two operators to both sides of the Schrödinger equation subject to the Hamiltonian of (4.306) yields

$$i\hbar \frac{\partial}{\partial t}(TS\psi) = \left(\frac{P^2}{2m} - \frac{e}{mc} \mathbf{A}(t) \cdot \mathbf{P} - \frac{e^2}{2mc^2} A^2(t)\right) TS\psi + V(\mathbf{x}) TS\psi$$

$$i\hbar \frac{\partial \psi}{\partial t} + i\hbar \left(i \frac{e}{mc\hbar} \mathbf{A}(t) \cdot \mathbf{P} - i \frac{e^2}{2mc^2} A^2(t)\right) \psi = \tag{4.323}$$

$$\left(\frac{P^2}{2m} - \frac{e}{mc} \mathbf{A}(t) \cdot \mathbf{P} - \frac{e^2}{2mc^2} A^2(t)\right) \psi + T^{-1} V(\mathbf{x}) T\psi$$

or, since T^{-1} is itself a translation operator,

$$i\hbar \frac{\partial \psi}{\partial t} = \frac{P^2}{2m} \psi + V\left(\mathbf{x} - \frac{e}{mc} \int^t \mathbf{A}(t') dt'\right) \psi. \tag{4.324}$$

If we follow the usual convention of the stabilization literature and denote the peak oscillation displacement of a free electron in a field of vector potential $\mathbf{A}(t)$ by α_0 (recall

Section 3.2.2 and note that $|\alpha_0| = x_{osc}$), we can rewrite the Hamiltonian in Eq. (4.324) as

$$H_{KH} = \frac{p^2}{2m} + V(\mathbf{x} - \alpha_0 \cos \omega t). \qquad (4.325)$$

This form of the Hamiltonian is known as the Kramers–Henneberger Hamiltonian (Henneberger 1968). The net effect of applying the translation operator is to yield a Hamiltonian of the form of pure atom but with an atomic binding potential which oscillates back and forth in a comoving frame with a classically oscillating electron. We can define the Kramers–Henneberger time-dependent periodic potential as

$$V_{KH} = V(\mathbf{x} - \alpha_0 \cos \omega t). \qquad (4.326)$$

Note that up to this point we have made no approximations. Equation (4.325) is a rigorously accurate Hamiltonian for the atom in an electromagnetic wave up to any incident field strength for a single electron atom (assuming that the single active electron approximation is a good description of the interaction with a multielectron atom). Because V_{KH} is periodic, it can be expressed as a Fourier expansion in time,

$$V_{KH} = \sum_\rho V_\rho \exp[i\rho\omega t], \qquad (4.327)$$

where the Fourier coefficients are given by

$$V_\rho = \frac{\omega}{2\pi} \int_0^{2\pi/\omega} V(\mathbf{x} - \alpha_0 \cos \omega t) \exp[-i\rho\omega t] dt. \qquad (4.328)$$

At this point we need to consider what effect the Kramers–Henneberger potential has on the ground state of the atom. It is helpful now to apply a high-frequency approximation. "High frequency" in this context means that the oscillatory velocity of the quivering KH potential must be much faster than an electron moves in the binding potential of the atom. If we define a characteristic bound electron frequency in terms of the atomic ionization potential $\omega_b = I_p/\hbar$ (which is, just equal to ω_a for the hydrogen atom), the high-frequency approximation is met when

$$\omega \gg I_p/\hbar. \qquad (4.329)$$

This is saying that the photon energy must exceed the binding energy of the bound electron, which means that the incident field must be at short enough wavelength where single-photon ionization of the atom is possible. For most gaseous atoms, this is in the VUV. In reality, this condition is somewhat relaxed since, as we will see, the binding energy of the ground state decreases as the strength of the high-frequency field increases.

Within the framework of this approximation we can ask, what is the shape of the binding potential if all terms in the Fourier decomposition of V_{KH} which oscillate as $e^{i\omega t}$ or faster, simply average to zero? Within such an approximation, we keep only the first term of the Fourier series of (4.333) which, from Eq. (4.328), yields for the time-averaged KH binding potential

$$V_{KH}^{(0)} \cong \frac{\omega}{2\pi} \int_0^{2\pi/\omega} V(\mathbf{x} - \alpha_0 \cos \omega t)\, dt. \tag{4.330}$$

The binding potential of the atom subject to the rapidly oscillating field can be thought of as a stationary potential dressed by the light field. This dressed potential will hold stationary bound states

$$\psi = \chi_i(\mathbf{x}) e^{-i\varepsilon_i t/\hbar} \tag{4.331}$$

that are solutions of the KH Schrödinger equation

$$H_{KH}\psi = i\hbar \frac{\partial \psi}{\partial t}, \tag{4.332}$$

which takes the time-independent form

$$\left[\frac{p^2}{2m} + V_{KH}^{(0)}(\mathbf{x}, \alpha_0)\right] \chi_i = \varepsilon_i \chi_i. \tag{4.333}$$

So the KH Hamiltonian in the high-frequency approximation is now stationary with a time-averaged "dressed" potential. This Hamiltonian supports the stationary bound states, χ_i.

The shape of the KH binding potential in the high-frequency limit takes on an interesting form when the field strength becomes large enough that the oscillation amplitude α_0 becomes comparable to or greater than the size of the initial, undressed ground state wavefunction. (This condition can be roughly expressed for neutral atoms as $\alpha_0 > a_B$.) Figure 4.48 shows the shape of the KH potential for linearly polarized light along the polarization axis for a pure Coulombic binding potential. The dressed potential is illustrated for two values of the oscillation amplitude along the direction of the driving light field. This plot is actually for the potential slightly away from the $y=z=0$ axis, as the dressed KH potential diverges logarithmically as this central axis is approached. Figure 4.48 shows that the dressed binding potential actually develops an elongated structure, taking on the character of a line of charge along the $y = z = 0$ axis, with the $r = 0$ singularities displaced along the polarization axis to locations at $\pm\alpha_0$.

4.8.2 Ground State Wavefunctions of the KH Hamiltonian and Reduction in Ionization Potential

The shape of the "quasi-stationary" ground state wavefunctions in the KH dressed potential take on a shape which is rather different than the field-free atom. The shape of these quasi-stationary wavefunctions yields an explanation for adiabatic stabilization. Figure 4.48 plots the shape of the ground state wavefunction near the central oscillation axis for the two oscillation amplitudes $5a_B$ and $30a_B$, as calculated by Gavrila (2002). Because of the elongation of the KH potential, the electron wavefunction becomes elongated along x and ultimately develops two peaks at locations near $\pm\alpha_0$. This delocalization of the electron wavefunction pushes the center of gravity of the probability distribution well away from the actual location of the binding ion at $x = 0$. The wavefunction at high fields is often said to exhibit "dichotomy" (Pont et al. 1988). The high-frequency field serves to

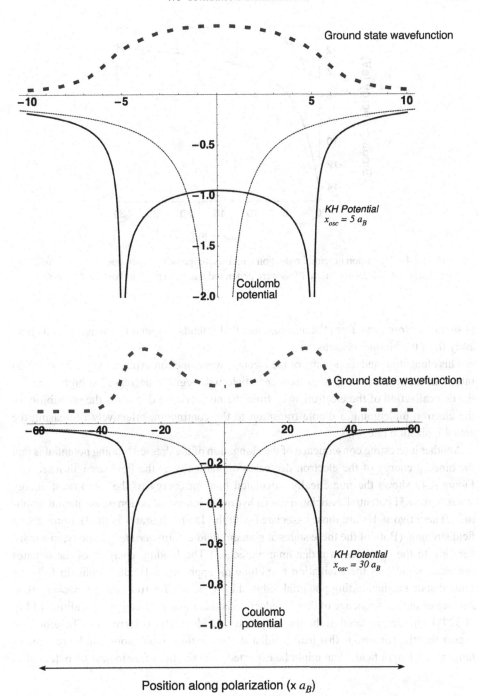

Figure 4.48 Plot of the typical first-order Kramers–Henneberger potential arising from a Coulomb potential along the polarization direction near the $r = 0$ axis. The KH potential for two electron oscillation amplitudes is shown, illustrating the formation of two, time-averaged, potential wells away from the nuclear center. Above each KH potential the ground state wavefunction is illustrated. (Wavefunction shapes were generated based on plots in (Gavrila 2002).)

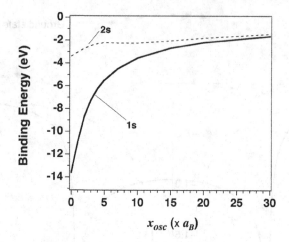

Figure 4.49 Ionization potential reduction which accompanies high-frequency strong-field irradiation of a hydrogen atom. Plots were generated based on plots in Pont et al. (1988).

push the electron away from the nucleus such that it tends to spend the majority of its time away from the binding nucleus.

This elongation and dichotomy of the electron wavefunction explains why the ionization rate of an atom in a strong, high-frequency field will begin to decrease. At high intensity, the delocalization of the electron away from the nuclear core decreases the probability of the electron undergoing a dipole transition to the continuum, effectively decreasing the actual ionization rate.

Another interesting consequence of the elongation of the dressed binding potential is that the binding energy of the electron decreases dramatically as the field strength increases. Figure 4.49 shows the numerically calculated binding energy of the two lowest energy states in the KH potential associated with a hydrogen ion as a function of oscillation amplitude. These two states are those associated with the 1s and 2s states in the H atom at zero field strength. (Both of the these states at nonzero field exhibit *gerade* symmetry, in a manner akin to the σ_g states in a diatomic molecule.) The binding energies of these states decreases rapidly as the oscillation amplitude α_0 approaches 10 Bohr radii. In fact, the ground state has its binding potential reduced by a factor of ~ 10 as α_0 approaches $30 a_B$. An interesting consequence of this I_p reduction is that the high-frequency condition of Eq. (4.329) is, in fact, reduced as the intensity of the incident field is increased. Therefore, we expect that the conditions that lead to adiabatic ionization stabilization can be reached in longer-wavelength fields than might be expected from the field-free ionization potential.

4.8.3 Ionization Rate in the High-Frequency Regime

In a field of infinitely high frequency, the approximation which led to the stationary KH Schrödinger equation, Eq. (4.339), would be exact, and the quasi-stationary states of the dressed potential would have infinite lifetime. The atom would undergo no ionization by the field. Of course, for any real field, this condition cannot hold and we expect some ionization to occur. The actual KH potential is time-dependent and will lead to finite ionization rate. We can estimate the ionization rate by considering the expansion of the KH potential. Consider the next term in the Fourier expansion of the KH potential (4.333), the first time-dependent term:

$$V_{KH} \cong V_{KH}^{(0)} + V_{KH}^{(1)} \exp[i\omega t]. \quad (4.334)$$

This second term acts as if it raises and lowers the quasi-stationary KH potential at the light frequency ω. We can calculate the ionization rate by treating this term as a perturbation to $V_{KH}^{(0)}$ and using first-order perturbation theory (Mostowski and Eberly 1991). Recalling the first-order perturbation theory transition amplitude, Eq. (4.17), and transition probability, Eq. (4.36), and combining these with the integration over a density state as given by Eqs. (4.41) and (4.42), we can write for the ionization rate resulting from the second term in the KH potential:

$$W^{(1)} = \frac{2\pi m}{\hbar^3} k_e \left| \left\langle \mathbf{k}_e \left| V_{KH}^{(1)} \right| \chi_i \right\rangle \right|^2. \quad (4.335)$$

Recall that this time-dependent perturbation can be written

$$V_{KH}^{(1)} = \frac{\omega}{2\pi} \int_0^{2\pi/\omega} V(\mathbf{x} - \alpha_0 \cos \omega t) \exp[-i\omega t] dt. \quad (4.336)$$

As a simple estimate, use the Born approximation, where we assume that the final continuum state is a plane-wave state of the form of Eq. (4.40). This permits us to write Eq. (4.355) for the transition matrix element

$$\left\langle \mathbf{k}_e \left| V_{KH}^{(1)} \right| \chi_i \right\rangle = \frac{\omega}{2\pi} \int \frac{1}{(2\pi)^{3/2}} e^{i\mathbf{k}_e \cdot \mathbf{x}} \times \\ \int_0^{2\pi/\omega} V(\mathbf{x} - \alpha_0 \cos \omega t) \exp[-i\omega t] dt \chi_i(\mathbf{x}) d\mathbf{x}. \quad (4.337)$$

Shift the coordinates in these integrals $\mathbf{x}' = \mathbf{x} - \alpha_0 \cos \omega t$, so that (4.337) becomes

$$\left\langle \mathbf{k}_e \left| V_{KH}^{(1)} \right| \chi_i \right\rangle = \frac{\omega}{(2\pi)^{5/2}} \int_0^{2\pi/\omega} \exp[i\mathbf{k}_e \cdot \alpha_0 \cos \omega t - i\omega t] \\ \int V(\mathbf{x}') \chi_i(\mathbf{x}' + \alpha_0 \cos \omega t) e^{i\mathbf{k}_e \cdot \mathbf{x}'} d\mathbf{x}' dt. \quad (4.338)$$

Because of the term $\exp[i\mathbf{k}_e \cdot \mathbf{x}']$, at large k_e (which is accessed if the photon energy is large compared to the ionization potential) the dominant contribution to the integral will occur near $\mathbf{x}' = 0$. So we can approximate the initial wavefunction by its value at $\mathbf{x}' = 0$ and pull it out of the integral. After we do this we recognize the second integral in (4.338) as the Fourier transform of the undressed atomic binding potential,

$$\tilde{V}(\mathbf{k}_e) = \frac{1}{(2\pi)^{3/2}} \int V(\mathbf{x}') e^{i\mathbf{k}_e \cdot \mathbf{x}'} d\mathbf{x}', \quad (4.339)$$

which, with $\varphi = \omega t$, lets us write the transition matrix element:

$$\left\langle \mathbf{k}_e \left| V_{KH}^{(1)} \right| \chi_i \right\rangle = \frac{\tilde{V}(\mathbf{k}_e)}{2\pi} \int_0^{2\pi} \chi_i \left(\alpha_0 \cos \varphi \right) \exp \left[i \mathbf{k}_e \bullet \alpha_0 \cos \varphi - i\varphi \right] d\varphi. \quad (4.340)$$

To get a rough sense of how this matrix element scales with incident field strength we can recall the shape of the ground state wavefunction for the dressed Coulombic potential (illustrated in Figure 4.43). We approximate χ_i as being constant over the region $\pm \alpha_0$. Consequently, over the limits of the integral in Eq. (4.340) we can say that

$$\chi_i \approx \frac{\text{const.}}{\alpha_0^{1/2}} \quad (4.341)$$

and pull that factor out of the integral. With that approximation we were left with an integral which we recognize as a Bessel function of the first kind:

$$\int_0^{2\pi} \exp \left[i \mathbf{k}_e \bullet \alpha_0 \cos \varphi - i\varphi \right] d\varphi = 2\pi i J_1 \left[\mathbf{k}_e \bullet \alpha_0 \right]. \quad (4.342)$$

The high-frequency condition mandates that the photon energies of the incident field are well above the ionization potential so $k_e \approx k$. The argument of the Bessel function in (4.342) scales as $k\alpha_0 \sim \alpha_0 \omega/c \sim I^{1/2}/\omega$, so this argument is large in intense pulses. If θ is the angle between the outgoing electron and the field polarization direction, the Bessel function can then be approximated

$$J_1 [k_e \alpha_0 \cos \theta] \approx \sqrt{\frac{2}{\pi k \alpha_0}} \cos \left(k\alpha_0 \cos \theta - \frac{3\pi}{4} \right). \quad (4.343)$$

This result interestingly implies that electrons are ejected preferentially at angles 90° to the field polarization.

Utilizing (4.341) and (4.343) in Eq. (4.340) yields the intensity scaling of the first-order matrix element:

$$\left\langle \mathbf{k}_e \left| V_{KH}^{(1)} \right| \chi_i \right\rangle \sim \frac{1}{\alpha_0} \sim \frac{1}{I^{1/2}}. \quad (4.344)$$

Equation (4.335) then indicates that the ionization rate scales as

$$W_{hf} \sim I^{-1}. \quad (4.345)$$

The ionization rate *decreases* with increasing intensity; ionization stabilization occurs. A more sophisticated analysis (Gavrila 2002) of the ionization rate in the high-frequency regime indicates that in linearly polarized light, the ionization rate deviates slightly from the scaling we have derived, decreasing as

$$W_{hf} \sim I^{-1} \ln I. \quad (4.346)$$

This is a remarkable finding. It indicates that photoionization rates increase linearly with intensity in the perturbative regime, but roll over when $\alpha_0 \approx 5-10$ and decrease as $\ln I / I$ as the intensity further increases. The implication for longer-wavelength fields

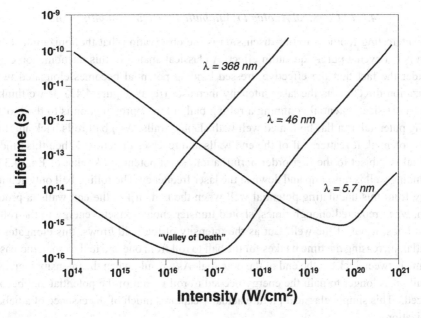

Figure 4.50 Plot of the predicted ionization rate and atomic lifetime of the ground state in a hydrogen atom subject to a high-frequency, (circularly polarized), high-intensity light pulse as a function of the instantaneous pulse intensity for three different wavelengths. Plots were generated based on calculations in Gavrila (2002).

is even more surprising. An atom subject to multiphoton ionization by a field with photon energy somewhat below the field-free ionization potential will experience nonlinear increase in ionization rate with increasing field strength. But at some point in intensity, the field-dressed reduction in the ionization potential leads to a fulfillment of the high-frequency condition and even these longer-wavelength fields will see an ultimate rollover in ionization rate and ionization stabilization.

The difficulty in observing experimentally this unexpected prediction is illustrated when the actual ionization rates are calculated. Figure 4.50 illustrates numerically calculated lifetimes of a hydrogen atom found by Gavrila (2002) as a function of intensity for three different wavelengths. (This particular calculation is for circularly polarized light but the trend for linearly polarized pulses will be virtually identical.) As expected, the lifetime of the atom decreases linearly at low intensity. At some critical intensity this decrease in lifetime bottoms out and the lifetime of the bound electron begins to increase in the stabilization regime. (The slightly "wavey" character of the lifetime increase seen in these curves is a manifestation of the Bessel function-type solution for the rate we found in our simple analysis.) The difficulty in experimental observation can be seen in the value of the "valley" in bound electron lifetime. Because this lifetime is 1 fs or less, any real pulse which has a finite temporal rise will almost always completely ionize this atom well before high enough intensity is reached in the pulse envelope to enter the stabilization regime. For this reason, this minimum in lifetime is often referred to as the "Valley of Death."

4.8.4 Classical Analog to Adiabatic Ionization Stabilization

An interesting footnote to our discussion is the observation that the ionization stabilization is not even a purely quantum effect. A classical analog of this phenomenon exists. Consider the fact that the effective dressed binding potential becomes elongated in the polarization direction as the laser intensity increases (recall Figure 4.43). If we think of this as a classical potential confining a rolling ball, a fair approximation is to think of the binding potential as a flat-bottomed well with slanted walls. As a ball rolls back and forth in this potential, it scatters off of the end walls and reverses direction. When this binding potential is subject to the first-order perturbation, say a potential like that of Eq. (4.334), the binding well is raised up and down at the laser frequency. The rolling ball only acquires energy from this undulating potential well when the ball strikes the end walls, a process which, when repeated enough times, should transfer enough kinetic energy to the rolling ball to boost it out of the well. But as the intensity of the field grows, this elongates the potential, increasing the time it takes for the ball to roll from one end to the other, increasing the time between kicks at the end of the potential. As a result, when the intensity increases, the ball takes longer to gain the energy needed to roll up out of the potential and become "ionized." This simple classical picture actually captures much of the essence of adiabatic stabilization.

4.9 Relativistic Effects

Before we conclude this chapter, a consideration of how relativity affects strong field interactions with atoms is in order. We argued in the introduction to this chapter that it was fair to ignore relativistic effects since most of the physics we were to consider involved bound electrons, which, for all but the most highly charged ions, are nonrelativistic. The models we have derived for strong field ionization of the atom have born out this assumption. However, some aspects of the interaction physics discussed, nonsequential double ionization and ATI in the tunneling regime in particular, have involved dynamics of the electron in the continuum, where we know relativistic effects play a role when a_0 approaches 1. It is reasonable to assume that relativistic effects play a role in these phenomena at high intensity and should be reconsidered (Salamin *et al.* 2006; Maquet *et al.* 2008).

4.9.1 Relativistic Suppression of Electron Rescattering

Rescattering of the electron subsequent to tunnel ionization was considered assuming that the electron oscillates on a trajectory parallel to the laser electric field once the electron has been released into the continuum by tunneling. Our discussions of Chapter 3 indicate that at relativistic intensity the electron trajectory deviates from this axis because of the laser's magnetic field (Prager and Keitel 2002). The forward motion of an electron in the continuum can prevent rescattering at sufficiently high intensity. The result is that rescattering NSDI will be suppressed at relativistic intensity (Dammasch *et al.* 2001).

We might expect this suppression of NSDI to occur when the forward drift of the electron becomes comparable to the spreading of the free electron wavepacket. Recalling the longitudinal trajectory of a free electron, Eq. (3.77), the forward drift during the course of the half cycle a tunnel-ionized electron is in the continuum is approximately

$$z_{cycle} \approx \frac{a_0^2 c\pi}{4\omega}. \tag{4.347}$$

Equation (4.238) tells us how much a wavepacket spreads (ignoring Coulomb focusing) over half an optical cycle, if we assume the electron emerged from an ion of residual charge Z with initial size a_B/Z

$$\delta r = \sqrt{\frac{a_B^2}{Z^2} + \frac{\pi^2 \hbar^2 Z^2}{4\omega^2 m^2 a_B^2}}$$
$$\approx \frac{\pi \hbar Z}{2m\omega a_B}. \tag{4.348}$$

The condition for suppression of NSDI occurs when $z_{cycle} \approx \delta r$, which can be written:

$$\frac{c\pi}{4\omega} \frac{e^2 E_0^2}{m^2 c^2 \omega^2} = \frac{\pi \hbar Z}{2m\omega a_B}, \tag{4.349}$$

leading to an estimate for the intensity at which we expect NSDI to be suppressed by relativistic effects:

$$I_{sup} = \frac{Zm^2 c^2 \omega^2}{4\pi \hbar}. \tag{4.350}$$

This formula implies, for example, that relativistic suppression of NSDI might occur in laser fields at a wavelength of 800 nm at $\sim Z \times 3.5 \times 10^{16}$ W/cm^2. If we use the scaling of the ionization potential of hydrogen-like ions with the BSI formula, Eq. (4.222), this result suggests that relativistic suppression of NSDI will occur in near-IR fields for $Z \approx 3$, an intensity of $\sim 1 \times 10^{17}$ W/cm^2. In reality, this relativistic drift is somewhat reduced by Coulomb focusing of the recolliding electron in the longitudinal direction. Experiments seem to indicate that in ions of Xe recollision ionization plays a role at intensities up to about 10^{18} W/cm^2, somewhat higher than our simple estimate (Dammasch et al. 2001).

4.9.2 ATI Electron Angular Distributions in the Relativistic Regime

Another interesting manifestation of relativistic effects arising in the energy spectrum and angular distributions of ATI electrons liberated at relativistic intensity (Meyerhofer et al. 1996; Keitel 1996; Hu and Starace 2002; Maltsev and Ditmire 2003). As we have seen, in the quasi-classical model for ATI, the electron residual energy is determined by the phase of birth of tunneling electrons in the field and, through conservation of canonical momentum in a plane-wave field, the subsequent propagation of the free electron in the laser pulse. So even if relativistic effects are unimportant in the tunnel ionization rate, they can play a role in the resulting energy of the electron and the angular distribution of ejected electrons. Conservation of momentum by absorption of many photons led to the

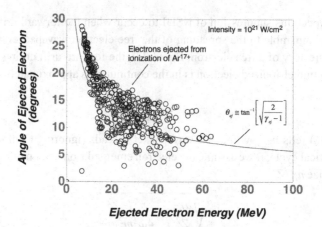

Figure 4.51 Monte Carlo simulation of relativistic ATI electrons ejected from a 5 μm focus of a 100 TW, 30 fs, 800 nm laser pulse. Probability of ionization in the focus and pulse were weighted by the PPT/ADK tunnel ionization rate. Each point is an ejected electron plotting its ejected ATI energy and its angle of ejection from the focus with respect to the laser propagation axis. Also plotted is the prediction of Eq. (4.351).

following formula for the angle of ejection by electrons acquiring $(\gamma_{ej} - 1)mc^2$ from the field:

$$\theta_{ej} \cong \tan^{-1}\left[\sqrt{\frac{2}{\gamma_{ej} - 1}}\right]. \tag{4.351}$$

Obviously, high-energy ATI electrons will be expected to have their ejection angle folded in a direction forward of the direction of field polarization, into an angle approximately given by the plane-wave formula, Eq. (4.351) (Moore et al. 1995).

In practice, ATI electrons produced in relativistic intensities have angular distributions which deviate significantly from Eq. (4.351). This arises from the fact that tight foci are often required to reach relativistic intensity, and the ultimately observed electron energy outside the laser focus is strongly influenced by the ponderomotive ejection of the electrons by the tightly focused field (which exhibit the longitudinal fields associated with real focus). Figure 4.51 shows the plot of electron ejection angle versus electron energy found from a Monte Carlo calculation of ATI electrons liberated by a 100 TW, 800 nm, 30 fs laser pulse focused to 5 μm, tunnel ionizing Ar^{17+} ions. These electrons clearly exit the focus at angles well forward of the polarization axis, though their distribution deviates significantly from Eq. (4.357) (also plotted in Figure 4.51) because of the interplay of electron motion with the tightly focused field.

4.9.3 Relativistic Corrections to the Tunnel Ionization Rate in Highly Charged Ions

Finally we ask, how is the tunnel ionization rate affected by relativity? Since the velocity of an electron prior to its emergence into the continuum is determined by its bound

4.9 Relativistic Effects

state energy, I_p, we expect that relativistic corrections to the ionization rate will only be important for ionization by very deeply bound ions (where I_p approaches mc^2). The intensities needed to ionize such deeply bound electrons is at the very upper limit of intensities experimentally available at the time of this writing. It is, in fact, possible to derive a relativistically correct semiclassical tunneling rate from the Dirac equation (Milosevic et al. 2002a; 2002b); but this result is of limited utility and we will not trouble ourselves with such a derivation here.

A qualitative intuition about the impact of relativistic effects on tunneling rates can be gleaned by noting that relativistic motion of the bound electron serves to increase the effective mass of the electron. Recalling that tunneling rates scale as

$$w_I \sim \exp\left[-\frac{2}{3}\frac{Z^3 m^2 e^5}{\hbar^4}\frac{1}{E_0}\right], \tag{4.352}$$

we can see that an increase in the effective mass of the tunneling electron will serve to lower the ionization rate. To slightly more quantitative, we can estimate the magnitude of this effect by writing an effective relativistic gamma factor for an electron bound with energy I_p

$$\gamma_{tun} = 1 + \frac{I_p}{mc^2}. \tag{4.353}$$

If we then say that the tunneling electron has effective mass $m_{eff} = \gamma_{tun} m$, then the tunneling rate might scale something like

$$w_I^{(Rel)} \sim \exp\left[-\frac{2}{3}\frac{Z^3 m^2 e^5}{\hbar^4}\frac{1}{E_0}\left(1 + I_p/mc^2\right)^2\right]. \tag{4.354}$$

This simple correction implies, for example, that a hydrogen-like ion with $Z = 50$ (a modestly relativistic ionization potential of 34 keV) will have its ionization rate at intensity 1×10^{24} W/cm^2 (one-half the BSI intensity for this charge state) reduced by a factor of ~ 5. The full tunneling formula derived from the Dirac equation indicates that the reduction in ionization rate for such an ion is around a factor of 3 (see e.g. Milosevic et al. 2002a, Eq. 25). So, relativistic effects will begin to play a role for very high charge states (say hydrogen-like ions with $Z > 30$).

5
Strong Field Interactions with Molecules

5.1 Introduction

In many ways, the physics of strong field interactions with small molecules is similar to interactions with single atoms. Multiphoton and tunnel ionization occur in a manner akin to those processes in atoms. Resonantly enhanced MPI, ATI and NSDI are also observed when molecules are irradiated at intensities similar to those in atoms. The most salient differences derive from the fact that the ground state wavefunctions in molecules are spatially extended and are not spherically symmetric. The dynamic localization of an electron near one of the nuclei of a molecule can also affect the character and rate of ionization in ways that are absent in atoms. Molecules also exhibit a new complication; the nuclei of the molecule can move during irradiation by the laser pulse. This added dynamic affects MPI and tunnel ionization and leads to new observables, such as the energy of ions ejected during the dissociation, or Coulomb explosion, of the ionized molecule.

In this chapter we will consider specifically strong field ionization and nuclear motion of "small" molecules in a strong laser field. What constitutes a small molecule depends to some extent on the physics under consideration, but practically speaking, most of the phenomena described in this chapter apply to molecules with two to, perhaps, twenty atoms. Larger structures begin to take on the character of clusters in the laser field and are often best described by the models presented in Chapter 7. In fact, much of our discussion will revolve around diatomic molecules, as these are the most widely studied in the literature.

5.2 Strong Field Ionization of Molecules

Qualitatively, strong field ionization of molecules is similar to the process in single atoms. Multiphoton ionization and tunnel ionization, along with the models we have developed in the previous chapter are accurate descriptions (DeWitt and Levis 1998). We will find that the models of the previous chapter are helpful in understanding ionization rates in molecules. Molecules do, however, have two very important differences with atoms in this regard (Nikura et al. 2008). First, because of the multinuclear structure of molecules and the fact that the ground state wavefunction will generally not be symmetric and is distributed spatially among the nuclei of the molecule, the orientation of the molecule with respect to the laser polarization will play a role in determining the ionization rate. Furthermore, the spatial spreading of electrons in the molecular orbitals leads to constructive and destructive interference in the outgoing ionized electron wave.

A second interesting aspect of molecular strong field ionization arises because the nuclei of the molecule can move during a laser pulse. Such motion results from previous ionization events (which can lead to Coulomb repulsion between the ions of the molecule) or alteration of the bonds of the molecule in the strong field. This nuclear motion can have a dramatic effect on the ionization of a molecule and can lead to transient enhancements of the ionization rate.

A particularly fascinating observation about strong field ionization of molecules is that, in general, for a given laser intensity, neutral molecules almost always exhibit lower ionization rates than atoms with nearly the same ionization potential (Muth-Böhm *et al.* 2000). This is at first puzzling, made even more mysterious by the fact that the magnitude of this ionization rate suppression depends on the molecular species and the wavelength employed (Hankin *et al.* 2000). In some molecules the observed rate suppression is only a factor of two or so (Niikura *et al.* 2008) (when, say, ionization of deuterium molecules by 800 nm pulses is compared to ionization of Ar atoms), while in others, the suppression can be over an order of magnitude (as is seen when ionization rates in the near-IR of O_2 are compared to Xe). This is an almost universal observation; even organic polyatomics exhibit a marked ionization rate suppression when compared to atomic ionization rates. In truth, at the time of the writing of this book, the detailed physics quantitatively explaining this is not completely understood; it would appear to result from a combination of factors. In this section we will call on some of the models we developed for single atoms in Chapter 4 to aid in explaining the trends describing ionization of small molecules.

5.2.1 Multiphoton Ionization of Diatomics

Ionization behavior of small molecules in intense laser fields falls into the same regimes as in that of multielectron atoms (Posthumus *et al.* 1996; Hankin *et al.* 2001). The ionization can be characterized as multiphoton in character when $\gamma_K > 1$ and can be described by tunneling ionization when $\gamma_K < 1$ (Corkum *et al.* 1997; Grasbon *et al.* 2001). As with atoms, molecular strong field ionization can be examined most easily in the context of the SAE approximation (though, as we shall see, this approximation is more suspect than atoms in small molecules and breaks down completely in large molecules). The most significant difference between molecules and atoms is that the ground state wavefunction in a molecule is typically not spherically symmetric. As mentioned, these rates also depend on the spacing between ions, though for the moment we shall consider ionization of a diatomic at its equilibrium nuclear separation.

In principle, LOPT can be applied to molecular systems if the ground state and a subset of the excited quantum states are known (Jaron-Becker *et al.* 2004). Orientational differences in the ionization of a diatomic can be seen in the context of LOPT when one considers that dipole coupling between the ground state and the excited states depends on the angle of the diatomic with the field direction, which will affect the strength of the multiphoton dipole matrix element, recalling Eq. (4.40). Quantitative calculations in the context of LOPT are complex and we will not delve into this method any further here.

On the other hand, let us consider implementation of the strong field approximation in a diatomic molecule. In particular, we would like to assess the effect that a distributed molecular wavefunction has on the differential ionization rate. Recall from Eq. (4.105) that the SFA differential ionization rate can be written (Milosevic 2006)

$$W_{fi}^{(SFA)}(\mathbf{k}_e) = 2\pi C_{Coul} \frac{m}{\hbar^3} \sum_{q=q_{min}}^{\infty} k_e (U_p - q\hbar\omega)^2 |\phi_i(\mathbf{k}_e)|^2 \\ \times \left[\sum_{n=-\infty}^{\infty} J_{q+2n}\left(\frac{eA_0 k_{ex}}{m\omega c}\right) J_n\left(\frac{U_p}{2\hbar\omega}\right) \right]^2, \quad (5.1)$$

where we have explicitly included the Coulomb correction derived in Section 4.4.6:

$$C_{Coul} = \left[2\left(\frac{I_p}{I_H}\right)^{3/2} \frac{E_a}{E_0} \right]^{2n^*}, \quad (5.2)$$

which uses the molecule's ionization potential, I_p, and the effective quantum number n^*, defined by Eq. (4.208). As noted in Chapter 4, within the SFA all of the physics of the quantum system being irradiated rests in the Fourier transform of the ground state wavefunction, $\phi_i(\mathbf{k}_e)$, reflecting conservation of momentum in a field described by the dipole approximation. This wavefunction will be the highest occupied molecular orbital (HOMO) of the diatomic molecule in question.

To derive a sense of how ionization of a diatomic differs from an atom, we can turn to the usual technique of approximating the HOMO by a linear combination of atomic orbitals (LCAO). The HOMO of the diatomic can be written in terms of an atomic orbital ψ_{at}

$$\Psi_{mol} = a\psi_{at}(\mathbf{r}, \mathbf{R}_a) + b\psi_{at}(\mathbf{r}, \mathbf{R}_b), \quad (5.3)$$

where $\mathbf{R}_{a,b}$ denote the positions of the two nuclei. As an illustration, consider a homonuclear diatomic whose HOMO wavefunction can be further simplified and written

$$\Psi_{mol} \cong \frac{1}{\sqrt{2}} \left(\psi_{at}(\mathbf{r} - \mathbf{R}/2) \pm \psi_{at}(\mathbf{r} + \mathbf{R}/2) \right), \quad (5.4)$$

if $\mathbf{R}$ is the spacing between the two nuclei. Here the positive sign is used for a bonding HOMO and the negative sign for an antibonding orbital, which will be antisymmetric. The Fourier transform of a molecular orbital of this form can be simply written in terms of the Fourier transform of the atomic orbital, $\phi_i^{(at)}$ since

$$\phi_i(\mathbf{k}_e) = \frac{1}{(2\pi)^{3/2}} \int \Psi_{mol} e^{i\mathbf{k}_e \cdot \mathbf{r}} d^3\mathbf{r} \\ = \frac{1}{\sqrt{2}} \phi_i^{(at)} \left(e^{i\mathbf{k}_e \cdot \mathbf{R}/2} \pm e^{-i\mathbf{k}_e \cdot \mathbf{R}/2} \right). \quad (5.5)$$

We can therefore conclude that the differential ionization rate for a homonuclear diatomic in the SFA will be closely related to the ionization rate of a free atom with ground state ψ_{at}

5.2 Strong Field Ionization of Molecules

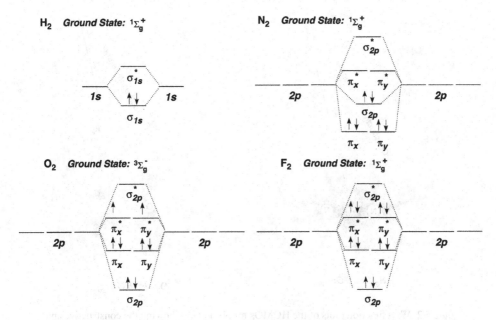

Figure 5.1 Molecular orbital diagrams and ground state term symbols for four common homonuclear diatomic molecules.

and depends on whether the HOMO of the molecule is a bonding or antibonding orbital such that

$$W_{\text{bond}}^{(SFA)}(\mathbf{k}_e) \cong 2N_e W_{at}^{(SFA)}(\mathbf{k}_e) \cos^2(\mathbf{k}_e \cdot R/2) \tag{5.6a}$$

$$W_{\text{anti-bond}}^{(SFA)}(\mathbf{k}_e) \cong 2N_e W_{at}^{(SFA)}(\mathbf{k}_e) \sin^2(\mathbf{k}_e \cdot R/2), \tag{5.6b}$$

where N_e is the number of electrons in that HOMO.

In both cases, the spatial extension of the wavefunction of the molecule has resulted in a modification of the molecular ionization rate over that of a single atom by the cosine^2 or sine^2 factor. Since the lowest multiphoton orders tend to dominate the total differential ionization rate, we note that $\hbar^2 k_e^2/2m \approx \hbar\omega$. This implies that for a diatomic with bond length $\sim 2a_B$ the magnitude of the argument of the cosine and sine factors is small: $k_e R/2 \approx (2m\omega a_B^2/\hbar)^{1/2} = 0.05$. Therefore, we expect that the ionization rate of a bonding orbital is not much different than that from a single atom while that of an antisymmetric, antibonding orbital is significantly reduced from this rate.

This can be understood when one considers the shape of bonding and antibonding orbitals. To draw on concrete examples, let us consider the electron structure of the outermost orbitals for the most common homonuclear diatomic molecules, illustrated in Figure 5.1. N$_2$ has electrons in a bonding σ_g orbital, while O$_2$ has its outermost electrons in an antibonding π_g^* orbital. These wavefunctions are plotted in Figure 5.2. An electron released by MPI from a bonding orbital, such as the σ_g orbital of N$_2$ plotted in Figure 5.2, will see the symmetric wavefunction of the ground state interfere constructively

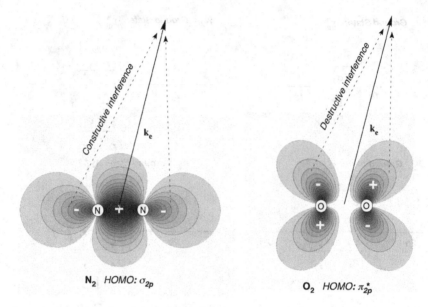

Figure 5.2 Wavefunction plots of the HOMOs for N_2 and O_2 showing the constructive and destructive interferences in the outgoing ionization channels, respectively.

in the continuum. On the other hand, the antisymmetric orbital of the oxygen molecule will destructively interfere in the outgoing electron wave. The cosine and sine cofactors that arise in the SFA differential ionization rates of Eqs. (5.6) reflect this constructive or destructive interference for bonding or antibonding orbitals (Hetzheim et al. 2007).

In truth, the experimental evidence for this quantum interference suppression of ionization in diatomics is, at the time of this writing, ambiguous. One way to assess the magnitude of molecular ionization is to compare ion yields from a molecule with the yields under identical conditions of an atom with nearly the same ionization potential. It is well established, for example, that the ionization rate of the N_2 molecule in a near-IR laser field (with 800 nm wavelength, say) with ionization potential of 15.58 eV is nearly identical to the rate of ionization of an Ar atom, which has a 15.76 eV ionization potential (Gibson et al. 1991). This is consistent with the N_2 molecule having bonding symmetry in its HOMO. It is also well established that the O_2 molecule, with ionization potential equal to 12.03 eV has ionization yields which tend to be over an order of magnitude less than yields in Xe atoms, which have nearly the same I_p of 12.17 eV (Talebpour et al. 1996). This seems to be a manifestation of the destructive interference we expect from the antibonding orbital of the oxygen molecule. However, this appears not to be a universal effect. The fluorine molecule ($I_p = 15.69$ eV), which we can see from Figure 5.1 to have its outer electrons in the same antibonding orbital as the oxygen molecule experimentally appears to have about the same ionization rate as Ar, at odds with the predictions of the SFA. The origin of this anomaly has at this time not been resolved theoretically, though it may have to do with the inadequacy of the SAE approximation for molecules, as we will discuss shortly.

5.2 Strong Field Ionization of Molecules

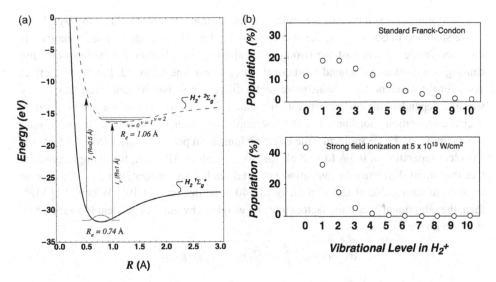

Figure 5.3 Potential energy curves for the ground states of H$_2$ and H$_2^+$ (a). Part (b) shows population distribution of vibrational levels in H$_2^+$ for ionization by electron impact and by a strong laser field. Figure was generated based on calculations from Walsh *et al.* (1997).

5.2.2 Vibrational Excitation During MPI of Diatomics

Another interesting aspect of the MPI of diatomics is the vibrational excitation that results from the ionization event (Urbain *et al.* 2004). Such vibrational level excitation is a well-known result of sudden ionization (i.e. ionization on a time scale much faster than the molecule's vibration period) of a diatomic because the equilibrium bond separation of the molecular ion AB$^+$ is typically different (usually longer) than the equilibrium separation of the neutral molecule (Posthumus 2004). For example, the ground state potential energy curves as a function of nuclear separation binding H$_2$ and H$_2^+$ are plotted in Figure 5.3 (the $^1\Sigma_g^+$ and $^2\Sigma_g^+$ states respectively) illustrating the spatial difference in the potential well minima (Wind 1965). As is well known from studies of ionization by single-photon ionization or electron impact ionization, a molecule in its ground vibrational state will be projected on to the vibrational levels of the molecular ion upon ionization. In the sudden approximation, the probabilities for populating the vibrational levels of the singly ionized molecule are given by the well-known Franck–Condon factors (Walsh *et al.* 1997):

$$M(v) = \left| \int \chi'_v(R)\chi_0(R)dR \right|^2, \tag{5.7}$$

where $\chi_0(R)$ is the ground state vibration wave function of the neutral molecule and $\chi'_v(R)$ is the v-th vibrational wavefunction of the molecular ion. These Franck–Condon factors are plotted in Figure 5.3 for sudden ionization of H$_2$ in its ground vibrational state to H$_2^+$, illustrating that the $v = 2$ state is the most likely populated.

The vibrational excitation of the molecular ion during MPI is altered in two ways. First, as we shall see in Section 5.4, the shape of the binding Born–Oppenheimer potentials is distorted in the presence of the strong field, altering the vibrational wavefunctions and causing some otherwise bound vibrational states to become unbound. Putting this effect aside for the moment, the second, more subtle difference is that the ionization rate is very strongly dependent on nuclear separation because the effective ionization potential varies over the separation coordinate R and the strongly nonlinear nature of the ionization rate. For example, in the H_2 molecule, the effective ionization potential varies from 18.5 eV for a nuclear separation of 0.5 Å to 14.8 eV for 1 Å separation. The SFA, for example, predicts that this small difference in ionization potential leads to an ionization rate difference in an 800 nm laser pulse of almost a factor of 140 at an intensity of 10^{14} W/cm^2. For MPI, then, the vibrational excitation factors must be weighted by the R-dependent ionization rate $W(R)$ to give

$$M_{MPI}(v) = \left| \int W^{1/2}(R) \chi'_v(R) \chi_0(R) dR \right|^2. \tag{5.8}$$

This weighting tends to push the maximum in the transition rate toward larger R separations (where I_p is lowest) and therefore pushes the excitation toward lower vibrational levels in the molecular ion. Figure 5.3, for example, shows the vibrational excitation distribution calculated for H_2^+ produced by tunnel ionization at an intensity of 5×10^{13} W/cm^2. Instead of populating vibrational levels out beyond $v = 8$ with peak at $v = 2$ as in the electron impact case, the distribution for strong field ionization is peaked at $v = 0$ with almost no population beyond $v = 4$. As the laser intensity increases, the ionization rate at smaller R increases and the vibrational state distribution in the molecular ion will evolve toward that of the pure Franck–Condon distribution.

5.2.3 Tunnel Ionization in Diatomics

As in atoms, the MPI regime of ionization for a diatomic gives way to a tunneling picture as the intensity or wavelength of the incident field increases (Yu *et al.* 1998; Posthumus and McCann 2001; Tong *et al.* 2002). As we have just discussed, ionization potentials of typical diatomics like N_2 are comparable to rare gas atoms and begin to ionize significantly when intensities exceed 10^{14} W/cm^2. At this point, in a near-IR field, the Keldysh parameter will drop below one and we would expect that tunneling models should be appropriate.

In fact, the PPT tunneling model that we developed for multielectron atoms in Section 4.4.7 can be applied to these diatomics as well. However, we must deal with the new complication that the molecular wavefunction is not spherically symmetric and cannot be simply described by a single spherical harmonic $Y_{lm_l}(\theta, \varphi)$. (For the moment, m refers to the angular momentum projection quantum number, not electron mass.) Instead, the angular distribution of the wavefunction in the asymptotic tunneling region (far from the nuclear cores) needs to be written in terms of a superposition of waves, expanded over the two-center molecular wavefunction. With this approach, the asymptotic wavefunction can be written

5.2 Strong Field Ionization of Molecules

$$\psi_i = f_i(r) \sum_{l,m_l} C_{lm_l} Y_{lm_l}(\theta,\varphi), \quad (5.9)$$

where the C_{lm_l}s are the expansion coefficients for the molecular wavefunction in the basis set of spherical harmonics. These coefficients are given for a variety of diatomic molecules aligned along the electric field direction in Kjeldsen et al. (2005). The radial part of the wavefunction in the asymptotic tunneling region can be approximated in the same way that we approximated this function for an atom, referring back to Eq. (4.207). The sum over C_{lm}s essentially replaces the C_{kl} factor in the derivation of the atomic tunneling rate. Note that for linear molecules (like the diatomics we presently consider), m_l is a good quantum number so no sum over m is needed. For molecules with a σ HOMO, we simply set $m_l = 0$ and for those molecules with π orbitals, $m_l = 1$.

As we did for complex atoms, we make the same approximation for the spherical harmonics that we did in Eq. (4.211) and say

$$Y_{lm_l}(\theta,\varphi) \approx \alpha_{lm_l} \sin^{|m_l|}\theta\, e^{im_l\varphi}. \quad (5.10)$$

We can follow the same derivation that we developed for complex atoms in Eqs. (4.210) to (4.218); however, now the atomic fitting factor is replaced by a sum over l:

$$C_{kl}\alpha_{lm_l} \to \sum_l C_l \left[\frac{(2l+1)(l+|m_l|)!}{(l-|m_l|)!}\right]^{1/2} \equiv \beta(m_l). \quad (5.11)$$

This leads to a formula for the tunnel ionization rate of the diatomic molecule AB very much like the PPT/ADK result of Eq. (4.219):

$$W_{AB}^{(tun)} = \frac{I_p}{\hbar}\beta^2(m_l)\frac{2^{2n^*-2|m_l|-1}}{|m_l|!}\sqrt{\frac{3}{\pi}}\left[\left(\frac{I_p}{I_H}\right)^{3/2}\frac{E_a}{E_0}\right]^{2n^*-|m_l|-3/2}$$

$$\times \exp\left[-\frac{2}{3}\left(\frac{I_p}{I_H}\right)^{3/2}\frac{E_a}{E_0}\right]. \quad (5.12)$$

This result highlights one of the significant differences between atoms and molecules: the ionization rate is dependent on orientation of a molecule with respect to the laser polarization (Zhao et al. 2003). Equation (5.12) yields the tunnel ionization rate for a linear molecule oriented along the laser field. For an arbitrary orientation of the molecule, it is necessary to express the asymptotic ground state wavefunction in terms of the expansion of Eq. (5.9) with a Wigner rotation operator, D, applied to the spherical harmonics:

$$\psi_i = f_i(r) \sum_{l,m_l'} C_l D^{(l)}_{m_l m_l'}(\varphi,\theta,\chi) Y_{lm_l'}(\hat{r}). \quad (5.13)$$

In this rotation, φ represents a rotation about the electric field axis, χ represents a rotation about the molecular bond axis and θ describes rotation of the molecular bond axis with respect to the field direction (and is equal to zero when the molecule is aligned along the field direction). For a linear molecule like a diatomic, the first two angles are irrelevant and we can simplify the notation by introduction of the "small-d" Wigner operator $D^{(1)}_{mm'}(0,\theta,0) = d^{(1)}_{mm'}(\theta)$. For example: $d^{(2)}_{20} = 6^{1/2}\sin^2\theta/4$ or $d^{(2)}_{21} = [(1+\cos\theta)/2]\sin\theta$.

We now carry through all possible m_l's, sum over them at the end, yielding a θ-dependent shape factor, β, for the rotated molecule

$$\beta(m'_l, \theta) = \sum_l C_l d^{(l)}_{m'_l m_l}(\theta)(-1)^{m'} \left[\frac{(2l+1)(l+|m'_l|)!}{(l-|m'_l|)!} \right]^{1/2}. \quad (5.14)$$

The tunnel ionization rate for the arbitrarily rotated linear molecule therefore must be written in terms of a sum:

$$W^{(tun)}_{AB}(\theta) = \frac{I_p}{\hbar} \sqrt{\frac{3}{\pi}} \exp\left[-\frac{2}{3} \left(\frac{I_p}{I_H} \right)^{3/2} \frac{E_a}{E_0} \right]$$

$$\times \sum_{m'_l} \beta^2(m'_l, \theta) \frac{2^{2n^* - 2|m'| - 1}}{|m'_l|!} \left[\left(\frac{I_p}{I_H} \right)^{3/2} \frac{E_a}{E_0} \right]^{2n^* - |m'_l| - 3/2}. \quad (5.15)$$

As an example, let us examine the tunnel ionization of the N_2 and O_2 molecules. Consideration of the shapes of the highest molecular orbitals in these molecules, reproduced in Figure 5.2, immediately suggests that the variation of the tunnel ionization rate with molecular rotation about the E-field should be quite different for the two molecules. As a simple illustrative approximation, assume as we did for atoms that the $m' = 0$ factor dominates the ionization rate. We can write the sum in Eq. (5.14) for an N_2 molecule (where $m = 0$) approximately:

$$\beta(0, \theta) \cong C_0 + 5^{1/2} C_2 d^{(2)}_{00}(\theta), \quad (5.16)$$

yielding for the θ dependence of tunnel ionization

$$W^{(tun)}_{N_2}(\theta) \sim \left(2.6 \cos^2 \theta + 1.2 \right)^2. \quad (5.17)$$

On the other hand, the O_2 molecule is described by $m_l = 1$, so

$$\beta(0, \theta) \cong 5^{1/2} C_2 d^{(2)}_{10}(\theta), \quad (5.18)$$

suggesting that the orientational dependence of tunneling in oxygen is

$$W^{(tun)}_{O_2}(\theta) \sim \sin^2 \theta \cos^2 \theta. \quad (5.19)$$

Not surprisingly, the tunneling rate is maximum when the N_2 molecule is aligned right along the field ($\theta = 0$). However, O_2 ionizes fastest when it is oriented roughly 45° with respect to the field. It is this orientation which aligns the π orbital lobes in the direction of the field, maximizing the tunneling rate (which we argued was why the $m_l = 0$ orbitals ionized fastest in atoms).

Figure 5.4 shows plots of tunnel ionization rates calculated from Eq. (5.15) for these two molecules. As we have ascertained in our approximate analysis, ionization of N_2 is about ten times faster when the molecule is oriented along the field. The ionization rate for O_2 peaks at around 40°, though the relative rate of ionization for perfectly aligned molecules does increase as intensity increases since the binding barrier is better suppressed at the higher intensity.

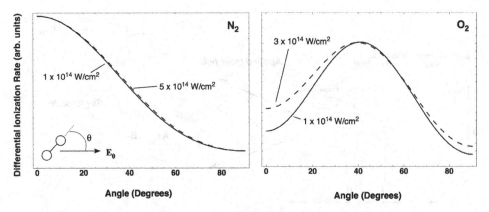

Figure 5.4 Tunnel ionization rate of N_2 and O_2 molecules as a function of molecular orientation angle at two different intensities.

5.2.4 Multielectron Effects in Ionization of Diatomics

Though the SAE approximation is often employed to describe strong field interactions with small molecules, it is, in fact, a poorer approximation than in atoms (Kono et al. 2006; Lezius et al. 2002). In atoms the characteristic time scale for electron motions is typically much faster than an optical oscillation time. This is not always true in a small molecule. The path length that a bound electron can traverse in the molecular potential is greater than in atoms, suggesting that driven dynamics by delocalized electrons inside the multinuclear molecular potential become important. We completely neglected such collective, bound electron motion in the deep, localized binding well of an atom. We might expect that motion of one or more bound electrons in the molecular potential might affect the ionization dynamics of another more weakly bound electron. Put differently, unlike atoms, molecules quite frequently have doubly excited states that exist below the ionization potential, indicating that there can likely be more than one active electron at play in a strong laser field. Understanding quantum multielectron effects is challenging. Short of numerically solving the multielectron Schrödinger equation including correlation effects, any consideration will need to be qualitative in nature. As an example of how two-electron effects might alter ionization in a diatomic, let us consider correlation effects in the ionization of an H_2 molecule aligned along a laser field. (Our discussion is just as relevant for other diatomics.)

If we accept that both electrons in the H_2 atom are driven by the incident laser field, a situation like that pictured in Figure 5.5 arises. We first consider H_2 in its ground state, a singlet state, $^1\Sigma_g^+$. In Figure 5.5 the two-electron wavefunction probability density is shown in a cartoon manner. In the strong E-field it is possible to push both electrons into the lower potential well of the atom, denoted well "B" in the figure. Localization of both electrons in one well is possible because of the anti-aligned spins of the two fermions in this singlet state. This leads to the transient formation of an $H^+ - H^-$ state. Qualitatively, we might expect that the H^- ion is easier to ionize by tunnel ionization than the H_2 molecule alone, leading to enhancement in the net ionization rate (Seideman et al. 1995).

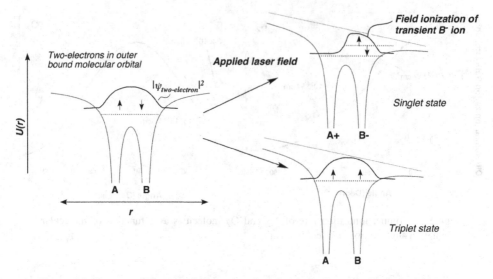

Figure 5.5 Cartoon of the two-electron wavefunction in a diatomic. The top right case illustrates how the electron density can be pushed to the lower nuclear well if the electrons are in a singlet state, while the bottom right illustration indicates that two electrons in a triplet state will remain spread out because of the Pauli exclusion principle.

To estimate if this transient H^+–H^- state can indeed be formed, we can make a simple energetics argument. We expect that this state forms if the laser field at its peak value can impart enough energy to an electron over the distance of the internuclear spacing, R, to overcome the difference in energy between a ground state H_2 molecule and the ionic state. This condition,

$$eE_0R \approx \varepsilon_{H^+H^-} - \varepsilon_{H_2}, \tag{5.20}$$

can be estimated in terms of the binding energy of both electrons and the attractive energy between the H^+ and H^- ions:

$$eE_0R \approx \left(-I_{H^-} - I_H - \frac{e^2}{R}\right) - \left(-I_{H_2} - I_{H_2^+}\right). \tag{5.21}$$

Since $I_{H^-} = 0.75$ eV, $I_{H_2} = 15.8$ eV, $I_{H_2^+} = 29$ eV, and the equilibrium spacing of H_2 is 0.74 Å, it follows that E_0 would need to be around 1.5×10^9 V/cm to create the ionic state. It is reasonable to expect that at fields within an order of magnitude of this value, say $\sim 10^8$ V/cm, should see some significant deformation of the two-electron wavefunction toward the H^+–H^- state. This field is reached at intensity of around 3×10^{13} W/cm^2.

What likely happens to the two-electron wavefunction (which has been confirmed in numerical solution of the two-electron Schrödinger equation (Harumiya et al. 2000)) is pictured in Figure 5.6. This schematically plots the probability density of two electrons along the internuclear axis of a diatomic molecule. The probability density in the field-free case peaks at $x_1 = +R/2$ and $x_2 = -R/2$ in the two-dimensional space describing the 1D wavefunction of two electrons. Simulations show that at the peak of the field, some

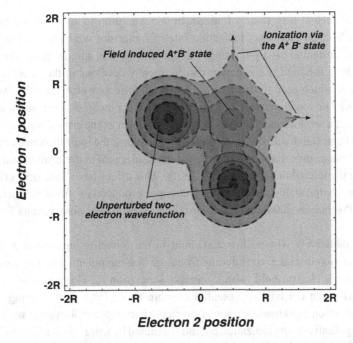

Figure 5.6 Schematic illustration of the probability density of a two-electron wavefunction along the internuclear axis. The contour plot ascribed by solid lines shows the ground state without a laser field. The dashed line contour plot illustrates how population density can form with both electrons centered on one nucleus at $+R/2$ with the ionization pathways along the respective electron axes.

significant population density can form at $x_1 = +R/2$ and $x_2 = +R/2$, which becomes a gateway state for tunnel ionization along the x_1 and x_2 axes. This two-electron phenomenon is likely to play a role in most neutral diatomics with singlet HOMOs, such as N_2.

The situation is quite different for electrons in a triplet state (Kono et al. 2006), as might be found in an excited state of H_2 or, say, the O_2 molecule. In this case, as illustrated in the cartoon of Figure 5.5, the Pauli exclusion principle will tend to keep both electrons separated on the two nuclear centers and strongly repels formation of an ionic state. For these molecules, we expect ionization rates will see no enhancement from multielectron effects. This might explain, in part, why O_2 seems to exhibit a much greater ionization suppression than the singlet state F_2 and N_2 molecules.

5.2.5 Tunnel Ionization as a Function of Nuclear Separation and the Critical Ionization Distance

So far, we have considered ionization of neutral diatomics with the two nuclei centered at or near their equilibrium separation. Of course, the nuclei can move during the laser pulse, driven apart through the process of bond softening (discussed in Section 5.4) or by the repulsive Coulomb forces between positively charged ions resulting after the molecule has been at least doubly ionized. In fact, variation in the nuclear separation has a dramatic effect on the tunnel ionization rate of molecules which are aligned along the laser field (Constant *et al.* 1996; Peng *et al.* 2003). (Again, alignment of the molecular bond axis along the field is a situation that frequently develops in the laser field because of field forces which tend to prealign the molecule, effects which we will discuss in the next section.) The ionization rate enhancement that accompanies increased nuclear separation peaks at a separation which has come to be known as the critical ionization distance (Chelkowski and Bandrauk 1995). At this critical distance the ionization rate can be higher by orders of magnitude than the molecule (or molecular ion) at the equilibrium separation of the neutral molecule (Nibarger *et al.* 2001). This effect has a dramatic effect on the observed ion charge states and ejected ion energies produced in strong field irradiation of diatomics (Posthumus 2004) and led to puzzling experimental observations for a number of years.

To see qualitatively why such a maximum in the tunneling rate exists at the critical nuclear separation distance, consider the Coulomb binding potential of a diatomic aligned along the laser's electric field, with a variable separation R. This potential is illustrated in Figure 5.7 when the binding potential is subject to a laser field pointing to the left. Clearly the electron must tunnel through a barrier to escape the molecular system, tunneling through the potential at the ionization potential denoted by a gray dashed line in Figure 5.7. Of course, as R varies, the ionization potential of the molecule will also vary. To understand this effect semi-quantitatively it is adequate to approximate the ionization potential in terms of the atomic ionization potentials of the two ionic constituents of the diatomic, I_1 and I_2, corrected for the attractive potential of the neighboring ion. If the ionization potentials for a diatomic of core charges Q_1 and Q_2 are averaged, we can approximate the ionization potential as a function of R by:

$$I_p(R) \cong \frac{1}{2}\left[I_1 + \frac{Q_2 e^2}{R} + I_2 + \frac{Q_1 e^2}{R}\right] \qquad (5.22)$$

or if both ions have the same charge the ionization potential is then approximately

$$I_p(R) \cong I_Q + Q e^2 / R. \qquad (5.23)$$

As R increases beyond the equilibrium separation of the neutral molecule, this ionization potential decreases, which suggests that the tunnel ionization rate should begin to increase rapidly. As R increases, something else very important occurs, the central potential barrier rises in the center of the diatomic. Eventually, this barrier will become high enough that the outermost bound electron will become localized in one well, illustrated in the second potential curve of Figure 5.7. Of course, if this happens with the electron trapped in the lower well, the situation is reversed one half cycle later. So every electron will find itself at some point in the upper well (the leftmost well in Figure 5.7). As the figure illustrates, when this happens, while the electron is in the upper well it now only needs to tunnel through the central barrier to be freed, a much narrower potential barrier than the electron encountered when it needed to tunnel through the outer barrier. At some critical distance, the central potential well will be at its narrowest; at this point the tunnel ionization rate

5.2 Strong Field Ionization of Molecules

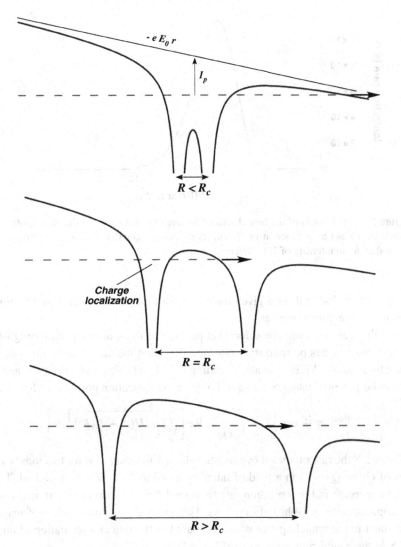

Figure 5.7 Plots of the binding potential of a diatomic aligned along an electric field with three different internuclear separations. The ionization potential of the outermost electron is illustrated as a dashed line.

reaches its maximum. An increase in R beyond this critical distance results in a broadening of the central potential barrier, and a dramatic decrease in the tunnel ionization rate ensues with further increases in nuclear separation. This is illustrated in the bottom curve of Figure 5.7.

Of course, at large enough separation, the ionization rate will necessarily approach that for a single free ion in the field. So at separations near R_{crit} the neighboring ion of the diatomic essentially enhances the field driving the tunnel ionization. Because of this enhanced tunneling, we therefore expect that if the nuclei of a molecule separate during a laser pulse,

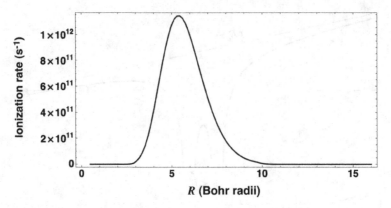

Figure 5.8 Calculation of the tunnel ionization rate of a field-aligned diatomic molecule using Eq. (5.24) as a function of R with core charges equal to $+2$ with electric field equivalent to an intensity of 10^{15} W/cm^2.

the charge states observed for a given intensity will tend to be higher from the molecule than from the lone atomic species.

We can illustrate the dramatic effect that passage of the molecular ion through R near the critical distance has on ionization rate by calculating the tunnel ionization rate of the molecule by a simple WKB estimate. Recalling Eq. (4.151), this formula can be applied to the molecular potential using our simple R-dependent ionization potential of Eq. (5.23):

$$W^{(tun)} \approx \frac{I_p}{\hbar} \exp\left[-\frac{2}{\hbar} \int_{-R/2}^{\infty} \mathrm{Re}\left[\sqrt{2m\left(U(r) - I_p(R)\right)}\right] dr\right]. \tag{5.24}$$

In Figure 5.8 the rate predicted by this formula as a function of R for two ions with core charges of $Q_1 = Q_2 = 2$ in a field of intensity equal to 10^{15} W/cm^2 is plotted. There is a dramatic increase in the ionization rate to around 1 ps^{-1} when the internuclear separation is approximately 6 Bohr radii ($\sim$3 Å). This is an almost three-orders-of-magnitude enhancement in the tunneling rate when compared to the rate at a separation of only half that (1.5 Å, the equilibrium separation of N_2 or O_2).

We can ask the question, at what separation does the electron begin to be localized in one well or the other. Roughly speaking, this should occur when the height of the central potential barrier reaches $I_p(R)$. The point at which this happens should also be near the separation where tunnel ionization is nearly maximized since the electron becomes localized and the width of the central barrier is near its thinnest. The localization condition, then, should give us a reasonable estimate for R_{crit}. It can be simply derived. First we find the location of the peak of the central barrier by noting that the 1D potential (for equally charged nuclear cores) is

$$U(r) = -\frac{Q'e^2}{|r + R/2|} - \frac{Q'e^2}{|r - R/2|} - eE_0 r. \tag{5.25}$$

We find the location of the peak in this potential, r_p, easily:

$$\left.\frac{\partial U}{\partial r}\right|_{r=r_p} = \frac{Q'e^2}{(r_p + R/2)^2} - \frac{Q'e^2}{(r_p - R/2)^2} - eE_0 = 0, \tag{5.26}$$

$$r_p \cong -\frac{E_0 R^3}{32eQ'}. \tag{5.27}$$

The condition for electron localization, where $R = R_l$, can be written with the aid of Eq. (5.23) as

$$U(r_p) \geq -I_p(R) \cong -I_Q - \frac{Qe^2}{R}, \tag{5.28}$$

$$U(r_p) \cong -\frac{4Qe^2}{R_l} + \frac{E_0^2 R_l^3}{32Q} = -I_Q - \frac{Qe^2}{R_l}, \tag{5.29}$$

which yields

$$R_l \cong \frac{3Qe^2}{I_Q} - \frac{81E_0^2 Q^3 e^8}{32 I_Q^3}. \tag{5.30}$$

This can be further simplified if we note that the ionization potential for many ions of charge Q is well approximated in terms of the neutral atom's ionization potential, $I_p^{(0)}$ simply by $I_Q \approx Q I_p^{(0)}$. (For example, this approximation is accurate to better than 10% in an atom like Cl, and better than 2% accurate in Cl when $Q > 2$.) The localization separation can be simply written

$$R_l \cong \frac{3e^2}{I_p^{(0)}} - \frac{81E_0^2 e^8}{32\left(I_p^{(0)}\right)^3}. \tag{5.31}$$

Remarkably, because the second term in Eq. (5.31) is small, this separation, which is a good approximation for the critical ionization distance, is only very weakly dependent on intensity and completely independent of ionic charge. If we ignore the small second term (which serves to decrease R_l slightly as intensity increases) and assume that $I_p^{(0)}$ is roughly equal to the ionization potential of the hydrogen atom, $I_H = e^2/2a_B$, this equation suggests that the critical ionization distance will be around 6 Bohr radii for almost all diatomic molecules, independent of charge state or intensity. In fact, that is reasonably close to what is observed experimentally. The implication of this fact is that if R varies in a molecule during a laser pulse, the ionization rate will tend to be rather slow until $R \approx 3-4$ Å, at which point the molecule rapidly ionizes to high charge states. (We will discuss this dynamic in greater detail when we consider Coulomb explosions in Section 5.5.)

Another way to look at the effects of nuclear separation on ionization rate is to consider the intensity at which barrier suppression ionization will occur as a function of R (Posthumus et al. 1996b). This approach was first presented by Posthumus (Posthumus et al. 1995; Posthumus 2004). The potential of the molecule is assumed to be of the form of Eq. (5.25). For cases when the two nuclear charges differ, we approximate Q' in this equation by $Q' = (Q_1 + Q_2)/2$. The intensity for barrier suppression ionization of the

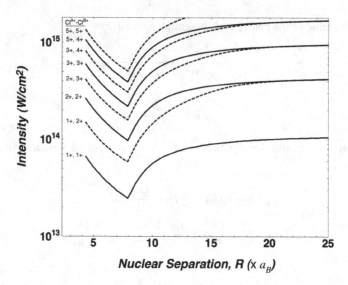

Figure 5.9 Plot of the BSI intensity in Cl_2 versus R needed to access the charge state pairs listed along the left of the plot. For the asymmetric pairs (q, q + 1), the BSI intensity is allowed to approach the intensity of the next-higher charge state pair (q + 1, q + 1) since at large R, the BSI intensity for both cases should approach the intensity needed to ionize a free ion q → q + 1.

Cl_2 molecule as a function of R is plotted in Figure 5.9. Since $I_{Cl} = 11.5$ eV, Eq. (5.30) predicts that the critical ionization distance should be near $7.1 a_B$. The effect of the critical ionization distance in the plot of Figure 5.9 is dramatic. Indeed, at R around $8 a_B$, the BSI intensity of all channels plotted drops precipitously.

5.2.6 Nonsequential Double Ionization of Diatomics

Recollision-driven NSDI occurs during the ionization of molecules in much the same manner it does in atoms (Guo et al. 1998). For example, the "knee" structure in the ion yields from N_2 ionization look very much like those in ionization of Ar. Not surprising, though, there are orientational differences in the magnitude of the NSDI yields when a diatomic is irradiated. The most significant effect is in the ellipticity dependence of the NSDI yield. As we noted in the last chapter, the recollision-driven double ionization rate falls off very rapidly as the ellipticity of the laser field departs from linear polarization. This is a result of the transverse velocity that the electron acquires upon ionization, which causes the electron to miss the parent core when it returns.

An elliptically polarized field can be expressed as

$$\mathbf{A}(\tau) = \frac{A_0}{\sqrt{1+\kappa^2}}(\sin \omega\tau \, \hat{\mathbf{x}} + \kappa \cos \omega\tau \, \hat{\mathbf{y}}), \tag{5.32}$$

with ellipticity parameter $\kappa = 1$ for circularly polarized fields and $\kappa = 0$ for linear polarization. This field will result in a transverse displacement of the electron in terms of birth

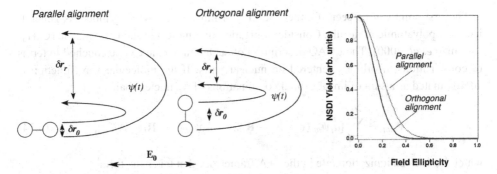

Figure 5.10 Illustration of how the difference in the transverse size of a wavepacket emerging from a diatomic can affect ellipticity dependence of NSDI depending on the orientation of the molecular axis with respect to the laser field polarization.

and return phases and can be easily calculated from conservation of canonical momentum in the y-direction:

$$y_{ret} = -\frac{\kappa}{\sqrt{1+\kappa^2}} \frac{eE_0}{m\omega^2} \left(\sin\varphi_r - \sin\varphi_b + \varphi_r \cos\varphi_b - \varphi_b \cos\varphi_b \right). \quad (5.33)$$

This transverse displacement will suppress NSDI if its magnitude is greater than the quantum diffusion of the wavepacket in the transverse direction while that tunnel ionized electron is in the continuum. In the absence of Coulomb focusing, this spreading is given by Eq. (4.245). How the orientation of the molecule affects this diffusing wavepacket is illustrated in Figure 5.10. If the molecule is aligned along the field, the initial extent of the wavepacket is small and the transverse spreading will be large. On the other hand, if the molecule is oriented perpendicularly to the field, we might expect that the wavepacket will have a larger spatial extent, of the order of the internuclear spacing, R. This leads to less transverse broadening. This results in an ellipticity dependence in the NSDI yield pictured in the right-hand panel of Figure 5.10. The aligned molecule will exhibit NSDI to a higher ellipticity parameter than the orthogonally aligned molecule. This effect has been established in experiments and confirms the supposition made earlier in this section that ionization of the diatomic occurs by electron probability density exiting over the entire extent of the HOMO occupied by the ionizing electron.

5.2.7 Ionization of Polyatomics

Trends that emerge in the ionization of diatomics persist in polyatomic molecules, though the dynamics can become vastly more complex in large molecules (Cornaggia 2001). In particular, two notable differences between ionization of diatomics and single atoms become even more pronounced in most polyatomics:

1) The extended, nonspherical nature of the molecular wavefunction affects ionization through orientational effects and quantum interference effects.
2) Multielectron dynamics become even more important in larger molecules.

The first point can be seen if either the SFA or molecular tunneling model is generalized to a polyatomic molecule. Consider ionization of a molecule such as benzene (C_6H_6) (Shimizu et al. 2000). The LCAO description of such a molecule can be couched in terms of combinations of orbitals centered on nuclear pairs. If the molecular wavefunction is approximated as a sum of linear orbital combinations of N_p nuclear pairs,

$$\Psi_{mol} \cong \sum_{i=1}^{N_p} [a_i \psi_{at}(\mathbf{r} - \mathbf{R}_i - \mathbf{R}/2) + b_i \psi_{at}(\mathbf{r} - \mathbf{R}_i + \mathbf{R}/2)], \qquad (5.34)$$

which leads to an ionization rate in the SFA framework that takes the form

$$W^{(tun)}(\mathbf{k}_e) \sim N_e N_p |\phi_i(\mathbf{k}_e)|^2 \begin{cases} \cos^2(\mathbf{k}_e \cdot \mathbf{R}/2) & \text{bonding pairs} \\ \sin^2(\mathbf{k}_e \cdot \mathbf{R}/2) & \text{anti-bonding pairs} \end{cases} \qquad (5.35)$$

(if the nuclear pairs are identical like they are in benzene). Again, quantum interferences can lead to ionization suppression.

Molecular shape and orientation effects can also be explored in polyatomics by generalizing the tunneling theory. Because a polyatomic is not, in general, linear like a diatomic, breaking of the axial symmetry demands that the expansion of the asymptotic wavefunction in the body-fixed frame of the molecule must now include a sum over m_l's. Following an analysis like that we performed in Eqs. (5.9) to (5.15) leads to a β factor given by a double sum, which is, in general, dependent on all three rotation angles.

$$\beta(m'_l, \varphi, \theta, \chi) = \sum_{lm_l} (-1)^{(|m'_l|+m'_l)/2} C_{lm_l} D^{(l)}_{m'_l m_l}(\varphi, \theta, \chi) \left[\frac{(2l+1)(l+|m'_l|)!}{(l-|m'_l|)!} \right]^{1/2}. \qquad (5.36)$$

C_{lm}'s of a few organic polyatomic molecules have been tabulated in Tong et al. (2002).

A potentially more fascinating aspect of strong field ionization in large polyatomics is the effect of multiple active electrons (Levis and DeWitt 1999). We have already discussed how polarization of the two degenerate electrons in an H_2 molecule can potentially affect ionization rates. Polarization of multiple delocalized electrons in a large polyatomic molecule will have an even greater impact on the ionization rates of that system. It is likely in large molecules that the SAE is completely inappropriate and the dynamics of many electrons, even those more deeply bound than the electrons in the HOMO, play a significant role (Lezius et al. 2002).

Any discussion of this physics must be qualitative. Let us consider an example molecule, such as naphthalene, to illustrate how delocalized electrons might affect tunnel ionization. The structure of naphthalene is illustrated in Figure 5.11. If the molecule is more or less oriented along the laser polarization, the actual extended potential seen by electrons bound in this system (illustrated schematically in the upper left of Figure 5.11) can very crudely be approximated as a box potential. In an SAE picture, application of a strong laser field deforms the box potential (as suggested in the upper right of Figure 5.11) and the outer active electron is easily ionized by tunneling through the suppressed outer confining barrier.

But this picture completely ignores the motion of the many other electrons confined in the delocalized orbitals of the naphthalene molecule. If the polarization of these electrons

5.2 Strong Field Ionization of Molecules

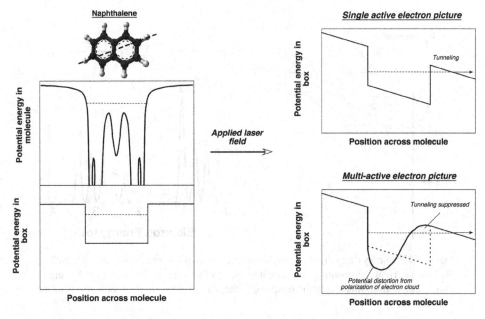

Figure 5.11 Multielectron effects in a large molecule such as naphthalene, explained by crudely approximating the binding potential of the aligned molecule (upper left) by a box potential.

is considered, the box model potential actually looks much more like the bottom right panel of Figure 5.11. The potential that results from the polarization of these electrons leads to a barrier that is thicker than the simple box model barrier of the SAE model, which would certainly greatly suppress the tunnel ionization rate of the outer electrons. Put differently, the field-induced "sloshing" of delocalized electrons toward the tunneling direction in the molecular potential serves to repel the electron trying to tunnel. We might conclude that the tunnel ionization rate of a large polyatomic might be significantly suppressed when compared to rate predictions made given simply the ionization potential of the molecule. This is exactly what is observed in experiments over a large number of organic molecules (Niikura *et al.* 2008).

5.2.8 ATI in Molecules

Generally, the photoelectron energy spectrum of ATI electrons emerging from strong field ionized molecules is qualitatively similar to that from single atoms. The most significant differences arise in the low-order peaks produced in the multiphoton regime. Complications in the ATI structure can occur as a result of the coupling between the outgoing electron energy and the energy left in the vibrational modes of the molecule. In single atoms, two-electron excitation is largely absent during ionization (with the exception of recollsions) and the atom has no other internal structure into which energy can be imparted. Consequently, low-order ATI peaks from atoms tend to be well separated into sharp peaks,

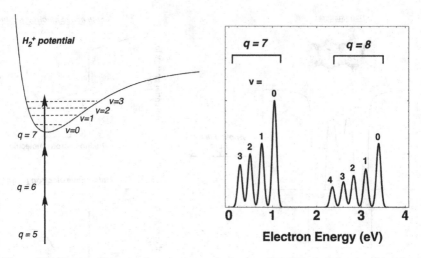

Figure 5.12 Energy diagram of seven-photon ionization of H_2 into the vibrational levels of H_2^+. On the right the resulting structure imparted to the seven-photon and first ATI eight-photon peaks from this electronic coupling to the various labeled vibrational levels of the molecular ion.

with structures only arising once the ponderomotive energy becomes large enough to shift the continuum significantly and shift intermediate states into transient resonance.

The situation is different in a molecule which has internal nuclear structure and vibrational modes. Consider, for example, multiphoton ionization of the H_2 molecule by 532 nm radiation. The minimum number of photons required for ionization is 7, though, as we have discussed, the ionization results in the population distribution in a range of vibrational levels in H_2^+. The energy diagram of this situation is illustrated in Figure 5.12. Energy coupled into excited vibrational levels must come out of the kinetic energy of the outgoing photo-electron. The result is structure in the ATI peaks like that shown in the right of Figure 5.12. As illustrated here, the $q = 7$ lowest-order peak and the first ATI peak at $q = 8$ become separated into a comb of peaks corresponding to the excitation of the residual molecular ion into various vibrational modes with amplitude given by the modified Franck–Condon factors (discussed in Section 5.2.2). As the intensity increases toward 10^{14} W/cm^2, ponderomotive shifts tend to wash out this vibrational structure.

As the intensity increases into the tunneling regime, ATI spectra from molecules look very much like those from atoms with similar ionization potentials. Molecular ATI spectra exhibit the characteristic direct tunneling electrons up to $2U_p$ as well as the rescattered electrons up to $10U_p$. One subtle difference does appear when ATI spectra from similar ionization potential molecules and atoms are directly compared. Shown schematically in Figure 5.13, ATI spectra from Xe and O_2 are plotted. As we derived in Eq. 5.6b, electrons ejected from the antibonding orbital of O_2 exhibit quantum destructive interference, as long as $k_e^\bullet R < \pi$. Consequently, low-order ATI peaks tend to have lower amplitude in O_2 spectra than in Xe spectra at similar intensities.

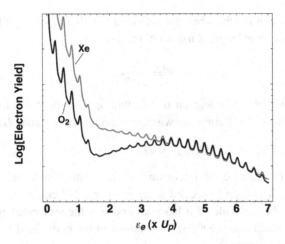

Figure 5.13 Schematic representation of ATI spectrum from O_2 and Xe in the tunneling regime.

5.3 Molecular Alignment in Strong Fields

At numerous points in the discussion of the past two sections we have considered specifically the ionization behavior of linear molecules aligned along the laser field polarization direction. This special consideration is appropriate not only because an ensemble of molecules will usually have random orientation so that some subset are prealigned with the field but is particularly relevant because linear molecules often have a propensity to align themselves along the laser polarization (Sakai et al. 1999; Seideman 2001; Dion et al. 2002; Stapelfeldt 2003). This effect, which can occur at modest intensity, is a consequence of the induced polarizability of a linear molecule (Seideman 2002). In fact, even complex molecules will typically have anisotropy in their polarizability and can be subject to some level of field-induced alignment (Larsen et al. 1999; Sakai et al. 1999). Though the intensities at which this tends to occur typically are below the threshold we have nominally set as the onset of "strong field" phenomena, we will consider this molecular alignment at least briefly because of the impact it often has on strong field molecular physics experiments (Litvinyuk et al. 2003).

Before proceeding, we need to review some of the characteristics of the rotation motion of molecules. We will limit our (somewhat superficial) considerations to linear molecules. A linear molecule can be described by a Hamiltonian written

$$H = H_0 + H_{int} + B\hat{J}^2 - \mathbf{d}_m \bullet \mathbf{E}(t) - \frac{1}{2}\mathbf{E}(t) \bullet \alpha \bullet \mathbf{E}(t) + \ldots \qquad (5.37)$$

where H_0 describes the electronic structure and vibrational contributions; H_{int} describes the laser's interaction with the electrons bound in the molecular structure. The third term is the energy arising from rigid rotor motion subject to the angular momentum operator with rotational constant B, which has units of energy. Within this convention, a diatomic of reduced mass μ and bond length R has rotational constant

$$B = \frac{\hbar^2}{2\mu R^2}. \quad (5.38)$$

Under this definition of B, the angular momentum operator in (5.37) has eigenfunctions given by the usual spherical harmonics with quantum numbers J and M_J:

$$\hat{J}^2 |J, M_J\rangle = J(J+1) |J, M_J\rangle. \quad (5.39)$$

The next two terms describe the interaction of the rotating molecule with the field, in a perturbation expansion, up to second order in the field strength. $\mathbf{d}_m$ is the permanent dipole moment of the molecule (if it has one) and $\boldsymbol{\alpha}$ is the molecule's polarizability tensor (Friedrich and Herschbach 1995b). This tensor in the body-fixed frame of the linear molecule can be diagonalized:

$$\alpha = \begin{pmatrix} \alpha_\perp & 0 & 0 \\ 0 & \alpha_\perp & 0 \\ 0 & 0 & \alpha_\parallel \end{pmatrix}, \quad (5.40)$$

where $\alpha_\perp$ is the polarizability of the molecule with field oriented perpendicular to the bond axis, and $\alpha_\parallel$ is the same for the molecule oriented parallel to the field. It is this fifth term in Eq. (5.37) and the difference between $\alpha_\perp$ and $\alpha_\parallel$ which lead to the alignment of the molecule in the strong field.

5.3.1 Pendular States in Long Laser Pulses

First we consider the quantum motion of a molecule in a laser pulse which is "long" and is turned on "slowly". What constitutes "long" pulse and "slow" turn on are driven to a large extent by the rotational time scale of the molecule (Ortigoso et al. 1999). The characteristic rotation time is

$$\tau_{rot} = \hbar/B. \quad (5.41)$$

For an H_2 molecule this time is 87 fs; for an iodine molecule it is 132 ps. We consider, then, laser pulses with duration and ramp-up time much longer than this quantity. This is often referred to as the adiabatic limit for molecular alignment. For such a situation, we can treat the electric field amplitude as constant and write the rotational part of the molecular Hamiltonian in a linearly polarized field as

$$H_{rot} = B\hat{J}^2 - d_m E_0 \cos \omega t \cos \theta \\ - \frac{1}{2} E_0^2 \cos^2 \omega t \left(\alpha_\parallel \cos^2 \theta + \alpha_\perp \sin^2 \theta \right) \quad (5.42)$$

if θ is the angle between the bond axis and the laser polarization.

A further simplification can be made if the laser field oscillates quickly with respect to molecular rotations (i.e. $\tau_{rot} \gg \omega^{-1}$), a condition easily met by visible and near-IR fields in all molecules. This allows us to average Eq. (5.42) over a laser period, yielding for the rotation Hamiltonian

$$H_{rot} = B\hat{J}^2 - \frac{1}{4}E_0^2 \left((\alpha_\parallel - \alpha_\perp)\cos^2\theta + \alpha_\perp\right). \quad (5.43)$$

For all linear molecules typically encountered, the molecule is more polarizable when the bond axis is oriented along the driving field so that $\alpha_\parallel > \alpha_\perp$. This means that the rotating molecule feels a potential well with a minimum at $\theta = 0$, oriented along the field. This double potential well with end-for-end symmetry will tend to attract the molecular rotors to an orientation along the field axis. The well depth of this conical potential is simply linear with the laser intensity and the difference between the polarizability tensor elements.

Quantum mechanically, the binding potential in the rotational Hamiltonian can support quantized states (some bound), which are termed pendular states. Because the field breaks the anisotropy of the rotational Hamiltonian, J is no longer a good quantum number, though M_J still is a good quantum number because of the rotational symmetry about the field axis. In reality, if the field ramps up slowly, the field-free rotation eigenstates will evolve adiabatically into pendular states described by quantum numbers $\tilde{J}$ and M_J. Consequently, an ensemble of molecules, initially in a randomly oriented distribution of rotational states, subject to a suitably intense field will evolve adiabatically to pendular states, resulting in a net, on average, orientation of the ensemble. Of course, the extent of this orientation depends on the well depth and will be greater as the laser intensity increases.

Figure 5.14 illustrates a few example pendular states for a sample rotational Hamiltonian. In this particular example, two bound states are supported. These two states are split by tunneling (one is symmetric about the two poles and one is antisymmetric). The number of bound states would increase with increasing intensity.

It is often informative to express the pendular eigenstates as a linear decomposition of field-free eigenstates

$$|J, M_J\rangle \rightarrow |\tilde{J}, M_J\rangle = \sum_J C_{JM}(E_0) |J, M_J\rangle \quad (5.44)$$

where, again,

$$|J, M_J\rangle = Y_{JM}(\theta, \varphi).$$

Figure 5.15 shows the first four coefficients for the ground pendular state as a function of intensity (with ordinate given in terms of the intensity needed for I_2 and O_2 alignment). Because of the symmetry of the ground state, only even J states contribute to the sum of Eq. (5.44). The adiabatic evolution of this pendular state as intensity increases can be seen in this plot. At low intensity, this state is dominated by the $J = 0$, $M_J = 0$ spherical harmonic state illustrating the adiabatic correspondence of this pendular state with the lowest-energy rotational state. As the laser intensity increases, higher-quantum-number rotational states play a greater role in the eigenstate decomposition. The two intensity scales illustrate how lighter, less polarizable molecules require somewhat higher intensity to align the molecule.

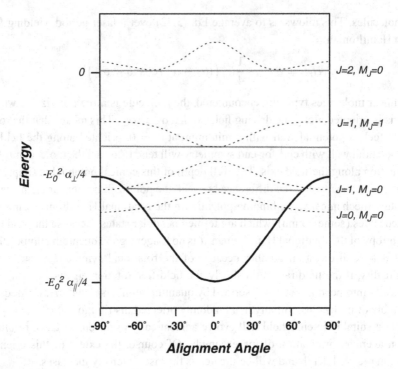

Figure 5.14 The laser-induced rotational potential well and some of the first few pendular states. Wavefunctions are denoted by dashed lines with quantum numbers describing those four states listed on the right. The parameters for this example are approximately those that would describe an O_2 molecule in a field at an intensity of 4×10^{11} W/cm^2. Figure was based on plots in (Friedrich and Herschbach 1995a).

The extent of alignment of a particular pendular state (or ensemble of states for that matter) can be quantified by calculating the expectation value of $\cos^2 \theta$. For a single quantum state

$$\left\langle \cos^2 \theta \right\rangle = \left\langle \psi(\theta, \varphi) \left| \cos^2 \theta \right| \psi(\theta, \varphi) \right\rangle. \quad (5.45)$$

In an isotropic distribution of molecules (or for a $J = 0$ rotational state) $<\cos^2 \theta> = 1/3$. Using the coefficients calculated in Figure 5.15, one finds that $<\cos^2 \theta>$ is greater than 0.9 for intensity greater than 2×10^{11} W/cm^2 for iodine molecules. This indicates a rather high degree of alignment for molecules initially in the lowest rotational state.

This long pulse alignment phenomenon can often be employed in experiments to produce prealigned molecules as a prepared target for a second, higher-intensity laser pulse. A pulse of picosecond to nanosecond duration at modest intensity can be focused into a gas of molecules, prealigning them adiabatically before the arrival of an ultrashort, high-intensity pulse. In practice, the molecules in the gas will initially be found in a range of rotational states. In that case, the extent of molecular alignment induced by a long

5.3 Molecular Alignment in Strong Fields

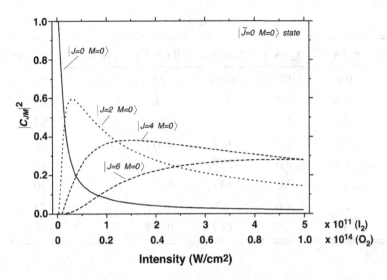

Figure 5.15 Calculation of the probability factors for the first four spherical harmonics in the eigenstate decomposition of the lowest-energy pendular state as a function of laser intensity. Intensity scale for an I_2 and an O_2 molecule are both given showing how the lighter, less-polarizable molecule requires higher intensity. Plot was generated based on plots in Larsen (2000).

pulse must be found by a double average over a weighted distribution of initial rotational states,

$$\langle\langle\cos^2\theta\rangle\rangle = \sum_{\tilde{J}} \rho_{\tilde{J}} \sum_{M_J=-J}^{J} \langle\cos^2\theta\rangle_{JM_J}, \tag{5.46}$$

with weighting coefficients usually given by the Boltzmann weighting factor for a gas with rotational temperature T_{rot},

$$\rho_J = \frac{e^{-BJ(J+1)/k_B T_{rot}}}{Z_{rot}}. \tag{5.47}$$

Here the rotational partition function is

$$Z_{rot} = \sum_J (2J+1)e^{-BJ(J+1)/k_B T_{rot}}. \tag{5.48}$$

At room temperature $k_B T_{rot} \approx 25$ meV. So for diatomic molecules where $\alpha_\| - \alpha_\perp \approx 1$–$5$ Å^3, laser intensities of >2–10×10^{12} W/cm^2 will be needed to prealign a significant population of room-temperature molecules (with the notable exception of H$_2$, which will typically need an order-of-magnitude-higher intensity). Table 5.1 gives some example numbers which can aid in the calculation of rotational orientation characteristics of a few diatomics (Friedrich and Herschbach 1995a). The final column of this table lists the depth

Table 5.1

Molecule	$\alpha_\parallel$ (Å^3)	$\alpha_\perp$ (Å^3)	τ_{rot} (ps)	B (meV)	ΔU_{rot}@10^{13} W/cm^2 (meV)
H_2	0.93	0.72	0.086	7.6	0.9
N_2	2.4	1.5	2.6	0.25	3.9
O_2	2.3	1.1	3.7	0.18	5.2
F_2	1.8	0.97	6.0	0.011	3.6
Cl_2	6.6	3.6	22.7	0.29	13
I_2	15	8.0	143	0.005	30
CO	2.6	1.6	2.7	0.24	43
CO_2	4.1	1.9	13.7	0.048	95

of the rotational potential well in a laser field at 10^{13} W/cm^2 to give a sense of how deeply bound low J levels are for that molecule.

5.3.2 Transient Alignment Driven by Short Laser Pulses

Rotational dynamics of a molecule are rather different when the aligning laser pulse is short in time (Stapelfeldt and Seidemann 2003). Roughly speaking, if the laser pulse duration, Δt_p, is significantly shorter than the characteristic rotation time of the molecule, such that

$$\Delta t_p \ll \hbar/B, \tag{5.49}$$

pendular states clearly cannot form. Instead, we think of the laser pulse as giving a fast rotational kick to the molecule. This kick will impart some angular momentum to the molecule, which will be a function of the angle between the molecular axis at the moment of irradiation and the laser polarization. Generally, if a linear molecule is nearly aligned along the laser field it receives a relatively small rotational impulse while those molecules nearly perpendicular to the field receive a significantly greater impulse. This problem, of course, must ultimately be solved quantum mechanically, and if a large ensemble of molecules in a range of initial rotational states is irradiated, as we might find in a molecular gas at finite rotational temperature, then a rather complicated quantum statistical mechanics treatment is needed.

Before we discuss this situation, some qualitative insight can be derived if we just examine the dynamics of a classical molecular rigid rotor subject to the impulse that an intense pulse might impart. If we consider the coordinate of motion of the molecule to be the angle between molecular axis and laser field, $\theta(t)$, the equation of motion can be written in terms of the angular momentum, L, and the Hamiltonian, Eq. (5.43):

$$\dot{L} = -\frac{\partial H}{\partial \theta}. \tag{5.50}$$

5.3 Molecular Alignment in Strong Fields

This yields an equation of motion in terms of the molecular polar moment of inertia, I_R,

$$I_R \frac{d^2\theta}{dt^2} = \frac{1}{2}E_0^2 \left(\alpha_\parallel - \alpha_\perp\right) \sin\theta \cos\theta$$

$$= \frac{1}{4}E_0^2 \Delta\alpha \sin 2\theta. \tag{5.51}$$

Since we consider the situation of a very short laser pulse, the impulsive angular momentum imparted to the molecule as a consequence of the torque applied momentarily by the laser field is

$$L_{im} = \int_{-\infty}^{\infty} \tau_\theta \, dt. \tag{5.52}$$

So, if θ_0 is the initial angle between the molecular axis and the laser field prior to irradiation, the impulse imparted to a molecule initially at rest by a pulse with peak intensity, I_0, is

$$L_{in}(\theta_0) = \frac{1}{4}\Delta\alpha \sin 2\theta_0 \int_{-\infty}^{\infty} |E_0|^2 dt$$

$$\cong \frac{2\pi}{c} \Delta\alpha I_0 \Delta t_p \sin 2\theta_0 t. \tag{5.53}$$

The rotational impulse is proportional to the fluence of the laser pulse (intensity times pulse duration) and depends on the initial angle between molecular axis and laser field. Using Eq. (5.53), the angular motion of our classical rigid rotor is simply

$$\theta(t) = \theta_0 + \frac{4\pi \Delta\alpha B I_0 \Delta t_p \sin 2\theta_0}{c\hbar^2} t. \tag{5.54}$$

What this simple model tells us is that when a collection of molecules is irradiated by an intense pulse, the distribution of impulses starts the molecules rotating, each with an angular velocity dependent on the initial rotor angle. Those molecules nearly aligned along the field rotate slowly (or not at all for $\theta_0 = 0$) while those 45° to the field rotate more quickly. If we use Eq. (5.54) to calculate the degree of alignment of a collection of many molecules, as defined by the average of $\cos^2\theta$, initially with random orientation with respect to the laser field, we find that there will be times at which the collection of molecules exhibit enhanced collective orientation along the field. Such a classical calculation with an ensemble of N_2 molecules irradiated by a short pulse with fluence of 1 J/cm² is illustrated in Figure 5.16. (This fluence would correspond to an intensity of 10^{13} W/cm² in a 100 fs pulse.) In this example, the molecules transiently increase their average orientation from a random distribution (with $\langle\cos^2\theta\rangle = 0.33$) to a more aligned distribution with $\langle\cos^2\theta\rangle \approx 0.7$. This simple classical model shows that such an impulse laser pulse can lead to delayed alignment of an initially unaligned ensemble of molecules.

The quantum treatment of a finite-temperature ensemble of molecules subject to "delta function" laser pulse also reveals periodic revivals of molecular alignment, though the periodicity is a result of quantum interference. When such an ensemble is irradiated by the

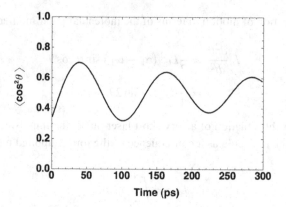

Figure 5.16 Rotational orientation of an initially randomly oriented collection of N_2 molecules subject to a fast pulse at 1 J/cm^2 calculated using the classical rigid rotor model.

delta function laser pulse, it creates a coherent wavepacket of rotational states, which can be written

$$\psi_{rot} = \sum_{J,M_J} a_{JM_J} e^{-iE_J t/\hbar} |J, M_J\rangle \qquad (5.55)$$

where the rotational energy levels are $E_J = BJ(J + 1)$. The absolute square of this wavepacket will inevitably lead to spikes in time at a repetition rate that is a result of the fact that the rotational levels are quantized.

As time evolves after the ensemble of molecules constituting this wavepacket have received the laser kick, any wavepacket formed from a large number of states will result in rapid cancellation, and the expectation value, Eq. (5.45), of the molecular ensemble is random. However, when the time passing reaches points where it is an integer, n, times a characteristic rotational time $T_{rot} = 2\pi\hbar/E_1$, the phases of the rotational states, $\varphi = E_J t/\hbar$, will be simply $\varphi_{rev} = n\pi J(J + 1)$. At these specific times ALL the phases of the rotational levels are an integer multiple of 2π and the wavefunction reflects a high degree of alignment. These times constitute "full revivals" and repeat at temporal intervals given by T_{rot}.

Revivals can also occur at specific fractions of T_{rot}, a situation typically called a "partial revival." For example, at times such that $t = nT_{rot}/2$ where n is an odd integer, a "half revival" takes place since $\varphi_{rev} = n\pi J(J + 1)/2$. States in which $J(J + 1)/2$ is even will be in phase and states where $J(J + 1)/2$ is odd will also be in phase. These two groups of rotation states will interfere with each other so the shape of the revival will be different than for full revivals. Partial revivals can occur with other fractional times of T_{rot}. Figure 5.17 shows a calculation where the alignment expectation value is plotted for an ensemble of nitrogen molecules irradiated by a short laser pulse at 1 J/cm^2. In this calculation, the full quantum statistical effects of a range of initial rotation states (for $T = 300$K) is calculated. The transient alignment that occurs at delayed times is evident. In this case, full and half revivals form, though as mentioned, the shape of the revival differs. At half revivals all

5.3 Molecular Alignment in Strong Fields

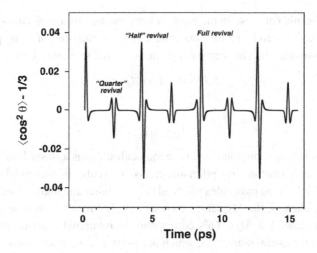

Figure 5.17 Calculated molecular alignment of a quantum ensemble of N_2 molecules with initial rotational temperature of 300K subject to a delta function pulse with fluence of 1 J/cm^2. Plot was generated based on plots in Poulsen (2005).

of the states have a π phase shift with respect to full revivals, so the same shape of the revival results with simply a change in sign. A small amplitude "quarter-revival" is also apparent at intervals between the half and full revivals. The upshot of this physics is that a moderately intense, ultrafast laser pulse (say one <1 ps in duration) can produce in a gas of linear molecules transient partial alignment at discrete time intervals after that pulse. This allows experiments to be conducted in which the pre-prepared gas of molecules can be interrogated at revival times so that the experiment is performed with some degree of alignment.

5.3.3 Rotational Pumping in Intense Laser Pulses

We are now left with the question of what is the rotational fate of molecules subject to even more intense pulses. Of course, most small molecules will begin to see significant ionization as the peak intensity approaches 10^{14} W/cm^2. However, even in fs pulses, the increasing field strength in the rising edge of a linearly polarized laser pulse will inevitably effect some degree of molecular alignment because of the torque that pulse applies (Schmidt et al. 1999). The simplest way to think of this strong field situation is similar to how we thought about atomic ionization when intensities reached the nonperturbative regime. This situation was depicted in Figure 4.14; we referred to this as strong coupling in the continuum. We argued that at intensities at which lowest-order perturbation theory failed, the atomic electrons could be thought of as absorbing and emitting by stimulated emission a large number of photons during the time of the laser pulse. The consequences of this for the atom was emission of many ATI electron energy peaks above the LOPT minimum energy peak (Zavriyev et al. 1990).

A similar dynamic can arise in the rotation wavefunction of a molecule subject to nonperturbative intensities and simultaneous absorption and emission of many photons. Each single photon transition in a molecule obeys the rotational selection rules

$$\Delta J = 0, \pm 1, \Delta M_J = 0 \qquad (5.56a)$$

with

$$J \geq |M_J|. \qquad (5.56b)$$

Since the rotational quantum number of the molecules, J, has a lower limit given by zero and $|M_J|$ after many photon absorption and emission events, the rotational distribution of the ensemble of irradiated molecules will tend toward higher and higher J values. However, as Eq. (5.56a) indicates, the M_J values do not grow. The result is an ensemble populated by high Js and rather low M_Js. This corresponds to rotational wavefunctions which are elongated along the quantization axis, which in this case is the laser polarization. This net alignment of the molecules is often called rotational pumping.

The consequence is that molecules subject to nonperturbative intensities will tend to be aligned along the laser polarization by the rising edge of the pulse. For example, if a gas of linear molecules is irradiated by a laser pulse at intensity of 10^{15} or 10^{16} W/cm^2, rotational pumping in the rising edge of the pulse when the intensity is around 10^{14} W/cm^2 will lead to significant molecular alignment by the time the pulse is intense enough to effect multiple ionization. This effect justifies our discussions where we assume that the molecule has a predilection toward alignment along the laser field prior to being ionized. It is interesting to note that the alignment from rotational pumping will tend to be more extreme in lasers of longer wavelength, because more photons are required to reach a given energy level, hence transferring more angular momentum to the molecule in the process.

5.4 Molecular Bonds in Strong Fields

In the rubric of the Born–Oppenheimer approximation, the dynamics of the nuclei in a molecular bond depend on the energy of the ground state electron as a function of nuclear separation. When the shape of this bound energy curve has a minimum at some ion separation, the molecular ions find themselves in potential energy minima, which support vibrational states. As we have already seen in the context of atoms, a strong laser field can quite significantly shift the energies of bound states through the AC Stark shift. It is not surprising then, that a strong laser field will have a major impact on the shape of the potential energy curves created by bound electron energies in molecular states, ultimately distorting them sufficiently to lead to dissociation (Zavriyev et al. 1990).

5.4.1 Floquet Theory of Strong Field Interactions

To ascertain how a strong laser field distorts the potential energy curves in a molecule, we require a quantitative theory for the shift of the molecular bound states as a function of internuclear separation, R. For this purpose, second-order perturbation theory will be

5.4 Molecular Bonds in Strong Fields

inadequate. Instead we turn to a powerful nonperturbative method for calculating energy levels (and, when included in the model properly, ionization rates) as a function of incident field intensity: Floquet theory (Colgan et al. 1998; Burke et al. 2000; Joachain 2007).

As usual, we treat the field classically and write the Hamiltonian in terms of the time-independent "atomic" or "molecular" Hamiltonian, H_0, which describes the binding quantum system, and the time-varying interaction Hamiltonian describing the strong field, H_{int} (Joachain et al. 2000):

$$H = H_0 + H_{int}(t). \tag{5.57}$$

Since we work in the SAE, we assume that the complete set of eigenstates and corresponding eigen-energies of H_0 are known and defined by

$$H_0|\alpha\rangle = \varepsilon_\alpha|\alpha\rangle. \tag{5.58}$$

Here α represents all quantum numbers describing the state, and two states, described by quantum numbers α and β are taken to be orthonormal $\langle \beta|\alpha\rangle = \delta_{\alpha\beta}$.

While not necessary, we assume at this stage that the interaction Hamiltonian is for a linearly polarized field. We must make a choice of gauge. The theory can be developed in either the Coulomb ($\mathbf{p} \cdot \mathbf{A}$) gauge or the length gauge without real physical distinction. The latter choice does permit easy treatment of the system with retardation and multipolar terms (Becker and Faisal 2005). The practical impact of choosing one gauge or the other is that the absolute value of the energy levels derived will be gauge dependent. In particular, choice of the length gauge in the dipole approximation eliminates the A_0^2 term from the Hamiltonian (compare Eqs. (3.335) and (3.340)). This amounts to neglecting a shift of the magnitude of the ponderomotive energy. For the problem of concern to us here, finding the shape of the potential energy curves for the molecular nuclei in the field, this shift is of little consequence, and we will find it easier to couch the Floquet equations in terms of a Hamiltonian in the length gauge. With this form we can write the interaction Hamiltonian as

$$H_I(t) = eE_0 x \cos \omega t. \tag{5.59}$$

For reasons that will become apparent, we will rewrite this as

$$H_I(t) = \frac{eE_0 x}{2} \left(e^{i\omega t} + e^{-i\omega t}\right)$$
$$= H_- e^{i\omega t} + H_+ e^{-i\omega t}. \tag{5.60}$$

These two terms in the interaction Hamiltonian will be associated with absorption and emission of a photon.

The next assumption that we must make is that the laser field is on for a long time; the length of the laser pulse must be much longer than an optical cycle. If it is, we can say that the interaction Hamiltonian is a periodic function with time, satisfying the relation:

$$H_I(t+T) = H_I(t), \tag{5.61}$$

where $T = 2\pi/\omega$ is the time for one complete laser cycle. Since H_0 is time independent we can make the same assertion for the entire Hamiltonian

$$H(t+T) = H(t). \tag{5.62}$$

The Schrödinger equation

$$H|\psi\rangle = i\hbar \frac{\partial}{\partial t}|\psi\rangle \tag{5.63}$$

can be compactly written if we define the operator

$$\hat{H} = H - i\hbar \frac{\partial}{\partial t} \tag{5.64}$$

as

$$\hat{H}|\psi\rangle = 0. \tag{5.65}$$

This is now a PDE with periodic coefficients. The Floquet theorem states that linear differential equations with such periodic coefficients have solutions (Chu and Telnov 2004) which can be expressed in the form

$$|\psi(t)\rangle = e^{-i\varepsilon t/\hbar}|\varphi(t)\rangle, \tag{5.66}$$

where $|\varphi(t)\rangle$ is itself periodic with the same periodicity of the PDE coefficients:

$$|\varphi(t+T)\rangle = |\varphi(t)\rangle. \tag{5.67}$$

Note that even though we have expressed the eigenfunction of the Schrödinger equation in terms of a periodic function, it is not itself necessarily periodic because ε is not commensurate with ω. (We know this must be true as there is no reason that the solutions of the Schrödinger equation subject to an oscillatory field will have periodicity of the laser or even be periodic in time, such as when a state is monotonically depleted by ionization.)

Substituting the solution of form (5.66) into Eq. (5.63) yields

$$He^{-i\varepsilon t/\hbar}|\varphi(t)\rangle = \varepsilon e^{-i\varepsilon t/\hbar}|\varphi(t)\rangle + i\hbar e^{-i\varepsilon t/\hbar}\frac{\partial}{\partial t}|\varphi(t)\rangle \tag{5.68}$$

or

$$\hat{H}|\varphi(t)\rangle = \varepsilon|\varphi(t)\rangle. \tag{5.69}$$

We have arrived at an eigenvalue equation for ε and $|\varphi(t)\rangle$. Use of the Floquet theorem has allowed us to transform the time-dependent Schrödinger equation, describing the interaction of the laser field with a quantum system into an eigenvalue problem with the operator $\hat{H}$. This means that we can draw on the usual techniques for solving such problems.

Equation (5.69) hints at the physical significance of the eigenvalues, ε. Since $\hat{H}$ includes terms for both the unperturbed binding Hamiltonian and the field, the ε's can be interpreted as the total time-averaged energy of the combined system, including both electron energy and energy of the electromagnetic field. These eigenvalues are often referred to as "quasi-energies" (Plummer and McCann 1996).

5.4 Molecular Bonds in Strong Fields

To solve Eq. (5.69) we need an appropriate decomposition of the functions $|\varphi(t)\rangle$. Because $|\varphi(t)\rangle$ is periodic we can express it as a Fourier series:

$$|\varphi(t)\rangle = \sum_{N=-\infty}^{\infty} \varphi_N e^{-iN\omega t}. \tag{5.70}$$

(Note that the $|\varphi(t)\rangle$ depend on position, $\mathbf{x}$, if expressed in configuration space, though we have not explicitly stated it to avoid clutter in the equations. This, of course, means that the Fourier coefficients $\varphi_N = \varphi_N(\mathbf{x})$ also depend on the position coordinates.) By expressing these functions in this way we see one of the important characteristics of the quasi-energies: they are not defined uniquely. Take any eigenstate solution of the Schrödinger equation case in Floquet–Fourier form:

$$|\psi_\varepsilon(t)\rangle = e^{-i\varepsilon t/\hbar} \sum_{N=-\infty}^{\infty} \varphi_N e^{-iN\omega t} \tag{5.71}$$

and shift the summation by an integer, s, so that the new summation variable is $N' = N - s$. The eigenstate becomes

$$|\psi_\varepsilon(t)\rangle = e^{-i\varepsilon t/\hbar} \sum_{N+s=-\infty}^{\infty} \varphi_{N+s} e^{-i(N+s)\omega t}$$

$$= e^{-i(\varepsilon + s\hbar\omega)t/\hbar} \sum_{N'=-\infty}^{\infty} \varphi_{N'} e^{-iN'\omega t} e^{is\omega t}$$

$$= e^{-i\varepsilon' t/\hbar} |\varphi'(t)\rangle. \tag{5.72}$$

So clearly the same eigenstate results under the transform

$$\varepsilon' = \varepsilon + s\hbar\omega, |\varphi'(t)\rangle = e^{is\omega t}|\varphi(t)\rangle. \tag{5.73}$$

Therefore, if we find a state $|\varphi(t)\rangle$, we also can write down an infinite number of states which are simply shifted in quasi-energy by increments of the photon energy.

Now we substitute the Fourier representation of $|\varphi(t)\rangle$ into the eigenvalue Eq. (5.69) and get

$$(H_0 - N\hbar\omega)\sum_{N=-\infty}^{\infty}\varphi_N e^{-iN\omega t} + H_+\sum_{N=-\infty}^{\infty}\varphi_N e^{-i(N+1)\omega t}$$
$$+ H_-\sum_{N=-\infty}^{\infty}\varphi_N e^{-i(N-1)\omega t} = \varepsilon \sum_{N=-\infty}^{\infty}\varphi_N e^{-iN\omega t} \tag{5.74}$$

or

$$(H_0 - N\hbar\omega)\varphi_N + H_+\varphi_{N-1} + H_-\varphi_{N+1} = \varepsilon\varphi_N. \tag{5.75}$$

Following Becker and Faisal (2005), we can further simplify this equation by introducing "shift" operators for the Fourier coefficients, $s_+\varphi_N = \varphi_{N+1}$ yielding

$$(H_0 - N\hbar\omega + H_+ s_- + H_- s_+)\varphi_N = \varepsilon\varphi_N. \tag{5.76}$$

We have transformed the time-dependent Schrödinger equation into an infinite set of *time-independent* equations for the Floquet–Fourier components of the periodic Floquet state $|\varphi(t)\rangle$. These operator equations can be expressed in matrix form if we introduce the infinite vector

$$\varphi = \begin{bmatrix} \vdots \\ \varphi_{N-1} \\ \varphi_N \\ \varphi_{N+1} \\ \vdots \end{bmatrix} \tag{5.77}$$

so that the system of Eq. (5.76) can be succinctly written

$$\mathbf{H}_F \varphi = \varepsilon \varphi. \tag{5.78}$$

The matrix $\mathbf{H}_F$ is referred to as the Floquet Hamiltonian. It is an infinite block tridiagonal matrix taking the form:

$$\mathbf{H}_F = \begin{bmatrix} \ddots & & & & & \\ & H_+ & H_0 - (N-1)\hbar\omega & H_- & 0 & 0 \\ \cdots & 0 & H_+ & H_0 - N\hbar\omega & H_- & 0 & \cdots \\ & 0 & 0 & H_+ & H_0 - (N+1)\hbar\omega & H_- \\ & & & & & \ddots \end{bmatrix} \tag{5.79}$$

Equation (5.79) tells us how we are to interpret the index N. It represents the number of photons in the system, and the operators H_+ and H_- are associated with absorption and emission of one photon respectively, shifting the number of photons present in the system down or up by one. The harmonic components, φ_N, of the Floquet state vector then have a clear interpretation: they can be seen as the states that the quantum system reaches from corresponding quasi-energy, ε, by exchanging a net N number of photons with the laser field. If $N > 0$, then this is the net number of photons that have been absorbed while if $N < 0$, than many photons have been emitted by stimulated emission. Such a state can be accessed in many ways; this Fourier component is accessed by absorbing and/or emitting a net N number of photons, reached by various combinations of absorption and emission. In the vocabulary of perturbation theory, the coupling of the quantum state to the Nth-order term is equivalent to summing all possible orders for a given N-photon process (say, absorption of N photons, plus absorption of $N+1$ photons followed by emission of one photon, plus absorption of $N+2$ photons with emission of two photons, etc.). The Floquet model is completely nonperturbative and accurate even in a regime when a perturbation expansion does not converge.

To find the quasi-energies we must diagonalize this matrix. The problem of solving the time-dependent Schrödinger equation has been transformed into a time-independent matrix eigenvalue problem without, to this point, making any approximations. It would seem, however, that we have not really improved the situation. We have simply traded one difficult problem for another: diagonalizing an infinite matrix. However, the problem can now be made tractable while remaining accurate by truncating the size of the matrix to a finite number of N. Since we now recognize N as the number of photons in the system, retaining the full matrix is the same as keeping an infinite number of photon orders in the interaction, and truncating the matrix means that we limit the number of multiphoton orders that participate in the interaction. The difficult time-dependent problem is now a virtually trivial numerical problem if the matrix elements in Eq. (5.60) can be evaluated. The solution can be checked for convergence by changing the size of the matrix (including greater or fewer photon orders). The eigenvectors that are then retrieved are so-called Floquet states. Again, each Floquet eigenstate has a quasi-energy which can be thought of as the total energy of the system; it is the sum of the electron energy and the energy in the field ($N\hbar\omega$). Put another way, the quasi-energies are the "field-dressed" energies of the AC Stark-shifted quantum states.

In practice, evaluation of the matrix elements in the Floquet Hamiltonian matrix means one must choose an orthonormal basis set of states to expand each Floquet–Fourier component. It makes sense to expand the Floquet states in terms of the unperturbed eigenstates of the quantum system being irradiated. Each Floquet–Fourier state, then, can be written

$$\varphi_N(\mathbf{x}) = \sum_\alpha a_\alpha^{(N)} |\alpha\rangle, \tag{5.80}$$

where the expansion coefficients $a_\alpha^{(N)}$ yield the weighting of each unperturbed state in the field-mixed Floquet states. Substitution into Eq. (5.76) yields for the Floquet equations

$$(H_0 - N\hbar\omega - \varepsilon) \sum_\alpha a_\alpha^{(N)} |\alpha\rangle + \sum_\alpha \left(a_\alpha^{(N-1)} H_+ + a_\alpha^{(N+1)} H_- \right) |\alpha\rangle = 0. \tag{5.81}$$

Recalling that $H_0|\alpha\rangle = \varepsilon_\alpha|\alpha\rangle$ we project these equations onto the orthonormal states $\langle\beta|$ (and use the fact that $\langle\beta|\alpha\rangle = \delta_{\alpha\beta}$) to derive finally the (in principle infinite) set of equations defining the Floquet state expansion coefficients:

$$(\varepsilon_\alpha - N\hbar\omega - \varepsilon) a_\alpha^{(N)} + H_+^{(\alpha\beta)} a_\alpha^{(N-1)} + H_-^{(\alpha\beta)} a_\alpha^{(N+1)} = 0. \tag{5.82}$$

In this set of equations, we have denoted the electromagnetic wave coupling matrix elements as

$$H_\pm^{(\alpha\beta)} = \langle\beta|H_\pm|\alpha\rangle. \tag{5.83}$$

With the form of the interaction Hamiltonian that we have chosen in Eq. (5.60), these are simply the dipole matrix elements between α and β. There are only as many independent solutions to the Eq. (5.82) as there are independent basis states included in the calculation. If there are N_α states included in the calculation, there will be this many equations for each photon number, N. The Floquet Hamiltonian, then, has dimensions equal to N_α times the number of photon orders retained in the calculation.

The Floquet states now take on the vectorial form:

$$\varphi_\varepsilon = \begin{bmatrix} \vdots \\ a_\alpha^{(N-1)} \\ a_\beta^{(N-1)} \\ \vdots \\ a_\alpha^{(N)} \\ a_\beta^{(N)} \\ \vdots \\ a_\alpha^{(N+1)} \\ a_\beta^{(N+1)} \\ \vdots \end{bmatrix}. \qquad (5.84)$$

Again, if the problem were solved exactly, with an infinite number of photon orders included, each of the a_αs would be the same ($a_\alpha^{(N)} = a_\alpha^{(N+1)} = a_\alpha^{(N+2)}$ etc.), each corresponding to a laser field–shifted eigenenergy itself shifted by N photons. The N_α independent eigenstates found by this procedure are often referred to as dressed states. The light field mixes the eigenstates of the unperturbed quantum system. There is a comb of these mixed states in energy, spaced by exactly one photon, representing a reduction or increase in the number of photons present around some large number of photons. The transfer of the electron from one to another Floquet state can be thought of as a transition from one eigenstate to another simultaneous with the absorption or emission of photons from or into the field. As we saw when we considered perturbation theory, even though we have treated the field classically, the field's sinusoidal periodicity led to an interpretation of the behavior in terms of absorption or emission of photons.

This representation of the Floquet states means that each element in the Floquet Hamiltonian matrix is itself a matrix with dimensionality equal to the number of orthonormal states utilized in the expansion of the Floquet state. Again, the dimensionality of these submatrices would in principle be infinite unless a finite number (N_α) of states is chosen. (For example, we will show that only two states will be needed to describe a significant amount of the physics involved in strong field dissociation of molecular bonds.) Once the matrix elements of the chosen basis states are evaluated and the dimension of the Floquet Hamiltonian matrix is truncated at a suitable dimension, the quasi-energies can be found from

$$\det |\mathbf{H}_F - \varepsilon \mathbf{I}| = 0. \qquad (5.85)$$

If only bound states are included in the basis state expansion of the Floquet states, it is easy to see that the off-diagonal elements in $\mathbf{H}_F$ satisfy the relation

$$H_-^{(\alpha\beta)} = \left(H_+^{(\beta\alpha)}\right)^* \qquad (5.86)$$

and the Floquet Hamiltonian matrix is Hermitian. We know that the eigenvalues of a Hermitian matrix are real, so the quasi-energies derived from Eq. (5.85) are real and correspond to the shifted energy of the dressed stationary states. In this way, we can calculate the AC Stark shift of the field-free quantum levels in the nonperturbative intensity regime (assuming that we have included enough photon orders in the matrix to assure accuracy).

If, however, continuum states are included in the expansion with outgoing (or so-called Siegert) boundary conditions imposed, the Floquet Hamiltonian matrix will be non-Hermitian (Chu and Telnov 2004). Solution of a Hamiltonian of this form leads to complex quasi-energy eigenvalues, which can be expressed as

$$\varepsilon = \varepsilon_R - i\frac{\Gamma}{2}. \tag{5.87}$$

The real part of the quasi-energy is $\varepsilon_R = \varepsilon_\alpha + \Delta\varepsilon_{Stark} + N\hbar\omega$. It includes the unperturbed energy of the dominant eigenstate, α, the shift of this level – the AC Stark shift, and the energy of the light field. The imaginary part has an interesting interpretation. That imaginary term means that the probability density of the electron in that state decays with time as $|\psi|^2 \sim e^{-\Gamma t}$. This can be associated with the ionization rate of that state in the field. Non-Hermitian Floquet theory is widely used as a numerically simple, but powerful tool for calculation of strong field ionization rates in the nonperturbative regime, for both complex atoms and molecules. It is particularly useful for finding ionization rates when intermediate resonances play an important role, where LOPT fails even at modest intensity. There is a widespread treatment of non-Hermitian techniques for calculating ionization rates in the literature; we will not consider them further here (Chu and Telnov 2004).

5.4.2 Floquet Treatment of a Two-Level System in a Strong Field

As we will illustrate in the next section, the fate of a bond, like that of the H_2^+ molecule can be treated in the context of two strongly dipole-coupled electronic levels. So let us examine the strong field coupling of two such levels with the Floquet formalism. Consider two levels, $|a\rangle$ and $|b\rangle$. The unperturbed energies of these two levels will presumably vary with some variable, such as interatomic spacing in a molecule, and will vary at differing rates. The two levels are assumed to be dipole coupled so that $\langle a|H_\pm|b\rangle \neq 0$. Of course, by symmetry considerations we also know that $\langle a|H_+|a\rangle = 0$.

Because of this latter fact, the off-diagonal elements in the Floquet Hamiltonian matrix elements will be zero between, say, the $|a, N\rangle$ and the $|a, N+1\rangle$ elements. It is therefore only necessary to retain coefficients for each unperturbed electronic state with even or odd numbers of photons in the Floquet state vectors. We can therefore write the Floquet Hamiltonian matrix for this two-state system as

$$H_F = \begin{bmatrix} \ddots & & & & & & \\ \cdots & \varepsilon_a - (N-2)\hbar\omega & H_-^{(ab)} & 0 & 0 & 0 & \\ & H_+^{(ab)} & \varepsilon_a - (N-1)\hbar\omega & H_-^{(ab)} & 0 & 0 & \\ & 0 & H_+^{(ab)} & \varepsilon_a - N\hbar\omega & H_-^{(ab)} & 0 & \cdots \\ & 0 & 0 & H_+^{(ab)} & \varepsilon_a - (N+1)\hbar\omega & H_-^{(ab)} & \\ & 0 & 0 & 0 & H_+^{(ab)} & \varepsilon_a - (N+2)\hbar\omega & \\ & & & & & & \ddots \end{bmatrix} \tag{5.88}$$

where
$$H_{\pm}^{(ab)} = \langle a | eE_0 x/2 | b \rangle \tag{5.89}$$

which means that
$$H_{-}^{(ab)} = H_{+}^{(ba)} = H_{ab} = \frac{eE_0}{2} x_{ab}. \tag{5.90}$$

An accurate solution for the shift of these two energy levels as a function of intensity (and parameters, like R) will mandate keeping a reasonable number of terms in the matrix and diagonalizing it. Doing this leads to a ladder of states, spaced by the photon energy.

The ladder of state energies plotted without the field-induced energy shift are usually referred to as *diabatic* states. These are the curves that would exist if the field does not couple strongly to the system. The way that a figure like this is to be interpreted is that wherever the two diabatic Floquet states $|a, N_1\rangle$ and $|b, N_2\rangle$ cross in such a diagram, these states are coupled in resonance by the field. If N_1 and N_2 differ by one at the crossing point, the two energy levels are separated by exactly one photon in energy. When N_1 and N_2 differ by three, the Floquet states are coupled by a three-photon resonance and so on. Diagonalization of the Hamiltonian finds the field-dressed energy states. Curves denoting these dressed state energies are referred to as *adiabatic* states. Well away from resonances, the mixing between unperturbed states will be weak; the Floquet states retain energies near the unperturbed energy of the field-free states, and the dressed state retains the primary characteristics of one of the initial unperturbed eigenfunctions. (Put differently, the Floquet state vector has one element with value near one.) However, near a resonance, the AC Stark shift of the unperturbed energies can be large and the two coupled states can be strongly mixed.

To get a sense of the magnitude of this energy shift near resonance, we can truncate the Floquet matrix to only two dimensions. Doing this is tantamount to saying that the dynamics are dominated by the absorption of one photon by the lower energy state (state a in our two-level system) or the emission of one photon from the upper level. This ignores higher-order processes and is equivalent to the well-known rotating wave approximation. In this approximation the Floquet Hamiltonian looks like

$$\mathbf{H}_F = \begin{bmatrix} \varepsilon_a & H_{ab} \\ H_{ab} & \varepsilon_b - \hbar\omega \end{bmatrix}. \tag{5.91}$$

Diagonalization of this matrix to find the quasi-energies is trivial. If we set the lower state to have zero energy, $\varepsilon_a = 0$, and introduce the detuning, $\Delta = \varepsilon_b - \hbar\omega$, we can employ Eq. (5.85)

$$\begin{vmatrix} -\varepsilon & eE_0 x_{ab}/2 \\ eE_0 x_{ab}/2 & \Delta - \varepsilon \end{vmatrix} = 0 \tag{5.92}$$

to find the two quasi-energies:

$$\varepsilon = \frac{\Delta}{2} \pm \frac{1}{2}\sqrt{\Delta^2 - (eE_0 x_{ab})^2}. \tag{5.93}$$

5.4 Molecular Bonds in Strong Fields

At resonance, $\Delta = 0$ and the state energies become, in terms of the Rabi frequency $\Omega_R = eE_0\, x_{ab}/\hbar$,

$$\varepsilon = \pm \frac{1}{2}\hbar\Omega_R. \tag{5.94}$$

The field-dressed Floquet states are then found to be

$$|\varphi_1\rangle = \frac{1}{\sqrt{2}}(|a\rangle - |b\rangle), \tag{5.95a}$$

$$|\varphi_2\rangle = \frac{1}{\sqrt{2}}(|a\rangle + |b\rangle). \tag{5.95b}$$

So when strongly coupled Floquet states cross, the field-dressed eigenstates are actually equal combinations of the two undressed states. Furthermore, the dressed state energies, as they approach resonance, are split, separated by $\Delta\varepsilon = \hbar\Omega_R$.

To be more explicit about the problem at hand, consider a molecule with two states that vary with internuclear separation, $|a(R)\rangle, |b(R)\rangle$. As the energy separation of these states approaches zero at some particular R, the diabatic states cross (see Figure 5.18). However, in a strong field, the states undergo an "avoided crossing" because of the dressing of the energies near that resonant point. The adiabatic states then look like the solid lines in Figure 5.18. To one side of the resonance a particular Floquet state has the characteristics of

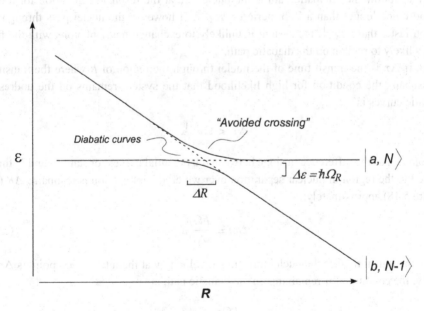

Figure 5.18 Schematic representation of two Floquet levels, a and b, varying with a parameter such as R near a single photon resonance. At R near this resonance, the two Floquet states, one shifted by one photon, cross. The diabatic states, characteristic of weak fields and weak coupling, are denoted by dashed lines. Solid lines denote the energy levels of the states when the coupling is strong enough to mix the states near the crossing. The energy shift of the coupled states leads to a separation in energy roughly equal to the Rabi frequency.

one of the unperturbed states. At the resonant point the two unperturbed states are strongly mixed and to the other side of that point the Floquet state takes on the character of the other basis state. The magnitude of the energy separation at the avoided crossing is roughly equal to the Rabi energy, and hence is proportional to the square root of the light intensity and the magnitude of the dipole matrix element coupling the two basis states.

If a molecule has nuclei that are moving near such a crossing of Floquet states, the system can transit the crossing either on the adiabatic curves or the diabatic curves depending on the rate at which it crosses the region of the intersection. Figure 5.18 describes such a crossing and the possible paths are shown. If the system remains on the diabatic curve, this is the same as saying that the molecule does not undergo any net photon absorption or emission at $R's$ near where the transition is in resonance; the nuclei traveled too quickly through this resonance region for a significant probability to undergo a transition to develop. If, on the other hand, the nuclei pass slowly thorough this region, there is sufficient time for the field near resonance to mix strongly the two states and there is a high probability that photons are interchanged with the molecule. In that case the molecular system will remain on the adiabatic (field-mixed) curve.

The probability for either of these two paths to occur can be estimated using the simple Landau–Zener semiclassical theory (Miret-Artes and Atabek 1994). We can simply estimate the conditions under which one or the other will occur if we note that the system is likely to follow the adiabatic path if the nuclei are at the resonant separation for a time period much longer than a Rabi period $\sim \Omega_R^{-1}$. If however, the nuclei pass through the region faster than Ω_R^{-1}, the system is unlikely to exchange many photons with the field and is likely to remain on the diabatic path.

So, if $\Delta \tau$ is the transit time of the nuclei through the region of R where the transition is resonant, the condition for high likelihood that the system remains on the undressed diabatic curves is

$$\Delta \tau < \Omega_R^{-1}. \tag{5.96}$$

Denote Δm as the difference in slope between the potential curves for states a and b; this is related to the region in nuclear separation over which the interaction is resonant, ΔR (see Figure 5.18) approximately:

$$\Delta m \cong \frac{\hbar \Omega_R}{\Delta R}. \tag{5.97}$$

Since the transit time of the nuclei traveling at velocity v at the intersection point is $\Delta \tau = \Delta R / v$, the condition for remaining on the diabatic path is

$$\frac{\hbar \Omega_R^2}{\Delta m v} < 1 \tag{5.98}$$

or in terms of intensity

$$I < \frac{\Delta m v c}{8\pi \hbar e^2 \langle x_{ab} \rangle^2}. \tag{5.99}$$

The light intensity must be low enough that the states are not strongly coupled during the period of passage; otherwise, the strong coupling is likely to push the system along the adiabatic, field-dressed path instead.

5.4.3 Bond Softening and Above-Threshold Dissociation

Armed with this Floquet technique, we are in a position to assess how a laser field distorts the Born–Oppenheimer binding potential of a molecular bond. We will consider mainly a diatomic molecule aligned along a linearly polarized laser field, a situation which we know likely arises from molecular alignment in the rising edge of an intense laser pulse. A simple picture can easily convince us that such a field will tend to weaken the molecular bond and drive dissociation (Yang et al. 1991; Giusti-Suzor et al. 1990; Jolicard and Atabek 1992; Levis et al. 2002). Considering a diatomic molecule of nuclear separation R oriented along the field polarization x axis we would expect during the half laser cycle when the field is pointing along $+x$ that the electron cloud of the bond will be pushed toward $-x$. This leaves a predominantly positive charge at $+R/2$ which is propelled away from the molecular center by that field. During the subsequent half cycle, when the field has reversed direction and points along $-x$, the electron cloud is pushed back in the $+x$-direction and now the other ion at $-R/2$ is pulled away from the molecular toward $-x$. A time-average of this effect over many cycles will clearly tend to pull both ions away from the molecule.

It turns out that this softening of the bond does not play a significant role in neutral diatomics in practice (Posthumus 2004). The reason for this can be seen if one asks how intense a field is likely needed for this bond softening to occur. It is reasonable to expect that for significant distortion of the electron cloud in the bond, the laser ponderomotive energy will need to be a reasonable fraction of the bond energy. In H_2 the bond energy is around 4.5 eV, so in visible or near-IR fields intensity of a few times 10^{13} W/cm^2 will be needed to soften this bond (Bucksbaum et al. 1990). At these intensities, ionization tends to be fairly rapid since ionization potentials for diatomics are around 12–15 eV. Consequently, ionization near the equilibrium nuclear separation tends to occur before the bond can be much weakened by the field.

The situation is rather different in singly ionized molecular ions. Consider the H_2^+ ion, which has been extensively studied as the simplest molecular system in a range of experiments (Lebedev et al. 2003). H_2^+ has an ionization potential of 29 eV, which easily survives even long (>1 ps) pulses to intensities of over 10^{14} W/cm^2. Given that the bond energy of this molecule is 2.8 eV, when intensities in excess of 10^{13} W/cm^2 occur, we expect significant distortion of the binding potential of the bond.

To quantify the shape of the potential energy surfaces which H_2^+ sees in the field we can use Floquet theory to calculate the ground and excited state energies as a function of nuclear separation. To perform such a calculation, a suitable set of electronic states must be chosen. It is well established that the H_2^+ bond can be described in a strong field by considering the ground state, $^2\Sigma_g^+$ ($1s\sigma_g$) and the first repulsive state, $^2\Sigma_u^+$ ($2p\sigma_u$), which is strongly coupled to the ground state by a dipole interaction. (In an actual

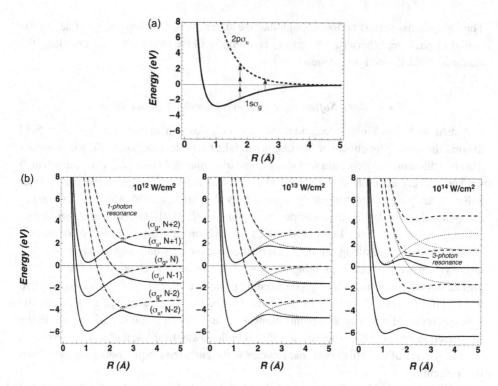

Figure 5.19 Panel (a) shows the ground state and first repulsive states as a function of nuclear separation in H_2^+. The vertical arrows illustrate the locations of one and three photon resonances for 800 nm light. The three panels in part (b) show the diabatic (thin dashed lines) and adiabatic Floquet states in H_2^+ for three different intensities subject to 800 nm radiation.

Floquet calculation, choice of gauge should be unimportant, though it has been shown that this two-state approximation for H_2^+ is most accurate in the length gauge (Sheehy and DiMauro 1996), the gauge we have developed in Section 5.4.1.) These two states are plotted as a function of internuclear separation, R, in Figure 5.19a. As this figure shows, at R near R_e three photon resonances occur between these two states in the visible and near-IR and single-photon transitions are possible at R near 2–3 Å.

The lower section of Figure 5.19 shows the diabatic Floquet states with these two electronic states vs. R plotted as thin dashed lines (plotted only in the left panel). Since the two states are coupled by one photon, only every other Floquet state is shown, anchored on the ground state $1s\sigma_g$ with N background photons. For this illustration, a laser wavelength of 800 nm is assumed. Recall that where these diabatic states cross, the two levels are in one-, three-, five- ... photon resonances. The adiabatic states, found by diagonalizing the Floquet Hamiltonian, are then plotted on top of these diabatic states. The three panels show these field-dressed states at three intensities, 10^{12}, 10^{13} and 10^{14} W/cm².

We see that even at low intensity the binding potential becomes distorted. The principal distortion in the potential at 10^{12} W/cm² occurs around the single-photon resonance near

$R = 2.5$ Å. Here, where the diabatic curves of the $|2s\sigma_g, N\rangle$ and $|2p\sigma_u, N-1\rangle$ Floquet states cross, an avoided crossing opens up. This leads to a binding potential which has a local maximum near this crossing. At higher intensity, 10^{13} W/cm², the magnitude of the gap in this avoided crossing increases. The depth of the binding potential well which forms the bond becomes more shallow. Finally, at high intensity, 10^{14} W/cm², the coupling at the three-photon resonance (which occurs near $R = 2$ Å) becomes significant and a gap between diabatic states opens up here. This further deforms the binding potential and depresses the binding barrier even more.

The consequences of the deformation of the dressed states in the field are obvious (Posthumus 2004). As the one-photon, and ultimately the three-photon coupling strength between the attractive ground state and the $2p\sigma_u$ repulsive state increase, fewer vibrational modes are bound. Any wavepacket population initially in high-lying vibrational states will be able to pass over the depressed barrier as the intensity irradiating the molecule increases. This process is commonly referred to as "bond softening" (Miret-Artes and Atabek 1994). In the example illustrated in Figure 5.19, only the $v = 0$ and $v = 1$ states are still bound in the H_2^+ molecule at intensity of 10^{14} W/cm². If there were any population in higher vibrational levels upon ionization from H_2 (which we know is likely from our discussion in Section 5.2.2) population in those states is likely to dissociate leading to ejection of H and H^+ fragments.

We can use the Floquet picture to get a better handle on how the H_2^+ molecule will dissociate at various intensities. This is illustrated in another example, shown in Figure 5.20, where the dressed Floquet states of the H_2^+ molecule in a 527 nm laser field at intensity

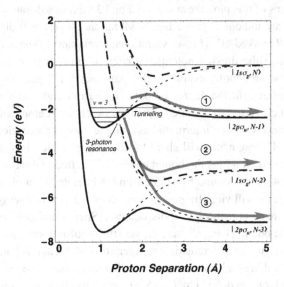

Figure 5.20 Diabatic and adiabatic Floquet states in H_2^+ as a function of nuclear separation for irradiation by 527 nm radiation at 2×10^{13} W/cm². The energy level locations of the first four vibrational levels are illustrated as are the three potential dissociation paths.

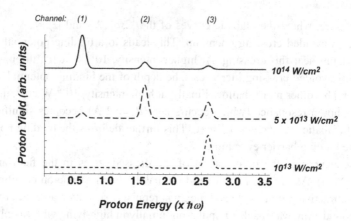

Figure 5.21 Schematic ATD proton spectra for three intensities. The proton energies corresponding to the three dissociation channels illustrated in Figure 5.20 are labeled above the corresponding proton peaks.

of 2×10^{13} W/cm^2 are plotted. (This is a wavelength near where many of the early bond-softening experimental studies were performed (Posthumus 2004).) Again, the thin dashed lines show the ladder of diabatic states for every other photon number state. The thick solid and dashed lines show the adiabatic states that result from field dressing and mixing of the two undressed states. At this intensity, an avoided crossing has developed at an internuclear separation near 2.2 Å and a significant gap has opened here.

There are essentially three pathways for dissociation (Ludwig et al. 1997). Under the particular conditions of this plot, the $v = 0, 1, 2$ and 3 vibrational states are bound; higher vibrational levels are unbound. These higher vibrational levels will dissociate along the path toward large R marked "1". (Lower vibrational levels can also pass along this pathway by tunneling through the dressed potential barrier, as illustrated in Figure 5.21, but of course, this is a low-probability event.) In this path the H–H$^+$ pair separates by passing over (or tunneling through) the depressed potential barrier and acquiring kinetic energy from this potential curve which is shared equally between the H atom and the proton. This bond-softening pathway can be interpreted as a single-photon absorption to the repulsive state. The H and H$^+$ fragments will share kinetic energy which is simply given by the energy of a single photon minus the bound energy of the freed vibrational state. (For this example, the $v = 4$ state is bound by 1.65 eV and is liberated by a photon of 2.35 eV so population in this level will yield fragments with about 0.35 eV of energy each).

However, there are other pathways for dissociation (Sheehy and DiMauro 1996). At an R of around 1.5 Å the bound $|1s\sigma_g, N\rangle$ state crosses the repulsive and $|2p\sigma_u, N - 3\rangle$ Floquet state. The system can undergo a transition between these states at this point (a three-photon absorption event) and the molecule can begin to dissociate along the repulsive state. If the intensity is not too large, as defined by Eq. (5.99), when the system passes through the one-photon coupling region near 2.2 Å, it will stay on the diabatic path. It remain in this Floquet state, traveling ultimately along the path labeled "3". This path is effectively a three-photon

5.4 Molecular Bonds in Strong Fields

event and results in fragments with total kinetic energy equal to $3\hbar\omega - \varepsilon_v$. If, on the other hand, the field is sufficiently intense that the coupling near the one-photon resonance is strong, the system will remain on the *adiabatic* curve and ultimately follow the path marked "2" in Figure 5.21. This path can be thought of as a four-photon order interaction, in which a three-photon absorption event is followed by a single-photon stimulated emission event to lead the system to evolve asymptotically to large R on the $|1s\sigma_g, N-2\rangle$ curve. In this case, the system has absorbed a net total of two photons and the kinetic energy of the fragments will be lower by the amount of a single photon than those molecules which dissociated along the "3" path.

The two dissociation pathways labeled "2" and "3" in Figure 5.20 are usually referred to as above-threshold dissociation or "ATD." This term is used in the same sense that ATI refers to the liberation of an electron during ionization with a net number of photons which is greater than the minimum needed to overcome the ionization potential. Above-threshold dissociation results in a proton and an H atom with kinetic energies which are greater by one or two photons (for paths "2" and "3," respectively) than are needed to dissociate the molecule. In experiments measuring the energy of ejected protons, indeed peaks in the proton energy spectrum separated by one-half of a photon in energy are observed, corresponding in kinetic energy to these three pathways.

The molecule's predilection for these various pathways actually depends quite sensitively on laser intensity. This we can see if we look back at the potential energy curves of Figure 5.19. A schematic of the observed proton energy spectrum at three intensities in light of this Floquet picture is illustrated in Figure 5.21. At low intensity (say at or below $\sim 10^{13}$ W/cm^2) the binding potential is not strongly deformed. As we have seen in Section 5.2.2, ionization of H$_2$ at moderate intensity leads to predominate population of the first four vibrational levels. At intensity below $\sim 10^{13}$ W/cm^2 all of these levels are bound and the only way to dissociate along the single-photon bond-softening pathway is to tunnel through the barrier. Consequently, at low intensity this single-photon path is very weak. Instead, the easiest dissociation path occurs by resonant three-photon excitation to the repulsive state. Because the nuclei gain kinetic energy from sliding down the repulsive $|2p\sigma_u, N-3\rangle$ state, by the time they reach the next-level crossing around 2.3 Å, they do not meet the strong coupling requirements of Eq. (5.99) and proceed along the diabatic passage of path 3. The result is the ejection of fragments from the three-photon absorption mechanism. Surprisingly, the strongest signal is observed in protons with the highest kinetic energy at *low intensity*.

At higher intensity, say 5×10^{13} W/cm^2, dissociation over the barrier along path 1 starts to become significant as high-lying vibrational states are freed, but the majority of the dissociation occurs via the three-photon absorption and passage along the adiabatic curve of path "2" which becomes more strongly favored over path "3" as the coupling strength increases with increasing intensity. At high intensity, near 10^{14} W/cm^2, the binding potential barrier is strongly depressed and most of the vibrational population exits along the single-photon bond-softening pathway, which manifests as an increase in the number of protons at that low kinetic energy (as shown in the uppermost proton spectrum of Figure 5.21). This picture leads to a rather surprising conclusion; in the strong

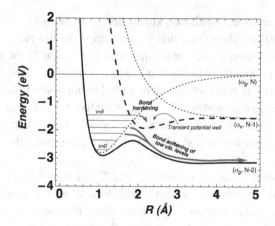

Figure 5.22 Diabatic and adiabatic Floquet states in H_2^+ as a function of nuclear separation for irradiation by 800 nm radiation at 1×10^{14} W/cm^2. The thick solid and thick dashed curves denote the adiabatic Floquet states, with the avoided crossing that develops near the three-photon resonance at ~2 Å. The avoided crossing causes a dip to form in the upper Floquet level. Higher vibrational levels such as $v = 4$ and 5 can be coupled into this transient binding potential well at 2.2 Å, temporarily stopping dissociation, while lower vibrational levels ($v = 2$ and 3 in this example) dissociate immediately by bond softening.

field-induced dissociation of the H_2^+ bond, the lower intensity actually favors the *higher* multiphoton order; only at high intensity does the single-photon absorption mechanism become significant (Plummer *et al.* 1998).

The physics described by this picture is likely to be a general effect in molecular ions, though it has not been well studied experimentally for species other than H_2^+. For other species, the competition between ATD and ionization of a second electron from the molecular ion is likely to vary in importance (also dependent on the wavelength employed) so that Coulomb explosion is the direct cause of molecular fragmentation. Bond softening is predicted to happen in neutral atoms such as H_2; however, simulations based on a Floquet state picture indicate that intensities over 2×10^{14} W/cm^2 are needed in H_2 to yield significant binding potential deformation. At these intensities, the molecule will ionize rather quickly (subpicosecond lifetimes) and ATD will have no time to occur.

5.4.4 Bond Hardening

Before concluding our discussion of bonds in strong laser fields, it is worth mentioning that the deformation of the potential energy surfaces which occurs by dressing the states can lead to more exotic phenomena (Posthumus 2004). These are, in practice, of minor consequence but do lead to some interesting dynamics. Looking back at the potential curves created by deformation of the repulsive state, one can see that there are regions around local minima which could, in principle, support a transient bound state in the field. Such a local minimum is illustrated in Figure 5.22, which shows a region of Floquet states in H_2^+

subject to 800 nm light at intensity of 10^{14} W/cm^2. In this case the field is strong enough to create an avoided crossing near the three-photon resonance at $R \approx 2$ Å. (At this point the $|1s\sigma_g, N\rangle$ and $|2p\sigma_u, N-3\rangle$ diabatic curves cross.) The trapping of population in such a state by a molecular ion that should otherwise dissociate is referred to as "bond hardening" (Sheehy and DiMauro 1996). Bond hardening is a transient phenomenon since such bound regions only occur in the field and only persist over a limited range of intensity.

In Figure 5.22, we see that at moderate intensity as the laser pulse ramps up, some population in a high-lying vibrational state can couple to the deformed repulsive state near that three-photon resonance. This population is then transiently trapped in this state while the laser is at high intensity (transient well at $R = 2.2$ Å). This temporary bound state (a "bond-hardened" state) disappears on the trailing edge of the pulse as the bound nucleus is freed, with kinetic energy characteristic of the asymptotic energy of the $|2p\sigma_u, N-3\rangle$ dressed state. The result is the ejection of a proton and an H atom with kinetic energy characteristic of three absorbed photons at an internuclear separation near that $R = 2.2$ Å potential well. There has been no definitive experimental confirmation of these bond-hardened states, though some experiments have been interpreted by claiming transient bond hardening.

5.5 Coulomb Explosion of Multiply Ionized Molecules

Like atoms, as the laser intensity increases much beyond 10^{14} W/cm^2, small molecules will undergo multiple ionization. After more than one electron is ejected by ionization, what remains are two or more charged nuclei, which repel each other by their Coulomb forces. The fragmentation that occurs subsequent to multiple ionization is referred to as a Coulomb explosion. The kinetic energies of the ejected ions are, more or less, given by the potential energies of the ions at the moment of ionization. For large molecules, this is not necessarily true, as light ions will tend to carry away more energy during the dynamics of the explosion, a process we will consider in more detail when we examine explosion of heteronuclear clusters in the next chapter. For small molecules, like diatomics, the kinetic energy of the ejected ions is a reflection of the internuclear reparation, R, at the moment of maximum ionization.

A naïve analysis of this situation might lead one to conclude that the kinetic energy release for a diatomic will be simply the potential energy of the observed charge state at the initial equilibrium bond separation of the molecule. This was the assumption in the strong field community in the early 1990s when the Coulomb explosions of diatomics in fs pulses at intensities $>10^{14}$ W/cm^2 were studied. It turns out that this is not the case. In general, kinetic energies of ion fragments are often 50–70% of what would be expected from the explosion of a molecule at its equilibrium nuclear separations. Even more puzzling at first look is that this behavior is observed over a rather large range of pulse durations, from 30 fs to 1 ps in a variety of molecules. While some pulse duration effects do exist, they tend to be rather weak effects for a given peak intensity. One might expect that fast pulses, namely ones that have a rise time faster than a characteristic Coulomb explosion time scale, might exhibit Coulomb explosion ion fragments with energies characteristic of R_e, while pulses longer than this time might result in ions with energies far below that resulting from an equilibrium explosion.

In fact, the tendency for diatomics to align along the field polarization and the strong dependence in tunnel ionization rate with nuclear separation that we explored in Section 5.2.5 make the situation more complex. The strong enhancement in ionization rate near the critical ionization distance derived in that section turns out to dominate the Coulomb explosion dynamics and explains the lion's share of observed Coulomb explosion kinetic energies.

5.5.1 Characteristic Coulomb Explosion Time

Before we consider the effects of R_c in Coulomb explosions, it is informative to assess what the characteristic time scales for a Coulomb explosion might be, to gauge the importance of the rise time of the laser pulse. Consider a diatomic in which the two ions are ionized to Q_1 and Q_2. The majority of the time, there is more or less equal ionization of the two ions, and $Q_1 = Q_2$ or they differ only by one. (Asymmetric fragmentation does, however, occur some of the time, where Q_1 and Q_2 differ by more than one (Posthumus et al. 1995; Gibson et al. 1998).) In this case, the ions evolve along the potential energy curve given by

$$U(R) = -\frac{Q_1 Q_2 e^2}{R(t)}, \tag{5.100}$$

with velocity

$$v(R) = \sqrt{\frac{2}{\mu} Q_1 Q_2 e^2 \left(\frac{1}{R_0} - \frac{1}{R(t)}\right)}. \tag{5.101}$$

Here μ is the reduced mass of a two-body system:

$$\mu = \frac{m_1 m_2}{m_1 + m_2}. \tag{5.102}$$

Integration of the equation of motion for the nuclear separation is straightforward, when starting from initial separation, R_0. The time to reach internuclear separation, R, is

$$t(R) = \int_{R_0}^{R} \frac{dR'}{v(R')}$$

$$= \sqrt{\frac{\mu R_0^3}{2 Q_1 Q_2 e^2}} \left[\sqrt{\frac{R}{R_0} \left(\frac{R}{R_0} - 1\right)} + \ln\left[\sqrt{\frac{R}{R_0}} + \sqrt{\frac{R}{R_0} - 1}\right]\right]. \tag{5.103}$$

If we say that the characteristic time for a diatomic to explode, τ_{Coul}, is the time it takes to expand to twice its initial nuclear separation, we can write that

$$\tau_{Coul} \cong 1.6 \sqrt{\frac{\mu R_0^3}{Q_1 Q_2 e^2}}. \tag{5.104}$$

5.5 Coulomb Explosion of Multiply Ionized Molecules

For example, a doubly ionized H_2 molecule at its equilibrium separation has a Coulomb explosion time of only 2 fs. On the other hand, a Cl_2 molecule ionized to 8^+ ($Q_1 = Q_2 = 4$) has a Coulomb explosion time from its 4 Å equilibrium separation of ~ 35 fs.

Clearly, Coulomb explosion occurs on a femtosecond time scale but, at least for heavier ions, potentially longer than an ultrafast laser pulse. If the pulse is indeed faster than Eq. (5.104), the simple model of the ionization process would argue that the kinetic energy of the fragments is

$$\varepsilon_{Coul} = -\frac{Q_1 Q_2 e^2}{R_0} \qquad (5.105)$$

5.5.2 Effect of the Critical Ionization Distance

Experiments consistently show that the ejected ions typically have kinetic energies characteristic of an explosion distance $\approx 1.4 - 2 \times R_e$. The answer rests in the interplay of the ionization rate with nuclear separation. Initial ionization of the molecule, even if only singly ionized, begins to push the nuclei apart. As the laser pulse ramps up in intensity, the ionization rate of the molecular ion is also increasing because the separation is approaching the critical ionization distance. What typically occurs is a multistep process which can be approximated by the following four-step model for diatomic A_2:

1) $A_2(R_0) + q\hbar\omega \rightarrow A_2^+(R_0)$
2) $A_2^+(R_0) \rightarrow A_2^+(R_c)$
3) $A_2^+(R_c) + s\hbar\omega \rightarrow A_2^{N+}(R_c)$
4) $A_2^{N+}(R_c) \xrightarrow[R \rightarrow \infty]{} A^{N_1+} + A^{(N-N_1)+}$

First (1), the laser field multiphoton ionizes the molecule. The molecule then begins to separate in the field, through bond softening or transient excitation to repulsive states (2). During this separation, the ionization rate is modest and removal of additional electrons is rare, or at least occurs only for a small number of additional electrons. Once the nuclei reach the critical separation, the tunneling barrier is greatly suppressed and the field rapidly ionizes the diatomic to a high charge state (3). This more highly charged system rapidly dissociates by Coulomb explosion, with kinetic energy release characteristic of R_c, not R_e (4).

The details of the explosion, the ultimate charge state reached and the ultimate Coulomb explosion ion kinetic energy depend on the temporal interplay of the pulse rise time and speed of ion separation, but the results are surprising similar for a range of molecular nuclear weights and laser pulse durations because of the pronounced nonlinear rise in ionization rate near R_c. A particularly illuminating way to visualize these explosion dynamics was first presented by Posthumus (2004) in which the trajectory of the exploding molecule is plotted in intensity-R space. Figure 5.23 reproduces the barrier suppression intensity thresholds that were plotted in Figure 5.9 for a Cl_2 molecule as a function of Cl-Cl separation. Superimposed on these plots is the trajectory that the molecule takes as its nuclei separate and the laser pulse intensity increases and then decreases in the pulse. The dashed

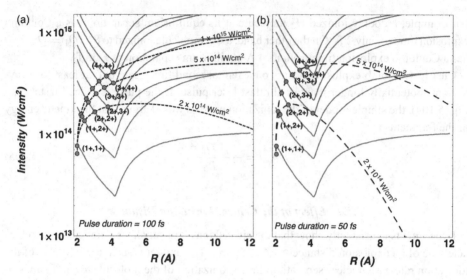

Figure 5.23 Plots of molecular separation for a Cl_2 molecule. The gray lines are the barrier suppression ionization intensity thresholds as a function of nuclear separation for ionization of the polarization-aligned molecule. The dashed lines are the paths taken by the molecule showing the separation of the molecule and the instantaneous intensity of the laser occurring when the molecule has that separation. The trajectories for different peak intensities are denoted by intensity labels. The plot in **(a)** is for Gaussian temporal envelopes with 100 fs pulse duration at FWHM. The plot in **(b)** is the same with a 50 fs pulse.

lines show numerical calculation for different peak intensities, the intensity that the molecule feels in the laser pulse at each internuclear separation. If the trajectory intercepts one of the barrier suppression curves, we can assume that the molecule undergoes BSI to the next-higher charge state and the subsequent evolution of the charge separation proceeds subject to the Coulomb forces of this next-higher charge state.

To approximate the repulsive curve of the first ionization state, we make the crude approximation that the ions repel as if each were charged to +1/2. Subsequent ionization stages are assumed to be distributed evenly ($Q_1 = Q_2$) or with only one ionization stage difference (so called symmetric ionization). The molecular history for two pulse durations is shown in the two plots (with pulse shapes given by Gaussian envelopes).

What these various trajectories illustrate is that the charge states accessed are very heavily determined by the deep "dip" in the BSI intensity at $R = R_c$. This dip occurs at about the same separation over a rather large variation in charge state accessed. As a result, the molecule tends almost always to stop further ionization once R passes R_c, which means that the observed Coulomb explosion kinetic energies will frequently be what is expected with explosion from $R \approx R_c$. For example, if we consider the 10^{15} W/cm^2 peak intensity curve for a 100 fs pulse (Figure 5.23a) the molecule quickly expands such that the chlorine molecule becomes eight times ionized, releasing two 4+ ions near R_c. Even though the intensity still rises as the molecule further expands, no further ionization occurs. This is compared to the trajectory that is only half the peak intensity at 5×10^{14} W/cm². A simple analysis would conclude that the strong nonlinearity in ionization rate would lead to significantly lower charge states. However, the passage of the separation trajectory through the R_c region leads to ionization up to 7+, at this separation. Again, the final ionization occurs near R_c and the Coulomb explosion energy will reflect this.

Comparison with the trajectories of the 50 fs pulse (Figure 5.23b) show explosion dynamics virtually identical to the 100 fs case. As we noted earlier, the characteristic Coulomb explosion time of Cl_2^{8+} is ~35 fs (Schmidt et al. 1994). We might, then, conclude that pulses with similar duration, like our 50 fs pulse, will lead to ionization and explosion at shorter separation than a pulse three times this explosion time. The trajectory for 5×10^{14} W/cm² peak intensity in the 50 fs case shows that this does not occur. The explosion occurs for the 4+–4+ ion pair right at R_c, just like the longer pulse case.

5.5.3 Asymmetric Fragmentation of a Diatomic

For the majority of Coulomb explosions the ions ejected have symmetric charge states ($Q_1 - Q_2 = 0$ or 1). However, asymmetric fragmentation where ($Q_1 - Q_2 = 2$) has been observed in explosion of molecules like I_2 (Codling and Frasinski 1993). The possible fragmentation channels can be written

$$I_2 + q\hbar w \rightarrow I_2^{2N+} \rightarrow I^{N+} + I^{N+}$$

or

$$\rightarrow I^{(N-1)+} + I^{(N+1)+}.$$

The essential reason for this asymmetric behavior, which can in some situations represent 20% of the explosion yield, is that the strong laser field can transfer an electron at small R.

We can make a rough estimate for the intensity required for this asymmetric charge transfer through a simple energetics argument. We equate the energy difference between the symmetric charge pair and the asymmetric pair with the energy that the laser field would transfer to an electron in its transfer across the nuclear separation distance:

$$eE_0R \approx \varepsilon_{20} - \varepsilon_{11}$$
$$\approx \left(I_p^{(2)} + I_p^{(1)} + U_{20}(R)\right) - \left(2I_p^{(1)} + e^2/R\right). \tag{5.106}$$

Here $I_p^{(2,1)}$ are the ionization potentials of the ions at the lower (1) and higher (2) ionization states. $U_{20}(R)$ is the repulsive energy between the two ions. For the asymmetric ionization situation described earlier, Eq. (5.106) predicts for the field strength needed:

$$eE_0R \approx \Delta\varepsilon_Q$$
$$\approx \left(I_p^{(Q+1)} + I_p^{(Q)} + \frac{(Q+1)(Q-1)e^2}{R}\right) - \left(2I_p^{(Q)} + \frac{Q^2e^2}{R}\right)$$
$$\approx I_p^{(Q+1)} - I_p^{(Q)} - \frac{e^2}{R}. \tag{5.107}$$

Further simplification can be made by making a reasonable approximation that the ionization potential of the Q-charged species is related to the ionization potential of the neutral species ($I_p^{(1)}$) by

$$I_p^{(Q)} \cong Q I_p^{(1)} \qquad (5.108)$$

(good to better than 20% for almost all of the first few charge states in a species like Cl). With this approximation, the field needed to drive asymmetric fragmentation is roughly

$$eE_0 R \approx I_p^{(1)} - e^2/R. \qquad (5.109)$$

Consider iodine: $I_p^{(1)} = 10.4$ eV. The intensity needed to fulfill Eq. (5.109) is only $\sim 2 \times 10^{13}$ W/cm^2. Not surprisingly, at the intensities needed for double ionization of I_2 ($\approx 10^{14}$ W/cm^2) the laser field is easily large enough to drive electron transfer for asymmetric explosions (Posthumus 2004).

5.5.4 Explosion of Multiply Charged Polyatomics

When more than two nuclei are present in the molecule, the dynamics of the Coulomb explosion quickly become complicated. At some point, in a large enough molecule (like C_{60}), the explosion is better treated with the models discussed in the next chapter. We can make, however, some qualitative observations about the Coulomb explosion of triatomics, educated by experimental observations (Cornaggia et al. 1991; Codling and Frasinski 1993; Cornaggia et al. 1994; Cornaggia 2010).

Like a diatomic, a linear triatomic (say CO_2) will tend to align along a linearly polarized laser field. Consequently, many of the observations we made about the ionization rate as a function of nuclear separation discussed earlier are relevant. Not surprisingly, Coulomb explosion energies reflect ionization at critical separations, not at the equilibrium bond distances. (As examples, it is observed that explosions of CO_2 into CO^+ and O^+ at intensity $\sim 10^{14}$ W/cm^2 yield fragments about 40% of the expected equilibrium distance Coulomb explosion (Cornaggia et al. 1994). The SO_2 molecule fragmenting into $O^{2+} + SO^{2+}$ exhibits 45% of the equilibrium Coulomb energy.)

It is often not obvious how a triatomic will fragment when it is doubly ionized at moderate intensity ($\sim 10^{14}$ W/cm^2). For example, for a triatomic composed of three nuclei ABC, two fragmentation channels should dominate:

$$ABC^{2+} \rightarrow AB^+ + C^+$$

or

$$\rightarrow A^+ + BC^+.$$

The branching ratio of these often depends on details of the potential energy surfaces of the dication. An interesting example is N_2O (with N–N–O structure), which has two possible dissociation pathways:

$$N_2O^{2+} \rightarrow N^+ + NO^+ \text{ (approx. 90\% yield)}$$

or

$$\rightarrow N_2^+ + O^+ \text{ (approx. 10\% yield)}.$$

5.5 Coulomb Explosion of Multiply Ionized Molecules

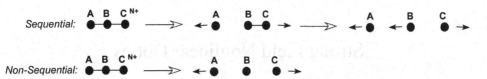

Figure 5.24 Illustration of sequential versus nonsequential explosion of a multiply ionized triatomic molecule in a strong field.

The fact that the bond energy of the N–N bond is about three times that of the N–O bond would suggest that N_2O^{2+} tends to prefer explosion into the second of these two channels. Experimental observations (listed next to the channels in parentheses) indicate that the opposite is true (Lezius et al. 2002). Simulations suggest that the potential energy barriers for N_2O^{2+} grow large for the separation of N_2^+ and O^+, pushing the molecular ion toward the first explosion path.

At higher intensity (say, $>10^{15}$ W/cm^2), triple and higher ionization occurs, which usually leads to fragmentation into three individual ions (though two-body fragmentation is frequently observed, as in the SO_2 example cited earlier). The fragmentation of ABC^{N+} could, in fact, proceed by two mechanisms, illustrated in Figure 5.24. The molecule could fragment sequentially, where the first step is the fragmentation into a diatomic ion (say B–C) and a single ion, followed by a second step where the diatomic then explodes. Alternatively, explosion could be nonsequential, where the three ions separate during the explosion nearly simultaneously. These, of course, represent two limiting cases for a range of possible explosion dynamics. However, experiments show that the most likely possibility in multiply ionized molecules like CO_2 and N_2O ionized at intensity 10^{15}–10^{16} W/cm^2 is the nonsequential pathway. In truth, the explanation behind the detailed physics of these triatomic explosion dynamics is still challenging.

6
Strong Field Nonlinear Optics

6.1 Introduction to High-Harmonic Generation

Second harmonic generation in an optical crystal was the first experimentally observed multiphoton phenomenon. It was seen very shortly after the demonstration of the laser, an effect famously observed first by Peter Franken and coworkers at the University of Michigan in 1961. Harmonic conversion of laser light in solid-state crystals represents a collection of phenomena. The vast majority of these effects are exploited in solid-state media at intensity well below that of the strong field regime (with most nonlinear optical applications occurring at intensities of MW/cm^2 to GW/cm^2, below the optical field ionization of the crystal). The underlying physics which makes all these effects possible is that the intense laser drives charge oscillations of electrons in the solid with great enough amplitude that the charge cloud motion becomes anharmonic. This leads to polarization terms in the material that oscillate nonlinearly which contribute to the production of radiation at harmonics of the laser frequency, as well as a host of other effects such an intensity-dependent change of the material's refractive index.

In the strong field regime, far more extreme driving fields can be incident on bound electrons in atoms, so, not surprisingly, much higher-order processes than are traditionally studied in nonlinear optics become observable. At the intensities considered in this book, ionization and optical damage prevent the propagation of the laser pulse through a crystal, so the nonlinear optical effects we consider will be produced in gases. The most famous and widely studied nonlinear optical effect in the strong field regime is the production of very high-order harmonic radiation from an intense laser pulse in a gas target. This chapter is devoted to a consideration of this widely explored phenomenon, typically referred to simply as high-harmonic generation or HHG.

The production of high harmonics by an intense laser pulse was greeted with some surprise when first observed in the late 1980s. This surprise arose because it was seen that beyond the first few harmonic orders, many orders could be produced with almost equal intensity; a perturbative view would suggest that higher orders should fall off with exponentially decreasing intensity (L'Huillier *et al.* 1992). It was soon shown that at intensities above 10^{14} W/cm^2 a "plateau" of very high harmonic orders of the laser radiation is produced (L'Huillier and Balcou 1993; Macklin *et al.* 1993), with orders beyond 100 possible under the right conditions (Chang et al. 1997). This triggered considerable excitement in the strong field community because of the prospect of producing coherent, ultrafast radiation pulses with wavelengths in the extreme ultraviolet (XUV) or soft X-ray region of the spectrum, with photon energies of many tens to hundreds of eV (Rundquist et al. 1998; Gibson et al. 2004). More recently, the possibility of generating trains of individual XUV pulses with temporal duration under 100 attoseconds ($\sim 10^{-16}$ s) has propelled research in HHG worldwide (Antoine et al. 1996; Scrinzi and Muller 2008).

In light of the discussions of Chapter 4, HHG to very high order with a plateau of intensities should not be surprising at all. High harmonic generation in nonperturbative intensities ultimately arises from the participation of a large number of photons in the multiphoton production of a harmonic. With strong coupling of many multiphoton pathways, it is natural that high harmonic orders might be produced with comparable intensity, just as above-threshold ionization can eject electrons above the ionization potential threshold with comparable yield among electrons with different numbers of absorbed photons. Indeed, HHG and ATI are closely related phenomena (Lewenstein and L'Huillier 2008).

6.1.1 The Strongly Driven Anharmonic Oscillator

At the core of all harmonic generation is radiating electrons made to oscillate anharmonically (nonsinusoidally) when driven harmonically by the laser. Fourier decomposition of reradiated fields from such electrons will have Fourier components at multiples of the drive frequency. We have already seen in Chapter 3 that single free electrons in vacuum will quiver in a strong laser field with anharmonic motion because of the dynamic relativistic mass increase of the electron driven to velocities near c.

Confinement of an electron in the potential of an atom can also be the source of sufficient anharmonic motion of the electron to yield reradiation at high harmonics. This effect leads to HHG at intensities as low as 10^{13} W/cm^2. High harmonic components can even be seen in the motion of a particle in a simple classical driven harmonic oscillator in which a basic, anharmonic potential term is introduced. Consider the simplest possible 1D classical anharmonic oscillating system in which a single quartic term is added to a parabolic confining potential:

$$U(x) \sim \frac{1}{2}\omega_0^2 x^2 + \frac{1}{4}\alpha x^4. \tag{6.1}$$

As a concrete example, let us examine the equation of motion of such an anharmonic system:

$$\ddot{x} + \beta\dot{x} + \omega_0^2 x + \alpha x^3 = F_0 \cos \omega t. \tag{6.2}$$

Numerical solution of this equation is illustrated in Figure 6.1 where the Fourier transform of the numerically calculated trajectory is shown. We have chosen $\omega = 1$, $\omega_0 = 5$, $\beta = 4.5$, $\alpha = 4$ and $F_0 = 550$. (As an aside, the choice of these parameters leads to a system which is not dramatically dissimilar to the conditions encountered in a real experimental situation. If one considers an 800 nm drive laser with photon energy of 1.55 eV, choice of $\omega_0 = 5$

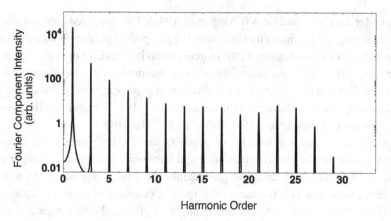

Figure 6.1 Fourier transform of the numerical solution of Eq. 6.2 where the following parameters have been used in the calculation: $\omega = 1$, $\omega_0 = 5$, $\beta = 4.5$, $\alpha = 4$ and $F_0 = 550$.

would correspond, in a general sense, to an atom with a resonance level at ~ 8 eV, which would imply an atom with ionization potential of ~ 11 eV, comparable to a large noble gas atom like Xe or Kr. Furthermore, with units normalized in this manner, choosing $F_0 = 550$ would be equivalent to a drive intensity of $\sim 3 \times 10^{14}$ W/cm^2.)

As can be seen in the spectrum of Figure 6.1, even this simple anharmonic system exhibits a comb of harmonics extending up the 29th order. Perhaps more interesting is that this harmonic spectrum already exhibits many of the features seen in actual strong field HHG: emission of odd-order harmonics over the first couple of orders (up to $q = 7$ or 9) with yield which falls off exponentially followed by a plateau of harmonic peaks with almost constant intensity out to high order, truncated by an abrupt cutoff at a well-defined harmonic.

6.1.2 Harmonic Radiation from an Ensemble of Strongly Driven Quantum Systems

Our interest is in the spectral character and yield of reradiated energy from an ensemble of strongly driven anharmonic dipoles. If we were to consider the trajectory, $\mathbf{x}(t)$, of a classical particle, the power reradiated would be given simply by Larmor's formula:

$$P = \frac{2}{3}\frac{e^2}{c^3}|\ddot{x}|^2. \tag{6.3}$$

This would suggest that the power radiated by a quantum system could be found using the quantum mechanical expectation value of $\mathbf{x}$ in the Larmor formula with the following replacement:

$$|\ddot{x}|^2 \rightarrow \frac{d^2}{dt^2} \langle \psi(t)|x|\psi(t)\rangle. \tag{6.4}$$

6.1 Introduction to High-Harmonic Generation

If we define the dipole moment in the usual manner,

$$\mathbf{d}(t) = -e\langle\psi(t)|x|\psi(t)\rangle, \tag{6.5}$$

the radiated power will be

$$P_{rad} = \frac{2}{3}\frac{e^2}{c^3}\left|\frac{d^2}{dt^2}\langle\psi(t)|x|\psi(t)\rangle\right|^2 = \frac{2}{3}\frac{1}{c^3}\left|\ddot{\mathbf{d}}(t)\right|^2. \tag{6.6}$$

Formally, this is not quite correct when considering the radiation from a single quantum system (see the discussion of this point by Mulser and Bauer (2010, p. 311)). However, when a large ensemble of radiators is considered (a situation well fulfilled by HHG experiments in which there might be 10^{11} to 10^{13} atoms in the focal volume), Eq. (6.6) is an excellent approximation. We can proceed then with the assumption that a calculation of the expectation value, Eq. (6.5), can serve as a source term in Maxwell's equations to ascertain the yield and properties of high harmonic radiation.

Frequently we will be concerned with the properties of radiation at a given harmonic order, q. Though in most experiments the drive pulse is tens of femtoseconds and therefore only a few cycles, because we are dealing with radiation at short wavelengths, it is still a reasonable approximation to assume that the harmonic generation is driven by a quasi-continuous wave. This enables easy calculations on the field radiated at a specific harmonic wavelength because we can deconstruct the time-dependent dipole in a Fourier series and derive the amplitude of the qth Fourier component of the dipole in the usual fashion:

$$\tilde{\mathbf{d}}_q = \frac{\omega}{2\pi}\int_0^{2\pi/\omega} \mathbf{d}(t)e^{iq\omega t}dt. \tag{6.7}$$

In this representation, the time-dependent dipole would be found from

$$\mathbf{d}(t) = \sum_q \tilde{\mathbf{d}}_q e^{-iq\omega t}, \tag{6.8}$$

and the total power radiated at the qth harmonic would be

$$P_{rad}^{(q)} = \frac{2}{3}\frac{q^4\omega^4}{c^3}\left|\tilde{\mathbf{d}}_q\right|^2, \tag{6.9}$$

proportional to the square of the Fourier component of the dipole and the fourth power of the harmonic order. So, the problem of finding the single-atom emitted harmonic spectrum reduces to performing a quantum mechanical calculation to find $\mathbf{d}(t)$ and Fourier decomposing it to find the power radiated at a specific harmonic wavelength.

The observed harmonic radiation from a real experiment is complicated by the fact that a large ensemble of radiators is driven coherently by the driving laser and their emission can then add coherently to form the observed radiation pattern, which is usually forward-directed and coherent (Bartels et al. 2002). This is the situation depicted schematically in Figure 6.2.

This demands using the single-atom dipole as a source in the wave equation, which can then be solved to find the coherent growth (and reconversion) of the harmonic field. If we define the polarization of the medium

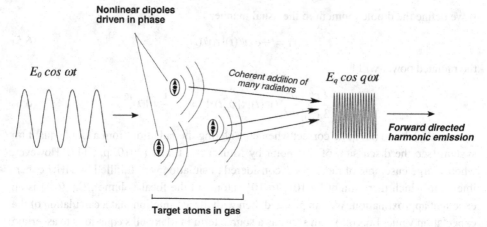

Figure 6.2 Atomic dipoles driven by the same coherent laser field will radiate in phase, leading to coherent addition of the emitted harmonic radiation in the forward direction, a process known as phase-matching.

$$\mathcal{P}(t) = n_a \mathbf{d}(t), \tag{6.10}$$

where n_a is the local atomic density of the target medium, the harmonic fields will grow according to the wave equation

$$\nabla^2 \mathbf{E} - \frac{1}{c^2} \frac{\partial^2 \mathbf{E}}{\partial t^2} = \frac{4\pi}{c^2} \frac{\partial^2 \mathcal{P}}{\partial t^2}. \tag{6.11}$$

As with the dipole, we can Fourier decompose the polarization,

$$\tilde{\mathcal{P}}_q = n_a \tilde{\mathbf{d}}_q, \tag{6.12}$$

and find the strength of the quasi-cw harmonic field amplitude, given by

$$\tilde{\mathbf{E}}_q = \frac{\omega}{2\pi} \int_0^{2\pi/\omega} \mathbf{E}(t) e^{iq\omega t} dt \tag{6.13}$$

via the equation for the qth harmonic component:

$$\nabla^2 \tilde{\mathbf{E}}_q - \frac{q^2 \omega^2}{c^2} \tilde{\mathbf{E}}_q = \frac{4\pi q^2 \omega^2}{c^2} \tilde{\mathcal{P}}_q. \tag{6.14}$$

6.1.3 Harmonic Generation Geometry and Typical Parameters

Now we are faced with two challenges: finding the single-atom response at the qth harmonic and using it as a source to find propagation effects on the harmonic yield and spatial structure. In the following two sections, we will consider each of these physical aspects of HHG in turn. However, before proceeding, it is worth giving a brief explanation of how actual harmonic generation experiments are conducted and considering some of the numerical parameters typically encountered.

6.2 Single-Atom Physics of High-Harmonic Generation

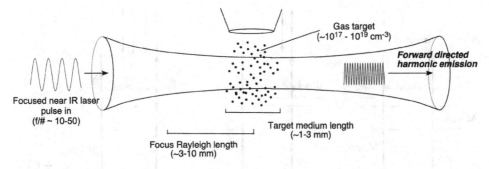

Figure 6.3 Typical experimental geometry of an ultrafast high-intensity laser-driven HHG experiment.

Figure 6.3 shows a cartoon of a generic harmonic generation experiment. Typically, a CPA laser such as an 800 nm Ti:sapphire laser will be employed with pulse duration of ~20–50 fs and pulse energy ranging from 0.1 µJ to 10 mJ (though ps pulses have been employed in many HHG experiments and wavelengths from 248 nm out to the mid-IR at a few µm wavelengths have been used). This laser pulse is focused weakly via a lens (or concave mirror) at f/# from ~10 to ~50 into a gaseous medium. The length of this medium will vary, but usually the interaction length is confined to between one and a few mm. This choice means that the Rayleigh length of the focused beam is typically a factor of a few times longer than the interaction length (a situation which is desirable from a phase-matching standpoint, as we will discuss in Section 6.3). The density of the target medium is usually chosen to be a fraction of atmospheric density with the majority of experiments performed with n_a between 10^{17} cm^{-3} and 10^{19} cm^{-3}. The target gas constituents vary, but the vast majority of experiments to date have been performed with rare gas atoms (He, Ne, Ar, Kr and Xe). Some work has been done on HHG in molecular gases (such as N_2 and CO_2) with some interesting twists, as we shall discuss.

6.2 Single-Atom Physics of High-Harmonic Generation

Many of the characteristics of observed HHG are determined by the physics of single atoms subject to the strong drive field. We will find it informative to develop a substantial theory of the single-atom response using the SFA (Lewenstein et al. 1994). Before we do that, it is useful to make some qualitative observations about the single-atom response and discuss what observable consequences those have.

6.2.1 Nonlinearities, Selection Rules and the Plateau Structure

Consider, first, the single-atom response for HHG in a semiperturbative picture. This view, which might be appropriate at modest intensity ($<10^{14}$ W/cm^2) would mandate that we view HHG as a multiphoton process like that pictured in Figure 6.4. We can see HHG

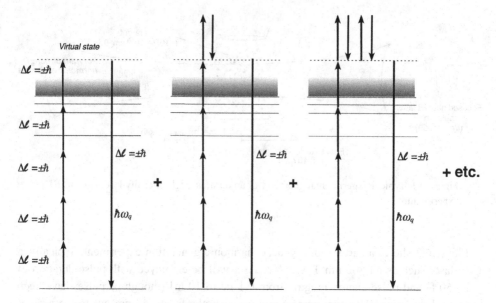

Figure 6.4 Semiperturbative, multiphoton view of high-order harmonic generation. An atom can absorb q photons to alight on a short-lived virtual state. Remission of this absorbed energy results in the emission of a harmonic photon. Higher-order multiphoton processes can and will contribute to the total probability amplitude of a harmonic photon emission. This picture illustrates that in linear polarization, only production of odd harmonics is possible because of conservation of angular momentum.

in this manner as the coherent sum of terms that involve the simultaneous absorption of q photons, which include terms involving q photons, $q+2$ photons, $q+4$ photons, and so on. These interactions drive an electron to a virtual state with energy $q\hbar\omega$ above the ground state. Reemission of a photon at $\omega_q = q\omega$ returns the electron to the ground state. If we consider atoms with centro-symmetric potentials (like all single atoms), angular momentum must be conserved. We see immediately that because of this, HHG in atoms demands that the field be nearly linearly polarized because conservation of angular momentum mandates that the virtual state have angular moment exactly $\pm\hbar$ different than the ground state to allow emission of one photon. This would not be possible if circularly polarized photons were used, which would drive the angular moment state up monotonically from the ground state with the absorption of each additional photon. Only linearly polarized photons allow $\pm\hbar$ with each absorption and permit the needed angular moment state for emission.

Secondly, we see by the same reasoning that only *odd* harmonics can be emitted by the linearly polarized drive field. The requirement that the virtual state have angular moment equal to $\pm\hbar$ of the ground state means that only an odd number of photons can be absorbed to reach the virtual state. HHG in some molecules breaks this conservation law and indeed even harmonics can be seen in these molecular media. Also, the presence of a second field (even a weak one) at a harmonic multiple of the drive (like the laser's second harmonic) is sufficient to yield harmonic emission on the "even" harmonic numbers.

6.2 Single-Atom Physics of High-Harmonic Generation

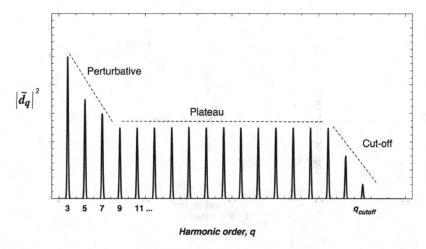

Figure 6.5 General morphology of the spectrum of high harmonics emitting in the strong field regime. The first few orders exhibit yield which falls off in intensity exponentially. However, many orders beyond this initial falloff can be emitted with constant intensity in the so-called plateau. Finally, the spectrum ends abruptly in a cutoff.

An interesting aspect of the single-atom response that became apparent from the very first experiments on HHG and was soon reproduced in numerical quantum simulations is that the emitted harmonic spectrum shows the character reproduced schematically in Figure 6.5. It is universally seen that the harmonics emitted fall off in strength exponentially over the first three or four orders (out to $q = 7$ or 9) as one might expect from a LOPT picture. The harmonic strength then levels off to a nearly constant intensity over many orders in what is called the plateau. This comb of nearly equal-strength harmonics is then abruptly truncated by a rapid falloff of harmonic yield in what is termed the cutoff. The extent of this plateau and the location of this cutoff can be predicted and will depend on the atomic ionization potential, the laser wavelength and laser intensity.

The emission strength of a single harmonic also tends to show generally reproducible trends. If one observes the strength of a high-order harmonic as a function of drive laser intensity, typically in the range of 10^{13} W/cm^2 to a few times 10^{15} W/cm^2 the harmonic yield frequently exhibits the scaling depicted in Figure 6.6. At modest intensity the yield increases quite rapidly with increasing intensity, typically with nonlinearity such that yield $\sim I^q$. This behavior tends to coincide with the regime in which the harmonic order in question rests in the cutoff region of the harmonic spectrum. After a small intensity window, the harmonic grows with intensity at a power, q' such that $q' \sim 3$–8. At higher intensity still, where field ionization becomes significant for the atom being irradiated, the yield saturates and grows slowly with intensity, mainly due to the volume increase of laser light in the focus above a threshold intensity.

In fact, this plateau and cutoff behavior in the harmonic spectrum can be easily explained using a simple model, the quasi-classical tunnel ionization and recollision model previously employed to explain nonsequential double ionization and the $10U_p$ extent of ATI peaks.

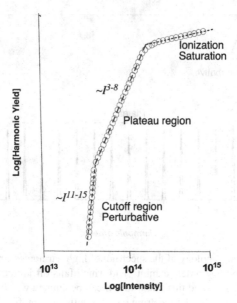

Figure 6.6 Generally observed yield of a high-order harmonic as a function of drive intensity.

6.2.2 The Quasi-Classical Three-Step Model of HHG and the Cutoff Law

Many of the most salient qualitative aspects of HHG can be explained by an exceedingly simple, largely classical model. This model, the so-called quasi-classical three-step model, has become one of the foundation stones in the strong field community for understanding HHG. Though simple, it predicts many aspects of HHG behavior, including the photon energy location of the high harmonic cutoff. It also serves as a convenient starting point to make predictions about other complicated behavior such as the ellipticity dependence of the HHG and attosecond pulse generation.

We considered the principal structural elements of this model earlier in the text when we investigated the origin of nonsequential double ionization in Chapter 4. The basis for the model is to consider interaction of the laser with the target atom in the tunneling regime (when $\gamma_K < 1$). In that case we can consider the driving field as an oscillatory deformation of the confining potential, as depicted in Figure 6.7. In the first step of the model, an electron tunnels through the deformed potential at a well-defined phase (the "birth" phase) and emerges in the continuum. In the second step, the electron propagates in the continuum and reverses direction as the drive field oscillates and reverses direction. The delay between tunneling and photon emission as the electron propagates in the continuum has a real effect on the phase of the emitted HHG phase, as we shall see. In the third step, the electron returns to the nucleus. At this point there is a probability that the electron can recombine, drop back to the ground state and emit a recombination photon. By conservation of energy, the energy of the outgoing photon will be the kinetic energy of the incoming laser-driven electron plus the ionization potential of the ground state.

6.2 Single-Atom Physics of High-Harmonic Generation

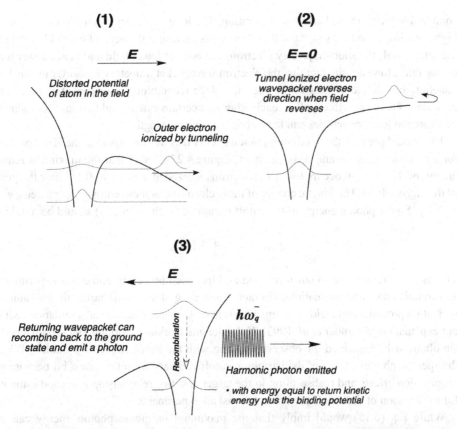

Figure 6.7 Description of the three-step, quasi-classical picture of high-harmonic generation. First, the electron tunnels from the binding potential of an atom near the peak of the laser field oscillation. The electron travels in the continuum out and then back toward the nucleus once the laser field reverses direction. Finally, back in the vicinity of the nucleus, the electron can recombine back to its ground state, resulting in the emission of a short-wavelength photon.

The kinematics of this recollision in the field have already been considered back in Chapter 4. Recall that the phase in the laser field at which the electron returns to the nucleus is directly related to the phase at which the electron was born into the continuum by tunneling. Solution of Eq. (4.242) yields the return phase as a function of birth phase, and Eq. (4.243) yields the kinetic energy of the returning electron. Analysis of these equations (see Figure 4.27) indicated that up to the maximum return energy, there are actually two solutions of birth phase that yield the same return energy. Those electrons born between the peak of the drive field and $0.1\,\pi$ off the peak (18°) yield return with "long trajectories," and those electrons born between $0.1\,\pi$ and $0.5\,\pi$ (18° and 90°) return with the same energies after undergoing a "short trajectory" in the continuum (see Figure 4.27). Therefore every harmonic photon energy would seem to be produced by two distinct families of recolliding electrons: the short- and long-trajectory electrons. In fact, both contribute comparable yields to the harmonic generation. The long-trajectory electrons are born with high tunneling probability because their birth phase is nearer the peak of the field cycle. On the other hand, the short-trajectory electrons are born at lower field and hence lower tunneling rate. However, the fact that the electron wavepacket trajectory is shorter means less quantum diffusion to the size of the packet, and the recombining wavepacket is more concentrated. These two effects offset each other to a certain extent, and harmonics produced by short and long trajectories can have about the same strength.

This model predicts the maximum photon energy that we can expect in the HHG process for a given target atom and drive intensity. Figure 4.27 showed that the maximum return energy of the electron occurs for those electrons ionized at a phase of 0.1π after the peak of the electric field. The kinetic energy of these electrons, as they return to the nucleus, was $3.17\, U_p$. So the photon energy of the cutoff harmonic (with order q_{cut}) should be roughly

$$q_{cut}\hbar\omega = I_p + 3.17 U_p. \tag{6.15}$$

This is an extremely well-known relation and has been proven in numerous experiments and complex quantum simulations. (In fact, this scaling of the cutoff harmonic with atomic ionization potential and incident intensity was first observed in numerical simulations without explanation (Kulander *et al.* 1993). Kulander and Corkum (Corkum 1993) both, nearly simultaneously, explained the observed scaling with the simple three-step model.) While the specific photon energy of the harmonic cutoff can vary from this law a bit because of propagation effects and reabsorption in the target gas, this relationship is a good estimate for the location of the cutoff harmonic in most all experiments.

While Eq. (6.15) would imply that the maximum harmonic photon energy can be increased without bound if we increase the focused intensity, in reality the target atom eventually ionizes by tunneling, depleting the population of neutral atoms producing the harmonic and increasing the electron density, an effect which clamps the harmonic generation because of phase-matching effects (see the next section). Therefore, a rather good estimate for the maximum cutoff photon energy can be made by calculating the ponderomotive energy at the ionization saturation intensity for the atom irradiated. Because of the λ^2 scaling of U_p, in fact the best way to reach shorter harmonic wavelengths is by using longer drive wavelengths, a seemingly nonintuitive result. (As we shall see, the price that is paid at longer drive wavelengths is a weaker nonlinear dipole and a reduction in harmonic yield.)

It might now be asked in the context of this model, why are a comb of harmonic peaks observed at all? The model would seem to suggest that a broad continuum of photons are emitted from the range of recollision energies acquired by electrons tunneling at a continuous range of phases in the field. However, in a quasi-CW drive field we must account for the fact that a burst of recollision recombination photons is emitted every half-cycle of the laser field. It is reasonable to assume that each of these recombination bursts is emitted coherently with well-defined phase. So, if $\tilde{E}_{coll}(\omega)$ is the continuous E-field spectrum that is emitted during a single half-cycle of the laser oscillation (like that depicted in Figure 6.8a), the observed E-field spectrum from repeated half-cycle bursts is

6.2 Single-Atom Physics of High-Harmonic Generation

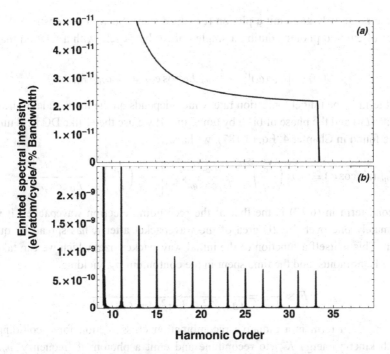

Figure 6.8 Phenomenological calculation of the shape of the harmonic spectrum expected from a hydrogen atom irradiated by 800 nm light at intensity of 2×10^{14} W/cm^2. The model contains the tunneling rate, wavepacket spreading in the continuum and radiative recombination, as described in the text. Panel (a) shows the spectrum from a single return of recombining electrons summed over short and long trajectories. Panel (b) shows the effects of 30 repeated coherently added returns from panel (a) showing the coalesence of the emitted radiation into distinct harmonic peaks at the odd harmonics of the drive laser.

$$\tilde{E}_{tot}(\omega) \sim \tilde{E}_{coll}(\omega) + \tilde{E}_{coll}(\omega)e^{-iq\omega(\pi/\omega)} + \tilde{E}_{coll}(\omega)e^{-2iq\omega(\pi/\omega)} + \ldots \quad (6.16)$$

This will lead to the coalescence of emitted spectral energy into a comb of peaks separated by 2ω in frequency space. Because emission occurs every half-cycle, the observed spectrum has peaks at odd harmonic numbers. Figure 6.8b shows the observed power spectrum when the continuous spectrum depicted in Figure 6.8a is repeated over 30 half-cycles of the drive laser (corresponding to about 50 fs of harmonic emission in an 800 nm drive field). The collapse of emitted energy into discrete harmonic peaks from this repeated coherent emission of short-wavelength bursts is dramatic.

Using the three-step model we can now construct a semiquantitative phenomenological model for the single-atom harmonic emission spectrum. The power spectrum emitted from an atom over a single half-cycle of the laser oscillation will scale something like a tunnel ionization rate (the first step) times a returning electron flux back at the nucleus which will be determined by the quantum diffusion of the free electron wavepacket while it is out in the continuum (the second step) times a cross section for that returning electron to recombine radiatively and emit a photon (the third step). So we can construct the power spectrum of emitted photons during a single optical half-cycle with a relation that looks like

$$S(\omega) \sim |\tilde{E}_{tot}(\omega)|^2 \sim W_{tun}(E_0 \cos\varphi) F_a(t_{ret}) \sigma_{rec}(K_{ret}). \tag{6.17}$$

The first term is the tunnel ionization rate, which depends on the magnitude of the E-field amplitude (E_0) and the phase of birth by tunneling. If we use the H-like DC field tunneling result we found in Chapter 4, Eq. (4.187), we have

$$W_{tun}(E_0 \cos\varphi) \cong 4\omega_a \left(\frac{I_p}{I_H}\right)^{5/2} \frac{E_a}{E_0 \cos\varphi} \exp\left[-\frac{2}{3}\left(\frac{I_p}{I_H}\right)^{3/2} \frac{E_a}{E_0 \cos\varphi}\right]. \tag{6.18}$$

The second term in (6.17) is the flux of the recolliding electron wavepacket. It will be approximately one over the 2D area of the wavepacket after it has spread by quantum diffusion. This is itself a function of the initial wavepacket size, which we can take to be about the Bohr radius, and the time spent in the continuum, τ_r, yielding

$$F \approx \frac{1}{\delta r^2} = \frac{1}{(a_B/Z)^2 + \hbar^2 \tau_r^2/4m^2(a_B/Z)^2}. \tag{6.19}$$

Finally, the third term is a radiative recombination cross section for a colliding electron with kinetic energy K_{ret} to recombine and emit a photon of frequency ω_q. From the atomic physics literature (Kotelnikov and Milstein 2018) such a cross section can be written approximately as

$$\sigma_{rec} \cong \frac{64}{3\sqrt{3}} \alpha \pi a_B^2 \frac{(\hbar\omega_q)^2}{mc^2 K_{ret}} \frac{1}{Z^2} \left(\frac{I_p}{\hbar\omega_q}\right)^3. \tag{6.20}$$

Here α is the fine structure constant (1/137). We know from the discussions of section 4.5.2 that the energy of the returning electron is related to the birth phase ωt_b and the return phase ωt_r by Eq. (4.243),

$$K_{ret} = 2U_p (\sin\omega t_r - \sin\omega t_b)^2. \tag{6.21}$$

These two phases are related by Eq. (4.242) and the fact that $t_r = t_b + \tau_r$.

As a reasonable approximation, it is appropriate to sum the spectral power of the long and short trajectories. The single-cycle emitted power spectrum calculated with the phenomenological formula Eq. (6.17) for a hydrogen atom irradiated at 2×10^{14} W/cm^2 by an 800 nm wavelength pulse summed over the two trajectories is illustrated in Figure 6.8a. Of course the abrupt cutoff in emission seen at a frequency just above 33ω is the 3.17 U_p maximum recollision energy. Coherent repeated emission of this spectrum over 30 half-cycles is then found from Eq. (6.16) and reproduced in Figure 6.8b. Remarkably all of the features observed in HHG experiments come out naturally from this simple model based on the three-step picture: an initial falloff in harmonic intensity in the lower orders followed by a nearly flat plateau out to a cutoff harmonic near $I_p + 3.2 U_p$.

As an aside we should note a fact highlighted by Lewenstein et al. (1994): detailed consideration of the quasi-classical recollision model indicates that Eq. (6.15) is not quite

6.2 Single-Atom Physics of High-Harmonic Generation

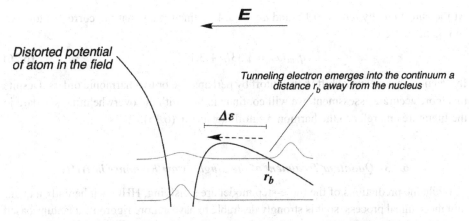

Figure 6.9 Shape of the distorted atomic potential during tunnel ionization showing that the electron is actually born not at the nuclear center but a distance r_b away from the nucleus. This results in a slightly additional boost in energy as the electron returns to the nuclear core because of the slightly longer path the electron must take. This boost in return energy has the practical effect of shifting the cutoff harmonic to a slightly higher order.

correct. Our analysis that led to this equation assumed that a tunnel-ionized electron is born into the continuum right at $x = 0$ (location of the nucleus) in the field. Strictly speaking, this is not really the case. Consider Figure 6.9. Deformation of the atomic Coulomb potential by the laser field will cause the tunneling electron wavepacket to emerge at a slight displacement from the center of the potential, denoted r_b in the figure. This offset causes a recolliding electron to acquire slightly more energy than predicted by Eq. (6.15) because of the slightly increased distance the electron must travel in the laser field to return to the core and radiate.

It is easy to calculate this additional energy acquired from the laser field. It is simply

$$\Delta \varepsilon \cong -eE_0 \cos \omega t_r r_b. \tag{6.22}$$

To evaluate this we need to find the birth radius of the ionized electron, which can be found simply by noting that

$$-\frac{e^2}{r_b} - eE_0 r_b \cos \omega t_b = -I_p \tag{6.23}$$

to yield

$$r_b \cong \frac{I_p}{eE_0 \cos \omega t_b}. \tag{6.24}$$

Inserting this into Eq. (6.22) tells us that the additional energy acquired is

$$\Delta \varepsilon \cong -I_p \frac{\cos \omega t_r}{\cos \omega t_b}. \tag{6.25}$$

At the cutoff energy $\omega t_b = 0.1\pi$ and $\omega t_t = 1.4\,\pi$, this means that the correct cutoff law should be written

$$q_{cut}\hbar\omega = 1.33 I_p + 3.17 U_p. \tag{6.26}$$

In practice, this means a shift in the cutoff by perhaps one or two harmonic orders. Despite this more accurate assessment, we will continue in line with the overwhelming standard in the literature and refer to the harmonic cutoff law as Eq. (6.15).

6.2.3 Quantum Treatment of the Single-Atom Response in HHG

While the predictions of the three-step model are satisfying, HHG is inherently a quantum mechanical process, so it is strongly desirable to have a more rigorous, quantum-based model for calculation of the atomic dipole. In this context, the problem becomes one of finding the time-dependent wavefunction of the single outer atomic electron (assuming that the SAE is appropriate), so that we can calculate

$$\mathbf{d}(t) = -e\,\langle\psi(t)|x|\psi(t)\rangle \tag{6.27}$$

and then use Eq. (6.7) to calculate the qth harmonic emission strength (and phase).

In practice, only a numerical integration of the Schrödinger equation can yield a detailed, quantitatively accurate answer to this problem. However, an analytic treatment is desirable for assessment of the underlying physics and scaling. An analytic nonperturbative treatment of the time-dependent wavefunction in a strong laser field is, in fact, possible using the previously introduced strong field approximation. Use of the SFA in treating HHG was first espoused by Lewenstein and coworkers in a classic paper (Lewenstein et al. 1994) and is generally called the "Lewenstein model" in this context. The SFA solution has subsequently been widely adapted in the HHG research community.

Before proceeding, recall the central assumptions of the SFA (as discussed in Chapter 4): (a) the SAE approximation is used and the ground state of that electron is assumed to remain undepleted; (b) the atomic potential of the outermost bound electron is treated as a perturbation. Again, the second assumption means that we effectively ignore any atomic structure and any intermediate bound states. The first assumption allows us to write the time-dependent wave function as a sum of the undepleted ground state and a perturbation sum over all possible final states at time t, weighted by the transition probability amplitude, A_{f0}. So in this model the time-dependent wave function is

$$\psi(t) = |\phi_0(t)\rangle + \int A_{f0}\,|\phi_f(\mathbf{k}_e)\rangle\,d^3\mathbf{k}_e. \tag{6.28}$$

In the SFA we find the time dependence of the ground state from the initial stationary state

$$|\phi_0(t)\rangle = e^{iI_p t/\hbar}|i\rangle, \tag{6.29}$$

and the final state is taken to be a Volkov state with canonical momentum $\hbar\mathbf{k}_e$

$$|\phi_f(t)\rangle = \left|\psi_V^{\mathbf{k}_e}(t)\right\rangle. \tag{6.30}$$

6.2 Single-Atom Physics of High-Harmonic Generation

In Chapter 4, we have already found the probability amplitude that we seek for Eq. (6.28) from the SFA within the derivation of ionization. This result, found in Eq. (4.96), is

$$A_{fi}^{(1)} = -\frac{i}{\hbar}\int_0^t dt' \left\langle \psi_V^{\mathbf{k}_e}(t') \left| H_I(t') \right| \psi_i(t') \right\rangle. \tag{6.31}$$

So, our desired wavefunction can be written

$$\psi(t) = e^{iI_i t/\hbar}|i\rangle - \frac{i}{\hbar}\int_0^t dt' \int d^3\mathbf{k}_e e^{iI_p t/\hbar} \left|\psi_v^{\mathbf{k}_e}(t)\right\rangle \left\langle \psi_V^{\mathbf{k}_e}(t') \left| H_I(t') \right| i \right\rangle. \tag{6.32}$$

For the moment, let us assume linear polarization and therefore restrict our calculation of Eq. (6.27) to the calculation of the x-component of $\mathbf{d}(t)$, which is with Eq. (6.32)

$$\begin{aligned}d_x(t) = & -e \langle i | x | i \rangle \\ & + i\frac{e}{\hbar}\int_0^t dt' \int d^3\mathbf{k}_e e^{iI_p(t'-t)/\hbar} \left\langle i|x|\psi_v^{\mathbf{k}_e}(t) \right\rangle \left\langle \psi_V^{\mathbf{k}_e}(t')|H_I(t')|i\right\rangle \\ & - i\frac{e}{\hbar}\int_0^t dt' \int d^3\mathbf{k}_e e^{-iI_p(t'-t)/\hbar} \left\langle \psi_v^{\mathbf{k}_e}(t)|x|i \right\rangle \left\langle \psi_V^{\mathbf{k}_e}(t')|H_I(t')|i\right\rangle^*. \\ & + i\frac{e}{\hbar}\int_0^t dt' \int_0^t dt'' \int d^3\mathbf{k}_e \int d^3\mathbf{k}_e' e^{iI_p(t'-t)/\hbar} \left\langle \psi_V^{\mathbf{k}_e'}(t)|x|\psi_v^{\mathbf{k}_e}(t)\right\rangle \\ & \times \left\langle \psi_V^{\mathbf{k}_e}(t')|H_I(t')|i\right\rangle \left\langle i|H_I(t'')|\psi_V^{\mathbf{k}_e''}(t'')\right\rangle\end{aligned} \tag{6.33}$$

The first term in this equation is zero by symmetry considerations. We also see that the third term is the complex conjugate of the second term. Assessing the importance of the fourth term requires us to look ahead a bit. We can generally associate a term such as $\langle 1 |x| 2\rangle$ as proportional to a radiative transition probability. We see that the second and third terms scale as

$$\sim \left\langle \psi_V^{\mathbf{k}_e}(t)|x|i\right\rangle,$$

which suggest that they are associated with transitions from a continuum Volkov state to the ground bound state. The fourth term, however, scales as

$$\sim \left\langle \psi_V^{\mathbf{k}_e'}(t)|x|\psi_v^{\mathbf{k}_e}(t)\right\rangle,$$

which implies that it describes transitions from one continuum state to another of different energy. Our intuition tells us that these kinds of transitions are weak when compared to free-bound transitions. We will therefore ignore the fourth term in Eq. (6.33).

The x-component of the dipole moment can then be written

$$d_x(t) = i\frac{e}{\hbar}\int_0^t dt' \int d^3\mathbf{k}_e e^{iI_p(t'-t)/\hbar} \left\langle i|x|\psi_v^{\mathbf{k}_e}(t)\right\rangle \left\langle \psi_V^{\mathbf{k}_e}(t')|H_I(t')|i\right\rangle + c.c. \tag{6.34}$$

For the moment we will suppress the c.c. for simplicity. Following Lewenstein, we will evaluate Eq. (6.34) in the length gauge, so the interaction Hamiltonian is

$$H_I = -e\mathbf{E}\cdot\mathbf{x} = -eE_0 x \cos\omega t, \tag{6.35}$$

and the Volkov state is

$$\psi_V^{k_e}(t) = \frac{1}{(2\pi)^{3/2}} \exp\left[-\frac{i}{\hbar}\int_0^t \frac{1}{2m}\left(\hbar k_e + \frac{e}{c}A_0 \sin\omega t''\right)^2 dt'' + i\frac{A_0 \cdot x}{\hbar c}\sin\omega t + ik_e \cdot x\right]. \quad (6.36)$$

With these insertions, the dipole moment becomes

$$d_x(t) = -i\frac{e^2}{\hbar}\int_0^t dt' \int d^3k_e E_0 \cos\omega t' \langle i|x| \exp[i(k_e + A_0 \sin\omega t/\hbar c)\cdot x]\rangle$$
$$\times \langle \exp[i(k_e + A_0 \sin\omega t'/\hbar c)\cdot x]|x|i\rangle$$
$$\times \frac{1}{(2\pi)^3}\exp\left[-\frac{i}{\hbar}\int_t^{t'}\frac{1}{2m}\left(\hbar k_e + \frac{e}{c}A_0 \sin\omega t''\right)^2 dt'' + \frac{i}{\hbar}I_p(t'-t)\right]. \quad (6.37)$$

This equation can be simplified considerably if we introduce a couple of definitions. First, we define

$$D_x^{k_e}(t) = \frac{1}{(2\pi)^{3/2}}\int d^3x\, x\psi_i(x)\exp\left[-i\left(k_e + \frac{A_0}{\hbar c}\sin\omega t\right)\cdot x\right]$$
$$= \langle p|x|i\rangle, \quad (6.38)$$

the dipole matrix element between the ground state and a plane-wave continuum state. We remind the reader that p is the kinetic momentum of the electron, related to canonical momentum in an EM field by Eq. (3.11). A plane-wave state with kinetic momentum p is written

$$|p\rangle \equiv \frac{1}{(2\pi)^{3/2}}\exp\left[\frac{i}{\hbar}p\cdot x\right]. \quad (6.39)$$

Second, we introduce the "modified" or "quasi-classical" action

$$S_{k_e}(t,t') = \int_{t'}^t \left[\frac{1}{2m}\left(\hbar k_e + \frac{e}{c}A_0\sin\omega t''\right)^2 + I_p\right]dt'', \quad (6.40)$$

which lets us rewrite Eq. (6.37) as

$$d_x(t) = -\frac{ie^2}{\hbar}\int_0^t dt'\int d^3k_e E_0 \cos\omega t' \left[D_x^{k_e}(t)\right]^* D_x^{k_e}(t') \exp\left[-iS_{k_e}(t,t')/\hbar\right]. \quad (6.41)$$

This is essentially the Lewenstein result. It describes the x-component of the time-dependent dipole moment. It is important to point out that S_{ke} is not the usual "action" as defined in Lagrangian mechanics, so

$$S_{k_e}(t,t') \neq \int_{t'}^t L(t'')dt''.$$

Looking ahead, we can already see in Eq. (6.41) correspondence with the quasi-classical three-step model for HHG described in Section 6.2.2. First recall that $D_x^{(ke)}(t)$ is the dipole matrix element between the ground state and a continuum plane-wave state of kinetic momentum **p**. It would appear then that, in Eq. (6.41),

$$eE_0 \cos \omega t \, D_x^{k_e}(t) = eE_0 \cos \omega t \, \langle \mathbf{p}(t)|x|i\rangle \tag{6.42}$$

is a term which is related to the probability amplitude of a laser-induced transition from the ground state into the continuum at time t'. The free electron then acquires a phase factor S_{k_e} over a time period of $\Delta t = t = t'$. Then $[D_x^{(k_e)}(t)]^*$ is associated with a probability amplitude for recombination from the plane-wave state back to the ground state. These terms appear to mimic the three steps in the quasi-classical model.

6.2.4 The SFA Result for Elliptically Polarized Light

It is worth noting at this point that this result can be easily generalized to elliptically polarized laser driving fields. If we define the ellipticity parameter, $\kappa = E_{0y}/E_{0x}$, to be the ratio of the field amplitudes in the x and y axes (equal to zero for linearly polarized light and one for purely circularly polarized light), we can write an elliptically polarized drive field

$$\mathbf{E}(t) = \frac{E_0}{\sqrt{1+\kappa^2}}(\hat{\mathbf{x}} \cos \omega t + \kappa \hat{\mathbf{y}} \sin \omega t). \tag{6.43}$$

The dipole moment now has both x and y components, which can be written

$$d_{x,y}(t) = -i\frac{e}{\hbar}\int_0^t dt' \int d^3\mathbf{k}_e E_0 \left(1+\kappa^2\right)^{-1/2} \cos \omega t' \left[D_{x,y}^{k_e}(t)\right]^* \\ \times \left[D_x^{k_e}(t') \cos \omega t' + \kappa D_y^{k_e}(t') \sin \omega t'\right] \\ \times \exp\left[-iS_{k_e}(t,t')/\hbar\right] \tag{6.44}$$

where

$$D_y^{k_e}(t) = \langle \mathbf{p}|y|i\rangle \tag{6.45}$$

and

$$S_{k_e}(t,t',\kappa) = \int_t^{t'} \left[\frac{1}{2m}\left(\hbar \mathbf{k}_e + \frac{e}{c}A_0\left(1+\kappa^2\right)^{-1/2}(\hat{\mathbf{x}} \sin \omega t'' + \kappa \hat{\mathbf{y}} \cos \omega t'')\right)^2 + I_p\right] dt''. \tag{6.46}$$

6.2.5 Evaluation of d(t) and Saddle Point Approximation for k_e

To evaluate the dipole moment in Eq. (6.41), we first need to evaluate the matrix element $D_x^{k_e}(t)$. As a concrete example, we proceed by evaluating this for a hydrogen-like ground state with nuclear charge Z. The wavefunction for this state is

$$\psi_{1s} = \frac{1}{\pi^{1/2}}\left(\frac{Z}{a_B}\right)^{3/2} e^{-Zr/a_B} = \frac{k_p^{3/2}}{\pi^{1/2}} e^{-k_p r}, \tag{6.47}$$

where k_p is defined with ionization potential

$$I_p = \frac{\hbar^2 k_p^2}{2m}. \tag{6.48}$$

The matrix element is calculated from Eq. (6.36):

$$D_{x1s}^{k_e}(t) = \frac{1}{(2\pi)^{3/2}} \int d^3x \, x\psi_{1s}(x) \exp[-i\mathbf{p} \cdot \mathbf{x}/\hbar]. \tag{6.49}$$

This is easily solved if we recall our previous result for the Fourier transform of the 1s state, Eq. (4.136):

$$\phi_i(\mathbf{p}) = \langle \mathbf{p} | 1s \rangle = -\frac{2^{3/2}}{\pi} \frac{k_p^{5/2}}{\left(p^2/\hbar^2 + k_p^2\right)^2}. \tag{6.50}$$

We can see that

$$D_{x1s}^{k_e}(t) = -\frac{\hbar}{i} \frac{\partial}{\partial p_x} \phi_i(\mathbf{p})$$

$$= i \frac{2^{7/2}}{\pi} \frac{k_p^{5/2}}{\hbar} \frac{p_x}{\left(p^2/\hbar^2 + k_p^2\right)^3}, \tag{6.51}$$

which yields our desired result:

$$D_{x1s}^{k_e}(t) = i \frac{2^{7/2}}{\pi} k_B^{5/2} \frac{k_{ex} + (eA_0/\hbar c) \sin \omega t}{\left[(k_{ex} + (eA_0/\hbar c) \sin \omega t)^2 + k_{ey}^2 + k_z^2 + k_B^2\right]^3}. \tag{6.52}$$

This formula can be easily generalized for calculating the matrix element in elliptically polarized fields with the substitution:

$$k_{ey}^2 \rightarrow \left(k_{ey} + \left(e\kappa A_0/\hbar c\sqrt{1+\kappa^2}\right) \cos \omega t\right)^2 \tag{6.53}$$

and division of A_0 by $(1+\kappa^2)^{1/2}$. The other component, $D_{y1s}^{k_e}$, is similar to Eq. (6.52) with the substitution of (6.53) but with

$$k_{ey} + \left(e\kappa A_0/\hbar c\sqrt{1+\kappa^2}\right) \cos \omega t \tag{6.54}$$

in the numerator.

To find the emission strength of the qth harmonic we Fourier decompose $d(t)$, yielding

$$\tilde{d}_q = \frac{\omega}{2\pi} \int_0^{2\pi/\omega} d_x(t) e^{iq\omega t} dt$$

$$= -i \frac{e^2 \omega}{2\pi \hbar} \int_0^{2\pi/\omega} dt \int_0^t dt' \int d^3k_e E_0 \cos \omega t' \left[D_x^{k_e}(t)\right]^* D_x^{k_e}(t')$$

$$\times \exp\left[-iS_{k_e}(t,t')/\hbar + iq\omega t\right] + c.c. \tag{6.55}$$

This is essentially our final result. It is a fully quantum description of the dipole strength of the qth harmonic in the framework of the SFA. In principle, it is straightforward, though

6.2 Single-Atom Physics of High-Harmonic Generation

computationally expensive, to calculate this five-dimensional integral numerically. In practice, it is beneficial for quick calculations to make further approximations to this equation using the saddle-point method. Not only will this make numerical calculations easier, but it will ultimately give us some additional physical insights.

In fact, all five integrals can be approximated using the saddle-point approach. Interestingly each saddle-point approximation will reveal a correspondence between the quantum SFA theory and the quasi-classical three-step model. Recall that one physical interpretation of the saddle-point approximation in this context is to consider the integrals as sums over all possible quantum paths and the saddle-point approximation as equivalent to finding the quantum paths that contribute the most. We will start by simplifying the three-dimensional integral over $\mathbf{k}_e$.

First, we recall that if the integral has a rapidly varying exponential argument, the integral can be approximated in the following way:

$$\int g(z)e^{if(z)}dz \cong \sum_{z_i} \sqrt{\frac{2\pi}{i|f''(z_i)|}} g(z_i)e^{if(z_i)}, \qquad (6.56)$$

where the sum is performed over all saddle points, z_i, found from the relation

$$f'(z)\big|_{z=z_i} = 0. \qquad (6.57)$$

Instituting this on the $d^3\mathbf{k}_e$ integrals in Eq. (6.55), relation (6.57) yields for the saddle-point momenta

$$\int_{t'}^{t} \frac{1}{m}\left(\hbar\mathbf{k}_e + \frac{e}{c}A_0\hat{\mathbf{x}}\sin\omega t''\right) dt'' = 0. \qquad (6.58)$$

This equation has a nice, intuitive physical interpretation. The significance can be seen by recalling the relationship of $\mathbf{k}_e$ to the kinetic momentum

$$\mathbf{p} = \hbar\mathbf{k}_e + \frac{e}{c}\mathbf{A}(t). \qquad (6.59)$$

This means that the saddle-point constraint in Eq. (6.58) corresponds to the condition that

$$\int_{t'}^{t} \frac{\mathbf{p}}{m} dt'' = \mathbf{x}(t) - \mathbf{x}(t') = 0. \qquad (6.60)$$

So, the main contributions to the integral are the paths with kinetic momentum such that an electron that is born into the field at time t' returns to the same place (the atomic nucleus) at a time t. Therefore, t' can be thought of as the tunnel birth time and $t = t' + \tau_r$ is the time of electron return. τ_r is the time spent by the electron wavepacket out in the continuum between birth and recombination.

The saddle-point condition in Eq. (6.58) then yields conditions for the saddle-point momenta $\mathbf{k}_{es}$

$$\hbar k_{exs}(t - t') - \frac{e}{c\omega} A_0(\cos\omega t - \cos\omega t') = 0, \qquad (6.61)$$

yielding for linear polarization

$$k_{exs} = \frac{e}{\hbar\omega^2} \frac{E_0}{\tau_r} (\cos\omega t - \cos[\omega(t - \tau_r)]) \qquad (6.62a)$$

$$k_{eys} = 0 \tag{6.62b}$$

$$k_{ezs} = 0. \tag{6.62c}$$

In elliptically polarized light, the y-component of $\mathbf{k}_{es}$ is

$$k_{eys} = -\frac{e}{\hbar\omega^2} \frac{E_0}{\tau_r} \frac{\kappa}{\sqrt{1+\kappa^2}} \left(\sin \omega t - \sin[\omega(t-\tau_r)]\right). \tag{6.63}$$

Note that $\mathbf{k}_{es}$ depends explicitly on t, and τ_r though we do not write this functional dependence to avoid clutter in the notation. With these saddle-point momenta and the fact that $f''(\mathbf{k}_{es}) = \hbar\tau_r/m$, the approximate result for the dipole moment of the qth harmonic can be written

$$\tilde{d}_q = \frac{e^2 \omega (2\pi m^3)^{1/2}}{i^{1/2} \hbar^{5/2}} \int_0^{2\pi/\omega} dt \int_0^\infty d\tau_r \frac{1}{\tau_r^{3/2}} E_0 \cos[\omega(t-\tau_r)]$$
$$\times \left[D_x^{kes}(t)\right]^* D_x^{kes}(t-\tau_r) \tag{6.64}$$
$$\times \exp\left[-iS_{k_e}(t,t')/\hbar + iq\omega t\right] + c.c.,$$

where we have transformed the integral over birth times, t', to one over all return times in the continuum, τ_r, and extended the integral to infinity to account for all possible return times. The quasi-action within this saddle-point approximation is

$$S_{k_{et}}(t,\tau_r) = \int_{t-\tau_r}^{t} \left\{ \frac{1}{2m} \left[\frac{e}{\omega^2} \frac{E_0}{\tau_r} (\cos \omega t - \cos[\omega(t-\tau_r)]) \right. \right.$$
$$\left. \left. + \frac{e}{\omega} E_0 \sin \omega t'' \right]^2 + I_p \right\} dt'', \tag{6.65}$$

$$S_{k_{es}}(t,\tau_r) = -\frac{2U_p}{\omega^2 \tau_r} (\cos \omega t - \cos[\omega(t-\tau_r)])^2 + (I_p + U_p) \tau_r$$
$$- \frac{U_p}{2\omega} (\sin 2\omega t - \sin[2\omega(t-\tau_r)]). \tag{6.66}$$

This result, Eqs. (6.64) and (6.66) combined with a calculation for the dipole matrix element like Eq. (6.52) with substitutions from Eq. (6.62), can be easily evaluated numerically. Such a calculation will yield a single-atom high harmonic spectrum if we plot the square of the dipole strength as a function of q. It can also yield an intensity scaling for the strength (and phase) of the qth harmonic. Such a calculation of the former using these equations is presented in Figure 6.10 for a helium atom irradiated by 800 nm light at intensities of 2×10^{14} W/cm^2 and 8×10^{14} W/cm^2. The plateau and cutoff structure described earlier emerges from this single-atom calculation, particularly evident at the higher intensity. The location of the cutoff as predicted by the quasi-classical three-step model delineated in Eq. (6.15) is noted on the figure. A calculation for the dipole strengths of the 25th and 55th harmonics as a function of intensity for 800 nm irradiation of He is presented in Figure 6.11 showing the strongly nonlinear increase in harmonic strength at first when the 25th harmonic is in the cutoff region followed by a more modest nonlinear growth at intensity above 3×10^{14} W/cm^2 when the harmonic is in the plateau region.

6.2 Single-Atom Physics of High-Harmonic Generation

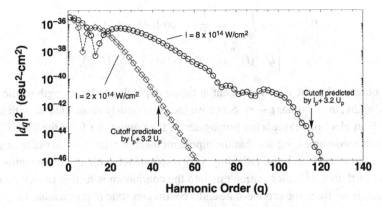

Figure 6.10 Strong field approximation calculation (using the saddle-point approximation for k_{ex}) of the square of the dipole for harmonics generated in He atoms. The spectral strength is shown for two different drive intensities assuming an 800 nm drive wavelength.

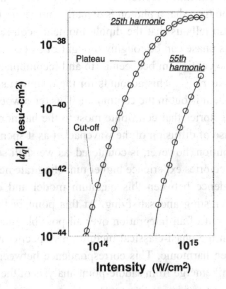

Figure 6.11 SFA calculation (using the saddle-point approximation for k_{ex}) of the square of the dipole for the 25th and 55th harmonics generated in He atoms as a function of intensity by 800 nm radiation. The cutoff and plateau regions of generation are clear from the differing nonlinear variation.

6.2.6 Classical Correspondence of d(t)

It is interesting at this point to examine the structure of our quantum model for the dipole moment. It reveals a close correspondence between our quantum model and the quasi-classical three-step model. Revert for the moment to a consideration of the time-dependent dipole (not its Fourier transform), which has a structure that looks like

$$d(t) \sim \text{const.} \int_0^\infty d\tau_r \frac{1}{\tau_r^{3/2}} E_0 \cos[\omega(t-\tau_r)]$$
$$\times \left[D_x^{k_{es}}(t)\right]^* D_x^{k_{es}}(t-\tau_r) \exp\left[-iS_{k_e}(t,\tau_r)/\hbar\right]. \tag{6.67}$$

First, we observe that the dipole moment at time t depends on the strength of the driving laser field at an *earlier* time, $t - \tau_r$. Since we have previously associated τ_r with the time over which an electron wavepacket propagates in the continuum from the time it tunnels to the time it recombines, we see that the dipole moment at t depends on the field strength at the time the electron is born, reflecting a correspondence with a field ionization rate at this earlier birth time. This ionization rate into the continuum is further related to a dipole matrix element between the ground state and a continuum state of momentum k_{es} evaluated at this earlier time, manifest in D_x.

Next, we see that $d(t)$ varies as the conjugate of such a dipole matrix element between the continuum and the ground state evaluated at the time t, a quantity that describes a recombination rate by a returning electron. Evaluation at time t is appropriate since the recombination time is the time that a photon is actually emitted in the quasi-classical picture. Equation (6.67) also tells us that the dipole moment acquires a phase factor given by the quasi-action. This phase can be roughly thought of as due to the time the electron wavepacket spends in the continuum between birth and recombination. Finally, the dipole moment is proportional to $\tau_r^{-3/2}$. This accounts for the quantum spatial spreading of the electron wavepacket while it is out in the continuum. This term would suggest that it is the shortest return time trajectories that contribute most to the harmonic emission (so-called short trajectories) because of diffusion of the wavepacket as it spends a longer time in the continuum. This contribution, however, is countered, as we shall see, by the fact that the second set of longer trajectories experience higher tunneling rate nearer to the field peak.

The close correspondence between this quantum model and our phenomenological three-step picture is interesting and satisfying. At this point in the analysis, the dipole moment strength is a result of an integration over all possible quantum return times, τ_r. We already know that in the quasi-classical picture only specific return times contribute to the strength of a given harmonic. This correspondence between specific return times and a given harmonic will emerge from saddle-point analysis of the remaining integrals in Eq. (6.64).

6.2.7 Saddle-Point Analysis of the Time Integrals for $\tilde{d}_q$

Now we apply the saddle-point approximation to the integrals over return times, τ_r, and emission time, t. To perform these analytic approximations, I follow closely the treatment of Ivanov *et al.* (1996). We look first at the rapidly varying exponential term in Eq. (6.64). Recalling our result for the quasi-action

$$S_{k_e}(t,\tau_r) = \int_{t-\tau_r}^t \left[\frac{1}{2m}\left(\hbar\mathbf{k}_e + \frac{e}{c}\mathbf{A}(t'')\right)^2 + I_p\right] dt'', \tag{6.68}$$

saddle-point analysis of the integral over τ_r demands finding the saddle-point return times from

$$\frac{\partial}{\partial \tau_r}(S_{k_e} - q\hbar\omega t) = \frac{1}{2m}\left(\hbar\mathbf{k}_e + \frac{e}{c}\mathbf{A}(t-\tau_r)\right)^2 + I_p = 0, \qquad (6.69)$$

and saddle-point approximation of the integral over emission time t requires finding saddle-point times from

$$\frac{\partial}{\partial t}(S_{k_e} - q\hbar\omega t) = \frac{1}{2m}\left(\hbar\mathbf{k}_e + \frac{e}{c}\mathbf{A}(t)\right)^2 \\ - \frac{1}{2m}\left(\hbar\mathbf{k}_e + \frac{e}{c}\mathbf{A}(t-\tau_r)\right)^2 - q\hbar\omega = 0. \qquad (6.70)$$

Inserting the first relation into the second yields

$$\frac{1}{2m}\left(\hbar\mathbf{k}_e + \frac{e}{c}\mathbf{A}(t_s)\right)^2 + I_p = q\hbar\omega, \qquad (6.71)$$

which, in terms of the kinetic momentum, is

$$\frac{\mathbf{p}(t_s)^2}{2m} + I_p = q\hbar\omega. \qquad (6.72)$$

We see that these final two approximations yield simply a statement of energy conservation in the quasi-classical model: the energy of the emitted qth-harmonic photon is equal to the kinetic energy of the electron at the time of return, t_s, plus the ionization potential.

Analytic solutions for the saddle-point return times, τ_{rs}, from Eqs. (6.69) and (6.70) cannot be derived, however, as Ivanov et al. (1996) have shown, approximate solutions can be found. We first need to find the saddle-point solutions for the return times, which we denote τ_{rs}, from the equation

$$\frac{1}{2m}\left(\hbar\mathbf{k}_e + \frac{e}{c}\mathbf{A}(t-\tau_{rs})\right)^2 + I_p = 0, \qquad (6.73)$$

which for linear polarization is

$$\frac{1}{2m}\left(\hbar k_{exs} + \frac{e}{c}A_0\sin[\omega(t-\tau_{rs})]\right)^2 + I_p = 0. \qquad (6.74)$$

Solutions of this equation for τ_{rs} must be complex. We can derive approximations after finding first-order approximate solutions, $\tau_{rs}^{(0)}$, by temporarily ignoring the ionization potential

$$\hbar k_{exs} + \frac{e}{c}A_0\sin\left[\omega(t-\tau_{rs}^{(0)})\right] \approx 0. \qquad (6.75)$$

Recalling Eq. (6.60), we already know that the k_{exs} correspond to canonical momenta that lead to the condition that the electron starts at the nucleus and returns to the nucleus after time τ_r. Equation (6.75) further tells us that

$$p_{xs}(t - \tau_{rs}^{(0)}) = 0; \qquad (6.76)$$

the momentum of the electron wavepacket is zero at the time it is born into the continuum. This confirms that we associate $\tau_{rs}^{(0)}$ with the classical recollision times of the three-step model. In other words, the greatest contributions to the integral over all τ_r are associated with electrons that start at rest and return right back to the origin, the location of the nucleus, after this time has elapsed. Inserting the previous result, Eq. (6.62a), for the saddle-point momentum into Eq. (6.75), we get the equation we need to solve for $\tau_{rs}^{(0)}$:

$$\frac{eA_0}{c\omega\tau_{rs}^{(0)}}\left(\cos\omega t - \cos\left[\omega(t - \tau_{rs}^{(0)})\right]\right) + \frac{e}{c}A_0 \sin\left[\omega(t - \tau_{rs}^{(0)})\right] = 0. \quad (6.77)$$

Rewriting (6.77) in terms of an emission or "return" phase, $\varphi_r = \omega t$, and a birth phase related to time $\varphi_b = \omega(t - \tau_{rs}^{(0)})$, this condition becomes

$$\cos\varphi_r - \cos\varphi_b + (\varphi_r - \varphi_b)\sin\varphi_b = 0. \quad (6.78)$$

This equation for $\tau_{rs}^{(0)}$ is the equation we derived in Chapter 4, Eq. (4.242), to calculate the classical return time of an electron born by tunneling at a well-defined phase in the field. This is not an unexpected result. Solutions found from Eq. (6.77) as a function of classical birth time for the quarter cycle after the peak of the laser field are plotted in Figure 6.12. Our previous analysis of Eq. (6.78) in Section 6.2.2 noted that over this quarter-cycle of possible tunneling birth times in the field there are two classes of trajectories with differing return times that yield the same returning kinetic energy: "long" trajectories produced by electrons liberated near the field oscillation peak between phase of 0 and 0.1π, and "short" trajectories followed by electrons tunnel-ionized at phase between 0.1π and 0.5π. These two classes of trajectories are noted on the plot of Figure 6.12. This plot also shows that for

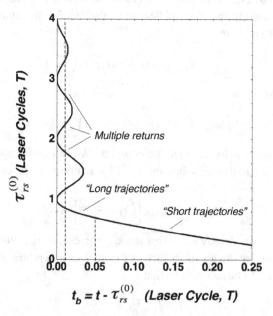

Figure 6.12 Solutions for the zeroth-order saddle-point return times as a function of the birth time $t - \tau_{rs}^{(0)}$.

6.2 Single-Atom Physics of High-Harmonic Generation

a small range of birth phases the electron can actually return to the nucleus more than once so that there are, in principle, multiple saddle-point contributions to the integral over τ_r. Quantum diffusion of the wavepacket tends to reduce significantly the emission strength of these multiple returns, so it is usually a good approximation to consider only the first return in calculating the harmonic dipole strength.

To find solutions for the full saddle-point equation, we assume that the imaginary part of τ_{rs} is small and perform a perturbation analysis starting with our zeroth-order solution, writing

$$\tau_{rs} = \tau_{rs}^{(0)} + \delta. \tag{6.79}$$

Defining $f(\tau_{rs})$ in terms of the equation we seek to solve,

$$f(\tau_{rs}) = \frac{1}{2m}\left(\hbar k_{exs} + \frac{e}{c}A_0 \sin[\omega(t - \tau_{rs})]\right)^2 + I_p = 0, \tag{6.80}$$

we write this function as a Taylor expansion in terms of the small (imaginary) parameter δ:

$$f(\tau_{rs}) = f_0 + f_1 \delta + \frac{1}{2} f_2 \delta^2 + \ldots. \tag{6.81}$$

The zeroth-order term of this function is found by substitution of the zeroth-order solution for τ_{rs}:

$$f_0 = f(\tau_{rs}^{(0)}) = I_p. \tag{6.82}$$

Taylor expansion of the sin function in Eq. (6.80) and use of Eq. (6.77) yields

$$f \cong \frac{1}{2m}[\frac{eA_0}{\omega \tau_{rs}^{(0)}}\left(\cos \omega t - \cos\left[\omega(t - \tau_{rs}^{(0)})\right]\right)$$
$$+ \frac{e}{c}A_0 \left(\sin\left[\omega(t - \tau_{rs}^{(0)})\right] + \omega \cos\left[\omega(t - \tau_{rs}^{(0)})\right]\delta\right)^2 + I_p$$
$$\cong f_0 + \frac{e^2 \omega^2 A_0^2}{2mc^2} \cos^2\left[\omega(t - \tau_{rs}^{(0)})\right]\delta^2 + \ldots, \tag{6.83}$$

which tells us that $f_1 = 0$ and

$$f_2 = \frac{e^2}{m} E_0^2 \cos^2 \omega t_b, \tag{6.84}$$

where we have reverted to use of the E-field and the birth time, t_b. Equation (6.80) rewritten to second order in δ is

$$I_p + \frac{e^2}{m} E_0^2 \cos^2 \omega t_b \delta^2 = 0, \tag{6.85}$$

yielding

$$\delta = \pm i \frac{(2m I_p)^{1/2}}{e E_0(t_b)}. \tag{6.86}$$

If we recall, in Chapter 4 the definition of the Keldysh parameter, we presented the tunneling time, t_{tun}, which we associated with the time it took for a bound electron to traverse a distance equal to the barrier width it must tunnel through to reach the continuum.

Our solution for the imaginary part of the recollision time takes on an interesting form if we use t_{tun}:

$$\gamma_K = \omega t_{tun} = \sqrt{\frac{I_p}{2U_p}} \Rightarrow t_{tun} = \frac{(2mI_p)^{1/2}}{eE_0} \tag{6.87}$$

$$\delta = \pm i t_{tun}. \tag{6.88}$$

The imaginary part of the saddle-point return time is the time interval that it takes the electron to tunnel through the barrier, a time during which *classically* the electron kinetic energy is negative. Much has been made in the strong field community about the physical significance of this imaginary time interval and whether this negative-energy time interval is observable. We would note simply that this result is an artifact of our approximate model and arises because of an attempt to apply a classical correspondence to a purely quantum effect: tunneling of the electron through a classically forbidden barrier.

Armed with this approximate result for the saddle-point return time we can find the modified action, noting the approximation that $\partial \tau_r = \partial \delta$, which lets us write

$$\frac{\partial S_{kes}}{\partial \tau_r} = \frac{\partial S_{kes}}{\partial \delta}, \tag{6.89}$$

and Eq. (6.69) with the definition of Eq. (6.80) can be written

$$\frac{\partial S_{kes}(t, \tau_r)}{\partial \delta} \cong f_0 + \frac{1}{2} f_2 \delta^2. \tag{6.90}$$

The approximate modified action under the saddle-point approximation for τ_r is then

$$\begin{aligned}
S_{kes}(t, \tau_r) &\cong S_{kes}(t, \tau_r^{(0)}) + f_0 \delta + \frac{1}{6} f_2 \delta^3 \\
&\cong S_{kes}(t, \tau_r^{(0)}) \pm i \frac{(2m)^{1/2} I_p^{3/2}}{eE(t_b)} \mp i \frac{2^{3/2} m^{1/2} I_p^{3/2}}{6eE(t_b)} \\
&\cong S_{kes}(t, \tau_r^{(0)}) \pm i \frac{2^{3/2} m^{1/2} I_p^{3/2}}{3eE(t_b)}.
\end{aligned} \tag{6.91}$$

To find the dipole strength we insert τ_{rs} and can use, to good approximation, $\tau_{rs}^{(0)}$ in all the prefactors of Eq. (6.64). We do, however, encounter one complication. Recall that the saddle-point condition is

$$k_{exs} + (eA_0/\hbar c) \sin[\omega(t - \tau_{rs})] \cong ik_p, \tag{6.92}$$

which means that the dipole matrix element

$$D_x^{kes}(t - \tau_r) = i \frac{2^{7/2}}{\pi} k_p^{5/2} \frac{k_{exs} + (eA_0/\hbar c) \sin[\omega(t - \tau_r)]}{\left[(k_{exs} + (eA_0/\hbar c) \sin[\omega(t - \tau_r)])^2 + k_p^2\right]^3} \tag{6.93}$$

6.2 Single-Atom Physics of High-Harmonic Generation

is singular at the saddle point for τ_r. To deal with this singularity, we Taylor-expand the denominator about $(\tau - \tau_{rs})$. The dipole matrix element has Taylor-expanded denominator

$$g(t, \tau_r - \tau_{rs}) \cong (\tau_r - \tau_{rs})^3 \left\{ (eA_0\omega/\hbar c) \cos\left[\omega(\tau_r - \tau_{rs}^{(0)})\right] \right.$$
$$\left. \times 2\left(k_{exs} + (eA_0/\hbar c)\sin\left[\omega(\tau_r - \tau_{rs}^{(0)})\right]\right) \right\}^3. \tag{6.94}$$

We are now faced with applying the saddle-point method to an integral with a singularity of the form $1/(\tau_r - \tau_{rs})^3$. Following the treatment found in Gribakin and Kuchiev (1997), we proceed by using the identity

$$\frac{1}{(\tau_r - \tau_{rs})^3} = \frac{1}{\Gamma(3)} \int_0^\infty d\xi\, \xi^2 \exp\left[-\xi(\tau_r - \tau_{rs})\right]. \tag{6.95}$$

From this formula, using the usual saddle-point formula and performing the secondary integral, it can be shown that the saddle-point approximation to a singular integral is

$$\int g(z) e^{if(z)} dz \cong \sum_{z_i} i^\zeta \frac{\Gamma(\eta/2)}{\Gamma(\eta)} \sqrt{\frac{2\pi}{i|f''(z_i)|}} \left[if''(z_i)\right]^{\zeta/2} \exp\left[if(z_i)\right] \tag{6.96}$$

if the pre-exponential factor has a singularity of the form

$$g(z) = \frac{g'(z)}{(z - z_i)^\zeta}. \tag{6.97}$$

We apply Eq. (6.96) to our specific case of $\zeta = 3$, noting that $\Gamma(3/2) = \pi^{1/2}/2$ and approximating τ_{rs} by $\tau_{rs}^{(0)}$ wherever we can.

For the moment, we revert to calculating the time-dependent dipole under these approximations:

$$d(t) \cong \frac{ie^2}{\hbar} \sum_{\tau_{rs}^{(0)}} \left\{ \left(\frac{2\pi m}{i\hbar\tau_{rs}^{(0)}}\right)^{3/2} E_0 \cos\left[\omega(t - \tau_{rs}^{(0)})\right] \left[D_x^{kes}(t)\right]^* \right.$$
$$\times i\frac{2^{7/2}}{\pi} k_p^{5/2} \frac{ik_p}{\left[(e\omega A_0/\hbar c)\cos\left[\omega(t - \tau_{rs}^{(0)})\right] 2(ik_p)\right]^3} \tag{6.98}$$
$$\times \exp\left[-iS_{kes}(t, \tau_{rs}^{(0)})/\hbar\right] \exp\left[-\frac{m^{1/2}(2I_p)^{3/2}}{3e\hbar E(t_b)}\right]$$
$$\left. \times i^3 \frac{\Gamma(3/2)}{\Gamma(3)} (2\pi)^{1/2} 2^{3/2} \left[if''(\tau_{rs}^{(0)})\right] + c.c. \right\}.$$

To evaluate (6.96) we use Eqs. (6.68) and (6.93) to find that

$$f''(\tau_{rs}) = \frac{1}{m\hbar}\left(\hbar k_{exs} + \frac{eA_0}{c}\sin\left[\omega(t - \tau_{rs})\right]\right)\frac{e\omega A_0}{c}\cos\left[\omega(t - \tau_{rs})\right]$$
$$\cong f_0 + \frac{e^2\omega^2 A_0^2}{2mc^2}\cos^2\left[\omega(t - \tau_{rs}^{(0)})\right]\delta^2 + \ldots. \tag{6.99}$$

We can say that

$$E_0 \cos[\omega(t - \tau_{rs})] = E_0(t_b), \quad (6.100)$$

which leaves us with the approximation for the time-dependent dipole:

$$d(t) \cong -\frac{1}{\sqrt{i}} 4\pi^{3/2} (m\hbar)^{1/2} \sum_{\tau_{rs}^{(0)}} \left\{ \frac{k_p^{3/2}}{(\tau_{rs}^{(0)})} \left[D_x^{k_{es}}(t) \right]^* \frac{1}{E_0(t_b)} \right.$$

$$\left. \times \exp\left[-iS_{k_{es}}(t, \tau_{rs}^{(0)})/\hbar \right] \exp\left[-\frac{m^{1/2}(2I_p)^{3/2}}{3e\hbar E(t_b)} \right] + c.c. \right\} \quad (6.101)$$

This result lets us write the dipole moment as the product of three amplitudes summed over all possible saddle-point return times:

$$d(t) \cong -\frac{1}{\sqrt{i}} \sum_{\tau_{ss}^{(0)}} \alpha_{rec}(t, \tau_{rs}^{(0)}) \alpha_{free}(t, \tau_{rs}^{(0)}) \alpha_{tun}(t, \tau_{rs}^{(0)}, E_0) + c.c. \quad (6.102)$$

The approximate dipole in this form is composed of a term

$$\alpha_{rec}(t, \tau_{rs}^{(0)}) = k_p^{3/2} \left[D_x^{k_{ss}}(t) \right]^*, \quad (6.103)$$

which is proportional to a recombination probability; a second term,

$$\alpha_{free}(t, \tau_{rs}^{(0)}) = 4\pi^{3/2} \frac{(m\hbar)^{1/2}}{\left(\tau_{rs}^{(0)}\right)^{3/2}} \exp\left[-iS_{k_{es}}(t, \tau_{rs}^{(0)})/\hbar \right], \quad (6.104)$$

which is a factor which accounts for wavepacket spreading in the continuum and a phase factor imparted to the electron from its propagation in the continuum; and a third term,

$$\alpha_{tun}(t, \tau_{rs}^{(0)}, E_0) = \frac{1}{E_0(t_b)} \exp\left[-\frac{m^{1/2}(2I_p)^{3/2}}{3e\hbar E(t_b)} \right], \quad (6.105)$$

which is proportional to a tunneling probability at $t_b = t - \tau_{rs}^{(0)}$.

The HHG spectrum, then, is found by summing over all possible free propagation return times which bring a tunnel-ionized electron back to the ion at time t and Fourier transforming the result. With Eq. (6.102), the correspondence between the fully quantum SFA model and the phenomenological quasi-classical three-step model for HHG is complete. This result is remarkable. It is virtually identical in structure to the HHG model that we assembled on a heuristic basis from the three-step model back in Eq. (6.17). It is the product of a recombination probability, a term accounting for quantum diffusion of the wavepacket and a tunnel ionization rate at an earlier time in the laser optical cycle.

6.2.8 Ellipticity Effects on HHG

Simple angular momentum considerations tell us that HHG is not allowed by circularly polarized light, at least in target media composed of atoms or molecules with centrosymmetric potentials. This can be easily seen in the multiphoton picture, as illustrated in

6.2 Single-Atom Physics of High-Harmonic Generation

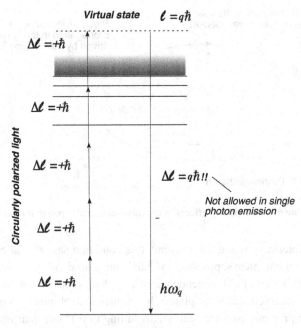

Figure 6.13 Schematic diagram of multiphoton driven HHG by a circularly polarized pulse, showing that HHG is not allowed by angular momentum conservation consideration.

Figure 6.13. Absorption of q circularly polarized photons promotes the electron to a virtual state with angular momentum qh above that of the ground state. It is not possible then to conserve angular momentum by the emission of a single harmonic photon.

There should be, then, a quick falloff of harmonic yield as the drive laser field becomes slightly elliptical. This effect can also be seen in the context of the quasi-classical three-step model, a view which will let us make a semiquantitative estimate for the amount of ellipticity that is tolerable before HHG is significantly reduced. Consider the trajectory of the tunnel-ionized wavepacket in the continuum in an elliptically polarized field, a situation shown schematically in Figure 6.14. The wavepacket will undergo a lateral displacement in the continuum from the elliptically polarized field. If this displacement is comparable to or larger than the quantum spreading of the wavepacket during its recollision trajectory, we expect that the harmonic yield drops significantly.

The condition for HHG suppression can be found by equating the size of wavepacket spreading with the lateral displacement in the field of ellipticity κ. This condition,

$$\delta r \approx \sqrt{\delta r_0^2 + \hbar^2 \tau_r^2 / 4m^2 \delta r_0^2}$$
$$\approx \frac{\kappa e E_0}{m \omega^2 \sqrt{1 + \kappa^2}} \sin \varphi_r - \sin \varphi_b + \varphi_r \cos \varphi_b - \varphi_b \cos \varphi_b, \quad (6.106)$$

can be solved easily depending on the trajectory in question. For higher-order harmonics approaching the cutoff we know that $\varphi_b \approx 0.1\pi$, $\varphi_r \approx 1.4\pi$ and $\tau_r \approx 1.3\pi/\omega$. So, in an

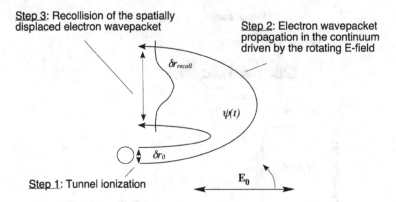

Figure 6.14 Ellipticity effects in the quasi-classical three-step model.

800 nm field at intensity around 10^{14} W/cm^2 this condition says that at ellipticity above $\kappa \approx 0.48$ we expect complete suppression of harmonic emission.

Experimentally the ellipticity dependence of high harmonics scales like the curves in Figure 6.15. The observed scaling of plateau harmonics with ellipticity is quantitatively in keeping with our previous estimate with yields falling to 0.1% of their level in a linearly polarized field at the same intensity when $\kappa \approx 0.4$. Lower-order harmonics (below about the 15th), which can thought to be more "perturbative" and closer to the multiphoton picture, tend to exhibit a slower roll-off in yield with increasing ellipticity (Budil *et al.* 1993). It is also interesting to note that harmonics in or very near the cutoff actually exhibit a dip in yield as the ellipticity of the field drops to zero (see thick dashed line in Figure 6.15). This has been attributed (Ivanov *et al.* 1996) to quantum interferences between the nearly equal time short and long trajectories that contribute to harmonics in this region.

To make quantitative calculations of the single-atom response in an elliptical field, one can use the SFA. We have already noted in Eq. (6.44) what the form of the dipole looks like in an elliptical field. Here we summarize the results for $\mathbf{d}(t)$ under the saddle-point approximations for the $\mathbf{k}_e$ integrals, noting that k_{eys} is no longer zero. The time-dependent dipole is

$$d_{x,y}(t) = \frac{ie^2}{\hbar} \int_0^\infty d\tau_r \left(\frac{2\pi m}{i\hbar\tau_r}\right)^{3/2} \frac{E_0}{\sqrt{1+\kappa^2}} \left[D_{x,y}^{k_{es}}(t)\right]^* \qquad (6.107)$$
$$\times \left[D_x^{k_{es}}(t-\tau_r)\cos[\omega(t-\tau_r)] + \kappa D_y^{k_{es}}(t-\tau_r)\sin[\omega(t-\tau_r)]\right]$$
$$\times \exp\left[-iS_{k_{es}}(t,\tau_r)/\hbar\right] + c.c.$$

with hydrogen-like dipole matrix elements reading

$$D_x^{k_{es}}(t) = i\frac{2^{7/2}}{\pi}k_p^{5/2} \qquad (6.108a)$$
$$\times \frac{k_{exs} + (eE_0'/\hbar\omega)\sin\omega t}{\left[(k_{exs} + (eE_0'/\hbar\omega)\sin\omega t)^2 + (k_{eys} + (eE_0'/\hbar\omega)\cos\omega t)^2 + k_p^2\right]^3}$$

6.2 Single-Atom Physics of High-Harmonic Generation

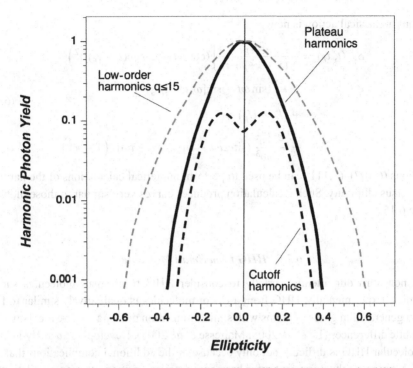

Figure 6.15 General experimentally observed scaling of high-order harmonic yields as a function of laser ellipticity. Thin dashed line is for low-order or nearly perturbative harmonics. The thick line is observed for harmonics in the plateau. The thick dashed line is the behavior seen for harmonics near the cutoff. The dip in yield for nearly zero ellipticity is a result of destructive interference between long and short trajectories. Plots were generated based on the data of Budil *et al.* (1993) and Ivanov *et al.* (1996).

$$D_x^{k_{es}}(t) = i\frac{2^{7/2}}{\pi}k_p^{5/2}$$

$$\times \frac{k_{exs} + (eE'_0\kappa/\hbar\omega)\cos\omega t}{\left[\left(k_{exs} + (eE'_0/\hbar\omega)\sin\omega t\right)^2 + \left(k_{eys} + (eE'_0/\hbar\omega)\cos\omega t\right)^2 + k_p^2\right]^3} \quad (6.108b)$$

where

$$E'_0 = \frac{E_0}{\sqrt{1+\kappa^2}} \quad (6.109)$$

and the saddle-point momenta are

$$k_{exs} = \frac{e}{\hbar\omega^2}\frac{E'_0}{\tau_r}\left(\cos\omega t - \cos[\omega(t-\tau_r)]\right), \quad (6.110a)$$

$$k_{eys} = -\frac{e}{\hbar\omega^2}\frac{E'_0\kappa}{\tau_r}\left(\sin\omega t - \sin[\omega(t-\tau_r)]\right). \quad (6.110b)$$

The quasi-classical action is now

$$S_{k_{es}}(t, \tau_r) = -\frac{2U_p}{(1+\kappa^2)\omega^2 \tau_r}\left[(\cos\omega t - \cos[\omega(t-\tau_r)])^2\right.$$
$$\left. + \kappa^2 (\sin\omega t - \sin[\omega(t-\tau_r)])^2\right]$$
$$+ (I_p + U_p)\tau_r$$
$$- \frac{e^2 E_0'^2}{8m\omega^3}(\sin 2\omega t - \sin[2\omega(t-\tau_r)])\left(1-\kappa^2\right). \tag{6.111}$$

Equations (6.107)–(6.111) can be used to perform numerical calculations of the harmonic yield versus ellipticity. Such a calculation produces curves very similar to those plotted in Figure 6.15.

6.2.9 HHG from Small Molecules

We now turn our attention briefly to consider HHG from small molecules such as diatomics. Experimentally, HHG from a gas of molecules is qualitatively similar to HHG spectra generated in gases of atoms with similar ionization potential. There are, however, a few subtle differences (Le *et al.* 2008; Mairesse *et al.* 2008). Developing an analytic model for molecular HHG is difficult, not only because of the additional complications that arise from having more than one center for harmonic reemission upon electron recollision, but also because multielectron effects start to play a role (McFarland *et al.* 2008). Therefore, we will confine our consideration to a few qualitative observations made in experiments which are more or less consistent with and explainable by the quasi-classical three-step model.

The first observation is that molecular orientation strongly influences the HHG yield (Kato *et al.* 2011). Consider, for example, a homonuclear molecule like N_2 (McFarland *et al.* 2008). The HOMO, as depicted in Figure 6.16, exhibits σ_g symmetry. As we observed in Chapter 5, the tunnel ionization rate is strongly enhanced when the laser polarization is oriented along the molecular nuclear axis. Consequently, HHG is greatest for this molecular orientation. Furthermore, the symmetry of the molecule ensures equal contributions to the nonlinear dipole for either half-cycle, so odd harmonics are solely emitted. We also note that in this orientation the σ_g orbital produces a large dipole for returning recolliding electrons.

The second important difference between molecules and atoms is that multielectron effects can play a much more significant role in the HHG process. Again, consider the N_2 molecule and consider electrons in the next-deeper bound HOMO-1 orbital, which has π_u symmetry. This orbital, depicted in Figure 6.17, plays almost no role in the HHG process when the N_2 molecule is oriented along the laser field. Looking at Figure 6.17, we see that not only should tunnel ionization be greatly suppressed because there is no limb of electron density extending out along the field as there is for the σ_g HOMO, but, in addition, the dipole generated by a recolliding electron will be nearly zero because of the +/− antisymmetry of the orbital's lobes.

6.2 Single-Atom Physics of High-Harmonic Generation

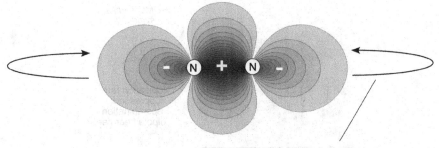

Tunnel ionized electron along molecular axis recolliding to produce HHG

Figure 6.16 High Harmonic Generation from an aligned N_2 molecule with plot of the HOMO. High Harmonic Generation by recolliding electrons occurs with highest efficiency when the molecule is oriented along the laser polarization, and the three-step HHG production is symmetric for both laser half-cycles.

The situation is quite different when the same molecule is oriented perpendicular to the laser field, as depicted in the bottom of Figure 6.17. Not only does the orbital have some electron density out along the field which can participate in tunneling, but a large recombination dipole is generated for a recolliding electron. The experimental consequence of this is that the more deeply bound electrons play a role in HHG in the perpendicularly oriented molecules. Because the ionization potential is greater for the HOMO-1 electrons, HHG from these electrons actually exhibits a cutoff at a significantly higher photon energy. The HHG spectrum observed in N_2 gas, then, has the usual strong plateau and cutoff characteristic of the outer electron like that in atomic HHG, but a lower-intensity plateau of harmonics extends beyond this cutoff as well to a much higher cutoff determined by the binding potential of those HOMO-1 electrons.

A third interesting aspect of HHG from diatomic molecules is observed when heteronuclear molecules such as CO are irradiated. Such molecules break the inversion symmetry of the ground state wavefunction and lead to even harmonic generation (Frumker et al. 2012). This can be seen when the HOMO of CO is considered, shown in Figure 6.18. Tunnel ionization is greatly enhanced when the molecule is more or less oriented along the laser polarization for the half-cycles that pull electrons from the carbon side of the molecule, where there is a more extended orbital structure. Half-cycles pulling from the oxygen side of the molecule will have reduced dipole contributions. This breaks the symmetry that causes only odd harmonic production and leads to harmonic spectra like the one plotted schematically in Figure 6.19. While there is a strong preponderance of odd harmonics from target gases like CO, there are observable emissions on the even harmonics at reduced intensity, interspersed with the strong odd peaks.

The final difference in molecular HHG that we will note is that interference effects between recombination from two nuclear centers in a diatomic molecule can shape the harmonic spectrum (Lein et al. 2005). Consider, for example, HHG from recolliding electrons in a molecule oriented like the one at the top of Figure 6.20. Because of the lateral

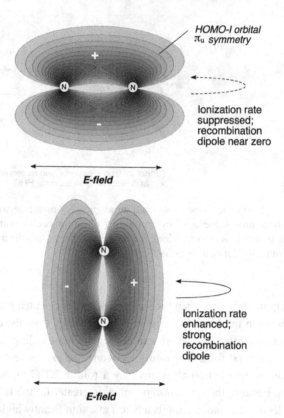

Figure 6.17 Rotational effects of the HOMO-I orbital of N_2, which has π_u symmetry. Though it is more deeply bound than the HOMO electron when the molecule is oriented perpendicular to the laser field, its enhancement in ionization rate causes it to contribute to the HHG. Consequently, HHG from these anti-oriented molecules will have a higher cutoff photon energy because of the more deeply bound electrons that participate in the generation process.

displacement of the two centers, denoted as δ in the figure, nearly simultaneous harmonic emission from the two nuclei can destructively interfere under certain conditions. Roughly speaking, when $k_q \delta \approx \pi$, this destructive interference will occur (Lin *et al.* 2018, p. 238). This leads to reductions of harmonic emission for certain wavelength bands and produces a "dip" in the HHG spectrum like the one illustrated in the spectrum at the bottom of Figure 6.20.

6.3 Propagation and Phase-Matching in High-Harmonic Generation

Single-atom physics plays an important role in determining the characteristics of HHG emission in a gas irradiated at high intensity. However, the actual observed yield, spatial pattern, and coherence properties of a high harmonic are the result of the coherent addition

6.3 Propagation and Phase-Matching in High-Harmonic Generation

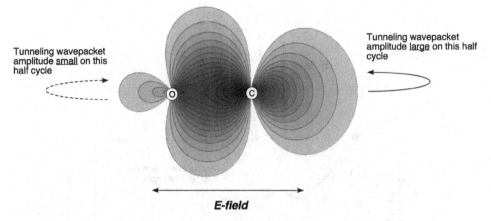

Figure 6.18 Molecular orbital shape of the CO molecule. Because of the asymmetry in the electron density distribution, when the molecule is oriented along the drive laser polarization axis there is a greater recolliding electron flux during one half-cycle than the other. This leads to even harmonic generation.

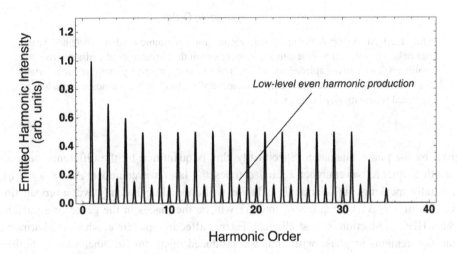

Figure 6.19 Schematic of the HHG spectrum observed from a CO target.

of many distributed single-atom radiators. We must, therefore, turn our attention to the coherent build-up of harmonic intensity as a laser pulse propagates through the harmonic generating medium (gas jet, cell, hollow core fiber, etc.), a process depicted in Figure 6.2 and governed by the wave equation, Eq. (6.14). Coherent buildup of a harmonic in any nonlinear process occurs as long as the radiators remain in phase along the driving laser's propagation path. The factors affecting this phase-matching will be the dominant focus of our consideration in this section.

As in classic low-order nonlinear optics, the rate at which the harmonic emission becomes dephased and stops adding coherently to previously produced harmonic light is

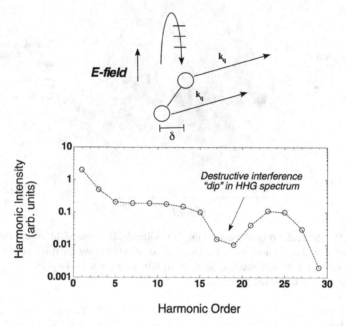

Figure 6.20 At the top is the recollision picture for a diatomic randomly oriented in the laser field. The interval δ is the lateral displacement of the two centers of emission from the recollision. These two displaced reemission points can destructively interfere for a certain range of harmonic wavelengths. This leads to "dip" features in the harmonic spectrum, as illustrated schematically in the lower panel.

given by the phase mismatch. Specifically, this is quantified by the difference between the qth harmonic wavenumber k_q and q times the laser wavenumber, $\Delta k = k_q - qk$. Generally speaking, high-harmonic radiation will only add coherently over a propagation distance of $\sim 1/\Delta k$, so regions of low Δk will be the places in the generating medium where HHG production is most efficient. Factors affecting the rate at which the harmonic emission remains in phase with emission produced upstream (in other words, the harmonic generation remains phase matched) will include geometric focus effects, refractive index dispersion in the gas or plasma produced by ionization, as well as the intrinsic phase imparted to the emission from the atomic physics of high-harmonic radiation.

6.3.1 Geometric Phase-Matching Effects: Phase-Matching at Moderate Intensity

In almost all real HHG experiments, the harmonics are produced by a focused laser beam. So, we will first consider the consequences of the phase and intensity variation of such a beam as it traverses the medium and passes through focus. A focused beam undergoes a π phase shift as it passes through the focal point, the well-known Gouy phase shift.

6.3 Propagation and Phase-Matching in High-Harmonic Generation

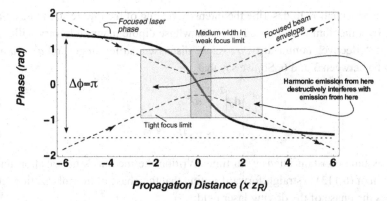

Figure 6.21 Plot of the phase of a focused Gaussian laser beam as it propagates through the focus. This is superimposed upon the beam waist as it travels through the focus. The Gouy phase shift causes a π shift of the beam through the focal region. This shows that if the HHG-generating medium is much longer than a Rayleigh-length harmonic radiation emitted before the focus will destructively interfere with emission after the focus, reducing the net harmonic yield. This effect is mitigated if the medium is comparable or shorter than the Rayleigh range.

This phase variation occurs in a region near the focal plane over a propagation region roughly $\pm z_R$. The general consequences on HHG can be seen in Figure 6.21.

If the medium used to generate the high harmonics is much longer than a Rayleigh range, HHG radiation is produced at points all along the laser's π phase shift. Harmonic light produced after the focus will be phase-shifted roughly $q\pi$ from harmonic light produced before focus. Since q is odd, destructive interference between light generated before and after focus will limit harmonic yield. This situation is generally called the "strong focus" regime because it tends to be an appropriate description when the laser is focused tightly and hence has a short Rayleigh range compared to the target medium depth. This geometric phase-matching clamp to the HHG yield can be circumvented in part by placing the medium somewhat before or after the focus. Even better yield can be achieved if the "weak" focus limit is used. As illustrated in Figure 6.21, weak focusing occurs when the laser Rayleigh range is much longer than the HHG medium. In this limit there is only a modest geometric contribution to the phase mismatch.

It is possible to quantify the effects of geometric phase-matching and laser propagation on the total harmonic photon yield and on the spatial output profile of the harmonic beam using a model introduced by L'Huillier *et al.* (1992). To explore the effects of the geometric phase mismatch we will, for the moment, ignore ionization and plasma formation and will also disregard the intrinsic atomic phase introduced in the strong field. We will accommodate these effects later in this section. Consequently, this model is a good approximation for intensity, say, between 10^{13} W/cm^2 and 10^{14} W/cm^2.

The basis of the model is to insert a single-atom source term with nonperturbative nonlinearity into an integral solution of the driven wave equation for the qth harmonic. We use a simple approximation to describe the intensity scaling of the dipole strength; we assume that the HHG medium is composed of atoms whose dipole magnitude vary with intensity with some "effective" nonlinearity, q'. As we discussed earlier, in general $q' < q$ and will typically be between 3 and 6. So, we say that

$$\left|\tilde{\mathbf{d}}_q\right|^2 \sim aI^{q'}, \tag{6.112}$$

$$\left|\tilde{\mathbf{d}}_q\right|^2 \sim \alpha_q |E_0|^{2q'}. \tag{6.113}$$

We will assume constant intensity in time. Writing the medium's polarization under the assumption of (6.113) is straightforward, noting that the phase of the polarization term will be q times the phase of the driving laser field:

$$\tilde{P}_q \cong \alpha_q n_a(\mathbf{x}) |E_0(\mathbf{x})|^{q'} \exp\left[iq Arg[E_0(\mathbf{x})]\right]. \tag{6.114}$$

Here $n_a(\mathbf{x})$ is the spatially varying atomic density of the target gas.

To solve for the qth harmonic field outside of the generating medium, we resort to integral formalism for solving the wave equation; see Jackson (1975, p. 224). The qth harmonic field $E_q(\mathbf{x}')$ at some point $\mathbf{x}'$ outside the generating medium is

$$E_q(\mathbf{x}') = \frac{q^2 \omega^2}{c^2} \iiint \tilde{P}_q(\mathbf{x}) \frac{e^{ik_q R}}{R} d^3\mathbf{x}. \tag{6.115}$$

The geometry of this integral is illustrated in Figure 6.22. Our principal concern is for the harmonic field in the far field of the interaction, well away from the HHG generating medium. Since the harmonic emission is predominantly in the forward direction (emitted along z), the distance from emission point to observation plane, $R = |\mathbf{x}' - \mathbf{x}|$, can be approximated as

$$R = |\mathbf{x}' - \mathbf{x}| \cong z' - z + \frac{(x' - x)^2 + (y' - y)^2}{2(z' - z)}. \tag{6.116}$$

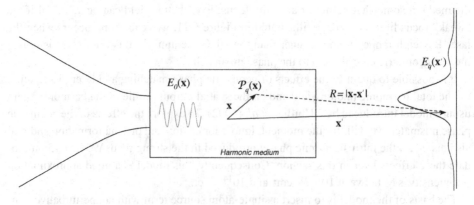

Figure 6.22 Geometry of the fields and polarizability in the integral of Eq. (6.115).

6.3 Propagation and Phase-Matching in High-Harmonic Generation

At this point we will introduce the slowly varying amplitudes for the harmonic field and medium polarization

$$E'_q(\mathbf{x}') = E_q(\mathbf{x}')e^{-ik_q z}, \tag{6.117}$$

$$\tilde{P}'_q(\mathbf{x}) = \tilde{P}_q(\mathbf{x})e^{-iqkz}. \tag{6.118}$$

Introducing these amplitudes and approximation (6.116) into (6.115), the slowly varying harmonic field can now be written as

$$E'_q(x') = \frac{q^2\omega^2}{c^2} \iiint \frac{\tilde{P}_q}{(z'-z)} e^{-i\Delta k_q t} \exp\left[ik_q \frac{[(x'-x)^2+(y'-y)^2]]}{2(z'-z)}\right] d^3\mathbf{x}. \tag{6.119}$$

Here we have introduced an intrinsic phase mismatch between the qth harmonic and laser wavenumbers:

$$\Delta k_q = k_q - qk. \tag{6.120}$$

This quantity will be nonzero if there is some dispersion in the refractive index between the driving wavelength and harmonic wavelength. Even a modest-density neutral gas will have a very slight variation in wavelength between a near-IR laser field and the XUV field of a high-order harmonic. This is usually negligible for most HHG calculations. However, field ionization of some of the medium's atoms by the drive laser will lead to plasma formation and the introduction of a substantial intrinsic phase mismatch between the near-IR and XUV field, as we shall see.

Let us now make the reasonable assumption that the focused laser field can be described by a TEM$_{00}$ Gaussian beam. For ease of computation we will introduce the Gaussian beam in complex notation, which is slightly different from the definition we have been employing from Chapter 3. We write the slowly varying envelope of a focused Gaussian laser beam field as

$$E_0(\mathbf{x}) = E_{00} \frac{1}{(1+z^2/z_R^2)^{1/2}} \exp\left[-\frac{r^2}{w_0^2(1+z^2/z_R^2)} - i\tan^{-1}\left(\frac{z}{z_R}\right) + i\frac{r^2}{w_0^2(1+z^2/z_R^2)}\frac{z}{z_R}\right]. \tag{6.121}$$

Under this definition, the local intensity in the beam is

$$I = \frac{c}{4\pi}|E_0(\mathbf{x})|^2. \tag{6.122}$$

Of course, the usual definitions remain, $r = \sqrt{x^2+y^2}$, and the Rayleigh range z_R is related to the focal spot size, w_0, in the usual way: $z_R = \pi w_0^2/\lambda$. We make the further assumption that the atomic medium density varies only along the propagation axis (and is constant over the radial extent of the focus). The harmonic E-field in the observation plane becomes

$$E'_q(\mathbf{x}') = \frac{q^2\omega^2}{c^2}\alpha_q E_{00}^{q'} \iiint n_a(z) \frac{e^{-i\Delta k_q z}}{(z'-z)} \frac{1}{(1+z^2/z_R^2)^{q'/2}}$$
$$\times \exp\left[ik_q \frac{[(x'-x)^2 + (y'-y)^2]}{2(z'-z)} - \frac{q'r^2}{w_0^2(1+z^2/z_R^2)}\right. \quad (6.123)$$
$$\left. - iq\tan^{-1}\left(\frac{z}{z_R}\right) + iq\frac{r^2}{w_0^2(1+z^2/z_R^2)}\frac{z}{z_R}\right)\right] dx\,dy\,dz$$

Though we have not written it explicitly, since the atomic medium density varies with z we should expect that the intrinsic phase mismatch Δk_q also varies with z. Furthermore, we should also note that the refractive index of the medium is, in all likelihood, complex, reflecting the fact that it is absorptive, particularly at the harmonic XUV wavelength which is likely to see substantial absorption by photoionization of the medium if the harmonic photon energy is above the medium atom's ionization potential. Though we will, for the moment, retain the simpler $\Delta k_q z$ notation to avoid clutter in the equations, in fact we can account in our final result for spatial variation of the gas density and the presence of harmonic absorption, κ_{abs}, with the following substitution for the intrinsic phase mismatch:

$$\Delta k_q z \rightarrow \int_{-\infty}^{z} \Delta k_q(z'')dz'' - \frac{i}{2}\int_{z}^{\infty} \kappa_{abs}(z'')dz'', \quad (6.124)$$

where the absorption coefficient κ is introduced in such a way that the intensity of a propagating beam falls by Beer's law, $I = I_0 e^{-\kappa_{abs} z}$. If the absorption in the medium is mainly by single-atom photoionization, it is convenient to couch the absorption in terms of a photoionization cross section and local neutral gas density:

$$\kappa_{abs}(z) = \sigma_{abs} n_a(z). \quad (6.125)$$

The integrals over x and y in Eq. (6.123) can be performed analytically. Consider, for example, the integral over transverse coordinate x,

$$I_x = \int_{-\infty}^{\infty} \exp\left[i\frac{k_q}{2z_R}\frac{(x'-x)^2}{\tilde{z}'-\tilde{z}} - \frac{q'x^2}{w_0^2(1+\tilde{z}^2)} + iq\frac{\tilde{z}x^2}{w_0^2(1+\tilde{z}^2)}\right] dx, \quad (6.126)$$

where we have normalized the propagation coordinate to a Rayleigh range: $\tilde{z} = z/z_R$. If we further introduce $r' = (x'^2 + y'^2)^{1/2}$, then the x and y integrals together become

$$I_x I_y = 2\pi \left[\frac{2(q' - iq\tilde{z})}{w_0^2(1+\tilde{z}^2)} - i\frac{k_q}{z_R(\tilde{z}'-\tilde{z})}\right]^{-1}$$
$$\times \exp\left[-\frac{k_q r'}{2iz_R(\tilde{z}'-\tilde{z}) + k_q w_0^2(1+\tilde{z}^2)/(q'-iq\tilde{z})}\right]^2. \quad (6.127)$$

6.3 Propagation and Phase-Matching in High-Harmonic Generation

The harmonic field is reduced to a single integral along the propagation axis,

$$E'_q(\mathbf{x}') = 2\pi \frac{q^2\omega^2}{c^2}\alpha_q E_{00}^{q'} \int_{-\infty}^{\infty} \frac{n_a(z)}{(1+\tilde{z}^2)^{q'/2}} \left[\frac{2(q'-iq\tilde{z})z_R(\tilde{z}'-\tilde{z})}{w_0^2(1+\tilde{z}^2)} - ik_q\right]^{-1}$$
$$\times \exp\left[-i\Delta k_q z_R\tilde{z} - iq\tan^{-1}(\tilde{z})\right] \qquad (6.128)$$
$$\times \exp\left[-\frac{k_q(q'-iq\tilde{z})r'^2}{2iz_R(\tilde{z}'-\tilde{z})(q'-iq\tilde{z}) + k_qw_0^2}\right] z_R d\tilde{z}.$$

We can now factor out a term of the form

$$i\frac{w_0^2(1+\tilde{z}^2)}{q'-iq\tilde{z}}$$

and use the fact that

$$\frac{1}{a+ib} = \frac{\exp[i\tan(a/b)]}{\sqrt{a^2+b^2}} \qquad (6.129)$$

to arrive at our final result for the harmonic field at the observation plane:

$$E'_q(\mathbf{x}') = i2\pi \frac{q^2\omega^2}{c^2} z_R \alpha_q w_0^2 E_{00}^{q'} \int_{-\infty}^{\infty} \frac{n_a(\tilde{z})}{(1+\tilde{z}^2)^{q'/2-1}(q'^2+q^2\tilde{z}^2)^{1/2}}$$
$$\times \frac{1}{\xi(\tilde{z},\tilde{z}')} \exp\left[-i\Delta k_q z_R\tilde{z} - iq\tan^{-1}(\tilde{z}) + i\tan^{-1}(q\tilde{z}/q')\right]. \qquad (6.130)$$
$$\times \exp\left[-\frac{k_q r'^2}{\xi(\tilde{z},\tilde{z}')}\right] d\tilde{z}$$

Here we have introduced the complex spot size function:

$$\xi(\tilde{z},\tilde{z}') = 2iz_R(\tilde{z}'-\tilde{z}) + \frac{k_q w_0^2(1+\tilde{z}^2)}{q'-iq\tilde{z}}. \qquad (6.131)$$

While cumbersome, numerical integration over propagation distance for arbitrary gas density profiles is trivial, yielding spatial profiles and absolute yields for the qth harmonic. (This assumes that suitable dipole moment coefficients α_q are supplied from a nonperturbative model, like the SFA, for the single-atom response.)

Figure 6.23 shows numerically calculated far-field profiles for the 21st harmonic of 800 nm light 0.5 m from a 1 mm long gas jet. The laser is assumed to be focused to a Gaussian spot size with $w_0 = 20$ μm and the medium is assumed to exhibit effective nonlinearity $q' = 5$. The solid lines in the two panels show calculations from Eq. (6.130) for two locations for the center of the gas jet with respect to the laser focal plane: right on the focal plane in the top of Figure 6.23 and one Rayleigh range (∼1.5 mm) past the focal plane in the bottom of Figure 6.23. The dip in the center of the distribution seen in the top panel of that figure is a consequence of the destructive interference that occurs directly on axis from the Gouy phase slip for a medium centered exactly on the focal plane. When the jet is moved down from the focus, the harmonic beam is broader at the observation plane because it is being produced by a diverging laser beam (of reduced intensity).

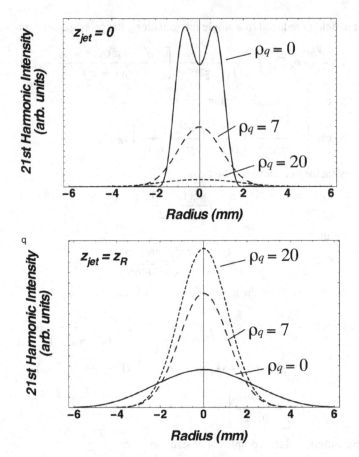

Figure 6.23 Calculation of the radial profile of the 21st harmonic of 800 nm light 0.5 m from a 40 μm focal waist diameter using Eq. (6.130) and Eq. (6.131). Assumed effective nonlinearity is $q' = 5$. The top panel shows harmonic yield with a 1 mm wide gas jet centered at the focal plane and the lower panel shows the same calculations but with the jet centered downstream of the focus by 1.5 mm (one Rayleigh range). Profiles without intensity-dependent phase are plotted (solid lines), as are profiles with intensity-dependent phase of two values (dashed lines).

Additional insight into the effects of the geometric phase mismatch can be gleaned from Eq. (6.130) if we consider this equation under the experimentally common situation of a weak focus, namely when $z_R \gg l$, when the Rayleigh range is much greater than the length of the generating medium. In that case the integral over z takes place within limits such that $|\tilde{z}| \ll 1$. We also have in the far field that $\tilde{z}' \gg \tilde{z}$. We can then make some simplifying approximations. First, we can say that (6.131) simplifies under these limits to

$$\xi(\tilde{z}, \tilde{z}') \cong 2iz_R\tilde{z}' + k_q w_o^2/q'. \qquad (6.132)$$

6.3 Propagation and Phase-Matching in High-Harmonic Generation

We can then see under the weak focus limit that the harmonic field takes on the spatial profile of a Gaussian beam:

$$E'_q(\mathbf{x}') \sim \exp\left[-\frac{k_q r'^2}{\xi(\bar{z}, \bar{z}')^2}\right]$$

$$\sim \exp\left[\frac{k_q r'^2}{4z'^2 + (k_q w_0^2/q')^2}\left(k_q w_0^2/q' - 2iz'\right)\right], \quad (6.133)$$

or considering only the real part to pull out the far field spatial profile we see that

$$E'_q(\mathbf{x}') \sim \exp\left[-\frac{k_q r'^2}{4z'^2 + (qz_R/q')^2}\frac{qz_R}{q'}\right]$$

$$\sim \exp\left[-\frac{r'^2}{w_0'^2\left(1 + z'^2/z_R'^2\right)}\right], \quad (6.134)$$

where we have introduced new quantities for the focal waist radius and Rayleigh range for the harmonic field given by

$$w_0' = w_0/\sqrt{q'}$$

$$z_R' = \frac{q}{q'} z_R. \quad (6.135)$$

Under weak focusing, the high-harmonic field propagates to the far field as a single-mode Gaussian beam with effective spot size reduced by a factor of $(q')^{-1/2}$ and Rayleigh range increased by a factor of q/q' (which might be a factor of roughly three to ten). The physical origin of this scaling is simple; in the weak focus limit, the harmonic is produced from a nearly plane-wave field at the laser focus, and the produced harmonic beam is this intensity profile taken to the power of q'. This exponentiation of the laser Gaussian profile results in an intensity profile reduction by $(q')^{-1/2}$ and, combined with the shorter wavelength of the harmonic, a commensurate increase in Rayleigh range.

Let us now calculate the intensity of the qth harmonic under the specific case of a harmonic-generating medium that is of uniform density over length l and is centered on the laser focus. If we consider the harmonic intensity, which under the definition of the fields used in this section is

$$I_q(\mathbf{x}') = \frac{c}{4\pi}\left|E'_q(\mathbf{x}')\right|^2, \quad (6.136)$$

we have that

$$|\xi|^2 = 4\left(z_R^2 \bar{z}'^2 + z_R'^2\right), \quad (6.137)$$

so we can write for the harmonic far-field intensity

$$I_q(\mathbf{x}') \cong \frac{\pi}{4} \frac{q^2\omega^4}{c^3} \alpha_q^2 E_{00}^{2q'} n_a^2 \frac{w_0^4}{(1+z'^2/z_R'^2)} \exp\left[-\frac{2r'^2}{w_0'^2(1+z'^2/z_R'^2)}\right]$$
$$\times \left|\int_{-l/2z_R}^{l/2z_R} \exp\left[-i\Delta k_q z_R \tilde{z} - iq\tilde{z} + i\frac{q}{q'}\tilde{z}\right] d\tilde{z}\right|^2 \quad (6.138)$$

This yields for the final far-field intensity of the qth harmonic in the weak focus limit

$$I_q(\mathbf{x}') \cong \frac{\pi}{4}(4\pi)^{q'} \frac{q^2\omega^4}{c^{q'+3}z_R^2} w_0^4 \alpha_q^2 n_a^2 l^2 I_0^{q'} \operatorname{sinc}^2\left[\left(\Delta k_q + \frac{q}{z_R} - \frac{q}{q'z_R}\right)\frac{l}{2}\right]$$
$$\times \frac{1}{(1+z'^2/z_R'^2)} \exp\left[-\frac{2r'^2}{w_0'^2(1+z'^2/z_R'^2)}\right]. \quad (6.139)$$

It is apparent that there is significant high-harmonic yield as long as the sinc^2 function has an argument that is much smaller than π, the point where the yield will drop to zero because of phase mismatch. It makes sense now to define the *total* phase mismatch,

$$\Delta k_q^{(tot)} = \Delta k_q + \frac{q}{z_R} - \frac{q}{q'z_R}, \quad (6.140)$$

which is a sum of the intrinsic phase mismatch and geometric terms. It is interesting to observe that as long at $\Delta k_q^{(tot)} l/2 \ll \pi$, the total harmonic yield scales as the square of the atomic density and the square of the harmonic generating medium length. This is a direct consequence of the coherent nature of harmonic growth as the laser propagates through the medium. The harmonic yield, of course, also scales as the peak laser intensity taken to the power of the effective nonlinearity.

This result also informs us about the length of generating medium that can be employed for HHG. Following the usual nonlinear optic practice, we can define a coherence length as the propagation length over which coherent growth of the harmonic field stops and starts to fall back to zero by reconversion from the phase mismatch. Since the harmonic field scales as $\sin^2 \Delta k l/2$ in (6.139), we can see that this coherence length is

$$l_{coh} = \left|\frac{\pi}{\Delta k_q^{(tot)}}\right| = \left|\frac{\pi z_R}{\Delta k_q z_R + q - q/q'}\right|. \quad (6.141)$$

If the intrinsic phase mismatch is zero (i.e. the refractive indices for the laser and harmonic field are essentially the same) then this coherence length is determined by the geometric limitations of the Gouy phase shaft,

$$l_{coh} \cong \frac{\pi z_R}{q - q/q'} \approx \pi z_R/q. \quad (6.142)$$

So we see that if operating at focus where the laser intensity is highest, one must limit the harmonic-generating medium length to something less than $2\pi/q$ times the focused laser Rayleigh range for efficient harmonic generation on the qth harmonic.

6.3 Propagation and Phase-Matching in High-Harmonic Generation

We can use Eq. (6.139) to estimate the total integrated harmonic yield if we have some knowledge of the qth-harmonic pulse duration $\Delta \tau_q$. The total number of photons produced on the qth harmonic is

$$N_q = \frac{\Delta \tau_q}{\hbar \omega_q} \int_0^\infty I_q(r', z' = \infty) 2\pi r' dr', \tag{6.143}$$

which yields

$$N_q = \frac{\pi^2}{8} (4\pi)^{q'} \frac{q}{q'} \frac{\omega^3 \Delta \tau}{c^{q'+3} z_R^2} w_0^6 \alpha_q^2 n_a^2 l^2 I_0^{q'} \,\text{sinc}^2[\Delta k l/2]. \tag{6.144}$$

In practice, this is not an extremely accurate formula but it does serve as a simple, convenient way to estimate the harmonic yield to an order of magnitude. As an example, consider an intense 527 nm pulse focused to a $1/e^2$ diameter of 140 μm into a He gas with a 1 mm length and density of 5×10^{18} cm^{-3}. In this case z_R is 3.3 cm, so we are in the weak focus regime. If we want to compare with experiment on the yield, say, of the 21st harmonic at 250Å, we need to make an estimate for the production intensity and emitted pulse duration. A reasonable assumption is that the production intensity is the intensity in the pulse near the ionization saturation intensity, which for He is about 3×10^{15} W/cm^2. For production by a longish pulse (say 100–500 fs) a production window of 30 fs is reasonable. Finally, we need some knowledge of the single-atom dipole strength. From numerical solution of the nonlinear Schrödinger equation for He (Ditmire et al. 1996) we have that the dipole strength is about 1.5×10^{-23} esu-cm with an effective nonlinear scaling around $q' = 3$. This means that $\alpha_{21} \approx 1.9 \times 10^{-41}$, which when inserted into (6.144) predicts a 21st harmonic yield around 10^9 photons. This would be about 7 nJ of harmonic light, which is pretty close to experimental observations under these conditions on this harmonic of 10 nJ (Ditmire et al. 1996).

6.3.2 Phase-Matching in Strong Fields and the Intensity-Dependent Phase

Now consider phase-matching of the HHG in fields at intensities above 10^{14} W/cm^2. Not only must we consider the geometric phase on the driving laser beam, but we must also consider any phase imparted to the harmonic field by the single-atom response. This dipole phase is imparted to the harmonic field through the polarization of the medium, recalling Eqs. (6.12) and (6.14).

We have already seen in our treatment of the single-atom harmonic strength by the SFA that the dipole element of a harmonic has some phase delay associated with it. Recall, for example, Eq. (6.104). In the quasi-classical picture, this phase is acquired by an electron as it propagates in the continuum in the time interval between its birth by tunnel ionization and its production of a harmonic photon by recombination. This single-atom-induced phase shift can have a significant impact on the macroscopic phase-matching of the harmonic field in the generating medium (Salieres et al. 1995).

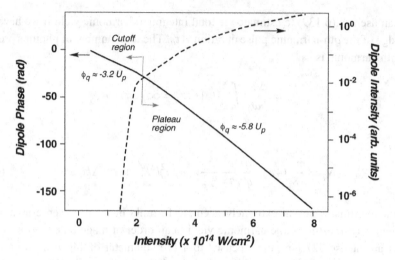

Figure 6.24 SFA calculation (using the saddle-point approximation for k_{ex}) conducted by Lewenstein et al. of the dipole strength and phase for the 45th harmonic generated in Ne atoms by long trajectories as a function of intensity by 800 nm radiation. Figure was generated based on plots in figure 1a of (Lewenstein et al. 1995).

The magnitude of the phase can be derived easily from a numerical analysis of the SFA result (Lewenstein et al. 1995). Figure 6.24 shows just such a calculation made by Lewenstein et al. for the 45th harmonic produced by long trajectories in Ne. This shows that the phase has a different scaling with intensity depending on whether the harmonic is in the cutoff region or the plateau region of the harmonic spectrum. We can quantify the origin of this phase scaling by examining the analytic results of the SFA under the saddle point approximation. Our treatment of the single-atom qth-harmonic dipole strength predicted a dipole phase

$$\varphi_q(t, \tau_r, E_0) = S_{k_e}(t, \tau_r, E_0)/\hbar$$

$$= \frac{1}{\hbar} \int_{t-\tau_r}^{t} \left[\frac{1}{2m} \left(\hbar \mathbf{k}_e + \frac{e}{c} \mathbf{A}_0 \sin \omega t'' \right)^2 + I_p \right] dt''. \quad (6.145)$$

A simple equation for this phase can be found recalling the result for the quasi-action under the saddle-point approximation for electron canonical momentum $\mathbf{k}_e$. Under this approximation, the single-atom phase is given by Eq. (6.111).

Consider the relative sizes of the terms in this equation. We know from subsequent saddle-point approximation to τ_r that we can associate this quantity with the propagation time in the continuum of a tunnel-ionized electron which is about three-fourths of an optical cycle. So, $\omega \tau_r \approx \pi - 1.5 \pi \approx 3 - 5$ (depending on whether we consider short or long return trajectories). The first term in Eq. (6.111) is smaller than the $U_p \tau_r$ term by a factor $\sim 2/(\omega \tau_r)^2$ while the second term is smaller by a factor $1/(2\omega \tau_r)$. So we can say that both terms will tend to be smaller than the $U_p \tau_r$ term by about an order of magnitude. It is also reasonable to ignore the $I_p \tau_r$ term as approximately constant. This leads to an interesting conclusion. The atomic phase then is approximately

6.3 Propagation and Phase-Matching in High-Harmonic Generation

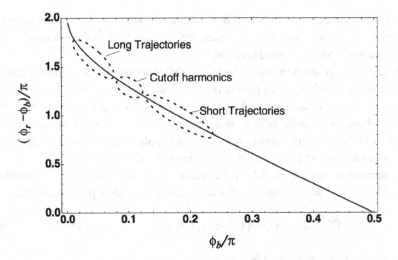

Figure 6.25 The calculated phase period a recolliding electron spends in the continuum as a function of birth phase.

$$\varphi_q \cong U_p \tau_r / \hbar. \tag{6.146}$$

It is more or less linearly proportional to the drive laser intensity.

Because of the variation of intensity in the focal region of the driving laser, this intensity-dependent phase can have a significant impact on the phase-matching of a given harmonic. If we rewrite the phase in the following manner,

$$\varphi_q \cong (\varphi_r - \varphi_b) \frac{U_p}{\hbar \omega}, \tag{6.147}$$

it is obvious that the phase takes on values up to maybe 200 rad for higher intensities (say $\sim 10^{15}$ W/cm^2).

It is also clear from Eq. (6.147) that, for a given harmonic, this phase scales with intensity differently for short and long trajectories. Figure 6.25 summarizes our previous calculations on the return and birth phase difference as a function of birth phase from the quasi-classical three-step model. Recall that the long trajectories for a given harmonic originate at birth phases below 0.1π and the short trajectories arise at birth phases longer than the cutoff phase. From this figure we can conclude that there are three regimes of this atomic phase scaling with intensity:

$$\text{Long Trajectories:} \quad \varphi_q(I) \cong 1.7\pi \frac{U_p}{\hbar \omega} \tag{6.148a}$$

$$\text{Near Cutoff:} \quad \varphi_q(I) \cong 1.3\pi \frac{U_p}{\hbar \omega} \tag{6.148b}$$

$$\text{Short Trajectories:} \quad \varphi_q(I) \cong \pi \frac{U_p}{\hbar \omega} \tag{6.148c}$$

Consider the relative sizes of the terms in this equation. We know from subsequent saddle-point approximation to τ_r that we can associate this quantity with the propagation time in the continuum of a tunnel-ionized electron.

We are now in a position to assess the effects that this intensity-dependent phase has on the spatial phase-matching of a given harmonic. The essence of this effect rests with the fact that the intensity in the focal region varies with position and hence imparts a spatially dependent phase on the coherently growing harmonic field. A useful way to see some of the impact of this atomic phase is to return to our simple integral model for the harmonic field with an addition made to account for this intensity-dependent phase.

The polarization with our model for the atomic response with an effective nonlinearity and an added term to account for this atomic intensity-dependent phase looks like

$$\tilde{\mathcal{P}}_q \cong \alpha_q n_a(\mathbf{x}) |E_0(\mathbf{x})|^{q'} \exp\left[iq \text{Arg}[E_0(\mathbf{x})] - i\rho_q G(\mathbf{x})\right]. \tag{6.149}$$

Here the atomic phase coefficient is

$$\rho_q = \omega \tau_r \frac{U_p^{(0)}}{\hbar \omega}, \tag{6.150}$$

where $U_p^{(0)}$ is the ponderomotive energy at the peak of the laser focus. This coefficient will range from, say, 20 for short trajectories at modest intensity to >200 for long-trajectory harmonics driven at intensity approaching 10^{15} W/cm^2. Again assuming a Gaussian focal profile, we have that the normalized focal intensity profile is a simple focused Gaussian:

$$G(\mathbf{x}) = \frac{1}{1+\tilde{z}^2} \exp\left[-\frac{2(x^2+y^2)}{w_0^2(1+\tilde{z}^2)}\right]. \tag{6.151}$$

Because of the strongly nonlinear response of the single-atom dipole to increasing intensity, it is a good approximation to say that the vast majority of harmonic generation occurs at radii much less than the $1/e^2$ focal radius. So, we Taylor-expand this Gaussian profile in the phase exponent

$$G(\mathbf{x}) \approx \frac{1}{1+\tilde{z}^2}\left(1 - \frac{2x^2}{w_0^2(1+\tilde{z}^2)} - \frac{2y^2}{w_0^2(1+\tilde{z}^2)}\right). \tag{6.152}$$

Following the same steps that led up to Eq. (6.126), we can rewrite the integral over the x coordinate with the intensity-dependent phase as

$$I_x = \int_{-\infty}^{\infty} \exp\left[i\frac{k_q}{2z_R}\frac{(x'-x)^2}{\tilde{z}'-\tilde{z}} - \frac{q'x^2}{w_0^2(1+\tilde{z}^2)} + iq\frac{\tilde{z}x^2}{w_0^2(1+\tilde{z}^2)} + i2\frac{\rho_q x^2}{w_0^2(1+\tilde{z}^2)}\right]dx \tag{6.153}$$

6.3 Propagation and Phase-Matching in High-Harmonic Generation

Some additional manipulation leads to our final result for the harmonic field

$$E'_q(\mathbf{x}') = i2\pi \frac{q^2\omega^2}{c^2} z_R \alpha_q w_0^2 E_{00}^q \int_{-\infty}^{\infty} \frac{n_a(\tilde{z})}{(1+\tilde{z}^2)^{q'/2-1}}$$
$$\times \frac{1}{\left[q'^2 + \left(q\tilde{z} + 2\rho_q(1+\tilde{z}^2)^{-1}\right)^2\right]^{1/2}} \frac{1}{\xi_s(\tilde{z},\tilde{z}')}$$
$$\times \exp\left[-i\Delta k_q z_R \tilde{z} - iq \tan^{-1}\tilde{z} + i\tan\left[\left(q\tilde{z} + 2\rho_q(1+\tilde{z}^2)^{-1}\right)/q'\right]\right]$$
$$\times \exp\left[-\frac{k_q r'^2}{\xi_s(\tilde{z},\tilde{z}')}\right] \exp\left[-i\rho_q(1+\tilde{z}^2)^{-1}\right] d\tilde{z}$$
(6.154)

Here the strong field version of the complex spot size function is

$$\xi_s(\tilde{z},\tilde{z}') = i2z_R(\tilde{z}' - \tilde{z}) + \frac{k_q w_0^2(1+\tilde{z}^2)}{q' - iq\tilde{z} + i2\rho_q(1+\tilde{z}^2)^{-1}}. \quad (6.155)$$

As before, this equation is very easily integrated numerically to derive a harmonic yield and spatial profile. In Figure 6.23 results from such numerical calculations for two values of the atomic phase coefficient characteristic of short trajectory harmonics at moderate peak intensity ($\rho_q = 7$ and 20). When the gas medium is placed at the focus we see that the intensity-dependent phase serves to lower the net harmonic yield. However, when the gas medium is placed after the focus, we see that in fact the harmonic yield is better than it was calculated without any atomic phase contribution ($\rho_q = 0$).

This can be understood in part if we plot the Gouy phase added to the atomic phase on axis as a function of propagation position (Figure 6.26). This plot shows that there is a region after the focal plane in which there is a relatively flat region of phase. This corresponds to a region where the harmonic field grows coherently in phase over the propagation distance. The presence of this flat phase region explains the increase in harmonic yield with increasing atomic phase coefficient when the gas was placed in this postfocus region. In this region the decreasing intensity leads to a phase that compensates the Gouy phase variation. In the region before the laser focus the opposite happens: the increasing intensity produces a phase that adds to the Gouy phase slip and further deteriorates phase-matching.

We can quantify this effect by examining the phase terms in Eq. (6.154). A simple approximation to the z variation of the atomic phase term near the focal plane is

$$\frac{\rho_q}{1+\tilde{z}^2} \approx \begin{cases} -\rho_q \tilde{z}/2 & \tilde{z} > 0 \\ \rho_q \tilde{z}/2 & \tilde{z} < 0 \end{cases}. \quad (6.156)$$

If the intrinsic phase mismatch is zero, the total phase mismatch in the weak focus limit with approximation (6.156) becomes

$$\Delta k_q^{(tot)} \cong \begin{cases} \frac{q}{z_R} + \frac{\rho_q}{2z_R} & \text{Before focus} \\ \frac{q}{z_R} - \frac{\rho_q}{2z_R} & \text{After focus} \end{cases}. \quad (6.157)$$

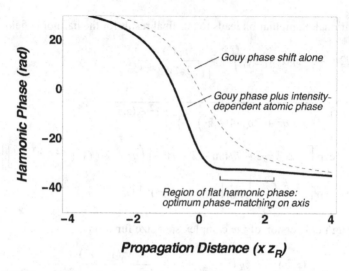

Figure 6.26 On-axis phase imparted to the 25th harmonic as a function of propagation distance through the laser focus. Dashed line is phase imparted only from the Gouy phase alone and the solid line is if the atomic phase proportional to intensity is included (in this case with $\rho_q = 20$).

The compensation to the geometric phase mismatch that the falling intensity just past the focus imparts through the intensity-dependent atomic phase is clear. Optimum phase-matching, then, will occur in the region just past the focus plane when $\rho_a \approx 2q$.

The effect that the intensity-dependent phase has on the far-field spatial profile of a harmonic is more complex. To get a handle on this physics we will use the excellent phase-matching visualization method first published by Balcou et al. (1997). First we must note that the direction of harmonic radiation production at a given point in the laser field is determined by the gradient of the total phase of the polarization source term. This phase is a sum of q times the laser phase and the single-atom phase. We denote this phase gradient as

$$\mathbf{k}'_q = \nabla \left(qkz + q\varphi_{Gaussian} + \varphi_{atomic} \right). \tag{6.158}$$

It is clear from this equation that the direction of harmonic emission, along $\mathbf{k}'_q$, will not necessarily be along the direction of laser propagation because of the introduction of the atomic phase in the nonlinear polarization.

The phase terms that enter into calculation of $\mathbf{k}'_q$ are

$$\varphi_{Gaussian} = \frac{r^2}{w_0^2 \left(1 + z^2/z_R^2\right)} \frac{z}{z_R} - \tan^{-1}\left[\frac{z}{z_R}\right] \tag{6.159}$$

and

$$\varphi_{atomic} \simeq \tau_r \omega \frac{U_p}{\hbar \omega}. \tag{6.160}$$

6.3 Propagation and Phase-Matching in High-Harmonic Generation

This suggests that the harmonic emission vector, $\mathbf{k}'_q$, can be thought of as composed of two components, a laser-directed component from the geometry of the focused laser beam, $\mathbf{K}_q$, and a component from the intensity-dependent phase of the single-atom response, $\mathbf{K}_p$:

$$\mathbf{k}'_q \cong \underbrace{qk\hat{\mathbf{z}} + q\frac{2r}{w_0^2}\frac{z}{z_R}\hat{\mathbf{r}} - \frac{q}{z_R}\frac{1}{1+z^2/z_R^2}\hat{\mathbf{z}}}_{\mathbf{K}_q} - \underbrace{\left(\frac{\tau_r}{\hbar}\right)\nabla U_p}_{\mathbf{K}_p}. \qquad (6.161)$$

Figure 6.27 shows vector plots of these two contributions to $\mathbf{k}'_q$. The laser field is propagates from left to right in these plots. The focal point and region of highest intensity is in the center of each plot. The laser component, $\mathbf{K}_q$, plotted in the top graph, is simply that of a focused laser beam. The k-vector converges and diverges after the focus. The atomic phase

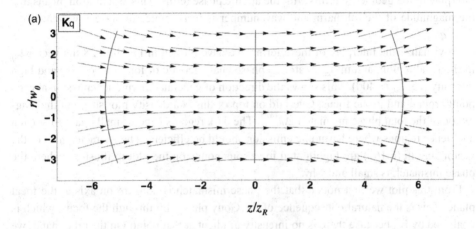

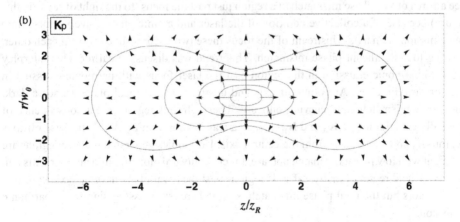

Figure 6.27 Vector plots of the two contributions to the harmonic wavenumber vector $\mathbf{k}_q$: the geometric phase from the focused laser, $\mathbf{K}_q$ (a) and the intensity-dependent atomic phase $\mathbf{K}_p$ (b).

component $\mathbf{K}_p$, on the other hand, reflects the shape of the intensity contours of the focused laser pulse near the focal plane (some of which are shown in the bottom plot). The intensity gradient is toward the center of the focal profile, leading to a set of **k**-vectors which point outward in a star-like pattern. The vector sum of these two contributions determines the complex local harmonic emission direction and the magnitude of the phase mismatch. (Also note that the size of the contribution from the $\mathbf{K}_p$ term will depend on whether the harmonics are being generated by long or short trajectories.)

Now the efficiency of harmonic production at any point in the focal profile will depend closely on the size of the total phase mismatch. This mismatch can be written

$$\Delta k_q^{(tot)} = k_q - \left|\mathbf{k}_q'\right|. \tag{6.162}$$

For the time being, we assume zero intrinsic phase mismatch so that we can assess the interplay of the geometric terms with the atomic phase terms. This assumption means that the magnitude of the qth-harmonic wavenumber is just q times the laser's wavenumber, $k_q = qk$.

To visualize the interplay of these terms, we have plotted in Figure 6.28 the vector $\mathbf{k}_q'$ in the focal region, assuming an atomic phase characteristic of long trajectories and high intensity (i.e. $\rho_a \approx 400$). This shows the direction of generated harmonic emission at each point around and in the focus. Overlaid on top of this is a density plot showing the magnitude of the total phase mismatch $\Delta k_q^{(tot)}$. The dark regions denote small phase mismatch and hence regions where harmonic emission should be efficient. The characteristics of the vector sum in (6.161) are highlighted in various points on the plot to illustrate where the phase mismatch is small and why.

From this plot we first notice that the phase mismatch is nonzero on axis at the focal plane. This is the natural consequence of the Gouy phase slip through the focus, which is unaltered by $\mathbf{K}_p$ because there is no intensity gradient at that point. On the other hand, we see an area of low phase mismatch in a region just past the focus (to the right of focus in this figure) because of a collective addition of the laser and atomic phase wave vectors. Here the emission is on axis. Upstream of the focus these two vectors subtract from each other, leading to a significant phase mismatch. This trend was discussed earlier. This interplay leads to harmonic emission in the region past the laser focus with harmonic emission in the forward direction. At points off axis and upstream of the focal plane we see a dark, low phase mismatch band curving outward. These off-axis regions are a consequence of noncollinear addition of $\mathbf{K}_q$ and $\mathbf{K}_p$ in such a way as to minimize $\Delta k_q^{(tot)}$. In these off-axis regions $\mathbf{k}_q'$ points out away from the central axis. Harmonics generated in these regions are efficient with low phase mismatch and are directed outward along $\mathbf{k}_q'$ into annular emission in the far field region. Finally, off-axis regions past the focal plane have off-axis emission wavevectors but the total phase mismatch is large here, suppressing significant harmonic emission.

We can see now what might be expected in harmonic emission as we move the generating medium through the focus. With the medium at the focus, phase mismatch serves to lower the harmonic yield. Moving the generating medium downstream serves to enhances the

6.3 Propagation and Phase-Matching in High-Harmonic Generation 369

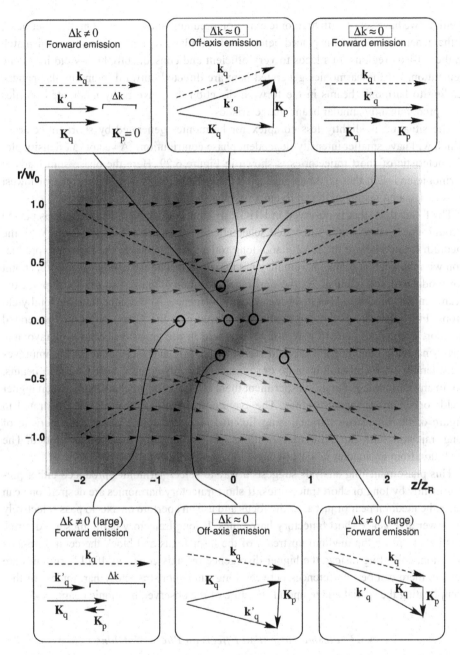

Figure 6.28 Vector plots of the harmonic wavenumber vector $\mathbf{k}_q$ direction in the focal region overlaid on a density plot of phase mismatch for a harmonic produced by *long* trajectories. White denotes large phase mismatch and black denotes small Δk and hence regions of efficient harmonic production.

yield (as we have seen) with harmonic emission predominantly on axis. On the other hand, if the generating medium is placed before the focal plane regions of low phase mismatch in the off-axis regions, this leads to very efficient and consequently high-yield harmonic generation, but the harmonics generated here are directed outward in an annular pattern leading to harmonic beams in the forward direction that have badly modulated profiles with predominantly annular beam characteristics.

The situation is slightly less complex for harmonics generated by short trajectories, which will have smaller intensity-dependent phase contributions. A vector/Δk density plot characteristic of short trajectories is shown in Figure 6.29. Here the phase sum leads to harmonic emission that is almost all in the forward direction with the regions of lowest phase mismatch resting in the region just past the focal plane.

The trend, then, that is observed in HHG emission as the generating medium is passed through the focal region is illustrated schematically in Figure 6.30a at the top. With the medium located before the focal plane, long trajectories dominate the harmonic production with the total integrated harmonic yield maximized. The far-field profiles at this point are modulated and annular. As the generating medium is moved to the focal point the beam remains modulated but does decrease in divergence. At the same time the total yield drops. Finally, as the generating medium is moved past the focus, harmonics generated by short trajectories start to play a dominant role in the emission because of favorable phase-matching conditions. The yield increases slightly and the spatial profile improves considerably with emission in more or less nicely low divergence near Gaussian beams. So, in any harmonic generation experiment there is a trade-off in choosing between higher yields or smoother spatial profiles. The evolution of the spatial profile is illustrated in Figure 6.30b where calculations using Eq. (6.154) with atomic phase characteristic of long trajectories for three different locations of the generating medium are plotted. The evolution from an annular to a near-Gaussian beam is evident.

This phase-matching analysis suggests a way to select harmonics produced either predominantly by long or short trajectories. If short trajectory harmonics are desired, one can place the medium near or just after the focus and use an aperture on axis to pass selectively the lower divergence short trajectory harmonics. If long trajectory harmonics are desired, it is best to place the medium upstream of the laser focus and block the central on-axis emission, collecting instead the higher-divergence off-axis emission. If all harmonics are passed, there can be interferences between long- and short-trajectory harmonics, further complicating the spatial and temporal structure of the observed harmonic beam.

6.3.3 Plasma Formation Effects on Phase-Matching

Up until now we have largely ignored the intrinsic phase mismatch under the assumption that it was near zero. This was the same as assuming that the refractive indices of the laser and the XUV harmonic pulse were the same. Of course, the fact is that the laser must propagate through a medium to generate the harmonics, so it is inevitable that the intrinsic phase mismatch is not exactly zero. In fact, the largest contributor to dispersion in the refractive index and the generation of an intrinsic phase mismatch is the formation of a

6.3 Propagation and Phase-Matching in High-Harmonic Generation

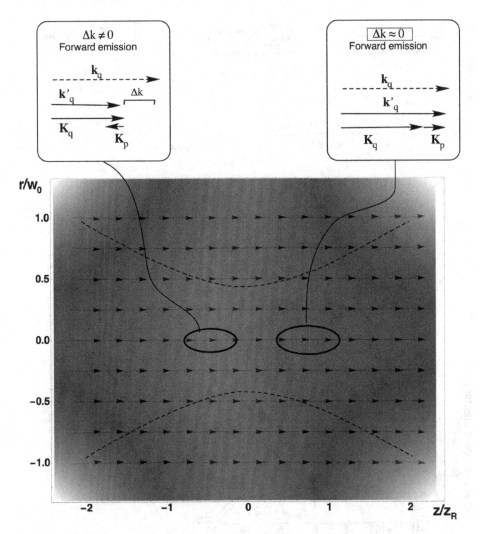

Figure 6.29 Vector plots harmonic wavenumber vector $\mathbf{k}_q$ direction in the focal region overlaid on a density plot of phase mismatch for a harmonic produced by *short* trajectories. White denotes large phase mismatch and black denotes small Δk.

plasma in the generating medium. Free electron production will become important in most gases when the peak laser intensity exceeds $\sim 10^{14}$ W/cm^2 where field ionization starts to become significant.

To assess the magnitude of the plasma-induced phase mismatch, recall the Drude model for the refractive index of a free electron plasma

$$\eta = \sqrt{1 - \omega_{pe}^2/\omega^2}, \tag{6.163}$$

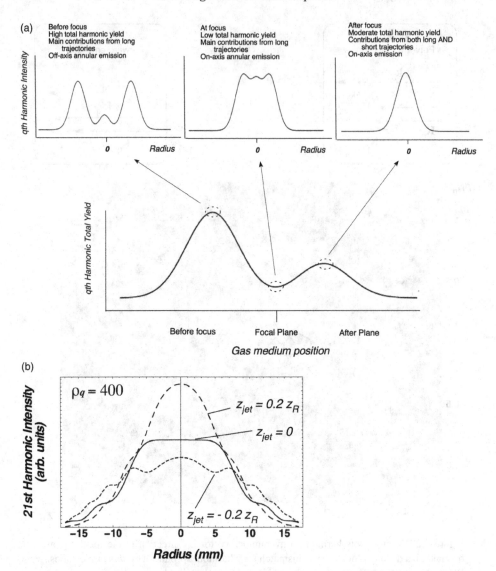

Figure 6.30 (a) Schematic plot of total harmonic yield as a function of gas medium position with respect to the laser focus with corresponding spatial profiles of the harmonics in the far field. (b) Calculation using Eq. (6.154) of 21st harmonic far-field profiles for three different gas jet positions under the same conditions as Figure 6.23 but with a larger atomic phase characteristic of a long trajectory harmonic showing the much wider spatial profile from the long trajectory.

where the electron plasma frequency is given by the usual formula:

$$\omega_{pe} = \sqrt{\frac{4\pi n_e e^2}{m}}. \quad (6.164)$$

6.3 Propagation and Phase-Matching in High-Harmonic Generation

We use this to calculate the intrinsic phase mismatch

$$\Delta k_q = k_q - qk, \tag{6.165}$$

$$\Delta k_{plas} = \frac{2\pi q}{\lambda}(\sqrt{1 - \omega_{pe}^2/\omega_q^2} - \sqrt{1 - \omega_{pe}^2/\omega^2}). \tag{6.166}$$

Since HHG almost always takes place in low-density gases, generally $\omega, \omega_q \gg \omega_p$, so we can Taylor-expand this expression around ω_p^2/ω^2 to yield for the plasma-induced phase mismatch

$$\Delta k_{plas} \cong \frac{\pi q}{\lambda} \frac{\omega_{pe}^2}{\omega^2}\left(1 - \frac{1}{q^2}\right), \tag{6.167}$$

which for large q reduces to

$$\Delta k_{plas} \cong \frac{\omega_{pe}^2}{4\pi c^2} q\lambda = \frac{e^2}{mc^2} n_e q\lambda. \tag{6.168}$$

So, the plasma phase mismatch is worse for longer laser drive wavelengths and high harmonic orders.

It turns out that typically only a modest level of ionization is sufficient to suppress completely the harmonic generation. For example, if we consider the coherence length from the plasma mismatch, $l_{coh} = \pi/\Delta k_{plas}$, we can ask what electron density is sufficient to reduce this length to only 1 mm, a typical length for the generating medium in many HHG experiments. If, for example, $q = 21$ and the drive wavelength is 800 nm, an electron density of $\sim 7 \times 10^{16}$ cm^3 yields a 1 mm coherence length. Such a plasma density may be only a few percent of the density of the generating medium. So some small amount of field ionization during the HHG process in the pulse serves to suppress the growth of a harmonic field.

The practical situation then for the temporal dynamics of HHG during an ultrafast pulse propagation is depicted in Figure 6.31. As the ultrafast pulse ramps up in intensity, high harmonic emission begins. Because this emission is accompanied by ionization, the plasma-induced phase mismatch quickly becomes dominant and suppresses further HHG even if the laser pulse continues to ramp up in intensity. This ionization-induced phase mismatch essentially acts as a temporal shutter to the HHG emission, causing the HHG pulse to come out in a window that is generally much shorter than the laser pulse itself. As a result, it is usually a good approximation to say that most HHG in an ionizing pulse gets generated at an intensity just below the ionization saturation intensity even in laser pulses that have a significantly higher peak intensity.

6.3.4 Neutral Gas Refractive Index and Waveguide Effects on Phase-Matching

Though not as dispersive as a free electron plasma, the neutral gas of the generating medium can also have refractive index dispersion, which affects high-harmonic phase-matching. Because the gas has virtually no effect on the refractive index of the harmonic XUV light, the intrinsic phase mismatch in the gas is

$$\Delta k_{gas} = \frac{2\pi q}{\lambda}\left(1 - \eta_{gas}(\lambda)\right). \tag{6.169}$$

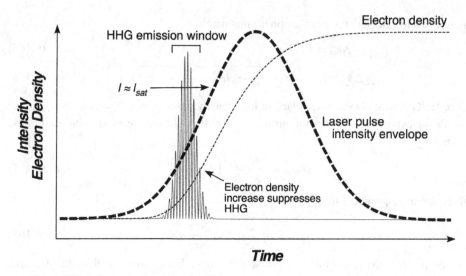

Figure 6.31 Schematic plot of the time history of harmonic emission during a laser pulse which is creating increasing electron density by field ionization.

Because $\eta_{gas} > 1$, this phase mismatch is of the opposite sign to the plasma-induced mismatch. The quantity $(1 - \eta_{gas})$ tends to scale linearly with gas density. As an example, consider Ar at a density of 10^{18} cm^{-3}. At 800 nm, $(n_{Ar} - 1) \approx 1.4 \times 10^{-5}$, which leads to a coherence length for the qth harmonic $l_{coh} \approx 2.9q$ cm. So, for example, for the 21st harmonic $l_{coh} \approx 1.5$ mm, which is about the length often used in experiments.

It is also interesting to note that while a gas slows the phase velocity of the laser, it is possible to speed up the drive field's phase velocity by propagating it down a single-mode hollow-core fiber, which can be back-filled with a variable-density gas to generate the harmonic (Durfee et al. 1999; Paul et al. 2006). In such a fiber of radius a, the wavenumber of the drive laser accounting for the presence of a gas and possible plasma is (Winterfeldt et al. 2008)

$$k = \frac{2\pi}{\lambda}\eta_{gas}(\lambda) - \frac{n_e e^2 \lambda}{mc^2} - \frac{u_{ab}\lambda}{4\pi a^2}. \quad (6.170)$$

The third term in this equation arises from the propagation down a fiber, and the factor u_{ab} is a constant depending on the propagation mode. For propagation by a zeroth-order Bessel mode, where $E(r) \sim J_0(r)$, $u_{11} = 2.4$ (Lin et al. 2018, p. 222). Because there is no Gouy phase shaft for a mode guided by the fiber, the total phase mismatch for the Bessel mode is

$$\Delta k_q^{(tot)} = \frac{2\pi q}{\lambda}\left(1 - \eta_{gas}(\lambda)\right) + \frac{n_e e^2}{mc^2}q\lambda + \frac{qu_{11}\lambda}{4\pi a^2}. \quad (6.171)$$

Because the first term has the opposite sign of the third term and varies roughly linearly with gas density, it is, in principle, possible to tune the total phase mismatch to zero with judicious choice of gas density. Even though fully phase matched, the harmonic yield is still ultimately clamped by absorption, as we shall see in the next section, so the yield

6.3.5 Atomic Absorption and Harmonic Yield Limitations

Our discussions of phase-matching would suggest that as long as the laser is weakly focused through the gas medium, the harmonic yield could be increased indefinitely as the medium length is increased. In practice, this is not the case. As we alluded to in Eq. (6.124), propagation of the harmonic field through the generating medium will often be accompanied by absorption of the VUV or XUV harmonic light. For photon energies above the ionization threshold of the atoms in the generating gas (usually true for harmonic orders above the 9th to 13th in most gases driven by near-IR pulses) photoionization is the predominant mechanism for absorption of the harmonic light. This absorption will ultimately limit the yield as the generating medium length is increased even if the coherence length is longer than the medium length (Constant et al. 1999).

We can think of this effect by defining an absorption length

$$l_{abs} = \kappa_{abs}^{-1}, \tag{6.172}$$

where, again, if photoionization is the predominant harmonic absorption mechanism, the absorption coefficient is in terms of the photoionization cross section and neutral gas density

$$\kappa_{abs} = n_a \sigma_{pi}. \tag{6.173}$$

For example, in Ar gas, $\sigma_{pi} \approx 3 \times 10^{-17}$ cm^2 for photons with energy in the range from 15 to 25 eV. This cross section drops to about 7×10^{-19} cm^2 for photons with energy up to 50 eV. So, the absorption length in an Ar medium at density around 10^{18} cm^{-3} is about 0.3 mm for lower-energy harmonic photons and up to 15 mm for higher-energy photons.

To quantify the effects of absorption on high harmonic yield we can fall back on our simple propagation model in the weak focus limit, Eq. (6.138). The output intensity is

$$I_q(\mathbf{x}') = I_q^{(0)} \left| \int \frac{1}{z_R} \exp[-i\Delta kz] dz \right|^2, \tag{6.174}$$

where we include the effects of absorption with the substitution

$$\Delta kz \rightarrow \int_{-\infty}^{z} \Delta k_q^{(tot)}(z'') \, dz'' - \frac{i}{2} \int_{z}^{\infty} \kappa_{abs}(z'') \, dz''. \tag{6.175}$$

Here the total phase mismatch can be approximated in the weak focus limit

$$\Delta k_q^{(tot)}(z) = \Delta k_q(z) + \frac{q}{z_R} - \frac{q}{q'z_R} \left\{ \pm \frac{\rho_a}{2z_R} \right\}, \tag{6.176}$$

where the final term in brackets can be inserted if the generating medium is located slightly before (+ sign) or after (−sign) the focal plane to approximate the effects of the atomic intensity-dependent phase. Again, if we assume a uniform medium of length l roughly centered on the focal plane, the output intensity is

$$I_q(\mathbf{x}') = I_q^{(0)} \left| \int_{-l/2}^{l/2} \exp\left[-i\Delta k_q^{(tot)} z\right] \exp\left[-\frac{\kappa}{2}\left(\frac{l}{2} - z\right)\right] dz \right|^2. \tag{6.177}$$

Evaluation of the integral yields for the output harmonic intensity

$$I_q = I_q^{(0)} \frac{4}{z_R^2 \left(\kappa^2 + 4\left(\Delta k_q^{(tot)}\right)^2\right)} \left(1 - 2e^{-\kappa l/2} \cos \Delta k_q^{(tot)} l + e^{-\kappa l}\right), \tag{6.178}$$

which in terms of l_{coh} and l_{abs} is

$$I_q = I_q^{(0)} \frac{4 l_{abs}^2 / z_R^2}{1 + 4\pi^2 l_{abs}^2 / l_{coh}^2} \left(1 + e^{-l/l_{abs}} - 2e^{-l/2l_{abs}} \cos\left[\frac{\pi l}{l_{coh}}\right]\right). \tag{6.179}$$

Note that when there is negligible absorption and $l_{abs} \to \infty$, then we retrieve our previous result

$$I_q \sim 1 - \cos\left[\pi l / l_{coh}\right]$$
$$\sim l^2 \operatorname{sinc}^2 \left[\pi l / 2 l_{coh}\right] \tag{6.180}$$

in which the harmonic yield oscillates between conversion and reconversion as the medium length increases. However, if l_{abs} is much shorter than the medium length, $l_{abs} \ll l$, and the coherence length is long, $l_{coh} \gg l$, then the output intensity $I_q \sim l_{abs}^2$ is independent of the actual medium length and varies as the square of the absorption length. Equation (6.179) is plotted as a function of medium length in Figure 6.32 for three values of l_{coh}/l_{abs}. These plots suggest that there is little to be gained by employing a generating medium that is much longer than three or four times the absorption length.

6.4 Attosecond Pulse Generation

Now we turn to a consideration of the temporal structure of the high harmonic emission. Though we have yet to address the question directly, we have implied many times during our invocation of the quasi-classical three-step model that the harmonic emission probably emerges in short bursts; we know that we can think of the high-order harmonics as arising from electron wavepacket recollisions with the nuclear core produced at a well-defined phase of the drive laser. Indeed, such bursts can have time duration of only a few tens of attoseconds (10^{-18} s) (Agostini and DiMauro 2004; Scrinzi et al. 2006; Krausz and Ivanov 2009). Using the models we have developed, we can glean considerable insight into the duration and temporal characteristics of these attosecond harmonic radiation bursts.

6.4 Attosecond Pulse Generation

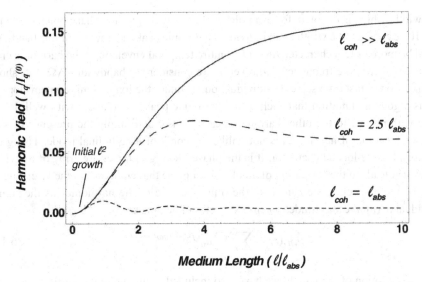

Figure 6.32 Plot of high harmonic yield as a function of generating medium length for three different cases of coherence length.

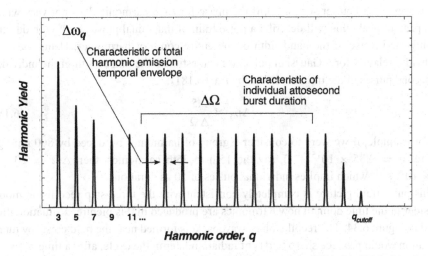

Figure 6.33 Harmonic spectrum and bandwidth. The spectral width of individual peaks reflects the duration of the entire harmonic emission envelope while the integrated coherent bandwidth of an entire group of harmonics is indicative of the duration of individual attosecond harmonic bursts from each half-cycle.

6.4.1 The Bandwidth of the Harmonic Spectrum and Attosecond Pulse Trains

That the high harmonic emission probably has attosecond temporal structure can be seen first from a consideration of the entire harmonic spectrum itself. Looking at Figure 6.33, it would seem that viewed on the whole, the high harmonic spectrum represents an enormous bandwidth, which is denoted, for example, $\Delta\Omega$ for a wide group of harmonics in the plateau. If we consider the width of an individual harmonic peak, the inverse of this bandwidth would be more or less characteristic of the entire temporal envelope of the harmonic emission at that harmonic frequency. However, if we consider the bandwidth $\Delta\Omega$, this should be linked to the fast time scale of individual bursts under the longer emission envelope.

It is a good assumption that each of the harmonic peaks is coherent with well-defined phase with respect to the other harmonic peaks. This would imply the presence of short pulses. The comb of harmonics is not unlike the comb of longitudinal modes lasing in a broadband mode-locked oscillator. It is the phase-locking of the entire array of individual modes that leads to the formation of an ultrashort pulse that circulates in the laser cavity.

To be more specific, we can write the time structure of the harmonics as the sum of individual harmonic amplitudes, each with well-defined phase φ_q:

$$E_{HHG}(t) = \sum_j \sum_q a_q^{(j)} e^{-i(q\omega t - \varphi_q^{(j)})}. \quad (6.181)$$

In this description of the E-field we have also included a sum over all contributing recollision trajectories, j, because we have already seen that the single-atom phase imparted to the harmonic radiation for a given harmonic is different for short and long trajectories. This sum represents a Fourier series. So, if the phase for each harmonic does not vary widely from peak to peak, this will describe a pulse train of individual pulses each with duration roughly the inverse of the bandwidth of the entire comb of harmonics. Using the time-bandwidth relation for a Gaussian pulse, we can estimate the pulsewidth of the individual attosecond pulses in the train described by Eq. (6.181):

$$\Delta t_{as} \approx \frac{2.5}{\Delta\Omega}. \quad (6.182)$$

So, for example, if we were to consider a group of harmonics produced by 800 nm light (where $\omega = 2.35 \times 10^{15}$ s^{-1}), say the 15th to 35th harmonics, then $\Delta\Omega = 20\omega = 4.7 \times 10^{16}$ s^{-1}, which implies individual pulses of 50 as duration.

This pulse train picture is completely consistent with the three-step recollision model. Consider in the time domain how harmonics are produced in this picture, a situation illustrated in Figure 6.34. The recollision of wavepackets formed near the field crests by tunnel ionization would produce sharp bursts of radiation later in the cycle, after a time τ_r (which is, of course, different for harmonic radiation produced by the long and short trajectories). These attosecond bursts come every half-cycle.

6.4.2 The Atto-Chirp

Our simple estimate for the time duration of individual attosecond pulses in the recollision pulse train assumes that the pulses are transform limited, or at least suggests what the best *potential* pulse duration could be if the phase across harmonic orders was reduced to a flat constant. In fact, these pulses will not be transform limited, and will be produced with a frequency chirp (i.e. a phase delay that varies quadratically with frequency in the

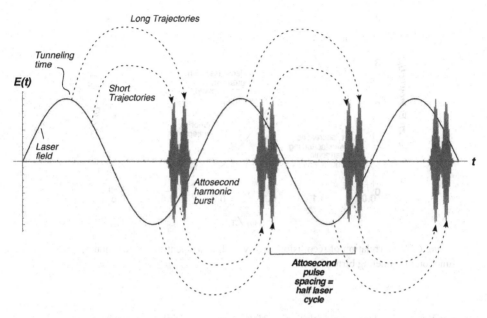

Figure 6.34 Attosecond pulse spacing in a constant-amplitude laser field is π/ω, every half-cycle.

spectral domain or a wavelength that varies linearly in the temporal domain). This chirp is a consequence of the single-atom phase associated with the harmonic production. It can be understood simply in the context of the three-step recollision model.

Recall the correlation between emission time for a given harmonic and that harmonic's photon energy. This relationship is summarized in Figure 6.35. Here the return energy, U_{ret}, of the recolliding wavepacket as a function of birth phase is plotted on top of the time interval spent in the continuum for the same birth phases. Clearly the temporal point of harmonic emission varies with the energy of that harmonic photon. Remember that the energy of the harmonic photon emitted from the recombination step is just $q\hbar\omega = I_p + U_{ret}$. As can be seen in Figure 6.35, if we consider first long trajectories, lower photon energies are emitted later in the laser cycle than higher photon energies emitted as one moves toward the cutoff. This implies a clear temporal chirp in each attosecond pulse. Figure 6.35 also indicates that such a chirp will also be present for the short trajectories but with opposite sign. This chirp is usually referred to as the "atto-chirp" (Agostini and DiMauro 2004; Varju et al. 2005).

Note that the pulse train of attosecond pulses will be composed of pulses from both short and long trajectories, each with a chirp of opposite sign. If this chirp is compensated, it can only be done for one sign. So, to derive sharp, short attosecond pulses it is critical to choose harmonics from one set of trajectories such as the short trajectories, which can be selected, as we have seen, by aperturing the harmonic beam and selecting radiation emitted predominantly with low divergence.

Compensating this chirp turns out to be critical for achieving the shortest pulses as the chirp lengthens the pulse considerably. As a concrete example, consider the grouping of

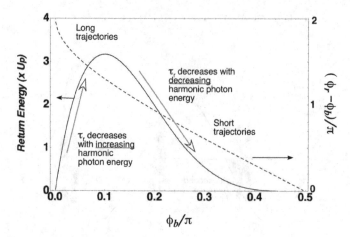

Figure 6.35 Correlation of recollision energy with time spent in the continuum, $\omega\tau_r$, as a function of tunneling birth phase.

800 nm harmonics from the 15th to the 35th. Assume that these harmonics are produced in Ne at an intensity of $\sim 2 \times 10^{14}$ W/cm² and that we are concerned with the chirp on the short trajectory radiation. The 15th harmonic is produced at a recollision phase of $\approx 0.72\pi$ while the 35th is produced at $\approx 0.6\pi$. This represents a temporal spread over this harmonic frequency range of nearly 550 as, far larger than the 50 as transform-limited duration we would expect from that bandwidth.

We can be a bit more quantitative and find the value of the spectral chirp we expect using the phase imparted to each harmonic as predicted by the SFA. In the time domain, this chirp can be written as an E-field phase, which varies quadratically with time:

$$\varphi_q(t) = -\left(\omega_{q0} t + b_c t^2\right) \tag{6.183}$$

or for a complicated time-varying phase

$$b_c = -\frac{1}{2}\frac{\partial^2 \varphi_q}{\partial t^2}. \tag{6.184}$$

For this analysis it will be convenient to think of the harmonic order, q, as a continuous variable instead of a discrete integer. (This is not necessarily a bad approximation. As we shall discuss, when experimental techniques are used to isolate a single attosecond pulse, the discrete peaks of individual harmonics coalesce into a single broad spectral feature.) We can for simplicity then, think of the harmonic spectrum as a continuous spectral feature with a harmonic band which we treat as a Gaussian spectrum to ease calculations.

A temporal phase like that of Eq. (6.183) will lead, in turn, to a spectral phase which varies quadratically with frequency. So the temporal structure of a given attosecond pulse is like

6.4 Attosecond Pulse Generation

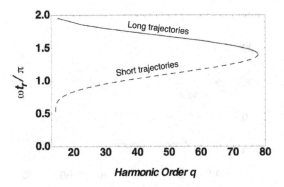

Figure 6.36 Plot of return time t_r as a function of harmonic for harmonics produced in Ne at 5×10^{14} W/cm².

$$E_q(t) \sim \int_{-\infty}^{\infty} \exp\left[-(\omega_q - \omega_{q0})^2 / \Delta\Omega^2\right] \\ \times \exp\left[i\alpha(\omega_q - \omega_{q0})^2\right] \exp\left[i\omega_q t\right] d\omega_q, \quad (6.185)$$

where ω_{q0} is the central frequency of the band of harmonics producing the attosecond pulses and $\Delta\Omega$ is interpreted as the spectral full width. This spectral chirp is derived from the phase

$$\alpha_c = \frac{1}{2} \frac{\partial^2 \varphi_q^{(j)}}{\partial \omega_q^2} = \frac{1}{2\omega^2} \frac{\partial^2 \varphi_q^{(j)}}{\partial q^2}. \quad (6.186)$$

The origin of this spectral chirp can be thought of as the consequence of the variation in recollision time as a function of harmonic order. As an example, this recollision time is plotted in Figure 6.36 as a function of q for harmonics in Ne produced at intensity of 5×10^{14} W/cm². The differing slope of the chirp for long and short trajectories is obvious.

To estimate the value of α for a harmonic pulse we can use our SFA-derived result for the single-atom dipole strength. Again, the phase can be calculated by finding the argument of the imaginary dipole strength from Eq. (6.55) or the phase itself found approximately under the saddle-point approximation for k_e from Eq. (6.66); but for the moment, to make basic estimates, we can use the simple approximation that

$$\varphi_q^{(j)} \cong q\omega t_r - \frac{U_p + I_p}{\hbar} t_r. \quad (6.187)$$

Using a calculation such as that shown in Figure 6.36 and Eq. (6.187) it is straightforward to calculate $\varphi_q^{(j)}$ and its second derivative with respect to harmonic order for both the long and short trajectories. Such a computation for harmonics produced in Ne at 5×10^{14} W/cm² is shown in Figure 6.37. We can see that the spectral chirp is considerably larger for long trajectory harmonics.

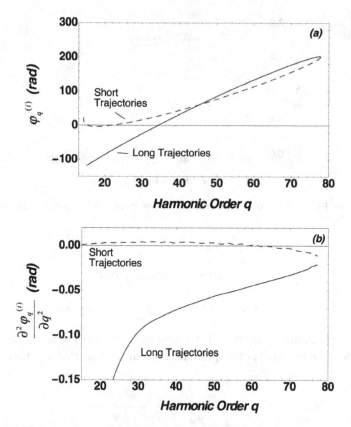

Figure 6.37 Plot of (a) harmonic phase and (b) second derivative of phase as a function of harmonic order for harmonics produced in Ne at 5×10^{14} W/cm^2.

With a calculation for α it is possible to make a more quantitative estimate for the chirped pulse duration. Fourier transforming Eq. (6.185) tells us that the temporal evolution of an individual attosecond pulse is

$$E_q(t) \sim \exp\left[-\frac{t^2}{4\left(1/\Delta\Omega^2 + \alpha_c^2 \Delta\Omega^2\right)}\right]$$
$$\times \exp\left[-i\frac{\alpha_c \Delta\Omega^2 t^2}{4\left(1/\Delta\Omega^2 + \alpha_c^2 \Delta\Omega^2\right)}\right] \exp\left[-i\omega_{q0} t\right] \quad (6.188)$$

so then we can say that the pulse duration is

$$\Delta t_{as} \cong 2\sqrt{1/\Delta\Omega^2 + \alpha_c^2 \Delta\Omega^2} \quad (6.189)$$

and the temporal chirp is

$$b_c = \frac{\alpha_c \Delta\Omega^2}{4\left(1/\Delta\Omega^2 + \alpha_c^2 \Delta\Omega^2\right)}. \quad (6.190)$$

For the example in Ne we see that for short trajectories $\alpha_c \approx 9 \times 10^{-34}$ s^2 and for the long trajectories $\alpha_c \approx 6 \times 10^{-33}$ s^2. (We would point out that our simple estimate for the single-atom phase underestimates the value of the chirp for the short trajectories. See more accurate calculations in Varju *et al.* (2005).) For a band of harmonics between the 31st and the 51st we would expect a transform-limited pulse duration of about 40 as but the short-trajectory harmonics are chirped to about 100 as (or even longer) and the long-trajectory harmonics are chirped to almost 600 as.

The differing emission times of the short- and long-trajectory harmonics, in fact, lead to complicated temporal patterns in each attosecond pulse. In experiments, then, it almost always critical to select one set of harmonics spatially with an aperture (Winterfeldt *et al.* 2008). The chirp of the short-trajectory harmonics is positive (i.e. higher frequencies come later in time). It is possible in experiments to compensate for such a positive chirp and recompress the attosecond pulses by introducing a thin filter medium which will impart a negative dispersion on the XUV pulse. With correct choice of material and thickness, the positive chirp of the short-trajectory harmonics can be removed. Typically, filters of Zr, Sn or Al of a few hundred nm thickness are employed in real experiments to retrieve sub-100 as pulses.

6.4.3 The Harmonic Chirp and Atto-Pulse Spacing In a Rising Laser Intensity

While it has been convenient to talk of high-harmonic generation in a constant-amplitude laser field, in reality most high harmonics are produced on the rising edge of a temporally short laser pulse. Consequently, the attosecond bursts are produced from laser cycles with increasing amplitude for each subsequent recollision. This has two consequences.

First, we know that the single-atom phase more or less scales linearly with intensity. In a linearly rising intensity envelope, each subsequent attosecond burst would acquire greater phase and this would lead to a slight shift in the center frequency of each individual harmonic peak. If the pulse has some curvature to its rising intensity, as does, say, a Gaussian intensity envelope of the form

$$I(t) = I_0 \exp\left[-2t^2/\tau_0^2\right], \tag{6.191}$$

this can lead to a temporal chirp on an individual harmonic frequency. With the definition of the temporal chirp parameter, b, in Eq. (6.184), this harmonic chirp is

$$b_c = -\frac{1}{2}\frac{\partial^2 \varphi_q^{(j)}}{\partial t^2} = -\frac{1}{2}\left(\frac{\partial \varphi_q^{(j)}}{\partial I}\frac{\partial^2 I}{\partial t^2} + \frac{\partial^2 \varphi_q^{(j)}}{\partial I^2}\left(\frac{\partial I}{\partial t}\right)^2\right)$$

$$\cong -\frac{1}{2}\frac{I_0}{\tau_0^2}\frac{\partial \varphi_q^{(j)}}{\partial I}. \tag{6.192}$$

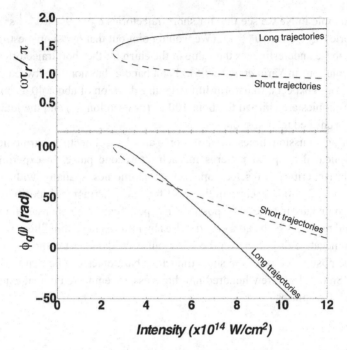

Figure 6.38 Plot of return times (top panel) and harmonic phase (bottom panel) for the 45th harmonic in Ne driven by 800 nm light as a function of intensity.

Figure 6.38 plots, as an example, in the top panel, the continuum propagation time of the 45th harmonic in Ne as a function of intensity showing that it does not vary much for harmonics once they have entered the plateau. So the principal variation of phase with intensity is the linear dependence of U_p with intensity. The single-atom phase estimated, then, from Eq. (6.187) for the 45th harmonic is plotted in the bottom panel of Figure 6.38. This shows that indeed the single-atom phase is nearly linear with intensity and has different slope for short and long trajectories. The chirp that this intensity variation of phase has with intensity on the rising curve of a laser pulse will broaden the spectral width of each harmonic peak so that they are not quite transform limited. This effect combined with ionization-induced blue shifting of the drive laser (discussed in Chapter 8) can significantly broaden the spectrum of each observed peak.

The second consequence of HHG in a time-varying intensity is that the spacing of the attosecond pulse bursts can be shifted slightly from one optical oscillation to the next. Considering first the short trajectories, for a given harmonic photon energy, the return time of the recolliding electron wavepacket decreases as the intensity of the drive light increases. Therefore, as the laser pulse ramps up in intensity, the attosecond pulse bursts arrive earlier in each subsequent cycle with the next effect being that these bursts are spaced closer than a half-cycle. The opposite is true for long-trajectory harmonics. This effect can be compensated if the laser pulse itself is chirped prior to the harmonic generation. For short-trajectory harmonics, if a negative chirp is induced, the frequency drops as the pulse progresses and the optical peaks spread out on each cycle compensating for the earlier returns in the higher-amplitude cycles. Of course, the spacing reverses on the falling edge of the pulse but generally, in experiments, ionization suppresses HHG in the trailing edge of the pulse.

6.4.4 Single Attosecond Pulse Generation

A train of attosecond pulses is the most likely output of an HHG experiment with most near-IR pulses of duration of a few tens of femtoseconds. For many applications of attosecond XUV pulses, however, such as time-resolved pump-probe experiments, it is strongly desirable to isolate a single attosecond pulse. At the time of this writing, there are generally three common experimental methods for producing or extracting a single attosecond pulse from HHG (Chini et al. 2014).

A) Temporal Gating

The first method of single attosecond pulse generation involves temporally gating the HHG process. With this method effectively only one optical cycle produces an attosecond burst, at least in the harmonic wavelength range of interest (Reider 2004).

The simplest method for temporal gating is *amplitude gating*, illustrated in Figure 6.39, in which an extremely short laser pulse, say only 5 fs or maybe 2 optical cycles, is used to generate the harmonics. Such short pulses can usually be produced by propagating a 15 or 20 fs near-IR pulse down a gas-filled hollow-core fiber to broaden the spectrum by self-phase modulation, and then compress this broadened bandwidth down to ∼5 fs. Carrier envelope phase stabilization is critical in this method so that a single optical half-cycle can be placed in the central, highest-intensity portion of the pulse envelope (Krausz and Ivanov 2009). As the top plot of Figure 6.39 illustrates, while the central optical half-cycle can produce a high-energy attosecond recollision, the two optical half-cycles on either side of this central cycle are reduced in amplitude and produce only lower-energy collisions. This manifests itself in the HHG spectrum, as shown in the bottom plot of Figure 6.39. Harmonic radiation at the high end of photon energies out to the cutoff is produced by only this central single cycle and have no individual harmonic peaks at all. Only at lower photon energies, where the other, lower-amplitude half-cycles produce these photon energies as well, do distinct harmonic peaks emerge. If a suitable metal filter is chosen to pass only the broad continuum part of the HHG spectrum, this is tantamount to passing only those higher-energy photons from the single high-energy recollision. Only a single attosecond pulse then traverses the metal filter.

A variation on this technique is known as ionization gating. As we have already discussed, the growth of a free electron plasma in the HHG generating gas on the rising edge of the laser pulse will quickly suppress the harmonic emission after a certain time by

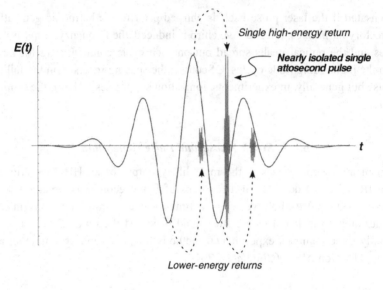
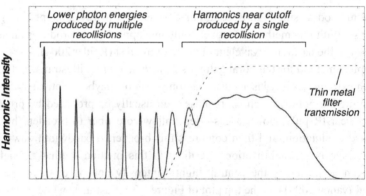

Figure 6.39 Single attosecond pulse generation by amplitude gating. A very short CEP stabilized pulse (~5 fs) is shown in the upper plot. Only the central field half-cycle produces high-energy returns. Lower-amplitude adjacent half-cycles produce electron returns with lower energies. The resulting harmonic spectrum is shown in the bottom plot. While at lower harmonic energies there are multiple returns leading to discrete harmonic peaks, at higher photon energy near the cutoff only the single high-energy return produces harmonic radiation. This single return generates a continuous spectrum. Use of a suitable metal filter will transmit only the higher-energy photons yielding a single attosecond pulse.

destroying the phase-matching. This process is illustrated for a very short CEP-stabilized pulse in Figure 6.40. In a short, few-femtosecond pulse focused to intensity near the saturation intensity of the generating medium atoms, the density of free electrons by field ionization will step up during each optical half-cycle. By a judicious choice of peak

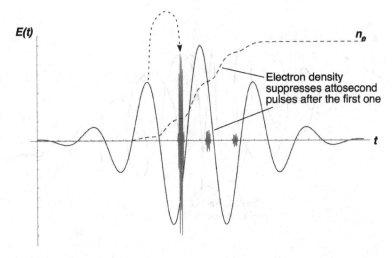

Figure 6.40 Single attosecond pulse generation by ionization gating. With appropriately chosen intensity, the rise in electron density, n_e, suppresses harmonic generation after the first main attosecond burst.

intensity, a situation can be created in which one of the half-cycles is high enough amplitude to produce a short-wavelength attosecond burst; but bursts from subsequent half-cycles, even if they are actually higher-intensity, have their harmonic radiation suppressed by the production of free electrons during the previous couple cycles. Again, in principle, a single attosecond burst emerges. This technique requires good shot-to-shot intensity stability of the drive laser.

B) Polarization Gating

A second method to produce a single attosecond pulse involves taking advantage of the strong suppression of HHG radiation by drive fields with some ellipticity (Shan *et al.* 2005). To do this, a short, near-IR pulse is split into two nearly identical pulses (equal in duration and intensity) but each is circularly polarized with opposite directions of field rotation. If one pulse is delayed by roughly one cycle with respect to the other pulse and recombined, a field of the form

$$E(t) = \frac{E_0}{\sqrt{2}} \exp\left[-\frac{(t + \Delta t/2)^2}{\tau_0^2}\right] (\cos[kz - \omega t]\hat{\mathbf{x}} + \sin[kz - \omega t]\hat{\mathbf{y}}) \\ + \frac{E_0}{\sqrt{2}} \exp\left[-\frac{(t - \Delta t/2)^2}{\tau_0^2}\right] (\cos[kz - \omega t]\hat{\mathbf{x}} - \sin[kz - \omega t]\hat{\mathbf{y}}) \quad (6.193)$$

results. This field is illustrated schematically in Figure 6.41. The rising and falling edges of the pulse exhibit some ellipticity. Only in the center, where the two counterrotating pulses have the same intensity, will the two pulses conspire to yield, for a short window, a

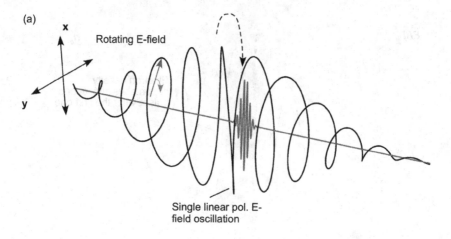

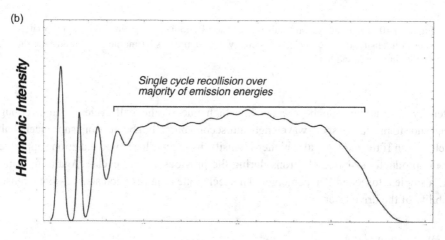

Figure 6.41 Single attosecond pulse generation by polarization gating. Two counterrotating temporally offset short pulses are shown in (a). Only the central linearly polarized field half-cycle produces a high-energy return. The resulting harmonic spectrum is shown in (b). This single return generates a continuous spectrum over nearly the entire harmonic spectrum.

linearly polarized field. With correct choice of pulse duration and delay between the two counterrotating fields, it is possible to produce nearly linear polarization over only a half-cycle. Because of the ellipticity in the majority of the pulse envelope, no recollisions are possible and no attosecond bursts are produced. Only in the single, linearly polarized half-cycle is a recollision possible, leading to a single attosecond burst. Experimentally, this situation can manifest itself in a spectrum that looks like Figure 6.41b: a nearly featureless continuum over the entirety of the plateau, indicative of only one electron wavepacket recollision.

6.4 Attosecond Pulse Generation

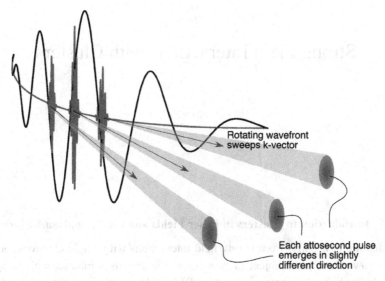

Figure 6.42 Scheme for separating single attosecond pulses by the attosecond lighthouse idea. A rotating wavefront is imparted onto a few-femtosecond pulse, which generates an attosecond pulse train. The rotating wavefront results in a **k**-vector which sweeps with optical cycle, sending each attosecond pulse into a slightly different direction so that an aperture can be used to select a single pulse.

C) Spatiotemporal Gating: The Attosecond Lighthouse

A third technique involves inducing a rotating wavefront on the focused laser pulse, which can be induced by inserting a thin misaligned wedge in the beam (Chini *et al.* 2014). This rotating wavefront situation is illustrated schematically in Figure 6.42. Because of the wavefront rotation, harmonic light emitted during each half-cycle emerges from the medium at a slightly different direction. Each angle will produce an attosecond pulse from a half-cycle of the drive laser. By using an aperture to choose a single beam of this "lighthouse" emission, it is possible to isolate a single attosecond pulse.

7

Strong Field Interactions with Clusters

7.1 Introduction to Clusters in Laser Fields and the Nanoplasma Model

Our attention now shifts from strong field interactions with single electrons, atoms or molecules toward the rich subject of how intense laser pulses interact with many body systems. As such we will enter the realm of laser-plasma physics, the main subject of the final two chapters of this book. Before we shift completely to plasma interactions, we shall consider the fascinating topic of how intense ultrafast lasers interact with atomic and molecular clusters. Clusters, as we consider them here, refer to agglomerations of atoms or molecules of anywhere from, say, ten to hundreds of thousands of atoms each. Such agglomerations are typically spherical or nearly spherical in shape and, even with the largest clusters considered, will be much smaller than a laser wavelength. Often they are loosely bound, as with van der Waals–bonded noble gas clusters, but some species can be tightly bound, as in a metallic cluster.

Clusters have seen particular focus within chemical physics research over the past fifty years as they represent an interesting bridge between the properties of single atoms and the bulk properties of a macroscopic solid. Often, much has been learned by tracing certain observables such as optical or thermodynamic properties as clusters become larger and larger, approaching the behavior of a bulk solid. From the standpoint of strong field interactions, this bridge spans the quantum models of ionization of atoms and small molecules that we have studied so far to the complex collective electron motion that dominates strong field interactions with plasmas. As we will see, collective motion of many electrons is often the principal manner in which the laser field is coupled to the cluster.

Experimentally, the first results on the interactions of intense laser pulses with clusters came as something of a surprise. Clusters, which turn out to be present in many high-intensity laser experiments utilizing gas jet targets in vacuum, yielded a number of unexpected and, in some ways, spectacular results when those gas jets were illuminated. In general these experimental studies are characterized by three significant observations: (1) Atoms in clusters are often ionized to much higher charge states than are the same atoms when irradiated individually, and frequently significantly higher than might be expected even in a solid-density plasma irradiated at very high intensity. (2) Ions are ejected from the cluster with high energy, suggesting that the absorption of laser energy by each cluster

7.1 Introduction to Clusters in Laser Fields and the Nanoplasma Model

is very high. (3) Gas jets composed of clusters, which are otherwise completely transparent to low-intensity light, become nearly 100% absorbing at high intensity. This third observation is usually accompanied by bright extreme ultraviolet (XUV) or X-ray emission which results from the high laser energy absorption and production of high-charge-states production. Gases of single atoms at the same density often show virtually no X-ray emission under identical irradiation conditions.

These three main experimental phenomena are augmented by a number of other fascinating observations. For example, it is seen that the absorption of laser light and energies of the subsequently emitted photons and ejected ions are strongly dependent on the laser pulse duration or on the irradiation of two suitably delayed pulses. The optimum pulse duration in these studies often appears at a length well longer and at subsequently lower intensity than the shortest pulses used in the experiment; sometimes pulses as long as a few hundred femtoseconds or even a few picoseconds are optimum for coupling energy into the cluster. This effect suggests that the evolution of the cluster during the laser pulse is important to the interaction. It is often seen that the electron spectra emitted from intensely irradiated clusters exhibit two peaks whose relative position and amplitude vary with pulse duration, suggesting that there is more than one mechanism for coupling laser energy into those electrons. Most of these effects can be understood within the context of the collective motions of many electrons in the cluster, to which we now turn our investigations.

7.1.1 Cluster Formation in Gas Jets and the Hagena Parameter

There are a number of ways that clusters can be formed for strong field experiments, depending on the nature of the cluster (metallic, liquid, gas, etc.). A significant majority of the cluster studies in the literature have focused on studies with van der Waals–bonded clusters. These clusters are formed from high-pressure reservoirs of noble gases (like Xe, Kr and Ar) or small molecules (like N_2, CO_2 or CH_4) which are expanded through a gas jet into vacuum. The cooling associated with the adiabatic expansion of the gas can cause the passage of a phase transition and the coalescence of clusters in the gas jet bonded by the van der Waals forces between the atoms or molecules. Strongly polarizable atoms or molecules (like Xe or CO_2) tend to form the largest clusters for a given gas expansion condition because of the greater strength of van der Waals forces among these species.

The structure of small clusters has been well studied and is quite frequently icosahedral with clusters of "magic numbers" corresponding to closed shells of icoahedrons exhibiting particular stability (such as clusters with 12, 55, 147 etc. atoms). Larger clusters (>1000 atoms/cluster) can usually be thought of as spherical liquid droplets. We will almost always assume that the initial atomic configuration is spherical with more or less uniformly and randomly distributed atoms within the cluster. This view is adequate for explaining most phenomena with a few exceptions (such as the appearance of shell structure in the explosions of small to moderate-sized clusters, discussed in Section 7.4.3).

While there are a number of analytic theories which predict an expected cluster size for given initial gas jet conditions (Krainov and Smirnov 2002), most experimentalists have found that the most accurate way to predict the average cluster size in a given gas

Table 7.1

Species	k_c	$\rho_l(g/cm^3)$
Xe	5550	2.94
Kr	2980	2.41
Ar	1650	1.39
Ne	185	1.21
H_2	184	0.07
He	4	0.13
CH_4	2360	0.42
N_2	528	0.81

expansion is by evaluating a purely empirically derived quantity, the "Hagana parameter" (Hagena and Obert 1972). This dimensionless number depends only on the gas jet's reservoir initial conditions and nozzle geometry, reading

$$\Gamma^* = k_c \frac{(d/\tan\alpha)^{0.85} p_0}{T_0^{2.29}}, \qquad (7.1)$$

where d is the gas jet's nozzle diameter in µm, α is the half-angle of a cone fitted to the jet orifice, p_0 is the initial gas backing pressure in mbar and T_0 is the backing gas's initial temperature in K. The k_c is a scaling parameter that varies with atomic species. This atomic clustering parameter is listed for various commonly encountered gases in Table 7.1. (The density of the liquid state of these gases is also listed in this table to enable easy estimation of cluster diameters for given cluster sizes.)

This table clearly shows that the large polarizable atoms like Xe and Kr along with polyatomic molecules like methane will tend to yield the highest Hagena parameter and will therefore cluster under the lowest backing pressure conditions.

Clustering in a jet tends to begin when $\Gamma^* \geq 300$ (Dorchies et al. 2003). Experiments usually show that the average cluster size in a jet scales with Γ^* in such a way that it follows a line roughly like that reproduced in Figure 7.1. Above the clustering threshold a good rule of thumb is that the average cluster size in the jet is roughly

$$\langle N \rangle \cong \left(\frac{\Gamma^*}{225}\right)^{2.35}. \qquad (7.2)$$

While this scaling gives a good indicator of the average cluster size, in fact there will be a wide range of cluster sizes in any clustering jet. Experimental observations indicate that the distribution of sizes in a clustering jet is more or less a log-normal distribution, characterized by a distribution function that can be written

$$P_c(N)dN = \frac{1}{N\sigma_c\sqrt{2\pi}} \exp\left[-(\ln[N/N_0])^2/2\sigma_c^2\right] dN. \qquad (7.3)$$

7.1 Introduction to Clusters in Laser Fields and the Nanoplasma Model

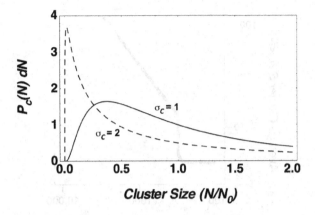

Figure 7.1 Typical scaling of observed average cluster size in a supersonic nozzle as a function of the Hagena parameter.

Here N denotes the number of atoms/molecules in a cluster. The N_0 and σ_c are parameters that characterize the average cluster size and width of the distribution. Two example distributions are plotted in Figure 7.2. This distribution is characterized by a long tail so most experiments with clustering jets have a small fraction of very large clusters present; this small fraction of large clusters can sometimes dominate the experimental results. While every gas jet is different, it is often seen that $\sigma_c \approx (\ln N_0)/2$.

The size moments of this distribution (Madison et al. 2003) are easily calculated:

$$N^p = \int_0^\infty N^p P_c(N)\, dN$$
$$= \exp\left[p \ln N_0 + p^2 \sigma_c^2 / 2\right] \quad (7.4)$$

which means that the average cluster size of the distribution is

$$N = N_0 e^{\sigma_c^2/2}. \quad (7.5)$$

7.1.2 Regimes of Strong Field Cluster Interactions and the Cluster Nanoplasma Model

Because of the complex interplay of ionization and cluster expansion during the laser interaction, the laser energy deposition and subsequent explosion can evolve through a number of regimes, even during a single laser pulse (Ditmire 1997; Ramunno et al. 2008; Fennel et al. 2010). Sometimes the cluster behaves as if it is two spherical charged fluids (the electron and ion fluids) while at other times single-particle interactions dominate. However, it is reasonable to view high-intensity irradiation of a cluster as falling into three possible regimes.

A) Coulomb Explosion Regime

If the situation arises in which all electrons that have been liberated from the cluster atoms by field or collisional ionization are removed rapidly from the cluster ions, the cluster

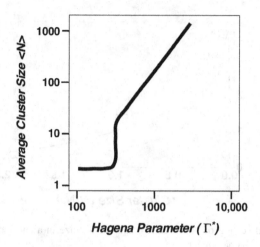

Figure 7.2 Log-normal size distributions characterizing many clustering jets plotted for two different size dispersions.

experiences a Coulomb explosion (this is the multibody analog of the Coulomb explosion of diatomic molecules discussed in Chapter 5). In this case "rapidly removed" means that the electrons are pulled from the ion core on a time scale much faster than the ion core expands. Such fast electron ejection leaves a charged sphere of ions near their initial equilibrium positions. Coulomb repulsion drives an explosion of the residual ion core. This regime is characterized by minimal interaction between the laser field and the electrons (minimal electron heating) and is most typical for small clusters, clusters of low Z (like hydrogen clusters) or in irradiation at very high intensity. It is commonly observed that this regime occurs when the laser ponderomotive energy significantly exceeds the electrostatic binding potential electrons feel by the bare ion core, written

$$U_p \gg \frac{\bar{Z} N_i e^2}{R_0}, \tag{7.6}$$

where $\bar{Z}$ represents the average ionization state of the cluster atoms, N_i is the number of ions in the cluster and R_0 is the initial radius of the cluster in its pre-ionized equilibrium. This condition can be expressed in terms of the laser field as

$$E_0 > \frac{Q_{tot}}{R_0^2}. \tag{7.7}$$

As a concrete example consider an H_2 cluster with initial radius of 2 nm, corresponding to $N_i \approx 700$ ions. In this case $N_i e^2/R_0 \approx 500$ eV, so in a near-IR field (say, wavelength of 800 nm) the condition in Eq. (7.6) is reached at an intensity exceeding 10^{16} W/cm^2.

B) Quasi-neutral Nanoplasma Regime

The opposite case to that described in the previous paragraph occurs when the space-charge forces of the cluster ions are strong enough to confine almost all of the electrons liberated by ionization of the cluster atoms. In this case the cluster can be viewed as a

7.1 Introduction to Clusters in Laser Fields and the Nanoplasma Model

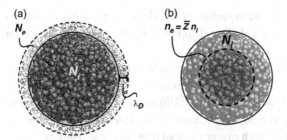

Figure 7.3 Schematic illustration of a cluster nanoplasma. On the left (a), the situation in which no outer-ionization has occurred (no free electrons have been removed from the cluster sphere), so the cluster is quasi-neutral. In that case the warm electrons will expand to neutralize the cluster ions with a very small fraction creating a Debye sheath outside the volume of the cluster ions. On the right (b), the situation in the cluster in which some electrons have been outer-ionized (removed by the laser field or thermal energy), and the density of the cluster is still high. The electron cloud contracts to leave an inner sphere of neutralized ions surrounded by a shell of electron-free ions.

nanoscopic plasma (i.e. a nanoplasma) that is more or less quasi-neutral within the cluster sphere (Ditmire et al. 1996). It is the opposite of the limit described in Eq. (7.6). Furthermore, it is only accurate to describe the cluster sphere as a plasma if the Debye radius of the cluster electrons given by the usual formula (ignoring ion mobility; see Appendix B),

$$\lambda_D = \sqrt{\frac{k_B T_e}{4\pi n_e e^2}}, \qquad (7.8)$$

is significantly less than the radius of the cluster (otherwise it would be more accurate to think of the electrons as forming a Debye cloud around the ion core). As a calibration point, the Debye radius is about 0.5 nm in a solid density plasma at an electron temperature of 1000 eV. In many experiments involving medium to large clusters (say a few hundred to a few thousand atoms per cluster) the cluster radius is many nm, so the situation depicted in Figure 7.3a develops: a quasi-neutral core is surrounded by a thin electron Debye sheath. In this regime the laser interacts predominantly with the collective motion of the electron cloud around the confining ion sphere. The expansion of the cluster is largely from the electron pressure. This regime has come to be called the nanoplasma regime and is associated in the literature with a rate equation model of the cluster dynamics (Ditmire et al. 1996; Micheau et al. 2008). I will use the term nanoplasma more generally to denote simply the regime in which an ionized cluster remains more or less quasi-neutral, whether or not a rate equation, "rigid oscillating electron cloud" picture is appropriate.

C) Partially Charged Nanoplasma Core Regime

A distinct regime exists between these two extremes, in which some but not all electrons are removed from the space-charge confinement of the cluster ions. This situation is depicted in Figure 7.3b. As in the nanoplasma regime, the laser still interacts with the collective oscillations of an electron cloud, but the body of the cluster is no longer quasi-neutral. The explosion of the cluster is a combination of Coulomb explosion (of an outer shell, as we shall see) followed by electron pressure–driven expansion of an inner core.

We shall consider each of these regimes in detail, examining the physics of the electron dynamics, the laser absorption mechanisms, hot electron production and ion expansion, among other details of the interaction. Ultimately, when we consider the entire interaction history of a cluster with an intense femtosecond laser pulse, we will see that, in some cases, the cluster evolves through two or even all three of these regimes.

7.1.3 Phases of the Laser Cluster Interaction

The concepts introduced in the last section are summarized in the chart of Figure 7.4. Some of the relevant physics at various points in the laser pulse (early during the rise time, during the bulk of the pulse and in the time after the pulse intensity falls) in these three regimes are noted.

Before we launch into discussions of strong field phenomena in the laser-cluster interaction, it is informative to consider interactions of a cluster in more modest fields, in which the interaction is nearly linear. At this point we shall introduce one powerful approximate picture of the cluster which will serve us well in understanding a whole range of phenomena in the strong field regime. In many cases we will encounter a situation in which atoms of the cluster have been ionized leading to free electron production, but in a regime in which some or all of these freed electrons are confined to the cluster body by space-charge forces (regimes "B" and "C" described earlier). In this case it is helpful to think of the cluster as composed of an incompressible rigid sphere of electrons which oscillates through the rigid shell of ions. The ions serve as a restoring force to a rigid charged sphere of electrons which is driven by the field of the laser. The shortcomings and potential failings of this simplistic picture are obvious: of course, the electron cloud is composed of individual particles which not only repel each other but which can scatter off of individual ions and the collective field of the ions. We would note, however, that numerous molecular dynamics simulations have confirmed that, at least for a time (tens or hundreds of femtoseconds), the electron cloud can retain its integrity and appears to oscillate as a collective rigid whole (Saalman and Rost 2003; Saalman et al. 2006; Ditmire 1998). Furthermore, multiple experimental observables are well described by the rigid electron cloud picture. Later in the chapter we will consider the conditions in which such a picture is most accurate and will consider dynamics when individual particle (i.e. kinetic) effects also play a role.

A second useful approximation arises by noting that the laser–cluster interaction is almost always in the dipole limit: when the laser wavelength is much greater than the cluster diameter. This allows us to treat the laser field as spatially uniform over the body of the cluster. The weak field limit can be described as the situation in which the rigid electron cloud oscillates as a dipole with oscillation amplitude much smaller than the

7.1 Introduction to Clusters in Laser Fields and the Nanoplasma Model

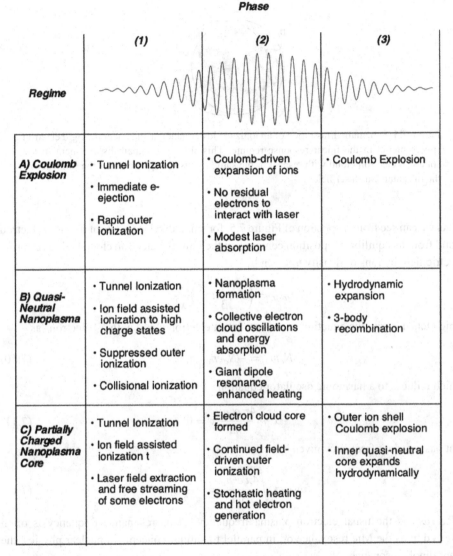

Figure 7.4 Chart describing three limits of cluster evolution and some description of what happens in the cluster in three phases of the interaction: (1) the rising edge of the laser pulse, (2) the high-intensity peak of the pulse, and (3) the falling edge of the pulse and the time after the pulse has passed.

cluster diameter (depicted in Figure 7.5). Here we consider the case when $x \ll R_0$. (We shall show that in many clusters, particularly high Z clusters like Xe and Ar of over a few hundred atoms, this "weak field" regime actually persists up to fairly high intensity, like $> 10^{16}$ W/cm^2.)

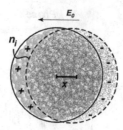

Figure 7.5 Schematic illustration of an ionized cluster nanoplasma under the rigid electron sphere model in the linear response regime. This shows the small displacement of the spherical electron cloud with respect to the same radius ion sphere under the influence of the instantaneous laser field.

As we can see from inspection of Figure 7.5, for small displacements of the rigid electron cloud from its equilibrium position centered on the ions, the electron cloud feels a restoring electric field by ions of density n_i given by

$$E_{rest} = \left(\bar{Z}n_i \frac{4}{3}\pi x^3\right) \frac{e}{x^2}, \qquad (7.9)$$

which lets us write the equation of motion for an electron cloud with N_e electrons as

$$N_e m \ddot{x} = -N_e e n_e \frac{4}{3}\pi x e, \qquad (7.10)$$

which reduces to a harmonic oscillator equation

$$\ddot{x} + \frac{4\pi e^2 n_e}{3m} x = 0 \qquad (7.11)$$

that has resonant frequency given by

$$\omega_M \equiv \sqrt{\frac{4\pi e^2 n_e}{3m}} = \frac{1}{\sqrt{3}} \omega_{pe}. \qquad (7.12)$$

(Here ω_{pe} is the usual electron plasma frequency.) This resonance frequency is often referred to as the Mie resonance or, in parallel to similar concepts in nuclear physics, the giant dipole resonance.

Next we ask, what is the field inside the cluster sphere in this linear dipole regime. Qualitatively the field will be a superposition of the uniform long-wavelength oscillating laser field and the field induced by the displacement of the rigid electron cloud with respect to the ion core. We know from basic plasma physics that the electrons in a dense plasma will tend to shield and lower the amplitude of the laser field inside the plasma, though when the plasma frequency approaches the laser frequency, the laser field amplitude can be enhanced.

We can treat the cluster nanoplasma as a material with dielectric constant

$$\varepsilon = 1 - \frac{\omega_p^2}{\omega(\omega + i\nu)}, \qquad (7.13)$$

7.1 Introduction to Clusters in Laser Fields and the Nanoplasma Model

where ν is some damping rate. This characterizes any process that damps the oscillations of the rigid electron cloud such as electron–ion collisions. Following the usual Mie scattering treatment, we work in the dipole approximation and treat the problem as a standard boundary value problem (Jackson 1975, p. 150). If θ is the angle with respect to the laser polarization, then we can write the potential inside and outside the cluster sphere in cylindrical coordinates:

$$\Phi_{out} = \sum_{l=0}^{\infty} \left[\beta_l r^l + \gamma_l r^{-(l+1)} \right] P_l(\cos\theta), \tag{7.14}$$

where the P_l are the usual Legendre polynomials. First, we note that at $r = \infty$ we must retrieve the uniform laser field:

$$\Phi_{out}(r = \infty) = -E_0 x = -E_0 r \cos\theta, \tag{7.15}$$

so all $\beta_l = 0$ except for $\beta_1 = -E_0$. Next, we use the usual boundary conditions on the cluster sphere surface, $\mathbf{D} = \varepsilon \mathbf{E}$. On the cluster surface the tangential E-field and normal D-field are continuous, yielding the following conditions:

$$E_{tan} = -\frac{1}{r}\frac{\partial \Phi}{\partial \theta} \tag{7.16}$$

$$-\frac{1}{R}\frac{\partial \Phi_{in}}{\partial \theta}\bigg|_{r=R} = -\frac{1}{R}\frac{\partial \Phi_{out}}{\partial \theta}\bigg|_{r=R}. \tag{7.17}$$

These equations then tell us that

$$\alpha_1 = -E_0 + \frac{\gamma_1}{R^3} \tag{7.18}$$

and

$$\alpha_l = \frac{\gamma_l}{R^{2l+1}}. \tag{7.19}$$

On the other hand, using the boundary conditions for $\mathbf{D}$ on the surface:

$$-\varepsilon \frac{\partial \Phi_{in}}{\partial r} = -\frac{\partial \Phi_{out}}{\partial r}, \tag{7.20}$$

$$\varepsilon \alpha_1 = -E_0 + 2\frac{\gamma_1}{R^3} \quad \text{for } l = 1, \tag{7.21}$$

$$\varepsilon l \alpha_l = -(l+1)\frac{\gamma_l}{R^{2l+1}} \quad \text{for } l \neq 1. \tag{7.22}$$

The $l \neq 1$ condition can only be satisfied if $\alpha_l = \gamma_l = 0$. Therefore, only $l = 1$ terms exist in the polynomial expansions. Substituting Eq. (7.18) into Eq. (7.21) yields

$$\gamma_1 = \left(\frac{\varepsilon - 1}{\varepsilon + 2}\right) R^3 E_0. \tag{7.23}$$

Therefore

$$\alpha_1 = -\left(\frac{3}{\varepsilon + 2}\right) E_0, \tag{7.24}$$

leading to the final solution for static potential inside and outside the cluster:

$$\Phi_{in} = -\left(\frac{3}{\varepsilon+2}\right) E_0 r \cos\theta, \tag{7.25}$$

$$\Phi_{out} = -E_0 r \cos\theta + \left(\frac{\varepsilon-1}{\varepsilon+2}\right) \frac{R^3}{r^2} \cos\theta. \tag{7.26}$$

This allows us to retrieve the field inside the cluster:

$$E_{in} = -\nabla \Phi_{in} = \left(\frac{3}{\varepsilon+2}\right) E_0. \tag{7.27}$$

We see that when $\varepsilon < -5$, in dense plasma (i.e. when $n_e > 6 n_{crit}$, where $n_{crit} = m\omega^2/4\pi e^2$ is the usual critical density of a plasma in a laser field), the field in the plasma is effectively shielded and the field strength in the cluster is less than the incident laser field. At very high density the electrons respond adiabatically to the laser field and, more or less, shield the field from the interior. Furthermore, what field does penetrate is shifted 180° out of phase with the drive laser field. On the other hand, as the electron density in the cluster drops, as would happen as the cluster expands, then ε approaches -2, where the field in the cluster is resonantly enhanced. This resonant enhancement of the field occurs when $n_e = 3 n_{crit}$, which corresponds to the Mie resonance derived in Eq. (7.12). In the absence of any oscillation damping the E-field in the cluster approaches infinity. In reality, the oscillating electron cloud is damped and the peak enhancement depends on this damping rate. Generally, the maximum enhanced field strength is roughly $E_{in}^{(max)} \approx (\omega/3\nu) E_0$.

This analysis suggests that one way to examine cluster-electron dynamics and laser absorption by the collective electron motion of the nanoplasma in weak to moderate intensities is to treat the ionized cluster as a driven harmonic oscillator. (Again, the quantitative accuracy of this picture has been confirmed by molecular dynamics (MD) simulations of clusters in strong laser fields (Saalman and Rost 2003; Jungreuthmayer et al. 2004).) Ignoring damping for the moment, the cluster as driven harmonic oscillator is subject to the usual simple HO equation of motion:

$$\ddot{x} + \frac{4\pi e^2 n_e}{3m} x = -\frac{e}{m} E_0 e^{i\omega t}, \tag{7.28}$$

which has a simple, steady-state solution,

$$x \sim x_0 e^{i\omega t}, \tag{7.29}$$

where the electron cloud oscillation amplitude is just

$$x_0 = \frac{eE_0}{m\left(\omega^2 - \omega_D^2\right)}, \tag{7.30}$$

with $\omega_D = (4\pi e^2 n_e/3m)^{1/2}$, or at high electron density (near solid density)

$$x_0 \cong \frac{3E_0}{4\pi e n_e}. \tag{7.31}$$

7.1 Introduction to Clusters in Laser Fields and the Nanoplasma Model

The linear dipole regime is valid when this oscillation amplitude is much smaller than the cluster radius, R, which leads to a condition on the peak laser field appropriate for the weak field approximation in a cluster

$$E_0 \ll \frac{4\pi e n_e R}{3}. \tag{7.32}$$

Eq. (7.32) means, for example, that in a 10 nm cluster with electron density around 10^{23} cm^{-3}, the linear regime occurs when the drive laser field is much less than, say, 10^8 statvolt/cm, corresponding to an intensity of 10^{18} W/cm^2. So, the "weak field" model with clusters is actually a pretty good approximation to reasonably high intensity in large clusters.

Next, we consider heating of the cluster. Because the dielectric constant is imaginary for this cluster nanoplasma, we expect some laser absorption and consequent electron heating. We calculate this absorption by considering a dielectric medium subject to an electromagnetic wave (Landau and Lifshitz 1984, p. 319). The wave has energy flux

$$\mathbf{S} = \frac{c}{4\pi} \mathbf{E} \times \mathbf{H}, \tag{7.33}$$

so the temporal change of the energy density, u, in the cluster medium is

$$\begin{aligned}\frac{\partial u}{\partial t} &= -\nabla \cdot \mathbf{S} \\ &= -\frac{c}{4\pi} [\mathbf{H} \cdot \nabla \times \mathbf{E} - \mathbf{E} \cdot \nabla \times \mathbf{H}],\end{aligned} \tag{7.34}$$

which, with the help of Maxwell's equations, is

$$\frac{\partial u}{\partial t} = \frac{1}{4\pi} \left[\mathbf{H} \cdot \frac{d\mathbf{B}}{dt} - \mathbf{E} \cdot \frac{d\mathbf{D}}{dt} \right]. \tag{7.35}$$

We write the linearly polarized laser field in the usual way,

$$\mathbf{E} = \frac{1}{2} E_0 \hat{x} e^{-i\omega t} + c.c, \tag{7.36}$$

where $\mathbf{D} = \varepsilon \mathbf{E}$. Within the dipole approximation used here we ignore the magnetic field heating; we can write the cluster heating rate as

$$\begin{aligned}\frac{\partial u}{\partial t} &= \frac{1}{4\pi} \frac{|E|^2}{4} \left(-i\omega\varepsilon e^{-2i\omega t} - i\omega\varepsilon + i\omega\varepsilon^* + i\omega\varepsilon^* e^{2i\omega t} \right) \\ &= \frac{\omega}{8\pi} Im[\varepsilon] |E|^2.\end{aligned} \tag{7.37}$$

In terms of the incident laser field amplitude, this heating rate becomes

$$\frac{\partial u}{\partial t} = \frac{9\omega}{8\pi} \frac{Im[\varepsilon]}{|\varepsilon + 2|^2} E_0^2. \tag{7.38}$$

Since we consider absorption by a small particle, we can rewrite this heating rate in terms of a laser absorption cross section. The cross section can be found by multiplying Eq. (7.38) by the cluster's volume and dividing by the incident electromagnetic power flux:

$$\sigma_{abs} = \frac{12\pi \omega R^3}{c} \frac{\text{Im}[\varepsilon]}{|\varepsilon+2|^2}. \tag{7.39}$$

In our simple nanoplasma model, we treat the cluster's dielectric constant as that of a simple free-electron Drude model:

$$\varepsilon = 1 - \frac{\omega_{pe}^2}{\omega(\omega+i\nu)}. \tag{7.40}$$

Here we see the usual electron plasma frequency,

$$\omega_{pe} = \sqrt{\frac{4\pi e^2 n_e}{m}}, \tag{7.41}$$

and ν is some damping rate for electron oscillations. We will discuss the detailed origins of this damping rate later in the chapter, though here we just note that it will be the sum of a damping resulting from electron–ion collisions in the plasma and damping from free-electron scattering from the surface of the cluster sphere. We now write the cluster heating rate explicitly in terms of the electron density in the cluster, noting that from (7.40)

$$\text{Im}[\varepsilon] = \frac{\omega_p^2}{\omega(\omega^2+\nu^2)}\nu, \tag{7.42}$$

so

$$\frac{\partial u}{\partial t} = \frac{9}{8\pi} \frac{\omega_p^2 \omega^2 \nu}{\left|3\omega(\omega+i\nu)-\omega_p^2\right|^2} E_0^2$$

$$= \frac{9}{8\pi} \frac{\omega_p^2 \omega^2 \nu}{9\omega^2(\omega^2+\nu^2)+\omega_p^2(\omega_p^2-6\omega^2)} E_0^2, \tag{7.43}$$

which means in terms of electron density and damping rate, the heating rate is

$$\frac{\partial u}{\partial t} = \frac{9\omega}{8\pi} \frac{n_e}{n_{crit}} \frac{\nu}{\omega} \frac{1}{9-6n_e/n_{crit}+n_e^2/n_{crit}^2+9\nu^2/\omega^2} E_0^2. \tag{7.44}$$

This rate is plotted in Figure 7.6 as a function of electron density for two values of the damping rate. This plot clearly shows the presence of a resonance in heating at density near $3n_{crit}$, the situation which we have already seen to correspond to the density at which the drive field frequency matches the giant dipole resonance frequency (though clearly shifted to higher density by large ν).

7.2 Electron Dynamics and Cluster Ionization

Now we turn to nonlinear aspects of the laser–cluster interactions. The discussion of the previous section should have illustrated that at the core of these interactions are dynamics of the ionizing and subsequently free but quasi-bound electrons.

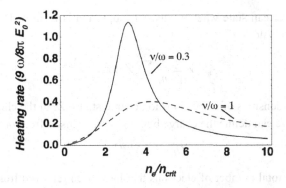

Figure 7.6 Heating rate of the electrons in a cluster nanoplasma in the linear harmonic oscillator regime. Here plotted for two values of the electron cloud damping frequency as a function of electron density.

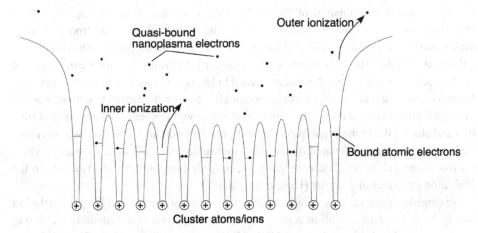

Figure 7.7 Cartoon representation of the confining potential felt by electrons in a cluster illustrating the difference between bound, inner-ionized and outer-ionized electrons.

7.2.1 Inner- and Outer-Ionization: Nanoplasma Formation and Continuum Lowering

In considering the ionization of a cluster in the strong laser field, it is helpful to think about electron ionization in two groups (Fennel *et al.* 2010). These are illustrated schematically in Figure 7.7 *(1) Inner-ionization* – refers to ionization of the clusters, atoms and ions. These ionized electrons are freed from the binding potential of the parent atom but are still confined by the space-charge forces of the cluster ion sphere to a spatial region within the body of the cluster. Inner ionized electrons are the electrons forming the cluster nanoplasma. *(2) Outer-ionization* – refers to the removal of free electrons from the cluster body and removing them from the cluster system altogether. Put differently, a delay between inner- and outer-ionization of a significant number of electrons leads to the transient nanoplasma formation.

Quantitatively, we can state when a cluster nanoplasma is formed. Define the average charge state in the cluster:

$$\bar{Z} = \frac{1}{n_i} \sum_{Z=1}^{Z_{max}} Z n_{Z+}, \quad (7.45)$$

where n_{Z+} are the densities of the various ion charge states, Z, in the cluster and n_i is the total ion density. We can define the charge buildup on the cluster from outer-ionization as

$$Q_{tot} = e N_e^{(OI)}, \quad (7.46)$$

where $N_e^{(OI)}$ is the total number of electrons that have been removed from the cluster via outer-ionization. We can say that a nanoplasma exists if

$$\bar{Z} N_i - Q_{tot}/e \approx N_i. \quad (7.47)$$

There is one significant consequence to the nanoplasma picture, which should be obvious from cursory examination of Figure 7.7. When we turn to calculating the rate of inner-ionization in the cluster (whether it be by the laser field or by electron collisional ionization), the ionization potential of the various ion charge states play a central role in calculating those rates. However the presence of cluster ions inside the cluster can distort the binding potential and reduce the energy needed to liberate an electron by inner-ionization. In essence, the usual tabulated ionization potentials represent the energy to remove a bound electron and move it to infinity. Inner-ionization, however, involves removing a bound electron and placing it into the bound potential of the cluster body, an energy that will inevitably be less than the energy denoted by the usual ionization potential. In dense-plasma physics this phenomenon is called continuum lowering. It can play an important role in the ionization dynamics of a cluster (Hilse et al. 2009).

We consider continuum lowering as a correction to the vacuum ionization potential of an ion: $I_p \rightarrow I_p - \Delta I_p(n_i, \bar{Z}, Te)$. In dense plasmas the magnitude of the continuum lowering is usually calculated using a theory published by Stewart and Pyatt (1966). The continuum lowering can, however, be thought of as occurring in two limits; considering these can be useful in understanding the magnitude of continuum lowering in a plasma. These two limits can be understood in the context of the strong coupling parameter

$$\Gamma_p = \frac{U_{Coul}}{k_B T_e} = \frac{Z e^2}{r_{WS}} \frac{1}{k_B T_e}, \quad (7.48)$$

which is essentially the ratio of the average potential energy between ions and their thermal kinetic energy. Here we have introduced the Wigner–Seitz radius

$$r_{WS} = \left(\frac{3}{4\pi n_i} \right)^{1/3}, \quad (7.49)$$

which is the average spacing between ions in a plasma. A "standard" plasma occurs when the parameter in (7.48) is much less than 1. In that case, the usual plasma picture of Debye shielding is appropriate and we can calculate the continuum lowering using the usual small parameter expansions of a Debye plasma. This regime is denoted in Figure 7.8.

7.2 Electron Dynamics and Cluster Ionization

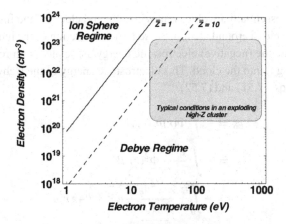

Figure 7.8 Regimes in electron temperature-density space for the validity of the ion sphere and Debye models for continuum lowering. The border lines for singly ionized and ten-times ionized ions are both shown.

To calculate continuum lowering in the weakly coupled limit, recall the usual derivation for Debye shielding in a weakly coupled plasma. The plasma density near an ion follows Boltzmann statistics and can be written

$$\rho_q(r) = -en_e \exp\left[\frac{e\Phi}{k_B T_e}\right] + \bar{Z}en_i \exp\left[\frac{\bar{Z}e\Phi}{k_B T_i}\right], \tag{7.50}$$

where Φ is the electrostatic potential around the ion and in the weakly coupled limit can be said to fulfill the conditions where $e\Phi \ll k_B T_e, k_B T_i$. Taylor expansion of Eq. (7.50) yields

$$\rho_q(r) \cong -en_e \left(\frac{e\Phi}{k_B T_e} + \frac{\bar{Z}e\Phi}{k_B T_i}\right). \tag{7.51}$$

We seek a solution to Poisson's equation, $\nabla^2 \Phi = -4\pi \rho_q$, which has the usual Debye shielding solution

$$\Phi(r) = \frac{\bar{Z}e}{r} e^{-r/\lambda_D}, \tag{7.52}$$

with shielding length given by the Debye radius

$$\lambda_D = \left(\frac{4\pi n_e e^2}{k_B T_e} + \frac{4\pi \bar{Z}^2 n_i e^2}{k_B T_i}\right)^{-1/2} \tag{7.53}$$

(confirmed by substitution of Eq. (7.51) and the solution (7.52) into Poisson's equation). In clusters, particularly of large atoms (like Ar and Xe), it is a good approximation to assume that the ions are immobile, which simplifies the Debye length to

$$\lambda_D \cong \sqrt{\frac{k_B T_e}{4\pi n_e e^2}}. \tag{7.54}$$

Inner-ionization essentially involves removing a bound electron and inserting it into the surrounding Debye cloud around the ion. So, to find the lowering of the ionization potential we need to calculate the (negative) electrostatic energy associated with removing a bound electron and placing it into the cloud. This electrostatic energy of the Debye cloud is easily calculated using Eqs. (7.51) and (7.52):

$$\begin{aligned}\varepsilon_{ES} &= -\int \Phi(r)\rho_e(r)dV \\ &= -\int \frac{e^2 n_e}{k_B T_e}\Phi^2(r)\,dV \\ &= -\int_0^\infty \frac{Z^2 e^4 n_e}{k_B T_e}\frac{1}{r^2}e^{-2r/\lambda_D}4\pi r^2 dr \\ &= -Z^2 e^3 \sqrt{\frac{\pi n_e}{k_B T_e}}.\end{aligned} \quad (7.55)$$

The ionization potential lowering then is the difference in the energy of the cloud before ionization and after ionization (which must include the energy of a cloud around an ion of one charge state higher AND the cloud around the newly injected electron)

$$\Delta I_p = \varepsilon_{ES}^{(Z-1)} - \left(\varepsilon_{ES}^{(Z)} + \varepsilon_{ES}^{(e-)}\right). \quad (7.56)$$

Therefore the continuum lowering in this limit is

$$\begin{aligned}\Delta I_p &= \left[-(Z-1)^2 + Z^2 + 1\right]e^3\sqrt{\frac{\pi n_e}{k_B T_e}} \\ &= 2Ze^3\sqrt{\frac{\pi n_e}{k_B T_e}} = \frac{Ze^2}{\lambda_D}.\end{aligned} \quad (7.57)$$

It is seen to be essentially the attractive energy between an ion of charge Z and an electron separated by a distance equal to the Debye radius. It is typical to replace Z with the average charge state in the plasma, Eq. (7.45), when conducting this calculation.

Finding the continuum lowering in the opposite, strongly coupled limit is a similar calculation. However, in this limit the Taylor expansion of Eq. (7.51) is inappropriate, so the usual approximation is to use what is called the "ion sphere model." The plasma around the ion is approximated as being a uniform cloud of Z electrons in a sphere around each ion of charge Z with radius given by the Wigner–Seitz radius. The calculation of attractive electrostatic energy then becomes trivial. Between the ion and the "muffin tin" electron cloud the electrostatic energy is

$$\varepsilon_{+-} = \int_0^{r_{WS}} \frac{Ze\rho_e}{r}4\pi r^2 dr, \quad (7.58)$$

where

$$\rho_e = \frac{Ze}{^4/_3\pi r_{WS}^3}, \quad (7.59)$$

so

$$\varepsilon_{+-} = -\int_0^{r_{WS}} \frac{3Ze^2}{r_{WS}^3} r\, dr$$

$$= -\frac{3}{2}\frac{Z^2 e^2}{r_{WS}}. \tag{7.60}$$

By the same token the repulsive energy associated with placing the ionized electron into the surrounding ion sphere electron cloud is

$$\varepsilon_{--} = \int_0^{r_{WS}} Ze\left(\frac{r}{r_{WS}}\right)^3 \frac{Ze}{r} \frac{4\pi r^2}{{}^4\!/_3 \pi r_{WS}^3} r\, dr$$

$$= \frac{3}{5}\frac{Z^2 e^2}{r_{WS}}. \tag{7.61}$$

Thus the total electrostatic energy is

$$\varepsilon_{Coul} = \varepsilon_{+-} + \varepsilon_{--}$$

$$= -\frac{9}{10}\frac{Z^2 e^2}{r_{WS}}, \tag{7.62}$$

which means that the ionization potential lowering, given by the difference in this energy before and after ionization, is

$$\Delta I_p = \varepsilon_{Coul}^{(Z-1)} - \varepsilon_{Coul}^{(Z)}, \tag{7.63}$$

$$\Delta I_p^{(IS)} = \frac{9}{10}(2Z-1)\frac{e^2}{r_{WS}}. \tag{7.64}$$

Note that in this strongly coupled plasma limit the ionization potential lowering is independent of electron temperature and is roughly equal to the attractive energy of the Z+ ion and an electron separated by half the Wigner–Seitz radius. These two intuitively derived formulas, Eqs. (7.57) and (7.64), are found to be limits to the well-known Stewart–Pyatt formula for continuum lowering (Stewart and Pyatt 1966):

$$\Delta I_p^{(SP)} = \frac{k_B T_e}{2(Z+1)}\left[\left(\frac{3(Z+1)}{k_B T_e}\frac{Ze^2}{\lambda_D} + 1\right)^{2/3} - 1\right], \tag{7.65}$$

where those authors show that the appropriate "Z" to use in the formula is the following ratio:

$$Z \to Z^* = \frac{\overline{Z^2}}{\overline{Z}} = \frac{\sum_j \hbar Z_j^2 n_j}{\sum_j Z_j n_j}. \tag{7.66}$$

In virtually all ionization calculations performed in a cluster, it is appropriate to include the ionization lowering predicted by one of these formulae. This will generally have the effect of accelerating the usual vacuum ionization rate of the cluster ions. This effect is one of a number of phenomena which accelerate ionization of ions in a laser-irradiated cluster.

7.2.2 Inner-Ionization by Laser-Driven and Ion-Field Assisted Tunneling

Turning now to ionization behavior in the early phases of the cluster interactions (phases 1 and 2 in Figure 7.4), we first consider inner-ionization. It is the process of inner-ionization that creates the free electrons forming a cluster nanoplasma and the resulting highly charged ions that are observed in experiments. It is the production of highly charged ions that is one of the salient experimental observables of intense laser–cluster interactions. As an aside, it is worth mentioning that at the time of this writing there is still some mystery surrounding the observation of the very highest charge states in high Z clusters. For example Xe charges as high as 40+ have been observed when large Xe clusters are irradiated at only 10^{16} W/cm^2 (Ditmire et al.1997d).

Generally inner-ionization proceeds by two mechanisms, (a) direct laser field ionization (i.e. tunneling) and (b) electron collisional ionization. We first consider the former mechanism. To the simplest approximation, the atoms and ions in the cluster are ionized by the laser field just as free atoms irradiated by such a laser field become sequentially ionized. Therefore a good starting place for understanding this form of inner-ionization is the tunneling models detailed in Chapter 4. In the cluster, however, there are a number of additional physical mechanisms that tend to enhance the laser-driven tunnel ionization rate, sometimes dramatically (Smirnov and Krainov 2004). In particular, there are four reasons why laser-driven tunneling is enhanced inside the cluster.

First, as we have discussed, the ionization potential is lowered when an atom is immersed inside the cluster. This mandates the substitution $I_p \to I_p - \Delta I_p$. Because of the very nonlinear nature of tunnel ionization, this continuum lowering can have a large impact on the ionization rate at a given laser intensity. Second, we have also seen that collective electron motion of nanoplasma electrons in a cluster retaining the free electrons can either shield the laser field inside the cluster or enhance it near the Mie resonance. Again, the nonlinear nature of tunneling can cause a dramatic increase in the ionization for a time while an expanding cluster has electron density near the Mie resonance ($\sim 3\, n_{crit}$). In a similar vein, in the regime where free electrons have been removed from the cluster (by outer-ionization) the field of the residual cluster ion sphere can aid in enhancing the laser field and increase the tunnel ionization rate.

The fourth mechanism for enhancing the laser-driven tunnel ionization rate in an ionizing cluster arises from local field enhancements that can occur when ions in the cluster are near the ionizing atom in question. This mechanism is akin to the peak in tunneling rate we observed in a diatomic molecule with its internuclear axis oriented along the laser polarization. As you will recall from Chapter 5, the presence of the second ion on an ionizing diatomic molecule when placed near the critical distance served to distort the binding potential of the up-field ion and enhance its ionization. Local field enhancements from ions in the cluster can play the same role if they venture to within the right distance of the ionizing atom. This tunnel ionization rate enhancement can be termed "ion-field assisted" tunneling (or IFA ionization). This effect is schematically illustrated in Figure 7.9.

We can derive a quantitative model for this field enhancement by realizing that this problem in the cluster is very similar to the well-known broadening of spectroscopic radiating

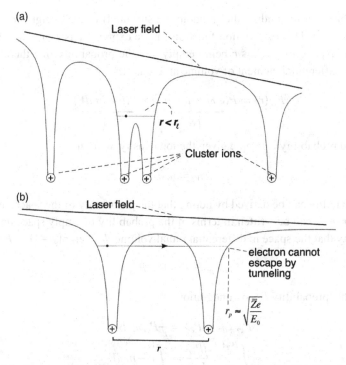

Figure 7.9 Cartoon of potential curves in a cluster with randomly distributed ions and an imposed laser field. Part (a) shows the situation in which the electron is delocalized between two closely spaced ions. The plot in (b) shows the weak field situation in which an ion-field assisted tunneling event has the tunneling frustrated by the confining potential of the down-field ion.

lines of atoms and ions in a plasma from quasi-static ion fields. The theory used to describe this ion-field line broadening is known as the Holtsmark ion field theory (Holtsmark 1919). We can adapt the approach of this theory to calculations of IFA ionization enhancement in the cluster (Gets and Krainov 2006). One basic approximation of the Holtsmark theory is the "Nearest Neighbor Approximation." This approximation assumes that only the field of the closest ion in the plasma contributes to the line broadening or, in our case, the tunneling enhancement. The simplest manifestation of the Holtsmark theory is to assume that the nearest neighbor ions are randomly distributed around the ionizing atom in question. This approximation ignores the effects of Coulomb repulsion between ions and, as a result, overestimates the probability of the nearest neighbor resting close to the ionizing atom. (Coulomb repulsion dramatically reduces the probability at close separations.) As we shall see, that has little consequence to our calculation. We shall also proceed by ignoring Debye shielding of the field of the nearest neighbor ion. This approximation is appropriate if the Debye radius is greater than the Wigner–Seitz radius, which will turn out to be close to the most probable nearest neighbor separation.

So we can amend the tunneling ionization rate formula by adding the fields of the randomly distributed nearest neighbor ions to the laser field. We first need a probability distribution function for finding the (randomly distributed) nearest neighbor ion at some radial separation, r. This distribution function can be expressed as a product of the probability of the space up to radius r being empty and the probability that there is one ion present in the differential element of volume at radius r:

$$F_H(r) = \underbrace{P(\text{empty up to } r)}_{P_{er}} \times \underbrace{P(\text{ion in } dV)}_{R_{dV}} \quad (7.67)$$

The second probability is simple given the ion density, written

$$P_{dV} = n_i \, dV. \quad (7.68)$$

The first probability can be derived by noting that the probability of the space being empty up to radius $r + dr$ can be written in terms of the probability of empty space up to r times the probability that the space in differential radial volume dr is empty $= (1 - P_{dV})$:

$$P_{er+dr} = P_{er}(1 - P_{dV}). \quad (7.69)$$

We can find the probability P_{er} by integration:

$$P_{er+dr} - P_{er} = -P_{er} n_i \, dV \quad (7.70)$$

$$\int_1^{P_{er}(r)} \frac{dP_{er}}{P_{er}} = \int_0^V -n_i \, dV \quad (7.71)$$

$$P_{er}(r) = e^{-n_i V}. \quad (7.72)$$

The probability distribution function we seek can be found using Eq. (7.69) and Eq. (7.72), written in terms of radial separation and Wigner–Seitz radius, r_{WS}, noting that $n_i = (4\pi r_{WS}/3)^{-1}$ to yield

$$F_H(r) dr = 3 \left(\frac{r}{r_{WS}}\right)^2 e^{-(r/r_{WS})^3} \frac{dr}{r_{WS}}. \quad (7.73)$$

It is well known in ion line broadening theory that this formula overestimates the number of ions at small distance because Coulomb repulsion between the nearest neighbor ions repels them. (Instead of falling off as r^2 as r approaches zero, the true function falls off exponentially for decreasing r.) This function is properly normalized.

Now we must add the fields from a nearest neighbor ion with separation probability described by Eq. (7.73) to the laser field in some static field ionization rate. We also assume that the nearest neighbor is oriented at random with the laser polarization axis. We can do this by using the PPT/ADK tunneling formula altered to calculate the tunneling rate in an instantaneous steady E-field. In Eq. (4.214) we found the cycle-averaged tunnel ionization rate for complex atoms. Furthermore, Eq. (4.186) shows how to relate this cycle-averaged rate to a static DC field ionization rate. This lets us write

$$W^{(stat)}(E_0) = \left(\frac{3}{\pi} \frac{\tilde{E}}{\tilde{I}_p^{3/2}}\right)^{-1/2} W^{(cycleave)}(E_0), \quad (7.74)$$

7.2 Electron Dynamics and Cluster Ionization

which means we can write a static complex atom tunneling rate as

$$W_{PPT}^{(stat)}(E) = \zeta_{ml} \left(\frac{\tilde{I}_p^{3/2}}{\tilde{E}}\right)^{2n-1} \exp\left[-\frac{2}{3}\frac{\tilde{I}_p^{3/2}}{\tilde{E}}\right], \qquad (7.75)$$

where the prefactor can be derived by comparison with Eq. (4.214). (Twiddled quantities represent dimensionless ionization potential and E-field. The I_p is normalized to the I_p of hydrogen and the E-field is normalized to the atomic E-field; see Chapter 4 for these definitions.)

We can now add in the effects of the (randomly oriented) nearest neighbor ion field. We shall assume that all ions in the cluster have the same charge state, given by the average charge state, Eq. (7.45). The ion field is simply

$$E_I = \frac{\bar{Z}e}{r^2}. \qquad (7.76)$$

At this point it is worth recalling the discussion about tunnel ionization of diatomic molecules. We noted that if the ion separation in the diatomic was too small in fact the down-field ion did not enhance the tunnel ionization; instead the electron was delocalized between the two ions. So we must say that there is a cut-off, short-range separation in which the ion field does not enhance ionization. This localization radius was found in a diatomic (recall Section 5.2.5):

$$R_l \cong \frac{3\bar{Z}e^2}{I_p(\bar{Z})}. \qquad (7.77)$$

We will therefore ignore ion-field enhancements for all nearest neighbor ions with separation less than Eq. (7.77). It is also useful to introduce the ion E-field at a separation equal to the Wigner–Seitz radius:

$$E_{I0} \equiv \frac{\bar{Z}e}{r_{WS}^2}, \qquad (7.78)$$

which allows us to rewrite the nearest neighbor separation probability distribution function as an ion-field-enhanced field distribution. Using

$$dE_I = -2\frac{\bar{Z}e}{r^3}dr = -\frac{2}{\sqrt{\bar{Z}e}}E_I^{3/2}dr, \qquad (7.79)$$

Eq. (7.73) becomes, in terms of the ion field,

$$F_H(E_I)\,dE_I = \frac{3}{2}\left(\frac{E_{I0}}{E_I}\right)^{5/2} e^{-(E_{I0}/E_I)^{3/2}}\frac{dE_I}{E_{I0}}. \qquad (7.80)$$

We must add only the component of the ion field that lies along the laser polarization axis. If θ is the angle between laser field direction and the separation axis of the nearest neighbor ion,

$$E_I^{\parallel} = E_I \cos\theta.$$

Now to find the IFA-enhanced tunneling rate we must take the static field tunneling rate, add the ion field and average over ion orientation angle and ion-field strength weighted by the probability function, Eq. (7.80):

$$W_{IFA}^{(stat)}(E) = \frac{1}{2}\int_0^\infty \int_0^\pi W_{PPT}^{(stat)}(E + E_I \cos\theta) F_H(E_I) \sin\theta \, d\theta \, dE_I. \qquad (7.81)$$

Remember that we say that the enhancing field is zero when $r < R_I$, or that there will only be an ion-field enhancement for ion fields less than a critical value:

$$E_I < E_{IC} = \frac{Ze^2}{r_I^2} = \frac{I_p^2}{9e^2\bar{Z}}. \qquad (7.82)$$

This breaks the average into two parts so that the IFA-enhanced tunneling rate is

$$W_{IFA}^{(stat)} = \int_{E_{IC}}^\infty W_{PPT}^{(stat)} F_H(E_I) dE_I$$

$$+ \frac{1}{2}\int_0^{E_{IC}} \int_0^\pi \zeta_{ml} \left(\frac{\tilde{I}_p^{3/2}}{\tilde{E} + \tilde{E}_I \cos\theta}\right)^{2n-1} \exp\left[-\frac{2}{3}\frac{\tilde{I}_p^{3/2}}{\tilde{E} + \tilde{E}_I \cos\theta}\right]$$

$$\times F_H(E_I)\sin\theta \, d\theta \, dE_I \qquad (7.83)$$

We can ignore the ion field in the prefactor and Taylor-expand around the small ion field in the exponent. This lets us pull the PPT/ADK ionization rate out of the integral and restore the cycle-averaged cofactor from Eq. (7.74). The IFA rate then becomes equal to the PPT/ADK tunneling rate times an enhancement factor.

At this point we do need to treat the field limits in the integral with a bit of care. In strong laser fields, our Taylor expansion and cutoff field is appropriate. However, in weak fields the Taylor expansion leads to the unphysical result that the ionization rate begins to grow exponentially with decreasing laser field. This is a consequence of treating the perturbing ion field as a constant in space. In weak fields, as illustrated in Figure 7.9b, an electron is bound by the down-field ion binding potential even though the tunneling formula predicts a rapid tunneling rate. We can fix this unphysical prediction by amending the critical cutoff field. The condition for frustrated tunneling (when the downfield ion can still confine the electron) is depicted in Figure 7.9b and is when the total potential energy from laser and ion equals the binding potential,

$$-\frac{\bar{Z}e^2}{r_p} - eE_0(r + r_p) = -I_p, \qquad (7.84)$$

where r_p is illustrated in the figure. So the critical separation for this frustrated tunneling occurs for a separation found by solving for r in (7.84):

$$r'_C = \left(I_p - eE_0 r_p - \frac{\bar{Z}e^2}{r_p}\right)\frac{1}{eE_0} = \frac{I_p}{eE_0} - 2\sqrt{\frac{\bar{Z}e}{E_0}}. \qquad (7.85)$$

7.2 Electron Dynamics and Cluster Ionization

In weak fields it is appropriate to ignore the second term in Eq. (7.85), which means that there is no IFA enhancement for ion fields greater than $\bar{Z}e/(r'_c)^2$, which is equal to

$$\frac{\bar{Z}e^3}{I_p^2}E_0^2 = \left(\frac{E_0}{4E_{BSI}}\right)E_0 \approx \frac{1}{4}E_0, \tag{7.86}$$

where E_{BSI} was the field calculated for barrier suppression ionization in Section 4.4.8. (We assumed in Eq. (7.86) that at the field of significant tunneling it is approximately equal to the BSI field.) So, we can account for this effect by amending the critical ion field used in Eq. (7.83):

$$E'_{IC} = Min[E_{IC}, \xi_{IC}E_0,],$$

where $\xi_{IC} \approx 1/4$.

With these considerations the IFA-enhanced tunneling rate Eq. (7.83) becomes

$$W_{IFA}^{(cyc\,ave)} \equiv W_{IFA} = W_{PPT}(E_0)\left\{\int_{E'_{IC}}^{\infty} F_H(E_I)dE_I\right.$$
$$\left.+\frac{1}{2}\int_0^{E'_{IC}}\int_0^{\pi} \exp\left[\frac{2}{3}\frac{\tilde{I}_p^{3/2}}{\tilde{E}_0^2}\tilde{E}_I\cos\theta\right] F_H(E_I)\sin\theta\,d\theta\,dE_I\right\}. \tag{7.87}$$

The first term can be evaluated with Eq. (7.80) and the fact that

$$\int_y^{\infty} \frac{3}{2}\frac{1}{x^{5/2}}\exp\left[-1/x^{3/2}\right]dx = 1 - \exp\left[-1/y^{3/2}\right].$$

The θ integral of the second term can be evaluated because

$$\int_0^{\pi} \exp[a\cos x]\sin x\,dx = \frac{(e^a - e^{-a})}{a},$$

yielding

$$W_{IFA} = W_{PPT}(E_0)\left\{\left(1 - \exp\left[-\left(\frac{E_{I0}}{E'_{IC}}\right)^{3/2}\right]\right) + \int_0^{E'_{IC}/E_a} \frac{3}{4}\frac{\tilde{E}_0^2}{\tilde{I}_p^{3/2}}\frac{1}{\tilde{E}_I}\right.$$
$$\left.\times\left(\exp\left[\frac{2}{3}\frac{\tilde{I}_p^{3/2}}{\tilde{E}_0^2}\tilde{E}_I\right] - \exp\left[-\frac{2}{3}\frac{\tilde{I}_p^{3/2}}{\tilde{E}_0^2}\tilde{E}_I\right]\right)F_H(\tilde{E}_I)d\tilde{E}_I\right\}, \tag{7.88}$$

where we have normalized electric field to an atomic field strength, E_a. With the help of Eq. (7.80), the integral in Eq. (7.88) can be written in the form

$$\int_0^{x_c} \frac{1}{x^{7/2}}\left(\exp\left[-\frac{1}{x^{3/2}} + \alpha x\right] - \exp\left[-\frac{1}{x^{3/2}} - \alpha x\right]\right)dx \tag{7.89}$$

if we define $x = E_I/E_{I0}$ and $x_c = E'_{IC}/E_{I0} = r_{WS}/r_l$, and write the constant from the tunneling formula as

$$\alpha \equiv \frac{2}{3}\frac{\tilde{I}_p^{3/2}}{\tilde{E}_0}\tilde{E}_{I0} = \frac{2}{3}\frac{\tilde{I}_p^{3/2}}{\tilde{E}_0}\bar{Z}\left(\frac{a_B}{r_{WS}}\right). \tag{7.90}$$

Both terms in Eq. (7.89) can be approximated by Laplace's Method (Zwillinger 1992, p. 221). (In brief, this method is applied to integrals of exponentials with strongly peaked arguments. The approximation argues that the integral is dominated by the region near the peak of the argument of the exponent, so for peaks internal to the range of integration, like the second term in (7.89), the argument is approximated by a Gaussian and for terms with maximum at the edge of the integration range, like the first term, the argument is approximated as a pure exponential.) Noting that

$$\tilde{E}_{I0} = \bar{Z}\left(\frac{a_B}{r_{WS}}\right)^2 \tag{7.91}$$

and introducing the definition

$$R'_I = \text{Max}\left[\frac{3\bar{Z}e^2}{I_p(Z)}, \sqrt{\frac{\bar{Z}}{\xi_{IC}\tilde{E}_0}}\, a_B\right], \tag{7.92}$$

Laplace integration of the two terms in Eq. (7.89) yields our final result for the IFA tunnel ionization rate:

$$\begin{aligned}
W_{IFA} = W_{PPT}(E_0) &\left\{\left(1 - \exp\left[-\left(\frac{R'_I}{r_{WS}}\right)^{3/2}\right]\right) + \frac{9}{8}\frac{\tilde{E}_0^2}{\tilde{I}_p^{3/2}}\frac{1}{\bar{Z}}\left(\frac{r_{WS}}{a_B}\right)^2\right. \\
&\times \left(\frac{R'_I}{r_{WS}}\right)^7 \left[\frac{3}{2}\left(\frac{R'_I}{r_{WS}}\right)^5 + \frac{2}{3}\frac{\tilde{I}_p^{3/2}}{\tilde{E}_0^2}\bar{Z}\left(\frac{a_B}{r_{WS}}\right)^2\right]^{-1} \\
&\times \exp\left[-\left(\frac{R'_I}{r_{WS}}\right)^3 + \frac{2}{3}\frac{\tilde{I}_p^{3/2}}{\tilde{E}_0^2}\bar{Z}\left(\frac{a_B}{r_{WS}}\right)^2\right] \\
&\left. - \sqrt{\frac{8\pi}{15}}\left[\frac{4}{9}\frac{\tilde{I}_p^{3/2}}{\tilde{E}_0^2}\bar{Z}\left(\frac{a_B}{r_{WS}}\right)^2\right]\exp\left[-\zeta_I\left(\frac{\tilde{I}_p^{3/2}}{\tilde{E}_0^2}\bar{Z}\left(\frac{a_B}{r_{WS}}\right)^2\right)^{3/5}\right]\right\}.
\end{aligned} \tag{7.93}$$

The numerical coefficient in the final exponential function is

$$\zeta_I = \left(\left(\frac{2}{3}\right)^{3/5} + \left(\frac{3}{2}\right)^{2/5}\right)\left(\frac{2}{3}\right)^{3/5} \cong 1.54. \tag{7.94}$$

While this formula is cumbersome, it is easily evaluated numerically. A slightly less cumbersome form can be used by noting that the third term in the brackets is usually negligible and can be dropped. (Physically, this term arose from integrating the angle from 0 to π and therefore represents the reduction to the tunneling rate for the half-cycle that the laser field is pointed in the direction opposite that of the enhancing ions field and is slightly reduced by the presence of a nearest neighbor ion.)

Plots of the enhancement factor (that factor multiplying the PPT/ADK rate in Eq. (7.93)) are shown in Figure 7.10a for ionization of Ar^{8+}. (Note that because α in Eq. (7.89) gets small with large r_{WS}, the Laplace method for integrating those terms breaks down and the equation derived therefore exhibits the nonphysical behavior of an enhancement factor that falls below 1 at large r_{WS}. A reasonable assumption then is to restrict the enhancement

7.2 Electron Dynamics and Cluster Ionization

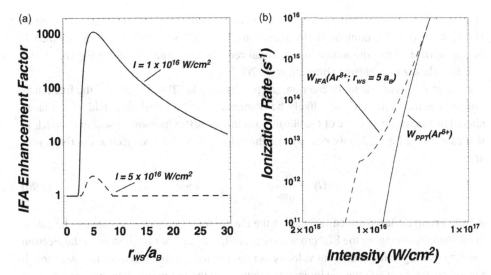

Figure 7.10 (a) Calculated IFA enhancement to the tunneling ionization rate of Ar^{8+} as a function of Wigner–Seitz radius in an Ar cluster nanoplasma for two different intensities. (b) Calculated IFA tunnel ionization rate for Ar^{8+} as a function of intensity with a Wigner–Seitz radius of five Bohr radii (2.5 Å). The weak field cutoff is discussed in the text. This IFA rate is compared to the standard PPT/ADK tunneling rate.

factor to 1.0 when it falls below 1 with large r_{WS}.) Here we see behavior that is very much like the tunnel ionization rate we calculated in a diatomic molecule as a function of the separation of the molecule's two ions (recall Figure 5.8). We see an enhancement, which can be quite large, when the average ion spacing, which is roughly equal to r_{WS}, is equal to a critical value. As in a diatomic, this critical ion spacing is ~5–6 Bohr radii.

The IFA ionization rate for the same ion is plotted as a function of laser intensity when $r_{WS} = 5a_B$ in Figure 7.10b and compared to the unaltered PPT/ADK tunneling rate for that ion. Clearly the appearance intensity for 9+ is reduced because of the IFA enhancement. This analysis suggests that as a cluster explodes it will pass through an ion density in which IFA tunneling will greatly increase the ionization rate and help to produce more highly charged ions.

7.2.3 Thermal and Laser-Driven Collisional Inner-Ionization

There is a second mechanism for inner-ionization in the cluster, namely ionization through collisions with other free electrons in the cluster nanoplasma, electron collisional ionization (ECI). The rate of ECI in a plasma can be calculated with a simple empirically derived formula for the cross section describing the ionization of an atom or ion of ionization potential I_p by an incident free electron of kinetic energy ε_e. This widely used formula, introduced by Lotz (Lotz 1968), reads

$$\sigma_{CI} \cong a_i q_i \frac{\ln[\varepsilon_e/I_P]}{\varepsilon_e I_P}. \tag{7.95}$$

Here $a_i = 4.5 \times 10^{-14}$ cm^2 – eV^2 = 1.15×10^{-37} cm^2 – erg^2 is an empirically derived constant and q_i is the number of electrons in the valence shell of the ion being ionized. It is, appropriate to use the ionization potential reduced by continuum lowering for cluster ions in a cluster nanoplasma (Bornath et al. 2007).

Early in the laser–cluster interaction, say phases (1) and (2) of Figure 7.4, the free inner-ionized electron velocities will likely be dominated by the oscillating field of the laser, altered by the restoring force of the cluster ions in the cluster harmonic oscillator model. In that case, using Eq. (7.30) we can say that the velocities of the free electrons in the cluster are

$$v(t) = \frac{eE_0}{m\omega} \frac{1}{1 - \omega_{pe}^2/3\omega^3} \sin\omega t. \tag{7.96}$$

A time-averaged, laser-driven ECI rate in the cluster for a given charge state can be calculated, then, by averaging the ECI cross section of (7.95) over the time history of the electron velocity. Noting that a minimum velocity is required for the free electrons to have enough kinetic energy to overcome the ionization potential of the ion in question, we can write the time-averaged ECI rate as

$$W^{(las)}_{ECI} = \langle n_e \sigma_{CI} v \rangle$$

$$= \frac{2n_e}{\pi} \int_{\varphi_{min}}^{\pi/2} \sigma_{CI} \frac{eE_0}{m\omega} \frac{1}{1 - \omega_p^2/3\omega^2} \sin\varphi \, d\varphi \tag{7.97}$$

It is convenient to transform this integral over phase to one over electron kinetic energy. If we define the ponderomotive energy of electrons oscillating in the cluster electron cloud, $U_p^{(c)}$:

$$\frac{eE_0}{m\omega} \frac{1}{1 - \omega_p^2/3\omega^2} = \sqrt{\frac{4U_p^{(c)}}{m}}, \tag{7.98}$$

we can write

$$d\varepsilon_e = 4U_p^{(c)} \cos\varphi \sin\varphi \, d\varphi \tag{7.99}$$

and recast the integral over energy

$$W^{(las)}_{ECI} = \frac{n_e a_i q_i}{\pi I_p} \frac{1}{\sqrt{mU_p^{(c)}}} \int_{I_p}^{2U_p^{(c)}} \frac{\ln[\varepsilon_e/I_p]}{\varepsilon_e \sqrt{1 - \varepsilon_e/2U_p^{(c)}}} \, d\varepsilon_e. \tag{7.100}$$

Actually, this integral can be performed analytically using a generalized hypergeometric function. However, it is almost as accurate to note that most of the ionization occurs for electrons in the range just above the ionization potential, so we can Taylor-expand the natural logarithm in (7.100) around 1:

$$\ln[\varepsilon_e/I_p] \cong \frac{3}{2} - 2\frac{I_p}{\varepsilon_e} + \frac{1}{2}\left(\frac{I_p}{\varepsilon_e}\right)^2, \tag{7.101}$$

which leads to an analytic formula for the cycle-averaged laser-driven ECI rate in the cluster

$$W^{(las)}_{ECI} = \frac{n_e a_i q_i}{\pi I_p} \frac{1}{\sqrt{mU_p^{(c)}}} \left\{ \left[3 + \frac{I_p}{U_p^{(c)}} + \frac{3}{32}\left(\frac{I_p}{U_p^{(c)}}\right)^2 \right] \ln\left[\frac{1+\sqrt{1-I_p/2U_p^{(c)}}}{\sqrt{I_p/2U_p^{(c)}}}\right] \right.$$
$$\left. - \left(\frac{7}{4} + \frac{3}{16}\frac{I_p}{U_p^{(c)}}\right)\sqrt{1-I_p/2U_p^{(c)}} \right\}. \tag{7.102}$$

Later in the interaction with the cluster (say Phase 3 as described in Section 7.1.3), the laser field amplitude has dropped and the laser-driven ECI rate falls. However, the interaction may have left behind a cluster nanoplasma with a heated electron plasma cloud. In that case further ECI ionization may occur from thermal electrons. To calculate this rate, we now have to perform the velocity average over a Maxwellian velocity distribution. Such a normalized Maxwellian distribution can be written

$$f(v)dv = \left(\frac{m}{2\pi k_B T_e}\right)^{3/2} \exp\left[-\frac{mv^2}{2k_B T_e}\right] 4\pi v^2 dv \tag{7.103}$$

or in terms of electron kinetic energy

$$f(\varepsilon_e)d\varepsilon_e = \frac{2}{\pi^{1/2}} \frac{1}{(k_B T_e)^{3/2}} \sqrt{\varepsilon_e} \exp[-\varepsilon_e/k_B T_e] d\varepsilon_e. \tag{7.104}$$

So, the thermal ECI rate is found, with an integration by parts:

$$W^{(kT)}_{ECI} = \langle n_e \sigma_{ECI} v \rangle_{T_e}$$
$$= \frac{n_e a_i q_i}{I_p} \frac{2\sqrt{2}}{\pi^{1/2} m^{1/2}} \frac{1}{(k_B T_e)^{3/2}} \int_{I_p}^{\infty} \ln[\varepsilon_e/I_p] \exp[-\varepsilon_e/k_B T_e] d\varepsilon_e$$
$$\frac{2\sqrt{2} n_e a_i q_i}{I_p \pi^{1/2} m^{1/2}} \frac{1}{(k_B T_e)^{1/2}} \int_{I_p/k_B T_e}^{\infty} \frac{e^{-x}}{x} dx. \tag{7.105}$$

$$W^{(kT)}_{ECI} = \frac{2\sqrt{2} n_e a_i q_i}{I_p \pi^{1/2} m^{1/2}} \frac{1}{(k_B T_e)^{1/2}} \Gamma[0, I_p/k_B T_e]. \tag{7.106}$$

Here $\Gamma[0,x]$ is the incomplete Gamma function. It has been shown in simulations that the ECI rate can be enhanced by the presence of a strong electric field from the cluster ion core if significant outer-ionization has occurred (Fennel et al. 2007). A reasonable approximation in numerical models of the cluster evolution is to continuum-lower the published ionization potentials.

Before concluding this section, it is worth noting that in a dense cluster the entire cast of free electrons can also recombine by three-body recombination and therefore reduce, to an extent, the level of inner-ionization. This well-known plasma recombination rate is strongly dependent on electron temperature and can be calculated roughly from the following formula of recombination rate per ion:

$$W_{3BR} = n_e^2 \frac{4\sqrt{2}\pi^{3/2}}{9} \frac{Z^3 e}{m^{1/2}(k_B T_e)^{9/2}} \ln \Lambda \tag{7.107}$$

Any numerical model of cluster dynamics should include this contribution which can play an important role later in the laser–cluster interaction. (Note that $\ln \Lambda$ is the well-known Coulomb logarithm, which is discussed in Appendix B.)

7.2.4 Cluster Ionization Dynamics and Ionization Ignition

Inner-ionization of cluster ions, then, can differ significantly from ionization in isolated atoms or small molecules. The four phenomena considered (ionization potential lowering, field shielding and enhancement from collective oscillations of the electron cloud, IFA tunneling, and electron collisional ionization) conspire to make the effective inner-ionization rate in the laser field higher than under identical intensity and pulse-duration conditions in isolated atoms. The clear observable is production of higher charge states in the laser pulse; some initial ionization aids further ionization. This phenomenon has generally been referred to as "ionization ignition" (Rose-Petruck et al. 1997; Bauer and Macchi 2003).

Another aspect to note is that two of these ionization enhancement mechanisms, IFA tunneling and collective electron cloud field enhancement, peak with an ion density (for the former) and electron density (for the latter) that are somewhat below solid density. Therefore, some cluster expansion is needed to enhance inner-ionization. An upshot of this observation is that laser pulse duration matters, since in some cases, depending on the cluster size and ion mass, longer laser pulses than the shortest used in experiments can actually lead to higher ionization states by initially ionizing the cluster in the rising edge of the pulse and then further ionizing it later in the pulse tens or hundreds of femtoseconds later when the cluster has expanded (Fukuda et al. 2003; Fennel et al. 2010). A semiquantitative extent of these dynamics can be illustrated with a few simple numerical examples.

Figure 7.11 shows the predicted ionization state history of an ensemble of isolated argon atoms calculated using a simple rate equation model and the standard PPT/ADK tunnel ionization rates assuming a 50 fs, 800 nm laser pulse irradiating the atoms at a peak intensity of 5×10^{15} W/cm^2. This kind of calculation is essentially identical to the ones presented in Chapter 4 (recall Figure 4.32). The population densities of the various charge states in argon are plotted versus time in the upper panel and the average charge is plotted in the lower panel. The usual tunneling theory predicts saturation of ionization to Ar^{4+} with some subsequent 5+ production before the pulse intensity starts to fall. For comparison, the same calculation performed using the IFA tunneling rate from Eq. (7.93) at a fixed Wigner–Seitz radius of $4.5 a_B$ is shown in Figure 7.12. Clearly higher charge states are produced in the latter case, with average charge at the end of the interaction increasing from $\sim$4.3 to $\sim$6.

The effects of expansion can be considered if the field in the cluster is affected by the collective electron oscillations as predicted by the rigid electron sphere harmonic oscillator model and by assuming that all inner-ionized electrons remain confined to the cluster body. Such a calculation assuming a 5 nm Ar cluster which expands as if the electron gas has pressure from a 100 eV temperature (which might be considered a "slow" expansion) is shown in Figure 7.13. In this case, with identical peak intensity and pulse duration as the

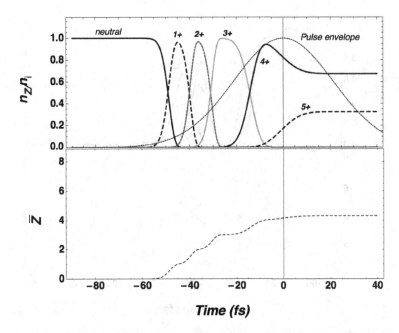

Figure 7.11 Calculation of sequential ionization history of individual Ar atoms irradiated by a 50 fs pulse with peak intensity of 5×10^{15} W/cm^2. This calculation was performed with the simple PPT/ADK tunneling rates of Eq. (4.214).

previous two examples, the average charge state reached is barely 3+, below that of single ions. The reason for this ionization rate suppression can be seen from the plots in the middle panel of Figure 7.13, which shows the cluster radius as a function of time (with its modest increase during the laser pulse) and the effective intensity in the cluster interior as calculated from Eq. (7.27). Electron cloud shielding reduces the intensity below the vacuum value for the entirety of the interaction (with a momentary exception on the rising edge of the pulse when initial ionization very briefly causes the cluster nanoplasma to pass through the Mie resonance.)

If, however, the cluster expansion is accelerated, a calculation presented in Figure 7.14, the situation is different. In this calculation the electron cloud is assumed to have a pressure associated with a 1000 eV plasma greatly accelerating the cluster expansion, which can be seen comparing the cluster radius history in the middle panels of Figure 7.13 and 7.14. The accelerated expansion serves to lower the electron density, plotted in the third panel of this figure to a value near 3 n_{crit} just after the peak of the laser pulse. The field enhancement associated with this Mie resonance boosts the final average ionization state to 8+.

Next, the manifestation of a behavior that could be viewed as ionization ignition appears when we perform the same simple rate equation calculation including the IFA tunneling rate from a shielded Ar cluster irradiated by a laser pulse that is ten times higher peak intensity. Here, in Figure 7.15, we see some initial tunnel ionization at −80 fs followed by a rather narrow time window in which ionization quickly tears through ionization stages 3+ to 8+ at around −40 fs.

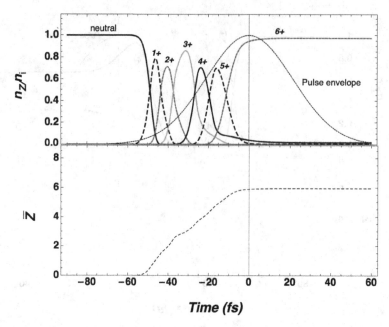

Figure 7.12 Calculation of sequential ionization history of Ar atoms in a cluster with a fixed Wigner–Seitz radius of 4.5 a_B irradiated by a 50 fs pulse with peak intensity of 5×10^{15} W/cm^2.

Finally, we consider the effects of ECI. In small to moderate-sized clusters (like the 5 nm clusters we have been examining) ECI plays a rather minor role in the ionization dynamics. Nonetheless, it can help to produce even in small clusters a small population of rather highly charged ions. Another simple rate equation calculation like those illustrated previously but for a 5 nm Xe cluster irradiated at 10^{17} W/cm^2 is presented in Figure 7.16. The upper panel shows average ionization state for isolated Xe atoms, the Xe atoms subject to IFA ionization in a "rapidly" expanding Xe cluster and the same situation in which the laser-driven ECI is added to the ionization rate equations. While isolated Xe is ionized to about 10+ at this intensity, IFA enhancement in the cluster increases the ion charge state a little, to 12+. ECI has a very minor effect on the average charge state, but, as the lower panel illustrates, which shows the population of ion charge states at the end of the laser pulse, ECI serves to produce a small number of more highly charged ions up to 18+, in this example.

It is worth noting that the trends seen here in this simple model are consistent with experiments as well as more sophisticated MD simulations (Saalman et al. 2006). In fact, the importance of cluster expansion because of IFA enhancements is seen in MD simulations even in very small clusters. Figure 7.17 illustrates an adaptation of MD simulations by Siedschlag and Rost (2002) which show that in a 16-atom Xe cluster higher average ionization is observed when the laser pulse is stretched to 1 ps even though the peak intensity falls with this stretching. This was attributed to IFA ionization by the authors. Similarly, high charge state production is consistently seen in MD simulations, like those of

7.2 Electron Dynamics and Cluster Ionization

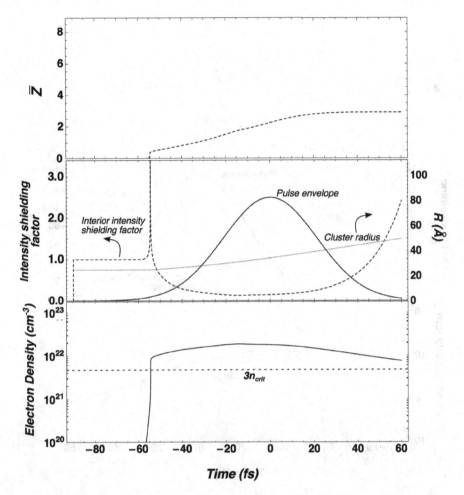

Figure 7.13 Example Ar cluster ionization calculation using the IFA tunneling formula and accounting for electron cloud field shielding inside the cluster. Again the pulse is 50 fs FWHM and has a peak intensity of 5×10^{15} W/cm^2. The cluster begins with a 5 nm diameter (~1400 atoms) and is assumed to expand slowly such that the Mie resonance occurs well after laser pulse. (Expansion rate is characteristic of a 100 eV electron temperature.)

Last and Jortner (2004). Charge state distributions they predict in ~1000-atom Xe clusters are shown in Figure 7.18.

On the whole, the current wisdom from numerical simulations including not only MD but other techniques like density functional theory, is that ionization ignition in moderate to large clusters results mainly from field enhancements by collective electron motion in the cluster coupled to passage through a Mie-type resonance during cluster expansion (Smirnov and Krainov 2004; Fennel et al. 2010).

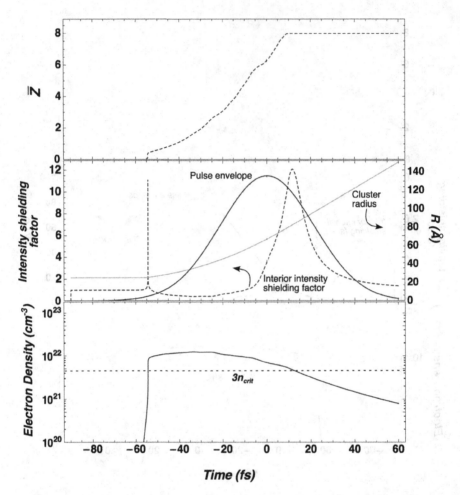

Figure 7.14 Example Ar cluster ionization calculation using the IFA tunneling formula under the same conditions as Figure 7.13; however, the cluster is now assumed to expand rapidly such that the Mie resonance occurs during laser pulse (when the electron density falls by cluster expansion 3 n_{crit}. (Expansion rate is characteristic of a 1000 eV electron temperature.)

7.2.5 Cluster Outer-Ionization by Laser Field–Driven Electron Ejection

We turn now to the question of cluster outer-ionization; how and how fast are electrons removed from the cluster as a whole leading to an electrostatic charging of the cluster sphere. In most macroscopic plasmas, space-charge forces enforce quasi-neutrality. The nanoscale of clusters means that space-charge forces can often be easily overcome by laser fields or the thermal energy of the cluster electrons. This suggests that we can think of

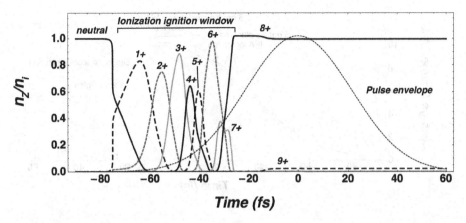

Figure 7.15 Calculation of sequential IFA ionization history of Ar atoms in a cluster with a varying Wigner Seitz due to fast expansion (1000 eV electron temperature). The laser pulse now is a 50 fs pulse with peak intensity of 5×10^{16} W/cm^2.

outer-ionization as occurring by two separate mechanisms: direct laser field extraction of free electrons and thermally activated escape of heated electrons from the cluster fields.

Considering the first mechanism, it is clear that significant outer-ionization will occur if the laser field transiently exceeds the space-charge field of the spherical ion assembly. Furthermore, the oscillation amplitude of the free electrons should probably exceed significantly the cluster radius; otherwise, we expect these electrons to remain localized to the cluster ion region. Therefore, the first simple constraint on laser field–induced outer-ionization is that in near-IR laser fields this condition is fulfilled in large clusters for even modest intensity:

$$\frac{eE_0}{m\omega^2} \geq R. \tag{7.108}$$

In truth, calculating the extent of outer-ionization in a cluster is a dynamic stochastic process and is therefore challenging, most accurately done with molecular dynamic (MD)-style particle calculations (Saalmann et al. 2006). However, simple estimates are possible. If we consider direct electron removal by the field of the laser, the most relevant electric field scale is the field at the edge of the cluster ion sphere with all electrons fully stripped, which we can write

$$E_{OI}^{(max)} = \frac{4}{3}\pi R \bar{Z} e n_i. \tag{7.109}$$

We would expect that the fraction of inner-ionized electrons that are extracted by outer-ionization will scale as the ratio of the laser peak field to the cluster ion space-charge field:

$$\xi_{OI} \equiv E_0 / E_{OI}^{(max)}. \tag{7.110}$$

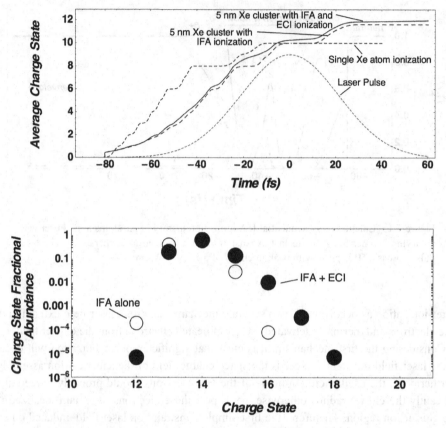

Figure 7.16 Sequential ionization in Xe calculated for a 50 fs pulse at 800 nm with peak intensity of 1×10^{17} W/cm^2. The top panel plots the average ionization level for three situations, individual Xe atoms, Xe atoms in a 5 nm cluster expanding with a 1000 eV electron temperature subject only to IFA tunnel ionization and Xe atoms in the cluster with the additional contribution of electron collisional ionization. The bottom panel shows the various charge states and their fraction of the ions observed when only IFA is used and when ECI is added.

Roughly speaking, the fraction of inner-ionized electrons that are stripped via outer-ionization, f_{OI}, is $\approx \xi_{OI}$. A somewhat more accurate estimate for f_{OI} can be found in two limits.

The first model, which we refer to as the "Breizman–Arefiev" model (Breizman et al. 2005), is appropriate early in the cluster expansion when the nanoplasma free-electron density is high. In this case we treat the laser field as uniform and quasi-static with an electron cloud which rearranges quickly to nullify the value of the E-field inside the electron cloud. This is the same as saying that the electron density is much higher than the laser critical electron density (i.e. $\gg 10^{21}$ cm^{-3}). If we consider the case in which no outer-ionization has yet to occur, the electron cloud simply fills the ion sphere volume and creates a quasi-neutral nanoplasma. If some outer-ionization has occurred (and we assume

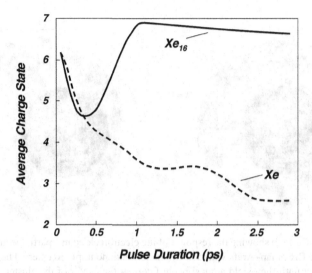

Figure 7.17 Illustration of pulse duration effects on the average charge state reached in small Xe clusters simulated through molecular dynamics simulations. The dashed line shows the average charge state of isolated Xe atoms, which decreases because the peak intensity decreases with increasing pulse duration, while the solid line shows that pulses of 1 ps duration maximize the ionization of the Xe atoms in a small Xe cluster even though the peak intensity drops as duration is increased. Plots were generated based on calculations of Siedschlag and Rost (2002).

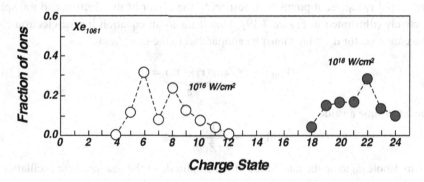

Figure 7.18 Plot of calculated charge state distribution in 1000 atom Xe clusters found from MD simulations at two peak intensities. Plot generated based on calculations of (Last and Jortner 2004).

uniform ion density), the remaining space-charge confined electrons will condense to a radius that yields zero internal E-field, namely an electron density equal to Zn_i, leaving a positively charged ion shell around this neutral region.

When an external laser field is imposed, which in this limit oscillates slowly compared to the response time of the free electron cloud, the electron cloud will respond adiabatically to the laser field and will be displaced in a direction opposite the laser E-field direction to cancel the laser field in the interior of the cloud. The magnitude of this lateral displacement can be easily found by noting that the E-field, interior to the cloud, is

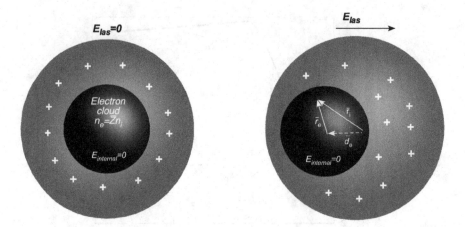

Figure 7.19 Cartoon showing the response of the electron cloud in a partially outer-ionized cluster in the Breizman–Arefiev model. On the left with no imposed external field the electrons contract until they yield a quasi-neutral core at the center of the cluster surrounded by a charged ion shell. When an external field from the laser is imposed, as shown on the right, the electron cloud adiabatically evolves to yield zero E-field in the now displaced cloud.

$$\mathbf{E}_{int} = \mathbf{E}_{las} + \frac{4}{3}\pi \bar{Z} e n_i \mathbf{r}_i - \frac{4}{3}\pi e n_e \mathbf{r}_e = 0, \tag{7.111}$$

where $\mathbf{r}_e$ and $\mathbf{r}_i$ represent points with respect to the center of the electron and ion sphere respectively (illustrated in Figure 7.19). This leads to an equation for the electron cloud displacement vector $\mathbf{d}_e$, which must be antiparallel to the laser field,

$$\mathbf{E}_{las} + \frac{4}{3}\pi \bar{Z} e n_i \underbrace{(\mathbf{r}_i - \mathbf{r}_e)}_{\mathbf{d}_e} = 0, \tag{7.112}$$

and must assume a value

$$\mathbf{d}_e = -\frac{3\mathbf{E}_0}{4\pi \bar{Z} e n_i} \cos\omega t. \tag{7.113}$$

(We can denote d_0 to be the magnitude of the amplitude of the displacement oscillations in the field.) If N_e is the number of free electrons remaining in the cluster, then the radius of the electron cloud is

$$R_e = (3N_e/4\pi \bar{Z} n_i)^{1/3}. \tag{7.114}$$

If R is the radius of the cluster ion sphere, clearly when $d_0 + R_e > R$ the laser field extracts free electrons from the confines of the cluster ions and increases the level of outer-ionization. The number of outer-ionized extracted electrons, N_{eI}, is then found from the volume of the ion sphere and the residual electron cloud sphere to be

$$\begin{aligned} N_{eI} &= \frac{4}{3}\pi R^3 \bar{Z} n_i - \frac{4}{3}\pi (R - d_0)^3 \bar{Z} n_i \\ &= \frac{4}{3}\pi \bar{Z} n_i \left[R^3 - \left(R - \frac{3E_0}{4\pi \bar{Z} e n_i} \right)^3 \right] \end{aligned} \tag{7.115}$$

7.2 Electron Dynamics and Cluster Ionization

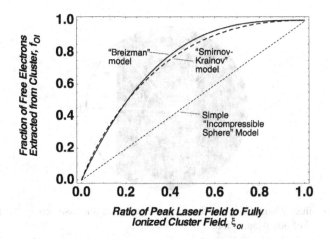

Figure 7.20 Plot of outer-ionized electron fraction as a function of peak laser field normalized to the space-charge confinement field of the cluster ion sphere. The simple estimate as well as the estimates of the two limiting models considered here are plotted.

or written in terms of the ratio of the laser field to surface field of the bare cluster ion sphere, defined in Eq. (7.110), the fraction of outer-ionized electrons $f_{OI} = N_{eI}/N_{e0}$, becomes simply

$$f_{OI} = 1 - (1 - \xi_{OI})^3, \tag{7.116}$$

noting the parameter ξ_{OI} can be written in terms of the number of cluster ions, averaged over charge state and cluster radius

$$\xi_{OI} = \frac{R^2 E_0}{\bar{Z} N_i e}. \tag{7.117}$$

Equation (7.116) is plotted in Figure 7.20 and tells us that nearly 90 percent of all free electrons are removed from the cluster in this high-density adiabatic limit when the laser field is just half the space-charge confining field of the ions (i.e. $\xi_{OI} = 0.5$). Note also that Eq. (7.117) shows that greater outer-ionization can occur in a constant-intensity laser field if the cluster radius increases by cluster explosion during the laser pulse.

When the electron density drops through cluster expansion and the rapidly responding electron cloud picture no longer works, we need another model to find the laser-driven outer-ionization fraction. In this regime (likely later in the interaction) the electrons cannot adjust adiabatically to the laser field. So, it is reasonable to assume that the electron response to the laser field is more like that depicted in Figure 7.21 (a configuration which we dub the "Krainov–Smirnov" model (Krainov and Smirnov 2002b)). In this picture the electrons do not have time to collapse to a spherical core and more or less fill the cluster ion sphere with an exception of a layer that is shifted away from the edges of the cluster in the direction of laser polarization.

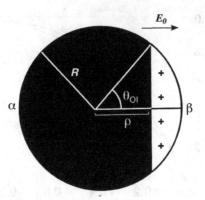

Figure 7.21 Illustration of the electron density configuration assumed in the Krainov–Smirnov model of outer-ionization.

In this model the number of outer-ionized electrons is found by integration over the volume of the ion sphere vacated by electrons. This volume integral can be cast in terms of the angle θ_{OI} defined in Figure 7.21:

$$
\begin{aligned}
N_{eI} &= \int \bar{Z} n_i dV \\
&= \bar{Z} n \int_0^{\theta_{OI}} \pi R^3 \sin^3\theta \, d\theta \\
&= \frac{4}{3}\pi R^3 \bar{Z} n_i \, (2 + \cos\theta_{OI}) \sin^4[\theta_{OI}/2],
\end{aligned} \qquad (7.118)
$$

which means that the fraction of electrons removed by laser-driven outer-ionization in this model is

$$ f_{OI} = (2 + \cos\theta_{OI}) \sin^4[\theta_{OI}/2]. \qquad (7.119) $$

To find the angle we note that electrons arranged in the configuration pictured in Figure 7.21 will be extracted from the cluster by the laser at point α. This point has some electric field E_α which will develop as electrons are removed until it equals the laser field. This field is

$$
\begin{aligned}
E_\alpha &= \bar{Z} e n_i \int_0^{2\pi} \int_{R\cos\theta_{OI}}^{R} \int_0^{\sqrt{R^2-\rho^2}} \frac{R+\rho}{\left[(R+\rho)^2 + r'^2\right]^{3/2}} r' dr' d\rho \, d\phi \\
&= \frac{2\pi}{3} \bar{Z} e n_i R \left[\sqrt{2}(1+\cos\theta_{OI})^{3/2} - 3\cos\theta_{OI} - 1\right].
\end{aligned} \qquad (7.120)
$$

Using the condition that $E_0 = E_\alpha$ we have that the ratio of the laser field to cluster confining field, Eq. (7.110), is

$$ \xi_{OI} = \frac{1}{2}\left[\sqrt{2}(1+\cos\theta_{OI})^{3/2} - 3\cos\theta_{OI} - 1\right]. \qquad (7.121) $$

7.2 Electron Dynamics and Cluster Ionization

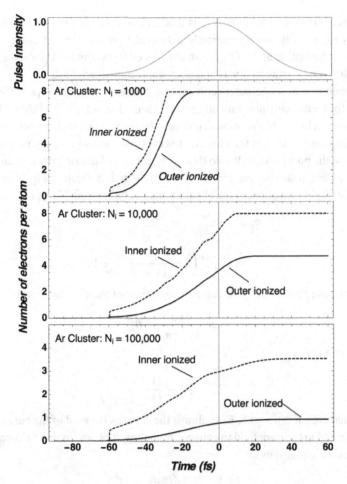

Figure 7.22 Rate equation calculation of the time history of inner- and outer-ionization in Ar clusters of three sizes in a 50 fs laser pulse at peak intensity of 5×10^{15} W/cm^2.

Equation (7.121) can be used with Eq. (7.119) to find the fraction of extracted outer-ionized electron as a function of normalized laser field. This curve is plotted in Figure 7.20. It turns out that the difference between this model and the Breizman–Arefiev model is scant.

The extent of outer-ionization depends on laser intensity through the ratio of the laser field to the electrostatic confining field of the cluster ions. This indicates that outer-ionization will happen much more easily for small clusters. This has been confirmed in MD simulations. To illustrate the interplay of inner- and outer-ionization with different cluster sizes, a rate equation calculation like that presented earlier is shown in Figure 7.22 for three different Ar cluster sizes in a 50 fs laser pulse at 5×10^{15} W/cm^2 assuming "fast" expansion with a 1000 eV electron temperature. In this simulation outer-ionization is calculated using Eq. (7.116), and the time history of the number of electrons per Ar atom that have been inner- and outer-ionized are plotted. We can see that in the case of a moderately sized Ar cluster (1000 atoms), even at this modest intensity, almost every electron created by inner-ionization is almost immediately extracted from the cluster; essentially there is no nanoplasma formation at all. Only with clusters of 10,000 and 100,000 atoms is there a significant difference between inner- and outer-ionization at any instant.

The development of some outer-ionization in a cluster suggests a possible additional mechanism for enhancement of tunneling inner-ionization: a bulk ion field addition to the laser field in tunneling. The situation is complicated by the fact that the bulk ion field will vary depending on position in the cluster. However, we can estimate the magnitude of the contribution at the pole point β. We do this in the context of the Breizman picture of an ion sphere surrounding a smaller electron cloud sphere which is displaced by a distance d_0 by the laser field. In this situation, at the cluster pole, the bulk ion field, E_I, is

$$E_I = \bar{Z}N_i \frac{e}{R^2} - \frac{N_e e}{(R+d_0)^2}$$

$$= E_{IO}^{(max)}\left(1 - \frac{1-f_{OI}}{(1+d_0/R)^2}\right), \quad (7.122)$$

where we have used Eq. (7.109). Recall the magnitude of the electron cloud displacement

$$d_0 = \frac{3E_0}{4\pi \bar{Z}en_i} = R\xi_{OI} \quad (7.123)$$

and use Eq. (7.116) to get

$$E_I = E_{IO}^{(max)}\left(1 - \frac{(1-\xi_{OI})^3}{(1+\xi_{OI})^2}\right). \quad (7.124)$$

So, we see that even if $\xi_{OI} = 0.5$, E_I is almost the same as the field of the bare cluster ions. This field can add to the laser field to enhance tunneling. The size of this field-enhancement felt by nearest neighbor ions is

$$\frac{E_I}{E_{IFA}} \approx \frac{4/3\pi \bar{Z}en_i R}{\bar{Z}e/r_{WS}^2} \approx \frac{R}{r_{WS}}. \quad (7.125)$$

In some regions of the cluster, if outer-ionization has become significant, the bulk ion field can be factors of ten or more larger than the IFA fields. Of course, ions inside the cluster within the electron cloud will not feel this full bulk ion field.

7.2.6 Evaporative and Free-Streaming Outer-Ionization

A second mechanism for outer-ionization can arise if the free inner-ionized electrons become significantly heated by the laser field. Then some of the electrons can exit the space-charge field of the charged cluster. This can be thought of as evaporative outer-ionization by free-streaming of hot electrons. The rate of ionization by this mechanism can be estimated if we assume that the electrons remain thermalized and have a Maxwellian energy distribution.

We note that only those electrons with kinetic energy sufficient to overcome the instantaneous space-charge field can free-stream and only those electrons within roughly one

7.2 Electron Dynamics and Cluster Ionization

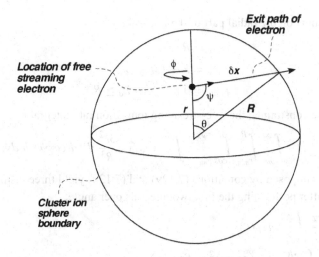

Figure 7.23 Geometry of the free-streaming outer-ionization rate calculation.

electron collisional mean free path will contribute to the outer-ionization rate. We need to integrate over the electron energy distribution and the cluster volume subject to the mean free path constraint to find the outer-ionization rate from the cluster. The geometry of this integral is shown in Figure 7.23.

To find the rate of electrons leaving the cluster sphere by free-streaming, we must integrate over the distribution of electrons with velocity v at radius r within the cluster traveling into solid angle dS:

$$W_{FS} = 4\pi R^2 \int v n_e \, d\Omega \, dr \, dv. \tag{7.126}$$

The constraints set on this integral are, first, that the electron velocity be greater than the velocity needed to overcome the space-charge potential of the partially outer-ionized cluster

$$v \geq v_{esc}. \tag{7.127}$$

Second, the electron must be within one collisional mean free path of the cluster's edge to exit without undergoing scattering,

$$\delta x_c \leq \lambda_e(v). \tag{7.128}$$

Based on the geometry of Figure 7.23, the constraints on the angular part of the integral can be written:

$$\cos \psi_{MAX} = -1 \tag{7.129a}$$

$$\cos \psi_{min} = \frac{r^2 + \lambda_e^2 - R^2}{2r\lambda_e} \quad \lambda_e < R + r \tag{7.129b}$$

$$\cos \psi_{min} = 1 \quad \lambda_e \geq R + r \tag{7.129c}$$

and the constraints on the radial part of the integral are

$$r_{max} = R \tag{7.130a}$$
$$r_{min} = R - \lambda_e \quad \lambda_e < R \tag{7.130b}$$
$$r_{min} = 0 \quad \lambda_e \geq R. \tag{7.130c}$$

Subject to these constraints, the free-streaming ionization rate integral is

$$W_{FS} = 4\pi R^2 \int_{v_{esc}}^{\infty} \int_{r_{min}}^{R} \int_{\cos\psi_{max}}^{\cos\psi_{max}} \int_{0}^{2\pi} v n_e f(v) \frac{3r^2}{R^3} d\varphi \, d(\cos\psi) \, dr \, dv. \tag{7.131}$$

The constraints imposed by conditions (7.129) and (7.130) yield three distinct regions for the r integral after performing the first two integrals over angle:

$$W_{FS} = \frac{6\pi}{R} \int_{v_{esc}}^{\infty} v n_e f(v)$$

$$\times \begin{cases} \int_{R-\lambda_e}^{R} r^2 \left(\frac{r^2 + \lambda_e^2 - R^2}{2r\lambda_e} + 1 \right) dr \, dv & \lambda_e < R \\ \int_{0}^{\lambda_e - R} 2r^2 \, dr \, dv & \\ + \int_{\lambda_e - R}^{R} r^2 \left(\frac{r^2 + \lambda_e^2 - R^2}{2r\lambda_e} + 1 \right) dr \, dv & R < \lambda_e < 2R \\ \int_{0}^{R} 2r^2 \, dr \, dv & \lambda_e > 2R \end{cases} \tag{7.132}$$

which becomes, after integrating over dr (and now noting the explicit electron velocity dependence of the mean free path),

$$W_{FS} = \begin{cases} \dfrac{\pi}{4R} \int_{v_{esc}}^{\infty} v n_e f(v) \lambda_e \left(12R^2 - \lambda_e^2 \right) dv & \lambda_e(v) < 2R \\ 4\pi R^2 \int_{v_{esc}}^{\infty} v n_e f(v) \, dv & \lambda_e(v) > 2R \end{cases}. \tag{7.133}$$

Finally, the integral over velocity can be recast as an integral over kinetic energy by noting that the electron kinetic energy must be at least equal to the electrostatic binding energy of the ionized cluster

$$\varepsilon_{esc} = (N_{eI} + 1) \frac{e^2}{R}. \tag{7.134}$$

We need an expression for the electron mean free path, which can be written in terms of the electron velocity and the electron–electron and electron–ion collision frequencies

$$\lambda_e = \frac{v_e}{v_{ee} + v_{ei}}. \tag{7.135}$$

Referring to the results of Appendix B (which address the question of collisions in plasmas), these collision frequencies can be inserted to yield for the electron mean free path in terms of electron energy, ε_e,

$$\lambda_e(\varepsilon_e) = \frac{\varepsilon_e^2}{\pi n_e(\bar{Z} + 1)e^4 \ln \Lambda_C} \equiv \alpha_e \varepsilon_e^2. \tag{7.136}$$

7.2 Electron Dynamics and Cluster Ionization

(Here $\ln \Lambda_C$ is the usual Coulomb logarithm, which is discussed in detail in Appendix B. We just note here that it is roughly equal to 2 in a solid-density cluster at an electron temperature of 100 eV.) Rewriting the free-streaming integral in terms of the escaping electron energy, we have

$$W_{FS} = \begin{cases} \dfrac{\pi}{4R} \displaystyle\int_{\varepsilon_{esc}}^{\varepsilon'_e} \sqrt{\dfrac{2\varepsilon_e}{m}} n_e f(\varepsilon_e) \lambda_e(\varepsilon_e) \left(12R^2 - \lambda_e(\varepsilon_e)^2\right) d\varepsilon_e \\ \quad + 4\pi R^2 \displaystyle\int_{\varepsilon'_e}^{\infty} \sqrt{\dfrac{2\varepsilon_e}{m}} n_e f(\varepsilon_e) d\varepsilon_e & \lambda_e(\varepsilon_{esc}) < 2R \\ 4\pi R^2 \displaystyle\int_{\varepsilon_{esc}}^{\infty} \sqrt{\dfrac{2\varepsilon_e}{m}} n_e f(\varepsilon_e) d\varepsilon_e & \lambda_e(\varepsilon_{esc}) > 2R \end{cases} \quad (7.137)$$

where

$$\varepsilon_e' = \sqrt{2R\pi n_e(\bar{Z}+2)e^4 \ln\Lambda_C} = \sqrt{2R/\alpha_e}. \quad (7.138)$$

Assuming a Maxwellian for the electron energy distribution in the cluster of temperature T_e, the integral is

$$W_{FS} = \sqrt{\dfrac{2}{\pi m}} \dfrac{2n_e}{(k_B T_e)^{3/2}} \begin{cases} \dfrac{\pi \alpha_e}{4R} \displaystyle\int_{\varepsilon_{esc}}^{\varepsilon'_e} \varepsilon^3 \left(12R^2 - \alpha_e^2 \varepsilon^4\right) e^{-\varepsilon/k_B T_e} d\varepsilon \\ \quad + 4\pi R^2 \displaystyle\int_{\varepsilon'_e}^{\infty} \varepsilon e^{-\varepsilon/k_B T_e} d\varepsilon & \lambda_e(\varepsilon_{esc}) < 2R \\ 4\pi R^2 \displaystyle\int_{\varepsilon_{esc}}^{\infty} \varepsilon e^{-\varepsilon/k_B T_e} d\varepsilon & \lambda_e(\varepsilon_{esc}) > 2R \end{cases}$$

$$(7.139)$$

It turns out that these integrals do have analytic solutions, but they are cumbersome. A more practical equation can be found by approximating the electron mean free path with its average over a Maxwellian yielding

$$\lambda_e \cong \dfrac{15}{4\pi} \dfrac{(k_B T_e)^2}{n_e(\bar{Z}+1)e^4 \ln\Lambda_C}. \quad (7.140)$$

This lets us pull the mean free path out of the integral, which is then trivial and evaluated to yield for the free-streaming ionization rate:

$$W_{FS} \cong 2n_e \sqrt{\dfrac{2\pi}{mk_B T_e}} (k_B T_e + \varepsilon_{esc}) e^{-\varepsilon_{esc}/k_B T_e} \begin{cases} \dfrac{\lambda_e}{4R}\left(12R^2 - \lambda_e^2\right) & \lambda_e < 2R \\ 4R^2 & \lambda_e > 2R \end{cases} \quad (7.141)$$

To illustrate the magnitude of this effect, we put in some numbers. In a large cluster with 5 nm radius, an average ionization state of 8 (which might characterize an Ar cluster), an electron temperature of 100 eV and an electron density near that of solid (1.6×10^{23} cm^{-3}) the electron mean free path is about 0.22 nm. The initial free-streaming rate (i.e. when $\varepsilon_{esc} = 0$) is 1.1×10^{19} s^{-1} or 10^4 electrons per fs. This is a very fast ionization rate. However, when 10% of the electrons have exited, the escape energy for this cluster has climbed to 2.3 keV and the free-streaming rate plummets to 3×10^{10} s^{-1}. So space-charge forces in a reasonably sized cluster tend to shut down free-streaming ionization effectively and most outer-ionization is field driven. However, as we discuss in the following section, the electron distribution can transiently pass through time windows of much higher thermal energy, so there can be spikes in the free-streaming ionization rate.

7.3 Laser Absorption and Heating of Cluster Nanoplasmas
7.3.1 Regimes of Cluster Electron Cloud Behavior in the Laser Field

In addition to depositing energy in the cluster by ionizing the cluster atoms, the laser field can be further absorbed by a cluster in the strong field regime by heating the free electron cloud confined by the cluster ion sphere. The laser deposits energy in these electrons through scattering of the electron in the oscillating field, converting the coherent oscillatory energy to thermal energy. The physics describing this deposition is complex and ultimately only calculated quantitatively by particle dynamics simulations. This complexity arises because of the interaction of the coherent collective oscillation of the electron cloud accompanied by kinetic effects of individual electrons, such as electron ion scattering.

To address then the question of how the laser field heats the electron nanoplasma, some simplification can be derived by thinking about the interaction as roughly falling into one of five regimes. The heating history of a cluster may, in fact, transit more than one of the these regimes as the laser intensity changes, the electron density changes and the cluster radius increases during the course of the interaction. Roughly speaking, the various heating regimes are governed by the relationship of three spatial scales in the interaction: the cluster radius $R(t)$, the free electron oscillation amplitude,

$$x_{osc} = eE_0/m\omega^2, \qquad (7.142)$$

and the oscillating displacement of the free electron cloud in the cluster, which at high density in a uniform ion density cluster is

$$d_0 = \frac{3E_0}{4\pi \bar{Z} e n_i} = \frac{3eE_0}{m\omega_p^2}. \qquad (7.143)$$

This relationship will change as the cluster expands and its electron density passes through the giant dipole resonance.

With these scales, we can think of the interaction and the nature of energy deposition as falling into one of these five regimes:

I. $d_0 \ll x_{osc} \leq R \rightarrow$ high density; large cluster and/or moderate intensity
   - The free-electron cloud dynamics are characterized by the linear harmonic oscillator model
II. $x_{osc} \leq R \leq d_0 \rightarrow$ low density/expanded cluster; possibly near resonance
   - The free-electron cloud dynamics are characterized by the nonlinear electron cloud oscillator model.
III. $d_0 \leq R < x_{osc} \rightarrow$ high density; high intensity
   - Harmonic motion of a reduced radius electron core; outer-ionization regime

IV. $R < d_0 \sim x_{osc} \rightarrow$ small cluster; high intensity/low charge
- Nearly complete outer-ionization; Coulomb explosion regime

V. $d_0 \sim x_{osc} \leq R \rightarrow \omega x_{osc} \sim c$; small cluster; high intensity/low charge
- Relativistic oscillations

Most cluster interactions during a laser pulse pass through a number of these regimes as the pulse intensity and cluster radius evolve. To illustrate this, Figure 7.24 discusses the phases that different clusters pass through under different irradiation conditions. One should note that the regimes seen in an interaction also depend strongly on pulse duration, as we shall illustrate.

7.3.2 Collective Effects and the Giant Dipole Resonance

It is now well understood that the effective absorption of intense laser radiation by atomic clusters is largely a result of the collective electron oscillations in an inner-ionized cluster nanoplasma (Ditmire et al. 1996). Generally this is complex and requires particle dynamic (often referred to as molecular dynamic or MD simulations) or kinetic codes to quantify. MD codes have shown (Saalmann and Rost 2003) that at moderate intensity the electron cloud exhibits a collective behavior that is indeed well approximated by the harmonic oscillator model that was previously presented. It is reasonably justified to examine the dynamics of the collective electron cloud in an ionized cluster with our simple harmonic oscillator (HO) model.

The basic assumptions of the linear cluster HO model are that the cluster has undergone minimal outer-ionization (most appropriate for phases 1 and 2 in the picture of Figure 7.24.) We assume a uniform spherical ion core where the ions and electrons are treated as an incompressible fluid. We also assume that the electron fluid retains a spherical shape (even though the spherical symmetry is strongly broken by the presence of the linearly polarized laser field). Finally, as we have noted previously, this model is appropriate when the electron cloud displacement is much less than the cluster radius.

Recalling the results of Section 7.1 for the electron cloud oscillation amplitude, Eq. (7.30), these assumptions put a constraint on the cluster radius for validity of the linear HO model:

$$R > 4.5\,nm \frac{\sqrt{I[/10^{18}\ W/cm^2]}}{n_e[/10^{23}\ cm^{-3}]}. \tag{7.144}$$

So, in moderate intensities ($\sim 10^{16}$ W/cm^2) where most experimental studies have been conducted, the HO model is applicable down to fairly small clusters, $\sim$1 nm.

We treat the dynamics of the electron cloud according to the simple harmonic oscillator equation

$$\ddot{x} + v_C \dot{x} + \omega_M^2 x = -\frac{e}{m} E_0 \cos \omega t, \tag{7.145}$$

where we have added a new term with v_c to account for damping of the electron cloud motion. This damping occurs because, the electron cloud is not just a fluid; it is composed

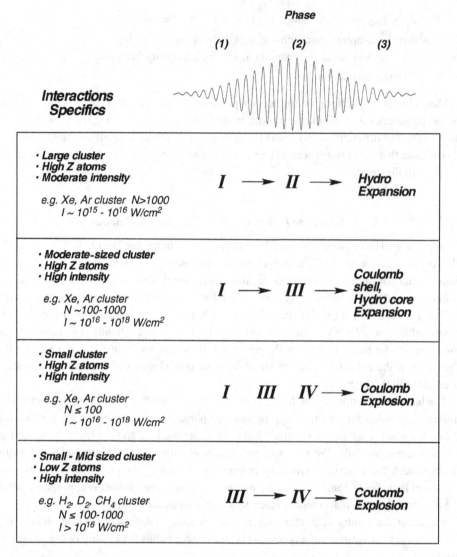

Figure 7.24 Illustration of how different cluster and pulse-intensity conditions can lead to different regimes of electron dynamics in an intense laser pulse.

of individual particles that can scatter, so v_C can be thought of as an effective electron-scattering frequency in the environment of the cluster.

This scattering frequency has two main contributions. First, there is electron–ion Coulomb scattering as is found in any plasma. This will generally lead to thermal heating of the electron population. Second, the driven electrons can scatter from the cluster surface, an effect which will tend to lead to a population of hot electrons. So, v_c is

$$v_C = v_{ei} + v_{eC}. \qquad (7.146)$$

7.3 Laser Absorption and Heating of Cluster Nanoplasmas

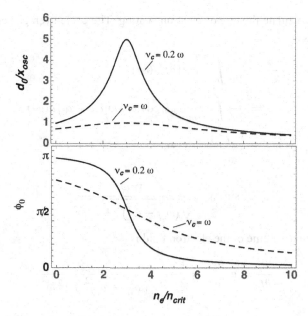

Figure 7.25 Plots of the peak oscillation amplitude and phase as a function of electron density in the cluster with different damping constants.

We will work to evaluate this later, but note that v_c can be as large as ω in some situations, meaning that the electron cloud is very strongly damped. Of course, the damped driven harmonic oscillator has well-known solutions from a freshman physics textbook. The electron cloud displacement $x(t)$ is

$$x(t) = d_0 \cos[\omega t - \varphi_0], \tag{7.147}$$

where the oscillation amplitude is

$$d_0 = \frac{-eE_0/m}{\sqrt{(\omega_M^2 - \omega^2)^2 + v_C^2 \omega^2}}, \tag{7.148}$$

and the phase of the electron oscillation compared to the drive field is

$$\varphi_0 = \arctan\left[\frac{v_C \omega}{\omega_M^2 - \omega^2}\right]. \tag{7.149}$$

As we have noted, the cluster response passes through a resonance with $\omega = \omega_M$ so that $n_e = 3 n_{crit}$. Plots of the peak oscillation amplitude and phase as a function of electron density in the cluster for different damping constants are shown in Figure 7.25. Note that the oscillation phase shifts as the cluster passes through resonance.

We can now calculate the rate of electron heating. The cycle-averaged electron heating rate in the HO is

$$\frac{\partial \varepsilon_e}{\partial t} = \frac{1}{2\pi} \int_0^{2\pi} F(t)\dot{x}(t)d(\omega t)$$

$$= \frac{1}{2\pi} \int_0^{2\pi} \frac{e}{m} E_0 d_0 \cos \omega t \, \sin[\omega t - \varphi_0] \, d(\omega t)$$

$$= \frac{1}{2} e E_0 d_0 \omega \sin \varphi_0, \quad (7.150)$$

and since

$$\sin \varphi_0 = \frac{\nu_C \omega}{\sqrt{(\omega_M^2 - \omega^2)^2 + \nu_C^2 \omega^2}} \quad (7.151)$$

the heating per unit volume of the electron fluid is

$$\frac{\partial u_\varepsilon}{\partial t} = n_e \frac{\partial \varepsilon_e}{\partial t} = \frac{1}{2} \frac{e^2 E_0^2}{m} \frac{\nu_C \omega^2}{(\omega_M^2 - \omega^2)^2 + \nu_C^2 \omega^2}. \quad (7.152)$$

Recall that

$$\omega_M = \sqrt{4\pi e^2 n_e / 3m} = \omega_{pe}/\sqrt{3} \quad (7.153)$$

so that we recover for the heating rate

$$\frac{\partial u_\varepsilon}{\partial t} = \frac{9\omega}{8\pi} \frac{n_e}{n_{crit}} \left(\frac{\nu_C}{\omega}\right) E_0^2 \frac{1}{9 - 6n_e/n_{crit} + n_e^2/n_{crit}^2 + 9\nu_C^2/\omega^2}. \quad (7.154)$$

This result is essentially identical to Eq. (7.44), but now we see that the damping constant is not just the electron–ion collision frequency, as was assumed in Eq. (7.44), but is a result of any anharmonic damping of the collective electron oscillation around the cluster ion core. As an example, note that in an 800 nm laser field far from resonance (say $n_e = 10^{23}$ cm^{-3}) we might expect $\nu/\omega \sim 0.1$ and a heating rate of 0.8 eV/fs at intensity of 10^{16} W/cm^2, whereas near resonance when $\nu/\omega \sim 1$ the heating rate is 3 keV/fs (!) at the same intensity.

7.3.3 Electron Dynamics in an Expanding Cluster: Effects of Passing Through a Dipole Resonance

With this background we can now see what happens at moderate intensity (say $\leq \sim 10^{16}$ W/cm^2) as a cluster is inner-ionized and then expands through the giant dipole Mie resonance during the laser pulse. These qualitative dynamics have been seen in MD-style simulations (Saalmann and Rost 2003). Figure 7.26 summarizes the time history that that is seen in these MD simulations in a high-Z cluster subject to a moderate-intensity femtosecond pulse. The intensity envelope of the laser pulse is shown in the top panel. As the intensity of the pulse ramps up, inner-ionization leads to an increase in the average ionization state of ions in the cluster. This ionization is accompanied by cluster expansion, so the ion sphere radius begins to increase with time (middle panel). Many of the inner-ionized

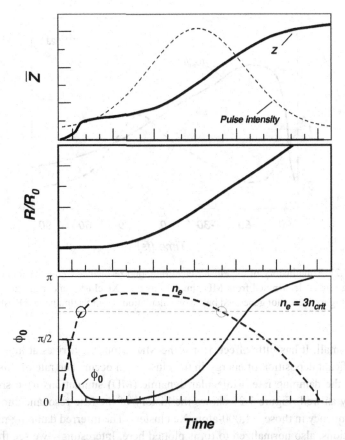

Figure 7.26 Time history plot of average charge state (top panel), cluster radius (middle panel), confined electron density and oscillation phase of the confined electron cloud in a ~1,000-atom Xe cluster in a near IR (780 nm) laser pulse, generated from molecular dynamics simulations of (Saalmann and Rost 2003). The laser peak intensity was 9×10^{14} W/cm^2, with pulse envelope shown in the top panel.

electrons remain confined to the cluster sphere, so the electron density quickly ramps up (bottom panel). This electron cloud passes quickly through the Mie resonance early in the pulse. The mean of the oscillating electron cloud has an oscillating phase which quickly falls into phase with the driving field (characteristic of a driven harmonic oscillator with increased spring constant passing through resonance). As the cluster expands, electron density drops and passes through the Mie resonance a second time. As illustrated here, this resonance (marked with a circle in the bottom plot) occurs not long after the peak of the laser pulse. When this resonance passage occurs depends on a whole host of factors (cluster size, ion species and mass, laser intensity, pulse duration, etc.)

It is the passage of the cluster through this Mie resonance during the laser pulse which can dominate the nature and energetics of the whole interaction. If the cluster expands slowly or the pulse is fast, this passage through a Mie resonance occurs when the laser

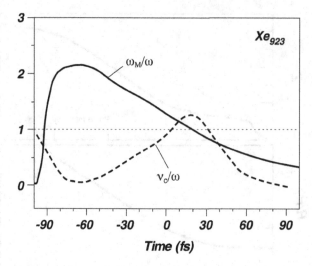

Figure 7.27 Time history of the cluster electron cloud resonance frequency and effective damping rate as determined from MD simulations on a Xe cluster under the same conditions of Figure 7.26. Plot generated based on calculations from Saalmann and Rost (2003).

intensity is small. It has little effect. But if the Mie resonance arrives at high intensity, a rather significant deposition of energy in the cluster can occur. The rate of this deposition depends on the damping rate. Molecular dynamic (MD) simulations have shed light on this quantity as well. Figure 7.27 shows the history of the Mie resonance normalized to the laser frequency in those ∼1,000-atom Xe clusters. The inferred damping rate from the MD simulations, also normalized to ω, is plotted here. Interestingly we see this damping rate maximize around $\nu/\omega \approx 1$ near the time that the cluster electron cloud comes into resonance with the laser field. We will discuss the likely reasons for this increased damping near the Mie resonance in later sections.

The interplay between the passage through a Mie resonance and the rise and fall of a laser pulse has some interesting and nonintuitive consequences in the cluster irradiation. To a simple approximation it would appear that energy deposition in the cluster will be maximum if a situation arises in which the expanding cluster nanoplasma passes through a Mie resonance with the driving laser field near the peak of the laser pulse. Consequently, optimum heating and ionization in the expanding cluster often occurs when pulses are stretched to a duration longer than their bandwidth limit, even though this stretching lowers the peak intensity of a fixed energy pulse (or when two pulses with delay matched to the expansion time of the cluster are employed (Springate et al. 2000)).

This effect is well known in experiments (Zweiback et al. 1999; Zweiback et al 2000a) and is seen in MD simulations as well. Figure 7.28, also adapted from the MD simulations of Saalmann and Rost (2003), illustrates this point. Here the average charge state per atom found from the MD simulations as a function of pulse duration is plotted. The corresponding peak intensity of the pulse is shown on the top axis. There is clearly an optimum pulse duration to maximize inner-ionization. What is more, the pulse duration of this optimum

7.3 Laser Absorption and Heating of Cluster Nanoplasmas

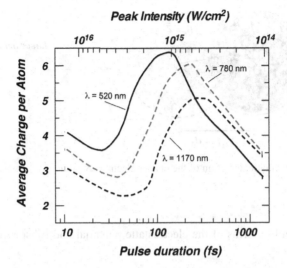

Figure 7.28 Calculated average charge state in a Xe cluster of ∼600 atoms subject to variable pulse widths (at constant pulse energy) of three different wavelength lasers. The changing peak intensity that results from changing the pulse duration is shown on the top axis.

changes with differing laser wavelength, with longer wavelengths requiring longer pulses for optimum ionization. This is completely understandable within the context of our HO cluster nanoplasma picture.

Optimum ionization occurs when the Mie resonance is passed through near the pulse peak. Since longer-wavelength pulses reach this Mie resonance at lower density and therefore greater cluster radii, the optimum occurs predictably at longer pulse duration in longer-wavelength fields.

7.3.4 Nonlinear Collective Electron Motion in the Cluster

So far we have relied on the linear harmonic oscillator picture of the cluster electron cloud. Near a Mie resonance or at sufficiently high intensity, the electron cloud oscillation amplitude becomes large and will approach the size of the cluster ion sphere. This regime deviates from the simple HO model and requires us to consider nonlinear motion of the inner-ionized electron cloud (Parks et al. 2001; Fomichev et al. 2003a).

If we continue modeling the inner-ionized electrons as a spherical incompressible cloud, the situation will look something like that depicted in Figure 7.29. To find the response of the system to an external laser field we need to find the restoring force, $\mathbf{F}_r$, of the electron cloud subject to this large amplitude oscillation. If $\mathbf{E}_i(\mathbf{x})$ is the electron field produced by the ion core, this restoring force can be written as a volume integral over the electron cloud:

$$\mathbf{F}_r = -e \int n_e \mathbf{E}_i(\mathbf{x}) \, dV. \tag{7.155}$$

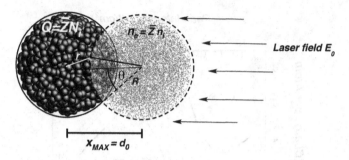

Figure 7.29 Schematic depiction of the cluster electron dynamics in the nonlinear oscillation regime.

Expressing the field in terms of the electrostatic potential and subsequent integration by parts yields

$$\mathbf{F}_r = e \int n_e(\mathbf{x}) \nabla \Phi_i(\mathbf{x}) \, dV, \tag{7.156}$$

$$\mathbf{F}_r = -e \int \nabla n_e(\mathbf{x}) \Phi_i(\mathbf{x}) \, dV + e \int \nabla (n_e(\mathbf{x}) \Phi_i(\mathbf{x})) \, dV. \tag{7.157}$$

Since the electron density is assumed uniform, the first term is zero and the second term can be expressed as a surface integral:

$$\mathbf{F}_r = e n_e \oint \nabla \Phi_i(\mathbf{x}) \, dS. \tag{7.158}$$

The electrostatic potential of the ion core is

$$\Phi_i(\mathbf{x}) = \bar{Z} N_i \frac{e}{r} \qquad \text{Outside ion sphere}$$

$$= \bar{Z} N_i \frac{e}{R} + \frac{1}{2} \bar{Z} N_i e \frac{R^2 - r^2}{R^3} \qquad \text{Inside ion sphere} \tag{7.159}$$

As Figure 7.29 shows, we must break the surface integral into two parts for the regions of the electron sphere inside and outside the ion core. This yields for the restoring force

$$F_{rx} = e n_e \int_1^{x/2R} \left[\frac{3}{2} \bar{Z} N_i \frac{e}{R} - \frac{1}{2} \bar{Z} N_i \frac{e}{R^3} \left(x^2 + R^2 - 2xR\cos\theta \right) \right]$$
$$\times \cos\theta \, 2\pi R^2 \, d(\cos\theta) \tag{7.160}$$
$$+ e n_e \int_{x/2R}^{-1} \bar{Z} N_i e \frac{1}{\sqrt{x^2 + R^2 - 2dR\cos\theta}} \cos\theta \, 2\pi R^2 \, d(\cos\theta),$$

which is easily integrated to yield

$$F_{rx} = 2\pi \bar{Z} N_i e^2 n_e R \left[\frac{2}{3} \frac{x}{R} - \frac{3}{8} \left(\frac{x}{R} \right)^2 + \frac{1}{48} \left(\frac{x}{R} \right)^4 \right] \qquad x \leq 2R$$
$$= \frac{4}{3} \pi \bar{Z} N_i e^2 n_e R^3 \frac{1}{x^2} \qquad x > 2R \tag{7.161}$$

7.3 Laser Absorption and Heating of Cluster Nanoplasmas

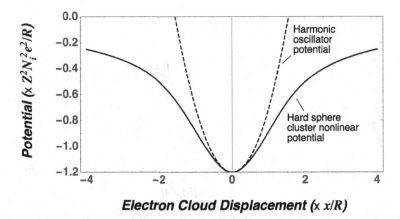

Figure 7.30 The one-dimensional confining potential felt by an incompressible electron sphere oscillating through a uniform ion sphere (plotted as a solid line). The confining potential of a harmonic oscillator is shown as a dashed line.

For $x \ll R$ this equation reduces to our previous HO result. The 1D potential felt by the electron cloud is plotted in Figure 7.30 and compared to the HO potential.

The large amplitude excursion of the electron cloud tends to soften the confining potential, pushing the effective Mie resonance to higher core ion density than that predicted by the simple HO model. For example, the Mie resonance in a 5 nm diameter cluster at $Z = 8$ in a laser field of intensity 10^{16} W/cm^2 is actually closer to 6 n_{crit} instead of the 3 n_{crit} resonance of a HO. The resonance frequency is shifted to longer frequencies and is amplitude dependent. The oscillation period in terms of the electron cloud oscillation amplitude d_0 is

$$T = 4 \int_0^{d_0} \frac{dx'}{\sqrt{2\left(U(d_0) - U(x')\right)/m}}, \qquad (7.162)$$

which is easily evaluated numerically. Such a calculation is shown in Figure 7.31. The free oscillation frequency begins to drop below that of the HO resonant frequency when the oscillation amplitude exceeds $\sim 0.5\,R$. At large enough amplitudes, the electron cloud large oscillates outside the volume of the ion core and the oscillation frequency scales like that expected from Kepler's third law. (As we will discuss a little later, the dashed line in this plot shows that there is an even greater deviation from the HO resonant frequency for large clusters because relativistic effects play a role and the electrons become effectively more massive.)

Note that the restoring force has a maximum. This is found from Eq. (7.161). The point that this force is maximum is found from

$$\left.\frac{\partial}{\partial x}\left(\frac{2}{3}\frac{x}{R} - \frac{3}{8}\left(\frac{x}{R}\right)^2 + \frac{1}{48}\left(\frac{x}{R}\right)^4\right)\right|_{x_{max}} = 0, \qquad (7.163)$$

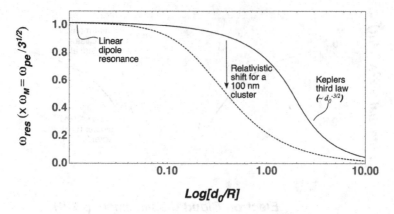

Figure 7.31 Numerically calculated cluster electron oscillation resonance frequency as a function of the maximum oscillation amplitude normalized to the cluster ion core radius (solid line). The dashed line is the same calculation when relativistic effects are considered for a cluster ion core that is 100 nm in diameter.

which indicates that $x_{MAX} = R$. The restoring force for a cloud at this amplitude is

$$F_{rx}^{(max)} = \frac{5}{8}\pi \bar{Z} N_i e^2 n_e R, \quad (7.164)$$

which corresponds to a force per electron in the cloud of

$$f_{rx}^{(max)} = \frac{15}{32}\bar{Z} N_i \frac{e^2}{R^2}. \quad (7.165)$$

This force is $15/32 \sim 1/2$ times the force needed to remove the last electron from the surface of a fully stripped ion core. This suggests that in a DC field the electrons of the cluster will be completely pulled free of the ions and will be outer-ionized at a field roughly half that of the field needed to outer-ionize the cluster sequentially. For reasons we will discuss, this collective stripping is probably not realized actually in a cluster but it does help to set an upper limit on the intensity regime of validity of this collective nonlinear oscillator model. This "breakdown intensity" is

$$I_{CO} \leq \frac{c}{32\pi}\bar{Z}^2 N_i^2 \frac{e^2}{R^4} \quad (7.166)$$

It is, for example, reached in a 5 nm Ar cluster ($N_i \approx 10^4$) inner-ionized to $Z = 8$ at an intensity of 7×10^{17} W/cm^2.

The consequences of the nonlinear oscillation are illustrated in Figure 7.32. Here the 1D equation of motion for an electron cloud is numerically solved by assuming an initially 5 nm cluster inner-ionized to $Z = 8$. The oscillation amplitude and the heating rate as the cluster expands and lowers its electron density subject to a driving 800 nm wavelength laser at 10^{17} W/cm^2 are plotted. Solutions assuming the nonlinear potential and the HO potential are shown (both with damping constant $\nu = \omega$). The most striking feature of these plots is that the peak in electron density shifts from the 3 n_{crit} resonance predicted by the HO

7.3 Laser Absorption and Heating of Cluster Nanoplasmas

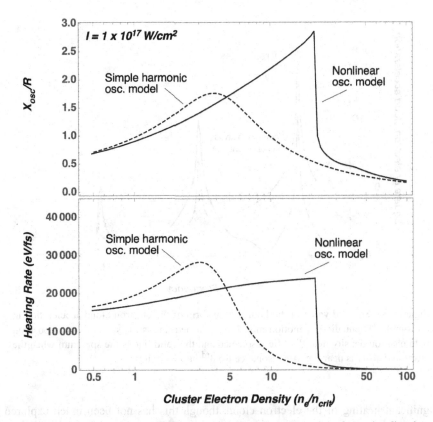

Figure 7.32 Numerical calculation of an initially 5 nm, $Z- = 8$ cluster irradiated at 10^{17} W/cm^2. Top panel shows oscillation amplitude normalized to cluster radius as a function of electron density (varied by increasing the radius of the cluster). The top panel shows the calculated oscillation amplitude and the bottom pane shows the heating rate per electron. The solid line results when the full nonlinear oscillator model is assumed and the dashed line is the result when a simple harmonic oscillator potential is used. (In both cases a damping constant of $v = \omega$ is used in the simulation.)

model to higher density, nearly 30 n_{crit}. Not surprisingly, the softer nonlinear potentially yields a greater oscillation amplitude of the electron cloud.

Another interesting consequence of the nonlinearity in the electron cloud motion is that the electrons can introduce a significant electric field at harmonics of the laser frequency (Fomyts'kyi et al. 2004). This field, particularly at the 3rd harmonic, can lead to additional electron heating (Fomichev et al. 2003b). Figure 7.33 shows a calculation of the squared Fourier transform of the electron cloud acceleration found from the numerical calculation of Figure 7.32. In one plot, that dipole spectrum generated near the linear Mie resonance is shown, where oscillation near the laser drive frequency obviously dominates. On the same plot, this dipole spectrum near the density at which the electrons are in resonance with the 3rd harmonic of the laser, at an order-of-magnitude higher ion density. Here we see a substantial polarization-induced field at 3ω. It is likely that this field leads

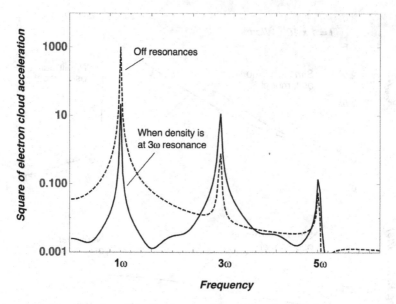

Figure 7.33 Squared value of the Fourier transform of the electron cloud acceleration in the simple 1D equation of motion model. The dashed line is the spectral decomposition with electron density near the Mie resonance and the solid line is the spectrum when the electron density is near the Mie resonance for 3rd harmonic light.

to significant heating of the electron cloud, though this has not been much explored in experiments.

Finally, briefly consider an ion density distribution that differs from the uniform density approximation. A good example might be a Gaussian density distribution:

$$n_i(r) = n_{i0} exp\left[-r^2/R^2\right]. \qquad (7.167)$$

Figure 7.34 shows the numerical equation of motion results for this profile. For illustrative purposes we have assumed a lower damping rate ($\nu = 0.15\,\omega$). At low intensity, the electron cloud oscillation amplitude in the Gaussian ion potential is nearly that of the harmonic oscillator. At higher intensity, the nonlinearity and softening of the restoring force leads to a shift in the resonance. We also see that the width of the resonance is also somewhat broader. We will return to this observation in a moment.

7.3.5 Electron Dynamics at High Intensity and Relativistic Effects

At even higher laser intensity, when $x_{osc} > R$, we enter region III described previously. In this regime we expect that the laser field will strip a significant number of electrons far from the cluster ion core and the outer-ionization fraction will increase rapidly (Breizman and Arefiev 2003). As we have seen in Section 7.2.5, field removal of electrons while the cluster is still at high density results in a contracted electron cloud which remains

7.3 Laser Absorption and Heating of Cluster Nanoplasmas

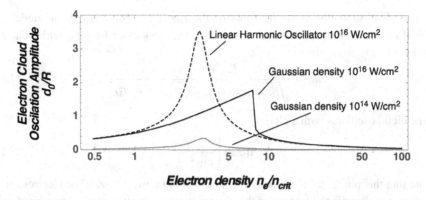

Figure 7.34 Numerically simulated oscillation amplitude versus electron density for electrons around a Gaussian ion spatial distribution compared to the same calculation for a harmonic oscillator.

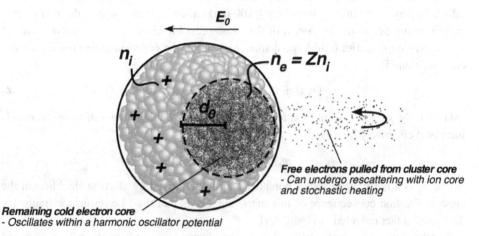

Figure 7.35 Electron dynamics in a high-density cluster at high intensity. The laser field strips electrons, pulling them into vacuum and leaving a contracted electron cloud that responds to the laser field inside the nearly harmonic confining field of the uniform ion sphere core.

in the confines of the ion core. The electron cloud in this case should adiabatically follow the oscillating field of the laser. This case is depicted in Figure 7.35. Inside the ion core sphere, the potential felt by the electrons is

$$\Phi_i(x) = \bar{Z} N_i \frac{e}{R} + \frac{1}{2} \bar{Z} N_i e \frac{R^2 - r^2}{R^3}, \tag{7.168}$$

so the restoring force on these remaining space-charge confined electrons is

$$F_r = -\bar{Z} N_i \frac{e}{R^3} r = m \omega_M^2 r, \tag{7.169}$$

which is identical to the restoring force of the linear harmonic oscillator model. The confined electron cloud will oscillate subject to the peak laser field E_0 with amplitude given by

$$d_0 \approx \frac{-eE_0/m}{\sqrt{(\omega_M^2 - \omega^2)^2 + v_C'^2 \omega^2}} \approx -\frac{eE}{m\omega_M^2}, \qquad (7.170)$$

and the cloud oscillates with a phase

$$\varphi_0 = \arctan\left[\frac{v_C'\omega}{\omega_M^2 - \omega^2}\right]. \qquad (7.171)$$

Note that this phase is determined by a modified damping rate v_C'. The electrons in the case do not scatter off the surface of the cluster, so the damping rate is determined solely by electron–ion collisions (a subject we consider next).

At this point we should say a few words about relativistic effects, though to date there has not been extensive data collected in the relativistic regime. We might expect relativistic effects to play a role in very large (say ≥ 100 nm) clusters which could be driven by very-high-intensity fields. In the context of the incompressible electron cloud oscillator model at relativistic intensities (and large-diameter clusters), the equation governing the cluster electron cloud is

$$\frac{d}{dt}(\gamma_e \dot{x}) + v_c \dot{x} + \omega_M^2 x = -\frac{e}{m} E_0 \cos\omega t. \qquad (7.172)$$

Qualitatively, we know from Chapter 3 that in the laser field the mass of each electron is increased effectively by

$$m_{eff} = m\sqrt{1 + a_0'^2/2}, \qquad (7.173)$$

where a_0' is a normalized vector potential amplitude altered by electron shielding in the cluster. The first consequence of this mass increase is that the Mie resonance frequency decreases, a fact reflected in Figure 7.31.

Another interesting relativistic effect in the cluster is that we might expect a second heating mechanism to become important. So far, we have only considered absorption by the damped-driven when driven harmonically by the electric field. But at high intensity our studies of single-electron motion in Chapter 3 inform us that the magnetic field can set up longitudinal oscillations at the second harmonic. The motion of the electron cloud in the z direction (along **k**) is

$$\frac{d}{dt}\left(\frac{\dot{z}}{\sqrt{1-(\dot{x}^2+\dot{z}^2)/c^2}}\right) + v_c \dot{z} + \omega_M^2 z = -\frac{e\dot{x}}{mc} B_0 \cos\omega t. \qquad (7.174)$$

The space-charge restoring force of the ions prevents the electrons from being pressed forward by the B-field and instead a longitudinal oscillation will be excited. The structure of this equation is like that of Eq. (7.145) with the driving term reduced by a factor of $\dot{x}/c$, so we might expect in mildly relativistic fields that this magnetic field heating will be smaller than the E-field heating by a factor of $(\dot{x}/c)^2$. So, in a 5 nm Ar cluster irradiated at 10^{17} W/cm^2 this will be about a 10 percent additional contribution to the cluster heating.

7.3.6 Electron Collisional Damping and Electron Thermal Heating

So far we have been nebulous about the specific mechanisms and value of the electron damping rate, noting that the electron oscillations will be damped by scattering from the cluster ions. This collisional scattering can really be thought of as two quite different mechanisms. The first mechanism for damping the cluster electron-driven oscillation is through Coulomb scattering off of individual ions. These electron–ion collisions lead to a conversion of coherent oscillation energy to randomized thermal energy and hence heat the cluster electron nano-plasma.

The physics of electron–ion collisions in classical plasmas is a very well-studied topic. A review of some of the most important results from Coulomb scattering in plasmas is presented in Appendix B. We will revisit the question of laser-driven heating of plasma electrons through electron–ion collisions (a process often called inverse Bremsstrahlung absorption) in the next chapter. For the purposes of finding a quantitative way to calculate this damping rate in a cluster we can make a heuristic calculation of the rate using a previously derived plasma electron–ion Coulomb scattering frequency. Since we seek the rate at which the oscillation momentum is damped we start with the result from Appendix B for the electron momentum loss collision frequency

$$v_{ei} = \frac{4\pi \bar{Z}^2 e^4 n_i}{m^2 v_e^3} \ln \Lambda_C. \tag{7.175}$$

Here v_e is the electron velocity in the cluster and $\ln \Lambda_C$ is the previously mentioned Coulomb logarithm. (See Appendix B for a discussion of how to calculate this term.)

Evaluation of the electron–ion Coulomb scattering contribution to the cluster electron cloud damping can be considered in two regimes. The first is when the laser-driven oscillation velocity is well below the thermal velocity of electrons, $v_{Te} = (2k_B T_e/m)^{1/2}$. In this case it is reasonable to evaluate Eq. (7.175) simply by substituting this thermal velocity for v_e. On the other hand, most of the time the laser-driven oscillation velocity of the electron cloud will dominate, and we require some suitable average over the sinusoidal velocity variation of the electrons. We will consider this problem with more rigor in the next chapter but for now it is adequate to take a heuristic approach to derive an appropriate collision frequency.

Consider an approximate heating rate, which we can estimate by a simple picture in which each electron acquires about the instantaneous oscillation energy in each collision, so the heating rate can be written as a product of the collision frequency and the instantaneous kinetic energy of a freely oscillating (nonrelativistic) electron

$$\frac{\partial \varepsilon_e}{\partial t} \cong v_{ei} \left[1/2 m \left(v_{osc} \sin \omega t \right)^2 \right]. \tag{7.176}$$

Here we must take the amplitude of electron velocity oscillations to be that inside the shielded cluster. (So, for example, at high density in the cluster $v_{osc} = \omega d_0$.) We can ask what the cycle-averaged heating rate of those electrons is. To do this, for the period in the cycle that the oscillation energy is below the thermal energy we will evaluate the velocity in the collision frequency with the thermal velocity (as noted earlier), but for the portion of the oscillation in which the electron velocity is higher than the thermal velocity we evaluate the velocity in the collision frequency with the instantaneous quiver velocity. Cycle-averaging with this approximation yields

$$\left\langle \frac{\partial \varepsilon_e}{\partial t} \right\rangle = \frac{m}{\pi} \left(\frac{4\pi \bar{Z}^2 e^4 n_i}{m^2} \ln \Lambda_C \right) \left[\int_{\varphi_{min}}^{\pi/2} \frac{d\varphi}{v_{osc} \sin \varphi} + \int_0^{\varphi_{min}} \frac{v_{osc}^2}{v_{Te}^3} \sin \varphi \, d\varphi \right] \quad (7.177)$$

where $\varphi_{min} = \sin^{-1}[v_{Te}/v_{osc}]$. We can evaluate this integral

$$\left\langle \frac{\partial \varepsilon_e}{\partial t} \right\rangle = \frac{4\bar{Z}^2 e^4 n_i}{m} \ln \Lambda_C \left\{ \frac{1}{v_{osc}} \ln \left[\frac{\cos(\varphi_{min}/2)}{\sin(\varphi_{min}/2)} \right] + \frac{v_{osc}^2}{v_T^3} \left[\frac{\varphi_{min}}{2} - \frac{1}{4} \sin(2\varphi_{min}) \right] \right\}. \quad (7.178)$$

Considering this cycle-averaged heating rate in the limit $v_{osc} \gg v_T$, we find

$$\frac{\partial \varepsilon_e}{\partial t} = \frac{4\bar{Z}^2 e^4 n_i}{m v_{osc}} \ln \Lambda_C \ln \left[\frac{2v_{osc}}{v_{Te}} \right]. \quad (7.179)$$

(By happy coincidence this is almost identical, to within a factor of 2, to the result in the high field limit we will derive in Chapter 8. The discrepancy rests with the simple approximate formula (7.176), which, as we will see in Chapter 8, would have been more accurate by multiplying the heating rate by 2×.) Now we pull out the cycle-averaged collision frequency by dividing by $mv_{osc}^2/2$ (see Section 8.2) to give

$$v_{ei} = \frac{8\bar{Z}^2 e^4 n_i}{m v_{osc}^3} \ln \Lambda_C \ln \left[\frac{2v_{osc}}{v_{Te}} \right]. \quad (7.180)$$

Note that this collision frequency essentially scales as E_0^{-3} and ω^3 through v_{osc}, so it falls off quickly at higher intensity (the hallmark of Coulomb collision-dominated processes in a plasma).

As an example, consider an Ar cluster, say at its initial liquid density $\sim 2 \times 10^{22}$ cm^{-3}, with electron temperature of 100 eV and internal (shielded) intensity of 10^{16} W/cm^2 (the field outside the cluster in that case could would have to be about an order of magnitude higher). Now $\ln \Lambda_C \approx 5$ and $\ln[2v_{osc}/v_T] \approx 1.6$. So $v_{ei} \approx 2 \times 10^{15}$ s$^{-1} \approx \omega$. This is a very collisional plasma with very strong damping of any driven electron cloud oscillation. On the other hand, in that cluster near resonance at $n_i \approx 6 \times 10^{20}$ cm^{-3} then $v_{ei} \approx 0.02 \, \omega$. Electron–ion collisions only weakly damp the driven electron cloud.

7.3.7 Collisionless Electron Damping by Interactions with the Cluster Edge

It was clear in early experiments on high field irradiation of noble gas clusters that the absorption near resonance and the width of the resonance with varying ion density as the cluster expanded was inconsistent with damping solely by microscopic electron–ion collisions (Zweiback et al. 1999; Milchberg et al. 2001). Clearly, early in the interaction during the phase when the cluster is at high density, Coulomb scattering probably is the primary damping mechanism, but near resonance and at large electron oscillation amplitudes it is

7.3 Laser Absorption and Heating of Cluster Nanoplasmas

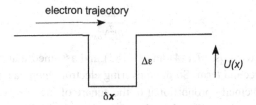

Figure 7.36 A driven free electron encountering a well potential, similar to the case of a cluster electron traversing a bare ion core.

obvious that something else must be at work. It is well known in studies of metallic clusters that spectral broadening of the Mie resonance is mainly by scattering of electrons off the macroscopic electrostatic potential drop at the edge of the metallic cluster. A similar mechanism plays an important role in strong field interactions with clusters in the large oscillation amplitude regime (say near the cluster resonance) (Saalmann and Rost 2008). This is often called "collisionless damping" since no collisions with individual ions are involved (Megi et al. 2003; Korneev et al. 2005; Krainov and Smirnov 2001).

The origin of the effect can be seen in two simple examples, which model the cluster in two limits, a fully stripped ion sphere in the first example and a fully internally field shielded sphere in the second. First consider a free electron oscillating in a laser field which encounters a sharp box potential well, like that depicted in Figure 7.36. Consider the case in which the electron encounters the potential well at $t = 0$ when its velocity in the field is at its maximum = v_{osc}. If the electron exits the well at t_e, integrating the equation of motion of the electron while in the well yields:

$$\int_{v_i}^{v(t_e)} dv = -\frac{eE_0}{m} \int_0^{t_e} \sin\omega t \, dt \qquad (7.181)$$

where v_i is the velocity of the electron immediately after it enters the well and is

$$v_i = \sqrt{v_{osc}^2 + 2\Delta\varepsilon/m} \approx v_{osc} + \frac{\Delta\varepsilon}{mv_{osc}}. \qquad (7.182)$$

We can estimate $t_e \approx \delta x / v_{osc}$. Integration of Eq. (7.181) yields

$$v(t_e) \approx v_{osc} \cos\omega t_e + \frac{\Delta\varepsilon}{mv_{osc}}. \qquad (7.183)$$

Upon emerging from the well, the electron has velocity

$$v_2 \approx v_e - \frac{\Delta\varepsilon}{mv_{osc}}. \qquad (7.184)$$

The subsequent velocity of the electron in the laser field is now (where $v_e = v(t_e)$)

$$v(t) \approx v_2 + \frac{eE_0}{m\omega}(\cos\omega t - \cos\omega t_e)$$

$$\approx v_{osc}\cos\omega t - \frac{\Delta\varepsilon}{m}\left(\frac{1}{v_e} - \frac{1}{v_{osc}}\right)$$

$$\approx v_{osc} \cos\omega t - \frac{\Delta\varepsilon^2}{m^2 v_{osc}^3}, \quad (7.185)$$

where we have inserted Eqs. (7.184) and (7.183) and assumed that $\cos^2 \omega t_e \approx 1$ in the denominator of the second term. So the quivering electron upon passage through the well has acquired a drift velocity proportional to the square of the energy depth of the square well (and inversely proportional to the cube of the free electron oscillation velocity). This single encounter with the potential well has served to absorb energy from the field into the electron kinetic energy just as an electron–ion collision does.

A second example, which more closely approximates the situation in which a driven cluster electron encounters a high-density cluster nanoplasma that is effectively shielding the field inside the cluster, is when a driven free electron encounters a region of width δx in which the laser field is 0. Again we assume that this region is encountered at $t = 0$ when the quivering electron is at its maximum velocity, v_{osc}. Now integrating the equation of motion for the electron leads to a lower limit in the force integration of the exit time of the electron from the field-free region, t_e, (in this case, $t_e = \delta x/v_{osc}$, is exactly true):

$$\int_{v_{osc}}^{v(t)} dv = -\frac{eE_0}{m} \int_{t_e}^{t} \sin\omega t' \, dt' \quad (7.186)$$

which yields

$$v(t) = v_{osc} \cos\omega t + v_{osc}(1 - \cos\omega t_c). \quad (7.187)$$

Once again we see that the electron has acquired a drift velocity by this encounter with the field-free region. The magnitude of the drift energy depends on the width of the field-free region and adds a drift energy between 0 and $2 U_p$ to the quivering electron.

In both scattering examples the encounter with the "cluster" potential leads to energy transfer to the electron. In principle, a given electron could scatter multiple times on subsequent returns. Conceptionally then, this "stochastic" scattering from the macroscopic cluster potentially is nearly the same as the effect of an electron scattering from an individual ion. Quantifying this effect to find an effective electron damping frequency is challenging. However, following the well-studied results from metallic clusters a very reasonable approximation can be made for the collisionless damping frequency. We can say that the damping frequency is the average electron velocity divided by the cluster radius. We take the electron velocity in this approximation to be the sum in quadrature of the oscillation velocity and the electron thermal velocity to yield for the stochastic scattering damping frequency

$$v_{ST} \approx \frac{\sqrt{v_{osc}^2 + v_{Te}^2}}{R}, \quad (7.188)$$

where again we have

$$v_{Te} = \sqrt{2k_B T_e/m}. \quad (7.189)$$

This simple estimate yields a remarkably good approximation to what is observed in particle dynamics simulations of strong field irradiated clusters (Megi et al. 2003). For

7.3 Laser Absorption and Heating of Cluster Nanoplasmas

example, note that early in the cluster interaction, when it is dense, we have that the electrons in the collective oscillating cloud have $v_{osc} \approx \omega d_0$. Since at this point $d_0 \ll R$, it follows for thermally cold electrons that $v_{ST} \ll \omega$, and damping is dominated by electron–ion collisions, at least until the cluster gets hot. For an $R = 5$ nm cluster the electron temperature must be elevated to about 400 eV to get $v_{ST} \approx \omega$ in an 800 nm wavelength laser field. On the other hand, near resonance $d_0 \approx R$, so $v_{ST} \approx \omega d_0/R \approx \omega$. Stochastic scattering will clearly dominate the electron cloud damping at this point in the interaction.

This is best illustrated in a simple simulation of an expanding Xe cluster in a 50 fs, 800 nm laser at peak intensity of 10^{16} W/cm^2. The calculated damping constant as a function of time is plotted in Figure 7.37a. Here the damping from the combination of stochastic cluster surface scattering as calculated with Eq. (7.188) is compared to the case if only damping from electron–ion collisions is considered. As this plot shows, there is a precipitous, two-order-of-magnitude drop in the damping rate near the Mie resonance late in the interaction if only Coulomb scattering is included. However, including cluster surface scattering leads to a damping rate $\sim \omega$ throughout the interaction. The consequence this has on the predicted field inside the cluster is compared in Figure 7.37b. The sharp spike in field strength seen at the Mie resonance with only Coulomb scattering is largely smoothed out by the much higher damping rate associated with the cluster surface scattering. Experiments are largely consistent with the high damping rate (Zweiback et al. 2000a).

We can view this cluster surface scattering as a heating mechanism, which leads to a slightly different way to look at and quantify this scattering. The heating from this collisionless scattering can be found from the Poynting theorem,

$$\langle \nabla \cdot \mathbf{S} \rangle = \langle \mathbf{j} \cdot \mathbf{E} \rangle, \quad (7.190)$$

where brackets denote cycle averaging. As long as either the electrons are completely free or are confined in a perfect harmonic oscillator potential, the driven electron current is purely sinusoidal,

$$\mathbf{j} = en_e v_e \sim sin\omega t, \quad (7.191)$$

and perfectly out of phase with the electric field. No net electron heating occurs. Another way to think of these electrons interacting with the cluster walls is to consider oscillations of the electron cloud that are comparable to or larger than the cluster radius. In the approximation of an incompressible driven electron cloud, this led to the model of a nonlinear confining potential, which served to dephase the oscillation of $\mathbf{j}$ and permit energy deposition into the electrons. So another way of modeling this cluster surface scattering is to consider the collective electron dynamics in the nonlinear HO potential of Eq. (7.161).

The effects of the anharmonic potential can be seen in some simple simulations. Figure 7.38 shows the numerical solution of the equation of motion for a damped driven harmonic oscillator characteristic of an initially $R = 5$ nm cluster ionized to an electron density of an eight-time-ionized Ar cluster. The drive field is that of an 800 nm pulse at 10^{16} W/cm^2. The peak oscillation amplitude calculated from simulations of an electron cloud in a purely harmonic oscillator as a function of cluster radius (as the cluster radius

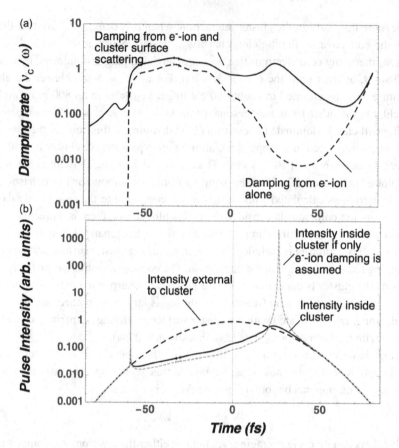

Figure 7.37 Simulation of a 5 nm Xe cluster in a 50 fs, 800 nm laser pulse at 10^{16} W/cm². Panel (a) plots the calculated damping rate when both electron–ion and cluster surface scattering are included (solid line) and when only Coulomb scattering damping is plotted (dashed line). Panel (b) shows the calculated effective intensity in the cluster for the two cases of panel (a) showing the initial shielding of the field in the cluster and the enhancement near the Mie resonance. The inclusion of cluster surface scattering effectively smooths out the unphysical spike in the field near the Mie resonance.

is increased, the electron density drops volumetrically). A damping constant of 0.1 ω or 0.3 ω is used in the numerical solutions. The damping rate here is the microscopic damping, as from electron–ion collisions. This oscillation amplitude is compared to the peak amplitude found when the anharmonic potential of Eq. (7.161) is numerically integrated, with a microscopic damping rate of 0.1 ω. The effective width with electron density of the amplitude peak near the Mie resonance of the anharmonic potential mimics the width and peak of the more strongly damped harmonic potential. This is even more obvious when the electron heating rate is found. This is plotted in Figure 7.39 as a function of electron density for the numerical integration of the harmonic oscillator EOM with $n = 1.0 \, \omega$ and the anharmonic potential with $\nu = 0.1 \, \omega$. The width and peak are about the same (though the

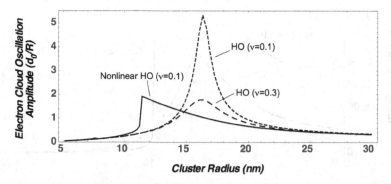

Figure 7.38 Plot of the peak amplitude from the numerical integration of the equation of motion for the electron cloud of an Argon cluster (with $Z = 8$) at intensity of 10^{16} w/cm^2 in an 800 nm wavelength field. The cluster radius is varied, decreasing the electron density, passing through the Mie resonance.

anharmonic potential has a shift in the peak position which is a consequence of the softening of the restoring force of the anharmonic potential at large oscillation amplitudes). These plots show that the anharmonicity of the confining cluster ion potential has an effect similar to increasing the damping rate.

The situation in reality is more complex and the incompressible electron cloud picture is greatly limited in its quantitative applicability. At large oscillation amplitudes near the Mie resonance, the electron cloud will disassemble from space-charge forces very quickly, and will hold the characteristics of an electron cloud sphere for maybe only a couple of laser cycles. So, individual electron trajectories are really the main contributor to the energy kinetics of the cluster. Only particle simulations can gain true quantitative answers on these dynamics.

Examining individual electron trajectories in such particle simulations of strong field irradiated clusters is illuminating. As an example we reproduce schematically some of the single-electron trajectory results of simulations published by Mulser and Kanapathipillai (2005) in Figure 7.40.

What becomes clear in these simulations is that quivering electrons in the cluster often are scattering strongly from the cluster potential near the Mie resonance driving them out of the cluster confines into free space. This serves as an effective damping mechanism to the remaining electron cloud as a whole

7.3.8 Stochastic Hot Electron Heating

Stochastic scattering from the cluster surface represents a heating mechanism for the bound cluster electrons (Kostyukov and Rax 2003). The situation we have described so far applies to regimes I and II of Section 7.3.1: the electron oscillation amplitude is smaller than the cluster, so we are describing occasional scattering of electrons as they encounter the cluster boundary. At higher laser intensity, characteristic of regime III, where the free

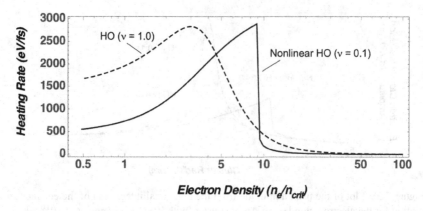

Figure 7.39 Plot of the electron heating rate from the numerical integration of the equation of motion for the electron cloud of an argon cluster (with $Z = 8$) at intensity of 10^{16} w/cm^2 in an 800 nm wavelength field. The cluster electron density is varied, passing through the Mie resonance.

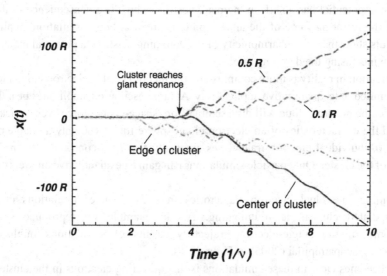

Figure 7.40 Single-electron trajectories in a cluster particle dynamics simulation. Plot generated based on work published in (Mulser and Kanapathipillai 2005).

electron oscillation amplitude is comparable to or larger than the cluster, the situation is qualitatively different. Laser-driven electrons can indeed scatter from the cluster boundary, but instead of rapidly thermalizing with the core cluster nanoplasma electron cloud, the strong laser field pulls these electrons free of the cluster in a small number of oscillations.

The consequence is that this stochastic scattering likely produces a population of "suprathermal" electrons with energy a few times the ponderomotive energy, well above the thermal energy of the confined electron nanoplasma (Chen et al. 2002). Two electron populations develop: a warm (100–1000 eV) thermalized inner electron core and a

7.3 Laser Absorption and Heating of Cluster Nanoplasmas

population of outer-ionized hot electrons. This situation is depicted in Figure 7.35. Experiments show some evidence of these hot electrons, mainly in a double-humped electron spectrum (Shao *et al.* 1996; Fukuda *et al.* 2003) but also some indirect evidence through high-photon-energy X-ray production (Deiss *et al.* 2006).

To quantify this effect, let us consider this hot electron production in two limits. First, consider a cluster nanoplasma still at high density and still nearly quasi-neutral (i.e. very little outer-ionization has yet taken place). The electron core cloud is driven by the laser field and those electrons at the edge momentarily emerge from the cluster ion boundary. Those electrons that are pulled from the edge will be liberated by the laser field and propagate freely in vacuum (like those electrons depicted in Figure 7.35). As we have discussed, the inner-core electron sphere oscillation is damped primarily by electron–ion collisions so the edge electrons pulled free by the field are injected into the vacuum field at a phase given by

$$\varphi_B = \arctan\left[\frac{v_{ei}\omega}{\omega_M^2 - \omega^2}\right]. \tag{7.192}$$

At high density $\omega_M \gg \omega$, so φ_B is small: electrons are born into the field at a phase in which the electron is pulled away from the cluster core. This will be true as long as $\varphi_B < \pi/2$; when $\pi/2 \le \varphi_B = \pi$, the electron cloud reaches the edge at a phase when the field just pushes the edge electrons back into the cluster. So, in this regime hot electron generation occurs only when the cluster is dense and unexpanded.

We can make an estimate for the number of electrons pulled from the cloud each half-cycle. If the cluster is nearly quasi-neutral at the peak of the electron cloud oscillation, a situation like that in Figure 7.41 arises. The number of free electrons liberated are those which emerge from the ion sphere volume and are, in terms of the electron cloud oscillation amplitude x_0,

$$\begin{aligned} N_{fe} &= n_e 2\pi R^2 \int_0^{\pi/2} x_0 \cos\theta \sin\theta \, d\theta \\ &= \frac{\pi^2}{2} R^2 x_0 n_e. \end{aligned} \tag{7.193}$$

From our simple HO model, this amplitude is

$$x_0 = \frac{-eE_0/m}{\sqrt{\left(\omega_M^2 - \omega^2\right)^2 + v_{ei}\omega}} = d_0. \tag{7.194}$$

This fraction of freed electrons propagate out into free space. In the regime that $x_{osc} \gg R$, we can treat the electron propagation as that of a free electron in the field. The electron then recollides with the cluster sphere. The energy of that recollision as a function of birth phase has already been calculated in Chapter 4 under Section 4.5.2 where we considered rescattering of tunnel-ionized electrons from atoms. This recollision energy was plotted in Figure 4.28; recall that it peaks at $3.17\,U_p$ at a birth phase near $0.1\,\pi$.

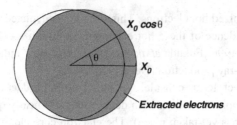

Figure 7.41 Geometry of edge-liberated electrons from the nearly quasi-neutral cluster nanoplasma sphere.

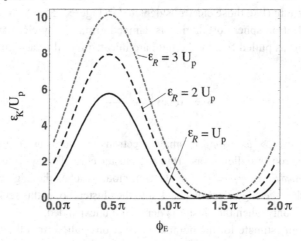

Figure 7.42 Calculation of the residual electron drift energy from recollided electrons in a cluster as a function of emergence phase for three different return energies.

Once these field-driven free electrons enter the cluster, they are shielded from the laser field and essentially propagate field-free through a chord of the cluster sphere. It is fair to assume that the electrons are high enough energy that collisional slowing is negligible.

Noting that θ is defined in Figure 7.41 and, if ε_R is the energy of the recolliding electron (as determined by its birth phase and the calculation of Figure 4.28), we can say that the phase at which the recollided (nonrelativistic) electron emerges from the other side of the cluster is

$$\varphi_E = \varphi_B + 2\omega R \sqrt{\frac{m}{2\varepsilon_R}} \cos\theta. \qquad (7.195)$$

The ultimate kinetic energy that the electron gains can then be calculated numerically. Figure 7.42 shows such a calculation of free-electron residual drift energy as a function of emergence phase for three different recollision energies. This figure shows that for those electrons that recollide with nearly the maximum possible recollision energy (3.2 U_p), if they are at a point in the cluster sphere such that they emerge at a phase near 0.5 π (near a zero in the laser field strength), the electrons can acquire almost 10 U_p of energy. So, in

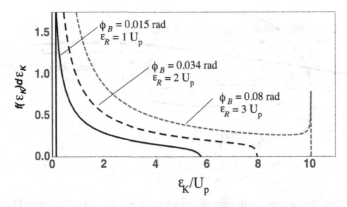

Figure 7.43 Hot electron spectra calculated for stochastic heating from 5 nm clusters for three different birth phases. The corresponding return energies are given below the birth phase.

a near-IR field with (still modest) intensity of 10^{17} W/cm^2, hot electrons with almost 100 keV energy can be produced.

Using the results of Figure 7.42 and Eq. (7.195), we can calculate the hot electron spectra for various birth phases (which, one will recall, depend on the density of the cluster nanoplasma through Eq. (7.192)). Such spectra for three different birth phases and their corresponding return energies in a 5 nm cluster are shown in Figure 7.43. In all three cases this stochastic hot electron production produces a significant hot electron tail extending to energy of a few times the laser ponderomotive energy.

The next logical question to ask then is: what do these electron dynamics look like at relativistic intensity? Recall that at high intensity the magnetic field of the laser causes a significant free-electron drift in the direction of laser propagation. If the drift velocity is, over the time of a single cycle, greater than the cluster size, namely if

$$v_d/\nu \geq R \tag{7.196}$$

or, using Eq. (3.79), if

$$\frac{a_0^2}{4+a_0^2} \geq \frac{\omega R}{2\pi}, \tag{7.197}$$

then the electron will not recollide with the cluster and this stochastic hot electron generation will be suppressed. So, it is fair to assume that in small clusters this mechanism is absent in strongly relativistic laser fields. For example, Eq. (7.197) predicts that in 5 nm clusters in 800 nm wavelength laser fields, the relativistic drift will suppress stochastic recollision hot electron generation at intensity above $\sim 5 \times 10^{16}$ W/cm^2. Numerous groups have experimentally observed two-humped electron spectra, which is consistent with the stochastic hot electron picture presented here. For example, Fukuda et al. (2003) observed an electron spectrum from Xe clusters of a few thousand atoms each irradiated at around 10^{16} W/cm^2 that had a dominant feature at energy between 0 and 1 keV and a second lower peak in the vicinity of 5 keV.

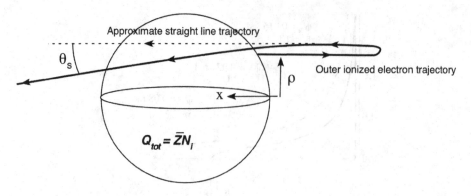

Figure 7.44 Rescattering geometry of a cluster in the limit of a fully stripped ion core.

Another regime in which we might consider stochastic hot electron production is regime IV, when the cluster is almost fully stripped and electrons produced by inner-ionization are almost immediately stripped by the laser field (Smirnov and Krainov 2003). This is the situation one might encounter in low-Z clusters (like H_2 clusters). In this case, electrons stripped from the cluster are pulled from the ion core and can then rescatter off this bare ion sphere. Now we must consider the kinematics of the electron in a slightly different limit. (Recall our two toy models of Section 7.3.7. The situation we describe now is more closely akin to the first of the two 1D scattering models we developed.) The situation we consider is an electron trajectory like that depicted in Figure 7.44. We can develop a simple analytic model if we make a couple of simplifying assumptions: (1) assume $x_{osc} \gg R$ with the free-electron oscillation centered on the cluster ion core, which means that the returning electron encounters the cluster sphere with its maximum oscillation velocity, v_0, namely $\varepsilon_K = 2U_p$, and (2) assume that the scattering angle can be calculated in the straight-line approximation, where the trajectory of the rescattering electron is assumed to be nearly a straight line back through the ion sphere.

The angle that the rescattering electron is deviated can be estimated by finding the perpendicular momentum imparted by the perpendicular force, assuming that the electron travels on a straight line at constant velocity:

$$p_\perp = \int F_\perp dt = \int_{-\infty}^{\infty} F_\perp(x) \frac{dx}{v_0}. \tag{7.198}$$

The Coulomb potential inside and outside the uniform ion density sphere is

$$\Phi_C = -\frac{\bar{Z}N_i e^2}{r} \qquad \text{outside}$$
$$= -\frac{3}{2}\frac{\bar{Z}N_i e^2}{R} + \frac{1}{2}\bar{Z}N_i e^2 \frac{r^2}{R^3} \qquad \text{inside} \tag{7.199}$$

which yields for the perpendicular force felt by the rescattering electron

7.3 Laser Absorption and Heating of Cluster Nanoplasmas

$$F_\perp = -\frac{\partial \Phi_C}{\partial \rho} = -\bar{Z} N_i e^2 \frac{\rho}{(x^2 + \rho^2)^{3/2}} \quad \text{outside}$$

$$= -\frac{\bar{Z} N_i e^2}{R^3} \rho \quad \text{inside} \quad (7.200)$$

We work with normalized quantities:

$$\tilde{\rho} \equiv \frac{\rho}{R}; \quad \tilde{x} \equiv \frac{x}{R} \quad (7.201)$$

in terms of which the perpendicular momentum kick is

$$p_\perp = -2\frac{\bar{Z} N_i e^2}{R v_0} \tilde{\rho} \left[\int_{\sqrt{1-\tilde{\rho}^2}}^\infty \frac{1}{(\tilde{x}^2 + \tilde{\rho}^2)^{3/2}} d\tilde{x} + \int_0^{\sqrt{1-\tilde{\rho}^2}} d\tilde{x} \right]$$

$$= -2\frac{\bar{Z} N_i e^2}{R v_0} \left[\frac{\tilde{\rho}}{1 + \sqrt{1-\tilde{\rho}^2}} + \tilde{\rho}\sqrt{1-\tilde{\rho}^2} \right]. \quad (7.202)$$

The scattering angle is

$$\theta_S \cong \left| \frac{p_\perp}{m v_0} \right|, \quad (7.203)$$

which can be expressed in terms of the ratio of impact kinetic energy to surface electrostatic potential energy

$$\chi_i = \frac{1}{2} m v_0^2 \frac{R}{\bar{Z} N_i e^2} \approx \frac{2 U_p R}{\bar{Z} N_i e^2} \quad (7.204)$$

as

$$\theta_S \cong \frac{\tilde{\rho}}{\chi_i} \left[\frac{2 - \tilde{\rho} + \sqrt{1-\tilde{\rho}^2}}{1 + \sqrt{1-\tilde{\rho}^2}} \right]. \quad (7.205)$$

(Note that Krainov and Smirnov (2002b) have shown that this rescattering problem can be solved exactly, albeit with a much more cumbersome formula that is quantitatively almost exactly the same as Eq. (7.205) at $\chi_I > 5$.)

To find a hot electron spectrum we need the fraction of electrons pulled from the cluster within impact parameter ρ to $\rho + d\rho$, which, if N_{ecyc} is the number of electrons pulled from the cluster each optical cycle, is

$$f(\rho)d\rho = \frac{4\pi \rho \sqrt{R^2 - \rho^2}}{4/3\pi R^3} d\rho N_{ecyc} \quad (7.206)$$

$$f(\tilde{\rho})d\tilde{\rho} = 3 N_{ecyc} \tilde{\rho} \sqrt{1 - \tilde{\rho}^2} \, d\tilde{\rho}, \quad (7.207)$$

with impact parameter normalized to radius. Note that electrons extracted with small ρ scatter with small angle and likely rescatter on subsequent optical oscillations. Only electrons scattered out of the cluster sphere cross section are extracted on the cycle considered. The condition for an electron to be scattered at a large enough angle to be freed from

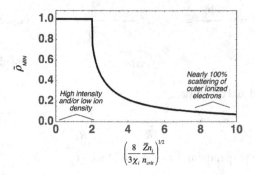

Figure 7.45 Numerical solution of Eq. (7.209) for the minimum impact parameter for which electrons are scattered out of the cluster.

rescattering is $2x_{osc}\theta_s \geq \rho + R$. Note that we can express the oscillation amplitude in terms of χ_i:

$$\frac{x_{osc}}{R} = \frac{1}{\omega}\sqrt{\frac{2}{m}\frac{\bar{Z}N_i e^2}{R^3}\chi_i}$$

$$= \left(\frac{2\chi_i}{3}\frac{\bar{Z}n_i}{n_{crit}}\right)^{1/2}. \quad (7.208)$$

The condition for electron extraction then is

$$\left(\frac{8}{3\chi_i}\frac{\bar{Z}n_i}{n_{crit}}\right)^{1/2}\frac{\tilde{\rho}}{\tilde{\rho}+1}\left[\frac{2-\tilde{\rho}+\sqrt{1-\tilde{\rho}^2}}{1+\sqrt{1-\tilde{\rho}^2}}\right] \geq 1. \quad (7.209)$$

The numerical solution for the minimum impact parameter for extraction is shown in Figure 7.45.

Finally we need to estimate the extent of heating of these extracted electrons. The (nonrelativistic) velocity of the freely oscillating electrons is, of course,

$$v_x(t) = \frac{eE_0}{m\omega}\sin\omega t - \frac{eE_0}{m\omega} + v_{ix}, \quad (7.210)$$

so the velocity of the scattered electron along the polarization axis emerging from the cluster is

$$v_{ix} \cong v_0 - \frac{1}{2}\frac{v_\perp^2}{v_0} = v_0 - \frac{1}{2}\theta_s^2 v_0, \quad (7.211)$$

which then propagates after scattering with velocity

$$v_x(t) = v_0 \sin\omega t - \frac{1}{2}\theta_s^2 v_0. \quad (7.212)$$

So, the residual total drift velocity of this scattered electron is

$$v_d^2 = \left(\frac{1}{2}\theta_s^2 v_0\right)^2 + v_\perp^2 = \left(\theta_s^2 + 1/4\theta_s^4\right)v_0^2, \quad (7.213)$$

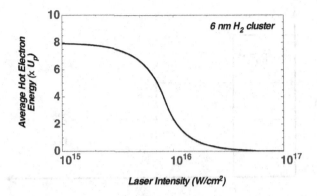

Figure 7.46 Calculated average hot electron energy in the bare ion core regime for a 6 nm hydrogen cluster subject to an 800 nm field.

which lets us conclude that the drift energy acquired by the electron is

$$\varepsilon_K \cong \left(2\theta_S^2 + 1/2\theta_S^4\right) U_p. \tag{7.214}$$

The average energy of scattered electrons then is this energy averaged over the distribution function (7.207):

$$\langle \varepsilon_K \rangle \cong U_p \int_{\tilde{\rho}_{min}}^{1} \left(2\theta_S(\tilde{\rho})^2 + 1/2\theta_S(\tilde{\rho})^4\right) 3\tilde{\rho}\sqrt{1 - \tilde{\rho}^2} d\tilde{\rho}. \tag{7.215}$$

This quantity is plotted for a 6 nm hydrogen cluster as a function of incident intensity in an 800 nm wavelength field in Figure 7.46. It is interesting to note that in this regime the scattered energy drops quickly with increasing intensity. This is more or less consistent with experimental observations that low-Z clusters like hydrogen do not show a significant suprathermal electron population.

7.4 Cluster Explosions

By now it is obvious that the expansion of a cluster nanoplasma during the irradiation by the laser pulse plays an important role in the electron dynamics, nanoplasma heating and inner-ionization of the cluster. We shall now look in more detail at the mechanisms driving this cluster expansion.

7.4.1 Hydrodynamic Expansion

The first expansion mechanism arises from the thermal pressure of the confined nanoplasma cloud (we typically assume that the ions in the cluster are cold and have no significant thermal pressure) (Ditmire et al. 1996; Milchberg et al. 2001). This hydrodynamic expansion can be thought of thermodynamically as the adiabatic expansion of a

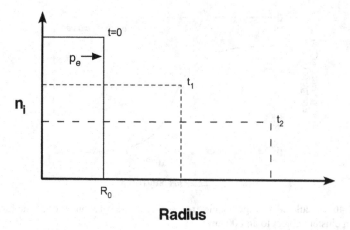

Figure 7.47 Cartoon showing the ion density of an expanding cluster ion sphere in the balloon approximation.

two-component gas whose pressure from the electrons performs work to accelerate the more massive ion gas, which provides the inertia to regulate the speed of expansion. Microscopically, what in fact happens is that the hot electron gas attempts to expand but the space-charge forces of the cold, slow ions confines this expansion. The electrons form a thin sheath around the edge of the ion sphere of a thickness comparable to the Debye length, given by Eq. (7.8). The sheath field set up by the charged layer sets up an ambipolar field which accelerates the ions. This picture is only accurate if the Debye length is much smaller than the cluster radius; often the two are comparable in the early stages of the expansion of small clusters. In that case a more appropriate model is the Coulomb explosion model we will consider later.

A simple approximation to estimate simply the rate at which a spherical cluster nanoplasma will explode under hydrodynamic forces is to assume that the cluster expands as a sphere with constant density over the entire sphere. In that case the ion density expands according to the picture illustrated in Figure 7.47; we might call this approximation the "expanding balloon" approximation. (This is a specific case of a self-similar expansion in which the radial profile does not change during the expansion, it simply increases in size with time.) In this case the ion density as a function of time is simply

$$n_i(t) = \frac{N_i}{4/3\pi R(t)^3}. \tag{7.216}$$

The radius of the ion sphere core, $R(t)$, can be calculated by considering the $p\,dV$ work performed on the ion sphere by the plasma electrons. First, we calculate the volume kinetic energy of such a uniform density sphere expanding with radial velocity $\dot{R}$:

$$\varepsilon_{VK} = \int_0^R \frac{1}{2} N_i m_i \frac{3r^2}{R^3} \left(\dot{R}\frac{r}{R}\right)^2 dr$$

7.4 Cluster Explosions

$$= \frac{3}{10} N_i m_i \dot{R}. \tag{7.217}$$

Conservation of energy demands that

$$d\varepsilon_{VK} = pdV. \tag{7.218}$$

This can be written in terms of the expanding cluster radius

$$p_e 4\pi R^2 dR = \frac{3}{5} N_i m_i \dot{R} d\dot{R} \tag{7.219}$$

$$p_e 4\pi R^2 = \frac{3}{5} N_i m_i \ddot{R} \tag{7.220}$$

or finally in terms of the time-varying pressure of the electron gas and varying ion density:

$$\ddot{R} = \frac{20}{3} \pi \frac{R(t)^2}{N_i m_i} p_e(t) = 5 \frac{p_e(t)}{m_i n_i(t) R(t)}. \tag{7.221}$$

There is, of course, a cooling of the electron temperature T_e as thermal energy is converted to directed expansion energy in the explosion. Again, conservation of energy in the conversion of thermal energy to work lets us write

$$p_e dV = n_e k_B T_e 4\pi R^2 dR \tag{7.222}$$

$$n_e k_B T_e 4\pi R^2 dR = -\frac{3}{2} n_e k_B dT_e \left(\frac{4}{3}\pi R^3 \right), \tag{7.223}$$

which yields an equation for the electron temperature cooling:

$$\frac{\partial T_e}{\partial t} = -2 \frac{T_e}{R} \dot{R}. \tag{7.224}$$

Strictly speaking, in numerical models of the cluster explosion, one must account for the heating of the electron fluid by the laser and account for energy transfer from the electrons to the confining cold ion background by electron–ion equilibration. So the equation governing the electron temperature is in fact

$$\frac{\partial T_e}{\partial t} = -2 \frac{T_e}{R} \dot{R} + \frac{2}{3} \frac{1}{k_B} \frac{\partial u_{las}}{\partial t} - \frac{T_e - T_i}{\tau_{ei}} \tag{7.225}$$

where the equilibration time can be written (see Appendix B for details)

$$\tau_{ei} = \frac{3 m_e m_i}{8(2\pi)^{1/2} n_i \bar{Z}^2 e^4 \ln \Lambda_C} \left(\frac{k_B T_e}{m_e} + \frac{k_B T_i}{m_i} \right). \tag{7.226}$$

This equilibration time is usually a few picoseconds for clusters with heavy ions (Ar, Kr Xe etc.), so it is usually negligible.

Using an ideal gas equation of state for the electrons, Eq. (7.221) becomes

$$\ddot{R} = 5 \frac{\bar{Z} k_B T_e}{m_i R}. \tag{7.227}$$

Taking the time derivative of this equation yields

$$\dddot{R} = 5 \frac{\bar{Z} k_B}{m_i} \left(\frac{\dot{T}_e}{R} - \frac{T_e \dot{R}}{R^2} \right). \tag{7.228}$$

Inserting Eq. (7.225) and ignoring equilibration gives

$$\ddot{R} = 5\frac{\bar{Z}k_B}{m_i}\left(3\frac{T_e R}{R^2} + \frac{2}{3}\frac{1}{Rk_B}\frac{\partial u_{las}}{\partial t}\right). \quad (7.229)$$

Noting that

$$\frac{\partial^3}{\partial t^3}R^2 = 6\dot{R}\ddot{R} + 2R\dddot{R} \quad (7.230)$$

and use of Eq. (7.227) yields finally a differential equation for the cluster radius expansion time history:

$$\frac{\partial^3}{\partial t^3}R^2 = \frac{20}{3}\frac{\bar{Z}}{m_i}\frac{\partial u_{las}}{\partial t}. \quad (7.231)$$

Combination of this equation with a model for the cluster heating allows simple numerical integration of the cluster radial expansion history.

If the energy deposition is much faster than the time it takes the cluster to expand significantly, we can integrate Eq. (7.224) directly to derive the electron temperature as a function of cluster radius and initial heated temperature T_{e0}:

$$\frac{\partial T_e}{T_e} = -2\frac{\partial R}{R}, \quad (7.232)$$

$$T_e(t) = T_{e0}\left(\frac{R_0}{R(t)}\right)^2. \quad (7.233)$$

We can also estimate, then, the final cluster expansion velocity as $R \to \infty$. The energy density of the electron ideal gas is

$$u_{tot} = {}^3/_2 k_B T_{e0}, \quad (7.234)$$

so integration of Eq. (7.231), assuming that the heating is a delta function and insertion of Eq. (7.234) yields

$$\frac{\partial^2}{\partial t^2}R^2 = 10\frac{\bar{Z}}{m_i}k_B T_{e0}. \quad (7.235)$$

The expansion velocity is constant at $t \to \infty$. So, since

$$2\dot{R}^2 + 2R\ddot{R} = 10\frac{\bar{Z}}{m_i}k_B T_{e0} \quad (7.236)$$

and the fact that $\ddot{R} = 0$ when $t \to \infty$ lets us write for the final explosion velocity of the cluster

$$\dot{R} = \sqrt{5\frac{\bar{Z}}{m_i}k_B T_{e0}} \quad (7.237)$$

$$= \sqrt{3}c_{sy},$$

7.4 Cluster Explosions

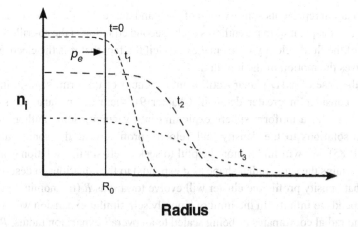

Figure 7.48 Schematic illustration of realistic radial density profiles as an initially uniform heated cluster plasma expands by hydrodynamic forces.

where $c_{s\gamma}$ is the adiabatic sound speed of a gas given in terms of the ratio of specific heats γ_c (= 5/3 for a monatomic ideal gas) as

$$c_{s\gamma} = \sqrt{\gamma_c \bar{Z} k_B T_e / m_i}. \tag{7.238}$$

This simple "expanding balloon" model can only give a semiquantitative estimate for the rate of cluster expansion; in reality, the time progression of a heated spherical cluster nanoplasma will progress more along the lines depicted in Figure 7.48. Even if the initial radial density profile is uniform, the leading edge of the expansion into vacuum will begin to smear and the density profile will evolve with time more toward an expanding super-Gaussian to Gaussian density profile. We expect the time scale for this smearing to occur to be about $R_0/c_{s\gamma}$. For a 10 nm Ar cluster eight-times ionized and heated to 100 eV, this time is ~150–300 fs. So, the uniform expanding sphere picture might be ok to assess the dynamics during a very short laser pulse but clearly we need to examine the hydro-expansion profile in more detail to assess what the dynamics are later and to determine what kind of ejected ion energy spectra might be observed.

In this regime we can describe the one-dimensional spherical expansion by the ion fluid equations. We assume that the electron gas responds very quickly, on the time scale of this ion hydrodynamic motion so the electrons merely serve as a neutralizing background that exerts a pressure on the ion fluid. In spherical coordinates the ion fluid equations for radial and time-dependent ion density and fluid velocity are

$$\frac{\partial}{\partial t} n_i(r, t) + \frac{1}{r^2} \frac{\partial}{\partial r} \left(r^2 n_i(r, t) v(r, t) \right) = 0 \tag{7.239}$$

$$n_i(r, t) m_i \left(\frac{\partial}{\partial t} v(r, t) + v(r, t) \frac{\partial}{\partial r} v(r, t) \right) = -\frac{\partial}{\partial r} p_e(r, t) \tag{7.240}$$

The first equation represents conservation of mass and the second describes conservation of momentum. (The quantity in parentheses in the second equation is the so-called convective derivative of the fluid velocity.) This makes explicit the fact that it is the electron pressure, p_e, that drives the motion of the ion fluid.

Unlike the case of a 1D planar outflow into vacuum of uniform, semi-infinite plasma (which we consider in greater detail in Chapter 9), there are no analytic solutions to these equations for a uniform sphere exploding into vacuum. Some authors have found asymptotic solutions to the density and velocity profiles near the sonic outflow point (Schmalz 1985). We will find it more helpful to seek a self-similar solution (one in which the profile retains the same radial shape as it expands) to the equations to describe approximately what density profile the cluster will evolve toward as R (the nominal radius of the cluster) expands to infinity. In the limit of a purely self-similar expansion we can write the velocity and radial coordinates as being scaled to an overall expansion radius, $R(t)$,

$$v(r,t) = \dot{R}(t)\frac{r}{R(t)}, \qquad (7.241)$$

and we can introduce the scaled variable $\tilde{r} \equiv r/R(t)$. The self-similar density profile, then, can be written as a product of the time-varying density at the center of the cluster times the self-similar profile between $r = 0$ and R. This is a function only of the scaled radial variable: $n_i(r,t) = n_{i0}(t)f(\tilde{r})$. This lets us write the electron pressure, assuming an ideal gas equation of state (EOS) and uniform electron temperature, as

$$p_e(r,t) = \bar{Z}n_i(r,t)k_B T_e(t)$$
$$= p_{e0}f(\tilde{r}) = \bar{Z}n_{i0}k_B T_e(t)f(\tilde{r}). \qquad (7.242)$$

As with the uniform density approximation, we assume that the expansion is adiabatic, so $p\,dV = d\varepsilon_K$, which for a differential radial element can be written

$$-\frac{\partial p_e}{\partial r'}dr'4\pi r^2 dr = n_i m_i 4\pi r^2\,dr\,\ddot{r}d\dot{r}. \qquad (7.243)$$

We divide this by dt and integrate over the plasma volume:

$$a_1 p_{e0}\frac{\dot{R}}{R} = a_2 n_{i0} m_i \frac{\partial}{\partial t}\left(\dot{R}\right)^2, \qquad (7.244)$$

where, with Eq. (7.242), we can see that the constants in this equation are

$$a_1 = -4\pi \int_0^1 \frac{\partial f}{\partial \tilde{r}}\tilde{r}^3\,d\tilde{r}, \qquad (7.245a)$$

$$a_2 = 2\pi \int_0^1 f(\tilde{r})\tilde{r}^4\,d\tilde{r}. \qquad (7.245b)$$

We seek the equation of motion for the front position of the self-similar expansion $R(t)$. Defining a third constant,

$$a_3 = 4\pi \int_0^1 f(\tilde{r})\tilde{r}^2\,d\tilde{r}, \qquad (7.245c)$$

7.4 Cluster Explosions

we use the fact that it is the electron thermal energy which does work on the system, so conservation of energy demands

$$\frac{3}{2}\bar{Z}n_i(r)\,d\,(k_B T_e) = -pdV + \bar{Z}n_i(r)\partial u, \tag{7.246}$$

where we have added a term to account for external heating of the cluster plasma by the laser (the second term on the right side). Dividing by dt and integrating over the plasma volume yields

$$\frac{3}{2}a_3\bar{Z}n_{i0}k_B\frac{\partial T_e}{\partial t} = -a_1\frac{p_{e0}}{R}\dot{R} + a_3\bar{Z}n_{i0}\frac{\partial u}{\partial t}. \tag{7.247}$$

We use Eq. (7.244),

$$a_1\frac{p_{e0}}{R} = 2a_2 n_{i0} m_i \ddot{R}, \tag{7.248}$$

recall the ideal gas EOS Eq. (7.242), and differentiate this equation with respect to time to get

$$\dddot{R} = \frac{a_1}{2a_2}\frac{1}{m_i}\left(\bar{Z}\frac{k_B}{R}\frac{\partial T_e}{\partial t} - \frac{k_B T_e}{R^2}\dot{R}\right). \tag{7.249}$$

Substitution of Eq. (7.247) to eliminate the time derivative of electron temperature gives us

$$\frac{3}{2}a_3\bar{Z}n_{i0}\left(\frac{a_1}{2a_2}\frac{m_i}{\bar{Z}}R\dddot{R} - \frac{k_B T_e}{R^2}\dot{R}\right) = -2a_2 n_{i0} m_i \dot{R}\ddot{R} + a_3\bar{Z}n_{i0}\frac{\partial u}{\partial t}. \tag{7.250}$$

This equation can be simplified by noting that if we integrate Eq. (7.245a) by parts and assume $f(1) = 0$, we see that

$$\frac{a_1}{a_3} = -\int_0^1 \frac{\partial f}{\partial \tilde{r}}\tilde{r}^3\,d\tilde{r} \Big/ \int_0^1 f\tilde{r}^2\,d\tilde{r} = 3. \tag{7.251}$$

We have, then, that

$$a_2 n_{i0} m_i R\ddot{R} + 3a_2 n_{i0} m_i \dot{R}\ddot{R} = a_3\bar{Z}n_{i0}\frac{\partial u}{\partial t}, \tag{7.252}$$

which reduces to our final result, the equation of motion for expanding cluster self-similar profile front:

$$\frac{\partial^3}{\partial t^3}\left(R^2\right) = 2\frac{a_3}{a_2}\frac{\bar{Z}}{m_i}\frac{\partial u}{\partial t}. \tag{7.253}$$

This is essentially the general form of Eq. (7.231), which we derived for a uniform cluster ion profile. It can be seen that if the integrals of Eqs. (7.245) are performed for a uniform density profile out to radius R, the earlier equation is retrieved. Our final task is to find the actual self-similar profile, $f(r)$. Rewriting the convective derivative of the momentum ion fluid, Eq. (7.240), instead as a complete derivative and using the self-similar definition of the velocity and pressure profiles, Eqs. (7.241) and (7.242), the ion momentum equation becomes

$$n_{i0} m_i f(\tilde{r})\tilde{r}\ddot{R} = p_{e0}\frac{\partial f}{\partial \tilde{r}}\frac{1}{R}. \tag{7.254}$$

From Eq. (7.248), the definitions of a_1 and a_2 and repeated integration by parts (assuming $f(1) = 0$ and $f'(0) = 0$) we can write

$$\frac{\partial f}{\partial \tilde{r}} \frac{1}{f(\tilde{r})\tilde{r}} = \frac{n_{i0} m_i}{p_{e0}} R\ddot{R} \qquad (7.255)$$
$$= \frac{a_1}{2a_2},$$

which trivially integrates to yield the self-similar density profile

$$f(\tilde{r}) = exp\left[-\frac{a_1}{4a_2} \frac{r^2}{R(t)^2}\right]. \qquad (7.256)$$

If we make the approximation that the limits in the integrals of Eqs. (7.245a–c) can be extended to infinity, we can perform the integrals in these definitions and algebraically solve for the constants, showing again that $a_1/a_3 = 3$ and $a_1/a_2 = 4$. So, the final solution for the self-similar expanding density profile of the expanding cluster sphere is a Gaussian:

$$f(r) = exp\left[-\frac{r^2}{R(t)^2}\right]. \qquad (7.257)$$

We must emphasize that this is an approximate solution; it does not obey the assumed property that the plasma density goes to zero at the front position $r = R$. A Gaussian density profile has density extending out to $r = \infty$, which is, of course, nonsense for a plasma profile that is initially well confined (like the radially uniform ion density of the initial cluster). Nonetheless, it gives us a good sense of the profile that an expanding plasma sphere will evolve toward at large R. Such a late-time spherical plasma evolution is well established in numerical hydrodynamic simulations (Haught and Polk 1970).

The equation describing the motion of this self-similar expansion comes from Eq. (7.253):

$$\frac{\partial^3}{\partial t^3}\left(R^2\right) = \frac{8}{3}\frac{\bar{Z}}{m_i}\frac{\partial u}{\partial t}. \qquad (7.258)$$

So, again with a short energy deposition time, as $R \to \infty$ the velocity of the expansion at the 1/e point of the density profile asymptotically approaches

$$\dot{R} = \sqrt{2\frac{\bar{Z}}{m_i} k_B T_e(0)} \qquad (7.259)$$

which means that the radially dependent velocity of the exploding cluster ions is

$$v(r) = \sqrt{2}c_s \frac{r}{R}, \qquad (7.260)$$

where c_s is the usual (isothermal) ion sound speed $(Zk_B T_e/m_i)^{1/2}$. Consequently, if the ions from this explosion were measured, they would exhibit a velocity distribution that is

$$f_i(v)dv = 4\pi v^2 \exp\left[-v^2/2c_s^2\right], \qquad (7.261)$$

7.4 Cluster Explosions

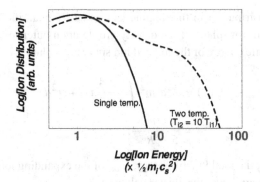

Figure 7.49 Solid line is a plot of the ion energy distribution from a hydrodynamically exploding cluster, as predicted by Eq. (7.266). The dashed line shows that a two-humped distribution is expected if there is a two-temperature initial electron distribution.

which means that the energy distribution of an ensemble of exploding clusters, in terms of the initial electron temperature is

$$f_i(\varepsilon_i)d\varepsilon_i = \alpha \sqrt{\varepsilon_i} \exp\left[-\varepsilon_i/\bar{Z}k_B T_{e0}\right]. \tag{7.262}$$

The expected ion energy distribution, then, which is observed will look something like the solid line in Figure 7.49. Such distributions from exploding clusters in the hydrodynamic regime are well established in experiment (Ditmire et al. 1997c; Ditmire et al. 1998b; Krishnamurthy et al. 2006). Note that the ion energy distribution in this regime is essentially independent of initial cluster size, R_0 (though the energy deposition dynamics that determine the initial cluster electron temperature might depend on this initial cluster size). So the observed ion energy spectra in this regime tend to be insensitive to the distribution of initial cluster sizes in the target gas.

Experimentally, real ion energy distributions often exhibit a two-humped shape like that shown by the dashed line of Figure 7.49. This is almost certainly the signature of cluster explosions with electron fluids that have, after laser heating, a two-temperature distribution. (Chapter 9 gives a much more detailed discussion of two-temperature expansions in the case of planar targets.) We have already seen that such a situation naturally occurs when the space-charge confined electrons are heated collisionally and a minority of "hot" electrons develop by stochastic heating.

7.4.2 Coulomb Explosion

A completely different mechanism for cluster explosions evolves in the opposite limit of the hydrodynamic regime. When most electrons are removed by outer-ionization, which is a situation encountered frequently in experiments with low-Z clusters like hydrogen clusters, or in small (say ≤ 100 atom) noble gas clusters, the cluster ion sphere will expand subject to the Coulomb repulsion forces between the ions. This situation is usually referred to as a Coulomb explosion (Last and Jortner 2000; Madison et al. 2004a; Islam et al. 2006).

The ion energy distribution in this regime can be easily quantified if we consider, as usual, an initially uniform spherical ion density profile upon outer-ionization by the laser. The stored electrostatic energy of the ionized ion sphere of radius R is

$$U(R) = \bar{Z}^2 e^2 n_i^2 \int_0^R \frac{1}{r} \frac{4}{3}\pi r^3 4\pi r^2 d$$

$$= \frac{3}{5} \frac{\bar{Z}^2 e^2 N_i}{R}. \tag{7.263}$$

By similar reasoning, the total kinetic energy, K_{tot}, of the expanding ions at some instant is given by a similar integral over the cluster volume:

$$K_{tot} = \int_0^R \tfrac{1}{2} m_i \left(\dot{R}\frac{r}{R}\right)^2 n_i 4\pi r^2 \, dr$$

$$= \frac{2}{5}\pi m_i n_i R^3 \dot{R}^2 = \frac{3}{10} m_i N_i \dot{R}^2. \tag{7.264}$$

We can find the instantaneous velocity of the expanding cluster radius by simple conservation of energy, equating the total kinetic energy of the exploding ion sphere plus the remaining electrostatic repulsion energy with the initial electrostatic energy

$$K_{tot} + U(R) = U(R_o), \tag{7.265}$$

$$\frac{3}{10} m_i N_i \dot{R}^2 = \frac{3}{5} \bar{Z}^2 e^2 N_i \left(\frac{1}{R_0} - \frac{1}{R(t)}\right), \tag{7.266}$$

which yields for the cluster radial velocity

$$\frac{dR}{dt} = \sqrt{\frac{2\bar{Z}^2 e^2 N_i}{m_i} \left(\frac{1}{R_0} - \frac{1}{R(t)}\right)}. \tag{7.267}$$

This equation can be integrated to yield

$$t(R) = \sqrt{\frac{m_i R_0^3}{2\bar{Z}^2 e^2 N_i} \left(\frac{1}{R_0} - \frac{1}{R(t)}\right)} \left\{ \frac{R}{R_0}\left(\frac{R}{R_0} - 1\right) + \ln\left[\left(\frac{R}{R_0}\right)^{1/2} + \left(\frac{R}{R_0} - 1\right)^{1/2}\right] \right\}. \tag{7.268}$$

This equation gives us a convenient analytic estimate for the time it takes a cluster to Coulomb explode: we can define this explosion time as the time it takes for the cluster sphere to double in size. From Eq. (7.268) this doubling time is

$$t_{2R_0} = 1.6 \sqrt{\frac{m_i R_0^3}{\bar{Z}^2 e^2 N_i}} = 2.8 \omega_{pi}^{-1}. \tag{7.269}$$

This time is about an inverse ion plasma period.

7.4 Cluster Explosions

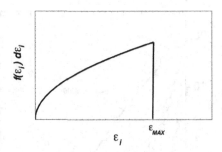

Figure 7.50 Ion energy distribution from a Coulomb exploding cluster.

In the context of this simple model, the ion energy distribution from a single exploding cluster is easy to calculate. From electrostatic energy considerations the velocity of a shell of ions at radius r (V_i) within the cluster asymptotically approaches

$$V_i(r) \underset{R \to \infty}{=} \sqrt{\frac{8\pi}{3} \frac{\bar{Z}^2 e^2 n_i}{m_i} r^2}. \tag{7.270}$$

The kinetic energy of an ion in this shell is

$$\varepsilon_i = \tfrac{1}{2} m_i V_i(r)^2 = \frac{4\pi}{3} \bar{Z}^2 e^2 n_i r^2. \tag{7.271}$$

The distribution of ion number in a shell at radius r is just

$$f(r)dr = 4\pi r^2 n_i dr. \tag{7.272}$$

Eq. (7.271) indicates that $\varepsilon_I \sim r^2$ and $d\varepsilon_I \sim r\, dr$, so it follows that the energy distribution of ions from this Coulomb exploding sphere scales like

$$\begin{aligned} f(\varepsilon_i)d\varepsilon_i &\sim \sqrt{\varepsilon_i}\, d\varepsilon_i && \text{for } \varepsilon_i \leq \frac{4\pi}{3}\bar{Z}^2 e^2 n_i R_0^2 \equiv \varepsilon_{MAX} \\ &\sim 0 && \text{for } \varepsilon_i > \varepsilon_{MAX}, \end{aligned} \tag{7.273}$$

where the maximum ion energy is just determined by the energy of the outermost ion shell. Defining $\tilde{\varepsilon}_i \equiv \varepsilon_i/\varepsilon_{MAX}$, normalization of the distribution leads to

$$\begin{aligned} f(\varepsilon_i)d\varepsilon_i &= \frac{3}{2}\sqrt{\tilde{\varepsilon}_i}\, d\tilde{\varepsilon}_i && \text{for } \tilde{\varepsilon}_i \leq 1 \\ &= 0 && \text{for } \tilde{\varepsilon}_i > 1 \end{aligned}. \tag{7.274}$$

This distribution is illustrated in Figure 7.50. For example, a 10 nm deuterium cluster will have a Coulomb explosion maximum energy of about 3 keV.

As an aside, note that this simple Coulomb explosion picture assumes a uniform ion fluid. This picture is not really very accurate for very small clusters (say <500 atoms per cluster). The ion energy distribution reflects the fact that these small clusters have some internal structure; picturing them as a uniform ion fluid is not very accurate (Rusek *et al.* 2001). The shell structure of these clusters can be seen experimentally (Erk *et al.* 2011).

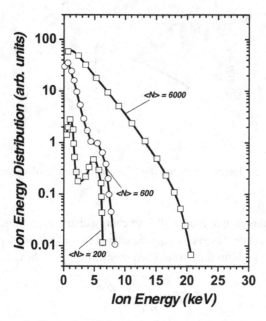

Figure 7.51 Schematic illustration of the kind of data seen in the Coulomb explosion of rather small clusters. A shelled structure becomes evident.

Figure 7.51 shows schematically the kind of ion energy data seen for small Ar clusters in the Coulomb explosion regime. The smallest sizes exhibit humps in the ion energy spectrum, which reflects the shell structure of the ions initially.

Unlike a plasma formed from hydrodynamically exploding clusters, the ion energy distribution observed depends on the initial size distribution of clusters in the target gas. Returning to rely on the simple uniform ion model of Eq. (7.274) and assuming that the laser intensity is sufficient to outer-ionize all the clusters in the size distribution, we can estimate the cluster ion energy distribution by recalling the log-normal cluster size distribution of Eq. (7.3). The overall ensemble ion energy distribution, then, of a plasma created from Coulomb exploding clusters can be written as an integral over the size distribution, weighted by Eq. (7.274):

$$g(\varepsilon_i)d\varepsilon_i = \int_{N_\varepsilon}^{\infty} P_c(N)dN\sqrt{\varepsilon_i}\,d\varepsilon. \tag{7.275}$$

Here the lower limit of the integration is the smallest cluster size that contributes to the distribution at ε_i. Written in terms of its initial radius, this cluster size is

$$N_\varepsilon = {}^4\!/\!_3\pi R_\varepsilon^3 n_{i0}. \tag{7.276}$$

Using Eq. (7.271), we can write this cluster size as

$$N_\varepsilon = \alpha_{Ci}\varepsilon_i^{3/2} \tag{7.277}$$

7.4 Cluster Explosions

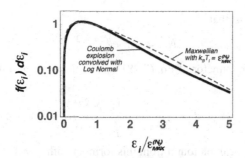

Figure 7.52 Plot of the convolved Coulomb explosion ion energy distribution compared to a Maxwellian distribution.

where

$$\alpha_{Ci} = \left(\frac{3}{4\pi}\right)^{1/2} \frac{1}{Z^3 e^3 n_{i0}^{1/2}}. \tag{7.278}$$

Integration of Eq. (7.275) yields for the ensemble cluster plasma ion energy distribution

$$g(\varepsilon_i)d\varepsilon_i = C\sqrt{\varepsilon_i}\, Erfc\left[\ln\left[\alpha_{Ci}\varepsilon_i^{3/2}/N_0\right]/\sqrt{2\sigma_c}\right] d\varepsilon. \tag{7.279}$$

Note that the factor in the natural log is

$$\alpha_{Ci}\varepsilon_i^{3/2}/N_0 = \left(\varepsilon_i/\varepsilon_{MAX}^{(N_0)}\right)^{3/2}. \tag{7.280}$$

This distribution is plotted in Figure 7.52. The shape of this distribution when $\sigma_c \approx (\ln N_0)/2$ is remarkably similar to a Maxwellian distribution with ion temperature equal to the maximum ion energy of a Coulomb exploding cluster of size N_0.

There is some interest in finding the kinetic energy moments of this ion energy distribution. The pth energy moment is written

$$\varepsilon_i^p = \int_0^\infty \varepsilon_i^p g(\varepsilon_i) d\varepsilon_i \bigg/ \int_0^\infty g(\varepsilon_i) d\varepsilon_i. \tag{7.281}$$

Looking back at Eq. (7.275), the numerator of this calculation can be expressed as a double integral:

$$Num = \int_0^\infty \int_{N_\varepsilon}^\infty \varepsilon_i^{p+1/2} P_c(N)\, dN\, d\varepsilon_i. \tag{7.282}$$

Reversing the order of integration:

$$Num = \int_0^\infty \int_0^{\varepsilon_{MAX}(N)} \varepsilon_i^{p+1/2} P_c(N)\, d\varepsilon_i dN$$

$$= \frac{2}{2p+3} \int_0^\infty \varepsilon_{MAX}^{p+1/2} P_c(N)\, dN$$

$$= \frac{2}{2p+3} \int_0^\infty \left(\frac{N}{\alpha_{Ci}}\right)^{(2p+3)/3} P_c(N)\, dN$$

$$= \frac{2}{2p+3} \frac{1}{\alpha_{Ci}^{(2p+3)/3}} N^{(2p+3)/3}. \tag{7.283}$$

Recalling from Eq. (7.4) that

$$N^{p'} = N_0^{p'} e^{p'^2 \sigma_c^2 / 2} \tag{7.284}$$

and noting that the denominator of Eq. (7.281) is just this with $p = 0$, we have for the pth energy moment

$$\varepsilon_i^p = \frac{2}{2p+3} \frac{1}{\alpha_{Ci}^{2p/3}} \frac{N^{(2p+3)/3}}{N}. \tag{7.285}$$

The average ion energy can be found from this formula with $p = 1$. It is

$$\varepsilon_i = \frac{3}{5} \frac{N_0^{2/3}}{\alpha_{Ci}^{2/3}} e^{(8/9)\sigma_c^2}. \tag{7.286}$$

So, the average energy of a gas from Coulomb exploding clusters is just a factor of $\exp[(8/9)\,\sigma_c^2]$ different from a monodisperse cluster distribution. For example, when $\sigma_c = 1$, this enhancement factor is ~ 2.4. The warm tail of the distribution greatly increases the average ion energy.

7.4.3 Explosions of Partially Outer-Ionized Cluster Plasmas

The intermediate case involves explosions of clusters that have only been partially ionized. If the outer-ionization is modest, and the fraction of inner-ionized electrons removed from the cluster is small, a reasonable way to treat this situation is to assume that the Coulomb forces act as an effective pressure which must be added to the thermal pressure of the space-charge confined electrons. For a uniform ion density sphere, this pressure is easily approximated by considering the cluster as a spherical capacitor charged to Q_c, with stored potential energy

$$U_Q = \frac{Q_c^2}{2R}. \tag{7.287}$$

The Coulomb contribution to the force acting per unit area on the cluster surface is

$$p_{Coul} = -\frac{1}{4\pi R^2} \frac{\partial u_Q}{\partial R}, \tag{7.288}$$

which means that the equation of motion, Eq. (7.221), for the expanding cluster is amended to read

$$\frac{\partial^2 R}{\partial t^2} = 5 \frac{p_e + p_{Coul}}{n_i m_i R(t)}. \tag{7.289}$$

This additional force to push the cluster explosion is typically most significant in the smaller clusters. For example, in Figure 7.53 is shown a schematic reproduction of particle simulations of exploding Xe clusters irradiated by 100 fs pulses at 10^{16} W/cm^2 (Last and Jortner 2004). The inner- and outer-ionization per atom is plotted as a function of time for Xe clusters of sizes ranging from 55 to 1061 atoms. The expansion rate of the smaller clusters is fastest because of significant outer-ionization early in the pulse and subsequent Coulomb pressure.

7.4 Cluster Explosions

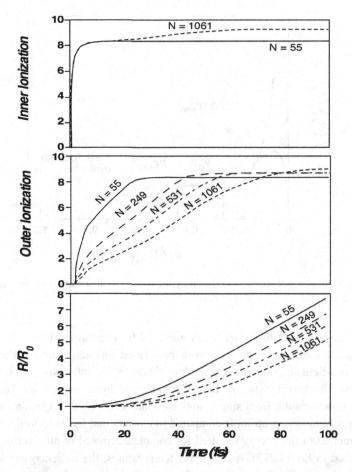

Figure 7.53 Schematic reproduction of particle simulations of exploding Xe clusters irradiated by 100 fs pulses at 10^{16} W/cm^2. The top two panels show inner- and outer-ionization fraction per atom in the cluster. The third panel shows the radius expanding as a function of time. Plot was generated using calculations from (Last and Jortner 2004).

When significant outer-ionization occurs before the cluster has much expanded, a situation encountered, say, when large, high-Z clusters are irradiated by high-intensity, very short pulses, a situation like that in Figure 7.19 likely develops. The remaining thermal electrons contract and form a core of radius R_e in the center of the ion sphere. In this case we might expect the outer shell of ions to explode by Coulomb explosion, with ejected kinetic energy at initial cluster radii of $R_e \leq r \leq R_0$ of

$$\varepsilon_i(r) = \frac{4}{3}\pi \bar{Z}^2 n_i e^2 \frac{(r^3 - R_e^3)}{r}. \tag{7.290}$$

The inner core then should explode by hydrodynamic expansion.

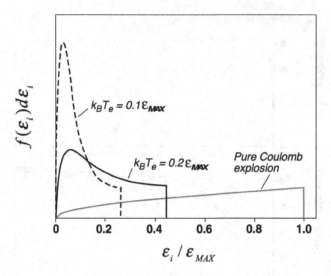

Figure 7.54 Ion energy spectra calculated by the two-fluid, nonneutral simulations. Plots generated from calculations of Peano *et al.* (2006).

In reality, these perfect limits are rarely achieved for experimentally realizable pulses. There has been some computational work performed on this situation. For example, nonneutral, two-fluid simulations of exploding clusters show interesting ion energy spectra which exhibit characteristics of both Coulomb and hydro-explosions. For example, Figure 7.54 shows results from simulations by Peano *et al.* (2006). Qualitatively, a pure Coulomb explosion ion spectrum is compared to clusters that have electron temperatures that are a fraction of the energy required for complete removal of all electrons from the cluster, defined in Eq. (7.273). At the lower temperatures, the ion energy spectrum still shows an abrupt cutoff like the pure Coulomb explosion, shifted to lower energy, with lower-energy ions having a spectrum more like a hydrodynamic expansion.

7.4.4 Coulomb Explosion of Heteronuclear Clusters

When clusters of mixed species Coulomb explode, there can be kinematic effects. Instead of ions simply acquiring energy based on their initial radial position in the cluster, the light ions can overtake the heavy ions during the expansion and derive a dynamic kick in kinetic energy from the slower moving large ions. A good example of where this occurs in experiments is in studies of exploding methane and deuterated methane clusters (Hohenberger *et al.* 2005). Whether such kinematic effects will occur can be characterized by the parameter (Last and Jortner 2002)

$$\eta_k = \frac{q_A m_B}{q_B m_A}, \tag{7.291}$$

where q_j and m_j represent the charge and mass of the two kinds of ions in the cluster ($j = A$ is for the lighter ion and B for the heavier).

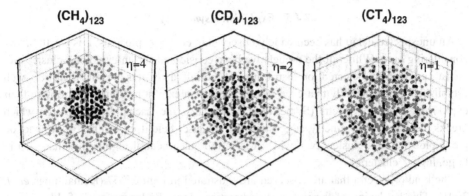

Figure 7.55 Illustration of particle dynamics simulations for three species of methane clusters with the carbon ions depicted as black dots and the hydrogen isotopes shown as gray dots. Figure was generated from calculations published in (Hohenberger et al. 2005).

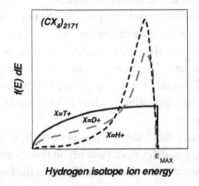

Figure 7.56 Ion energy spectra from particle simulations of 2,171-molecule CX_4 clusters showing the spectra of the light ion. Plot was generated from calculations in (Last and Jortner 2002).

If $\eta_k > 1$, then the light ions will overtake the heavy ions and acquire a kinematic enhancement in their asymptotic ion energy. There is no longer a direct relationship between final ion energy and initial radial position of the ion like there is in a homonuclear cluster. This can be seen in particle simulations of exploding CX_4 clusters (where X = H, D or T). Snapshots of ions from exploding $(CX_4)_{123}$ clusters calculated by MD simulations are illustrated in Figure 7.55. In standard methane clusters with fully ionized carbon and hydrogen ions we have $\eta = 4$. The light-colored H+ ions outrun the carbon core and pick up additional energy from the more highly charged core.

Of course, the maximum energy of the light ions does not increase. These ions reside on the cluster surface and they are accelerated to the usual ε_{MAX}. It is the inner light ions that can receive the kinematic boost in energy. This can be seen in the ion energy spectra found in particle simulations like those shown in Figure 7.55. Figure 7.56 shows light ion energy spectra from the three methane species. The tritiated methane simulation has $\eta_k = 1$, and that spectrum is just like that of a Coulomb exploding homonuclear cluster. The lighter species (H+ and D+) show a boost of the lower-energy ions up toward ε_{MAX}.

7.4.5 Explosion Asymmetry

An unexpected detail has been widely observed in cluster explosions; it is worth briefly considering here. While all the models we have considered for an exploding cluster predict a pure isotropic ejected ion distribution if the initial clusters are spherical, it is well established experimentally that even spherical clusters exhibit a slight anisotropy in their exploded ion energy spectra (Symes *et al.* 2007). It is the laser polarization axis which breaks the spherical symmetry, suggesting that the laser field itself can have an impact on the explosions of a cluster, even when it is in the quasi-neutral regime like that seen in large, high-Z clusters.

The kind of spectra that are observed are illustrated in Figure 7.57 (Kumarappan *et al.* 2001). This hardening of the ions in the direction along the laser electric field axis has been seen in a variety of clusters (H_2, Ar, N_2 and Xe). This effect cannot be simply acceleration from the laser field itself because that field time averages to zero during the many-femtosecond time scale of the cluster explosion.

There are two contributions to this anisotropy. First, the polarization field set by the laser-driven electron oscillation could enhance inner-ionization near the poles and give rise to somewhat higher charge states at the poles oriented along the laser field. This effect, however, does not explain the observed anisotropy in proton energies from exploding hydrogen clusters (Symes *et al.* 2007). A more important contribution is most likely a result of the polarization field set up by the oscillating electron cloud. This is particularly important for large amplitude oscillations.

A simple model to quantify this second effect is possible within the scope of the non-linear harmonic oscillator picture discussed already in this chapter. Consider a cluster ion sphere that has undergone some outer-ionization. The ξ_e is the fraction of electrons remaining in the cluster after some outer-ionization. In that case the number of electrons in the electron cloud is

$$N_e = \xi_e \bar{Z} N_i \tag{7.292}$$

and its radius is

$$R_e = \xi_e^{1/3} R. \tag{7.293}$$

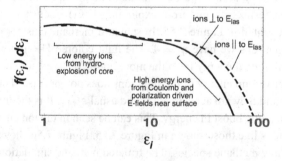

Figure 7.57 Illustration of anisotropy observed in many cluster explosion experiments.

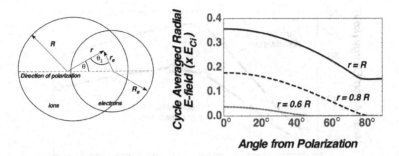

Figure 7.58 On the left is the geometry of the polarization-induced field. On the right is the calculated cycle-averaged E-field as a function of angle from the polarization axis in terms of the ion-induced field at the ion sphere surface for $x_{osc} = 0.5\,R$.

The geometry of the interaction is shown at left in Figure 7.58. The electric field from the ion sphere is directed radially with magnitude

$$E_i = \frac{\bar{Z}N_i e}{r^2} \frac{r^3}{R^3} \equiv E_{Ci} \frac{r}{R}. \tag{7.294}$$

Similarly, the E-field from the electron cloud is

$$E_e = \xi_e \frac{\bar{Z}N_i e}{r_e^2} \frac{r_e^3}{R_e^3} \quad \text{inside electron cloud}$$

$$= \xi_e \frac{\bar{Z}N_i e}{r_e^2} \quad \text{outside electron cloud}. \tag{7.295}$$

It is convenient to write the fields in terms of the fields from the ion and electron spheres separately at the surface of the spheres respectively:

$$E_{Ci} = \frac{\bar{Z}N_i e}{R^2}; \quad E_{Ce} = \frac{N_e e}{R_e^2} = \xi_e^{1/3} E_{Ci}. \tag{7.296}$$

The field experienced by an ion is the laser field plus the field of the ion sphere and the oscillating electron cloud

$$E_r(\mathbf{x}) = E_{laser}\cos\theta + E_i - E_e\cos\theta'. \tag{7.297}$$

The position of this ion from the center of the electron cloud is

$$r_e = \sqrt{r^2 + [x_0\cos(\omega t + \varphi_0)]^2 - 2rx_0\cos(\omega t + \varphi_0)\cos\theta}, \tag{7.298}$$

which is related to the oscillation amplitude and the angles defined in Figure 7.58 by the cosine rule

$$\cos\theta' = \frac{r - x_0\cos(\omega t + \varphi_0)\cos\theta}{r_e}. \tag{7.299}$$

Therefore, the total radial electric field felt by an ion is

$$E_r(\mathbf{x}) = E_0\cos\omega t\,\cos\theta + E_{Ci}\frac{r}{R} - E_{Ce}\frac{R_e^2\,(r - x_0\cos(\omega t + \varphi_0)\cos\theta)}{r_e^3} \quad r_e < R_e$$

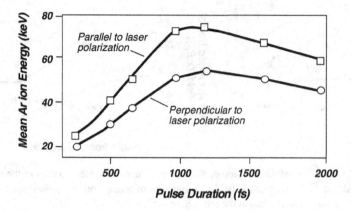

Figure 7.59 Schematic of data showing mean Ar ion energy from exploding clusters parallel and perpendicular to the laser polarization as a function of pulse duration.

$$= E_0 cos\omega t\ cos\theta + E_{Ci}\frac{r}{R} - E_{Ce}\frac{(r - x_0 cos(\omega t + \varphi_0))cos\theta)}{R_e} \qquad r_e \geq R_e.$$
(7.300)

Ignoring the tangential field, averaging over one optical cycle yields the time-averaged field as a function of angle, which is plotted for three different depths within the cluster ion core is shown as a function of polar angle in Figure 7.58 when $x_{osc} = 0.5\ R$. The time-averaged field is higher along the laser polarization, and this will boost the energy of ions expanding in this direction.

Because this effect is particularly significant when the electron cloud oscillation amplitude is large (i.e. $x_{osc} \sim R$) it would suggest that the anisotropy observed should be enhanced if the cluster passes through resonance during the laser pulse. This has been confirmed in experiments where the mean energy of Ar ions parallel and perpendicular to the polarization were measured in Kumarappan *et al.* (2001) as a function of laser pulse duration. A schematic of the kind of data observed is shown in Figure 7.59 (Kumarappan *et al.* 2002). At the shortest pulses, the mean ion energies are almost identical along the two axes. However, when some stretching of the pulse occurs, which likely allows for the expanding cluster to pass through resonance during the laser pulse, the ion energy enhancement along the polarization is maximized.

7.5 Laser Absorption Dynamics in an Exploding Cluster

Before concluding this chapter, full consideration of the physics discussed here really demands that we take a step back and assess the dynamics of the high-intensity laser irradiation of a cluster in an integrated sense. So far in this chapter we have examined various aspects of the interaction: inner- and outer-ionization, electron heating, oscillation dynamics, and cluster expansion; in each case we have examined these effects in various simplified limits that allowed us to develop basic analytic toy models and derive some

7.5 Laser Absorption Dynamics in an Exploding Cluster

quantitative insight into what and why they occur. It should be no surprise that virtually no real experiment closely approximates during the entire interaction just one of these simple limits; in fact, the interaction with a single cluster can enter various regimes and sample different limits during and after the interaction with the irradiating pulse. We have already alluded to how one effect, such as cluster expansion, affects the dynamics of another, such as inner-ionization and ionization ignition. In this section let us take a look at how the various effects interplay to lead to experimental observables.

7.5.1 Electron Heating and Ionization Dynamics at High Intensity During Cluster Explosions

To bring some order to the chaos, in Section 7.3.1 we argued that the laser-cluster interaction loosely fell into five different regimes, each characterized more or less by one limit of the effects we have considered (e.g. hydro versus Coulomb expansion, small or large outer-ionization, and so on). We then suggested that the interaction fell into these regimes based on an interplay of the magnitude of three spatial scales: the instantaneous radius of the cluster, the oscillation amplitude of free electrons in vacuum at the instantaneous intensity, and the deflection amplitude of the confined electron cloud (if there is one) in the cluster. This method of deconstructing the interaction suggests that it might be informative to plot these various regimes in a phase space of cluster radius versus instantaneous laser intensity, and plot the trajectory a given cluster of some initial size and species might take in the phase space for some laser pulse of a given wavelength, pulse duration and peak intensity.

Figure 7.60 illustrates just such a phase space plot. In this case the plot describes the interaction of an 800 nm wavelength laser pulse with an Ar cluster that has an initial radius (R_0) of 2.5 nm, corresponding to a moderately sized cluster of about 1,400 atoms. The intensity that the laser pulse might exhibit is delineated along the bottom of the plot. The vertical axis denotes the cluster size. To ascertain the limits of the various regimes it is assumed that the cluster becomes ionized to 8+ on the rising edge of the laser pulse, and the electron plasma density is assumed to drop with the plasma volume as the cluster size increases in the vertical direction of the plotted phase space. Under these assumptions, the horizontal dashed line denotes the cluster radius at which the cluster nanoplasma comes into Mie resonance with 800 nm light. The solid lines denote the point in intensity/plasma density space at which either $x_{osc} = R$ or $d_0 = R$, and, as such, delineate roughly the boundaries of the various interaction regimes described in Section 7.3.1. The five regions are annotated on this plot with a short description reminding the reader what characteristics are expected of the interaction in that regime.

Any real pulse will begin at zero intensity off the left axis of the plot and trace a trajectory that rises up in intensity, carrying the trajectory to the right until the pulse reaches its peak intensity and then falls, and then plotting a trajectory that returns to the left axis. The cluster is presumed to start at its initial radius, but during the interaction the cluster will expand. Therefore, the history of the interaction can be thought of as a trajectory that starts

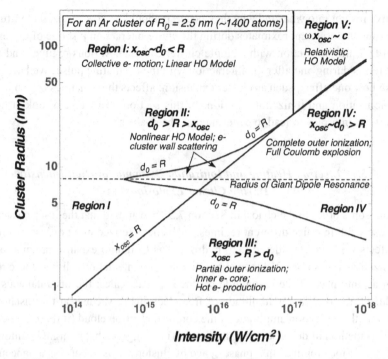

Figure 7.60 Phase space plot describing the regimes of interaction in which an 800 nm laser pulse might enter when irradiating a 5 nm diameter Ar cluster.

at the left axis on R = 2.5 nm which sweeps out to the right and increases upward until it returns to the left axis as the pulse falls in intensity at some increased radius.

Depending on the laser pulse parameters, the cluster trajectory will enter various regions on this plot, letting us visualize which physical effects play a role and when in the pulse they are important. Different size and species clusters will have the regime boundaries located at different places in the phase space, based on expected ionization level, initial radius, and so on.

Figure 7.61 reproduces the phase space plot for a 2.5 nm Ar cluster and plots four notional potential trajectories to show how laser parameters could drive the same cluster into quite different regimes of interaction. The two trajectories labeled "S1" and "S2" illustrate what we might expect to happen to the cluster in a very short (say $\sim$30 fs) pulse. The S1 trajectory denotes a low-intensity pulse, with peak intensity of only a few times 10^{14} W/cm^2. In this case the interaction is solely in region I, where we expect virtually no outer-ionization and a more or less linearly driven cluster nanoplasma cloud. The pulse is so fast that it falls in intensity before the cluster can expand to the Mie resonance. For such an interaction we expect some thermal heating of the cluster electrons and a purely hydrodynamic expansion.

On the other hand, "S2" describes a short pulse with high intensity, having a peak near 10^{17} W/cm^2. The rapid rise of the pulse while the cluster is still at high density pushes the interaction into region III where we expect some outer-ionization and stochastic heating.

7.5 Laser Absorption Dynamics in an Exploding Cluster

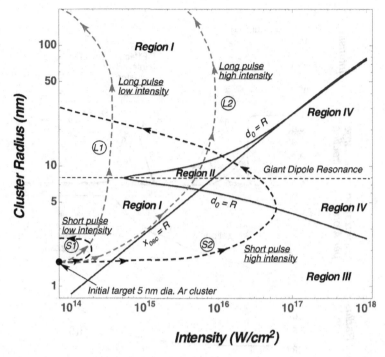

Figure 7.61 Cluster phase space plot like that of Figure 7.60 with four notional interaction trajectories plotted.

We therefore expect a population of hot electrons from this cluster and high-energy X-ray emission (see the next section). The Ar cluster now expands more rapidly, entering region IV and passing through the Mie resonance. This phase of the interaction likely removes most of what is left of the Ar cluster electrons, and the final phase of the explosion is a Coulomb explosion (though the initial acceleration occurred in Region III, so we expect a two-humped ion energy distribution like that of Figures 7.49 and 7.57).

The other two trajectories plotted in Figure 7.61, labeled "L1" and "L2," show what we expect from "long pulses" (say a few hundred fs). The low-intensity long pulse, like the short pulse, remains in region I for the duration of the interaction, though, unlike the short pulse cluster expansion, does lead to the cluster nanoplasma passing through the Mie resonance near the peak intensity of the pulse. So, in this case we expect to see ions accelerated by hydrodynamic expansion but with greater mean energy resulting from the greater electron heating that accompanies passing through the Mie resonance near the pulse intensity peak. The trajectory explains what is seen in experiments like that of Figure 7.59 or the absorption measurements described later; some lengthening of the laser pulse leads to enhanced heating, greater absorption and higher ion energies than seen in measurements with very short pulses. Finally "L2" denotes a longer pulse at higher intensity. In the case of the trajectory plotted, the cluster enters the regime of the nonlinear harmonic oscillator regime around the time period that the cluster approaches the Mie resonance. Here we

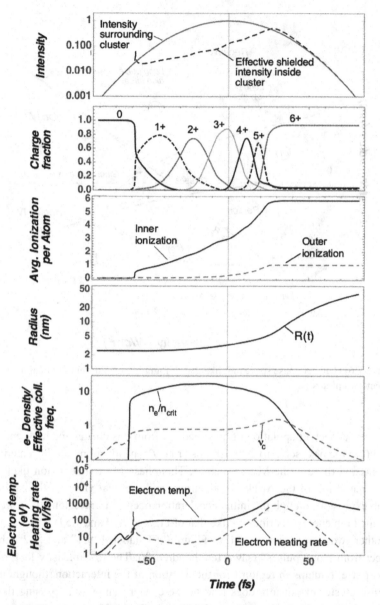

Figure 7.62 Illustration of 5 nm diameter Ar cluster parameters during irradiation by a 10^{16} W/cm^2, 50 fs 800 nm pulse.

expect the plasma oscillation damping to be dominated by cluster ion wall scattering and not by electron–ion collisions.

We can generate these trajectories using the 1D cluster nanoplasma model which we have developed through this chapter. Figure 7.62 shows the results of just such a calculation for a 2.5 nm initial radius Ar cluster subject to a 50 fs, 800 nm pulse at peak intensity of 10^{16} W/cm^2. The model calculates the radial expansion of the cluster in the uniform "expanding balloon" model, accounting for both hydro and Coulomb pressures. The electrons are inner-ionized by IFA tunneling and electron impact ionization. Outer-ionization by field extraction and thermal free streaming is calculated. The electron cloud is heated using the linear electron cloud harmonic oscillator model. The field interior to the cluster is shielded as predicted by the HO model. Electron cloud damping is modeled assuming electron–ion collisions and the simple wall-scattering model for the damping rate of Eq. (7.188).

The time history of various quantities is plotted in this figure. The top panel shows the pulse envelope and the effective shielded intensity in the cluster. We see that the electron cloud shielding lowers the intensity that the internal ions see by almost an order of magnitude during the rising part of the pulse. Only on the falling edge does the field see some enhancement, heavily damped. The next panel shows the history of the fractional populations of various charges from inner-ionization of the Ar constituents. In this case we see that nanoplasma field shielding has limited inner-ionization to about 6+. The third panel plots the average extent of inner-ionization and outer-ionization per atom. Though some fraction of the electrons are ejected by outer-ionization, the majority are not and the nanoplasma picture is appropriate.

The cluster radius as a function of time is shown in the fourth panel, the electron density and effective damping frequency are plotted in the fifth panel and the electron heating rate and temperature are plotted in the final panel. These show the interplay of cluster expansion and the passage through the Mie resonance. Some slow cluster expansion brings the cluster into Mie resonance on the pulse's falling side. This boosts the electron heating rate and increases the electron temperature to over 1 keV. This broad window of enhanced heating leads to an acceleration of the cluster radial expansion and its eventual disintegration by the end of the laser pulse.

This is to be contrasted with the same kind of calculation of the same 2.5 nm radius Ar cluster irradiated at ten times higher intensity, reproduced in Figure 7.63. In this case the Ar ions are ionized to 8+. Passage of the cluster through the Mie resonance occurs near the peak of the pulse, spiking the electron temperature to tens of keV. This drives rapid outer-ionization, which evacuates nearly 100 percent of the inner-ionized electron, leaving a bare ion core. We might expect in this case to see significant hot electron production (in the 10–100 keV range) with significant keV X-ray emission in a gas composed of these clusters. We would also expect to see Ar ions ejected with energies well in excess of 100 keV. The trends seen in these two calculations are completely consistent with experimental observations.

As a final note, it is interesting to take these calculations and map them onto the phase space of Figure 7.60. Using the parameters plotted in Figures 7.62 and 7.63, the phase space trajectories of these calculated interactions are plotted and labeled in Figure 7.64. We see that the lower-intensity pulse briefly enters region III, explaining the modest outer-ionization, and then the passage through the Mie resonance late in the pulse, boosting the

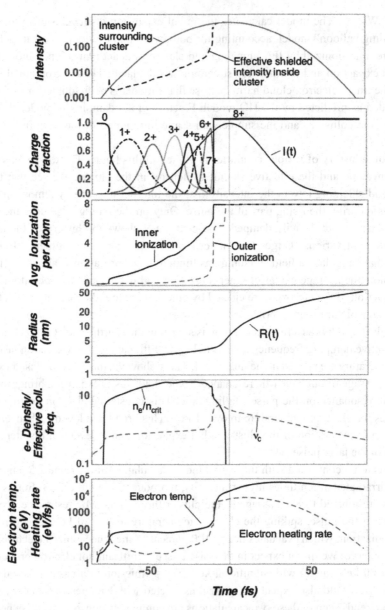

Figure 7.63 Illustration of ionization dynamics in a 5 nm diameter Ar cluster; 50 fs with various pulse peak intensities (at 800 nm) and pulse durations. The labeled curves show the history of cluster radius as a function of intensity as the laser pulse ramps up and then back down in intensity.

cluster expansion rate. In comparison, the higher-intensity interaction drives the cluster well into region IV, where complete outer-ionization occurs and significant hot electron generation is expected.

7.6 Bulk Cluster Plasma Target Effects

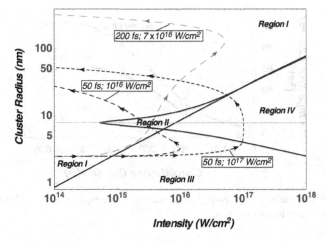

Figure 7.64 Illustration of sequential versus nonsequential explosion of a multiply ionized triatomic molecule in a strong field.

For comparison, the same kind of calculation was conducted for a 200 fs pulse at high intensity and plotted here. Clearly this cluster remains a nearly quasi-neutral nanoplasma throughout the interaction with a boost in expansion coming as the cluster passes through resonance. In such an interaction we expect high ion energies but minimal hot electron production.

7.5.2 Pulse Duration and Chirp Effects

From these discussions it is clear that there are significant pulse-duration effects in the ion energy. Optimum pulse duration brings the cluster nanoplasma into resonance with the laser field. Furthermore, it is interesting to note that the chirp, or direction of the frequency sweep, of the pulse also plays a role (Parra *et al.* 2000; Madison *et al.* 2004a; Fennel *et al.* 2010; Magunov *et al.* 2001). Figure 7.65 shows the kind of mean ion energy data that are seen when a cluster target containing, say large Xe clusters, is irradiated with fs laser pulses which are deliberately chirped and lengthened (Kumarappan *et al.* 2002). As already seen, there is a peak in the ion energy with longer pulses; however, the magnitude of this peak depends on the chirp direction of the pulse. When the pulse is negatively chirped, its frequency sweeps down in time. As a result, as the cluster passes through resonance the expanding cluster has a Mie resonance which decreases in time. The negatively chirped pulse has a frequency which tracks this downward sweep in Mie resonance and therefore stays in resonance longer. This leads to a greater ion energy enhancement than the opposite chirp case.

7.6 Bulk Cluster Plasma Target Effects

Before concluding this chapter, we should briefly examine some of the physical phenomena that are observed when an intense, ultrafast laser pulse irradiates a high-density

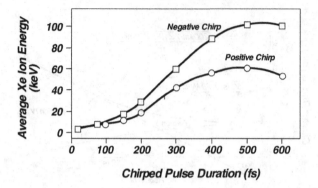

Figure 7.65 Illustration of sequential versus nonsequential explosion of a multiply ionized triatomic molecule in a strong field.

gas jet containing clusters. The presence of the clusters leads to plasma conditions that can be quite interesting after the clusters disassemble. Here we look at three effects seen in irradiation of such clustering gas jets.

7.6.1 Intense Laser Pulse Absorption in Clustering Gas Jets

Typically when an intense pulse is focused at ionizing intensity into a gas jet or gas cell with density of under 10^{20} cm^{-3}, the plasma that results is underdense. The level of laser absorption over a typical Rayleigh range, say $\sim$1–5 mm, is minimal. This situation is dramatically different when that target gas contains mid-sized to large ($>$1,000-atom) clusters (Ditmire et al. 1997b). Figure 7.66 shows an example of the kind of data that are observed when a subpicosecond pulse irradiates a gas jet of $\sim$5 mm in extent and gas jet density $\sim 10^{19}$ cm^{-3}. When the jet contains Ne, which does not form clusters under the experimental conditions considered, the jet is virtually transparent to the pulses incident on it at 10^{17} W/cm^2. On the other hand, as the backing pressure in the same jet backed with Ar or Xe is irradiated, the absorption climbs to nearly 100%. This is a consequence of the strong absorption mechanisms of a nanoplasma.

The expansion dynamics of the cluster scan be seen in such absorption measurements. Figure 7.67 shows a schematic of laser absorption in an Ar jet as a function of the pulse duration for three different cluster sizes. The absorption goes through a peak with pulse lengths that have been increased from the original 35 fs pulses. This is clearly a signature of the Mie resonance.

7.6.2 X-ray Emission From Cluster Plasmas

Another experimental observable which came to the fore in the first experiments on cluster gas jet targets was anomalously bright X-ray emission. The high absorption discussed

7.6 Bulk Cluster Plasma Target Effects

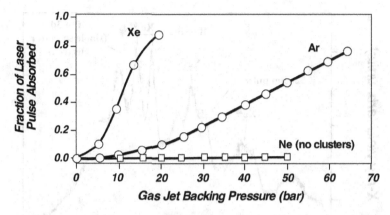

Figure 7.66 Schematic of absorption observed when an intense subpicosecond pulse irradiates jets as a function of backing pressure.

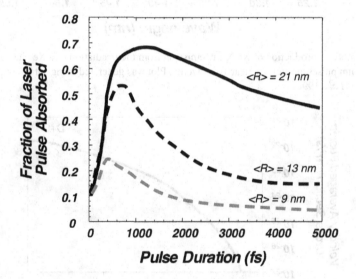

Figure 7.67 Schematic of observed absorption at high intensity in a clustering gas jet as a function of laser duration.

earlier explains this in part; however, it is not uncommon to observe keV to multi-keV photons emitted from Ar, Kr and Xe cluster targets even at quite modest intensity (say $\geq 10^{16}$ W/cm^2) (Junkel-Vives et al. 2001; Hansen et al. 2002). The photon energies and fluxes are not explainable by a simple thermal plasma given the level of laser absorption, even when it approaches 100%. In fact, this hard X-ray emission is probably due, at least in part, to the hot electron population that is formed from stochastic heating. We will not attempt to survey the breadth of X-ray emission characteristics from cluster plasmas other than to note that the emission is consistent with the models and physics we have discussed in this chapter.

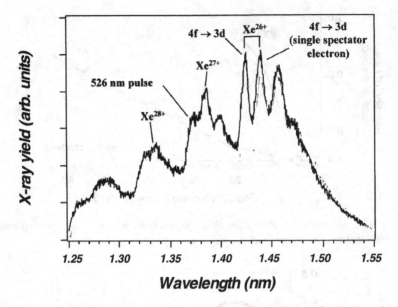

Figure 7.68 Reproduction of Xe X-ray emission from the irradiation of a Xe clustering jet by 526 nm pulses at intensity near 10^{17} w/cm². Plot was generated using data published in Ditmire et al. (1998a).

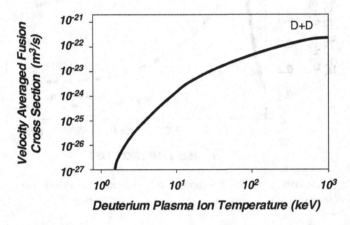

Figure 7.69 Plot of velocity averaged cross section for DD fusion versus plasma ion temperature.

As an example, Figure 7.68 shows X-ray emission when a Xe clustering jet is illuminated by 526 nm pulse at intensity $\sim 10^{17}$ W/cm². Very strong M-band emission is observed at photon energies around 1 keV (Ditmire et al. 1998a). Time-resolved studies of this kind of emission shows a sharp spike in the emission at $t = 0$ followed by longer-scale (~ 100 ps – ns) lower brightness emission (Ditmire et al. 1995). The initial spike results from emission by the dense cluster during its disassembly.

7.6.3 High Ion Temperature Plasma Production and Nuclear Fusion in Cluster Plasmas

Given that exploding clusters eject high-energy ions, it is natural then that a plasma created from the irradiation of a gas of clusters likely exhibits a high average ion energy. This fact has been harnessed to drive nuclear fusion in a gas of Coulomb exploding deuterium clusters (Ditmire et al. 1999; Zweiback et al. 2000b; Madison et al. 2004b). It turns out that the energies of these Coulomb exploding clusters is well matched to the energies needed in a thermonuclear plasma. Figure 7.69 reproduces the velocity-averaged cross section for DD fusion. The two pathways for this fusion are $D(D,n)^3He$ and $D(D,p)T$. This shows that substantial reactivity occurs once a plasma with an ion temperature above about

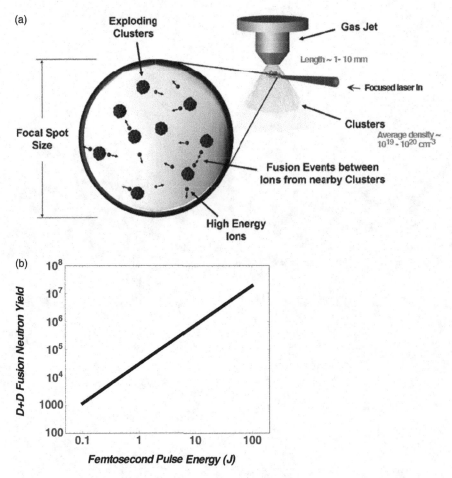

Figure 7.70 (a) A cartoon of how a cluster fusion experiment works. (b) Plot of the approximate fusion yield that is observed in these experiments as a function of input laser energy (typically performed with sub-100 fs pulses weakly focused into a deuterium clustering jet).

10 keV is achieved. We have seen that a few-nanometer D_2 cluster Coulomb explodes with a few keV of maximum energy. So, a gas of deuterium clusters irradiated at intensity of around 10^{17} W/cm² ($U_p \sim 10$ keV) produces a burst of fusion neutrons from the first of the two channels that deuterium can fuse. (Similar results are seen when deuterated methane clusters are irradiated (Bang *et al.* 2013).)

Figure 7.70 shows how such an experiment works. At the right of this illustration is a plot showing approximately how many fusion neutrons are observed in these experiments as a function of pulse energy (for ≤ 100 fs pulses) when weakly focused ($\sim$f/10–20) into a deuterated gas jet.

8

Strong Field Interactions with Underdense Plasmas

8.1 Introduction to Light Propagation in Underdense Plasma

At this point we change the thrust of our investigations into strong field interactions. Until now we have considered intense laser interactions with microscopic systems much smaller than the laser wavelength, like single electrons, atoms, molecules or small clusters. Now we shift our considerations to interactions with plasmas, where the macroscopic collective response of the ionized medium leads to new, fascinating effects. We will consider how the target interaction affects the propagation of the focused laser beam and temporal pulse, a feedback which leads to a wealth of interesting dynamics. We will first consider plasmas of density low enough that the pulse can propagate through the plasma, so-called underdense plasmas.

8.1.1 Propagation of an Electromagnetic Wave in a Collisional Plasma

Before proceeding, it is worth spending a little time recalling the linear optical properties of an underdense plasma. Appendix B gives a brief summary of a few of the most important concepts in plasma physics relevant to our discussion. We consider in this chapter plasmas in a realm where an electromagnetic wave has real solutions of the dispersion relation, meaning that the laser frequency is greater than the electron plasma frequency:

$$\omega_{pe} = \sqrt{\frac{4\pi n_{e0} e^2}{m}} < \omega \qquad (8.1)$$

This condition, put differently, means that the background electron density is lower than the critical density, the density at which the electron plasma frequency equals the laser frequency,

$$n_{e0} < n_{crit} \equiv \frac{m\omega^2}{4\pi e^2}. \qquad (8.2)$$

An easy mnemonic is to note that the critical electron density is about 10^{21} cm^{-3} for laser light with wavelength near 1 μm and varies as $1/\lambda^2$. We will adopt a phenomenological approach to describe the optics of such a plasma, and think of the plasma as exhibiting some effective, frequency-dependent dielectric constant, ε (which can be complex), such that we can throughout rewrite Maxwell's equations in the plasma as (Jackson 1975, p. 218)

$$\nabla \cdot (\varepsilon \mathbf{E}) = 0, \tag{8.3}$$

$$\nabla \cdot \mathbf{B} = \frac{1}{c} \frac{\partial}{\partial t} (\varepsilon \mathbf{E}). \tag{8.4}$$

To find this dielectric constant, and the effective refractive index of the plasma, we consider explicitly the charged particle motion driven by the laser field. The first assumption is that on the time scale of laser light oscillations, the ions can be considered immobile and serve as a neutralizing background to the motion of the electrons. In this limit the optical response is given by the motion of the electrons, which we can describe as a fluid with fluid velocity, $\mathbf{u}_e$. The equation of motion of the electron fluid, assuming (for the moment) that the intensity is low enough that the magnetic field can be ignored, can be written

$$m \frac{\partial \mathbf{u}_e}{\partial t} + \nu_{ei} m \mathbf{u}_e = -e \mathbf{E}(t). \tag{8.5}$$

In this equation we have added a term, proportional to the electron–ion collision frequency, ν_{ei}. This term essentially accounts for a slowing down of the electron fluid velocity because of collisions between the driven electrons and the immobile heavy ions. This collisional term damps the laser-driven electron oscillations. The addition of such a term in a fluid equation for electrons at this point is heuristic, but it is justified in the context of plasma kinetic theory. Strictly speaking, the appropriate collision frequency to use in Eq. (8.5) is the electron–ion longitudinal momentum-stopping collision frequency, discussed in Appendix B.

We describe the plasma by the charge density, ρ_c, and electron current density, $\mathbf{J}_e$, which can be written in terms of the average ionization state of the plasma atoms, $\bar{Z}$, as

$$\rho_c = -e n_e + \bar{Z} e n_i, \tag{8.6}$$

$$\mathbf{J}_e = -e n_e \mathbf{u}_e. \tag{8.7}$$

Accounting explicitly for charge motion in the plasma, we use Gauss's law and Ampere's law,

$$\nabla \cdot \mathbf{E} = 4\pi \rho_c, \tag{8.8}$$

$$\nabla \cdot \mathbf{B} = \frac{1}{c} \frac{\partial \mathbf{E}}{\partial t} + \frac{4\pi}{c} \mathbf{J}_e, \tag{8.9}$$

and assume harmonic oscillations of the plane-wave (quasi-monochromatic) light of the form $\mathbf{E} = \mathbf{E}_0 e^{-i\omega t}$, $\mathbf{B} = \mathbf{B}_0 e^{-i\omega t}$. This Fourier decomposition implies that the electron fluid also oscillates harmonically, $\mathbf{u}_e = \mathbf{u}_{e0} e^{-i\omega t}$, which, when substituted into Eq. (8.5), yields

$$-i\omega m \mathbf{u}_{e0} + \nu_{ei} m \mathbf{u}_{e0} = -e \mathbf{E}_0, \tag{8.10}$$

$$\mathbf{u}_{e0} = \frac{-ie\mathbf{E}_0}{m(\omega + i\nu_{ei})}. \tag{8.11}$$

8.1 Introduction to Light Propagation in Underdense Plasma

This allows us to write an AC conductivity for the plasma, σ_e, by combining Eqs. (8.7) and (8.11),

$$\mathbf{J}_e = \sigma_e \mathbf{E} = -en_e \mathbf{u}_e = \frac{ie^2 n_e}{m(\omega + i\nu_{ei})} \mathbf{E}, \tag{8.12}$$

$$\sigma_e = \frac{ie^2 n_e}{m(\omega + i\nu_{ei})} = \frac{i\omega_{pe}^2}{4\pi(\omega + i\nu_{ei})}. \tag{8.13}$$

This can be used in Eq. (8.9) and compared to Eq. (8.4),

$$\nabla \cdot \mathbf{B} = \frac{-i\omega}{c} \mathbf{E} + \frac{i\omega_{pe}^2}{c(\omega + i\nu_{ei})} \mathbf{E} = -i\frac{\omega}{c}(\varepsilon \mathbf{E}), \tag{8.14}$$

to give us the complex optical dielectric function of a collisional plasma, which is

$$\varepsilon = 1 - \frac{\omega_{pe}^2}{\omega(\omega + i\nu_{ei})}, \tag{8.15}$$

and can be written in terms of the AC conductivity, σ_e as

$$\varepsilon = 1 + i\frac{4\pi\sigma_e}{\omega}. \tag{8.16}$$

This dielectric function is complex because we have retained electron–ion collisions in Eq. (8.5). This will give rise to laser light absorption and plasma heating. If this collision frequency is low, as it is in many underdense plasmas, we can ignore it for the moment and retrieve the usual, optical dielectric function,

$$\varepsilon = 1 - \frac{\omega_{pe}^2}{\omega^2} = 1 - \frac{n_{e0}}{n_{crit}}, \tag{8.17}$$

written here in terms of the background, quiescent electron density n_{e0}. We often work with the refractive index of an optical medium (which we denote in the chapter by η_R instead of the more commonly used "n" to avoid confusion with our frequent employment of n in the coming pages to denote various gas or plasma densities). This refractive index, from basic optics, is related to ε by

$$\eta_R = \sqrt{\varepsilon}. \tag{8.18}$$

Since many of the plasmas considered in this chapter involve electron densities that are well below the critical density, Taylor expansion of the refractive index around n_{eo}/n_{crit} yields an expression we will use often:

$$\eta_R \cong 1 - \frac{n_{e0}}{2n_{crit}}. \tag{8.19}$$

This expression, ignoring collisions, returns the dispersion function found in Appendix B, since

$$k^2 = \frac{\omega^2}{c^2} \varepsilon, \tag{8.20}$$

giving

$$\omega^2 = c^2 k^2 + \omega_{pe}^2. \tag{8.21}$$

We finally remind ourselves that since the dielectric constant is <1, the phase velocity of the laser light in the plasma is faster than c,

$$v_\varphi = \frac{\omega}{k} = \frac{c}{\sqrt{\varepsilon}}$$

$$= \frac{c}{\sqrt{1-\omega_{pe}^2/\omega^2}} > c, \qquad (8.22)$$

while the group velocity in the plasma is necessarily less than the speed of light,

$$v_g = \frac{\partial \omega}{\partial k}$$

$$= c\sqrt{1-\omega^2{}_{pe}/\omega^2}$$

$$= c\sqrt{\varepsilon} \leq c. \qquad (8.23)$$

8.2 Plasma Formation and Heating Mechanisms

8.2.1 Underdense Plasma Formation by Multiphoton and Tunnel Ionization

First, we should answer: how are underdense plasmas typically formed in most strong field experiments? This is closely tied to the next question, which is, to what temperature will the electron plasma be heated by its formation and during the subsequent interaction of the intense laser pulse with that plasma? This second question is actually complex since, at the densities we consider, the electron distribution function may become strongly deformed from a Maxwellian, and asking simply what temperature arises is an ill-defined question. We shall for the moment consider interactions that are sufficiently well behaved, often meaning interactions in plasmas with pulse intensities that are well below relativistic intensity (say $\ll 10^{18}$ W/cm^2). Even at higher intensity, since the plasma is created and heated on the lower-intensity rising edge of a laser pulse, this question of plasma heating and electron temperature is still important.

From an experimental standpoint, with a few exceptions in which a plasma is preformed as a target, for example, by some kind of high-voltage electrical discharge in a gas, most strong field experiments with underdense plasma are initiated by focusing the intense laser pulse into a low-density gas such as a gas-filled cell with a small entrance pinhole or into a gas jet puffed into vacuum. It can also mean, in the case of long-range plasma filamentation considered in Section 8.3, focusing the pulse into air. This means that the initial, unionized gas density, n_0, is at most a few atmospheres, $\sim 10^{20}$ cm^{-3}, and is typically at least 10^{16} cm^{-3}. Plasma formation in this case is almost always initiated by multiphoton ionization in the rising edge of the laser pulse. This means in practice that intensities of at least 10^{13} W/cm^2 are needed to reach appreciable ionization rates in gases used in most experiments (like the noble gases or atmospheric gases like N_2 or O_2). In most ultrafast laser pulses the only significant ionization mechanism is laser field ionization, so the average charge state, $\bar{Z}$, and commensurate electron density, $n_{e0} = \bar{Z} n_0$, can usually be calculated using tunnel ionization rates and single-atom ionization rate equations along the lines discussed in Chapter 4.

Longer, lower-intensity pulses may have the plasma formation and subsequent final electron density increased by electron collisional ionization. This process was discussed in a cluster in Chapter 7 and the appropriate equations from that chapter, Eqs. (7.102) and (7.106), can be used to calculate this collisional ionization rate. As in the cluster nanoplasma, there can be two contributions, ionization driven by the kinetic energy of the oscillating electrons in the laser field and ionization by collisions from the thermal energy of the electrons. Though often not completely negligible, in the majority of high-intensity laser experiments, the duration of the pulse is too short to see much collisional ionization, and field ionization is by far the dominant mechanism for the plasma formation.

8.2.2 ATI Heating

For the most part, the plasma formation and subsequent heating by the intense laser pulse deposits negligible amounts of energy in the plasma ions, because of their large mass and the short time scales involved. We can almost always assume that any plasma heating is directed into the electrons. Furthermore, in the underdense plasmas considered in this chapter, the electron–ion collisional equilibration time is typically quite long on the time scale of many experiments. For example, Eq. (B.85) tells us that a helium plasma, ionized to $Z = 2$, at atmospheric density with electrons heated to 100 eV will equilibrate their temperatures over about 40 μs. This means that in almost all experiments, the electron temperature is substantially higher than the ion temperature, and we can estimate the electron temperature by ignoring any substantial transfer of energy to the ions.

To ascertain what electron temperature is reached, we have to consider two avenues in which the laser can heat. First, there is heating during the ionization process itself, and, second, the laser can further heat the ionized electrons in the plasmas by collisional absorption. The first process is a consequence of mechanisms we have already considered in the context of single-atom ionization, that is, the transfer of energy to electrons during ionization, namely above-threshold ionization, or ATI.

In Chapter 4 we considered the magnitude of electron energy gain during ATI in some detail. In the tunneling regime, we derived a simple analytic formula for the energy spectrum of electrons undergoing tunnel ionization at a given intensity, described by Eq. (4.289). To determine the overall heating imparted during the plasma formation in a gas, the energy put into such an ATI spectrum has to be calculated over the time history and ionized charge state history within the laser pulse. This is best done by performing a rate equation calculation utilizing an appropriate ionization rate formula (such as the PPT/ADK formula) and summing the electron production rate with the ATI spectrum. This means that the actual energy given to the electron plasma is dependent on the intensity and laser wavelength (through the λ^2 scaling of the ponderomotive energy).

Inspection of a typical tunneling ionization ATI spectrum like that reproduced in Figure 4.41 immediately illustrates that the resulting electron energy distribution is significantly different from a Maxwellian. However, unlike electron–ion equilibration, which

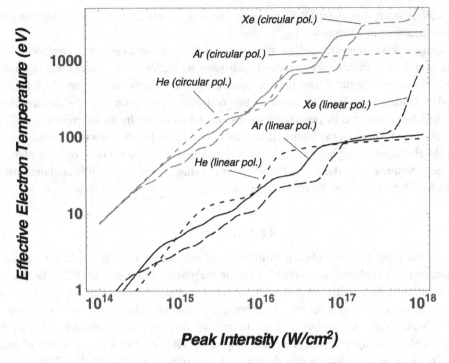

Figure 8.1 Effective electron temperature of a plasma in three different gases produced by tunneling ionization and ATI heating as a function of peak intensity. Calculation was performed for a 30 fs pulse with 800 nm wavelength. Temperatures for linear polarization (black lines) and circular polarization (gray lines) are plotted.

is slow on experimental time scales because of the large difference between electron and ion masses, electron–electron thermal equilibration typically occurs fairly quickly. It is a fair approximation to find an effective ATI electron plasma temperature by assuming fast equilibration to a Maxwellian and defining the ATI heated temperature, $k_B T_e$, simply in terms of the average energy imparted to the plasma electrons,

$$k_B T_e = \frac{2}{3} \langle \varepsilon_e \rangle . \qquad (8.24)$$

Using this definition, it is straightforward to calculate the ATI temperature of various gases at the end of an ionizing pulse. Figure 8.1 shows the calculated temperature of various gases ionized by a 30 fs, 800 nm laser pulse as a function of the peak intensity. Steps in the temperature as peak intensity increases result from higher-intensity pulses accessing higher charge states, leading to ionization of a group of electrons at higher intensity and resulting in a contribution of electrons with higher ATI energy. The temperatures for linearly polarized light and circularly polarized light are shown. Circularly polarized light results in significantly hotter plasmas because of the ATI physics discussed in Chapter 4; circular polarization yields electrons with about U_p of energy while linear polarization imparts about 0.1 U_p of energy.

It is interesting to note that the final plasma temperature does not vary much with gas for a given peak intensity. A good rule of thumb is that the final plasma temperature from ATI heating is about 3 percent the peak ponderomotive energy of a linearly polarized pulse and about $0.5 \times$ peak-U_p in circularly polarized pulses.

8.2.3 Strong Field Inverse Bremsstrahlung Heating

Once a plasma has formed during the laser pulse from field ionization (or collisional ionization) and the electrons have acquired some thermal energy in the ionization process, the plasma electrons can be further heated by the laser field through collisions with plasma ions, a process often called inverse bremsstrahlung (IB) heating. The origin of this heating is illustrated in two limits in Figure 8.2. The way to picture this absorption of laser energy by a free electron is implied by the term inverse bremsstrahlung, depicted on the left in Figure 8.2; this is the opposite process of the well-known bremsstrahlung process in which an electron emits a photon as a result of the acceleration it feels during a Coulomb scattering encounter with an ion. In a quantum mechanical sense, free electrons in the plasma can only absorb photons from the laser field when a dipole is present and some ion is close. In the high intensity regime, where we can treat the laser field classically, a picture like that on the right of Figure 8.2 is more informative. We have seen in Chapter 3 that a free electron oscillating in the laser field will return its quiver energy back to the laser pulse if the intensity ramps down adiabatically. If, however, the adiabaticity of the electron's oscillation motion is interrupted by scattering off an ion, the net trajectory of the electron after the Coulomb collision will involve further laser-driven oscillation in the field combined with some drift velocity resulting from the redirection of the electron by scattering from the ion. The electron will return the coherent oscillatory motion to the laser field but will retain the drift velocity; averaged over many electrons and many collisions the result is a net increase in the thermal energy by the electrons in the plasma.

This process can be trivially illustrated through a simple calculation like that in Figure 8.3. The velocity of an electron is calculated here subject to an oscillating electric field that ramps to zero. The dashed line shows the same calculation when the electron has its velocity suddenly set to zero during one of the oscillations, simulating a 90° scattering event that rapidly shifts the velocity of the electron along the field's polarization direction to velocity in an orthogonal coordinate. Subsequent integration of the equation of motion for this electron results in it retaining some nonzero velocity in the polarization direction.

This plasma heating mechanism by laser radiation has been very well studied in the plasma physics literature back to shortly after the first employment of lasers in fusion research and has been considered by a number of authors over the years (Silin 1965; Pert 1972; Seely and Harris 1973; Langdon 1980; Jones and Lee 1982). Most of these early studies were performed for interactions in the "weak field" regime where the laser ponderomotive energy is smaller than the thermal energy of the electron plasma. We will instead require a strong field theory to quantify the heating rate. It is, in fact, possible to develop a complete analytic theory spanning both regimes through a full collisional kinetic theory treatment using the kinetic equation with an appropriate Fokker–Planck collision

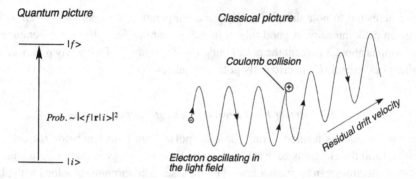

Figure 8.2 Schematic illustration of two pictures for inverse bremsstrahlung laser absorption in a plasma. On the left shows laser heating in a quantum picture in which an electron absorbs one photon in the presence of an ion. On the right is a classical picture for collisional heating in which a free electron oscillating in the field scatters off an ion at a well-defined time and the scattering event breaks the adiabaticity of the electron's sinusoidal motion imparting a residual drift velocity.

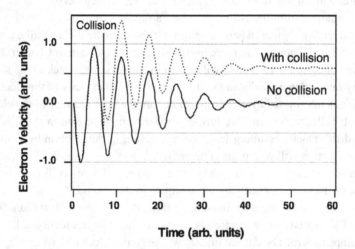

Figure 8.3 Numerical solution for the motion of an electron in an oscillating electric field which has a slowly decreasing amplitude. The solid line shows the electron velocity decreasing to zero as the field amplitude ramps down in time, while the dashed line illustrates the velocity in which the electron velocity is set to zero during an oscillation cycle to simulate a Coulomb scattering event. This interruption of the electron oscillation leads to a residual velocity and kinetic energy after the field has ramped down to zero.

term incorporating an oscillating external electric field (and assuming cold, immobile ions) (Silin 1965). This derivation was performed most rigorously in a classic paper by Jones and Lee (Jones and Lee 1982).

We will forgo this rigorous kinetic equation treatment and will instead develop a formula for IB heating in the strong field regime using a more intuitive approach, first developed by

Pert (1972). This approach requires us to assume a Maxwellian electron distribution without specific justification. Our approach returns the same heating rate as the Jones and Lee derivation. It turns out that the assumption of a Maxwellian distribution by electrons during this heating is an excellent approximation anyway. As we have already noted, electron–electron equilibration toward a Maxwellian is often very fast (few-femtosecond time scale) in the plasmas encountered in most strong field laser-plasma experiments. Furthermore, as Jones and Lee show, the self-similar solution to the electron distribution function during electron–ion collisional heating in the strong field regime is also a Maxwellian.

Before we dive into a detailed calculation of the collisional heating rate, let us first look at the implications to the laser absorption rate in a plasma within the context of the phenomenological introduction of the electron–ion collision frequency that we made to the plasma optical dielectric function, Eq. (8.15). Recalling this form of ε, and Taylor-expanding, assuming $n_e \ll n_{crit}$, the complex refractive index for the laser light in the plasma is

$$\eta_r = \sqrt{\varepsilon}$$
$$\cong 1 - \frac{n_{e0}}{2n_{crit}} + \frac{i}{2}\frac{\omega_{pe}^2}{\omega^2}\frac{\nu_{ei}}{\omega}. \tag{8.25}$$

The attenuation of the laser intensity from the imaginary part of η_r is

$$I(z) = I_0 \exp\left[-\kappa_{plas} z\right], \tag{8.26}$$

with absorption coefficient of

$$\kappa_{plas} = \frac{2\pi}{\lambda}\frac{\omega_{pe}^2}{\omega^2}\frac{\nu_{ei}}{\omega} = \frac{1}{c}\frac{\omega_{pe}^2}{\omega^2}\nu_{ei}. \tag{8.27}$$

The change of average electron thermal energy ε_e and plasma heating rate per unit volume w_{IB} is derived by noting that

$$\frac{d}{dt}(n_e \langle \varepsilon_e \rangle) \equiv w_{IB},$$

which, from Eq. (8.27), is seen to be

$$w_{IB} = -\frac{dI}{dz} = \alpha I$$
$$= \frac{\omega_{pe}^2}{\omega^2} \nu_{ei} \frac{I}{c}. \tag{8.28}$$

This equation leads to a nice intuitive interpretation of the collision frequency in this case if we rewrite it in terms of the laser ponderomotive energy:

$$w_{IB} = \frac{4\pi n_e e^2}{m}\frac{\nu_{ei}}{\omega^2}\frac{E_0^2}{8\pi}$$
$$= 2n_e \nu_{ei} U_p. \tag{8.29}$$

The average rate that the laser deposits thermal energy in the plasma electrons is twice the ponderomotive energy times the frequency at which those electrons undergo momentum-changing collisions.

This phenomenological formulation gives a pretty good estimate for the heating rate. However, a visit to Appendix B will illustrate that the appropriate electron–ion collision frequency varies as the electron velocity $\sim 1/v^3$. So a strong field application of this simple formula will beg the question, what is the appropriate velocity to insert here? A reasonable estimate in the weak field limit would be a thermal velocity, while in the strong field limit we would expect that the dominant ponderomotive quiver velocity is most appropriate. A more precise theory is desirable; I opted here to develop a heating rate in a treatment that closely follows that of Pert (1972).

The basis for our derivation of the IB heating rate is to utilize what is known as the "sudden approximation" for describing Coulomb collisions between electrons and ions in the laser field. Consider an electron in the plasma with a velocity that we can decompose into a thermal velocity and a laser-driven velocity $\mathbf{v} = \mathbf{v}_T + \mathbf{v}_L$. The laser-driven component (in a nonrelativistic field) is simply

$$\mathbf{v}_L = \frac{e\mathbf{E}_0}{m\omega} \sin \omega t \qquad (8.30)$$

We make two well-founded assumptions about Coulomb collisions in the plasma: (1) Collisions occur on a time scale much faster than the laser oscillation period, the essence of the sudden approximation, and (2) every collision is completely elastic, occurring with infinitely heavy, immobile ions. Denoting quantities after a collision with primes and those before that collision as unprimed, the first assumption, the sudden approximation, lets us say that the laser component of the velocity is unchanged in the collision:

$$\mathbf{v}'_L(t') = \mathbf{v}_L(t). \qquad (8.31)$$

The second assumption allows us to state conservation of energy over the collision as

$$\tfrac{1}{2} m \mathbf{v}'(t')^2 = \tfrac{1}{2} m \mathbf{v}(t). \qquad (8.32)$$

The acquisition of thermal energy during a collision, then, is

$$\Delta \varepsilon_T = \tfrac{1}{2} m \left(\mathbf{v}_T'^2 - \mathbf{v}_T^2 \right). \qquad (8.33)$$

Expanding the total kinetic energy of the electron, ε_K, gives

$$\varepsilon_K = \tfrac{1}{2} m |\mathbf{v}|^2 = \tfrac{1}{2} m \left(\mathbf{v}_T^2 + \mathbf{v}_L^2 + 2 \mathbf{v}_T \cdot \mathbf{v}_L \right), \qquad (8.34)$$

and, if we assume that the collision frequency is much smaller than the laser frequency, the time average of the electron kinetic energy causes the third term in (8.34) to average to zero, so ε_K can be approximated by

$$\varepsilon_K \cong \tfrac{1}{2} m \left(\mathbf{v}_T^2 + \mathbf{v}_L^2 \right). \qquad (8.35)$$

From Eq. (8.33) the gain in thermal energy by the electron in this collision is

$$\Delta \varepsilon_T = \tfrac{1}{2} m \left[\left(\mathbf{v}'^2 - \mathbf{v}_L'^2 - 2 \mathbf{v}' \cdot \mathbf{v}'_L \right) - \left(\mathbf{v}^2 - \mathbf{v}_L^2 - 2 \mathbf{v} \cdot \mathbf{v}_L \right) \right].$$

8.2 Plasma Formation and Heating Mechanisms

Assumptions (1) and (2) tell us that $\mathbf{v}'_L = \mathbf{v}_L$ and $v'^2 = v^2$, so

$$\begin{aligned}\Delta \varepsilon_T &= m\left(\mathbf{v} - \mathbf{v}'\right) \cdot \mathbf{v}_L \\ &= m\left(\mathbf{v}_T + \mathbf{v}_L - \mathbf{v}'_T - \mathbf{v}'_L\right) \cdot \mathbf{v}_L \\ &= m\left(\mathbf{v}_T - \mathbf{v}'_T\right) \cdot \mathbf{v}_L.\end{aligned} \qquad (8.36)$$

We will now define a differential collision frequency, v'_{ei}, which is the rate of Coulomb scattering into solid angle $d\Omega$ for electrons with thermal energy between $\mathbf{v}_T$ and $\mathbf{v}_T + d\mathbf{v}_T$. We need to find the rate of gain of thermal energy by electrons in the plasma over a number of Coulomb scattering events. This will require an average of the thermal energy gain, (8.36), over all scattering angles, Ω, and over all angles between the electron velocity, $\mathbf{v}$, and its laser-driven component of velocity, $\mathbf{v}_L$, angles we denote with Ω'. Furthermore, to find the overall plasma heating rate we must average these angle-averaged heating events over the distribution of electron thermal velocities, denoted in the usual way by an average of the distribution function $f(\mathbf{v}_T) d\mathbf{v}_T$. This leads us to write for the heating rate per electron

$$\frac{d\varepsilon_T}{dt} = \iiint v'_{ei} \Delta \varepsilon_T f(\mathbf{v}_T) d\Omega \frac{d\Omega'}{4\pi} d\mathbf{v}_T. \qquad (8.37)$$

The differential collision frequency can be expressed in terms of the differential Coulomb scattering cross section, the electron velocity and background ion density as

$$v'_{ei} = \frac{d\sigma}{d\Omega} v\, n_i. \qquad (8.38)$$

Combining Eqs. (8.36), (8.37) and (8.38) yields the (seven-dimensional) integral for the heating rate per electron

$$\frac{d\varepsilon_T}{dt} = \frac{n_i m}{4\pi} \iiint f(\mathbf{v}_T) \frac{d\sigma}{d\Omega} v\left(\mathbf{v} - \mathbf{v}'\right) \cdot \mathbf{v}_L\, d\Omega\, d\Omega'\, d\mathbf{v}_T. \qquad (8.39)$$

To perform this integral, we need to introduce appropriate coordinates and angles, which are summarized in Figure 8.4. Referring to this figure, we can say that the dot product term in Eq. (8.39) is

$$(\mathbf{v} - \mathbf{v}') \cdot \mathbf{v}_L = v_L v \cos \theta' - v_L v \left(\sin \theta' \sin \theta \cos \varphi + \cos \theta' \cos \theta\right), \qquad (8.40)$$

or, since $v_L v \cos \theta' = \mathbf{v}_L \cdot \mathbf{v}$, is

$$(\mathbf{v} - \mathbf{v}') \cdot \mathbf{v}_L = \mathbf{v} \cdot \mathbf{v}_L (1 - \cos \theta), \qquad (8.41)$$

and so the heating rate is

$$\frac{d\varepsilon_T}{dt} = \frac{n_i m}{4\pi} \iiint f(\mathbf{v}_T) \frac{d\sigma}{d\Omega} v_L \cdot \mathbf{v} (1 - \cos \theta)\, d\Omega\, d\Omega'\, d\mathbf{v}_T. \qquad (8.42)$$

We introduce the usual Coulomb scattering cross section (Goldstein 1980, p. 109)

$$\frac{d\sigma}{d\Omega} = \frac{Z^2 e^4}{4 m^2 v^4 \sin^4(\theta/2)}, \qquad (8.43)$$

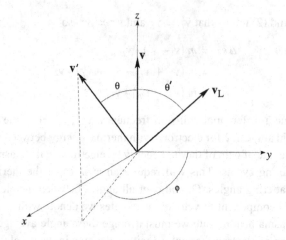

Figure 8.4 Geometry of the electron velocities in the Coulomb scattering event.

for collisions with ions of charge Z. In real plasmas the ions will consist of a range of charges and we would, in principle, have to average the heating rate over all these charges weighted by the density of each particular charge. This is an unnecessary complication, so a good approximation in this formula is to replace the charge Z with the average charge of ions in the plasma, $\bar{Z}$. This cross section with this substitution gives for the heating rate

$$\frac{d\varepsilon_T}{dt} = \frac{n_i \bar{Z}^2 e^4}{16\pi m} \int f(\mathbf{v}_T) \int \frac{\mathbf{v}_L \cdot \mathbf{v}}{v^3} \int_0^\pi \frac{(1-\cos\theta)}{\sin^4(\theta/2)} 2\pi \sin\theta \, d\theta \, d\Omega' \, d\mathbf{v}_T. \quad (8.44)$$

At this point we encounter the usual problem in considerations of collisional processes in plasma physics: the divergence of the integral over all scattering angles, θ, arising from the long-range nature of the Coulomb scattering cross section. This integral diverges because scattering angles of 0° correspond to infinite impact parameter. Referring the reader to Appendix B to get a more detailed discussion of how to deal with this divergence, we merely note that we limit the integration over θ to a minimum scattering angle to yield

$$\int_{\theta_{min}}^\pi \frac{(1-\cos\theta)}{\sin^4(\theta/2)} \sin\theta \, d\theta \cong 8 \ln\left[\frac{2}{\theta_{min}}\right]. \quad (8.45)$$

This result is the well-known Coulomb logarithm:

$$\ln \Lambda'_C \equiv \ln\left[\frac{2}{\theta_{min}}\right]. \quad (8.46)$$

Referring to the discussion of Appendix B, and noting that in underdense plasmas, the Coulomb logarithm is best evaluated with maximum impact parameter given by the Debye length in the plasma, yielding for Eq. (8.46)

$$\theta_{min} \approx v_\perp/v_0 = \frac{2\bar{Z}e^2}{mv_0^2 \lambda_D}, \quad (8.47)$$

where λ_D is the usual plasma Debye length derived in Appendix B.1 and v_0 is some appropriately chosen average electron velocity, such as the ponderomotive quiver velocity, v_{osc}, in the strong field regime or the electron thermal velocity in weak fields. (Using the latter yields the Coulomb logarithm derived in Appendix B). With the usual formula for the Debye length in the plasma of immobile ions this term is

$$\Lambda'_C \cong \frac{mv_0^2}{\bar{Z}e^2}\sqrt{\frac{k_B T_e}{4\pi e^2 \bar{Z} n_0}}. \tag{8.48}$$

(We have retained the primed notation for this Coulomb logarithm as a reminder that if its velocity is evaluated in the strong field limit, it is different from the usual thermal Coulomb logarithm found in most plasma physics texts and that found in Appendix B.)

We now have for the heating rate an average over all angles between the electron velocity and its laser-driven velocity and average over the thermal distribution

$$\frac{d\varepsilon_T}{dt} = \frac{n_i \bar{Z}^2 e^4}{m} \int f(\mathbf{v}_T) \int \frac{\mathbf{v}_L \cdot \mathbf{v}}{v^3} \ln \Lambda'_C d\Omega' \, d\mathbf{v}_T. \tag{8.49}$$

As is usually done in calculations of this sort, we pull the slowly varying logarithmic term out of the integration over $\mathbf{v}_T$. Considering first the integration over angles, the term in the denominator can be written, with reference to Figure 8.4, as

$$v^3 = \left(v_T^2 + v_L^2 + 2v_T v_L \cos\theta'\right)^{3/2}, \tag{8.50}$$

letting us evaluate the angle integral to derive a compact result:

$$\int \frac{\mathbf{v}_L \cdot \mathbf{v}}{v^3} \ln \Lambda'_C d\Omega' = 2\pi \int_0^\pi \frac{1}{v^3} \left(\mathbf{v}_T \cdot \mathbf{v}_L + v_L^2\right) \sin\theta' \, d\theta'$$

$$= 2\pi \int_{-1}^1 \frac{(v_T v_L \cos\theta' - v_L^2)}{(v_T^2 + v_L^2 + 2v_T v_L \cos\theta')^{3/2}} d(\cos\theta')$$

$$= \begin{cases} \dfrac{4\pi}{v_L} & v_L > v_T \\ 0 & v_L \le v_T \end{cases}. \tag{8.51}$$

The heating rate becomes

$$\frac{d\varepsilon_T}{dt} = \begin{cases} \dfrac{4\pi n_i \bar{Z}^2 e^4}{m} \ln \Lambda'_C \int \dfrac{1}{v_L} f(\mathbf{v}_T) d\mathbf{v}_T & v_L > v_T \\ 0 & v_L \le v_T \end{cases}. \tag{8.52}$$

Before proceeding, we note that what we really seek is not necessarily an instantaneous heating rate of the electrons but instead a heating rate averaged over a laser cycle. Since the direction of the quivering electron's velocity is irrelevant, it is adequate to average over a quarter of the laser cycle. Accounting for the fact that the heating rate is zero for values of the instantaneous laser-driven velocity that are less than the thermal velocity, this cycle average takes the form

$$\left\langle \frac{d\varepsilon_T}{dt} \right\rangle = \frac{4\pi \bar{Z}^2 e^4 n_i}{m} \ln \Lambda'_C \int \left[\frac{2}{\pi} \int_{v_{osc} \sin\varphi > v_T} \frac{1}{v_{osc} \sin\varphi} d\varphi \right] f(\mathbf{v}_T) d\mathbf{v}_T, \qquad (8.53)$$

where v_{osc} is the maximum oscillation velocity in the laser cycle. Since we have yet to average over the thermal distribution, we can assume that there will always be some population of electrons that have thermal velocity greater than the maximum quiver velocity, so evaluation of the phase integral in Eq. (8.53) is straightforward if we denote the phase in the oscillation where $v_L = v_T$ (the lower limit of the integral) as φ_1, giving

$$\int_{\varphi_1}^{\pi/2} \frac{1}{\sin\varphi} d\varphi = \ln[\cos(\varphi_1/2)] - \ln[\sin(\varphi_1/2)]. \qquad (8.54)$$

A reasonable (if not completely justified in high-temperature plasma) approximation to these terms can be made if we assume that for most of the thermal electron distribution $v_{osc} \gg v_T$, so

$$\ln[\cos(v_T/2v_{osc})] \approx \ln[1] = 0 \qquad (8.55)$$

and

$$-\ln[\sin(v_T/2v_{osc})] \approx \ln\left[\frac{2v_{osc}}{v_T}\right], \qquad (8.56)$$

yielding for the cycle-averaged heating rate

$$\left\langle \frac{d\varepsilon_T}{dt} \right\rangle = \begin{cases} \dfrac{8\bar{Z}^2 e^4 n_i}{m v_{osc}} \ln \Lambda_C \int \ln\left[\dfrac{2v_{osc}}{v_T}\right] f(\mathbf{v}_T) d\mathbf{v}_T & v_{osc} > v_T \\ 0 & v_{osc} \leq v_T \end{cases}. \qquad (8.57)$$

(This last averaging step is not accurate at low intensity when the electron temperature is significantly higher than the laser's ponderomotive energy. But since the term derived varies slowly and is logarithmic, we can assume that it is sufficiently accurate.)

At this point we need an appropriate electron velocity distribution function for the final integral. As we have mentioned, a full kinetic theory treatment of this problem shows that in the strong field limit electron heating pushes the distribution toward a Maxwellian. This effect combined with the fact that electron equilibration times are often faster than or at least comparable to a typical femtosecond pulse duration allows us, to a good approximation, to assume a Maxwellian distribution of the form

$$f(\mathbf{v}_T) d\mathbf{v}_T = \sqrt{\frac{2}{\pi}} \left(\frac{m}{k_B T_e}\right)^{3/2} \exp\left[-m v_T^2 / 2k_B T_e\right] d\mathbf{v}_T. \qquad (8.58)$$

Evaluation of Eq. (8.57) can be performed numerically. We can, however, derive analytic formulas in two limits, the weak and strong field limits, which in this case mean that v_{osc} is either much less (weak field) or much greater (the strong field case) than $\sqrt{2k_B T_e/m}$. In the weak field limit, most of the contribution to the heating rate will occur for thermal

8.2 Plasma Formation and Heating Mechanisms

velocities comparable to v_{osc}, so we set $v_T = v_{osc}$ in the log term within the integral over $d\mathbf{v}_T$. With that approximation, the heating rate can then be evaluated analytically in terms of the standard error function,

$$\left\langle \frac{d\varepsilon_T}{dt} \right\rangle_{WF} = \frac{8\ln[2]\bar{Z}^2 e^4 n_i}{mv_{osc}} \ln \Lambda'_C \int_0^{v_{osc}} \sqrt{\frac{2}{\pi}} \left(\frac{m}{k_B T_e}\right)^{3/2} \exp\left[-mv_T^2/2k_B T_e\right] v_T^2 dv_T$$

$$= \frac{8\ln[2]\bar{Z}^2 e^4 n_i}{mv_{osc}} \ln \Lambda'_C \left(\mathrm{Erf}\left[v_{osc}\sqrt{\frac{m}{2k_B T_e}}\right] - \sqrt{\frac{2m}{\pi k_B T_e}} v_{osc} \right.$$

$$\left. \exp\left[-\frac{mv_{osc}^2}{2k_B T_e}\right] \right). \tag{8.59}$$

Further simplification in this weak field limit can be made by Taylor-expanding the error function around the small parameter $v_{osc}/(2k_B T_e/m)^{1/2}$,

$$\mathrm{Erf}[x] \cong \frac{2}{\sqrt{\pi}}\left(x - \frac{x^3}{3} + \ldots\right), \tag{8.60}$$

yielding a nice succinct formula for the heating rate

$$\left\langle \frac{d\varepsilon_T}{dt} \right\rangle_{WF} \cong \frac{8\ln[2]}{\sqrt{2\pi}} \frac{\bar{Z}^2 e^4 n_i}{m} \left(\frac{m}{k_B T_e}\right)^{3/2} v_{osc}^2 \ln \Lambda'_C. \tag{8.61}$$

The more relevant case is the one in which $v_{osc} \gg (2k_B T_e/m)^{1/2}$. In this limit it is a reasonable approximation to set $v_T = v_{Te} \equiv (2k_B T_e/m)^{1/2}$ in the logarithmic term and pull it out of the integral, which gives

$$\left\langle \frac{d\varepsilon_T}{dt} \right\rangle_{SF} = 8\frac{\bar{Z}^2 e^4 n_i}{m} \frac{1}{v_{osc}} \ln \Lambda'_C \ln \left[\frac{2v_{osc}}{v_{Te}}\right]$$

$$\times \left(\mathrm{Erf}\left[v_{osc}\sqrt{\frac{m}{2k_B T_e}}\right] - \sqrt{\frac{2}{\pi}} \sqrt{\frac{m}{k_B T_e}} v_{osc} \exp\left[-\frac{mv_{osc}^2}{2k_B T_e}\right] \right). \tag{8.62}$$

This limit of the heating rate per electron can be simplified by noting that the argument to the Erf term is large and $\mathrm{Erf}[\infty] \to 1$, while the exponential term falls to zero in the same limit. This results in a very useful formula:

$$\left\langle \frac{d\varepsilon_T}{dt} \right\rangle_{SF} = 8\frac{\bar{Z}^2 e^4 n_i}{m} \frac{1}{v_{osc}} \ln \left[\frac{2v_{osc}}{v_{Te}}\right] \ln \Lambda'_C. \tag{8.63}$$

Our strong field IB heating rate is essentially identical to the rate derived by Silin and almost identical to that of Jones and Lee with only a slight difference in the numerical cofactor.

The scaling of both weak field and strong field heating rate with v_{osc} and $k_B T_e$ is easily understood. We expect in both instances the heating rate to reflect that electrons acquire something on the order of the field's ponderomotive energy in every Coulomb scattering event. So it is natural that the heating rate scales like $U_p \sim v_{osc}^2$ times a collision frequency,

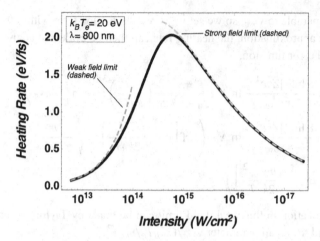

Figure 8.5 Inverse bremsstrahlung heating rate as a function of laser intensity for 800 nm light propagating in a plasma of $Z = 8$ with ion density of 2×10^{19} cm^{-3} and electron temperature of 20 eV found from numerical integration of Eq. (8.57). The scaling of the heating rate from the analytic results in the weak field and strong field limits are shown in dashed gray lines for comparison.

which should scale something like a Coulomb scattering cross section ($\sim 1/v^4$) times an electron velocity, v. In the weak field limit, the collision rate is dominated by the thermal velocity of the electrons $\sim (k_B T_e)^{1/2}$, so in that case the heating rate varies as $v_{osc}^2/(k_B T_e)^{3/2}$ in Eq. (8.61). On the other hand, in the strong field case, the collision rate will be dominated by the oscillation velocity, so the total heating rate we derived in Eq. (8.63) scales simply as $1/v_{osc}$. At low intensities the heating rate increases linearly with laser intensity but then peaks when $v_{osc} \approx v_{Te}$ (when the ponderomotive energy is comparable to the electron temperature) and then drops as the square root of intensity as the laser intensity further increases. Higher-intensity light is less efficient at heating the plasma. This scaling also suggests that shorter-wavelength light is less effective at heating the plasma at low intensity but is *more* effective at heating in the strong field regime.

We would also remark that while our strong field formula matches those formulae published in the literature, our weak field formula has the same scaling as the low-intensity heating rate predicted by Jones and Lee or Silin but with a numerical cofactor which is a little higher (a numerical factor of 2.2 in our Eq. (8.61), whereas the cofactor in Silin's formula is 1.7 and that of Jones and Lee is 0.74). The difference is of no real significance and is a result of the assumptions in the cycle-averaging and in assuming a Maxwellian distribution, while the kinetic theory predicts a non-Maxwellian distribution in the low field limit, an effect predicted by Langdon (1980).

The trends predicted earlier and the magnitude of heating is illustrated in Figure 8.5, where the heating rate of a near-atmospheric-density ($n_0 = 2 \times 10^{19}$ cm^{-3}), eight-times-ionized, 20 eV plasma subject to 800 nm laser light is plotted as a function of intensity as calculated by numerical integration of Eq. (8.57). The weak and strong field formulae are plotted as dashed lines for comparison. This plot illustrates that in a typical experiment at this density, the plasma will be heated on the order of an electron volt per femtosecond. Comparison with Figure 8.1 suggests that plasmas created by ~30–100 fs pulses will have roughly equal contributions of ATI and collisional heating. A good rule of thumb, given the interplay of ATI heating (which increases with intensity) and IB heating (which tends to become less important at higher intensity), is that gas targets irradiated by linearly polarized, near-IR, femtosecond pulses at intensity of $\sim 10^{16}$ to 10^{18} W/cm^2 will typically produce plasma electron temperature around 100 to 300 eV.

8.3 Beam Propagation of Intense Laser Pulses through Underdense Plasma

Now consider the effect that a plasma has on the propagation of an intense focused laser pulse. We review the propagation of a Gaussian beam through focus in vacuum within the paraxial wave approximation in Appendix A, a framework we will rely upon as a starting point in this section. First, I make a few observations about the general nature of propagation of intense light in the plasma. In essence, we need to consider two overall effects of the plasma on the laser: (1) on the evolution of the beam's spatial profile as it propagates and (2) on the temporal structure of the pulse, both on its amplitude and its spectrum.

The first of these effects results from the plasma imparting some radial phase variation on the pulse, leading to focusing or defocusing, often varying over the temporal envelope of the pulse. This radial phase imprint can be the result of radial variation in electron density from, say, radially dependent ionization or the excitation of plasma waves with amplitude that varies with radius. It can also be the result of nonlinear optical effects, such as the variation in refractive index with the variation of intensity at various spatial points in the beam. This second effect can result from atomic effects in the target gas or more exotic strong field plasma effects such as the variation of refractive index by intensity-dependent change of the electron mass at relativistic intensity. This and the next section of the chapter are devoted to these plasma-induced effects on the laser's beam profile as it propagates and the consequences these effects have, such as on resonant excitation of plasma waves or expulsion of electrons or ions from the laser focus.

The second of these effects, the impact that plasma evolution has on the temporal and phase structure of the pulse, will be examined in Section 8.5. This discussion will encompass two aspects of plasma impact on the pulse envelope: temporal variation in the phase velocity of the light in the evolving plasma and group velocity variation in time. Phase-modulation affects the pulse spectrum; group velocity variation modulates the amplitude through energy bunching. Again, contributions to these effects include temporal variation in plasma electron density through time-dependent ionization or through the excitation of longitudinal plasma waves. Nonlinear refractive index effects can also play a role.

First we consider propagation by intense pulses at subrelativistic intensity ($< 10^{17}$ W/cm^2). The optical propagation can be fairly summarized as being influenced by two impacts on the target gas's refractive index, varying electron density

$$\eta_R \cong 1 - \frac{n_e(r,t)}{2n_{crit}}, \tag{8.64}$$

and through a so-called nonlinear optical refractive index, typically treated as a correction to the medium's refractive index by a term proportional to intensity

$$\eta_R = \eta_0 + \gamma_{NL} I(r, t). \tag{8.65}$$

8.3.1 Radial Envelope Model for Propagation of Intense Pulses in Ionizing Plasma

The various phenomena we will consider in the next couple of sections can only be treated accurately by numerical solution of the wave equation. We can, however, derive intuition and some useful analytic results by utilizing a rather simple model for the propagation of the beam. A real laser beam, even if it enters the plasma with spatial profile well approximated by a cylindrically symmetric Gaussian, will experience refraction and nonlinear optical effects which lead to deviation from a Gaussian. However, a simple analytic model can be developed assuming the beam is roughly cylindrically symmetric during its propagation. We merely track an appropriate radial point in such a beam as it propagates. This permits us to develop a model for the beam as it propagates, employing a single parameter, its effective beam radius as a function of propagation distance, $w(z)$.

This model for the evolution of the radial beam envelope has been employed by a number of authors (Fill 1994; Sprangle *et al.* 1996) but has been most extensively developed by Esarey and coworkers (Esarey *et al.* 1994; 1997). They utilize a somewhat more rigorous source-dependent expansion method for describing the pulse radial envelope propagating under the paraxial wave equation. We chose a slightly simpler way of developing the model by assuming that the beam remains Gaussian and that the $1/e^2$ radius, $w(z)$, can be used to describe the beam throughout its evolution. This simplification leads to some loss of accuracy in a few results we will derive, but it is intuitive and essentially captures all of the physics in a model that is necessarily approximate. The essential idea is illustrated schematically in Figure 8.6.

We begin by recalling the radial intensity profile of a typical Gaussian beam,

$$I(r, z) = I_0 \left(\frac{w_0}{w(z)^2} \right)^2 \exp\left[-\frac{2r^2}{w(z)^2} \right], \tag{8.66}$$

where w is the usual $1/e^2$ intensity point, which evolves in vacuum according to the scaling summarized in Appendix A:

$$w(z) = w_0 \sqrt{1 + z^2/z_R^2}. \tag{8.67}$$

Here z_R is the usual Rayleigh range, describing the characteristic propagation distance over which a beam of initial size w_0, diffracts. For a Gaussian beam focused by a lens or curved mirror of focal length f, with input beam radius, w_i, the vacuum focus of the beam can be characterized by $f/\# = f/(2w_i)$, which yields a diffraction limited focal spot of

$$w_0 = 2\frac{\lambda}{\pi} f/\#. \tag{8.68}$$

8.3 Beam Propagation of Intense Laser Pulses through Underdense Plasma

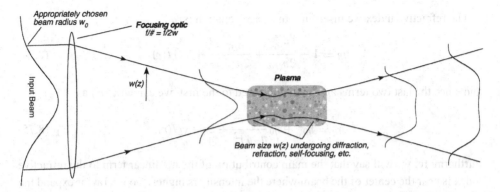

Figure 8.6 Schematic illustration showing how beam propagation through a plasma can be approximately modeled by tracking an effective beam radius as a function of propagation position, $w(z)$.

We turn to the paraxial wave equation for the slowly varying amplitude envelope of the laser's oscillating electric field, which we describe as

$$E(r, z, t) = \hat{E}(r, z) e^{i(kz - \omega t)}, \tag{8.69}$$

so that the field envelope, propagating in a medium with refractive index η_R, evolves according to the paraxial wave equation of the form

$$\nabla_\perp^2 \hat{E} + 2ik \frac{\partial \hat{E}}{\partial z} = k^2 (1 - \eta_R^2) \hat{E}. \tag{8.70}$$

We seek to develop a differential equation of the fields' radius, described by the usual Gaussian form:

$$\hat{E} = E_0 \left(\frac{w_0}{w(z)^2} \right) \exp\left[-\frac{r^2}{w(z)^2} \right] \exp\left[i\alpha_G(z) r^2 / w(z)^2 + i\varphi(z) \right]. \tag{8.71}$$

(See Appendix A for definitions of the terms in the exponentials.) Our main assumption is that the beam evolves in a gas and/or plasma that has two contributions to the refractive index: electron density and some possible nonlinear refractive index so that

$$\eta_R \cong 1 - \frac{n_e}{2 n_{crit}} + \gamma_{NL} I(r). \tag{8.72}$$

This form assumes slow variation of electron density with z. We want to consider the refractive effects of radially varying plasma density. We will approximate these effects by assuming that the electron density varies parabolically with beam radius,

$$n_e = n_{e0} \left(1 - \frac{r^2}{r_0^2} \right). \tag{8.73}$$

This radial shape for electron density has the useful consequence that a Gaussian beam propagating in such a plasma remains Gaussian in radial profile. In the equations that result, if we desire uniform electron density, we can simply take r_0 to go to infinity.

The refractive index we insert into the wave equation is

$$\eta_R \cong 1 - \frac{n_{e0}}{2n_{crit}} + \frac{n_{e0}}{2n_{crit}} \frac{r^2}{r_0^2} + \gamma_{NL} I(r), \qquad (8.74)$$

and since the last two terms are small compared to the first, we can write η_R^2 as

$$\eta_R^2 \cong \eta_0^2 + \eta_0 \frac{n_{e0}}{n_{crit}} \frac{r^2}{r_0^2} + 2\eta_0 \gamma_{NL} I(r). \qquad (8.75)$$

Furthermore, we will say that the main contribution of the nonlinear term to the refractive index is near the center of the beam where the intensity is highest, so we Taylor expand the intensity, Eq. (8.66), with r, and insert into Eq. (8.70),

$$1 - \eta_R^2 \cong -\eta_0 \frac{n_{e0}}{n_{crit}} \frac{r^2}{r_0^2} + \frac{4\eta_0 \gamma_{NL}}{w^2} I_0 r^2 - 2\eta_0 \gamma_{NL} I_0. \qquad (8.76)$$

We use Eq. (8.76) in Eq. (8.70) and put in the E-field of the form written in Eq. (8.71) to get

$$\left[-4\frac{i\alpha_G - 1}{w^2} + 4\frac{(i\alpha_G - 1)^2}{w^4} r^2 \right] \hat{E}$$

$$- 2ik \left[2r^2 \frac{w'}{w^3} + i\frac{\alpha_G' r^2}{w^2} - 2i\frac{\alpha_G w'}{w^3} r^2 + i\varphi - \frac{w'}{w} \right] \hat{E} \qquad (8.77)$$

$$= k^2 \left[-\eta_0 \frac{n_{e0}}{n_{crit}} \frac{r^2}{r_0^2} + \frac{4\eta_0 \gamma_{NL}}{w^2} I_0 r^2 - 2\eta_0 \gamma_{NL} I_0 \right] \hat{E}$$

To derive an equation for $w(z)$ we equate the real terms and equate the imaginary terms in (8.77), then further equate terms with no r dependence and terms that scale as r^2 to get four equations. These four equations yield solutions for α_G and its derivative in terms of w,

$$\alpha_G = \frac{-kw'w}{2} \qquad (8.78a)$$

$$\alpha_G' = \frac{-kw''w}{2} - \frac{kw'^2}{2}. \qquad (8.78b)$$

They can then be used to derive a second-order ordinary differential equation for $w(z)$, our desired result:

$$w''(z) = \frac{4}{k^2 w^3} + \eta_0 \frac{n_{e0}}{n_{crit}} \frac{w}{r_0^2} - \frac{4\eta_0 \gamma_{NL}}{w} I_0. \qquad (8.79)$$

The three terms on the right side of this equation each have a clear physical significance: the first term describes diffraction of the beam, the second term describes refraction by a radially varying plasma density profile, and the third describes nonlinear self-focusing. In the absence of plasma refraction or nonlinear effects, as we would encounter for vacuum propagation of the beam, the equation describing w reduces to

8.3 Beam Propagation of Intense Laser Pulses through Underdense Plasma

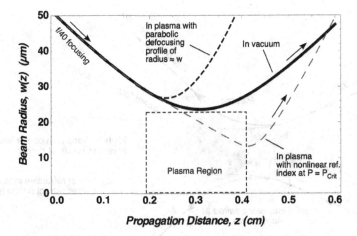

Figure 8.7 Numerical solutions of Eq. (8.79). The solid line shows typical Gaussian focusing of 800 nm light in vacuum with an f/40 focusing geometry. The short dashed line shows the beam trajectory from refraction at a plasma ∼2 mm long and a parabolic radial density profile with plasma radius equal to the focused spot size. The gray long dashed line shows the beam trajectory undergoing nonlinear self-focusing if the laser power in the plasma region is equal to the critical power.

$$w''(z) = \frac{4}{k^2 w^3}. \tag{8.80}$$

It can be seen easily by substitution that the vacuum propagation solution for the spot size given in Eq. (8.67) satisfies this equation.

Equation (8.79) can be easily solved numerically for various conditions and peak intensities. Figure 8.7 illustrates three such numerical solutions for 800 nm wavelength light. The thick black line is the spot size evolution of a beam focused by an f/40 optic as the beam passes through focus. The thick dashed line shows the same beam entering a 2 mm long plasma centered on the vacuum focal point, located between 0.2 and 0.4 cm with a radial variation such that the electron density is highest on axis and falls off with r characterized by r_0 equal to the vacuum spot size, w_0. Clearly, strong refractive defocusing is evident in this solution and the beam diverges before it can come to its vacuum focus. Such plasma-induced defocusing will be considered in the next subsection. The third solution of (8.79) shown in the figure, the thin, long dashed gray line, is the solution of that f/40 focused beam in plasma with the nonlinear refractive index term included with a beam power equal to the so-called critical power, described in Section 8.3.3. This beam undergoes further focusing once it enters the 2 mm length of plasma and the beam size collapses to a smaller spot (an effect we will discuss in detail later).

8.3.2 Ionization-Induced Plasma Defocusing

Now consider a laser pulse focused into a gas at nonrelativistic intensity. The target might be a gas jet, a gas cell or the pulse might be focused in air. Typical experimen-

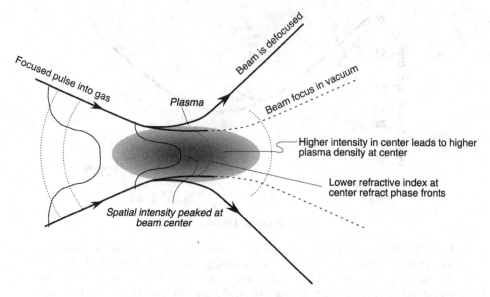

Figure 8.8 Schematic illustration showing how an intense laser pulse focused in a gas can produce a plasma with electron density peaked on axis and therefore lead to refraction and defocusing of the pulse as it propagates in that plasma.

tal targets often involve low-Z gases for wakefield acceleration experiments (like H_2) but frequently they involve higher-Z elements like N_2, Ar or Xe, which will undergo multiple stages of ionization during the pulse. As we have discussed, on the rising edge of the laser pulse multiphoton and tunnel ionization will create a plasma. Since the center of the focused beam is almost always at higher intensity, it will ionize the gas first and will produce the highest charge states and highest electron density.

The consequences of creating a radial electron density gradient during focusing are illustrated in Figure 8.8. Because of the drop in refractive index with increasing electron density, the phase velocity of the focused beam will be higher in the center of the beam leading to increased divergence and defocusing of the beam. This plasma-induced refraction is a well-known phenomenon in experiments when weakly focused intense pulses are focused in higher-density gases, particularly at density above about 10^{18} cm^{-3} (Rankin et al. 1991; Monot et al. 1992; Auguste et al. 1992; Leemans et al. 1992a; Mackinnon et al. 1996; Chessa et al. 1999). It is generally considered a deleterious effect because it prohibits the pulse from focusing to its highest intensity. Because of the increase of critical density with decreasing wavelength, this plasma-induced defocusing tends to be greater with IR pulses than with visible or near-UV lasers.

The three main experimental observables of this refraction are as follows: (1) The beam reaches its minimum focal spot size upstream of its expected focus in vacuum. (2) The peak intensity is clamped at a lower value than the focus in vacuum. (3) The ionization state and hence the maximum produced electron density are lower than would be expected for the focused intensity achieved in vacuum. A simple analysis can estimate the limit that

8.3 Beam Propagation of Intense Laser Pulses through Underdense Plasma

refraction sets to achieving the maximum electron density that can be produced by field ionization in a gas target. The phase accumulated by the center of the beam as it propagates along the z-axis is

$$\varphi_0 = \frac{2\pi}{\lambda} \int \eta_R(z) dz \approx \frac{2\pi}{\lambda} \eta_R(0) l, \quad (8.81)$$

if l is the length of the plasma. We expect significant defocusing when the center of the beam has its phase advanced $\sim \pi/2$ with respect to the edge of the beam. The phase difference between beam center and edge is approximately

$$\Delta\varphi_0 \cong \frac{2\pi}{\lambda}(1 - \eta_R) l$$

$$= \frac{\pi}{\lambda} \frac{n_e}{n_{crit}} l. \quad (8.82)$$

The defocusing will clamp the peak intensity if the plasma length needed to yield that $\pi/2$ defocusing phase is comparable to or less than the focused beam's Rayleigh range. So we can write the condition in which the peak intensity will be clamped as

$$\frac{\pi}{\lambda} \frac{n_e}{n_{crit}} z_R \geq \frac{\pi}{2}, \quad (8.83)$$

which implies that the maximum electron density that could be achieved by field ionization will be of the order of

$$n_e^{(MAX)} \approx \frac{\lambda}{2 z_R} n_{crit}. \quad (8.84)$$

Using the usual formula for the Rayleigh range and the vacuum focal spot size for a given $f/\# = f/2w_{in} \equiv f_\#$, the maximum possible electron density estimated in terms of the focal geometry is

$$n_e^{(MAX)} \approx \frac{\pi}{8} \frac{1}{f_\#^2} n_{crit}. \quad (8.85)$$

For a focused pulse with wavelength near 1 μm, the critical density is 10^{21} cm^{-3}, which means that focusing a near-IR pulse with weak f/20 focusing would allow production of plasma with electron density up to about 10^{18} cm^{-3}, while tighter f/6 focusing would be needed to ionize a gas up to electron density of $\sim 10^{19}$ cm^{-3} (assuming, of course, that the gas density and accessible ionization states are high enough to yield that electron density). This implies that a weakly focused, near-IR pulse in air (with gas density 2.7×10^{19} cm^{-3}) would likely not even fully ionize the air's atoms to a 1+ charge state.

We can develop a few other approximate scaling laws for focusing limits in an ionizing gas by following a treatment similar to that presented by Fill (1994) and using our radial beam envelope model. The appropriate form of the envelope propagation equation to use, Eq. (8.79), is one in which we retain the diffraction term and the plasma refraction term for some appropriate radial density gradient, characterized by r_0,

$$w''(z) = \frac{4}{k^2 w^3} + \frac{n_{e0}}{n_{crit}} \frac{w}{r_0^2}. \quad (8.86)$$

To utilize this equation to derive any analytic results we will need some kind of analytic model to estimate the peak electron density and r_0, which will be a function of the local peak intensity and the focal radius as the beam propagates. The electron density will be a result of the sequential ionization of the gas's atoms during the time history of the laser pulse and would normally need to be found by a solution of the ionization rate equations using a suitable ionization rate model from Chapter 4.

We can dispense with this more complicated numerical approach which relies on integration of ionization rates over an intensity temporal envelope and derive an approximate estimate for the local ionized electron density by recalling that the strongly nonlinear nature of MPI and tunnel ionization rates with intensity leads to a situation in which a particular ionization state will quickly saturate once an intensity is reached. It is, therefore, a fair approximation to say that the local charge state accessed by ionization depends only on the local, instantaneous intensity. We can develop an analytic estimate for the charge state reached using the barrier suppression ionization equation, Eq. (4.221), which gives us an estimate of intensity at which a particular charge state Z is produced. This BSI intensity estimate is

$$I_{BSI}(Z) = \frac{cI_p^4}{128\pi e^6} \frac{1}{Z^2}$$
$$= \kappa_{BSI} I_p^4 Z^{-2}. \tag{8.87}$$

Next, we make the further approximation that the charge state accessed is a continuous variable with intensity so the average charge state produced in the plasma is a simple equation depending on the constants in Eq. (8.87), the instantaneous intensity, and some functional estimate for how the ionization potential increases in the target atom as this continuous variable Z increases. This average ionization state can be written

$$\bar{Z} \cong \sqrt{\frac{\kappa_{BSI}}{I}} I_p(Z)^2. \tag{8.88}$$

As we have already discussed in Chapter 5, for most of the mid to high-Z atoms that typically constitute gaseous targets in many experiments (N_2, O_2, Ar etc.) the first few ionization potentials tend to increase roughly linearly with charge state. A reasonable estimate would be to say that $I_p(Z) \approx I_{p0}\bar{Z}$, if I_{p0} is the first ionization potential of the atom. With Eq. (8.88) this lets us write a very simple analytic estimate for the average charge state as a function of instantaneous local intensity

$$\bar{Z} \cong \frac{1}{\kappa_{BSI}^{1/2} I_{p0}} I^{1/2} \equiv \alpha_I I^{1/2}. \tag{8.89}$$

Since most of the atoms we have mentioned have $I_{p0} \approx 12\text{-}15$ eV, α_I is in the range of $5 - 7 \times 10^{-8}$ if the intensity is expressed in terms of W/cm^2. The predictions of this simple formula for the peak ionized charge state can be compared to the predictions of a rate equation calculation using PPT/ADK tunnel ionization rates. Such a comparison for 30 fs pulses as a function of peak intensity is made in Figure 8.9 for nitrogen and argon, illustrating that this simple formula gives a pretty reasonable estimate for ionized charge state.

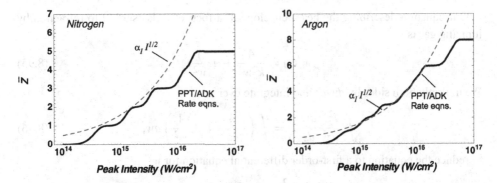

Figure 8.9 Average ionization states in nitrogen atoms (left) and argon atoms (right) as a function of the peak intensity of a 30 fs long pulse. These average ionization states were calculated with a rate equation model using the tunnel ionization rate predicted by the PPT/ADK model, Eq. (4.214). Our simple approximate scaling model is overlaid on these plots.

This equation lets us write a simple analytic equation for the electron density profile for a Gaussian beam with local peak intensity I_0,

$$n_e = \bar{Z} n_0 \tag{8.90a}$$

$$= \alpha_I n_0 I_0^{1/2} \left(\frac{w_0}{w}\right) \exp\left[-r^2/w^2\right]$$

$$\cong \alpha_I n_0 I_0^{1/2} \frac{w_0}{w} \left(1 - \frac{r^2}{w^2}\right), \tag{8.90b}$$

if we Taylor-expand the Gaussian for radii less than w. Inspection of the parabolic form for electron density we inserted in the envelope propagation equation, Eq. (8.73), tells us that our simple ionization model predicts that the peak electron density is

$$n_{e0} = \alpha_I n_0 I_0^{1/2} \frac{w_0}{w}, \tag{8.91}$$

and $r_0 = w$. The plasma refraction term in Eq. (8.86) can be written

$$\frac{n_{e0}}{n_{crit}} \frac{w}{r_0^2} = \alpha_I n_0 I_0^{1/2} \frac{w_0}{w^2} \equiv A_I(P_\lambda) \frac{1}{w^2}. \tag{8.92}$$

This term varies only with the instantaneous power P_λ of a temporal point in the pulse,

$$P_\lambda = I_0 \frac{\pi w_0^2}{2} \quad \Rightarrow \quad I_0^{1/2} w_0 = \sqrt{2P_\lambda/\pi}, \tag{8.93}$$

so the coefficient A_I in the refraction term is

$$A_I(P_\lambda) = \alpha_I \frac{n_0}{n_{crit}} \sqrt{2P_\lambda/\pi}. \tag{8.94}$$

Our equation describing the beam envelope of a focused pulse subject to refraction by ionizing gas is

$$w''(z) = \frac{4}{k^2 w^3} + A_I \frac{1}{w^2}. \tag{8.95}$$

We multiply both sides by dw/dz and integrate over z,

$$\int \frac{d^2 w}{dz^2} \frac{dw}{dz} dz = \int \left(\frac{4}{k^2 w^3} + A_I \frac{1}{w^2} \right) dw, \tag{8.96}$$

to reduce the equation to a first-order differential equation for w:

$$\frac{1}{2} \left(\frac{dw}{dz} \right)^2 = -\frac{2}{k^2 w^2} - \frac{A_I}{w} + C, \tag{8.97}$$

where C is merely a constant of integration to be determined from the focal geometry. We now have this equation to work with:

$$\frac{dw}{dz} = -\left(2C - \frac{4}{k^2 w^2} - \frac{2A_I}{w} \right)^{1/2}, \tag{8.98}$$

from which we can determine C by noting that far from the focus, ionization is negligible and therefore

$$\left. \frac{dw}{dz} \right|_{w=\infty} = -\sqrt{2C} = -w_{lens}/f = -\frac{1}{2f_\#}. \tag{8.99}$$

Inserting this solution for C,

$$\frac{dw}{dz} = -\left(\frac{1}{4f_\#^2} - \frac{4}{k^2 w^2} - \frac{2A_I}{w} \right)^{1/2}, \tag{8.100}$$

lets us find the minimum spot size the laser beam reaches subject to plasma refraction, which will occur when $dw/dz = 0$, so

$$0 = \frac{1}{4f_\#^2} - \frac{4}{k^2 w_{min}^2} - \frac{2A_I}{w_{min}}, \tag{8.101}$$

$$w_{min} = 4A_I f_\#^2 + \sqrt{16 A_I^2 f_\#^4 + \frac{4\lambda^2 f_\#^2}{\pi^2}}. \tag{8.102}$$

Inspection of this equation tells us that refraction by the plasma dominates the focusing when $8A_I \gg \lambda/f_\#$ (for example, in a gas with $n_0 \sim 10^{18}$ cm^{-3} with weak focusing $8A_I \sim 10^{-3}$ cm while $\lambda/f_\# \sim 10^{-5}$ cm). In the absence of refraction, Eq. (8.102) yields the usual vacuum focal spot

$$w_{min} = \frac{2\lambda f_\#}{\pi}, \tag{8.103}$$

but in the presence of strong plasma-induced defocusing, the minimum spot increases to

$$w_{min} \cong 8A_I f_\#^2, \tag{8.104}$$

8.3 Beam Propagation of Intense Laser Pulses through Underdense Plasma

which is determined by the pulse's peak power and the focusing f/#. The maximum focused intensity that can be produced in the strongly defocusing plasma is

$$I_{MAX} \cong \frac{2P_\lambda}{\pi w_{min}^2} = \frac{P_\lambda}{32\pi A_I^2 f_\#^4}, \qquad (8.105)$$

which, inserting for A_I, yields a basic formula for the refraction-clamped maximum focused intensity

$$I_{MAX} \cong \frac{1}{\alpha_I^2} \left(\frac{n_{crit}}{n_0}\right)^2 \frac{\pi}{2P} \frac{P_\lambda}{32\pi f_\#^4}$$

$$\cong \frac{1}{64} \frac{1}{\alpha_I^2} \left(\frac{n_{crit}}{n_0}\right)^2 \frac{1}{f_\#^4}. \qquad (8.106)$$

This is, surprisingly, completely independent of the pulse's peak power.

This indicates that at 10% atmospheric density, where singly ionized gas has density about 10^{-3} n_{crit}, if fairly tight f/10 focusing is employed, the maximum intensity possible is only $\sim 3 \times 10^{14}$ W/cm^2. In reality, this is something of an underestimate but does illustrate the dramatic clamping effect that ionization in a high-density gas can have on the focusing of a high-power pulse.

It is also worth noting that the actual f/# used in experiments is a bit different from the Gaussian formula we have used, because in most real experiments the beam is typically closer to flattop in profile at the input to the focusing optic. For these cases, it is more accurate to define an effective f/#, using the formula of an Airy pattern focus we expect from a flattop input beam (Jackson 1975, p. 441). This means that the effective f/# as the beam approaches the focus is

$$f_\#^{(eff)} = \frac{\pi}{2} 1.22 f_\# = 1.9 f_\#, \qquad (8.107)$$

which suggests we can rewrite our estimates in terms of a realistic experimental f/# to yield a more realistic estimate for the maximum focused intensity,

$$I_{MAX} \cong \frac{13}{64} \frac{1}{\alpha_I^2} \left(\frac{n_{crit}}{n_0}\right)^2 \frac{1}{\left(f_\#^{(eff)}\right)^4}, \qquad (8.108)$$

and the maximum charge state and ionized electron density,

$$\bar{Z}_{MAX} = \alpha_I I_{MAX}^{1/2}$$

$$= \frac{1}{8} \left(\frac{n_{crit}}{n_0}\right) \frac{1}{f_\#^2}, \qquad (8.109)$$

$$n_e^{(MAX)} = \frac{1}{8} \frac{n_{crit}}{f_\#^2} \approx 0.45 \frac{n_{crit}}{\left(f_\#^{(eff)}\right)^2}. \qquad (8.110)$$

This last equation is actually rather close (different only by a factor π) to our simple estimate for maximum electron density made earlier in this section, in Eq. (8.85).

Finally, we can use these equations to make an estimate for at what gas density refraction becomes important. Referring back to Eq. (8.102), we can write the condition for refraction as

$$16 A_I^2 f_\# \cong \frac{4\lambda^2 f_\#^2}{\pi^2}, \qquad (8.111)$$

which yields for the gas density in which refraction is a problem

$$n_0^{(thresh)} = \frac{1}{\alpha_I} \frac{\lambda}{2\sqrt{2}\pi} \frac{1}{P_\lambda^{1/2}} \frac{n_{crit}}{f_\#}. \qquad (8.112)$$

For a 1 TW peak power pulse this says that a density $\sim 10^{-4}\, n_{crit}/f_\#$ will produce serious refraction.

Recall that these results depend on the target gas being composed of atoms with multiple ionization states, like N_2, Ar or Xe, and that the plasma interaction length is long enough for refraction to play a significant role. Experiments in hydrogen or helium targets are far less susceptible to plasma refraction, so much higher peak intensities are possible than would be implied by the preceding equations. Also, it is possible to design experiments with gas jets or gas cells such that the laser is brought close to focus in vacuum before it enters the gas region, allowing greater peak intensity in the gas than if the pulse comes to focus in the gas. Furthermore, these equations ignore the possibility of relativistic self-focusing, an effect considered in the next section.

Consequently, these scaling equations should be utilized with some care. However, the trends that they predict are well established in experiments (Rae 1993; Monot et al. 1992). For example, the experimental trends observed in maximum electron density and peak focused intensity are illustrated in Figure 8.10. In the left plot, the trends observed in sub-picosecond 1 μm wavelength pulses of a few-terawatt power in the maximum electron density observed as a function of target gas pressure are plotted. The electron density produced tracks the increase in target gas density until a rollover density is observed. The point at which this rolls over increases in density when lower $f_\#$ focal geometries are employed. The trends observed in similar experiments for the maximum peak intensity achieved as a function of the density of Ar gas are shown in part (b) of Figure 8.10. When the vacuum peak intensity should be 10^{18} W/cm^2, the peak intensity actually observed drops precipitously to under 10^{15} W/cm^2, when the argon target is increased in density to $\sim 10^{19}$ cm^{-3}. The peak intensity model of Eq. (8.108) is plotted as well to show how observed data more or less follow the trend predicted by our simple envelope propagation model.

The experimentally observed trends in the upstream shift of the focal position from refraction as a function of target gas density are plotted in Figure 8.11 for various peak powers. This can be a significant effect if appropriately designed gas cell or jet targets with sharp density input points are not used. Presaging our discussion of filamentation in the next section, we would point out that the shift of focal position from plasma refraction is actually dependent on the temporal point in the pulse, leading to rather complicated focal plane intensity distributions, as illustrated in Figure 8.12. The leading edge of the pulse sees little ionization and therefore focuses near the vacuum focal point, but later in the

8.3 Beam Propagation of Intense Laser Pulses through Underdense Plasma

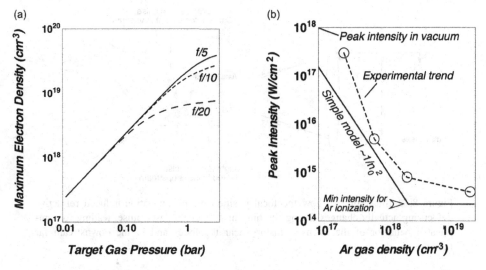

Figure 8.10 Experimentally observed trends in plasma-induced defocusing of pulses with wavelength near 1 μm and duration under 1 ps. In (a) the maximum electron density produced in a gas as a function of the initial atomic gas pressure for various focal geometries is shown. Plot generated based on data shown in Rae (1993). In (b) is shown the observed trend of decreasing peak intensity obtainable in Ar as its gas density is increased (for ~f/10 focusing of a 1 μm wavelength pulse. Plot generated based on data from (Monot et al. 1992). Our simple analytic model prediction is also plotted here.

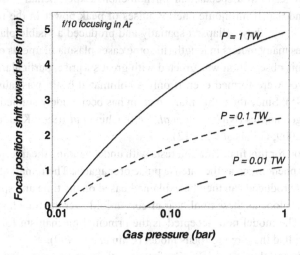

Figure 8.11 Plot of the observed trends in the shift of the focal position upstream toward the focusing optic in a gas like Ar as the gas density is increased. Various peak laser powers are plotted, showing how higher ionization by a higher-power pulse leads to greater refraction and focal position shift. Plot was generated using data from (Monot et al. 1992).

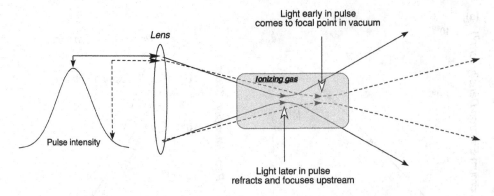

Figure 8.12 Illustration of how the focal position during ionization-induced refractive defocusing actually changes during the time history of the laser pulse, leading to complicated behavior of the intensity history near the focus and in the downstream far field.

pulse, after significant ionization has occurred, the pulse energy there focuses upstream from the vacuum focus.

8.3.3 Laser Filamentation at Moderate Intensity in Atmosphere

Sometimes the propagation of collimated or weakly focused, few-millijoule, femtosecond pulses in air leads to a spectacular phenomenon. Experimental observations in the early 1990s found that if millijoule energy pulses of peak power in the range of 10–100 GW propagated in air, they collapsed spatially and produced a visible plasma filament in the air, sometimes many meters in length (in some cases plasma filaments of 10–100 m can be produced). This observation was greeted with great surprise, particularly since the first filaments observed were formed even if only a collimated beam was launched in the air (Braun *et al.* 1995). Since then, this filamentation has been widely studied over a range of parameters and geometries (Nibbering *et al.* 1996; Chin *et al.* 1999; Kandidov *et al.* 2009; Bernstein *et al.* 2009; Geints *et al.* 2012).

The explanation for this filamentation rests with understanding the interplay of two competing nonlinear phenomena as the intense pulse propagates. The remarkable length of the plasma filaments produced can then be explained based on the time and space evolution of the laser pulse envelope as these nonlinear effects unfold over the time history of the laser pulse envelope. The model now accepted as the principal mechanism for these filaments has come to be called the moving focus model (Chin *et al.* 1999).

The first important physical mechanism involves the nonlinear correction to the refractive index of the air by the high intensity of the pulse. In short, in most polarizable media, the refractive index increases at high light intensity. As already mentioned, this correction to the refractive index, essentially a manifestation of nonlinearity in the third-order optical susceptibility of a medium when the light field is strong, is usually written:

$$\eta = \eta_0 + \gamma_{NL} I. \tag{8.113}$$

8.3 Beam Propagation of Intense Laser Pulses through Underdense Plasma

This correction has a number of consequences, one of which is depicted in Figure 8.13. As an intense pulse is launched in a medium, spatial regions in the beam where the intensity is highest, say in the center of a Gaussian-like shaped beam profile, will experience phase retardation which is higher than the low-intensity regions in the beam because of the shift in refractive index. This retardation causes wavefront curvature, which focuses the beam, called nonlinear self-focusing. If the beam's intensity (actually instantaneous power as we shall see) is high enough, this nonlinear self-focusing can overcome the divergence resulting from diffraction, and this can lead to a sharp focusing of the beam. In air, the nonlinear refractive index can lead to the spatial collapse of a propagating femtosecond pulse and initiate ionization, the first steps toward plasma filament formation.

We can derive the conditions for self-focusing to overcome diffraction by considering a simple, single-mode Gaussian beam, whose radial phase can be written

$$\text{Gaussian} \sim E_0 \exp\left[i\frac{kr^2}{2R(z)}\right], \quad (8.114)$$

where $R(z)$ is the local radius of curvature of the beam's phase fronts and is (as summarized in Appendix A) related to the beam's Rayleigh range z_R and distance z from the waist of the beam as $R = z + z_R^2/z$. Self-focusing counteracts diffraction when the rate of phase advance in the middle of the beam from diffraction is cancelled by the phase retardation from the nonlinear refractive index. Since from Eq. (8.114) we see that the phase on the beam from diffraction is

$$\varphi_D = -\frac{kr^2}{2(z + z_R^2/z)}, \quad (8.115)$$

and, since the phase retardation from the second term of Eq. (8.113) is

$$\varphi_{NL} = \frac{2\pi}{\lambda} \eta_0 \gamma_{NL} I z, \quad (8.116)$$

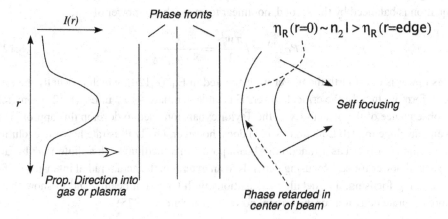

Figure 8.13 Illustration of how a medium with a finite nonlinear refractive index can cause phase retardation at the center of a beam where intensity is highest, leading to self-focusing.

we can find the condition for self-focusing by asking, when does $d\varphi_D/dz + d\varphi_{NL}/dz = 0$? We can write

$$\frac{d\varphi_D}{dz} = \frac{4\pi}{\lambda} w_0^2 \left[\left(2 - 2\frac{z_R^2}{z^2}\right) \frac{1}{(2z + 2z_R^2/z)^2} \right], \quad (8.117)$$

and evaluate it at the beam waist, giving

$$\left.\frac{d\varphi_D}{dz}\right|_{z=0} = -\frac{2}{\pi}\frac{\lambda}{w_0^2}. \quad (8.118)$$

Therefore, the condition on the peak central intensity, I_0, which would drive self-focusing is

$$\frac{d\varphi_D}{dz} + \frac{d\varphi_{NL}}{dz} = 0 = -\frac{2}{\pi}\frac{\lambda}{w_0^2} + \frac{2\pi}{\lambda} \eta_0 \gamma_{NL} I_0. \quad (8.119)$$

Since the instantaneous power in the laser pulse is the intensity times the effective beam area, which for a Gaussian beam is $P_L = I_0 \pi w_0^2/2$, we see that the condition for self-focusing depends only on the laser pulse power. The condition delineated in Eq. (8.119) says then that light above some critical power, given as

$$P_C = \frac{\lambda^2}{2\pi \eta_0 \gamma_{NL}}, \quad (8.120)$$

will undergo self-focusing if left to propagate long enough in the nonlinear medium.

The consequences on pulse propagation at power around this critical power can be nicely seen by examining the radial beam envelope equation we derived in Section 8.3.1. Including the nonlinear refractive index, this equation reads

$$\frac{d^2w}{dz^2} = \frac{4}{k^2 w^3} - \frac{4\eta_0}{w} \gamma_{NL} I. \quad (8.121)$$

Divergence from the diffraction modeled by the first term on the right-hand side of this equation is balanced by the second, nonlinear term at a beam power of

$$P_C^{(en)} = I\frac{\pi w^2}{2} = \frac{\lambda^2}{8\pi \eta_0 \gamma_{NL}}. \quad (8.122)$$

This power is one-fourth the power we estimated in Eq. (8.120), which is usually the textbook form written for the critical power. This underestimate for P_c in Eq. (8.121) is merely a consequence of the inaccuracy in the Taylor expansion used to develop this approximate beam envelope model. Equation (8.121) does, however, nicely illustrate how the evolution of a beam is affected as a function of beam power in a medium with a finite γ_{NL}. In fact, the actual power for self-focusing depends to an extent on the actual radial intensity profile undergoing focusing, and detailed calculations with Gaussian beams tend to show that a more accurate equation for the critical power is (Marburger 1975)

$$P_C \cong \frac{3.77}{8}\frac{\lambda^2}{\pi \eta_0 \gamma_{NL}} \approx \frac{\lambda^2}{2\pi \eta_0 \gamma_{NL}} \leftarrow \text{by convention.} \quad (8.123)$$

8.3 Beam Propagation of Intense Laser Pulses through Underdense Plasma

Convention in the literature is usually to adapt Eq. (8.120) for the analytic form of the P_c. For numerical computations, the best approach, when plasma is present, is to recast the envelope propagation equation in terms of an experimentally observed P_c to yield

$$\frac{d^2 w}{dz^2} = \frac{4}{k^2 w^3}\left(1 + \frac{k^2 \eta_0}{4}\frac{n_{e0}}{n_{crit}}\frac{w^4}{r_0^2} - \frac{P}{P_c}\right). \tag{8.124}$$

The situation in a medium such as air is more complicated than this simpler picture suggests. Equation (8.113) is an approximation that implies that the nonlinear correction to the refractive index is only a function of the instantaneous intensity. This means that any nonlinear response of the medium is very fast compared to any temporal variation in intensity. For media in which the contribution to the polarizability is only determined by the electronic structure of the atoms, this is usually an excellent approximation. However, some media exhibit polarizability, which is slow, meaning that the polarization, and hence the γ_{NL}, are a function of the prior intensity history. The dipole in molecules like N_2 and O_2 in the air falls into this category. Their polarizability is a function of the angle of orientation of the axis of the molecule with the laser field direction. Molecules oriented along the direction of linearly polarized light are more polarizable and will therefore exhibit a higher refractive index. We have seen in Chapter 5 that linearly polarized light can orient diatomic molecules, though the rotation of the molecule into the polarization direction takes some time. Air, therefore exhibits noninstantaneous nonlinear refractive index, and this complicates the dynamics of an intense pulse propagating in air. For very short pulses, however, like those $<\sim$ 50 fs, this molecular rotation effect is not large and it is a fair approximation to treat the nonlinear response as instantaneous. Longer pulses, however, do experience a higher effective n_2 for this reason.

The actual experimentally observed n_2 in air, then, depends on a number of laser parameters, like pulse duration, wavelength, and so on. Air's γ_{NL} is typically observed to be $\approx 2-6 \times 10^{-19}$ cm^2/W (Mlejnek et al. 1999; Kandidov et al. 2009) which corresponds to critical power of $\approx 1.5-5$ GW. Sub-100 fs pulses of only a few millijoules have peak power of tens to hundreds of gigawatts, which means that the power over most of the temporal pulse envelope will exceed the critical power for self-focusing in air.

We see, then, the first element contributing to the formation of plasma filaments in air: when a multi-GW femtosecond pulse is launched to propagate in air, self-focusing causes energy throughout the pulse envelope to collapse to a small spot. This intensity can then exceed that needed for significant multiphoton ionization (MPI) of the air molecules and plasma is formed. This triggers the second phenomenon; plasma-induced refractive defocusing. If the instantaneous power is not badly depleted, that time slice remains above the critical power and can refocus further downstream in the pulse's propagation. Clearly, the propagation length required for self-focusing collapse of the light to an ionizing small spot will get shorter with increasing power. Consequently, different temporal slices in the propagating pulse will collapse at different points along the propagation, contributing to plasma formation at different lengths along the propagation axis. This qualitative view of plasma filament formation in air has come to be called the "dynamic moving focus model."

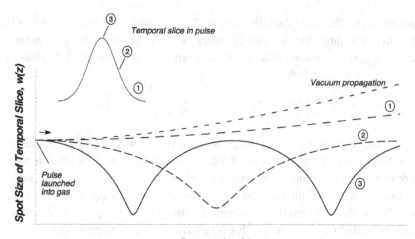

Figure 8.14 Illustration of the dynamics of the spot size of a pulse propagating in a medium with nonlinear refractive index. Various temporal time slices, each with different intensity will undergo self-focusing at different rates. In the low intensity, front edge of the pulse, the spatial spot size grows by diffraction at a rate only slightly lower than propagation of the bam in vacuum. Later time slices, however, have power above the critical power and undergo self-focusing until ionization-induced refraction clamps the intensity and causes the spot to diverge. Different intensity time slices undergo self-focusing at different rates and therefore collapse to a small size at different points during the beam's propagation.

A qualitative picture leading to filamentation can be drawn, illustrated in Figure 8.14. Different temporal slices self-focus in the air, collapsing at different points along the propagation axis. The low-intensity leading and trailing edges of a laser pulse simply expand by diffraction; however, later in the pulse a moving focal point develops over the pulse temporal envelope. Refocusing of the light after plasma refraction can occur for some slices in the pulse, prolonging the observed filament (like that illustrated for the beam envelope trajectory for time slice "3" located near the peak of the pulse).

The real dynamics of filamentation are far more complex for a range of reasons:

1) γ_{NL} varies in time, so the rear of the pulse tends to see a higher nonlinear refractive index.
2) MPI on axis increases the plasma density in the center of the beam at later times in the pulse. Consequently, the rising edge of the pulse, which propagates in the air prior to any significant plasma formation, undergoes very little plasma refraction and simply continues to self-focus along the propagation axis. The rear of the pulse in the center of the beam is strongly refracted by plasma produced by ionization earlier in the pulse. This leads to the radial 'wings" developing in the intensity profile late in the pulse. These wings can subsequently refocus by the delayed enhancement in the γ_{NL} produced early in the pulse.

8.3 Beam Propagation of Intense Laser Pulses through Underdense Plasma

3) The energy of various slices in the pulse will be depleted as they deposit energy into the plasma. This absorption into MPI ultimately stops multiple bounces from repeated self-focusing and sets the limit on the observed plasma filament length.

The ultimate length of the observed plasma filament varies with conditions, but generally speaking, if a collimated pulse of a few mm in diameter is launched into air, the first plasma is generally observed after the pulse has propagated a fraction of the beam's Rayleigh range and the observed plasma is a few Rayleigh lengths. For example, if a 5 mm diameter 800 nm pulse is launched in air, its Rayleigh length is about $\pi\,(0.25\text{ cm})^2/(800\text{ nm}) \approx 2.5$ m. Note that the energy deposited into MPI over such a range will significantly deplete the pulse energy. As we have seen, plasma-induced refraction limits the peak density a weakly focused pulse can create, quantified in Eq. (8.110). This is confirmed in experiments which show that electron density of $\sim 10^{16} - 10^{17}$ cm^{-3} are produced in the plasma filament.

We can estimate the power consumed by ionization in terms of the ionization rate W_{ion}, the ionization potential of the nitrogen molecules (around 14.5 eV), the gas density n_0, the self-focused beam area and the propagation length, l as

$$P_{consumed} \cong W_{ion} I_p^{(1)} n_0 \frac{\pi w_0^2}{2} l. \tag{8.125}$$

Estimating the focused spot size in terms of the self-focus focal length, this estimate becomes

$$P_{consumed} \cong W_{ion} I_p^{(1)} n_0 \frac{\lambda^3}{2\pi^2} \frac{f_{NL}^4}{w_i^4}. \tag{8.126}$$

Finding the self-focal length is usually complex, but it can be estimated using an empirical formula developed by Marburger (1975) based on numerical simulations and some experimental observations. This focal length is

$$f_{NL} \cong \frac{0.37 k w_i^2}{\left\{[(P/P_C)^{1/2} - 0.85]^2 - 0.022\right\}^{1/2}}, \tag{8.127}$$

which suggests that for slices in the pulse which are a few times the critical power, they have a focal length $\approx z_R/2 \approx 1$ m. Equation (8.126) then predicts that for propagation over a Rayleigh range, ~ 2 GW of power is consumed by ionization. As a consequence, energy absorption of a pulse ~ 10–100 GW limits observed filaments to ~ 5–30 m (Chin et al. 1999).

We can articulate some of the more complex aspects of the dynamics in filamentation. Figures 8.15 and 8.16 attempt to summarize these complex dynamics. Figure 8.15 presents a slightly more sophisticated view of the focusing envelope trajectories of various time slices in the laser pulse in air, and Figure 8.16 attempts to illustrate the complex spatio-temporal dynamics that result by schematically illustrating the temporal and radial intensity profiles of the pulse at various stages of filamentation. First, as shown at the time of Figure 8.16, by the intensity profile labeled $t = 0$ a pulse is launched into air. The most intense time slice of the pulse undergoes the fastest self-focusing (time slice labeled (4) in Figure 8.15). The distance from the point that the pulse is compressed and launched into

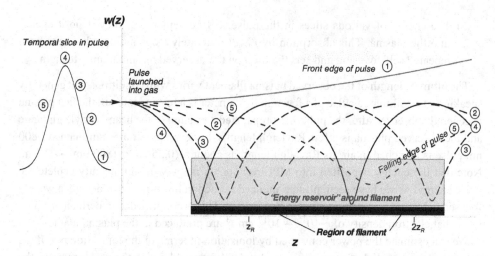

Figure 8.15 Schematic plot of the spot size of various time slices in a femtosecond pulse in air, illustrating how self-focusing (and refocusing) of different intensity slices in the pulse at different propagation points leads to the experimental observation of a long plasma filament. The range of focal lengths also leads to the observation of an apparent halo of pulse energy around the high-intensity core on axis that leads to filament ionization.

the air to the point that the plasma filament is first formed is determined by the focal length of the power at the peak of the pulse, estimated by Eq. (8.127). This is usually roughly half the Rayleigh range, as illustrated by the focus trajectory of time slice (4) in Figure 8.15. As shown by the profile labeled $t = 1$ in Figure 8.16, the peak of the pulse has become spatially pinched and begins then to diverge from plasma refraction.

Midway along the filament, as shown by the profile at $t = 2$, two effects start to manifest. The leading edge of the pulse, which has lower power than the peak and therefore self-focuses more slowly, continues to self-focus, but, because it sees very little plasma formation that early in the pulse, does not refract much. Consequently the front edge of the pulse builds a sharp spike in intensity. At the rear of the pulse, the light directly on axis has encountered significant plasma formation and therefore refracts away before it can self-focus. This leads to "wings in the spatial profile late in the pulse, as shown by the spatiotemporal profile labeled $t = 3$. These wings, however, see an enhanced γ_{NL}, as illustrated in Figure 8.16 at the bottom left, and they then undergo refocusing. So, the filament looks to regenerate late in the propagation because of the refocusing of this initially plasma-diffracted energy early in the filament formation.

These dynamics lead to an interesting experimental observation, shown at the bottom right of Figure 8.16. When a beam image is taken along the filament, a bright core is observed, formed by that portion of the pulse that is undergoing its focus from self-focusing. That bright core is surrounded by a halo of energy. This can be explained in Figure 8.15 by considering the trajectory of light at points in the pulse that have already come to a focus or are coming down to a focus. This halo is often called an "energy reserve"

8.3 Beam Propagation of Intense Laser Pulses through Underdense Plasma 531

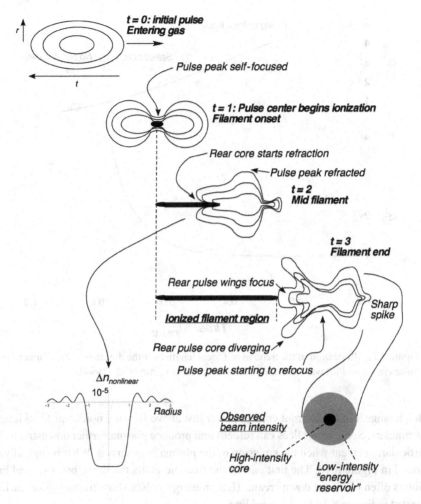

Figure 8.16 Here the evolution of the intensity profile of a pulse in time and radius is shown as that pulse produces a long plasma filament in air. The initial pulse has near-Gaussian intensity profiles both in spatial radius and over time. Subsequent evolution of the spatial temporal profile of the pulse as it undergoes self-focusing in the air, refraction from ionization and subsequent refocusing are shown at three time shots along the pulse's filament formation and propagation. The inset at bottom shows the nonlinear contribution to the refractive index in the late time spatial wings of the pulse leading to refocusing and ionization downstream in the filament. The image at bottom right shows the experimentally observed fluence profile of a filamenting pulse with hot central focused core and wings of energy around this core.

that feeds the filament. It is merely a manifestation of the moving focus of various time slices in the pulse.

The spatial scale of the filament can be understood on the basis of these spatiotemporal dynamics. Figure 8.17 plots the power of the front half of a pulse and nonlinear focal length at various points in this temporal envelope. At the peak, the pulse focuses in under a

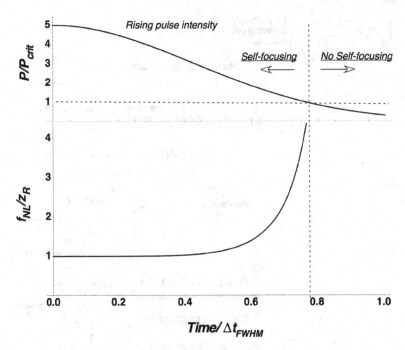

Figure 8.17 Illustration of the trend in self focal length over the duration of the filamenting pulse envelope. Plot was generated based on data from Chin et al. (1999).

Rayleigh range, but in the front edge at power just above P_{crit} the nonlinear focal length is a few times z_R. Some time slices can refocus and produce plasma further downstream. This is particularly evident when the emission of the plasma is observed. What is typically seen is plotted in Figure 8.18. The first part of the filament emits the most, but a second bright section is often observed downstream. Higher-energy pulses show filamentation earlier as illustrated in Figure 8.18 by the solid line.

Numerical simulations confirm this refocus picture (Sprangle et al. 1996; Meijnek et al. 1999). Figure 8.19 shows trends typically seen in simulations. The peak intensity on axis undergoes two peaks, which is matched by the electron density seen on axis, plotted in the bottom panel of Figure 8.19. The temporal envelope of light on axis in simulations often shows the development of a sharp spike in intensity on the leading edge of the pulse, illustrated by the three temporal envelopes in Figure 8.20 (labeled to correspond roughly to the three points labeled on the peak intensity plot of Figure 8.19). Finally the radial integrated fluence observed in simulations shows the development of spatial wings from refraction, as shown in Figure 8.21. The refocusing of these wings leads to the second point of bright plasma emission shown in Figures 8.18 and 8.19.

This complex spatiotemporal dance leads to a remarkable effect in the lab which to witness is quite spectacular. At this point it is worth noting a few qualitative observations about filamentation. Employing a weak focus lens ($\sim$f/100–f/300) will often aid in the initiation of filamentation. Tight focusing in air tends, however, to eliminate filament formation, as refraction and diffraction are not easily overcome by self-focusing in this geometry. Second, it is observed that filamentation generally requires laser pulses that are shorter than

8.3 Beam Propagation of Intense Laser Pulses through Underdense Plasma 533

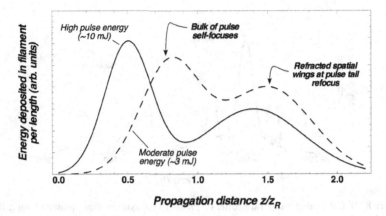

Figure 8.18 Trend of often observed pulse energy deposition as a function of length in the filament for two incident femtosecond pulse energies.

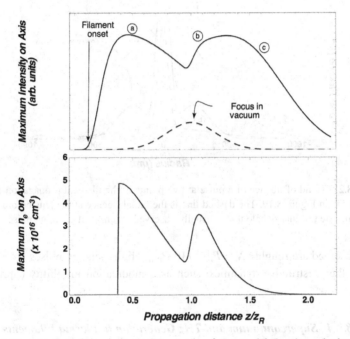

Figure 8.19 Adaption of simulations of the moving focus model showing calculated maximum intensity on axis for a pulse (whose location in time varies as different time slices come to focus on axis) focused at ~f/1000 in the top panel. The bottom panel shows the corresponding maximum electron density observed on axis in these simulations. Plots were generated based on calculations of Mlejnek et al. (1999).

about 200 fs. Longer pulses are more susceptible to the effects of plasma refraction over the longer pulse envelope and therefore filamentation is less robust. Finally, when particularly high peak-power pulses are employed, say $P \gg 10\, P_{crit}$, it is usually observed that instead of the formation of one long filament, the beam typically breaks up into many filaments, seeded by spatial inhomogeneities on the beam. As a rule of thumb, the number of small

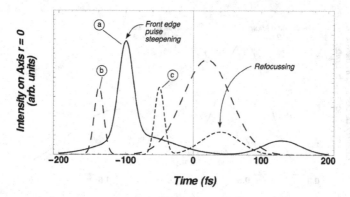

Figure 8.20 Calculated temporal profile of intensity on axis for three points along a filament, denoted by three letters in Figure 8.19 assuming an initial 200 fs Gaussian temporal profile pulse. Plot adapted from calculations of (Mlejnek *et al.* 1999).

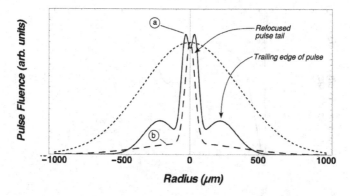

Figure 8.21 Trend of fluence of a pulse at two points in the filament propagation labeled "a" and "b" in Figure 8.19. The dashed line is the radial fluence at the lens before. These plots illustrate the energy halo that is usually observed around a filamenting pulse.

filaments observed are roughly $N \approx P_{peak}/10\, P_{crit}$. Higher-energy pulses are also subject to various other destructive dynamics, such as a modulation instability (Sprangle *et al.* 1996).

8.3.4 Supercontinuum and THz Generation in Plasma Filaments

A notable aspect of laser filamentation is the production of a forward cone of bright white light and THz radiation (Nibbering *et al.* 1996). The first effect, often referred to as conical supercontinuum generation can be understood simply as a result of extreme temporal phase-modulation of the self-focused pulse. We will consider self-modulation again in the context of relativistic intensity pulses later in this chapter. In brief, the ultrashort pulse undergoes a phase shift which shifts the spectrum both into the red and into the blue regions around the near-IR peak of the spectrum. The local frequency in the pulse can be thought of as the derivative of the temporal phase imparted on the beam by a temporally evolving refractive index, $\omega_1 = -d\varphi/dt$, where $\varphi(t) = \omega\eta_R(t)z/c - \omega t$. For our filamenting pulse this leads to three sources of frequency shift:

$$\omega_1 \cong \frac{\omega}{c}z\left(-\gamma_{NL}\frac{\partial I}{\partial t} - I\frac{\partial n_2}{\partial t} + \frac{1}{2n_{crit}}\frac{\partial n_e}{\partial t}\right). \tag{8.128}$$

The first two terms result from the nonlinear refractive index in the air; the first leads to a red shift mainly on the rising edge of the pulse, the second term from the molecular rotation delay in the γ_{NL} leads to an overall red shift on the pulse and the third term leads to a blue shift as the air undergoes ionization.

The spectrum that results often looks something like that pictured in Figure 8.22. Red shifting predominates, however, so the plasma ionization–induced blue-shifting can shift near-IR light into the visible. Since greater shift toward the blue is a consequence of the highest temporal gradient in ionized electron density, these time slices of the laser pulse also see the highest radial gradient in electron density. This leads to a radial refraction of the light by an amount

$$\Delta k_r \sim \frac{\partial n_e(r)}{\partial r}, \tag{8.129}$$

and white light emission angle

$$\text{emission angle} \sim \frac{\Delta k_r}{k_z}. \tag{8.130}$$

Since the most blue-shifted light sees the greatest refraction by Eq. (8.129), the resulting blue-shifted visible light tends to be emitted into a visually spectacular cone with blue light at the highest angle at the outer edges of the cone and a rainbow of supercontinuum moving to longer wavelengths at lower emission angle. The data that are typically measured in such a conical rainbow tend to look like that plotted in Figure 8.23.

Finally we note that a rather interesting cone of THz radiation is usually observed from the filament. The geometry of the observations is illustrated in Figure 8.24. An annulus of radially polarized THz radiation is typically observed, resulting from plasma wave currents excited in the plasma filament as the laser pulse passes (Kandidov et al. 2009).

8.4 Propagation in Underdense Plasma in the Relativistic Regime

At higher intensity, when a_0 approaches 1 ($\sim 10^{18}$ W/cm^2), two effects become important: the laser-driven oscillations increase the mass of the electrons, and the laser ponderomotive force can drive significant motion of the electron fluid, leading to large-amplitude plasma generation and cavitation of electrons around the pulse. These have important consequences on propagation of laser pulses at these intensities (Umstadter 2003).

8.4.1 Relativistic Dispersion Relation, Nonlinear Refractive Index and Self-Focusing

Relativistic mass increase of the electrons in a strong field can be accounted for with a simple correction to the dispersion relation for propagation of the EM wave in the plasma.

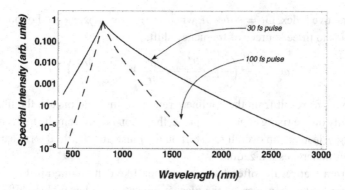

Figure 8.22 Typical shape of the supercontinuum spectrum observed by filamentation of few-terawatt, 800 nm pulses of duration ~100 fs and ~30 fs. Plot was generated from data presented by (Kandidov et al. 2009).

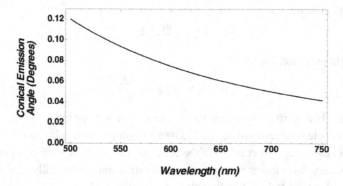

Figure 8.23 General trend observed in broadband, "white light" emission observed from filamenting pulses, showing the general variation of observed wavelength as a function of angle with respect to the filament axis. Plot was generated based on data presented in (Chin 1999).

Forgoing, for the moment, a more rigorous derivation, we account for this effect by a simple phenomenological amendment to Eq. (8.21), noting that the mass in the electron plasma frequency increases with some relativistic gamma factor

$$\omega^2 = c^2 k^2 + \frac{\omega_{pe}^2}{\gamma_{eff}}, \qquad (8.131)$$

which means that we need to rewrite our Drude model result for the plasma refractive index with this relativistic correction

$$\eta = \sqrt{1 - \omega_{pe}^2/\omega^2} \qquad (8.132a)$$

$$\cong 1 - \frac{1}{2}\frac{\omega_{pe}^2}{\omega^2} = 1 - \frac{1}{2\omega^2}\frac{4\pi n_{e0} e^2}{\gamma_{eff} m} \qquad (8.132b)$$

8.4 Propagation in Underdense Plasma in the Relativistic Regime

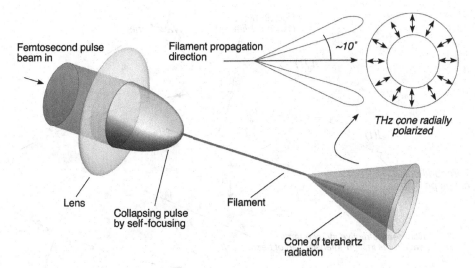

Figure 8.24 Illustration of the geometry of experimentally observed conical, radially polarized terahertz emission from plasma currents generated in a plasma filament in air.

$$\cong 1 - \frac{n_{e0}}{2\gamma_{eff} n_{crit}}. \tag{8.132c}$$

Here n_{e0} is the quiescent equilibrium electron density in the plasma (which we usually take to be equal to the average charge state times the background ion density, n_i, and may differ from the local electron density n_e if there is electron fluid movement as in, say, an electron plasma wave). γ_{eff} accounts for the cycle-averaged mass increase in the strong field which we discussed in Chapter 3 and can be reasonably taken to be given by Eq. (3.120). The refractive index is now a function of laser intensity,

$$\eta \cong 1 - \frac{1}{2} \frac{n_{e0}}{n_{crit}} \frac{1}{\sqrt{1 + a_0^2/2}}. \tag{8.133}$$

We can also explicitly write the refractive index in a way to acknowledge that the local electron density can now be altered by the ponderomotive force of the laser pulse, both longitudinally and radially, which we do by adding to the background electron density an electron density perturbation, $\delta n_e = n_e - n_{e0}$ (positive for electron density bunching or negative for electron cavitation):

$$\eta(\mathbf{x}) \cong 1 - \frac{(n_{e0} + \delta n_e(\mathbf{x}))}{2\gamma_{eff} n_{crit}}. \tag{8.134}$$

The effects of driving electron plasma waves by the laser pulse are manifested in this term. Generically, the consequences of these two effects, the intensity dependence of the refractive index and the ponderomotive-driven motion of the electron density, are illustrated in Figure 8.25.

538 *Strong Field Interactions with Underdense Plasmas*

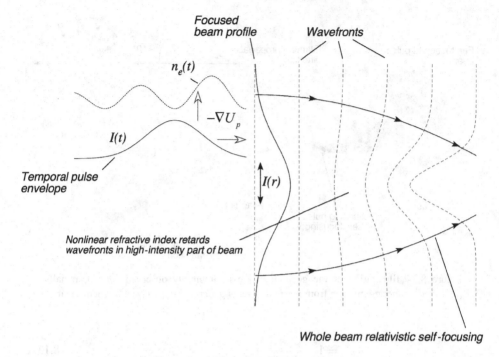

Figure 8.25 Schematic illustration of how a laser pulse in plasma can excite plasma waves by the rise in pulse intensity and can also undergo self-focusing in the plasma by relativistic mass increase of the plasma electrons in the most intense part of the beam.

8.4.2 *Relativistic Self-Focusing in the Weak Focus, Quasi-CW Regime*

Let us first look at the effects of the intensity dependence of the refractive index, specifically examining what effect this has as a consequence of the radial intensity distribution of a focused intense laser pulse. As Figure 8.25 illustrates, because the refractive index in the plasma in regions of higher intensity is higher, as indicated by Eq. (8.133), in the center of a focused beam phase fronts are retarded. This leads to a positive radius of curvature of phase fronts in the beam and causes it to focus. We encountered this already when we looked at propagation of moderate intensity pulses in air, an effect which led to the formation of a moving focus plasma filament. In that case, the intensity dependence to the refractive index arose from the nonlinearity in the polarizability of the bound electrons in the air's molecules; in that lower-intensity regime, plasma formation served to defocus the beam by refraction from higher electron density on axis. Now we see that at higher, relativistic intensity, specifically when a_0 approaches and exceeds 1, the relativistic mass increase of the electrons leads to the opposite effect (Sprangle *et al.* 1987; Sprangle *et al.* 1992; Brandi *et al.* 1993; Ritchie 1994).

We will need to look at the consequences of this relativistic self-focusing in various regimes (such as tight or weakly focused pulses, and pulses that are temporally long or short). Let us start with the most simple case, when the laser pulse is weakly focused into

8.4 Propagation in Underdense Plasma in the Relativistic Regime

the plasma and the pulse duration is long enough that we can essentially think of it as a quasi-CW wave. Practically this means the situation is characterized by two conditions. That the pulse is quasi-CW means that its duration is much longer than the oscillation period of any electron plasma wave. This condition is not yet rigorously justified yet, but later in this chapter it will become apparent why we require this condition. Appendix B describes the physics of basic linear longitudinal electron-density plasma waves which exhibit oscillations at the usual electron plasma frequency, Eq. (8.1), (or at least they do when they are linear and small amplitude). Since, for example, an inverse plasma frequency in a plasma of electron density $\sim 10^{19}$ cm^{-3} is $\omega_{pe}^{-1} \sim 5$ fs, this condition for "CW" pulses is often fulfilled by pulses of a few hundred femtoseconds or longer.

The condition that the pulse is weakly focused, is constrained by similar physics, namely that we assume the focal spot is much larger than any radial plasma electron-density oscillations, mandating that the focal spot radius, w, is much larger than an electron plasma oscillation wavelength, λ_{pe}. For electron plasma waves traveling at a phase velocity near c this means that $w \gg 2\pi c/\omega_{pe}$. (We shall always assume w represents the $1/e^2$ radius of a Gaussian intensity profile, and w_0 is that radius at best focus in vacuum.) Again, in a plasma of $n_{e0} \approx 10^{19}$ cm^{-3}, the spot size would need to be much larger than ~ 10 μm, (easily satisfied by focal geometries of $>\sim$f/20). Finally, since in the following analysis we assume that the electron density remains constant at $\sim n_{e0}$, we will also require that the focused pulse does not expel a significant amount of the plasma electrons radially because of the ponderomotive force developed by a radial gradient in the focused laser intensity. This is essentially a constraint on peak intensity, and the condition can be roughly stated by saying that the ponderomotive force must be much less than any electrostatic restoring force that electrons would feel if they were ponderomotively expelled from a focal volume. Roughly speaking, for cylindrical electron expulsion this mandates that

$$\nabla U_p \approx \frac{U_p}{w} \ll e^2 n_e w. \tag{8.135}$$

In an $n_{e0} \approx 10^{19}$ cm^{-3} plasma with a focal spot radius of ~ 100 μm that would require an intensity of under about 10^{17} W/cm^2, which is the intensity ($a_0 \sim 0.3$) at which we expect relativistic self-focusing to begin playing an important role in the pulse propagation.

Experimentally, these conditions are fulfilled when a near-IR pulse with about a joule of energy and a duration of a half picosecond or longer is weakly focused at >f/20 into a gas cell or jet of close to atmospheric density, assuming that refractive defocusing has been circumvented by appropriate gas geometry or preionization of the gas. (We will address the propagation of a pulse in other relevant experimental conditions in later sections.) Within this experimental regime, the electron motion is essentially dominated by the quiver motion in the field, and the refractive index governing the pulse propagation is fairly accurately given by Eq. (8.133).

Recall that we describe the laser field as continuous, with

$$a(t) = a_0 \sin \omega t, \tag{8.136}$$

and recall that this normalized field parameter is related to the other observable parameters like intensity and electric field amplitude by

$$I = \frac{c}{8\pi}E_0^2 = \frac{\omega^2}{8\pi c}A_0^2 = \frac{\omega^2 m^2 c^3}{8\pi e^2}a_0^2. \qquad (8.137)$$

If we are in a weakly relativistic intensity field, we can Taylor-expand about a_0 and say that the refractive index is

$$\eta_R \cong 1 - \frac{n_{e0}}{2n_{crit}}\left(1 - \frac{a_0^2}{4}\right)$$

$$\cong \eta_0 + \frac{n_{e0}}{8n_{crit}}a_0^2$$

$$\cong \eta_0 + \frac{\omega_{pe}^2}{8\omega^2}a_0^2. \qquad (8.138)$$

It takes on the classic, linearly scaling intensity-dependent refractive index of the form we have considered previously,

$$\eta = \eta_0 + \gamma_{NL}I. \qquad (8.139)$$

The relativistic nonlinear refractive index intensity coefficient, then, can be read from Eq. (8.138) using relation (8.137) giving,

$$\gamma_{NL}^{(rel)} = \frac{\pi e^2 \omega_{pe}^2}{m^2 c^3 \omega^4}. \qquad (8.140)$$

We know from our previous discussions that when the medium exhibits a nonlinear refractive index of the form of Eq. (8.139) we can couch our beam envelope size propagation equation in terms of a beam critical power,

$$\frac{d^2 w}{dz^2} = \frac{4}{k^2 w^3}\left(1 - \frac{P}{P_c}\right). \qquad (8.141)$$

Recalling the critical self-focusing power from Eq. (8.120),

$$P_c = \frac{2\pi c^2}{\omega^2 \eta_0 \gamma_{NL}} \qquad (8.142)$$

the relativistic critical power would be, using relation (8.140),

$$P_c^{(rel)} = \frac{2m^2 c^5}{e^2 \eta_0}\frac{\omega^2}{\omega_{pe}^2}. \qquad (8.143)$$

This can be related to the peak intensity and normalized vector potential in the center of the beam, a_{00}, by

$$P = \frac{\pi w^2}{2}I_0$$

$$= \frac{\omega^2 m^2 c^3}{8\pi e^2}\left(\frac{\pi w^2}{2}\right)a_{00}^2. \qquad (8.144)$$

8.4 Propagation in Underdense Plasma in the Relativistic Regime

Since the background refractive index of most underdense plasmas is close to 1 for laser light, this expression for the relativistic P_{crit} can be written in a form often reproduced in the high-intensity laser-plasma literature, $P_{crit} = 17\, GW \times (\omega/\omega_{pe})^2 = 17\, GW \times (n_{crit}/n_e)$. In other words, near-IR laser pulses (with $n_{crit} \sim 10^{21}$ cm^{-3}) in plasma at $n_{e0} \sim 10^{19}$ cm^{-3} will undergo relativistic self-focusing with peak power above a couple of terawatts.

While Eqs. (8.140) and (8.141) were helpful in giving us a form for the relativistic critical power, they are not very useful in calculating the real propagation dynamics of an intense beam. Employing Eq. (8.141) to find the evolution of the beam size will ultimately be inadequate; it, in fact, predicts that the beam size collapses to zero if it propagates far enough in the plasma. This unphysical development, of course, is not realized in any real experiment because this model was based on a Taylor expansion and the fact that $a_0^2 \ll 1$, a situation which obviously breaks down when the beam collapses to a very small beam size. (This equation also fails us eventually because it was based on the assumption that the beam shape remained Gaussian, also incorrect under extreme self-focusing but a fact we will ignore to make progress.) To get a slightly better picture, let us revisit the envelope propagation equation retaining the original form of the refractive index at relativistic a_0. Looking back at the derivation for this equation we see that the refractive term in the second-order ODE for the beam radius was of the form

$$\frac{d^2 r_B}{dz^2} = \left.\frac{d\eta_R(r)}{dr}\right|_{r=r_B}. \qquad (8.145)$$

The refractive index in this weak-focus, quasi-CW regime is written

$$\eta_R \cong 1 - \frac{n_{e0}}{2 n_{crit}} \frac{1}{\sqrt{1 + a_0(r)^2/2}}, \qquad (8.146)$$

and, as usual, we assume a Gaussian focal profile,

$$a_0(r)^2 = a_{00}(w)^2 \exp\left[-2r^2/w(z)^2\right]. \qquad (8.147)$$

Now the nonlinear refractive term governing the envelope equation takes the form

$$\frac{d\eta_R(r)}{dr} = -\frac{n_{e0}}{2 n_{crit}} \left[\frac{a_{00}(w)^2 r \exp[-r^2/w^2]}{w^2 \left(1 + a_{00}(w)^2 \exp[-r^2/w^2]/2\right)^{3/2}}\right]. \qquad (8.148)$$

We need to evaluate this at some appropriate radius. Evaluating it at the $1/e^2$ radius might be the obvious choice at first, but out at that radius in the Gaussian the radial intensity variation is slow compared to the inner radial parts of the focal profile. Therefore, using this radius will underestimate the actual rate of self-focusing. A slightly better estimate would be the half-width at half maximum (HWHM) radius, which describes a point on the beam with a faster intensity gradient:

$$r_{HWHM} = \sqrt{\ln 2/2}\, w. \qquad (8.149)$$

Using this radius (which is 58% of w) and our previous definition for P_c in Eq. (8.143) gives

$$\left.\frac{d\eta_R(r)}{dr}\right|_{r=r_{FWHM}} = -\frac{n_{e0}}{2n_{crit}} \left[\frac{\sqrt{\ln 2/2}\, a_{00}^2}{2w} \frac{1}{(1+a_{00}^2/4)^{3/2}}\right], \quad (8.150)$$

$$\frac{d\eta_R}{dr} = -4\sqrt{2\ln 2}\frac{c^2}{e^2}\frac{P}{P_c}\frac{1}{w^3}\frac{1}{(1+a_{00}^2/4)^{3/2}}. \quad (8.151)$$

So now a more accurate description of the envelope beam evolution during relativistic self-focusing is

$$\frac{d^2w}{dz^2} = \frac{4}{k^2 w^3} - \frac{4\sqrt{2\ln 2}}{k^2 w^3}\frac{P}{P_c}\frac{1}{(1+a_{00V}^2 w_0^2/4w^2)^{3/2}}, \quad (8.152)$$

where we have introduced the peak focused normalized vector potential amplitude on axis for the beam in vacuum (absent self-focusing) with a_{00V}. This equation has used the fact that the peak a_{00} at the beam center varies with beam size as $a_{00} = a_{00V}(w_0/w)$. (More rigorous derivations of this equation have additional logarithmic terms; see, for example, Esarey et al. (1997), but given the approximate nature of the entire model, this equation is adequate for illustrating the beam dynamics.) In the weak field limit this equation reduces to

$$\frac{d^2w}{dz^2} \cong \frac{4}{k^2 w^3}\left(1 - \sqrt{2\ln 2}\frac{P}{P_c}\right), \quad (8.153)$$

which is different by a factor of 1.17 from our previous result of Eq. (8.141). This is a result of the beam radius we chose, but is of no real consequence.

The kinds of beam envelope propagation predicted by Eq. (8.152) are illustrated in Figure 8.26. The focus of a 1 μm wavelength laser focused by an f/40 optic in vacuum is shown as a function of propagation distance by the long dashed line. If the vacuum peak $a_{00} = 1$ and plasma density is increased to move the power of the laser to three times the critical power, the solid black line is the predicted beam radius. Self-focusing causes the beam to collapse to a small spot size about a millimeter upstream of the vacuum focus. The beam then diverges from this tight focus. If the plasma density is increased by a factor of three to push the laser's power to nine times P_{crit}, a beam size envelope illustrated by the thick gray dashed line evolves. The self-focusing is more extreme, moving the focus further upstream by a millimeter, resulting in an even smaller minimum focal spot. The laser beam then undergoes further self-focusing and refocuses 8 mm downstream.

The propagation described by Eq. (8.152) follows a predictable trend. The beam undergoing self-focusing will have a radius that oscillates along the propagation axis, with period and amplitude dictated by the beam power and the focusing trajectory of the initial converging beam (set by the f/# of the focusing optic). An interesting way to view this propagation dynamic and derive further intuition is to follow the development of Esarey et al. (1997) and realize that Eq. (8.152) is actually in the form of an equation-of-motion of a particle moving in a potential. In this case the "particle position" is the

8.4 Propagation in Underdense Plasma in the Relativistic Regime

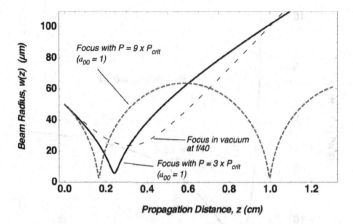

Figure 8.26 Solutions off the beam envelope equation for a 1 µm wavelength beam focused with an f/40 optic. The long dashed line is the focus in vacuum while the short dashed and solid lines show the effects that the relativistic nonlinear refractive index has on the propagation of the beam. In the case that the power is only three times the critical power, the beam self-focuses to a tighter spot in the plasma but then diverges, but if the plasma density is reduced so that the beam is now nine times P_{crit}, the self-focusing causes refocusing of the beam within a few millimeters of its first focal point.

beam size which is subject to an equation like Newton's law. In this view the propagation position, z, can be seen as "time" so that the beam-size envelope equation describes an acceleration of the beam-size quasi-particle. Developing this idea, the "potential" in which the beam-size quasi-particle is propagating is

$$g(w) \equiv - \int \frac{d^2 w}{dz^2} dw, \qquad (8.154)$$

which, when using Eq. (8.152), and ignoring the factor of 1.17 in front of P, gives for the beam size potential

$$g(w) = \frac{2}{k^2 w^2} + \frac{4}{k^2} \frac{P}{P_c} \frac{4}{a_{00V}^2 w_0^2} \frac{1}{\left(1 + a_{00V}^2 w_0^2 / 4w^2\right)^{1/2}} + const$$

$$= \frac{2}{k^2 w^2} - \frac{16}{a_{00V}^2 w_0^2} \frac{P}{P_c} \left(1 - \frac{1}{\sqrt{1 + a_{00V}^2 w_0^2 / 4w^2}} \right). \qquad (8.155)$$

Plots of a few possible beam size potentials are illustrated in Figure 8.27.

The initial "velocity" of the beam size particle is its initial convergence angle ($f/\# = dw/dz$) as it enters the plasma. If dw/dz is too negative (focal geometry is very tight), the beam will self-focus to a spot smaller than the vacuum focus, turn around in the potential and diverge. If the beam is weakly focused and dw/dz is not too negative, the beam will bounce back and forth in the confining potential $g(w)$ and refocus multiple times (at least until it exits the plasma). This behavior is akin to that which we saw in filamentation in

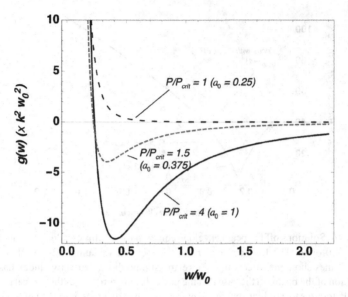

Figure 8.27 Plots of the beam size quasi-particle potential $g(w)$ for three values of the beam power and a_0.

which the tail of the femtosecond pulse could refocus downstream and extend the length of the plasma filament.

We can use this picture and $g(w)$ to ascertain how tightly focused a beam can be in the plasma and still refocus, hence self-channel. The trick is to use "conservation of energy" in this quasi-potential. The condition for self-channeling (repetitive refocusing) is the same as saying that the beam-size "particle" is confined in the potential so that at the entry to the plasma, the sum of kinetic and potential energy is less than zero:

$$\frac{1}{2}\left(\frac{dw}{dz}\right)_i^2 + g(w_i) < 0. \qquad (8.156)$$

The focused beam at the entry point to the plasma (at propagation point z_i) has a trajectory of a Gaussian beam in vacuum which is

$$\left.\frac{dw}{dz}\right|_i = \frac{w_0}{z_R}\left(\frac{z_i}{z_R}\right)\frac{1}{\sqrt{1+z_i^2/z_R^2}}. \qquad (8.157)$$

If the beam at this entry point is well away from the vacuum focus point, it is reasonable to say that

$$\left.\frac{dw}{dz}\right|_i \cong \frac{w_0}{z_0} = \frac{1}{2f_\#}, \qquad (8.158)$$

so the condition for self-channeling by the beam is

$$f_\# > \frac{1}{\sqrt{-2g(w_i)}}. \qquad (8.159)$$

8.4 Propagation in Underdense Plasma in the Relativistic Regime

As we have discussed previously, a more experimentally realistic situation is for a flattop beam to be focused so we can say that $f_\#^{(eff)} \approx 2 f_\#$ and a more useful estimate for having a self-channeling beam in the plasma is

$$f_\#^{(eff)} \gtrsim \sqrt{\frac{-2}{g(w_i)}}. \tag{8.160}$$

Putting the beam propagation potential in terms of P/P_{crit},

$$\begin{aligned}
g(w) &= \frac{2}{k^2 w^2} - \frac{1}{4}\frac{\omega_{pe}^2}{\omega^2}\left(1 - \frac{1}{\sqrt{1 + 16(P/P_c)(k_{pe}w)^{-2}}}\right) \\
&= \frac{1}{2\pi^2}\left(\frac{\lambda}{w}\right)^2 - \frac{1}{4}\frac{n_e}{n_{crit}}\left(1 - \frac{1}{\sqrt{1 + (4/\pi^2)(P/P_c)(n_e/n_{crit})(\lambda/w)^2}}\right),
\end{aligned} \tag{8.161}$$

it is fairly easy to estimate the conditions in which relativistic self-channeling might occur.

As a concrete example, consider a 1 μm wavelength petawatt power pulse focused into a plasma of electron density of 10^{18} cm^{-3} to a spot radius ~100 μm at the plasma entrance. Here $P/P_{crit} \approx 60$ and Eq. (8.161) says that the potential $g(w_i) \approx -0.00011$, which means that a flattop focused beam would have to be focused by an f/# at least >135. Tighter f/#s can produce channeling if the electron density is higher. This stringent condition on how weak the focal geometry must be to make the pulse self-channel is one of the reasons that it is notoriously challenging to observe this relativistic self-channeling experimentally. For example, if the pulse beam does not focus to a near-diffraction limited spot, that f/135 focus would not produce a small enough spot for the beam to undergo relativistic refocusing and self-channeling. Finding experimental conditions in which dynamics like that predicted in Figure 8.26 manifest is one of the reasons that predictions of self-channeling are hard to realize in laboratory experiments. The consequences of a finite pulse driving plasma waves is another reason, as we shall investigate in the next section.

8.4.3 Relativistic Self-Focusing and Propagation of Short Pulses in the Weak Focus Regime

If the laser pulse is short enough that the quasi-CW condition does not apply, we are faced with a more challenging task. Now we must assess how the longitudinal ponderomotive force of the finite rise time of the intense pulse envelope moves the plasma and affects the pulse's propagation. To simplify the task for the time being, we will still assume that the beam is in the weak focus limit described earlier. As we mentioned, this is not particularly hard to achieve experimentally, typically requiring the focal diameter to be at least a few tens of micrometers as long as the intensity is low enough that radial electron expulsion is also a small effect. Under these conditions it is possible to treat the problem of dealing with plasma motion in 1D; we only need be concerned with the longitudinal motion of the plasma. We also assume that the heavy ions are stationary on the time scale of the laser pulse, typically an excellent approximation. We treat the ions as a uniform positively charged background that provides quasi-neutrality to the electron plasma fluid if it is quiescent in equilibrium and a restoring force to any electron plasma waves that get excited. (As an aside, this approximation also demands that the ion quiver velocity is negligible, which is certainly true for pulses at intensity below about 10^{22} W/cm^2.)

A qualitative picture of what we expect to happen in this short pulse regime is illustrated in Figure 8.28. In 1D we ignore any lateral (radial) motion of the plasma, but in the axis of propagation, the pulse can be thought of as a disturbance with longitudinal force from the ponderomotive force of a rising temporal envelope. The laser pulse will travel through the plasma at the group velocity. The gradient in ponderomotive energy, U_p, associated with the pulse envelope intensity rise produces a ponderomotive force on the electrons that is forward on the rising edge and backward on the falling edge of the pulse. The rising edge of the laser intensity envelope will push the plasma electrons like a snowplow and cause a rise in electron density. The extent and magnitude of the plasma density rise will depend on the pulse intensity but will also depend on the ion density because the stationary ions produce an electrostatic electric field that wants to pull the electron density back to equilibrium. The combination of the reverse direction of the ponderomotive force on the falling edge of the laser pulse and the restoring force of the positively charged ions causes the electron density increase to push back, overshoot and yield an electron-density trough at the back edge of the pulse. This sets up longitudinal plasma oscillations and an electron plasma wave behind the pulse. Appendix B discusses the physics of such plasma waves. These oscillations persist after the pulse has passed, producing oscillating electron-density peaks and troughs as well as longitudinal electric fields. This wave travels with a phase velocity along behind the laser at the speed of the pulse's group velocity, which is just below c.

Later in this chapter we will discuss how the electric fields arising from the plasma wave produced by passage of the laser pulse can accelerate high-energy electrons. For the moment, however, let us assess the impact this could have on the propagation of the laser. We know that the laser's increasing intensity leads to a relativistic increase in the refractive index. However, we now see that the finite rise time of a pulse can lead to a plasma density increase, with longitudinal spatial scale of the order of an electron plasma wave wavelength, $\lambda_{pe} = 2\pi v_{\varphi e}/\omega_{pe} \approx 2\pi c/\omega_{pe}$, where the second expression is true since the electron wave phase velocity, $v_{\varphi e}$, is given by the group velocity of the passing laser pulse, which is close to c in an underdense plasma. The increase in electron density leads to a *decrease* in refractive index, offsetting the increase from the relativistic correction to the electron mass. Recall Eq. (8.134) for the refractive index. This will frustrate relativistic self-focusing. It can be expected that pulses that are very short, meaning of the same time scale or faster than the plasma oscillation period, $2\pi/\omega_{pe}$, will not undergo significant relativistic self-focusing, and longer pulses will have a rising edge that will not self-focus. This second situation will lead to longer pulses at relativistic intensity, with diffraction eating away at the leading edge of the pulse.

8.4 Propagation in Underdense Plasma in the Relativistic Regime

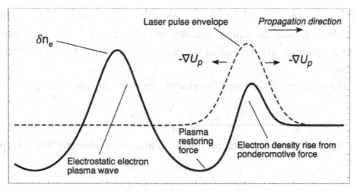

Figure 8.28 Schematic illustration of how the longitudinal ponderomotive force of a pulse propagating through a plasma can excite 1D longitudinal plasma waves as it passes through the plasma. The rising intensity drives a rise in electron density and the restoring force of the stationary ions, augmented by the opposite-direction ponderomotive force of the falling pulse intensity, creating a residual electron density oscillation in the pulse's wake.

8.4.4 The Electron Plasma Self-Consistent Response and the Quasi-Static Approximation in 1D

To describe propagation of short pulses in the plasma it is now apparent that it is not sufficient to consider only the relativistic mass correction. We must include the effects of the self-consistent response of the plasma to the passing laser pulse. Because we consider high-intensity pulses, it is not sufficient to treat the waves as we have in Appendix B (and as they are treated in most plasma physics textbooks); we must consider the nonlinear motion of the plasma if the electron-density oscillation amplitude becomes large. Nonlinear plasma waves were first considered in a classic paper by Akhiezer and Polovin (1956). The best way to handle these challenges and get a quantitative handle on the effect pictured in Figure 8.28 is to consider the plasma motion within what is known as the quasi-static approximation, or QSA. This model has been extensively developed by a number of authors, most substantially by Sprangle, Esarey and collaborators as well as by Feit and coworkers (Sprangle *et al.* 1990a; Sprangle *et al.* 1990b; Sprangle *et al.* 1992; Feit *et al.* 1996).

To tackle this problem, which in 1D can be described by four variables, we are best to work with the fields in terms of their potentials, $\Phi(\mathbf{x}, t)$ and $\mathbf{A}(\mathbf{x}, t)$, as well as the two variables describing the electron fluid, the local electron density $n_e(\mathbf{x}, t)$ and the electron fluid velocity, $\mathbf{u}(\mathbf{x}, t)$. (Strictly speaking, since we are working in 1D, the only velocity component we ultimately need is the z-component but we will retain the vector notation for both electron fluid velocity and vector potential for the moment because the laser field is transverse to z, so the x-component must still be retained.) We require four equations to solve for these four variables. We can employ the wave equation for the field, Poisson's equation, the electron fluid continuity equation and the electron fluid momentum equation. (These last two equations are derived in the nonrelativistic limit from the kinetic equation in Appendix B).

Recalling the relationship between the electric field and the potentials,

$$\mathbf{E} = -\frac{1}{c}\frac{\partial \mathbf{A}}{\partial t} - \nabla \Phi, \tag{8.162}$$

the wave equation for the vector potential in a plasma comes directly from Ampere's law,

$$\nabla \times \mathbf{B} = \frac{4\pi \mathbf{J}}{c} + \frac{1}{c}\frac{\partial \mathbf{E}}{\partial t}. \tag{8.163}$$

Since the heavy ions are effectively immobile on the oscillation time scale of the laser, the current density arises only from electron motion It can be written in terms of the electron fluid velocity, $\mathbf{J} = -e n_e \mathbf{u}$, so Ampere's law is

$$\nabla \times \nabla \times \mathbf{A} = -\frac{4\pi e n_e}{c}\mathbf{u} - \frac{1}{c^2}\frac{\partial^2 \mathbf{A}}{\partial t^2} - \frac{1}{c}\frac{\partial}{\partial t}\nabla \Phi, \tag{8.164}$$

or when using the vector identity $\nabla \times \nabla \times = \nabla(\nabla \cdot) - \nabla^2$ and the Coulomb gauge condition $\nabla \cdot \mathbf{A} = 0$, this becomes

$$\nabla^2 \mathbf{A} - \frac{1}{c^2}\frac{\partial^2 \mathbf{A}}{\partial t^2} = \frac{4\pi e n_e}{c}\mathbf{u} + \frac{1}{c}\frac{\partial}{\partial t}\nabla \Phi. \tag{8.165}$$

Poisson's equation is

$$\nabla^2 \Phi = 4\pi e \left(n_e - \bar{Z} n_i \right), \tag{8.166}$$

which can be written in terms of the equilibrium electron density $n_{e0} = \bar{Z} n_i$ to become

$$\nabla^2 \Phi = 4\pi e (n_e - n_{e0}). \tag{8.167}$$

To describe the motion of the plasma subject to the laser fields and the electrostatic longitudinal fields, we will work with the electron fluid equations generalized for relativistic motion of the electron fluid (see Appendix B). The continuity equation for the electron fluid is

$$\frac{\partial n_e}{\partial t} + \nabla \cdot (n_e \mathbf{u}) = 0, \tag{8.168}$$

and the relativistic fluid momentum equation is

$$\frac{\partial}{\partial t}(\gamma m \mathbf{u}) + \mathbf{u} \cdot \nabla (\gamma m \mathbf{u}) = e \nabla \Phi + \frac{e}{c}\frac{\partial \mathbf{A}}{\partial t} - \frac{e}{c}\mathbf{u} \times (\nabla \times \mathbf{A}). \tag{8.169}$$

Here γ is the relativistic factor for an electron fluid element $\gamma = (1 - u^2/c^2)^{-1/2}$. Following the bulk of the literature on this topic, we will work with normalized, dimensionless quantities:

$$\mathbf{a} = e\mathbf{A}/mc^2, \tag{8.170a}$$

$$\phi = e\Phi/mc^2, \tag{8.170b}$$

8.4 Propagation in Underdense Plasma in the Relativistic Regime

$$\boldsymbol{\beta}_u = \mathbf{u}/c, \tag{8.170c}$$

$$\tilde{\mathbf{p}} = \gamma \mathbf{u}/c = \gamma \boldsymbol{\beta}_u. \tag{8.170d}$$

In terms of these normalized quantities and the electron plasma frequency, the four relevant equations are

$$\nabla^2 \mathbf{a} - \frac{1}{c^2} \frac{\partial^2 \mathbf{a}}{\partial t^2} = \frac{\omega_{pe}^2}{c^2} \frac{n_e}{n_{e0}} \boldsymbol{\beta}_u + \frac{1}{c} \frac{\partial}{\partial t} \nabla \phi, \tag{8.171}$$

$$\nabla^2 \phi = \frac{\omega_{pe}^2}{c^2} \left(\frac{n_e}{n_{e0}} - 1 \right), \tag{8.172}$$

$$\frac{1}{c} \frac{\partial n_e}{\partial t} + \nabla \cdot (n_e \boldsymbol{\beta}_u) = 0, \tag{8.173}$$

$$\frac{1}{c} \frac{\partial}{\partial t} (\gamma \boldsymbol{\beta}_u) + \boldsymbol{\beta}_u \cdot \nabla (\gamma \boldsymbol{\beta}_u) = \nabla \phi + \frac{1}{c} \frac{\partial \mathbf{a}}{\partial t} - \boldsymbol{\beta}_u \times (\nabla \times \mathbf{a}). \tag{8.174}$$

Since we consider the situation in which the plasma motion is in 1D, we need not worry about magnetic fields generated by plasma motion; therefore, the vector potential is only that of the laser field, which we will assume to be linearly polarized along x in the usual way: $\mathbf{a} = a_0 \sin(kz - \omega t) \hat{\mathbf{x}}$. At high intensity, though we know there is plasma motion in the z-direction, the restoring force of the ions in the plasma inhibits the longitudinal drift that a free electron in such a pulse would exhibit. We can then make a significant approximation to the wave equation describing propagation of the laser. The electron fluid motion is dominated by the *transverse* oscillation motion of the electrons in Eq. (8.171). With this consideration, to a good approximation, we can ignore the second term on the right-hand side of (8.171) (though we must retain the electrostatic potential in Poisson's equation to describe the longitudinal motion of the plasma), and note the first term on the right can be rewritten

$$\boldsymbol{\beta}_u \cong \boldsymbol{\beta}_\perp. \tag{8.175}$$

By conservation of transverse canonical momentum and the fact that canonical momentum, $\mathbf{P}$, must be zero before the arrival of the laser pulse in the quiescent plasma, we write

$$\mathbf{P} = \mathbf{p} - \frac{e}{c} \mathbf{A} = 0, \tag{8.176}$$

and develop that transverse first term,

$$\gamma m \dot{\mathbf{u}}_\perp = \frac{e}{c} \mathbf{A}_\perp, \tag{8.177}$$

$$\gamma \boldsymbol{\beta}_\perp = \mathbf{a}. \tag{8.178}$$

This gives for the wave equation describing propagation of the laser field, breaking out the transverse and longitudinal terms of the ∇^2 operator,

$$\nabla_\perp^2 \mathbf{a} + \frac{\partial^2 \mathbf{a}}{\partial z^2} - \frac{1}{c^2} \frac{\partial^2 \mathbf{a}}{\partial t^2} = \frac{\omega_{pe}^2}{c^2} \frac{n_e}{n_{e0}} \frac{\mathbf{a}}{\gamma}. \tag{8.179}$$

The remaining equations need only describe the longitudinal z motion of the plasma, so Poisson's equation and the continuity equation can be written in simplified form as

$$\frac{\partial^2 \phi}{\partial z^2} = \frac{\omega_{pe}^2}{c^2}\left(\frac{n_e}{n_{e0}} - 1\right), \qquad (8.180)$$

$$\frac{1}{c}\frac{\partial n_e}{\partial t} + \frac{\partial}{\partial z}(n_e \beta_z) = 0. \qquad (8.181)$$

To write the z-component of the electron fluid momentum equation, we note that since the laser field is transverse to the plasma fluid motion in 1D, there is no slowly varying magnetic field (only the longitudinal restoring force of the electrostatic electric field), so there is no time-varying longitudinal component of $\mathbf{a}$:

$$\left(\frac{\partial \mathbf{a}}{\partial t}\right)_z = 0. \qquad (8.182)$$

Again, using the fact that the electron fluid velocity will be dominated by the transverse laser-driven oscillation, the third term on the right-hand side of the momentum equation can be approximated:

$$(\boldsymbol{\beta}_u \times \nabla \times \mathbf{a})_z \cong \beta_\perp \frac{\partial a}{\partial z}$$

$$= \frac{a}{\gamma}\frac{\partial a}{\partial z} = \frac{1}{2\gamma}\frac{\partial}{\partial z}(a^2). \qquad (8.183)$$

Eq. (8.174) becomes

$$\frac{1}{c}\frac{\partial}{\partial t}(\gamma \beta_z) + \beta_z \frac{\partial}{\partial z}(\gamma \beta_z) = \frac{\partial \phi}{\partial z} - \frac{1}{2\gamma}\frac{\partial}{\partial z}a^2. \qquad (8.184)$$

The relativistic factor for the plasma fluid motion is

$$\gamma^2 = \left(1 - \beta_\perp^2 - \beta_z^2\right)^{-1}$$
$$= 1 + \gamma^2 \beta_\perp^2 + \gamma^2 \beta_z^2, \qquad (8.185)$$

and since from Eq. (8.178) $\beta_\perp = a/\gamma$, this can be recast as

$$\gamma^2 \left(1 - \beta_z^2\right) = 1 + a^2 \qquad (8.186)$$

$$\gamma = \sqrt{\frac{1 + a^2}{1 - \beta_z^2}}. \qquad (8.187)$$

More progress can be made by reformulating these equations in a coordinate frame that moves along with laser pulse (and hence moves with the phase velocity of the incited plasma wave). This frame will need to move at the group velocity of the laser pulse in the plasma, given by Eq. (8.23). We introduce a new set of coordinates describing this moving frame

$$\zeta = z - v_g t, \qquad (8.188a)$$
$$\tau = t. \qquad (8.188b)$$

8.4 Propagation in Underdense Plasma in the Relativistic Regime

The ζ describes the position in the frame moving along z at the group velocity and consequently maps out the pulse-intensity envelope, the fast oscillation of the laser field and the plasma variables in the region inside and behind the laser pulse. The τ describes the slow variation of any of these parameters in time as the frame moves along. (Please do not confuse the use of τ in this context with the proper time introduced in Chapter 3, which we denoted with the same variable. I have chosen the variable τ for this derivation to mirror the majority of the literature on the QSA.) The derivatives are

$$\frac{\partial}{\partial z} = \frac{\partial}{\partial \zeta}; \quad \frac{\partial^2}{\partial z^2} = \frac{\partial^2}{\partial \zeta^2};$$

$$\frac{\partial}{\partial t} = \frac{\partial}{\partial \tau} - v_g \frac{\partial}{\partial \zeta}; \quad \frac{\partial^2}{\partial t^2} = \frac{\partial^2}{\partial \tau^2} + v_g^2 \frac{\partial^2}{\partial \zeta^2} - 2 v_g \frac{\partial^2}{\partial \tau \partial \zeta}. \quad (8.189)$$

Now our four equations as reformulated in the comoving frame are

$$\nabla_\perp^2 \mathbf{a} + \frac{1}{\gamma_g^2} \frac{\partial^2 \mathbf{a}}{\partial \zeta^2} + 2 \frac{\beta_g}{c} \frac{\partial^2 \mathbf{a}}{\partial \tau \partial \zeta} - \frac{1}{c^2} \frac{\partial^2 \mathbf{a}}{\partial \tau^2} = \frac{\omega_{pe}^2}{c^2} \frac{n_e}{n_{e0}} \frac{\mathbf{a}}{\gamma}, \quad (8.190)$$

$$\frac{\partial^2 \phi}{\partial \zeta^2} = \frac{\omega_{pe}^2}{c^2} \left(\frac{n_e}{n_{e0}} - 1 \right), \quad (8.191)$$

$$\frac{1}{c} \frac{\partial n_e}{\partial \tau} + \frac{\partial}{\partial \zeta} \left(n_e (\beta_z - \beta_g) \right) = 0, \quad (8.192)$$

and the momentum fluid equation can be simplified

$$\frac{1}{c} \frac{\partial}{\partial \tau} (\gamma \beta_z) - \beta_g \frac{\partial}{\partial \zeta} (\gamma \beta_z) + \beta_z^2 \frac{\partial \gamma}{\partial \zeta} + \beta_z \gamma \frac{\partial \beta_z}{\partial \zeta} = \frac{\partial \phi}{\partial \zeta} - \frac{1}{2\gamma} \frac{\partial}{\partial \zeta} \left(\gamma^2 \left(1 - \beta_z^2\right) - 1 \right) \quad (8.193)$$

$$\frac{1}{c} \frac{\partial}{\partial \tau} (\gamma \beta_z) - \frac{\partial \phi}{\partial \zeta} + (1 - \beta_g \beta_z) \frac{\partial \gamma}{\partial \zeta} - \beta_g \gamma \frac{\partial \beta_z}{\partial \zeta} = 0 \quad (8.194)$$

$$\frac{1}{c} \frac{\partial}{\partial \tau} (\gamma \beta_z) + \frac{\partial}{\partial \zeta} \left[\gamma \left(1 - \beta_g \beta_z\right) - \phi \right] = 0. \quad (8.195)$$

These four equations are, within the assumptions already stated, a complete, nonlinear description of the plasma in one dimension.

Further analytic progress is stymied without additional approximation. A powerful simplification can be made now, called the quasi-static approximation or QSA. The trick is to look at how quickly the laser envelope varies with τ. Looking at the laser wave equation, Eq. (8.190), ignore the effects of diffraction for the moment and assume that the group velocity is close to c. That means we can ignore the first two terms. The fourth term on the left will vary more slowly than the third because the ζ derivative in the third term varies rapidly as k. So, noting that the oscillating laser field could be written $\mathbf{a} \approx \mathbf{a}_{env} \, e^{-ik\zeta}$, the wave equation describing the slow variation of the field amplitude looks something like

$$2i \frac{\omega}{c^2} \frac{\partial a_{env}}{\partial \tau} \approx \frac{\omega_{pe}^2}{c^2} \frac{n_e}{n_{e0}} \frac{a_{env}}{\gamma}. \quad (8.196)$$

This implies that the time scale on which the amplitude of the field changes as it propagates through the plasma is something like

$$\tau_{env} \approx 2\gamma \frac{\omega}{\omega_{pe}^2} \frac{n_{e0}}{n_e}. \tag{8.197}$$

Since γ is realistically $\sim 1\text{--}10$ for most laser intensities, the amplitude will change on a time scale $\sim (2-20) \times \omega_{pe}^{-1} (\omega/\omega_{pe})$. In underdense plasmas of most experiments, $\omega/\omega_{pe} \gg 1$, so the amplitude of the laser field varies slowly on a time scale of the plasma oscillation period, which is the time scale the plasma density will vary in the ζ frame. A similar conclusion is reached if diffraction is reconsidered. Diffraction and expansion of the beam size will cause variation of the laser field amplitude, but it is too slow, on a time scale of $z_R/c (\sim 10$ ps$)$. So, in any event it is likely that the amplitude of the laser in the comoving frame will only change on a many-picosecond time scale, considerably longer than the tens or hundreds of femtoseconds time scale we must worry about in looking at the plasma motion in the moving frame or the time scale of the laser-pulse envelope itself. We can therefore set all derivatives with respect to τ equal to zero. This is the essence of the QSA.

The two fluid equations, Eqs. (8.195) and (8.195), take on a beautiful, simple form:

$$\frac{\partial}{\partial \zeta} \left[n_e (\beta_z - \beta_g) \right] = 0, \tag{8.198}$$

$$\frac{\partial}{\partial \zeta} \left[\gamma (1 - \beta_g \beta_z) - \phi \right] = 0, \tag{8.199}$$

and can be trivially integrated

$$n_e(\zeta) (\beta_z(\zeta) - \beta_g) = n_e (\beta_z - \beta_g)|_{\zeta=-\infty}, \tag{8.200}$$

$$\gamma(\zeta) (1 - \beta_g \beta_z(\zeta)) - \phi = (\gamma (1 - \beta_g \beta_z) - \phi)|_{\zeta=-\infty}. \tag{8.201}$$

The integration constants are evaluated well before the arrival of the laser pulse, which is presumed to be in equilibrium, where $n_e = n_{e0}$, there is zero plasma motion, the electrostatic potential is zero and $\gamma = 1$. The differential equations for the fluid variables have been reduced to algebraic equations:

$$n_e (\beta_g - \beta_z) = n_{e0} \beta_g, \tag{8.202}$$

$$\gamma (1 - \beta_g \beta_z) - \phi = 1. \tag{8.203}$$

In most plasmas encountered in real experiments, the group velocity is close enough to the speed of light that it is an excellent approximation to set $\beta_g = 1$ and $\gamma_g^{-2} = 0$. We end up with two elegant, easily solvable equations for the plasma fluid variables.

$$n_e (1 - \beta_z) = n_{e0} \tag{8.204}$$

$$\gamma (1 - \beta_z) - \phi = 1 \tag{8.205}$$

The QSA then leaves us with one, simple ordinary differential equation for the electrostatic potential as a function of z, which is very easily solved numerically for a given laser vector potential

8.4 Propagation in Underdense Plasma in the Relativistic Regime

$$\frac{\partial^2 \phi}{\partial \zeta^2} = \frac{\omega_{pe}^2}{2c^2}\left(\frac{1+a^2}{(1+\phi)^2} - 1\right) \tag{8.206}$$

and, with the help of the expression for γ in Eq. (8.187), three algebraic equations for the plasma state variables:

$$\frac{n_e}{n_{e0}} = \frac{2+a^2+2\phi+\phi^2}{2(1+\phi)^2} = \frac{1}{2} + \frac{1+a^2}{2(1+\phi)^2}, \tag{8.207}$$

$$\beta_z = \frac{a^2 - 2\phi - \phi^2}{2+a^2+2\phi+\phi^2}, \tag{8.208}$$

$$\gamma = \frac{2+a^2+2\phi+\phi^2}{2(1+\phi)}. \tag{8.209}$$

The effects of the plasma motion on the propagation of the laser can then be expressed by the wave equation for the light's vector potential (8.190) in terms of any of these plasma variables. It is conveniently expressed in terms of the electrostatic potential as

$$\nabla_\perp^2 \mathbf{a} + \frac{2}{c}\frac{\partial^2 \mathbf{a}}{\partial \tau \partial \zeta} - \frac{1}{c^2}\frac{\partial^2 \mathbf{a}}{\partial \tau^2} = \frac{\omega_{pe}^2}{c^2}\frac{\mathbf{a}}{(1+\phi)}. \tag{8.210}$$

The QSA result gives a nice bit of insight into how the electron plasma motion excited by the ponderomotive force of a rising laser pulse affects the laser's propagation. This approximation has reduced a coupled set of nonlinear PDEs to a single ODE and three algebraic equations. It is a tool that we can now use to examine the nonlinear response of a plasma to a short intense pulse and see how this plasma response will affect the laser pulse's focusing and its spatiotemporal intensity profile.

Perhaps the easiest way to discern how the plasma motion is affecting propagation is to think in terms of an effective refractive index. Thinking in terms of normal linear optics, recalling the wave equation for the light wave, say Eq. (6.11), the polarizability term on the right-hand side (in our medium the polarizability arises from electron motion) can be written $\omega^2(1-n_R^2)\mathbf{a}/c^2$. Inspection of the wave equation resulting from the QSA, Eq. (8.210) would suggest that we can think of the effective refractive index of the plasma being altered by electron fluid motion as

$$\eta_R(\zeta) = \sqrt{1 - \frac{\omega_{pe}^2}{\omega^2}\frac{1}{1+\phi(\zeta)}} \tag{8.211}$$

$$\cong 1 - \frac{\omega_{pe}^2}{2\omega^2}\frac{1}{1+\phi(\zeta)}. \tag{8.212}$$

What form this takes in the quasi-CW limit, when the pulse duration is much longer than a plasma wave period, or in other words the variation of the pulse in ζ is much slower than $\lambda_{pe} = 2\pi c/\omega_{pe}$, can be seen by setting the derivative term on the left side of Eq. (8.206) equal to zero. Then

$$1 + \phi \cong \sqrt{1+a^2}, \tag{8.213}$$

or under in terms of time-averaged vector potential is

$$1 + \phi \cong \sqrt{1 + a_0^2/2}, \tag{8.214}$$

which, when inserted into Eq. (8.212), recovers our familiar result for the relativistic correction to the refractive index,

$$\eta_R(\varsigma) \cong 1 - \frac{\omega_{pe}^2}{2\omega^2} \frac{1}{\sqrt{1 + a_0^2/2}}. \tag{8.215}$$

We also know from our previous qualitative discussion that the passage of the laser incites the formation of electron plasma waves. This can be seen explicitly if we consider the equation for the variation of electrostatic potential with z in the limit of small ϕ. Taylor expansion of the right-hand side yields

$$\frac{\partial^2 \phi}{\partial \varsigma^2} \cong \frac{\omega_{pe}^2}{2c^2} \left((1 + a^2)(1 - 2\phi) - 1\right). \tag{8.216}$$

If we ignore the product of a^2 and ϕ as a small quantity, this rearranges into the form of a driven harmonic oscillator, with driving term proportional to the laser pulse intensity:

$$\frac{\partial^2 \phi}{\partial \varsigma^2} + \frac{\omega_{pe}^2}{c^2} \phi \cong \frac{\omega_{pe}^2}{c^2} \frac{a^2}{2}. \tag{8.217}$$

This is a satisfying if not unexpected result. As we have already discussed, the ponderomotive force from the passage of the laser pulse acts like a snowplow in the plasma and excites electron plasma waves.

Numerical solution of the QSA equations, (8.206)–(8.209), for a given laser pulse-intensity envelope is quite instructive. Two such numerical calculations are illustrated in Figure 8.29. In the top, a rather modest-intensity pulse, with peak $a_0 = 0.7$ (corresponding to peak intensity $\sim 10^{18}$ W/cm^2 at Ti:sapphire wavelength) passing through plasma with electron density 10^{18} cm^{-3}, is illustrated in a frame as a function of ς. The 30 fs long pulse envelope and the resulting electron density are shown in the top frame and the longitudinal electron field is shown in the second panel. It is clear that the pulse has plowed the electron plasma up in density in the rising part of the pulse and the restoring force of the ion leaves an oscillating wave behind the pulse (to the left of the pulse in ς-space). In this case the wave is more or less linear, taking on a sinusoidal oscillation (though a small amount of nonlinear deviation can be barely discerned in the electron density and electric field behind the laser). Because we plot this wave in a frame moving with the group velocity of the laser pulse, we see that the incited wave has phase velocity equal to the speed of the moving frame. The wavelength of this wave closely matches that expected for a linear electron plasma wave with phase velocity near c, $\lambda_{pe} = 2\pi c/\omega_{pe}$. The fast oscillation seen in the electron density during the pulse is a consequence of the fast longitudinal motion impacted by the laser field through the $\mathbf{E} \times \mathbf{B}$ component of force along z of the EM wave, discussed at length in Chapter 3.

A more interesting situation is calculated in the bottom two panels of the figure, in which that 30 fs, 800 nm pulse is 20× higher peak intensity, with peak $a_0 = 3$. Now the

8.4 Propagation in Underdense Plasma in the Relativistic Regime

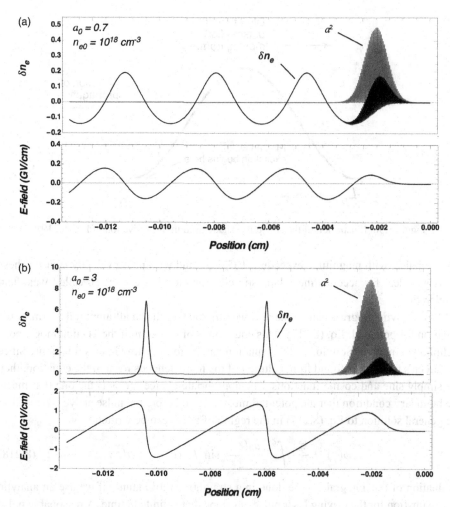

Figure 8.29 QSA solutions for the electron density and electric fields in and behind a 30 fs, 800 nm wavelength laser pulse in plasma of electron density 10^{18} cm^{-3} for weakly relativistic intensity, 10^{18} W/cm^2 (top, a) and strongly relativistic intensity, 2×10^{19} W/cm^2 (bottom, b). This illustrates how the lower-intensity pulse generates a nearly linear, sinusoidal electron plasma wave while the high-intensity pulse generates a highly nonlinear wave.

wave excited has considerably higher amplitude and has become highly nonlinear. The electron wave after passage of the pulse exhibits large spikes in density, nearly 7× the initial equilibrium density, with long density troughs between these spikes. The wavelength of this disturbance is considerably longer than the wavelength of the nearly linear oscillation. The longitudinal electric field exhibits a sawtooth profile, with nearly linear variation in the low-density trough parts of the wave as we would naturally expect from Gauss's law. Both of these calculations presage an effect that we will consider in some detail later in the chapter, the formation of very large longitudinal electric fields in the plasma behind the

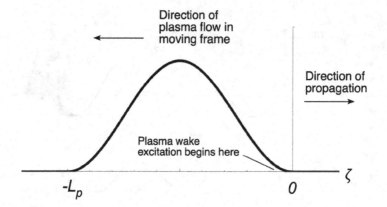

Figure 8.30 Illustration of the $\sin^2$ pulse shape used in the derivation of Eq. (8.219).

intense pulse, with magnitude exceeding 1 GV/cm and with phase velocity near c. These fields are ideal for accelerating relativistic electrons to high-energy, so-called wakefield acceleration.

More analytic progress can be made if we turn back to the small-amplitude form of the equation for potential, Eq. (8.217). This equation is of the form of the 1D inhomogeneous Helmholtz equation. Solution to this equation can be found with Green's functions. Since the Green's function is found from solution of the homogeneous form of the equation, they are simply sine and cosine functions. The appropriate choice for us is the one that meets the boundary condition that the potential must be zero before the pulse arrives at $z = 0$ so the general solution to Eq. (8.217) in the region of $z < 0$ can be written

$$\phi(\zeta) = \frac{k_{pe}}{2} \int_{-\infty}^{\zeta} \frac{a_0(\zeta')^2}{2} \sin\left[k_{pe}(\zeta - \zeta')\right] d\zeta'. \tag{8.218}$$

Evaluation of this integral can be done in a straightforward manner if we use an analytic approximation for the driving laser pulse envelope that is finite in time. A reasonable pulse shape to use is illustrated in Figure 8.30, where we take the intensity envelope of the form of a $\sin^2$ profile of total length L_p. The field envelope would then be written $a_0(\zeta) = a_0 \sin(\pi \zeta / L_p)$. Such a pulse has a full width at half maximum which is exactly half the total pulse duration, so $\Delta t_{FWHM} = L_p/2c$. Using a pulse of this form in (8.218) leads to a solution for the potential during the laser pulse of

$$\phi(\zeta) = \frac{a_0^2}{8} \left[1 + \frac{k_{pe}^2 L_p^2 \cos(2\pi \zeta / L_p) - 4\pi^2 \cos k_{pe}\zeta}{4\pi^2 - k_{pe}^2 L_p^2} \right]. \tag{8.219}$$

The formation of waves oscillating in z with wavenumber $k_{pe} = \omega_{pe}/c$ can easily be seen in this solution. First, if the pulse is much longer than a plasma wavelength, so $k_{pe}L_p \gg 1$, then the excited amplitude of potential oscillations is small, justifying our neglect of plasma waves in the long pulse limit.

Consider instead a short laser pulse, when pulse length is shorter than the plasma oscillation wavelength, so the opposite is true, $k_{pe}L_p \ll 1$. The approximate solution for the potential during that short pulse then is

$$\phi(\zeta) \cong \frac{k_{pe}^2 \zeta^2 a_0^2}{16}. \tag{8.220}$$

This means that in the short pulse we can write the effective refractive index, from Eq. (8.212), as

$$\eta_R(\zeta) \cong 1 - \frac{\omega_{pe}^2}{2\omega^2} \frac{1}{1 + k_{pe}^2 \zeta^2 a_0^2/16}. \tag{8.221}$$

The relativistic correction to the refractive index is reduced with the a_0^2 term, reduced by a factor $\sim k_{pe}^2 \zeta^2 \ll 1$. This greatly reduces relativistic self-focusing for a short pulse. This result also suggests that the front edge of the longer pulse will see reduced self-focusing and will tend to diffract away faster than the rest of the pulse. We can also use this solution to quantify the amplitude of the wave on the rising edge by writing the solution for electron density, Eq. (8.207), in terms of the short-time estimate for electrostatic potential. This tells us that

$$\frac{n_e(\zeta)}{n_{e0}} \cong \frac{1}{2} + \frac{1 + a_0^2/2}{2(1 + k_{pe}^2 \zeta^2 a_0^2/16)^2}, \tag{8.222}$$

or in the rising edge the electron snowplow increases the density by a factor

$$\frac{n_e}{n_{e0}} \cong 1 + a_0^2/4. \tag{8.223}$$

8.4.5 Plasma Nonlinear Response and the QSA in the 3D Tight Focus Regime

It is also possible to apply the QSA to the more general case, plasma motion in three dimensions. This derivation is a bit more complex than the 1D case and the equations that result are more involved. The resulting equations are of utility in numerical modeling of the more general problem.

First, we will assume that the problem is axisymmetric, so we are concerned about plasma motion in z and in r. For this analysis we have to be a bit more careful about the slow and fast components of the potentials and plasma state variables. For example, we have assumed in the previous section that the relativistic gamma factor could be taken to be slowly varying with average over fast laser oscillations, letting us employ the cycle-averaged version, $\gamma \cong (1 + a_0^2/2)^{1/2}$. We need now to pay more attention to fast time scales, which are the time scales associated with the laser oscillating field, $\sim 1/\omega$, and the slow time scales, which are the times of plasma fluid motion, $\sim 1/\omega_{pe}$. In this 3D case we again assume that the laser fields are purely transverse, but now we must consider that the plasma-generated fields can be both longitudinal and transverse. (The derivation that follows draws heavily from the development by Feit et al. (1996)).

As before, we start with the equations describing the laser field and plasma motion. Those include the four we started with in the 1D case. In addition to the normalized quantities we used in the previous section, Eqs. (8.170a–d), we further introduce the normalized plasma fluid momentum:

$$\tilde{\mathbf{p}} = \frac{\mathbf{p}_e}{mc} = \gamma \frac{\mathbf{u}_e}{c} = \gamma\, \boldsymbol{\beta}_u. \qquad (8.224)$$

First, we need the wave equation for the laser field,

$$\nabla^2 \mathbf{a} - \frac{1}{c^2}\frac{\partial^2 \mathbf{a}}{\partial t^2} = \frac{\omega_{pe}^2}{c^2}\frac{n_e}{n_{e0}}\boldsymbol{\beta}_u + \frac{1}{c}\frac{\partial}{\partial t}\nabla\phi, \qquad (8.225)$$

then we also require Poisson's equation for electrostatic potential,

$$\nabla^2 \phi = \frac{\omega_{pe}^2}{c^2}\left(\frac{n_e}{n_{e0}} - 1\right). \qquad (8.226)$$

The plasma continuity and relativistic plasma fluid momentum equations read

$$\frac{1}{c}\frac{\partial n_e}{\partial t} + \nabla \bullet (n_e \tilde{\mathbf{p}}/\gamma) = 0, \qquad (8.227)$$

$$\frac{1}{c}\frac{\partial \tilde{\mathbf{p}}}{\partial t} + \frac{\tilde{\mathbf{p}}}{\gamma} \bullet \nabla \tilde{\mathbf{p}} = \nabla\phi + \frac{1}{c}\frac{\partial \mathbf{a}}{\partial t} - \frac{\tilde{\mathbf{p}}}{\gamma} \times \nabla \times \mathbf{a}. \qquad (8.228)$$

Now we introduce the rapidly varying relativistic factor in terms of normalized momentum:

$$\gamma = \sqrt{1 + \tilde{\mathbf{p}}^2}. \qquad (8.229)$$

Finally, we will need to use a gauge condition for vector potential, which we take to be

$$\nabla \bullet \mathbf{a} = 0. \qquad (8.230)$$

In the 3D case we now need to care for magnetic fields generated by plasma currents, so it makes sense at this point to separate the laser field, $\mathbf{a}_0$, from the plasma-generated vector potential $\mathbf{a}_p$, noting that $\mathbf{a}_p \ll \mathbf{a}_0$. We still employ all the usual optical approximations, that the field is linearly polarized and is purely transverse to the direction of propagation, and that the pulse can be described by a slowly varying envelope, $\mathbf{a}_0$, such that

$$\mathbf{a}_{laser} = \frac{a_0(\mathbf{x}, t)}{2} \exp\left[i\,(kz - \omega t)\right] \hat{\mathbf{x}} + c.c. \qquad (8.231)$$

As before, we make the same approximation that the fast component of plasma motion is dominated by the laser field oscillation,

$$(\boldsymbol{\beta}_u)^{fast} = \frac{\mathbf{a}_{laser}}{\gamma}, \qquad (8.232)$$

and as before we can say that on the fast time scale

$$\frac{1}{c}\frac{\partial}{\partial t}\nabla\phi \rightarrow negligible. \qquad (8.233)$$

The rapidly varying part of the wave equation describes the laser field

$$\nabla^2 \mathbf{a}_{laser} - \frac{1}{c^2}\frac{\partial^2}{\partial t^2}\mathbf{a}_{laser} = \frac{\omega_{pe}^2}{c^2}\frac{n_e}{n_{e0}}\frac{\mathbf{a}_{laser}}{\gamma}. \qquad (8.234)$$

This can be transformed as before into the frame moving with the laser pulse. For simplicity, we will just assume that the group velocity of the laser pulse and hence the phase

8.4 Propagation in Underdense Plasma in the Relativistic Regime

velocity of any plasma wave produced is simply c. (This assumption is not needed and the group velocity can be retained in the derivation, but this is an unnecessary complication for our purposes. See Esarey et al. (1997) for equations with v_g included.) The wave equation becomes

$$\nabla_\perp^2 \mathbf{a}_{laser} + \frac{2}{c} \frac{\partial^2}{\partial \zeta \partial \tau} \mathbf{a}_{laser} - \frac{1}{c^2} \frac{\partial^2}{\partial \tau^2} \mathbf{a}_{laser} = \frac{\omega_{pe}^2}{c^2} \tilde{n}_e \mathbf{a}_{laser}, \quad (8.235)$$

where we have introduced the proper normalized electron density,

$$\tilde{n}_e \equiv \frac{n_e}{n_{e0}} \frac{1}{\gamma}. \quad (8.236)$$

In this moving frame the laser field can be written

$$\mathbf{a}_{laser} = \frac{a_0}{2} \exp[ik\zeta] \hat{\mathbf{x}} + c.c. \quad (8.237)$$

As usual, we employ the slowly varying envelope approximation for pulses that have envelope much longer than the laser wavelength so

$$k a_0 \gg \frac{\partial a_0}{\partial \zeta}, \quad (8.238)$$

which yields the equation governing the laser propagation

$$\nabla_\perp^2 a_0 + 2i \frac{k}{c} \frac{\partial a_0}{\partial \tau} = \frac{\omega_{pe}^2}{c^2} \tilde{n}_e a_0. \quad (8.239)$$

Now let us look at the slowly varying plasma motion equations. The slowly varying part of the wave equation after transforming into the (ζ, τ) frame reads

$$\nabla_\perp^2 \mathbf{a}_p + \frac{2}{c} \frac{\partial^2 \mathbf{a}_p}{\partial \zeta \partial \tau} - \frac{1}{c^2} \frac{\partial^2 \mathbf{a}_p}{\partial \tau^2} = \frac{\omega_{pe}^2}{c^2} \tilde{n}_e \tilde{\mathbf{p}} + \nabla \left(\frac{1}{c} \frac{\partial}{\partial \tau} - \frac{\partial}{\partial \zeta} \right) \phi. \quad (8.240)$$

Likewise the slowly varying electrostatic potential is governed by

$$\nabla^2 \phi = \frac{\omega_{pe}^2}{c^2} (\gamma \tilde{n}_e - 1). \quad (8.241)$$

It turns out that the continuity equation and the gauge condition are redundant. We will, at this point, then, diverge from the treatment of Sprangle (Sprangle et al. 1990b) and utilize Eq. (8.230) instead of (8.227). The grad operator can be broken out into transverse and longitudinal derivatives in the moving frame so that

$$\nabla = \nabla_\perp + \frac{\partial}{\partial \zeta} \hat{\mathbf{z}}. \quad (8.242)$$

Finally, we need to look at the plasma force equation, which we can write

$$\frac{1}{c} \frac{\partial}{\partial t} (\tilde{\mathbf{p}} - \mathbf{a}_p) = \nabla \phi - \frac{\tilde{\mathbf{p}}}{\gamma} \cdot \nabla \tilde{\mathbf{p}} - \frac{\tilde{\mathbf{p}}}{\gamma} \times \nabla \times \mathbf{a}_p. \quad (8.243)$$

The last two terms in this equation can be rearranged using the vector identity

$$\nabla (\mathbf{A} \cdot \mathbf{B}) = (\mathbf{A} \cdot \nabla) \mathbf{B} + (\mathbf{B} \cdot \nabla) \mathbf{A} + \mathbf{A} \times \nabla \times \mathbf{B} + \mathbf{B} \times \nabla \times \mathbf{A} \quad (8.244)$$

to give for those two terms

$$\tilde{\mathbf{p}} \cdot \nabla \tilde{\mathbf{p}} - \tilde{\mathbf{p}} \times \nabla \times \tilde{\mathbf{p}} = \frac{1}{2} \nabla \tilde{\mathbf{p}} \cdot \tilde{\mathbf{p}}. \tag{8.245}$$

The relativistic gamma factor in the plasma can be expressed in terms of the normalized fluid momentum in the form

$$\nabla \gamma = \nabla \sqrt{1 + \tilde{\mathbf{p}} \cdot \tilde{\mathbf{p}}} = \frac{\nabla \tilde{\mathbf{p}} \cdot \tilde{\mathbf{p}}}{2\gamma}, \tag{8.246}$$

so we recast the momentum equation in terms of a ponderomotive force as

$$\frac{1}{c}\frac{\partial}{\partial t}\left(\tilde{\mathbf{p}} - \mathbf{a}_p\right) = \nabla \phi - \nabla \gamma + \frac{\tilde{\mathbf{p}}}{\gamma} \times \nabla \times \left(\tilde{\mathbf{p}} - \mathbf{a}_p\right). \tag{8.247}$$

The last term in this equation has an element that looks like $\nabla \times (\tilde{\mathbf{p}} - \mathbf{a}_p)$, which is often referred to as the "plasma vorticity." The laser forces on the plasma are axi-symmetric: radial and longitudinal, so there is no way that a plasma initially at rest could develop any rotational vorticity from the passing laser. Put more rigorously, canonical angular momentum in the plasma must be conserved in a purely radial force. This means that this last term must be zero throughout the interaction with the laser. This equation takes on a simple interpretation: the slowly varying force on the electron plasma fluid is a result of slowly varying electric fields, and a ponderomotive force, which manifests itself in the $\nabla \gamma$ term. In the moving frame this equation is, then,

$$\left(\frac{1}{c}\frac{\partial}{\partial \tau} - \frac{\partial}{\partial \zeta}\right)\left(\tilde{\mathbf{p}} - \mathbf{a}_p\right) = \nabla(\phi - \gamma). \tag{8.248}$$

As in the 1D case, we make further progress by employing the QSA and dropping terms containing derivatives of τ (with the exception of the laser propagation equation in which we retain the time derivative so that we can find how the pulse evolves as it propagates). We have for the relevant equations now the equation describing laser pulse propagation,

$$\nabla_\perp^2 a_0 + 2i\frac{k}{c}\frac{\partial a_0}{\partial \tau} = \frac{\omega_{pe}^2}{c^2}\tilde{n}_e a_0, \tag{8.249}$$

the equations for the perpendicular and longitudinal components of the slowly varying part of the vector potential,

$$\nabla_\perp^2 \mathbf{a}_{p\perp} = \frac{\omega_{pe}^2}{c^2}\tilde{n}_e \tilde{\mathbf{p}}_\perp - \nabla_\perp \frac{\partial \phi}{\partial \zeta}, \tag{8.250}$$

$$\nabla_\perp^2 a_{pz} = \frac{\omega_{pe}^2}{c^2}\tilde{n}_e \tilde{p}_z - \frac{\partial^2 \phi}{\partial \zeta^2}, \tag{8.251}$$

Poisson's equation,

$$\nabla_\perp^2 \phi + \frac{\partial^2 \phi}{\partial \zeta^2} = \frac{\omega_{pe}^2}{c^2}(\gamma \tilde{n}_e - 1), \tag{8.252}$$

8.4 Propagation in Underdense Plasma in the Relativistic Regime

the electron fluid force equation,

$$\frac{\partial}{\partial \zeta}(\tilde{\mathbf{p}} - \mathbf{a}_p) = \nabla(\gamma - \phi) \tag{8.253}$$

and the gauge condition on the vector potential,

$$\nabla_\perp \cdot \mathbf{a}_{p\perp} + \frac{\partial \mathbf{a}_{p\perp}}{\partial \zeta} = 0. \tag{8.254}$$

This is a complete set of equations describing three-dimensional (axisymmetric) plasma and laser pulse dynamics in the QSA.

As in the 1D case, the longitudinal component of the force equation, (8.253), can be directly integrated

$$\frac{\partial}{\partial \zeta}(\tilde{p}_z - a_{pz}) = \frac{\partial}{\partial \zeta}(\gamma - \phi), \tag{8.255}$$

yielding a constant of motion since $\gamma = 1$ and $\phi = 0$ before arrival of the laser pulse

$$\tilde{p}_z - a_{pz} - \gamma + \phi = -1. \tag{8.256}$$

We introduce at this point what we can call the plasma wake potential,

$$\psi_p \equiv \phi - a_{pz}, \tag{8.257}$$

so Eq. (8.256) can be written

$$1 + \psi_p = \gamma - \tilde{p}_z. \tag{8.258}$$

The physical significance of ψ becomes quickly apparent. In the 1D case, we did not need to worry about magnetic field generation of electric fields by Faraday's law; but now, in 3D, we must consider how transverse plasma currents can produce magnetic fields that contribute to longitudinal electric fields. Since the E-field is

$$\mathbf{E} = -\nabla \Phi - \frac{1}{c}\frac{\partial \mathbf{A}}{\partial t}, \tag{8.259}$$

its longitudinal component in the moving frame is

$$E_z = -\frac{\partial \Phi}{\partial \zeta} - \frac{1}{c}\left(\frac{\partial A_z}{\partial \tau} - v_{e\varphi}\frac{\partial A_z}{\partial \zeta}\right), \tag{8.260}$$

which we see is simply proportional to the ζ derivative of ψ:

$$\begin{aligned} E_z &= -\frac{\partial \Phi}{\partial \zeta} + \frac{\partial A_z}{\partial \zeta} \\ &= -\frac{mc^2}{e}\frac{\partial}{\partial \zeta}(\phi - a_z) \\ &= -\frac{mc^2}{e}\frac{\partial \psi_p}{\partial \zeta}. \end{aligned} \tag{8.261}$$

When we consider plasma electron acceleration, this quantity will play a central role in our calculations.

The slowly varying part of the relativistic gamma factor will be needed, so it makes sense to separate explicitly the slow and fast parts of the normalized electron fluid momentum, the latter of which we assumed to be dominated by the fast transverse oscillations driven by the laser field

$$\gamma = \sqrt{1 + \left(\tilde{\mathbf{p}}_s + \mathbf{a}_{laser}\right)^2}. \tag{8.262}$$

The slow part of γ is defined in terms of its cycle average as

$$\gamma_s \equiv \sqrt{\langle \gamma^2 \rangle},$$

so we know it can be expressed

$$\gamma_s = \sqrt{1 + \tilde{p}_s^2 + a_0^2/2}. \tag{8.263}$$

A little algebraic manipulation on this quantity,

$$2\gamma_s^2 - 2\gamma_s \tilde{p}_{sz} = 1 + \tilde{p}_s^2 + a_0^2/2 - 2\gamma_s \tilde{p}_{sz} + \gamma_s^2,$$

leads, with the insertion of Eq. (8.258), to an expression for the slowly varying gamma as

$$\gamma_s \to \gamma = \frac{1 + \tilde{p}_\perp^2 + a_0^2/2 + (1 + \psi_p)^2}{2(1 + \psi_p)}. \tag{8.264}$$

Going forward, we drop the "s" subscript on γ so the relativistic factor is implicitly meant to signify a cycle-averaged quantity.

Next, we need to find the normalized electron density in terms of the wake potential by eliminating $\partial^2 \phi / \partial \zeta^2$ using Eqs. (8.251) and (8.252):

$$\nabla_\perp^2 \phi + \frac{\omega_{pe}^2}{c^2} \tilde{n}_e \tilde{p}_z - \nabla_\perp^2 a_{pz} = \frac{\omega_{pe}^2}{c^2} (\gamma \tilde{n}_e - 1), \tag{8.265}$$

which is, when written in terms of the wake potential

$$\nabla_\perp^2 \psi_p + \frac{\omega_{pe}^2}{c^2} \tilde{n}_e \left[\gamma - (1 + \psi_p)\right] = \frac{\omega_{pe}^2}{c^2} (\gamma \tilde{n}_e - 1),$$

delivering us the equation we seek for $\tilde{n}_e$,

$$\tilde{n}_e = \frac{1 + \nabla_\perp^2 \psi_p / k_{pe}^2}{1 + \psi_p}. \tag{8.266}$$

This equation, in large radial gradients of the wake potential, can yield negative numbers for the electron density, obviously physically nonsense. In that case, we need simply to set $\tilde{n}_e = 0$. This situation corresponds to that in which the laser pulse is intense enough to drive complete electron cavitation.

Further progress can be made by eliminating $\tilde{p}_\perp$. Employing Eq. (8.251), the gauge condition (8.254) and the definition for ψ_p, Eq. (8.257), we can write

8.4 Propagation in Underdense Plasma in the Relativistic Regime

$$\nabla_\perp^2 \mathbf{a}_{p\perp} = \frac{\omega_{pe}^2}{c^2} \tilde{n}_e \tilde{\mathbf{p}}_\perp - \nabla_\perp \frac{\partial \psi_p}{\partial \zeta} + \nabla_\perp (\nabla_\perp \cdot \mathbf{a}_{p\perp}). \tag{8.267}$$

Since we assume cylindrical symmetry, we note that

$$\nabla_\perp^2 \mathbf{a}_{p\perp} - \nabla_\perp (\nabla_\perp \cdot \mathbf{a}_{p\perp}) = 0, \tag{8.268}$$

which yields for the transverse electron fluid normalized momentum

$$\tilde{\mathbf{p}}_\perp = \frac{c^2}{\omega_{pe}^2 \tilde{n}_e} \nabla_\perp \frac{\partial \psi_p}{\partial \zeta}. \tag{8.269}$$

Now we do see that if the electron density goes to zero, as we have seen it can now in the 3D case, this transverse momentum becomes indeterminate. That is alright, however, because there are no electrons in that region anyway.

Our last task is to find an equation that solves for the wake potential. This goal can be achieved by taking the divergence of Eq. (8.253):

$$\frac{\partial}{\partial \zeta} \nabla \cdot \tilde{\mathbf{p}} = \nabla^2 (\gamma - \phi), \tag{8.270}$$

and then using Poisson's equation to eliminate $\nabla^2 \phi$ along with the longitudinal constant of motion (8.258) to eliminate the longitudinal momentum, leading to

$$\frac{\partial^2 \psi_p}{\partial \zeta^2} = \frac{\omega_{pe}^2}{c^2} (\gamma \tilde{n}_e - 1) - \nabla_\perp^2 \gamma + \frac{\partial}{\partial \zeta} \nabla_\perp \cdot \tilde{\mathbf{p}}_\perp. \tag{8.271}$$

This can be recognized as the 3D generalization of the ODE found for the electrostatic potential in the 1D QSA, Eq. (8.206), to which it reduces if the transverse derivatives are set to zero. The equations that yield the other plasma state variables are then

$$\tilde{n}_e = \frac{1 + \nabla_\perp^2 \psi_p / k_{pe}^2}{1 + \psi}, \tag{8.272}$$

$$\tilde{\mathbf{p}}_\perp = \frac{c^2}{\omega_{pe}^2 \tilde{n}_e} \nabla_\perp \frac{\partial \psi}{\partial \zeta}, \tag{8.273}$$

$$\gamma = \frac{1 + \tilde{p}_\perp^2 + a_0^2/2 + (1 + \psi_p)^2}{2(1 + \psi)}. \tag{8.274}$$

These are a complete set of equations which can be solved numerically in a fairly straightforward manner. The laser propagates, then, according to the equation

$$\nabla_\perp^2 \mathbf{a}_{laser} + 2i \frac{k}{c} \frac{\partial}{\partial \tau} \mathbf{a}_{laser} = \frac{\omega_{pe}^2}{c^2} \tilde{n}_e \mathbf{a}_{laser}. \tag{8.275}$$

Equations (8.271)–(8.275) constitute a complete, self-consistent description of the three-dimensional relativistic dynamics of an electron plasma and laser field within the QSA. Equation (8.272) combined with Eq. (8.275) suggest that we can write the effective refractive index of the plasma in terms of the wake potential,

$$\eta_R(\mathbf{x}) \cong 1 - \frac{\omega_{pe}^2}{2\omega^2} \frac{1 + c^2 \nabla_\perp^2 \psi_p / \omega_{pe}^2}{1 + \psi}, \tag{8.276}$$

which also reduces to the previous 1D result when the transverse derivatives are removed.

The principal utility of these equations rests with numerical solutions for laser propagation in the plasma. They can, however, give us some analytic insight into the transverse motion of the electron fluid subject to the transverse ponderomotive force of a focused beam. This ponderomotive self-channeling is the next topic we look at.

8.4.6 Ponderomotive Electron Expulsion and Electron Channel Formation

We already have seen in the dynamics of single electrons in Chapter 3, that the radial gradient in intensity of a focused laser pulse sets up an effective ponderomotive force that pushes electrons out of the focus. In the plasma this expulsion is countered by the electrostatic attraction of the (mostly) immobile positive ions. In the 3D QSA equations, this ponderomotive expulsion manifests itself in Eq. (8.272) in the $\nabla_\perp^2 \psi_p / k_{pe}^2$ term. A significant consequence of this is that it leads to a radial electron-density profile during the pulse that is hollow on axis and therefore serves to focus the laser beam. This effect is known as ponderomotive channeling (Kar *et al.* 2007). A second effect, considered in a subsequent section, is that the electron cavitation produces a transient radial force on the ions, leading potentially to radial ion acceleration on a many-picosecond time scale.

To grapple with the first effect, it is helpful to consider first the situation of a tightly focused beam in the long pulse limit $\tau_p \gg \omega_{pe}^{-1}$, a regime, we have seen, that does not generate significant longitudinal electron-density oscillations. In this case we can to a good approximation find the wake potential by substituting the rapidly varying relativistic factor, γ, with the slowly evolving cycle-averaged version, γ_s, which from Eq. (8.258) says

$$\psi_p + 1 = \gamma - \tilde{p}_z \cong \gamma_s - \tilde{p}_z, \tag{8.277}$$

where γ_s takes its familiar form in terms of the normalized vector potential amplitude, a_0,

$$\gamma_s = \sqrt{1 + \tilde{p}_z^2 + a_0^2/2}. \tag{8.278}$$

In the long pulse regime, we can ignore longitudinally excited plasma waves and electron fluid motion; we approximate the wake potential

$$\psi_p + 1 \cong \sqrt{1 + a_0^2/2}. \tag{8.279}$$

The 3D QSA result for the refractive index can be written

$$\eta_R(\mathbf{x}) \cong 1 - \frac{\omega_{pe}^2}{2\omega^2} \frac{1 + c^2 \nabla_\perp^2 \psi_p / \omega_{pe}^2}{1 + \psi_p}$$

$$\cong 1 - \frac{\omega_{pe}^2}{2\omega^2} \frac{1 + \left(c^2/\omega_{pe}^2\right) \nabla_\perp^2 \sqrt{1 + a_0^2/2}}{\sqrt{1 + a_0^2/2}}. \tag{8.280}$$

This equation, we admit, could have been written with less formalism and instead with basic intuition. In the long pulse regime it is a fair approximation to consider the electron

8.4 Propagation in Underdense Plasma in the Relativistic Regime

expulsion in steady state such that the electron-density profile adjusts so that the ponderomotive force is exactly canceled by the electrostatic potential created by this expulsion and the constant-ion-density background. The electron density then can be found by writing

$$-e\nabla_\perp \Phi = -\nabla_\perp U_p = -mc^2 \nabla_\perp \gamma_\perp. \tag{8.281}$$

In terms of normalized potential this is

$$\nabla_\perp \phi = \nabla_\perp \gamma = \nabla_\perp \sqrt{1 + a_0^2/2}, \tag{8.282}$$

which when combined with Poisson's equation for the cylindrically symmetric potential

$$\nabla_\perp^2 \phi = \frac{\omega_{pe}^2}{c^2}(\gamma \tilde{n}_e - 1) = \frac{\omega_{pe}^2}{c^2}\left(\frac{n_e - n_{e0}}{n_{e0}}\right)$$

$$= \frac{\omega_{pe}^2}{c^2}\frac{\delta n^2}{n_{e0}} \tag{8.283}$$

gives for the quasi-steady state response of the electron density

$$\delta n_e = n_{e0}\frac{c^2}{\omega_{pe}^2}\nabla_\perp^2\sqrt{1 + a_0^2/2}. \tag{8.284}$$

This is essentially the same result as that derived from the 3D QSA, Eq. (8.280). The physical situation that is evolving is illustrated in Figure 8.31.

The depth of the electron-density depression on the central axis of laser propagation can be calculated easily. That density drop in the center of a cylindrically symmetric beam is

$$\left.\frac{\delta n_e}{n_{e0}}\right|_{center} = \frac{c^2}{\omega_{pe}^2}\nabla_\perp^2\sqrt{1 + a_0^2/2}\bigg|_{r=0}. \tag{8.285}$$

This gives

$$\frac{\delta n_e}{n_{e0}} = \frac{c^2}{\omega_{pe}^2}\left[\frac{1}{2(1+a_0^2/2)^{1/2}}\nabla_\perp^2\left(\frac{a_0^2}{2}\right) - \frac{1}{4(1+a_0^2/2)^{3/2}}\left(\nabla_\perp\frac{a_0^2}{2}\right)^2\right], \tag{8.286}$$

which at $r = 0$ has $\nabla_\perp a_0^2 = 0$ for a cylindrically symmetric beam, so

$$\left.\frac{\delta n_e}{n_{e0}}\right|_{r=0} = \frac{c^2}{2\omega_{pe}^2}\frac{1}{(1+a_0^2/2)^{1/2}}\nabla_\perp^2\left(\frac{a_0^2}{2}\right)\bigg|_{r=0}. \tag{8.287}$$

If the intensity of the laser beam is Gaussian, we can write

$$\nabla_\perp^2 a_0^2 = \frac{8a_{00}^2}{w^2}\exp\left[-2r^2/w^2\right]\left(\frac{2r^2}{w^2} - 1\right), \tag{8.288}$$

which means that the electron-density reduction at the beam center is

$$\left.\frac{\delta n_e}{n_{e0}}\right|_{r=0} = -\frac{2c^2 a_{00}^2}{\omega_{pe}^2 w^2}\frac{1}{\sqrt{1 + a_{00}^2/2}}. \tag{8.289}$$

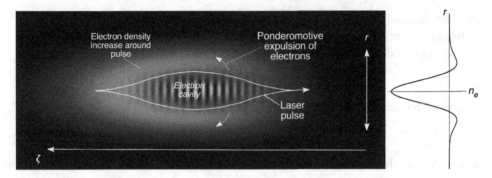

Figure 8.31 Schematic illustration of how electron density can be expelled radially by the radial ponderomotive force of an intense pulse passing through a plasma.

It is interesting to ask what the condition is for complete electron cavitation at the beam center. The condition for complete electron expulsion can be written

$$\frac{\delta n_e}{n_{e0}} \leq -1. \tag{8.290}$$

The peak intensity needed to reach this cavitation condition is

$$\frac{a_{00}^2}{\sqrt{1+a_{00}^2/2}} \geq \frac{\omega_{pe}^2 w^2}{2c^2} = \frac{n_{e0}}{n_{crit}} \frac{\omega^2 w^2}{2c^2}$$

$$\geq \left(\frac{n_{e0}}{n_{crit}}\right) 2\pi^2 \frac{w^2}{\lambda^2}. \tag{8.291}$$

For example, a $\lambda = 1$ μm laser focused to a spot radius of $w_0 = 10$ μm in an electron plasma $10^{-3}\, n_{crit} \approx 10^{18}$ cm^{-3} would require $a_0 \geq 1.8$, an intensity just above 10^{18} W/cm^2.

To get a little more insight into how this electron cavity affects the laser propagation, it is informative to consider the electron-density profile formed in weakly relativistic foci, so that Eq. (8.287) can be written

$$\delta n_e \approx \frac{c^2}{\omega_{pe}^2} n_{e0} \nabla_\perp^2 \frac{a_0^2}{4}. \tag{8.292}$$

Recall the envelope beam propagation equation in the presence of a radial electron density profile that is nearly parabolic, Eq. (8.124), which reads

$$\frac{d^2 w}{dz^2} = \frac{4}{k^2 w^3}\left(1 - \frac{P}{P_{crit}}\right) + \frac{\delta n_{e0}}{n_{crit}} \frac{w}{r_0^2} \tag{8.293}$$

when the electron-density profile is approximated as

$$n_e = \delta n_{e0}\left(1 + \frac{r^2}{r_0^2}\right) \tag{8.294}$$

8.4 Propagation in Underdense Plasma in the Relativistic Regime

$$\delta n_e = \delta n_{e0} \frac{r^2}{r_0^2}. \tag{8.295}$$

Employing the moderate field approximation for electron density, Eq. (8.292), with the radial derivatives of a Gaussian intensity profile, Eq. (8.288), lets us approximate, near the axial center of the beam, a ponderomotively created electron-density profile of

$$\delta n_e \approx \frac{c^2}{\omega_{pe}^2} n_{e0} \frac{2a_{00}^2}{w^2} \left(2\frac{r^2}{w^2} - 1 \right). \tag{8.296}$$

The depth of the electron density depression created by the laser pulse is

$$\delta n_e(r=0) \approx -2 \frac{c^2}{\omega_{pe}^2} \frac{2a_{00}^2}{w^2} n_{e0} \tag{8.297}$$

so the envelope propagation equation takes the form

$$\frac{d^2 w}{dz^2} = \frac{4}{k^2 w^3} \left(1 - \frac{P}{P_{crit}} \right) - \frac{2}{w} \frac{\delta n_{e0}}{n_{crit}}. \tag{8.298}$$

Obviously, the final term on the right side, which arose from the electron-density cavity, is negative and aids in focusing the beam. Consequently, the beam can be self-guided, an effect which supplements guiding from relativistic self-focusing as the pulse power approaches P_{crit}. As an example, consider a modest-intensity laser pulse, a 1 μm wavelength pulse of 1 TW peak power focused in a plasma of electron density 10^{18} cm^{-3} to $w = 25$ μm, which would yield a peak intensity of $\approx 10^{17}$ W/cm^2, or an $a_{00} = 0.27$. The magnitude of the density depression predicted by Eq. (8.297) would be fairly small, about 6×10^{15} cm^{-3}. The critical power for relativistic self-focusing at this density is 17 TW. Therefore this term in the propagation equation (the second term on the right-hand side) $\sim$.0038 while the ponderomotive channeling term, the third on the right is a bit higher, $\sim$.0048. Generally speaking, ponderomotive channeling typically dominates beam propagation at weak intensities but relativistic self-focusing assumes a dominant role once the pulse power approaches P_{crit}. As a general rule of thumb, the effective critical power for self-focusing in the plasma at relativistic intensity is reduced from the usual value discussed in the previous sections of $\sim(n_{crit}/n_e) \times 17$ GW to around $(n_{crit}/n_e) \times 16$ GW (Esarey et al. 1997).

At very high intensities, however, the situation illustrated schematically in Figure 8.32 can develop. The strong ponderomotive expulsion of a very high intensity pulse leads to a region of complete cavitation the size of the laser beam and pulse length. In that case nonlinear self-focusing is suppressed (there are no electrons at all to interact with the beam) and the pulse propagates as if in a hollow waveguide. This regime is complex, and factors such as the modulation instability likely play a significant role in the propagation.

8.4.7 Electron Channel Oscillations and Radial Beam Self-Modulation

The final propagation regime we need to examine is when a short pulse (one comparable to the electron plasma oscillation period) is tightly focused (so that its focal spot size is

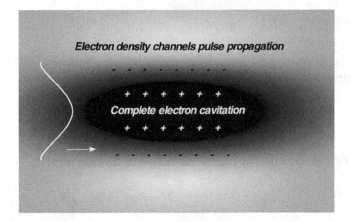

Figure 8.32 At very high intensity the radial ponderomotive force can expel virtually all the electrons in the plasma, creating its own electron plasma waveguide. The heavy ions are largely stationary during an ultrashort pulse, but their Coulomb repulsion can cause them to explode radially if the pulse is long enough.

comparable to an electron plasma wavelength). It is challenging to divine what happens in this regime from the fully nonlinear, 3D QSA equations. Qualitative insight can be derived in the case of moderately relativistic laser fields and small electron-density perturbations. Recall that the rising intensity of a short pulse drives longitudinal electron plasma waves whose amplitude is related to the peak intensity at the radial position the longitudinal wave is driven.

Return to the plasma fluid equations. The electron fluid momentum equation reads

$$\frac{\partial}{\partial \zeta}(\tilde{\mathbf{p}} - \mathbf{a}_p) = \nabla(\gamma - \phi). \tag{8.299}$$

Since, for illustrative purposes, we confine ourselves to moderately intense fields ($a_0 < 1$) and weak perturbations to the plasma response, let us ignore the time-varying plasma-induced vector potential and keep only the electrostatic restoring force in Eq. (8.299). To simplify matters, we revert to lab coordinates and, since we have ignored $\partial/\partial \tau$ terms in the QSA and can therefore take $\partial/\partial \zeta \approx -(1/c)\partial/\partial t$, we can write the gradient of Eq. (8.299) as

$$\frac{\partial}{\partial t}\nabla \cdot \tilde{\mathbf{p}} = c\left(\nabla^2 \gamma - \nabla^2 \phi\right). \tag{8.300}$$

The continuity equation is

$$\frac{1}{c}\frac{\partial n_e}{\partial t} + \nabla \cdot \left(n_e \tilde{\mathbf{p}}/\gamma\right) = 0. \tag{8.301}$$

In the spirit of a perturbation treatment in which the fluid momentum is a small quantity, linearizing this equation lets us set $n_e = n_{e0}$ in the second term and set $\gamma \approx 1$. Taking the time derivative of this equation then and using it to eliminate $\tilde{\mathbf{p}}$ yields

8.4 Propagation in Underdense Plasma in the Relativistic Regime

$$\frac{\partial^2 n_e}{\partial t^2} + n_{e0} c^2 \left(\nabla^2 \phi - \nabla^2 \gamma \right) = 0. \tag{8.302}$$

Finally, elimination of the normalized potential with Poisson's equation in the nonrelativistic limit

$$\nabla^2 \phi = \frac{\omega_{pe}^2}{c^2} \left(\frac{n_e}{n_{e0}} - 1 \right)$$

gives us a driven harmonic oscillator equation, the generalization of the equation for small-amplitude electron-density perturbations derived in Appendix B with a drive term now present that is proportional to the gradient in the ponderomotive force of the laser:

$$\frac{\partial^2 n_e}{\partial t^2} + \omega_{pe}^2 (n_e - n_{e0}) = n_{e0} c^2 \nabla^2 \gamma. \tag{8.303}$$

In the weakly relativistic intensity limit we take

$$\gamma \cong \sqrt{1 + a_0^2/2} \cong 1 + a_0^2/4,$$

which yields the sibling equation to the one we derived previously for the potential in the weak field limit, Eq. (8.217), a driven harmonic oscillator equation for small electron-density perturbations

$$\frac{\partial^2}{\partial t^2} \delta n_e + \omega_{pe}^2 \delta n_e = \frac{n_{e0} c^2}{4} \nabla^2 a_0^2(\mathbf{x}, t). \tag{8.304}$$

(The earlier equation for ϕ was found in the 1D case; the equation we just derived in principle is accurate in 3D, subject to the approximations inherent in linearizing the plasma fluid equations.)

We can, again, use a Green's function to solve this equation, which has formal solution

$$\delta n_e(\mathbf{x}, t) = \frac{n_{e0} c^2}{4 \omega_{pe}} \int_{-\infty}^{t} \sin \omega_{pe}(t - t') \nabla^2 a_0^2(t') \, dt'. \tag{8.305}$$

As we have previously, we can take the beam shape to be Gaussian and use the half-sine-wave function of Figure 8.30 for the temporal profile,

$$a_0(\mathbf{x}, t) = a_{00} \exp\left[-r^2/w^2\right] \sin\left[\pi t/2\tau_p\right], \tag{8.306}$$

which, when inserted into Eq. (8.304), yields the solution for the electron-density perturbations produced during the pulse in the focal spot area:

$$\delta n_e(\mathbf{x}, t) = \frac{n_{e0} c^2 a_{00}^2}{8 \omega_{pe}} \left[\left(1 + \frac{\omega_{pe}^2 \tau_p^2 \cos(\pi t/\tau_p) - \pi^2 \cos \omega_{pe} t}{\pi^2 - \omega_{pe}^2 \tau_p^2} \right) \right.$$

$$\times \left[\frac{8}{w^2} \left(2 \frac{r^2}{w^2} - 1 \right) \right] \exp\left[-2r^2/w^2\right] \tag{8.307}$$

$$\left. + \frac{\pi^2 \omega_{pe}^2}{c^2} \left(\frac{\cos \omega_{pe} t - \cos(\pi t/\tau_p)}{\pi^2 - \omega_{pe}^2 \tau_p^2} \right) \exp\left[-2r^2/w^2\right] \right].$$

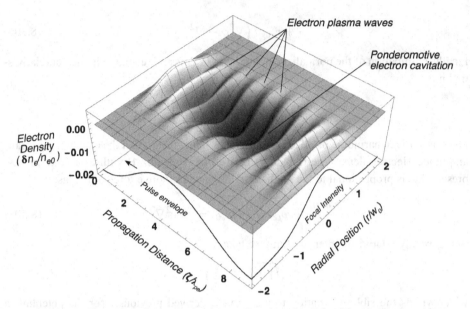

Figure 8.33 Illustration of electron density as a function of radial and longitudinal position around a laser pulse in the regime in which the rising edge of the pulse excites a longitudinal electron plasma wave as well as expelling some of the electrons radially. This plot was made for conditions including an electron density of 4×10^{19} cm^{-3}, a peak a_0 of 0.5, a $1/e^2$ focal radius of 20 μm and a pulse duration FWHM of 80 fs.

A plot of this solution is shown in Figure 8.33. The conditions were chosen, with peak $a_0 = 0.5$, such that the laser pulse width was 4.5 plasma periods and a focal $1/e^2$ radius was 3.75 plasma wavelengths. The electron cavitation from the radial ponderomotive expulsion is clear, with about 1% of the electrons expelled at the beam center. Superimposed on this cavitation are electron plasma waves excited by the rise of the laser pulse. The consequences that such an electron fluid response can have on the propagation of the pulse is clear if we consider the effects of the alternating electron density troughs and crests, seen most clearly in the electron-density plot of Figure 8.34.

Though certainly there will be overall focusing and guiding of the laser pulse as a result of the electron cavitation, the troughs produced by the electron plasma wave will enhance the pulse focusing and the crests will reduce the focusing. As the pulse propagates, the regions of the pulse in a plasma density trough will focus to higher intensity than the temporal slices of the pulse in the plasma crests in which the focusing is reduced. The result is that the pulse develops a "sausage"-like modulation of the beam as it propagates through the plasma, much like that illustrated in Figure 8.35. This spatial modulation leads to intensity modulation in time.

An interesting follow-on consequence of this modulation is that, because it is at the spatial modulation frequency of a plasma wavelength, the intensity modulation leads to a ponderomotive force modulation which can in turn resonantly drive the plasma wave to large amplitude. The increase in plasma wave amplitude enhances the spatial focusing

8.4 Propagation in Underdense Plasma in the Relativistic Regime

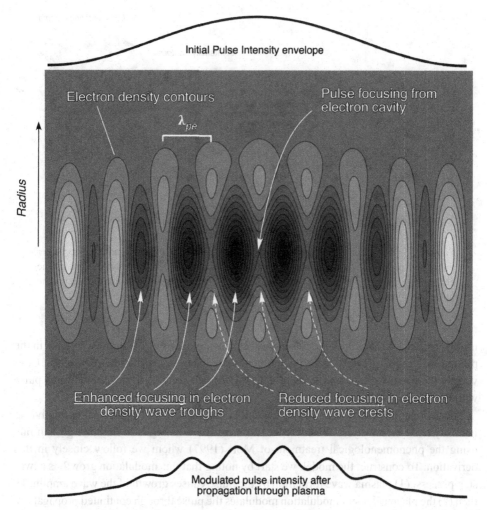

Figure 8.34 Illustration of the propagation effects that the excitation of a plasma wave can have on the focused beam envelope of the pulse as a function of time. Because the amplitude of the electron plasma wave is higher in the center where the intensity of the driving laser pulse is greater, hills and troughs of electron density radially peaked in the center of the beam are created. The high-electron-density regions of the peaks of the plasma wave defocus the beam (or reduce focusing from the overall electron cavitation) while the troughs in density enhance the focusing. The result is that after some distance of propagation the intensity of the laser pulse is modulated in time.

modulation and a positive feedback results. This "envelope self-modulation instability" can lead to exponential growth of the plasma wave amplitude and the size of intensity modulation on the laser pulse (Guerin et al. 1995).

A word about the relationship of the phase of the plasma wave to the phase of the intensity modulation is in order here. Essentially the feedback of this process is such that the laser-intensity modulation is in phase with the electron plasma wave (or strictly speaking it

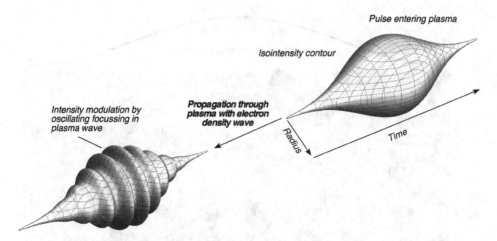

Figure 8.35 Plot of an iso-intensity contour in radius and time of a focused ultrashort laser pulse as it propagates in a plasma which has had an electron plasma wave excited by the early rising edge of the pulse. Alternate focusing and defocusing in time modulates the pulse at a spatial interval equal to the plasma wavelength.

is 180° shifted in phase with the plasma wave). The positions of the deepest troughs in the plasma lead to the greatest focusing and hence the locations of the intensity peaks. Conversely the plasma crests lead to the greatest defocusing and yield the points in the pulse with the intensity valleys.

The exponential growth rate of this instability can be derived (in the weakly relativistic intensity case we have just considered). The most illuminating is to find this growth rate using the phenomenological treatment of Mori (1997) whom we follow closely in this derivation. To construct the model, we start by noting that this modulation growth is a two-step process: (1) plasma wave formation by the pulse causes growth of the wave amplitude, then (2) the plasma density modulation modulates the pulse through continued propagation. Consider the second step first, which we can treat with the beam-size envelope equation, Eq. (8.79), employing only the plasma focusing/defocusing term:

$$\frac{d^2 w}{dz^2} = -\frac{\delta n_{e0}}{n_{crit}} \frac{w}{r^2}, \tag{8.308}$$

which results from a parabolic plasma density

$$\delta n_e = \delta n_{e0} \frac{r^2}{r_0^2}. \tag{8.309}$$

To model the first step, we once again work in the frame moving with the laser pulse which we approximate as moving at c, and for this treatment consider only the longitudinal ponderomotive force of the pulse, so the equation describing the electron plasma wave is

$$\frac{\partial^2}{\partial \zeta^2} \delta n_e + \frac{\omega_{pe}^2}{c^2} \delta n_e = \frac{n_{e0}}{4} \frac{\partial^2}{\partial \zeta^2} a_0^2. \tag{8.310}$$

8.4 Propagation in Underdense Plasma in the Relativistic Regime

Now we posit that the electron-density perturbation and the modulation in a^2 vary as $e^{ik_{pe}\zeta}$, so we write for the electron density and square of the field amplitude

$$\delta n_e = \delta n_{e1} e^{ik_{pe}\zeta} + c.c. \tag{8.311}$$

$$a_0^2 = (a_0^{(0)})^2 + \delta(a_0^2) e^{ik_{pe}\zeta} + c.c. \tag{8.312}$$

Employing the usual perturbation analysis in which (8.311) is considered a small perturbation and the second term in Eq. (8.312) is also considered small, insertion into Eq. (8.310) and suppression of the c.c. terms to reduce clutter, we get

$$\frac{\partial^2}{\partial \zeta^2}\delta n_{e1} + ik_{pe}\frac{\partial}{\partial \zeta}\delta n_{e1} - k_{pe}^2 \delta n_{e1} + \frac{\omega_{pe}^2}{c^2}\delta n_{e1}$$
$$= \frac{n_{e0}}{4}\left(\frac{\partial^2}{\partial \zeta^2}\left[(a_0^{(0)})^2 + \delta(a_0)^2\right] + ik_{pe}\frac{\partial}{\partial \zeta}\delta(a_0)^2 - k_{pe}^2 \delta(a_0)^2\right). \tag{8.313}$$

Since the pulse temporal profile likely increases over many plasma periods, we ignore the first term on the right side as small. A similar argument can be made to drop the first term on the left-hand side. Furthermore, we drop the ζ derivatives of $\delta(a_0)^2$ also as slowly varying (and use $k_{pe} = \omega_{pe}/c$) to give a simple, slowly varying envelope-style equation for the variation of the electron-density perturbation:

$$\frac{\partial}{\partial \zeta}\delta n_{e1} = i\frac{n_{e0}k_{pe}}{4}\delta(a_0)^2. \tag{8.314}$$

The growth of δn_{e1} is directly related to the amplitude of the intensity modulation.

Returning now to consider the second step of the instability, say that the plasma radial trough size is about the focal radius, so that $r_0^2 \approx w^2$, in $\zeta - \tau$ space, since $z \to c\tau$, Eq. (8.308) reads

$$\frac{d^2 w}{d\tau^2} = c^2 \frac{\delta n_{e1}}{n_{crit}} \frac{1}{w(\tau)}. \tag{8.315}$$

The modulation in beam size around the initial focal radius, w_0, modulates the intensity

$$a_0^2 = a_0^2(w_0)\left(\frac{w_0}{w(\tau)}\right)^2, \tag{8.316}$$

allowing us to state

$$(a_0^{(0)})^2 + \delta(a_0^2) = (a_0^{(0)})^2 - (a_0^{(0)})^2\frac{2}{w}\delta w. \tag{8.317}$$

Now differentiate this twice with respect to the propagation time

$$\frac{\partial^2}{\partial \tau^2}\delta(a_0^2) = (a_0^{(0)})^2 \frac{2}{w}\frac{\partial^2 w}{\partial \tau^2} \tag{8.318}$$

and insert Eq. (8.315) to get

$$\frac{\partial^2}{\partial \tau^2}\delta(a_0^2) = (a_0^{(0)})^2 \frac{c^2}{w^2}\frac{\delta n_{e1}}{n_{crit}}. \tag{8.319}$$

Finally, take one more derivative of this equation with respect to ζ and eliminate δn_{e1} with Eq. (8.314) to deliver, for the evolution of laser wave intensity perturbations during beam self-modulation:

$$\frac{\partial^3}{\partial \zeta \partial \tau^2} \delta(a_0^2) = -\frac{c^2 k_{pe}}{4} \frac{1}{w(\tau)^2} \frac{n_{e0}}{n_{crit}} (a_0^{(0)})^2 \delta(a_0^2). \tag{8.320}$$

This equation describes both the evolution of the intensity perturbations over the envelope of the pulse through the ζ dependence as well as the growth of those perturbations as the pulse propagates on through the plasma via the τ variation. The underlying intensity, proportional to $(a_0^{(0)})^2$, varies with the local beam size, by conservation of energy, as

$$(a_0^{(0)})^2 \sim (a_0^{(0)})^2(w_0) \left(\frac{w_0}{w(\tau)}\right)^2. \tag{8.321}$$

Unfortunately, there is no analytic solution to Eq. (8.320), even for a square intensity-pulse shape, but inspection lets us surmise the asymptotic exponential scaling of the amplitude growth of intensity modulations. A constant-intensity pulse starting at $\zeta = 0$ should have modulations that grow like

$$\delta(a_0^2) \sim \exp\left[3i^{1/3} \left(\frac{c^2 k_{pe}}{4w^2} \frac{n_{e0}}{n_{crit}}\right)^{1/3} (a_0^{(0)})^{2/3} \tau^{2/3} |\zeta|^{1/3}\right], \tag{8.322}$$

or since $i^{1/3} = e^{i\pi/6} = i/2 + \sqrt{3}/2$ we can say that the exponential growth factor of these modulations, growing like $\exp[\Gamma_{SM}\tau]$, has the value

$$\Gamma_{SM}\tau = \frac{3^{3/2}}{2} \left(\frac{c^2 k_{pe}}{4w^2} \frac{n_{e0}}{n_{crit}} (a_0^{(0)})^2\right)^{1/3} |\zeta|^{1/3} \tau^{2/3}. \tag{8.323}$$

As expected intuitively, early in the laser pulse, where $\zeta \to 0$, there is little modulation growth. It is the front of the laser pulse that drives the formation of the plasma wave in the first place as the pulse plows through the plasma, so it is only later in the pulse that the wave-amplitude growth can be driven resonantly by pulse modulation of the wave created by the pulse front. The growth rate increases later in the pulse as $|\zeta|$ gets bigger and modulation growth in time scales like $\exp[\tau^{2/3}]$. The modulation growth rate is also largest in regions of the highest pulse intensity as this growth rate scales like $\exp[(a_0^{(0)})^{2/3}]$.

Also note, interestingly, that because the solution of Eq. (8.322) also produces an imaginary term in the exponent, there is a phase shift on the modulation that increases with pulse intensity and later in the pulse as $|\zeta|$ increases. This phase term scales like a prefactor, C, which itself increases with pulse intensity times $|\zeta|^{1/3}$. The consequence of this is that the phase velocity of the resonantly excited plasma wave shifts during the pulse. This can be observed when one notes that the plasma wave in the moving frame oscillates like

$$\delta n_e \sim \exp\left[ik_{pe}\zeta + iC|\zeta|^{1/3}\right]. \tag{8.324}$$

This implies that the local wavenumber of the resonantly driven plasma wave is $k_e \approx k_{pe} + (C/3)|\zeta|^{-2/3}$, yielding a shifting phase velocity, $v_{\varphi e} = \omega_{pe}/k_e$, during the evolution

8.4 Propagation in Underdense Plasma in the Relativistic Regime

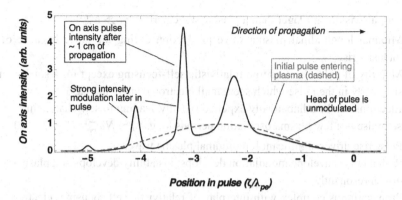

Figure 8.36 Plot of the intensity modulation seen in many simulations of pulse propagating in the plasma wave beam modulation regime. This shows the nonlinear growth of intensity modulations later in the laser pulse. Plot was generated from simulations by Sprangle et al. (1992).

of the pulse. This can have implications for the efficacy of the plasma wave generated in this manner in accelerating high-energy electrons.

This simple analytic treatment can only give some intuition about the growth of the beam self-modulation instability. The treatment has many weaknesses; not only was it a perturbation treatment that quickly breaks down as the beam modulation becomes comparable to the base intensity, but, also, the model assumes Gaussian spot propagation. In fact, the spatial modulation imposed on the beam by the plasma wave quickly distorts the beam shape, leading to formation of higher-order spatial modes and, consequently, scatter of laser light to large angles. Ultimately treating this problem with more accuracy demands at least numerical solution of the 3D QSA equations coupled to the wave equation for the laser pulse, or simulations using particle in cell (PIC) codes. The kind of beam intensity modulations that such simulations often predict is shown schematically in a plot of pulse intensity in Figure 8.36. The nonlinearities of the process quickly result in nonlinear, non-sinusoidal modulations of the pulse intensity with varying period along the pulse. This kind of beam modulation is known to play a significant role in the modulation of pulses in experiments.

At this point it is worth summarizing our discussions of the past few subsections. It should be clear that high-intensity short-pulse propagation in these underdense plasmas can be thought of as occurring in one of three regimes.

(1) Laser pulse is shorter than a plasma period, $\tau_p \lesssim 2\pi \omega_{pe}^{-1}$, characterized by
   - Minimal radial ponderomotive electron expulsion during the pulse; no time for the plasma to respond
   - Longitudinal (and at high intensity in small spots, radial) plasma waves excited behind the laser pulse
   - Relativistic self-focusing is frustrated by electron density ramp-up during the short pulse

(2) Laser pulse is much longer than a plasma wave period, $\tau_p \gg \omega_{pe}^{-1}$
- Minimal longitudinal plasma wave production during the pulse because of slow intensity rise
- Majority of pulse can undergo relativistic self-focusing except for a small temporal slice early in the pulse which sees small electron density increase
- Electrons can be adiabatically expelled radially, leading to self-channeling

(3) Laser pulse is a few plasma oscillation periods long, $\tau_p \sim N\omega_{pe}^{-1}; N \approx 2 - 10$
- Pulse rise drives significant longitudinal plasma waves
- Radial pulse envelope modulation develops; instability develops and plasma wave is driven resonantly
- Propagation is complex with interplay of relativistic self-focusing, electron cavitation and plasma wave production

8.4.8 Electron Cavitation-Driven Ion Motion, Ion Channel Formation, and Ion Acceleration

Until now we have assumed that ions are heavy and immobile, so we have simply ignored any ion motion. This is, for the most part, an excellent approximation, particularly for subpicosecond pulses in gases of relatively heavy ions (say nitrogen, argon, etc.). This assumption of stationary ions is not necessarily correct; however, if the pulse is longer and if lighter ions, say helium or hydrogen, compose the target gas. Ion motion should also be examined if we consider the longer time period after the passage of the laser pulse. In long laser pulses (say, nanosecond-duration pulses) ion acoustic waves can develop and instabilities such as stimulated Brilloiun scattering can occur (Kruer 2003, p. 87). These waves are of little consequence for most experiments in the strong field regime driven by short laser pulses. What can happen, however, is that the radial electrostatic field set up during the laser pulse by ponderomotive electron cavitation can be given a radial kick to the ions, and radial ion motion results.

There are two main experimentally observable consequences of this. First, at moderate intensities, the radial electric field sets ions into motion and they eventually form an ion density hole on the laser axis. Over the time scale of many picoseconds, this radial ion wave travels out at the ion sound speed and forms an ion density channel (followed and neutralized adiabatically by the light electrons), which persists for many tens or hundreds of picoseconds after the passage of the laser. Such ion channel formation can, under some circumstances such as in light ions like He and with longer-picosecond pulses, affect the propagation of the pulse itself, leading to a plasma waveguide that can self-channel the trailing part of a longer pulse. The second consequence occurs with higher-intensity pulses, such as PW-class few-hundred-femtosecond or picosecond pulses that drive complete electron cavitation for a significant period of time. This leads to a Coulomb explosion of ions and the radial expulsion of ions that can sometimes reach MeV kinetic energies.

The second effect has typically resulted in more spectacular experimental results (Krushelnick et al. 1999) so we will consider it first. We will consider electron expulsion in the long pulse regime (regime #2 described in the last subsection) so we ignore electron plasma oscillations and use the concepts developed in Section 8.4.6 to describe the size of the electron-cavitated region. As we discussed in this section, we can find the electron response with the approximation that the electrons adiabatically respond to the radial ponderomotive force of the laser pulse, setting up an electrostatic potential that balances this ponderomotive push. This condition is expressed

$$e\nabla_\perp \Phi = mc^2 \nabla_\perp \gamma, \quad (8.325)$$

which, with Poisson's equation to eliminate Φ and the usual form for the cycle averaged gamma factor, gives for the electron density

$$\frac{\delta n_e}{n_{e0}} = \frac{c^2}{\omega_{pe}^2} \nabla_\perp^2 \sqrt{1 + a_0(r)^2/2}. \quad (8.326)$$

As before, assume a Gaussian focus to derive an equation for the electron radial density profile

$$\frac{\delta n_e}{n_{e0}} = \frac{2c^2}{\omega_{pe}^2} \frac{a_{00}^2}{w^2} \left\{ \frac{[(8r^2/w^2 - 2) + a_{00}^2 (2r^2/w^2 - 1)] \exp[-4r^2/w^2]}{(1 + a_{00}^2 \exp[-4r^2/w^2]/2)^{3/2}} \right\}. \quad (8.327)$$

This profile is illustrated in Figure 8.37. At high enough peak intensity, complete cavitation of the electrons occurs. As peak intensity increases in the pulse, this cavitation will occur out to some radius we will denote as the critical radius, r_c. The radial boundary of this complete cavitation is dynamic; as the pulse ramps up, r_c moves outward, then, as the laser intensity falls back down, the electrons adiabatically come back in to smaller radii. The situation schematically illustrated in Figure 8.38 develops. The ions inside the cavitated region experience a nearly cylindrical radial electric field, which accelerates them. Something akin to a Coulomb explosion develops, at least over the time scale of complete radial electron cavitation (Tripathi et al. 2005; Rajouria et al. 2015). At densities above 10^{18} cm^{-3} this radial field can become quite significant, accelerating ions to MeV energies.

The critical radius can be estimated if we assume no motion of the ions internal to the cavity with Eq. (8.327) and the condition for complete cavitation given by

$$\frac{\delta n_e(r_c)}{n_{e0}} = -1. \quad (8.328)$$

This condition imposed on Eq. (8.327) yields a transcendental equation for r_c, which can be easily solved numerically. Such solutions for r_c in terms of the focal spot $1/e^2$ radius, w_0, are plotted for electron densities of 10^{18} and 10^{19} cm^{-3} in Figure 8.39. The upshot is that at these typical plasma densities, pulses at intensity above $\sim 10^{18}$ W/cm^2 will drive electron cavitation out to radii roughly half w_0.

Using this kind of estimate for r_c and a typical pulse time history, it is possible to estimate the ion energy spectrum, assuming Coulomb expulsion in the cylindrical electron cavity. As the geometry of Figure 8.38 suggests, we can write the field felt by an ion

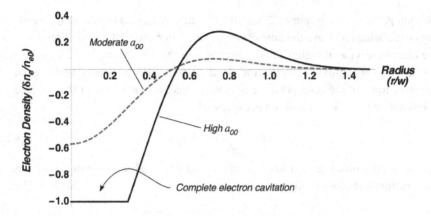

Figure 8.37 Plots of the radial electron density profiles generated by the radial ponderomotive force of a moderate- and high-intensity pulse found from Eq. (8.327). Complete electron cavitation occurs at high intensity.

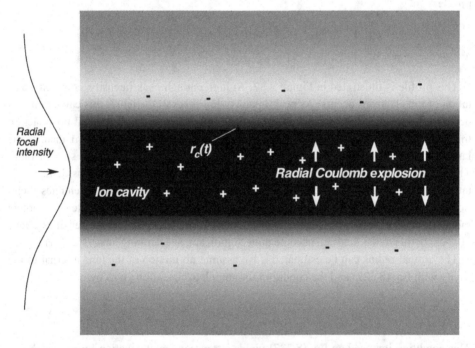

Figure 8.38 Schematic illustration of the critical radius to which complete electron cavitation occurs at high intensity leading to ion Coulomb explosion during the laser pulse.

initially located in the cavity channel at r_i as it expands by Coulomb explosion to greater radius r as

$$E(r) = \frac{Zen_i}{2}\frac{r_i^2}{r}. \tag{8.329}$$

8.4 Propagation in Underdense Plasma in the Relativistic Regime

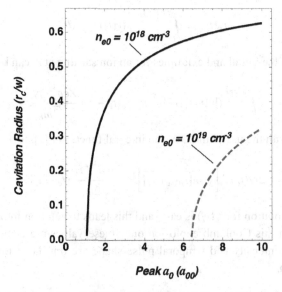

Figure 8.39 Calculation of the location of the critical electron cavitation radius as a function of laser peak a_0 for two values of the electron density. This shows that electrons are pushed to about half the $1/e^2$ radius of the focused beam at sufficiently high intensity.

(This estimate can be made in the regime of a Coulomb explosion because the ions originating at greater r_i always feel a greater radial field and are never overtaken by ions at inner starting radii.) In this regime the energy of an ion when it has reached radial point r is

$$\varepsilon_i(r) = Ze \int_{r_i}^{r} E(r)dr$$
$$= \frac{Z^2 e^2 n_i r_i^2}{2} \ln[r/r_i] \qquad (8.330)$$

We can make a reasonable estimate for the total energy achieved by an ion by finding its energy when it reaches the critical radius, at which point the electric field is greatly reduced by the presence of electrons. The location of this radius defining the end point of the ion acceleration in this simple model is dynamic, moving outward then inward as the laser pulse intensity ramps up and then drops back down. We estimate the beginning of the ion acceleration as occurring at the point in time that the critical radius expands past the ion's initial radius. So to calculate the ion acceleration from starting point r_i, the time an ion begins to accelerate is

$$r_i = r_C(t_i), \qquad (8.331)$$

which can be solved numerically with condition Eq. (8.328) and the temporal pulse shape. The time that the ion exits the cavitated region, t_e, is given by finding when the electron reaches r_c, found from solution of

$$r_c(t_e) = \int_{t_i}^{t_e} \left[\frac{2}{m_i} \varepsilon_i \left(r(t) \right) \right]^{1/2} dt. \tag{8.332}$$

Using Eq. (8.330), the initial and exit times for an ion starting at r_i can be related:

$$\int_{r_i}^{r_c(t_e)} (\ln[r/r_i])^{-1/2} dr = (t_e - t_i) \frac{Z e n_i^{1/2} r_i}{\sqrt{2 m_i}}, \tag{8.333}$$

which evaluates, with the use of the Dawson integral function F[x], to

$$2 r_c(t_e) \, F \left[\sqrt{\ln[r_c(t_e)/r_i]} \right] = \frac{Z e n_i^{1/2} r_i}{\sqrt{2 m_i}} (t_e - t_i). \tag{8.334}$$

The numerical solution for $t_e(r_i)$ is easy, and this leads to solution for $r_c(t_e)$. Finally, the ion spectrum from this Coulomb explosion once these values are computed for a given focal radius, peak intensity and temporal pulse shape are translated to an observed ion energy distribution,

$$dn_i = f(\varepsilon_i) d\varepsilon_i = 2\pi r_i n_i dr_i, \tag{8.335}$$

in which the ejected ion energy, ε_i, for an ion at initial radius r_i is given by

$$\varepsilon_i(r_c(t_e)) = \frac{Z^2 e^2 n_i r_i^2}{2} \ln[r_c(t_e)/r_i]. \tag{8.336}$$

Two ion energy spectra calculated using this simple model are plotted in Figure 8.40. The two spectra plotted are for identical plasma and peak intensity conditions: a helium plasma with electron density of 5×10^{18} cm^{-3} irradiated with a 20 μm diameter focal spot at peak intensity of 6×10^{20} W/cm^2 (corresponding to $a_{00} = 20$ for 1 μm wavelength pulses). The difference between the spectra is that one was generated with a 100 fs pulse and the second generated with a 500 fs pulse. This illustrates the dramatic enhancement in ejected ion energy when longer pulses are employed, because such pulses for given peak intensity produce a cavitated region for longer and allow more time for greater ion acceleration in the Coulomb explosion phase. The prediction of MeV He ions at these kinds of intensities with half-picosecond pulses is completely consistent with the ion energies observed in experiments (Krushelnick et al. 1999) as is the $\sim$100 keV estimate for sub-100 fs pulses (Kahaly et al. 2016).

As a final note, a simple analytic approximation can be made to derive estimates for the ion energies expected in experiment. For long pulses, one can make the approximation that the pulse is constant in intensity, so that r_c can be approximated as a constant in time and that the pulse is long enough for all ions in the cavitated region to Coulomb explode to the critical radius. In this simplified picture, the maximum ion energy can be estimated by noting that the maximum ion energy occurs for ions with $r_i \approx r_c/2$. The maximum ion energy expected then is

$$\varepsilon_{MAX} \approx \frac{\ln 2}{8} Z^2 e^2 n_i r_c^2. \tag{8.337}$$

8.4 Propagation in Underdense Plasma in the Relativistic Regime

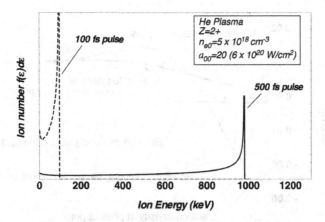

Figure 8.40 Calculation of the ion Coulomb explosion energy spectrum from electron cavitation of a He plasma at high intensity with two different pulse durations. This shows that a longer pulse leads to a longer cavity duration and significantly higher ion energies even when at the same intensity as a short (100 fs) pulse.

This equation, for example, predicts that the maximum ion energy in the helium plasma of the spectrum in Figure 8.40 ($n_e = 5 \times 10^{18}$ cm^{-3}) would be 1.4 MeV, right in line with the spectrum plotted and experiments.

The other effect seen in experiments has to do with longer-term evolution of the ion density. The previous discussion should make evident that the passage of the laser pulse through the plasma imparts radial momentum to the ions whose inertia will cause them to expand away from the axis even if the mobile electrons move back in to restore quasi-neutrality once the pulse intensity has dropped. This electron push will cause a hole in the ion density to develop, as illustrated in Figure 8.41. If the drive laser pulse is long enough, its temporal tail will be guided by the ion density hole left on axis, and this hole will expand on a tens- or hundreds-of-picosecond time scale. In fact, this will form something of a waveguide that a subsequently injected pulse can propagate down. As Figure 8.41 illustrates, the ponderomotive radial injection of electrons set up a radial electric field during the peak intensity of the pulse, and it is this electric field which accelerates the ions.

To get a handle on the ion density evolution, we will follow the treatment developed by Esarey *et al.* (1997). Since we wish to track the ion acoustic response to this ponderomotive kick, we cannot assume a cold plasma and must presume that the electron fluid has finite temperature, T_e. We can still assume that the ions are cold and set ion temperature $T_i = 0$. Assume an ideal gas equation of state with electron pressure (which is the determining pressure in setting the sound speed of the ions) to $p_e = n_e k_B T_e$. The plasma electron force equation now has an extra term; there is the usual ponderomotive force, but there is also now included a force term from any gradient in electron pressure. For the purposes of this analysis we will consider pulses at moderate intensity, in which complete electron cavitation does not occur. In that case the force equation, once again ignoring the inertia of the light electrons, relates the electrostatic potential to the radial ponderomotive force and force from gradient in electron pressure:

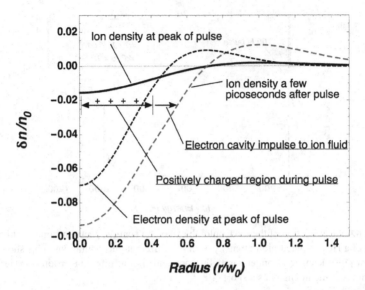

Figure 8.41 Plot of radial electron and ion densities in a laser focus during and after a short laser pulse passes through the plasma. During the pulse, electrons are largely expelled radially, producing a transient radial impulse to the heavy, slow ions. Once this electron expulsion is released with the passage of the short laser pulse, the impulse given the ions leads to a longer (few-picosecond) time scale expansion of the ions to produce a lingering density cavity in the plasma after the laser has passed.

$$e\nabla_\perp \Phi = mc^2 \nabla_\perp \gamma + \nabla_\perp p_e/n_e. \tag{8.338}$$

In this approximation we are ignoring the possibility of electron plasma waves (which is fine on the many-picosecond time scale of plasma motion that we now consider).

For the ion motion we use the cold ion temperature form of the ion force and continuity equations. As with the electrons we work with the equilibrium ion density, n_{i0}, and the perturbed ion density δn_i, related in the usual way:

$$\delta n_i = n_i(t) - n_{i0}. \tag{8.339}$$

Since we consider only radial motion of the ions from the laser propagation axis, the continuity equation is

$$\frac{\partial n_i}{\partial t} + \nabla_\perp \cdot (n_i \mathbf{v}_\perp) = 0, \tag{8.340}$$

which becomes for small ion-density perturbations

$$\frac{\partial}{\partial t}\delta n_i \cong -n_{i0} \nabla_\perp \cdot \mathbf{v}_\perp. \tag{8.341}$$

The ion momentum equation reflects that the ions are accelerated by an electrostatic potential set up by motion of the electrons in Eq. (8.338),

$$m_i \frac{\partial \mathbf{v}_\perp}{\partial t} \approx -Ze\nabla_\perp \Phi. \tag{8.342}$$

8.5 Temporal Phase and Amplitude Modulation of Intense Pulses in Plasma

Combining Equations (8.338), (8.341) and (8.342) yields

$$\frac{m_i}{n_{i0}}\frac{\partial^2}{\partial t^2}\delta n_i = Z\left(mc^2\nabla_\perp^2 \gamma + \nabla_\perp^2 p_e/n_e\right), \qquad (8.343)$$

where here γ is the usual cycle-averaged electron gamma $\gamma = (1+a_0^2/2)^{1/2}$. We can relate electron and ion density, assuming that perturbations to n_i are small so $n_i \approx n_{i0}$:

$$p_e/n_e = Zn_i k_B T_e/Zn_i$$
$$= \frac{n_i}{n_{i0}} k_B T_e, \qquad (8.344)$$

which gives for the ion density equation

$$\frac{m_i}{n_{i0}}\frac{\partial^2}{\partial t^2}\delta n_i = Zmc^2\nabla_\perp^2\left(1+a_0^2/2\right)^{1/2} + \frac{Zk_B T_e}{n_{i0}}\nabla_\perp^2 \delta n_i. \qquad (8.345)$$

Introducing here, the isothermal ion sound speed

$$c_{si} \equiv \sqrt{Zk_B T_e/m_i}, \qquad (8.346)$$

gives finally a driven wave equation for ion-density perturbations:

$$\frac{m_i}{n_{i0}}\frac{\partial^2}{\partial t^2}\delta n_i - c_{si}^2 \nabla_\perp^2 \delta n_i = \frac{Zmc^2 n_{i0}}{m_i}\nabla_\perp^2\left(1+a_0^2/2\right)^{1/2}. \qquad (8.347)$$

This equation illustrates that the ions are set in motion by the ponderomotive force of the laser pulse, transmitted through the electrostatic potential set up by the radial expulsion of the electrons. Put in terms of the frame moving with the drive laser pulse defined by the moving coordinate ζ, this equation can be formally solved with use of a 2D Green's function of the wave equation, which delivers us the explicit equation for evolution of a small ion-density perturbation:

$$\delta n_i(\zeta) = \frac{Zmc^2 n_{i0}}{4\pi m_i c_{si}^2}\int_{-\infty}^{\infty} d\zeta' \int_{-\infty}^{\infty} dx' \int_{-\infty}^{\infty} dy' \frac{\nabla_\perp^2 \left(1+a_0^2/2\right)^{1/2}}{\left(c_{si}^2(\zeta-\zeta')/c^2 - (\mathbf{r}_\perp - \mathbf{r}'_\perp)^2\right)^{1/2}}. \qquad (8.348)$$

Numerical solution of these integrals for a 500 fs pulse of peak $a_0 = 0.25$ (intensity of just under 10^{17} W/cm^2) passing through a helium plasma with 100 eV electron temperature is shown in Figure 8.42 for three times. Right at the end of this pulse a small axial dip in ion density has developed. After 10 ps have elapsed, the hole in the ion density has grown by almost 8×. After 300 ps, the ion momentum has set up a radially expanding ion-density wave, propagating outward at approximately the ion sound speed. This shows that if a pulse was injected 10 ps after the first pulse had passed, this second pulse would experience a channel to guide it.

8.5 Temporal Phase and Amplitude Modulation of Intense Pulses in Plasma

Discussions in the previous section essentially focused on the effects that the plasma has on the radial profile of the propagating laser pulse. It was these radial variations in

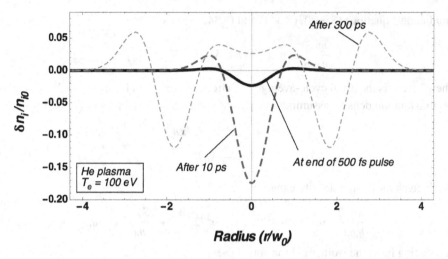

Figure 8.42 Radial ion density plots showing the consequences of ion cavitation during and after a 500 fs laser pulse has passed through a He plasma of 100 eV electron temperature. The laser for this case had a peak a_0 of 0.25. A small dip in ion density has developed after the end of the pulse, but this launches an ion-density wave that further deepens the cavity ~10 ps after the laser has passed and ultimately sends a radial ion plasma wave traveling radially outward at roughly the ion sound speed over the next few hundred picoseconds.

phase induced by ionization, relativistic mass shift of the plasma electrons or plasma wave motion that affected how the beam propagated. Now we should consider complementary effects: how these same three physical phenomena impact the temporal phase of the pulse envelope and lead to wavelength shifting and spectral broadening of the pulse. We will need to consider not only the temporal variation of the phase velocity of the EM wave in the plasma but also variation in group velocity.

8.5.1 Ionization-Induced Spectral Blue Shifting

As we started in the last section, we begin here by examining how the actual plasma formation by field ionization affects pulse temporal phase. As illustrated in Figure 8.43, propagation of the intense pulse in an unionized gas leads to plasma formation. The temporal increase in electron density during the pulse leads to an effective drop in refractive index during the rising edge of the pulse and a phase sweep imparted to the laser field. Because this plasma-induced phase sweep has negative derivative, the result is that it induces a shift of some of the laser spectrum toward higher frequencies: a blue shift (Penetrante et al. 1992; Rau et al. 1997; Giulietti et al. 2013).

A simple heuristic model is adequate to describe this process and predict the magnitude of observed blue shift. The phase accumulated along the pulse envelope is, in terms of the plasma-modified refractive index

$$\varphi(z,t) = \frac{\omega}{c} \int_0^z \eta_r(z',t) dz'. \tag{8.349}$$

8.5 Temporal Phase and Amplitude Modulation of Intense Pulses in Plasma

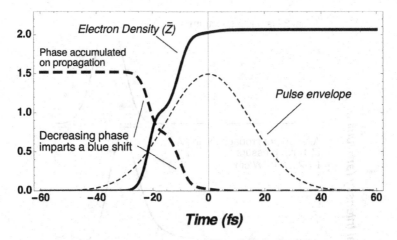

Figure 8.43 Plot of the time history of electron density during a 40 fs laser pulse in an example high-Z gas (such as Ar or Xe). The temporal phase imparted to the pulse as a result of the electron density increase from tunnel ionization is also plotted. The negative derivative in phase leads to a blue shift in the pulse spectrum.

The instantaneous shift in frequency can be written as a temporal derivative of wave phase

$$\Delta\omega = -\frac{\partial}{\partial t}\varphi(z,t). \tag{8.350}$$

Since any observed wavelength shift in an experiment is related to frequency shift by $\Delta\lambda = -(2\pi c/\omega^2)\Delta\omega$, we estimate the extent of blue shifting to be

$$\Delta\lambda = -\frac{\lambda}{2cn_{crit}}\int_0^z \frac{\partial}{\partial t}n_e(z',t)dz'. \tag{8.351}$$

A back-of-the-envelope estimate for the extent of blue shifting is to estimate the highest achieved ionization state, Z, and the ionization time Δt_{ionize}, which is typically a few tens of femtoseconds in an ultrashort pulse, to yield for the magnitude of blue shifting

$$\Delta\lambda \approx -(\lambda L/2cn_{crit})(Zn_0/\Delta t_{ionize}). \tag{8.352}$$

This equation suggests that a 1 μm wavelength pulse transiting 1 mm of plasma ionized up to 10^{18} cm^{-3} over 10 fs will experience nearly 30 nm of blue shifting. This is likely comparable to or larger than the pulse's initial bandwidth (which would be of the order of 10–30 nm for a 30–100 fs pulse).

Figure 8.44 shows the calculated spectral blue shifting that a moderate-intensity bandwidth-limited 30 fs pulse would experience propagating though just 10 μm of Ar gas at density around 2×10^{19} cm^{-3}. Even at only 10^{15} W/cm^2 the spectrum will be strongly distorted. In fact, the actual observed spectrum of a pulse transiting an ionizing gas is heavily dependent on the details of the experiment, since the peak intensity varies with focal radius, and most measurements are made in the far field of the focus. What is more, the blue shifting is further complicated by propagation effects such as plasma-induced refraction. Consequently, it is rare that experimental observation nicely follows any simple model.

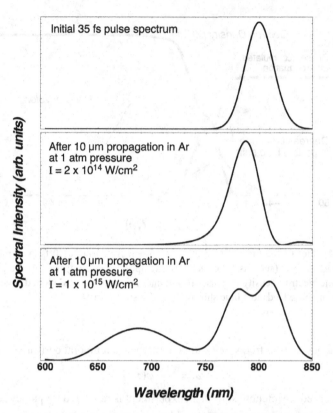

Figure 8.44 Calculation of the spectrum of a 35 fs, 800 nm laser pulse (top panel) as it enters an argon gas target at density of 1 atm. The second and third panels show the blue-shifted spectrum calculated after propagation through 10 μm of that Ar target when the pulse has peak intensity of 2× and 10× 10^{14} W/cm^2.

8.5.2 Relativistic Self-Phase Modulation

At higher peak intensity another phase-modulation effect appears in the plasma, brought about by the now-familiar intensity-dependent shift of the plasma refractive index because of the relativistic mass shift of the oscillating electrons. This relativistic self-phase modulation results from the refractive index shift with intensity over the time history of the pulse envelope. Because the pulse intensity both rises and falls, this phase-modulation leads to both red and blue shifting of the spectrum (Watts et al. 2002; Umstadter et al. 2005; Panwar and Sharma 2009).

As usual, we can employ Eq. (8.133) for the refractive index of the plasma at relativistic intensity, using the usual cycle-averaged approximation for the gamma factor, $\gamma \cong (1 + a_0(t)^2/2)^{1/2}$. As we estimated in the last subsection, the frequency shift is the derivative of the imparted phase from this refractive index telling us that

$$\Delta\omega = -\frac{\omega}{4c}\frac{a_0(t)}{\gamma^3}\frac{\partial a_0(t)}{\partial t}\int_0^z \frac{n_e(z')}{n_{crit}}dz', \qquad (8.353)$$

8.5 Temporal Phase and Amplitude Modulation of Intense Pulses in Plasma

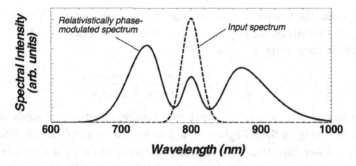

Figure 8.45 Calculation of the spectrum of a 30 fs, 800 nm laser pulse compared to that pulse's spectrum after it has undergone relativistic self-phase modulation. The pulse had peak $a_0 = 1$ and it has propagated through 1 mm of plasma with $n_e = 0.01\, n_{crit}$.

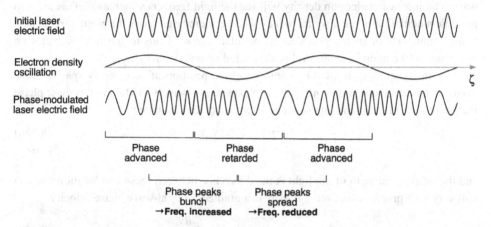

Figure 8.46 Schematic illustration of how the electron-density modulation of an electron plasma wave can lead to phase bunching and spreading, usually referred to as photon acceleration and deceleration respectively.

which yields a back-of-the-envelope estimate for the extent of spectral broadening in the pulse of duration τ_p,

$$\Delta\lambda \approx (\lambda L n_e / 4 c n_{crit})(a_0^2 / \gamma^3 \tau_p). \tag{8.354}$$

A calculation for the relativistic self-phase modulation deformation of the spectrum of a 30 fs transform-limited pulse after propagation through 1 mm of plasma at 1 percent of critical density with peak $a_0 = 1$ is shown in Figure 8.45. The broadening in both blue and red directions is clear, with nearly a quadrupling of bandwidth.

8.5.3 Photon Acceleration and Plasma Wave Phase Modulation

The third effect we had to consider in plasma pulse propagation in the last section was the excitation of plasma waves by the finite rise time of the pulse. This has interesting consequences in the time domain: it leads to periodic variations of both phase and group velocity. Consider the impact of phase velocity modulation first. (In these next two subsections we lean heavily on the elegant intuitive treatment published by Mori (1997).) The phase velocity in the plasma is

$$v_\varphi = \frac{c}{\eta_r} \cong c\left(1 + \frac{n_{e0}}{2\gamma n_{crit}} + \frac{\delta n_e}{2\gamma n_{crit}}\right). \tag{8.355}$$

An electron plasma-density oscillation will then have an effect on a monochromatic light wave illustrated in Figure 8.46. Regions of higher-than-background electron density in the wave will have lower refractive index and cause the phase of the light wave copropagating with this part of the plasma wave to advance. The light wave traveling along with the trough of the wave will see its phase retarded. In the regions with falling electron density, the light wave peaks will be spread out, the frequency is decreased, and the regions in the plasma wave with increasing electron density will see the light frequency increased. This leads to periodic modulation of the laser frequency with periodicity given by the plasma frequency. Fourier transform of such a periodically modulated phase leads to spectral sidebands on either side of the main laser spectral peak, shifted up and down by ω_{pe}.

This effect can be quantified by considering the position of two closely spaced phase fronts in the laser wave comoving with the plasma wave-density modulation. If these phase fronts are located in the moving frame at locations ζ_1 and ζ_2, they propagate as

$$\zeta_1 = \zeta_{10} + \Delta\tau v_{\varphi 1}, \tag{8.356}$$

$$\zeta_2 = \zeta_{20} + \Delta\tau v_{\varphi 2}, \tag{8.357}$$

and the local wavelength of the light is the difference between these two locations, which will vary with propagation time τ if there is a gradient in light-wave phase velocity:

$$\zeta_2 - \zeta_1 = \lambda_\ell = \lambda - \Delta\tau\lambda\frac{\partial v_\varphi}{\partial \zeta}. \tag{8.358}$$

The shift of wavelength, then, at this point in the light wave is

$$\Delta\lambda = \lambda_\ell - \lambda = \Delta\tau\lambda\frac{\partial v_\varphi}{\partial \zeta}, \tag{8.359}$$

or, using the usual refractive index ignoring spatial variation of a_0, yields the local wavelength shift

$$\frac{\partial \lambda}{\partial \tau} = -\frac{\lambda c}{2\gamma n_{crit}}\frac{\partial n_e}{\partial \zeta}. \tag{8.360}$$

Regions of rising electron density blue-shift the light, and falling density regions yield a red shift.

This process is sometimes called photon acceleration, because electron-density variations in the plasma wave can upshift (or downshift) the energy of the laser photons. This has the practical consequence of modulating the ponderomotive energy of electrons in this wave because U_p varies as $I\lambda^2$. Since this modulation is at periodicity given by the plasma wavelength, the modulation can in turn resonantly drive the plasma wave to higher amplitude. We will return to this feedback phenomenon in a bit.

8.5.4 Plasma Wave-Induced Longitudinal Laser Energy Bunching

Density oscillations in the plasma wave also temporally modulate the laser's local group velocity. Now regions of higher electron density slow down energy transport speed in the wave, and lower-density regions speed it up. This will modulate the laser field's energy density and give rise to oscillations in the light wave's intensity at the periodicity of the plasma electron wavelength. This process is illustrated in Figure 8.47. If we think in terms of photon density, the photon density will tend to spread out in regions of the plasma wave with falling electron density and bunch up where electron density rises.

The change in local intensity (i.e. a_0^2) can be thought of as a change in photon "spacing," ΔL, in the light wave. With this phenomenological view, the local intensity can be said to change by

$$\Delta(a_0^2) = -\frac{\Delta L}{L}(a_0^2). \tag{8.361}$$

Given a difference of group velocity at two points moving in the frame of the plasma wave, this photon spacing changes with time as

$$\Delta L = -\left(v_{g2} - v_{g1}\right)\Delta t$$
$$= L\frac{\partial v_g}{\partial \zeta}\Delta t. \tag{8.362}$$

The photon density will then change as a segment of length containing those photons changes with propagation time, which from (8.362) can be written

$$\frac{1}{L}\frac{\partial L}{\partial \tau} = \frac{c}{2\gamma n_{crit}}\frac{\partial}{\partial \zeta}\delta n_e. \tag{8.363}$$

It is natural to relate this segment of photons to the intensity in that region by the change of that segment length, yielding a relation for the rate of change of intensity, given the gradient in electron density:

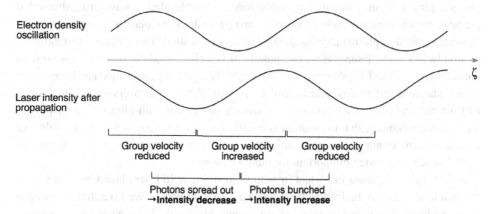

Figure 8.47 Schematic illustration of how the electron-density modulation of an electron plasma wave leads to group velocity modulation, bunching and spreading of the energy density of the laser wave.

$$\frac{\partial}{\partial \tau}(a_0^2) = -\frac{1}{L}\frac{\partial L}{\partial \tau}(a_0^2)$$
$$= -\frac{c(a_0^2)}{2\gamma n_{crit}}\frac{\partial}{\partial \zeta}\delta n_e. \quad (8.364)$$

As predicted, the rate of intensity modulation varies with the derivative of electron density. Like the photon acceleration mechanism, this group velocity bunching also modulates the ponderomotive energy at the plasma frequency, in this case through modulation of intensity. We shall see in the next section how these two effects, phase velocity and group velocity modulation in a plasma wave, conspire to drive an important plasma instability.

8.6 Plasma Instabilities Driven by Intense Laser Pulses

We have seen that if an intense laser pulse has a sufficiently fast rise time, say, comparable to the inverse of the electron plasma frequency, then the ponderomotive force associated with the longitudinal gradient in intensity can excite an electron plasma wave. Our considerations up to this point have assumed that "long" pulses essentially see no significant plasma wave excitation because their intensity rise is effectively adiabatic from the standpoint of the plasma electron response. This view is, in principle, correct. However, it is not completely true that a long pulse will not, or cannot excite plasma waves to any significant amplitude.

We have overlooked an important aspect of how electromagnetic waves can interact with the plasma. In the previous sections we have seen that if a plasma wave has been excited, it can modulate the intensity and frequency of the light wave. Since this modulation is at the plasma frequency, the light-wave modulation could, in turn, resonantly drive the plasma wave. This leads to the kind of positive feedback which drives exponential growth of the amplitude of an oscillating system, an instability in the plasma.

This kind of plasma instability, involving electron plasma waves growing in otherwise uniform plasma, is only one of a veritable zoo of possible plasma wave instabilities that are possible when intense light propagates through underdense plasma. There is an entire thriving subfield of plasma physics devoted to these so-called "Laser Plasma Instabilities" or simply LPI in the laser-plasma community. It is well beyond the scope of this book to consider most of the LPI phenomena. (The reader is urged to peruse what has become the classic standard text in this field by Bill Kruer (2003) to get an overview of the various LPI interactions.) In strong field physics, usually conducted with ultrashort laser pulses, we have the advantage that ion motion is usually too slow to matter on the time scales we consider, so the group of LPI instabilities that excite ion wave motion, such as stimulated Brilloiun scattering, are unimportant for our considerations.

In fact, by far the most important instability in strong field laser plasma experiments is more or less related to the feedback process we discussed earlier. We have already encountered one form of electron plasma wave–driven instabilities, the beam envelope modulation instability. However, a finite rise time of some pulse is not necessarily needed to excite electron plasma wave instabilities. In any plasma or real pulse there will inevitably be small

8.6 Plasma Instabilities Driven by Intense Laser Pulses

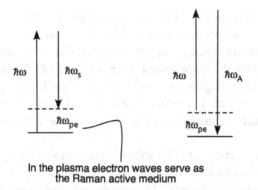

Figure 8.48 Illustration of Raman scattering of laser light in a plasma. The laser photon couples to electron plasma-wave plasmons. Downshifted photons result when a plasmon is created (Stokes radiation) and frequency-upshifted light is produced when a plasmon is absorbed (anti-Stokes radiation).

variations in electron density, perhaps driven by thermal fluctuations, or variations in the laser intensity. Because the instabilities we consider grow exponentially, a small initial fluctuation may be all that is needed to start off the feedback process that can lead to rapid growth of the plasma-wave amplitude. What this means is that an intense light wave, even if it is essentially constant in amplitude and with uniform spatial profile is still able to generate electron plasma waves which grow exponentially (at least initially).

This class of instabilities, in which quasi-CW light drives electron plasma waves seeded from noise in the plasma or in the light wave, are usually referred to as stimulated Raman scattering, or SRS. The reason for this is because the process is akin to the widely studied phenomena of Raman shifting in a molecular medium. This nonlinear optical process in a Raman active medium results because polarization oscillations in a molecule can serve to absorb or give energy to the incident laser wave, resulting in scattered light shifted in frequency by the resonant frequency of the Raman-active molecule. In a plasma, electron plasma waves take on the role of the polarization oscillation of the Raman-active molecules. This can be qualitatively described in a quantized photon picture like that of Figure 8.48 in which we can think of plasma oscillations at ω_{pe} acting as plasmons with quantized energy equal to $\hbar\omega_{pe}$. The energy of a laser photon can interact with the plasma, giving energy to a plasmon, resulting in a scattered photon of energy downshifted by $\hbar\omega_{pe}$, called a Stokes-shifted wave. If plasma oscillations are already present, the laser photon can receive one plasmon worth of energy during propagation and result in scattering of a photon with energy upshifted, an anti-Stokes wave. Because in a plasma the feedback we discussed can occur, we therefore refer to the process as stimulated.

8.6.1 Nonrelativistic Theory of Stimulated Raman Scattering in Plasma

Our principal goal at this point is find at what rate a plasma wave will grow subject to SRS. For now, we will restrict ourselves to driving laser waves at nonrelativistic intensity (say, $< 10^{17}$ W/cm^2). The equations we develop in this regime will actually be adapted fairly easily to the relativistic case. This can be developed rigorously for linearly polarized light, a convenience we do not have in the relativistic regime. In fact, there are a couple of ways to attack the problem of finding the growth rates and electron plasma mode frequency. We will, at first, employ the method most common in the LPI literature, Fourier decomposition of the various driving and scattered waves combined with a perturbation analysis on the wave and electron fluid equations. Later, we will revisit Raman Forward Scattering (FRS) using a different, intuitive approach to give us more insight into the underlying physics.

Our goal in this subsection is to find a dispersion relation, relating the laser frequency, ω, the Stokes wave frequency ω_S, the anti-Stokes frequency, ω_A, and the electron plasma wave frequency ω_e, which we expect will be complex since we consider a situation in which the plasma wave is expected to grow exponentially in time. This dispersion relation will relate these frequencies to the wavenumbers of the waves in question and the intensity of the driving laser wave. The Stokes and anti-Stokes frequencies are related by conservation of energy, $\omega_S = \omega - \omega_{pe}$, and $\omega_A = \omega + \omega_{pe}$ shifted by ω_{pe} because we assume that $\text{Re}[\omega_e] \cong \omega_{pe}$. The wave numbers must be related by conservation of momentum. Once we find this dispersion relation we can solve for the imaginary part of ω_e which yields the exponential growth rate in time of the plasma wave amplitude, $\text{Im}[\omega_e] = \Gamma_{SRS}$.

To match much of the original literature on SRS (Drake et al. 1974; Forslund et al. 1975; Kruer 2003) we will work, for the time being, with vector potential, **A**. The first equation we need is the wave equation for **A**, Eq. (8.165), neglecting the electrostatic potential,

$$\nabla^2 \mathbf{A} - \frac{1}{c^2} \frac{\partial^2 \mathbf{A}}{\partial t^2} = \frac{4\pi e n_e}{c} \mathbf{u}_e. \tag{8.365}$$

As has been our usual practice in the strong field regime, we assume that the zeroth-order electron plasma fluid motion is dominated by the laser oscillation,

$$\mathbf{u}_e \cong \mathbf{u}_{osc} = \frac{e\mathbf{A}}{mc}, \tag{8.366}$$

which, when inserted into Eq. (8.365), gives

$$\nabla^2 \mathbf{A} - \frac{1}{c^2} \frac{\partial^2 \mathbf{A}}{\partial t^2} = \frac{4\pi e^2 n_e}{mc^2} \mathbf{A}. \tag{8.367}$$

Working in the 1D limit, so that we assume that the light fields have transverse polarization and the electron plasma wave motion is purely longitudinal, the electron fluid velocity can be broken down

$$\mathbf{u}_e = u_{osc}\hat{\mathbf{x}} + u_z\hat{\mathbf{z}}, \tag{8.368}$$

where here we consider the longitudinal motion to be a small perturbation on the transverse laser-driven oscillations. In this limit the three equations we need, the wave equation, the electron density continuity equation (again, assuming that ions are stationary and provide a neutralizing background), and electron fluid momentum equation, can be written

$$\frac{\partial^2 \mathbf{A}}{\partial z^2} - \frac{1}{c^2} \frac{\partial^2 \mathbf{A}}{\partial t^2} = \frac{4\pi e^2 n_e}{mc^2} \mathbf{A}, \tag{8.369}$$

8.6 Plasma Instabilities Driven by Intense Laser Pulses

$$\frac{\partial n_e}{\partial t} + n_e \frac{\partial u_z}{\partial z} + u_z \frac{\partial n_e}{\partial z} = 0, \tag{8.370}$$

$$\frac{\partial u_z}{\partial t} + \frac{1}{2}\frac{\partial}{\partial z}\left| u_z \hat{z} + \frac{e\mathbf{A}}{mc} \right|^2 = \frac{e}{m}\frac{\partial \Phi}{\partial z}. \tag{8.371}$$

Now we linearize these equations by performing a perturbation analysis. We treat the light fields as a sum of large laser amplitude, $\mathbf{A}_0$, and small-perturbation Raman scattered wave

$$\mathbf{A} = \mathbf{A}_0 + \mathbf{A}_R. \tag{8.372}$$

We then consider small perturbations to the quiescent electron density in the usual way

$$n_e = n_{e0} + \delta n_e. \tag{8.373}$$

Insertion of these perturbation expansions into the wave equation and keeping terms first order in small quantities gives us an equation describing growth of the Raman-scattered light waves,

$$\frac{\partial^2 \mathbf{A}_R}{\partial z^2} - \frac{1}{c^2}\frac{\partial^2 \mathbf{A}_R}{\partial t^2} - \frac{4\pi e^2 n_{e0}}{mc^2}\mathbf{A}_R = \frac{4\pi e^2}{mc^2}\delta n_e \mathbf{A}_0, \tag{8.374}$$

a continuity equation for the small perturbations to electron density,

$$\frac{\partial}{\partial t}\delta n_e + n_{e0}\frac{\partial u_z}{\partial z} = 0, \tag{8.375}$$

and an equation for the small longitudinal component of electron fluid velocity,

$$\frac{\partial u_z}{\partial t} + \frac{1}{2}\frac{e^2}{m^2 c^2}\frac{\partial}{\partial z}\left(\mathbf{A}_0 \bullet \mathbf{A}_R^* + \mathbf{A}_0^* \bullet \mathbf{A}_R\right) = \frac{e}{m}\frac{\partial \Phi}{\partial z}. \tag{8.376}$$

Addition of Poisson's equation completes the set of equations we require:

$$\frac{\partial^2 \Phi}{\partial z^2} = -4\pi e \delta n_e. \tag{8.377}$$

Taking the time derivative of Eq. (8.375), the z derivative of (8.376) and combining these equations to eliminate u_z, yields a driven harmonic oscillator equation for the electron density perturbation,

$$\frac{\partial^2}{\partial t^2}\delta n_e + \omega_{pe}^2 \delta n_e = \frac{n_{e0}e^2}{2m^2 c^2}\frac{\partial^2}{\partial z^2}\left(\mathbf{A}_0 \bullet \mathbf{A}_R^* + \mathbf{A}_0^* \bullet \mathbf{A}_R\right). \tag{8.378}$$

If we assume constant drive-laser intensity (treating $\mathbf{A}_0$ as a constant, which means we ignore pump depletion as laser energy is deposited into the Raman field and electron plasma oscillation energy), we have two coupled equations, Eqs. (8.374) and (8.378), that describe the Raman scattering process in the plasma. These two equations also confirm our intuition about the SRS feedback process. Equation (8.378) illustrates that it is the beating of the laser field and the scattered, shifted Raman field that drives oscillations in δn_e at the resonance frequency ω_{pe}, and then Eq. (8.374) says that it is the beating of oscillations of the electron plasma wave with the laser field that serves as the source term to create the scattered Raman field. Clearly, such a two-step feedback process will lead to unstable growth of the plasma and Raman waves (making sure not to forget that these equations are only valid in the perturbation limit of small wave amplitudes).

At this point it is worth noting that these two equations describe undamped growth of both Raman and electron plasma waves. We know that, in fact, both waves will experience damping in a real plasma. The Raman scattered field can undergo inverse bremsstrahlung absorption and the electron plasma wave can be damped by electron–ion collisions and by the collisionless Landau damping mechanism (a process discussed in detail in many plasma physics texts and which is briefly derived and explained in Appendix B). So, a more accurate form of these two equations should include at least a phenomenologically motivated insertion of damping terms, which we denote as v_I for the scattered Raman light wave (mainly due to IB absorption) and v_P for the electron plasma wave, which can be calculated to good approximation as a sum of the electron–ion collision frequency and the Landau damping rate derived in Appendix B. Furthermore, we neglected the effects of force from the gradient of a finite electron temperature when we wrote down the electron fluid momentum equation, Eq. (8.371). (See Appendix B, section B.3 for the formal origin and form of that term). If we were to include it and assume an adiabatic ideal gas equation of state with electron fluid compression in one dimension, we would have an additional thermal correction term in the electron density equation, usually called the Bohm–Gross frequency (see Appendix B). With reinclusion of this thermal correction and the damping terms, the two coupled SRS equations take the form

$$\frac{\partial^2 \mathbf{A}_R}{\partial z^2} - \frac{1}{c^2}\frac{\partial^2 \mathbf{A}_R}{\partial t^2} - 2v_I \frac{\partial \mathbf{A}_R}{\partial t} - \frac{\omega_{pe}^2}{c^2}\mathbf{A}_R = \frac{4\pi e^2}{mc^2}\delta n_e \mathbf{A}_0, \tag{8.379}$$

$$\frac{\partial^2}{\partial t^2}\delta n_e + \omega_{pe}^2 \delta n_e + 2v_P \frac{\partial}{\partial t}\delta n_e + \frac{3k_B T_e}{m}\frac{\partial^2}{\partial z^2}\delta n_e = \frac{n_{e0}e^2}{2m^2 c^2}\frac{\partial^2}{\partial z^2}\left(\mathbf{A}_0 \cdot \mathbf{A}_R^* + c.c.\right). \tag{8.380}$$

The development of the Raman scatter growth rates can proceed with inclusion of these terms; this is done by a number of authors in the original treatments of SRS (Drake et al. 1974; Forslund et al. 1975). The terms add no insight for us and are of little importance in the strong field regime, where drive field intensities and growth rates can be much higher than those considered in traditional nanosecond LPI interactions. So we will neglect them in the next steps of our derivation.

Now, we Fourier-decompose the waves in question. We will treat the drive laser field as quasi-continuous with constant intensity and the Raman field as having a slowly varying amplitude; these two fields can be written

$$\mathbf{A}_0 = \frac{\mathbf{A}_{00}}{2}e^{i(kz-\omega t)} + c.c.; \quad \mathbf{A}_R = \frac{\mathbf{A}_{R0}}{2}e^{i(k_R z-\omega_R t)} + c.c., \tag{8.381}$$

while the plasma wave amplitude can be written

$$\delta n_e = \frac{\delta n_{e0}}{2}e^{i(k_e z - \omega_e t)} + c.c. \tag{8.382}$$

We retain the vector notation for the amplitudes of the light fields so that polarization information could be included.

8.6 Plasma Instabilities Driven by Intense Laser Pulses

The frequencies of these waves are related by conservation of energy, and I point out that the Raman field actually has two terms, oscillating at the Stokes and anti-Stokes frequencies so those frequencies are related by

$$\omega_S \equiv \omega - \omega_e; \quad \omega_A \equiv \omega + \omega_e. \tag{8.383}$$

In these relations it is implicit that ω_e refers to the real part of the electron plasma-wave oscillation frequency. The conditions on conservation of momentum of the scattered Raman field and excited plasma wave should read

$$\mathbf{k}_S \equiv \mathbf{k} - \mathbf{k}_e; \quad \mathbf{k}_A \equiv \mathbf{k} + \mathbf{k}_e. \tag{8.384}$$

A rigorous treatment would retain these wavenumbers as vectors to account for the possibility of scattering at various angles. For the most important effects, it is sufficient to consider scatter of the light along the laser propagation direction or counter to the laser axis, so the wavenumbers employed in Eq. (8.381) and (8.382) are implied to be along the z-direction. The scattered Raman light field then is

$$A_R = \frac{A_{S0}}{2} e^{i(k_S z - \omega_S t)} + \frac{A_{A0}}{2} e^{i(k_A z - \omega_A t)} + c.c. \tag{8.385}$$

We can insert this into (8.379) and since, as we will see in a moment, δn_e varies as A_{S0}^*, we group and keep terms that vary as $e^{i\omega_S t}$ to get

$$\left(-k_S^2 + \frac{\omega_S^2}{c^2} - \frac{\omega_{pe}^2}{c^2}\right) A_{S0}^* = \frac{\omega_{pe}^2}{2n_{e0}c^2} \delta n_{e0} A_{00}^*. \tag{8.386}$$

A similar exercise, grouping terms that vary as $e^{-i\omega_A t}$, gives

$$\left(-k_A^2 + \frac{\omega_A^2}{c^2} - \frac{\omega_{pe}^2}{c^2}\right) A_{A0} = \frac{\omega_{pe}^2}{2n_{e0}c^2} \delta n_{e0} A_{00}. \tag{8.387}$$

Next, substitution of the Fourier-decomposed waves into Eq. (8.380) and keeping terms oscillating like $e^{-i\omega_e t}$ yields

$$\left(-\omega_e^2 + \omega_{pe}^2\right) \delta n_{e0} = \frac{-k_e^2 n_{e0} e^2}{2m^2 c^2} \left(\mathbf{A}_{00}^* \cdot \mathbf{A}_{A0} + \mathbf{A}_{00} \cdot \mathbf{A}_{S0}^*\right). \tag{8.388}$$

This equation illustrates that the electron plasma wave is actually driven by the complex conjugate component of the Stokes field, A_{S0}^*, so we expect that the scattered Stokes wave, if the drive laser were circularly polarized, will actually have polarization with the opposite sense of rotation, though the anti-Stokes wave would retain the polarization of the drive laser wave.

Now it is a straightforward algebraic exercise to combine these three equations and eliminate δn_e, A_{S0}^* and A_{A0}. To simplify the final dispersion relation that results from this elimination, it is convenient to introduce a Stokes and an anti-Stokes dispersion term:

$$D_S \equiv \frac{\omega_S^2}{c^2} - \frac{\omega_{pe}^2}{c^2} - k_S^2, \tag{8.389}$$

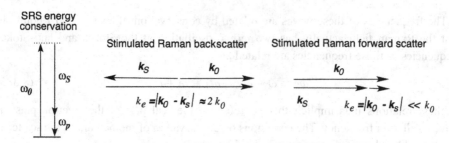

Figure 8.49 Illustration of how photon and plasma-wave momentum is conserved in Raman scattering of laser light in a plasma. When a Stokes wave is backscattered, it produces a plasma wave with large wavenumber (small plasma wavelength). However, in Raman Forward Scattering, the plasma wave produced from production of a Stokes wave that copropagates in the same direction as the driving laser wave must have a small wavenumber.

$$D_A \equiv \frac{\omega_A^2}{c^2} - \frac{\omega_{pe}^2}{c^2} - k_A^2. \tag{8.390}$$

Insertion of these definitions with combination of Eqs. (8.386)–(8.388) gives us our desired SRS dispersion relationship:

$$\omega_e^2 - \omega_{pe}^2 = \frac{k_e^2 \omega_{pe}^2 e^2}{4m^2 c^4} A_{00}^2 \left(\frac{1}{D_S} + \frac{1}{D_A} \right). \tag{8.391}$$

This is an equation that can be used to find the complex electron oscillation frequency, which will depend on the ponderomotive energy of the laser, as the relation includes a term proportional to A_{00}^2.

Numerical solution of this equation is straightforward, to find both the imaginary part of ω_e, the instability growth rates we seek, as well as solutions for the real parts of ω_e, telling us what the actual oscillation frequency of the excited plasma waves can be. This we will present in a moment, but first it is informative to explore this equation in a couple of important limits.

8.6.2 Stimulated Raman Backscattering

Our principal goal at this point is find at what rate a plasma wave will grow. Consider first the growth of the instability-generating SRS light that is backscattered, (Raman Backscatter or RBS). Look first at the implications that momentum conservation demands, illustrated in the middle wavenumber diagram of Figure 8.49. In the modest-density plasmas usually encountered in experiments, the frequency shift of the scattered Raman light is modest, so, $k_S \approx k_A \approx k$. This means that the wavenumber of the excited plasma wave during SRS backscatter is large, $k_e \approx 2k$, so fairly slow-phase-velocity ($v_{\varphi e} = \omega_e/k_e \approx \omega_{pe}/2k \ll c$), short-wavelength electron plasma waves are excited. One practical consequence of this is that because these waves have low phase velocity, they can pick up electrons from the background plasma with thermal velocity comparable to the wave-phase velocity and accelerate

8.6 Plasma Instabilities Driven by Intense Laser Pulses

them. This leads to multi-keV hot electron generation, a process we will look at in greater detail in the next section.

The physical explanation for RBS in a plasma is that the backscattered radiation beats with the forward-propagating laser light. Because the backscattered radiation has a slight frequency shift, the beating does not set up a stationary standing wave (as one produced, say, from the reflection of monochromatic light from a mirror) but a beating pattern that moves with velocity $\sim \Delta\omega/2k = \omega_{pe}/2k$. This beating in intensity drives the plasma wave. The electron-density oscillations act as a grating which backscatters more of the laser light. Because the electron wave is moving, the backscattered light is Doppler shifted (down for Stokes radiation) and builds the amplitude of the backscattered field. This sets up the exponentially growing feedback process.

To derive a growth rate for RBS, first look at the two dispersion denominators, (8.389) and (8.390). Since $\omega_e \approx \omega_{pe}$ and $\omega_A \approx \omega_S \approx \omega$, these two terms are approximately

$$D_S \approx \frac{\omega^2}{c^2} - \frac{2\omega\omega_{pe}}{c^2} - k_S^2 \approx -\frac{2\omega\omega_{pe}}{c^2} + 2kk_e - k_e^2, \tag{8.392}$$

$$D_A \approx \frac{\omega^2}{c^2} + \frac{2\omega\omega_{pe}}{c^2} - k_A^2 \approx \frac{2\omega\omega_{pe}}{c^2} - 2kk_e - k_e^2. \tag{8.393}$$

Since the last two terms are almost equal in both equations, they nearly cancel in D_S and add in D_A; the D_S term is clearly much smaller and therefore dominates the dispersion relation. The upshot is that the Stokes field is the resonant field. The anti-Stokes field is not resonant and yields damped solutions. Therefore, only the downshifted Stokes wave grows in backscattering. RBS is often referred to as a three-wave process, since the laser, Stokes and plasma waves are the important participants.

Now ignore the $1/D_A$ term and treat the plasma oscillation frequency as complex: $\omega_e = \omega_{eR} + i\Gamma_R$. The plasma-wave amplitude, then, grows like $\delta n_e \sim e^{\Gamma t}$. First, consider the solution for weak fields, such that the growth rate is small compared to the plasma frequency. Ignore the Γ_R^2 term in D_S and the left side of Eq. (8.391) to give

$$2i\Gamma_R \omega_{pe} = \frac{k_e^2 \omega_{pe}^2}{m} \left(\frac{e^2 A_{00}^2}{4mc^2}\right) \frac{1}{(\omega - \omega_{eR} - i\Gamma_R)^2 - \omega_{pe}^2 - c^2(k - k_e)^2}. \tag{8.394}$$

The maximum growth rate occurs for the resonance condition when the denominator goes to zero,

$$(\omega - \omega_{eR} - i\Gamma_R)^2 - \omega_{pe}^2 - c^2(k - k_e)^2 = 0, \tag{8.395}$$

which gives for the instability growth rate

$$\Gamma_{RBS} = \sqrt{\frac{k_e^2 \omega_{pe}}{4m} \left(\frac{e^2 A_{00}^2}{4mc^2}\right)\left(\frac{1}{\omega - \omega_{eR}}\right)}. \tag{8.396}$$

Since $\omega \gg \omega_{eR}$ (and $k_e \approx 2k$), the growth rate reduces to

$$\Gamma_{RBS} \cong \sqrt{\frac{k^2}{m}\frac{\omega_{pe}}{\omega}\left(\frac{e^2 A_{00}^2}{4mc^2}\right)}$$

$$= \sqrt{\frac{k^2}{m}\frac{\omega_{pe}}{\omega}U_p} \cong \sqrt{\omega\omega_{pe}\frac{U_p}{mc^2}}, \quad (8.397)$$

or in terms of the normalized vector potential can be succinctly written

$$\Gamma_{RBS} = \sqrt{\omega\omega_{pe}}\left(\frac{a_0}{2}\right). \quad (8.398)$$

Remember, this growth rate is valid in the so-called weak-coupling regime in which the laser-drive intensity and plasma-electron density are low enough to insure that $\Gamma_{RBS} \ll \omega_{pe}$. This condition is met when $(a_0/2)(n_{crit}/n_e) \ll 1$, which suggests that RBS is weakly coupled in plasma of $n_e \sim 10^{18}$ cm^{-3} when $a_0 \lesssim 0.4$, intensity up to about 10^{17} W/cm^2. What is usually observed in experiments, particularly those up to this intensity with pulse durations of hundreds of femtoseconds focused into a moderate-density gas jet, is backscattered light with a well-defined spectral feature at the Stokes-shifted wavelength (Darrow et al. 1992). Note that the RBS growth rate, $\Gamma_{RBS} \sim n_{e0}^{1/4}\lambda^{1/2}I^{1/2}$, scales weakly with intensity and is faster for longer-wavelength laser pulses.

At higher density and higher laser intensities, RBS can enter a regime in which $\Gamma_{RBS} \gg \omega_{pe}$, which is usually termed the strong-coupling regime. In that case the dispersion relation gives for the growth rate

$$-\left(\Gamma_{RBS}^{(SC)}\right)^2 \cong \frac{k_e^2 \omega_{pe}}{m}\left(\frac{e^2 A_{00}^2}{4mc^2}\right)\left(\frac{-1}{2i\Gamma_{RBS}^{(SC)}\omega}\right), \quad (8.399)$$

which means it becomes

$$\Gamma_{RBS}^{(SC)} \cong i^{-1/3}\left(\frac{k_e^2 \omega_{pe}}{2\omega}U_p\right)^{1/3} \quad (8.400)$$

$$\cong i^{-1/3}\left(\frac{\omega_{pe}^2 \omega}{2}\right)^{1/3} a_0^{2/3}. \quad (8.401)$$

The real part of this is the strongly coupled exponential growth rate, which, with use of the fact that $i^{-1/3} = e^{-i\pi/6} = \cos 30° - i\sin 30°$, is in terms of a_0,

$$\Gamma_{RBS}^{(SC)} = \sqrt{3}\left(\frac{\omega_{pe}^2 \omega}{16}\right)^{1/3} a_0^{2/3}. \quad (8.402)$$

Now, interestingly, the plasma oscillation frequency is not at the usual plasma frequency but is shifted by the intense driving laser field. These oscillation frequencies, sometimes called quasi-modes, are at the real part of ω_e:

8.6 Plasma Instabilities Driven by Intense Laser Pulses

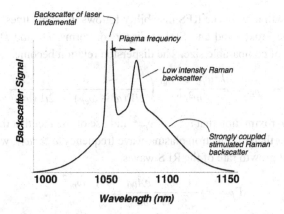

Figure 8.50 Schematic illustration of the spectrum of stimulated Raman backscatter in the intensity and density regime in which the process is strongly coupled (instability growth rate is comparable to the electron plasma frequency). The sharp downshifted Stokes wave from the lower-intensity parts of the pulse rides on top of a broad spectral feature from the strongly coupled backscatter. Figure was generated using data reported in Darrow et al. (1992).

$$\omega_e \approx \omega_{pe} + \left(\frac{\omega_{pe}^2 \omega}{16}\right)^{1/3} a_0^{2/3}. \tag{8.403}$$

Such strongly coupled backscatter is seen in experiments with subpicosecond pulses at intensity above 10^{18} W/cm^2. The kind of backscattered spectra seen in such experiments is schematically depicted in Figure 8.50. In addition to the scattered laser light frequency, Stokes-shifted light at $\omega - \omega_{pe}$ from the low-intensity parts of the pulse is seen as a sharp spectral feature riding on top of a broad peak of light, characteristic of strongly coupled SRS growth.

8.6.3 Raman Forward Scattering

Now consider a similar process, forward scatter (or RFS). In this situation the Raman-shifted light copropagates with the laser field and the shifted radiation beats with the laser wave to create ponderomotive energy modulations that can resonantly drive the plasma wave (McKinstrie and Bingham 1992; Wilks et al. 1992; Mori et al. 1994; Sakharov and Kirsanov 1994). Unlike RBS, RFS creates a beating that travels near c, so the plasma waves set up have phase velocity near c. If there are plasma electrons with thermal speed near c, they can be picked up in the wave and accelerated to many mega-electron volts of energy. In fact, RBS can serve as a way to boost thermal electrons to high enough energy that they can be picked up by the much faster-moving RFS scatter wave. The difference between RBS and RFS can be seen in the momentum conservation diagram of Figure 8.49. The right-hand vector diagram for RFS shows that the plasma wave will have wave number at k_{pe}, so the phase velocity of the plasma wave is $v_{\varphi e} = \omega_{pe}/k_{pe} \cong c$.

Finding the growth rate for the RFS instability follows the same lines as RBS; however, examination of the Stokes and anti-Stokes dispersion terms, D_S and D_A, illustrates that both must be kept of comparable size. The dispersion relation becomes

$$\omega_{eR}^2 - 2i\Gamma\omega_{eR} - \omega_{pe}^2 = \frac{k_e^2 \omega_{pe}^2 e^2 A_{00}^2}{4m^2 c^6}\left(\frac{-1}{2i\Gamma(\omega - \omega_{eR})} + \frac{1}{2i\Gamma(\omega + \omega_{eR})}\right), \quad (8.404)$$

which, with the approximation that $\omega_{eR}^2 \approx \omega_{pe}^2$ and use of the fact that the laser frequency is now much bigger than the electron plasma-wave frequency $\omega \gg \omega_{er}$, we can easily solve for the exponential growth rate of the RFS waves:

$$\Gamma_{RFS} = \frac{1}{\sqrt{8}}\frac{\omega_{pe}^2}{\omega}\frac{eA_{00}}{mc^2} = \frac{1}{\sqrt{8}}\frac{\omega_{pe}^2}{\omega}a_0. \quad (8.405)$$

Note that this growth rate is for both Stokes and anti-Stokes waves. So the forward scatter process is often spoken of as a resonant four-wave process.

It is instructive to compare the RFS growth rate to the backscatter growth rate, and we see that

$$\Gamma_{RBS}/\Gamma_{RFS} = \sqrt{2}(\omega/\omega_{pe})^{3/2} \gg 1. \quad (8.406)$$

Backscatter grows much faster than forward scatter, independent of intensity (at least in the nonrelativistic intensity regime). This is why in nanosecond pulse interactions RBS dominates. However, because of the convective nature of back scattering (with the growing scattered field propagating backward, out of the laser pulse), ultrashort pulses tend to exhibit RFS as the dominant mechanism.

8.6.4 Stimulated Raman Scattering at Relativistic Intensity

Let us now revisit the SRS equations in the regime of relativistic intensity ($> \sim 10^{18}$ W/cm^2). We will, to be consistent with the literature on this subject, recast the relevant equations in terms of normalized potentials (Barr et al. 1995; Decker et al. 1996b). In the relativistic limit, the 1D wave equation for normalized vector potential can be written as a slight simplification of Eq. (8.171) and reads

$$\frac{\partial^2 \mathbf{a}}{\partial z^2} - \frac{1}{c^2}\frac{\partial^2 \mathbf{a}}{\partial t^2} = \frac{\omega_{pe}^2}{c^2}\frac{n_e}{n_{e0}}\frac{\mathbf{a}}{\gamma}. \quad (8.407)$$

Recall that, at this point, the relativistic gamma factor still must be written in its most general rapidly varying form, accounting for both transverse laser-field oscillations and longitudinal plasma oscillation momentum as

$$\gamma = \sqrt{1 + \tilde{\mathbf{p}}^2} = \sqrt{1 + \tilde{p}_z^2 + a^2}. \quad (8.408)$$

This leads to the primary complication of these equations in the relativistic regime: Fourier decomposition of the right-hand side of the wave equation, if linear polarization is considered, now has an infinite series of harmonics of the laser field because of the $1/\gamma$ term. This should be clear if one recalls the solution for free-electron motion in the field,

8.6 Plasma Instabilities Driven by Intense Laser Pulses

which becomes highly anharmonic. The usual way to work around this complication is to assume that the laser is circularly polarized, which eliminates the higher-order harmonics in the $1/\gamma$ term, because a^2 is now just a constant $= a_0^2/2$. Decker et al. have shown that the growth rate of the Raman instability for linear polarization is close to that of circular polarization (Decker et al. 1996b), so the SRS growth derived in this way is still essentially applicable to relativistic linearly polarized laser experiments.

The other equations needed are the relativistic versions of the 1D continuity equation, plasma momentum equation and Poisson's equation:

$$\frac{1}{c}\frac{\partial n_e}{\partial t} + \frac{\partial}{\partial z}\frac{n_e \tilde{p}_z}{\gamma} = 0, \tag{8.409}$$

$$\frac{1}{c}\frac{\partial \tilde{p}_z}{\partial t} + \frac{\tilde{p}_z}{\gamma}\frac{\partial \tilde{p}_z}{\partial z} = \frac{\partial \phi}{\partial z} - \frac{1}{2\gamma}\frac{\partial}{\partial z}a^2, \tag{8.410}$$

$$\frac{\partial^2 \phi}{\partial z^2} = \frac{\omega_{pe}^2}{c^2}\left(\frac{n_e}{n_{e0}} - 1\right) = \frac{\omega_{pe}^2}{c^2}\frac{\delta n_e}{n_{e0}}. \tag{8.411}$$

As in the nonrelativistic case, we will linearize these equations, so, as before, we break the vector potential into the laser field and scattered Raman field, $\mathbf{a} = \mathbf{a}_L + \mathbf{a}_R$. The relativistic factor is

$$\gamma^2 = 1 + a_0^2 + 2\mathbf{a}_L \cdot \mathbf{a}_R + a_R^2 + \tilde{p}_z^2 \tag{8.412}$$

so its inverse can be Taylor-expanded to first order in small quantities,

$$\frac{1}{\gamma} \cong \frac{1}{\gamma_0}\left(1 - \frac{\mathbf{a}_L \cdot \mathbf{a}_R}{\gamma_0^2}\right), \tag{8.413}$$

with the usual definition of the laser-driven relativistic factor for circular polarization, $\gamma_0 = (1 + a_0^2/2)^{1/2}$. The wave equation becomes

$$\frac{\partial^2 \mathbf{a}}{\partial z^2} - \frac{1}{c^2}\frac{\partial^2 \mathbf{a}}{\partial t^2} = \frac{\omega_{pe}^2}{c^2}\frac{n_e}{n_{e0}}\mathbf{a} - \frac{\omega_{pe}^2}{2c^2\gamma_0^3}\frac{n_e}{n_{e0}}(\mathbf{a}_L \cdot \mathbf{a}_R)\mathbf{a}, \tag{8.414}$$

and the linearized continuity and momentum equations are

$$\frac{1}{c}\frac{\partial}{\partial t}\delta n_e + \frac{n_{e0}}{\gamma_0}\frac{\partial \tilde{p}_z}{\partial z} = 0, \tag{8.415}$$

$$\frac{1}{c}\frac{\partial \tilde{p}_z}{\partial t} = \frac{\partial \phi}{\partial z} - \frac{1}{2\gamma_0}\frac{\partial}{\partial z}\left(\mathbf{a}_L \cdot \mathbf{a}_R^* + \mathbf{a}_L^* \cdot \mathbf{a}_R\right). \tag{8.416}$$

We proceed as before, so we will skim quickly over the details. We use Poisson's equation to eliminate ϕ, and combine (8.415) with (8.416) to eliminate $\tilde{p}_z$. The wave equation for the electron-fluid perturbation can be written

$$\frac{\partial^2}{\partial t^2}\delta n_e + \frac{\omega_{pe}^2}{\gamma_0}\delta n_e = \frac{n_{e0}c^2}{\gamma_0}\frac{\partial^2}{\partial z^2}\left(\mathbf{a}_L \cdot \mathbf{a}_R^* + \mathbf{a}_L^* \cdot \mathbf{a}_R\right). \tag{8.417}$$

Writing the Stokes and anti-Stokes fields explicitly, and Fourier-decomposing, transforms the two equations for electron-fluid motion and scattered Stokes field as

$$\left(-\omega_e^2 + \frac{\omega_{pe}^2}{\gamma_0}\right)\delta n_e = \frac{-k_e^2 n_{e0} c^2}{2\gamma_0^2}\left(\mathbf{a}_0^* \cdot \mathbf{a}_{A0} + \mathbf{a}_0 \cdot \mathbf{a}_{S0}^*\right), \tag{8.418}$$

$$\left(-k_e^2 - \frac{\omega_e^2}{c^2} - \frac{\omega_{pe}^2}{c^2\gamma_0}\right)\mathbf{a}_{S0}^* = \frac{\omega_{pe}^2}{2n_{e0}c^2\gamma_0}\delta n_e \mathbf{a}_0$$

$$- \frac{\omega_{pe}^2}{c^2\gamma_0^3}\left(\frac{\mathbf{a}_0 \cdot \mathbf{a}_{S0}^*}{4} + \frac{\mathbf{a}_0^* \cdot \mathbf{a}_{A0}}{4}\right)\mathbf{a}_0$$

$$= \frac{\omega_{pe}^2}{2n_{e0}c^2\gamma_0}\delta n_e \mathbf{a}_{00}\left(1 - \frac{\omega_e^2}{k_e^2 c^2} + \frac{\omega_{pe}^2}{\gamma_0 k_e^2 c^2}\right). \tag{8.419}$$

Similarly for the anti-Stokes field

$$\left(-k_e^2 + \frac{\omega_e^2}{c^2} - \frac{\omega_{pe}^2}{c^2\gamma_0}\right)\mathbf{a}_{A0} = \frac{\omega_{pe}^2}{2n_{e0}c^2\gamma_0}\delta n_e \mathbf{a}_0\left(1 - \frac{\omega_e^2}{k_e^2 c^2} + \frac{\omega_{pe}^2}{\gamma_0 k_e^2 c^2}\right). \tag{8.420}$$

Introducing, finally, the relativistically corrected Stokes and anti-Stokes dispersion factors

$$D'_{S,A} \equiv \frac{\omega_{S,A}^2}{c^2} - \frac{\omega_{pe}^2}{c^2\gamma_0} - k_{S,A}^2 \tag{8.421}$$

leads to the relativistic dispersion relation for SRS:

$$\omega_e^2 + \frac{\omega_{pe}^2}{\gamma_0} = \frac{k_e^2 \omega_{pe}^2}{4\gamma_0^3}\left(1 - \frac{\omega_e^2}{k_e^2 c^2} + \frac{\omega_{pe}^2}{\gamma_0 k_e^2 c^2}\right)\left(\frac{1}{D'_S} + \frac{1}{D'_A}\right)a_0^2. \tag{8.422}$$

Compare this result with Eq. (8.391). If we accept that in the middle set of parentheses, the second and third terms are almost equal and therefore nearly cancel, then we see that we can find the relativistic corrections to our previously derived growth rates merely by making the following substitutions:

$$\omega_{pe} \to \omega_{pe}/\gamma_0^{1/2} \quad a_0^2 \to a_0^2/\gamma_0^2.$$

The Raman backscatter growth rate becomes

$$\Gamma'_{RBS} = \frac{\sqrt{\omega \omega_{pe}}}{\gamma_0^{5/4}}\left(\frac{a_0}{2}\right), \tag{8.423}$$

which, in the strong coupling limit, can be written

$$\Gamma'^{(SC)}_{RBS} = \frac{\sqrt{3}}{\gamma_0}\left(\frac{\omega_{pe}^2 \omega}{16}\right)^{1/3} a_0^{2/3}. \tag{8.424}$$

The RFS growth rate becomes

$$\Gamma'_{RFS} = \frac{1}{\sqrt{8}}\frac{\omega_{pe}^2}{\omega}\frac{a_0}{\gamma_0^2}. \tag{8.425}$$

8.6 Plasma Instabilities Driven by Intense Laser Pulses

The most significant thing about these corrected growth rates is that they actually decrease at very high intensity. At high a_0, we see that $\Gamma_{RBS}^{'(SC)} \sim a_0^{-1/3}$, and the RFS growth rate decreases with drive field as $\Gamma_{RFS}' \sim a_0^{-1}$. In fact, RFS peaks with $a_0^{(RFS\,Max)} = \sqrt{2}$, an intensity of just above 10^{18} W/cm^2, and backscatter peaks at an even lower field $a_0^{(RBS\,Max)} = 1/\sqrt{2}$. The RFS growth rate at this maximum field amplitude is $\omega_{pe}^2/4\omega$. This is $0.1\times$ to $0.01\times \omega_{pe}$ for typical underdense plasma experiments. So clearly RFS never enters the strongly coupled regime.

The next question that we might ask is, when might one expect to see significant backscatter in an experiment? This likely occurs when a few e-foldings of growth are possible as the Stokes field propagates back through the pulse. This condition would imply a threshold of $\Gamma_{RBS}\tau_p \sim 10$ if τ_p is the drive laser pulse duration. So a 1 ps long, 1 μm wavelength pulse in a plasma of 10^{18} e/cm^3 meets this condition when $a_0 \approx 0.03$ or an intensity of a few times 10^{16} W/cm^2. A sub-100 fs pulse would require intensity closer to 10^{18} W/cm^2. Shorter wavelengths require higher intensity, as the threshold intensity would scale roughly as $\sim 1/\lambda$. These estimated threshold intensities are consistent with experimental observations (Blyth et al. 1995). Typical Raman backscattered laser energy fraction for an experiment as a function of peak intensity is illustrated in Figure 8.51 (in this illustration with 248 nm light having 400 fs duration in atmospheric-density plasma).

The threshold for Raman forward scatter depends on different conditions than RBS. Because of the copropagation of the shifted light with the laser pulse, the time scale for growth depends on the length of propagation in a plasma. In typical experiments this can usually be between 1 mm and 1 cm. The condition for 10 e- foldings implies that the field amplitude threshold for RFS in a 10^{18} cm^{-3} plasma would be $a_0 \geq 3 \times 10^4 \, c/(\omega L_{prop})$ which suggests that intensity around 10^{18} W/cm^2 is needed, again in agreement with experiments.

As a final observation, now that we have considered RFS and RBS in the various limits, it is interesting to plot the instability growth rates as a function of plasma wavenumber by numerical solution of the relativistic dispersion relation, Eq. (8.422). This is done in Figure 8.52 for both forward scatter and backscatter for a few intensities. In Figure 8.52a, the growth rate of RBS versus electron wavenumber for three field values is plotted, showing the transition to the strongly coupled regime at high intensity. At low a_0, the growth rate is modest and centered in a region around the wavenumber expected for linear plasma waves. At a_0 of ~ 1 this spectrum of growth wavenumbers broadens as the instability enters the strongly coupled regime and finally at very high intensity, when $a_0 = 10$, while peak growth rate is reduced by relativistic mass increase, waves over a very large range of wavenumbers would be driven. In Figure 8.52b the same numerical solutions of the RBS growth rate are shown, plotted as a function of the real part of electron oscillation frequency, showing why experiments observe a broadened Stokes backscatter spectrum in the strongly coupled regime. Finally, in Figure 8.52c, the RFS growth rates are plotted, illustrating that forward scattering is largely confined to shifts of well-defined frequency and that the strongly coupled regime is never reached even at very high intensity.

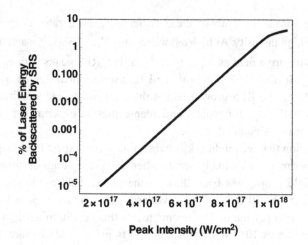

Figure 8.51 Typical data observed for SRS backscatter percentage of incident laser light for few-hundred-femtosecond pulses in electron density $\sim$1% critical. Plot was generated from data published in Blyth et al. (1995).

8.6.5 Conceptual Picture of Stimulated Raman Forward Scattering

The Fourier technique that we have employed in the previous two sections is convenient for deriving dispersion relations and growth rates in various regimes, but it does not yield much insight into the physical reasons for something like the growth of Raman Forward Scattering. We have made qualitative explanations for the feedback process but the real physical effects leading to the growth rates we have derived are still not obvious from the previous derivations. In fact, in the case of RFS, the growth rates can be derived in a much more intuitive way. This derivation was elegantly described by Mori (1997), and we revisit here his enlightening treatment of RFS.

We have seen from Figure 8.49 that RFS excites waves with small wavenumber and hence large wavelength. The phase velocity of these waves, $v_\varphi \cong \omega_{pe}/k_e \cong c$, is such that they copropagate with the laser pulse. The growth of the plasma wave can modulate a light field in two ways: (1) Longitudinal bunching of photons leads to intensity modulation and (2) photon acceleration leads to frequency modulation of the light. Both modulate the ponderomotive energy U_p at the (relativistic) plasma frequency $\omega'_{pe} = \omega_{pe}/\gamma_0^{1/2}$. So let's consider modulation by these two mechanisms. We can write the local change in the amplitude of the vector potential of the laser field in terms of the longitudinal bunching and frequency modulation as

$$\Delta \langle a^2 \rangle = -\frac{\Delta L}{L} \langle a^2 \rangle - \frac{\Delta \omega}{\omega} \langle a^2 \rangle. \tag{8.426}$$

Variation of a^2 with proper time then can be written

$$\frac{\partial}{\partial \tau} \langle a^2 \rangle = -\frac{1}{L}\frac{\partial L}{\partial \tau} \langle a^2 \rangle - \frac{1}{\omega}\frac{\partial \omega}{\partial \tau} \langle a^2 \rangle, \tag{8.427}$$

8.6 Plasma Instabilities Driven by Intense Laser Pulses

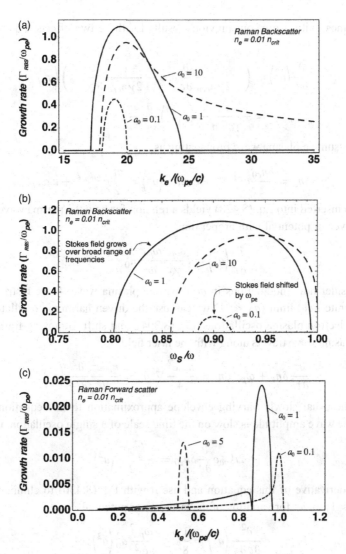

Figure 8.52 (a) Numerical solution of the relativistic dispersion relation for the imaginary part of the electron plasma frequency (the instability growth rate) in stimulated Raman backscatter as a function of induced plasma wavenumber. The solution is for plasma with n_e that is 1% n_{crit} and is plotted for three different values of the light a_0. (b) Shows the same calculation plotting instability growth rate as a function of the real part of the electron plasma wave frequency. These plots show the transition from backscattering of a well-defined Stokes frequency in the weakly coupled regime to the broad range of frequencies and wavenumbers that can grow in the strongly coupled regime at higher intensity. (c) Numerical solution of the same dispersion relation for the case of Raman Forward Scattering. This illustrates that RFS growth rate is considerably lower than backscatter under the same conditions and that strong coupling never occurs for RFS. The instability growth is defined to a sharp peak in electron wavenumber.

which becomes, with use of our previous results for these two effects, Eqs. (8.360) and (8.363),

$$\frac{\partial}{\partial \tau}\langle a^2\rangle = \left(-\frac{c}{2\gamma n_{crit}}\frac{\partial}{\partial \zeta}\delta n_e - \frac{c}{2\gamma n_{crit}}\frac{\partial}{\partial \zeta}\delta n_e\right)\langle a^2\rangle$$

$$= -\frac{c}{\gamma_0 n_{e0}}\frac{\omega_{pe}^2}{\omega^2}\langle a^2\rangle \frac{\partial}{\partial \zeta}\frac{\delta n_e}{n_{e0}}. \quad (8.428)$$

As before, assume a plasma wave propagating as

$$\delta n_e = \frac{\delta n_{e0}}{2}e^{-ik_{pe}\zeta} + c.c.; \quad \langle a^2\rangle = \frac{a_0^2}{2}e^{-ik_{pe}\zeta} + c.c., \quad (8.429)$$

which when inserted into Eq. (8.428) yields a relation for how the plasma wave affects the laser field's vector potential with proper time

$$\frac{\partial}{\partial \tau}\langle a^2\rangle = i\frac{ck_{pe}}{\gamma_0}\frac{\omega_{pe}^2}{\omega^2}a_0^2\frac{\delta n_e}{n_{e0}}. \quad (8.430)$$

Now consider how modulations in a_0 drive the plasma wave. If we restrict ourselves to the moderate field limit ($a_0 \ll 1$) we can use the driven harmonic oscillator equation describing electron plasma oscillations, Eq. (8.304), and shift to moving frame variables since this plasma wave travels along with the laser field,

$$\frac{\partial^2}{\partial \tau^2}\delta n_e + \omega_{pe}^2 \delta n_e = \frac{n_{e0}c^2}{4}\nabla^2 a_0^2 = \frac{n_{e0}c^2}{2}\frac{\partial^2}{\partial \zeta^2}\langle a^2\rangle. \quad (8.431)$$

We apply the usual slowly varying envelope approximation to this equation, since the growth of the wave amplitude is slow on the time scale of a single oscillation, to give

$$-2ik_{pe}c\frac{\partial}{\partial \zeta}\delta n_{e0} = -\frac{k_{pe}^2 c^2}{2}\langle a^2\rangle. \quad (8.432)$$

Take the τ derivative of this equation and use it with Eq. (8.430) to eliminate δn_e (and assume $\gamma_0 \approx 1$) to get for variation of the laser vector potential

$$\frac{\partial^2}{\partial \zeta \partial \tau}\langle a^2\rangle = \frac{1}{8}k_{pe}^2 c^2 \frac{\omega_{pe}^2}{\omega^2}a_0^2\langle a^2\rangle. \quad (8.433)$$

This we can intuitively associate with the Raman forward scatter growth rate, which we note, affects the growth of a^2 modulations as

$$\frac{\partial^2}{\partial \zeta^2}\langle a^2\rangle = \Gamma_{RFS}^2\langle a^2\rangle. \quad (8.434)$$

Comparison with Eq. (8.433) delivers a result for the FRS growth rate

$$\Gamma_{RFS} = \frac{1}{\sqrt{8}}\frac{\omega_{pe}^2}{\omega}a_0, \quad (8.435)$$

which is identical to the nonrelativistic FRS growth rate derived in Eq. (8.405).

What this essentially tells us is that the growth of FRS is a result of equal contributions of photon acceleration and longitudinal bunching of laser energy. This analysis

8.6 Plasma Instabilities Driven by Intense Laser Pulses

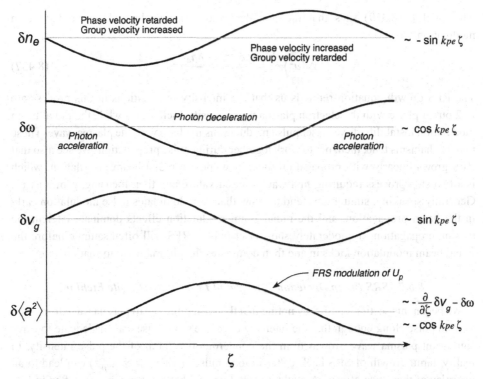

Figure 8.53 Schematic illustration of how plasma waves excited by Forward Raman Scattering produce modulations in the ponderomotive energy of the laser pulse through modulation of both field amplitude by group velocity modulation and frequency modulation through photon acceleration. The modulations of ponderomotive energy on the pulse envelope that result are $\pi/2$ out of phase with the density modulations.

and Eq. (8.430) also illustrate that the modulations imposed on a are $\pi/2$ out of phase with the modulations in the electron density. This interplay of wave phase is illustrated in Figure 8.53.

8.6.6 Pulse Modulation by SRS Compared to Radial Beam Modulation

We see that RFS has a consequence on the pulse-intensity envelope that is much like that seen when the radial beam modulation instability grows. Both modulate the pulse U_p at the plasma frequency and lead to further resonant excitation of plasma waves with phase velocity moving with the pulse. However, they do have one subtle difference: The phase relationship between the plasma wave and the intensity modulation on the laser pulse are different.

Equation (8.430) tells us that in RFS the amplitude of modulation of the vector potential envelope grows as

$$\Delta \langle a^2 \rangle \sim ick_{pe}\frac{\omega_{pe}^2}{\omega^2}\frac{a_0^2}{2}\frac{\delta n_e}{n_{e0}}\tau, \tag{8.436}$$

whereas Eq. (8.319) says that the modulation of this envelope during radial beam modulation grows as

$$\Delta \langle a^2 \rangle \sim \frac{c^2}{w^2} \frac{\omega_{pe}^2}{\omega^2} a_{00}^2 \frac{\delta n_e}{n_{e0}} \tau^2. \tag{8.437}$$

The RFS growth equation reminds us that the intensity modulations in this process are $\pi/2$ out of phase with the electron plasma wave-density oscillations while the radial beam modulation instability drives laser pulse modulations in phase with the plasma wave. These two mechanisms tend, then, not to work together during pulse propagation. We see also that RFS grows linearly with propagation time, τ, while the radial beam modulation, which is a two-step process requiring first wavefront curvature and then focusing, grows as τ^2. Generally speaking, simulations tend to show that RFS dominates pulse modulation early in the beam propagation, and the radial beam modulation effects dominate later in the plasma propagation. In moderately short laser pulses, RFS will often saturate before the radial beam modulation kicks in and then dominates the plasma wave formation.

8.6.7 SRS Pump Depletion Effects and Laser Pulse Profile Etching

We began this section on SRS by noting that this instability led to plasma wave formation even in pulses long enough that the intensity envelope's slow rise was inadequate to drive significant plasma wave growth from the ponderomotive ramp of the pulse intensity. In reality, rapid growth of SRS backscatter in long pulses (when $\tau_p \gg \omega_{pe}^{-1}$) can lead to an interplay of these two effects. A qualitative picture of what can happen, seen often in PIC simulations (Decker et al. 1996b), is illustrated in Figure 8.54.

When a long pulse enters the plasma, its slow rise does not excite a plasma wave initially. Raman backscatter, however, can grow rapidly and will saturate in a fairly short distance. This saturation depletes the energy of the laser pulse in a localized region in the front part of the pulse envelope, leading to the situation depicted in the upper right of Figure 8.54. This "notch" in the pulse temporal envelope can then ponderomotively excite a plasma wave to greater amplitude. This process further depletes the energy of the laser pulse at this front as the energy of the laser wave is transferred to the energy of the longitudinal plasma wave, leading to a sharp front in the pulse as shown in the lower left of the figure. The energy depletion naturally is localized at this sharp front where the plasma wave is excited in the pulse. In essence the snowplow effect of this sharp-intensity front drives a sharp spike in the electron density which further depletes the energy of the pulse in a narrow window around and just behind the sharp front. This energy depletion then "etches" the pulse envelope backward in time.

The consequences of this pulse etching will be explored in the context of electron acceleration in the next section. In short, because it is this sharp front which ends up being the location in the pulse where the plasma wave is excited, the net phase velocity of the excited plasma wave is actually slightly lower than the group velocity of the pulse traveling through the plasma. The net plasma wave excitation velocity can be estimated from conservation of energy arguments, but numerical simulations have been the best empirical predictor of this etching velocity backward through the pulse. The upshot is that the phase velocity of the

8.7 Laser Plasma Electron Acceleration

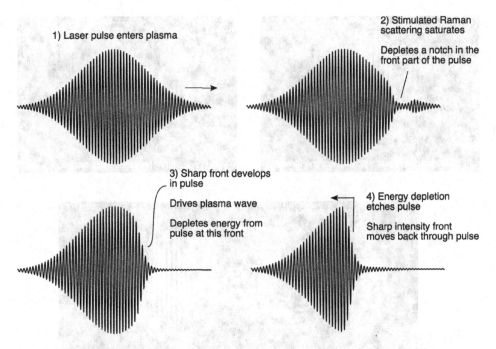

Figure 8.54 Schematic illustration of laser pulse energy depletion dynamics when an intense pulse ($>\sim 10^{19}$ W/cm^2) enters a plasma. When the pulse is moderately long, its rise does not efficiently excite plasma waves as it first enters the plasma. However, SRS instability develops early in the pulse envelope, which quickly saturates and depletes a notch in the pulse. This intensity notch can then locally excite plasma waves, which deplete the pulse at a sharp front. This front etches its way back into the pulse as it propagates.

excited plasma wave is reduced from the linear group velocity of the pulse in the plasma (more on that in the next section) from $v_{\varphi e} = v_g \cong 1 - \omega_{pe}^2/2\omega^2$ to a velocity closer to $v_{\varphi e} \cong 1 - 3\omega_{pe}^2/2\omega^2$ (Decker et al. 1994).

8.7 Laser Plasma Electron Acceleration

Understanding from the past few sections puts us now in a position to explore one of the principal reasons for modern strong field laser–plasma interaction research: the possibility of employing laser-driven plasma waves to accelerate electrons to high energy. It was the proposal of this idea in the classic paper by Tajima and Dawson (1979) that spurred much of the worldwide community to perform research on high-intensity lasers in gaseous plasmas (Leemans et al. 1992b; Pukhov et al. 1999; Esarey et al. 2009; Malka 2012; Hooker 2013). The technological motivation for this field of study is driven by a desire to accelerate electrons to many GeV (or even ultimately TeV) over very short distances (Malka et al. 2008). Traditional radio-frequency electron accelerators are limited to acceleration fields of about 100 MV/m; higher fields lead to breakdown of the metallic structures

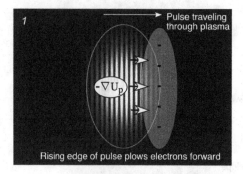
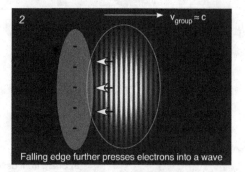
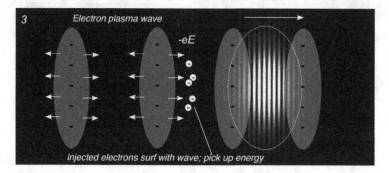

Figure 8.55 Cartoon of how a laser pulse excites a plasma wave, which can then accelerate electrons in the wake of the passing pulse.

used to form the fields and result in catastrophic failure of the accelerator. Plasmas (which, it is often joked, are a material to which one can do no further damage) can support electric fields which are far greater than this, possibly 1000× greater than fields in RF accelerators. This field arises from the longitudinal oscillations left behind by the passage of an intense laser pulse and has therefore come to be called laser–plasma wakefield acceleration.

8.7.1 The Plasma Wakefield Generation Concept

Discussions in the last few sections make it apparent how such an accelerating plasma wakefield is generated by an intense pulse in underdense plasma. A generic illustration of the mechanism is depicted in Figure 8.55.

If the pulse intensity rise is fast enough, or other mechanisms like stimulated Raman scattering impose a steep rise on the pulse, the pulse exerts a longitudinal force on the plasma electrons as it passes through the plasma. The pulse pushes the electron fluid in an accordion-like manner to regions of higher and lower intensity. The ions, whose large mass keeps them largely immobilie on these time scales, set up a restoring Coulomb force as the pulse passes. A large longitudinal electric field, E_z, develops and the electrons overshoot their initial position on their return, leading to accordion-like oscillations in electron density, accompanied by the large restoring electric field. The oscillation is, of course, near the electron plasma frequency (though deviations from this simple picture become significant with large plasma wave amplitudes).

The phasing of these electron oscillations is at a phase velocity given by the group velocity of the passing laser pulse $v_{\varphi e} = v_g$. Because the longitudinal electric field associated with this plasma wave moves along with the wave phase velocity, which is near c, any relativistic electrons that can manage to be injected to copropagate with this wave will get accelerated, picking up energy as they surf along with the wave, just as a surfer on the crest of a wave in the ocean picks up gravitation energy as he surfs along at the velocity of the wave.

A slightly more sophisticated view of this plasma wakefield accelerating structure is depicted in Figure 8.56. This plot, in the frame moving with the laser pulse to the right, is by now familiar from Section 8.4, and shows the comoving electron-density wave with the laser pulse field from QSA calculations. On top of the pulse an electron-density wave is plotted the amplitude of the comoving longitudinal electric field force on an electron in the wave. We can see that in the plasma wave of wavelength λ_{pe}, there is a longitudinal region of about half this plasma wavelength in which there is a longitudinal electric field force on an electron in the positive ζ-direction. If an electron were injected just ahead of the first crest of the electron plasma wave, the electron feels the maximum accelerating field. If the electron is relativistic, it moves at velocity near c, a speed near that of the propagating laser pulse, so it surfs with the field and can pick up substantial energy. In reality, strongly relativistic electrons propagate with velocity very close to c while the laser pulse propagates with a group velocity in the plasma which is a bit below c (refer to Appendix B), so eventually after some propagation length the accelerated electron will overtake the wave and enter a region in which the longitudinal field of the wave is reversed, ending the electron's acceleration and beginning to take energy back from it. It is this so-called dephasing length which typically sets the limit on the maximum acceleration an electron can gain in a wakefield. As we shall see, the longitudinal field can often have peak value of 1–10 GV/cm and dephasing length L_d of 1–10 cm (depending on plasma density), so it is common to see electrons accelerated to energies of many hundreds of MeV up to a few GeV (Wang et al. 2013; Gonsalves et al. 2019).

Such a plasma wakefield also has another desirable property from the standpoint of electron acceleration. If we consider the structure of the electric field in the plasma wake in two dimensions (ζ and r), as illustrated in Figure 8.57, there are also radially directed electric fields, pointing away from the positively charged electron density trough of the plasma wave. This is a result of the finite spot size of the laser pulse, and the fact that the highest-amplitude plasma waves are produced in the radial center of the pulse where the intensity is highest. In the positively charged regions, the field force on electrons points toward the propagation axis (and therefore serves to focus copropagating electrons) and the field points away from the radial center in the crest regions of the electron wave (defocusing any copropagating electrons). The plasma wave, then, has four regions, two in the electron-density crest regions of defocusing fields, half of which has an accelerating-oriented longitudinal field and half of which is decelerating, and two in the positively charged region where electrons are focused, again split into accelerating

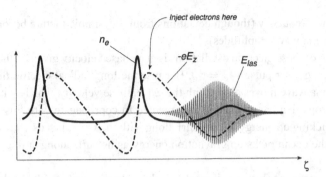

Figure 8.56 Plot of typical nonlinear plasma wave excited by passage of the laser pulse, plotted in the frame moving with the laser pulse. A nonlinear plasma wave is formed with sharp electron-density spikes. The electric field develops into a sawtooth shape. The optimum location in this wave for electron acceleration is to inject electrons at the peak of the electric field just ahead of the first electron wave-density spike following passage of the laser pulse.

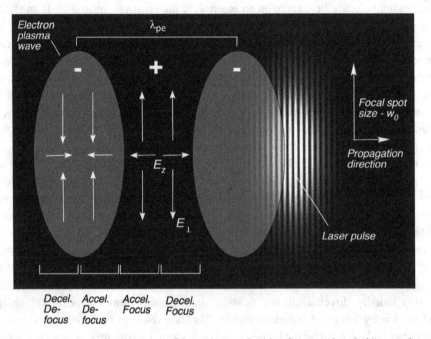

Figure 8.57 Schematic illustration of the various regions in a laser-produced plasma wake, showing the direction of the wave-produced electric field in these regions. This shows that half of the wave phase accelerates electrons along the direction of laser propagation and half of the phase decelerates injected electrons. Of the accelerating phase, half of that phase has radial electric fields that focus any injected electron beam and the other quarter-phase of the accelerating part of the wave has fields that defocus an injected electron bunch.

and decelerating regions. The ideal region for producing high-energy, collimated electron bunches is to accelerate them in the wave region of accelerating and focusing fields. This quarter-wave region is just ahead of the electron density crest and one of the challenges of wakefield electron acceleration is to inject electrons into this favorable accelerating and focusing phase.

There are various experimental realizations of a wakefield plasma accelerator, varying with the method employed to excite the wave by the laser. These various schemes more or less fall into four general catagories, whose optimum implementattion depends on the parameters of the laser (such as pulse duration or intensity) and plasma density (plasma wavelength). Schematic illustrations of these four approaches are illustrated in Figure 8.58. The most often implemented concept is the single-pulse wakefield accelerator in which the laser pulse length is roughly matched to the plasma wavelength. A variation on this is the multiple-pulse approach in which a train of pulses spaced to excite a plasma wave resonantly are injected in the plasma. The first proposal for wakefield acceleration, which uses long laser pulses, is the beatwave accelerator where a resonant drive of the plasma wave comes from the coherent beats of two laser pulses with slightly differently wavelengths. Finally, when plasma density is high and the drive pulse is moderately long, plasma instabilities can modulate the pulse intensity, leading to resonantly driven plasma waves, a scheme called the self-modulated wakefield accelerator. We will consider each of these schemes in the sections to come.

8.7.2 Wavebreaking and the Maximum Accelerating Fields

Before considering the detailed physics of these various schemes it is useful to look at some general aspects of the physics governing plasma wakefield acceleration. The first question that arises is: what electric fields are produced and what is the maximum longitudinal accelerating field that could be produced in a plasma if we are able to employ extremely intense driving pulses (or a long enough resonant modulation of intensity). A simple heuristic model yields a reasonable estimate for the maximum field attainable.

The maximum field in the plasma will be limited in some way by the fact that any plasma wave amplitude is limited by initial plasma density and at most the plasma can be driven to 100 percent electron cavitation. Though we know that large-amplitude plasma waves become highly nonlinear, a reasonable first estimate on the maximum field supported in a plasma wave is to assume the electron density wave, and the associated longitudinal electric field is sinusoidal, which takes the form

$$n_e = n_{e0} + \delta n_{e0} \sin(k_{pe}z - \omega_{pe}t); \quad E_z = E_{z0}\cos(k_{pe}z - \omega_{pe}t), \tag{8.438}$$

since the electric field is $\pi/2$ out of phase with the electron density. Gauss's law is

$$\frac{\partial E_z}{\partial z} = 4\pi e\left(\bar{Z}n_i - n_e\right) = 4\pi e\left(n_{e0} - n_e\right), \tag{8.439}$$

which becomes, with insertion of the wave form in Eq. (8.438),

$$k_{pe}E_{z0} = 4\pi e\,\delta n_e. \tag{8.440}$$

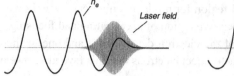

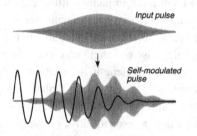

Figure 8.58 Illustration of four different schemes typically employed experimentally to generate wakefields in plasmas to accelerate electrons. Grey lines illustrate the laser field, and the black lines are plasma-wave amplitude.

Recalling that the phase velocity of a linear plasma wave is a $v_{\phi e} = \omega_{pe}/k_{pe}$, the maximum field amplitude that can be achieved in this sinusoidal wave is when $\delta n_e = n_{e0}$. This yields for the maximum plasma-generated field

$$E_{z0}^{(MAX)} = \frac{4\pi e n_e v_{\varphi e}}{\omega_{pe}}$$
$$= \frac{v_{\varphi e} m \omega_{pe}}{e}. \qquad (8.441)$$

Since, in a laser-driven wakefield $v_{\varphi e} \cong c$, this field is

$$E_{WB0} = \frac{mc\omega_{pe}}{e}. \qquad (8.442)$$

This field strength is usually called the cold, nonrelativistic wavebreaking field. It is an estimate for the maximum field that can be generated in a plasma. Driving a plasma to higher amplitude will cause the electron wave to break up incoherently, or "wavebreak," much like a wave in water whose amplitude becomes high enough that the crest of the wave spills over and creates a turbulent froth. To be more quantitative, wavebreaking occurs when the plasma-wave amplitude becomes high enough that the local electron fluid velocity exceeds the phase velocity of the wave. Such a situation would lead to electron plasma fluid elements overtaking upstream plasma and results in a shocklike region in electron density. The wavelike periodic structure assumed in our model for the plasma wave breaks down. Now, in reality, plasma waves acquire nonlinear structure (are not sinusoidal) at large

8.7 Laser Plasma Electron Acceleration

amplitude, so the actual wavebreaking field in a plasma is somewhat larger than (8.442). This expression, however, sets a convenient scale for fields in a plasma. In practical terms, it can be expressed in an easily memorized formula, $E_{WB0}[V/cm] = 0.96(n_e[cm^{-3}])^{1/2}$. So, a plasma at $n_e = 10^{18}$ cm^{-3} would support an accelerating field in the vicinity of 10^9 V/cm.

To find a more accurate equation describing the wavebreaking field, we need to consider the nonlinear behavior of a large-amplitude plasma wave (Kim and Umstadter 1999). This can be tackled by using the plasma wave equations in the quasi-static approximation if we associate the moving frame velocity, taken in the initial QSA derivation, to be the laser's group velocity, with the phase velocity of the plasma wave. In this frame the comoving space variable is $\zeta = z - v_{\varphi e} t$. From the QSA result, Eq. (8.202), the local electron density is, in terms of the wave phase velocity and plasma fluid velocity,

$$\frac{n_e}{n_{e0}} = \frac{\beta_{\varphi e}}{\beta_{\varphi e} - \beta_z}. \tag{8.443}$$

These velocities are related to the normalized electrostatic potential and plasma fluid relativistic factor by

$$\gamma \left(1 - \beta_{\varphi e}\beta_z\right) - \phi = 1. \tag{8.444}$$

In a plasma wave behind the laser pulse, $a = 0$ and $\gamma = (1 - \beta_z^2)^{-1/2}$. Following the same steps as previously, we solve for the electron plasma density,

$$\frac{n_e}{n_{e0}} = \beta_{\varphi e} \left[\beta_{\varphi e} - \left(2\beta_{\varphi e} - \sqrt{4\beta_{\varphi e}^2 + 4\phi(2+\phi)(1+\beta_{\varphi e}^2 + 2\phi + \phi^2)}\right) \right.$$
$$\left. \times \left(2(1 + \beta_{\varphi e}^2 + 2\phi + \phi^2)\right)^{-1} \right]^{-1}, \tag{8.445}$$

and define the relativistic factor for the plasma-wave phase velocity $\gamma_{\varphi e} = (1 - \beta_{\varphi e}^2)^{-1/2}$. After some considerable (hairy) algebra, we can write the electron plasma density in terms of the local potential

$$\frac{n_e}{n_{e0}} = \gamma_{\varphi e}^2 \beta_{\varphi e} \left[\left(1 - \frac{1}{\gamma_{\varphi e}^2(1+\phi)^2}\right)^{-1/2} - \beta_{\varphi e} \right]. \tag{8.446}$$

Poisson's equation, then, is

$$\frac{\partial^2 \phi}{\partial \zeta^2} = \frac{\omega_{pe}^2}{c^2} \gamma_{\varphi e}^2 \left[\beta_{\varphi e} \left(1 - \frac{1}{\gamma_{\varphi e}^2(1+\phi)^2}\right)^{-1/2} - 1 \right]. \tag{8.447}$$

Wavebreaking at the maximum electric field sustainable in the plasma occurs at the point that the electron fluid overtakes the wave and $\beta_z \to \beta_{\varphi e}$. Obviously from Eq. (8.443), n_e/n_{e0} becomes singular at that point. Eq. (8.446) says that that singularity in electron density occurs when $(1 + \phi) = 1/\gamma_{\varphi e}$. We just now need to find the E-field when this condition is met in the plasma wave. The electrostatic field is

$$E = -\frac{\partial \Phi}{\partial \zeta} = -\frac{mc^2}{e}\frac{\partial \phi}{\partial \zeta}. \tag{8.448}$$

For simplicity, we normalize the field to the cold nonrelativistic wavebreaking field, $\hat{E} \equiv E/E_{WB0}$, define the parameter $\chi \equiv 1 + \phi$ and normalize the spatial coordinate $\zeta' \equiv \zeta(\omega_{pe}/c)$. The E-field is

$$\hat{E} = -\frac{c}{\omega_{pe}} \frac{\partial \chi}{\partial \zeta}, \qquad (8.449)$$

and Poisson's equation takes the form

$$\frac{\partial^2 \chi}{\partial \zeta'^2} = \gamma_{\varphi e}^2 \left[\beta_{\varphi e} \left(1 - \frac{1}{\gamma_{\varphi e}^2 \chi^2}\right)^{-1/2} - 1 \right] \equiv \chi''. \qquad (8.450)$$

This equation can be integrated once with limits taken from the peak potential χ_{MAX} to the zero in potential. Note that the maximum electric field in the wave occurs at the zero crossing point of electrostatic potential where $\phi = 0$ or $\chi = 1$.

Integration yields

$$(\chi')^2 = 2\gamma_{\varphi e}^2 \left[(\chi_{MAX} - \chi) + \beta_{\varphi e} \left((\chi^2 - 1/\gamma_{\varphi e}^2)^{1/2} - (\chi_{MAX}^2 - 1/\gamma_{\varphi e}^2)^{1/2} \right) \right]. \qquad (8.451)$$

Since we determined that the maximum E-field occurs when $(1 + \phi) = 1/\gamma_{\varphi e}$, Eq. (8.449) and Eq. (8.451) can be combined to yield

$$\left(\hat{E}_{MAX}\right)^2 = 2\gamma_{\varphi e}^2 \left[1/\gamma_{\varphi e} - 1 + \beta_{\varphi e}(1 - 1/\gamma_{\varphi e}^2)^{1/2} - 0 \right]$$
$$= 2(\gamma_{\varphi e} - 1). \qquad (8.452)$$

The maximum wavebreaking field then is

$$E_{WB} = \sqrt{2(\gamma_{\varphi e} - 1)} E_{WB0} = \sqrt{2(\gamma_{\varphi e} - 1)} \frac{mc\omega_{pe}}{e}. \qquad (8.453)$$

We see that in a nonlinear wave, the maximum field is a factor of a few larger than the cold nonrelativistic result found with the heuristic model. The maximum value depends on plasma density in a real laser-driven wakefield because the wave phase velocity depends on laser pulse group velocity, which is directly related to plasma density.

What is the actual plasma wave phase velocity? As we have already discussed, plasma wave production and energy depletion lead to an etching away of the laser pulse front, and the actual plasma-wave phase velocity is slightly lower than the pulse-group velocity. Ignoring this for the time being, the plasma-wave phase velocity is close to v_g. Finding the laser field's group velocity at low intensity is a trivial exercise (which is laid out in Appendix B), but it turns out that it is nontrivial to find this group velocity in the strong field regime. The low-intensity, linear result for pulse group velocity is the usual result:

$$v_g = \frac{\partial \omega}{\partial k} = c\sqrt{1 - \frac{\omega_{pe}^2}{\omega^2}}. \qquad (8.454)$$

At high intensity, with a_0 approaching 1, the field shifts the plasma frequency such that $\omega_{pe} \to \omega_{pe}/(1 + a_0^2/2)^{1/4}$. This leads to the complicated situation that in a pulse envelope

8.7 Laser Plasma Electron Acceleration

of varying intensity temporally, the group velocity varies during the pulse, getting faster in regions of high intensity. Also recall, however, that the ponderomotive pressure of the rising laser pulse leads to a rise in electron density, which pushes the group velocity down as the pulse envelope rises and cancels the mass shift effect. We have already seen how this interplay largely prohibits relativistic self-focusing in short laser pulses. This complicated intercombination of effects on the pulse group velocity has been examined in an interesting sequence of papers by Decker and Mori (1994). We will not delve into these issues further other than to note that their conclusions were that, to a good approximation in short pulses (with duration comparable to or shorter than the plasma wavelength), the cancellation of these effects lets one use the linear result for pulse group velocity, even at high intensity.

With this in mind, we say that the electron phase velocity relativistic factor of a wave produced by a passing laser pulse is

$$\gamma_{\varphi e} = \frac{1}{\sqrt{1 - v_{\varphi e}^2/c^2}} = \left(1 - (1 - \omega_{pe}^2/\omega^2)\right)^{-1/2}$$

$$= \frac{\omega}{\omega_{pe}} = \sqrt{\frac{n_{crit}}{n_e}}. \tag{8.455}$$

So, for example, in a plasma driven by a laser at $\lambda = 1$ μm where $n_{crit} = 10^{21}$ cm^{-3} and a plasma has electron density $\sim 10^{18}$ cm^{-3}, then $\gamma_{\varphi e} \approx 30$, and $E_{z0}^{(MAX)} \approx 8 E_{WB0} \sim 8$ GeV/cm. This number is consistent with experimental observations (Gonsalves et al. 2019).

We note that there are other effects that alter the laser group velocity. In longer pulses, the relativistic mass correction must be accounted for and in the subrelativistic regime,

$$\gamma_g = \gamma_{\varphi e} = \frac{\omega}{\omega_{pe}} \rightarrow \frac{\omega}{\omega_{pe}} \left(\frac{\sqrt{1 + a_0^2/2} + 1}{2}\right)^{1/2}$$

$$\approx \frac{\omega}{\omega_{pe}} \left(1 + \frac{a_0^2}{16}\right). \tag{8.456}$$

This will cause some pulse steepening. The net group velocity of the laser pulse is also reduced by 3D focusing effects. A focused pulse has wavenumber components that are converging, which can be thought of as having rays that travel at an angle with respect to the propagation axis of the order of $\theta_d \approx w_0/\sqrt{2}z_R$. So the net longitudinal group velocity (in vacuum) looks something like

$$v_g \approx c \cos \theta_d$$

$$\approx c \left(1 - \frac{w_0^2}{4z_R^2}\right) = c \left(1 - \frac{\lambda^2}{4\pi^2 w_0^2}\right), \tag{8.457}$$

which means that the phase velocity gamma factor of the excited plasma wave is actually slightly modified from its plane-wave plasma value to be

$$\gamma_{\varphi e} \cong \left[1 - \left(1 - \frac{\omega_{pe}^2}{\omega^2}\right)\left(1 - \frac{\lambda^2}{4\pi^2 w_0^2}\right)\right]^{-1/2}$$

$$\cong \left[1 - \left(1 - \frac{\omega_{pe}^2}{\omega^2} - \frac{\lambda^2}{4\pi^2 w_0^2} + \ldots\right)\right]^{-1/2}$$

$$\cong \left(\frac{\omega_{pe}^2}{\omega^2} + \frac{\lambda^2}{4\pi^2 w_0^2}\right)^{-1/2}. \tag{8.458}$$

The focusing effect, then, is a nonnegligible correction to the wave-phase velocity when

$$\sqrt{2}\pi \frac{w_0}{\lambda} \lesssim \frac{\omega}{\omega_{pe}} = \sqrt{\frac{n_{crit}}{n_e}}. \tag{8.459}$$

So a 1 μm wavelength laser in a lower-density plasma like one with $n_e \sim 10^{17}$ cm^{-3}, will have significant focus-slowing of the net wave-phase velocity when the focal spot radius is smaller than about 40 μm.

As a final note, we would remark that the wavebreaking field in Eq. (8.453) was derived assuming a "cold" plasma in which the electron temperature was taken to be zero and plasma-pressure effects are ignored. Generally speaking, thermal effects of a finite-electron-temperature plasma will reduce the maximum wave-breaking field. Electron thermal pressure resists the fluid compression of very large-amplitude waves and a hotter plasma will tend to seed more thermal electrons into the wave, destroying the wave and electric fields below the cold wavebreaking limit. We will not delve into this in detail; there are many subtleties in defining what actually constitutes wavebreaking in a finite-temperature plasma and there continues to be some controversy in the literature about the definition of wavebreaking in a warm plasma. For the sake of estimating the magnitude of the thermal correction, it is reasonable to use an equation derived by Trines and Norreys (2006), which alters the wavebreaking field with a term that varies with thermal energy $\beta_{Th} = 3k_B T_e/mc^2$:

$$E_{WB} = \sqrt{2(\gamma_{\varphi e} - 1)} E_{WB0} \left[1 - \frac{\gamma_{\varphi e}}{2(\gamma_{\varphi e} - 1)}\left(\gamma_{\varphi e}^{1/2}\beta_{Th}^{1/4} - \gamma_{\varphi e}\beta_{Th}^{1/2}\right)\right]^{1/2}. \tag{8.460}$$

This equation implies that in a 100 eV plasma at a density of 10^{18} cm^{-3}, thermal effects lower the wavebreaking field by about 5 percent.

8.7.3 Maximum Electron Energy Gain and Dephasing Length

Now that we have found the maximum accelerating field that a plasma can support, the next obvious question is: what is the limit to the amount of energy gain an electron can achieve in single stage of wakefield acceleration? The first limitation rests with the maximum length over which a laser pulse can create a high-amplitude plasma wave. The other limit is more subtle and determined by how long an accelerating electron remains within the accelerating phase of the plasma wave, given the difference between the speed

8.7 Laser Plasma Electron Acceleration

of the relativistic electron and the phase velocity of the wave. Typically, it is the latter that sets the limits on acceleration, though in some experimental configurations it can be the former.

The length of plasma wave that can be excited is itself potentially limited for two reasons: (1) the Rayleigh range of the focused laser pulse and (2) energy depletion of the laser pulse from deposition of its energy in the plasma wave as it propagates (Bulanov et al. 1992). The first limitation, in which the plasma length $L \leq \pi w_0^2/\lambda$, can be mitigated by some kind of channeling, either though self-focusing or, more effectively, by preforming a plasma channel and injecting the pump pulse into the core of this channel (Esarey et al. 2009).

Calculating the pump depletion length L_{pd} is more challenging and depends on a lot of specific aspects of how the plasma wave is produced. A back-of-the-envelope estimate for the pump depletion length can be made simply by asking: at what propagation distance does the energy in the plasma wave approach the initial pump laser-pulse energy? A simple estimate involves integrating the electric field energy over a plasma wavelength, multiplying by the area of the focused beam, A_B, and the pump depletion length, and set it equal to the injected pump pulse energy:

$$\varepsilon_{pw} = L_{pd} A_B \frac{1}{\lambda_{pe}} \int_0^{\lambda_{pe}} \frac{E_{wave}^2}{8\pi} dz. \tag{8.461}$$

If we assume that the plasma wave electric field is close to the wavebreaking limit, this yields

$$\varepsilon_{pw} \cong L_{pd} \frac{\pi w_0^2}{2} \frac{E_{WB}^2}{16\pi}$$

$$\cong L_{pd} \frac{w_0^2}{16} \left(\gamma_{\varphi e} - 1 \right) \frac{m^2 c^2 \omega_{pe}^2}{e^2}$$

$$\approx \frac{1}{16 e^2} L_{pd} w_0^2 m^2 c^2 \omega \omega_{pe}, \tag{8.462}$$

which leads to an estimate for the pump depletion length in terms of laser power, P_L,

$$L_{pd} = 8\pi P_L \tau_L m^2 c^2 \omega \omega_{pe}/e^2. \tag{8.463}$$

We see from this that the pump depletion length scales as the square root of the plasma electron density. This sublinear scaling with electron density would seem to argue for using higher electron density for a larger net electron energy gain.

This reasoning fails because eventually, as one increases electron density, the accelerated electrons will have their net energy gain limited by the second mechanism, dephasing with the plasma wave. A strongly relativistic electron is traveling very close to c, but the plasma wave has phase velocity something below c because of the reduced group velocity of the pump pulse in the plasma. Indeed acceleration only occurs in the plasma wave over a wavelength roughly one half of the plasma wave wavelength, $\Delta L = \lambda_{pe}/2$. To find the propagation distance over which the accelerating electrons move in the plasma wave by this amount, we note that the speed difference between a highly relativistic electron and the plasma wave is $\Delta v = c - v_{\varphi e}$, so the time which it takes an electron injected at the optimum phase to become dephased and stop accelerating is

$$\frac{\lambda_{pe}}{2(c - v_{\varphi e})} = \Delta t. \tag{8.464}$$

The dephasing length of propagation in the lab frame is simply $L_{de} = \Delta t v_{\varphi e}$, which can be written

$$L_{de} = \frac{\lambda_{pe}}{2} \frac{v_{\varphi e}}{c - v_{\varphi e}}$$
$$= \frac{\lambda_{pe}}{2} \frac{1}{1 - v_{\varphi e}/c} = \frac{\lambda_{pe}}{2} \frac{1}{1 - (1 - 1/\gamma_{\varphi e}^2)^{1/2}},$$

so the dephasing length is

$$L_{de} \cong \gamma_{\varphi e}^2 \lambda_{pe} \cong 2\pi c \frac{\omega^2}{\omega_{pe}^3} \sim \frac{1}{n_{e0}^{3/2}}. \tag{8.465}$$

This scaling suggests that it is best for large electron acceleration to work at *low* electron density. At the end of the day, the optimum set of parameters for maximum electron acceleration are determined by compromising between dephasing length and laser pump depletion (or guiding). The effective group velocity of a tightly focused pulse is also lower than c, so focal geometry can also serve to lower the dephasing length in some cases (Esarey et al. 1995).

Eq. (8.465) is the dephasing length that is often quoted in the literature. In reality, because only half of the accelerating phase of the plasma wave also has electron radial focusing, the actual realistic dephasing length is around one-half of this value. More sophisticated simulations show that, because accelerating electrons do not acquire much energy from the plasma wave near the ends of the accelerating phase, a realistic estimate for the effective accelerating phase is reduced from $\pi/4$ to about $5\pi/32$, which means that the dephasing length is somewhat shorter than in Eq. (8.465) and about

$$L_{de} \cong \frac{5}{16} \gamma_{\varphi e}^2 \lambda_{pe}. \tag{8.466}$$

This estimate for the maximum propagation distance over which acceleration can occur is complicated by the fact that the actual wavelength of the plasma wave in the high-amplitude, nonlinear regime is longer than the linear plasma wavelength. These factors all vary the actual energy gain that can be achieved.

The maximum electron energy that we can expect, then, in a wakefield accelerator, if the dephasing length is the limit on energy gain, is

$$\Delta \varepsilon_{WF} \cong e E_{z0} L_{de}$$
$$= \frac{5}{8} \pi e c \frac{\omega^2}{\omega_{pe}^3} \frac{mc\omega_{pe}}{e} \frac{E_{z0}}{E_{WB0}}$$
$$= \frac{5}{8} \pi \gamma_{\varphi e}^2 mc^2 \left(\frac{E_{z0}}{E_{WB0}}\right), \tag{8.467}$$

or, if we assume that the accelerating field is close to the wavebreaking limit, can be expressed as

$$\Delta\varepsilon_{WF} = \frac{5\pi}{4\sqrt{2}}\gamma_{\varphi e}^2(\gamma_{\varphi e} - 1)^{1/2} mc^2. \quad (8.468)$$

At plasma density of 10^{18} cm^{-3}, for example, this estimate predicts an energy gain of 7 GeV.

8.7.4 Nonlinear Effects: Plasma Wavelength Shift and Wavefront Curvature

When the electron plasma-wave amplitude is driven to large values and the accelerating field approaches the wavebreaking limit, the wavelength of the plasma wave gets longer. At low amplitude, the wavelength for a wave traveling near c is just $\lambda_{pe} = 2\pi c/\omega_{pe}$. The QSA calculation in the top panel of Figure 8.59 illustrates this. A plasma wave driven to 20 percent of the cold nonrelativistic wavebreaking limit is shown at top; the plasma wave is linear, sinusoidal and has wavelength near λ_{pe}. Below that, a calculation of a wave driven to $2 \times E_{WB0}$ is shown. The wave shows the signature of a highly nonlinear plasma oscillation and the wavelength has stretched to $1.6 \times \lambda_{pe}$. The electron plasma-wave wavelength in this nonlinear regime can be easily estimated. We know from Eq. (8.207) that in a 1D highly nonlinear wave up to half of the electrons in the trough of one plasma wavelength get swept up into a very thin electron-density peak. Treating this plasma crest as an infinite plane sheet of charge of surface-charge density $\sigma_e = en_{e0}\lambda_e/2$, we know from Gauss's law that the peak longitudinal electric field near this sheet of charge is $E_{z0} = 2\pi\sigma_e = \pi en_{e0}\lambda_e$. This implies that the wavelength of the highly nonlinear wave will be stretched from $\lambda_{pe} = c/\omega_{pe}$ to a value scaling with the plasma wave's electric field as $\lambda_e \approx (2/\pi)(E_{z0}/E_{WB0})\lambda_{pe}$. So, a good estimate for the plasma wavelength is a quadrature interpolation from the linear to the nonlinear regime of the form

$$\lambda_e \approx \lambda_{pe}\left[1 + \frac{4}{\pi^2}\frac{E_{z0}^2}{E_{WB0}^2}\right]^{1/2}. \quad (8.469)$$

A 1D QSA calculation of the plasma wavelength as a function of the electron plasma-wave amplitude (left) and the maximum longitudinal electric field (right) are plotted at the bottom of Figure 8.59. This illustrates how the plasma wavelength begins to lengthen once E_Z starts to exceed E_{WB0}. The approximate formula of Eq. (8.469) is compared (as a gray dashed line) to the QSA calculation in this figure.

This plasma wavelength lengthening has some consequences on the various wakefield acceleration schemes we explore:

(1) In the plasma beatwave scheme with a periodically modulated light intensity, as the plasma wave amplitude grows and its length increases, the driving beatwave will grow out of resonance with the plasma wave, suppressing its further excitation and ultimately dampening its amplitude.

(2) In schemes using multiple, spaced short pulses, the pulse spacing must be increased later in the pulse train to keep the laser drive resonant with the lengthening plasma wavelength.
(3) In a wakefield scheme the plasma is almost always driven by a focused pulse which is higher intensity in the center of the beam. This means that the plasma-wave amplitude is higher in the beam center and will have longer wavelength than the waves driven further out in the beam's radius. The consequence of this is illustrated in Figure 8.60. The plasma wave fronts will become curved, and ultimately radial wavebreaking can occur, destroying the coherence of the whole plasma-accelerating structure (Esarey et al. 2009).

8.7.5 Single- and Multiple-Pulse Resonant Excitation of Wakefields

With the development of high-energy lasers with pulse durations under 100 fs in the late 1990s, the most commonly employed scheme for wakefield production is single-pulse resonant excitation of the wake (Andreev et al. 1992). If the pulse duration and plasma electron density are judiciously chosen, the ponderomotive force of the fast rise of the pulse drives the plasma wave, and the ponderomotive force in the opposite direction of the pulse's trailing half reinforces the ion-driven restoring force, resonantly driving the plasma wave to high amplitude. Intuitively, it makes sense that the highest-amplitude plasma wave that can be excited for a given peak intensity would be when the pulse envelope resonantly drives the wave and the FWHM pulse duration is about half of a plasma wavelength.

We can estimate analytically the magnitude of the excited plasma wave by such a pulse in the weak field limit ($a_0 \ll 1$) by recalling the Green's function solution for the normalized electrostatic potential from the 1D QSA, Eq. (8.218). Again, we use a half sine wave for the pulse shape,

$$a_0 = a_{00} \sin(\pi \zeta / 2L_{HW}), \tag{8.470}$$

where $L_p = 2L_{HW}$ is the spatial length of the entire pulse envelope and L_{HW} is close to the pulse's full width at half maximum ($\tau_{FWHM} \approx L_{HW}/c$). Since we desire the plasma wave amplitude after the entire pulse has passed, we set the limits in Eq. (8.218) to the endpoints of the pulse envelope and write

$$\phi(\zeta) = \frac{k_{pe}}{4} \int_{-2L_{HW}}^{0} a_{00}^2 \sin^2(\pi \zeta / 2L_{HW}) \sin[k_{pe}(\zeta - \zeta')] \, d\zeta', \tag{8.471}$$

giving

$$\phi(\zeta) = \frac{1}{4} \frac{a_{00}^2 \pi^2}{\pi^2 - k_{pe}^2 L_{HW}^2} \sin k_{pe} L_{HW} \sin k_{pe}(L_{HW} + \zeta). \tag{8.472}$$

This leads to the greatest field amplitude when the wave is driven resonantly with $k_{pe} L_{HW} = \pi$, namely with the FWHM length of the pulse equal to half of a plasma wavelength. Since the longitudinal field in the 1D limit is just

8.7 Laser Plasma Electron Acceleration

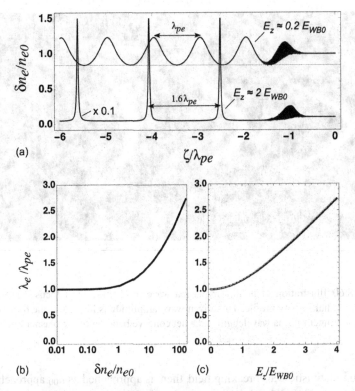

Figure 8.59 Panel (a) shows QSA calculations of plasma waves in electron density of 10^{18} cm^{-3} produced in two regimes, the top when the wave is essentially linear with longitudinal electric field on 20 percent of the cold plasma wavebreaking field and the second, lower plot when the wave is driven much harder such that the peak E_z is $2 \times E_{WB0}$. In panel (a) the electron density oscillation is sinusoidal and the wave peak spacing is the usual plasma wavelength $\lambda_{pe} \approx \omega_{pe}/(2\pi c)$. Panels (b) and (c) show that the highly nonlinear wave has wavelength that is stretched and is considerably longer than λ_{pe}. These panels show QSA calculations of actual plasma wavelength as a function of electron density amplitude and longitudinal electric field. (The gray dashed line in the panel (c) is the analytic formula given in the text.)

$$E_z = -\frac{mc^2}{e}\frac{\partial \phi}{\partial \zeta}, \qquad (8.473)$$

we see that the peak accelerating field produced by such a pulse is

$$E_{z0} = \frac{\pi^2}{4}\frac{\sin k_{pe}L_{HW}}{\pi^2 - k_{pe}^2 L_{HW}^2} a_{00}^2 E_{WB0}. \qquad (8.474)$$

This field scales linearly with peak intensity (a_{00}^2), and, at the optimum pulse duration, using L'Hospital's rule to evaluate (8.474), we see that the optimum peak field produced is

$$E_{z0}^{(opt)} = \frac{\pi a_{00}^2}{8} E_{WB0}. \qquad (8.475)$$

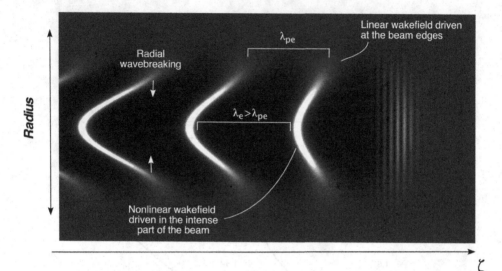

Figure 8.60 Illustration of nonlinear plasma wave production by a focused pulse with Gaussian radial density profile. The driven wave amplitude is highest in the beam center, leading to longer plasma wavelength and a net concave bending of the plasma wavefronts.

The cold nonrelativistic wavebreaking field, then, is approached as a_{00} approaches 1.

Numerical solution of the maximum accelerating field for a more realistic Gaussian-shaped pulse using the 1D QSA with realistic experimental parameters in density and field strength is shown in Figure 8.61. This shows that generally the optimum pulse duration is slightly below half a plasma wavelength, near $0.39\lambda_{pe}$ with these parameters (a result which is largely insensitive to density and peak intensity). The peak longitudinal excited field for the same plasma density and this optimum Gaussian pulse duration $L_{HW} = 0.39\lambda_{pe}$ as a function of peak laser normalized vector potential from the same kind of calculation is illustrated in Figure 8.62. As predicted earlier, the wakefield scales linearly with peak laser intensity at low a_0 and then transitions to a linearly scaling with a_0 at high peak fields. As a simple estimation of the field possible with an optimum pulse width, one can utilize an analytic formula for the maximum field derived for a square laser pulse by Berezhiani and Murusidze (1990),

$$E_{z0}^{(opt)} = \frac{a_0^2/2}{\sqrt{1+a_0^2/2}} E_{WB0}, \qquad (8.476)$$

which is also plotted in Figure 8.62 to compare to the numerical result for a Gaussian pulse.

Optimization of a single-pulse resonant wakefield accelerator, then, involves something of a trade-off in the experimental parameters chosen. The electron dephasing length increases as one decreases the electron density, scaling as $L_{de} \sim 1/n_{e0}^{3/2}$. On the other hand, the peak accelerating field increases with *increasing* electron density, growing as

8.7 Laser Plasma Electron Acceleration

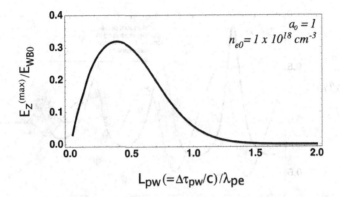

Figure 8.61 Calculation of the peak longitudinal electric field excited in a plasma by a Gaussian temporal pulse envelope as a function of pulse full width at half maximum. The maximum excited E-field for this pulse shape occurs with pulse length near 0.4× the plasma wavelength.

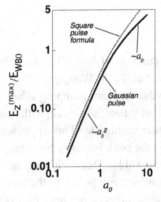

Figure 8.62 Plot of the peak longitudinal electric field excited in a plasma by a Gaussian temporal pulse envelope with FWHM equal to the optimum 0.39 λ_{pe} as a function of peak a_0. The gray dashed line is the analytic formula for a square pulse cited in the text.

$E_{z0} \sim \omega_{pe} \sim n_{e0}^{1/2}$. Since we would expect the total electron energy gain to scale as something like $L_{de} \times E_{z0}$, these scalings would suggest that optimum electron acceleration is best at lower electron density. This is true to a limit, because, to realize these higher electron energies at lower density requires longer propagation lengths. This is ultimately limited by the ability to channel a laser pulse at high intensity over long distances and can also be limited by laser pulse pump depletion. Lower density also requires longer laser pulses to excite the wave resonantly, which can lower the peak drive intensity for a given laser pulse energy. The optimum depends on the parameters of the laser available and differs among the various published experiments. Generally speaking, to now (the time of this writing) the highest electron energy gains seen in experiments have been demonstrated with laser

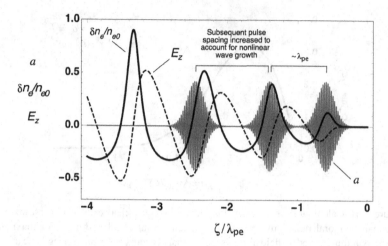

Figure 8.63 QSA calculation of an electron plasma wave and resulting longitudinal E-field generated by a train of three pulses. The third pulse has its spacing from the second pulse increased to account for nonlinear increase in the plasma wavelength at high wave amplitude.

pulses of 30–130 J and pulse duration of 40–130 fs with plasma density in the regime of $10^{17} - 10^{18}$ cm^{-3} (Gonsalves et al. 2019; Wang et al. 2013).

A variation on this single-pulse resonant pumping scheme is illustrated in Figure 8.63. Instead of a single pulse, a train of pulses is injected, with spacing such that they resonantly drive the plasma wave to higher amplitude. This approach is best used when the laser energy available is limited and the peak intensity for strong plasma-wave excitation with a single pulse is not easily achievable. This technique has the advantage that the pulse spacing can be increased with subsequent pulses in the train to account for the increase in λ_e as the plasma wave field grows and becomes nonlinear (Umstadter et al. 1994).

8.7.6 Acceleration in the Blow-Out Regime

A calculation such as that reproduced in Figure 8.59 illustrates that at very high intensities, even in 1D, the region in the plasma wave after the pulse has passed sees a large depletion of the electrons, leaving a region of positively charged ions. Radial expulsion of the electrons by the ponderomotive force of the focused intense beam reinforces this effects, and, at high enough intensity with a properly chosen pulse duration, nearly complete cavitation of the electrons behind the pulse occurs (Lu et al. 2006). Electron acceleration in this regime is often referred to as the blowout regime, or is sometimes called "bubble acceleration" (Gahn et al. 1999). This latter term is often used in the literature because the situation depicted in Figure 8.64 arises in this regime. The complete cavitation of the electrons behind the laser pulses produces a comoving, nearly spherical bubble of ions. This region can accelerate electrons with very high gradients.

8.7 Laser Plasma Electron Acceleration

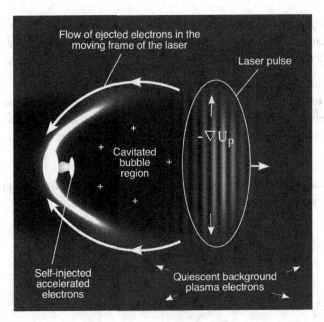

Figure 8.64 Illustration of the two-dimensional nonlinear electron plasma dynamics driven by an intense pulse in the bubble regime. Nearly complete cavitation of electrons behind the pulse produces an ion bubble propagating along behind the laser pulse.

The condition for complete radial expulsion of the electrons in the quasi-static limit was derived previously in Section 8.4.6. Though acceleration in the bubble regime is usually enacted with a short pulse and the longitudinal ponderomotive force works in conjunction with the radial force to form the ion cavity, this quasi-static radial expulsion condition can be used as a rough estimate for the intensity needed to reach the bubble regime. From this section the condition for peak a_0 to reach complete cavitation is

$$a_{00} \gtrsim \sqrt{2}\pi^2 \frac{w_0^2}{\lambda^2} \frac{n_{e0}}{n_{crit}}. \tag{8.477}$$

The ion bubble formation at these fields can be described qualitatively in the picture of Figure 8.64, undergoing the following steps:

(1) The passing short laser pulse expels the electrons radially to nearly complete cavitation.
(2) The expelled electrons move back around the bubble in the comoving frame of the propagating electrons. The expulsion produces electrostatic electric fields in the ion bubble region and the expelled electron current produces an azimuthal magnetic field.
(3) The electrons traveling backward in the comoving frame feel a return force from the ion bubble; these electrons collapse and stagnate at the rear of the bubble.
(4) The collision of high currents of electrons and the bubble end point leads to wave-breaking and injection of some of these electrons into the region at the back of the bubble.

(5) This self-injection of electrons at the bubble rear, the optimum point for electron acceleration, leads to high electron-energy acceleration as the relativistic injected electrons acquire energy in the ion cavity.

We first should estimate the bubble size by matching the ponderomotive expulsion force to the ion-restoring force. The actual dynamics of the bubble formation are complex, since a short pulse kicks the electrons radially and the expulsion is dynamic. Proceeding, nonetheless, with the static approximation, the laser's field at best focus is

$$a_0(r) = a_{00} \exp\left[-r^2/w_0^2\right]. \tag{8.478}$$

The actual dynamics must account for currents and magnetic field generation, but the approximate restoring force is estimated from Gauss's law to be

$$E_r \approx 2\pi r e n_e. \tag{8.479}$$

Equating this with the radial ponderomotive force yields

$$eE_r = -\nabla_\perp \gamma mc^2 = \frac{\partial}{\partial r}\left(1 + a_{00}^2 \exp\left[-2r^2/w_0^2\right]/2\right)^{1/2}, \tag{8.480}$$

which becomes, with introduction of the bubble radius, r_B,

$$2\pi e^2 n_e r_B = \frac{1}{2}\frac{mc^2}{\left(1 + a_{00}^2 \exp\left[-2r_B^2/w_0^2\right]/2\right)^{1/2}} a_{00}^2 \frac{2r_B}{w_0^2} \exp\left[-2r_B^2/w_0^2\right]. \tag{8.481}$$

This is a transcendental equation for the bubble radius. We estimate the condition for the optimum peak-focused normalized vector potential by setting $r_B \approx w_0$ and making the approximation that $e^{-2} \approx 1/8$ to reduce Eq. (8.481) to give

$$2\pi e^2 n_e w_0 \cong \frac{1}{8}\frac{mc^2}{\left(1 + a_{00}^2/16\right)^{1/2}} a_{00}^2 \frac{1}{w_0}, \tag{8.482}$$

yielding for the optimum focused field

$$a_{00} \approx \frac{4\pi e^2 n_e}{mc^2} w_0^2 = \frac{\omega_{pe}^2 w_0^2}{c^2}. \tag{8.483}$$

We see then that the optimum focused field for bubble formation is

$$k_{pe} w_0 \approx k_{pe} r_B \approx \sqrt{a_{00}}. \tag{8.484}$$

Numerical simulations of bubble formation (Lu et al. 2007) confirm that the optimum conditions for bubble formation are close to this simple estimate, suggesting that the optimum focused field with a short laser pulse is $k_{pe} r_B \approx 2 a_{00}^{1/2}$.

To understand better the dynamics of electrons injected into this co-moving ion bubble, we need to derive the electric and magnetic fields inside the bubble. This derivation follows that of Kostyukov et al. (2004). Since we have a freedom in choosing gauge, we will, for the moment, use a gauge in which $a_z = -\phi$. We start with Ampere's law, which is, in terms of normalized vector potential,

$$\nabla \times \nabla \times \mathbf{a} = -\frac{\omega_{pe}^2}{c^2}\frac{n_e}{n_{e0}}\frac{\mathbf{p}}{\gamma mc} - \frac{1}{c^2}\frac{\partial^2 \mathbf{a}}{\partial t^2} - \frac{1}{c}\frac{\partial}{\partial t}\nabla\phi. \tag{8.485}$$

8.7 Laser Plasma Electron Acceleration

We also need Poisson's equation, and, as with the discussion on the 3D QSA, we work with the wake potential, $\psi = \phi - a_z$, so this equation can be written

$$\nabla^2 \psi_p + \nabla^2 a_z = \frac{\omega_{pe}^2}{c^2}\left(\frac{n_e}{n_{e0}} - 1\right). \tag{8.486}$$

Using the z-component of Ampere's law and the gauge condition ($\psi_p = 2\phi$), this equation becomes

$$\nabla^2 \psi_p = -\frac{\partial}{\partial z}(\nabla \cdot \mathbf{a}) + \frac{\omega_{pe}^2}{c^2}\left(\frac{n_e}{n_{e0}} - 1 - \frac{n_e}{n_{e0}}\frac{p_z}{\gamma mc}\right) - \frac{1}{2c}\frac{\partial}{\partial t}\left(\frac{1}{c}\frac{\partial}{\partial t} - \frac{\partial}{\partial z}\right)\psi_p. \tag{8.487}$$

Now we do the usual transformation into the moving frame of the bubble, to the coordinate $\zeta = z - v_{\varphi e}t$, and use the fact that $v_{\varphi e} \approx c$ and $\gamma_{\varphi e} \gg 1$. Then we apply the QSA, and separate $\mathbf{a}$ into two components, $\mathbf{a}_\perp$ and a_z, to transform Eqs. (8.485) and (8.487) into

$$\nabla^2 \psi_p = -\frac{3}{2}\frac{\omega_{pe}^2}{c^2} + \frac{1}{2}\frac{\partial}{\partial \zeta}(\nabla_\perp \cdot \mathbf{a}_\perp), \tag{8.488}$$

and

$$\nabla_\perp^2 \mathbf{a}_\perp = \nabla_\perp(\nabla_\perp \cdot \mathbf{a}_\perp) - \frac{1}{2}\nabla_\perp \frac{\partial \psi_p}{\partial \zeta}. \tag{8.489}$$

We can find the solution to these equations by assuming spherical symmetry in the bubble, setting $\zeta = 0$ to the bubble center and introducing the radius variable in the bubble $r^2 = x^2 + y^2 + \zeta^2$. The solutions for the wake and vector potentials are

$$\psi_p(r) = -\frac{\omega_{pe}^2}{c^2}\frac{r^2}{4} + const \tag{8.490}$$

$$\mathbf{a}_\perp = 0 \tag{8.491}$$

$$a_z = -\phi = -\frac{\psi_p}{2} = \frac{\omega_{pe}^2}{c^2}\frac{r^2}{8} + const. \tag{8.492}$$

From these potentials it is now straightforward to find the fields in the bubble cavity. The electric field components are

$$E_\zeta = -\frac{mc^2}{e}\frac{\partial \psi_p}{\partial \zeta}$$

$$= \frac{m}{e}\omega_{pe}^2\frac{\zeta}{2} = 2\pi n_{e0}e\zeta, \tag{8.493}$$

and

$$E_r = E_x = E_y = -\frac{\partial}{\partial x}\frac{mc^2}{e}(\phi - a_x) = -\frac{mc^2}{e}\frac{\partial}{\partial x}(\psi_p + a_z)$$

$$E_x = \frac{m\omega_{pe}^2}{4e}x; \quad E_y = \frac{m\omega_{pe}^2}{4e}y. \tag{8.494}$$

The radial focusing field is cylindrically symmetric and is, in terms of cylindrical radius, $r' = (x^2 + y^2)^{1/2}$,

$$E'_r = \pi n_{e0}er'. \tag{8.495}$$

We also see that inside the bubble there is an azimuthal magnetic field, found from

$$\mathbf{B} = \frac{mc^2}{e} \nabla \times \mathbf{a} \tag{8.496}$$

to be

$$B_\varphi = -\pi e n_{e0} r'. \tag{8.497}$$

Now we can estimate the accelerating field. Experiments in the bubble regime tend to be performed with sub-100 fs pulses in plasma of density in some cases up to $\sim 10^{19}$ cm^{-3} with bubble radius in the vicinity of $r_b \approx 20$ μm. At the back edge of the bubble, $E_z \cong 2\pi n_{e0} e r_B$, which is up to 18 GV/cm for these typical parameters. (Longer pulse experiments have been performed (Wang et al. 2013) at somewhat reduced electron density, so fields of ~ 2 GV/cm are more typical in those studies.) The intensity required to reach the bubble regime is often high. For optimized bubble formation in the preceding parameter range one typically would require $a_0 > 10$ or 15. The field structure in the bubble is illustrated in Figure 8.65. The radial electric field is almost linear with radius, which is optimum for limiting emittance growth of an electron bunch accelerating near the central axis (Joshi 2017). We also note that the tail of the laser pulse can be spatially guided by the electron-density profile at the front edge of the bubble. During propagation, the front edge of the laser pulse is etched away by pump-depletion from the energy needed to form the bubble, which we have noted previously, and further lowers the phase velocity of the traveling bubble wave. Based on the discussion in the previous section, the phase velocity of the bubble can be well approximated by $v_{\varphi e} = c(1 - 3n_e/2n_{crit})$ (Lu et al. 2007).

Because electrons tend to be self-seeded at a localized region at the back of the bubble, experiments very frequently exhibit electron energy spectra that are quasi-monoenergetic (Joshi 2017) and of the form schematically illustrated in Figure 8.66. The effective energy gain is frequently set by the dephasing length, so we can find the likely energy gain as

$$\Delta \varepsilon \cong e \frac{E_z}{2} L_{de}. \tag{8.498}$$

Since the dephasing length set by the bubble's phase velocity limited by the etching laser pulse

$$L_{de} \cong \frac{c r_b}{c - v_\varphi} \cong \frac{2}{3} r_B \frac{\omega^2}{\omega_{pe}^2} \tag{8.499}$$

and the optimum bubble radius is

$$r_B^{(opt)} \approx \frac{2c\sqrt{a_0}}{\omega_{pe}}, \tag{8.500}$$

we can use Eq. (8.493) to express the likely electron-energy gain in the bubble as

$$\Delta \varepsilon \cong \frac{2}{3} mc^2 \frac{\omega^2}{\omega_{pe}^2} a_0. \tag{8.501}$$

At plasma density of 10^{18} cm^{-3} and typical a_0 of 2–4, this implies electron energy around 1 GeV.

8.7 Laser Plasma Electron Acceleration

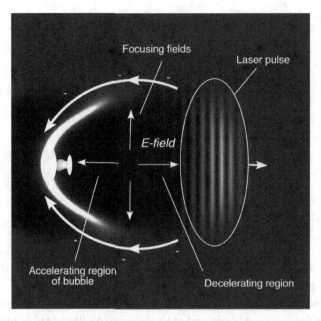

Figure 8.65 Illustration of how the ion cavity in the bubble acceleration regime produces a half-cavity region in which the electric fields are accelerating and focusing. Injection of electrons at the rear of the bubble occurs by nonlinear wavebreaking from the electrostatic collapse of the radially ejected plasma electrons at the rear of the bubble.

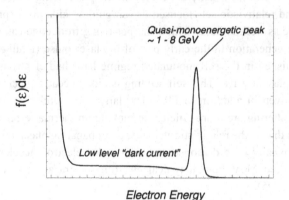

Figure 8.66 Schematic illustration of the multi-GeV, quasi-monoenergetic electron spectra that are often observed in experiments performed with pulses in the bubble acceleration regime.

8.7.7 Self-Modulated Plasma Wakefield Generation

When the electron density is higher and the pulses are longer, say 100 fs or many hundreds of femtoseconds, the resonant drive of a plasma wakefield is quite inefficient. This regime is entered when the pulse length, L_p, is much longer than the plasma wavelength,

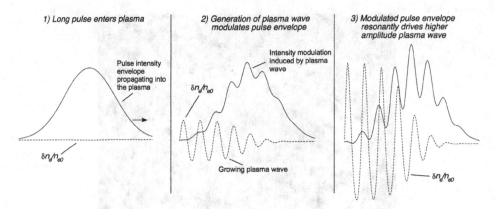

Figure 8.67 Principal steps in self-modulated wakefield acceleration.

λ_{pe}. Then, as we have seen, there are instabilities which can modulate the intensity of the long laser-pulse envelope. The two most prominent are the beam modulation instability and forward Raman scattering. Which one dominates depends on a number of factors such as focusing conditions and pulse duration. As illustrated in Figure 8.67, such instability-driven modulations will resonantly drive a plasma wave to high amplitude, which can accelerate the electrons. This process is called self-modulated wakefield acceleration.

This acceleration process usually benefits from the pulse being above the self-focusing critical power, and relativistic self-focused driven guiding aids in the process. The pulse shape plays a role as well. A pulse with a sharper-rising front edge can help seed the initial plasma wave generation in the early part of the laser pulse (Krall *et al.* 1993). Most experiments published in the self-modulated regime have had electrons self-seeded into the accelerating plasma wave. This self-seeding is aided because backward Raman scattering is often driven in addition to FRS. The large-wavenumber plasma waves driven in Raman backscattering are more efficient at picking up thermal electrons and initiating their acceleration than is the relativistic-velocity, copropagating plasma wave. Experiments in this regime typically have demonstrated self-seeded electron acceleration to energies up to a few hundred MeV, though typical electron spectra are rather broad in energy (Coverdale *et al.* 1995).

8.7.8 Beatwave Acceleration

The final wakefield acceleration scheme we will examine is the beatwave accelerator (Noble 1985). Though this scheme was the first one proposed and explored after the first publication on wakefield electron acceleration (Tajima and Dawson 1979), it is not really employed much anymore. The development of high-energy, femtosecond lasers has essentially made the beatwave scheme obsolete; it was initially proposed because long pulses could be used to excite the wave. The beat-wave approach also has other physics limitations. Nonetheless, there is some interesting plasma physics involved in describing and modeling this scheme, so we will briefly examine it in this section.

8.7 Laser Plasma Electron Acceleration

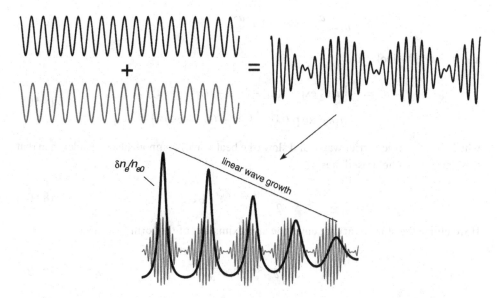

Figure 8.68 Beatwave production by adding laser waves of slightly different frequency, leading to a beating in the intensity. If the frequency difference is chosen to be equal to the plasma frequency, the beating intensity modulation will resonantly drive a plasma wave whose amplitude will grow nearly linearly (at least until the wave amplitude becomes nonlinear).

The essential idea, explained in Figure 8.68, is to copropagate two laser pulses at two wavelengths. The beating of these two laser fields effectively leads to a train of equally spaced pulses which resonantly drive the amplitude of the plasma wave. This occurs when the difference in frequencies of the two laser pulses is equal to the plasma frequency, resulting in an intensity beatwave pattern which oscillates at ω_{pe}. This condition can be achieved experimentally by tuning the electron density in the target gas. Pulse intensity typically employed is in the vicinity of 10^{15} W/cm^2. Two obvious choices for experimentally realizing the two-color beat wave are to utilize the two gain lines in a CO_2 laser, or to use phosphate and silicate glass lasers with wavelengths at 1053 nm and 1064 nm respectively. The latter case, for example, mandates using an electron density of 1.1×10^{17} cm^{-3}.

To model this process, which utilizes fairly modest laser intensity, it is reasonable to consider excitation in the small amplitude regime and recall Eq. (8.217) for the wave potential,

$$\frac{\partial^2 \phi}{\partial \zeta^2} + \frac{\omega_{pe}^2}{c^2}\phi \cong \frac{\omega_{pe}^2}{c^2}\frac{a^2}{2}, \tag{8.502}$$

which can be recast in terms of the longitudinal accelerating field as

$$\frac{\partial^2 E_z}{\partial \zeta^2} + k_{pe}^2 E_z \cong -E_{WB0}\frac{\omega_{pe}}{2e}\frac{\partial}{\partial z}\left(a^2\right). \tag{8.503}$$

The two-color beating field is

$$a = \frac{a_1}{2}e^{i(k_1z-\omega_1 t)} + \frac{a_2}{2}e^{i(k_2z-\omega_2 t)} + c.c. \tag{8.504}$$

and its intensity looks like

$$a^2 = \frac{1}{2}[a_1 a_2^* \exp[i((k_1-k_2)z - (\omega_1-\omega_2)t)] \\ + a_1 a_2 \exp[i((k_1+k_2)z - (\omega_1+\omega_2)t)] + c.c.], \tag{8.505}$$

which, has fast (the carrier wave) and slow (the beat-wave) components. Consider then that the plasma wakefield oscillates as

$$E_z = \frac{E_{z0}}{2}e^{i(k_{pe}z-\omega_{pe}t)} + c.c. \tag{8.506}$$

If we utilize the slowly varying envelope approximation of the form

$$\frac{\partial^2 E_{z0}}{\partial \zeta^2} \ll 2ik_{pe}\frac{\partial E_{z0}}{\partial \zeta} \tag{8.507}$$

and find the terms in Eq. (8.503) that vary as $e^{i(\omega_1-\omega_2)t}$, we get an equation for the growth of the plasma field amplitude,

$$\frac{\partial E_{z0}}{\partial \zeta} = \frac{1}{4}\frac{\omega_{pe}}{c}E_{WB0}a_1 a_2^*. \tag{8.508}$$

This equation is trivially integrated to yield

$$E_{z0} = \frac{1}{4}k_{pe}E_{WB0}a_1 a_2^* |\zeta|, \tag{8.509}$$

showing that the plasma wake electric field grows linearly back through the beating laser pulses. The time scale for the growth of the beatwave can be estimated from this result and is

$$\tau_{BW} \approx \frac{4}{a_1 a_2 \omega_{pe}}. \tag{8.510}$$

For example, in a plasma of density $\sim 10^{17}$ cm^{-3} with drive-laser intensity of $\sim 10^{15}$ W/cm^2 (meaning that $a_1 = a_2 = 0.03$), this equation implies that a beat-wave-driven wakefield will grow on a time scale of ~ 250 ps, illustrating that long, near-nanosecond pulses are needed for this approach.

The linear growth of the plasma-wave amplitude will typically saturate at a fairly modest value, well below the wavebreaking maximum field desired for short-scale acceleration. There are two principal reasons for this saturation in the plasma wave's growth: (1) Practically speaking, it is experimentally difficult to set the plasma electron density precisely to match ω_{pe} to $\omega_1 - \omega_2$, so some detuning from resonant excitation will always exist, and (2) detuning from resonant excitation will become more severe as the plasma-wave amplitude grows because of the wavelength increase of the plasma wave as it becomes nonlinear.

Consider the second effect. Recalling the plasma wavelength shift with increasing field, approximated in Eq. (8.469), it is fair to write for the plasma wavenumber the approximate formula

$$k_e \approx k_{pe} - \frac{3}{16}\left(\frac{E_{z0}}{E_{WB0}}\right)^2 k_{pe}. \tag{8.511}$$

The plasma-wave amplitude will stop growing once the beat-wave wavelength falls $\pi/2$ out of phase with the lengthening plasma wave. The condition for the point in the laser propagation where this detuned saturation occurs, ζ_{sat}, can be written

$$\int_0^{\zeta_{sat}} (k_{pe} - k_e)\,d\zeta = \frac{\pi}{2}, \tag{8.512}$$

which, with the approximation of Eq. (8.511), becomes

$$\int_0^{\zeta_{sat}} k_{pe} \frac{3}{16}\frac{E_{z0}^2}{E_{WB0}^2}\,d\zeta = \int_0^{\zeta_{sat}} \frac{3}{256} k_{pe}^2 a_1 a_2 \zeta^2\,d\zeta = \frac{\pi}{2} \tag{8.513}$$

and evaluates to yield for the saturation propagation length

$$\zeta_{sat} = \frac{(128\pi)^{1/3}}{k_{pe}(a_1 a_2)^{2/3}}. \tag{8.514}$$

With Eq. (8.509) we use this to estimate the maximum accelerating field that can realistically be achieved:

$$E_{z0}^{(sat)} \cong E_{WB0}\,(2\pi a_1 a_2)^{1/3}. \tag{8.515}$$

This field is typically well below the maximum wavebreaking field, and this nonlinear detuning is one of the principal limitations to the beatwave acceleration technique. For example, using the typical experimental parameters noted previously, this equation predicts that the maximum accelerating field that might be produced is only about 35 MV/cm, quite small compared to the kinds of fields that can now be produced by resonant short-pulse excitation by femtosecond pulses ($\sim$1–10 GV/cm). This beat-wave accelerating field prediction is completely consistent with experiments (Esarey et al. 2009). The nonlinear wavelength detuning can be compensated to some extent in experiments by slightly increasing the plasma density so that the plasma wave will come into resonance with the laser beat wave as the plasma-wave amplitude increases. This trick can lead to about a 50 percent increase in the attainable accelerating field (Bingham et al. 2004). In practice, it is very challenging to tune the plasma density with the precision that this technique demands.

8.7.9 Electron Trapping and Injection Seeding Techniques

A number of times I have made reference to the seeding of electrons into the plasma wave so that they can be accelerated, and I have mentioned that electrons can be self-seeded into the wave from the background electrons in the plasma. While most published experiments at the time of this writing have been conducted with self-seeded electrons, a practical, near monoenergetic plasma-based electron accelerator will likely require some method of injecting a bunch of electrons into the optimum accelerating phase of the plasma wave at a high enough electron energy that those electrons will be trapped in the wave and copropagate long enough to see substantial acceleration.

An obvious approach might be to inject electrons that have been preaccelerated to relativistic energy in a traditional compact RF accelerator. However, there are a number of all-optical and plasma-based approaches to electron seeding, and we will explore them briefly in this subsection. Before we do that, we need to look at what it means to meet the conditions for an electron to be injection-seeded effectively into the plasma wave in the first place. The easiest way to do this is to employ a common plasma physics tool, a 1D phase space plot for electron trajectories in a plasma wave. One such plot for a short pulse-driven plasma wave is shown in Figure 8.69. This plot shows a window in the comoving frame behind the driving laser pulse and plots test electrons' momenta in the z-direction as a function of longitudinal position in the wave. Above the main plot the electron-density wave that results in this window from the passage of a short laser pulse (whose field is shown in gray on this part of the figure) illustrates what the structure of the wave is in the phase-space plot below it. The gray lines in the phase-space plot show possible trajectories of a test electron at various positions in the plasma wave. If an electron is at a phase of the wave that accelerates it, the electron will move up one of the gray curves, ascending to higher longitudinal momentum. The thick black line is what is commonly called the separatrix; it denotes the boundary in this phase space between electron trajectories that are effectively trapped in the wave and those that are untrapped. Those electrons inside the separatrix have sufficient longitudinal mometum to remain comoving with the wave, climbing up or falling in momentum depending on which phase of the wave the electron is in.

Electrons above the separatrix are too fast to remain trapped in the wave and essentially outrun the wave. Those electrons will have momentum that rises and falls as they zoom along, passing through the plasma wave (which we recall always have phase velocity somewhat below c). Electrons below the separatrix are too slow to move with the plasma wave and essentially see the wave move past them as they move from right to left in this comoving plasma wave window. It is in this region where the vast majority of the background, thermal electrons of the plasma, reside.

One can see, then, that the trick to injection-seeding into the wave, to see substantial acceleration, is to promote some bunch of background electrons up in momentum, at the right spatial point in the wave, to get them on a trajectory inside the separatrix that accelerates them. This plot also shows that an electron inside the separatrix will oscillate in momentum. If it is in the accelerating phase of the wave, the electron moves up one of the bound trajectories until it reaches the top of the gray curve, after which it starts to drop in momentum to the bottom of the curve, where the process repeats. This is a reflection of the dephasing we have already discussed. Because the plasma wave's phase velocity is slightly less than c, as an electron gets accelerated it will move forward in the frame of the wave until it reachs a phase in the wave where it starts to decelerate (corresponding to

8.7 Laser Plasma Electron Acceleration

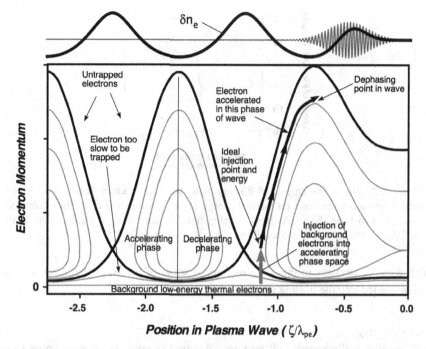

Figure 8.69 One-dimensional phase-space plot of possible electron trajectories in position-momentum space when moving in an electron plasma wave. The structure of such a laser-driven plasma wave is plotted at the top of the figure.

the top of one of the bound gray trajectories). The idea of injection-seeding and optimum acceleration, is annotated in this plot. Some thermal electrons must be promoted into the separatrix and then the plasma wave should end after a propagation distance corresponding to the time needed for the seeded electrons to rise to the maximum momentum point before they dephase.

Because the phase velocity of the wave is relativistic, substantial energy is needed for an electron to be seeded into the wave. Thermal energy is usually insufficient, so there are usually a negligible number of electrons in a Maxwellian tail to self-seed into the wave for typical plasma temperatures (which are typically of the order of 50–1000 eV), though the actual probability for such a thermal electron to be trapped depends on the wave amplitude. Waves with higher field amplitude can trap lower-energy electrons To illustrate this, the fraction of electrons with energy in the tail of a Maxwellian with sufficient velocity to be trapped in a plasma wave is plotted in Figure 8.70 as a function of plasma temperature for two different waves, a moderate field wave that has been driven by a pulse with peak $a_0 = 1$ and a second, higher field wave driven at 25× the peak laser intensity with $a_0 = 5$. This plot shows that multi-keV to tens of keV plasma temperatures are needed for any substantial seeding.

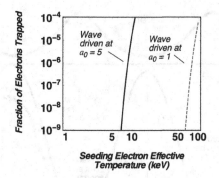

Figure 8.70 Plot of the calculated fraction of background electron density that can be seeded into a plasma wave on thermal energy alone as a function of electron temperature. The two lines show calculated trapping fractions for waves at electron density of 10^{18} cm^{-3} driven by 30 fs pulses of two different peak a_0s, 1 and 5.

With these constraints in mind, let us look at some of the ways that electrons can be injection-seeded into the plasma wave.

(A) Self-Trapping

Self-trapping really means that no targeted action is taken in an experiment to inject electrons. Because thermal distributions of the background electrons are usually incapable of injecting any appreciable number of electrons into the accelerating plasma wave, self-trapping usually requires triggering some kind of nonthermal mechanism to produce the 10–100 keV electrons needed to become seeded into the relativistic plasma wave. There are two mechanisms common in self-seeded acceleration experiments.

In bubble acceleration, background electrons are ponderomotively ejected radially, and those electrons recollapse at the back side of the bubble. This stagnation of electrons that have been ejected with up to MeV of energy leads to an ejection of some significant number of electrons with MeV energy into the back section of the bubble, self-seeding the acceleration. As we have noted, this injection at a well-defined phase of the bubble wave structure leads to quasi-monoenergetic electron energy distributions in the GeV range.

The second common self-seeding mechanism is primarily observed with longer drive pulses in the self-modulated acceleration regime. This seeding process relies on a two-step, "boot-strap" injection mechanism. A longer pulse (say, many tens or hundreds of femtoseconds) can drive Raman backscattering (RBS). As the middle vector diagram of Figure 8.49 illustrates, RBS excites large-wavenumber, small-wavelength and low-phase-velocity, forward-propagating plasma waves. These waves *are* slow enough to pick up a large number of thermal electrons and the wave can accelerate these subrelativistic electrons to energies of hundreds of keV. Such preaccelerated electrons acquire enough momentum that they *can* be subsequently trapped in the relativistic high-phase-velocity plasma wave excited by Raman forward scattering. Because of the nonlocalized, two-step nature of this mechanism, electrons tend to be seeded into the relativistic wave at all locations in the phase of the wave. This results in a broad energy spectrum, often with

8.7 Laser Plasma Electron Acceleration

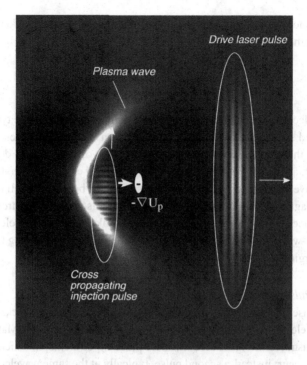

Figure 8.71 Schematic illustration of ponderomotive injection.

a distribution resembling a high-temperature Maxwellian with a high energy (>100 MeV) cutoff.

(B) Ponderomotive Injection

A second laser pulse propagating perpendicular to the main plasma wave-driving pulse can be employed to "kick" background plasma electrons, injecting them into the accelerating wave (Umstadter *et al.* 1996). This process is illustrated in Figure 8.71. A tightly focused second laser pulse will accelerate background plasma electrons by the radial ponderomotive force of the focused beam. This can easily yield MeV-energy electrons that are then trapped in the wave. This permits a choice of where in the plasma the electrons are injected by choosing where in the wave the second pulse is to be injected. The energy of the seed electrons can be easily estimated.

Recall that the ponderomotive force of the seeding laser pulse can be written

$$F_p = -\frac{mc^2}{\bar{\gamma}} \nabla \frac{a^2}{2}, \tag{8.516}$$

which leads to a z-directed force by the perpendicularly injected seed pulse with spot size w_0 of

$$F_Z \approx \frac{mc^2}{2\bar{\gamma}} \frac{a_1^2}{w_0}. \tag{8.517}$$

The injection momentum from a seed pulse of duration τ_p and peak vector potential amplitude a_1 is then roughly

$$\Delta p_Z \approx F_z \tau_p = \frac{mc^2}{\sqrt{1 + a_1^2/2}} \frac{a_1^2}{2w_0} \tau_p. \tag{8.518}$$

To achieve electron injection at well-defined plasma wave phase, one typically requires a tightly focused spot and a fairly short pulse duration, as well as femtosecond time synchronization of the seed pulse with the drive pulse. One of the practical downsides of this approach is that, as Eq. (8.518) illustrates, relativistic seed electron generation requires the seed pulse to have a_1 of at least 1. Such an intense focused beam will itself generate its own transverse propagating plasma wake. The presence of this wake can destroy the coherence of the primary accelerating wave, and the secondary wake can even seed electrons into subsequent plasma-wave buckets of the primary wakefield structure, leading to a broad range of electron energies.

(C) Colliding Pulse Injection

Colliding pulse injection relies on the same principle as the SRS backscatter sequence that self-seeds electrons in longer-pulse self-modulated experiments (Malka 2012). This technique is illustrated in Figure 8.72. In short-pulse experiments, in which SRS backscattering does not occur; instead, a second pulse, typically at the same wavelength as the drive pulse, is injected in the opposite direction to the main driving laser pulse. This collision of coherent pulses leads to a beating pattern that has low phase velocity and can bootstrap thermal electrons to the energy needed for injection into the main plasma wave.

(D) Ionization Injection

In mid to longer-duration drive pulses, it is possible to seed electrons into the accelerating plasma wave by ionization, since we know that field ionization produces ATI electrons with energy up to $2U_p$. The typical approach is to mix low- and high-Z gases in the target gas cell with resulting temporal dynamics illustrated in Figure 8.73. A low-Z gas such as H_2 or He ionizes early in the laser pulse, producing a plasma that can have a wave driven in it early in the pulse envelope. A small fraction of higher-Z gas, like N_2, Ar or Kr is mixed with the majority low-Z background gas. Field ionization occurs near the peak of the laser pulse in more deeply bound shells (such as the K-shell in N), and ATI ejection produces a population of electrons with high enough momenta to be seeded into the plasma wave. Because the phase of electron ejection in the plasma wave is not controlled, this technique, while it can supply quite a large number of seeded electrons, results in a fairly broad accelerated electron-energy spread.

(E) Density Ramp Injection

Another possibility for injecting a large number of electrons into a wakefield plasma is to set up the target with a large downward density ramp along the axis of laser propagation (Bulanov et al. 1998). As electron density drops, the wavelength of the excited plasma

8.7 Laser Plasma Electron Acceleration

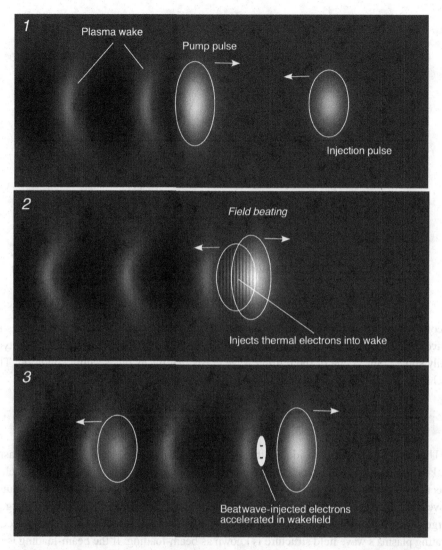

Figure 8.72 The three steps in involved in the colliding pulse injection technique. The main drive pulse propagates to the right, producing the plasma wave. Then the counter-propagating pulse collides with the main pulse and sets up a beating wave, which produces a low-phase-velocity plasma wave. This beat-wave-produced plasma wave picks up thermal electrons and accelerates them to sufficient velocity for seeding into the main relativistic comoving plasma wake.

wave increases. So, in a decreasing electron density gradient (assuming that the gradient scale length is $\gg \lambda_{pe}$), the effective phase velocity is greatly reduced because of the retarding peak of the wave as plasma wavelength increases. The essence of the idea is to slow the phase velocity down to the point that it almost matches the fluid velocity of the oscillating electron wave. This will result in a dramatic pickup of background electrons and seeding

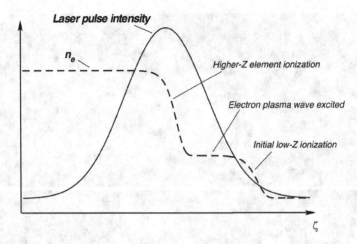

Figure 8.73 Ionization dynamics in ionization injection seeding schemes showing the electron density history in the pulse resulting from mixing low- and high-Z gases in the target cell.

a considerable charge into the wave. Because the reduction in phase velocity increases for waves many oscillations behind the driver for a sufficiently sharp gradient, there is typically some point in one of the trailing wave crests that meets this seeding condition. This technique has been studied in simulations but is not much utilized in experiments.

8.7.10 Beam-Loading Effects

Until now, we have considered the acceleration of a test electron in some plasma wakefield structure. Real acceleration experiments attempt to accelerate a large bunch of electrons, so now we need to consider what impact the fields from this accelerated bunch have back on the structure of the plasma wave, a significant effect if the bunch's charge is large enough. This back action of the fields of an accelerated electron bunch on the accelerating plasma's wave field structure is known as beam-loading. If the beam-loading is too large, the accelerated bunch can destroy the coherence of the accelerating plasma wave. This effect tends to set a limit on the maximum number of electrons that can be accelerated in a wakefield plasma (Tzoufras et al. 2007).

In a modest-amplitude, linear plasma wave, when beam-loading is a problem can be estimated by asking, at what bunch charge does the bunch itself produce its own wakefield of amplitude comparable to the laser-generated wave. In more relevant situations that are highly nonlinear, such as the blow-out regime, the beam-loading limit on how large a bunch charge can be supported can be estimated by asking, when is the accelerated bunch energy Comparable to the energy in the cavitated electrons, roughly equal to the energy in the plasma wake's fields? In the bubble regime, while there will indeed be energy imparted to the expelled electrons, the energy in the wave can be fairly estimated by integrating over the stored field energy in the cavitated bubble (Lu et al. 2007). This bubble field energy, ε_B, takes the form of an integral over the volume of the approximately spherical bubble of radius, r_b, such that it is

$$\varepsilon_B \cong \int \frac{|E|^2 + |B|^2}{8\pi} dV$$

$$= \int_0^{r_B} \int_0^{\pi} \int_0^{2\pi} \frac{1}{8\pi} \left[\pi^2 n_{e0}^2 e^2 \left(4r^2 \cos^2 \theta + 2r^2 \sin^2 \theta\right) r^2 \sin \theta \right] d\varphi d\theta dr$$

$$= \frac{4}{15} \pi^2 n_{e0}^2 e^2 r_B^5. \tag{8.519}$$

In tightly focused experiments, the acceleration length is often set by the Rayleigh range of the laser so the beam-loading condition on the maximum number of electrons that can be accelerated is roughly

$$N_e \langle E_z \rangle z_R \approx \varepsilon_B. \tag{8.520}$$

Since the accelerating field ramps up linearly in the bubble, the average field felt by the electron bunch as it moves forward in the bubble is roughly

$$\langle E_z \rangle \cong \frac{E_z^{(max)}}{2} = \pi n_{e0} e r_B, \tag{8.521}$$

which, combined with Eqs. (8.519) and (8.520), suggest that the maximum number of electrons that can be accelerated is

$$N_e \approx \frac{4}{15} n_{e0} r_B^2 \lambda. \tag{8.522}$$

This formula implies that a 1 μm wavelength laser pulse producing a 20 μm bubble in a plasma of $\sim 10^{18}$ cm^{-3} could support about 20 pC, or ~ 200 pC in an order-of-magnitude higher plasma density. This is more or less consistent with what is observed in experiments, which typically observe about 100 pC accelerated to energies of 1–10 GeV.

As an interesting side note, in bubble acceleration, the self-generated fields of the self-injected bunch at the rear of the bubble tend to flatten the gradient in the electron field in the bubble (Tzoufras et al. 2009), as illustrated in Figure 8.74. The fortunate consequence is that a spatially distributed charge bunch will see a more uniform accelerating field, and this aids in producing the near monoenergetic electron energy spectra seen in experiments illustrated schematically in Figure 8.66.

8.8 X-ray Production by Betatron Oscillations in Wakefield Ion Cavities

The final topic we will consider in this chapter is an interesting effect which is a consequence of the very strong fields that can be produced in highly nonlinear plasma wakes. It centers on the production of highly collimated X-rays from the trajectories that electrons develop in the fields of the plasma wave. These X-rays are emitted in the direction of laser propagation and exhibit considerable high peak brightness (Phuoc et al. 2005; Thomas and Krushelnick 2009; Cipiccia et al. 2011). Because they are often emitted in

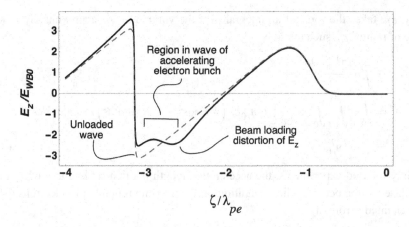

Figure 8.74 Plot of electric field as a function of distance behind the excitation pulse in a strongly nonlinear plasma wave. The electric field in the absence of beam-loading, shown as a solid black line, exhibits the typical sawtooth gradient in the field strength. The insertion of a high charge bunch (in this case of Gaussian density profile and diameter about 25% of the plasma wavelength) distorts the field and flattens it in space, serving to aid in uniform field acceleration over the entire bunch and helping lead to monoenergetic energy spectra.

photon energy ranges that are of great utility in a variety of applications, such as the 5–50 keV photon energy range, they have been widely explored by a number of groups for the possibility of building compact, high-brightness sources.

8.8.1 Electron Trajectory in a Wakefield Ion Cavity: Betatron Oscillations

The essence of the effect is schematically shown in Figure 8.75. We have focused so far on the longitudinal acceleration of electrons injected in the back area of a nonlinear plasma wave bubble of spherical structure like that shown in the figure. However, we have also noted in the derivation of Eqs. (8.494) and (8.497) that there are also transverse electric and azimuthal magnetic fields in the cavitated region. Both the electric and magnetic fields will exert a force on any electron traveling the same direction as the laser pulse on a trajectory that is off the central axis toward the central axis. The consequence, as shown by the thick white trajectory in Figure 8.75, is that these off-axis electrons will undergo oscillations about the central cavity axis, known as betatron oscillations. The large transverse accelerations that such relativistic electrons can undergo in the ion cavity can be large, leading the electron to radiate X-ray photons. Because the comoving electron is also undergoing longitudinal acceleration along z, it can become highly relativistic (with energy of hundreds of MeV) so the radiated photons are folded into a forward-directed cone of divergence of the order of $1/\gamma$.

8.8 X-ray Production by Betatron Oscillations in Wakefield Ion Cavities

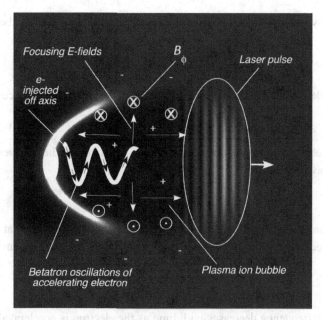

Figure 8.75 Schematic illustration of an electron undergoing betatron oscillations in the ion-cavity region of a nonlinear 3D plasma wake as a result of the off-axis restoring force of the radial electric fields and the azimuthal magnetic fields.

The motion of an electron undergoing betatron oscillations is easily calculated from the relativistic equations of motion. The motion of a test electron in the bubble is determined by

$$\frac{d\mathbf{p}}{dt} = -e(\mathbf{E} + \boldsymbol{\beta} \times \mathbf{B}), \quad (8.523)$$

which, with insertion of the fields from the previous section on bubble acceleration, Eqs. (8.493), (8.495) and (8.497), yields for the EOM

$$\frac{d\mathbf{p}}{dt} = -\frac{1}{2}m\omega_{pe}^2 \zeta \,\hat{\zeta} - \frac{1}{4}m\omega_{pe}^2 (1+\beta_z) r\,\hat{r}. \quad (8.524)$$

The value of $\beta_z \approx 1$ for the accelerated comoving electrons. Breaking the motion down into longitudinal and transversion accelerations, we write

$$\frac{dp_z}{dt} = \frac{1}{2}m\omega_{pe}^2 |\zeta|, \quad (8.525)$$

$$\frac{dp_\perp}{dt} = -\frac{1}{2}m\omega_{pe}^2 r. \quad (8.526)$$

Consider the transverse motion of the electron. If the relativistic factor of the electron

$$\gamma = \sqrt{1 + \frac{p_\perp^2}{m^2 c^2} + \frac{p_z^2}{m^2 c^2}} \quad (8.527)$$

is dominated by its longitudinal acceleration, it can be approximated by the z-component of the factor

$$\gamma \cong \gamma_z = \sqrt{1 + p_z^2/m^2 c^2}. \tag{8.528}$$

The equation describing the transverse oscillations of the electron then becomes

$$\frac{d}{dt}(m\gamma_z v_r) = -\frac{1}{2} m\omega_{pe}^2 r, \tag{8.529}$$

which nicely reduces to the equation of a harmonic oscillator for the transverse position of the forward-traveling electron

$$\ddot{r} + \frac{\omega_{pe}^2}{2\gamma_z} r = 0. \tag{8.530}$$

Clearly, such an electron oscillates about the central axis of the ion cavity at what is called the betatron frequency, which is related to the usual electron plasma frequency of the initial plasma by

$$\omega_\beta = \frac{\omega_{pe}}{\sqrt{2\gamma_z}}. \tag{8.531}$$

Obviously, this frequency decreases with time as the electron is accelerated in the cavity because of the electron's relativistic mass increase.

We can further develop this simple model if we assume that the electron remains near the back edge of the ion cavity with $r \approx r_b$. The longitudinal equation of motion is then

$$\frac{d}{dt}(m\gamma_z v_z) = \frac{1}{2} m\omega_{pe}^2 r_b, \tag{8.532}$$

which yields for highly relativistic electrons that are initially injected into the cavity with energy $\gamma_{z0} mc^2$,

$$\gamma_z(t) \cong \frac{\omega_{pe}^2 r_b}{2c} t + \gamma_{z0}. \tag{8.533}$$

The transverse equation of motion then is

$$\gamma_z(t) \frac{d^2 r}{dt^2} + \frac{d\gamma_z}{dt}\frac{dr}{dt} + \frac{\omega_{pe}^2}{2} r = 0. \tag{8.534}$$

We see that the oscillation frequency is time dependent and the electrons' oscillations are effectively damped by a factor $\dot{\gamma}_z/\gamma_z$. Defining the factor, $\delta_\beta = 2c\gamma_{z0}/\omega_{pe}^2 r_b$, this equation can be recast in a simple form,

$$(t + \delta_\beta)\frac{d^2 r}{dt^2} + \frac{dr}{dt} + c\frac{r}{r_b} = 0, \tag{8.535}$$

which predicts an oscillation amplitude, as shown in Figure 8.76, which decays as $1/\gamma^{1/4}$.

A slightly more accurate picture is to account for the forward motion of the electrons in the ion cavity as they propagate from mismatch in electron speed and plasma-wave phase velocity. In that case, the equations of motion are

$$\frac{d}{dt}(m\gamma_z \dot{r}) = -\frac{1}{2} m\omega_{pe}^2 r, \tag{8.536}$$

8.8 X-ray Production by Betatron Oscillations in Wakefield Ion Cavities

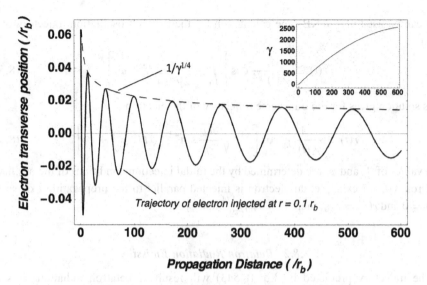

Figure 8.76 Solution for the transverse motion of an electron injected into an ion bubble cavity offset 10 percent of the bubble radius from the central axis as a function of propagation distance. This shows the decay of the betatron oscillation amplitude as γ grows during the acceleration, which is shown in the inset plot for a 20 μm radius bubble in electron plasma density of 10^{18} cm^{-3}.

$$\frac{d}{dt}(m\gamma_z \dot{z}) = -\frac{1}{2} m\omega_{pe}^2 \zeta, \tag{8.537}$$

and we know that in the comoving frame the electrons move forward from an injection point at the rear of the bubble, where $\zeta = r_b$, to the center of the bubble at $\zeta = 0$ over a dephasing propagation time

$$t_{de} \cong \frac{L_{de}}{c} \cong \frac{2}{3} \frac{r_b}{c} \frac{\omega^2}{\omega_{pe}^2} = \frac{2}{3} \frac{r_B}{c} \frac{n_{crit}}{n_e}. \tag{8.538}$$

Then the time evolution of the ζ position in the ion cavity with time is

$$\zeta = -\zeta_{init} + \frac{t}{t_{de}}\zeta_{init} = r_B\left(\frac{t}{t_{de}} - 1\right). \tag{8.539}$$

The electron acceleration with time then is

$$\frac{d}{dt}(m\gamma_z v_z) \cong \frac{m\omega_{pe}^2 r_B}{2}\left(1 - \frac{t}{t_{de}}\right), \tag{8.540}$$

which yields a quadratic time correction to Eq. (8.533),

$$\gamma_z(t) = \frac{\omega_{pe}^2 r_B}{2c}\left(t - \frac{t^2}{t_{de}}\right) + \gamma_{z0}, \tag{8.541}$$

that can be used in a WKB-type calculation for the betatron oscillation trajectory of the form

$$r(t) \cong \frac{A_r}{\gamma_z(t)^{1/4}} \cos\left[\int_0^t \frac{\omega_{pe}}{\sqrt{2\gamma_z(t')}} dt' + \varphi_r\right]. \tag{8.542}$$

This solution gives for the transverse trajectory of the electron

$$r(t) \cong \frac{A_r}{\gamma_z(t)^{1/4}} \cos\left[2^{3/2}\sqrt{\frac{ct_{de}}{r_b}} \sin^{-1}\left(\sqrt{t/2t_{de}}\right) + \varphi_r\right]. \tag{8.543}$$

The values of A_r and φ_r are determined by the initial injection conditions of the oscillating electron. (If, for example, an electron is injected parallel to the propagation axis so that $\dot{r}(0) = 0$ and $r(0) = r_{\beta 0}$, then $A_r = r_{\beta 0}\gamma_{z0}^{1/4}$ and $\varphi_r = 0$.)

8.8.2 Betatron Radiation Emission

The trajectory predicted by Eq. (8.543) will result in betatron radiation emission. Because the emitted photons are radiated by strongly relativistic electrons, the photons will come out in a rather collimated cone of divergence $\sim 1/\gamma_z$ and with Doppler upshift in frequency into the X-ray photon energy range. In the lab frame the electron oscillates at the betatron frequency, Eq. (8.531), and, because phase is a Lorentz invariant, the electron oscillates in its rest frame at $\gamma_z\omega_\beta = \omega_e$. If the electron radiates, then, in its frame at this frequency, we know from the transform of Eq. (3.134) that the frequency of radiation back in the lab frame, ω_R, is

$$\omega_R = \gamma_z\omega_e(1+\beta_z)$$
$$= 2\gamma_z^2\omega_\beta = \sqrt{2}\gamma_z^{3/2}\omega_{pe}, \tag{8.544}$$

which is, in terms of the incident laser photon energy,

$$\hbar\omega_R = \left(2\gamma_z^3 \frac{n_e}{n_{crit}}\right)^{1/2} \hbar\omega, \tag{8.545}$$

It becomes apparent that for laser photons of energy in the range of 1–1.5 eV that the reradiated betatron photons, for electrons in the 100s of MeV energy range ($\gamma \approx 200-2000$) in plasma of $n_e/n_{crit} \sim 10^{-2}$ would be $\sim 0.5 - 13$ keV, well into the X-ray spectral range.

The actual radiated spectrum can be found from the electron trajectory from Eq. (3.185). It is evident from this equation that the observed radiation in the lab frame must be highly directed along the electron propagation axis because of the $(1 - \hat{\beta}\cdot\hat{n})^{-1}$ term in this equation. Some further intuition and estimates for the radiation spectrum can be made by examining the simple picture for the betatron radiation depicted in Figure 8.77. In principle, the betatron radiation can be thought of as occurring from the wiggling electron in two regimes. The first is one with small-amplitude oscillations in which the electron radiates in more or less the same direction over its entire oscillation cycle. This could be called the "undulator regime," an analog to the situation in an undulator insertion device in a synchrotron. This regime is rarely found in real experiments, so in practice most observed

8.8 X-ray Production by Betatron Oscillations in Wakefield Ion Cavities

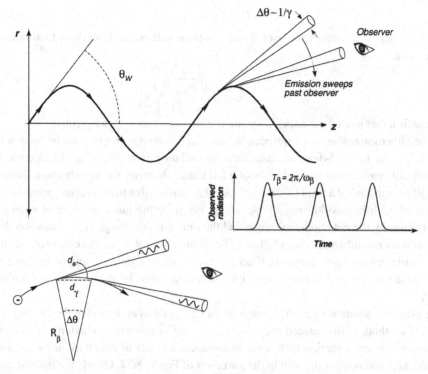

Figure 8.77 Geometry of betatron radiation, showing that an observer in the forward direction would see short pulses of radiation in time as the cone of radiation from the oscillating electron sweeps past the observer periodically. The lower left panel shows that each sweep of observed radiation can be approximated by the synchrotron radiation from an electron following a circular trajectory.

betatron radiation is in the second regime, one characterized by large-amplitude oscillations and could, by the same reasoning, be termed the "wiggler regime."

In this context, large-amplitude oscillations mean that the instantaneous radiation emission cone angle $\Delta\theta \sim 1/\gamma_z$ is much smaller than the deviation angle of the wiggling electron, denoted θ_W in the figure. The second regime is entered, then, when the betatron undulator parameter, K_β, is

$$K_\beta \equiv \gamma_z \theta_W \geq 1. \tag{8.546}$$

The wiggler angle of the electron is

$$\theta_W = \left.\frac{dr}{dz}\right|_{max} = \left.\frac{d}{dz} r_\beta \cos k_\beta z\right|_{max} = r_\beta k_\beta, \tag{8.547}$$

where the betatron wavenumber is

$$k_\beta = \omega_\beta/c = \omega_{pe}/c\sqrt{2\gamma_z}, \tag{8.548}$$

so the betatron undulator parameter can be written

$$K_\beta = \gamma_z k_\beta r_\beta. \tag{8.549}$$

Making realistic estimates for normal experimental values for these three factors in this parameter,

$$K_\beta = \underbrace{\pi\sqrt{2\gamma_z}}_{\sim 100-200} \underbrace{\sqrt{\frac{n_e}{n_{crit}}}}_{\sim 0.1-0.3} \underbrace{\frac{r_\beta}{\lambda}}_{\sim 5-20} \tag{8.550}$$

we confirm that indeed the wiggler regime is most common in experiments.

The rather interesting consequences of being in this wiggler regime can be seen in Figure 8.77. The forward-directed, instantaneous radiation from the wiggling electron will periodically pass across the sight axis of a lab frame observer. So the betatron radiation is actually composed of a train of short pulses. This implies that the radiation spectrum must be broad, with bandwidth, roughly speaking, given by the pulse duration of each pass, determined by the radiation cone angle and the time rate-of-change of the emission direction of the sinusoidally moving electron. This picture indicates that the observed spectrum is a Fourier transform of a train of short pulses, which will yield a comb of harmonics of the fundamental emission frequency with width determined by the duration of each short burst.

A bit more quantitative analytic progress can be made in ascertaining the emitted spectrum if we think of the emitted spectrum in terms of synchrotron radiation of an electron undergoing circular motion with some instantaneous radius of curvature in its sinusoidal dance. This geometry is detailed in the lower left of Figure 8.77. Clearly the tightest radius of curvature will occur when its emission is directed straight down the propagation axis, so a good estimate of the observed spectrum can be articulated by writing the synchrotron emission spectrum of an electron on a circular path at this extreme. Such synchrotron radiation is a textbook solution (Jackson 1975, p. 672).

The instantaneous radius of curvature of any trajectory, $y(x)$, is

$$R = \left| \frac{(1+y'(x)^2)^{3/2}}{y''(x)} \right|, \tag{8.551}$$

so, since the electron traces a trajectory of the form

$$r(z) = r_\beta \sin k_\beta z, \tag{8.552}$$

its radius of curvature is

$$R_\beta(z) = \frac{\left(1 + k_\beta^2 r_\beta^2 \cos^2 k_\beta z\right)^{3/2}}{k_\beta^2 r_\beta \sin k_\beta z}. \tag{8.553}$$

The distance traveled by the electron on a circular arc during the duration that the burst of radiation can be observed, d_e, is determined by

$$\frac{d_e}{2\pi R_\beta} = \frac{\Delta\theta}{2\pi}, \tag{8.554}$$

$$d_e = \frac{2R_\beta}{\gamma}. \tag{8.555}$$

8.8 X-ray Production by Betatron Oscillations in Wakefield Ion Cavities

On the other hand, the distance that the radiated light has traveled in this time along the chord of this radial trajectory, d_γ, is, with Taylor expansion about $1/\gamma$,

$$d_\gamma = 2R_\beta \sin \Delta\theta/2$$
$$= 2R_\beta \sin(1/\gamma)$$
$$\approx 2R_\beta \left(\frac{1}{\gamma} - \frac{1}{6\gamma^3}\right). \tag{8.556}$$

The duration of the wiggler burst, then, is

$$\Delta t_W = \left(\frac{d_e}{c} - \frac{d_\gamma}{c}\right) = \frac{R_\beta}{3c\gamma^3}. \tag{8.557}$$

The uncertainty principle for this short radiation burst, $\Delta\omega \Delta t_W \approx 1$, yields the frequency up to which we would expect significant radiated energy by this train of short pulses,

$$\Delta\omega \equiv \omega_c = \frac{3c\gamma^3}{R_\beta}, \tag{8.558}$$

where we have now defined the betatron wiggler critical frequency, ω_c. We expect, to observe betatron radiation up to the critical frequency evaluated at the minimum radius of curvature from Eq. (8.553), which yields for the approximate maximum betatron radiation frequency

$$\hbar\omega_R = 3\hbar c \gamma^3 k_\beta^2 r_\beta$$
$$= \frac{3}{2}\hbar \gamma^2 \frac{r_\beta}{c} \omega_{pe}^2. \tag{8.559}$$

If we estimate for the amplitude of the betatron oscillations about a tenth of the ion cavity radius, set by the following optimization condition,

$$r_\beta \approx \frac{r_b}{10} = \frac{1}{5}\sqrt{a_0}\frac{c}{\omega_{pe}}, \tag{8.560}$$

we derive a compact analytic estimate for the expected photon energy we might expect from betatron oscillations in terms of laser and plasma parameters,

$$\hbar\omega_R = \left(\frac{3}{10}\gamma^2 \sqrt{a_0}\sqrt{\frac{n_e}{n_{crit}}}\right)\hbar\omega. \tag{8.561}$$

Again we see that photons $\sim 10^4 \times$ the laser photon energy are expected, and we see that higher electron densities yield harder X-ray photons. Consequently, most betatron radiation production experiments are performed at somewhat higher densities ($\sim 10^{19}$ cm^{-3}) than experiments on high-quality wakefield electron acceleration (Corde et al. 2013).

It is possible to derive an analytic expression for the betatron radiation harmonic spectrum from Eq. (3.134), using a procedure similar to that used to find the spectrum for nonlinear Thomson scattering in Chapter 3. This is done in another excellent paper by Esarey and coworkers (Esarey et al. 2002). But for our illustrative purposes, it is adequate to estimate the observed spectral shape of the betatron emission using the classic textbook solution to synchrotron radiation from a circularly moving electron, famously laid out in

Jackson (1975, chapter 14). From this well-known solution, the power radiated per unit frequency, per steradian at angle θ from the instantaneous electron propagation direction on its arc as a function of radiation frequency, ω_R, is

$$\frac{d^2 P_\gamma(t)}{d\omega d\Omega} = \frac{e^2 R_\beta \omega_R^2}{6\pi^3 c^2} \left(\frac{1}{\gamma^2} + \theta^2\right) \left[K_{2/3}(\xi_\beta)^2 + \frac{\theta^2}{(1/\gamma^2) + \theta^2} K_{1/3}(\xi_\beta)^2\right]. \qquad (8.562)$$

Here $K_j(x)$ is the jth-order modified Bessel function of the second kind, and

$$\xi_\beta = \frac{\omega R_\beta}{3c} \left(\frac{1}{\gamma^2} + \theta^2\right)^{3/2}. \qquad (8.563)$$

This equation for our electron's emission is time dependent through the variation in time of R_β and γ. Because the actual observed spectrum comes from the electron oscillating through a range of emission angles as it wiggles in the x-y plane, the more appropriate equation for explaining experiments is the angle-averaged version of this spectrum, which can be expressed (Jackson 1975, p. 677) as

$$\frac{dP_\gamma}{d\omega} = \frac{\sqrt{3}}{\pi} \frac{e^2}{R_\beta} \gamma \frac{\omega_R}{\omega_c} \int_{2\omega_R/\omega_c}^{\infty} K_{5/3}(\chi) d\chi. \qquad (8.564)$$

At low photon energy, this equation predicts a spectral shape that varies as $\omega_R^{1/3}$. As shown in Figure 8.78, where this spectral shape is plotted, this integral equation can be fairly well approximated by an analytic formula shown in the figure. The observed spectrum can be found from this formula by integrating over all times, accounting for the time dependences of R_β, γ and ω_c.

8.8.3 Betatron Radiation Properties: Divergence, Duration and Photon Yield

Actual observed betatron radiation arises not from a single electron but from an injected bunch of electrons, which can have a spread in initial injection radii, energy, angle and other parameters. The properties of such a radiating electron charge bunch typically set the range of X-ray properties observed in experiments. Because the electrons are randomly distributed in the radiating bunch, their emission is incoherent and the observed X-ray properties are the incoherent sum of emission from individual electrons with the range of electron parameters found in the bunch. Typical observed X-ray parameters include the following:

(1) X-ray Pulse Duration

This quantity is difficult to measure experimentally, but it is a fair assumption that the duration of the emitted pulse is roughly equal to the bunch length injected at the back of the bubble. This length must be a fraction of the radius of the ion cavity so it is expected that most betatron x-ray pulses are only a few tens of femtoseconds in duration.

8.8 X-ray Production by Betatron Oscillations in Wakefield Ion Cavities

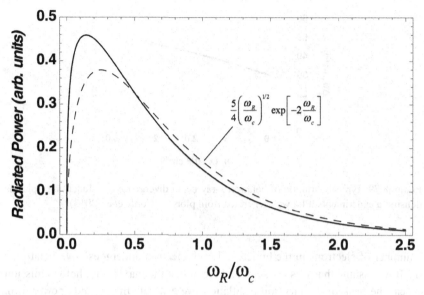

Figure 8.78 Plot of the spectral shape predicted by Eq. (8.564). A simple analytic approximation to this formula is overlaid.

(2) X-ray Beam Divergence

A naïve estimate for the X-ray beam divergence might be simply the radiated cone angle of a single electron, $\theta_W \sim 2/\gamma$. Using the usual condition for optimum bubble formation for a given laser a_0, we can estimate this beam divergence as

$$\Delta\theta_{beam} \cong \theta_W = r_\beta k_\beta$$

$$= \sqrt{\frac{1}{2\gamma}} r_\beta \frac{\omega_{pe}}{c} = \sqrt{\frac{2a_0}{\gamma}}. \quad (8.565)$$

This implies a typical beam divergence of ~50 mrad, in the ballpark of what is experimentally observed (Albert and Thomas 2016). Typical X-ray beam divergence as a function of plasma density is illustrated in Figure 8.79. Higher electron density tends to produce a more diverging X-ray beam because of the reduction in betatron oscillation period (increase in k_β) at higher density. The typical X-ray source size (an important parameter is determining the imaging resolution that can be obtained with such an X-ray source) is usually quite small, typically about the spatial size of the injected electron bunch, a fraction of r_b or a few μm.

(3) Photon Yield

How many X-rays are actually produced is an important parameter in many X-ray applications. This quantity can be estimated by temporal integration of the emitted power, estimated from Eq. (8.564). A fairly simple and reasonable estimate can be made if N_e

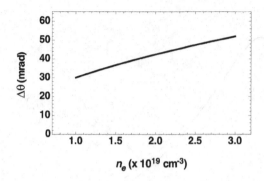

Figure 8.79 Typical variation of betatron x-ray beam divergence as a function of plasma density in experiments. Plot was generated from plots in (Corde et al. 2013).

is the number of electrons in the bunch and each electron undergoes $\sim N_\beta$ betatron oscillations. If we assume that γ is constant and recall that the maximum photon emission will occur near the peak of the electron oscillation, we estimate the radiated power using the usual result for the total power radiated by an accelerating electron, Eq. (3.215). Since the electron velocity is $v = (ck_\beta/\gamma)\sin\omega_\beta t$, if we ignore any energy gain from acceleration in Eq. (3.215) we can use this equation to estimate the total energy radiated by the bunch during its propagation to be

$$\varepsilon_{R\omega} \approx N_e N_\beta \frac{2\pi e^2 \gamma^4 \omega_\beta v_\perp^2}{3c^3}. \tag{8.566}$$

With inspection of the synchrotron spectrum, we see that the average energy of each emitted photon is, in terms of the critical frequency $\omega_c = 3\gamma^2 \omega_\beta K_\beta$,

$$\langle \hbar \omega_R \rangle \cong \frac{2}{3}\hbar\omega_c, \tag{8.567}$$

which, when divided into (8.566), gives us an estimate for the total number of X-ray photons that might be expected in terms of the undulator parameter and the fine structure constant ($\alpha_{FS} = 1/137$),

$$N_\gamma \approx \frac{\pi}{3} N_e N_\beta \alpha_{FS} K_\beta. \tag{8.568}$$

Since $K_\beta \approx 5 - 10$ in many experiments, this equation suggests that each electron radiates about 0.1 photons per betatron oscillation. In a normal plasma with dephasing length (or plasma length) of ~ 1 cm, there are a few (~ 2–5) betatron oscillations while the electrons are near the peak of their accelerated energy. Therefore, electron bunches with 10–50 pC would be expected to radiate about $2 - 10 \times 10^7$ X-ray photons, a very good estimate for what is indeed observed in most experiments (Albert and Thomas 2016). (This corresponds to a very high peak spectral brightness, in the range of $10^{21} - 10^{22}$ photons/s-mrad^2-mm^2-0.1% BW.) The typical observed scaling of emitted X-ray pulse energy as a function of electron density is shown schematically in Figure 8.80. What is seen is that

8.8 X-ray Production by Betatron Oscillations in Wakefield Ion Cavities

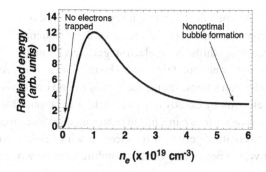

Figure 8.80 Schematic representation of the trend often observed in the betatron X-ray pulse energy as plasma density is varied. Plot was generated from plots in Corde *et al.* (2013).

there is an optimum in electron density; at low density, no electrons are trapped to radiate, while at high density nonoptimal bubble formation stunts electron acceleration.

8.8.4 Direct Laser Acceleration of Electrons at the Betatron Resonance

Until now we have ignored the presence of the laser field itself in affecting the dynamics of electrons undergoing betatron oscillations. The laser is typically just a light bullet that creates the traveling ion cavity, and ignoring it in the electrons' equations of motion is well justified. However, when the laser pulse is a bit longer (say comparable in length to the plasma wavelength or the bubble radius) the tail end of the laser pulse can interact with the electron bunch, particularly as it nears the ion cavity center near the dephasing limit. For example, a 50 fs pulse will extend back $\sim$20 μm into the ion bubble, comparable to the bubble's radius in many experiments. This situation is particularly important in regimes in which the plasma wavelength is much smaller than the bubble (the "forced laser wakefield" regime).

The main consequence of this situation is that electrons oscillating in the plane of the laser's polarization can have their betatron oscillations dephased by the laser field, or, in a special set of parameters and circumstances, can oscillate in phase with the laser frequency (or a harmonic of that frequency) and in fact have their betatron oscillation amplitude increased. The former situation reduces betatron X-ray emission; the latter can dramatically enhance the emission. This resonant enhancement of the electron's oscillation and kinetic energy by the laser field is known as "Direct Laser Acceleration" (DLA). This effect tends to be more important in experiments with modest peak power ($\sim$10 TW) where electrons reach moderate energies ($\sim$10–50 MeV) (Cipiccia *et al.* 2015).

The main experimental observables are:

(1) Increase of the electron energy
(2) Broadening of the usual quasi-monoenergetic electron energy spectrum
(3) Enhanced betatron X-ray output
(4) Shift of betatron X-ray spectrum to higher photon energy.

These occur specifically when the laser frequency, in the frame of the electron, is equal to that electron's betatron frequency, or a harmonic of that frequency. In fact, the dynamics are complicated because ω_β shifts as the electron gets accelerated.

Consider when this resonance condition would be met. Specifically resonant acceleration occurs when, in the electron's frame during the period of one betatron oscillation, the laser phase overtakes the electron by exactly one laser-field oscillation. This occurs because the laser's phase velocity is faster than c in a plasma and can, therefore, overtake the relativistic electron. If T is the oscillation period of the electron with velocity component along the laser propagation of v_z, the betatron resonance condition can be written

$$\left(v_{l\varphi} - v_z\right) T = \lambda, \tag{8.569}$$

which can be cast in terms of the shifted laser frequency, ω_D, in the electron frame

$$\left(1 - v_z/v_{l\varphi}\right) \frac{1}{\omega_D} = \frac{1}{\omega}. \tag{8.570}$$

For a highly relativistic electron we can say

$$v_z/c \cong 1 - \gamma^2/2, \tag{8.571}$$

and we have the usual relationship for light wave-phase velocity in a plasma,

$$v_{l\varphi}/c \cong 1 + \omega_{pe}^2/2\omega^2, \tag{8.572}$$

to give for the laser frequency experienced by the electron in the frame of its z motion

$$\omega_D = \omega \left(1 - \frac{1 - \gamma^2/2}{1 + \omega_{pe}^2/2\omega^2}\right) \tag{8.573}$$

$$= \omega \left(\frac{1}{2\gamma^2} + \frac{\omega_{pe}^2}{2\omega^2}\right). \tag{8.574}$$

Typically, the second term in this expression is much greater than the first for a highly relativistic electron, so we can say

$$\omega_D \cong \frac{\omega}{2} \frac{n_e}{n_{crit}}. \tag{8.575}$$

The betatron resonance condition occurs when $\omega_\beta = \omega_{pe}/(2\gamma)^{1/2} = \omega_D$, so DLA at a betatron resonance is expected for a plasma in which

$$\omega_{pe} = \sqrt{\frac{2}{\gamma}} \omega, \tag{8.576}$$

which means we would need an electron density of $n_e = 2n_{crit}/\gamma$. So, for example, when 800 nm pulses are employed, if electrons reach $\sim$50 MeV energies, the electron density needed for betatron resonance is about 3×10^{19} cm^{-3}.

The nature of DLA can be illuminated if we return to the transverse equations of motion for the electrons and add terms for the laser electric and magnetic fields. This gives for the radial oscillatory motion of an electron

8.8 X-ray Production by Betatron Oscillations in Wakefield Ion Cavities

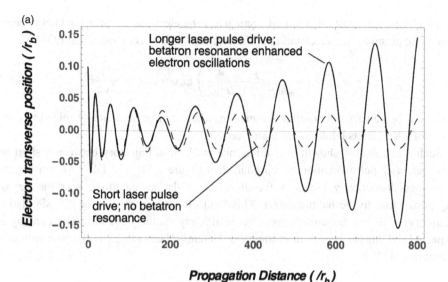

Figure 8.81 Solution of the transverse electron equation of motion in a plasma bubble in a plasma with $n_e = 10^{19}$ cm^{-3} for short (30 fs – dashed line) and long (100 fs – solid line) pulses.

$$\gamma \ddot{r} + \dot{\gamma}\dot{r} + \frac{\omega_{pe}^2}{2} r = -\frac{eE_0}{m}\cos\omega_D t + \frac{e}{m}\beta_z B_0 \cos\omega_D t$$

$$\cong -\frac{e}{m}(1-\beta_z)E_0 \cos\omega_D t$$

$$\cong -\frac{e}{2m\gamma^2}E_0 \cos\omega_D t. \tag{8.577}$$

Now the energy gain of the electron must include the **v • E** term from the laser field, so this gives

$$\dot{\gamma} = \frac{\omega_{pe}^2 r_B}{2c}(1 - t/t_{de}) - \frac{e}{mc^2}\dot{r}E_0 \cos\omega_D t - \frac{\omega_{pe}^2}{2c^2}\dot{r}r. \tag{8.578}$$

If we assume that most DLA would occur when the electrons are near the bubble center, where the plasma-wave acceleration of the electrons falls to zero, the radial EOM is

$$\ddot{r} + \omega_\beta^2 r = \left(\frac{e}{\gamma_m mc^2} \dot{r}^2 - \frac{e}{2m\gamma_m^3} \right) E_0 \cos \omega_D t + \frac{\omega_\beta^2}{c^2 \gamma_m} \dot{r}^2 r. \tag{8.579}$$

This is a forced harmonic oscillator equation, which can be numerically solved to give a sense of what effects DLA can have on electrons.

Such a calculation, showing the difference in electron trajectory between a short and a longer laser pulse situation is reproduced in Figure 8.81. The DLA from a betatron resonance dramatically increases the electron oscillation amplitude near the dephasing point close to the bubble center. This kind of dynamic is seen in PIC simulations (Curcio et al. 2015). Enhancements in betatron X-ray photon yield and photon energy are expected and this has been seen in some experiments (Hazra et al. 2018), along with other signatures of DLA.

9
Strong Field Interactions with Overdense Plasmas

Irradiating a solid piece of material with an intense laser is one of the most natural things to do with an intense laser. In practice, while this experiment is almost trivial to execute, it is among the most difficult of all the interactions we have attempted so far to understand and explain. The plasma that naturally occurs when an intense laser pulse begins to ionize and heat a target initially at or near solid density undergoes a rather convoluted interaction. Consequently, this kind of strong field interaction is among the most difficult to explain with analytic or simple numerical models. This is the challenge we will tackle in this final chapter. Because of the interplay of multiple time and spatial scales, from microscopic, subfemtosecond single-particle interactions up to hydrodynamic motion over hundreds of microns and many nanoseconds, much of our discussion to explain this physics will necessarily be more qualitative than in many of the explanations of the previous chapters.

The interaction of a high-intensity pulse with a solid target leads to a veritable zoo of physical phenomena, many of them interrelated and of varying importance (Pukhov 2003). Many of these are illustrated in the cartoon of Figure 9.1. When the intense laser light impinges on the surface, a plasma is rapidly created by a combination of tunnel and collisional ionization. This has two immediate consequences: the plasma absorbs laser light by collisional absorption and begins to heat. A thermal wave is driven into the cold underlying bulk material and the pressure associated with the heated electron plasma begins to drive a hydrodynamic expansion outward toward the illuminating laser. At sufficiently high intensity, the light pressure can stop or reverse this expansion, boring a hole in the expanding plasma. Further interactions with the target surface can lead to driven nonlinear oscillations of that surface, generating high-order harmonics of the laser light. At the highest intensity, the laser field can shift the plasma frequency through the relativistic increase of the mass of the laser-driven electrons to convert an opaque, overdense plasma to one that is momentarily transparent to the incident light field.

The laser's electromagnetic field can then interact with the expanding plasma, leading to further absorption by driving plasma waves or accelerating electrons at a sharp plasma surface. Most of these so-called collisionless absorption mechanisms lead to the production of a population of electrons with kinetic energy well above the background thermal energy,

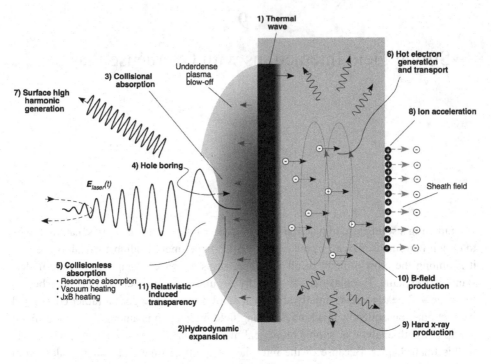

Figure 9.1 Illustration of some of the various physical effects in the interaction of an intense laser pulse with a solid target.

usually called "hot electrons." These electrons penetrate into the target, producing high peak currents which generate transient magnetic fields of extremely high field strength. The hot electrons will also interact with the underlying matter, producing X-rays through atomic inner shell processed (K-alpha radiation) or, at higher electron energy, hard gamma rays via bremsstrahlung emission. If the target is sufficiently thin, these fast electrons will exit the back side of the target and set up an extremely high sheath field which will then accelerate ions, such as protons, adhered to the back of the target. In fact, the pressure of the laser pulse can in some cases itself produce a large population of fast ions, in a process known as radiation pressure acceleration.

In the sections to come in this chapter we will look in detail at most of these phenomena (along with a few other, even more exotic effects). We will begin this consideration by looking at some of the processes that affect the structure of the plasma, such as temperature and density profiles, which then impact the various strong field effects that come into play once the laser pulse intensity rises toward its peak.

9.1 The Structure of Solid Target Plasmas

9.1.1 Short- and Long-Pulse Interaction Regimes

Plasma formation on the surface of a solid target, in fact, occurs at fairly modest intensity. A small number of seed electrons will be produced by multiphoton ionization and

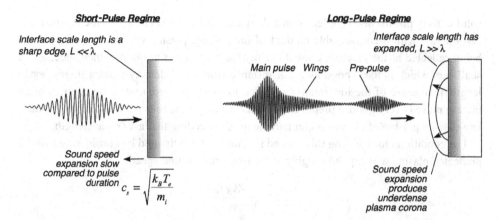

Figure 9.2 Illustration of some of the various physical effects in the interaction of an intense laser pulse with a solid target.

then these electrons are heated to temperature sufficient to ionize other atoms by collisional ionization. Because of the high density of a solid, this collisional ionization, avalanche ionization, quickly leads to many more ionization events. Plasma formation typically starts at intensity as low as 10^{11} W/cm^2; we will not spend any time detailing this physics, as it falls a bit outside the confines of strong field physics, but it has been well characterized over the years (see, for example, Strafe et al. (2014, chapters. 4 and 5); Mulser and Bauer (2010, chapter 1). For our purposes it is sufficient to realize that a high-density plasma is formed very early in the rising edge of any incident intense pulse.

Because the critical electron density for most IR or optical pulses occurs at electron density of $\sim 10^{21}$ cm^{-3}, even a few percent ionization of a solid target will lead to plasma formation which is opaque and at least partially reflective of the laser pulse. Determining how much laser pulse energy gets absorbed at this reflection point and where that energy is deposited represents a considerable portion of the focus of this chapter. While there is a continuum of situations, the structure of the plasma that then evolves can be characterized by two regimes, illustrated schematically in Figure 9.2: (1) The short-pulse regime, in which the rise of the laser is so fast from the time of initial plasma formation that the bulk of the intense pulse interacts with a sharp plasma edge, whose scale length is much less than a single wavelength. (2) A long-pulse regime, in which the front temporal edge of the pulse arrives sufficiently long before the arrival of the majority of the laser-pulse energy that the initial plasma has had time to expand by hydrodynamic pressure, so the main pulse interacts with a cloud of underdense plasma.

Generally, in most strong field experiments, in which the main laser pulse is subpicosecond or only a few tens of femtoseconds, the latter regime is entered because of the presence of a low-intensity prepulse before the main pulse, sometimes irradiating the solid many picoseconds or even nanoseconds before the arrival of the main pulse. The origins and removal of these prepulses are discussed in Chapter 2. The two regimes usually result in rather different kinds of interactions. In case (1) a strong laser field interacts with solid-density plasma and is reflected in a sharp interface with vacuum. In case (2) the pulse can interact with a considerable amount of underdense plasma out in front of the solid, being reflected at the electron critical density (or even lower at oblique incidence, as we shall see) which is itself possibly a fair distance from solid-density plasma (many wavelengths). Because of the low intensity needed to ionize a solid, unless the contrast of an intense pulse is very good, a preplasma may have a long time to expand and might produce low-density plasma of thickness comparable to or exceeding the laser's wavelength.

The conditions for entering this second regime can be estimated by considering that the preheated plasma will expand roughly at the isothermal sound speed:

$$c_{si} = \sqrt{\frac{Zk_B T_e}{m_i}} \tag{9.1}$$

given in terms of the heated electron temperature T_e and the inertia set by the ion mass m_i. So, for example, if an aluminum target ($m_i = 4.5 \times 10^{-23}$ g) is heated to 100 eV (a temperature fairly easy to reach as intensity ramps up toward $\sim 10^{14}$ W/cm^2) the Al plasma will have expanded by about a laser wavelength (~ 1 μm) in roughly 50 ps.

9.1.2 Surface Temperature and Thermal Waves

Frequently it will be necessary to find the electron plasma temperature at or near the target surface, or we may wish to know how long a hot plasma persists near the surface before it cools. This heating may be from some long-time-scale ($\sim$ns) prepulse or it may be from the main laser pulse, which, in most strong field experiments, is quite short (<ps) on the time scale of the plasma cooling. In the latter case a rough estimate for the front surface temperature can be derived simply by estimating the volume heating done by a given laser-deposited fluence in a layer the depth of the optical skin depth (discussed in Subsection 9.1.5).

Generally speaking, the deposited fluence depends on some surface absorption by the solid-density (or critical-density) plasma. It is usually necessary to think of laser energy deposition in two terms:

$\eta_{abs}^{(T)}$ = fraction of laser energy absorbed into thermal energy of the target plasma
$\eta_{abs}^{(NT)}$ = fraction of laser energy absorbed by nonthermal plasma effects such as plasma waves and hot electrons.

It is the first which contributes principally to the rise in electron temperature at and near the target surface. As in the previous chapter, we note that generally most absorbed laser energy is deposited into the electrons and so we concern ourselves mainly with the elevation and evolution of the electron plasma temperature profile. With this in mind, the first relevant quantity is the absorbed energy fluence (absorbed energy per area) into the thermal plasma defined in terms of the laser pulse intensity history $I(t)$:

$$F_{abs} = \eta_{abs}^{(T)} \int I(t) dt. \tag{9.2}$$

9.1 The Structure of Solid Target Plasmas

Before proceeding, recall the definition of the plasma's specific heat (see Appendix B.8) at constant volume (the relevant quantity for finding plasma heating at constant density),

$$C_V = \left(\frac{\partial u}{\partial T}\right)_V, \tag{9.3}$$

where u is the internal energy density of the plasma. From Appendix B.8 we can think of this quantity in two limits, a plasma of low-Z atoms, in which most deposited energy goes into electron thermal energy, the ideal gas case, giving

$$C_V = \frac{3}{2} n_e k_B \quad \text{ideal gas} \tag{9.4}$$

and the limit of high-Z, where a significant fraction of the deposited energy must go into ionization, in which case the simple estimate of Appendix B.8 yields

$$C_V \cong \frac{17}{4} n_e k_B \quad \text{ionizing high} - Z \text{ gas.} \tag{9.5}$$

(For purposes of proceeding, simply think of $C_V = \alpha_T C_V(\text{ideal gas})$ where $\alpha_T = 1$ for low-Z targets and $\approx 17/6 \approx 3$ in high-Z plasma.) For short pulses, then, we might estimate the surface temperature assuming volume heating in a skin depth, δ_{SD}, which is in a plasma $\delta_p \approx c/\omega_{pe}$, implying that a short pulse heats the plasma to

$$k_B T_e \approx \frac{F_{abs}}{C_V \delta_p}. \tag{9.6}$$

Such an estimate would say, for example, that even a 100 mJ laser pulse focused to a 20 µm diameter spot on solid density ($n_e \sim 10^{23}$ cm^{-3}) would yield an electron temperature in the vicinity of 100 keV. This is a gross overestimate. It illustrates that even for ultrashort pulses we must consider plasma cooling mechanisms. Generally there are two: cooling associated with the adiabatic expansion of the plasma, and cooling by heat conduction into underlying cold, unheated buried matter in the target. The former is not very effective on subpicosecond time scales because of the inertia of the heavy ions (though in thin foil targets it can be important even on a $\sim$1 ps time scale). The latter effect, however, can very quickly cool the surface because of the very high-temperature gradients that get set up when the laser heats only a skin layer (which may be only a few tens of nm).

If we look at heat conduction into the bulk of a target, we need the plasma heat conductivity, which from Appendix B can be written in the so-called "Spitzer" regime as

$$\kappa_e = \frac{20\sqrt{2}}{\pi^{3/2}} \frac{k_B (k_B T_e)^{5/2}}{\bar{Z} e^4 m^{1/2} \ln \Lambda_C}. \tag{9.7}$$

$\bar{Z}$ can be interpreted to be the average ionization state in the plasma and $\ln \Lambda_C$ is the usual Coulomb logarithm. For most high-intensity experiments, it is a good approximation to treat the heat conduction in one dimension since the focal spot size ($\sim$10 µm or larger) is usually significantly bigger than the depth of heat penetration, at least on the time scale during and immediately after the laser pulse. Assuming that the heat conduction is diffusive (not always a good assumption in many short-pulse laser plasma experiments), the electron temperature is governed by the heat equation, which, if $\mathbf{Q}_e$ is the energy flux in the electron plasma, can be written

$$C_V \frac{\partial T_e}{\partial t} = -\nabla \cdot \mathbf{Q}_e = \frac{\partial}{\partial z} \kappa_e \frac{\partial}{\partial z} T_e. \quad (9.8)$$

As is usual in the literature, we define the thermal diffusivity χ_e:

$$\chi_e \equiv \frac{\kappa_e}{C_V} = a T_e^b, \quad (9.9)$$

and, in the interest of generality, define it in terms of a power law with temperature to allow for possible heat diffusion mechanisms other than electron thermal conduction (such as diffusive radiation transport) which may have a different power-law dependence than $b = 5/2$ from Eq. (9.7).

The 1D heat equation of the form

$$\frac{\partial T_e}{\partial t} = a \frac{\partial}{\partial z} T_e^b \frac{\partial}{\partial z} T_e \quad (9.10)$$

can be solved in terms of a self-similar solution for the temperature profile into the target. Such a heat wave solution has the same functional shape in depth as time advances; only the spatial scale of the self-similar temperature profile evolves. That means that we seek a solution for the electron temperature $T_e(z, t)$ in terms of a function of a single dimensionless variable. (This approach is well developed by Zel'dovich and the reader is encouraged to consult this classic text for more details on these self-similar heat waves (Zeldovich and Raizer 2002, pp. 652–681).)

To make progress, we couch the boundary conditions for heat deposition by an ultrashort pulse by treating it as a delta function in time, in a region very close to the target surface. In that case the relevant parameter to express the initial conditions for solution of Eq. (9.10) is the integral of the temperature over the depth of energy deposition at the target surface, which we call G and write in terms of the absorbed laser fluence as

$$G \equiv \int_{-\infty}^{\infty} T_e(z) dz = \frac{2}{3} \frac{F_{abs}}{k_B n_e}. \quad (9.11)$$

(Here we have assumed an ideal gas specific heat. This equation can be divided by α_T when the target plasma is high-Z.) We then construct a dimensionless spatial variable from a, G, z and t:

$$\tilde{z} = \frac{z}{a^{1/(b+2)} G^{b/(b+2)} t^{1/(b+2)}}. \quad (9.12)$$

Since G/z has units of temperature, it makes sense to seek a solution of the form

$$T_e(z, t) = \left(\frac{G^2}{at}\right)^{1/(b+2)} g(\tilde{z}). \quad (9.13)$$

Substitution of this trial solution into the heat equation yields

$$(b+2) \frac{\partial}{\partial \tilde{z}} \left(g^b \frac{\partial g}{\partial \tilde{z}}\right) + \tilde{z} \frac{\partial g}{\partial \tilde{z}} + g = 0, \quad (9.14)$$

9.1 The Structure of Solid Target Plasmas

which has a solution found in Zel'dovich and Kompaneets (1959) that reads

$$g(\tilde{z}) = \left[\frac{b}{2(b+2)}\left(\tilde{z}_0^2 - \tilde{z}^2\right)\right]^{1/b} \quad \tilde{z} < \tilde{z}_0 \qquad (9.15)$$
$$= 0 \qquad \tilde{z} > \tilde{z}_0$$

The integration constant can be found from the fact that the derivative of the temperature must be zero at the target surface (since no heat can flow into or from the vacuum just outside the target) and insertion of the solution (9.13) into the definition, Eq. (9.11), to yield

$$G = \int_{-\infty}^{\infty} \left(\frac{G^2}{at}\right)^{1/(b+2)} a^{1/(b+2)} G^{b/(b+2)} t^{1/(b+2)} g(\tilde{z}) \, d\tilde{z}, \qquad (9.16)$$

which reduces to

$$\int_{-\infty}^{\infty} g(\tilde{z}) \, d\tilde{z} = 1, \qquad (9.17)$$

from which the constant $\tilde{z}_0$ is found to be

$$\tilde{z}_0 = \left[\frac{(b+2)^{b+1} 2^{1-b}}{b\pi^{b/2}}\right]^{1/(b+2)} \left[\frac{\Gamma(1/2 + 1/b)}{\Gamma(1/b)}\right]^{b/(b+2)}. \qquad (9.18)$$

When $b = 5/2$, this is simply $\tilde{z}_0 = 1.01$. Inspection of Eq. (9.15) tells us that there will be a temperature profile that has a front moving along z into the target with front location $\tilde{z} = \tilde{z}_0$ so the front depth z_f is

$$z_f = \tilde{z}_0 a^{1/(b+2)} G^{b/(b+2)} t^{1/(b+2)}. \qquad (9.19)$$

We see, then, that the short-pulse energy deposition on the target surface drives a heat wave which penetrates, if heat conduction is by electrons, as $z_f \sim t^{2/9}$.

An example of a heat wave driven by a modest laser energy in a solid-density Al target is illustrated in Figure 9.3a. The heat front at early time penetrates very rapidly, as the thermal gradient is initially very high right after the laser heating. The front velocity slows significantly once the deposited energy has spread into the target bulk on a 10 ps time scale. Figure 9.3b shows the anatomy of such an impulse-driven self-similar heat wave as a function of depth. The shape of the heat front as time evolves is shown in the top panel and the heat flux associated with the heat wave after 100 ps is shown below that. The time derivative of the electron temperature at the 100 ps point is shown in the bottom panel, illustrating how the cold plasma out in front of the wave is heated while the plasma nearer the surface is cooling as its energy propagates into the target. The most important takeaway point of this simple illustration is that heat conduction very quickly brings the electron temperature in the region near the target surface to a temperature of 100–500 eV. This tends to be a good rule of thumb in most high-intensity ultrafast solid target experiments: heat conduction tends to push the target into a temperature of a few hundred eV.

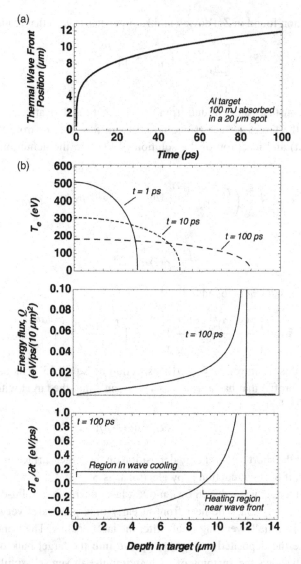

Figure 9.3 (a) Position of the thermal wavefront as a function of time from a delta function heat impulse of 100 mJ of energy in a 20 μm laser spot, calculated in Al as an example. (b) Three panels showing the spatial structure into the target of the heat wave plotted in 9.3(a). Panel (b1) shows electron temperature at three times after heat deposition. The first panel (b2) in (b) illustrates the energy flux and Panel (b3) the time rate of temperature in the heat wave at 100 ps showing regions of heating at the wavefront and cooling behind the front.

We can infer the surface temperature from the self-similar solution, which tells us that at the target surface,

$$T_e(0,t) = \left(\frac{b}{2(b+2)}\right) \tilde{z}_0^{2/b} \frac{G^{2/(b+2)}}{(at)^{1/(b+2)}}, \qquad (9.20)$$

in which for electron heat conduction, where $b = 5/2$, the surface temperature falls like $T_e(0,t) \propto t^{-2/9}$. So, instead of Eq. (9.6), a better estimate for the surface temperature at the end of a laser pulse is to use Eq. (9.20) and evaluate it at a time equal to the pulse duration. This suggests that in our previous example, a 100 mJ pulse in a 20 µm spot with pulse duration of 30 fs will see surface temperature more like ~1 keV, much more like what is inferred from experiments.

When the heating is in the long-pulse regime, say by some long pedestal in front of the high-intensity main ultrashort pulse, we can use similar techniques to deduce the surface temperature. A constant heat input by absorption on the surface will also drive a thermal wave whose position can be estimated simply by conservation of energy. If the thermal wave penetrates z_f in a time t subject to I_{abs} constant absorbed power per area on the surface, the location of the front is roughly found from

$$I_{abs} \frac{t}{z_f} \approx \frac{3}{2} \alpha_T n_e k_B T_e. \tag{9.21}$$

The thermal gradient in the heat equation can be simply approximated as

$$I_{abs} = a T_e^b \frac{\partial T_e}{\partial z} \approx \frac{a T_e^{b+1}}{z_f}, \tag{9.22}$$

which can be used to eliminate T_e in (9.21) to give

$$z_f = \left(\frac{2}{3\alpha_{ion} n_e k_B}\right)^{(b+1)/(b+2)} a^{1/(b+2)} I_{abs}^{b/(b+2)} t^{(b+1)/(b+2)}. \tag{9.23}$$

The same reasoning allows us to estimate the surface temperature:

$$C_V T_e z_f \cong I_{abs} t, \tag{9.24}$$

which grows with time (with $b = 5/2$) as

$$T_e(z=0,t) \cong \frac{2 I_{abs} t}{3 \alpha_{ion} n_e k_B z_f} \sim t^{2/9}. \tag{9.25}$$

The shape of the thermal wave can be found, again assuming a self-similar solution in terms of the quantity:

$$H \equiv \frac{I_{abs}}{C_V}. \tag{9.26}$$

The appropriate self-similar variable is

$$\tilde{z}' = \frac{z}{a^{1/(b+2)} H^{b/(b+2)} t^{(b+1)/(b+2)}} \tag{9.27}$$

and the solution for the temperature takes the form

$$T_e(z,t) = \left(\frac{H^2 t}{a}\right)^{1/(b+2)} h(\tilde{z}'). \tag{9.28}$$

Substituting this solution into the heat equation yields a differential equation for the shape of the heat wave,

$$\frac{\partial}{\partial \tilde{z}'}\left(h^b \frac{\partial h}{\partial \tilde{z}'}\right) + \frac{b+1}{b+2}\tilde{z}'\frac{\partial h}{\partial \tilde{z}'} - \frac{1}{b+2}h = 0, \qquad (9.29)$$

which is, for electron heat conduction,

$$\frac{\partial}{\partial \tilde{z}'}\left(h^{5/2} \frac{\partial h}{\partial \tilde{z}'}\right) + \frac{7}{9}\tilde{z}'\frac{\partial h}{\partial \tilde{z}'} - \frac{2}{9}h = 0. \qquad (9.30)$$

The appropriate boundary conditions are set by the heat flux from laser irradiation on the target surface such that

$$Q(z=0) = \eta_{abs}^{(Th)} I_0 = -aC_V T_e^b \left.\frac{\partial T_e}{\partial z}\right|_{z=0}, \qquad (9.31)$$

which yields a boundary condition for h, on the target surface

$$\left. h^b \frac{\partial h}{\partial \tilde{z}'}\right|_{z=0} = -1, \qquad (9.32)$$

that can be combined with the fact that deep into the target the material remains unheated, so

$$h(\infty) = 0. \qquad (9.33)$$

This equation does not have an analytic solution, but numerical solution yields for the position of the heat front an equation almost identical to our simple estimate in Eq. (9.23),

$$z_f = 1.05 a^{1/(b+2)} H^{b/(b+2)} t^{(b+1)/(b+2)}, \qquad (9.34)$$

and a solution for the target surface temperature:

$$T_e(z=0) = 1.35 \left(\frac{H^2 t}{a}\right)^{1/(b+2)}. \qquad (9.35)$$

A numerical solution to the heat equation subject to a low-intensity constant heat source, as might be characteristic of a low-intensity nanosecond-scale prepulse before a high-intensity femtosecond pulse, is illustrated in Figure 9.4 (using as input a constant absorbed heat flux on the surface of 10^{13} W/cm^2 and assuming an Al target). The buildup of temperature to a couple of hundred electron volts on a nanosecond time scale is seen. This is illustrated in Figure 9.5 by a plot of surface temperature as a function of time for two absorbed heat fluxes.

9.1.3 Hydrodynamic Expansion of the Target Surface

Very frequently in high-intensity experiments on solid target surfaces, some heat deposition during the pulse or by a prepulse well before the main laser pulse will set up a pressure

9.1 The Structure of Solid Target Plasmas

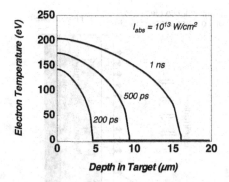

Figure 9.4 Numerical simulations of the heat equation in Al for a constant heat deposition at the target surface showing electron temperature profiles into the target at three times after the initiation of heat deposition.

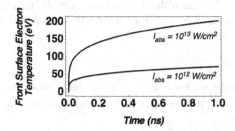

Figure 9.5 Plot of front surface temperature as a function of time in Al at two heat deposition rates.

that drives hydrodynamic expansion of the plasma into vacuum (Gurevich et al. 1966; Crow et al. 1975). This has two practical consequences: It leads to a lower-density plasma that the main pulse has to traverse on its way to the critical density surface, affecting absorption and hot electron generation. It also leads to production of fast ions from the front surface. Therefore, understanding the structure of the plasma in front of the target with which the main laser pulse interacts will mandate a brief consideration of this physics.

On a microscopic spatial scale, the origin of this expansion in a two-component, electron–ion plasma can be understood by the simple picture illustrated in Figure 9.6. Since typically the laser deposits most of its energy, at least initially, into the electrons, the thermal motion of these light particles causes the electron plasma fluid to want to expand quickly. This fluid is quickly confined by the space-charge attraction of the more slowly moving, massive ions. A so-called ambipolar sheath electric field is set up (Piel 2010, p. 86) which accelerates the ions. Consequently, the characteristic expansion sound speed is set by the thermal energy of the electrons and the inertia of the heavy ions.

To examine this analytically, we assume that the expansion is in 1D (planar) and treat the problem as an expansion in the standard two-fluid regime in which the ion temperature is zero. The conservation of mass and momentum equations for the ions subject to

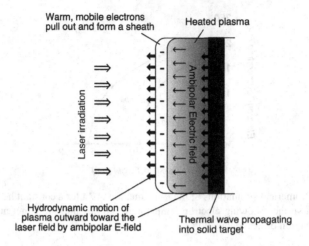

Figure 9.6 Schematic illustration showing how an ambipolar field set up by the expansion of the light, hot electron plasma accelerates the plasma into vacuum and drives hydrodynamic expansion.

an ambipolar electric field are now familiar as we have encountered them frequently in Chapters 7 and 8. They read for the ion density n_i and ion fluid velocity u_i:

$$\frac{\partial n_i}{\partial t} + \frac{\partial}{\partial z}(n_i u_i) = 0, \tag{9.36}$$

$$m_i n_i \frac{\partial u_i}{\partial t} + m_i n_i u_i \frac{\partial u_i}{\partial z} - Z e n_i E = 0. \tag{9.37}$$

On the time scale of the expansion, we treat the electrons as a light fluid which responds instantaneously to the motion of the slower ions. Consequently, though we must include the effect of finite electron temperature through inclusion of the electron pressure in the electron fluid equation, we set $m = 0$ and write for the electron fluid and its pressure p_e

$$\frac{\partial p_e}{\partial z} = -e n_e E. \tag{9.38}$$

The expansion of the warm electron fluid sets up an electric field in the gradient present at the target surface. As usual, we assume an ideal gas equation of state for the electrons

$$p_e = n_e k_B T_e, \tag{9.39}$$

and recall the usual definition for the isothermal sound speed of an ion fluid subject to electron pressure:

$$c_{si}^2 = \left(\frac{\partial p}{\partial \rho}\right)_S = \frac{1}{m_i}\frac{\partial p_e}{\partial n_i} = \frac{Z k_B T_e}{m_i}. \tag{9.40}$$

The ion fluid momentum equation can be rewritten:

$$\frac{\partial u_i}{\partial t} + u_i \frac{\partial u_i}{\partial z} + c_{si}^2 \frac{\partial n_i}{\partial z} = 0. \tag{9.41}$$

The solution to this equation can be found, assuming a self-similar solution for the ion density and fluid velocity in terms of a self-similar variable, z/t, such that $n_i = N(z/t)$, $u_i = U(z/t)$ and $\tilde{z}_0 = z/t$ (Schmalz 1985). The continuity equation (9.36) reads, in terms of these functions,

$$-\frac{z}{t^2}N' + \frac{U}{t}N' + \frac{N}{t}U' = 0, \quad (9.42)$$

$$N'(U - z/t) + NU' = 0. \quad (9.43)$$

Likewise, the ion momentum equation is

$$-\frac{z}{t^2}U' + \frac{U}{t}U' = c_s^2 \frac{1}{N}\frac{1}{t}N', \quad (9.44)$$

$$U'(U - z/t) = c_s^2 \frac{N'}{N}. \quad (9.45)$$

Inserting (9.43) into (9.45) to eliminate N'/N yields for the ion velocity

$$U = c_s + z/t. \quad (9.46)$$

This reduces Eq. (9.43) to

$$\frac{N'}{N} = -\frac{1}{c_s}, \quad (9.47)$$

which is trivially integrated to give the ion density profile as a function of time:

$$n_i(z, t) = n_0 \exp\left[-\frac{z + c_s t}{c_s t}\right]. \quad (9.48)$$

We see that the hydrodynamic expansion leads to an exponential density profile out in front of the solid target with a scale-length increasing linearly with time at a rate equal to the isothermal sound speed and a rarefaction wave traveling back into the target at the same speed. The size of the electric ambipolar field, E_A, is found from Eq. (9.38):

$$E_A = -\frac{k_B T_e}{e n_e}\frac{\partial n_e}{\partial z}$$

$$= \frac{k_B T_e}{e c_{si} t}. \quad (9.49)$$

The field is constant in space, yielding a potential out in the plasma expansion which varies linearly with distance from the rarefaction front:

$$\Phi = -\frac{k_B T_e}{e c_{si} t}z. \quad (9.50)$$

A plot of these quantities in such a 1D, single-temperature expansion is shown in Figure 9.7. This analysis indicates, for example, that an Al target heated to 100 eV by a prepulse will expand to $\sim$5 μm if heated on a $\sim$100 ps time scale, a scale-length that is many laser wavelengths. This low-density preplasma can have a dramatic effect on the nature of how the intense pulse interacts with the target.

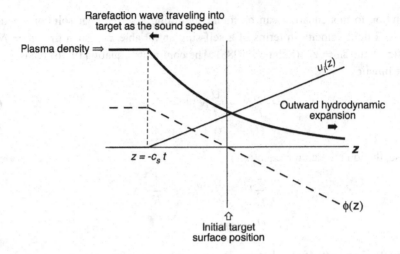

Figure 9.7 Plot of the ion density, ion velocity and electron potential in an ambipolar hydrodynamic expansion of plasma driven by a warm electron fluid.

9.1.4 Two-Electron Temperature Hydrodynamic Expansion

The assumption that the target electron plasma can be characterized by a Maxwellian of a single temperature is a fair approximation for plasma heated by modest intensity, as say by a prepulse. This simple approximation often breaks down in real laser-plasma experiments because of the presence of two (or more) absorption mechanisms. Collisional absorption can lead to a 'warm" electron plasma component, with temperature of $\sim$100 eV, and collisionless absorption (explained in the next section) can boost the energy of a minority component of the target electrons to well above this. Though the second electron component is typically far from thermal equilibrium, it is not a bad first approximation to describe this situation as if the electron fluid was composed of two Maxwellian components of different electron temperature. This has an interesting, observable impact on the target interaction.

We will consider the plasma expansion of such a two-temperature plasma by briefly examining the analytic formalism published in a set of papers by Wickens, Allen and Rumsby (Wickens et al. 1978; Wickens and Allen 1979; Wickens and Allen 1981). Though the analytic result has a rather limited range of applicability, it does nicely illustrate the qualitative trend evinced by this situation, a trend which is well documented in experiments. We start by noting that the electron fluid is composed of two fluids with cold and hot electron temperatures, T_{ec} and T_{eh} respectively, of densities n_{ec} and n_{eh}. Quasi-neutrality in the plasma mandates that

$$n_{ec} + n_{eh} = n_e = \bar{Z} n_i. \qquad (9.51)$$

At this point the analysis follows similar lines as the single-temperature analysis. Once again the ion fluid is assumed to have zero temperature and its density and velocity evolve by the continuity and momentum equation driven by an ambipolar potential:

9.1 The Structure of Solid Target Plasmas

$$\frac{\partial n_i}{\partial t} + \frac{\partial}{\partial z}(n_i u_i) = 0, \tag{9.52}$$

$$\frac{\partial u_i}{\partial t} + u_i \frac{\partial u_i}{\partial z} = -\frac{Ze}{m_i} \frac{\partial \Phi}{\partial z}. \tag{9.53}$$

The hot and cold components of the electron fluid are considered to distribute themselves instantaneously according to the plasma potential,

$$n_{ec}(z) = n_{ec0} \exp[e\Phi/k_B T_{ec}], \tag{9.54}$$

$$n_{eh}(z) = n_{eh0} \exp[e\Phi/k_B T_{eh}]. \tag{9.55}$$

Using these in Eq. (9.51) and taking the z derivative lets us eliminate the potential in (9.53) to give

$$\frac{\partial u_i}{\partial t} + u_i \frac{\partial u_i}{\partial z} = -\frac{Z}{m_i} \frac{n_{ec} + n_{eh}}{(n_{ec}/k_B T_{ec}) + (n_{eh}/k_B T_{eh})} \frac{\partial n_i}{\partial z}. \tag{9.56}$$

We can now define a sound speed which is not a constant, as in the single-temperature case, but varies with z through the distributions of the hot and cold electron components:

$$c_{s2}(z) = \sqrt{\frac{Z}{m_i} \frac{n_{ec} + n_{eh}}{(n_{ec}/k_B T_{ec}) + (n_{eh}/k_B T_{eh})}}. \tag{9.57}$$

This lets us write the ion fluid moment equation in a familiar form:

$$\frac{\partial u_i}{\partial t} + u_i \frac{\partial u_i}{\partial z} = -\frac{c_{s2}^2}{n_i} \frac{\partial n_i}{\partial z}. \tag{9.58}$$

As before, we can assume that the solution takes on a self-similar form in terms of the variable $\tilde{z} = z/t$. This gives for our two ion fluid equations

$$(u_i - \tilde{z}) \frac{\partial n_i}{\partial \tilde{z}} + n_i \frac{\partial u_i}{\partial \tilde{z}} = 0, \tag{9.59}$$

$$(u_i - \tilde{z}) \frac{\partial u_i}{\partial \tilde{z}} + \frac{c_{s2}^2}{n_i} \frac{\partial n_i}{\partial \tilde{z}} = 0. \tag{9.60}$$

Following steps similar to the one-temperature derivation, we can arrive at the equation

$$c_{s2}^2 = (u_i - \tilde{z})^2, \tag{9.61}$$

resulting in a form for the ion velocity similar to our previous result.

$$u_i(z) = c_{s2}(z) + z/t. \tag{9.62}$$

In this case, however, the z variation of the sound speed means that the ion velocity is not simply linear with position out in the expansion. The presence of a hot electron component gives a "boost" to the velocity of ions out in the leading part of the expansion.

To make further progress, we have to assume that the potential is monotonic and quasi-neutrality is maintained (this condition will break down for very high-temperature hot electron components). We need to relate n_i to Φ. From the z derivative of (9.51) we can write that

$$\frac{c_{s2}^2}{n_i}\frac{\partial n_i}{\partial \tilde{z}} = \frac{Ze}{m_i}\frac{\partial \Phi}{\partial \tilde{z}}. \tag{9.63}$$

Using the continuity equation for the ion fluid (9.59)

$$\frac{c_{s2}^2}{n_i}\frac{\partial n_i}{\partial \tilde{z}} = -c_{s2}\frac{\partial u_i}{\partial \tilde{z}} \tag{9.64}$$

and Eq. (9.63), we have that

$$\frac{\partial u_i}{\partial \tilde{z}} = -\frac{1}{c_{s2}}\frac{Ze}{m_i}\frac{\partial \Phi}{\partial \tilde{z}}, \tag{9.65}$$

which, with the definition for the two-temperature sound speed, can be written in terms of an integrable equation:

$$\frac{m_i}{Ze}\partial u_i = -\sqrt{\frac{(n_{ec}/k_B T_{ec}) + (n_{eh}/k_B T_{eh})}{n_{ec} + n_{eh}}}\frac{m_i}{Z}\partial \Phi. \tag{9.66}$$

Integration yields

$$u_i(\Phi) = \left(\frac{1}{T_{ec}} - \frac{1}{T_{eh}}\right)\left\{\frac{c_{sc}}{T_{ec}}\ln\left[\frac{(c_{s2} - c_{sc})(c_{s20} + c_{sc})}{(c_{s2} + c_{sc})(c_{s20} - c_{sc})}\right]\right.$$
$$\left. + \frac{c_{sh}}{T_{eh}}\ln\left[\frac{(c_{sh} + c_{s2})(c_{sh} - c_{s20})}{(c_{sh} - c_{s2})(c_{sh} + c_{s20})}\right]\right\}, \tag{9.67}$$

where we have introduced the following constants:

$$c_{s20} = c_{s2}(\Phi = 0); \quad c_{sc} = \sqrt{Zk_B T_{ec}/m_i}; \quad c_{sh} = \sqrt{Zk_B T_{eh}/m_i}. \tag{9.68}$$

This result can be coupled with Eq. (9.62) along with Eqs. (9.51), (9.54) and (9.55), using the potential simply as a parameter to plot the ion density and velocity as a function of position in the expansion.

Such a plot with a hot electron temperature component $9\times$ that of the cold component is shown in Figure 9.8. Examination of the ion density shows a deviation from the exponential profile of the single-temperature expansion, with a fast ion component racing out in front of the slower bulk expansion of the main plasma. Such two-component ion expansions are well known in strong field experiments and can, in fact, be used to diagnose the magnitudes of the two electron temperatures (Meyerhofer et al. 1993). We will revisit this effect when we consider ion acceleration in a subsequent section.

In a postscript to this analysis, we should note that it can be shown that the potential profile with z predicted by the Di model reflected in Eq. (9.67) becomes multivalued when $T_{eh}/T_{ec} > 9.9$. This unphysical result and breakdown of the model is a consequence of the fact that at such high hot-electron temperatures the assumption of quasi-neutrality breaks down and a more complex model is needed to describe the expansion of the fast

9.1 The Structure of Solid Target Plasmas

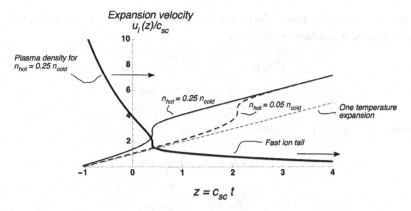

Figure 9.8 Plot of the ion density and ion velocity in an ambipolar hydrodynamic expansion of plasma driven by a two-temperature warm electron fluid of temperatures in which the hot component is 9× higher temperature than the cold component. The solid lines show the plasma profile and ion velocity when the hot component is 25% the density of the cold component. The long dashed line shows the velocity profile for a 5% hot component.

ion component. This will mandate a hot-electron transport model, which usually requires numerical simulation.

9.1.5 Laser Field Profile Approaching a Solid Target in Finite Plasma Scale Length

Now that we understand how the plasma-density profile in front of a solid target evolves as it has been heated by the incident laser, we now seek to find the electric field profile at the plasma surface so that we can understand the various absorption mechanisms and hot-electron generation mechanisms that come into play at high intensity. This topic has been treated by numerous textbook authors, as far back as the classic text of Ginzburg (Ginzburg 1960, pp. 319–376; Kruer 2003, pp. 27–39; Eliezer 2002, pp. 79–95; Liu et al. 2019, pp. 35–42; Larsen 2017, pp. 631–650). It is a topic that is fraught with subtleties, which have been widely discussed in the laser-plasma literature. We will examine this physics because of its importance in understanding the various collisional and collisionless absorption mechanisms that manifest in strong field irradiation of solid targets.

For reasons that will become apparent, we will follow the usual treatment and solve not for **A** but for the electric and magnetic fields. For this purpose we will treat the fields as imaginary and harmonic:

$$\mathbf{E} \sim \mathbf{B} \sim e^{-i\omega t}. \tag{9.69}$$

We derive the wave equation governing these fields in the usual manner, but now taking account of the fact that the plasma dielectric function, ε, is a function of position because of gradients in the plasma density approaching the solid target surface. Faraday's law reads

$$\nabla \times \mathbf{E} = -\frac{1}{c}\frac{\partial \mathbf{B}}{\partial t} = i\frac{\omega}{c}\mathbf{B}, \tag{9.70}$$

and Ampere's law is, in terms of the plasma electron current $\mathbf{J}_e$ (assume ions are stationary on the time scale of the laser oscillation):

$$\nabla \times \mathbf{B} = \frac{1}{c}\frac{\partial \mathbf{E}}{\partial t} + \frac{4\pi \mathbf{J}_e}{c}$$

$$= \frac{1}{c}\frac{\partial}{\partial t}(\varepsilon \mathbf{E}) = \frac{-i\omega \varepsilon}{c}\mathbf{E}. \quad (9.71)$$

As before, we write the plasma dielectric function

$$\varepsilon = 1 - \frac{\omega_{pe}^2}{\omega(\omega + iv_d)} \quad (9.72)$$

$$\cong 1 - \frac{n_e}{n_{crit}} + i\frac{n_e}{n_{crit}}\frac{v_d}{\omega} \quad (9.73)$$

Spatial variation in ε arises from the spatial variation of the electron plasma frequency because of nonuniformity in electron density. We take the curl of Eq. (9.70) and the time derivative of (9.71), combining them to yield a wave equation for $\mathbf{E}$:

$$\nabla^2 \mathbf{E} - \nabla(\nabla \cdot \mathbf{E}) + \frac{\omega^2}{c^2}\varepsilon \mathbf{E} = 0 \quad (9.74)$$

and, reversing the operations and using the fact that $\nabla \times (\varepsilon \mathbf{E}) = \varepsilon \nabla \times \mathbf{E} + \nabla \varepsilon \times \mathbf{E}$, getting a wave equation for $\mathbf{B}$:

$$\nabla^2 \mathbf{B} + \frac{\omega^2}{c^2}\varepsilon \mathbf{B} + \frac{1}{\varepsilon}\nabla \varepsilon \times \nabla \times \mathbf{B} = 0. \quad (9.75)$$

As an aside, we note that, unlike EM wave propagation in vacuum, the inhomogeneous plasma density can lead to situations in which $\nabla \cdot \mathbf{E} \neq 0$.

We consider, at the moment, the experimental situation in which there is some plasma expansion, either from the rise of a "long" pulse (say >1 ps) or because of some prepulse ahead of the main laser short pulse. This situation is illustrated in Figure 9.9. The geometry of an obliquely incident wave approaching a solid target plasma at angle of incidence of θ_0 is shown on the right in that figure. This figure illustrates that it is easiest if we now abandon

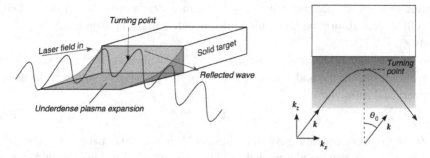

Figure 9.9 Schematic illustration of an EM wave incident on a target surface with some plasma expansion out in front of the solid (left). On the right the trajectory that the ray of this wave will take as it approaches the target critical surface at oblique angle of incidence.

9.1 The Structure of Solid Target Plasmas

the convention of defining the z-axis as the direction of the laser wave propagation and x- and y-axes as along the E and B fields respectively. Instead we will now let the target geometry set the axes, defining z as the direction directly into the target and the plasma expansion in the $-z$-direction.

We will, in turn, consider the field structure for three cases:

1) Normal Incidence ($\theta_0 = 0$)
2) Oblique incidence, S-pol. (**E** out of the plane of incidence)
3) Oblique incidence, P-pol. (**E** in the incidence plane)

Each case presents some unique, interesting physics.

1) Normal Incidence Obviously, in this case **E** is perpendicular to the direction of the plasma density gradient, so electron oscillations in the field occur in a direction of uniform density, which means that $\nabla \cdot \mathbf{E} = 0$. The wave propagates along the z-axis and the wave equation for the electric field, Eq. (9.74) reduces to a simple Helmholtz equation:

$$\nabla^2 \mathbf{E} + \frac{\omega^2}{c^2}\varepsilon(z)\mathbf{E} = 0. \tag{9.76}$$

This equation can be easily solved numerically for some known, arbitrary density profile. Such a calculation for a wave incident on an exponential plasma density profile with scale length (1/e length) of 5λ is shown in Figure 9.10. The main takeaway is that the field

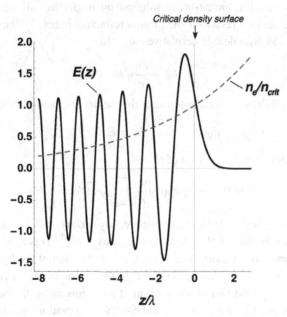

Figure 9.10 Numerical solution of the Helmholtz equation Eq. (9.76) for an electromagnetic wave incident normally on target with an exponential plasma profile with 1/e scale length of 5λ.

oscillates as it travels to the critical density point where it is reflected and an evanescent wave penetrates into the overdense region over a spatial scale of a skin depth (see Appendix B, section B.4). Of more interest is the fact that the electric field amplitude swells in magnitude as the wave approaches this reflection point at density $n_e = n_{crit}$.

A more quantitative estimate for this swelling effect can be seen analytically if we assume that the plasma-density variation is slow on the spatial scale of a laser wavelength, a condition that can be stated as

$$\frac{\varepsilon}{d\varepsilon/dz} = \frac{n_e}{dn_e/dz} \equiv L \gg \lambda. \tag{9.77}$$

In this case, we can employ a WKB solution to estimate the variation of the E-field amplitude (an approach that clearly has to break down at electron density near the critical density). Proceeding with a formal WKB-type solution, we seek a solution to Eq. (9.76) of the form

$$E(z) = E_0(z) \exp\left[i\frac{\omega}{c} \int \eta(z) dz\right], \tag{9.78}$$

where $E_0(z)$ and $\eta(z)$ are unknown functions (though it should be apparent that $\eta(z)$ assumes the role of a refractive index). Substituting into (9.76) this solution yields

$$E_0'' + \frac{2i\omega}{c}\eta E_0' - \frac{\omega^2}{c^2}\eta^2 E_0 + \frac{i\omega}{c}\eta' E_0 + \frac{\omega^2}{c^2}\varepsilon E_0 = 0, \tag{9.79}$$

where primes denote derivative with respect to z. Since we assume slow density variation over z, we find the lowest-order solution by neglecting all derivatives yielding $\eta(z) = \varepsilon(z)^{1/2}$, an obvious result for the plasma refractive index. To the next order retain single derivatives and drop double derivatives to yield

$$\frac{2i\omega}{c}\eta E_0' + \frac{i\omega}{c}\eta' E_0 = 0, \tag{9.80}$$

which gives for the field amplitude in terms of the incident field amplitude in vacuum, E_{00}:

$$E_0(z) = E_{00}/\sqrt{\eta} = E_{00}/\varepsilon^{1/4} \tag{9.81}$$

and a formal solution for the variation of E:

$$E(z) \cong \frac{E_{00}}{\varepsilon^{1/4}} \exp\left[i\frac{\omega}{c} \int_{-\infty}^{z} \sqrt{\varepsilon(z')} dz'\right]. \tag{9.82}$$

This WKB solution is singular at the critical density surface, but does show that the field amplitude swells as the laser light approaches n_{crit} as $1/\varepsilon^{1/4}$. This result is in fact, to be expected based simply on conservation of energy and variation of the light group velocity through the changing plasma density. The energy flux is the electric field energy density times a group velocity, and this must be equal at all points along the propagation in the absence of absorption, which means then that energy conservation mandates

$$\frac{E_{00}^2}{8\pi}c = \frac{E^2}{8\pi}c\sqrt{\varepsilon} \quad \rightarrow \quad E = \frac{E_{00}}{\varepsilon^{1/4}}. \tag{9.83}$$

9.1 The Structure of Solid Target Plasmas

The magnetic field is found easily from the WKB result and Ampere's law, which is for normal incidence

$$\frac{i\omega}{c} B = \frac{\partial E}{\partial z}, \qquad (9.84)$$

yielding for the magnetic field

$$B(z) \cong E_{00} \left(\varepsilon^{1/4} + \frac{ic}{4\omega} \frac{1}{\varepsilon^{5/4}} \frac{\partial \varepsilon}{\partial z} \right) \exp\left[i \frac{\omega}{c} \int_{-\infty}^{z} \sqrt{\varepsilon(z')} dz' \right]. \qquad (9.85)$$

Since the second term in parentheses is small in long plasma-density gradients, we see that, while the electric field amplitude swells as the light approaches the critical density (as $\varepsilon^{-1/4}$), the magnetic field amplitude decreases (as $\varepsilon^{1/4}$).

A bit more insight (particularly for the oblique incidence cases we consider next) can be derived analytically by using a plasma profile that is a linear density ramp, (a situation not really found experimentally but one that does yield a reasonable approximation to a plasma-density expansion). Such a plasma profile is depicted in Figure 9.11 and can be expressed analytically as

$$n_e = n_{crit}(1 + z/L). \qquad (9.86)$$

The dielectric function for such a plasma profile is easily expressed in terms of z

$$\begin{aligned} \varepsilon &= -z/L & z &> -L \\ &= 1 & z &< -L \end{aligned} \qquad (9.87)$$

which yields for the Helmholtz equation describing E

$$\frac{\partial^2 E}{\partial z^2} - \frac{\omega^2}{c^2} \frac{z}{L} E = 0. \qquad (9.88)$$

This equation is of the form of Stoke's differential equation:

$$E''(\alpha) - \alpha E(\alpha) = 0, \qquad (9.89)$$

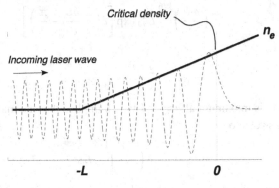

Figure 9.11 Geometry of a linear plasma-density ramp and the electric field (dashed line) coming into that plasma.

where

$$\alpha = \tilde{z}_L \equiv (\omega^2/c^2 L)^{1/3} z. \quad (9.90)$$

This equation can be solved in terms of Bessel functions of order 1/3, functions more commonly known as the Airy functions Ai(z) and Bi(z). Since the laser is reflected at the critical density surface ($z = 0$) we require that the solution to (9.89) go to zero at $z \to \infty$. Only the function Ai(z) fulfills this condition, and it can be written in terms of the Airy integral as

$$\text{Ai}(\tilde{z}_L) = \frac{3}{\pi} \int_0^\infty \cos\left(\frac{x^3}{3} - \tilde{z}_L x\right) dx. \quad (9.91)$$

The solution then is, in terms of a constant, a, found from boundary conditions

$$E(z) = a\, \text{Ai}(\tilde{z}_L). \quad (9.92)$$

We will need to make use of the asymptotic expression for Ai when $|\tilde{z}_L| \gg 1$,

$$\text{Ai}(-\tilde{z}_L) \approx \frac{1}{\tilde{z}_L^{1/4} \sqrt{\pi}} \cos\left(\frac{2}{3}\tilde{z}_L^{3/2} - \frac{\pi}{4}\right) \quad (9.93)$$

and note that, in the absence of absorption (which we assume for the moment) that outside the plasma the field is a standing wave formed from 100% reflection of the laser field and some phase shift from this plasma reflection:

$$E(z = -L) = E_{00} + E_{00} \exp[i\varphi_s]. \quad (9.94)$$

To find a, and the phase of the reflected light, we set the two solutions (9.93) and (9.94) equal to each other at the entrance to the plasma ramp at $z = -L$ to get

$$\frac{a}{\sqrt{\pi}} \left(\frac{c}{\omega L}\right)^{1/6} \frac{1}{2} \left(\exp\left[i\left(\frac{2}{3}\frac{\omega L}{c} - \frac{\pi}{4}\right)\right] + \exp\left[i\left(-\frac{2}{3}\frac{\omega L}{c} - \frac{\pi}{4}\right)\right]\right),$$
$$= E_{00}\left(1 + e^{i\varphi_s}\right) \quad (9.95)$$

so we can deduce that

$$a = 2\sqrt{\pi} \left(\frac{\omega L}{c}\right)^{1/6} E_{00} \exp\left[i\left(\frac{2}{3}\frac{\omega L}{c} - \frac{\pi}{4}\right)\right] \quad (9.96)$$

and

$$\varphi_s = \frac{2}{3}\frac{\omega L}{c} - \frac{\pi}{4}. \quad (9.97)$$

The Airy function is maximum at $a = -1$ which implies that the maximum amplitude of the electric field swells to

$$E_{MAX} \cong 1.9 \left(\frac{\omega L}{c}\right)^{1/6} E_{00}$$
$$\cong 2.6 \left(\frac{L}{\lambda}\right)^{1/6} E_{00}. \quad (9.98)$$

9.1 The Structure of Solid Target Plasmas

So, if the plasma has a scale length of ~10 λ (say 10 μm in IR irradiation), the field at the turning point is about 3.8 times larger than the vacuum laser field amplitude, a significant effect. The magnetic field is easily found from Ampere's law and the solution of (9.92),

$$B = -\frac{ic}{\omega}\frac{\partial E}{\partial z} = -i2\sqrt{\pi}\left(\frac{c}{\omega L}\right)E_{00}Ai'(\tilde{z}_L). \tag{9.99}$$

2) Oblique Incidence with S-Polarization

Next, we consider the case in which the E-field is still perpendicular to the plasma density gradient but now the light is incident at nonzero incidence angle at θ_0. The light now has **k** components in both the z- and x-directions and, since the light wave direction will be deviated by refraction and reflection in the plasma, we write the wavenumber in terms of a spatially varying angle,

$$\mathbf{k} \cdot \mathbf{x} = kx\sin\theta(z) + kz\cos\theta(z). \tag{9.100}$$

Because the plasma density varies only in z, conservation of momentum demands that k_x must be conserved, so we write for the E field

$$E = E_y(z)e^{ik_0 x \sin\theta_0}, \tag{9.101}$$

which, when substituted into the wave equation, gives

$$\frac{\partial^2 E}{\partial z^2} - k_0^2 \sin^2\theta_0 E_y + \frac{\omega^2}{c^2}\varepsilon(z)E_y = 0. \tag{9.102}$$

Considering the z-component of **k**, we can write

$$\sqrt{\varepsilon(0)}\sin\theta_0 = \sqrt{\varepsilon(z)}\sin\theta, \tag{9.103}$$

which is just a statement of Snell's law. If we recast Eq. (9.102) using (9.103), we get

$$\frac{\partial^2 E}{\partial z^2} + \frac{\omega^2}{c^2}\left(\varepsilon(z) - \sin^2\theta_0\right)E_y = 0, \tag{9.104}$$

which can be compared to the normal incidence equation (9.76), illustrating that the equation governing the E-field evolution at oblique incidence has merely resulted in this substitution: $\varepsilon(z) \to \varepsilon(z) - \sin^2\theta_0$. So now the reflection point (or turning point for the light rays) has been displaced from the critical density surface, where $\varepsilon(z_{turn}) = 0$, to a point in which $\varepsilon(z_{turn}) = \sin^2\theta_0$.

Using the result for a linear density ramp at normal incidence, we can just now write the solution for E_y as

$$E_y(z) = a\,\mathrm{Ai}\left[\left(\frac{\omega^2}{c^2 L}\right)^{1/3}(z - \sin^2\theta_0 L)\right]. \tag{9.105}$$

Since **B** is oriented along x, as before it can be found from E_y, using Ampere's law. We can then also find the trajectory of a ray of laser light entering the plasma by noting that the ray direction is governed by the ratio of the components of **k**, which is

$$\frac{dz}{dx} = \frac{k_z}{k_x} = \frac{\sqrt{\varepsilon(z) - \sin^2\theta_0}}{\sin\theta_0}. \tag{9.106}$$

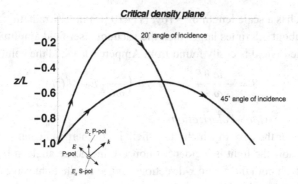

Figure 9.12 Calculation of two light ray trajectories in a linear plasma density ramp up to the critical density.

This can be easily integrated numerically to get the ray trajectory:

$$\int_{-L}^{-z} \frac{dz'}{\sqrt{\varepsilon(z) - \sin^2 \theta_0}} = \frac{x}{\sin \theta_0}. \quad (9.107)$$

Figure 9.12 shows the calculation of ray trajectories for two angles in incidence in a linear plasma density ramp. One can see that the plasma density gradient refracts and bends the light trajectory as the light approaches the critical density surface, causing reflection at a density out in front of the critical density surface, a point that moves out further as the angle of incidence is increased.

3) Oblique Incidence with P-Polarization Finally we consider the most interesting irradiation configuration, P-polarization, in which the electric field is oriented in the plane of incidence. The greatest consequence of this incidence geometry is that the laser now has a component of the E-field in both x- and z-directions. That means that the z-component of electric field drives electron plasma oscillations in and out of the plasma density gradient. This situation leads to two complications in calculating the fields. First, because of plasma refraction, the orientation of the electric field changes during propagation. Second, of even greater consequence, the z-component of E field no longer drives electron oscillations in homogeneous density. This means that the electric field of the laser can now directly couple to plasma waves with oscillation along the plasma density gradient.

This can be easily seen if one considers variation in electron density perturbations in the z direction produced by the laser-driven oscillation of the electron fluid over a distance z_{osc} which lead to nonzero-density perturbation:

$$\delta n_e = n_e(z + z_{osc}) - n_e(z)$$
$$\cong z_{osc} \frac{\partial n_e}{\partial z}. \quad (9.108)$$

This means that $\nabla \cdot \mathbf{E} \neq 0$. We see this by considering electric displacement $\mathbf{D} = \varepsilon \mathbf{E}$, and writing Gauss's law in quasi-neutral plasma:

9.1 The Structure of Solid Target Plasmas

$$\nabla \cdot \mathbf{D} = 4\pi n_q = 0, \tag{9.109}$$

$$\nabla \cdot (\varepsilon \mathbf{E}) = \varepsilon \nabla \cdot \mathbf{E} + \mathbf{E} \cdot \nabla \varepsilon, \tag{9.110}$$

$$\nabla \cdot \mathbf{E} = -\frac{1}{\varepsilon} \mathbf{E} \cdot \nabla \varepsilon. \tag{9.111}$$

We find the equation for E_z from the wave equation, again assuming conservation of k_y,

$$\frac{\partial^2 E_z}{\partial y^2} + \frac{\partial^2 E_z}{\partial z^2} + \frac{\omega^2}{c^2}\varepsilon(z)E_z - \frac{\partial}{\partial z}\nabla \cdot \mathbf{E} = 0, \tag{9.112}$$

which yields

$$\frac{\partial^2 E_z}{\partial z^2} + \frac{\omega^2}{c^2}\left(\varepsilon(z) - \sin^2 \theta_0\right) E_z - \frac{1}{\varepsilon^2}\left(\frac{\partial \varepsilon}{\partial z}\right)^2 E_z + \frac{1}{\varepsilon}\frac{\partial^2 \varepsilon}{\partial z^2} E_z + \frac{1}{\varepsilon}\frac{\partial \varepsilon}{\partial z}\frac{\partial E_z}{\partial z} = 0. \tag{9.113}$$

It is apparent that the turning point is no longer when $\varepsilon(z) = \sin^2 \theta_0$. However, in slow density gradients the spatial derivatives of ε are small, so the turning point is close to the location when $\varepsilon(z) \approx \sin^2 \theta_0$. We can, however, find the location of the turning point by inserting a transformation in this equation inspired by our WKB solution (Liu et al. 2019, p. 38):

$$E_z \equiv \frac{G(z)}{\varepsilon^{1/2}}. \tag{9.114}$$

Inserting this into (9.113) gives

$$\frac{\partial^2 G}{\partial z^2} + \left[\frac{\omega^2}{c^2}\left(\varepsilon - \sin^2 \theta_0\right) - \frac{3}{4}\frac{1}{\varepsilon^2}\left(\frac{\partial \varepsilon}{\partial z}\right)^2 + \frac{1}{2}\frac{1}{\varepsilon}\frac{\partial^2 \varepsilon}{\partial z^2}\right] G = 0. \tag{9.115}$$

The condition for the turning point becomes

$$\frac{\omega^2}{c^2}\left(\varepsilon - \sin^2 \theta_0\right) - \frac{3}{4}\frac{1}{\varepsilon^2}\left(\frac{\partial \varepsilon}{\partial z}\right)^2 + \frac{1}{2}\frac{1}{\varepsilon}\frac{\partial^2 \varepsilon}{\partial z^2} = 0 \tag{9.116}$$

In a linear density gradient of length L this becomes

$$-\frac{z}{L} - \sin^2 \theta_0 - \frac{3}{4}\frac{1}{(kz)^2} = 0 \tag{9.117}$$

$$z^3 + z^2 L \sin^2 \theta_0 - \frac{3}{16\pi^2}L\lambda^2 = 0. \tag{9.118}$$

This must be solved numerically but does show that the turning point for P-pol is further from the critical density surface than for S-pol light.

To make further progress, at this point it makes more sense to work with the B-field since it is transverse to the density gradient. Using the wave equation for $\mathbf{B}$ in Eq. (9.75) and using the conservation of $\mathbf{k}$ in the y direction, the equation for magnetic field becomes

$$\frac{\partial^2 B_y}{\partial z^2} + \frac{\omega^2}{c^2}\left(\varepsilon - \sin^2 \theta_0\right) B_y - \frac{1}{\varepsilon}\frac{\partial \varepsilon}{\partial z}\frac{\partial B_y}{\partial z} = 0. \tag{9.119}$$

In gentle density gradients, we can ignore the third term and we recover the Airy function solutions, now for magnetic field. This solution, however, is not accurate at the critical density plane where $\varepsilon \to 0$ because the third term cannot be ignored there. From Ampere's law we have

$$\nabla \times \mathbf{B} = -i\frac{\omega}{c}\varepsilon \mathbf{E}, \qquad (9.120)$$

which gives for the z-component of E

$$E_z = i\frac{c}{\varepsilon\omega}\frac{\partial B_y}{\partial x} \qquad (9.121)$$

or

$$E_z = -\frac{1}{\varepsilon}\sin\theta_0 B_y. \qquad (9.122)$$

This is an interesting result. It says that the z-component of the electric field is strongly peaked at the critical density surface, even though the turning point for the laser field is at density below n_{crit}. (E_x also diverges logarithmically; see Eliezer (2002, p. 93).)

So the question is, what is E_z at the critical density? The reason for the peaking of the field there, even though the laser does not propagate to that point, is that E is in resonance with plasma waves there because $\omega = \omega_{pe}$. So, to calculate E_z there we need B_y at $z = 0$. As an approximation, we can assume the usual Airy function solution in the region before the turning point from the S-polarized case which says that

$$B(z_t) \approx -i2\sqrt{\pi}\left(\frac{c}{\omega L}\right)^{1/6} E_{00} Ai'(0)$$

$$\approx 0.9i\left(\frac{c}{\omega L}\right)^{1/6} E_{00}. \qquad (9.123)$$

We then assume that the B-field "tunnels" through the region of plasma in which propagation is not allowed. We can, therefore, estimate B at the critical density to be

$$B(z=0) \cong B(z_t)\exp\left[\int_{z_t}^{0} i\frac{\omega}{c}\sqrt{\varepsilon_{eff}}\,dz'\right]$$

$$\cong B(z_i)\exp\left[\int_{-L\sin^2\theta_0}^{0} i\frac{\omega}{c}\sqrt{-\frac{z'}{L}n_{crit} - \sin^2\theta_0}\,dz'\right]$$

$$\cong B(z_t)\exp\left[-\frac{2}{3}\frac{\omega}{c}L\sin^3\theta_0\right]. \qquad (9.124)$$

Eq. (9.122) then gives the electric field at n_{crit},

$$|E_z(0)| \cong \frac{1}{\varepsilon(0)} 0.9\left(\frac{c}{\omega L}\right)^{1/6}\sin\theta_0 \exp\left[-\frac{2}{3}\frac{\omega}{c}L\sin^3\theta_0\right] E_{00}$$

$$\cong \frac{0.9}{\varepsilon(0)}\left(\frac{c}{\omega L}\right)^{1/2}\alpha \exp\left[-\frac{2}{3}\alpha^3\right] E_{00}, \qquad (9.125)$$

where

$$\alpha = \left(\frac{\omega L}{c}\right)^{1/3}\sin\theta_0 = (2\pi)^{1/3}\left(\frac{L}{\lambda}\right)^{1/3}\sin\theta_0. \qquad (9.126)$$

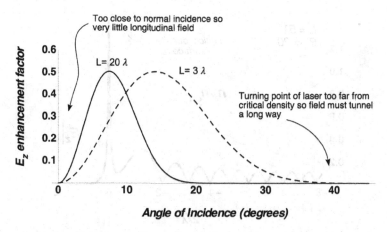

Figure 9.13 Plot of the longitudinal electric field at the critical density surface in p-polarized irradiation as a function of incidence angle for two different plasma scale lengths.

A plot of this function is illustrated in Figure 9.13, showing the longitudinal field as a function of incidence angle for two plasma scale lengths: three wavelengths and 20 waves (with the terms for E_{00} and $\varepsilon(0)$ factored out). These curves illustrate the essence of the physics at work. Near normal incidence, the field is small, because the laser does not have much field component in the z-direction at normal incidence. On the other hand, at glancing incidence approaching 10° the field must tunnel through a considerable amount of plasma to the critical density plane, particularly in the long scale length case, so the field is small as well. There is some optimum angle for resonant enhancement at n_{crit}. This will lead to an important collisionless absorption mechanism, resonance absorption.

To find the actual value of the E-field at this resonant point, we need the initially incident field E_{00} and knowledge of the value of the dielectric constant at the critical density. This factor will not be zero because there will have to be some damping of plasma oscillations, either by collisional damping or, more likely, by Landau damping. (See Appendix B for a discussion of Landau damping.) This means that if the dielectric function can be written in terms of some phenomenological damping rate ν_d

$$\varepsilon = 1 - \frac{\omega_{pe}^2}{\omega(\omega + i\nu_d)} \cong 1 - \frac{n_e}{n_{crit}} + i\frac{\nu_d}{\omega}, \tag{9.127}$$

then the E-field at resonance is

$$E_z \approx \frac{\omega}{\nu_d} \sin\theta_0 B_y. \tag{9.128}$$

Figure 9.14 shows a numerical solution of the wave equation for E_z in an exponential plasma density profile with $1/e$ scale length of 5λ and incidence angle of 20°. Here a damping rate of $10^{-4}\,\omega$ is assumed. This shows the dramatic effect that the resonance has at the critical density surface. This sharp spike can have an important effect on hot electron generation on a solid target in the strong field regime.

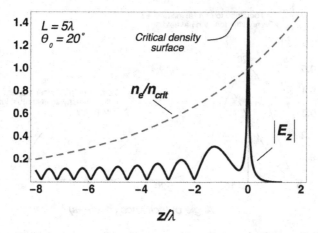

Figure 9.14 Numerical solution of the wave equation for the z-component of electric field as a laser approaches the critical density surface in an exponential density profile (shown as a dashed line).

9.1.6 Laser Field Profile at a Sharp Interface

In the limit of small or no plasma expansion, we have a more straightforward problem facing us in finding the field distributions. In this limit we can essentially think of the plasma interface as a metal mirror, a problem that has been well treated in many textbooks (see e.g. Born and Wolf 1980, pp. 615–624; Larsen 2017, pp. 659–664). This is the limit encountered when very short pulses (say <30 fs) with very good temporal contrast are employed. In practice it is quite difficult to access this regime, though at high intensity, extreme ponderomotive steepening can create local conditions that approximate a sharp metal mirror-like plasm interface, even for longer pulses.

We treat the plasma, as usual, as a Drude metal described by the dielectric function of Eq. (9.127). Outside the target we have a standing wave and inside the overdense target is an evanescent wave. For illustrative purposes let us consider S-polarization, meaning that we can treat the surface as having an effective dielectric constant $\varepsilon_{\mathit{eff}} = \varepsilon_0 - \sin^2\theta_0$. The skin depth of this target is

$$\delta_p = \frac{1}{\sqrt{\varepsilon}} \frac{c}{\omega} \tag{9.129}$$

$$= \frac{c}{\omega_{pe}} \frac{1}{\sqrt{1 - (\omega^2/\omega_{pe}^2)\cos^2\theta_0}} \tag{9.130}$$

$$\cong \frac{c}{\omega_{pe}}. \tag{9.131}$$

Outside the target we have the usual solution of a standing wave for the electric field,

$$E_y(x,z) = 2E_0 \sin(kz\cos\theta_0 + \varphi_R), \tag{9.132}$$

9.1 The Structure of Solid Target Plasmas

and inside the plasma the E-field is described by an evanescent wave with field at the surface E_S, and skin depth δ_p:

$$E_y(x,z) = E_S \exp\left[-z/\delta_p\right]. \tag{9.133}$$

These two solutions and their derivatives must match at $z = 0$, the plasma surface, which gives us two equations:

$$2E_0 \sin(\varphi_R) = E_S, \tag{9.134}$$

$$2k \cos\theta_0 E_0 \sin(\varphi_R) = -\frac{1}{\delta_p} E_S. \tag{9.135}$$

Taking the ratio of these two equations gives us

$$\tan\varphi_R = -k\delta_p \cos\theta_0, \tag{9.136}$$

which lets us write

$$\frac{E_S^2}{4E_0^2} = \frac{\delta_p^2 k^2 \cos^2\theta_0}{\delta_p^2 k^2 \cos^2\theta_0 + 1} \cong \delta_p^2 k^2 \cos^2\theta_0, \tag{9.137}$$

from which we can solve for the electric field on the plasma surface,

$$E_S = \pm 2E_0 \left(\frac{\delta_p^2 k^2 \cos^2\theta_0}{\delta_p^2 k^2 \cos^2\theta_0 + 1}\right)^{1/2} \cong -2\frac{\omega}{\omega_{pe}} \cos\theta_0 E_0. \tag{9.138}$$

The magnetic field for this solution is found from Faraday's law:

$$B_y = -i\frac{c}{\omega}\frac{\partial E_x}{\partial z} = -2iE_0 \cos\theta_0 \cos(kz\cos\theta_0 + \varphi_R) \quad \text{(outside)} \tag{9.139}$$

$$= i\frac{c}{\omega}\frac{E_S}{\delta_p} \exp\left[-z/\delta_p\right]. \quad \text{(inside)} \tag{9.140}$$

Examples of these solutions are shown in Figure 9.15. One can see that the derivative of the magnetic field has an inflection because of surface currents driven by the electric field.

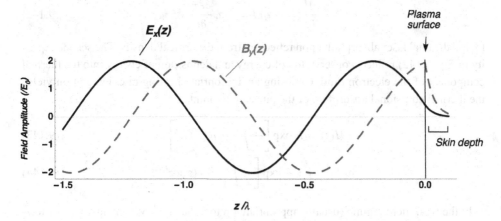

Figure 9.15 Plot of E and B fields at a sharp plasma interface.

Similar solutions can be derived for P-polarization. In that case tangential E and B fields must be matched at the plasma interface and normal B and D components matched. These solutions can be found in Born and Wolf (1980, p. 620).

9.1.7 Inverse Bremsstrahlung Absorption at a Plasma Surface with Finite Scale-Length

Now that we understand how the laser field propagates as it approaches the critical density surface, we turn to the question of how the laser energy is absorbed. From this point forward, we will have to think of the laser absorption by two avenues, one which deposits energy into the thermal distribution of electrons, resulting in overall heating of the plasma, and absorption mechanisms which deposit energy into a small number of high-energy electrons, so-called non-thermal absorption. We can write that the absorption of the laser is $\eta_{abs} = \eta_{abs}^{(Th)} + \eta_{abs}^{(NT)}$, explicitly separating absorption leading to bulk thermal heating of the plasma electrons (superscript "Th"), and that absorption that deposits energy into fast electrons or the acceleration of the plasma ions ("NT"). We then denote the power absorbed per unit area $I_{abs} = \eta_{abs} I_0$.

Thermal absorption in the underdense plasma as the pulse propagates toward the critical density surface is largely a result of inverse bremsstrahlung (i.e. collisional) heating, a topic we covered in Chapter 8. Consider collisional absorption in the nonrelativistic regime, the regime of validity of Eqs. (8.61) and (8.63). At relativistic intensity, we shall see that absorption is almost completely dominated by collisionless absorption mechanisms.

We need to consider thermal absorption in the weak and strong field regimes; the latter is when

$$v_{osc} > \sqrt{2k_B T_e/m}. \tag{9.141}$$

Unlike the situation considered in Chapter 8, the laser now propagates through a varying density profile and we must integrate

$$\frac{\partial I}{\partial z} = -\alpha_{IB} I = -n_e \frac{\partial \varepsilon_T}{\partial t} \tag{9.142}$$

to find the total laser absorption upon reflection from the critical density. The second equality in Eq. (9.142) reflects conservation of energy and deposition of energy into the thermal component of the electron fluid. Choosing an appropriate heating rate, integration yields the thermal collisional absorption of the pulse up to point z:

$$I(z) = I_0 \exp\left[-\int_{-\infty}^{z} \alpha_{IB}(z')dz'\right], \tag{9.143}$$

$$\eta_{abs}^{(T)} = 1 - \exp\left[-\int_{-\infty}^{z} \alpha_{IB}(z')dz'\right]. \tag{9.144}$$

In the weak field regime (usually appropriate up to about 10^{16} W/cm^2 in a typical few-hundred-eV plasma at IR wavelengths), we can refer to Eq. (8.61) and use Eq. (9.142) to

9.1 The Structure of Solid Target Plasmas

retrieve a weak field absorption coefficient, which becomes, when the group velocity in the plasma is considered ($I = c\varepsilon^{1/2}E_0^2/8\pi$),

$$\alpha_{IB} = \frac{8\ln 2}{\sqrt{2\pi}} \frac{\bar{Z}e^6 m^{1/2}}{(k_B T_e)^{3/2}} \ln\Lambda_C v_{osc}^2 \frac{n_e^2}{I} \qquad (9.145)$$

$$= \tilde{\alpha}_{IB} \left(\frac{n_e}{n_{crit}}\right)^2 \frac{1}{I_0 \sqrt{1 - n_e/n_{crit}}}, \qquad (9.146)$$

where we have grouped all cofactors (including the incident vacuum intensity) to the electron density normalized to n_{crit} into the coefficient $\tilde{\alpha}_{IB}$. Our goal is to find the total laser absorption as it propagates through the underdense plasma to the turnaround point near the critical density and then back out after reflection. As before, we consider the underdense region of the solid target to be in the negative z region. We must assume something for the plasma density profile, so we will assume an exponential density, defined, in the region $z < 0$ as

$$n_e = n_{crit} \exp[z/L]. \qquad (9.147)$$

Accounting for the possibility of oblique incidence we will use the solution for S-polarized light, recalling the ray trajectory detailed in Fig. 9.12, to find the deepest penetration of the laser light, the turnaround point, which happens at

$$\varepsilon(z) - \sin^2\theta_0 = 0, \qquad (9.148)$$

so that the reflection point is

$$z_T = L\ln\left[\cos^2\theta_0\right]. \qquad (9.149)$$

Using Eq. (9.144), we write the laser collisional absorption as

$$\eta_{abs}^{(T)} = 1 - \exp\left[-2\tilde{\alpha}_{IB}\int_{-\infty}^{L\ln[\cos^2\theta_0]} \frac{\exp[2z/L]}{(1-\exp[z/L])^{1/2}} \frac{dz}{\cos\theta}\right]. \qquad (9.150)$$

To evaluate this, we must use Snell's law,

$$\sin\theta_0 = \varepsilon^{1/2}\sin\theta, \qquad (9.151)$$

(where θ_0 is the initial incident angle) giving $\cos\theta = (1 - \sin^2\theta_0/\varepsilon)^{1/2} = (\varepsilon - 1 + \cos^2\theta_0)^{1/2}/\varepsilon^{1/2}$, and since $\varepsilon = 1 - \exp[z/L]$ (9.150) yields a result for the total absorption of the laser:

$$\eta_{abs}^{(T)} = 1 - \exp\left[-2\tilde{\alpha}_{IB}\int_{-\infty}^{L\ln[\cos^2\theta_0]} \frac{\exp[2z/L]}{(\cos^2\theta_0 - \exp[z/L])^{1/2}} dz\right] \qquad (9.152)$$

$$= 1 - \exp\left[-\frac{8}{3}\tilde{\alpha}_{IB} L \cos^3\theta_0\right]. \qquad (9.153)$$

A similar calculation can be conducted for a linear plasma gradient (Kruer 2003, p. 51). In that case $\varepsilon = -z/L$ in the region of z from $-L$ to 0. Integration of Eq. (9.144) for this case yields a slightly different result,

$$\eta_{abs}^{(T)} = 1 - \exp\left[-2\tilde{\alpha}_{IB} \int_{-L}^{L\sin^2\theta_0} \frac{(1+z/L)^2}{(\cos^2\theta_0 - 1 - z/L)^{1/2}} dz\right] \quad (9.154)$$

$$= 1 - \exp\left[-\frac{32}{15}\tilde{\alpha}_{IB} L \cos^5\theta_0\right]. \quad (9.155)$$

Note that these equations predict that the absorption drops to zero at high angle of incidence. The laser light merely glances from the plasma at low density and deposits little thermal energy. At angles near normal incidence the thermal collisional absorption can be a significant fraction of the incident laser light. Equation (9.153) predicts that a target with a 1 μm wavelength laser incident on it with a temperature of 500 eV and $Z = +5$ having a 10 μm density gradient will absorb 47% of the incident laser energy.

At higher intensity, it becomes necessary to use the strong field limit of the inverse bremsstrahlung absorption coefficient. Using Eq. (8.63), we can write the propagation differential equation in the strong field regime as

$$\frac{\partial I}{\partial z} = -8\bar{Z}\frac{e^4}{m}\frac{1}{v_{osc}} \ln\left[2\frac{v_{osc}}{v_{Te}}\right] \ln \Lambda'_C n_e^2 \quad (9.156)$$

$$= -8^{3/2}\pi^{1/2}\bar{Z}e^3\omega c^{1/2} \ln\left[2\frac{v_{osc}}{v_{Te}}\right] \ln \Lambda'_C n_e^2 \varepsilon^{1/4} \frac{1}{I^{1/2}}. \quad (9.157)$$

The integration of this equation is a bit more complex. Consider the case of a linear density gradient. Integration of (9.157), again using Snell's law, yields

$$\frac{2}{3}\left(I(z)^{3/2} - I_0^{3/2}\right) = -32\sqrt{2\pi}c\bar{Z}e^3\omega \ln\left[2\frac{v_{osc}}{v_{Te}}\right] \ln \Lambda'_C n_{crit}^2$$

$$\times \underbrace{\int_{-L}^{-L\sin^2\theta_0} (1+z/L)^2 \frac{(-z/L)^{3/4}}{(\cos^2\theta_0 - 1 - z/L)^{1/2}} dz}_{\beta_{SF}(L,\theta_0)} \quad (9.158)$$

$$= -\tilde{\alpha}_{SF}\beta_{SF}(L,\theta_0). \quad (9.159)$$

The strong field absorption function $\beta_{SF}(L,\theta_0)$ has analytic solutions with hypergeometric functions. Solving for the absorption gives

$$\eta_{abs}^{(SF)} = 1 - \left(1 - \frac{3}{2}\frac{\tilde{\alpha}_{SF}\beta_{SF}(L,\theta_0)}{I_0^{3/2}}\right)^{2/3}. \quad (9.160)$$

This equation is only appropriate for weak absorption because, if the laser light is strongly absorbed (say in long density gradients), the intensity drops out of the strong field regime and the weak field absorption takes over, lowering the intensity ultimately to zero. At normal incidence,

$$\beta_{SF}(L,0) = \frac{128}{585}L. \quad (9.161)$$

Plots of the predicted absorption in the weak field regime and strong field regime are reproduced in Figures 9.16 and 9.17 respectively. The weak field regime shows that

9.1 The Structure of Solid Target Plasmas

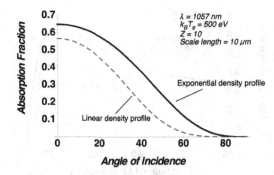

Figure 9.16 Plot of predicted laser light absorption as a function of the angle of incidence in the weak field regime for two density profiles in a 500 eV multiply ionized plasma with gradient of ~10 wavelengths.

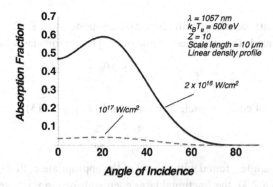

Figure 9.17 Plot of predicted laser light absorption as a function of the angle of incidence in the strong field regime for two laser intensities in a 500 eV multiply ionized plasma with gradient of ~10 wavelengths.

collisional absorption is maximum (and a large fraction) at normal incidence. Absorption in the strong field limit shows a dip in the absorption at normal incidence, a consequence of the field swelling and collisional absorption reduction when the laser propagates at low incidence angles. Figure 9.17 illustrates absorption at two intensities, showing that the thermal absorption drops significantly when intensity, in the example 500 eV plasma, increases to $\sim 10^{17}$ W/cm^2. In this regime absorption is predominantly by collisionless mechanisms leading to significant nonthermal electron production.

9.1.8 Collisional Absorption at a Sharp Plasma Interface: Fresnel Equations

In the short pulse regime, with minimal plasma expansion, thermal collisional absorption essentially happens by an electromagnetic wave incident on a flat conducting surface. The solution is treated in many textbooks and will not be repeated in detail here. Essentially, the solution is found by matching the normal component of the electric displacement

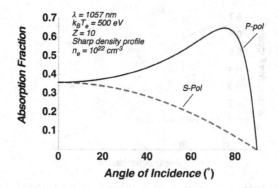

Figure 9.18 Plot of the Fresnel equation solutions for laser absorption by P- and S-polarized light at a sharp, solid density plasma interface in the weak field limit.

D at the conductor surface, along with tangential components of **E**. One uses the complex form of Snell's law inside the overdense region of the plasma for a laser incident at angle θ_0,

$$\theta_t = \text{Arcsin}\left[\frac{\sin\theta_0}{\sqrt{\varepsilon}}\right], \qquad (9.162)$$

where we use the usual complex dielectric function from Eq. (8.15),

$$\varepsilon \cong 1 - \frac{n_e}{n_{crit}} + i\frac{\nu_{ie}}{\omega}. \qquad (9.163)$$

With these complex angles, found with the aid of the appropriate collision frequency in the plasma (see Section 8.2.3), the fractional laser energy absorption, η_{abs}, can be found from the Fresnel equation solutions,

$$\eta_{abs}^{(S-pol)} = 1 - \left|\frac{\sin(\theta_0 - \theta_t)}{\sin(\theta_0 + \theta_t)}\right|^2, \qquad (9.164)$$

$$\eta_{abs}^{(P-pol)} = 1 - \left|\frac{\tan(\theta_0 - \theta_t)}{\tan(\theta_0 + \theta_t)}\right|^2. \qquad (9.165)$$

Figure 9.18 plots these solutions for a solid density plasma at 500 eV in the weak field limit as a function of incidence angle for both P- and S-polarizations. The P-polarization curve exhibits typical behavior with a peak in absorption at some large incidence angle.

9.2 Ponderomotive Force Effects on Solid Target Plasmas

Until now we have considered the structure of a solid target plasma subject only to the motion of plasma from pressure exerted by the thermal energy of the electron fluid. This led us to consider two limits, one in which the time scale after ionizing the plasma is long enough that hydrodynamic motion has led to plasma expansion and the other in which the time scales are so short that the target surface has not much expanded. While this is a fair picture at modest intensity, at higher intensity, the pressure of the laser light itself must be

9.2 Ponderomotive Force Effects on Solid Target Plasmas

included in determining plasma motion during the interaction (Lindl and Kaw 1971). It is to these ponderomotive force effects on the plasma surface that we now turn.

9.2.1 Ablation Pressure

Before considering these ponderomotive effects, we should first mention that even in the absence of significant ponderomotive pressure, the deposition of energy by a laser pulse can exert a significant pressure on a solid target. This pressure results from the conservation of momentum at the target surface resulting from the hydrodynamic expansion of the heated plasma. This "rocket action" results in a pressure directed into the target and is termed the ablation pressure. The structure of an ablating plasma is illustrated in Figure 9.19. Expansion of the plasma outward back toward the laser leads to pressure exerted on the stationary, underlying dense plasma near a depth which is termed the ablation front.

A simple phenomenological model can be developed to describe the ablation pressure. Ignore the heat conduction into the cold solid and consider only thermal absorption by collisional heating. In that case, all of the absorbed laser energy must be transferred (on a hydrodynamic time scale) into the kinetic energy of expanding plasma. Making the simple approximation that the ablated plasma expands at constant velocity, v, with density ρ, conservation of absorbed laser energy, ε, per unit area, A, demands that

$$\frac{1}{A}\frac{\partial \varepsilon}{\partial t} = \eta_{abs}^{(T)} I = \frac{1}{2}\rho v^2. \tag{9.166}$$

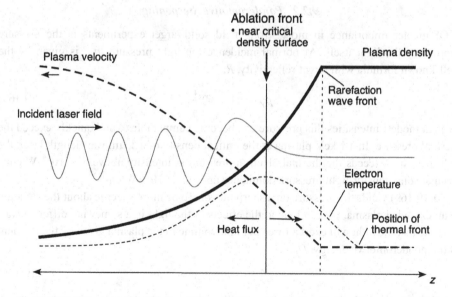

Figure 9.19 Structure of an ablating plasma showing the ablation front and heat conduction near the ablation surface.

Conservation of momentum by the ablated plasma mass must lead to a force per unit area, the ablation pressure, p_a,

$$p_a = \frac{1}{A}\frac{\partial p}{\partial t} = \rho v^2. \tag{9.167}$$

A reasonable assumption is to take the ablated plasma density to be that at the critical density surface, $\rho \approx n_{crit} m_i/\bar{Z}$. Equations (9.166) and (9.167) then yield a simple formula for ablation pressure at the ablation front

$$p_a = 2^{2/3} \rho^{1/3} \left(\eta_{abs}^{(T)} I\right)^{2/3}. \tag{9.168}$$

This equation suggests that the ablation pressure increases as the two-thirds power of incident intensity. In reality, this well-known result has been found not to be particularly accurate in experiments and a $I^{2/3}$ scaling of ablation pressure is rarely observed in real experiments. Generally Eq. (9.168) overestimates actual ablation pressure, in large part because of the neglect of heat conduction. Mulser and Bauer (2010, p. 82) have presented a more sophisticated model that includes heat conduction and find an ablation pressure one-half that of Eq. (9.168), $p_a = 2^{-2/3} \rho^{1/3} (\eta_{abs} I)^{2/3}$. While this analysis is not particularly accurate, it does illustrate the physical origin of the ablation pressure and is reasonable to make order-of-magnitude estimates. For example, this formula predicts that light at intensity of 10^{14} W/cm^2 incident on a 10× ionized Al plasma will experience an ablation pressure of ∼30 Mbar, an estimate consistent with many experiments.

9.2.2 Ponderomotive Steepening

Of greater importance in most strong field solid-target experiments is the pressure exerted by the light itself. At normal incidence the light pressure, p_γ, is given by the well-known formula with target reflectivity, R,

$$p_\gamma = (1+R)\frac{I}{c}. \tag{9.169}$$

Even at modest intensities this pressure can become comparable to and quickly exceed the thermal pressure. In a 1 keV plasma at the critical density of a 1 μm wavelength laser, the light pressure exceeds the thermal electron pressure at intensity above 5×10^{15} W/cm^2. Even at solid density, light pressure dominates above 5×10^{17} W/cm^2.

Eq. (9.169) yields the pressure on a sharp mirror. To be more specific about the situation in an extended plasma, we can look at the effects of the light in a somewhat different way. We note that the light will exert a force per unit volume in the plasma given by the gradient of the ponderomotive energy, U_p,

$$f_p = -n_e \nabla U_p. \tag{9.170}$$

A CW wave in an inhomogeneous plasma will exert this force. Consider the time-averaged electron ponderomotive energy in a continuous propagating light field,

$$U_p \sim \left\langle |E_0 \sin(kz - \omega t)|^2 \right\rangle_t, \tag{9.171}$$

9.2 Ponderomotive Force Effects on Solid Target Plasmas

which implies that the ponderomotive force in an extended, *homogeneous* plasma is zero:

$$\nabla U_p \sim 2kE_0^2 \langle \cos(kz - \omega t) \sin(kz - \omega t) \rangle_t = 0. \tag{9.172}$$

Ponderomotive force in homogeneous plasma requires a finite laser pulse rise time (it is this effect which gives rise to wakefield generation in underdense homogenous plasma).

The presence of a plasma-density gradient changes the situation. Consider instead a standing wave produced by reflection from a plasma surface, so that now

$$U_p \sim \langle |E_0 \sin kz \sin \omega t|^2 \rangle_t. \tag{9.173}$$

Now, upon averaging over a laser cycle the ponderomotive force is

$$\nabla U_p \sim 2kE_0^2 \langle \cos kz \sin kz \sin^2 \omega t \rangle_t, \tag{9.174}$$

$$\nabla U_p \sim kE_0^2 \cos kz \sin kz \sim kE_0^2 \sin 2kz. \tag{9.175}$$

The force is now considerable, because of the presence of the standing wave, and the force oscillates in space. A standing wave produces a large light pressure because the field has a gradient of the order of $\lambda/4$.

This ponderomotive force in a plasma with a density gradient is, in fact, merely a manifestation of the light pressure, which can be seen intuitively if we consider a normally incident light wave on a sharp plasma interface. The ponderomotive force in this case arises from the $\mathbf{v} \times \mathbf{B}$ force of the light wave, as illustrated in Figure 9.15. We have previously found the solution for the electric field at a sharp plasma interface; recalling Eqs. (9.133) and (9.140), inside the plasma we have

$$E = 2\frac{\omega}{\omega_{pe}} E_0 e^{-z/\delta_p} \cos \omega t, \tag{9.176}$$

$$B = 2\frac{c}{\omega \delta_p} \frac{\omega}{\omega_{pe}} E_0 e^{-z/\delta_p} \cos \omega t, \tag{9.177}$$

which means that the ponderomotive force on each electron is

$$F_p = 4\frac{e^2}{m\omega_{pe}^2 \delta_p} E_0^2 e^{-2z/\delta_p} \cos^2 \omega t. \tag{9.178}$$

Since the cycle-averaged ponderomotive force per unit volume inside the plasma is $f_p = n_e \langle F_p \rangle$, the net pressure on the plasma surface is an integration of this force per volume over the depth inside the plasma,

$$P_p = -\int_0^\infty f_p(z) dz \tag{9.179}$$

$$= -\int_0^\infty 2n_e \frac{e^2}{m\omega_{pe}^2 \delta_p} E_0^2 e^{-2z/\delta_p} dz. \tag{9.180}$$

Integration leads to the natural result that the pressure on a sharp plasma interface is

$$P_p = \frac{E_0^2}{4\pi} = \frac{2I}{c}. \tag{9.181}$$

This conclusion is generally true in an extended plasma as well. Following Mulser and Baue (2010, p. 210), we can write the Helmholtz equation for the laser electric field in the plasma in front of a solid target,

$$\frac{\partial^2 \mathbf{E}}{\partial z^2} + \frac{\omega^2}{c^2}\left(1 - \frac{n_e}{n_{crit}}\right)\mathbf{E} = 0, \tag{9.182}$$

and recall the nonrelativistic ponderomotive force per unit volume in this plasma

$$f_p = -n_e \nabla U_p = \frac{n_e e^2}{4m\omega^2}\frac{\partial}{\partial z}|\mathbf{E}|^2. \tag{9.183}$$

The total pressure is just this force integrated over all space,

$$P_p = -\int_{-\infty}^{\infty} f_p(z) dz \tag{9.184}$$

which yields

$$P_p = -\int_{-\infty}^{\infty} \frac{n_e(z) e^2}{4m\omega^2}\frac{\partial}{\partial z}\mathbf{E}\cdot\mathbf{E}^* dz \tag{9.185}$$

$$P_p = -\int_{-\infty}^{\infty} \frac{1}{16\pi}\frac{n_e}{n_{crit}}\left(\mathbf{E}^*\cdot\frac{\partial \mathbf{E}}{\partial z} + c.c.\right)dz. \tag{9.186}$$

Multiply Eq. (9.182) by $\partial \mathbf{E}^*/\partial z$,

$$\frac{c^2}{\omega^2}\frac{\partial \mathbf{E}^*}{\partial z}\cdot\frac{\partial^2 \mathbf{E}}{\partial z^2} + \frac{\partial \mathbf{E}^*}{\partial z}\cdot\mathbf{E} = \frac{n_e}{n_{crit}}\frac{\partial \mathbf{E}^*}{\partial z}\cdot\mathbf{E} \tag{9.187}$$

and substitute this into (9.186)

$$P_p = -\int_{-\infty}^{\infty} \frac{1}{16\pi}\frac{\partial}{\partial z}\left(\mathbf{E}\cdot\mathbf{E}^* + \frac{c^2}{\omega^2}\frac{\partial \mathbf{E}}{\partial z}\cdot\frac{\partial \mathbf{E}^*}{\partial z}\right)dz. \tag{9.188}$$

The integral can be evaluated noting that at $z = +\infty$ the plasma is over dense and the electric field is zero while at $z = -\infty$ the field can be written in terms of the incident laser field and a reflected field giving

$$P_p = \frac{1}{16\pi}\left[(\mathbf{E}_0 + \mathbf{E}_R)\cdot(\mathbf{E}_0^* + \mathbf{E}_R^*) + (\mathbf{E}_0 - \mathbf{E}_R)\cdot(\mathbf{E}_0^* - \mathbf{E}_R^*)\right] \tag{9.189}$$

$$= \frac{1}{8\pi}\left(|\mathbf{E}_0|^2 + |\mathbf{E}_R|^2\right), \tag{9.190}$$

which can be written in terms of the plasma reflectivity, R,

$$P_p = (1 + R)\frac{I_{inc}}{c}, \tag{9.191}$$

illustrating that the forward ponderomotive force is identical to the well-known result for light pressure.

This light pressure has an interesting consequence on strong field laser interactions with solid targets at intensity above about 10^{17} W/cm^2 because the light pressure becomes comparable to or exceeds the electron pressure at solid density. The light pressure can slow or even reverse the hydrodynamic expansion driven by the thermal electron pressure

9.2 Ponderomotive Force Effects on Solid Target Plasmas

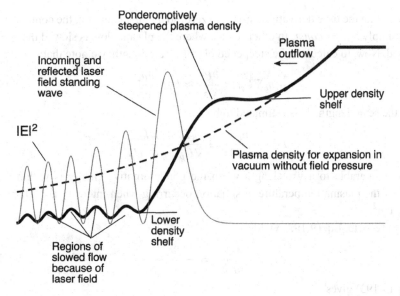

Figure 9.20 Plot of ponderomotively steepened plasma and standing wave field in the steady-state regime.

(Lee *et al.* 1977). We can consider this in two distinct regimes. The first, which is not usually relevant in strong field experiments but is useful to examine, is a "long pulse" regime in which the plasma expansion slows and comes into equilibrium with the light pressure. The time scale for this to happen is approximately $\lambda/4c_s$, where c_s is the usual plasma sound speed describing its expansion rate. This is the time scale over which a plasma flows over a wavelength of the laser's standing wave. This time scale is of the order of a few to tens of picoseconds for plasmas in the 100 to 1000 eV temperature range. For example, it is 4 ps for an Al plasma at 1 keV.

This steady-state description is not entirely accurate, even for pulses longer than this time scale – see Mulser and Bauer (2010, p. 210) – but in this steady-state approximation, a situation like that illustrated in Fig. 9.20 develops. The plasma outflow is slowed by the standing wave of the reflected laser field. This leads to a plasma density shelf in front of the reflection point, resulting in an increased density gradient. The plasma that continues to flow is slowed in regions of the standing wave where the ponderomotive pressure is higher. This leads to oscillations in the plasma density.

Following Kruer (2003, pp. 117–121) we can develop an analytic theory to estimate the scale length of plasma that develops in this regime. We start with the usual plasma fluid equations for the ions which are amended to include a term for the ponderomotive force, $f_p - n_e \nabla U_p$:

$$\frac{\partial n_i}{\partial t} + \frac{\partial}{\partial z}(n_i u_i) = 0 \tag{9.192}$$

$$\frac{\partial u_i}{\partial t} + u_i \frac{\partial u_i}{\partial z} + c_s^2 \frac{1}{n_i}\frac{\partial n_i}{\partial z} = -\frac{n_e}{m_i n_i}\frac{\partial U_p}{\partial z}. \tag{9.193}$$

In steady-state the time derivatives are set to zero, so the first equation, the continuity equation, is simply $n_i u_i = \text{const}$. In other words, where the plasma flow is slowed the density is increased. Now, to calculate the steepened plasma scale-length, we note that

$$n_e \nabla U_p \approx \frac{\partial P_e}{\partial z} \approx k_B T_e \frac{\partial n_e}{\partial z}, \tag{9.194}$$

so that the scale length, L, is estimated to be

$$L = \frac{n_e}{\partial n_e/\partial z} \cong \frac{k_B T_e}{\nabla U_p} \approx \frac{\lambda}{4} \frac{k_B T_e}{U_p}. \tag{9.195}$$

This is comparable to a wavelength when the ponderomotive energy is within a factor of four of the plasma temperature, a situation occurring when intensity is perhaps above 10^{15} W/cm^2.

In steady-state, Eq. (9.192) yields

$$\frac{u_i}{n_i} \frac{\partial n_i}{\partial z} = -\frac{\partial u_i}{\partial z}, \tag{9.196}$$

and Eq. (9.193) gives

$$\left(1 - \frac{u_i^2}{c_s^2}\right) \frac{1}{n_i} \frac{\partial n_i}{\partial z} + \frac{\bar{Z}}{m_i c_s^2} \frac{\partial U_p}{\partial z} = 0. \tag{9.197}$$

We see from (9.197) that the sonic point, the point in the flow where the plasma outflow velocity is c_s (an appropriate sound speed), is where $\partial U_p/\partial z = 0$, namely at the maxima of the standing wave. If the density at the sonic point is n_s, then from Eq. (9.192) we have that $n_i u_i = n_s c_s$, and we can rewrite Eq. (9.193):

$$\left(\frac{1}{n_i} - \frac{n_s^2}{n_i^3}\right) \frac{\partial n_i}{\partial z} = -\frac{\bar{Z}}{k_B T_e} \frac{\partial U_p}{\partial z} = 0. \tag{9.198}$$

This equation is easily integrated, defining U_{p0} as the ponderomotive energy at the wave peak, to yield

$$\ln\left[\frac{n_i(z)}{n_s}\right] + \frac{n_s^2}{n_i(z)^2} - \frac{1}{2} = \frac{\bar{Z}}{k_B T_e} (U_{p0} - U_p(z)). \tag{9.199}$$

To quantify the extent of the ponderomotive steepening, we seek the plasma density of the upper shelf, which we denote as n_2 and the density at the bottom of the gradient, denoted n_1. With these definitions and reference to the density profile shown in Fig. 9.20, we can say that the plasma-density gradient at the steepened point is

$$L \approx \frac{n_{crit}}{(n_2 - n_1)} \frac{\lambda}{4}. \tag{9.200}$$

A simple approximation is to assume that a standing wave from a sharp interface is formed. At the two shelf points (upper and lower), we have that $U_p \approx 0$, so Eq. (9.199) becomes an implicit algebraic equation that can be solved numerically for n_1/n_s and n_2/n_s versus U_{p0}:

$$2\ln\left[\frac{n}{n_s}\right] + \frac{n_s^2}{n^2} - 1 = 2\bar{Z} \frac{U_{p0}}{k_B T_e}. \tag{9.201}$$

9.2 Ponderomotive Force Effects on Solid Target Plasmas

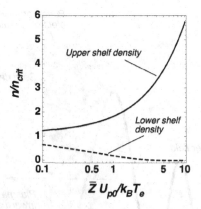

Figure 9.21 Plot of numerical solutions for the shelf densities in a steady-state ponderomotively steepened plasma.

Also recall that $n_i = n_{crit}/\bar{Z}$ at the interface, so from the solution of the field at a sharp interface, Eq. (9.138), we can say that the laser field at the steepened surface is

$$E_s \cong 2\sqrt{\frac{n_{crit}}{n_2}} E_0 \qquad (9.202)$$

and

$$U_p \cong \frac{n_{crit}}{n_2} U_{p0}. \qquad (9.203)$$

Note that U_{p0} is 4× the vacuum wave ponderomotive energy because the reflection sets up a standing wave. We can write then for n_2,

$$2\ln\left[\frac{n_{crit}}{n_s}\right] + \frac{n_s^2}{n_{crit}^2} - 1 = \bar{Z}\left(1 - \frac{n_{crit}}{n_2}\right)\frac{U_{p0}}{k_B T_e}. \qquad (9.204)$$

These equations can be numerically solved to derive the steady-state plasma shelf densities at the steepened surface as a function of $U_{p0}/k_B T_e$. Figure 9.21 shows such a solution for n_1 and n_2. This plot shows that when the standing wave peak ponderomotive energy is about 5× the plasma temperature, a large degree of steepening occurs with a steepened gradient that is about five times the critical density. This suggests that significant ponderomotive steepening occurs in the steady-state regime when the intensity of a 1 μm laser exceeds about 2×10^{15} W/cm^2.

9.2.3 Hole Boring

At higher intensities a more severe situation occurs at the reflection front. In this case we can have $-n_e \nabla U_p \gg \nabla P_e$, so the laser pressure far exceeds the thermal pressure of the electron fluid. In this case the laser pressure can drive the plasma inward and we enter what is commonly called the hole boring regime (Fuchs et al. 1999). In this case a situation at the plasma surface like that illustrated in Fig. 9.22 develops. The ponderomotive pressure

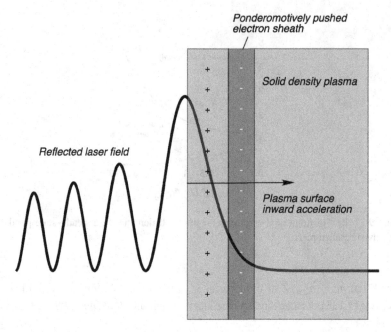

Figure 9.22 Schematic illustration of a ponderomotively driven double layer at a sharp overdense plasma front in the hole boring regime.

of the incident laser pushes a layer of electrons into the plasma, leaving behind a sheath of unneutralized ions. Consequently, a large longitudinal electric field develops, pointed into the plasma surface, which can accelerate the ions at the surface inward, driving the entire plasma front surface in at high velocity. This double layer results in an electron sheath of the order of the laser skin depth and can accelerate the front and the ions composing the front to high velocity.

Some simple estimates can be made by noting that, in the nonrelativistic limit, the longitudinal electric field, E_z, balances the ponderomotive force at the reflection point, which is

$$eE_z = -\nabla U_p \approx 4U_{p0}/(\lambda/4). \tag{9.205}$$

Using Gauss's law, this implies that the field is

$$eE_z \cong 4\pi \bar{Z}e^2 n_i \delta_i \approx 16\frac{U_{p0}}{\lambda}, \tag{9.206}$$

where δ_i is the thickness of the ion layer produced by pushing the electron sheath. From this we can estimate the thickness of the ion layer to be

$$\delta_i \cong \frac{4}{\pi} \frac{U_{p0}}{\bar{Z}e^2 n_i \lambda}. \tag{9.207}$$

On the other hand, the electron sheath layer will be comparable to the skin depth of the laser in the compressed electron plasma, which is

$$\delta_e \cong \frac{c}{\omega_{pe}} = \frac{c}{\omega}\sqrt{\frac{n_{crit}}{n_e}}. \tag{9.208}$$

So, for example, in an Al target illuminated by a 1 μm wavelength laser at 10^{18} W/cm^2, assuming $Z \approx 10$ and $n_i \approx 6 \times 10^{22}$ cm^{-3}, the ion layer thickness is $\delta_i \approx .14$ nm and the electron layer thickness would be $\delta_e \approx 6$ nm. At higher intensities the ion layer can be much thicker. At above 10^{20} W/cm^2, we have that $\delta_I \gg \delta_e$.

The velocity at which the plasma surface is driven inward by this longitudinal electric field, which we will call the hole boring velocity, u_s, can be estimated from the ion fluid equation. Ignoring the electron fluid pressure, the ion fluid equation is now familiar, reading

$$n_i m_i \frac{\partial u_i}{\partial t} + n_i m_i u_i \frac{\partial u_i}{\partial z} = -n_e \frac{\partial U_p}{\partial z}. \tag{9.209}$$

Using the usual relation for ponderomotive pressure, P_p, at a partially absorbing, reflecting interface, we can say

$$\frac{\partial p_p}{\partial z} = \frac{\partial}{\partial z}(2 - \eta_{abs})\frac{I}{c}\cos\theta_0, \tag{9.210}$$

where θ_0 is the angle of incidence. Equation (9.209) in steady-state (ignoring time derivatives) can be rewritten

$$\frac{n_i m_i}{2} u_i^2 = (2 - \eta_{abs})\frac{I}{c}\cos\theta_0. \tag{9.211}$$

If we finally say that the ions in the ion layer bounce off the electrostatic potential set up by the longitudinal electric field, then to a good approximation $u_s \approx u_i/2$, so the hole boring surface velocity, in this nonrelativistic steady-state approximation, is

$$u_S \cong \sqrt{\frac{(2 - \eta_{abs})I\cos\theta_0}{2n_i m_i c}}. \tag{9.212}$$

This equation predicts that, for the above Al target example at 10^{18} W/cm^2, the hole boring velocity might be (at normal incidence) something like $\approx 2 \times 10^7$ cm/s or $u_s/c \approx 0.0007$. A Doppler shift from this surface (a good experimental observable of this phenomena) would be

$$\omega_R = \omega \frac{1 - u_S/c}{1 + u_S/c}, \tag{9.213}$$

suggesting about a 0.2% shift or about a 2 nm shift. This is comparable to the bandwidth of a subpicosecond 1 μm laser but is observable.

9.2.4 Relativistic Hole Boring at High Intensity

At higher intensities a more dynamic situation occurs at the reflection front. When intensity approaches 10^{20} W/cm^2, the front velocity u_s can become relativistic. In the treatment of this high-intensity situation we closely follow the treatment of Schlegel et al. (2009).

Consider the laser and plasma dynamics in a frame that is receding from the incoming laser field at hole boring velocity, $u_s/c = \beta_S$. The plasma profiles are quasi-stationary in this frame. We seek the fields, laser intensity and densities in this frame, which we denote with a $\hat{}$. From Jackson (1975, p. 552) we use the usual field Lorentz transforms so that

$$\hat{E}_0 = \gamma_S(E_0 - \beta_S B_0) = \gamma_S(1 - \beta_S)E_0 \tag{9.214}$$

(where we use the fact that $E_0 = B_0$). This indicates that the intensity on the plasma surface in the moving plane is

$$\hat{I} = \gamma_S^2(1 - \beta_S)^2 I. \tag{9.215}$$

More generally, at oblique incidence the radiation pressure p_{rad} would read

$$p_{rad} = \frac{2I_0}{c} \frac{(\cos\theta_0 - \beta_S)^2}{(1 - \beta_S^2)}. \tag{9.216}$$

We also point out that because of length contraction, the ion density in the receding frame is increased to $\hat{n}_{i0} = n_{i0}\gamma_S$. The structure of the relevant quantities in the receding frame is illustrated in Figure 23 (Naumova et al. 2009).

A standing wave is set up by the laser. Its ponderomotive pressure creates an ion separation layer of thickness δ_i along z with an accompanying longitudinal electric field E_z (which is the field accelerating the ions). The electrons are pressed into a sheath layer of thickness δ_e out in front of the ions. The acceleration of the ions into the stationary background ions leads to a density jump in the ion density by a factor of 2; the jump moving at the hole boring velocity in the quasi-stationary frame.

Assume near 100% reflection of the light in the moving frame. In this frame ions stream into the sheath layer and are reflected by the longitudinal electric field, sending them back into the plasma (to the right in Fig. 23). This is what sets up the electrostatic ion density shock moving into the target at u_s with density $\hat{n}_i = 2\hat{n}_{i0}$. The ion momentum flux back away from the incident field must equal the light pressure. This lets us write

$$\frac{2\hat{I}}{c} = 2\gamma_s m_i \hat{n}_i u_s^2 = 2m_i c^2 \hat{n}_{i0} \gamma_s^2 \beta_s^2$$

$$= \frac{2I}{c} \frac{1 - \beta_s}{1 + \beta_s}. \tag{9.217}$$

We will solve for β_s and introduce the dimensionless parameter

$$\Gamma_i = \frac{I}{n_{i0} m_i c^3}, \tag{9.218}$$

which is

$$\Gamma_i^2 = \frac{1 + \beta_s}{1 - \beta_s} \frac{\beta_s^2}{(1 - \beta_s)(1 + \beta_s)}, \tag{9.219}$$

$$\beta_s = \frac{\Gamma_i}{1 + \Gamma_i}. \tag{9.220}$$

At low intensity note that $\beta_s \cong \Gamma_i$ and we recover the result of Eq. (9.212). We recover the ion recession velocity in the lab frame by another Lorentz transform where $\beta = -\beta_s$, $\hat{\beta}_i = \beta_s$,

$$\beta_i = \frac{\hat{\beta}_i - \beta}{1 - \beta \hat{\beta}_i}$$
$$= \frac{2\beta_s}{1 + \beta_s^2}, \qquad (9.221)$$

so the expected forward accelerated ion energy is

$$\varepsilon_i = (\gamma_i - 1) m_i c^2$$
$$= 2 m_i c^2 \gamma_s^2 \beta_s^2, \qquad (9.222)$$

which does not depend on the charge of the ions.

Finally we again observe that this ion acceleration represents a collisionless absorption mechanism in which laser energy is converted to ion energy. We find this absorption coefficient by noting that the number of reflected photons is conserved, so from the Doppler shift of the reflected photons we can say

$$\frac{\varepsilon_{refl}}{\varepsilon_{inc}} = \frac{\omega_r}{\omega} = \frac{1 - \beta_s}{1 + \beta_s}. \qquad (9.223)$$

At high intensity we have that

$$\gamma_s = \frac{1}{\left(1 - (1 + 1/\Gamma_i)^{-2}\right)^{1/2}} \cong \sqrt{\frac{\Gamma_i}{2}} \sim I^{1/4}. \qquad (9.224)$$

For example, at 10^{21} W/cm^2 in a CH target, $\Gamma_i \approx 0.02$.

9.2.5 Structure and Dynamics of the Hole Boring Piston

Before concluding the discussion of hole boring, it is interesting to look in a bit more detail into the structure illustrated in Figure 9.23 and consider its implication for experimental observables. First, we illustrate some of the predictions of hole boring velocity and resulting ion energy as a function of target ion density for a range of laser peak intensities in Figure 9.24. This shows that in most standard, solid target experiments, the hole boring velocity is only weakly relativistic and the ion energies expected are in the 1–10 MeV range, even at very high intensity.

This ion acceleration essentially occurs in the sheath field from the ponderomotively produced pressure separation of electrons and ions. Ions swept into the sheath region reflect off of E_z. This means the ions are initially reflected at velocity β_s at the back of the sheath region and then are further accelerated to β_i by the longitudinal field. The structure is such that usually $\delta_i \gg \delta_e$.

We can find the ion density structure in the charge-depleted sheath region. In the quasi-stationary frame, the ion density in the upstream counterpropagating shock region is

$$\hat{n}_i = 2\gamma_s n_{i0}. \qquad (9.225)$$

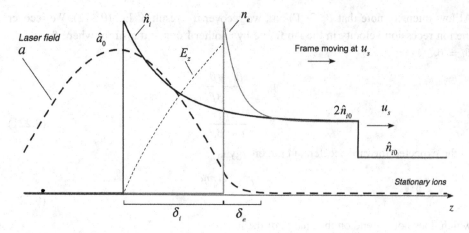

Figure 9.23 Plot of the incident laser field, plasma density and electrostatic longitudinal field structure in the quasi-stationary frame receding at the hole boring velocity.

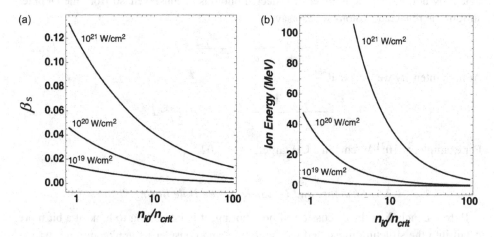

Figure 9.24 Plot of predicted hole boring velocity (left) and accelerated ion energy as a function of target ion density for three different laser intensities.

In the charge separation layer, mass conservation demands that

$$\hat{n}_i = 2\gamma_s n_{i0} \frac{\beta_s}{\hat{\beta}_i}. \tag{9.226}$$

Gauss's law in 1D is

$$\frac{\partial E_z}{\partial \hat{z}} = 4\pi \bar{Z} e \hat{n}_i, \tag{9.227}$$

and the equation of motion for the ions is

$$\bar{Z} e E_z = m_i c^2 \frac{\partial \hat{\gamma}_i}{\partial z}. \tag{9.228}$$

Substituting in Gauss's law and using mass conservation, we can write

$$\frac{\partial^2 \hat{\gamma}_i}{\partial z^2} = \frac{4\pi \bar{Z}^2 e^2}{m_i c^2}\hat{n}_i = \frac{2\omega_{pi}^2 \gamma_s}{c^2}\frac{\beta_s}{\hat{\beta}_i}. \qquad (9.229)$$

We integrate this once with the following boundary conditions, which reflect the turning point location and the fact that the electron sheath cancels the ion layer field

$$\hat{\gamma}_i(-\delta_i) = 1, \qquad (9.230a)$$

$$E_z(-\delta_i) = 0, \qquad (9.230b)$$

$$\frac{\partial \hat{\gamma}_i(-\delta_i)}{\partial z} = 0. \qquad (9.230c)$$

Using

$$\int_1^{\hat{\gamma}_i} \frac{1}{\sqrt{1 - 1/\hat{\gamma}_i^2}} d\hat{\gamma}_i = \sqrt{\hat{\gamma}_i^2 - 1}, \qquad (9.231)$$

this leads to

$$\frac{\partial \hat{\gamma}_i}{\partial \hat{z}} = \frac{2\omega_{pi}}{c}\sqrt{\gamma_s \beta_s}\left(\hat{\gamma}_i^2 - 1\right)^{1/4}. \qquad (9.232)$$

We use this to find E_z in the ion sheath region:

$$E_z = \frac{2 m_i c}{\bar{Z} e}\omega_{pi}\sqrt{\gamma_s \beta_s}\left(\hat{\gamma}_i^2 - 1\right)^{1/4}. \qquad (9.233)$$

The maximum electric field in the sheath is at $\hat{z} = 0$; we ignore the ion energy change of the ions transiting the thin electron sheath layer, which means that ions enter the ion sheath region at $\hat{z} = 0$ with velocity β_s, so $\hat{\gamma}_i(0) = \gamma_s$. We can therefore write for the maximum longitudinal electric field

$$E_z^{(MAX)} = E_z(0) = \frac{2 m_i c}{\bar{Z} e}\omega_{pi}\sqrt{\gamma_s \beta_s}\left(\gamma_s^2 - 1\right)^{1/4}$$

$$= \frac{2 m_i c}{\bar{Z} e}\gamma_s \beta_s. \qquad (9.234)$$

Putting this in terms of the laser field and the previously introduced dimensionless intensity parameter

$$\Gamma_i = \gamma_s \beta_s \left(\frac{1+\beta_s}{1-\beta_s}\right)^{1/2}, \qquad (9.235)$$

we find the satisfying conclusion that

$$E_z^{(MAX)} = \sqrt{2}E_0. \qquad (9.236)$$

This is a remarkable result: the longitudinal field is comparable to the laser field itself. This interaction converts an oscillating laser field to a quasi-static longitudinal accelerating field.

Now we can find the thickness of the ion sheath region. We integrate Eq. (9.232) from $-\delta_I$ to 0 to solve for the sheath thickness:

$$\hat{\delta}_i = \frac{c}{2\omega_{pi}\sqrt{\gamma_s \beta_s}} \int_1^{\gamma_s} \frac{1}{(\hat{\gamma}_i^2 - 1)^{1/4}} d\hat{\gamma}_i. \qquad (9.237)$$

This integral can be evaluated analytically in terms of a hypergeometric function $_aF_b$ and the Euler Gamma function Γ, yielding

$$\hat{\delta}_i = \frac{c}{2\omega_{pi}\sqrt{\gamma_s \beta_s}} \left[2\gamma_s^{1/2} {}_2F_1\left(-\frac{1}{4}; \frac{1}{4}; \frac{3}{4}; \frac{1}{\gamma_s^2}\right) - \frac{2}{\sqrt{\pi}} \Gamma\left(\frac{3}{4}\right)^2 \right]. \qquad (9.238)$$

Back in the lab frame, the sheath thickness is $\delta_i = \gamma_s \hat{\delta}_i$.

The ion energy distribution can be found from Eq. (9.237) and mass conservation, Eq. (9.225), to give

$$\frac{\hat{n}_i}{n_{i0}} = \frac{2\gamma_s \beta_s}{(1 - 1/\hat{\gamma}_i^2)^{1/2}}, \qquad (9.239)$$

where

$$\hat{\beta}_i = \left(1 - 1/\hat{\gamma}_i^2\right)^{1/2}. \qquad (9.240)$$

Figure 9.25 shows the ion sheath thickness from this formula.

The electron sheath structure is a bit more complicated. It can be calculated in the moving frame with the quasi-steady-state condition of the equalization of the ponderomotive force with the longitudinal electric field such that $F_p = eE_z$. As usual, we can write

$$\frac{\partial E_z}{\partial z} = 4\pi e \left(\tilde{n}_e - \bar{Z}\hat{n}_i\right), \qquad (9.241)$$

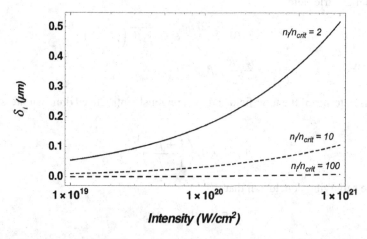

Figure 9.25 Plot of calculated ion sheath thickness as a function of laser intensity for three target ion densities.

9.2 Ponderomotive Force Effects on Solid Target Plasmas

where the relativistic ponderomotive force is

$$F_p = \nabla mc^2 \sqrt{1 + a_0^2/2}. \quad (9.242)$$

This lets us write a differential equation for the electron density:

$$\frac{\partial^2}{\partial z^2} \frac{mc^2}{e} \sqrt{1 + a_0^2/2} = 4\pi e (n_e - n_{e0})$$

$$\frac{n_e}{n_{e0}} = 1 + \frac{c^2}{\omega_{pe}^2} \frac{\partial^2}{\partial z^2} \sqrt{1 + a_0^2/2}. \quad (9.243)$$

This equation must be solved numerically in combination with the wave equation for the incident laser field. A simplified calculation can be performed if the E-field of the laser at a sharp interface is assumed,

$$a_0(z) \cong 2a_0(-\infty) \frac{\omega}{\omega_{pe}} e^{-z/\delta_p}, \quad (9.244)$$

and one assumes that the electron sheath layer is then about the skin depth $\delta_e \approx \delta_p$.

The reality of hole boring in real experiments is even more complicated. The 1D approximation for the hole boring piston structure is almost impossible to realize in experiments. In actuality, variations in intensity across a focused laser pulse lead to Rayleigh–Taylor-type instabilities. The result, seen in 2D PIC simulations (Wilks and Kruer 1997) and apparent in experiments, is a hole boring front that looks more like that pictured in Figure 9.26. Ripples in the ion front of the order of a few times the laser wavelength appear (Pukhov and Meyer-ter-Veyn 1997). Figure 9.27 schematically illustrates two of the trends that are

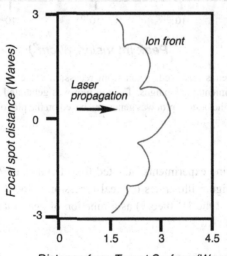

Figure 9.26 Schematic illustration of the rippling at the ion sheath front that is seen in PIC simulations of hole boring in 2D. The figure was generated using simulation results presented in Wilks and Kruer (1997, figure 5) and Pukhov and Meyer-ter-Veyn (1997, figure 1).

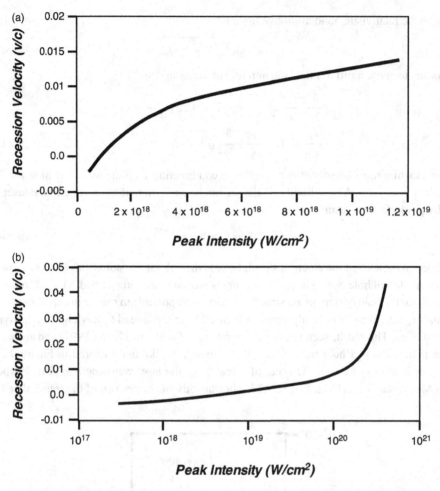

Figure 9.27 Plot of trends observed in ion front recession velocity versus intensity as observed in two experimental campaigns. The top plot was generated based on a plot in Zepf et al. (1996), and the bottom plot was generated based on the plot of Ping et al. (2008, figure 3).

observed in real hole boring experiments, adapted from the data of Zepf et al. (1996) and Ping et al. (2008). This figure illustrates the real recession velocity seen (not too different from the predictions of the 1D theory) as a function of peak intensity from these two experimental campaigns.

9.3 Collisionless Absorption

At modest intensity, the principal mechanism for the plasma surface to absorb laser light is by collisional absorption. At high intensity, the laser field drives large-amplitude collective plasma oscillations and the energy absorption shifts toward collisionless mechanisms.

9.3 Collisionless Absorption

This new regime leads to two major consequences. First, the absorption, which drops with increasing intensity in the collisional regime because of the decrease in collisionality, begins to increase again with rising intensity. The second consequence is that the driving of large-amplitude plasma oscillations can lead to acceleration of a lot of hot electrons.

There are a whole host of collisionless absorption mechanisms, some of which are merely different manifestations of the same physical effect at different laser intensity, plasma density or plasma scale length (Liseykina et al. 2015). In fact we have already encountered one collisionless mechanism: hole boring. In this case laser energy is transferred to directed ion energy through collective space-charge fields set up by the ponderomotive pressure of the laser. This effect turns out to be a rather minor mechanism in terms of laser energy absorbed. In fact, most collisionless absorption gets transferred to hot electrons. Indeed, collisionless absorption into electrons tends to reduce the actual observed hole boring velocities (Ping et al. 2008).

9.3.1 Trends in Laser Absorption as a Function of Intensity

At higher intensities, a more severe situation occurs at the reflection front. The experimental trends observed in the laser absorption as a function of peak intensity for both near-normal and oblique incidence with P-polarization are schematically illustrated in Figure 9.28. This plot is the amalgamation of a number of experiments and only illustrates the trends typically observed; of course, the actual absorption depends on a number of other factors, like laser wavelength, pulse duration, temporal contrast and so on. What is generally observed (Price et al. 1995) is a decrease in absorption up to around 10^{16} W/cm^2 as the plasma collisionality decreases with increasing intensity in the strong field collisional absorption regime; recall Eq. (8.63). Another effect that lowers collisional absorption is the ponderomotive steepening which reduces the density gradient and amount of plasma through which the pulse propagates. Above this intensity, as peak intensity increases above

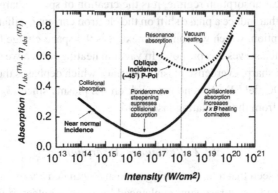

Figure 9.28 Plot of trends observed in laser absorption at a plasma surface as a function of laser intensity for a couple of typical configurations, normal incidence, and oblique P-polarized irradiation. Plot was generated from data of Ping et al. (2008, figure 1) and Price et al. (1995).

$\sim 10^{17}$ W/cm^2, collisionless processes cause the absorption to rise quickly. At normal incidence the absorption is dominated by $\mathbf{J} \times \mathbf{B}$ heating. At oblique incidence with P-pol, the additional absorption mechanisms of resonance absorption occur, which transitions at high intensity with severe ponderomotive steeping to vacuum (Brunel) absorption. These three processes will be considered in the following discussion, but essentially represent conversion of the laser oscillating field into collective longitudinal oscillations of electrons.

Before considering these mechanisms, it is informative to consider the nature of collisionless absorption formally (Listykina et al. 2015). Poynting's theorem gives the energy flux, $\mathbf{S}$, at some point in the plasma:

$$\langle \nabla \cdot \mathbf{S} \rangle = \langle \mathbf{J}_e \cdot \mathbf{E} \rangle. \tag{9.245}$$

Since the laser field varies as $\sin \omega t$, it is natural to assume that the collective electron current density will scale like

$$\mathbf{J}_e \sim \mathbf{v}_e \sim \cos(\omega t + \varphi), \tag{9.246}$$

where φ is the dephasing produced by fields or friction in the plasma, so

$$\langle \nabla \cdot \mathbf{S} \rangle \sim \langle \cos(\omega t + \varphi) \sin \omega t \rangle = -\frac{1}{2} \sin \varphi. \tag{9.247}$$

Clearly, some form of dephasing ($\varphi \neq 0$) of the collective electron current is needed to have nonzero energy flow into the plasma. For example, recall that in the collisional regime of an underdense plasma, this dephasing is caused by the friction from electron–ion collisions. Recalling Eq. (8.12), the effect of finite collision frequency, ν_{ei}, on the electron current is

$$\mathbf{J}_e = i \frac{e^2 n_e}{m} \frac{1}{\omega^2 + \nu_{ei}^2} (\omega - i\nu_{ei}) \mathbf{E} \tag{9.248}$$

so that the dephasing angle is nonzero only with finite collision frequency $\varphi \cong \arctan(\nu_{ei}/\omega)$.

In the collisionless absorption regime, it is the creation of space-charge fields with ponderomotive forces that induce a phase shift on the electron current oscillation. For example, in resonance absorption, which we consider next, it is the space-charge field at the critical density surface which serves this purpose. In vacuum heating, which we consider later, it is the sheath field at a sharp, overdense plasma interface which dephases the electron current. However, note, a DC field $\mathbf{E}_s$ does no net work on the plasma since $\langle \mathbf{J}_e \cdot \mathbf{E}_s \rangle = 0$, so the energy must come from the oscillating laser field.

9.3.2 Resonance Absorption

We have already seen that a laser entering a plasma with density gradient greater than a few wavelengths produces a resonantly enhanced longitudinal electric field at the critical density surface. Figure 9.29 illustrates how that z-directed electric field can couple to a plasma wave. In a completely uniform underdense plasma the oscillating laser field in the linear regime (namely at modest intensity) can drive no plasma waves; the electron current

9.3 Collisionless Absorption

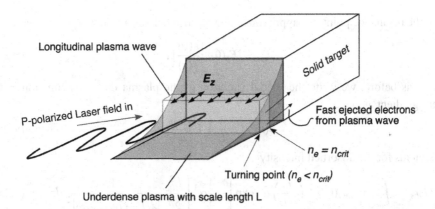

Figure 9.29 Schematic illustration of how the resonantly enhanced electric field in the direction of a plasma gradient can lead to plasma oscillations at n_{crit}.

is spatially uniform. In a plasma-density gradient the situation is different. The displacement of electrons in the direction of the gradient creates a restoring field because of the variation in ion density. This restoring force gives rise to a Langmuir wave. The net result is that laser energy gets coupled into the energy of this wave (Forslund et al. 1975).

The situation depicted in Figure 9.29 is one in which the plasma-density gradient is greater than a laser wavelength, namely that $L \cong n_e/\nabla n_e \gg \lambda$, and a situation in which the laser field is not strong enough to drive electron oscillations with amplitude comparable to the density gradient. This limits this regime typically to laser interactions with somewhat longer pulses, in which a plasma-density gradient can develop, and at moderate intensity, typically below about 10^{17} W/cm^2. This absorption of the laser by the field enhancement at the critical density is called resonance absorption (RA) and it has been a widely explored phenomenon for many years (Pert 1978; Rajouria et al. 2013; Palastro et al. 2018).

The absorption process is essentially a two-step process: the laser field energy is converted to electron wave energy, then this plasma wave is itself damped by various mechanisms, the most important being Landau damping (see Appendix B for a discussion of Landau damping). The result is that laser energy is converted to fast electrons (many keV), which propagate into the target. We can make a quantitative estimate, following the well-known treatment first presented in Ginzburg (1960, p. 379), but articulated well in Kruer (2003, p. 42). We assume some phenomenological damping rate for the excited plasma wave, ν_{damp}, and calculate the absorbed energy, per time per area, I_{abs}, which is done by integrating over all depths:

$$I_{abs} = \int_{-\infty}^{\infty} \nu_{damp} \frac{E_z^2}{8\pi} dz. \qquad (9.249)$$

Assume a simple, linear-density plasma gradient, so that E_z is given by Eq. (9.122):

$$E_z = \frac{ic}{\varepsilon \omega} \frac{\partial B_y}{\partial x} = -\frac{1}{\varepsilon} \sin \theta_0 B_y. \qquad (9.250)$$

Near the resonance point, we approximate the electric field as a function of depth as

$$E_z \cong |E_z(0)| \left|\frac{\varepsilon(0)}{\varepsilon(z)}\right|, \qquad (9.251)$$

where, as before, we write the spatial variation of the plasma dielectric function in the linear gradient as

$$\varepsilon(z) = -\frac{z}{L} + i\frac{v_{damp}}{\omega}. \qquad (9.252)$$

This yields for the absorbed intensity

$$I_{abs} = \int_{-\infty}^{\infty} v_{damp}(0.9)^2 \left(\frac{c}{\omega L}\right) \alpha^2 \exp\left[-\frac{4}{3}\alpha^3\right] \frac{E_{00}^2}{8\pi} \frac{1}{|z/L - iv_{damp}/\omega|^2} dz. \qquad (9.253)$$

As usual, the incident laser intensity is defined in terms of the incident vacuum laser field by

$$I_{inc} = \frac{cE_{00}^2}{8\pi}. \qquad (9.254)$$

In Eq. (9.253) we have introduced the definition

$$\alpha = \left(\frac{\omega L}{c}\right)^{1/3} \sin\theta_0, \qquad (9.255)$$

a number that interrelates the plasma-density scale length and incident angle. Performing the integral yields for the absorption a very well-known result:

$$\eta_{abs}^{(RA)} = \frac{I_{abs}}{I_{inc}} = 0.8\pi\alpha^2 \exp\left[-\frac{4}{3}\alpha^3\right]. \qquad (9.256)$$

Interesting to note is that the damping rate falls out of the result, justifying our ignoring the details of just how the wave is damped. Instead, this formula predicts laser absorption as a function of incidence angle for a given plasma-density scale length. The general wisdom is that this formula predicts absorption that is a bit too high. More sophisticated theories have been published (Pert 1978). One often quoted result that is a little more accurate is a theory-inspired approximation formula for the absorption in terms of Airy functions:

$$\eta_{Pert}^{(RA)} = \frac{18\alpha^2 Ai'^3(\alpha^2)}{Ai(\alpha^2)}. \qquad (9.257)$$

A comparison of these two absorption results as a function of α is illustrated in the left-hand panel of Figure 9.30. One sees that these formulae predict an optimum in the absorption, at some optimum incidence angle for a given scale length. The absorption as a function of incidence angle for a 1057 nm laser wavelength in two different plasma-density gradients is illustrated at the right in Figure 9.30. One sees that the optimum angle for absorption moves closer to normal incidence as the plasma scale length increases.

The presence of the optimum absorption angle can be explained by two competing effects. First, the simple geometric amount of laser field directed along the z-axis increases with increasing angle of incidence, so absorption increases with increasing angle. However,

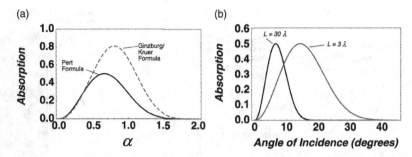

Figure 9.30 In part (a) is a plot of the two formulas for RA. The plot in (b) shows the predicted absorption by RA as a function of incidence angle for a 1057 nm wavelength laser for a long, 30-wavelength density gradient, and a relatively sharp, three-wavelength gradient.

at higher angle of incidence, the laser turns around at a lower density, and the E_z component must tunnel further through the plasma to excite a wave at n_{crit}. These two effects lead to an optimum angle.

The principal take-away of this analysis is that RA leads to about 50% absorption at an optimum angle, say $10° - 20°$. While it might be expected that RA disappears near normal incidence, the actual reality is that in most experiments there is some rippling of the plasma surface, so some RA almost always occurs. As mentioned, the plasma-wave energy is mostly transferred to hot electrons (Estabrook and Kruer 1978). These electrons usually have a distribution of kinetic energies that looks very much like a Maxwellian distribution with a temperature T_{hot}. A reasonable rule of thumb for estimating this hot electron temperature in the strong field regime is given by Wilks et al. (1992) and Estabrook and Kruer (1978)

$$T_{hot} \approx 10 \left[\frac{I[W/cm^2]}{10^{15}} \frac{T_e[eV]}{1000} \lambda[\mu m] \right]^{1/3} \text{keV}, \tag{9.258}$$

which illustrates that temperatures of over 100 keV are produced even at moderate ($\sim 10^{16}$ W/cm^2) intensities. This leads to a significant population of MeV electrons, which in turn produce hard X-rays (an effect we briefly discuss later).

As a final comment, it should be noted that ponderomotive steepening by the enhanced resonant field at n_{crit} can occur (Albritton and Langdon 1980). The consequence is that the range of angles in a real experiment with a focal spot intensity variation broadens the range of angles that efficient RA occurs.

9.3.3 Vacuum Heating

When the incident laser pulse is fast and high contrast, or of such high intensity that ponderomotive steepening and hole boring occur, the nature of collisionless absorption at the critical density surface changes. Specifically, this situation is reached when the density scale length is much shorter than a wavelength; in this case resonance absorption and

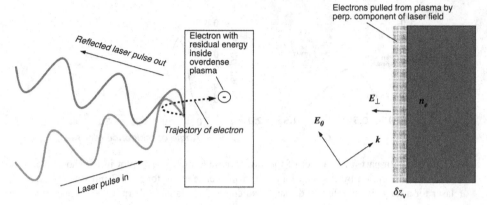

Figure 9.31 Illustration of how electrons are accelerated at a sharp plasma interface in vacuum heating.

plasma waves in a gradient cannot occur and the regime of so-called vacuum heating is entered. This regime is often referred to as "Brunel absorption" after the classic paper by Brunel (1987) who first analyzed this kind of absorption.

The essence of the effect is illustrated in Figure 9.31. When laser light of electric-field amplitude E_0 illuminates the sharp interface of a plasma with vacuum with p-polarization and incidence angle θ_0, there is a normal component of this field of magnitude $E_\perp \cong 2E_0 \sin \theta_0$ (ignoring absorption for the moment). This component of the field can pull plasma electrons from the surface into the vacuum, and, when the field reverses direction, it accelerates the electrons back into the plasma surface. Inside the overdense plasma, the field is shielded and the accelerated electron retains the kinetic energy acquired during the half-cycle it was out in the vacuum. Thought of differently, this vacuum heating is a result of the fact that the presence of the plasma interface breaks the adiabaticity of the oscillating field, and an electron will acquire roughly the ponderomotive energy in the process.

The interplay of the space-charge of a real plasma makes this situation complicated; while further analytic progress can be made (Cai et al. 2006b), PIC simulations are the best way to quantify the effect. We can, however, develop a simple 1D model to find the laser absorption, following along similar lines to Brunel's (1987) original paper. Essentially we ignore any transverse motion of the electrons and treat the plasma interface as oscillating sheets of electrons, as is illustrated in the right-hand figure of Figure 9.31. We seek the fraction of incident power absorbed by the accelerated electrons, which we here denote η_V. We estimate the normal field to be

$$E_\perp \cong \left(1 + (1 - \eta_V)^{1/2}\right) E_0 \sin \theta_0. \tag{9.259}$$

Now we treat the electron sheet like a rigid body (akin to how we treated the oscillating electron cloud in a cluster discussed in Chapter 7). This electron sheet will acquire velocity normal to the surface $v_\perp$, and will have an excursion distance in the vacuum δz_V.

9.3 Collisionless Absorption

A space-charge field E_{sc} will build up from the amount of electrons that the laser has pulled off this rigid electron fluid, which we, for the moment, assume is the density of the overdense target n_{e0},

$$E_{sc} = 4\pi e n_{e0} \delta z_V. \tag{9.260}$$

We assume the laser field is strong and we ignore this space-charge field for the moment, allowing us to write for the equation of motion of the relativistic electron sheet fluid element

$$\frac{dp_\perp}{dt} = eE_\perp \cos \omega t, \tag{9.261}$$

$$\gamma_\perp m v_\perp = \frac{eE_\perp}{\omega} \sin \omega t, \tag{9.262}$$

from which we can write an equation for the normal component of electron fluid velocity returning to the surface, $\beta_{\perp R}$, in terms of the normal component of laser vector potential $a_\perp$:

$$\frac{\beta_{\perp R}}{(1-\beta_{\perp R}^2)^{1/2}} = \frac{eE_\perp}{mc\omega} \equiv a_\perp. \tag{9.263}$$

The kinetic energy of the returning electrons is

$$\varepsilon_R \cong (\gamma_{\perp R} - 1) mc^2 = \left(\sqrt{1+a_\perp^2} - 1\right) mc^2, \tag{9.264}$$

which implies an absorbed power per area over one optical cycle, T, of

$$I_{abs} = \frac{n_{e0} \delta z_V}{T} \varepsilon_R. \tag{9.265}$$

The absorption fraction is the ratio of power per area of energy in fast electrons to the incident laser intensity $\eta_V = I_{abs}/I_{inc}$. This allows us to combine Equations (9.263), (9.264) and (9.265) to yield an equation for the absorbed fraction of incident energy,

$$\eta_V = \frac{\omega}{2\pi} \frac{E_\perp}{4\pi e} \left(\sqrt{1+a_\perp^2} - 1\right) mc^2 \frac{8\pi}{cE_0^2 \cos\theta_0}, \tag{9.266}$$

or with Eq. (9.259) is written

$$\eta_V = \frac{1}{\pi a_0} \left(1 + (1-\eta_V)^{1/2}\right) \left(\sqrt{1+a_\perp^2} - 1\right) \frac{\sin\theta_0}{\cos\theta_0}. \tag{9.267}$$

Writing in terms of laser-normalized vector potential yields an implicit equation for the absorption fraction:

$$\eta_V = \frac{1}{\pi a_0} \left(1 + (1-\eta_V)^{1/2}\right) \left\{\left[1 + \left(1 + (1-\eta_V)^{1/2}\right)^2 \sin\theta_0 a_0^2\right]^{1/2} - 1\right\} \frac{\sin\theta_0}{\cos\theta_0}, \tag{9.268}$$

which is easily solved numerically (see Figure 9.32).

This heuristic model gives some easy insight to the rough scaling of laser absorption by vacuum heating. Brunel offered a slightly more accurate model which does not assume

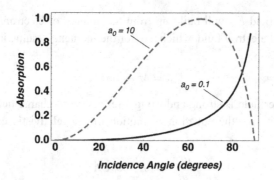

Figure 9.32 Numerical solutions for amount of laser light absorption as a function of incident angle for *P*-polarized light. The solid line is 1053 nm light with a_0 of 0.1 (around 10^{16} W/cm²) and the dashed line is for a_0 of 10 (around 10^{20} W/cm²).

a rigid electron sheet and accounts for the space-charge field. In this approach we again consider a laser field which oscillates, giving field in the normal direction (*z*-axis):

$$E_z = -E_\perp \sin \omega t. \quad (9.269)$$

A sheath field develops, which in terms of the areal density of electrons, $\hat{N}_e$, is, from Eq. (9.260),

$$-e\hat{N}_e(t) = \sigma_e = -\frac{E_\perp}{4\pi} \sin \omega t. \quad (9.270)$$

During the pulling of electrons from the surface, a space-charge field dynamically builds up on the target surface, and, since electrons do not pass each other as they are pulled from the surface, for a given sheet pulled from the surface at time τ_R, the areal charge that has been pulled from the surface up to that point is $-e\hat{N}_e(\tau_R)$. The field felt by that particular sheet of charge is a combination of the laser field and the space-charge built up to the time of release τ_R:

$$E_{tot_e} = -E_\perp \sin \omega t + E_\perp \sin \omega \tau_R. \quad (9.271)$$

Now the equation of motion for the charge sheets is

$$\frac{d}{dt}\gamma_\perp m u_\perp = eE_\perp (\sin \omega t - \sin \omega \tau_R), \quad (9.272)$$

which can be integrated to yield

$$\gamma_\perp \beta_\perp = a_\perp (\cos \omega t - \cos \omega \tau_R) - \omega a_\perp (t - \tau_R) \sin \omega \tau_R \equiv \Gamma_V(t). \quad (9.273)$$

Solving for the return velocity of the charge sheets $\beta_{\perp R}(t)$,

$$\beta_{\perp R}(t) = \frac{\Gamma_V}{\sqrt{1 + \Gamma_V^2}}, \quad (9.274)$$

9.3 Collisionless Absorption

which integrates to give the trajectory of the returning charge sheet, $z_{\perp R}(t)$, while it is out in the vacuum (noting that the location of the sheet when it is initially pulled is $z_{\perp R}(\tau_R) = 0$):

$$z_{\perp R}(t) = c \int_{\tau_R}^{t} \frac{\Gamma_V(t')}{\sqrt{1+\Gamma_V(t')^2}} dt. \quad (9.275)$$

This is a complicated integral; in the nonrelativistic limit, it yields

$$\frac{z_{\perp R}(t)}{c} = -\frac{a_\perp}{\omega}(\sin \omega t - \sin \omega \tau_R) + a_\perp (t - \tau_R) \cos \omega \tau_R \\ - \frac{\omega a_\perp}{2}(t - \tau_R)^2 \sin \omega \tau_R. \quad (9.276)$$

Now we want to find the charge density as a function of location in the vacuum, z. The electron sheets, which travel in the region of negative z in the vacuum, that were released at time $\tau_R + d\tau_R$ are located at $z_R - dz_R$, so

$$-n_e(z_{\perp R}(t)) dz_{\perp R} = \frac{d\hat{N}_e}{d\tau_R}, \quad (9.277)$$

and the electron density in the vacuum is

$$n_e = -\frac{d\hat{N}_e/d\tau_R}{dz_{\perp R}/d\tau_R}. \quad (9.278)$$

From Eq. (9.270), we can write

$$\frac{d\hat{N}_e}{d\tau_R} = \frac{\omega}{4\pi e} E_\perp \cos \omega \tau_R = \frac{mc\omega^2}{4\pi e^2} a_\perp \cos \omega \tau_R. \quad (9.279)$$

The value of $dz_{\perp R}/d\tau_R$ can be calculated in the relativistic regime from Equation (9.275). In the simpler, nonrelativistic limit, it is

$$\frac{dz_{\perp R}(t)}{d\tau_R} = -\frac{\omega^2 a_\perp c}{2}(t - \tau_R)^2 \cos \omega \tau_R, \quad (9.280)$$

so at nonrelativistic intensity the temporal history of electron density is

$$n_{eR}(t) = \frac{m}{2\pi e^2} \frac{1}{(t-\tau_R)^2} = \frac{2n_{crit}}{\omega^2(t-\tau_R)^2}. \quad (9.281)$$

These equations give us a recipe for numerical calculation of the absorbed intensity. Eq. (9.275) can be used to calculate the return time, $t_R(\tau_R)$, of the sheet released at τ_R,

$$z_\perp(t_R, \tau_R) = c \int_{\tau_R}^{t_R} \frac{\Gamma_V(t')}{\sqrt{1+\Gamma_V^2}} dt' = 0, \quad (9.282)$$

and then Eq. (9.278) can be employed to calculate the electron density of the returning sheet $n_{eR} = n_e(t_R)$. Finally, the absorbed intensity is found by integration over a half-cycle:

$$I_{abs} = \frac{\omega}{\pi} \int_0^{\pi/\omega} \left(\frac{1}{\sqrt{1-\beta_{\perp R}(t_R)^2}} - 1 \right) mc^3 n_{eR} \beta_{\perp R}(t_R) d\tau_R, \quad (9.283)$$

where the integration limits were chosen because electrons are only pulled out toward the vacuum for half the cycle. Inspired by the simple form of absorbed intensity in Eq. (9.265), we can parameterize the intensity absorption with a similar equation, inserting some numerically calculated factor α_V, a space-charge amendment to Eq. (9.265):

$$I_{abs} = \alpha_V \frac{n_{e0} \delta z}{T} \varepsilon_R. \qquad (9.284)$$

Brunel numerically calculated this cofactor in the nonrelativistic limit to give $\alpha_V \approx 1.57$, and then performed PIC simulations with a finite thermal temperature plasma to get a correction to this factor (Brunel 1987)

$$\alpha_V \approx 1.75 \, (1 + 2v_{Te}/ca_\perp) \qquad (9.285)$$

(where $v_{Te} = (k_B T_e/m)^{1/2}$). Other authors have found slightly different correction factors using PIC simulations as a function of various parameters. For example, Kato et al. (1993) found a density-dependent correction factor of

$$\alpha_V \approx \frac{1}{1 - n_{crit}/n_{e0}}. \qquad (9.286)$$

With such a parameterization, finding the absorption is once again a simple numerical exercise with the implicit equation for η_V now including this correction factor α_V:

$$\eta_V = \frac{\alpha_V}{\pi a_0} \left(1 + (1-\eta_V)^{1/2}\right) \left\{ \left[1 + \left(1 + (1-\eta_V)^{1/2}\right)^2 \sin\theta_0 a_0^2\right]^{1/2} - 1 \right\} \frac{\sin\theta_0}{\cos\theta_0}. \qquad (9.287)$$

In the low-intensity limit, this equation simplifies to

$$\eta_V \cong \frac{\alpha_V}{2\pi} \left(1 + (1-\eta_V)^{1/2}\right)^3 \frac{\sin^3\theta_0}{\cos\theta_0} a_0, \qquad (9.288)$$

which illustrates that before absorption is large, the magnitude of the absorption,

$$\eta_V \cong \frac{4\alpha_V}{\pi} \frac{\sin^3\theta_0}{\cos\theta_0} a_0, \qquad (9.289)$$

increases as the square root of laser intensity.

On the other hand, in the high-intensity (large a_0) limit, Equation (9.287) indicates that

$$\eta_V \cong \frac{\alpha_V}{\pi} \left(1 + (1-\eta_V)^{1/2}\right)^2 \frac{\sin^2\theta_0}{\cos\theta_0} \qquad (9.290)$$

and absorption is essentially independent of intensity:

$$\eta_V \cong \frac{4\alpha_V}{\pi} \frac{\sin^2\theta_0}{\cos\theta_0} \left(1 + \frac{\alpha_V}{\pi} \frac{\sin^2\theta_0}{\cos\theta_0}\right)^{-2}. \qquad (9.291)$$

This result for the absorption peaks at an incidence angle of 73°. Figure 9.32 plots numerical solutions for absorption based on this equation as a function of incidence angle for two intensities (a_0 of 0.1 and 10) From an experimental standpoint, not a lot should be taken from this quantitative prediction, but two takeaway points can be gleaned from

9.3 Collisionless Absorption

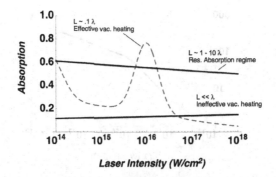

Figure 9.33 Schematic illustration of representative PIC simulation results for collisionless absorption as a function of laser intensity in various plasma scale length regimes. Figure was generated based on simulations in Gibbon and Bell (1992, figure 4).

this analysis. First, we see that absorption at lower intensity is nonlinear with $\eta_V \sim I_{inc}^{1/2}$; $I_{abs} \sim I_{inc}^{3/2}$. Second, at relativistic intensity ($a_0 > 1$) absorption approaches unity. So, vacuum heating is a significant absorption process and is important in many high-intensity solid target experiments. Experimentally, it has been found at modest intensity, around 10^{15} W/cm^2 in 800 nm Ti:sapphire pulses, that the absorption more or less scales like this heuristic Brunel model predicts with $\eta_V \sim I_{inc}^{0.64}$ (Grimes et al. 1999).

Nonideal effects arise in real situations because of even small density gradients at the target surface (Gibbon and Bell 1992). Two trends that are seen in simulations with real density gradients are (1) absorption is $\sim 3\times$ in small gradients because the laser field does penetrate into more plasma and more electrons can be pulled out each cycle than our simple model predicts with a sharp interface, and (2) some electrons do not return in one cycle, so some stochastic heating leads to a hot electron tail and increased absorption. PIC simulations like those of Gibbon and Bell (1992) find more interesting intensity scaling of vacuum heating absorption with laser intensity than this simple model suggests. A schematic illustration of simulations results like these is shown in Figure 9.33 showing that in certain regimes of small but finite density scale-lengths ($\sim 0.1\,\lambda$), there can be intensity ranges where the vacuum heating is very efficient.

In addition to efficient absorption of laser light, vacuum heating has another important experimental signature, which is the production of high-energy electrons (Cui et al. 2013). This heating analysis suggests that, at high intensity, absorption approaches 1 and virtually ALL of the absorbed laser energy is transferred to high-energy electrons. Experiments have often shown over 50% of the incident laser light will be converted to hot electrons. Given our considerations, it makes sense that many of these hot electrons have a kinetic energy which is comparable to the laser ponderomotive energy (many MeV above 10^{18} W/cm^2). A sense of the trend in hot electron temperatures from experiments and simulations in the vacuum heating-dominated regime is illustrated in the plot of Figure 9.34. At modest intensity, like absorption, hot electron temperature scales as the square root of intensity, folding over to a $I^{1/3}$ scaling as intensity reaches into the relativistic regime (Malka and Miquel 1996).

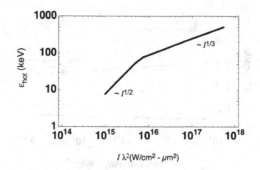

Figure 9.34 Schematic illustration of the trend observed in experiments and simulations of collisionless absorption as a function of laser peak intensity. The commonly observed transition of hot electron temperature increase with intensity from $I^{1/2}$ to $I^{1/3}$ is shown.

As one final note, one might suspect that vacuum heating does not play a role in plasmas that have long density gradients (perhaps because of prepulse) or in experiments that employ near-normal incidence or S-polarized light. The first is certainly true as lower intensity, where resonance absorption certainly dominates in long scale length plasmas. However, when intensity reaches, say, 10^{19} or 10^{20} W/cm^2, it is likely that vacuum heating dominates almost all absorption of the laser, even in these other cases. The reason is that at high intensity, the laser ponderomotively steepens long scale lengths to sharper ones. In addition, hole boring occurs, leading to regions of P-polarized incidence, even when the initial experimental configuration uses normal incidence or S-polarization.

9.3.4 J × B Heating

Absorption and hot electron production by vacuum heating considered in the previous section probably dominate collisionless absorption in most solid target experiments. This was an effect resulting from longitudinal forces of the laser's oscillating electric field. We did not consider the effects of the laser's magnetic field, which we know from considerations in Chapter 3 leads to electron motion in the field along the axis of laser propagation (the often-referenced "figure-8" motion of an electron). In fact, at relativistic intensity, this longitudinal motion can be a significant fraction of the quiver amplitude and can lead to absorption by electrons at a plasma interface just as a longitudinal component of the electric field caused absorption. This magnetic field absorption effect is usually termed **J × B** heating. Its importance was first considered by Kruer and Estbrook in the 1980s (Kruer and Estabrook 1985).

The essence of this absorption mechanism is that, even at normal incidence on a solid target interface, where, once again, we consider plasma with sharp interfaces, the longitudinal oscillation of the **J × B** term leads to vacuum heating. This only happens when the laser is not circularly polarized (recall that in pure circular polarization the electrons acquire longitudinal velocity but do not oscillate longitudinally).

9.3 Collisionless Absorption

This physics can be easily seen from examination of the ponderomotive force (Macchi 2013, p. 74). Consider, for simplicity, the nonrelativistic intensity regime. The usual ponderomotive force is

$$F_p = -\nabla U_p, \tag{9.292}$$

which gives rise to longitudinal force given by

$$F_z = -\frac{mc^2}{2}\frac{\partial}{\partial z}|\mathbf{a}|^2. \tag{9.293}$$

Further consider, to keep polarization general, elliptically polarized light with ellipticity parameter κ, which at the surface of a sharp plasma interface has some skin depth of penetration, δ_s,

$$\mathbf{a}(z,t) = a_{00} e^{-z/\delta_s} \frac{1}{\sqrt{1+\kappa^2}} \left(\hat{\mathbf{x}} \cos \omega t + \kappa \hat{\mathbf{y}} \sin \omega t\right). \tag{9.294}$$

As in previous chapters, $\kappa = 0$ is pure linear polarization and $\kappa = 1$ gives circular polarization. The longitudinal component of the ponderomotive force is

$$F_z = \frac{mc^2}{\delta_s} a_{00}^2 \frac{1}{1+\kappa^2} \left(\cos^2 \omega t + \kappa^2 \sin^2 \omega t\right) e^{-2z/\delta_s}, \tag{9.295}$$

or, with use of a simple trigonometric identity, becomes

$$F_z = \frac{mc^2}{2\delta_s} a_{00}^2 \left(1 + \frac{1-\kappa^2}{1+\kappa^2} \cos 2\omega t\right) e^{-2z/\delta_s}. \tag{9.296}$$

So, as long as the field is not circularly polarized ($\kappa = 1$), the longitudinal force has an oscillatory component, oscillating at twice the frequency of the laser field. This force can perform vacuum heating just as $E_\perp$ did in the Brunel heating case.

Now consider a more general case in which the laser is normally incident on a plasma with a finite scale length and the intensity may be relativistic. This analysis closely follows that of Cai et al. (2006a). The usual relativistic ponderomotive energy is

$$U_p = mc^2 (\gamma_{las} - 1). \tag{9.297}$$

The electron fluid velocity $\mathbf{u}_e$ can be considered in terms of its transverse and longitudinal (z-axis) components. As usual, the transverse velocity, $\mathbf{u}_T$, is dominated by the strong laser electric field,

$$\mathbf{u}_e = \mathbf{u}_T + \mathbf{u}_z. \tag{9.298}$$

Conservation of transverse canonical momentum yields the equations

$$\mathbf{p}_T = \frac{e}{c} \mathbf{A}_T \tag{9.299}$$

$$m \frac{1}{\sqrt{1 - u_e^2/c^2}} \mathbf{u}_T = \frac{e}{c} \mathbf{A}, \tag{9.300}$$

where the laser-driven relativistic factor has the usual form in terms of instantaneous laser-normalized vector potential (not cycle-averaged), $\mathbf{a} = e\mathbf{A}/mc^2$,

$$\gamma_{las} = \sqrt{1 + a^2}. \tag{9.301}$$

Note at this point, in the relativistic case, the longitudinal ponderomotive force, which is given by the z-derivative of the relativistic ponderomotive energy, $\partial[mc^2(\gamma_{las} - 1)]/\partial z$, unlike in the low-intensity case considered earlier, actually has *many* harmonics.

We proceed with Ampere's law for the transverse fields,

$$\nabla^2 \mathbf{A} - \frac{1}{c^2}\frac{\partial^2 \mathbf{A}}{\partial t^2} = \frac{4\pi e n_e}{c} \mathbf{u}_T. \tag{9.302}$$

Ampere's law for the longitudinal field is

$$\frac{\partial E_z}{\partial t} = -4\pi J_z, \tag{9.303}$$

which, in terms of electrostatic potential, is

$$\frac{\partial^2 \Phi}{\partial t \partial z} = -4\pi e n_e u_z, \tag{9.304}$$

and Poisson's equation is

$$\frac{\partial^2 \Phi}{\partial z^2} = 4\pi e \left(n_e - \bar{Z} n_i\right). \tag{9.305}$$

We will, at this point, assume linear polarization to simplify things a bit, so

$$\mathbf{a} = a_{00}(z)\hat{\mathbf{x}} \cos \omega t. \tag{9.306}$$

Now Fourier-expand all the relevant quantities including the laser relativistic factor

$$\gamma_{las} = \gamma_0(z) + \gamma_1(z) \cos \omega t + \gamma_2(z) \cos 2\omega t + ..., \tag{9.307}$$

which has zeroth order Fourier coefficient

$$\gamma_0 = \frac{\omega}{2\pi} \int_0^{2\pi/\omega} \sqrt{1 + a^2}\, dt. \tag{9.308}$$

We can approximate the first term in the Fourier expansion and replace it with the cycle-averaged form of the relativistic factor:

$$\gamma_0 \cong \sqrt{1 + a_{00}^2/2}. \tag{9.309}$$

The next coefficient is zero:

$$\gamma_1 = \frac{\omega}{2\pi} \int_0^{2\pi/\omega} \sqrt{1 + a^2} \cos \omega t\, dt = 0. \tag{9.310}$$

The term for the second harmonic factor is not zero:

$$\gamma_2 = \frac{\omega}{2\pi} \int_0^{2\pi/\omega} \sqrt{1 + a^2} \cos 2\omega t\, dt. \tag{9.311}$$

9.3 Collisionless Absorption

We integrate this by parts and approximate with the cycle-averaged term

$$(1+a^2)^{-1/2} \cong (1+a_{00}^2/2)^{-1/2}, \qquad (9.312)$$

pull it out of the integral, and use this integral,

$$\int_0^{2\pi} \sin 2x \frac{\partial}{\partial x} \cos^2 x \, dx = -\pi, \qquad (9.313)$$

to find the coefficient for the relativistic factor oscillating at the second harmonic of the laser:

$$\gamma_2 \cong \frac{a_{00}^2}{4\gamma_0}. \qquad (9.314)$$

If the ions are immobile, space-charge demands that $u_{z0} = 0$ in a quasi-steady-state situation. Also, in quasi-steady state, an electrostatic potential at the plasma surface must grow to counteract the steady ponderomotive pressure of the laser, such that

$$e\frac{\partial \Phi}{\partial z} = mc^2 \frac{\partial \gamma_0}{\partial z}. \qquad (9.315)$$

Note that we can write the wave equation for the normalized vector potential, recalling Eq. (8.179) as

$$\frac{\partial^2 a}{\partial z^2} - \frac{\omega^2}{c^2} a = \frac{4\pi e^2}{c^2} \frac{n_e}{\gamma_0} a. \qquad (9.316)$$

Again, electron oscillations can be driven by the γ_2 term of the ponderomotive force, which means that the equation of motion for the longitudinal component of the electron fluid is

$$\frac{\partial}{\partial t} m\gamma u_{ez} = e\frac{\partial \Phi}{\partial z} - mc^2 \frac{\partial \gamma}{\partial z} \gamma_2. \qquad (9.317)$$

Taking the time derivative of this and substituting it into (9.304) gives

$$\frac{m}{e} \frac{\partial^2}{\partial t^2} (\gamma u_{ez}) + \frac{mc^2}{e} \frac{\partial^2 \gamma}{\partial z \partial t} + 4\pi n_e u_{ez} = 0. \qquad (9.318)$$

The first two components of the electron fluid velocity Fourier expansion are zero: u_{z0} and $u_{z1} = 0$. (There is no steady flow of electron fluid because of ion immobility, and there is no oscillating component at the frequency of the laser in the longitudinal direction.) The first nonzero component is the second harmonic electron oscillation given by this equation of motion,

$$-\frac{4m\gamma_0 \omega^2}{e} u_{z2} + 4\pi n_{e0} e u_{z2} - \frac{2mc^2 \omega}{e} \frac{\partial \gamma_z}{\partial z} = 0, \qquad (9.319)$$

which we solve for the electron second harmonic fluid velocity:

$$u_{z2} = \frac{1}{n_{e0} - 4\gamma_0 n_{crit}} \frac{mc^2 \omega}{2\pi e^2} \frac{\partial \gamma_z}{\partial z}. \qquad (9.320)$$

The second harmonic component of the longitudinal electric field is then

$$E_{z2} = -\frac{mc^2 n_{e0}}{e} \frac{1}{n_{e0} - 4\gamma_0 n_{crit}} \frac{\partial \gamma_z}{\partial z}. \qquad (9.321)$$

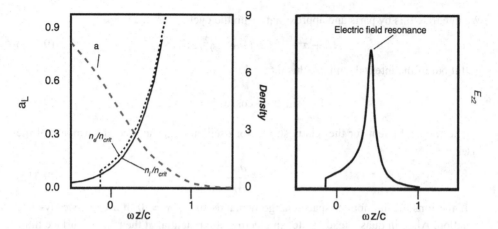

Figure 9.35 Electron and ion densities also shown with the laser field in the modest intensity case ($\sim 10^{18}$ W/cm^2) case on the left. The calculated longitudinal electric field is shown in the right panel, illustrating the clear resonance at four times the critical density. Plots were generated using simulations presented in (Cai et al. 2006a, Fig. 3).

We see that the longitudinal field-driving electron oscillations actually peak in a resonance at an electron density of $4\gamma_0 n_{crit}$. We have seen this physics before, in a slightly different situation. This is the $\mathbf{J} \times \mathbf{B}$ heating analog of resonance absorption at the critical density surface. In this case the resonant density point in the plasma is now at (relativistically corrected) four times critical density, the resonant density for the 2ω oscillating field. This is an interesting result: it says that a form of $\mathbf{J} \times \mathbf{B}$ absorption also occurs when there is a significant plasma gradient, just as the resonant absorption effect occurred (Lefebvre and Bonnaud 1997). This resonance enhances absorption even when the laser intensity is not very high. To find the rate of absorption, as with RA we introduce a phenomenological damping term on E_z. To determine where this resonance occurs (how deep and at what gradient) we have to find the self-consistent plasma density profile taking into account the steepening of the electron front by the laser pressure. We considered this in the context of hole boring in Sections 9.2.4 and 9.2.5.

The field at the interface can be found numerically with the help of the Helmholtz equation for the laser field,

$$\frac{\partial^2 a_T}{\partial z^2} + \frac{\omega^2}{c^2}\varepsilon(z)a_T = 0, \qquad (9.322)$$

Eq. (9.315), and the appropriate boundary conditions for the field at the steepened interface. The result are profiles like those illustrated in Figure 9.35, which shows a modest intensity case as calculated by Cai et al. (2006a). A small amount of steepening occurs from the ponderomotive pressure of the laser, leading to a small electron density shelf at the outer edge of the plasma gradient. A resonance in the longitudinal field, shown at the right in the figure, is clear, leading to the $\mathbf{J} \times \mathbf{B}$ version of resonance absorption at 2ω. Numerical calculations (Cai et al. 2006a) indicate that absorption at normal incidence in this regime is modest, increasing from about 1% with a scale length 0.01 λ up to around 6% with longer scale length of 0.05 λ.

9.3.5 $J \times B$ Heating at High Intensity

At high intensity the situation changes somewhat. Because of the large ponderomotive pressure that steepens the electron density profile, as illustrated in Fig. 9.36a, the laser sees no density gradient and instead experiences a sharp overdense profile. At these intensities, the electron density shelf is pushed to a density above the critical density and no 2ω resonance occurs. In this regime $J \times B$ heating is akin to vacuum heating with electron excursions larger than the steepened gradient, in this case by the magnetic-field-driven figure-8 motion of the electrons, which are accelerated every half-cycle into the target. The longitudinal second harmonic electric field just decays away exponentially and has no resonance, as the field did in Eq. (9.321).

To find the intensity at which this regime is reached and the 2ω resonance is quenched, we need to find when the electron density shelf exceeds $4 \times n_{crit}$. We can approximate the laser vector potential as that at a sharp surface with no absorption, a_s, recalling Eq. (9.138):

$$a_S \approx 2\frac{\omega}{\omega_{pe}}a_0. \tag{9.323}$$

We then define the critical field for reaching this sharp interface regime, $a_S \approx a_C = 2(1/4)^{1/2}a_0$, and assume for the profile of the electrons in the density

$$n_e - \bar{Z}n_i \cong \delta n_e e^{-2z'/\delta_p}, \tag{9.324}$$

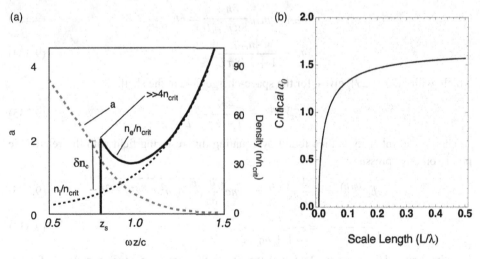

Figure 9.36 (a) Schematic illustration of ponderomotive steepening in the high-intensity regime with the development of a large overcritical shelf in the electron density and a laser field profile like that at the interface of a sharp-edged conductor. (b) Calculation of the critical field–normalized vector potential needed to reach this overdense, resonance-quenched high-intensity $J \times B$ regime.

with density scale approximated by the skin depth at 4× critical density:

$$\delta_p \approx \frac{c}{\omega_{pe}} = \frac{c}{2\omega} = \frac{\lambda}{4\pi}. \qquad (9.325)$$

So, the longitudinal restoring force at the shelf position, z_s, from the space-charge separation pulling the electrons back against the ponderomotive pressure has a field which is, from Gauss's law,

$$E_z^{(shelf)} \cong 8\pi e \int_{-\infty}^{z_s} n_{crit} e^{z/L} dz = 4\pi e \int_0^{\infty} \delta n_e e^{-8\pi z'/\lambda} dz' \qquad (9.326)$$

$$\cong e\lambda \delta n_e. \qquad (9.327)$$

Conservation of charge mandates that

$$\int_0^{\infty} \delta n_e e^{-8\pi z'/\lambda} dz' = \int_{-\infty}^{z_s} n_{crit} e^{z/L} dz \qquad (9.328)$$

$$-\delta n_e \frac{\lambda}{8\pi} e^{-8\pi z'/\lambda} \Big|_0^{\infty} = n_{crit} L e^{z/\lambda} \Big|_{-\infty}^{z_s} \qquad (9.329)$$

$$\delta n_e \frac{\lambda}{8\pi} = n_{crit} L e^{z_s/\lambda}. \qquad (9.330)$$

At the shelf surface, from Eq. (9.324) we can say

$$n_e - n_{e0}(z_s) = \delta n_e, \qquad (9.331)$$

and since $n_e = 4\, n_{crit}$ at the critical laser intensity, this means

$$4n_{crit} - n_{crit} e^{z_s/L} = \delta n_e \qquad (9.332)$$

$$4n_{crit} - n_{crit} \frac{\delta n_e \lambda}{8\pi n_{crit} L} = \delta n_e \qquad (9.333)$$

$$\delta n_e = \frac{4n_{crit}}{1 + \lambda/8\pi L}, \qquad (9.334)$$

which, with Eq. (9.327), gives, for the space-charge field at the shelf,

$$E_z^{(shelf)} \cong \frac{4e\lambda n_{crit}}{1 + \lambda/8\pi L}. \qquad (9.335)$$

The critical intensity is then found by equating this restoring field with the relativistic ponderomotive pressure:

$$eE_z^{(shelf)} = -\frac{\partial}{\partial z} \gamma_0^{(crit)} = -mc^2 \frac{\partial}{\partial z} \sqrt{1 + a_0^2 e^{-8\pi z'/(\lambda/2)}}, \qquad (9.336)$$

$$\frac{2}{1 + \lambda/8\pi L} = \frac{a_0^2}{\sqrt{1 + a_0^2/2}}. \qquad (9.337)$$

For simplicity, define $\alpha_L \equiv 2/(1 + \lambda/8\pi L)$, so this equation gives a solution for the normalized vector potential for reaching this resonance-quenched $\mathbf{J} \times \mathbf{B}$ regime:

$$a_0 = \sqrt{\frac{\alpha_L^2/4 + \sqrt{\alpha_L^4/4 + 4\alpha_L^2}}{2}}. \qquad (9.338)$$

9.3 Collisionless Absorption

The field calculated from this formula is shown in Fig. 9.36b. This illustrates that the intensity needed when the scale length is less than a tenth of a wavelength increases sharply, but with longer scale lengths the intensity is about constant and is at a_0 of 1.5, an intensity of a few times 10^{18} W/cm^2 (depending on laser wavelength). This simple estimate is consistent with numerical calculations of the profile steepening in Cai et al. (2006a).

The consequences of entering this high-intensity vacuum heating-like $\mathbf{J} \times \mathbf{B}$ regime is that absorption increases significantly above the resonant low-intensity regime. Simulations indicate that absorption at normal incidence in this regime is between 10% and 20% (Lefebvre and Bonnaud 1997). Because the absorption is by ponderomotively driven electrons from vacuum back into the overdense target, significant hot electron production accompanies this regime. For example, some experiments have observed $\sim$1 MeV electrons at intensity of 10^{19} W/cm^2.

As a final comment, note that this effect does not occur when the laser field is circularly polarized, because there is no "figure-8" motion in such a field. This implies that there will be some threshold in the ellipticity of the laser polarization in which $\mathbf{J} \times \mathbf{B}$ heating is suppressed. Macchi (2013, pp. 75–76) has estimated this ellipticity threshold in terms of ellipticity parameter, κ, (Macchi 2013, pp. 75–76) to be $\kappa \leq 2(n_{crit}/n_e)^{1/2}$, which is a fairly small threshold for ellipticity suppression with a highly overdense plasma target.

9.3.6 Sheath Inverse Bremsstrahlung and Anomalous Skin Effect Absorption

To summarize our understanding of the regimes where various collisionless absorption mechanisms occur: (1) when a significant density gradient develops out in front of the target and oblique incidence resonance absorption occurs; (2) at normal incidence $\mathbf{J} \times \mathbf{B}$ absorption occurs and becomes significant at high intensity because of electron density steepening; (3) in sharp density profiles at oblique incidence and high-intensity vacuum, heating becomes significant. Are these the only regimes in which collisionless absorption matters? For example, at moderate intensities on a sharp plasma interface, is the absorption only by collisional absorption? It turns out there are a couple of additional collisionless mechanisms that can contribute, even at moderate intensity through the interaction of thermal electrons with the sharp plasma interface and the evanescent laser field. It is usually difficult to observe these mechanisms definitively in experiments because they do not produce the tell-tale sign of hot electrons, but they are well established in theory and simulation (Catto and More 1977; Yang et al. 1995; Yang et al. 1996).

As we have discussed in the context of collisional absorption, the way a laser-driven electron absorbs energy from the light field (classically) is when its coherent oscillation motion is disturbed on a nonadiabatic time scale, faster than a laser oscillation time, $\sim \omega^{-1}$. This is what happens when a laser-driven electron scatters from an ion. However, at a sharp plasma interface, such nonadiabatic interruption of electron oscillation motion can also occur. We consider two such absorption mechanisms which are closely related:

(1) The Anomalous Skin Effect

Electrons oscillating within the skin depth of the plasma surface will also have thermal motion. If the thermal velocity of the electron is sufficient, it can potentially exit the skin

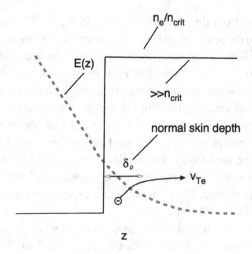

Figure 9.37 Simple illustration of absorption by the anomalous skin effect.

depth region on a time scale faster than a single oscillation. This serves the purpose of breaking the electron oscillation adiabaticity and will lead to some absorption of energy from the laser field. This effect is depicted in Fig. 9.37 and is known as the anomalous skin effect (Rozmus and Tikhonchuk 1990; Gibbon 2005, p. 163).

This effect is significant only when the temperature of the overdense plasma is high enough to lead to a significant population of electrons with thermal velocity high enough to traverse a skin depth on a time scale faster than an electron oscillation. Therefore, since thermal velocity is

$$v_{Te} = \sqrt{2k_B T_e/m}, \qquad (9.339)$$

the time for a thermal electron to transit the skin depth is roughly

$$\Delta t_{SD} \approx \delta_p/v_{Te} \approx c/\omega_{pe} v_{Te}. \qquad (9.340)$$

So, the condition for the anomalous skin effect is

$$c/v_{Te} \ll \omega_{pe}/\omega \qquad (9.341)$$

or namely that

$$n_e/n_{crit} \gg mc^2/2k_B T_e. \qquad (9.342)$$

This implies that for short-pulse laser experiments, where heat conduction typically limits thermal electron temperatures to $\sim$100 eV – 500 eV, this absorption is only significant when $n_e \gg 10^3 \, n_{crit} \sim 10^{24}$ cm^{-3}. This high electron density is rather uncommon except possibly in very high-Z targets (like Au) with high ionization. If for some reason higher thermal temperatures are achieved such as in longer laser pulses at shorter wavelength, then anomalous skin effect should become more important.

9.3 Collisionless Absorption

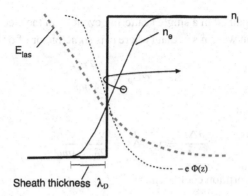

Figure 9.38 Simple illustration of absorption by sheath inverse bremsstrahlung.

(2) Sheath Inverse Bremsstrahlung

A closely related absorption mechanism occurs when electrons can acquire enough longitudinal velocity that they bounce off the sheath potential barrier at the surface of the plasma on a time scale faster than a laser oscillation, interrupting the transverse motion, as depicted in Fig. 9.38. This electron heating process is akin to the stochastic heating in a cluster that we encountered in Chapter 7 and is known as sheath inverse bremsstrahlung (SIB). It is probably more relevant in ultrafast laser–solid interactions than is the anomalous skin effect (Gibbon 2005, p. 164).

To find the conditions for this to occur, let us look a little at the interaction of an electron with the plasma surface electric potential structure. The condition for SIB is, in terms of the Debye length of the sheath field, λ_D,

$$\Delta t_{sheath} \approx \lambda_D / v_{Te} \ll \omega^{-1}, \tag{9.343}$$

which, if we introduce the Debye length of the sheath (see Appendix B), is

$$\frac{k_B T_e}{4\pi e^2 n_{e0}} \frac{m}{2 k_B T_e} \ll 1/\omega^2 \rightarrow \frac{n_{e0}}{n_{crit}} \gg 1. \tag{9.344}$$

This situation is common in short-pulse solid-target experiments. Note, at normal incidence, because the "collision" occurs in the longitudinal direction, the transverse canonical momentum of an electron in the field is still preserved, and it acquires no energy from that component of the laser field. So, in that case, SIB would involve heating by the $\mathbf{v} \times \mathbf{B}$ component of its oscillatory motion (Yang et al. 1995).

We can make a heuristic estimate of the magnitude of SIB absorption at normal incidence. We define an effective SIB collision frequency,

$$\nu_{SIB} \approx \frac{v_{Te}}{2\delta_p}, \tag{9.345}$$

and note that the rate of absorption is

$$\eta_{Abs}^{(SIB)} I = n_e \delta_p \frac{\partial \Delta \varepsilon}{\partial t}, \tag{9.346}$$

$\Delta\varepsilon$ being the energy gained in a single scattering event from the electron off the sheath. In a rough approximation we can say that (where α is some factor of order unity)

$$\Delta\varepsilon \cong \alpha U_p = \alpha \frac{e^2 E_0^2}{4m\omega^2}. \tag{9.347}$$

The heating rate is then

$$\frac{\partial \Delta\varepsilon}{\partial t} \cong \nu_{SIB}\Delta\varepsilon = \frac{v_{Te}}{2\delta_p}\alpha\frac{e^2 E_0^2}{4m\omega^2}, \tag{9.348}$$

giving for the SIB absorption coefficient

$$\eta_{Abs}^{(SIB)} = \frac{8\pi}{c}\frac{v_{Te}}{2}\frac{e^2}{4m\omega^2}n_e\alpha$$
$$= \frac{\alpha}{4}\frac{v_{Te}}{c}\frac{\omega_{pe}^2}{\omega^2} = \frac{\alpha}{4}\frac{v_{Te}}{c}\frac{n_e}{n_{crit}}. \tag{9.349}$$

With a plasma temperature of $\sim$500 eV and a low-Z, target where n_e/n_{crit} might be $\sim$10, the absorption fraction into SIB is $\sim\alpha/12$, or on the order of a few percent. With higher-density targets, this fraction increases. Our simple estimate is consistent with kinetic theory in Yang et al. (1995). At oblique incidence the SIB collision frequency increases because of the addition of laser-driven velocity to the thermal velocity; however, the interaction is not instantaneous, so this tends to cancel, and the absorption fraction is about the same. This entire effect can be thought of in some sense as the low-intensity limit of vacuum heating and is sometimes called in the literature "sheath transit heating."

9.3.7 Laser Plasma Instabilities in Density Gradients

In Section 8.6 we considered laser plasma instabilities (LPI) in underdense, uniform-density plasmas, in particular stimulated Raman scattering (SRS). In experiments on solid targets, where there is a shallow intensity gradient (say, one produced by some significant, early-time-scale prepulse) LPI can also occur. These drive plasma waves and will be a source of hot electrons. As illustrated in Fig. 9.39, SRS can occur in the region in front of the target where density is less than quarter critical density. In addition, a related instability can occur at the quarter critical point, two-plasmon decay (Kruer 2003, pp. 81–83; Singh et al. 2020). Recall that SRS is a resonant process with plasma waves, so density inhomogeneity limits the region over which the three-wave instability of backward scatter SRS will grow.

We can determine a phenomenological threshold in intensity at which SRS will grow in a plasma-density gradient of scale length L. Here we follow the treatment of Kruer (2003, p. 79). First, recall Eq. (8.391), the dispersion relationship for SRS growth, and further introduce in Eq. (8.389) a damping term

$$D_s \to \frac{\omega_S}{c^2}(\omega_S + i\gamma_S) - \frac{\omega_{pe}^2}{c^2} - k_S^2. \tag{9.350}$$

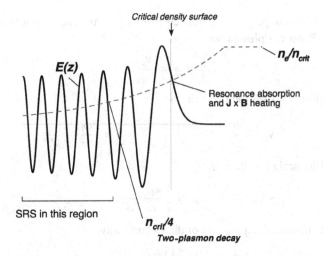

Figure 9.39 Structure of LPI in long-scale-length solid-target plasmas with laser field incident from the left.

Putting this into Eq. (8.391) and keeping only the Stokes term D_s, the SRS growth rate, Γ_{RBS}, is positive only if

$$\Gamma_{RBS} > \frac{1}{2}\sqrt{\gamma_e\gamma_s}, \qquad (9.351)$$

related to the damping rates of the Stokes (γ_S) and electron waves (γ_e) waves. This can be considered the threshold for SRS growth. In a plasma with varying density, the convection of waves out of resonance as the laser propagates up the plasma density ramp can be thought of as an effective damping rate.

To find this effective damping rate, we consider the wavenumber of the plasma wave as a function of position, and, as the laser propagates along, the phase mismatch of the laser, Stokes and plasma wave is $\Delta k(z) = k(z) - k_s(z) - k_e(z)$. SRS instability growth stops in a plasma gradient when the waves shift out of phase, namely when

$$\int_0^{l_{SRS}} \Delta k\, dz \approx \pi/2. \qquad (9.352)$$

Assuming a near linear variation in density, and using the fact that the laser and Stoke wave number will not vary much over a density gradient in underdense plasma, we approximate the phase mismatch

$$\Delta k(z) \approx \Delta k(0) + \Delta k' z \approx k - k_s - k_e - \Delta k'_e z \qquad (9.353)$$

and find from Eq. (9.352) the length over which SRS can grow,

$$l_{SRS} \approx \sqrt{\frac{\pi}{\Delta k'_e}}. \qquad (9.354)$$

The effective SRS damping rate is related to this growth length in terms of the group velocity of the Stokes or plasma wave v_g as

$$v_{eff} \approx \frac{v_g}{l_{SRS}}. \tag{9.355}$$

The electron plasma wave group velocity is

$$v_{ge} \approx \frac{\partial \omega_e}{\partial k_e} \approx \frac{\partial \omega_{pe}}{\partial z}\left(\frac{\partial k_e}{\partial z}\right)^{-1} \approx \frac{\partial \omega_{pe}}{\partial z}\frac{1}{\Delta k_e'} \tag{9.356}$$

and, if the density scale length is L,

$$v_{ge} \approx \frac{1}{2}\frac{\omega_{pe}}{L}\frac{1}{\Delta k_e'}, \tag{9.357}$$

giving, for the effective damping rate of the electron wave,

$$v_{eff}^{(e)} \cong \frac{1}{2}\frac{\omega_{pe}}{L}\frac{1}{\sqrt{\pi \Delta k_e'}}. \tag{9.358}$$

The group velocity of the Stokes wave in underdense plasma is close to c, so its convective damping rate in the plasma density gradient is

$$v_{eff}^{(s)} \cong c\sqrt{\frac{\Delta k_e'}{\pi}}. \tag{9.359}$$

These two give an estimate for the threshold for growth of SRS in the preplasma gradient:

$$\Gamma_{RBS} > \frac{1}{2}\sqrt{\frac{\omega_{pe}c}{\pi L}}. \tag{9.360}$$

Using our result for the growth of SRS in terms of laser-normalized vector potential, a_0, from Eq. (8.398) the intensity threshold for SRS growth is

$$a_0 \geq \sqrt{\frac{c}{\pi \omega L}} = \frac{1}{2^{1/2}\pi}\sqrt{\frac{\lambda}{L}}. \tag{9.361}$$

So, in long-scale-length plasmas, say $L \sim 10\,\lambda$, SRS will grow significantly with $a_0 \geq 0.07$, an intensity of around 10^{16} W/cm². This will be a source of moderately hot electrons. It is well known in experiments that at moderate intensities (up to $\sim 10^{17}$ W/cm²), hot electron production is greatly enhanced with the production of a long plasma scale length with a prepulse, suggesting that SRS growth is important.

As one more note, in a density gradient, it is also possible for a related plasma instability to develop right at the electron density which is at $n_{crit}/4$, the so-called two-plasmon decay (TPD) or the $2\,\omega_{pe}$ instability. We will not delve into this because it is likely less important in short-pulse laser interactions than it is for nanosecond lasers because of the short time scales over which there is laser intensity at the quarter critical surface. The phenomenon is similar to the SRS except, instead of a laser photon coupling to a plasma phonon and a Stokes photon, the laser photon couples to 2× plasma phonons. As with SRS, this mechanism produces plasma waves and hot electrons, and, in nanosecond laser experiments, can dominate the LPI. By conservation of momentum these plasma waves are produced at 45° with respect to the laser propagation. The growth rate for this instability can be found though a similar instability analysis as that performed for SRS in Section 8.6, see Kruer (2003, p. 83) for details. From that analysis it can be shown (at nonrelativistic intensity) that the TPD growth rate is

$$\Gamma_{2\omega_{pe}} \cong \frac{\omega}{c}\frac{a_0}{4}, \tag{9.362}$$

which incidentally is the growth rate of SRS backscatter at density $n_{crit}/4$. At subrelativistic intensity we see that the growth rate is a fraction of an inverse laser period, so many exponential growth factors will only develop for pulses that are many-picosecond or longer time scale.

9.4 Hot Electron Production and Transport

It should now be clear that there is a veritable zoo of collisionless absorption mechanisms that lead to suprathermal hot electron production particularly as a_0 approaches or exceeds 1 ($>10^{17}$ W/cm^2). In fact, in an actual experiment it is likely that a range of mechanisms contribute at the same time and the actual dynamics are very complicated. So, at this point, we should make some observations about real experiments and observed hot electron production trends.

9.4.1 Hot Electron Spectra and Temperature Scaling

The hot electron energy spectra observed across various experiments is usually quite similar (see, for example, Zheng et al. 2004; Cowan et al. 1999; Chen, H. et al. 2009; and Taylor et al. 2013 among many other examples). Although the hot electrons are so fast to be completely collisionless and cannot thermalize to thermodynamic equilibrium and a Maxwellian energy distribution, what is usually observed are electron energy spectra that are exponential with energy, a quasi-Maxwellian, characterized by energy distribution scaling like $\sim \exp[-\varepsilon_e/k_B T_{eh}]$. The typical shape of electron spectra is illustrated in Fig. 9.40. Often a "two-temperature" distribution is observed, likely the result of two (or more) collisionless absorption mechanisms at work.

An often-cited formula to estimate the temperature of the hottest component as a function of laser $I\lambda^2$ is one developed empirically from fitting a number of experiments, typically performed with few-hundred-femtosecond to few-picosecond pulses at oblique incidence (Beg et al. 1997):

$$T_{hot} \cong 250 \left(\frac{I\lambda^2[\mu m]}{10^{15} W/cm^2 - \mu m^2}\right)^{1/3} keV \tag{9.363}$$

This formula is usually referred in the literature as the Beg hot electron temperature formula. It is remarkably accurate in many experiments up to intensity of 10^{21} W/cm^2 (Chen et al. 2009).

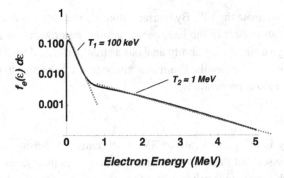

Figure 9.40 Typical shape of observed hot electron spectra in the forward direction at intensities of $10^{19} - 10^{21}$ W/cm^2 showing the often-observed two-temperature hot electron distribution.

While this formula is purely an empirical law fit to data, it turns out to be rather consistent with a simple "black box" model of laser conversion to hot electrons due to Haines (Haines *et al.* 2009). This model is worth examining, as it gives some general physics insight through simple energy and momentum conversion considerations to why the Beg law is so accurate.

Consider a thin region of plasma a few collisional skin depths thick. We seek a scaling of the hot electron temperature using relativistic equations for conservation of energy and momentum between the laser and the hot electrons. Assume near 100% laser absorption (reasonably consistent with experiments which observe ~50%) and hot electron density n_h, hot electron relativistic factor γ_h, relativistic forward moment $p_z = m\gamma_h v_z$, and usual nonrelativistic critical density n_{crit}. Electrons are accelerated at the laser turning point, so they are produced near n_{crit}. In the lab frame, due to Lorentz contraction,

$$n_h \cong \gamma_h n_{crit}. \tag{9.364}$$

Energy flux conservation in the thin layer where laser energy is deposited into electrons demands that energy carried away by hot electrons must be equal to incoming energy flux from the laser, so we can say

$$I = n_h m (\gamma_h - 1) v_z c^2. \tag{9.365}$$

On the other hand, momentum conservation in this thin layer says that

$$\frac{I}{c} = n_h p_z v_z = n_{crit} p_z^2 / m. \tag{9.366}$$

Combining these two equations, eliminating I, gives

$$n_{crit} p_z (\gamma_h - 1) c = n_{crit} p_z^2 / m$$

$$\frac{p_z}{mc} = (\gamma_h - 1). \tag{9.367}$$

Now transform to the axial rest frame of the streaming electrons to get their total energy ε_0:

$$\varepsilon_0^2 = \varepsilon^2 - p_z^2 c^2, \tag{9.368}$$

where $\varepsilon = \gamma_h mc^2$. Then, with Eq. (9.367), we write this as

$$\varepsilon_0^2 = \gamma_h^2 m^2 c^4 - p_z^2 c^2$$
$$= m^2 c^4 \left(1 + \frac{2p_z}{mc}\right). \qquad (9.369)$$

If $\varepsilon_0 = \gamma_0 mc^2$ and we approximate the hot electron temperature with this hot electron energy, then $k_B T_h \approx \gamma_0 mc^2$, meaning that the hot electron temperature (in the lab frame) is

$$k_B T_h \cong mc^2 \left[\left(1 + \frac{2p_z}{mc}\right)^{1/2} - 1\right] \qquad (9.370)$$

or, from Eq. (9.366), the hot electron temperature, in terms of laser intensity, is

$$k_B T_h \cong mc^2 \left[\left(1 + \frac{2}{mc}\sqrt{\frac{mI}{n_{crit}c}}\right)^{1/2} - 1\right]. \qquad (9.371)$$

In terms of a_0 this equation is

$$k_B T_h \cong mc^2 \left[\left(1 + 2^{1/2} a_0\right)^{1/2} - 1\right]. \qquad (9.372)$$

This equation illustrates the expected hot electron temperature scaling with intensity at weak and strong fields. At $a_0 \ll 1$, the temperature is

$$k_B T_h \cong mc^2 a_0 / 2^{1/2} \sim I^{1/2}, \qquad (9.373)$$

and at large a_0 it scales as

$$k_B T_h \cong mc^2 2^{1/4} a_0^{1/2} \sim I^{1/4}. \qquad (9.374)$$

Note, that these scalings fall on either side of the $I^{1/3}$ empirically deduced scaling of the Beg law. The Haines equation is, in fact, remarkably close to the Beg scaling over a wide intensity range, as illustrated in Fig. 9.41, where the Beg experimental and Haines theoretical formula are plotted over five orders of magnitude in intensity. This analysis confirms the intuitive conclusion that the hot electron temperature, independent of the actual collisionless absorption mechanism at work, is more or less the result of energy and momentum scaling in the conversion of laser light to hot electron current.

With this said, it should be noted that experimental evidence shows that, while the specific absorption mechanism may not strongly affect temperature, it does affect emission angle. As illustrated in Fig. 9.42 and confirmed in a number of experiments (Santala et al. 2000; Cho et al. 2009), electrons will exhibit a distinct emission direction depending on mechanism; for example, $\mathbf{J} \times \mathbf{B}$ electrons are emitted in the direction along the laser axis while resonant absorption and vacuum heated electrons emerge normal to the target surface. As discussed in Section 9.4.4, this is observable by the frequency of the hot electron bunches emerging, with $\mathbf{J} \times \mathbf{B}$ electrons coming every laser half-cycle and vacuum-heated electrons emerging every laser cycle.

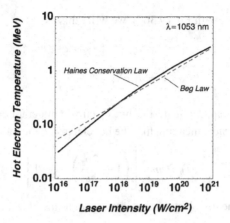

Figure 9.41 Comparison of hot electron temperature scaling with peak intensity from both the Beg, experimental law and the Haines conservation law.

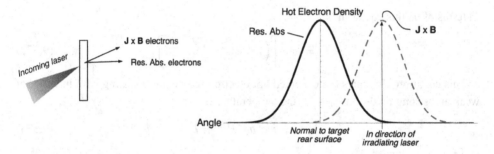

Figure 9.42 Schematic illustration of the direction of emission of hot electrons deduced from experiments and PIC simulations.

The conversion efficiency into electrons varies somewhat with experiment but in the relativistic intensity regime it is usually observed to be quite high, 10 to 50 percent of the total incident laser energy (Wharton *et al.* 1998). The total number of electrons tends to scale roughly linearly with laser energy (Zheng *et al.* 2004).

9.4.2 Hot Electron Transport

Once hot electrons are produced near the target surface, they transport into the cold solid-density target (McKenna and Quinn 2013). Studies of how this occurs and what affects the hot electron transport have been extensively motivated by the importance of transport to the fast ignition concept described in Section 9.10.3 (Chen, C. D. *et al.* 2009; Atzeni *et al.* 2009). Some of the physics affecting hot electron transport away from the laser are illustrated in Fig. 9.43. To the simplest approximation, hot electron transport is affected by two factors, self-generated magnetic fields and electrostatic fields built up by charge separation from the hot electron current. Both of these effects are strongly shaped

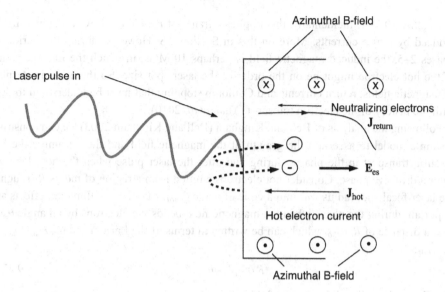

Figure 9.43 Schematic illustration of the structure of hot electron transport within the target.

by so-called return currents, which flow in the opposite direction of the hot electron current to attempt to neutralize the charge buildup that resulted from the flow of hot electrons away from the target surface. On top of these two effects, the hot electrons will lose some energy by Coulomb collisions and will exhibit an effective Coulomb scattering mean free path that is also important.

The temporal dynamic of the hot electron transport can be more or less characterized by five temporal phases (Norreys et al. 2014):

(1) Early transient regime in which the electrons fly ballistically into the target. This sets up an electrostatic E-field which begins to induce a return current.
(2) Heating regime as the electrostatic E-field slows the hot electrons and the induced return current ohmically heats the underlying cold solid.
(3) Conductivity increase of the underlying cold material as it is heated and it becomes less collisional. This allows an increase in the return current and freer penetration of the hot electrons with the neutralization of the electrostatic field.
(4) A drag phase in which the fast electrons slow and start contributing, along with the return currents, to the target heating.
(5) Finally a diffusive phase in which the hot electrons have been slowed and penetrate diffusively in the target.

A simple estimate for the magnitude of the hot electron current can be made. If a modest-energy, 10 J pulse of 500 fs duration is focused to 10^{19} W/cm^2, then the likely hot electron temperature by the Beg formula is ∼0.5 MeV, so with 30 percent conversion into hot electrons this implies roughly 5×10^{13} electrons over a time scale of 500 fs, a current of roughly 10 MA. Incidentally, this implies a transient B field of 1 GGs, though quickly reduced by return currents. More on this in Section 9.8. However, during the period of phases 2–5, the induced magnetic field is perhaps 10 MGs, in which the Larmor radius of the hot electron might be on the order of the laser spot size. So there is an interplay of magnetic fields, return currents and Coulomb stopping that must be understood to deal with the history of the electron transport (Yuan et al. 2010).

Following the analysis of Bell and Kingham (Bell and Kingham 2003) we can construct a simple model to ascertain the growth of the magnetic field and then examine the hot electron transport in the phases during and after the laser pulse when the hot electrons enter a diffusive phase. Consider hot electrons emitted from a region of radius R (roughly the laser focal spot radius w_0) into a cone of angle θ_{ehot}. Magnetic collimation effects are important during the phase when the magnetic field bends the electrons by an angle θ_{ehot} over a distance of R/θ_{ehot} which can be written in terms of the Larmor radius, $r_{Lar}\theta_{ehot} < R/\theta_{ehot}$:

$$R/r_{Lar} > \theta_{ehot}^2. \tag{9.375}$$

This demands finding the magnetic field in a regime in which return currents are attempting to neutralize the hot electron current. These return currents depend strongly on the conductivity of the underlying cold plasma, which is itself increasing with time as that underlying plasma gets heated.

Let us start by estimating the magnetic field from Faraday's law:

$$\frac{\partial \mathbf{B}}{\partial t} = -\nabla \times \mathbf{E}. \tag{9.376}$$

Quasi-neutrality demands that $\mathbf{J}_{hot} \cong -\mathbf{J}_{ret}$ although they will not be exactly equal because of the finite conductivity of the plasma. This means that the electrostatic E-field set up is, from Ohm's law

$$E = J_{ret}/\sigma_C \cong J_{hot}/\sigma_C, \tag{9.377}$$

which means that Faraday's law gives

$$\frac{\partial \mathbf{B}}{\partial t} \cong \nabla \times \mathbf{J}_{hot}/\sigma_C. \tag{9.378}$$

Simple ohmic heating of the plasma by the return current is (since power dissipated is $I^2 R$):

$$3/2 n_e k_B \frac{\partial T}{\partial t} = J_{ret}^2/\sigma_C, \tag{9.379}$$

or, with the quasi-neutrality condition,

$$3/2 n_e k_B \frac{\partial T_{cold}}{\partial t} = J_{hot}^2/\sigma_C. \tag{9.380}$$

To include the variation of the conductivity with heating, we can use the usual Spitzer scaling for plasma conductivity (see Eq. (B.109) in Appendix B for a quantitative formula):

$$\sigma_C = \sigma_{C0} \left(\frac{T_{cold}}{T_0}\right)^{3/2}. \tag{9.381}$$

9.4 Hot Electron Production and Transport

Equation (9.378) is, for a hot electron beam radius of R,

$$\frac{\partial B}{\partial t} \cong \frac{J_{hot}}{\sigma_C R}, \tag{9.382}$$

and the temperature increases according to the equation

$$3/2 n_e k_B \frac{\partial T_c}{\partial t} = \frac{J_{hot}^2}{\sigma_{C0}} \frac{T_0^{3/2}}{T_c^{3/2}}. \tag{9.383}$$

Integration

$$\int_0^{T_c} T_c^{3/2} \, dT_c = \int_0^t \frac{2}{3} \frac{J_{hot}^2 T_0^{3/2}}{n_e k_B \sigma_{C0}} dt \tag{9.384}$$

yields

$$\frac{2}{5} T_c^{5/2} = \frac{2}{3} J_{hot}^2 \frac{T_0^{3/2}}{n_e k_B \sigma_{C0}} t. \tag{9.385}$$

Grouping all the constants, this with Eq. (9.381) means that the conductivity scales with hot electron current density and time as

$$\sigma_C = \kappa_c J_{hot}^{6/5} t^{3/5}. \tag{9.386}$$

We write Eq. (9.382) as

$$\frac{\partial B}{\partial t} \cong \frac{1}{\kappa_c R} J_{hot}^{-1/5} t^{-3/5}, \tag{9.387}$$

yielding the following scaling for magnetic field with hot electron current and time

$$B \cong \frac{5}{2} \frac{1}{\kappa_c R} J_{hot}^{-1/5} t^{2/5}. \tag{9.388}$$

(We will revisit this mechanism in Section 9.8 when we consider magnetic fields specifically and make more quantitative estimates.) A good estimate for the time scale of the B-field growth is the laser pulse duration. PIC simulations indicate that the effects of magnetic field collimation on the electron transport tend to last for about the laser-pulse duration and a few tens of femtoseconds afterward (Huang et al. 2019).

On longer time scales, space-charge effects dominate the hot electron transport. In fact, space-charge fields will almost always reduce the hot electron penetration into a target to much shorter distances than a collisional mean free path. To assess the effect of space-charge we resort to an intuitive 1D model by Bell et al. (1998). We consider space-charge effects in two phases: (1) during the laser pulse, when hot electron current is continuously produced by the laser at the target surface ($t < \tau_p$), and (2) the few picoseconds after the pulse ($t \geq \tau_p$).

In this 1D model, assume a spot of uniform intensity, with α_{he} hot electron conversion efficiency:

$$I_{abs} = \alpha_{he} I. \tag{9.389}$$

The total current density with hot electrons and return current is almost zero (but not exactly because of resistivity effects)

$$J_{tot} = J_{hot} + J_{therm} \approx 0. \tag{9.390}$$

The continuity equation for the hot electron density is

$$\frac{\partial n_{eh}}{\partial t} = \nabla \cdot \frac{\mathbf{J}_{hot}}{e}. \tag{9.391}$$

As above, we define the conductivity of the thermal plasma and the hot electron-induced space-charge field

$$\mathbf{J}_{therm} = \sigma_C \mathbf{E}_{eh}, \tag{9.392}$$

which lets us write

$$\frac{\partial n_{eh}}{\partial t} = \nabla \cdot \frac{\sigma_C}{e} \mathbf{E}_{eh}$$

$$= \frac{\partial}{\partial z} \frac{\sigma_C}{e} E_{eh}. \tag{9.393}$$

The potential formed by this space-charge is

$$E_{eh} = -\frac{\partial}{\partial z} \Phi_{eh}, \tag{9.394}$$

so we assume that the hot electron density then responds to this potential as

$$n_{eh} = n_{0h} e^{e\Phi_{eh}/k_B T_{eh}}. \tag{9.395}$$

If, indeed, the hot electrons can be described by a quasi-Maxwellian of temperature T_{eh}, the space-charge field with Eqs. (9.394) and (9.395) can be written

$$E_{eh} = -\frac{\partial}{\partial z}\left(\frac{k_B T_{eh}}{e} \ln\left[\frac{n_{eh}}{n_{0h}}\right]\right), \tag{9.396}$$

$$E_{eh} = -\frac{k_B T_{eh}}{e n_{eh}} \frac{\partial n_{eh}}{\partial z}, \tag{9.397}$$

which, with Eq. (9.393) is

$$\frac{\partial n_{eh}}{\partial t} = \frac{\partial}{\partial z}\left[\frac{\sigma_C k_B T_{eh}}{e^2 n_{eh}} \frac{\partial n_{eh}}{\partial z}\right]. \tag{9.398}$$

We see that this is a diffusion equation for the hot electron density with a diffusion constant $\sim 1/n_{eh}$. Now we consider this equation in the two time regimes:

(1) During the laser pulse: Transport phases (4) to (5).

Assume the conductivity is constant in time and position (which we know is a rather crude approximation because, as we discussed earlier in the context of magnetic field growth, heating changes conductivity with time and focal spot sizes gives it spatial variation, but this nonetheless yields an intuitive picture for the short period during the laser pulse). Also assume T_{eh} is constant. The solution to this equation in terms of the initial hot electron density n_0, pulse duration, τ_p, and some effective hot electron penetration depth (yet to be found) z_p, is

$$n_{eh} = n_0 \left(\frac{t}{\tau_p}\right) \frac{z_p^2}{(z+z_p)^2}, \qquad (9.399)$$

Which, with Eq. (9.398), we write

$$\frac{n_0}{\tau_p} \frac{z_p^2}{(z+z_p)^2} = \frac{\sigma_C k_B T_{eh}}{e^2} \frac{2}{(z+z_p)^2}. \qquad (9.400)$$

To find the constant z_p we write in terms of the total number of hot electrons produced per unit area N_{eh},

$$\int_0^\infty n_{eh}(\tau_p) dz = N_{eh} = n_0 z_p. \qquad (9.401)$$

This areal density of hot electrons is just found from the absorbed intensity and conservation of energy:

$$I_{abs} \tau_p = N_{eh} \frac{3}{2} k_B T_{eh}. \qquad (9.402)$$

From Eq. (9.400) we can say at the surface of the target ($z = 0$)

$$z_p^2 \frac{n_0}{\tau_p} = \frac{2\sigma_c k_B T_{eh}}{e^2}. \qquad (9.403)$$

Algebraically solving these equations for the constants in Eq. (9.399) gives us

$$n_0 = \frac{2e^2 I_{abs}^2 \tau_p}{9(k_B T_{eh})^2 \sigma_c} \qquad (9.404)$$

and the penetration depth of the hot electrons *during the laser pulse*:

$$z_p = \frac{3\sigma_c (k_B T_{eh})^2}{e^2 I_{abs}}. \qquad (9.405)$$

What this tells us about space-charge mediated electron transport during a laser pulse is that the depth of hot electron penetration into the target increases as the square of the hot electron temperature. It also depends on the plasma conductivity (i.e. hotter plasma and/or lower-Z plasmas allow greater penetration and inversely with the intensity of the hot electron flux (a result from the fact that more hot electrons require more return current to neutralize them).

After the laser pulse: Later times in transport phase (5)

We cannot assume that the hot electron temperature remains constant after the laser has turned off, so we assume that $T_{eh}(t)$ decreases by expansion of the space-charge confined hot electron cloud. We consider a hot electron penetration length $L_h(t)$. Addressing the issue of the hot electron cloud cooling by expansion (a problem we have addressed a number of times before such as with clusters in Section 7.4.1), we write

$$\frac{dT_{eh}(t)}{T_{eh}(t)} = -\frac{2dz}{3z}, \tag{9.406}$$

which integrates to yield temperature in terms of the initial penetration depth during the pulse, the time-dependent penetration length and the hot electron temperature that emerged from the laser T_{eh0}:

$$T_{eh}(t) = T_{eh0} \frac{z_p^{2/3}}{L_h(t)^{2/5}}. \tag{9.407}$$

So we now seek a self-similar expanding hot electron solution to Eq. (9.398). Such a solution exists; it is of the form

$$n_{eh}(t,z) = \frac{2n_0 z_p}{\pi} \frac{L_h(t)}{z^2 + L_h(t)^2}, \tag{9.408}$$

which conserves the total number of electrons such that

$$\int_0^\infty n_{eh}\, dz = N_{eh} = n_0 z_p. \tag{9.409}$$

Inserting into Eq. (9.398) and imposing the constraint $L(\tau_p) = z_p$ yields finally the hot electron diffusive penetration depth versus time of

$$L(t) = z_p \left[\frac{5\pi}{6\tau_p}(t - \tau_p) + 1 \right]^{3/5}. \tag{9.410}$$

Figure 9.44 plots this equation for two times after the laser pulse, one full pulse duration and three times the pulse duration, illustrating the diffusion of the cloud. To put in a few numbers to get a sense of the expected hot electron transport depth from this model: with a

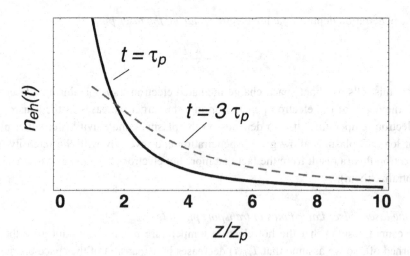

Figure 9.44 Plot of space-charge limited hot electron depth penetration at two times.

500 fs laser pulse, using the hot electron temperature from the Beg formula at 10^{19} W/cm^2, a thermal temperature of 100 eV and a $\bar{Z} = 10$, the initial penetration depth z_p would be 3 μm. So in experiments, from the standpoint of hot electron transport, "thick" targets are those greater than say 10 μm, while "thin" targets, those in which hot electrons penetrate through the far side, would be thinner than a few μm. This penetration depth will be important when we consider ion acceleration by the target normal sheath acceleration mechanism.

9.4.3 Coherent Transition Radiation from Hot Electrons

If the hot electrons penetrate to the back side of the target, they can induce a transient polarization on this back surface which will emit radiation, so-called transition radiation (Bae and Cho 2015). A schematic of this happening is shown in Fig. 9.45. If we recall the mechanisms for hot electron production like vacuum heating or $\mathbf{J} \times \mathbf{B}$ heating, we realize that the hot electrons will be emitted in bunches. For example, if produced by vacuum heating (or resonance absorption) the bunches are emitted every laser cycle. If they are produced by the $\mathbf{J} \times \mathbf{B}$ mechanism, they are launched every half-cycle. These bunches will spread out longitudinally as they propagate, because there is an energy spread in the electrons and their velocities are slightly less than c. However, if the target is sufficiently thin that there are still distinct bunches at the rear surface, the periodicity of the induced polarization has an interesting consequence. The transition radiation will be produced periodically (coherently) and the emission will be nearly monochromatic with wavelength

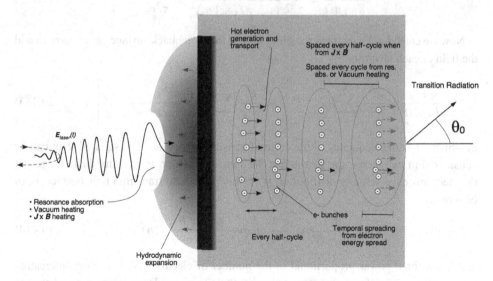

Figure 9.45 Illustration of how hot electron bunching leads to coherent light emission at the rear surface. The electron bunches spread longitudinally as they propagate because of their finite energy spread.

determined by the period of the bunches. This coherent transition radiation (CTR) is an important observable in experiments in diagnosing hot electron generation and transport (Cho et al. 2009).

The electromagnetic theory of CTR is complex so we will give merely the outlines of how it is calculated; in this we follow the treatment of Bae and Cho (2015) and conceptually consider the electron transport in a 1D sense. To start, if we consider the transition radiation of a single hot electron traveling normal toward the back surface of the conducting target, the electron can be thought of as having a mirror charge traveling toward the surface from the other side. This means that the radiation is a textbook dipole radiation calculation.

Treat this as two stationary charges at $t = 0$ and start moving at velocity v. The radiated spectrum (with transition radiation frequency ω_T) as a function of angle, from this well-known EM result (Jackson 1975, p. 670), is summed over the electron and its mirror charge

$$\frac{d^2\varepsilon_T}{d\omega_T d\Omega} = \frac{1}{4\pi^2 c^3} \left[\sum_i q_i \frac{\mathbf{v}_i \times \hat{\mathbf{n}}}{1 - \hat{\mathbf{n}} \cdot \boldsymbol{\beta}_i} \right]^2$$

$$= \frac{1}{4\pi^2 c^3} \left[-e\frac{v\hat{\mathbf{z}} \times \hat{\mathbf{n}}}{1 - \beta\hat{\mathbf{n}} \cdot \hat{\mathbf{z}}} + e\frac{(-v\hat{\mathbf{z}}) \times \hat{\mathbf{n}}}{1 + \beta\hat{\mathbf{n}} \cdot \hat{\mathbf{z}}} \right]^2, \quad (9.411)$$

where $\hat{\mathbf{z}}$ is the direction of propagation and $\hat{\mathbf{n}}$ is the direction to the observer, who is at angle θ_0 with respect to the normal of the back surface. Doing the geometry yields

$$\frac{d^2\varepsilon_T}{d\omega_T d\Omega} = \frac{e^2}{\pi^2 c} \frac{\beta^2 \sin^2 \theta_0}{\left(1 - \beta^2 \cos^2 \theta_0\right)^2} \equiv \frac{e^2}{\pi^2 c} A_j^2. \quad (9.412)$$

Now we consider N electrons traveling toward the target back surface, so we have to add the field of each electron:

$$\frac{d^2\varepsilon_{tot}}{d\omega_T d\Omega} = \frac{e^2}{\pi^2 c} \left| \sum_{j=1}^N A_j e^{-i\omega_T t_j} \right|^2 = \frac{e^2}{\pi^2 c} \left| \sum_{j=1}^N \frac{\beta_j \sin \theta_0}{1 - \beta_j^2 \cos^2 \theta_0} e^{-i\omega_T t_j} \right|^2, \quad (9.413)$$

in which ω_T is the frequency of the emitted transition radiation. The squared term can be considered in two terms, the first is just the sum of radiation from individual charges. If the electrons have velocity distribution $f(\beta)$, this incoherent transition radiation term can be written

$$\frac{d^2\varepsilon_I}{d\omega_T d\Omega} = \frac{e^2 N}{\pi^2 c} \int dv \frac{\beta^2 \sin^2 \theta}{1 - \beta^2 \cos^2 \theta} f(\beta) \quad (9.414)$$

and has emitted power proportional to the number of electrons N. The more interesting term is that arising from the cross terms in (9.413). Considering cross terms between two electrons, the ith and jth electrons, which are emitted at time separation $\tau = t_j - t_i$, and energy and time separation distribution $f(\beta, \tau) = f(\beta)g(\tau)$ (assuming no correlation between energy and separation), we have for this term

9.4 Hot Electron Production and Transport

$$\frac{d^2\varepsilon_C}{d\omega_T d\Omega} = \frac{e^2}{\pi^2 c} \sum_{i \neq j} A_i A_j^* e^{i\omega_T(t_i - t_j)} \tag{9.415}$$

$$= \frac{e^2 N(N-1)}{\pi^2 c} \left| \int d\tau dv \frac{\beta^2 \sin^2\theta}{1 - \beta^2 \cos^2\theta} f(\beta, \tau) e^{i\omega_T \tau} \right|^2, \tag{9.416}$$

which can be rewritten in terms of the incoherent term as

$$\frac{d^2\varepsilon_C}{d\omega_T d\Omega} = \frac{d^2\varepsilon_I}{d\omega_T d\Omega}(N-1) \left| \int d\tau\, g(\tau) e^{i\omega_T \tau} \right|^2. \tag{9.417}$$

The coherent, CTR emission is $N-1$ times greater than the incoherent emission and usually dominates the experimentally observed transition radiation.

These equations can be used to calculate numerically the CTR angular emission and spectrum. A particularly interesting consequence of this analysis is that since the electrons arriving are bunched, their emission is peaked at well-defined frequencies. This is illustrated in Figure 9.46 – adapted from Bae and Cho (2015) – which shows the electron current as a function of time. The energy smearing of the bunches has occurred. The same

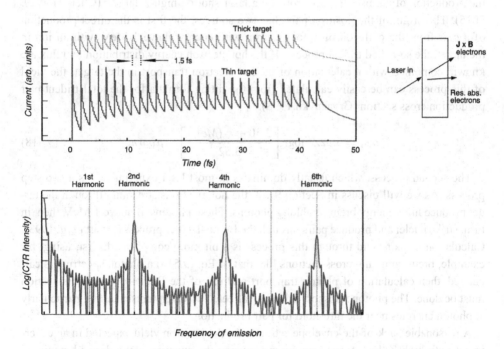

Figure 9.46 Illustration of the electron current at the rear target surface that a typical experiment might generate (upper panel) and the sort of CTR spectrum this generates is shown (bottom panel). The energy spread smearing of the electron bunches when they must propagate through a thick target is illustrated by the dotted line in the top panel. Plots were generated based on (Bae and Cho 2015, figure 3).

current one might expect for a thicker target is also shown in gray; here the longer propagation length has further smeared the bunches at the back surface, so the periodic structure is not as pronounced. Nonetheless, if these bunches arrive every half-cycle of the laser, as they would if produced by $\mathbf{J} \times \mathbf{B}$ heating, the CTR calculated from this periodic bunching leads to a peak in the emission at 2ω of the laser wavelength (with higher harmonics visible with reduced intensity). On the other hand, if they arrive every cycle, the CTR emission is predominantly at the laser wavelength. This aspect of the CTR can be used to differentiate hot electron production mechanisms by observing the color of the CTR. In fact, as might be expected, at oblique incidence, experiments show that 2ω CTR is emitted from a spot on the back side of the target consistent with $\mathbf{J} \times \mathbf{B}$ electron emitted in the laser propagation direction, while a spot of CTR is observed at the laser wavelength in a position consistent with electrons traveling normal to the target surface, as would indeed be expected from vacuum heating.

9.4.4 Positron Production by Hot Electrons

Another interesting byproduct of hot electron production is the generation of positrons in the solid target (Liang et al. 1998). Experiments with kJ-class PW lasers have observed the production of almost 10^{12} positrons on a laser shot on high-Z targets (Chen, H. et al. 2015). The origin of the positron pairs has two sources, the first is the direct production of a pair from the collision of a hot electron with energy above 1 MeV with an ion in the target, the so-called trident process. If the hot electron energy distribution and flux are known, combined with a calculation of the hot electron transport in the target, the yield of this process can be easily calculated from the fitting formula for the total trident pair production cross section (Gryaznykh 1998):

$$\sigma_{Tri} = 5.22 Z^2 \log^3 \left[\frac{2.30 + \varepsilon_e(MeV)}{3.52} \right] \, \mu barn. \quad (9.418)$$

The second process, which usually dominates in most laser experiments, is a two-step process. As we will discuss in Section 9.7.3, the hot electrons streaming through the target produce high-energy bremsstrahlung photons. These photons, if above 1 MeV, then, in turn, strike nuclei and produce pairs through the Bethe–Heitler process (Myatt et al. 1993). Calculation of pair yield through this process is a bit more complicated; first, using, for example, bremsstrahlung cross sections like that of Eq. (9.515) a photon spectrum is calculated, then calculation of photon transport and use of the Bethe–Heitler cross section must be done. The photon transport is heavily affected by Compton scattering, particularly at photon energies near the threshold for pair production.

A reasonable back-of-the-envelope estimate for the positron yield expected in an experiment with high-Z thick targets can be made using the experimentally derived empirical formula of Chen, H. et al. (2015)). They found for pulses around 1 ps in duration the positron yield scaled with laser energy as $Y_{e+} \cong 4 \times 10^5 \varepsilon_{laser}^{2.3} [J]$.

9.5 High-Harmonic Generation from Solid-Density Plasma

Another unique effect in strong field interactions with solids has to do with the nonlinearity of the plasma oscillations at the plasma surface and how these can lead to the emission of high order harmonics of the laser frequency, just as high harmonics are generated in gas from the nonlinear oscillations of the atomic dipole. High harmonics from solid targets were observed very early on with CO_2 lasers in the 1970s; even at modest intensity, high orders were observed because the CO_2 laser is at 10 μm and the emission was still just in the optical and near-UV (Carman et al. 1980). High-intensity lasers now produce harmonics from solid targets to very high orders well into the soft X-ray region. Unlike gas-phase high-order harmonic generation (HHG), HHG from solids is not a phase-matched process. So there is no coherent addition of radiation over an extended medium. However, because of the high density and the coherent motion of the electron fluid at the plasma surface, HHG from solid targets is shown to be at least as efficient as HHG in gases (Teubner and Gibbon 2009).

9.5.1 Basic Mechanisms

Harmonic generation from a solid surface is not surprising, and, for example, second harmonic light is easily seen from a gold surface with extremely modest intensity. High harmonic generation (HHG) from solids in the strong field regime, however, does take on some unique characteristics. The general wisdom is that there are two main mechanisms for this HHG (Thaury and Quere 2010).

a) Coherent Wake Emission

At modest intensity ($a_0 \ll 1$), when a plasma density gradient exists, this is thought to be the most important mechanism. It results from the excitation of plasma waves in the overdense region of the plasma gradient. The laser pulls electrons in and out of the plasma when irradiated at oblique incidence. These ultrashort bunches spaced at the laser frequency excite plasma waves at harmonics of this bunching frequency. Different local plasma frequencies in the plasma gradient are excited by these bunches and radiate. This mechanism must have a maximum harmonic frequency given by the maximum plasma frequency. So, the highest harmonic order that is likely to be produced is something like $q_{MAX} \cong (n_e/n_{crit})^{1/2}$ which implies that perhaps orders up to $q \sim 15$ will be produced by this mechanism.

b) Oscillating Mirror Model

At higher intensity ($a_0 > 1$), a more interesting process takes place. When the laser pulse is incident on a sharp plasma interface, or the ponderomotive pressure steepens the density gradient, the oscillatory forces of the laser, whether they be the electric field at oblique incidence with P-polarization or the $\mathbf{J} \times \mathbf{B}$ force at near-normal incidence, exceed the plasma pressure and drive the electron fluid in and out about the (stationary) ion front. This acts as a mirror that moves in and out, which Doppler-shifts the reflected wave. This oscillating mirror induces nonlinearities on the reflected wave and the production of high harmonics. Unlike the coherent wake emission mechanism, there is, in principle, no upper limit to the harmonic order, and, as a consequence, experiments have observed quite high harmonics (Teubner and Gibbon 2009). This mechanism also has the interesting consequence of producing attosecond bursts from the emission of light during the large-amplitude motions of the electron fluid mirror in the direction of reflection (Tarasevitch and von der Linde 2009). This is the mechanism we will explore further in this section.

9.5.2 Doppler Shift from a Receding/Advancing Mirror

Before considering HHG from the solid target, let us first review the Doppler shift expected on light reflected from a mirror that is advancing toward or receding away from the laser pulse at relativistic velocity. Consider a mirror moving with velocity

$$\beta_M = v_M/c \tag{9.419}$$

with relativistic factor $\gamma_M = 1/(1 - \beta_M^2)^{1/2}$. In the frame of the mirror the laser frequency is $\omega_1 = \omega(1 + \beta_M)$. The reflected wave in the lab frame has frequency

$$\omega_R = \frac{\omega_1}{1 - \beta_M}, \tag{9.420}$$

which means that the total Doppler shift of the reflected laser is

$$\omega_R = \frac{1 + \beta_M}{1 - \beta_M}\omega = (1 + \beta_M)^2 \gamma_M^2 \omega, \tag{9.421}$$

which leads to the simple formula as the mirror velocity approaches c,

$$\omega_R = 4\gamma_M^2 \omega \quad v_M \to c. \tag{9.422}$$

9.5.3 The Oscillating Mirror Model

Armed with this Doppler shift, we are in a position to see how high harmonics are generated at the plasma surface. Consider normal incidence. The situation like that pictured in Fig. 9.47 arises (Macchi 2013, pp. 110–114; Gibbon 2005, pp. 200–212). The $\mathbf{J} \times \mathbf{B}$ force

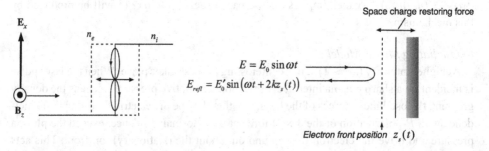

Figure 9.47 Illustration of the motion of a sharp plasma surface subject to normal irradiation by an intense laser field.

9.5 High-Harmonic Generation from Solid-Density Plasma

drives longitudinal oscillations of the electron–vacuum interface (the heavy ions remain stationary), and the reflected wave has time-varying reflection position.

To see how this affects the reflected laser light, assume purely harmonic motion for the electron fluid surface position z_s, so that

$$z_s = z_0 \sin(\omega t' + \varphi_s) \tag{9.423}$$

where φ_s is the phase shift of the electron fluid with respect to the laser field. (In reality the reflection position motion is not harmonic nor is it even symmetric but that is of no consequence in this simple model. PIC simulations are needed to find the real motion of the reflection surface (Lichters *et al.* 1996).) At high plasma densities (many times n_{crit}) the space-charge restoring force is strong and the plasma is relatively "stiff," which means that $\varphi_s \approx 0$. (For a lower-density plasma, at resonance when $\omega_{pe} \cong 2\omega$, then $\varphi_s \approx \pi/2$.) Taking $\varphi_s = 0$ for simplicity, the Doppler-shifted electric field takes the form

$$E_{ref} = E_0 \sin(\omega t + 2kz_0 \sin 2\omega t'), \tag{9.424}$$

where the retarded time is

$$t' = t + z_s(t')/c. \tag{9.425}$$

For small-amplitude oscillations $t' \approx t$, so we can write for the reflected wave

$$E_{ref} \sim E_0 \exp\left[i\omega t + i2k \sin 2\omega t\right]. \tag{9.426}$$

As we have done a number of times previously, we can expand a function of this form using the Jacobi–Anger identity

$$e^{ib \sin \alpha} = \sum_{s=-\infty}^{\infty} J_s(b) e^{is\alpha}, \tag{9.427}$$

yielding for the reflected field,

$$E_{ref} \sim E_0 \sum_{s=-\infty}^{\infty} J_s(2k) \exp\left[i(2s+1)\omega t\right]. \tag{9.428}$$

The high-order harmonic content of the reflected field is immediately apparent, resulting from the oscillatory motion of the critical density surface. We see in this case the reflected radiation has components at $(2s+1)\omega$, namely only odd harmonics are produced.

The physical origin of these high harmonics can be seen from the plots in Fig. 9.48 (Baeva *et al.* 2006). The parallel electron momentum is plotted above plots for the surface velocity and relativistic gamma factor:

$$v_s = dz_s/dt \quad \gamma_s = 1/(1 - v_s^2/c^2)^{1/2}. \tag{9.429}$$

The periodic large Doppler shift that arises and which produces the short-wavelength high harmonics, has strong peaks in time, a structure that is largely independent of the subtle shape of the surface velocity. Consequently the high frequency high harmonics come out in periodic attosecond bursts at the peak of the Doppler shift.

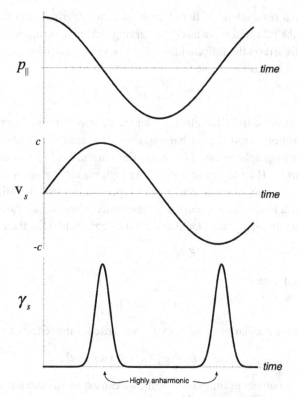

Figure 9.48 Schematic plots of the cyclical evolution of the surface velocity and relativistic factor. Plots were generated based on plots of (Baeva et al. 2006, figure 2).

Quantifying the spectrum in this simple model reduces to finding the amplitude of the electron "mirror," z_0. We know that the maximum velocity amplitude is

$$v_s = 2\omega z_0 \sin 2\omega t' < c \tag{9.430}$$

so

$$z_0^{(MAX)} \cong \frac{c}{2\omega} = \frac{\lambda}{4\pi}. \tag{9.431}$$

A slightly better estimate can be made based on the analysis of Tsakiris et al. (2006), where the surface velocity can be expressed as

$$\beta_s^{MAX} = (1 - \gamma_s^{-2})^{1/2} \quad \gamma_s = (1 + a_0^2)^{1/2} \tag{9.432}$$

so that the mirror amplitude is

$$z_0^{(MAX)} \cong \frac{\lambda}{4\pi} \frac{a_0}{\sqrt{1 + a_0^2}} \tag{9.433}$$

which is essentially Eq. (9.431) at intensity above $\sim 10^{19}$ W/cm^2. Furthermore, we have that

9.5 High-Harmonic Generation from Solid-Density Plasma

$$z_s(t) = z_0 \sin(2\omega t'), \qquad (9.434)$$

$$t' = t + \frac{z_0}{c} \sin(2\omega t'), \qquad (9.435)$$

which is an implicit equation for t', allowing one to solve for the reflected electric field numerically with

$$E_{ref} = E_0 \sin(\omega t + 2kz_0 \sin 2\omega t'). \qquad (9.436)$$

The reflected electric field found from these equations is illustrated in Figure 9.49. The sawtooth motion of the mirror surface leads to anharmonic reflected electric field and a

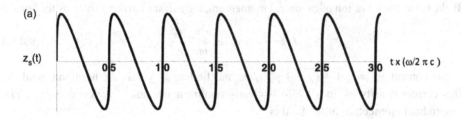

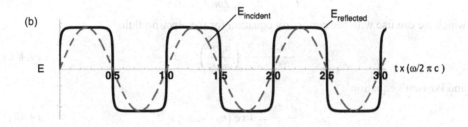

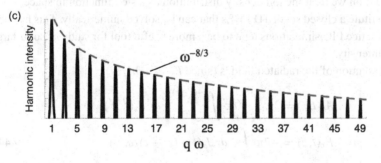

Figure 9.49 Numerical solution of Eqs. (9.434)–(9.436) illustrating the nonlinear motion of the relativistic plasma-moving mirror. The plot in (a) is of the nonlinear motion of the mirror position, z_s. The plot in (b) is of the incident and Doppler-shifted reflected laser electric fields. The Fourier transform and high harmonic spectrum that results from the reflected anharmonic electric field is shown in the plot in (c).

comb of harmonics to high order. The fall of harmonics as $\omega_R^{-8/3}$ is illustrated, a scaling that does a reasonable job of predicting experimental observations.

9.5.4 Nonlinear Fluid Formulation of HHG from Solids

A more formal solution is possible by considering the nonlinear motion of the plasma fluid. Again for simplicity consider normally incident laser light. We need to find the surface currents to calculate the emitted HHG radiation. In the 1D model of normal incidence, we consider the transverse laser vector potential A_x, with Ampere's law:

$$\frac{1}{c^2}\frac{\partial^2 A_x}{\partial t^2} - \nabla^2 A_x = \frac{4\pi}{c}J_x. \tag{9.437}$$

By the usual conservation of canonical momentum, the surface current driven by the laser is

$$J_x = -en_e v_x = \frac{e^2 n_e}{mc}\frac{A_x}{\gamma_e}. \tag{9.438}$$

In this context $\gamma_e = \sqrt{1 + (p_x^2 + p_z^2)/mc}$ and finding J_x yields the harmonic field A_x. This current density in Eq. (9.438) is clearly nonlinear because it scales as n_e/γ_e. The longitudinal momentum of the fluid is

$$\frac{\partial p_z}{\partial t} = e\frac{\partial \Phi}{\partial z} - \frac{e^2}{2mc^2 \gamma_e}\frac{\partial^2 A_x}{\partial x^2}, \tag{9.439}$$

which we can use with the continuity equation for the electron fluid

$$\frac{\partial n_e}{\partial t} + \frac{\partial}{\partial z}\left(\frac{n_e p_z}{m\gamma_e}\right) = 0 \tag{9.440}$$

and Poisson's equation

$$\frac{\partial^2 \Phi}{\partial z^2} = 4\pi e\left(n_e - \bar{Z}n_i\right). \tag{9.441}$$

In this last equation we treat the ion density distribution as a step function in space. These equations constitute a closed set of 1D PDEs that can be solved numerically. This has been done, but in practice PIC simulations tend to be a more useful tool for study of solid-target HHG at high intensity.

The formal solution of the radiated field is (since $E_x = -\partial A/\partial t$)

$$A_x(t,z) = 2\pi \int_z^\infty dz' \int_{-\infty}^{t-(z'-z)/c} dt' \, J_x(t',z')$$

$$E_x(t,z) = -2\pi \int_z^\infty dz' J_x(t - (z'-z)/c, z'). \tag{9.442}$$

9.5.5 Oblique Incidence and the Moving Mirror Model

This formalism, developed at normal incidence, can be used for oblique incidence as well by performing a simple transform, usually known as the moving-mirror model. The

9.5 High-Harmonic Generation from Solid-Density Plasma

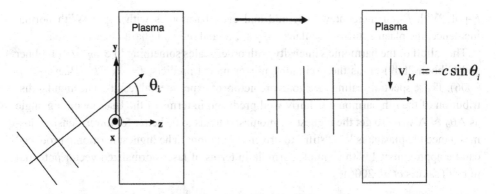

Figure 9.50 Geometry of oblique incidence and the transform to the M frame.

geometric idea is shown in Figure 9.50. We transform into what we will call the M frame, which is moving parallel to the target surface at velocity

$$\beta_M = -\sin\theta_i, \tag{9.443}$$

which yields the relativistic factor in this frame of

$$\gamma_M = \frac{1}{\cos\theta_i}. \tag{9.444}$$

In the lab frame, the transformed laser frequency is

$$\omega_M = \omega \cos\theta_i, \tag{9.445}$$

so we can insert this frequency in for the wavenumber in the M frame and lab (L) frames:

$$k_M = (0, 0, \omega \cos\theta_i/c) \tag{9.446}$$

$$k_L = (0, \omega \sin\theta_i/c, \omega \cos\theta_i/c). \tag{9.447}$$

This leads, in the M frame, to radiation at normal incidence with plasma densities

$$n_e^{(M)} = \gamma_M n_e = \frac{n_e}{\cos\theta_i} \tag{9.448}$$

$$n_i^{(M)} = \frac{n_i}{\cos\theta_i}. \tag{9.449}$$

This transform allows use of any models developed at normal incidence to be employed for oblique incidence.

9.5.6 HHG Orders, Polarization Selection Rules and Scaling

A few comments about the scaling and polarization of the reflected harmonics: at oblique incidence plasma oscillations at the first harmonic are set up, so even harmonics are possible. More specifically, with S-pol incident, the 1ω component leads to odd harmonics radiated with S-pol and, because the 2ω component is along P-pol, even harmonics with

P-pol. With P-pol incident we get odd and even harmonics with P-pol. With normal incidence, the plasma mirror oscillates only at 2ω and only odd harmonics are radiated.

The falloff of the harmonic's intensity with order scales something like $\omega_R^{-5/2}$ (Teubner and Gibbon 2009) or, as the oscillating mirror model predicts, $I \sim \omega_R^{-8/3}$ (Baeva et al. 2006). These spectral scalings are good predictors of experimental results. The angular distribution of the qth harmonic is fairly well predicted in terms of the laser incoming angle as $\Delta\theta_q \approx \Delta\theta_0/q$. To get the highest harmonics, it tends to be better for lower-density plasmas, since the plasma is less "stiff" to driven oscillations. The highest harmonic produced can be approximated with a simple formula in terms of laser-normalized vector potential to be (Tsakiris et al. 2006)

$$\omega_{MAX} \approx 4\gamma_{MAX}^2 \omega \cong 4a_0^2 \omega, \qquad (9.450)$$

which implies that intensities of 10^{19} W/cm^2 will yield up to $q \sim 40$.

9.6 Ion Acceleration

One of the most intensely studied phenomena in high-intensity laser–plasma physics is the acceleration of energetic ions from laser irradiation of solid foils (Snavely et al. 2000; Cowan et al. 2000b; Henig et al. 2009; Roth and Schollmeier 2013). It turns out that, depending on the intensity, target thickness, pulse contrast and many other factors, there are various mechanisms that lead to fast ion ejection (Daido et al. 2012). Typically ions with many 10s of keV up to $\sim$100 MeV are produced. We will examine a couple of the most common mechanisms for ion acceleration though we will not present an exhaustive survey.

9.6.1 Energetic Ions Produced by Hydrodynamic Expansion

We have already encountered and discussed one of the reasons that fast ions are produced. This involves ejection from the front side of the surface and results without particularly high-intensity irradiation. The production of a high-temperature thermal plasma at the target surface by the laser produces an electron pressure which drives a fast hydrodynamic expansion, the result being ejection from the front target surface of ions with tens of keV energy. Recalling our discussion in Section 9.1.3 and Eq. (9.48), we know that the expansion of the plasma from the front surface produces an ion distribution in space given by

$$n_i(z,t) = n_{i0} \exp\left[\frac{z + c_s t}{c_s t}\right], \qquad (9.451)$$

where the (isothermal) sound speed is the usual form,

$$c_{si} = \sqrt{\bar{Z} k_B T_e / m_i}, \qquad (9.452)$$

and the ion velocity distribution in space and time is

$$u_i = c_{si} + z/t. \qquad (9.453)$$

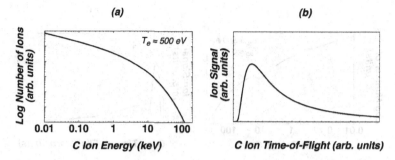

Figure 9.51 (a) Ion spectrum of $Z = 6+$ carbon ions resulting from the hydrodynamic expansion of a solid density 500 eV carbon plasmas. (b) The shape of the ion time-of-flight spectrum that one would expect to see in an experiment with such an exploding carbon plasma.

It naturally follows that the ion energy distribution from the front surface hydrodynamic expansion is simply

$$n_i(\varepsilon_i) = n_{i0} \exp\left[-\sqrt{\frac{\varepsilon_i}{\bar{Z}k_B T_e}}\right]. \quad (9.454)$$

As an example, the ion distribution from a fully ionized 500 eV carbon target is shown in Figure 9.51(a) and the resulting time-of-flight ion distribution that would be observed in an experiment is illustrated in Figure 9.51 (b). One sees that ions with energy a few times $\bar{Z}k_B T$ are produced, meaning that even with modest electron temperatures of a few hundred electron volts, a small fraction of ions with ~100 keV will be ejected. The number of ions is limited by the finite depth of the heated layer, with higher energy pulses producing more ions.

As an aside, Eq. (9.454) predicts some small number of ions with energy up to infinity. This is a mathematical consequence of the quasi-neutrality assumption that led to this result. We will need to address this shortcoming when we look at more energetic ion production from the back side of the target by target normal sheath acceleration.

9.6.2 Ions from Two-Temperature Hydrodynamic Expansion

As we have seen, the view of the laser-plasma as a single-electron thermal temperature is too simple, given the various absorption mechanisms at work. Quite commonly it is appropriate to view the front surface as having a two-temperature electron thermal energy distribution, a fact that has an effect on the observed fast ion energy distribution from the target surface. Recalling our discussion of a two-temperature ion hydro-expansion in Section 9.1.4, we can find the ion velocity distribution from

$$\frac{dn_i}{du_i} = \frac{\partial n_i}{\partial z}\frac{\partial z}{\partial \tilde{z}_0}\frac{\partial \tilde{z}_0}{\partial u_i} \quad (9.455)$$

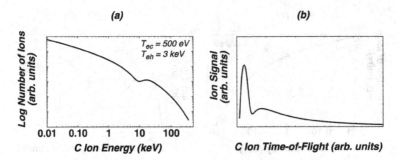

Figure 9.52 (a) The 6+ carbon ion spectrum that now results from a two-temperature electron distribution in which 75% of the electrons are at a temperature of 500 eV and there is a minor 25% electron component at 3 keV. (b) The two-peaked ion time-of-flight signal that is observed in experiments that produce such a two-electron-temperature plasma.

with the sound speed given by Eq. (9.57) to yield

$$\frac{\partial \tilde{z}_0}{\partial u_i} = \frac{\partial}{\partial u_i}(u_i - c_{sz}) = 1 - \frac{\partial c_{sz}}{\partial \Phi}\frac{\partial \Phi}{\partial \tilde{z}_0}\frac{\partial \tilde{z}_0}{\partial u_i}$$

$$= 1 + \frac{\partial c_{sz}}{\partial \Phi}\frac{\partial \Phi}{\partial \tilde{z}_0}\frac{c_{sz}m_i}{Ze}. \qquad (9.456)$$

This gives an ion velocity distribution which can be combined with Eq. (9.67) to derive an energy distribution

$$\frac{dn_i}{du_i} = \text{const.}\left[1 - \frac{1}{2}\frac{n_{eh}n_{ec}}{(n_{eh}/T_{eh} + n_{ec}/T_{ec})^2}\left(\frac{1}{T_{ec}} - \frac{1}{T_{eh}}\right)^2\right] \qquad (9.457)$$

The interesting upshot is that the ions observed in this case have a double-peaked energy distribution, well documented in experiments (Meyerhofer et al. 1993) as shown by the calculation in Figure 9.52.

9.6.3 Target Normal Sheath Acceleration

A more spectacular ion acceleration, which is akin to the acceleration in a hydrodynamic expansion, is one that instead arises from the back surface of the target. In this process, as in hydro-expansion, a sheath field is set up by mobile electrons, though in this case the field is set up by energetic, hot suprathermal electrons that have transited a thin target. Because of the much greater effective temperature of hot electrons, the sheath fields can be enormous, and ions with up to ~100 MeV result (Snavely et al. 2000). This process is known as Target Normal Sheath Acceleration or TNSA (Wilks et al. 2001). The ions accelerated are predominantly the light ones, and since, in most experiments, no matter the composition of the target, a thin layer of hydrocarbons (originating from oil impurities) rests on the target surface donating hydrogen ions to the acceleration, in most experiments, unless some technique is employed to remove actively this hydrocarbon layer, the accelerated ions observed in TNSA are overwhelmingly protons (Borghesi et al. 2006; Roth and Schollmeier 2013).

9.6 Ion Acceleration

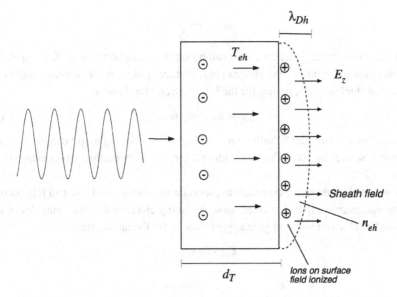

Figure 9.53 Schematic illustration of how a sheath field develops in a solid-target experiment leading to TNSA acceleration.

A schematic of how this TNSA process occurs is illustrated in Figure 9.53. If the target is thin (say less than a few μm) then the hot electrons generated at the surface, by the various mechanisms we have discussed (resonance absorption, vacuum heating, SRS, etc.) will transit the target with small energy losses. When they emerge, they set up a space-charge field, forming a Debye sheath of extent given by a hot-electron-temperature Debye length (λ_{Dh}). This field first field-ionizes ions at the back surface and then accelerates a thin layer of these ions. Their energy will be determined by the size of the sheath field which is, in turn, determined by the hot electron density and temperature.

It is straightforward to estimate this sheath by first estimating the hot electron density at the rear target surface. We employ an empirical scaling formula for hot electron conversion efficiency from laser energy deduced from experiments by Fuchs (2006)

$$\eta_{eh} = 1.2 \times 10^{-15} \left(I[W/cm^2]\right)^{0.74}. \tag{9.458}$$

Generally, it is believed this efficiency maxes out at around 50% at intensity ~6 × 10^{19} W/cm². We desire an estimate of the hot electron cloud volume at the back side in this case. Conservation of energy demands that the number of hot electrons from a laser pulse of energy ε_{las} be roughly

$$N_{eh} \approx \frac{2\eta_{eh}\varepsilon_{las}}{3k_B T_{eh}}, \tag{9.459}$$

where T_{eh} can be estimated from the Beg formula (Beg et al. 1997). Making the reasonable assumption that the hot electrons are produced over a time duration roughly equal to the pulse FWHM duration, their density is

$$n_{eh0} \approx \frac{N_{eh}}{\pi r_{eh}^2 c\tau_p}. \tag{9.460}$$

The lateral width of the electron cloud can be estimated to be the laser focal spot size (of radius w_0) plus blooming of the electron rear surface spot by some opening angle over the target transit thickness d_T, giving for the hot electron cloud radius

$$r_{eh} \approx w_0 + d_T \tan\theta_{op}. \tag{9.461}$$

Roth has argued (Roth and Schollmeier 2013) that this opening angle can be estimated by hot electron scattering from the target atoms, but experimentally a good estimate is that $\theta_{op} \approx 10 - 25°$.

Now we need to use these estimates to ascertain the sheath field. We can rely on our previous hydrodynamic expansion discussion, replacing electron densities and Debye lengths with their hot electron value to give approximately for the sheath field

$$E_{sh} \approx en_{eh}\lambda_{Dh} \tag{9.462}$$

where

$$\lambda_{Dh} = \sqrt{\frac{k_B T_{eh}}{4\pi e^2 n_{eh}}}. \tag{9.463}$$

Our usual self-similar solution to this problem assumes quasi-neutrality everywhere, but, as we shall discuss, this approximation has an important inconsistency. Nonetheless, proceeding with this assumption, we can say that at some initial time, when the hot electrons emerge from the back surface, they will have longitudinal density profile given by the usual Boltzmann distribution:

$$n_{eh}(z) = n_{eh0} \exp\left[e\Phi(z)/k_B T_{eh}\right]. \tag{9.464}$$

The structure of the fields and potential at the back surface of a target are illustrated in Figure 9.54. These are numerical solutions of the equations we present in what follows, but at this point we illustrate the structure at the target back surface. The top panel in Figure 9.54 shows a rather sharp ion density drop at the target back surface plotted on top of the hot electron-density cloud which distributes itself at this ion–vacuum interface. This sets up an electric potential gradient (shown in the bottom plot), which leads to a peaked longitudinal electric field, E_z, at the ion–vacuum interface (plotted on top of the ion density in the top plot).

After this sharp electric field is formed, ions are accelerated and their density distribution evolves toward a self-similar solution. The problem with this picture is that it implies ions with density and energy out to infinity, which is obviously not possible. The reality is depicted in Figure 9.55 (Mora 2003). As the ions are pulled from the back surface by the electron sheath field, they evolve from their sharp density shelf (n_0) toward the exponential spatial distribution, n_i, already discussed, but necessarily must develop a front, whose expanding location we denote in Figure 9.55 as z_f. At this point the longitudinal electric field develops a sharp peak because of the electron Debye sheath at the ion front. The formation of this ion front means that experimentally, a maximum ion energy will be observed at the high-energy end of the ion spectrum.

9.6 Ion Acceleration

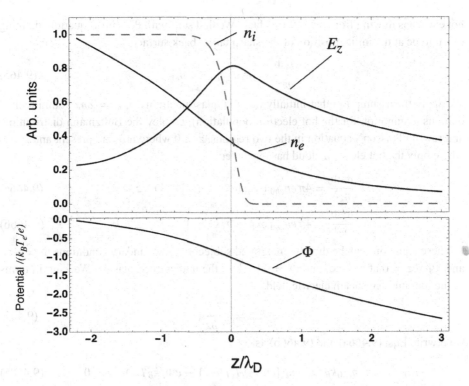

Figure 9.54 Numerical solution showing the structure of the longitudinal field (top) and potential (bottom) as a function of position for a realistic ion density profile (shown in the top panel) at the back side of the surface. The original position of the sharp backside of the target is at $z = 0$. And distance is shown in units of the hot electron Debye length. The structure of the hot electron cloud density, n_e, is also plotted in the top panel.

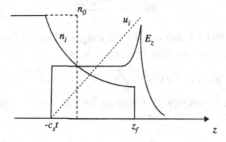

Figure 9.55 Schematic illustration of the spatial structure of the longitudinal electric sheath field after the initial sharp ion density, n_0, has experienced sum acceleration, with expanded density n_i, and an ion front has developed at z_f, and an ion velocity distribution u_i has formed. Plots were generated based on figures in Mora (2003).

To calculate the energy of this ion spectrum high energy cutoff we follow the treatment of Mora (2003). For this backside expansion we assume 1D density evolution and assume that the hot electron temperature remains isothermal (constant temperature). This second assumption is based on the idea that during the laser pulse, there is a continuous supply of hot electrons flowing from the front surface. We first start with Poisson's equation, defining $z = 0$ to be at the initial position of the sharp target back surface:

$$\frac{\partial^2 \Phi}{\partial z^2} = 4\pi e \left(n_{eh} - \bar{Z} n_i \right). \tag{9.465}$$

We apply the assumption that initially there is quasi-neutrality $n_{eh0} = \bar{Z} n_{i0}$ (admittedly a dubious assumption for the hot electron population), employ the Boltzmann distribution, and rewrite Poisson's equation in the two regions, $z < 0$ where ions are present and $z > 0$ where only the hot electron cloud has expanded:

$$\frac{\partial^2 \Phi}{\partial z^2} = 4\pi e n_{eh0} \left(\exp\left[e\Phi / k_B T_{eh} \right] - 1 \right) \quad z < 0, \tag{9.466a}$$

$$\frac{\partial^2 \Phi}{\partial z^2} = 4\pi e n_{eh0} \exp\left[e\Phi / k_B T_{eh} \right] \quad z > 0. \tag{9.466b}$$

These equations can be directly integrated subject to the boundary conditions $\Phi = -\infty$ and $\partial \Phi / \partial z = 0$ at $z = \infty$, and $\Phi = 0$ back in the undisturbed plasma. Working in terms of the longitudinal sheath electric field,

$$E_{sh} = -\frac{\partial \Phi}{\partial z}, \tag{9.467}$$

we rewrite Eqs. (9.466a) and (9.466b) as

$$\frac{1}{2} E_{sh}^2 = 4\pi e n_{eh0} k_B T_{eh} \left(\exp\left[e\Phi / k_B T_{eh} \right] - 1 - e\Phi / k_B T_{eh} \right) \quad z < 0, \tag{9.468a}$$

$$\frac{1}{2} E_{sh}^2 = 4\pi e n_{eh0} k_B T_{eh} \exp\left[e\Phi / k_B T_{eh} \right] \quad z > 0. \tag{9.468b}$$

The functions E_s and Φ must be continuous at $z = 0$. We can then determine from (9.466a) and (9.466b) that at the surface the potential must be

$$\Phi(0) = \frac{-k_B T_{eh}}{e}, \tag{9.469}$$

which, when inserted into (9.468a) tells us the magnitude of the sheath field at the surface which launches the acceleration of the outermost ions:

$$E_{sh0} = E_{sh}(0) = \sqrt{8\pi n_{eh} k_B T_{eh} \exp[-1]}. \tag{9.470}$$

We can find the electron density distribution in the sheath region by integrating Eq. (9.466b)

$$\frac{\partial \Phi}{\partial z} = E_{sh0} \exp[1/2] \exp\left[e\Phi / 2k_B T_{eh} \right] \tag{9.471}$$

$$-\frac{2k_B T_{eh}}{e} \exp\left[-e\Phi / 2k_B T_{eh} \right] \Big|_{-2k_B T_{eh}}^{\Phi} = E_{sh0} \exp[1/2] z \tag{9.472}$$

$$\Phi = -\frac{2k_B T_{eh}}{e} \ln\left[-\frac{e}{2k_B T_{eh}} E_{sh0} \exp[1/2] z + \exp[1/2] \right]$$

$$= -\frac{2k_B T_{eh}}{e} \left(\ln\left[1 + \sqrt{\frac{2\pi n_{eh0} e^2}{k_B T_{eh}}} \exp[1/2] z + \exp[1/2] \right] + \frac{1}{2} \right), \tag{9.473}$$

9.6 Ion Acceleration

or in terms of the hot electron Debye length, λ_{Dh},

$$\Phi = -\frac{2k_B T_{eh}}{e} \ln\left[1 + \sqrt{\frac{1}{2\exp[1/2]}\frac{z}{\lambda_{Dh}}}\right] - \frac{k_B T_{eh}}{e}. \tag{9.474}$$

The electron-density distribution in the sheath is found from Boltzmann's law:

$$n_{eh}(z) = n_{eh0}\exp[e\Phi/k_B T_{eh}]$$

$$= n_{eh0}\exp[-1]\left(1 + \sqrt{\frac{1}{2\exp[1/2]}\frac{z}{\lambda_{Dh}}}\right)^{-2}. \tag{9.475}$$

Not surprisingly, the hot electrons expand out into a region of size roughly equal to the hot electron Debye length, though the actual distribution shape is not exponential.

Once expansion has started, an ion front will develop and propagate outward with a hot electron sheath out ahead of this ion front. So, this electron sheath distribution of Eq. (9.475) still applies out ahead of the ion front though the electron density at the front position, z_f, will drop. So we can define the Debye length at the moving front position $\lambda_{Dhf} \equiv (k_B T_{eh}/4\pi e^2 n_{ehf})^{1/2}$ in terms of the falling hot electron density, n_{ehf}, and write the sheath distribution in the moving front as

$$n_{eh}^{(sheath)} \cong n_{ehf}\exp[-1]\left(1 + \sqrt{\frac{1}{2\exp[1/2]}\frac{(z-z_f)}{\lambda_{Dhf}}}\right)^{-2}. \tag{9.476}$$

We now can tackle the ion expansion. Well behind the ion front, we can rely on our previous hydrodynamic self-similar solutions to give us an approximate ion distribution solution. We can employ Eqs. (9.48)–(9.50), provided that we define a hot electron sound speed $c_{sh} = (\bar{Z}k_B T_{eh}/m_i)^{1/2}$. These solutions are only approximate because, as we have discussed, they predict a tail of ions extending to infinity with infinite velocity, a consequence of the assumption of quasi-neutrality, an assumption that clearly breaks down near the ion front. However, inside the bulk of the expansion, away from the front we know that the sheath field is constant in position and decays with time, given as $E_s^{(interior)} \cong k_B T_{eh}/ec_{sh}t$. At large times after the beginning of the expansion, we can then estimate the location of the front, z_f, as being where the local Debye length is equal to the extent of the expansion, namely

$$\lambda_{Dhf} \cong c_{sh}t \tag{9.477}$$

or, using Eq. (9.48),

$$\lambda_{Dhf} \cong \lambda_{Dh}\left(\frac{n_{eh0}}{n_{ehf}}\right)^{1/2} = \lambda_{Dh}\exp\left[\frac{1}{2}\left(1 + \frac{z}{c_{sh}t}\right)\right] \tag{9.478}$$

yields for the ion front position

$$z_f \cong c_{sh}t\,(2\ln[c_{sh}t/\lambda_{Dh}] - 1)$$
$$\cong c_{sh}t\,(2\ln[\omega_{pi}t] - 1) \tag{9.479}$$

in terms of the ion acoustic frequency $\omega_{pi} = (4\pi e^2 n_{i0}/m_i)$.

This implies an ion front velocity of

$$u_f \cong 2c_{sh} \ln[c_{sh}t/\lambda_{Dh}]. \qquad (9.480)$$

From this we can get the sheath field at the moving ion front

$$eE_{shf} = m_i \frac{\partial u_f}{\partial t} \qquad (9.481)$$

$$E_{shf} = \frac{2m_i c_{sh}}{et} = 2E_{Ah}, \qquad (9.482)$$

where we have looked back at Eq. (9.49) and introduced

$$E_{Ah} = \frac{k_B T_{eh}}{ec_{sh}t}. \qquad (9.483)$$

So, at the ion front, the accelerating sheath field peaks with a magnitude about twice the field in the bulk expansion. Alternatively, using $\lambda_{Dh}/c_{sh} = \omega_{pi}^{-1}$, we can also write

$$E_{shf} \cong \sqrt{2\exp[1]} E_{sh0} \frac{\lambda_{Dh}}{c_{sh}t}. \qquad (9.484)$$

The front sheath accelerating field we see evolves from

$$E_{shf}(t=0) = E_{sh0} \qquad (9.485)$$

initially to

$$E_{shf}(t > \omega_{pi}^{-1}) = \sqrt{2\exp[1]} \frac{\lambda_{Dh}}{c_{sh}t} E_{sh0} \qquad (9.486)$$

late in time (specifically at time scales longer than an ion acoustic time). Mora suggests a simple interpolation formula to give a continuous function for the accelerating field in time (Mora 2003):

$$E_{shf}(t) \cong \frac{E_{sh0}}{\left(1 + c_{sh}^2 t^2 / \sqrt{2\exp[1]} \lambda_{Dh}^2\right)^{1/2}}. \qquad (9.487)$$

Using this approximation formula, we can integrate to get the time history of the ion velocity at the front, and use this to estimate the maximum TNSA ion energy in an experiment. Introducing

$$\omega_{hot} = \frac{c_{sh}}{\lambda_{Dh}\sqrt{2\exp[1]}} = \frac{\omega_{pi}}{\sqrt{2\exp[1]}}, \qquad (9.488)$$

integration yields

$$u_f(t) = \frac{eE_{s0}}{m_i} \int_0^t \frac{1}{\left(1 + \omega_{hot}^2 t^2\right)^{1/2}} dt$$

$$= \frac{eE_{s0}}{m_i \omega_{hot}} \ln\left[\omega_{hot} t + \sqrt{\omega_{hot}^2 t^2 + 1}\right] \qquad (9.489)$$

$$\approx 2c_{sh} \ln[2\omega_{hot} t] \quad \text{late in time} \qquad (9.490)$$

9.6 Ion Acceleration

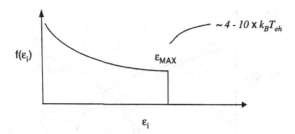

Figure 9.56 Schematic of the shape of the fast ion spectrum that develops during TNSA acceleration illustrating the sharp ion energy cutoff that forms from the development of the ion front shown in Figure 9.55.

The maximum ion energy that would then be observed in an experimental ion energy spectrum would then be estimated as

$$\varepsilon_i^{(MAX)} \cong 1/2 m_i u_f^2 = 2k_B T_{eh} \ln[2\omega_{hot} t]^2. \tag{9.491}$$

Clearly, this estimate diverges at infinite time. This is a consequence of our iso-thermal hot electron assumption, which we used during a period when there is a constant supply of hot electrons from the laser. This can only be true during the irradiation of the laser pulse, so a reasonable estimate for finding an expected maximum ion energy would be to set the time in Eq. (9.491) equal to the laser pulse duration. Often in the literature, this time is treated as a fitting parameter to experimental observation. This exercise suggests that a slightly better estimate for finding the maximum ion energy from Eq. (9.491) is to insert $t \cong 1.6\tau_{pulse}$ (Fuchs et al. 2006).

In TNSA experiments, ion spectra like that illustrated in Figure 9.56 are observed, with a fairly well-defined cutoff ion energy. So, as an example, consider a 1 ps, 1 = 1 µm laser focused to $\sim 10^{21}$ W/cm^2 (numbers consistent with many Nd:glass PW-class CPA system experiments on TNSA). The Beg formula implies this would produce a hot electron temperature of $\sim$25 MeV. Given proton acceleration at initial density of 3×10^{22} cm^{-3}, Eq. (9.491) predicts a maximum cutoff ion energy of $\sim$70 MeV. This is consistent with what is observed in experiments (Snavely et al. 2000) where up to 100 MeV protons are often observed.

It is also interesting to point out that this maximum ion energy estimate suggests that higher ion energies are increased linearly with hot electron temperature. With the weak intensity dependence of hot electron temperature predicted by the Beg scaling formula ($T_{eh} \sim I^{1/3}$) this equation predicts that maximum ion energy in TNSA weakly increases with increasing laser intensity. Also, note that this result implies that, at a given intensity, longer laser pulses will produce higher ion energies, a simple consequence that such longer pulses have more energy and produce more hot electrons. Both predictions are indeed observed in experiments. It has become well established that in TNSA acceleration, higher ion energies are seen with high-energy picosecond Nd:glass lasers than with lower-energy Ti:sapphire lasers.

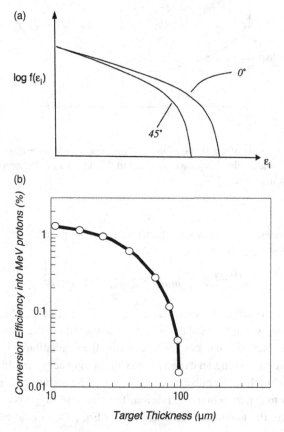

Figure 9.57 In part (a) is a schematic plot showing the qualitative differences seen in TNSA ion spectra taken normal to the target (0°) and at 45° from the normal. Adapted from data of Snavely et al. (2000). In part (b) is a schematic illustration of the experimentally observed trend in conversion efficiency of the laser energy into fast ions as a function of target thickness. Plots were generated based on plots in (Fuchs et al. 2006).

This simple theory has many limitations. For example, it is established in studies that hot electron recirculation plays a role in enhancing the ion energies, an effect neglected in our simple isothermal electron density solution. Other interesting experimental observables are illustrated schematically in Figure 9.57. For example, as might be expected with the acceleration by the sheath normal to the target, the majority of ions are ejected directly away from the rear target surface, independent of laser irradiation direction. Those ions emitted directly normal to the target show a somewhat "harder" ion energy spectrum than those emitted off normal. (Data like that illustrated in Figure 9.57a are often observed (Snavely et al. 2000).) The conversion efficiency from laser energy into fast (>1 MeV) protons in TNSA experiments tends to vary a fair amount with experimental conditions (Zimmer et al. 2021). With thinner targets (say less than 10 μm thick) at high-intensity conversion efficiency can be a few percent, perhaps as high as 10 percent into fast protons.

As the target thickness increases, this conversion efficiency rapidly rolls off. The plot in Figure 9.57b is adapted from data of Fuchs *et al.* (2006) who found conversion efficiency dropping quickly once the target was thickened to ~100 μm. Finally I would comment that the simple picture of a hot electron sheath accelerating the ions necessarily leads to an ion spectrum that is broad. There are some experimental techniques that can help to narrow the energy width of the TNSA ions (Yin *et al.* 2011).

9.6.4 Radiation Pressure Acceleration

Another way that an intense laser can accelerate ions is suggested by our previous discussions of laser-light-pressure hole boring in Section 9.2.4. The ponderomotive pressure at high intensity can be considerable and we have seen how it accelerates ions forward. If this technique is coupled to use of a thin target for purposes of ion acceleration, it is termed Radiation Pressure Acceleration or RPA. The attraction of this technique is that, when thin targets are used, it promises to produce nearly monoenergetic ions. When such thin targets are employed, say of thickness only 10s to a few hundred nm in thickness, it is said that it is RPA in the "light sail regime" (Macchi *et al.* 2009). This is in distinction to the thick target hole boring regime we have already considered. For reasons we shall explain, experimental realization of RPA and monoenergetic ions in the light sail regime is experimentally challenging (or practically impossible). Nonetheless, consideration of the physics is still illuminating.

We treat the thin target as a rigid sheet of ions, in what is often called the "Accelerated Mirror Model." We consider acceleration by light intensity I_0 of this rigid ion mirror target (with quantities subscripted with "T") to relativistic normalized velocity, β_T. We recall from Section 9.2.2 that the ponderomotive pressure is formally equally to the light pressure at a surface of reflectivity R, with $p_{rad} = (1 + R)I_0/c$. We proceed with a relativistic treatment following Macchi (2013, p. 98) though, in reality, experimentally achievable ion velocities are nonrelativistic. As usual, we define $\gamma_T = 1/(1 - \beta_T^2)^{1/2}$.

We transform intensity and laser electric field into the frame moving with the accelerated target. The usual Lorentz transform of the fields is

$$E_0' = \gamma_T (E_0 - \beta_T B_0)$$
$$= \gamma_T (1 - \beta_T) E_0 \qquad (9.492)$$

and since $I_0' \sim (E_0')^2$, the intensity in the frame of the target is

$$I_0' = \frac{1 - \beta_T}{1 + \beta_T} I_0. \qquad (9.493)$$

The same goes for the Doppler-shifted frequency of the laser in this moving frame:

$$\omega' = \sqrt{\frac{1 - \beta_T}{1 + \beta_T}} \omega. \qquad (9.494)$$

The radiation pressure in the frame of the target is

$$P_{rad} = (1 + R) \frac{1 - \beta_T}{1 + \beta_T} \frac{I_0}{c}. \qquad (9.495)$$

This is consistent with a picture of conservation of momentum by a bundle of reflected photons. The frequency of the Doppler-shifted reflected photons back in the lab frame is

$$\omega_r = \frac{1 - \beta_T}{1 + \beta_T}\omega. \tag{9.496}$$

For a laser pulse of duration τ_p, we write the intensity in terms of number of incident photons, N_ω, as

$$I = N_\omega \hbar \omega / \tau_p \tag{9.497}$$

and the duration of the reflected pulse

$$\tau_{ref} = \frac{1}{1 - \beta_T}\tau_p, \tag{9.498}$$

so the impulse imparted per unit area by these reflected Doppler-shifted photons is

$$P_{rad} = \frac{\Delta p}{\Delta t} = \frac{N_\omega \hbar \, \omega + \omega_r}{c \quad \tau_{ref}} = \frac{N_\omega \hbar}{c}\left(1 + \frac{1 - \beta_T}{1 + \beta_T}\right)1 - \beta_T, \tag{9.499}$$

consistent with Eq. (9.495).

We need to write an equation of motion for the rigid light sail, in terms of retarded time in the frame of the moving target, $\tau = t - z/c$. We denote the mass per unit area of the rigid ion sail σ_m. The equation of motion is, using Eq. (9.495),

$$\frac{dp}{dt} = F_{rad}, \tag{9.500}$$

$$\frac{d}{dt}(\gamma_T \sigma_m v_T) = (1 + R)\frac{1 - \beta_T}{1 + \beta_T}\frac{I_0(\tau)}{c}, \tag{9.501}$$

and since, recalling Eq. (3.73),

$$d\tau = (1 - \beta_T)dt, \tag{9.502}$$

the equation of motion in terms of retard time is

$$\frac{d}{dt}(\gamma_T \beta_T) = (1 + R)\frac{1 - \beta_T}{1 + \beta_T}\frac{I_0(\tau)}{\sigma_m c^2}, \tag{9.503}$$

$$(1 + \beta_T)\frac{d}{d\tau}\left(\frac{\beta_T}{\sqrt{1 - \beta_T^2}}\right) = (1 + R)\frac{I_0(\tau)}{\sigma_m c^2}, \tag{9.504}$$

$$\frac{\gamma_T}{1 - \beta_T}\frac{\partial \beta_T}{\partial \tau} = (1 + R)\frac{I_0(\tau)}{\sigma_m c^2}. \tag{9.505}$$

This can be integrated to find the velocity of the target versus retarded time:

$$\int_0^{\beta_T}\frac{1}{(1 - \beta_T)\sqrt{1 - \beta_T^2}}d\beta_T = \left(\frac{1 + \beta_T}{1 - \beta_T}\right)^{1/2} - 1 = \frac{1 + R}{\sigma_m c^2}\int_0^\tau I(\tau)d\tau. \tag{9.506}$$

The integral over the time history of the laser pulse intensity merely yields the total laser fluence on target. If we say that the total pulse fluence is $F_0 = F(\tau = \infty)$, then we can write the maximum target velocity at the end of the pulse as

$$\beta_T^{(MAX)} = \frac{(1+\tilde{F}_0)^2 - 1}{(1-\tilde{F}_0)^2 + 1}, \tag{9.507}$$

where we have introduced a normalized total fluence:

$$\tilde{F}_0 = \frac{1+R}{\sigma_m c^2} F_0. \tag{9.508}$$

The ion energy that would then result from the RPA of the thin foil is

$$\varepsilon_i = m_i c^2 (\gamma_T^{(MAX)} - 1)$$
$$= m_i c^2 \frac{\tilde{F}_0^2}{1+\tilde{F}_0}. \tag{9.509}$$

So, at lower fluence, namely the nonrelativistic case, the RPA ion energy scales as $\varepsilon_i \sim F_0^2$, while at high fluence, when the ion foil motion would become relativistic, $\varepsilon_i \sim F_0$. As a numerical example, consider a 100 nm thick Al foil. Such a foil has areal density $\sigma_m = 2.7 \times 10^{-5}$ g/cm². If irradiated by a 500 fs pulse at 10^{21} W/cm² (a typical Nd:glass PW laser parameter) at normal incidence with high reflectivity (R ≈ 1) then $\tilde{F} = 0.4_0$, and the ion energies driven reach 1.5 GeV. This exceeds what is generally possible via TNSA with such a laser and would, in principle allow easy choice of ion and, more attractive, the ion energy spectrum would be nearly monoenergetic.

The experimental reality is that realizing such an idealized light sail as an accelerator is far more difficult, for a number of reasons. First, the production of hot electrons represents a strong absorption mechanisms for the laser and reduces its effective pressure. This could in an idealized situation be circumvented by employing circularly polarized pulse normally incident on the foil (Robinson *et al.* 2008). Normal incidence eliminates Brunel absorption and the use of circular polarization eliminates **J** × **B** heating. So most all RPA experiments have been performed with circularly polarized light (Henig *et al.* 2009). As we will explain in a moment, the practical reality of this situation is almost impossible to achieve.

The next problem to achieve monoenergetic ions is that the ion foil must be accelerated as a rigid light sail. Of course, this is not the actual situation; the electrons and ions must, at the very least, be considered as two fluids. Recall the structure of the electron fluid subject to the light's ponderomotive pressure when we considered hole boring, Figure 9.23. In the thin-target, light sail regime, this ponderomotively pushed ion fluid develops a structure that looks more like that depicted in Figure 9.58. The laser field presses the electrons in a compressed layer at the back side of the target. A longitudinal E-field, peaked near the front side of the compressed electron layer, like that depicted in the figure, develops. So there will be a region near this peaked E-field where the ions are accelerated fastest. Those overrun the remaining ions, which witness a range of lower-accelerating fields. The result, as shown in PIC simulations (Eliasson *et al.* 2009), is an actual ion spectrum with a peaked

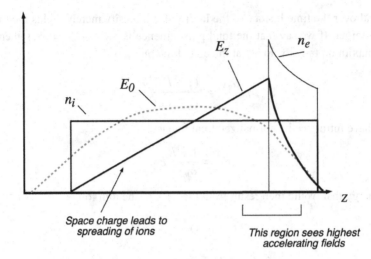

Figure 9.58 Illustration of the structure of light sail acceleration. Shown are laser field reflected at the surface, E_0, and the longitudinal field, E_z, that consequently develops as the electron density, n_e, is bunched up at the back side of the thin foil of ions from the light pressure.

fast layer followed by a diffuse range of slower ions coming from the "space-charge tail" region of the foil. This defeats the ability to produce a monoenergetic beam.

However, the most significant obstacle to experimental realization is the classic Rayleigh–Taylor instability (Ott 1972). This hydrodynamic instability occurs when there is a sharp density gradient and there is a pressure gradient in the opposite direct (such as when a glass of water is held upside down; density increases upward, but because of the force of gravity, pressure effectively increases downward). The situation we consider in the RPA is a classic example of this situation, as illustrated in Figure 9.59. There is a sharp increase in density at the front surface of the foil (from zero density to solid density), however, the pressure, coming from the incident light, goes from high (in the laser field) down to zero inside the electron-shielded portion of the foil.

It is beyond the scope of the chapter here to consider the theoretic development of this effect; the general case of the interface of two unequal-density fluids is well covered in many textbooks, such as Drake (2006, p. 170). Generally, we can say that at the interface of these two fluids, in our case the heavy foil and the light field, ripples in the surface will develop and their amplitude will grow exponentially. In an idealized situation, such as that depicted in Figure 9.59, we describe ripples by a spatial mode number, k_{RT}, and the amplitude of the ripple as a_{RT}. The foil is subject to some instantaneous acceleration by the light pressure, g_{rad}. Classic linear Rayleigh–Taylor theory (Drake 2006, p. 181) says that the amplitude of some spatial mode ripple at the accelerated interface between a heavy and light fluid will grow exponentially as

$$a_{RT}(t) = a_{RT}(0)e^{\Gamma_{RT}t}, \qquad (9.510)$$

9.6 Ion Acceleration

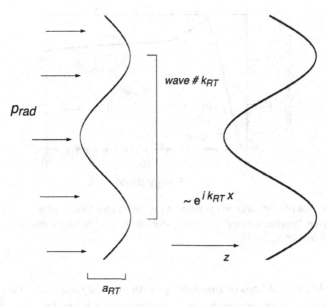

Figure 9.59 Structure of the development of ripples in thin-foil RPA as a result of the Rayleigh–Taylor instability and the pressure of the laser light.

where the exponential growth rate is given by

$$\Gamma_{RT} = \sqrt{Ak_{RT}g_{rad}}. \quad (9.511)$$

Here A is the Atwood number, given in terms of the densities of the two fluids at the interface in terms of their mass density as

$$A = \frac{\rho_{heavy} - \rho_{light}}{\rho_{heavy} + \rho_{light}}. \quad (9.512)$$

For our RPA situation, the light field has no mass density, so we can take $A = 1$.

To make a semiquantitative estimate for how fast ripples in the foil will grow, look at the acceleration of that 100 nm Al foil considered above. The acceleration that it experiences at 10^{21} W/cm^2 can be estimated as

$$g_{rad} = \frac{P_{rad}}{\sigma_m} \simeq \frac{2I_0}{\sigma_m c}, \quad (9.513)$$

which implies that the Al foil is accelerated at something like 2.5×10^{23} cm/s^2. We then need to determine which spatial ripple wavelength will tend to grow fastest. Equation (9.511) describes growth in the absence of viscosity or strength of the foil, so it suggests that infinitesimally small wavenumbers grow fastest. This is nonphysical. We can say, however, that because of diffraction of the laser EM wave, spatial scales much smaller than a laser wavelength are unlikely to grow. So, a reasonable estimate for k_{RT} of the fastest mode is that of the laser wavelength $\sim k$. Equation (9.511) predicts that ripples of this modenumber on our accelerated Al foil will grow with growth rate around $\approx 4 \times 10^{13}$ s^{-1}.

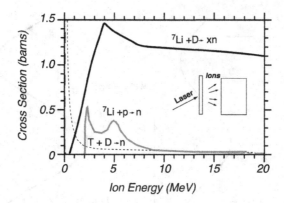

Figure 9.60 Plot of cross sections of neutron production for three typically used reactions in laser-driven "pitcher-catcher" neutron production experiments. Cross sections taken from (Higginson et al. 2011).

This would yield 20 e-foldings of amplitude growth over the time scale of our 500 fs drive pulse (!). Therefore, even extremely small spatial perturbations (of the scale of atomic spacing) on the foil will quickly build when subject to the extreme accelerations of such high light pressure. We would expect that the nice initial flat foil will break up quickly by Rayleigh–Taylor. This completely destroys any possible monoenergetic ion production. The situation is somewhat ameliorated if very short ($\sim$30 fs) pulses are employed, though they impart less impulse to a foil and result in much reduced ion energies.

What is more, as Figure 9.59 suggests, even modest growth of perturbations leads to regions of the target no longer irradiated normally; and hot electron production occurs even with circularly polarized light. This rapid breakup of the foil has been seen in simulations (Pegararo and Bulanov 2007). Experiments on RPA tend to confirm that true thin-foil, light sail acceleration is not possible (Henig et al. 2009), though these experiments do show a difference in the shape of the ion energy spectrum between linear and circularly polarized.

9.6.5 Neutron Production

Among the motivations for study of high-intensity acceleration of ions is in the possibility of a high-flux, laser-driven neutron source. The general idea is shown as the inset to Figure 9.60. A thin target foil is illuminated by the laser, producing MeV ions, most commonly by TNSA. A "catcher" target is placed immediately behind the ion-producing foil with material chosen such that nuclear reactions yield one or more neutrons. Commonly Li or Be catchers are employed (Roth et al. 2013), and with petawatt-class Nd:glass lasers it is possible to produce $10^{10} - 10^{11}$ MeV neutrons per shot. Commonly employed nuclear reactions are based on the fact that typically TNSA yields bare protons, though in some experiments D^+ ions are accelerated. Common nuclear reactions and energy release, Q, used in such pitcher-catcher experiments include (Petrov et al. 2012)

$$p + {}^7Li \rightarrow {}^7Be + n \quad Q = -1.65 \text{ MeV}$$
$$D + {}^7Li \rightarrow {}^8Be + n \quad Q = 15.03 \text{ MeV}$$
$$D + D \rightarrow {}^3He + n \quad Q = 3.25 \text{ MeV}$$

The cross sections of some of these reactions are plotted versus incoming p+ or D+ ion energies in Figure 9.60, showing that they are well suited to the multiple tens of MeV ions produced by laser-driven TNSA. Some experiments have employed Be catchers using ^{9}Be(d,n) and ^{9}Be(p,n) reactions while others employed LiF foils using ^{7}Li(d,xn) reactions (Higginson *et al.* 2011).

9.7 Solid Target X-ray Production

Vast efforts have been devoted in the high-intensity laser-plasma community to the study of X-ray production from intense irradiation of solids. This has been motivated both for the possibility of compact laser-driven ultrafast sources of X-rays for applications and also as a diagnostic on the state of dense plasmas or hot electrons produced by the laser. It is well beyond the scope of this book to discuss the radiative properties of laser-irradiated plasmas, but there is extensive material in the literature (Kieffer *et al.* 1993). Given the important role that experiments on X-ray production have played in the strong field laser plasma community, it is worth spending some time in this section discussing the most prominent X-ray production mechanisms. While there are many ways X-ray can be produced from a solid target, I will discuss the three most prevalent. Here we consider X-rays to be photons with a few hundred electron volts (wavelengths around 1 nm) out to the $\sim$10 MeV photons produced by hot electrons.

9.7.1 Thermal X-ray Emission

A high-temperature ($k_B T_e >$ 100 eV) dense plasma, like that formed on the surface of a solid irradiated at high intensity, will radiate copious amounts of soft X-rays. This thermal X-ray emission results because, in the plasma, Saha equilibrium (see Appendix B, Section B.7 for a discussion of Saha equilibrium plasmas) strips atoms down to high charge states. Then warm electrons can recollide, be captured in an upper quantum state and then drop to the ground state by radiative deexcitation. Because the ions will typically be ionized a few times, say 5+ to 20+, these quantum transitions produce photons in the 1 – 10 Å range. In short-pulse-produced plasmas, rapid thermal conduction can quickly cool the hot plasma, quenching the thermal X-ray emission and the burst of soft X-rays may last only a few picoseconds (Murnane *et al.* 1989).

The details of the X-ray spectrum produced are extremely complex, shaped by many factors such as atomic composition of the target, radiative opacity effects, the heating and cooling time history, and many other factors, so a generalized characterization of what is seen in a high-intensity laser–solid target experiment is not possible. To give a sense of the kind of X-ray spectra observed, a schematic illustration of (time-integrated) X-ray emission from Al target is shown in Figure 9.61 (Cobble *et al.* 1989; Kieffer *et al.* 1993). This

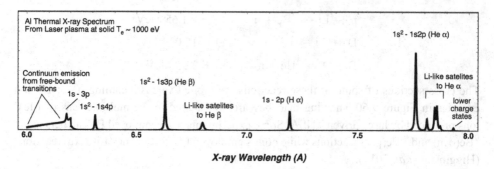

Figure 9.61 An example of the form of a typical thermal X-ray spectrum observed in solid-target interactions, in this case an Al spectrum from solid-target, short-pulse irradiation at intensity of $\sim 10^{16} - 10^{17}$ W/cm^2. Formed from typical data published in Keifer et al. (1993) and Cobble et al. (1989).

figure was formed from an amalgam of data published by a variety of authors, performing experiments on solid Al targets at intensity of $10^{16} - 10^{17}$ W/cm^2. As discussed earlier in the chapter, typical subpicosecond pulses will produce near-solid-density plasma at thermal electron temperature up to about 1 keV. At this temperature, Al in Saha equilibrium is predominantly stripped down to its He-like ionization state ($\sim$Al^{11+}). Therefore, electrons get collisionally excited (or populated by recombination) to upper states of He-like Al, and then radiative decay back to the 1s^2 ground state, producing strong line emission from this radiative decay. This includes the strong emission from the 1s2p $\rightarrow$ 1s^2 line (the so-called He-α line) at around 7.76 Å and slightly weaker emission from the 1s3p $\rightarrow$ 1s^2 (He-β) line near 6.63Å. Similar emission from the 2p $\rightarrow$ 1s line of hydrogen-like Al is produced, but with reduced intensity because H-like Al is less prevalent than the stable closed-shell of He-like Al.

Weaker emission from transitions from higher-lying levels is typically observed. In addition, because short-pulse laser–solid target plasmas are typically very dense (near solid density), there is a strong possibility for ions, such as He-like Al, to have additional electrons populating higher levels, giving excited Li-like ions. These ions, with so-called spectator electrons, produce weaker, but still significant emission at slightly longer wavelength than the main line. Examples of these Li-like "satellite lines" are shown as a grouping of lines between 7.8 and 7.9Å. Finally, note that weak emission is often observed from the recombination of electrons from the continuum into bound states of the Al ions, yielding a shorter-wavelength broad "continuum" X-ray emission at shorter wavelengths. All experiments are different, but this figure gives a sense of what kind of thermal X-rays are often produced.

9.7.2 K-α X-Ray Production

An X-ray production mechanism which has served the community often in the characterization of hot electron yields is the production of K-α X-rays (Chen, C. D. et al. 2009).

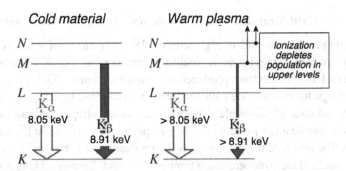

Figure 9.62 Quantum energy structure showing the origin of K-α emission along with weaker K-β emission at higher photon energy (in this case with a Cu target). The production of these X-rays in a dense plasma causes some photon energies to shift higher because ionization depletes electrons in upper shells, as illustrated on the right.

This mechanism has also been used to produce femtosecond pulses of multi-keV photons for use in ultrafast X-ray experiments, such as transient diffraction (Reich et al. 2000). Unlike the thermal X-rays discussed in the previous subsection, K-α X-rays result from hot, suprathermal electron production, through the various mechanisms we have discussed in this chapter.

The origin of K-α X-rays is shown with the quantum level diagram in Figure 9.62. The photon energy listed for the various transitions are, in this case, from a Cu target, though all sorts of X-ray wavelengths have been produced with different target materials. When hot electrons (typically of 10–100 keV energy) from the intense laser irradiation transport into the bulk solid, they can knock out electrons from the innermost shell, the K-shell. This leads to radiative decay of an electron from a higher-energy shell to fill the vacancy left in the K-shell leading to emission of an X-ray photon. If the decay is from the L-shell, the most probable, a K-α is emitted (of photon energy near 8 keV in Cu). A lesser, but significant amount of photons from decay of a higher shell, such as from the M-shell, is usually also observed in experiments (the K-β line). With Cu this photon has energy closer to 8.9 keV. When mid to high-Z atoms are used (like Cu, Sn, Ta, etc.) a range of photon energies is possible. Typically the conversion efficiency from laser light K-α to photons increases with intensity, but it is typically around 10^{-5} at modest intensity (say $\sim 10^{17}$ W/cm^2 in Cu). This tends to increase and reaches as high as $\sim 5 \times 10^{-4}$ at intensity above 10^{19} W/cm^2 (Nilson et al. 2008). Obviously, higher intensities are required with higher-Z targets as higher-energy electrons are needed to produce the K-shell vacancy in that case.

A side effect fairly unique to ultrafast high-intensity solid target is a shift in the K-α photon energy when high-contrast pulses are employed. In this case, the keV electrons produce K-shell vacancies in the atoms of the laser-heated dense plasma. As illustrated in the right hand side of Figure 9.62, the thermal plasma depopulates some of the upper shells of the target's atoms in collisional equilibrium. This serves to shift the K-α photons to slightly higher photon energy.

9.7.3 Hard X-ray and Gamma Production through Bremsstrahlung

Another unique characteristic of high-intensity laser irradiation of solids is the production of a burst of hard X-rays (or some would say gamma rays). This is the natural result of the very hot (>10 MeV) electrons produced at relativistic intensity (Galy et al. 2007). As a result, rather significant multi-MeV photon bursts are observed, typically emitted from the target in a broad cone of half-angle ~30–40°. Conversion efficiencies into hard gammas above 100 keV can be a few percent or even ~10 percent of the incident laser energy (Key et al. 1998). The origin of these gammas is the production of bremsstrahlung photons by the hot suprathermal electrons streaming through the target. Bremsstrahlung occurs when a fast electron scatters from the nuclei of an atom, and the acceleration of the electron by the scattering nucleus leads to production of a photon (Zel'dovich and Raizer 2002, p. 113).

Because of the Z^2 scaling of the bremsstrahlung cross section, where Z is the charge of the scattering nuclei, high-Z targets, such as gold or lead, produce the highest gamma yields. Because of its use as a diagnostic on hot electron temperatures, it is informative to look at the gamma energy spectrum for a given hot electron energy distribution. Though not thermalized, experiments often show that a laser-driven hot electron spectrum can be characterized by a Maxwellian. So, we take the hot electron energy distribution to be

$$f_e(\varepsilon_e) = \frac{\varepsilon_e}{k_B T_{eh}} \exp[-\varepsilon_e/k_B T_{eh}], \quad (9.514)$$

with hot electron distribution characterized by effective temperature T_{eh}. A reasonable approximate formula for the angle-integrated energy-differential bremsstrahlung cross section is (Findley 1989)

$$\sigma_b(\varepsilon_\gamma, \varepsilon_e) = aZ^2 \left(\frac{1}{\varepsilon_\gamma} - \frac{b}{\varepsilon_e} \right). \quad (9.515)$$

Here Z is the charge of the target atom nuclei, ε_γ is the energy of the gamma produced, and the fitting constants are $a = 11$ Mbarn, and $b = 0.83$.

The observed bremsstrahlung spectrum, $f_\gamma(\varepsilon_\gamma)$, is found by convolving this cross section with the hot electron energy distribution:

$$f_\gamma(\varepsilon_\gamma) = n_0 l \int_{\varepsilon_\gamma}^{\infty} \sigma(\varepsilon_\gamma, \varepsilon_e) f_e(\varepsilon_e) \, d\varepsilon_e \quad (9.516)$$

$$= n_0 l a Z^2 \int_{\varepsilon_\gamma}^{\infty} \frac{\varepsilon_e}{k_B T_{eh}} \left(\frac{1}{\varepsilon_\gamma} - \frac{b}{\varepsilon_e} \right) \exp[-\varepsilon_e/k_B T_{eh}] \, d\varepsilon_e \quad (9.517)$$

$$= n_0 l a Z^2 \left(1 - b - \frac{k_B T_{eh}}{\varepsilon_\gamma} \right) \exp[-\varepsilon_\gamma/k_B T_{eh}]. \quad (9.518)$$

So, we see that the gamma energy spectrum has roughly an exponential shape with exponential falloff given by the hot electron temperature (ignoring the small third term in the parentheses out in the high-energy tail of the distribution). Consequently, measuring the slope of the logarithm of measured high-energy photon spectra is an excellent

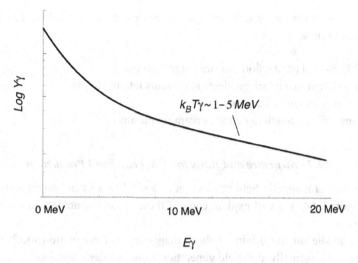

Figure 9.63 Shape of a typical bremsstrahlung hard X-ray spectrum observed in subpicosecond solid-target experiments at intensity $\sim 10^{20}$ W/cm^2.

experimental diagnostic on the effective temperature of the hot electron spectrum produced in an experiment.

In reality, it is quite common to observe gamma spectra in experiments at intensity $\geq 10^{20}$ W/cm^2 that look like that in Figure 9.63 (Key *et al.* 1998). In the few-MeV photon energy range the distribution falls off quickly with increasing photon energy, but above about 5 MeV the spectrum usually hardens with a characteristic photon (hot electron) temperature of 1 – 5 MeV, out to photon energies of a few tens of MeV. Such a distribution results from multiple hot electron temperatures produced by more than one collisionless absorption mechanism during the interaction.

9.8 Strong Magnetic Field Production

With the production and inward flow of laser-generated hot electrons, it is not surprising that strong magnetic fields will be generated at the surface inside the solid target (Sudan 1993; Belyaev *et al.* 2008). Indeed, very strong fields are observed experimentally, though these are notoriously difficult to characterize experimentally. At modestly relativistic intensity, it is believed that fields in the vicinity of 10 MGs are produced (Borghesi *et al.* 1998), and at intensity above 10^{20} W/cm^2 B-fields in the vicinity of a GGs are likely produced (Wagner *et al.* 2004, Huang *et al.* 2019; Tatarakis *et al.* 2002).

One consequence of this field production is that it, in turn, affects the propagation of hot electrons into the target, pinching their flow into a tighter, more collimated beam (Mason and Tabak 1998). As with all these solid-target, high-intensity laser phenomena, the interplay of physics leading to such high magnetic fields is complex. However, it is more or less possible to isolate four principal mechanisms for B-field production, each producing fields in somewhat different spatial regions in and around the target. We will consider each of these in turn:

1) $\nabla n \times \nabla T_e$ B-field production near the target surface
2) Direct ponderomotively driven electron currents into the target
3) Ohmic effects on return currents
4) Weibel instability growth in counterstreaming plasmas

9.8.1 Structure and Basis for Magnetic Field Production

The structure of magnetic field production in a solid target can be schematically illustrated in Figure 9.64. We will explain each of these B-field structures in the subsequent subsections.

To get a handle on the origin of these magnetic field production mechanisms it is instructive to look formally at B-field generation from Faraday's law:

$$\frac{\partial \mathbf{B}}{\partial t} = -c \nabla \times \mathbf{E}. \tag{9.519}$$

We account for the B-field generation by examining the various contributions to the electric field in Faraday's law (Bell et al. 1997; Huang et al. 2019), accounting explicitly for plasma currents. Formally, the sources of the B-field can be written with five contributing terms:

$$\mathbf{E} = \underbrace{\eta_C \mathbf{J}}_{Ohmic} + \underbrace{\frac{1}{en_e}\mathbf{J} \times \mathbf{B}}_{Hall\ effect} - \underbrace{\frac{1}{en_e}\nabla p_e}_{Ambipolar\ field} - \underbrace{\frac{\mathbf{v}}{c} \times \mathbf{B}}_{Field\ convection} - \underbrace{\frac{1}{2ecn_e}\nabla I_0}_{Ponderomotive\ current}. \tag{9.520}$$

The first term is just a statement of Ohm's law for current flow in the plasma, with $\eta_C = 1/\sigma_C$ equal to the underlying plasma resistivity, and the second is the usual Hall

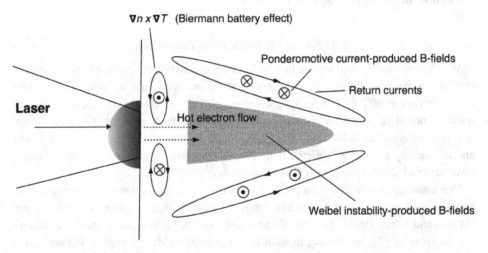

Figure 9.64 Structure of magnetic field production in short-pulse laser interactions at the surface of a solid target.

effect. The third term is one we have encountered previously in discussions of hydrodynamic expansion, the space-charge sheath E-field produced by a pressure gradient in the plasma. A term accounting for field convection and finally a current term driven by the ponderomotive force of the laser, which we here write in terms of the gradient of the laser-intensity profile. We will examine the contributions of some of these terms explicitly next. Note that the current term has to include both the ponderomotively driven fast electron current and the slow return current which neutralizes the region of the target from which the hot electrons have been expelled, two currents which do not completely cancel each other because of ohmic resistive effects on the return current.

9.8.2 $\nabla n \times \nabla T$ Magnetic Field Generation

First, we will consider the weakest contribution, the one arising from the third term in Eq. (9.520). A pressure gradient and a sheath field form in the hydrodynamic blow-off at the target surface, so we expect this to generate magnetic fields very close to the surface, as depicted in Figure 9.64. This generation mechanism is often called the thermoelectric effect or the Biermann battery effect. It will generate a magnetic field proportional to $\nabla n_e \times \nabla T_e$ at the target surface (Eliezer 2002, p. 145). At the surface there is a gradient of electron density normal to the target surface as a result of some hydrodynamic expansion, and there is a gradient in temperature radially, parallel to the surface of the target because the laser spot is most intense in its center and consequently produces the hottest plasma at the center and cooler plasma at the spot size edges. These perpendicularly oriented gradients give rise to the thermoelectric contribution.

Ignoring other terms and considering only ambipolar sheath field, and, as usual, ignoring electron inertia, we can write from Eq. (9.520)

$$\mathbf{E} = -\frac{1}{en_e}\nabla p_e, \qquad (9.521)$$

which is essentially Eq. (9.38). Using the ideal gas equation of state for the electron fluid and inserting into Eq. (9.519) gives

$$\frac{\partial \mathbf{B}}{\partial t} = \frac{ck_B}{e}\nabla \times \frac{1}{n_e}\nabla n_e T_e \qquad (9.522)$$

$$= \frac{ck_B}{e}\nabla \times \left(\frac{T_e}{n_e}\nabla n_e + \nabla T_e\right). \qquad (9.523)$$

Since the curl of the gradients in temperature and density is zero, this yields

$$\frac{\partial \mathbf{B}}{\partial t} = -\frac{ck_B}{en_e}\nabla n_e \times \nabla T_e. \qquad (9.524)$$

This contribution, for typical hydro-expansions and laser spot sizes, develops magnetic fields with toroidal shape close to the target surface with magnitudes of tens of megagauss. These B-fields persist as long as the thermal and density gradients persist, which are on a typical hydrodynamic expansion time scale and, consequently, typically last typically longer than the laser-pulse duration.

9.8.3 Ponderomotively Accelerated Electron Magnetic Field Production

Suprathermal electron currents driven by the ponderomotive force of the laser will produce transient currents that are large; after a short time, space-charge forces will start return currents flowing which (partially) neutralize the fast electron currents, but for some short period of time we would expect there to be very large B-field from the flow of fast electrons into the target. We can make a heuristic estimate for the time scale and magnitude of field generated this way. We do this by estimating the time scale over which fast electron flow before the space-charge field drives a significant return current.

Assume that the flow of fast electrons builds up a surface-charge density, which we estimate in terms of the hot electron density and the fast electron velocity as

$$\sigma_q = en_{eh}v_{fast}t. \tag{9.525}$$

We can assume $v_{fast} = c$. The space-charge field, E_{sc}, then also builds up in time as

$$E_{sc} = 2\pi\sigma_q = 2\pi en_{eh}ct. \tag{9.526}$$

The laser ponderomotive force will continue to drive the fast electron current over a time τ_h, at which time the space-charge force is equal and opposite to the ponderomotive force. So, with (9.526), we can write for the time at which these forces equalize as

$$F_p = -\nabla U_p = eE_{sc}(\tau_h) \tag{9.527}$$

$$2\pi e^2 n_{eh} c\tau_h = \nabla U_p, \tag{9.528}$$

which means that the time over which free-streaming ponderomotive electron current will flow is approximately

$$\tau_h \cong \frac{\nabla U_p}{2\pi e^2 n_{eh} c}. \tag{9.529}$$

To estimate this time, estimate the ponderomotive force as a gradient over the spatial scale of the skin depth, $\delta_{pe} \cong c/\omega_{pe}$

$$\nabla U_p \cong \frac{U_p}{\delta_{pe}}. \tag{9.530}$$

This build-up time is

$$\tau_h \cong \frac{U_p \omega_{pe}}{2\pi e^2 n_{eh} c^2}. \tag{9.531}$$

Assuming that the hot electron density is about the critical density for 1 μm, a ponderomotive energy of 1 MeV (around 10^{21} W/cm^2) yields a time scale of ~25 fs for which the fast electron current free streams. The magnetic field then, if the fast electron current area is about the size of the laser focal radius, w_0, is

$$B \cong 2\pi en_{crit}w_0. \tag{9.532}$$

A petawatt laser focused to a 10 μm spot at 10^{21} W/cm^2 would then be expected to produce a transient field of over 3 GGs. This is an enormous magnetic field of ~300,000 T.

9.8.4 Ohmic Current Magnetic Fields

On time scales longer than the fast electron free-streaming time, Eq. (9.531), return currents will eventually build up because of the space-charge field produced by exiting fast electrons. This return current will attempt to neutralize the fast electron current, but cannot completely neutralize it because that current is inhibited by the finite resistivity of the underlying solid target. Consequently the hot electron current is only approximately equal to the return current and their difference leads to a net current that can still generate a rather strong magnetic field. This is usually referred to as Ohmic current B-field production.

We can write for these two currents and the space-charge electric field

$$\mathbf{J}_{hot} \approx -\mathbf{J}_{therm} = -\mathbf{E}/\eta. \tag{9.533}$$

Ampere's law is

$$\nabla \times \mathbf{B} = \frac{4\pi}{c}(\mathbf{J}_{hot} + \mathbf{J}_{therm}). \tag{9.534}$$

Solving (9.534) for $\mathbf{J}_{therm}$, using its equality with the Ohm's equation, Eq. (9.533), and inserting into Faraday's law, Eq. (9.519), we get the ohmic contribution of hot electron flow to the buildup of magnetic field:

$$\frac{\partial \mathbf{B}}{\partial t} = -c\nabla \times \left(\frac{\eta c c}{4\pi}\right)\mathbf{B} + c\nabla \times (\eta_C \mathbf{J}_{hot}). \tag{9.535}$$

Ignoring magnetic field convection, this reduces to

$$\frac{\partial \mathbf{B}}{\partial t} = c\nabla \eta_C \times \mathbf{J}_{hot} + c\eta_C \nabla \times \mathbf{J}_{hot} \tag{9.536}$$

$$\cong c\eta_C \nabla \times \mathbf{J}_{hot} \quad \text{with uniform resistivity.} \tag{9.537}$$

Obviously, if the resistivity falls to zero, no B-field is produced; the return current can completely neutralize the hot electron current. As illustrated in Figure 9.64, the ohmic current magnetic fields tend to form in a toroidal geometry in a direction within the bulk of the solid target, in a direction opposite to the thermoelectric-produced fields near the target surface. These ohmic fields are thought to be in the vicinity of 100 MGs in petawatt-class laser experiments. The resistivity of the underlying target does matter in this case with metallic targets yielding lower magnetic fields than nonconducting targets.

9.8.5 Magnetic Fields from the Weibel Instability

The final B-field production mechanism is one which has been well known for many years in the context of beam–plasma interactions. Small-scale magnetic fields are formed in regions of counterstreaming currents as a result of an instability, first discussed by Weibel (1959). The Weibel instability can be qualitatively understood by considering two counterpropagating currents, as depicted in Figure 9.65. If there is some small displacement of one current with respect to the other, a magnetic field develops. This field will be oriented such that it will repel the two counter currents, leading to further separation and

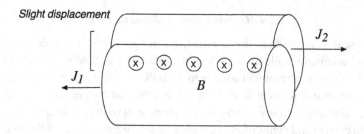

Figure 9.65 Basic current structure for the origin of the Weibel instability in plasmas.

further growth of the B-field. The field and displacement initially grow exponentially until the beams break up from this B-field growth.

This situation is more or less what happens in the central part of the hot electron flow in the laser solid target. The fast electrons are countered by return current in the central region. Small inhomogeneities in the current densities can lead to chaotic, small-scale magnetic fields due to the Weibel instability. These small-spatial-scale B-fields are observed in PIC simulations (Huang et al. 2019) and lead to the experimental observable of the hot electron current breaking up into filaments.

We will not develop the rigorous theory of the Weibel instability (see the initial paper by Weibel or Califano et al. (1997)). Though we can estimate how quickly these B-fields should grow using Weibel's initial result for the instability exponential growth rate for plasma inhomogeneities of spatial mode number k_{eW}. Weibel presents this growth rate which, in the linear regime, is independent of the current density, as

$$\Gamma_W \approx \frac{c\omega_{pe}k_{eW}}{\left(\omega_{pe}^2 + k_{eW}^2 c^2\right)^{1/2}}, \tag{9.538}$$

or for small-scale perturbations (large k_{eW}) is simply

$$\Gamma_W \approx \omega_{pe} \tag{9.539}$$

(though for very large wavenumbers the instability is damped). So in solid-density plasma inside the target, it appears that this growth happens in a just a few tens of femtoseconds. More sophisticated models for this growth rate can be found, for example, in Califano et al. (1997). The Weibel instability is a well-known limit to the extent of free hot electron propagation in a solid target.

9.9 Relativistic-Induced Transparency

Our definition in the chapter of "solid targets" was that they are overdense, and hence the laser does not propagate through the target plasma. However, at high-enough intensity, an overdense plasma can, in principle, become underdense because of the mass shift of the electron in the strong laser field, shifting the plasma frequency below the laser frequency. This effect is known as relativistic-induced transparency (RIT) or sometimes referred to as "self-induced transparency" (Macchi 2013, pp. 49–53).

9.9.1 Plasma Transparency With Electron Mass Shifts

First, let us consider the simple (perhaps naïve) view of the interaction. Consider the relativistic dispersion relation:

$$-k^2 c^2 + \omega^2 - \frac{\omega_{pe}^2}{\gamma_l} = 0, \qquad (9.540)$$

where the cycle-averaged relativistic factor in a linear polarized field is written in the usual way in terms of normalized vector potential in linear polarization:

$$\gamma_l = \sqrt{1 + a_0^2/2}. \qquad (9.541)$$

From an analytic standpoint in consideration of RIT, linear polarization is complicated by the presence of longitudinal currents in the plasma induced by $\mathbf{J} \times \mathbf{B}$ terms, considering circular polarization in the derivation is cleaner and analytic theory is more rigorous. Consequently much of the theory literature considers only circular polarization (Macchi *et al.* 2010). But since our considerations will be semiheuristic, we will approximate the extent of the effect for the experimentally more interesting linear polarized case.

Simple consideration of the relativistic dispersion relation indicates that an overdense plasma has its effective plasma frequency shifted down by the relativistic gamma factor (in other words, by increasing the cycle-averaged mass of the electrons). The overdense plasma becomes underdense at an a_0 such that

$$\omega > \frac{\omega_{pe}}{\sqrt{\gamma_l}} \to \omega > \frac{\omega_{pe}}{(1 + a_0^2/2)^{1/4}}. \qquad (9.542)$$

At high a_0, the condition for RIT then can be simply stated as

$$a_0 > \sqrt{2} \frac{n_e}{n_{crit}}, \qquad (9.543)$$

which implies that a solid-density target will become transparent at a_0 around 100, or intensity around 10^{22} W/cm^2 in near-IR fields.

9.9.2 Relativistic-Induced Transparency with Ponderomotive Pressure

In reality, the picture is more complicated than implied by the simple analysis of the last section (Cattani *et al.* 2000). The intensity threshold for RIT is, in fact, considerably higher. This is because the ponderomotive pressure piles up a layer of electrons on the rising edge of the pulse (an effect we have already discussed in the context of hole boring and RPA.) This ponderomotively pushed "wall" of electrons increases the electron density. This effect is sometimes called "self-induced opacity." We need to account for this ponderomotive snowplowing of electrons to get a more accurate estimate for the intensity threshold for RIT. (We will find that it is considerably higher than that predicted by (9.543).)

Given our previous discussions, we know that the actual situation at the solid target surface is like that shown in Figure 9.66. With a stationary shelf of ions, a layer is created

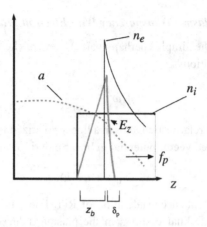

Figure 9.66 Plot of the laser field, longitudinal field and electron and ion densities in the situation of relativistically-induced "self-opacity."

where electrons have been pushed out and then form a sharp spike, with width of the order of the skin depth (as this is the depth over which the incident field ponderomotive force can act). A longitudinal space-charge electric field forms, with a cavitated region of ions of depth z_b (Eremin et al. 2010).

Relativistic-Induced Transparency theories have been developed to account for this electron density structure (Macchi 2013, p. 49). Here we will develop a simpler, toy model that will give us some intuitive insight into the physics scaling. To simplify the picture in Figure 9.66, assume that the electrons are pushed into a square hat of increased density over a depth equal to the skin depth:

$$\delta_p \cong \frac{c}{\omega_{pe}} = c\sqrt{\frac{m}{4\pi n_e e^2}}. \tag{9.544}$$

In front of this region of increased electron density, the electron density is zero, containing only ions in a layer of thickness z_b. In the square hat region, the increased electron density is n_e, and past the square hat of electrons at depth δ_p beyond the electron front, the electron density is at its initial quiescent density n_{c0}. We take the laser field skin depth of Eq. (9.544) to be from the increased, piled-up electron density.

From Eq. (9.138) we know that at normal incidence, the amplitude of the laser field at the sharp electron front is reduced from the peak field E_0 by

$$E_S = 2\sqrt{\frac{n_e}{n_{crit}}} E_0, \tag{9.545}$$

which is the same for the normalized vector potential

$$a_S = 2\sqrt{\frac{n_e}{n_{crit}}} a_0. \tag{9.546}$$

9.9 Relativistic-Induced Transparency

The ponderomotive force pushing on this electron shelf is

$$F_p = mc^2 \nabla \gamma_l = \frac{\partial}{\partial z} mc^2 \sqrt{1 + a_0^2/2}, \qquad (9.547)$$

which is at high a_0

$$F_p \cong \sqrt{2} mc^2 k a_0 = \sqrt{2} mc\omega a_0 \qquad (9.548)$$

since we know from Eqs. (9.132) and (9.136) that

$$a(z) = 2a_0 \sin(kz + \varphi_R) \qquad (9.549)$$

and $\varphi_R = \arctan(-k\delta_p) \approx 0$ for a skin depth much thinner than the laser wavelength.

In quasi-equilibrium, the longitudinal space-charge force from the evacuated ion layer balances the ponderomotive force on the surface of the electron shelf, so we write

$$eE_z = F_p \qquad (9.550)$$

with Eq. (9.548) and the space-charge field from the stationary ion layer of thickness z_b, to give

$$4\pi n_{e0} e^2 z_b = \sqrt{2} m\omega c a_0. \qquad (9.551)$$

From this we can find the thickness of the evacuated ion layer, at high a_0

$$z_b \cong \sqrt{2} \frac{\omega c}{\omega_{pe0}^2} a_0, \qquad (9.552)$$

which we have written in terms of the electron plasma frequency of the initial quiescent electron fluid, ω_{pe0}.

We now need to calculate the density of the snow-plowed electron shelf, which we can say is

$$n_e \cong n_{e0} + n_{e0} \frac{z_b}{\delta_p}. \qquad (9.553)$$

Using the evacuated layer thickness result of (9.552), this electron shelf density is

$$n_e \cong n_{e0} + \sqrt{2} n_e^{1/2} n_{crit}^{-1/2} a_0. \qquad (9.554)$$

This density is the actual density that the laser needs to penetrate through by RIT, a rather more difficult task than to pass through the initial quiescent solid-target electron density. Solving Eq. (9.554) for n_e gives

$$n_e \cong n_{e0} + n_{crit} a_0^2 + \sqrt{2 n_{e0} n_{crit} a_0^4 + n_{crit}^2 a_0^4} \qquad (9.555)$$

or for $n_e \gg n_{crit}$ gives

$$n_e \cong n_{e0} + \sqrt{2} n_{e0}^{1/2} n_{crit}^{1/2} a_0. \qquad (9.556)$$

Now the true condition for RIT, in terms of this snow-plowed electron density shelf, is

$$a_0 > \sqrt{2} \frac{n_e}{n_{crit}}, \qquad (9.557)$$

which is, with Eq. (9.556),

$$a_0 > \sqrt{2}\left(\frac{n_{e0}}{n_{crit}} + a_0^2 + \sqrt{2\frac{n_{e0}}{n_{crit}}a_0^2 + a_0^4}\right). \quad (9.558)$$

This equation has no real solution. This toy model implies that the snowplow effect *completely* counteracts the relativistic transparency. More sophisticated treatments indicate that there are some regimes in which the plasma becomes transparent, but that accessing these regimes is experimentally extremely difficult. For example, Macchi (2013, p. 51) derived a formula that says that RIT is accessible for $a_0 >\sim 0.6(n_e/n_{crit})^2$. This scaling implies that, with a solid-density target, intensity approaching 10^{26} W/cm^2 would be required for RIT, way beyond experimental capabilities at the time of this writing. Other factors may come into play, however. For example, we have completely ignored transverse ponderomotive pushing of electrons which will serve to lower their density. Nonetheless, this simple analysis indicates that it is far more difficult to enter the RIT regime than would be thought.

9.10 High-Intensity Laser Solid-Target Effects and Applications

To conclude this chapter on strong field irradiation of solid targets, I will discuss a few topics which have arisen in studies. They perhaps fall into the category of "applications" so start to go beyond the scope of this book. However, there is still some interesting physics encompassed by these topics, so a cursory understanding is relevant.

9.10.1 Electromagnetic Pulse Generation

Quite often in experiments, particularly when high-Z targets, like gold, are irradiated by 100 TW to petawatt-class lasers, a very significant electromagnetic pulse (EMP) is observed. This pulse, which contains EM wave frequencies from gigahertz to a few terahertz can have rather strong field strengths, up to 1 kV/cm at ~ 1 m from the target (Consoli *et al.* 2020). The specifics of what is emitted depend rather strongly on the configuration of the target and the way a target is mounted in the target chamber.

The essence of the effect is illustrated in Figure 9.68. When hot electrons are ejected from a target, ultimately neutralization of the target must come from currents flowing along the wire or stalk that mounts the target to a target chamber. What is observed is an EMP pulse with frequency distribution like that shown at the right in Figure 9.67. The largest emission arises from the resonance of the currents flowing along the mounting stalk, which is length l_s. Ejection of hot electrons produces a fast positive charge, Q_e, on the target. The ground plate of the chamber is the plane of symmetry of this charge, so there is Q_e at l_s and a mirror charge, $-Q_e$ at $-l_s$. These two charge meet at the ground plane at a time $t = l_s/c$ and invert their motion after $t = 2l_s/c$. A full oscillation occurs after time $4l_s/c$. So the strong EMP observed tends to be at frequency

$$\nu_{EMP} \approx \frac{c}{4l_s}. \quad (9.559)$$

9.10 High-Intensity Laser Solid-Target Effects and Applications

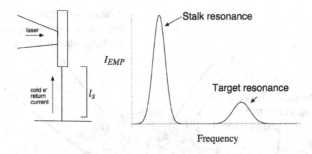

Figure 9.67 Cartoon illustration of the target geometry that leads to EMP production along with a schematic of the multi-GHz EMP radiation spectra typically observed.

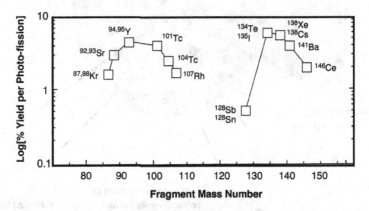

Figure 9.68 Schematic illustration of the photo-fission fragment distribution typically observed in petawatt laser solid-target experiments with uranium targets.

Typically higher frequencies are present in observed EMP bursts, presumably from resonances with smaller structures such as the metal target itself.

9.10.2 Photo-fission

An interesting consequence of the hard X-ray and gamma-ray photons produced by an intense petawatt laser is the photo-fission of the target nuclei if the target is very high-Z. For example, petawatt experiments using a ^{238}U backing on a thin gold target leads to production of fission nuclei of the uranium (Cowan et al. 2000a). The distribution of isotopes in such an experiment mirror the isotopes produced in electron accelerator–based bremsstrahlung measurements, producing a two-humped distribution of nuclear fragments.

Figure 9.68 schematically illustrates an adaption of data observed in such a photo-fission experiment with uranium backing. The various isotopes observed are labeled on the plot. With a 500 J PW pulse, approximately 2×10^7 photo fission events can be produced (though at the time of this writing there are limited published data on this effect). These yields are consistent with roughly 50 MeV effective bremsstrahlung gamma temperature and few percent conversion efficiency from laser energy to gammas.

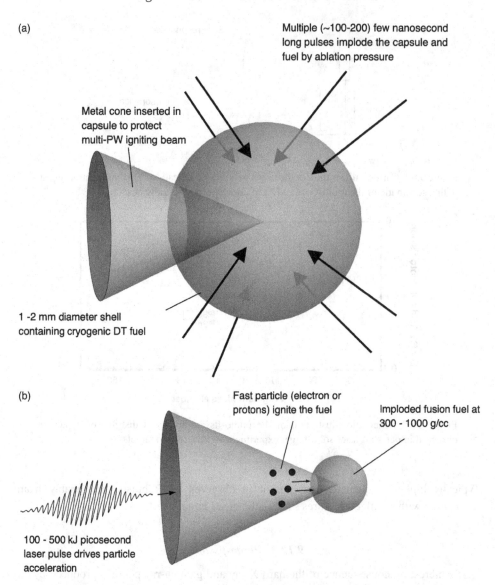

Figure 9.69 Illustration of a cone-in-shell target that is envisioned for inertial confinement fusion fast ignition. In (a) the first phase, where the DT fuel-filled capsule is imploded by ablation pressure from an array of long-pulse lasers. In (b), in the second phase after compression, a multi-100 kJ picosecond laser focuses in the metal cone, driving either fast electron or proton beams that propagate in, depositing their energy and igniting the fusion fuel.

9.10 High-Intensity Laser Solid-Target Effects and Applications

9.10.3 Fast Ignition

Because an enormous amount of research in high-intensity laser–solid interactions has been motivated by the so-called fast ignition (FI) approach to laser-driven inertial confinement fusion (ICF), it seems essential that we at least discuss this approach briefly (Tabak et al. 1994; Roth et al. 2001; Ditmire et al. 2023). So I conclude this book by examining FI with a cursory discussion. A detailed consideration of the physics of ICF is way beyond the scope of this text and the reader is encouraged to refer to the excellent text by Atzeni and Meyer-ter-Vehn (2004) on the subject.

In ICF the intent is to use the ablation pressure of lasers on a small ($\sim$1 mm) capsule containing fusion fuel, most typically a 50–50 mix of deuterium and tritium. This is accomplished in two possible ways: with direct irradiation of the capsule by nanosecond lasers to ablate the capsule surface (typically made from diamond or some plastic) in what is known as direct drive, or by placing the capsule in a small metal can, known as a hohlraum, and converting the nanosecond lasers to X-ray by irradiation of the inside of the can which then irradiate and compress the capsule. The goal is to assemble the fuel with enough density that it can burn by fusion reactions and deplete the fuel before the compressed fuel can disassemble by its own pressure. Once the fuel is assembled, it is necessary to ignite it by creating a hot spot with temperature sufficient to start burning the fuel by fusion reactions. In the traditional approach to ICF, that hot spot is created at the center by the compression itself. The $P\,dV$ work done is intended to create the hot spot in a low-density center of the compressed fuel.

This is challenging and mandates very uniform compression with no hydrodynamic instabilities mixing cold fuel from the outer part of the imploding capsule cooling the hot spot. With the advent of CPA and petawatt lasers in the 1990s, combined with the physics we have discussed at length in this chapter, namely the production of hot electrons and the subsequent acceleration of protons or other ions, it was proposed that such a laser could inject enough energy to ignite a previously compressed ICF fusion capsule. This proposal sparked 20 years of interest in strong field laser–solid interaction physics (Norreys 2009), research that we have drawn upon in much of the discussions of this chapter. The physics is complex and cannot be discussed here. I will just conclude by illustrating the kind of ICF target that is likely to be used should FI be made to work.

This kind of target is known as a cone-in-shell target, illustrated in Figure 9.69. The fusion fuel is contained in the shell which is imploded by the traditional ablation pressure approach. Then a high-energy intense laser is injected into the cone, producing hot electrons at the tip or, if a foil is inserted as illustrated, protons can be produced by TNSA. In either case these fast electrons would propagate into a compressed target (in principle) and ignite it. Generally, it is believed that $\sim$20 kJ must be deposited in a hot spot to ignite it, so given the discussions in the preceding sections about conversion efficiency of laser light to hot electrons or fast protons, it is believed that a $\sim$100–300 kJ short-pulse laser is needed. At the time of this writing, this is an active area of work and the topic of how to build such a laser is being researched (including by the author of this book).

Appendix
List of Symbols

The following is a *partial* list of symbols used throughout this book.

Greek symbols

α	=	angle between beam electron and incoming laser
$\boldsymbol{\alpha}$	=	molecular polarizability tensor
α_c	=	high harmonic chirp from recollision time variation
α_{Ci}	=	Coulomb exploding cluster ion energy coefficient
α_{Coul}	=	collected constant coefficients in the Coulomb differential scattering formula
α_{FS}	=	fine structure constant
α_{he}	=	hot electron conversion efficiency
α_{IB}	=	inverse bremsstrahlung absorption coefficient
$\tilde{\alpha}_{IB}$	=	inverse bremsstrahlung absorption coefficient density prefactor
α_{lm_l}	=	spherical harmonic normalization constant
α_L	=	plasma scale length parameter for resonance quenched $\mathbf{J} \times \mathbf{B}$ heating
α_q	=	atomic polarizability at the qth harmonic
α_V	=	Brunel co-factor in vacuum heating rate
β	=	relativistic velocity
$\beta_{\phi e}$	=	electron plasma wave relativistic phase velocity
β_ζ	=	ground state wavefunction Fourier transform sub-shell coefficient
β_d	=	relativistic drift velocity
β_i	=	relativistic ion velocity in hole boring in lab frame
β_M	=	relativistic velocity of the moving mirror
β_S	=	relativistic hole boring velocity
β_T	=	relativistic velocity of radiation pressure accelerated target surface
$\beta_T^{(MAX)}$	=	maximum moving target relativistic velocity at the end of a laser pulse

List of Symbols

β_u	=	relativistic velocity of a plasma fluid element
$\beta_{\perp R}$	=	target surface normal relativistic return velocity of charge sheets in vacuum heating
$\beta(m'_l, \theta)$	=	molecular tunneling shape-dependent factor
γ	=	relativistic factor
$\bar{\gamma}$	=	cycle-averaged relativistic factor
γ_0	=	first order term in Taylor expanded relativistic factor
$\gamma_{\phi e}$	=	electron plasma wave phase velocity relativistic factor
γ_C	=	adiabatic or polytropic index
γ_d	=	relativistic factor for electron drift velocity
γ_{eff}	=	relativistic factor for the cycle-averaged mass increase of an electron in a laser field
γ_{ej}	=	relativistic factor of electron ejected from focus
γ_h	=	hot electron relativistic factor
γ_i	=	relativistic factor for moving ions in hole boring in lab frame
$\hat{\gamma}_i$	=	relativistic factor for moving ions in hole boring in frame of accelerated target surface
γ_K	=	Keldysh parameter
γ_l	=	cycle-averaged relativistic factor in a laser field
γ_{las}	=	instantaneous (not cycle-averaged) relativistic factor in a laser field
γ_{Lit}	=	Littrow angle
γ_M	=	moving mirror relativistic factor
γ_{NL}	=	nonlinear refractive index intensity coefficient
γ_{OPA}	=	optical parametric amplification gain coefficient
γ_s	=	slowly varying, cycle-averaged plasma relativistic factor
γ_S	=	hole boring surface velocity relativistic factor
γ_T	=	relativistic factor of radiation pressure accelerated target surface
γ_{tun}	=	tunneling electron relativistic gamma factor
γ_z	=	relativistic factor of the z-component of electron motion
$\Gamma(a)$	=	gamma function
Γ^*	=	Hagana parameter
$\Gamma_{2\omega_{pe}}$	=	two-plasma decay growth rate
Γ_i	=	hole boring parameter
Γ_p	=	plasma strong coupling parameter
Γ_{SM}	=	radial beam self-modulation growth rate
Γ_R	=	Raman instability growth rate
Γ_{RBS}	=	Raman backscatter growth rate
$\Gamma_{RBS}^{(SC)}$	=	Raman backscatter growth rate in the strongly coupled regime
Γ'_{RBS}	=	relativistically corrected Raman backscatter growth rate
Γ_{RFS}	=	Raman forward scatter growth rate
Γ'_{RFS}	=	relativistically corrected Raman forward scatter growth rate
Γ_{RT}	=	Rayleigh Taylor instability growth rate

Γ_W	=	Weibel instability growth rate
δ_e	=	electron sheath layer thickness
δ_{gr}	=	diffraction grating line spacing
δ_i	=	thickness of an ion layer produced by ponderomotive pushing of an electron sheath
δ_p	=	skin depth in an overdense plasma
δ_s	=	general surface skin depth
δ_{SD}	=	collisionless skin depth
δ_{SH}	=	Spitzer–Harm thermal conductivity correction factor
δn_e	=	electron density perturbation on the quiescent electron density
δn_{e0}	=	electron density perturbation on axis at peak focused intensity
δn_i	=	ion density perturbation on the background quiescent ion density
$\Delta \varepsilon_s$	=	Stark energy shift
$\Delta \varepsilon_{WF}$	=	maximum electron energy gain in a wakefield
$\Delta \lambda$	=	spectral shift of phase modulated laser; also, laser pulse bandwidth
$\Delta \phi_{CEP}$	=	carrier envelope phase offset
ΔI_p	=	plasma continuum lowering energy of the ionization potential
$\Delta I_p^{(SP)}$	=	Stewart–Pyatt continuum lowering term
Δk	=	phase mismatch
Δk_{gas}	=	neutral gas contribution to the phase mismatch
Δk_{plas}	=	plasma induced phase mismatch
Δk_q	=	phase mismatch at the qth harmonic
$\Delta \lambda_{amp}$	=	gain narrowed amplified spectral width
$\Delta \nu_{FWHM}$	=	frequency spectrum full width at half maximum
$\Delta \tau$	=	laser temporal pulse duration; also, group delay spread in pulse spectrum
Δt_{as}	=	attosecond pulse duration
Δt_b	=	tunneling time
Δt_{SD}	=	thermal electron skin depth transit time
ε	=	dielectric constant; also, energy of a particle
ε_B	=	plasma bubble field energy
ε_{Coul}	=	ejected particle kinetic energy in a Coulomb explosion
ε_e	=	free electron kinetic energy
ε_{esc}	=	electron energy needed to escape a space charge field
ε_i	=	ion kinetic energy
$\varepsilon_i^{(MAX)}$	=	maximum ion energy in target normal sheath acceleration
ε_i^p	=	pth energy moment of an ion energy distribution
ε_j	=	energy of eigenstate j; also, energy of the jth particle
ε_K	=	electron kinetic energy
ε_{MAX}	=	maximum ion energy in ion channel cavitation

List of Symbols

ε_{pw}	=	total electrostatic energy in a single wavelength of a plasma wave
ε_R	=	kinetic energy of a returning electron during vacuum heating
ε_{res}	=	residual drift energy of the electron
ε_T	=	thermal energy of heated electrons; also, transition radiation energy
ε_{VK}	=	volume kinetic energy of an exploding cluster plasma
$\hat{\varepsilon}$	=	polarization direction unit vector
ζ	=	spatial coordinate of frame moving with laser pulse
ζ_{sat}	=	beat-wave saturation propagation length
$\varsigma(t)$	=	temporal variation phase factor in the Volkov wavefunction
η	=	refractive index
η_0	=	refractive index of a media at low intensity
η_2	=	nonlinear refractive index
$\eta_{abs}^{(SF)}$	=	collisional absorption fraction in the strong field intensity regime
$\eta_{abs}^{(RA)}$	=	fraction of laser absorption from resonance absorption
$\eta_{Abs}^{(SIB)}$	=	fraction of laser absorption from sheath inverse bremsstrahlung
$\eta_{abs}^{(T)}$	=	fraction of laser energy absorbed on a solid target into plasma thermal energy
$\eta_{abs}^{(T)}$	=	fraction of laser energy absorbed on a solid target by non-thermal, collisionless mechanisms
η_C	=	plasma resistivity
η_{eh}	=	hot electron conversion efficiency from laser energy
η_V	=	fraction of laser absorption by vacuum heating
η_k	=	kinematic two-species explosion parameter
η_R	=	refractive index of a material
θ_{ej}	=	ejection angle of the electron
θ_r	=	residual drift angle
θ_L	=	lab frame emission angle
θ_{min}	=	minimum scattering angle when calculating the Coulomb logarithm
θ_R	=	R-frame emission angle
θ_S	=	electron scattering angle
θ_W	=	wiggler angle
κ	=	ellipticity parameter
κ_{abs}	=	absorption coefficient
κ_{BSI}	=	BSI intensity coefficient
κ_f	=	tight focusing parameter
κ_e	=	electron thermal heat conductivity
κ_{plas}	=	plasma absorption coefficient
κ_R	=	radiative thermal heat conductivity
κ_{SH}	=	Spitzer–Harm electron thermal heat conductivity
λ	=	wavelength of the laser field in vacuum

List of Symbols

λ_D	=	plasma Debye length
λ_{Dh}	=	Debye length of hot electron cloud
λ_{Dhf}	=	Debye length of hot electron cloud at expanding ion front position
λ_e	=	electron collisional mean free path; also, electron plasma wave wavelength
λ_{pe}	=	wavelength associated with the electron plasma frequency
λ_R	=	Rossland mean free path
Λ_C	=	Coulomb logarithm argument
μ	=	reduced mass
ν_C	=	cluster electron cloud damping rate
ν_d	=	frequency of plasma wave damping due to collisional and collisionless mechanisms
ν_{damp}	=	damping frequency
ν_ε	=	energy loss collision frequency
ν_{eff}	=	effective damping rate for SRS in a plasma gradient
ν_{ei}	=	electron momentum loss collision frequency
ν_{EMP}	=	frequency of EMP emission
ν_j	=	frequency of oscillator jth longitudinal mode
ν_p	=	momentum loss collision frequency
ν_{SIB}	=	sheath inverse bremsstrahlung effective collision frequency
$\nu_{\parallel}^{ei}$	=	electron–ion slowing down collision frequency
$\nu_{\parallel}^{ee}$	=	electron–electron slowing down collision frequency
ξ	=	barrier penetration factor in WKB approximation; also, differential nonlinear Thomson emitted intensity;
ξ_e	=	fraction of electrons left in a cluster after outer ionization
ξ_{OI}	=	ratio of peak laser field to fully stripped cluster surface field
ρ_c	=	total net charge density
ρ_e	=	electron fluid charge density
ρ_{mn}	=	density matrix elements
ρ_q	=	atomic phase coefficient; also, charge density
$\tilde{\rho}$	=	electron radial position normalized to the cluster radius
$\sigma^{(q)}$	=	qth order multiphoton ionization cross section
$\sigma_0(\lambda)$	=	laser frequency dependent cross section
σ_{abs}	=	cluster nanoplasma absorption cross section
σ_b	=	bremsstrahlung cross section
σ_c	=	width of a cluster size distribution
σ_C	=	electrical conductivity
σ_{CI}	=	Lotz collisional ionization cross section
σ_e	=	electron plasma AC conductivity
σ_m	=	mass per unit area
σ_{pi}	=	photoionization cross section

σ_q	=	surface charge density
σ_{rec}	=	radiative recombination cross section
σ_{SB}	=	Stefan Boltzmann constant
σ_{WF}	=	root mean square spatial wavefront
τ	=	proper time
$\tau_\perp$	=	transverse momentum relaxation time
$\tau_\parallel$	=	parallel momentum relaxation time
$\tau_{12}^{(kT)}$	=	two-species temperature relaxation time
τ_{BW}	=	time scale for growth of a beat-wave
τ_{Coul}	=	Coulomb explosion time
τ_{ee}	=	electron temperature equilibration time
τ_{ei}	=	electron–ion temperature equilibration time
τ_{env}	=	variation time of laser amplitude as a pulse propagates in plasma
τ_h	=	hot electron drive time
τ_p	=	laser pulse duration
τ_r	=	recolliding electron return time
τ_{ref}	=	duration of pulse reflected from a moving target
τ_{rs}	=	recolliding electron saddle point return times
τ_R	=	extraction time for a sheet of electrons in vacuum heating
τ_{rot}	=	characteristic rotation time of a molecule
Υ	=	electric field amplitude normalized to the Schwinger field
ϕ	=	phase; also polar angle
$\phi(t)$	=	Floquet quantum states
ϕ_1	=	group delay
ϕ_2	=	group delay dispersion
ϕ_{atomic}	=	atomic/intensity-dependent contribution to high harmonic phase
ϕ_b	=	birth phase of a tunnel-ionized electron
ϕ_B	=	phase of field an electron is injected into the vacuum from a cluster
ϕ_D	=	phase accumulation on propagating beam from diffraction
ϕ_E	=	phase of electron reemergence after transiting a cluster
ϕ_G	=	Gaussian focal phase term; also denoted $\phi_{Gaussian}$
ϕ_j	=	jth order dispersion
ϕ_N	=	Floquet state's Fourier coefficients
ϕ_{NL}	=	phase accumulation on propagating beam in a medium due to a non-linear refractive index
ϕ_q	=	dipole phase of the qth harmonic
ϕ_r	=	return phase of a recolliding tunnel-ionized electron
ϕ_{WF}	=	spatial wavefront phase
$\bar{\phi}_{WF}$	=	average spatial wavefront phase over the beam
φ	=	normalized electrostatic potential
φ_i	=	ground state wavefunction Fourier transform

Φ	=	electrostatic potential
Φ_C	=	Coulomb potential inside and outside a uniform ion density sphere
$\Phi_{\hbar\omega}$	=	photon flux at frequency ω
$\Phi_{in/out}$	=	electrostatic potential inside/outside a cluster
χ	=	power radiated by an electron; also, gauge transformed quantum wavefunction
χ_e	=	electron thermal diffusivity
χ_i	=	ratio of electro impact energy to surface electrostatic potential energy
χ'_v	=	vth state vibrational wavefunction
ψ	=	general symbol for a quantum wavefunction
ψ_p	=	plasma wake potential
ψ_V	=	Volkov state wavefunction
$\psi_V^{\mathbf{k}_e}$	=	Volkov state with canonical momentum $\hbar\mathbf{k}_e$
ψ_V^{Coul}	=	Coulomb correct Volkov state
$\Psi(\mathbf{x})$	=	slowly varying field amplitude of a propagating laser wave in the paraxial approximation
Ψ_0	=	peak slowly varying amplitude in the paraxial approximation
Ψ_{mol}	=	highest occupied molecular orbital wavefunction
ω	=	angular frequency of the laser field
ω_β	=	betatron frequency
ω_a	=	atomic unit of frequency
ω_A	=	anti-Stokes wave frequency
ω_c	=	betatron critical frequency
ω_D	=	Doppler-shifted laser frequency in plasma-accelerating electron frame
ω_e	=	electron fluid oscillation frequency
ω_{hot}	=	hot electron ion plasma frequency
ω_M	=	Mie resonance frequency; also, laser frequency in a moving mirror frame
ω_{pe}	=	electron plasma frequency
ω_{pi}	=	ion plasma frequency
ω_q	=	qth harmonic angular frequency
ω_r	=	frequency of Doppler-shifted reflected photons from a moving target surface
ω_R	=	radiated angular frequency; also, radiated frequency of betatron radiation
ω_s	=	Stokes wave frequency; also, angular frequency of photon scattered from electron beam
ω_T	=	transition radiation angular frequency
Ω_R	=	Rabi frequency

List of Symbols

Alphanumeric symbols

a	=	normalized vector potential
a	=	magnitude of the normalzed vector potential; also, thermal diffusivity coefficient
a_0	=	normalized vector potential amplitude
a_{00}	=	peak normalized vector potential at best focus
a_B	=	Bohr radius
$a_\alpha^{(N)}$	=	weighting of unperturbed quantum states in field-mixed Floquet states
$\mathbf{a}_L$	=	laser component of the normalized vector potential
$\mathbf{a}_p$	=	plasma-generated component of the normalized vector potential
$\mathbf{a}_R$	=	Raman scattered component of the normalized vector potential
a_S	=	normalized vector potential at a reflecting target surface
A	=	vector potential
A_{00}	=	peak amplitude of the laser field vactor potential
$\mathbf{A}_{A0}$	=	anti-Stokes wave slowly varying vector potential amplitude
A_{fi}	=	transition amplitude from initial to final quantum state
$\mathbf{A}_R$	=	Raman scattered vector potential field
$\mathbf{A}_{S0}$	=	Stokes wave slowly varying vector potential amplitude
$Ai(a)$	=	Airy function of the first kind
$\mathcal{A}_q$	=	nonlinear Thomson scatter intensity factor for the qth harmonic
$\mathcal{B}_q$	=	nonlinear Thomson scatter intensity factor for the qth harmonic
b	=	impact parameter; also, normal separation of the planar faces of gratings
b_c	=	chirp parameter
B	=	molecular rotational constant
c	=	speed of light
$c_{s\gamma}$	=	adiabatic sound speed
c_{sh}	=	hot electron sound speed
c_{si}	=	isothermal sound speed
C_{Coul}	=	Coulomb correction coefficient
C_V	=	specific heat at constant volume
C_{kl}	=	asymptotic wavefunction normalization coefficient
C_{lm_l}	=	expansion coefficients for molecular wavefunction in basis set of spherical harmonics
d	=	dipole moment
d_0	=	magnitude of the peak displacement of a field-driven cluster nanoplasma electron cloud
$\mathbf{d}_e$	=	cluster nanoplasma electron cloud field induced spatial displacement
d_{eff}	=	effective nonlinear polarization of a nonlinear optical crystal

List of Symbols

$\mathbf{d}_m$	=	permanent dipole moment of a molecule
$d^{(l)}_{m_l m'_l}$	=	small-d Wigner operator
$\tilde{\mathbf{d}}_q$	=	Fourier component of the dipole moment at the qth harmonic
d_T	=	target transit thickness
$D^{k_e}_x$	=	dipole matrix element between the ground state and a continuum state
$D_{S,A}$	=	Stokes or anti-Stokes wave dispersion term
$D'_{S,A}$	=	relativistically corrected Stokes or anti-Stokes wave dispersion term
D_{pe}	=	plasma dispersion function
$D^{(l)}_{m_l m'_l}$	=	Wigner rotation operator
e	=	charge of the electron
$\hat{E}$	=	oscillating electric field amplitude
$\tilde{E}$	=	electric field normalized to the atomic electric field
E_0	=	electric field amplitude
E_a	=	atomic unit of electric field
E_{BSI}	=	atomic barrier suppression field
E_I	=	electric field at an ion during ion field enhanced ionization
E_{IC}	=	ion field enhancement critical electric field
$E^{(max)}_{OI}$	=	space charge electric field at the surface of a fully stripped cluster
E_S	=	magnitude of laser electric field at a sharp solid target surface
E_{sc}	=	space charge field build up at a target surface
E_{sh}	=	hot electron-produced sheath field
E_{sh0}	=	hot electron-produced sheath field at target rear surface
E_{shf}	=	hot electron-produced sheath field at the moving ion front
E_{WB}	=	relativistic phase velocity wave breaking field
E_{WB0}	=	cold plasma wave breaking field
E_z	=	in most contexts, the longitudinal electric field induced in a plasma
E_{z0}	=	longitudinal plasma wave electric field peak amplitude
$E^{(MAX)}_z$	=	peak longitudinal ion accelerating field in hole boring
$E^{(sat)}_{z0}$	=	maximum saturated accelerating field in a beat-wave
f	=	focal length of a focusing optic (lens or curved mirror)
$f_\#$	=	focusing optic $f/\#$
$f_\gamma(\varepsilon_\gamma)$	=	bremsstrahlung photon spectrum
f_e	=	electron kinetic distribution function
$f(\varepsilon)$	=	electron energy distribution
$f_i(\varepsilon_i)$	=	ion energy distribution
$f_i(r)$	=	radial variation of the molecular wavefunction
f_{NL}	=	self-focal length from nonlinear self-focusing
f_{OI}	=	fraction of outer ionized electrons
$\mathbf{F}_p$	=	cycle-averaged ponderomotive force
f_p	=	ponderomotive force per unit volume on an electron fluid element

List of Symbols

$f/\#$	=	focal number of a focusing optic
F	=	thermal flux limiter
F_0	=	total pulse fluence on target
$\tilde{F}_0$	=	normalized total fluence
F_{abs}	=	absorbed laser energy fluence
$F_H(r)$	=	nearest neighbor ion probability distribution
F_p	=	ponderomotive force
$\mathbf{F}_r$	=	electron cloud restoring force
$g(v_z)$	=	1D averaged electron distribution function
g_j	=	degeneracy of the jth state
$g(w)$	=	beam size potential
$g(\tilde{z})$	=	self-similar spatial profile of a thermal wave
G	=	integral of temperature over the depth of energy deposition
$G(\mathbf{x})$	=	Gaussian spatial profile
G_{OPA}	=	optical parametric amplification gain
G_0	=	peak small signal gain
$\hbar$	=	Planck's constant dived by 2π
$h(\tilde{z}')$	=	spatial profile of a thermal wave driven by a long pulse laser
H	=	Hamiltonian
H_0	=	unperturbed Hamiltonian
H_I	=	interaction Hamiltonian
$\hat{H}_I$	=	modified interaction Hamiltonian
$H_{-,+}$	=	interaction Hamiltonian associated with absorption and emission of a photon, respectively
H_{KH}	=	Kramers–Henneberger Hamiltonian
$\mathbf{H}_F$	=	Floquet Hamiltonian
I	=	intensity
$\hat{I}$	=	relativistically transformed intensity on a moving target surface
I'_0	=	Lorenz-transformed intensity in the frame of a moving target surface
I_a	=	atomic unit of intensity
I_{abs}	=	absorbed laser power per unit area per time on a solid target
I_{BSI}	=	barrier suppression intensity
I_{CO}	=	breakdown intensity of the cluster collective oscillator model
I_H	=	ionization potential of hydrogen
I_{inc}	=	intensity in vacuum incident on a plasma-density gradient
I_p	=	ionization potential
I_q	=	intensity of the qth harmonic
I_{MAX}	=	maximum peak intensity of a pulse undergoing plasma defocusing
I_Q	=	ionization potential of a molecular atom stripped to charge state Q
I_p	=	ionization potential
I_{sat}	=	ionization saturation intensity
$\tilde{I}_p$	=	ionization potential normalized to the ionization potential of hydrogen

j	=	current density
J	=	rotation angular momentum quantum number
J_s	=	Bessel function of order s
J	=	current density
J$_e$	=	electron current density
J$_{hot}$	=	hot electron current density
J$_{ret}$	=	thermal electron return current density
J$_{therm}$	=	thermal electron current density
$\hat{J}$	=	angular momentum operator
k	=	wavenumber of the laser field in vacuum
k_β	=	betatron wavenumber
k$_A$	=	anti-Stokes field wave vector
k_B	=	Boltzmann constant
k_c	=	scaling constant in the Hagana parameter based on atomic species
k_e	=	electron fluid oscillation wave number
k$_e$	=	quantum wave vector of the electron
k$_{es}$	=	saddle point momenta (wave number)
k_{eW}	=	mode number of plasma inhomogeneities seeding Weibel instabilities
k$_S$	=	Stokes field wave vector
$k_{x,y}$	=	x or y component of the saddle-point momentum
k_p	=	ionization potential wavenumber
k_{pe}	=	electron plasma wavenumber
K_β	=	betatron undulator parameter
K$_q$	=	focused laser-generated component of harmonic wavenumber vector
K$_p$	=	intensity-dependent phase component of harmonic wavenumber vector
K_{res}	=	residual kinetic energy imparted to electron
K_{ret}	=	recolliding electron kinetic energy
K_{tot}	=	total kinetic energy of expanding ions
K_0	=	modified Bessel function of the zeroth order
l_{abs}	=	absorption length
l_{coh}	=	coherence length
l_s	=	length of the target stalk
l_{SRS}	=	length for growth of stimulated Raman scattering in a density gradient
L	=	Lagrangian; also, plasma scale length
L_C	=	effective oscillator cavity length
L_{de}	=	dephasing length of an electron in a plasma wave
L_h	=	hot electron penetration length
L_{HW}	=	half length of a $\sin^2$ pulse
L_p	=	total length of a $\sin^2$ pulse
L_{pd}	=	pump depletion length of laser pulse driving a plasma wave
m	=	mass of the electron
m_d	=	grating diffraction order

m_i	=	ion mass
m_l	=	orbital (magnetic) quantum number
$M(v)$	=	Franck–Condon factor of the vth state
n^*	=	effective principal quantum number
n_a	=	atomic density
n_{crit}	=	plasma critical density
n_e	=	electron density
$n_e^{(MAX)}$	=	maximum electron density produced by field ionization in a defocusing plasma
n_{e0}	=	quiescent electron density
$\tilde{n}_e$	=	proper (relativistically corrected) normalized electron density
n_{ec}	=	density of the cold electron component of a two-temprature plasma
n_{ec}	=	density of the hot electron component of a two-temprature plasma
n_0	=	average atomic density
n_{eh}	=	quasi-Maxwellian hot electron density
n_h	=	hot electron density
n_i	=	ion density
$\hat{n}_i$	=	ion density in upstream shock region during hole boring
n_{i0}	=	quiescent background plasma ion density
n_s	=	plasma density at the sonic point in a plasma outflow
N	=	number of particles or photons
N_β	=	number of betatron oscillations undergone by an electron
N_γ	=	number of emitted photons
N_ω	=	number of photons incident on target
N_e	=	number of electrons
N_{eh}	=	number of hot electrons
N_q	=	number of photons at the qth harmonic
p	=	pressure; also, magnitude of the kinetic momentum; also used as the longitudinally varying phase in the paraxial approximation
$\mathbf{p}$	=	kinetic momentum
$\tilde{\mathbf{p}}$	=	relativistic momentum normalized to c
$\tilde{\mathbf{p}}_\perp$	=	transverse electron fluid normalized momentum
p_0	=	gas jet backing pressure
p_γ	=	light pressure
p_a	=	ablation pressure
p_{Coul}	=	Coulomb explosion pressure in a cluster
p_p	=	ponderomotive pressure
$\mathbf{p}_T$	=	kinetic electron momentum transverse to laser propagation
$\mathcal{P}$	=	polarization of a medium
$\tilde{\mathcal{P}}_q$	=	Fourier component at the qth harmonic of the medium's polarization
p_e	=	electron fluid pressure
$\mathbf{P}$	=	canonical momentum
P_C	=	critical power for self-focusing

$P_c^{(rel)}$	=	critical power for self-focusing resulting from relativistic mass increase of electrons in a plasma
$P_c(N)$	=	cluster size distribution
P_{fi}	=	probability of a transition from the initial state to a final state
P_λ	=	instantaneous power at some point in a laser pulse
$P_l(\theta)$	=	Legendre polynomials
P_L	=	laser pulse power
$P_L^{(q)}$	=	power radiated in the lab frame
$P_R^{(q)}$	=	power radiated in the R-frame
P_{rad}	=	radiated power
q	=	harmonic order; also, particle charge; also, complex radius of beam curvature
q_{cut}	=	cutoff harmonic order
q_i	=	charge of the ith particle
q_{min}	=	minimum number of photons needed to ionize an atom
Q_i	=	electric charge of the ith core ion in a molecule
$\mathbf{Q}$	=	thermal energy flux
$\mathbf{Q}_e$	=	electron thermal energy heat flux
Q_{tot}	=	total charge buildup on a cluster from outer ionization
r_b	=	tunnel-ionized electron birth radius
r_c	=	critical radius for electron cavitation
r_i	=	initial ion radial position in a cavitating channel
r_B	=	plasma bubble radius
r_{eh}	=	hot electron cloud radius
r_{Lar}	=	Larmor radius
r_{WS}	=	Wigner–Seitz radius
R	=	radius of a cluster; also Gaussian focal beam (real) radius of curvature; also, distance from emitter to observer
$R_{1,2}$	=	radius of curvature of end mirrors in a stable resonator
R_e	=	cluster electron cloud radius
R_l	=	separation distance between ions in which a bound electron is localized
s	=	temporal and spatial envelope of a laser pulse
s_{in}	=	slowly varying laser pulse envelope
$\tilde{s}_{in}$	=	initial pulse spectrum
$\tilde{s}_{fin}$	=	final phase-modulated pulse spectrum
S	=	pulse temporal envelope
$\mathbf{S}$	=	Poynting vector
$S_{k_e}(t, t')$	=	modified/quasi-classical action
$S_{k_e s}$	=	modified action at saddle-point times
S_{seed}	=	seed pulse spectrum
S_{amp}	=	pulse spectrum after amplification

List of Symbols

t'	=	retarded time
t_{2R_0}	=	time for a Coulomb exploding cluster to double its radius
t_{de}	=	electron dephasing propagation time
t_e	=	exit time of an electron from the field-free region in a cluster
t_{tun}	=	tunneling time
T	=	finite time interval for emission; also oscillation period
T_c	=	background cold electron temperature in a solid target
T_e	=	electron temperature
T_{ec}	=	electron temperature of the cold component of a two-temperature distribution
T_{ec}	=	electron temperature of the hot component of a two-temperature distribution
T_h	=	hot electron temperature
T_{hot}	=	hot electron temperature from collisionless absorption on a solid target
T_i	=	ion temperature
T_{rot},	=	molecular ensemble rotation temperature
u	=	energy density
$\mathbf{u}_e$	=	electron fluid velocity
u_ε	=	electron fluid internal energy density
u_f	=	TNSA ion front velocity
u_i	=	ion fluid velocity
u_{las}	=	energy density deposited by the laser
$\mathbf{u}_{osc}$	=	laser-driven oscillatory component of the electron fluid velocity
u_R	=	plasma photon energy density
u_S	=	hole boring surface velocity
$\hat{u}(t)$	=	evolution operator
U	=	total energy
U_p	=	ponderomotive energy
U_r	=	electron recollision energy
v_d	=	electron forward drift velocity
v_g	=	group velocity
v_ϕ	=	phase velocity
v_{osc}	=	electron oscillation velocity amplitude
v_s	=	moving mirror surface velocity
v_T	=	thermal velocity
v_{Te}	=	electron thermal velocity
V	=	potential energy
$\hat{V}$	=	modified binding potential
w	=	Gaussian focal beam $1/e^2$ intensity radius as a function of propagation distance
w_0	=	Gaussian focal beam smallest focal spot $1/e^2$ intensity radius
w_{CA}	=	cycle-averaged total ionization rate

w_i	=	total ionization rate
w_{tun}	=	tunnel ionization rate
w	=	electron randomly distributed thermal velocity
$w^{(SFA)}$	=	integrated ionization rate from the strong field approximation
w_{IB}	=	heating rate per unit volume in a plasma from inverse bremsstrahlung
w_{min}	=	minimum focal spot radius of a pulse undergoing plasma defocusing
W_{3BR}	=	three-body recombination rate
$W^{(stat)}$	=	static DC field tunnel-ionization rate
$W^{(tun)}$	=	general tunnel-ionization rate
$W_{AB}^{(tun)}$	=	differential tunnel-ionization rate of the diatomic molecule AB
W_{FS}	=	rate of electron free-streaming from a cluster
W_i	=	ionization rate for the ith charge state
W_{IFA}	=	ion field–enhanced ionization rate
$W_{fi}^{(q)}$	=	q-photon ionization rate
$W_{PPT/ADK}$	=	Perelomov, Popov and Teren'tev/ Ammosov, Delone and Krainov complex atom tunnel-ionization rate
$W_{fi}^{(SFA)}$	=	differential-ionization rate found from the strong field approximation
W_{PPT}	=	Perelomov, Popov and Teren'tev tunnel-ionization rate
W_{SFA}	=	strong field approximation ionization rate
W_{SFA}^{Coul}	=	Coulomb-corrected SFA ionization rate
$x_{fi}^{(q)}$	=	q-photon multiphoton dipole matrix element
x_0	=	oscillation amplitude of the electron cloud in a cluster
Y_{lm_l}	=	spherical harmonics
$z_0^{(MAX)}$	=	maximum possible amplitude of the oscillating mirror excursion
z_b	=	ion-layer thickness at a ponderomotively pushed target surface
z_f	=	position of a thermal wavefront
z_p	=	effective hot electron penetration depth
z_R	=	Rayleigh range
z_s	=	electron fluid surface position in the oscillating mirror model
z_T	=	turning point of an obliquely incident laser wave on a plasma-density gradient
$\tilde{z}$	=	propagation position normalized to the Rayleigh range; also, self-similar
Z	=	charge state of an ion
$\bar{Z}$	=	average charge state of a mixed-ionization-species plasma
$\bar{Z}_{MAX}$	=	maximum average charge state of high-Z atoms in a gas subject to a laser pulse undergoing plasma defocusing
$Z_p(T_e)$	=	partition function
Z_{rot}	=	rotational partition function

Appendices A and B appear online at www.cambridge.org/ditmire

Bibliography

Important Books Relevant to Strong Field Physics

(Referenced throughout this book with page numbers)

Borovsky, A. V., A. L. Galkin, O. B. Shiryaev and T. Auguste (2003), *Laser Physics at Relativistic Intensities* (Berlin: Springer).

Boyd, R. W. (1992), *Nonlinear Optics* (Boston: Academic Press).

Eliezer, S. (2002), *The Interaction of High-Power Lasers with Plasmas* (Bristol: Institute of Physics Publishing).

Faisal, F. H. M. (1987), *Theory of Multiphoton Processes* (New York: Plenum Press).

Fedorov, M. V. (1997), *Atomic and Free Electrons in a Strong Light Field* (Singapore: World Scientific).

Gibbon, P. (2005), *Short Pulse Laser Interactions with Matter: An Introduction* (London: Imperial College Press).

Ginzburg, V. L. (1960), *Propagation of Electromagnetic Waves in Plasma*, Trans. Royer and Roger, Inc. (New York: Gordon and Breach).

Joachain, C. J., N. J. Kylstra and R. M. Potvliege (2012), *Atoms in Intense Laser Fields* (Cambridge: Cambridge University Press).

Kruer, W. L. (2003), *The Physics of Laser Plasma Interactions* (Boulder, CO: Westview Press).

Larsen, J. (2017), *Foundations of High-Energy-Density Physics* (Cambridge: Cambridge University Press).

Lin, C. D., A.-T. Le, C. Jin and H. Wei (2018), *Attosecond and Strong-Field Physics: Principles and Applications* (Cambridge: Cambridge University Press).

Liu, C. S., V. K. Tripathi and B. Eliasson (2019), *High-Power Laser–Plasma Interaction* (Cambridge: Cambridge University Press).

Liu, J. (2014), *Classical Trajectory Perspective of Atomic Ionization in Strong Laser Fields: Semiclassical Modeling* (Heidelberg: Springer).

Macchi, A. (2013), *A Superintense Laser–Plasma Interaction Theory Primer* (Dordrecht: Springer).

Mulser, P. and D. Bauer (2010), *High Power Laser–Matter Interaction* (Berlin: Springer).

Reintjes, J. F. (1984), *Nonlinear Optical Parametric Processes in Liquids and Gases* (Orlando, FL: Academic Press).

Wegener, M. (2005), *Extreme Nonlinear Optics: An Introduction* (Berlin: Springer).

Other General Reference Books Helpful in the Study of Strong Field Physics

Atzeni, S. and J. Meyer-ter-Vehn (2004), *The Physics of Inertial Fusion* (Oxford: Oxford University Press).

Born, M. and Wolf, E. (1980), *Principles of Optics: Electromagnetic Theory of Propagation, Interference and Diffraction of Light*, 6th Edition (Oxford: Pergamon Press).

Boyd, T. J. M. and J. J. Sanderson (2003), *The Physics of Plasmas* (Cambridge: Cambridge University Press).

Drake, R. P. (2006), *High-Energy-Density Physics: Fundamentals, Inertial Fusion, and Experimental Astrophysics* (Heidelberg: Springer).

Goldstein, H. (1980), *Classical Mechanics* (Reading: Addison-Wesley).

Jackson, J. D. (1975), *Classical Electrodynamics* (New York: John Wiley & Sons).

Koechner, W. (2006), *Solid-State Laser Engineering* (New York: Springer).

Landau, L. D. and E. M. Lifshitz (1984), *Electrodynamics of Continuous Media*, Trans. J. B. Sykes, J. S. Bell and M. J. Kearsley (Oxford: Pergamon Press).

Landau, L. D. and E. M. Lifshitz (1991), *Quantum Mechanics: Non-Relativistic Theory*, 3rd Edition (London: Butterworth and Heinemann).

Landau, L. D. and E. M. Lifshitz (1994), *The Classical Theory of Fields*, Vol. 2. (New York: Pergamon Press).

Milonni, P. W. and J. H. Eberly (2010), *Laser Physics* (Hoboken, NJ: John Wiley and Sons).

Miyamoto, K. (1989), *Plasma Physics for Nuclear Fusion* (Cambridge: The MIT Press).

Nicholson, D. R. (1983), *Introduction to Plasma Theory* (New York: John Wiley and Sons).

Piel, A. (2010), *Plasma Physics: An Introduction to Laboratory, Space, and Fusion Plasmas* (Heidelberg: Springer).

Schiff, L. I. (1968), *Quantum Mechanics* (New York: McGraw Hill).

Siegman, A. E. (1986), *Lasers* (Melville, NY: University Science Books).

Stafe, M., A. Marcu and N. N. Puscas (2014), *Pulsed Laser Ablation of Solids: Basics, Theory and Applications*, (Heidelberg: Springer).

Zel'dovich, Y. B. and Y. P. Raizer (2002), *Physics of Shock Waves and High-Temperature Hydrodynamic Phenomena*, W. D. Hayes and R. F. Probstein, eds. (Mineola, NY: Dover Publications).

Zwillinger, D. (1992), *Handbook of Integration* (Boston, MA: Jones and Bartlett Publishers).

References in Individual Chapters

Chapter 1 Introduction to Strong Field Physics

Abu-Shawareb, H. *et al. (NIF Team)*, (2022), "Lawson Criterion for Ignition Exceeded in an Inertial Fusion Experiment," *Physical Review Letters* **129**, 075001.

Agostini, P., F. Fabre, G. Mainfray, G. Petite and N. K. Rahman (1979), "Free-Free Transitions Following 6-Photon Ionization of Xenon Atoms," *Physical Review Letters* **42**, 1127.

Augst, S., D. D. Meyerhofer, D. Strickland and S. L. Chin (1991), "Laser Ionization of Noble-Gases by Coulomb-Barrier Suppression," *Journal of the Optical Society of America B-Optical Physics* **8**, 858.

Braun, A., G. Korn, X. Liu, D. Du, J. Squier and G. Mourou (1995), "Self-Channeling of High-Peak-Power Femtosecond Laser-Pulses in Air," *Optics Letters* **20**, 73.

Brown, L. S. and T. W. B. Kibble (1964), "Interaction of Intense Laser Beams with Electrons," *Physical Review A-General Physics* **133**, A705.

Burnett, N. H. and P. B. Corkum (1989), "Cold-Plasma Production for Recombination Extreme-Ultraviolet Lasers by Optical-Field-Induced Ionization," *Journal of the Optical Society of America B-Optical Physics* **6**, 1195.

Chang, Z. H., A. Rundquist, H. W. Wang, M. M. Murnane and H. C. Kapteyn (1997), "Generation of Coherent Soft X Rays at 2.7 nm Using High Harmonics," *Physical Review Letters* **79**, 2967.

Chelkowski, S. and A. D. Bandrauk (1995), "Two-Step Coulomb Explosions of Diatoms in Intense Laser Fields," *Journal of Physics B-Atomic Molecular and Optical Physics* **28**, L723.

Corkum, P. B. (1993), "Plasma Perspective on Strong-Field Multiphoton Ionization," *Physical Review Letters* **71**, 1994.

Cowan, T. E., M. D. Perry, M. H. Key, T. Ditmire, S. P. Hatchett, E. A. Henry, J. D. Moody, M. J. Moran, D. M. Pennington, T. W. Phillips, T. C. Sangster, J. A. Sefcik, M. S. Singh, R. A. Snavely, M. A. Stoyer, S. C. Wilks, P. E. Young, Y. Takahashi, B. Dong, W. Fountain, T. Parnell, J. Johnson, A. W. Hunt and T. Kuhl (1999), "High Energy Electrons, Nuclear Phenomena and Heating in Petawatt Laser-Solid Experiments," *Laser and Particle Beams* **17**, 773.

Cowan, T. E., A. W. Hunt, T. W. Phillips, S. C. Wilks, M. D. Perry, C. Brown, W. Fountain, S. Hatchett, J. Johnson, M. H. Key, T. Parnell, D. M. Pennington, R. A. Snavely and Y. Takahashi (2000), "Photonuclear Fission from High Energy Electrons from Ultraintense Laser–Solid Interactions," *Physical Review Letters* **84**, 903.

Demaria, A. J., D. A. Stetser and H. Heynau (1966), "Self Mode-Locking of Lasers with Saturable Absorbers - (Regenerative Pulse Oscillator Bleachable Dyes E)," *Applied Physics Letters* **8**, 174.

Ditmire, T., T. Donnelly, R. W. Falcone and M. D. Perry (1995), "Strong X-Ray-Emission from High-Temperature Plasmas Produced by Intense Irradiation of Clusters," *Physical Review Letters* **75**, 3122.

Ditmire, T., J. Zweiback, V. P. Yanovsky, T. E. Cowan, G. Hays and K. B. Wharton (1999), "Nuclear Fusion from Explosions of Femtosecond Laser-Heated Deuterium Clusters," *Nature* **398**, 489.

Fabre, F., G. Petite, P. Agostini and M. Clement (1982), "Multi-Photon Above-Threshold Ionization of Xenon at 0.53 and 1.06-Mu-M," *Journal of Physics B-Atomic Molecular and Optical Physics* **15**, 1353.

Fittinghoff, D. N., P. R. Bolton, B. Chang and K. C. Kulander (1992), "Observation of Nonsequential Double Ionization of Helium with Optical Tunneling," *Physical Review Letters* **69**, 2642.

Franken, P. A., A. E. Hill, C. W. Peters and G. Weinreich (1961), "Generation of Optical Harmonics," *Physical Review Letters* **7**, 118.

Gallagher, T. F. (1992), "Rydberg Atoms in Strong Microwave Fields," in *Atoms in Intense Laser Fields*, M. Gavrila, ed. (Boston: Academic Press), p. 67.

Goeppert-Mayer, M. (1931), "Über Elementarakte mit zwei Quantensprüngen," *Annals of Physics* **9**, 273.

Gonsalves, A. J., K. Nakamura, J. Daniels, C. Benedetti, C. Pieronek, T. C. H. de Raadt, S. Steinke, J. H. Bin, S. S. Bulanov, J. van Tilborg, C. G. R. Geddes, C. B. Schroeder, Cs. Toth, E. Esarey, K. Swanson, L. Fan-Chiang, G. Bagdasarov, N. Bobrova, V. Gasilov, G. Korn, P. Sasorov and W. P. Leemans (2019), "Petawatt Laser Guiding and Electron Beam Acceleration to 8 GeV in a Laser-Heated Capillary Discharge Waveguide," *Physical Review Letters* **122**, 084801.

Hentschel, M., R. Kienberger, C. Spielmann, G. A. Reider, N. Milosevic, T. Brabec, P. Corkum, U. Heinzmann, M. Drescher and F. Krausz (2001), "Attosecond Metrology," *Nature* **414**, 509.

Keldysh, L. V. (1965), "Ionization in Field of a Strong Electromagnetic Wave," *Soviet Physics JETP-USSR* **20**, 1307.

Kmetec, J. D., C. L. Gordon, J. J. Macklin, B. E. Lemoff, G. S. Brown and S. E. Harris (1992), "Mev X-Ray Generation with a Femtosecond Laser," *Physical Review Letters* **68**, 1527.

Krausz, F. and P. Corkum (2002), "Research Supports Observation of Attosecond Pulses," *Laser Focus World* **38**, 7.

Kulander, K. C., K. J. Schafer and J. L. Krause (1993), in *Super-Intense Laser Atom Physics*, NATO ASI Ser., B. Piraux, and K. Rzazewski, eds. (New York: Plenum Press).

L'Huillier, A., L. A. Lompre, G. Mainfray and C. Manus (1982), "Multiply Charged Ions Formed by Multi-Photon Absorption Processes in the Continuum," *Physical Review Letters* **48**, 1814.

L'Huillier, A. and P. Balcou (1993), "High-Order Harmonic-Generation in Rare-Gases with a 1-Ps 1053-Nm Laser," *Physical Review Letters* **70**, 774.

McClung, F. J. and R. W. Hellwarth (1962), "Giant Optical Pulsations from Ruby," *Journal of Applied Physics* **33**, 828.

McPherson, A., G. Gibson, H. Jara, U. Johann, T. S. Luk, I. A. McIntyre, K. Boyer and C. K. Rhodes (1987), "Studies of Multiphoton Production of Vacuum Ultraviolet-Radiation in the Rare-Gases," *Journal of the Optical Society of America B-Optical Physics* **4**, 595.

McPherson, A., T. S. Luk, B. D. Thompson, K. Boyer, and C. K. Rhodes (1993), "Multiphoton-Induced X-Ray-Emission and Amplification from Clusters," *Applied Physics B-Photophysics and Laser Chemistry* **57**, 337.

Maine, P., D. Strickland, P. Bado, M. Pessot, and G. Mourou (1987), "Towards the Development of Petawatt Power Pulses," *Revue De Physique Appliquee* **22**, 1657.

Perelomov, A. M., V. S. Popov and M. V. Terentev (1966), "Ionization of Atoms in an Alternating Electric Field," *Soviet Physics JETP-USSR* **23**, 924.

Perry, M. D., D. Pennington, B. C. Stuart, G. Tietbohl, J. A. Britten, C. Brown, S. Herman, B. Golick, M. Kartz, J. Miller, H. T. Powell, M. Vergino and V. Yanovsky (1999), "Petawatt Laser Pulses," *Optics Letters* **24**, 160.

Posthumus, J. H., L. J. Frasinski, A. J. Giles and K. Codling (1995), "Dissociative Ionization of Molecules in Intense Laser Fields: A Method of Predicting Ion Kinetic Energies and Appearance Intensities," *Journal of Physics B-Atomic Molecular and Optical Physics* **28**, L349.

Reiss, H. R. (1980), "Effect of an Intense Electromagnetic-Field on a Weakly Bound System," *Physical Review A* **22**, 1786.

Roth, M., D. Jung, K. Falk, N. Guler, O. Deppert, M. Devlin, A. Favalli, J. Fernandez, D. Gautier, M. Geissel, R. Haight, C. E. Hamilton, B. M. Hegelich, R. P. Johnson, F. Merrill, G. Schaumann, K. Schoenberg, M. Schollmeier, T. Shimada, T. Taddeucci, J. L. Tybo, F. Wagner, S. A. Wender, C. H. Wilde and G. A. Wurden (2001), "Fast Ignition by Intense Laser-Accelerated Proton Beams," *Physical Review Letters* **86**, 436.

Schmidt, M., D. Normand, and C. Cornaggia (1994), "Laser-Induced Trapping of Chlorine Molecules with Picosecond and Femtosecond Pulses," *Physical Review A* **50**, 5037.

Shank, C. V. and E. P. Ippen (1974), "Subpicosecond Kilowatt Pulses from a Mode-Locked CW Dye Laser," *Applied Physics Letters* **24**, 373.

Snavely, R. A., M. H. Key, S. P. Hatchett, T. E. Cowan, M. Roth, T. W. Phillips, M. A. Stoyer, E. A. Henry, T. C. Sangster, M. S. Singh, S. C. Wilks, A. MacKinnon, A. Offenberger, D. M. Pennington, K. Yasuike, A. B. Langdon, B. F. Lasinski, J. Johnson, M. D. Perry and E. M. Campbell (2000), "Intense High-Energy Proton Beams from Petawatt-Laser Irradiation of Solids," *Physical Review Letters* **85**, 2945.

Strickland, D. and G. Mourou (1985), "Compression of Amplified Chirped Optical Pulses," *Optics Communications* **55**, 447.

Tabak, M., J. Hammer, M. E. Glinsky, W. L. Kruer, S. C. Wilks, J. Woodworth, E. M. Campbell, M. D. Perry and R. J. Mason (1994), "Ignition and High-Gain with Ultrapowerful Lasers," *Physics of Plasmas* **1**, 1626.

Tajima, T. and J. M. Dawson (1979), "Laser Electron-Accelerator," *Physical Review Letters* **43**, 267.

Talebpour, A., C. Y. Chien and S. L. Chin (1996), "The Effects of Dissociative Recombination in Multiphoton Ionization of O-2," *Journal of Physics B-Atomic Molecular and Optical Physics* **29**, L677.

Walker, B., B. Sheehy, L. F. Dimauro, P. Agostini, K. J. Schafer and K. C. Kulander (1994), "Precision-Measurement of Strong-Field Double-Ionization of Helium," *Physical Review Letters* **73**, 1227.

Chapter 2 Strong Field Generation by High-Intensity Lasers

Adachi, S., N. Ishii, T. Kanai, A. Kosuge, J. Itatani, Y. Kobayashi, D. Yoshitomi, K. Torizuka and S. Watanabe (2008), "5-Fs, Multi-mJ, CEP-Locked Parametric Chirped-Pulse Amplifier Pumped by a 450-Nm Source at 1 kHz," *Optics Express* **16**, 14341.

Ahmad, I., S. A. Trushin, Z. Major, C. Wandt, S. Klingebiel, T. J. Wang, V. Pervak, A. Popp, M. Siebold, F. Krausz and S. Karsch (2009), "Frontend Light Source for Short-Pulse Pumped OPCPA System," *Applied Physics B-Lasers and Optics* **97**, 529.

Aoyama, M., K. Yamakawa, Y. Akahane, J. Ma, N. Inoue, H. Ueda, and H. Kiriyama (2003), "0.85-Pw, 33-Fs Ti : Sapphire Laser," *Optics Letters* **28**, 1594.

Backus, S., R. Bartels, S. Thompson, R. Dollinger, H. C. Kapteyn, and M. M. Murnane (2001), "High-Efficiency, Single-Stage 7-kHz High-Average-Power Ultrafast Laser System," *Optics Letters* **26**, 465.

Backus, S., C. G. Durfee, M. M. Murnane, and H. C. Kapteyn, "High Power Ultrafast Lasers," *Review of Scientific Instruments* **69**, 1207 (1998).

Bahk, S. W., P. Rousseau, T. A. Planchon, V. Chvykov, G. Kalintchenko, A. Maksimchuk, G. A. Mourou and V. Yanovsky (2004), "Generation and Characterization of the Highest Laser Intensities (10^{22} W/cm^2)," *Optics Letters* **29**, 2837.

Banks, P. S., M. D. Perry, V. Yanovsky, S. N. Fochs, B. C. Stuart and J. Zweiback (2000), "Novel All-Reflective Stretcher for Chirped-Pulse Amplification of Ultrashort Pulses," *IEEE Journal of Quantum Electronics* **36**, 268.

Barty, C. P. J., C. L. Gordon and B. E. Lemoff (1994), "Multiterawatt 30-Fs Ti-Sapphire Laser System," *Optics Letters* **19**, 1442.

Barty, C. P. J., T. Guo, C. LeBlanc, F. Raksi, C. RosePetruck, J. Squier, K. R. Wilson, V. V. Yakovlev and K. Yamakawa (1996), "Generation of 18-Fs, Multiterawatt Pulses by Regenerative Pulse Shaping and Chirped-Pulse Amplification," *Optics Letters* **21**, 668.

Bonlie, J. D., F. Patterson, D. Price, B. White and P. Springer (2000), "Production of $>10^{21}$ W/cm^2 from a Large-Aperture Ti: Sapphire Laser System," *Applied Physics B-Lasers and Optics* **70**, S155.

Brabec, T., C. Spielmann, P. F. Curley and F. Krausz (1992), "Kerr Lens Mode-Locking," *Optics Letters* **17**, 1292.

Caird, J. A., A. J. Ramponi and P. R. Staver (1991), "Quantum Efficiency and Excited-State Relaxation Dynamics in Neodymium-Doped Phosphate Laser Glasses," *Journal of the Optical Society of America B-Optical Physics* **8**, 1391.

Cerullo, G. and S. De Silvestri (2003), "Ultrafast Optical Parametric Amplifiers," *Review of Scientific Instruments* **74**, 1.

Chambaret, J. P., C. LeBlanc, G. Cheriaux, P. Curley, G. Darpentigny, P. Rousseau, G. Hamoniaux, A. Antonetti and F. Salin (1996), "Generation of 25-TW, 32-fs Pulses at 10 Hz," *Optics Letters* **21**, 1921.

Chanteloup, J. C., F. Druon, M. Nantel, A. Maksimchuk and G. Mourou (1998), "Single-Shot Wave-Front Measurements of High-Intensity Ultrashort Laser Pulses with a Three-Wave Interferometer," *Optics Letters* **23**, 621.

Cheriaux, G., P. Rousseau, F. Salin, J. P. Chambaret, B. Walker, and L. F. Dimauro (1996), "Aberration-Free Stretcher Design for Ultrashort-Pulse Amplification," *Optics Letters* **21**, 414.

Corkum, P. B. (1985), "Amplification of Picosecond 10-μm Pulses in Multiatmosphere CO_2-Lasers," *IEEE Journal of Quantum Electronics* **21**, 216.

Cundiff, S. T. and J. Ye (2005), "Phase Stabilization of Mode-Locked Lasers," *Journal of Modern Optics* **52**, 201.

Cundiff, S. T., F. Krausz and T. Fuji (2008), "Carrier-Envelope Phase of Ultrashort Pulses," in *Strong Field Laser Physics*, T. Brabec, ed. (New York: Springer), p. 61.

Danson, C. N., P. A. Brummitt, R. J. Clarke, J. L. Collier, B. Fell, A. Frackiewicz, S. Hancock, S. Hawkes, C. Hernandez-Gomez, P. Holligan, M. H. R. Hutchinson, A. Kidd, W. J. Lester, I. O. Musgrave, D. Neely, D. R. Neville, P. A. Norreys, D. A. Pepler, C. J. Reason, W. Shaikh, T. B. Winstone, R. W. W. Wyatt and B. E. Wyborn (2004), "Vulcan Petawatt: An Ultra-High-Intensity Interaction Facility," *Nuclear Fusion* **44**, S239.

Ditmire, T. and M. D. Perry (1993), "Terawatt Cr-LiSrAlF$_6$ Laser System," *Optics Letters* **18**, 426.

Divall, E. J., C. B. Edwards, G. J. Hirst, C. J. Hooker, A. K. Kidd, J. M. D. Lister, R. Mathumo, I. N. Ross, M. J. Shaw, W. T. Toner, A. P. Visser and B. E. Wyborn (1996), "Titania: A 10^{20} W cm^{-2} Ultraviolet Laser," *Journal of Modern Optics* **43**, 1025.

Dubietis, A., G. Jonusauskas and A. Piskarskas (1992), "Powerful Femtosecond Pulse Generation by Chirped and Stretched Pulse Parametric Amplification in BBO Crystal," *Optics Communications* **88**, 437.

Eimerl, D., L. Davis, S. Velsko, E. K. Graham and A. Zalkin (1987), "Optical, Mechanical, and Thermal-Properties of Barium Borate," *Journal of Applied Physics* **62**, 1968.

Ell, R., U. Morgner, F. X. Kartner, J. G. Fujimoto, E. P. Ippen, V. Scheuer, G. Angelow, T. Tschudi, M. J. Lederer, A. Boiko and B. Luther-Davies (2001), "Generation of 5-Fs Pulses and Octave-Spanning Spectra Directly from a Ti:Sapphire Laser," *Optics Letters* **26**, 373.

Fernmann, M. E. and I. Hartl (2009), "Ultrafast Fiber Laser Technology," *IEEE Journal of Selected Topics in Quantum Electronics* **15**, 191.

Fernmann, M. E. and I. Hartl (2013), "Ultrafast Fibre Lasers," *Nature Photonics* **7**, 868.

Fittinghoff, D. N., B. C. Walker, J. A. Squier, C. S. Toth, C. Rose-Petruck and C. P. J. Barty (1998), "Dispersion Considerations in Ultrafast CPA Systems," *IEEE Journal of Selected Topics in Quantum Electronics* **4**, 430.

Frantz, L. M. and J. S. Nodvik (1963), "Theory of Pulse Propagation in a Laser Amplifier," *Journal of Applied Physics* **34**, 2346.

Fu, Q., F. Seier, S. K. Gayen and R. R. Alfano (1997), "High-Average-Power Kilohertz-Repetition-Rate Sub-100-Fs Ti:Sapphire Amplifier System," *Optics Letters* **22**, 712.

Gaul, E. W., M. Martinez, J. Blakeney, A. Jochmann, M. Ringuette, D. Hammond, T. Borger, R. Escamilla, S. Douglas, W. Henderson, G. Dyer, A. Erlandson, R. Cross, J. Caird, C. Ebbers and T. Ditmire (2010), "Demonstration of a 1.1 Petawatt Laser Based on a Hybrid Optical Parametric Chirped Pulse Amplification/Mixed Nd:Glass Amplifier," *Applied Optics* **49**, 1676.

Haberberger, D., S. Tochitsky and C. Joshi (2010), "Fifteen Terawatt Picosecond CO_2 Laser System," *Optics Express* **18**, 17865.

Hardy, J. W. (1978), "Active Optics: A New Technology for the Control of Light," *Proceedings of IEEE* **66**, 651.

Hariharan, A., M. E. Fermann, M. L. Stock, D. J. Harter and J. Squier (1996), "Alexandrite-Pumped Alexandrite Regenerative Amplifier for Femtosecond Pulse Amplification," *Optics Letters* **21**, 128.

Hunt, J. T., J. A. Glaze, W. W. Simmons and P. A. Renard (1978), "Suppression of Self-Focusing through Low-Pass Spatial-Filtering and Relay Imaging," *Applied Optics* **17**, 2053.

Jain, S. C., V. V. Rampal and U. C. Joshi (1992), "Mode-Locked Lasers," *Optical Engineering* **31**, 1287.

Jones, D. J., S. A. Diddams, J. K. Ranka, A. Stentz, R. S. Windeler, J. L. Hall and S. T. Cundiff (2000), "Carrier-Envelope Phase Control of Femtosecond Mode-Locked Lasers and Direct Optical Frequency Synthesis," *Science* **288**, 635.

Jullien, A., C. G. Durfee, A. Trisorio, L. Canova, J. P. Rousseau, B. Mercier, L. Antonucci, G. Cheriaux, O. Albert and R. Lopez-Martens (2009), "Nonlinear Spectral Cleaning of Few-Cycle Pulses via Cross-Polarized Wave (XPW) Generation," *Applied Physics B-Lasers and Optics* **96**, 293.

Jung, I. D., F. X. Kartner, N. Matuschek, D. H. Sutter, F. Morier Genoud, G. Zhang, U. Keller, V. Scheuer, M. Tilsch and T. Tschudi (1997), "Self-Starting 6.5-fs Pulses from a Ti:Sapphire Laser," *Optics Letters* **22**, 1009.

Kane, S. and J. Squier (1997), "Fourth-Order-Dispersion Limitations of Aberration-Free Chirped-Pulse Amplification Systems," *Journal of the Optical Society of America B-Optical Physics* **14**, 1237.

Keller, U., K. J. Weingarten, F. X. Kartner, D. Kopf, B. Braun, I. D. Jung, R. Fluck, C. Honninger, N. Matuschek and J. A. der Au (1996), "Semiconductor Saturable Absorber Mirrors (Sesam's) for Femtosecond to Nanosecond Pulse Generation in Solid-State Lasers," *IEEE Journal of Selected Topics in Quantum Electronics* **2**, 435.

Keppler, S., R. Bodefeld, M. Hornung, A. Savert, J. Hein and M. C. Kaluza (2011), "Pre-pulse Suppression in a Multi-10-TW Diode-Pumped Yb:Glass Laser," *Applied Physics B-Lasers and Optics* **104**, 11.

Klingebiel, S., C. Wandt, C. Skrobol, I. Ahmad, S. A. Trushin, Z. Major, F. Krausz and S. Karsch (2011), "High Energy Picosecond Yb:Yag Cpa System at 10 Hz Repetition Rate for Pumping Optical Parametric Amplifiers," *Optics Express* **19**, 5357.

Kmetec, J. D., J. J. Macklin and J. F. Young (1991), "0.5-TW, 125-Fs Ti-Sapphire Laser," *Optics Letters* **16**, 1001.

Leblanc, C., G. Grillon, J. P. Chambaret, A. Migus and A. Antonetti (1993), "Compact and Efficient Multipass Ti-Sapphire System for Femtosecond Chirped-Pulse Amplification at the Terawatt Level," *Optics Letters* **18**, 140.

Lemoff, B. E. and C. P. J. Barty (1993), "Quintic-Phase-Limited, Spatially Uniform Expansion and Recompression of Ultrashort Optical Pulses," *Optics Letters* **18**, 1651.

Limpert, J., S. Hadrich, J. Rothhardt, M. Krebs, T. Eidam, T. Schreiber and A. Tunnermann (2011), "Ultrafast Fiber Lasers for Strong-Field Physics Experiments," *Laser & Photonics Reviews* **5**, 634.

Lureau, F., O. Chalus, G. Matras, S. Laux, C. Radier, O. Casagrande, C. Derycke, S. Ricaud, G. Rey, T. Morbieu, A. Pellegrine, L. Boudjemaa, C. Simon-Boisson, A. Baleanu, R. Banici, A. Gradinariu, C. Caldararu, P. Ghenuche, A. Naziru, C. Caldaruru, P. Ghenuche,

A. Naziru, S. Kolliopoulos, L. Neagu, B. De Boisdeffre, D. Ursescu and I. Dancus (2020), in *Solid State Lasers XXIX: Technology and Devices*, Proceedings of the SPIE Vol. **11259**, 112591J.

Maine, P., D. Strickland, P. Bado, M. Pessot and G. Mourou (1988), "Generation of Ultra-high Peak Power Pulses by Chirped Pulse Amplification," *IEEE Journal of Quantum Electronics* **24**, 398.

Martinez, O. E. (1987), "3000 Times Grating Compressor with Positive Group-Velocity Dispersion: Application to Fiber Compensation in 1.3-1.6 Mu-M Region," *IEEE Journal of Quantum Electronics* **23**, 59.

Moulton, P. F. (1986), "Spectroscopic and Laser Characteristics of Ti-Al_2O_3," *Journal of the Optical Society of America B-Optical Physics* **3**, 125.

Murray, J. E. and W. H. Lowdermilk (1980), "Nd - YAG Regenerative Amplifier," *Journal of Applied Physics* **51**, 3548.

Nees, J., S. Biswal, F. Druon, J. Faure, M. Nantel, G. A. Mourou, A. Nishimura, H. Takuma, J. Itatani, J. C. Chanteloup and C. Honninger (1998), "Ensuring Compactness, Reliability, and Scalability for the Next Generation of High-Field Lasers," *IEEE Journal of Selected Topics on Quantum Electronics* **4**, 376.

Noom, D. W. E., S. Witte, J. Morgenweg, R. K. Altmann and K. S. E. Eikema (2013), "High-Energy, High-Repetition-Rate Picosecond Pulses from a Quasi-CW Diode-Pumped Nd:Yag System," *Optics Letters* **38**, 3021.

Norris, T. B. (1992), "Femtosecond Pulse Amplification at 250 kHz with a Ti-Sapphire Regenerative Amplifier and Application to Continuum Generation," *Optics Letters* **17**, 1009.

Ortac, B., J. Limpert and A. Tunnermann (2007), "High-Energy Femtosecond Yb-Doped Fiber Laser Operating in the Anomalous Dispersion Regime," *Optics Letters* **32**, 2149.

Payne, S. A., L. K. Smith, R. J. Beach, B. H. T. Chai, J. H. Tassano, L. D. Deloach, W. L. Kway, R. W. Solarz and W. F. Krupke (1994), "Properties of Cr:$LiSrAlF_6$ Crystals for Laser Operation," *Applied Optics* **33**, 5526.

Perry, M. D., R. D. Boyd, J. A. Britten, D. Decker, B. W. Shore, C. Shannon and E. Shults (1995), "High-Efficiency Multilayer Dielectric Diffraction Gratings," *Optics Letters* **20**, 940.

Perry, M. D., O. L. Landen, J. Weston and R. Ettlebrick (1989), "Design and Performance of a High-Power, Synchronized Nd-YAG Dye Laser System," *Optics Letters* **14**, 42.

Perry, M. D., T. Ditmire and B. C. Stuart (1994), "Self-Phase Modulation in Chirped Pulse Amplification," *Optics Letters* **19**, 2149.

Perry, M. D., D. Pennington, B. C. Stuart, G. Tietbohl, J. A. Britten, C. Brown, S. Herman, B. Golick, M. Kartz, J. Miller, H. T. Powell, M. Vergino and V. Yanovsky (1999), "Petawatt Laser Pulses," *Optics Letters* **24**, 160.

Pessot, M., J. Squier, P. Bado, G. Mourou and D. J. Harter (1989), "Chirped Pulse Amplification of 300-Fs Pulses in an Alexandrite Regenerative Amplifier," *IEEE Journal of Quantum Electronics* **25**, 61.

Pittman, M., S. Ferre, J. P. Rousseau, L. Notebaert, J. P. Chambaret and G. Cheriaux (2002), "Design and Characterization of a Near-Diffraction-Limited Femtosecond 100-TW 10-Hz High-Intensity Laser System," *Applied Physics B-Lasers and Optics* **74**, 529.

Planchon, T. A., J. P. Rousseau, F. Burgy, G. Cheriaux and J. P. Chambaret (2005), "Adaptive Wavefront Correction on a 100-TW/10-Hz Chirped Pulse Amplification Laser and Effect of Residual Wavefront on Beam Propagation," *Optics Communications* **252**, 222.

Ple, F., M. Pittman, G. Jamelot and J. P. Chambaret (2007), "Design and Demonstration of a High-Energy Booster Amplifier for a High-Repetition Rate Petawatt Class Laser System," *Optics Letters* **32**, 238.

Powell, H. T., A. C. Erlandson, K. S. Jancaitis and J. E. Murray (1990), "Flashlamp Pumping of Nd-Glass Disk Amplifiers," in *High-Power Solid State Lasers and Applications*, SPIE Proceedings Vol. **1277**.

Qiao, J., A. Kalb, T. Nguyen, J. Bunkenburg, D. Canning and J. H. Kelly (2008), "Demonstration of Large-Aperture Tiled-Grating Compressors for High-Energy, Petawatt-Class, Chirped-Pulse Amplification Systems," *Optics Letters* **33**, 1684.

Ricci, A., A. Jullien, J. P. Rousseau, Y. Liu, A. Houard, P. Ramirez, D. Papadopoulos, A. Pellegrina, P. Georges, F. Druon, N. Forget and R. Lopez-Martens (2013), "Energy-Scalable Temporal Cleaning Device for Femtosecond Laser Pulses Based on Cross-Polarized Wave Generation," *Review of Scientific Instruments* **84**, 043106.

Rimmer, M. P. and J. C. Wyant (1975) "Evaluation of Large Aberrations Using a Lateral-Shear Interferometer Having Variable Shear," *Applied Optics* **14**, 142.

Ross, I. N., J. L. Collier, P. Matousek, C. N. Danson, D. Neely, R. M. Allot, D. A. Pepler, C. Hernandez-Gomez and K. Osvay (2000), "Generation of Terawatt Pulses by Use of Optical Parametric Chirped Pulse Amplification," *Applied Optics* **39**, 2422.

Ross, I. N., A. R. Damerell, E. J. Divall, J. Evans, G. J. Hirst, C. J. Hooker, J. R. Houliston, M. H. Key, J. M. D. Lister, K. Osvay and M. J. Shaw (1994), "A 1 TW KrF Laser Using Chirped Pulse Amplification," *Optics Communications* **109**, 288.

Ross, I. N., P. Matousek, M. Towrie, A. J. Langley and J. L. Collier (1997a), "The Prospects for Ultrashort Pulse Duration and Ultrahigh Intensity Using Optical Parametric Chirped Pulse Amplifiers," *Optics Communications* **144**, 125.

Ross, I. N., M. Trentelman and C. N. Danson (1997b), "Optimization of a Chirped-Pulse Amplification Nd:Glass Laser," *Applied Optics* **36**, 9348.

Rouyer, C., E. Mazataud, I. Allais, A. Pierre, S. Seznec, C. Sauteret, G. Mourou and A. Migus (1993), "Generation of 50-TW Femtosecond Pulses in a Ti-Sapphire/Nd-Glass Chain," *Optics Letters* **18**, 214.

Rudd, J. V., G. Korn, S. Kane, J. Squier, G. Mourou and P. Bado (1993), "Chirped-Pulse Amplification of 55-Fs Pulses at a 1-kHz Repetition Rate in a Ti-Al_2O_3 Regenerative Amplifier," *Optics Letters* **18**, 2044.

Saraceno, C. J., C. Schriber, M. Mangold, M. Hoffmann, O. H. Heckl, C. R. E. Baer, M. Golling, T. Sudmeyer and U. Keller (2012), "Sesams for High-Power Oscillators: Design Guidelines and Damage Thresholds," *IEEE Journal of Selected Topics in Quantum Electronics* **18**, 29.

Sasnett, M. W. (1989), "Propagation of Multimode Laser Beams – M^2 Factor," in *The Physics and Technology of Laser Resonators*, ed. D. R. Hall and P. E. Jackson (Bristol: Adam Hilger), p. 132.

Schafer, F. P., W. Schmidt and J. Volze (1966), "Organic Dye Solution Laser," *Applied Physics Letters* **9**, 306.

Schneider, W., A. Ryabov, C. Lombosi, T. Metzger, Z. Major, J. A. Fulop and P. Baum (2014), "800-Fs, 330-µJ Pulses from a 100-W Regenerative Yb:Yag Thin-Disk Amplifier at 300 kHz and THz Generation In Linbo3," *Optics Letters* **39**, 6604.

Shore, B. W., M. D. Perry, J. A. Britten, R. D. Boyd, M. D. Feit, H. T. Nguyen, R. Chow, G. E. Loomis and L. F. Li (1997), "Design of High-Efficiency Dielectric Reflection Gratings," *Journal of the Optical Society of America A-Optics Image Science and Vision* **14**, 1124.

Spielmann, C., P. F. Curley, T. Brabec and F. Krausz (1994), "Ultrabroadband Femtosecond Lasers," *IEEE Journal of Quantum Electronics* **30**, 1100.

Strickland, D. and G. Mourou (1985), "Compression of Amplified Chirped Optical Pulses," *Optics Communications* **55**, 447.

Sullivan, A., J. Bonlie, D. F. Price and W. E. White (1996), "1.1-J, 120-fs Laser System Based on Nd:Glass-Pumped Ti:Sapphire," *Optics Letters* **21**, 603.

Tavella, F., K. Schmid, N. Ishii, A. Marcinkevicius, L. Veisz and F. Krausz (2005), "High-Dynamic Range Pulse-Contrast Measurements of a Broadband Optical Parametric Chirped-Pulse Amplifier," *Applied Physics B-Lasers and Optics* **81**, 753.

Tiwari, G., E. Gaul, M. Martinez, G. Dyer, J. Gordon, M. Spinks, T. Toncian, B. Bowers, X. Jiao, R. Kupfer, L. Lisi, E. McCary, R. Roycroft, A. Yandow, G. D. Glenn, M. Donovan, T. Ditmire and B. M. Hegelich (2019), "Beam Distortion Effects upon Focusing an Ultrashort Petawatt Laser Pulse to Greater Than 10^{22} W/cm^2," *Optics Letters* **44**, 2764.

Tochitsky, S. Y., J. J. Pigeon, D. J. Haberberger, C. Gong and C. Joshi (2012), "Amplification of Multi-Gigawatt 3 Ps Pulses in an Atmospheric CO_2 Laser Using AC Stark Effect," *Optics Express* **20**, 13762.

Tournois, P. (1997), *Optics Communications* **140**, 245.

Treacy, E. B. (1969), "Optical Pulse Compression with Diffraction Gratings," *IEEE Journal of Quantum Electronics* **QE 5**, 454.

Wall, K. F. and A. Sanchez (1990), "Titanium Sapphire Lasers," *The Lincoln Laboratory Journal* **3**, 447.

Walling, J. C., H. P. Jenssen, R. C. Morris, E. W. Odell and O. G. Peterson (1979), "Tunable-Laser Performance in $BeAl_2O_4$-Cr^{3+}," *Optics Letters* **4**, 182.

White, W. E., J. R. Hunter, L. Vanwoerkom, T. Ditmire and M. D. Perry (1992), "120-fs Terawatt Ti-Al_2O_3/Cr-$LiSrAlF_6$ Laser System," *Optics Letters* **17**, 1067.

Yamakawa, K. and C. P. J. Barty (2000), "Ultrafast, Ultrahigh-Peak, and High-Average Power Ti : Sapphire Laser System and Its Applications," *IEEE Journal of Selected Topics in Quantum Electronics* **6**, 658.

Yamanaka, C., Y. Kato, Y. Izawa, K. Yoshida, T. Yamanaka, T. Sasaki, M. Nakatsuka, T. Mochizuki, J. Kuroda and S. Nakai (1981), "Nd-Doped Phosphate-Glass Laser Systems for Laser-Fusion Research," *IEEE Journal of Quantum Electronics* **17**, 1639.

Zhang, M., E. J. R. Kelleher, S. V. Popov and J. R. Taylor (2014), "Ultrafast Fibre Laser Sources: Examples of Recent Developments," *Optical Fiber Technology* **20**, 666.

Zheng, J. and H. Zacharias (2009), "Non-Collinear Optical Parametric Chirped-Pulse Amplifier for Few-Cycle Pulses," *Applied Physics B-Lasers and Optics* **97**, 765.

Chapter 3 Strong Field Interactions with Free Electrons

Bardsley, J. N., B. M. Penetrante and M. H. Mittleman (1989), "Relativistic Dynamics of Electrons in Intense Laser Fields," *Physical Review A* **40**, 3823.

Bituk, D. R. and M. V. Fedorov (1999), "Relativistic Ponderomotive Forces," *Journal of Experimental and Theoretical Physics* **89**, 640.

Brown, L. S. and T. W. B. Kibble (1964), "Interaction of Intense Laser Beams with Electrons," *Physical Review A-General Physics* **133**, A705.

Bucksbaum, P. H., M. Bashkansky and T. J. McIlrath (1987), "Scattering of Electrons by Intense Coherent Light," *Physical Review Letters* **58**, 349.

Castillo-Herrera, C. I. and T. W. Johnston (1993), "Incoherent Harmonic Emission from Strong Electromagnetic-Waves in Plasmas," *IEEE Transactions on Plasma Science* **21**, 125.

Chen, S., A. Maksimchuk and D. Umstadter (1998), "Experimental Observation of Relativistic Nonlinear Thomson Scattering," *Nature* **396**, 653.

Cicchitelli, L., H. Hora and R. Postle (1990), "Longitudinal-Field Components for Laser-Beams in Vacuum," *Physical Review A* **41**, 3727.

Davis, L. W. (1979), "Theory of Electromagnetic Beams," *Physical Review A* **19**, 1177.

Di Piazza, A. (2008), "Exact Solution of the Landau-Lifshitz Equation in a Plane Wave," *Letters on Mathematical Physics* **83**, 305.

Eberly, J. H. and A. Sleeper (1968), "Trajectory and Mass Shift of a Classical Electron in a Radiation Pulse," *Physical Review* **176**, 1570.

Esarey, E., S. K. Ride and P. Sprangle (1993), "Nonlinear Thomson Scattering of Intense Laser-Pulses from Beams and Plasmas," *Physical Review E* **48**, 3003.

Gunn, J. E. and J. P. Ostriker (1971), "Motion and Radiation of Charged Particles in Strong Electromagnetic Waves: 1. Motion in Plane and Spherical Waves," *Astrophysical Journal*, **165**, 523.

Hartemann, F. V. (1998), "High-Intensity Scattering Processes of Relativistic Electrons in Vacuum," *Physics of Plasmas* **5**, 2037.

Hartemann, F. V., S. N. Fochs, G. P. Lesage, N. C. Luhmann, J. G. Woodworth, M. D. Perry, Y. J. Chen and A. K. Kerman (1995), "Nonlinear Ponderomotive Scattering of Relativistic Electrons by an Intense Laser Field at Focus," *Physical Review E* **51**, 4833.

Holstein, B. R. (1999), "Strong Field Pair Production," *American Journal of Physics* **67**, 499.

Kaplan, A. E. and A. L. Pokrovsky (2005), "Fully Relativistic Theory of the Ponderomotive Force in an Ultraintense Standing Wave," *Physical Review Letters* **95**, 053601.

Kaplan, A. E. and A. L. Pokrovsky (2005), "Fully Relativistic Theory of the Ponderomotive Force in an Ultraintense Standing Wave," *Physical Review Letters* **95**, 053601.

Kibble, T. (1966), "Mutual Refraction of Electrons and Photons," *Physical Review* **150**, 1060.

Maltsev, A. and T. Ditmire (2003), "Above Threshold Ionization in Tightly Focused, Strongly Relativistic Laser Fields," *Physical Review Letters* **90**, 053002.

Melissinos, A. C. (2008), "Tests of QED with Intense Lasers," in *Strong Field Laser Physics*, T. Brabec, ed. (New York: Springer), p. 497.

Meyerhofer, D. D. (1997), "High-Intensity-Laser-Electron Scattering," *IEEE Journal of Quantum Electronics* **33**, 1935.

Monot, P., T. Auguste, L. A. Lompre, G. Mainfray and C. Manus (1993), "Energy Measurements of Electrons Submitted to an Ultrastrong Laser Field," *Physical Review Letters* **70**, 1232.

Quesnel, B. and P. Mora (1998), "Theory and Simulation of the Interaction of Ultraintense Laser Pulses with Electrons in Vacuum," *Physical Review E* **58**, 3719.

Sarachik, E. S. and G. T. Schappert (1970), "Classical Theory of Scattering of Intense Laser Radiation by Free Electrons," *Physical Review D* **1**, 2738.

Schmidt, G. and T. Wilcox (1973), "Relativistic Particle Motion in Nonuniform Electromagnetic Waves," *Physical Review Letters* **31**, 1380.

Schwinger, J. (1951), "On Gauge Invariance and Vacuum Polarization," *Physical Review* **82**, 664.

Startsev, E. A. and C. J. McKinstrie (1997), "Multiple Scale Derivation of the Relativistic Ponderomotive Force," *Physical Review E* **55**, 7527.

Chapter 4 Strong Field Interactions with Single Atoms

Agostini, P. and L. F. DiMauro (2008), "Atoms in High Intensity Mid-Infrared Pulses," *Contemporary Physics* **49**, 179.

Agostini, P., F. Fabre, G. Mainfray, G. Petite and N. K. Rahman (1979), "Free-Free Transitions Following 6-Photon Ionization of Xenon Atoms," *Physical Review Letters* **42**, 1127.

Ammosov, M. V., N. B. Delone and V. P. Krainov (1986), "Tunnel Ionization of Complex Atoms and Atomic Ions in a Varying Electromagnetic-Field," *Zhurnal Eksperimentalnoi I Teoreticheskoi Fiziki* **91**, 2008 [*Soviet Physics – JETP* **64**, 1191 (1986)].

Augst, S., D. D. Meyerhofer, D. Strickland and S. L. Chin (1991), "Laser Ionization of Noble-Gases by Coulomb-Barrier Suppression," *Journal of the Optical Society of America B-Optical Physics* **8**, 858.

Bauer, J. H. (2008), "Simple Proof of Gauge Invariance for the S-Matrix Element of Strong-Field Photoionization," *Physica Scripta* **77**, 015303.

Bebb, H. B. and A. Gold (1966), "Multiphoton Ionization of Hydrogen and Rare-Gas Atoms," *Physical Review* **143**, 1.

Becker, A. and F. H. M. Faisal (1999), "Interplay of Electron Correlation and Intense Field Dynamics in the Double Ionization of Helium," *Physical Review A* **59**, R1742.

Becker, A. and F. H. M. Faisal (2005), "Intense-Field Many-Body S-Matrix Theory," *Journal of Physics B-Atomic Molecular and Optical Physics* **38**, R1.

Becker, A., L. Plaja, P. Moreno, M. Nurhuda and F. H. M. Faisal (2001), "Total Ionization Rates and Ion Yields of Atoms at Nonperturbatve Laser Intensities," *Physical Review A* **64**, 023408.

Becker, A., R. Dorner and R. Moshammer (2005), "Multiple Fragmentation of Atoms in Femtosecond Laser Pulses," *Journal of Physics B – Atomic Molecular and Optical Physics* **38**, S753.

Becker, W. and H. Rottke (2008), "Many-Electron Strong-Field Physics," *Contemporary Physics* **49**, 199.

Becker, W., A. Lohr and M. Kleber (1994), "Effects of Rescattering on Above-Threshold Ionization," *Journal of Physics B – Atomic Molecular and Optical Physics* **27**, L325.

Becker, W., F. Grasbon, R. Koplod, D. B. Milosevic, G. G. Paulus and H. Walther (2002), "Above-Threshold Ionization: From Classical Features to Quantum Effects," in *Progress in Optics Volume XLVIII* (Amsterdam: Elsevier Science Publishers), p. 35.

Bisgaard, C. Z. and L. B. Madsen (2004), "Tunneling Ionization of Atoms," *American Journal of Physics* **72**, 249.

Brabec, T., M. Y. Ivanov and P. B. Corkum (1996), "Coulomb Focusing in Intense Field Atomic Processes," *Physical Review A* **54**, R2551.

Brandi, H. S., L. Davidovich and N. Zagury (1981), "High-Intensity Approximations Applied to Multi-Photon Ionization," *Physical Review A* **24**, 2044.

Burnett, K., V. C. Reed and P. L. Knight (1993), "Atoms in Ultra-Intense Laser Fields," *Journal of Physics B – Atomic Molecular and Optical Physics* **26**, 561.

Burnett, N. H. and P. B. Corkum (1989), "Cold-Plasma Production for Recombination Extreme-Ultraviolet Lasers by Optical-Field-Induced Ionization," *Journal of the Optical Society of America B – Optical Physics* **6**, 1195.

Chang, B., P. R. Bolton and D. N. Fittinghoff (1993), "Closed-Form Solutions for the Production of Ions in the Collisionless Ionization of Gases by Intense Lasers," *Physical Review A* **47**, 4193.

Chen, Z. J., A. T. Le, T. Morishita and C. D. Lin (2009), "Quantitative Rescattering Theory for Laser-Induced High-Energy Plateau Photoelectron Spectra," *Physical Review A* **79**, 033409.

Chirila, C. C. and R. M. Potvliege (2005), "Low-Order Above-Threshold Ionization in Intense Few-Cycle Laser Pulses," *Physical Review A* **71**, 021402.

Colosimo, P., G. Doumy, C. I. Blaga, J. Wheeler, C. Hauri, F. Catoire, J. Tate, R. Chirla, A. M. March, G. G. Paulus, H. G. Muller, P. Agostini and L. Dimauro (2008), "Scaling Strong-Field Interactions Towards the Classical Limit," *Nature Physics* **4**, 386.

Corkum, P. B., N. H. Burnett and F. Brunel (1989), "Above-Threshold Ionization in the Long-Wavelength Limit," *Physical Review Letters* **62**, 1259.

Corkum, P. B., N. H. Burnett and F. Brunel (1992), "Multiphoton Ionization in Large Ponderomotive Potentials," in *Atoms in Intense Laser Fields*, M. Gavrila, ed. (Boston: Academic Press), p. 109.

Dammasch, M., M. Dorr, U. Eichmann, E. Lenz and W. Sandner (2001), "Relativistic Laser-Field-Drift Suppression of Nonsequential Multiple Ionization," *Physical Review A* **64**, 061402.

Dörner, R., Th. Weber, M. Weckenbrock, A. Staudte, M. Hattass and H. Schmidt-Böcking (2002), "Multiple Ionization in Strong Laser Fields," in *Progress in Optics Volume XLVIII* (Amsterdam: Elsevier Science Publishers).

Eberly, J. H., J. Javanainen and K. Rzazewski (1991), "Above-Threshold Ionization," *Physics Reports – Review Section of Physics Letters* **204**, 331.

Faisal, F. H. M. (1973), "Collisions of Electrons with Laser Photons in a Background Potential," *Journal of Physics B* **6**, L312.

Fittinghoff, D. N., P. R. Bolton, B. Chang and K. C. Kulander (1992), "Observation of Nonsequential Double Ionization of Helium with Optical Tunneling," *Physical Review Letters* **69**, 2642.

Freeman, R. R. and P. H. Bucksbaum (1991), "Investigations of Above-Threshold Ionization Using Subpicosecond Laser-Pulses," *Journal of Physics B – Atomic Molecular and Optical Physics* **24**, 325.

Freeman, R. R., P. H. Bucksbaum, W. E. Cooke, G. Gibson, T. J. McIlrath and L. D. Van Woerkom (1992), "Photoionization with Ultrashort Laser Pulses," in *Atoms in Intense Laser Fields*, M. Gavrila, ed. (Boston: Academic Press), p. 43.

Gavrila, M. (2002), "Atomic Stabilization in Superintense Laser Fields," *Journal of Physics B – Atomic Molecular and Optical Physics* **35**, R147.

Grasbon, F., G. G. Paulus, H. Walther, P. Villoresi, G. Sansone, S. Stagira, M. Nisoli and S. De Silvestri (2003), "Above-Threshold Ionization at the Few-Cycle Limit," *Physical Review Letters* **91**, 173003.

Henneberger, W. C. (1968), "Perturbation Method for Atoms in Intense Light Beams," *Physical Review Letters* **21**, 838.

Hu, S. X. and A. F. Starace (2002), "GeV Electrons from Ultraintense Laser Interaction with Highly Charged Ions," *Physical Review Letters* **88**, 245003.

Karnakov, B. M., V. D. Mur, S. V. Popruzhenko and V. S. Popov (2009), "Strong Field Ionization by Ultrashort Laser Pulses: Application of the Keldysh Theory," *Physics Letters A* **374**, 386.

Keitel, C. H. (1996), "Ultra-Energetic Electron Ejection in Relativistic Atom–Laser Field Interaction," *Journal of Physics B – Atomic Molecular and Optical Physics* **29**, L873.

Keldysh, L. V. (1965), "Ionization in Field of a Strong Electromagnetic Wave," *Soviet Physics JETP-USSR* **20**, 1307.

Kling, M. F., J. Rauschenberger, A. J. Verhoef, E. Hasovic, T. Uphues, D. B. Milosevic, H. G. Muller and M. J. J. Vrakking (2008), "Imaging of Carrier-Envelope Phase Effects in Above-Threshold Ionization with Intense Few-Cycle Laser Fields," *New Journal of Physics* **10**, 025024.

Krainov, V. P. (1997), "Ionization Rates and Energy and Angular Distributions at the Barrier-Suppression Ionization of Complex Atoms and Atomic Ions," *Journal of the Optical Society of America B-Optical Physics* **14**, 425.

Lambropoulos, P. and X. Tang (1992), "Resonances in Multiphoton Ionization," in *Atoms in Intense Laser Fields*, M. Gavrila, ed. (Boston: Academic Press), p. 335.

L'Huillier, A., L. A. Lompre, G. Mainfray and C. Manus (1982), "Multiply Charged Ions Formed by Multi-Photon Absorption Processes in the Continuum," *Physical Review Letters* **48**, 1814.

Liu, X. and C. Figueira de Morisson Faria (2004), "Nonsequential Double Ionization with Few-Cycle Laser Pulses," *Physical Review Letters* **92**, 133006.

Lotz, W. (1968), "Electron-Impact Ionization Cross-Sections and Ionization Rate Coefficients for Atoms and Ions from Hydrogen to Calcium," *Zeitschrift Fur Physik* **216**, 241.

Mainfray, G. and C. Manus (1991), "Multiphoton Ionization of Atoms," *Reports on Progress in Physics* **54**, 1333.

Maltsev, A. and T. Ditmire (2003), "Above Threshold Ionization in Tightly Focused, Strongly Relativistic Laser Fields," *Physical Review Letters* **90**, 053002.

Maquet, A., R. Taïeb and V. Véniard (2008), "Relativistic Laser-Atom Physics," in *Strong Field Laser Physics*, T. Brabec, ed. (New York: Springer), p. 477.

Meyerhofer, D. D., J. P. Knauer, S. J. McNaught and C. I. Moore (1996), "Observation of Relativistic Mass Shift Effects During High-Intensity Laser Electron Interactions," *Journal of the Optical Society of America B – Optical Physics* **13**, 113.

Milonni, P. W. and B. Sundaram (1993), "Atoms in Strong Fields: Photoionization and Chaos," in *Progress in Optics Volume XXXI*, E. Wolf, ed. (Amsterdam: Elsevier Science Publishers) p. 3.

Milosevic, N., V. P. Krainov and T. Brabec (2002a), "Relativistic Theory of Tunnel Ionization," *Journal of Physics B – Atomic Molecular and Optical Physics* **35**, 3515.

Milosevic, N., V. P. Krainov and T. Brabec (2002b), "Semiclassical Dirac Theory of Tunnel Ionization," *Physical Review Letters* **89**, 193001.

Milosevic, D. B., G. G. Paulus, D. Bauer and W. Becker (2006), "Above-Threshold Ionization by Few-Cycle Pulses," *Journal of Physics B – Atomic Molecular and Optical Physics* **39**, R203.

Milosevic, D. B., E. Hasovic, M. Busuladzic, A. Gazibegovic-Busuladzic and W. Becker (2007), "Intensity-Dependent Enhancements in High-Order Above-Threshold Ionization," *Physical Review A* **76**, 053410.

Milosevic, D., B., W. Becker, M. Okunishi, G. Prumper, K. Shimada and K. Ueda (2010), "Strong-Field Electron Spectra of Rare-Gas Atoms in the Rescattering Regime: Enhanced Spectral Regions and a Simulation of the Experiment," *Journal of Physics B – Atomic Molecular and Optical Physics* **43**, 015401.

Moore, C. I., J. P. Knauer and D. D. Meyerhofer (1995), "Observation of the Transition from Thomson to Compton Scattering in Multiphoton Interactions with Low-Energy Electrons," *Physical Review Letters* **74**, 2439.

Mostowski, J. and J. H. Eberly (1991), "Approximate Atomic Ionization Rates in the Stabilization Regime of the Kramers–Henneberger Picture," *Journal of the Optical Society of America B – Optical Physics* **8**, 1212.

Muller, H. G., P. Agostini and G. Petite (1992), "Multiphoton Ionization," in *Atoms in Intense Laser Fields*, M. Gavrila, ed. (Boston: Academic Press), p. 1.

Paulus, G. G., F. Grasbon, H. Walther, R. Kopold and W. Becker (2001), "Channel-Closing-Induced Resonances in the Above-Threshold Ionization Plateau," *Physical Review A* **64**, 021401.

Perelomov, A. M., V. S. Popov and M. V. Terentev (1966), "Ionization of Atoms in an Alternating Electric Field," *Soviet Physics JETP-USSR* **23**, 924.

Perry, M. D. and O. L. Landen (1988), "Resonantly Enhanced Multiphoton Ionization of Krypton and Xenon with Intense Ultraviolet-Laser Radiation," *Physical Review A* **38**, 2815.

Perry, M. D., O. L. Landen, A. Szoke and E. M. Campbell (1988a), "Multiphoton Ionization of the Noble-Gases by an Intense 10^{14}-W/cm^2 Dye-Laser," *Physical Review A* **37**, 747.

Perry, M. D., A. Szoke, O. L. Landen and E. M. Campbell (1988b), "Nonresonant Multiphoton Ionization of Noble-Gases: Theory and Experiment," *Physical Review Letters* **60**, 1270.

Pont, M., N. R. Walet, M. Gavrila and C. W. McCurdy (1988), "Dichotomy of the Hydrogen-Atom in Superintense, High-Frequency Laser Fields," *Physical Review Letters* **61**, 939.

Popov, V. S. (2004), "Tunnel and Multiphoton Ionization of Atoms and Ions in a Strong Laser Field (Keldysh Theory)," *Physics-Uspekhi* **47**, 855.

Popruzhenko, S. V. (2014), "Keldysh Theory of Strong Field Ionization: History, Applications, Difficulties and Perspectives," *Journal of Physics B – Atomic Molecular and Optical Physics* **47**, 204001.

Popruzhenko, S. V., V. D. Mur, V. S. Popov and D. Bauer (2008), "Strong Field Ionization Rate for Arbitrary Laser Frequencies," *Physical Review Letters* **101**, 193003.

Potvliege, R. M. and R. Shakeshaft (1992), "Nonperturbative Treatment of Multiphoton Ionization within the Floquet Framework," in *Atoms in Intense Laser Fields*, M. Gavrila, ed. (Boston: Academic Press), p. 373.

Potvliege, R. M. and S. Vucic (2009), "Freeman Resonances in High-Order Above-Threshold Ionization," *Journal of Physics B – Atomic Molecular and Optical Physics* **42**, 055603.

Prager, J. and C. H. Keitel (2002), "Laser-Induced Nonsequential Double Ionization Approaching the Relativistic Regime," *Journal of Physics B – Atomic Molecular and Optical Physics* **35**, L167.

Protopapas, M., C. H. Keitel and P. L. Knight (1997), "Atomic Physics with Super-High Intensity Lasers," *Reports on Progress in Physics* **60**, 389.

Reiss, H. R. (1980), "Effect of an Intense Electromagnetic-Field on a Weakly Bound System," *Physical Review A* **22**, 1786.

Salamin, Y. I., S. X. Hu, K. Z. Hatsagortsyan and C. H. Keitel (2006), "Relativistic High-Power Laser–Matter Interactions," *Physics Reports: Review Section of Physics Letters* **427**, 41.

Sheehy, B., R. Lafon, M. Widmer, B. Walker, L. F. DiMauro, P. A. Agostini and K. C. Kulander (1998), "Single- and Multiple-Electron Dynamics in the Strong-Field Tunneling Limit," *Physical Review A* **58**, 3942.

Usachenko, V. I., V. A. Pazdzersky and J. K. McIver (2004), "Reexamination of High-Energy Above-Threshold Ionization (ATI): An Alternative Strong-Field ATI Model," *Physical Review A* **69**, 013406.

van der Hart, H. W. and K. Burnett (2000), "Recollision Model for Double Ionization of Atoms in Strong Laser Fields," *Physical Review A* **62**, 013407.

Walker, B., B. Sheehy, L. F. Dimauro, P. Agostini, K. J. Schafer and K. C. Kulander (1994), "Precision-Measurement of Strong-Field Double-Ionization of Helium," *Physical Review Letters* **73**, 1227.

Walker, M. A., P. Hansch and L. D. Van Woerkom (1998), "Intensity-Resolved Multiphoton Ionization: Circumventing Spatial Averaging," *Physical Review A* **57**, R701.

Wassaf, J., V. Veniard, R. Taieb and A. Maquet (2003), "Strong Field Atomic Ionization: Origin of High-Energy Structures in Photoelectron Spectra," *Physical Review Letters* **90**, 013003.

Weber, T., M. Weckenbrock, A. Staudte, L. Spielberger, O. Jagutzki, V. Mergel, F. Afaneh, G. Urbasch, M. Vollmer, H. Giessen and R. Dörner (2000), "Recoil-Ion Momentum Distributions for Single and Double Ionization of Helium in Strong Laser Fields," *Physical Review Letters* **84**, 443.

Yang, B. R., K. J. Schafer, B. Walker, K. C. Kulander, P. Agostini and L. F. DiMauro (1993), "Intensity-Dependent Scattering Rings in High-Order Above-Threshold Ionization," *Physical Review Letters* **71**, 3770.

Yudin, G. L. and M. Y. Ivanov (2001), "Physics of Correlated Double Ionization of Atoms in Intense Laser Fields: Quasistatic Tunneling Limit," *Physical Review A* **63**, 033404.

Chapter 5 Strong Field Interactions with Molecules

Becker, A. and F. H. M. Faisal (2005), "Intense-Field Many-Body S-Matrix Theory," *Journal of Physics B – Atomic Molecular and Optical Physics* **38**, R1.

Bucksbaum, P. H., A. Zavriyev, H. G. Muller and D. W. Schumacher (1990), "Softening of the H_2^+ Molecular-Bond in Intense Laser Fields," *Physical Review Letters* **64**, 1883.

Burke, P. G., J. Colgan, D. H. Glass and K. Higgins (2000), "R-Matrix-Floquet Theory of Molecular Multiphoton Processes," *Journal of Physics B – Atomic Molecular and Optical Physics* **33**, 143.

Chelkowski, S. and A. D. Bandrauk (1995), "Two-Step Coulomb Explosions of Diatoms in Intense Laser Fields," *Journal of Physics B – Atomic Molecular and Optical Physics* **28**, L723.

Chu, S. I. and D. A. Telnov (2004), "Beyond the Floquet Theorem: Generalized Floquet Formalisms and Quasienergy Methods for Atomic and Molecular Multiphoton Processes in Intense Laser Fields," *Physics Reports: Review Section of Physics Letters* **390**, 1.

Codling, K. and L. J. Frasinski (1993), "Dissociative Ionization of Small Molecules in Intense Laser Fields," *Journal of Physics B – Atomic Molecular and Optical Physics* **26**, 783.

Colgan, J., D. H. Glass, K. Higgins and P. G. Burke (1998), "The Calculation of Molecular Multiphoton Processes Using the R-Matrix-Floquet Method," *Computer Physics Communications* **114**, 27.

Constant, E., H. Stapelfeldt and P. B. Corkum (1996), "Observation of Enhanced Ionization of Molecular Ions in Intense Laser Fields," *Physical Review Letters* **76**, 4140.

Corkum, P. B., M. Y. Ivanov and J. S. Wright (1997), "Subfemtosecond Processes in Strong Laser Fields," *Annual Review of Physical Chemistry* **48**, 387.

Cornaggia, C. (2001), "Small Polyatomic Molecules in Intense Laser Fields," in *Molecules and Clusters in Intense Laser Fields*, J. H. Posthumus, ed. (Cambridge: Cambridge University Press), p. 84.

Cornaggia, C. (2010), "Enhancements of Rescattered Electron Yields in Above-Threshold Ionization of Molecules," *Physical Review A* **82**, 053410.

Cornaggia, C., J. Lavancier, D. Normand, J. Morellec, P. Agostini, J. P. Chambaret and A. Antonetti (1991), "Multielectron Dissociative Ionization of Diatomic-Molecules in an Intense Femtosecond Laser Field," *Physical Review A* **44**, 4499.

Cornaggia, C., M. Schmidt and D. Normand (1994), "Coulomb Explosion of CO_2 in an Intense Femtosecond Laser Field," *Journal of Physics B – Atomic Molecular and Optical Physics* **27**, L123.

DeWitt, M. J. and R. J. Levis (1998), "Observing the Transition from a Multiphoton-Dominated to a Field-Mediated Ionization Process for Polyatomic Molecules in Intense Laser Fields," *Physical Review Letters* **81**, 5101.

Dion, C. M., A. Ben Haj-Yedder, E. Cances, C. Le Bris, A. Keller and O. Atabek (2002), "Optimal Laser Control of Orientation: The Kicked Molecule," *Physical Review A* **65**, 063408.

Friedrich, B. and D. Herschbach (1995a), "Alignment and Trapping of Molecules in Intense Laser Fields," *Physical Review Letters* **74**, 4623.

Friedrich, B. and D. Herschbach (1995b), "Polarization of Molecules Induced by Intense Nonresonant Laser Fields," *Journal of Physical Chemistry* **99**, 15686.

Gibson, G. N., R. R. Freeman and T. J. McIlrath (1991), "Dynamics of the High-Intensity Multiphoton Ionization of N_2," *Physical Review Letters* **67**, 1230.

Gibson, G. N., M. Li, C. Guo and J. P. Nibarger (1998), "Direct Evidence of the Generality of Charge-Asymmetric Dissociation of Molecular Iodine Ionized by Strong Laser Fields," *Physical Review A* **58**, 4723.

Giusti-Suzor, A., X. He, O. Atabek and F. H. Mies (1990), "Above-Threshold Dissociation of H_2^+ in Intense Laser Fields," *Physical Review Letters* **64**, 515.

Grasbon, F., G. G. Paulus, S. L. Chin, H. Walther, J. Muth-Bohm, A. Becker and F. H. M. Faisal (2001), "Signatures of Symmetry-Induced Quantum-Interference Effects Observed in Above-Threshold-Ionization Spectra of Molecules," *Physical Review A* **63**, 041402.

Guo, C., M. Li, J. P. Nibarger and G. N. Gibson (1998), "Single and Double Ionization of Diatomic Molecules in Strong Laser Fields," *Physical Review A* **58**, R4271.

Hankin, S. M., D. M. Villeneuve, P. B. Corkum and D. M. Rayner (2000), "Nonlinear Ionization of Organic Molecules in High Intensity Laser Fields," *Physical Review Letters* **84**, 5082.

Hankin, S. M., D. M. Villeneuve, P. B. Corkum and D. M. Rayner (2001), "Intense-Field Laser Ionization Rates in Atoms and Molecules," *Physical Review A* **64**, 013405.

Harumiya, K., I. Kawata, H. Kono and Y. Fujimura (2000), "Exact Two-Electron Wave Packet Dynamics of H_2 in an Intense Laser Field: Formation of Localized Ionic States H+H," *Journal of Chemical Physics* **113**, 8953.

Hetzheim, H., C. Faria and W. Becker (2007), "Interference Effects in Above-Threshold Ionization from Diatomic Molecules: Determining the Internuclear Separation," *Physical Review A* **76**, 023418.

Jaron-Becker, A., A. Becker and F. H. M. Faisal (2004), "Ionization of N_2, O_2 and Linear Carbon Clusters in a Strong Laser Pulse," *Physical Review A* **69**, 023410.

Joachain, C. J., M. Dörr and N. Kylstra (2000), "High-Intensity Laser-Atom Physics," in *Advances in Atomic, Molecular and Optical Physics Vol 42*, B. Bederson and H. Walther eds. (San Diego: Academic Press) p. 225.

Joachain, C. J. (2007), "R-Matrix-Floquet Theory of Multiphoton Processes: Concepts, Results and Perspectives," *Journal of Modern Optics* **54**, 1859.

Jolicard, G. and O. Atabek (1992), "Above-Threshold-Dissociation Dynamics of H_2^+ With Short Intense Laser-Pulses," *Physical Review A* **46**, 5845.

Kjeldsen, T. K., C. Z. Bisgaard, L. B. Madsen and H. Stapelfeldt (2005), "Influence of Molecular Symmetry on Strong-Field Ionization: Studies on Ethylene, Benzene, Fluorobenzene and Chlorofluorobenzene," *Physical Review A* **71**, 013418.

Kono, H., Y. Sato, M. Kanno, K. Nakai and T. Kato (2006), "Theoretical Investigations of the Electronic and Nuclear Dynamics of Molecules in Intense Laser Fields: Quantum Mechanical Wave Packet Approaches," *Bulletin of the Chemical Society of Japan* **79**, 196.

Larsen, J. J. (2000), "Laser Induced Alignment of Neutral Molecules," PhD Thesis University of Aarhus.

Larsen, J. J., H. Sakai, C. P. Safvan, I. Wendt-Larsen and H. Stapelfeldt (1999), "Aligning Molecules with Intense Nonresonant Laser Fields," *Journal of Chemical Physics* **111**, 7774.

Lebedev, V. S., L. P. Presnyakov and I. I. Sobel'man (2003), "Radiative Transitions in the Molecular H_2^+ Ion," *Physics-Uspekhi* **46**, 473.

Levis, R. J. and M. J. DeWitt (1999), "Photoexcitation, Ionization and Dissociation of Molecules Using Intense Near-Infrared Radiation of Femtosecond Duration," *Journal of Physical Chemistry A* **103**, 6493.

Lezius, M., V. Blanchet, M. Y. Ivanov and A. Stolow (2002), "Polyatomic Molecules in Strong Laser Fields: Nonadiabatic Multielectron Dynamics," *Journal of Chemical Physics* **117**, 1575.

Litvinyuk, I. V., K. F. Lee, P. W. Dooley, D. M. Rayner, D. M. Villeneuve and P. B. Corkum (2003), "Alignment-Dependent Strong Field Ionization of Molecules," *Physical Review Letters* **90**, 233003.

Ludwig, J., H. Rottke and W. Sandner (1997), "Dissociation of H_2^+ and D_2^+ in an Intense Laser Field," *Physical Review A* **56**, 2168.

Milosevic, D. B. (2006), "Strong-Field Approximation for Ionization of a Diatomic Molecule by a Strong Laser Field," *Physical Review A* **74**, 063404.

Miret-Artes, S. and O. Atabek (1994), "Isotope Effects and Bond Softening in Intense-Laser-Field Multiphoton Dissociation of H_2^+," *Physical Review A* **49**, 1502.

Muth-Böhm, J., A. Becker and F. H. M. Faisal (2000), "Suppressed Molecular Ionization for a Class of Diatomics in Intense Femtosecond Laser Fields," *Physical Review Letters* **85**, 2280.

Nibarger, J. P., S. V. Menon and G. N. Gibson (2001), "Comprehensive Analysis of Strong-Field Ionization and Dissociation of Diatomic Nitrogen," *Physical Review A* **63**, 053406.

Niikura, H., V. R. Bhardwaj, F. Légaré I. V. Litvinyuk, P. W. Dooley D. M. Rayner, M. Y. Ivanov, P. B. Corkum and D. M. Villeneuve (2008), "Ionization of Small Molecules by Strong Laser Fields," in *Strong Field Laser Physics*, T. Brabec, ed. (New York: Springer), p. 185.

Ortigoso, J., M. Rodriguez, M. Gupta and B. Friedrich (1999), "Time Evolution of Pendular States Created by the Interaction of Molecular Polarizability with a Pulsed Nonresonant Laser Field," *Journal of Chemical Physics* **110**, 3870.

Peng, L. Y., D. Dundas, J. F. McCann, K. T. Taylor and I. D. Williams (2003), "Dynamic Tunneling Ionization of H_2^+ in Intense Fields," *Journal of Physics B – Atomic Molecular and Optical Physics* **36**, L295.

Plummer, M. and J. F. McCann (1996), "Field-Ionization Rates of the Hydrogen Molecular Ion," *Journal of Physics B – Atomic Molecular and Optical Physics* **29**, 4625.

Plummer, M., J. F. McCann and L. B. Madsen (1998), "The Calculation of Multiphoton Ionization Rates of the Hydrogen Molecular Ion," *Computer Physics Communications* **114**, 94.

Posthumus, J. H. (2004), "The Dynamics of Small Molecules in Intense Laser Fields," *Reports on Progress in Physics* **67**, 623.

Posthumus, J. H., L. J. Frasinski, A. J. Giles and K. Codling (1995), "Dissociative Ionization of Molecules in Intense Laser Fields: A Method of Predicting Ion Kinetic Energies and Appearance Intensities," *Journal of Physics B – Atomic Molecular and Optical Physics* **28**, L349.

Posthumus, J. H., A. J. Giles, M. R. Thompson and K. Codling (1996), "Field-Ionization, Coulomb Explosion of Diatomic Molecules in Intense Laser Fields," *Journal of Physics B – Atomic Molecular and Optical Physics* **29**, 5811.

Posthumus, J. H., A. J. Giles, M. Thompson, W. Shaikh, A. J. Langley, L. J. Frasinski and K. Codling (1996b), "The Dissociation Dynamics of Diatomic Molecules in Intense Laser Fields," *Journal of Physics B – Atomic Molecular and Optical Physics* **29**, L525.

Posthumus, J. H. and J. F. McCann (2001), "Diatomic Molecules in Intense Laser Fields," in *Molecules and Clusters in Intense Laser Fields*, J. H. Posthumus, ed. (Cambridge: Cambridge University Press), p. 27.

Poulsen, M. D. (2005), "Alignment of Molecules Induced by Long and Short Laser Pulses," PhD Thesis, University of Aarhus, Denmark.

Sakai, H., C. P. Safvan, J. J. Larsen, K. M. Hilligsoe, K. Hald and H. Stapelfeldt (1999), "Controlling the Alignment of Neutral Molecules by a Strong Laser Field," *Journal of Chemical Physics* **110**, 10235.

Schmidt, M., S. Dobosz, P. Meynadier, P. D'Oliveira, D. Normand, E. Charron and A. Suzor-Weiner (1999), "Fragment-Emission Patterns from the Coulomb Explosion of Diatomic Molecules in Intense Laser Fields," *Physical Review A* **60**, 4706.

Schmidt, M., D. Normand and C. Cornaggia (1994), "Laser-Induced Trapping of Chlorine Molecules with Picosecond and Femtosecond Pulses," *Physical Review A* **50**, 5037.

Seideman, T. (2001), "On the Dynamics of Rotationally Broad, Spatially Aligned Wave Packets," *Journal of Chemical Physics* **115**, 5965.

Seideman, T. (2002), "Time-Resolved Photoelectron Angular Distributions: Concepts, Applications and Directions," *Annual Review of Physical Chemistry* **53**, 41.

Seideman, T., M. Y. Ivanov and P. B. Corkum (1995), "Role of Electron Localization in Intense-Field Molecular Ionization," *Physical Review Letters* **75**, 2819.

Sheehy, B. and L. F. DiMauro (1996), "Atomic and Molecular Dynamics in Intense Optical Fields," *Annual Review of Physical Chemistry* **47**, 463.

Shimizu, S., J. Kou, S. Kawato, K. Shimizu, S. Sakabe and N. Nakashima (2000), "Coulomb Explosion of Benzene Irradiated by an Intense Femtosecond Laser Pulse," *Chemical Physics Letters* **317**, 609.

Stapelfeldt, H. and T. Seideman (2003), "Colloquium: Aligning Molecules with Strong Laser Pulses," *Reviews of Modern Physics* **75**, 543.

Talebpour, A., C. Y. Chien and S. L. Chin (1996), "The Effects of Dissociative Recombination in Multiphoton Ionization of O_2," *Journal of Physics B – Atomic Molecular and Optical Physics* **29**, L677.

Tong, X. M., Z. X. Zhao and C. D. Lin (2002), "Theory of Molecular Tunneling Ionization," *Physical Review A* **66**, 033402.

Urbain, X., B. Fabre, E. M. Staicu-Casagrande, N. de Ruette, V. M. Andrianarijaona, J. Jureta, J. H. Posthumus, A. Saenz, E. Baldit and C. Cornaggia (2004), "Intense-Laser-Field Ionization of Molecular Hydrogen in the Tunneling Regime and Its Effect on the Vibrational Excitation of H_2^+," *Physical Review Letters* **92**, 163004.

Walsh, T. D. G., F. A. Ilkov and S. L. Chin (1997), "The Dynamical Behaviour of H_2 and D_2 in a Strong, Femtosecond, Titanium:Sapphire Laser Field," *Journal of Physics B – Atomic Molecular and Optical Physics* **30**, 2167.

Wind, H. (1965), "Vibrational States of Hydrogen Molecular Ion," *Journal of Chemical Physics* **43**, 2956.

Yang, B., M. Saeed, L. F. Dimauro, A. Zavriyev and P. H. Bucksbaum (1991), "High-Resolution Multiphoton Ionization and Dissociation of H_2 and D_2 Molecules in Intense Laser Fields," *Physical Review A* **44**, R1458.

Yu, H. T., T. Zuo and A. D. Bandrauk, (1998), "Intense Field Ionization of Molecules with Ultra-Short Laser Pulses: Enhanced Ionization and Barrier-Suppression Effects," *Journal of Physics B – Atomic Molecular and Optical Physics* **31**, 1533.

Zavriyev, A., P. H. Bucksbaum, H. G. Muller and D. W. Schumacher (1990), "Ionization and Dissociation of H_2 in Intense Laser Fields at 1.064 μm, 532 nm and 355-nm," *Physical Review A* **42**, 5500.

Zhao, Z. X., X. M. Tong and C. D. Lin (2003), "Alignment-Dependent Ionization Probability of Molecules in a Double-Pulse Laser Field," *Physical Review A* **67**, 043404.

Chapter 6 Strong Field Nonlinear Optics

Agostini, P. and L. F. DiMauro (2004), "The Physics of Attosecond Light Pulses," *Reports on Progress in Physics* **67**, 813.

Antoine, P., A. L'Huillier and M. Lewenstein (1996), "Attosecond Pulse Trains Using High-Order Harmonics," *Physical Review Letters* **77**, 1234.

Balcou, P., P. Salieres, A. L'Huillier and M. Lewenstein (1997), "Generalized Phase-Matching Conditions for High Harmonics: The Role of Field-Gradient Forces," *Physical Review A* **55**, 3204.

Bartels, R. A., A. Paul, H. Green, H. C. Kapteyn, M. M. Murnane, S. Backus, I. P. Christov, Y. W. Liu, D. Attwood and C. Jacobsen (2002), "Generation of Spatially Coherent Light at Extreme Ultraviolet Wavelengths," *Science* **297**, 376.

Budil, K. S., P. Salieres, A. L'Huillier, T. Ditmire and M. D. Perry (1993), "Influence of Ellipticity on Harmonic-Generation," *Physical Review A* **48**, R3437.

Chang, Z. H., A. Rundquist, H. W. Wang, M. M. Murnane and H. C. Kapteyn (1997), "Generation of Coherent Soft X Rays at 2.7 nm Using High Harmonics," *Physical Review Letters* **79**, 2967.

Chini, M., K. Zhao and Z. Chang (2014), "The Generation, Characterization and Applications of Broadband Isolated Attosecond Pulses," *Nature Photonics* **8**, 178.

Constant, E., D. Garzella, P. Breger, E. Mevel, C. Dorrer, C. Le Blanc, F. Salin and P. Agostini (1999), "Optimizing High Harmonic Generation in Absorbing Gases: Model and Experiment," *Physical Review Letters* **82**, 1668.

Corkum, P. B. (1993), "Plasma Perspective on Strong-Field Multiphoton Ionization," *Physical Review Letters* **71**, 1994.

Ditmire, T., K. Kulander, J. K. Crane, H. Nguyen and M. D. Perry (1996), "Calculation and Measurement of High-Order Harmonic Energy Yields in Helium," *Journal of the Optical Society of America B – Optical Physics* **13**, 406.

Durfee, C. G., A. R. Rundquist, S. Backus, C. Herne, M. M. Murnane and H. C. Kapteyn (1999), "Phase Matching of High-Order Harmonics in Hollow Waveguides," *Physical Review Letters* **83**, 2187.

Frumker, E., N. Kajumba, J. B. Bertrand, H. J. Worner, C. T. Hebeisen, P. Hockett, M. Spanner, S. Patchkovskii, G. G. Paulus, D. M. Villeneuve, A. Naumov and P. B. Corkum (2012), "Probing Polar Molecules with High Harmonic Spectroscopy," *Physical Review Letters* **109**, 233904.

Gibson, E. A., A. Paul, N. Wagner, R. Tobey, S. Backus, I. P. Christov, M. M. Murnane and H. C. Kapteyn (2004), "High-Order Harmonic Generation up to 250 eV from Highly Ionized Argon," *Physical Review Letters* **92**, 033001.

Gribakin, G. F. and M. Y. Kuchiev (1997), "Multiphoton Detachment of Electrons from Negative Ions," *Physical Review A* **55**, 3760.

Ivanov, M. Y., T. Brabec and N. Burnett (1996)., "Coulomb Corrections and Polarization Effects in High-Intensity High-Harmonic Emission," *Physical Review A* **54**, 742.

Kato, K., S. Minemoto and H. Sakai (2011), "Suppression of High-Order-Harmonic Intensities Observed in Aligned CO_2 Molecules with 1300-nm and 800-nm Pulses," *Physical Review A* **84**, 021403.

Kotelnikov, I. A. and A. I. Milstein (2018), "Electron Radiative Recombination with a Hydrogen-like Ion," *Physica Scripta* **94**, 05503.

Krausz, F. and M. Ivanov (2009), "Attosecond Physics," *Reviews of Modern Physics* **81**, 163.

Kulander, K. C., K. J. Schafer and J. L. Krause (1993), "Dynamics of Short-Pulse Excitation, Ionization and Harmonic Conversion," in *Super-Intense Laser-Atom Physics*, B. Piraux, A. Lhuillier and K. Rzazewski, eds., NATO ASI Series: Physics Vol. **316** (New York: Plenum Press), p. 95.

Le, A. T., R. Della Picca, P. D. Fainstein, D. A. Telnov, M. Lein and C. D. Lin (2008), "Theory of High-Order Harmonic Generation from Molecules by Intense Laser Pulses," *Journal of Physics B – Atomic Molecular and Optical Physics* **41**, 081002.

Lein, M., R. de Nalda, E. Heesel, N. Hay, E. Springate, R. Velotta, M. Castillejo, P. L. Knight and J. P. Marangos (2005), "Signatures of Molecular Structure in the Strong-Field Response of Aligned Molecules," *Journal of Modern Optics* **52**, 465.

Lewenstein, M., P. Balcou, M. Y. Ivanov, A. L'Huillier and P. B. Corkum (1994), "Theory of High-Harmonic Generation by Low-Frequency Laser Fields," *Physical Review A* **49**, 2117.

Lewenstein, M. and A. L'Huillier (2008), "Principles of Single Atom Physics," in *Strong Field Laser Physics*, T. Brabec, ed. (New York: Springer), p. 147.

Lewenstein, M., P. Salieres and A. L'Huillier (1995), "Phase of the Atomic Polarization in High-Order Harmonic-Generation," *Physical Review A* **52**, 4747.

L'Huillier, A. and P. Balcou (1993), "High-Order Harmonic-Generation in Rare-Gases with a 1-ps 1053-nm Laser," *Physical Review Letters* **70**, 774.

L'Huillier, A., L-A. Lompré, G. Mainfray and C. Manus (1992), "High-Order Harmonic Generation in Rare Gases," in *Atoms in Intense Laser Fields*, M. Gavrila, ed. (Boston: Academic Press), p. 139.

Macklin, J. J., J. D. Kmetec and C. L. Gordon (1993), "High-Order Harmonic-Generation Using Intense Femtosecond Pulses," *Physical Review Letters* **70**, 766.

Mairesse, Y., J. Levesque, N. Dudovich, P. B. Corkum and D. M. Villeneuve (2008), "High Harmonic Generation from Aligned Molecules: Amplitude and Polarization," *Journal of Modern Optics* **55**, 2591.

McFarland, B. K., J. P. Farrell, P. H. Bucksbaum and M. Guhr (2008), "High Harmonic Generation from Multiple Orbitals in N_2," *Science* **322**, 1232.

Paul, A., E. A. Gibson, X. S. Zhang, A. Lytle, T. Popmintchev, X. B. Zhou, M. M. Murnane, I. P. Christov and H. C. Kapteyn (2006), "Phase-Matching Techniques for Coherent Soft X-Ray Generation," *IEEE Journal of Quantum Electronics* **42**, 14.

Reider, G. A. (2004), "XUV Attosecond Pulses: Generation and Measurement," *Journal of Physics D – Applied Physics* **37**, R37.

Rundquist, A., C. G. Durfee, Z. H. Chang, C. Herne, S. Backus, M. M. Murnane and H. C. Kapteyn (1998), "Phase-Matched Generation of Coherent Soft X-rays," *Science* **280**, 1412.

Salieres, P., A. L'Huillier and M. Lewenstein (1995), "Coherence Control of High-Order Harmonics," *Physical Review Letters* **74**, 3776.

Scrinzi, A., M. Y. Ivanov, R. Kienberger and D. M. Villeneuve (2006), "Attosecond Physics," *Journal of Physics B – Atomic Molecular and Optical Physics* **39**, R1.

Scrinzi, A. and H.-G. Muller (2008), "Attosecond Pulses: Generation, Detection and Applications," in *Strong Field Laser Physics*, T. Brabec, ed. (New York: Springer), p. 281.

Shan, B., S. Ghimire and Z. Chang (2005), "Generation of the Attosecond Extreme Ultraviolet Supercontinuum by a Polarization," *Journal of Modern Optics* **52**, 277.

Varju, K., Y. Mairesse, B. Carre, M. B. Gaarde, P. Johnsson, S. Kazamias, R. Lopez-Martens, J. Mauritsson, K. J. Schafer, P. H. Balcou, A. L'Huillier and P. Salieres (2005), "Frequency Chirp of Harmonic and Attosecond Pulses," *Journal of Modern Optics* **52**, 379.

Winterfeldt, C., C. Spielmann and G. Gerber (2008), "Colloquium: Optimal Control of High-Harmonic Generation," *Reviews of Modern Physics* **80**, 117.

Chapter 7 Strong Field Interactions with Clusters

Bang, W., M. Barbui, A. Bonasera, G. Dyer, H. J. Quevedo, K. Hagel, K. Schmidt, F. Consoli, R. De Angelis, P. Andreoli, E. Gaul, A. C. Bernstein, M. Donovan, M. Barbarino, S. Kimura, M. Mazzocco, J. Sura, J. B. Natowitz and T. Ditmire (2013), "Temperature Measurements of Fusion Plasmas Produced by Petawatt-Laser-Irradiated D_2-He^3 or CD_4-He^3 Clustering Gases," *Physical Review Letters* **111**, 055002.

Bauer, D. and A. Macchi (2003), "Dynamical Ionization Ignition of Clusters in Intense Short Laser Pulses," *Physical Review A* **68**, 033201.

Bornath, T., P. Hilse and M. Schlanges (2007), "Ionization Dynamics in Nanometer-Sized Clusters Interacting with Intense Laser Fields," *Laser Physics* **17**, 591.

Breizman, B. N. and A. V. Arefiev (2003), "Electron Response in Laser-Irradiated Microclusters," *Plasma Physics Reports* **29**, 593.

Breizman, B. N., A. V. Arefiev and M. V. Fomyts'kyi (2005), "Physics of Laser-Irradiated Microclusters," *Physics of Plasmas* **12**, 056706.

Chen, L. M., J. J. Park, K. H. Hong, I. W. Choi, J. L. Kim, J. Zhang and C. H. Nam (2002), "Measurement of Energetic Electrons from Atomic Clusters Irradiated by Intense Femtosecond Laser Pulses," *Physics of Plasmas* **9**, 3595.

Deiss, C., N. Rohringer, J. Burgdorfer, E. Lamour, C. Prigent, J. P. Rozet and D. Vernhet (2006), "Laser-Cluster Interaction: X-Ray Production by Short Laser Pulses," *Physical Review Letters* **96**, 013203.

Ditmire, T. (1997), "Atomic Clusters in Ultrahigh Intensity Light Fields," *Contemporary Physics* **38**, 315.

Ditmire, T. (1998), "Simulation of Exploding Clusters Ionized by High-Intensity Femtosecond Laser Pulses," *Physical Review A* **57**, R4094.

Ditmire, T., T. Donnelly, R. W. Falcone and M. D. Perry (1995), "Strong X-Ray-Emission from High-Temperature Plasmas Produced by Intense Irradiation of Clusters," *Physical Review Letters* **75**, 3122.

Ditmire, T., T. Donnelly, A. M. Rubenchik, R. W. Falcone and M. D. Perry (1996), "Interaction of Intense Laser Pulses with Atomic Clusters," *Physical Review A* **53**, 3379.

Ditmire, T., P. K. Patel, R. A. Smith, J. S. Wark, S. J. Rose, D. Milathianaki, R. S. Marjoribanks and M. H. R. Hutchinson (1998a), "KeV X-Ray Spectroscopy of Plasmas Produced by the Intense Picosecond Irradiation of a Gas of Xenon Clusters," *Journal of Physics B – Atomic Molecular and Optical Physics* **31**, 2825.

Ditmire, T., R. A. Smith, J. W. G. Tisch and M. H. R. Hutchinson (1997b), "High Intensity Laser Absorption by Gases of Atomic Clusters," *Physical Review Letters* **78**, 3121.

Ditmire, T., E. Springate, J. W. G. Tisch, Y. L. Shao, M. B. Mason, N. Hay, J. P. Marangos and M. H. R. Hutchinson (1998b), "Explosion of Atomic Clusters Heated by High-Intensity Femtosecond Laser Pulses," *Physical Review A* **57**, 369.

Ditmire, T., J. W. G. Tisch, E. Springate, M. B. Mason, N. Hay, R. A. Smith, J. Marangos and M. H. R. Hutchinson (1997c), "High-Energy Ions Produced in Explosions of Superheated Atomic Clusters," *Nature* **386**, 54.

Ditmire, T., J. Zweiback, V. P. Yanovsky, T. E. Cowan, G. Hays and K. B. Wharton (1999), "Nuclear Fusion from Explosions of Femtosecond Laser-Heated Deuterium Clusters," *Nature* **398**, 489.

Dorchies, F., F. Blasco, T. Caillaud, J. Stevefelt, C. Stenz, A. S. Boldarev and V. A. Gasilov (2003), "Spatial Distribution of Cluster Size and Density in Supersonic Jets as Targets for Intense Laser Pulses," *Physical Review A* **68**, 023201.

Erk, B., K. Hoffmann, N. Kandadai, A. Helal, J. Keto and T. Ditmire (2011), "Observation of Shells in Coulomb Explosions of Rare-Gas Clusters," *Physical Review A* **83**, 043201.

Fennel, T., K. H. Meiwes-Broer, J. Tiggesbaumker, P. G. Reinhard, P. M. Dinh and E. Suraud (2010), "Laser-Driven Nonlinear Cluster Dynamics," *Reviews of Modern Physics* **82**, 1793.

Fennel, T., L. Ramunno and T. Brabec (2007), "Highly Charged Ions from Laser-Cluster Interactions: Local-Field-Enhanced Impact Ionization and Frustrated Electron-Ion Recombination," *Physical Review Letters* **99**, 233401.

Fomichev, S. V., S. V. Popruzhenko, D. F. Zaretsky and W. Becker (2003a), "Laser-Induced Nonlinear Excitation of Collective Electron Motion in a Cluster," *Journal of Physics B – Atomic Molecular and Optical Physics* **36**, 3817.

Fomichev, S. V., S. V. Popruzhenko, D. F. Zaretsky and W. Becker (2003b), "Nonlinear Excitation of the Mie Resonance in a Laser-Irradiated Cluster," *Optics Express* **11**, 2433.

Fomyts'kyi, M. V., B. N. Breizman, A. V. Arefiev and C. Chiu (2004), "Harmonic Generation in Clusters," *Physics of Plasmas* **11**, 3349.

Fukuda, Y., K. Yamakawa, Y. Akahane, M. Aoyama, N. Inoue, H. Ueda and Y. Kishimoto (2003), "Optimized Energetic Particle Emissions from Xe Clusters in Intense Laser Fields," *Physical Review A* **67**, 061201.

Gets, A. V. and V. P. Krainov (2006), "The Ionization Potentials of Atomic Ions in Laser-Irradiated Ar, Kr and Xe Clusters," *Journal of Physics B – Atomic Molecular and Optical Physics* **39**, 1787.

Hagena, O. F. and W. Obert (1972), "Cluster Formation in Expanding Supersonic Jets: Effect of Pressure, Temperature, Nozzle Size and Test Gas," *Journal of Chemical Physics* **56**, 1793.

Hansen, S. B., A. S. Shlyaptseva, A. Y. Faenov, I. Y. Skobelev, A. Magunov, T. A. Pikuz, F. Blasco, F. Dorchies, C. Stenz, F. Salin, T. Auguste, S. Dobosz, P. Monot, P. D'Oliveira, S. Hulin, U. I. Safronova and K. B. Fournier (2002), "Hot-Electron Influence on L-Shell Spectra of Multicharged Kr Ions Generated in Clusters Irradiated by Femtosecond Laser Pulses," *Physical Review E* **66**, 046412.

Haught, A. F. and D. H. Polk (1970), "Formation and Heating of Laser Irradiated Solid Particle Plasmas," *Physics of Fluids* **13**, 2825.

Hilse, P., M. Moll, M. Schlanges and T. Bornath (2009), "Laser-Cluster-Interaction in a Nanoplasma-Model with Inclusion of Lowered Ionization Energies," *Laser Physics* **19**, 428.

Hohenberger, M., D. R. Symes, K. W. Madison, A. Sumeruk, G. Dyer, A. Edens, W. Grigsby, G. Hays, M. Teichmann and T. Ditmire (2005), "Dynamic Acceleration Effects in Explosions of Laser-Irradiated Heteronuclear Clusters," *Physical Review Letters* **95**, 195003.

Holtsmark, J. (1919), "Uber die Verbreiterung von Spektrallinien," *Annalen der Physik* **363**, 577–630.

Islam, M. R., U. Saalmann and J. M. Rost (2006), "Kinetic Energy of Ions after Coulomb Explosion of Clusters Induced by an Intense Laser Pulse," *Physical Review A* **73**, 041201.

Jungreuthmayer, C., M. Geissler, J. Zanghellini and T. Brabec (2004), "Microscopic Analysis of Large-Cluster Explosion in Intense Laser Fields," *Physical Review Letters* **92**, 133401.

Junkel-Vives, G. C., J. Abdallah, F. Blasco, C. Stenz, F. Salin, A. Y. Faenov, A. I. Magunov, T. A. Pikuz and I. Y. Skobelev (2001), "Observation of H-Like Ions within Argon Clusters Irradiated by 35-Fs Laser via High-Resolution X-Ray Spectroscopy," *Physical Review A* **64**, 021201.

Korneev, P. A., S. V. Popruzhenko, D. F. Zaretsky and W. Becker (2005), "Collisionless Heating of a Nanoplasma in Laser-Irradiated Clusters," *Laser Physics Letters* **2**, 452.

Kostyukov, I. and J. M. Rax (2003), "Collisional versus Collisionless Resonant and Autoresonant Heating in Laser-Cluster Interaction," *Physical Review E* **67**, 066405.

Krainov, V. P. and M. B. Smirnov (2001), "Heating of Deuterium Clusters by a Superatomic Ultra-Short Laser Pulse," *Journal of Experimental and Theoretical Physics* **92**, 626.

Krainov, V. P. and M. B. Smirnov (2002b), "Cluster Beams in the Super-Intense Femtosecond Laser Pulse," *Physics Reports: Review Section of Physics Letters* **370**, 237.

Krishnamurthy, M., J. Jha, D. Mathur, C. Jungreuthmayer, L. Ramunno, J. Zanghellini and T. Brabec (2006), "Ion Charge State Distribution in the Laser-Induced Coulomb Explosion of Argon Clusters," *Journal of Physics B – Atomic Molecular and Optical Physics* **39**, 625.

Kumarappan, V., M. Krishnamurthy and D. Mathur (2001), "Asymmetric High-Energy Ion Emission from Argon Clusters in Intense Laser Fields," *Physical Review Letters* **87**, 085005.

Kumarappan, V., M. Krishnamurthy and D. Mathur (2002), "Two-Dimensional Effects in the Hydrodynamic Expansion of Xenon Clusters under Intense Laser Irradiation," *Physical Review A* **66**, 033203.

Last, I. and J. Jortner (2000), "Dynamics of the Coulomb Explosion of Large Clusters in a Strong Laser Field," *Physical Review A* **62**, 013201.

Last, I. and J. Jortner (2002), "Nuclear Fusion Driven by Coulomb Explosion of Methane Clusters," *Journal of Physical Chemistry A* **106**, 10877.

Last, I. and J. Jortner (2004), "Electron and Nuclear Dynamics of Molecular Clusters in Ultraintense Laser Fields. I. Extreme Multielectron Ionization," *Journal of Chemical Physics* **120**, 1336.

Last, I. and J. Jortner (2004), "Electron and Nuclear Dynamics of Molecular Clusters in Ultraintense Laser Fields. II. Electron Dynamics and Outer Ionization of the Nanoplasma," *Journal of Chemical Physics* **120**, 1348.

Lotz, W. (1968), "Electron-Impact Ionization Cross-Sections and Ionization Rate Coefficients for Atoms and Ions from Hydrogen to Calcium," *Zeitschrift für Physik* **216**, 241.

Madison, K. W., P. K. Patel, M. Allen, D. Price and T. Ditmire (2003), "Investigation of Fusion Yield from Exploding Deuterium-Cluster Plasmas Produced by 100-TW Laser Pulses," *Journal of the Optical Society of America B – Optical Physics* **20**, 113.

Madison, K. W., P. K. Patel, M. Allen, D. Price, R. Fitzpatrick and T. Ditmire (2004a), "Role of Laser-Pulse Duration in the Neutron Yield of Deuterium Cluster Targets," *Physical Review A* **70**, 053201.

Madison, K. W., P. K. Patel, D. Price, A. Edens, M. Allen, T. E. Cowan, J. Zweiback and T. Ditmire (2004b), "Fusion Neutron and Ion Emission from Deuterium and Deuterated Methane Cluster Plasmas," *Physics of Plasmas* **11**, 270.

Magunov, A. I., T. A. Pikuz, I. Y. Skobelev, A. Y. Faenov, F. Blasco, F. Dorchies, T. Caillaud, C. Bonte, F. Salin, C. Stenz, P. A. Loboda, I. A. Litvinenko, V. V. Popova, G. V. Baidin, G. C. Junkel-Vives and J. Abdallah (2001), "Influence of Ultrashort Laser Pulse Duration on the X-Ray Emission Spectrum of Plasma Produced in Cluster Target," *JETP Letters* **74**, 375.

Megi, F., M. Belkacem, M. A. Bouchene, E. Suraud and G. Zwicknagel (2003), "On the Importance of Damping Phenomena in Clusters Irradiated by Intense Laser Fields," *Journal of Physics B – Atomic Molecular and Optical Physics* **36**, 273.

Micheau, S., H. Jouin and B. Pons (2008), "Modified Nanoplasma Model for Laser-Cluster Interaction," *Physical Review A* **77**, 053201.

Milchberg, H. M., S. J. McNaught and E. Parra (2001), "Plasma Hydrodynamics of the Intense Laser-Cluster Interaction," *Physical Review E* **64**, 056402.

Mulser, P. and M. Kanapathipillai (2005), "Collisionless Absorption in Clusters Out of Linear Resonance," *Physical Review A* **71**, 063201.

Parks, P. B., T. E. Cowan, R. B. Stephens and E. M. Campbell (2001), "Model of Neutron-Production Rates from Femtosecond-Laser-Cluster Interactions," *Physical Review A* **63**, 063203.

Parra, E., I. Alexeev, J. Fan, K. Y. Kim, S. J. McNaught and H. M. Milchberg (2000), "X-Ray and Extreme Ultraviolet Emission Induced by Variable Pulse-Width Irradiation of Ar and Kr Clusters and Droplets," *Physical Review E* **62**, R5931.

Peano, F., F. Peinetti, R. Mulas, G. Coppa and L. O. Silva (2006), "Kinetics of the Collisionless Expansion of Spherical Nanoplasmas," *Physical Review Letters* **96**, 175002.

Ramunno, L., T. Brabec and V. Krainov (2008), "Intense Laser Interactions with Noble Gas Clusters," in *Strong Field Laser Physics*, T. Brabec, ed. (New York: Springer), p. 225.

Rose-Petruck, C., K. J. Schafer, K. R. Wilson and C. P. J. Barty (1997), "Ultrafast Electron Dynamics and Inner-Shell Ionization in Laser Driven Clusters," *Physical Review A* **55**, 1182.

Rusek, M., H. Lagadec and T. Blenski (2001), "Cluster Explosion in an Intense Laser Pulse: Thomas-Fermi Model," *Physical Review A* **63**, 013203.

Saalmann, U. and J. M. Rost (2003), "Ionization of Clusters in Intense Laser Pulses through Collective Electron Dynamics," *Physical Review Letters* **91**, 223401.

Saalmann, U. and J. M. Rost (2008), "Rescattering for Extended Atomic Systems," *Physical Review Letters* **100**, 133006.

Saalmann, U., C. Siedschlag and J. M. Rost (2006), "Mechanisms of Cluster Ionization in Strong Laser Pulses," *Journal of Physics B – Atomic Molecular and Optical Physics* **39**, R39.

Schmalz, R. F. (1985), "New Self-Similar Solutions for the Unsteady One-Dimensional Expansion of a Gas into a Vacuum," *Physics of Fluids* **28**, 2923.

Shao, Y. L., T. Ditmire, J. W. G. Tisch, E. Springate, J. P. Marangos and M. H. R. Hutchinson (1996), "Multi-keV Electron Generation in the Interaction of Intense Laser Pulses with Xe Clusters," *Physical Review Letters* **77**, 3343.

Siedschlag, C. and J. M. Rost (2002), "Electron Release of Rare-Gas Atomic Clusters under an Intense Laser Pulse," *Physical Review Letters* **89**, 173401.

Smirnov, M. B. and V. P. Krainov (2003), "Hot Electron Generation in Laser Cluster Plasma," *Physics of Plasmas* **10**, 443.

Smirnov, M. B. and V. P. Krainov (2004), "Ionization of Cluster Atoms in a Strong Laser Field," *Physical Review A* **69**, 043201.

Springate, E., N. Hay, J. W. G. Tisch, M. B. Mason, T. Ditmire, J. P. Marangos and M. H. R. Hutchinson (2000), "Enhanced Explosion of Atomic Clusters Irradiated by a Sequence of Two High-Intensity Laser Pulses," *Physical Review A* **61**, 044101.

Stewart, J. C. and K. D. Pyatt (1966), "Lowering of Ionization Potentials in Plasmas," *Astrophysical Journal* **144**, 1203.

Symes, D. R., M. Hohenberger, A. Henig and T. Ditmire (2007), "Anisotropic Explosions of Hydrogen Clusters under Intense Femtosecond Laser Irradiation," *Physical Review Letters* **98**, 123401.

Zweiback, J., T. Ditmire and M. D. Perry (1999), "Femtosecond Time-Resolved Studies of the Dynamics of Noble-Gas Cluster Explosions," *Physical Review A* **59**, R3166.

Zweiback, J., T. Ditmire and M. D. Perry (2000a), "Resonance in Scattering and Absorption from Large Noble Gas Clusters," *Optics Express* **6**, 236.

Zweiback, J., R. A. Smith, T. E. Cowan, G. Hays, K. B. Wharton, V. P. Yanovsky and T. Ditmire (2000b), "Nuclear Fusion Driven by Coulomb Explosions of Large Deuterium Clusters," *Physical Review Letters* **84**, 2634.

Chapter 8 Strong Field Interactions with Underdense Plasmas

Akhiezer, A. I. and R. V. Polovin (1956), "Theory of Wave Motion of an Electron Plasma," *Soviet Physics JETP-USSR* **3**, 696.

Albert, F. and A. G. R. Thomas (2016), "Applications of Laser Wakefield Accelerator-Based Light Sources," *Plasma Physics and Controlled Fusion* **58**, 103001.

Andreev, N. E., L. M. Gorbunov, V. I. Kirsanov, A. A. Pogosova and R. R. Ramazashvili (1992), "Resonant Excitation of Wakefields by a Laser-Pulse in a Plasma," *JETP Letters* **55**, 571.

Auguste, T., P. Monot, L. A. Lompre, G. Mainfray and C. Manus (1992), "Defocusing Effects of a Picosecond Terawatt Laser-Pulse in an Underdense Plasma," *Optics Communications* **89**, 145.

Barr, H. C., T. J. M. Boyd, F. I. Gordon and S. J. Berwick (1995), "Stimulated Raman Scattering in Plasmas Produced by Short Intense Laser Pulses," *Laser and Particle Beams* **13**, 525.

Berezhiani, V. I. and I. G. Murusidze (1990), "Relativistic Wake-Field Generation by an Intense Laser-Pulse in a Plasma," *Physics Letters A* **148**, 338.

Bernstein, A. C., M. McCormick, G. M. Dyer, J. C. Sanders and T. Ditmire (2009), "Two-Beam Coupling between Filament-Forming Beams in Air," *Physical Review Letters* **102**, 123902.

Bingham, R., J. T. Mendonca and P. K. Shukla (2004), "Plasma Based Charged-Particle Accelerators," *Plasma Physics and Controlled Fusion* **46**, R1.

Blyth, W. J., S. G. Preston, A. A. Offenberger, M. H. Key, J. S. Wark, Z. Najmudin, A. Modena, A. Djaoui and A. E. Dangor (1995), "Plasma Temperature in Optical-Field Ionization of Gases by Intense Ultrashort Pulses of Ultraviolet-Radiation," *Physical Review Letters* **74**, 554.

Brandi, H. S., C. Manus, G. Mainfray, T. Lehner and G. Bonnaud (1993), "Relativistic and Ponderomotive Self-Focusing of a Laser-Beam in a Radially Inhomogeneous Plasma: 1. Paraxial Approximation," *Physics of Fluids B-Plasma Physics* **5**, 3539.

Braun, A., G. Korn, X. Liu, D. Du, J. Squier and G. Mourou (1995), "Self-Channeling of High-Peak-Power Femtosecond Laser-Pulses in Air," *Optics Letters* **20**, 73.

Bulanov, S. V., I. N. Inovenkov, V. I. Kirsanov, N. M. Naumova and A. S. Sakharov (1992), "Nonlinear Depletion of Ultrashort and Relativistically Strong Laser-Pulses in an Underdense Plasma," *Physics of Fluids B-Plasma Physics* **4**, 1935.

Bulanov, S. V., N. Naumova, F. Pegoraro and J. Sakai (1998), "Particle Injection into the Wave Acceleration Phase due to Nonlinear Wave Wave-Breaking," *Physical Review E* **58**, R5257.

Chessa, P., E. De Wispelaere, F. Dorchies, V. Malka, J. R. Marques, G. Hamoniaux, P. Mora and F. Amiranoff (1999), "Temporal and Angular Resolution of the Ionization-Induced Refraction of a Short Laser Pulse in Helium Gas," *Physical Review Letters* **82**, 552.

Chin, S. L., A. Brodeur, S. Petit, O. G. Kosareva and V. P. Kandidov (1999), "Filamentation and Supercontinuum Generation during the Propagation of Powerful Ultrashort Laser Pulses in Optical Media (White Light Laser)," *Journal of Nonlinear Optical Physics & Materials* **8**, 121.

Cipiccia, S., M. R. Islam, B. Ersfeld, R. P. Shanks, E. Brunetti, G. Vieux, X. Yang, R. C. Issac, S. M. Wiggins, G. H. Welsh, M. P. Anania, D. Maneuski, R. Montgomery, G. Smith, M. Hoek, D. J. Hamilton, N. R. C. Lemos, D. Symes, P. P. Rajeev, V. O. Shea,

J. M. Dias and D. A. Jaroszynski (2011), "Gamma-Rays from Harmonically Resonant Betatron Oscillations in a Plasma Wake," *Nature Physics* **7**, 867.

Cipiccia S., M. R. Islam, B. Ersfeld, G. H. Welsh, E. Brunetti, G. Vieux, X. Yang, S. M. Wiggins, P. Grant, D. R. Gil, D. W. Grant, R. P. Shanks, R. C. Issac, M. P. Anania, D. Manueski, R. Montgomery, G. Smith, M. Hoek, D. Hamilton, D. Symes, P. P. Rajeev, V. O'Shea, J. M. Dias, N. R. C. Lemos and D. A. Jaroszynski (2015), "Gamma-Ray Production from Resonant Betatron Oscillations in Plasma Wakes," in *Advances in X-Ray Free-Electron Lasers Instrumentation*, Proceedings of SPIE, ed. S. G. Biedron, Vol **9512**.

Corde, S., K. T. Phuoc, G. Lambert, R. Fitour, V. Malka, A. Rousse, A. Beck and E. Lefebvre (2013), "Femtosecond X Rays from Laser-Plasma Accelerators," *Reviews of Modern Physics* **85**, 1.

Coverdale, C. A., C. B. Darrow, C. D. Decker, W. B. Mori, K. C. Tzeng, K. A. Marsh, C. E. Clayton and C. Joshi (1995), "Propagation of Intense Subpicosecond Laser-Pulses through Underdense Plasma," *Physical Review Letters* **74**, 4659.

Curcio, A., D. Giulietti, G. Dattoli and M. Ferrario (2015), "Resonant Interaction between Laser and Electrons Undergoing Betatron Oscillations in the Bubble Regime," *Journal of Plasma Physics* **81**, 495810513.

Darrow, C. B., C. Coverdale, M. D. Perry, W. B. Mori, C. Clayton, K. Marsh and C. Joshi (1992), "Strongly Coupled Stimulated Raman Backscatter from Subpicosecond Laser–Plasma Interactions," *Physical Review Letters* **69**, 442.

Decker, C. D. and W. B. Mori (1994), "Group-Velocity of Large-Amplitude Electromagnetic-Waves in a Plasma," *Physical Review Letters* **72**, 490.

Decker, C. D., W. B. Mori, K. C. Tzeng and T. Katsouleas (1996b), "Evolution of Ultra-Intense, Short-Pulse Lasers in Underdense Plasmas," *Physics of Plasmas* **3**, 2047.

Drake, J. F., P. K. Kaw, Y. C. Lee and G. Schmidt, C. S. Liu and M. Rosenbluth (1974), "Parametric-Instabilities of Electromagnetic-Waves in Plasmas," *Physics of Fluids* **17**, 778.

Esarey, E., J. Krall and P. Sprangle (1994), "Envelope Analysis of Intense Laser-Pulse Self-Modulation in Plasmas," *Physical Review Letters* **72**, 2887.

Esarey, E., P. Sprangle, M. Pilloff and J. Krall (1995), "Theory and Group-Velocity of Ultrashort, Tightly Focused Laser-Pulses," *Journal of the Optical Society of America B – Optical Physics* **12**, 1695.

Esarey, E., P. Sprangle, J. Krall and A. Ting (1997), "Self-Focusing and Guiding of Short Laser Pulses in Ionizing Gases and Plasmas," *IEEE Journal of Quantum Electronics* **33**, 1879.

Esarey, E., B. A. Shadwick, P. Catravas and W. P. Leemans (2002), "Synchrotron Radiation from Electron Beams in Plasma-Focusing Channels," *Physical Review E* **65**, 056505.

Esarey, E., C. B. Schroeder and W. P. Leemans (2009), "Physics of Laser-Driven Plasma-Based Electron Accelerators," *Reviews of Modern Physics* **81**, 1229.

Feit, M. D., J. C. Garrison and A. M. Rubenchik (1996), "Short Pulse Laser Propagation in Underdense Plasmas," *Physical Review E* **53**, 1068.

Fill, E. E. (1994), "Focusing Limits of Ultrashort Laser-Pulses: Analytical Theory," *Journal of the Optical Society of America B – Optical Physics* **11**, 2241.

Forslund, D. W., J. M. Kindel and E. L. Lindman, "Theory of Stimulated Scattering Processes in Laser-Irradiated Plasmas," *Physics of Fluids* **18**, 1002 (1975).

Gahn, C., G. D. Tsakiris, A. Pukhov, J. Meyer-ter-Vehn, G. Pretzler, P. Thirolf, D. Habs and K. J. Witte (1999), "Multi-MeV Electron Beam Generation by Direct Laser Acceleration in High-Density Plasma Channels," *Physical Review Letters* **83**, 4772.

Geints, Y. E., A. D. Bulygin and A. A. Zemlyanov (2012), "Model Description of Intense Ultra-Short Laser Pulse Filamentation: Multiple Foci and Diffraction Rays," *Applied Physics B – Lasers and Optics* **107**, 243.

Giulietti Andre, A., A. Dufrenoy, S. Dobosz Giulietti, D. Hosokai, T. Koester, P. Kotaki, H. Labate, L. Levato, T. Nuter, R. Pathak, N. C. Monot, P. Gizzi, L. A. Gizzi and A. Leonida (2013), "Space- and Time-Resolved Observation of Extreme Laser Frequency Upshifting during Ultrafast-Ionization," *Physics of Plasmas* **20**, 082307.

Gonsalves, A. J., K. Nakamura, J. Daniels, C. Benedetti, C. Pieronek, T. C. H. de Raadt, S. Steinke, J. H. Bin, S. S. Bulanov, J. van Tilborg, C. G. R. Geddes, C. B. Schroeder, C. Toth, E. Esarey, K. Swanson, L. Fan-Chiang, G. Bagdarov, N. Bobrova, V. Gasilov, G. Korn, P. Sasorov and W. P. Leemans (2019) "Petawatt Laser Guiding and Electron Beam Acceleration to 8 GeV in a Laser-Heated Capillary Discharge Waveguide," *Physical Review Letters* **122**, 084801.

Guerin, S., G. Laval, P. Mora, J. C. Adam, A. Heron and A. Bendib (1995), "Modulational and Raman Instabilities in the Relativistic Regime," *Physics of Plasmas* **2**, 2807.

Hazra, D., A. Moorti, B. S. Rao, A. Upadhyay, J. A. Chakera and P. A. Naik (2018), "Betatron Resonance Electron Acceleration and Generation of Relativistic Electron Beams Using 200fs Ti: Sapphire Laser Pulses," *Plasma Physics and Controlled Fusion* **60**, 085015.

Hooker, S. M. (2013), "Developments in Laser-Driven Plasma Accelerators," *Nature Photonics* **7**, 775.

Jones, R. D. and K. Lee (1982), "Kinetic-Theory, Transport, and Hydrodynamics of a High-Z Plasma in the Presence of an Intense Laser Field," *Physics of Fluids* **25**, 2307.

Joshi, C. (2017), "Laser-Driven Plasma Accelerators Operating in the Self-Guided, Blowout Regime," *IEEE Transactions on Plasma Science* **45**, 3134.

Kahaly, S., F. Sylla, A. Lifschitz, A. Flacco, M. Veltcheva and V. Malka (2016), "Detailed Experimental Study of Ion Acceleration by Interaction of an Ultra-Short Intense Laser with an Underdense Plasma," *Scientific Reports* **6**, 31647.

Kandidov, V. P., S. A. Shlenov and O. G. Kosareva (2009), "Filamentation of High-Power Femtosecond Laser Radiation," *Quantum Electronics* **39**, 205.

Kar, S., M. Borghesi, C. A. Cecchetti, L. Romagnani, F. Ceccherini, T. V. Liseykina, A. Macchi, R. Jung, J. Osterholz, O. Willi, L. A. Gizzi, A. Schiavi, M. Galimberti and R. Heathcote (2007), "Dynamics of Charge-Displacement Channeling in Intense Laser–Plasma Interactions," *New Journal of Physics* **9**, 402.

Kim, J. K. and D. Umstadter (1999), "Cold Relativistic Wavebreaking Threshold of Two-Dimensional Plasma Waves," in *Advanced Accelerator Concepts, Eighth Workshop*, W. Lawson, C. Bellamy and D. F. Brosius, eds., pp. 404.

Kostyukov, I., A. Pukhov and S. Kiselev (2004), "Phenomenological Theory of Laser–Plasma Interaction in Bubble Regime," *Physics of Plasmas* **11**, 5256.

Krall, J., A. Ting, E. Esarey and P. Sprangle (1993), "Enhanced Acceleration in a Self-Modulated-Laser Wake-Field Accelerator," *Physical Review E* **48**, 2157.

Krushelnick, K., E. L. Clark, Z. Najmudin, M. Salvati, M. I. K. Santala, M. Tatarakis, A. E. Dangor, V. Malka, D. Neely, R. Allott and C. Danson (1999), "Multi-MeV Ion Production from High-Intensity Laser Interactions with Underdense Plasmas," *Physical Review Letters* **83**, 737.

Langdon, A. B. (1980), "Non-Linear Inverse Bremsstrahlung and Heated-Electron Distributions," *Physical Review Letters* **44**, 575.

Leemans, W. P., C. E. Clayton, W. B. Mori, K. A. Marsh, P. K. Kaw, A. Dyson, C. Joshi and J. M. Wallace (1992a), "Experiments and Simulations of Tunnel-Ionized Plasmas," *Physical Review A* **46**, 1091.

Leemans, W. P., C. Joshi, W. B. Mori, C. E. Clayton and T. W. Johnston (1992b), "Nonlinear Dynamics of Driven Relativistic Electron-Plasma Waves," *Physical Review A* **46**, 5112.

Lu, W., C. Huang, M. Zhou, W. B. Mori and T. Katsouleas (2006), "Nonlinear Theory for Relativistic Plasma Wakefields in the Blowout Regime," *Physical Review Letters* **96**, 165002.

Lu, W., M. Tzoufras, C. Joshi, F. S. Tsung, W. B. Mori, J. Vieira, R. A. Fonseca and L. O. Silva (2007), "Generating Multi-GeV Electron Bunches Using Single Stage Laser Wakefield Acceleration in a 3D Nonlinear Regime," *Physical Review Special Topics: Accelerators and Beams* **10**, 061301.

Mackinnon, A. J., M. Borghesi, A. Iwase, M. W. Jones, G. J. Pert, S. Rae, K. Burnett and O. Willi (1996), "Quantitative Study of the Ionization-Induced Refraction of Picosecond Laser Pulses in Gas-Jet Targets," *Physical Review Letters* **76**, 1473.

Malka, V. (2012), "Laser Plasma Accelerators," *Physics of Plasmas* **19**, 055501.

Malka, V., J. Faure, Y. A. Gauduel, E. Lefebvre, A. Rousse and K. T. Phuoc (2008), "Principles and Applications of Compact Laser-Plasma Accelerators," *Nature Physics* **4**, 447.

Marburger, J. H (1975), "Self-focusing: Theory," *Progress in Quantum Electronics* **4**, 35.

McKinstrie, C. J. and R. Bingham (1992), "Stimulated Raman Forward Scattering and the Relativistic Modulational Instability of Light Waves in Rarefied Plasma," *Physics of Fluids B – Plasma Physics* **4**, 2626.

Mlejnek, M., E. M. Wright and J. V. Moloney (1999), "Moving-Focus versus Self-Waveguiding Model for Long-Distance Propagation of Femtosecond Pulses in Air," *IEEE Journal of Quantum Electronics* **35**, 1771.

Monot, P., T. Auguste, L. A. Lompre, G. Mainfray and C. Manus (1992), "Focusing Limits of a Terawatt Laser in an Underdense Plasma," *Journal of the Optical Society of America B – Optical Physics* **9**, 1579.

Mori, W. B. (1997), "The Physics of the Nonlinear Optics of Plasmas at Relativistic Intensities for Short-Pulse Lasers," *IEEE Journal of Quantum Electronics* **33**, 1942.

Mori, W. B., C. D. Decker, D. E. Hinkel and T. Katsouleas (1994), "Raman Forward Scattering of Short-Pulse High-Intensity Lasers," *Physical Review Letters* **72**, 1482.

Nibbering, E. T. J., P. F. Curley, G. Grillon, B. S. Prade, M. A. Franco, F. Salin and A. Mysyrowicz (1996), "Conical Emission from Self-Guided Femtosecond Pulses in Air," *Optics Letters* **21**, 62.

Noble, R. J. (1985), "Plasma-Wave Generation in the Beat-Wave Accelerator," *Physical Review A* **32**, 460.

Panwar, A. and A. K. Sharma (2009), "Self-Phase Modulation of a Laser in Self-Created Plasma Channel," *Laser and Particle Beams* **27**, 249.

Penetrante, B. M., J. N. Bardsley, W. M. Wood, C. W. Siders and M. C. Downer (1992), "Ionization-Induced Frequency-Shifts in Intense Femtosecond Laser-Pulses," *Journal of the Optical Society of America B – Optical Physics* **9**, 2032.

Pert, G. J. (1972), "Inverse Bremsstrahlung Absorption in Large Radiation Fields During Binary Collisions: Classical Theory," *Journal of Physics A: General Physics* **5**, 506.

Phuoc, K. T., F. Burgy, J. P. Rousseau, V. Malka, A. Rousse, R. Shah, D. Umstadter, A. Pukhov and S. Kiselev (2005), "Laser Based Synchrotron Radiation," *Physics of Plasmas* **12**, 023101.

Pukhov, A., Z. M. Sheng and J. Meyer-ter-Vehn (1999), "Particle Acceleration in Relativistic Laser Channels," *Physics of Plasmas* **6**, 2847.

Rae, S. C. (1993), "Ionization-Induced Defocusing of Intense Laser-Pulses in High-Pressure Gases," *Optics Communications* **97**, 25.

Rajouria, S. K., H. K. Malik, V. K. Tripathi and P. Kumar (2015), "Step Density Model of Laser Sustained Ion Channel and Coulomb Explosion," *Physics of Plasmas* **22**, 023104.

Rankin, R., C. E. Capjack, N. H. Burnett and P. B. Corkum (1991), "Refraction Effects Associated with Multiphoton Ionization and Ultrashort-Pulse Laser Propagation in Plasma Wave-Guides," *Optics Letters* **16**, 835.

Rau, B., C. W. Siders, S. P. LeBlanc, D. L. Fisher, M. C. Downer and T. Tajima (1997), "Spectroscopy of Short, Intense Laser Pulses due to Gas Ionization Effects," *Journal of the Optical Society of America B-Optical Physics* **14**, 643.

Ritchie, B. (1994), "Relativistic Self-Focusing and Channel Formation in Laser–Plasma Interactions," *Physical Review E* **50**, R687.

Sakharov, A. S. and V. I. Kirsanov (1994), "Theory of Raman-Scattering for a Short Ultrastrong Laser-Pulse in a Rarefied Plasma," *Physical Review E* **49**, 3274.

Seely, J. F. and E. G. Harris (1973), "Heating of a Plasma by Multiphoton Inverse Bremsstrahlung," *Physical Review A* **7**, 1064.

Silin, V. P. (1965), "Nonlinear High-Frequency Plasma Conductivity," *Soviet Physics JETP-USSR* **20**, 1510.

Sprangle, P., C. M. Tang and E. Esarey (1987), "Relativistic Self-Focusing of Short-Pulse Radiation Beams in Plasmas," *IEEE Transactions on Plasma Science* **15**, 145.

Sprangle, P., E. Esarey and A. Ting (1990a), "Nonlinear-Interaction of Intense Laser-Pulses in Plasmas," *Physical Review A* **41**, 4463.

Sprangle, P., E. Esarey and A. Ting (1990b), "Nonlinear-Theory of Intense Laser–Plasma Interactions," *Physical Review Letters* **64**, 2011.

Sprangle, P., E. Esarey, J. Krall and G. Joyce (1992), "Propagation and Guiding of Intense Laser-Pulses in Plasmas," *Physical Review Letters* **69**, 2200.

Sprangle, P., E. Esarey and J. Krall (1996), "Self-Guiding and Stability of Intense Optical Beams in Gases Undergoing Ionization," *Physical Review E* **54**, 4211.

Tajima, T. and J. M. Dawson (1979), "Laser Electron-Accelerator," *Physical Review Letters* **43**, 267.

Thomas, A. G. R. and K. Krushelnick (2009), "Betatron X-Ray Generation from Electrons Accelerated in a Plasma Cavity in the Presence of Laser Fields," *Physics of Plasmas* **16**, 103103.

Trines, R. and P. A. Norreys (2006), "Wave-Breaking Limits for Relativistic Electrostatic Waves in a One-Dimensional Warm Plasma," *Physics of Plasmas* **13**, 123102.

Tripathi, V. K., T. Taguch and C. S. Liu (2005), "Plasma Channel Charging by an Intense Short Pulse Laser and Ion Coulomb Explosion," *Physics of Plasmas* **12**, 043106.

Tzoufras, M., W. Lu, F. S. Tsung, C. Huang, W. B. Mori, T. Katsouleas, J. Vieira, R. A. Fonseca and L. O. Silva (2009), "Beam Loading by Electrons in Nonlinear Plasma Wakes," *Physics of Plasmas* **16**, 056705.

Tzoufras, M., W. Lu, F. S. Tsung, C. Huang, W. B. Mori, J. Vieira, R. A. Fonseca and L. O. Silva (2007), "The Physical Picture of Beam Loading in the Blowout Regime," 2007 *IEEE Particle Accelerator Conference*, Vols 1–11.

Umstadter, D. (2003), "Relativistic Laser–Plasma Interactions," *Journal of Physics D – Applied Physics* **36**, R151.

Umstadter, D., E. Esarey and J. Kim (1994), "Nonlinear Plasma-Waves Resonantly Driven by Optimized Laser-Pulse Trains," *Physical Review Letters* **72**, 1224.

Umstadter, D., S. Y. Chen, A. Maksimchuk, G. Mourou and R. Wagner (1996), "Nonlinear Optics in Relativistic Plasmas and Laser Wake Field Acceleration of Electrons," *Science* **273**, 472.

Umstadter, D., S. Sepke and S. Y. Chen (2005), "Relativistic Nonlinear Optics," in *Advances in Atomic Molecular and Optical Physics*, Vol. **52**, P. R. Berman and C. C. Lin, eds., p. 331.

Wang, X. M., R. Zgadzaj, N. Fazel, Z. Y. Li, S. A. Yi, X. Zhang, W. Henderson, Y. Y. Chang, R. Korzewkwa, H. E. Tsai, C. H. Pai, H. Quevedo, G. Dyer, E. Gaul. M. Martinez, A. C. Bernstein, T. Borger, M. Spinks, M. Donovan, V. Khudik, G. Shvets, T. Ditmire and M. C. Downer (2013), "Quasi-monoenergetic Laser-Plasma Acceleration of Electrons to 2 GeV," *Nature Communications* **4**, 1988.

Watts, I., M. Zepf, E. L. Clark, M. Tatarakis, K. Krushelnick, A. E. Dangor, R. Allott, R. J. Clarke, D. Neely and P. A. Norreys (2002), "Measurements of Relativistic Self-Phase-Modulation in Plasma," *Physical Review E* **66**, 036409.

Wilks, S. C., W. L. Kruer, K. Estabrook and A. B. Langdon (1992), "Theory and Simulation of Stimulated Raman Scatter at Near-Forward Angles," *Physics of Fluids B – Plasma Physics* **4**, 2794.

Chapter 9 Strong Field Interactions with Overdense Plasmas

Albritton, J. R. and A. B. Langdon (1980), "Profile Modification and Hot-Electron Temperature from Resonant Absorption at Modest Intensity," *Physical Review Letters* **45**, 1794.

Atzeni, S., A. Schiavi and J. R. Davies (2009), "Stopping and Scattering of Relativistic Electron Beams in Dense Plasmas and Requirements for Fast Ignition," *Plasma Physics and Controlled Fusion* **51**, 015016.

Bae, L. J. and B. I. Cho (2015), "Coherent Transition Radiation from Thin Targets Irradiated by High Intensity Laser Pulses," *Current Applied Physics* **15**, 242.

Baeva, T., S. Gordienko and A. Pukhov (2006), "Theory of High-Order Harmonic Generation in Relativistic Laser Interaction with Overdense Plasma," *Physical Review E* **74**, 046404.

Beg, F. N., A. R. Bell, A. E. Dangor, C. N. Danson, A. P. Fews, M. E. Glinsky, B. A. Hammel, P. Lee, P. A. Norreys and M. Tatarakis (1997), "A Study of Picosecond Laser-Solid Interactions up to 10^{19} W cm^{-2}," *Physics of Plasmas* **4**, 447.

Bell, A. R., J. R. Davies and S. M. Guerin (1997), "Magnetic Field in Short-Pulse High-Intensity Laser-Solid Experiments," *Physical Review E* **58**, 2471.

Bell, A. R., J. R. Davies, S. Guerin and H. Ruhl (1998), "Fast-Electron Transport in High-Intensity Short-Pulse Laser-Solid Experiments," *Plasma Physics and Controlled Fusion* **39**, 653.

Bell, A. R. and R. J. Kingham (2003), "Resistive Collimation of Electron Beams in Laser-Produced Plasmas," *Physical Review Letters* **91**, 035003.

Belyaev, V. S., V. P. Krainov, V. S. Lisitsa and A. P. Matafonov (2008), "Generation of Fast Charged Particles and Superstrong Magnetic Fields in the Interaction of Ultrashort High-Intensity Laser Pulses with Solid Targets," *Physics-Uspekhi* **51**, 793.

Borghesi, M., J. Fuchs, S. V. Bulanov, A. J. Mackinnon, P. K. Patel and M. Roth (2006), "Fast Ion Generation by High-Intensity Laser Irradiation of Solid Targets and Applications," *Fusion Science and Technology* **49**, 412.

Borghesi, M. J., A. J. MacKinnon, A. R. Bell, R. Gaillard and O. Willi (1998), "Megagauss Magnetic Field Generation and Plasma Jet Formation on Solid Targets Irradiated by an Ultraintense Picosecond Laser Pulse," *Physical Review Letters* **81**, 112.

Brunel, F. (1987), "Not-So-Resonant, Resonant Absorption," *Physical Review Letters* **59**, 52.

Cai, H. B., W. Yu, S. P. Zhu and C. Y. Zheng (2006a), "Short-Pulse Laser Absorption via J x B Heating in Ultrahigh Intensity Laser Plasma Interaction," *Physics of Plasmas* **13**, 113105.

Cai, H. B., W. Yu, S. P. Zhu, C. Y. Zheng, L. H. Cao and W. B. Pei (2006b), "Vacuum Heating in the Interaction of Ultrashort, Relativistically Strong Laser Pulses with Solid Targets," *Physics of Plasmas* **13**, 063108.

Califano, F., F. Pegoraro and S. V. Bulanov (1997), "Spatial Structure and Time Evolution of the Weibel Instability in Collisionless Inhomogeneous Plasmas," *Physical Review E* **56**, 963.

Carman, R. L., D. W. Forslund and J. M. Kindel (1980), "Visible Harmonic Emission as a Way of Measuring Profile Steepening," *Physical Review Letters* **46**, 29.

Cattani, F., A. Kim, D. Anderson and M. Lisak (2000), "Threshold of Induced Transparency in the Relativistic Interaction of an Electromagnetic Wave with Overdense Plasmas," *Physical Review E* **62**, 1234.

Catto, P. J. and R. M. More (1977), "Sheath Inverse Bremsstrahlung in Laser-Produced Plasmas," *Physics of Fluids* **20**, 704.

Chen, C. D., P. K. Patel, D. S. Hey, A. J. Mackinnon, M. H. Key, K. U. Akli, T. Bartal and F. N. Beg (2009), "Bremsstrahlung and K-alpha Fluorescence Measurements for Inferring Conversion Efficiencies into Fast Ignition Relevant Hot Electrons," *Physics of Plasmas* **16**, 082705.

Chen, H., A. Link, Y. Sentoku, P. Audebert, F. Fiuza, A. Hazi, R. F. Heeter, M. Hill, L. Hobbs, A. J. Kemp, G. E. Kemp, S. Kerr, D. D. Meyerhofer, J. Myatt, S. R. Nagel, J. Park, R. Tommasini and G. J. Williams (2015), "The Scaling of Electron and Positron Generation in Intense Laser–Solid Interactions," *Physics of Plasmas* **22**, 056705.

Chen, H., S. C. Wilks, W. L. Kruer, P. K. Patel and R. Shepherd (2009), "Hot Electron Energy Distributions from Ultraintense Laser Solid Interactions," *Physics of Plasmas* **16**, 020705.

Cho, B. I., J. Osterholz, A. C. Bernstein, G. M. Dyer, A. Karmakar, A. Pukhov and T. Ditmire (2009), "Characterization of Two Distinct, Simultaneous Hot Electron Beams in Intense Laser–Solid Interactions," *Physical Review E* **80**, 055402.

Cobble, J. A., G. A. Kyrala, A. A. Hauer, A. J. Taylor, C. C. Gomez, N. D. Delamater and G. T. Schappert (1989), "Kilovolt X-Ray Spectroscopy of a Subpicosecond-Laser-Excited Source," *Physical Review A* **39**, 454.

Consoli, F., V. T. Tikhonchuk, M. Bardon, P. Bradford, D. C. Carroll, J. Cikhardt, M. Cipriani, R. J. Clarke, T. E. Cowen, C. N. Danson, R. De Angelis, M. De Marco, J. L. Dubois, B. Etchessahar, A. L. Garcia, D. I. Hillier, A. Honsa, W. M. Jiang, V. Kmetik, J. Krasa, Y. T. Li, F. Lubrano, P. McKenna, J. Metzkes-Ng, A. Poye, I. Principe, P. Raczka, R. A. Smith, R. Vrana, N. C. Woolsey, E. Zemaiyte, Y. H. Zhang, Z. Zhang, B. Zielbauer and D. Neely (2020), "Laser Produced Electromagnetic Pulses: Generation, Detection and Mitigation," *High Power Laser Science and Engineering* **8**, e22.

Cowan, T. E., M. D. Perry, M. H. Key, T. R. Ditmire, S. P. Hatchett, E. A. Henry, J. D. Moody, M. J. Moran, D. M. Pennington, T. W. Phillips, T. C. Sangster, J. A. Sefcik, M. S. Singh, R. A. Snavely, M. A. Stoyer, S. C. Wilks, P. E. Young, Y. Takahashi, B. Don, W. Fountain, T. Parnell, J. Johnson, A. W. Hunt and T. Kuhl (1999), "High Energy Electrons, Nuclear Phenomena and Heating in Petawatt Laser-Solid Experiments," *Laser and Particle Beams* **17**, 773.

Cowan, T. E., A. W. Hunt, T. W. Phillips, S. C. Wilks, M. D. Perry, C. Brown, W. Fountain, S. Hatchett, J. Johnson, M. H. Key, T. Parnell, D. M. Pennington, R. A. Snavely and Y. Takahashi (2000a), "Photonuclear Fission from High Energy Electrons from Ultraintense Laser–Solid Interactions," *Physical Review Letters* **84**, 903.

Cowan, T. E., M. Roth, J. Johnson, C. Brown, M. Christl, W. Fountain, S. Hatchett, E. A. Henry, A. W. Hunt, M. H. Key, A. MacKinnon, T. Parnell, D. M. Pennington, M. D. Perry, T. W. Phillips, T. C. Sangstewr, M. Singh, R. Snavely, M. Stoyer, Y. Takahashi, S. C. Wilks and K. Yasuike (2000b), "Intense Electron and Proton Beams from Petawatt Laser–Matter Interactions," *Nuclear Instruments & Methods in Physics Research Section A: Accelerators, Spectrometers, Detectors and Associated Equipment* **455**, 130.

Crow, J. E., P. L. Auer and J. E. Allen (1975), "Expansion of a Plasma into a Vacuum," *Journal of Plasma Physics* **14**, 65.

Cui, Y. Q., W. M. Wang, Z. M. Sheng, Y. T. Li and J. Zhang (2013), "Laser Absorption and Hot Electron Temperature Scalings in Laser Plasma Interactions," *Plasma Physics and Controlled Fusion* **55**, 085008.

Daido, H., M. Nishiuchi and A. S. Pirozhkov (2012), "Review of Laser-Driven Ion Sources and Their Applications," *Reports on Progress in Physics* **75**, 056401.

Ditmire, T., M. Roth, P. K. Patel, D. Callahan, G. Cheriaux, P. Gibbon, D. Hammond, A. Hannasch, L. C. Jarrot, G. Shaumann, W. Theobald, C. Therrot, O. Turianska, X. Vaisseau, F. Wasser, S. Zähter, M. Zimmer and W. Goldstein (2023), "Focused Energy, A New Approach Towards Inertial Fusion Energy," *Journal of Fusion Energy* **42**, 27.

Eliasson, B., C. S. Liu, Z. Shao, R. Z. Sagdeev and P. K. Shukla (2009), "Laser Acceleration of Monoenergetic Protons via a Double Layer Emerging from an Ultra-thin Foil," *New Journal of Physics* **11**, 073006.

Eremin, V. I., A. V. Korzhimanov and A. V. Kim (2010), "Relativistic Self-Induced Transparency Effect during Ultraintense Laser Interaction with Overdense Plasmas: Why it Occurs and its Use for Ultrashort Electron Bunch Generation," *Physics of Plasmas* **17**, 043102.

Estabrook, K. and W. L. Kruer (1978), "Properties of Resonantly Heated Electron Distributions," *Physical Review Letters* **40**, 42.

Findley, D. J. S. (1989), "Analytic Representation of Bremsstrahlung Spectra from Thick Radiators as a Function of Photon Energy and Angle," *Nuclear Instruments and Methods A: Accelerators, Spectrometers, Detectors and Associated Equipment* **276**, 598.

Forslund, D. W., J. M. Kindel, K. Lee, E. L. Lindman and R. L. Morse (1975), "Theory and Simulation of Resonant Absorption in a Hot Plasma," *Physical Review A* **11**, 679.

Fuchs, J., J. C. Adams, F. Amiranoff, S. D. Baton, N. Blanchot, P. Gallant, L. Gremillet, A. Heron, J. C. Kieffer, G. Laval, G. Malka, J. L. Miquel, P. Mora, H. Pepin and C. Rousseaux (1999), "Experimental Study of Laser Penetration in Overdense Plasmas at Relativistic Intensities. I: Hole Boring through Preformed Plasmas Layers," *Physics of Plasmas* **6**, 2563.

Fuchs, J., P. Antici, E. D'Humieres, E. Lefebvre, M. Borghesi, E. Brambrink, C. A. Cecchetti, M. Kaluza, V. Malka, M. Manclossi, S. Meyroneinc, P. Mora, J. Schreiber, T. Toncian, H. Pepin and T. Audebert (2006), "Laser-Driven Proton Scaling Laws and New Paths Towards Energy Increase," *Nature Physics* **2**, 48.

Galy, J., M. Maucec, D. J. Hamilton, R. Edwards and J. Magill (2007), "Bremsstrahlung Production with High-Intensity Laser Matter Interactions and Applications," *New Journal of Physics* **9**, 23.

Gibbon, P. and A. R. Bell (1992), "Collisionless Absorption in Sharp-Edged Plasmas," *Physical Review Letters* **68**, 1535.

Grimes, M. K., A. R. Rundquist, Y. S. Lee and M. C. Downer (1999), "Experimental Identification of 'Vacuum Heating' at Femtosecond-Laser-Irradiated Metal Surfaces," *Physical Review Letters* **82**, 4010.

Gryaznykh, D. A. (1998) "Cross Section for the Production of Electron–Positron Pairs by Electrons in the Field of a Nucleus," *Physics of Atomic Nuclei* **61**, 394.

Gurevich, A. V., L. V. Pariiska and L. P. Pitaevsk (1966), "Self-Similar Motion of Rarefied Plasma," *Soviet Physics JETP-USSR* **22**, 449.

Haines, M. G., M. S. Wei, F. N. Beg and R. B. Stephens (2009), "Hot-Electron Temperature and Laser-Light Absorption in Fast Ignition," *Physical Review Letters* **102**, 045008.

Henig, A., S. Steinkie, M. Schnurer, T. Sokollik, R. Horlein, D. Kiefer, D. Jung, J. Schreiber, B. M. Hegelich, X. Q. Yan, J. Meyer-ter-Vehn, T. Tajima, P. V. Nickles, W. Sandner and D. Habs (2009), "Radiation-Pressure Acceleration of Ion Beams Driven by Circularly Polarized Laser Pulses," *Physical Review Letters* **103**, 245003.

Higginson D. P., J. M. McNaney, D. C. Swift, G. M. Petrov, J. Davis, J. A. Frenje, L. C. Jarrott, R. Kodama, K. L. Lancaster, A. J. MacKinnon, H. Nakamura, P. K. Patel, G. Tynan and F. N. Beg (2011), "Production of Neutrons up to 18 MeV in High-Intensity, Short-Pulse Laser Matter Interactions," *Physics of Plasmas* **18**, 100703.

Huang, L. G., H. Takabe and T. E. Cowan (2019), "Maximizing Magnetic Field Generation in High Power Laser–Solid Interactions," *High Power Laser Science and Engineering* **7**, e22.

Key, M. H., M. D. Cable, T. E. Cowan, K. G. Estabrook, B. A. Hammel, S. P. Hatchett, E. A. Hentry, D. E. Hijnkel, J. D. Kilkenny, J. A. Koch, W. L. Kruer, A. B. Langdon, B. F. Lasinski, R. W. Lee, B. J. MacGowan, A. MacKinnon, J. D. Kieffer, J. C., M. Chaker, J. P. Matte, H. Pepin, C. Y. Cote, Y. Beaudoin, T. W. Johnston, C. Y. Chien, S. Coe, G. Mourou and O. Peyrusse (1998), "Ultrafast X-Ray Sources," *Physics of Fluids B – Plasma Physics* **5**, 2676.

Kruer, W. L. and K. Estabrook (1985), "JxB Heating by Very Intense Laser-Light," *Physics of Fluids* **28**, 430.

Lee, K., D. W. Forslund, J. M. Kindel and E. L. Lindman (1977), "Theoretical Derivation of Laser-Induced Plasma Profiles," *Physics of Fluids* **20**, 51.

Lefebvre, E. and G. Bonnaud (1997), "Nonlinear Electron Heating in Ultrahigh-Intensity-Laser–Plasma Interaction," *Physical Review E* **55**, 1011.

Liang, E. P., S. C. Wilks and M. Tabak (1998), "Pair Production by Ultraintense Lasers," *Physical Review Letters* **81**, 4887.

Lichters, R., J. Meyer-ter-Vehn and A. Pukhov (1996), "Short-Pulse Laser Harmonics from Oscillating Plasma Surfaces Driven at Relativistic Intensity," *Physics of Plasmas* **3**, 3425.

Lindl, J. D. and P. K. Kaw (1971), "Pondermotive Force on Laser-Produced Plasmas," *Physics of Fluids* **14**, 371.

Liseykina, T., P. Mulser and M. Murakami (2015), "Collisionless Absorption, Hot Electron Generation, and Energy," *Physics of Plasmas* **22**, 03302.

Listykina et al. 2015 Scaling in Intense Laser–Target Interaction," *Physics of Plasmas* **22**, 033302.

Macchi, A., S. Veghini, T. V. Liseykina and F. Pegoraro (2010), "Radiation Pressure Acceleration of Ultrathin Foils," *New Journal of Physics* **12**, 045013.

Macchi, A., S. Veghini and F. Pegoraro (2009), "'Light Sail' Acceleration Reexamined," *Physical Review Letters* **103**, 085003.

Malka, G. and J. L. Miquel (1996), "Experimental Confirmation of Ponderomotive-Force Electrons Produced by an Ultrarelativistic Laser Pulse on a Solid Target," *Physical Review Letters* **77**, 77.

Mason, R. J. and M. Tabak (1998), "Magnetic Field Generation in High-Intensity-Laser–Matter Interactions," *Physical Review Letters*, **80**, 524.

McKenna, P. and M. N. Quinn (2013), "Energetic Electron Generation and Transport in Intense Laser–Solid Interactions," in *Laser–Plasma Interactions and Applications*, P. McKenna, D. Neely, R. Bingham and D. A. Jaroszynski, eds. (Heidelberg: Springer).

Meyerhofer, D. D., H. Chen, J. A. Delettrez, B. Soom, S. Uchida and B. Yaakobi (1993), "Resonance-Absorption in High-Intensity Contrast, Picosecond Laser–Plasma Interactions," *Physics of Fluids B – Plasma Physics* **5**, 2584.

Mora, P. (2003), "Plasma Expansion into a Vacuum," *Physical Review Letters* **90**, 185002.

Murnane, M. M., H. C. Kapteyn and R. W. Falcone (1989), "High-Density Plasmas Produced by Ultrafast Laser Pulses," *Physical Review Letters* **62**, 155.

Myatt, J., J. A. Delettrez, A. V. Maximov, D. D. Meyerhofer, R. W. Short, C. Stoeckl and M. Storm (1993) "Optimizing Electron–Positron Pair Production on Kilojoule-Class High-Intensity Lasers for the Purpose of Pair-Plasma Creation," *Physical Review E* **79**, 066409.

Naumova, N., T. Schlegel, V. T. Tikhonchuk, C. Labaune, I. V. Sokolov and G. Mourou (2009), "Hole Boring in a DT Pellet and Fast-Ion Ignition with Ultraintense Laser Pulses," *Physical Review Letters* **102**, 025002.

Nilson, P. M., W. Theobald, J. Myatt, C. Stoeckl, M. Storm, O. V. Gotchev, J. D. Zuegel, T. Betti, D. D. Meyerhofer and T. C. Sangster (2008) "High-Intensity Laser–Plasma Interactions in the Refluxing Limit," *Physics of Plasmas* **15**, 056308.

Norreys, P. A. (2009), "Progress in Fast Ignition," in *Laser–Plasma Interactions*, D. A. Jaroszynski, R. Bingham, R. A. Cairns, eds. (Boca Raton: CRC Press), p. 361.

Norreys, P., D. Batani, S. Baton, F. N. Beg, R. Kodama, P. M. Nilson, P. Patel, F. Perez, J. J. Santos, R. H. Scott, V. T. Tikhonchuk, M. Wei and J. Zhang (2014) "Fast Electron Energy Transport in Solid Density and Compressed Plasma," *Nuclear Fusion* **54**, 054004.

Ott, E. (1972), "Nonlinear Evolution of Rayleigh-Taylor Instability of a Thin-Layer," *Physical Review Letters* **29**, 1429.

Palastro, J. P., J. G. Shaw, R. K. Follett, A. Colaitis, D. Turnbull, A. V. Maximov, V. N. Goncharov and D. H. Froula (2018), "Resonance Absorption of a Broadband Laser Pulse," *Physics of Plasmas* **25**, 123104.

Pegoraro, F. and S. V. Bulanov (2007), "Photon Bubbles and Ion Acceleration in a Plasma Dominated by the Radiation Pressure of an Electromagnetic Pulse," *Physical Review Letters* **99**, 065002.

Pert, G. J. (1978), "Analytic Theory of Linear Resonant Absorption," *Plasma Physics and Controlled Fusion* **20**, 175.

Petrov, G. M., D. P. Higginson, J. Davis, Tz. B. Petrova, J. M. McNaney, C. McGuffey, B. Qiao and F. N. Beg (2012), "Generation of High-Energy (>15 MeV) Neutrons Using Short Pulse High Intensity Lasers," *Physics of Plasmas* **19**, 093106.

Ping, Y., R. Shepherd, B. F. Lasinski, M. Tabak, H. Chen, H. K. Chung, K. B. Fournier, S. B. Hansen, A. Kemp, D. A. Liedahl, K. Widmann, S. C. Wilks, W. Rozmus and M. Sherlock (2008), "Absorption of Short Laser Pulses on Solid Targets in the Ultrarelativistic Regime," *Physical Review Letters* **100**, 085004.

Price, D. F., R. M. More, R. S. Walling, G. Guethlein, R. L. Shepherd, R. E. Stewart and W. E. White (1995), "Absorption of Ultrashort Laser-Pulses by Solid Targets Heated Rapidly to Temperatures 1–1000 eV," *Physical Review Letter* **75**, 252.

Pukhov, A. (2003), "Strong Field Interaction of Laser Radiation," *Reports on Progress in Physics* **66**, 47.

Pukhov, A. and J. Meyer-ter-Veyn (1997), "Laser Hole Boring into Overdense Plasma and Relativistic Electron Currents for Fast Ignition of ICF Targets," *Physical Review Letters* **79**, 2686.

Rajouria, S. K., K. K. M. Kumar and V. K. Tripathi (2013), "Nonlinear Resonance Absorption of Laser in an Inhomogeneous Plasma," *Physics of Plasmas* **20**, 083112.

Reich, C., P. Gibbon, I. Uschmann and E. Forster (2000), "Yield Optimization and Time Structure of Femtosecond Laser Plasma K Alpha Sources," *Physical Review Letters* **84**, 4846.

Robinson, A. P. L., M. Zepf, S. Kar, R. G. Evans and C. Bellei (2008), "Radiation Pressure Acceleration of Thin Foils with Circularly Polarized Laser Pulses," *New Journal of Physics* **10**, 013021.

Roth, M., D. Jung, K. Falk, N. Guler, O. Deppert, M. Devlin, A. Favalli, J. Fernandez, D. Gautier, M. Geissel, R. Haight, C. E. Hamilton, B. M. Hegelich, R. P. Johnson, F. Merrill, G. Schaumann, K. Schoenberg, M. Schollmeier, T. Shimada, T. Taddeucci, J. L. Tybo, F. Wagner, S. A. Wender, C. H. Wilde and G. A. Wurden (2001), "Fast Ignition by Intense Laser-Accelerated Proton Beams," *Physical Review Letters* **86**, 436.

Roth, M. and M. Schollmeier (2013), "Ion Acceleration: TNSA," in *Laser–Plasma Interactions and Applications*, P. McKenna, D. Neely, R. Bingham and D. A. Jaroszynski, eds. (Heidelberg: Springer).

Roth, M., D. Jung, K. Falk, N. Guler, O. Deppert, M. Devlin, A. Favalli, J. Fernandez, D. Gautier, M. Geissel, R. Haight, C. E. Hamilton, B. M. Hegelich, R. P. Johnson, F. Merrill, G. Schaumann, K. Schoenberg, M. Schollmeier, T. Shimada, T. Taddeucci, J. L. Tybo, F. Wagner, S. A. Wender, C. H. Wilde and G. A. Wurden (2013), "Bright Laser-Driven Neutron Source based on the Relativistic Transparency of Solids," *Physical Review Letters* **110**, 044802.

Rozmus, W. and V. T. Tikhonchuk (1990), "Skin Effect and Interaction of Short Laser-Pulses with Dense-Plasmas," *Physical Review A* **42**, 7401.

Santala, M. I. K., M. Zepf, I. Watts, F. N. Beg, E. Clark, M. Tatarkis, K. Kruschelnick, A. E. Dangor, T. McCanny, I. Spencer, R. P. Singhal, K. W. D. Ledingham, S. C. Wilks, A.

C. Machacek, J. Wark, R. Allott, R. J. Clarke and P. A. Norreys (2000), "Effect of the Plasma Density Scale Length on the Direction of Fast Electrons in Relativistic Laser–Solid Interactions," *Physical Review Letters* **84**, 1459.

Schlegel, T., N. Naumova, V. T. Tikhonchuk, C. Labaune, I. V. Sokolov and G. Mourou (2009), "Relativistic Laser Piston Model: Ponderomotive Ion Acceleration in Dense Plasmas Using Ultraintense Laser Pulses," *Physics of Plasmas* **16**, 083103.

Schmalz, R. F. (1985), "New Self-Similar Solutions for the Unsteady One-Dimensional Expansion of a Gas into a Vacuum," *Physics of Fluids* **28**, 2923.

Singh, P. K., A. Adak, A. D. Lad, G. Chatterjee and G. R. Kumar (2020), "Two-Plasmon-Decay Induced Fast Electrons in Intense Femtosecond Laser–Solid Interactions," *Physics of Plasmas* **27**, 083105.

Snavely, R. A., M. H. Key, S. P. Hatchett, T. E. Cowan, M. Roth, T. W. Phillips, M. A. Stoyer, E. A. Henry, T. C. Sangster, M. S. Singh, S. C. Wilks, A. MacKinnon, A. Offenberger, D. M. Pennington, K. Yasuike, A. B. Langdon, B. F. Lasinski, J. Johnson, M. D. Perry and E. M. Campbell (2000), "Intense High-Energy Proton Beams from Petawatt-Laser Irradiation of Solids," *Physical Review Letters* **85**, 2945.

Sudan, R. N. (1993), "Mechanism for the Generation of 10^9 G-Magnetic Fields in the Interaction of Ultraintense Short Laser-Pulse with an Overdense Plasma Target," *Physical Review Letters* **70**, 3075.

Tabak, M., J. Hammer, M. E. Glinksky, W. L. Kruer, S. C. Wilks, J. Woodworth, E. M. Campbell, M. D. Perry and R. J. Mason (1994), "Ignition and High-Gain with Ultrapowerful Lasers," *Physics of Plasmas* **1**, 1626.

Tarasevitch, A. and D. von der Linde (2009), "High Order Harmonic Generation from Solid Targets: Towards Intense Attosecond Pulses," *European Physical Journal-Special Topics* **175**, 35.

Tatarakis, M., A. Gopal, I. Watts, F. N. Beg, A. E. Dangor, K. Krushelnick, U. Wagner, P. A. Norreys, E. L. Clark, M. Zepf and R. G. Evans (2002), "Measurements of Ultrastrong Magnetic Fields during Relativistic Laser–Plasma Interactions," *Physics of Plasmas* **9**, 2244.

Taylor D., E. Liang, T. Clarke, A. Henderson, P. Chaguine, X. Wang, G. Dyer, K. Serratto, N. Riley, M. Donovan and T. Ditmire (2013), "Hot Electron Production Using the Texas Petawatt Laser Irradiating Thick Gold Targets," *High Energy Density Physics* **9**, 363.

Teubner, U. and P. Gibbon (2009), "High-Order Harmonics from Laser-Irradiated Plasma Surfaces," *Reviews of Modern Physics* **81**, 44.

Thaury, C. and F. Quere (2010), "High-Order Harmonic and Attosecond Pulse Generation on Plasma Mirrors: Basic Mechanisms," *Journal of Physics B – Atomic Molecular and Optical Physics* **43**, 213001.

Tsakiris, G. D., K. Eidmann, J. Meyer-ter-Vehn and F. Krausz (2006), "Route to Intense Single Attosecond Pulses," *New Journal of Physics* **8**, 19.

Wagner, U., M. Tatarkis, A. Gopal, F. N. Beg, E. L. Clark, A. E. Dangor, R. G. Evans, M. G. Haines, S. P. D. Mangles, P. A. Norrieys, M. S. Wei, M. Zepf and K. Krruschelnick (2004), "Laboratory Measurements of 0.7 GG Magnetic Fields Generated during High-Intensity Laser Interactions with Dense Plasmas," *Physical Review E* **70**, 026401.

Weibel, E. S. (1959), "Spontaneously Growing Transverse Waves in a Plasma Due to an Anisotropic Velocity Distribution," *Physical Review Letters* **2**, 83.

Wharton, K. B., S. P. Hatchett, S. C. Wilks, M. H. Key, J. D. Moody, V. Yanovsky, A. A. Offenberger, B. A. Hammel, M. D. Perry and C. Joshi (1998), "Experimental Measurements of Hot Electrons Generated by Ultraintense ($>10^{19}$ W/cm^2) Laser–Plasma Interactions on Solid-Density Targets," *Physical Review Letters* **81**, 822.

Wickens, L. M. and J. E. Allen (1979), "Free Expansion of a Plasma with 2 Electron Temperatures," *Journal of Plasma Physics* **22**, 167.

Wickens, L. M. and J. E. Allen (1981), "Ion Emission from Laser-Produced, Multi-Ion Species, 2-Electron Temperature Plasmas," *Physics of Fluids* **24**, 1894.

Wickens, L. M., J. E. Allen and P. T. Rumsby (1978), "Ion Emission from Laser-Produced Plasmas with 2 Electron Temperatures," *Physical Review Letters* **41**, 243.

Wilks, S. C. and W. L. Kruer (1997), "Absorption of Ultrashort, Ultra-Intense Laser Light by Solids and Overdense Plasmas," *IEEE Journal of Quantum Electronics* **33**, 1954.

Wilks, S. C., W. L. Kruer, M. Tabak and A. B. Langdon (1992), "Absorption of Ultra-Intense Laser-Pulses," *Physical Review Letters* **69**, 1383.

Wilks, S. C., A. B. Langdon, T. E. Cowan, M. Roth, M. Singh, S. Hatchett, M. H. Key, D. Pennington, A. MacKinnon and R. A. Snavely (2001), "Energetic Proton Generation in Ultra–Intense Laser–Solid Interactions," *Physics of Plasmas* **8**, 542.

Yang, T. Y. B., W. L. Kruer, A. B. Langdon and T. W. Johnston (1996), "Mechanisms for Collisionless Absorption of Light Waves Obliquely Incident on Overdense Plasmas with Steep Density Gradients," *Physics of Plasmas* **3**, 2702.

Yang, T. Y. B., W. L. Kruer, R. M. More and A. B. Langdon (1995), "Absorption of Laser-Light in Overdense Plasmas by Sheath Inverse Bremsstrahlung," *Physics of Plasmas* **2**, 3146.

Yin, L., B. J. Albright, D. Jung, K. J. Bowers, R. C. Shah, S. Palaniyappan, J. C. Fernandez and B. M. Hegelich (2011), "Mono-energetic Ion Beam Acceleration in Solitary Waves during Relativistic Transparency Using High-Contrast Circularly Polarized Short-Pulse Laser and Nanoscale Targets," *Physics of Plasmas* **18**, 053103.

Yuan, X. H., A. P. L. Robinson, M. N. Quinn, D. C. Carroll, M. Borghesi, R. J. Clarke, R. G. Evans, J. Fuchs, P. Gallegos, L. Lancia, D. Neely, K. Quinn, L. Romagnani, G. Sarri, P. A. Wilson and P. McKenna (2010), "Effect of Self-Generated Magnetic Fields on Fast-Electron Beam Divergence in Solid Targets," *New Journal of Physics* **12**, 063018.

Zel'dovich, Ya. B. and A. S. Kompaneets (1959), "On the Propagation of Heat for Nonlinear Heat Conduction," in *Collection Dedicated to the Seventieth Birthday of Academician A. F. Ioffe*, ed. P. I. Lukirskii (Moscow: Izdat Akad. Nauk SSSR).

Zepf, M., M. CastroColin, D. Chambers, S. G. Preston, J. S. Wark, J. Zhang, C. N. Danson, D. Neely, P. A. Norreys, A. E. Dangor, A. Dyson, P. Lee. A. P. Fews, P. Gibbon, S. Moustaizis and M. Key (1996), "Measurements of the Hole Boring Velocity from Doppler Shifted Harmonic Emission from Solid Targets," *Physics of Plasmas* **3**, 3242.

Zheng, J., K. A. Tanaka, T. Sato, T. Yabuuchi, T. Kurahashi, Y. Kitagawa, R. Kodama, T. Norimatsu and T. Yamanaka (2004), "Study of Hot Electrons by Measurement of Optical

Emission from the Rear Surface of a Metallic Foil Irradiated with Ultraintense Laser Pulse," *Physical Review Letters* **92**, 165001.

Zimmer, M., S. Scheuren, T. Ebert, G. Schaumann, B. Schmitz, J. Hornung, V. Bagnoud, C. Rödel and M. Roth (2021), "Analysis of Laser-Proton Acceleration Experiments for Development of Empirical Scaling Laws," *Physical Review E* **104**, 045210.

Appendix B Summary of Basic Plasma Physics

Braginskii, S. I. (1965), "Transport Processes in a Plasma," in *Reviews of Plasma Physics, Vol. 1*, M. A. Leontovich, ed. (New York: Consultants Bureau).

Spitzer, L. and R. Härm (1953), "Transport Phenomena in Completely Ionized Gas," *Physical Review* **89**, 977.

Index

aberrations
  chromatic, 39
  in stretcher, 40
  optical, 38
  wavefront, 53, 62
ablation
  plasma density, 694
  pressure, 693
above-threshold dissociation, 307
above-threshold ionization, 3, 167, 223, 307, 317, 323, 499
  angular distributions, 223, 239, 246, 259
  by few-cycle pulses, 244
  carrier envelope phase effects, 244
  critical intensity for onset of nonperturbative character, 230
  cutoff, 237
  ellipticity effects, 240
  Freeman resonance, 233
  from tunneling, 234
  heating, 499
  in molecules, 281
  nonperturbative, 224, 230
  peak suppression, 230, 231
  perturbative, 226
  plateau, 225, 237
  quasi-classical, 234
  rescattering effects, 225, 236
  resonant enhancement, 232
  strong field approximation, 223, 238
absorption
  at a plasma surface, 662, 667, 709
  atomic, 140, 147, 326, 375
  Brunel. *See* vacuum heating
  collisional. *See also* inverse bremsstrahlung heating. *See* inverse bremsstrahlung heating 501
  collisional at sharp interface, 691
  collisionless, 659, 672, 675, 685, 703, 708
  fraction, 715
  in the continuum, 235
  multiphoton, 144, 149, 170, 180, 204, 223
  nonthermal, 688
  of circularly polarized light, 240
  photoionization, 356, 375
  resonance. *See* resonance absorption
  thermal, 688
  two-photon, 149
accelerated frame. *See* Kramers–Henneberger frame
accelerated mirror model, 765
acceleration
  beam loading effects, 642
  beatwave, 632
  blow-out regime, 626
  bubble, 6, 626, 638, 642
  direct laser, 655
  electron, 562, 609
  electron injection seeding, 635
  hydrodynamic expansion, 754
  ion, 703, 754
  photon, 587
  radiation pressure, 765
  target normal sheath, 8, 743, 755, 756, 767, 770, 787
  wakefield, 6, 610, 621, 624
  wakefield, maximum energy gain, 618
acousto-optic
  modulation, 29
  programmable dispersive filters, 46
action
  quasi-classical, 332, 336, 338, 348
adaptive optics, 65
adiabatic
  curve, 302, 307
  evolution of pendular states, 285
  expansion, 391, 463, 468, 663
  limit, 184
  limit for molecular alignment, 284
  limit in a cluster, 427
  sound speed, 467
  stabilization, 249, 252, 258

states, 300, 301, 304
adiabatically
  varying field, 142
adiabaticity
  electron in a field, 237
  of electron oscillations, 75, 87, 501, 577, 714, 728
ADK formula. *See* PPT/ADK formula
Airy function, 680, 683, 712
Airy pattern, 63
Akheizer–Polovin equations, 547
alexandrite, 22
alignment
  molecular, 274, 283, 310, 348
  transient, 288
ambipolar
  potential, 672
  sheath field, 464, 669, 671, 777
Ammosov–Delone–Krainov formula. *See* PPT/ADK formula
Ampere's law, 496, 548, 628, 676, 679, 681, 684, 722, 752, 779
amplified spontaneous emission, 27, 58
amplitude modulation, 589
anharmonic
  damping, 438
  dipoles, 318
  motion, 71, 83, 92, 97
  potential, 453
anharmonic motion, 316
anomalous skin effect, 727
anti-Stokes wave, 591
ASE. *See* amplified spontaneous emission
ATD. *See* above-threshold dissociation
ATI. *See* above-threshold ionization
atomic photoionization. *See* ionization, photoionization
atomic unit
  electric field, 2, 142, 153
  frequency, 153
  intensity, 153
atto-chirp. *See* attosecond pulse, attochirp
attosecond pulse
  lighthouse, 389
  atto-chirp, 378
  atto-pulse spacing, 383
  generation, 5, 376
  polarization gating, 387
  single pulse generation, 385
  temporal gating, 385
  trains, 377
Atwood number, 769
avoided crossing, 301

backscatter
  Raman, 596
barrier suppression ionization. *See* ionization, barrier suppression
BBO. *See* beta-barium borate

beam loading, 642
beam quality, 21, 41, 50, 65
beatwave acceleration. *See* acceleration, beatwave
Bebb and Gold, method, 157, 242, 246
Beg formula, 733, 737, 743, 757, 763
benzene, 280
beta-barium borate, 24, 35, 62
betatron
  electron trajectory, 644
  oscillations, 643
  radiation emission, 7, 648
  resonance, 655
  X-ray properties, 652
Bethe–Heitler process, 746
B-integral, 20, 55, 59, 62
Bloch equations, 163
blow-off plasma, 668, 777
blow-out regime, 626
blue shift
  ionization-induced, 535, 584
  relativistic, 586
Bohm–Gross frequency, 594
Bohr radius, 144, 153, 187, 212
Boltzmann
  distribution, 758, 760
  statistics, 761
bond hardening, 309
bond softening, 5, 273, 303, 311
Born approximation, 151, 255
Born–Oppenheimer approximation, 268, 292, 303
Breizman–Arefiev model, 424, 429
bremsstrahlung
  inverse. *See* ierse bremsstrahlung heating, 501
  X-rays, 660, 746, 774, 785
Brunel absorption/heating. *See* vacuum heating
BSI. *See* ionization, barrier suppression
bubble acceleration. *See* acceleration, bubble

canonical momentum, 73
  conservation, 78, 85, 110, 121, 131, 549
carrier envelope phase, 245
  control, 32
cavitation
  electron plasma, 535, 562, 564
  ion, 576
CEP. *See* carrier envelope phase channel
channel
  electron plasma, 564
  oscillation, 567
channel closing, 180, 231
channel, ion, 576
charge resonant enhanced ionization. *See* critical ionization distance, molecules
chirped pulse amplification, 4, 13
cluster, 6
  Coulomb explosion. *See* Coulomb, explosion
  formation in gas jets, 390
  nanoplasma, 393

CO_2 laser, 67
coherence length, 360, 373, 375
coherent transition radiation, 743
coherent wake emission, 747
colliding pulse injection, 640
collision
  Coulomb, 449, 501, 504
  electron–ion, 449, 453, 497
  frequency, 496, 497, 505, 692
  frequency, electron–ion, 449, 496, 503, 504, 594
  term, 501
collisional
  absorption, 688
  damping, 399, 457, 486, 496, 594, 685
  damping, in cluster, 449
  equilibration time, electron–ion, 499
  excitation, 214
  ionization. See ionization, collisional
  ionization, in cluster, 415
  plasma, 450, 497
collisional absorption. See absorption, collisional, 672
collisionality, 709
collisionless
  absorption, 688, 708, 710
  damping, 594
  skin depth, 678
collisionless damping
  cluster, 450
  frequency, 452
collisionless scattering, 453
compressor. See pulse compressor
conductivity
  effect on magnetic field generation, 779
  electrical, 737, 738, 741
  thermal, 663
continuity equation, 548, 559, 568, 582, 592, 601, 671, 674, 698, 740
continuum lowering, 404, 408, 416
continuum state, 145, 154, 170, 228, 255
contrast
  temporal pulse, 57
Coulomb
  collision, 237, 504, 737
  continuum state, 156
  correction, 194, 199, 264
  explosion, 5, 577
  explosion time, 310
  explosion, cluster, 396, 435, 471, 477, 478
  explosion, ion channel, 576
  explosion, molecular, 309, 314
  focusing, 211, 259
  force, 169
  force neglect in the strong field approximation, 170
  gauge, 72, 293, 548
  logarithm, 418, 433, 506, 663
  potential distortion in laser field, 183, 200, 274
  pressure, 476

scattering, 237, 248, 436, 449, 450, 453, 501, 505, 509
scattering cross section, 505
CPA. See chirped pulse amplification
Cr:LiSAF, 22
critical density, 495, 662
critical ionization distance, 5, 310
  cluster, 408
  molecule, 273, 311
critical power. See self-focusing, critical power
critical radius, 577
critical radius, electron cavitation, 577
cross-polarized wave generation, 60
CTR. See coherent transition radiation
cutoff law, high harmonic, 318, 323, 324, 326

Debye
  cloud, 395, 406
  length, 405, 506, 729, 761
  radius, 395
  sheath, 395, 757
  shielding, 404
defocusing
  ionization-induced, 515
deformable mirror, 65
density of states, 134, 151
density profile modification. See ponderomotive, steepening
density ramp injection, 640
dephasing length
  in wakefield acceleration, 619
diabatic
  curve, 302
  states, 300, 304
dielectric
  constant, 495, 498
  effective, angle-dependent, 686
  function, 686
  function, complex, 692
  function, complex optical, 497
  function, plasma, 675
  function, plasma in density gradient, 712
  function, plasma optical, 503, 685
  function, plasma profile, 679
diffraction limit, 53, 63
diffusivity, thermal, 664
dipole approximation, 137, 144, 168, 174
dipole moment
  molecular, 284
dipole resonance. See giant dipole resonance
direct laser acceleration. See acceleration, direct laser
dispersion, 41
  compensation, 27, 41
  group-delay, 43
  stretcher, 45
  third-order, 43, 46
dispersion relation, 592
DLA. See acceleration, direct laser

Doppler shift, 93, 97, 597, 701, 703, 748
double ionization. *See* ionization, nonsequential double
dressed states, 298, 306
drift velocity
   electron, 77, 78, 234
   electron longitudinal, 83, 86, 94, 98, 126
Drude model, 371, 402, 536, 686
dye laser, 68
dynamic localization, 262

ECI. *See* ionization, collisional
effective principal quantum number, 198
electromagnetic pulse generation, 784
ellipticity. *See* polarization, ellipticity
   efffect on high harmonics, 324
ellipticity. *See* polarization, ellipticity
emittance, electron beam, 132, 630
EMP. *See* electromagnetic pulse generation
EOS. *See* equation of state
equation of state
   ideal gas, 465, 468, 581, 594, 670, 777
etching, laser pulse, 608
Euler–Lagrange equations, 74, 80
evolution operator, 146
excimer laser, 23

Faraday's law, 561, 675, 687, 738, 776, 779
fast ignition, 7, 787
Fermi's golden rule, 140, 147
figure-8 motion, 96, 720, 725
filamentation
   laser in air, 7, 524
   length, 529
fission, photo, 785
flashlamp pumping, 19, 22, 48, 67
Floquet
   diabatic states, 304
   Fourier component, 297
   Hamiltonian, 294, 296, 300
   quasi-energies, 295
   state, 296, 298, 301
   theory, 292
   theory for two-level system, 299
   theory of bonds in a strong field, 303
fluid equations, 548, 568, 752
Franck–Condon factors, 267, 282
free-streaming. *See* ionization, free-streaming
Freeman resonance, 233, 240
Fresnel equations, 691
fusion
   in cluster plasmas, 6, 493
   inertial confinement, 787

gain narrowing, 15, 21, 27, 48, 53
gain saturation, 53
gas jet, 390, 490, 494

gauge
   Coulomb, 72, 137
   length, 139
   radiation, 72, 137
   transformation, 137
   velocity, 172
gauge invariance, in strong field approximation, 168
Gauss's law, 496, 555, 613, 621, 628, 682, 700, 704, 726
Gaussian beam, 111, 512, 519, 525, 541, 565
   efffective beam area, 526
   focus, 126, 219
   longitudinal fields in focus, 115
   refraction, 512
Gaussian pulse shape, 75, 88, 244
generalized Bessel function, 172
gerade symmetry, 254
giant dipole resonance, 398, 402, 434, 438
Gouy phase shift, 352, 357, 360
grating, 8, 14, 23, 35
   equation, 37, 45
   fiber Bragg, 31
   gold, 41
   in stretcher, 38
Green's function, 556, 569
group delay dispersion, 41
group velocity, in plasma, 498

Hagena parameter, 391
Haines hot electron conversion law, 734
Hamiltonian, 73, 137, 283, 288
   Floquet, 294
   interaction, 137, 144, 146, 238, 293
   interaction operators, 148
   interaction, velocity form, 171
   Kramers–Henneberger, 249
   length form, 137, 144, 167
   molecular rotational, 285
   relativistic, 80
harmonic oscillator
   plasma wave, 569
heat capacity
   plasma. *See* specific heat
heat conduction, 663
heat equation, 664, 668
Helmholtz equation, 556, 677, 679, 696
hermal velocity, 452
Hermitian, matrix, 298
HHG. *See* high-harmonic generation
high harmonic chirp, 383
high harmonic cutoff law, 326
high-harmonic generation, 4, 316
   ellipticity effects, 344
   from solids, 747
   in molecules, 348
   phase-matching, 350
   single-atom model, 321
   three-step model, 324

highest occupied molecular orbital, 264, 270
hole boring, 699, 709, 713
    relativistic intensity, 701
    surface velocity, 701
HOMO. *See* highest occupied molecular orbital
hot electron
    generation, 660. *See also* collisionless, absorption, 733
    production, 7
    recirculation, 764
    temperature, 672, 713, 733
    transport, 736
hydrodynamic expansion, 659, 668, 693, 696
    cluster, 463
    two-electron temperature, 672

IFA. *See* ionization, tunneling, ion-field assisted
inelastic excitation, 214
intensity
    relativistic, 1
inverse bremsstrahlung heating, 499, 501, 594, 688
    in finite scale length, 688
inverse Compton scattering, 128
ion acceleration. *See* acceleration, ion
ion acoustic waves, 576
ion temperature, 499, 581
ion-field assisted tunneling. *See* ionization, tunneling, ion-field assisted
ionization
    absorption in a filament, 529
    avalanche, 661
    average state, 663
    barrier suppresion, 518
    barrier suppression, 200
    collective tunneling, 206
    collisional, 212, 499, 659, 661
    collisional, cross section, 212
    evaporative, 430
    free-streaming, 430
    ignition, 418
    impact, 487
    injection, 640
    knee structure, 222
    multiphoton. *See* multiphoton ionization
    nonsequential double, 204
    photoionization, 140, 206, 256, 356, 375
    polyatomic molecule, 279
    potential, 140, 264, 266, 268, 274, 303, 518
    saturation, 180, 212, 418
    sequential, 141
    SFA rate in tunneling limit, 193
    shake-off, 206
    stabilization, 248
    state, 496, 516, 518, 585
    time-dependent, 511
    tunnel, 142, 498, 499, 518
    tunneling, 4, 143, 167, 183, 184, 205, 212, 323, 345, 361, 378
    tunneling in circular polarization, 203
    tunneling in complex atoms, 196
    tunneling in hydrogen atom, 186
    tunneling, Coulomb correction, 194
    tunneling, cycle-averaged, 191
    tunneling, in clusters, 408
    tunneling, in molecules, 262, 268, 348
    tunneling, ion-field assisted, 408
    tunneling, molecular, 263
    tunneling, relativistic correction, 260
ionization-induced plasma defocusing, 515
isothermal
    expansion, 759
    sound speed, 470, 583, 662, 670

J × B heating, 720
Jacobi–Anger formula, 105, 171

KDP. *See* potassium dihydrogen phosphate
Keldysh
    approximation, 167, 238
    parameter, 185, 207, 234, 247, 268, 341
Keldysh–Faisal–Reiss theory. *See* strong field approximation
Kepler's third law, 443
Kerr effect, 60, 67
Kerr-lens mode-locking, 30
KFR theory. *See* strong field approximation
KH frame. *See* Kramers–Henneberger, potential
kinetic equation, 501, 548
Krainov–Smirnov model, 427
Kramers–Henneberger
    ground state wavefunction, 252
    Hamiltonian, 249
    potential, 251
KrF laser, 23
Kummer confluent hypergeometric function, 156

L'Hospital's rule, 623
Lagrangian, 72
    cyclic variables, 74, 77
    relativistic, 79
Landau damping, 594, 685
Landau–Zener theory, 302
Langmuir wave, 711
Laplace's Method, 414
Larmor
    formula, 91, 109, 318
Larmor formula, 91
laser energy bunching, 589
laser plasma instabilities, 590
    in plasma density gradients, 730
LBO. *See* lithium triborate
LCAO. *See* linear combination of atomic orbitals
Lewenstein model. *See also* strong field approximation, high-harmonic generation, 330
Liénard–Wiechert potentials, 100

light pressure. *See* pressure, light
linear combination of atomic orbitals, 264
lithium triborate, 25, 68
Littrow angle, 37, 45
localization
   electron in molecular tunneling, 276
longitudinal fields
   effect on ponderomotive scattering, 118
   in focus, 114
LOPT. *See* lowest-order perturbation theory
Lorentz
   factor, 108
   frame, 80
   invariant, 95, 134, 648
   transform, 93, 98, 128, 135
Lotz formula, 415
lowest-order perturbation theory, 3, 4, 143, 144, 156, 180. *See also* multiphoton ionization, 263
   breakdown, 164
   intensity validity range, 152
   multiphoton cross section, 154
   relation to the strong field approximation, 167
LPI. *See* laser plasma instabilities

magnetic field generation, 775
Maxwell's equations, 115, 117, 496
Maxwellian
   distribution, 500, 508
Mie
   resonance, 398, 400, 408, 419, 438, 441, 445, 451, 453, 483
   resonance, at relativistic intensity, 448
   resonance, with large amplitude electron oscillation, 443
   scattering, 399
mode-locking, 27
molecular
   alignment, 303
   bond, 303
   bond hardening, 308
MPI. *See* multiphoton ionization multielectron atoms
multielectron atoms
   multiple ionization, 216
multielectron effects, molecular, 6, 271
multiphoton ionization, 3, 143, 498, 527, 660
   above-threshold ionization regime, 224
   cross section, 152, 154
   diatomic molecules, 263
   diatomic molecules, vibrational excitation, 267
   hydrogen molecule, 282
   perturbation expansion, 144
   resonantly enhanced, 158
   resonantly enhanced, in ATI, 233
   transition rate, 149
multiphoton process, 321
multiphoton processes, 142
multiple ionization, 204

nanoplasma, 6, 395, 398, 403, 408, 424, 440, 457, 463, 483
   heating, 434, 456
nanoplasma model, 390, 393, 400, 402, 404, 415, 441
Nd:glass, 15, 19
Nd:YAG, 67
NDSI. *See* ionization, nonsequential double
neutron production
   by laser-driven ion acceleration, 770
   cross sections, 770
nonlinear refractive index, 56, 524
nonlinear Thomson scattering
   angular distribution, 100
nonperturbative
   nonlinear optics, 4
nonperturbative effects, 2, 3, 166, 167, 234
nonsequential double ionization, 5, 204
   Coulomb focusing effect, 211
   inelastic excitation, 214
   recollision dynamics, 207
   wavepacket spreading, 211
normalized vector potential, 79, 82, 94
   definition, 548
NSDI. *See* nonsequential double ionization

Offner triplet, 40
Ohm's law, 738
OPCPA. *See* optical parametric chirped pulse amplification
optical parametric chirped pulse amplification, 15, 24
   nanosecond, 21
   picosecond, 61
   with Nd
      glass amplifiers, 21
orbital
   molecular, 263, 264
   antibonding, 264, 265
   atomic, 264
   bonding, 265
   molecular, 6
   pi, 269
organic molecule, 263, 280
oscillating mirror model, 748
over the barrier ionization. *See* ionization, barrier suppresion
overdense plasma, 659, 692, 710, 715

pair production, 746
parabolic coordinates, 187, 196
paraxial wave equation. *See* wave equation, paraxial
Parseval's theorem, 101
partition function
   rotational, 287
pendular states, 284
perturbation
   expansion, 146
   theory, 140, 143

theory breakdown, 1, 153
phase control, CPA, 41
phase velocity, 511
   light in plasma, 498, 511, 516
   plasma wave, 539
phase-matching, 320
phase-matching, in high-harmonic generation, 350
photoelectric effect, 140
photoionization. *See* ionization, photoionization
photon acceleration. *See* acceleration, photon
plasma frequency, 398, 402, 495, 549, 588, 590, 597, 616, 676, 747
   relativistic, 536, 604
plasma wave
   nonlinear, 547
plasma wave, electron, 511
plateau, high-order harmonic, 4, 317, 321, 323, 328, 336, 346, 349, 362, 378
Poisson's equation, 548, 549, 558, 560, 563, 569, 577, 593, 601, 615, 616, 629, 722, 752, 760
polar moment of inertia, 289
polarizability
   atomic, 391
polarization
   atomic, 316, 319, 354, 361, 527
   circular, 77, 96, 203, 500, 601, 720, 767
   cluster, 445, 480
   effects in HHG selection rules, 322
   eliptically, 727
   ellipical, 721
   ellipticity, 278, 333
   ellipticity effects on high-harmonic generation, 344
   gating, 387
   laser, 71, 75, 82, 86, 90, 92, 102, 114, 127, 137, 263, 269, 280, 283, 284, 288, 292, 303, 331, 348, 388, 399, 408, 410, 427, 462, 480, 482, 501, 600, 655, 781
   molecular, 280, 591
   P-polarization, 682, 688, 692, 709, 720, 747
   S-polarization, 686, 692, 720
   scattered emission, 102
   Stokes wave, 595
   surface, 743
   surface high harmonic, 753
pole approximation, 229
polyatomic molecule, 279, 280
   Coulomb explosion, 314
ponderomotive
   channeling, 564
   continuum energy shift, 174, 226, 232, 282
   driven electron current, 776–778
   ejection from focus, 118, 226, 231, 260, 539, 564
   electron density driven motion, 537
   electron expulsion, 564, 570, 575, 626, 628, 638
   electron injection, 581
   energy, 76, 78, 85, 139, 170, 175, 210, 224, 231, 237, 243, 303, 326, 394, 456, 499, 509, 588, 604, 719

   energy compared to thermal, 501
   energy gradient, 114, 546
   energy level shift, 160, 282, 293
   energy modulation, 599
   energy of cluster electron cloud, 416
   energy phase shift in high-harmonic generation, 363
   force, 111, 124, 535, 539, 553, 554, 560, 565, 569, 577, 581, 639, 692, 695, 697, 721
   force approximation breakdown, 124
   force, longitudinal, 545
   force, relativistic, 120, 707, 783
   hole boring. *See* hole boring
   injection, 639
   ion kick, 581
   potential, 76
   pressure, 617, 695, 702, 723, 726, 747, 765, 781
   quiver velocity, 75, 504, 507
   scattering, 110
   shift of resonance. *See* Freeman resonance
   steepening, 686, 694, 709, 713, 720, 724
   term in Hamiltonian, 168, 250
   wakefield excitation. *See* wakefield generation
positron production, 8, 746
potassium dihydrogen phosphate, 25, 46, 68
Poynting theorem, 453
PPT/ADK formula. *See also* ionization, tunneling
prepulse, 49, 57, 61, 662, 668, 671, 732
pressure
   ablation, 693
   electron, 395, 396, 418, 419, 463, 465, 468
   gas backing, 392
   light. *See* pressure, ponderomotive, 696, 701
   ponderomotive. *See* ponderomotive, pressure, 697, 701, 726, 765
   radiation, 765
profile steepening, 727
propagation operator, 164, 168, 238
proper time, 80, 95, 103
proton acceleration. *See* acceleration, ion
pulse compressor, 35, 50
pulse stretcher, 14, 38
pulse temporal modulation
   radial beam, 607
   stimulated Raman, 599, 604
pump depletion length
   wakefield generation, 619

Q-switch, 7, 19, 26
QSA. *See* quasi-static approximation
quantum interference
   in ionization of diatomic molecules, 266
quantum interferences
   in the strong field approximation, 179
quasi-classical model, 5
quasi-classical three-step model, 4, 324
quasi-energy. *See* Floquet, quasienergies

quasi-neutrality, 422, 546, 581, 673, 674, 738, 755, 760
quasi-static approximation, 547, 611, 615, 621, 629
  3D, 557, 564, 575
quasi-stationary
  potential, 255
  wavefunction, 252
quiver
  amplitude, 75
  energy, 2
  velocity, 75

Rabi frequency, multiphoton, 163
Rabi oscillations, multiphoton generalization, 162
radial beam envelope model, 512, 526
radial beam modulation, 607
radiation pressure acceleration. *See* acceleration, radiation pressure
radiation reaction
  free electron, 109
Raman forward scattering, 599
Raman scattering
  forward, 7, 599
  stimulated, 591
  stimulated, backscatter, 591
  stimulated, in density gradient, 730
  stimulated, relativistic, 600
  stimulated, threshold in density gradient, 732
rarefaction wave, 671
Rayleigh range, 88, 111, 126, 321, 353, 355, 357
Rayleigh–Taylor instability, 707, 768
recollision
  atomic field ionization, 207
  dynamics in high-harmonic generation, 324
  dynamics in ionization, 207
  in cluster, 457
  ion momentum distribution, 215
  relativistic suppression, 258
recombination, 772
  three-body, 417
refraction
  at a plasma surface, 681, 682
  during super-continuum generation, 535
  in filamentation, 528
  plasma induced, 512, 514, 515, 538, 585
  plasma, clamped maximum intensity, 521
refractive index
  complex, plasma, 503
  nonlinear, 55, 512, 513, 526
  nonlinear in air, 524, 528
  nonlinear, phase-modulation, 535
  plasma, 497
  relativistic nonlinear, 536, 540
regen. *See* regenerative amplifier
regenerative amplifier, 25, 48, 50, 55, 58, 68
relativistic induced transparency, 9
relativistic intensity
  definition, 79

radiation emission, 92, 94
relativistic-induced transparency, 780
  ponderomotive pressure effect, 781
REMPI. *See* multiphoton ionization, resonantly enhanced
rescattering. *See also* recollision. *See* recollision, 225
  cluster, 456
  in above-threshold ionization. *See* above-threshold ionization, rescattering effects
  in nonsequential double ionization, 204
residual drift
  angle, 89
  distance, in finite pulse, 88
  energy, 89
  motion, 87
residual drift energy, electron, 77
resistivity
  effect on return current, 740, 779
  influence on magnetic field generation, 776
resonance absorption, 710, 713
resonances
  in multiphoton ionization, 158
retarded time, 91, 99–101
return current, 737, 776, 778
R-frame, 94
RIT. *See* relativistic-induced transparency
rotating wave approximation, 300
rotational constant, molecular, 284
rotational partition function, 287
rotational pumping, 291
rotational revival, 290
rotational state, 286
RPA. *See* acceleration, radiation pressure

saddle point
  approximation, 176, 180
  approximation, in high-harmonic generation, 333
  method, 176
SAE. *See* single active electron approximation
saturable absorber, 30, 31, 68
saturation intensity, 217, 220, 326, 361, 373, 386
  barrier suppression ionization, 217
  knee structure, 222
  multiphoton ionization, 157, 180
Schrödinger equation, 136, 144, 164, 169, 250, 294
  length form, 187
Schwinger field, 134
selection rules
  atomic high-harmonic generation, 321
  solid target high-harmonic generation, 753
self-channeling. *See also* ilamentation. *See also* ilamentation
self-focusing, 13, 514
  critical power, 526, 540, 567, 632
  in air during filamentation, 524
  in laser amplifiers, 55, 64
  Kerr-lens mode-locking, 30
  refractive reduction in plasma, 557

relativistic, 7, 538, 567, 575
self-focal length, 529
self-induced opacity, 781
self-induced transparency. *See* relativistic-induced transparency
self-modulated wakefield generation, 613
self-modulation
  radial beam, 567
self-phase modulation, 7, 385
  fiber, 34
  laser amplifier, 41, 55
  plasma wave, 587
  relativistic, 586
self-similar
  density profile, 468, 469, 671
  expansion, 464
  heat wave, 665
  solution, 468, 664, 666, 671, 673, 758, 761
  solution, expanding hot electron, 742
  temperature profile, 664
  variable, 667
self-trapping
  in wakefields, 637
semiconductor saturable absorber mirror (SESAM), 30
separatrix, 636
SFA. *See* strong field approximation
shake-off double ionization, 206
sheath inverse bremsstrahlung, 729
Siegert boundary conditions, 299
single active electron approximation, 141, 204, 251
skin depth
  collisional, 734
  collisionless, 700, 726, 782
skin depth, collisionless, 663, 678
slowly varying amplitude approximation, 87, 112, 513
Snell's law, 690, 692
sound speed
  adiabatic, 467
  hot electron, 761
  hydrodynamic expansion, 669, 697
  isothermal, 470, 662, 670, 671, 754
  two-temperature, 673
SP-OPCPA. *See* optical parametric chirped pulse amplification, picosecond
space-charge, 423, 430, 464, 709, 714, 716, 726, 740, 757, 778, 779
  amendment, 718
  confinement, 741
  force, 403
  potential, 431
  separation, 726
spatial filter, 51, 56
specific heat, 663
  ideal gas, 663
  ionizing plasma, 663
spectral shaping, 19
SRS. *See* Raman scattering, stimulated

stabilization
  dynamic, 249
  ionization, 248
  quasistationary, 249
standing wave, 680, 686, 695, 698
Stark shift
  AC, 159, 163, 249
  effect on resonant enhanced ionization, 224, 231
steepest descent, method. *See* saddle point, method
stimulated Brillouin scattering, 576, 590
stimulated Raman scattering. *See* Raman scattering, stimulated
stochastic heating, 423, 455
stochastic scattering, 452
Strehl ratio, 63
stretcher. *See* pulse stretcher
strong coupling in the continuum, 164
strong field approximation, 3, 6, 166, 167, 171
  above-threshold ionization prediction, 223
  Coulomb correction, 194
  diatomic molecule, 264
  high-harmonic generation, 330
  ionization rate, 177
  limit of many photons, 175
  weak field limit, 175
  with complex atoms, 196
strong field physics
  definition, 1
sudden approximation, 504
supercontinuum generation, 534
suprathermal electrons, 756, 774
synchrotron radiation, 651

target normal sheath acceleration. *See* acceleration, target normal sheath
thermal velocity, 449, 504, 507, 508
thermal wave, 662
Thomson scattering, 110
  linear, 91, 97, 128
  nonlinear relativistic, 92
three step model, quasi-classical, 324, 344, 363
  quantum correspondence, 332, 335, 337
THz radiation generation
  plasma filament, 534
  solid target, 784
Ti:sapphire, 15, 16
time-bandwidth product, 43
TNSA. *See* acceleration, target normal sheath
trajectory, electron, 317, 345
  long, 325, 364, 370, 384
  short, 325, 370, 380, 385
transformation
  wavefunction, 145
trapping, electron, 635
truncated summation method, 154, 157, 242, 246
tunnel ionization. *See* ionization, tunneling
tunneling time, 184

turning point, in a plasma density gradient, 681, 683, 705, 734
two-photon absorption, 3
two-plasmon decay, 732

unitary transform, 250

vacuum heating, 713
valley of death, 257
vector potential, 74, 77, 547, 592, 606
  at sharp interface, 725
  Coulomb gauge, 72
  electron momentum, 121
  gauge condition, 558
  in electron Lagrangian, 72
  normalized, 79, 88, 540, 624, 715
  normalized, in ponderomotive force, 122
  relation to fields, 72
  time-averaged, 554
velocity distribution, 508, 744
  ion, 754
velocity distribution, Maxwellian, 417, 470
vibrational
  excitation, 267
  levels, 268
Volkov state, 138, 168, 172, 330
  Coulomb correction, 194
  in rescattering, 238
vorticity, plasma, 560

wake potential, 561, 564, 629
wakefield
  generation, resonant excitation, 622
wakefield acceleration. *See* acceleration, wakefield
wakefield generation. *See also* acceleration, wakefield
  multipulse, 626
  self-modulated, 613, 631
wave equation, 558, 681
  electric field, 676, 683
  Helmholz equation, 677
  ion density perturbation, 583
  magnetic field, 676, 683
  paraxial, 111, 513
  vector potential, 548, 549, 553, 592, 600, 723
wavebreaking, 613, 619, 621, 624, 635
  radial, 622
wavefront curvature, plasma, 621
wavefunction
  plane-wave, 138, 151, 172, 239, 255
wavelength shift, plasma, 621
wavepacket
  barrier penetration, 183
  electron, 326
  in the continuum, 335, 338, 339
  nuclear in bond softening, 305
  of rotational states, 290
  periodicity in above-threshold ionization, 235, 244
  recollision, 211, 214
  recombining in *high-harmonic* generation, 326
  spreading, 211, 239, 259, 279, 327, 341, 344, 345
  tunneling, 185, 329
Weibel instability, 776, 779
white light emission, 535
Wigner rotation operator, 269
Wigner–Seitz radius, 404, 410, 411
WKB
  continuum states, 228
WKB approximation, 171, 185, 188, 194, 228, 276, 648, 678, 683

X-ray production, 660, 771
  betatron, 643
  cluster, 457, 490
  hard, 774
  inverse Compton, 129
  K-alpha, 772
  soft, 317, 747
  thermal, 771
XPW. *See* crossed-polarized wave generation
XUV radiation, 317

Yb-based laser materials, 21

Printed in the United States
by Baker & Taylor Publisher Services